Hyperspectral Imaging and Applications

Hyperspectral Imaging and Applications

Special Issue Editors

Chein-I Chang
Meiping Song
Junping Zhang
Chao-Cheng Wu

MDPI • Basel • Beijing • Wuhan • Barcelona • Belgrade

Special Issue Editors
Chein-I Chang
Dalian Maritime University
China
University of Maryland
Baltimore County, USA

Meiping Song
Dalian Maritime University
China

Junping Zhang
Harbin Institute of Technolgy
China

Chao-Cheng Wu
National Taipei University
of Technology
Taiwan

Editorial Office
MDPI
St. Alban-Anlage 66
4052 Basel, Switzerland

This is a reprint of articles from the Special Issue published online in the open access journal *Remote Sensing* (ISSN 2072-4292) from 2017 to 2019 (available at: https://www.mdpi.com/journal/remotesensing/special_issues/Hyperspectral_Imaging_Applications)

For citation purposes, cite each article independently as indicated on the article page online and as indicated below:

LastName, A.A.; LastName, B.B.; LastName, C.C. Article Title. *Journal Name* **Year**, *Article Number*, Page Range.

ISBN 978-3-03921-522-5 (Pbk)
ISBN 978-3-03921-523-2 (PDF)

Contents

About the Special Issue Editors

Chein-I Chang

Chein-I Chang is a Professor with Department of Computer Science and Electrical Engineering at the University of Maryland, Baltimore County and is currently holding a Chang Jiang Scholar Chair Professorship at Dalian Maritime University, China. Dr. Chang has piublsihed over 183 referred SCI publications including more than 60 journal articles in IEEE *Transaction on Geoscience and Remote Sensing*. He authored three books, Hyperspectral Imaging (Kluwer, 2003), Hyperspectral Data Processing (Wiley, 2013), Real Time Progressive Hyperspectral Image Processing (Springer, 2016) and Real-Time Recursive Hyperspectral Sample and Band Processing (Springer, 2017). In addition, He edited two books, Recent Advances in Hyperspectral Signal and Image Processing (Trasworld Research Network, India, 2006) and Hyperspectral Data Exploitation (Wiley, 2007) and co-edited with A. Plaza a book on High Performance Computing in Remote Sensing (CRC Press, 2007). Dr. Chang is a Life Fellow of IEEE and a Fellow of SPIE with contributions to hyperspectral image processing.

Meiping Song

Meiping Song received her Ph.D. degree in the College of Computer Science and Technology from the Harbin Engineering University in 2006. From 2013–2014, she was a visiting associate research scholar at Remote Sensing Signal and Image Processing Laboratory, University of Maryland, Baltimore County. She is currently an Associate Professor in the College of Information Science and Technology, Dalian Maritime University. Her research includes remote sensing and hyperspectral image processing.

Junping Zhang

Junping Zhang received the B.S. degree in biomedical engineering and instrument from Harbin Engineering University and Harbin Medical University, Harbin, China, in 1993, and the M.S. and Ph.D. degrees in signal and information processing from the Harbin Institute of Technology (HIT), Harbin, China, in 1998 and 2002, respectively. Since 2008, she has been a Professor in Department of Information Engineering, School of Electronics and Information Engineering, HIT. Between Aug. 2011 and August 2012, she was a Visiting Scholar at the Center for Global Change & Earth Observations, Michigan State University, USA. Her research interests include hyperspectral data analysis and image processing, multisource information fusion and applications, multitemporal image analysis, pattern recognition and machine learning.

Chao-Cheng Wu

Chao-Cheng Wu received the B.S. degree in electrical engineering from the Tamkang University, Taipei, Taiwan in 2002, the M.S and Ph.D. in electrical engineering from the University of Maryland, Baltimore County (UMBC), Baltimore, MD in 2006 and 2009 respectively. Dr. Wu is currently an associate professor in Department of Electrical engineering, Taipei Tech. His research interests include image and signal processing, detection and estimation, and pattern recognition algorithms in in multispectral, hyperspectral and medical fields. He is a reviewer for several international journals. Dr. Wu is a member of the honor society of IEEE.

remote sensing

MDPI

Editorial

Editorial for Special Issue "Hyperspectral Imaging and Applications"

Chein-I Chang [1,2,*], Meiping Song [1], Junping Zhang [3] and Chao-Cheng Wu [4]

1 Center for Hyperspectral Imaging in Remote Sensing (CHIRS), Information and Technology College, Dalian Maritime University, Dalian 116026, China
2 Remote Sensing Signal and Image Processing Laboratory, Department of Computer Science and Electrical Engineering, University of Maryland, Baltimore, MD 21250, USA
3 Harbin Institute of Technology, Harbin 150001, China
4 Department of Electrical Engineering, National Taipei University of Technology (Taipei Tech), Taipei 10608, Taiwan
* Correspondence: cchang@umbc.edu

Received: 22 August 2019; Accepted: 23 August 2019; Published: 27 August 2019

Abstract: Due to advent of sensor technology, hyperspectral imaging has become an emerging technology in remote sensing. Many problems, which cannot be resolved by multispectral imaging, can now be solved by hyperspectral imaging. The aim of this Special Issue "Hyperspectral Imaging and Applications" is to publish new ideas and technologies to facilitate the utility of hyperspectral imaging in data exploitation and to further explore its potential in different applications. This Special Issue has accepted and published 25 papers in various areas, which can be organized into 7 categories, Data Unmixing, Spectral variability, Target Detection, Hyperspectral Image Classification, Band Selection, Data Fusion, Applications.

Keywords: band selection; data fusion; data unmixing; hyperspectral image classification; hyperspectral imaging; spectral variability; target detection

1. Introduction

Over the past years hyperspectral imaging has received considerable interests [1] such as parallel processing [2], real-time processing [3,4]. It deviates from traditional spatial domain-based image processing and multispectral imaging in many different ways. It has attracted many people from different disciplinary areas to explore new ideas and new applications [5]. In recent years, a significant increase in publications in hyperspectral imaging has provided evidence that hyperspectral image processing has broken away from traditional spatial domain analysis-based remote sensing and successfully branched out to stand alone as a potential and promising research area. Most importantly, hyperspectral imaging have also changed many ways in which algorithms are designed and developed. As a consequence, many problems such as subpixels and mixed pixels that are generally encountered in hyperspectral imaging have become major issues for traditional spatial domain-based techniques [6]. Also, the traditional concept of "*seeing-is-believing*" by visual inspection may no longer true in hyperspectral imaging since targets of interest may be completely embedded in a single pixel or partially but not fully occupy a single pixel in which case only spectral properties that can be used to characterize such targets for data analysis. Therefore, this Special Issue "Hyperspectral Imaging and Applications" is devoted to topics which can demonstrate the utility of hyperspectral imaging in data exploitation and to further explore its potential in different applications. This Special Issue has accepted and published 25 papers in various areas, which can be organized into 7 categories with the number of papers published in every category included in its open parenthesis.

1. Data Unmixing (2 papers)

2. Spectral variability (2 papers)
3. Target Detection (3 papers)
4. Hyperspectral Image Classification (6 papers)
5. Band Selection (2 papers)
6. Data Fusion (2 papers)
7. Applications (8 papers)

Under every category each paper is briefly summarized by a short description in the following section so that readers can quickly grab its content to find what they are interested in.

2. Overview of Published Papers

Part I: Data Unmixing (2 papers)

09-001074
Nonnegative Matrix Factorization with Data-Guided Constraints For Hyperspectral Unmixing
Risheng Huang, Xiaorun Li and Liaoying Zhao
This paper proposes a nonnegative matrix factorization with data-guided constraints (DGC-NMF) for hyperspectral unmixing where DGC-NMF imposes on the unknown abundance vector of each pixel with either an L1/2 constraint or an L2 constraint to enforce sparseness and evenness, respectively.

09-01224
Joint Local Abundance Sparse Unmixing for Hyperspectral Images
Mia Rizkinia and Masahiro Okuda
This paper proposes propose an algorithm that exploits the low-rank local abundance by applying the nuclear norm to the abundance matrix for local regions of spatial and abundance domains where the local abundance regularizer is collaborated with the L2,1 norm and the total variation for sparsity and spatial information, respectively.

Part II: Spectral Variability (2 papers)

09-00748
Criteria Comparison for Classifying Peatland Vegetation Types Using In Situ Hyperspectral Measurements
Thierry Erudel, Sophie Fabre, Thomas Houet, Florence Mazier and Xavier Briottet
This paper develops various criteria to inventory and evaluating the performance of discrimination techniques for peatland habitats based on in situ spectra. Two main methods are applied for vegetation discrimination: the similarity measurement techniques and the supervised classification methods with sometimes application of a preliminary spectral band reduction technique.

09-00884
Reducing the Effect of the Endmembers' Spectral Variability by Selecting the Optimal Spectral Bands
Omid Ghaffari, Mohammad Javad Valadan Zoej and Mehdi Mokhtarzade
This paper proposes an approach, called Prototype Space (PS) to simultaneously mitigate spectral variability and reduce correlation among different endmembers in hyperspectral datasets. The idea is to utilize the angular discrepancy of bands where each band is treated as a space point and the proposed method is to identify independent bands according to their angles.

Part III: Target Detection (3 papers)

10-00103
Recursive Local Summation of RX Detection for Hyperspectral Image Using SlidingWindows
Liaoying Zhao, Weijun Lin, Yulei Wang and Xiaorun Li

This paper develops a recursive local summation RX anomaly detection approach using sliding windows which can avoid the calculation of historical information and thus speed up the processing. In addition, a background suppression algorithm is also proposed to remove the current under test pixel from the recursively update processing.

10-00434

A Sliding Window-Based Joint Sparse Representation (SWJSR) Method for Hyperspectral Anomaly Detection

Seyyed Reza Soofbaf, Mahmod Reza Sahebi and Barat Mojaradi

This paper develops a new sliding window-based joint sparse representation (SWJSR) anomaly detector for hyperspectral data which can improve the detection probability of anomaly presence in signals using the integration of information gathered during transition of sliding window for each pixel.

00-516

A Deep Pipelined Implementation of Hyperspectral Target Detection Algorithm on FPGA Using HLS

Jie Lei, Yunsong Li, Dongsheng Zhao, Jing Xie, Chein-I Chang, Lingyun Wu, Xuepeng Li, Jintao Zhang and Wenguang Li

This paper uses a deep pipelined background statistics (DPBS) approach to optimizing and implementing a well-known subpixel target detection algorithm, called constrained energy minimization (CEM) on FPGA by using high-level synthesis (HLS) and offers significant benefits in terms of increasing data throughput and improving design efficiency.

Part IV: Hyperspectral Image Classification (6 papers)

09-00872

Multiscale Union Regions Adaptive Sparse Representation for Hyperspectral Image Classification

Fei Tong, Hengjian Tong, Junjun Jiang and Yun Zhang

This paper proposes an approach, called Multiscale Union Regions Adaptive Sparse Representation (MURASR) for hyperspectral image classification to make full use of the advantages of two recentky developed classifiers Multiscale Adaptive Sparse Representation (MASR) classifier and Multiscale Superpixel-Based Sparse Representation (MSSR) classifier and overcome their weakness.

09-00924

Hyperspectral Image Classification Based on Semi-Supervised Rotation Forest

Xiaochen Lu, Junping Zhang, Tong Li and Ye Zhang

This paper develops a semi-supervised rotation forest (SSRoF) algorithm as an improved Rotation forest (RoF) algorithm to take advantage of both the discriminative information and local structural information provided by the limited labeled and massive unlabeled samples, thus providing better class separability for subsequent classifications.

10-00515

Semi-Supervised Classification of Hyperspectral Images Based on Extended Label Propagation and Rolling Guidance Filtering

Binge Cui, Xiaoyun Xie, Siyuan Hao, Jiandi Cui and Yan Lu

This paper proposes a semi-supervised classification method based on extended label propagation (ELP) and a rolling guidance filter (RGF) called ELP-RGF, in which ELP is designed to take advantage of unlabeled samples to improve classification performance.

09-001094

Hashing Based Hierarchical Feature Representation for Hyperspectral Imagery Classification

Bin Pan, Zhenwei Shi, Xia Xu and Yi Yang

This paper combines the advantages of a multiple feature fusion (MFF) and spectral-spatial feature extraction (FE), and further proposes an ensemble based feature representation method for hyperspectral imagery classification, which aims at generating a hierarchical feature representation for the original hyperspectral data.

10-00441

Classification of Hyperspectral Images by SVM Using a Composite Kernel by Employing Spectral, Spatial and Hierarchical Structure Information

Yi Wang and Hexiang Duan

This paper introduce a classification framework for hyperspectral images (HSIs) by jointly employing spectral, spatial, and hierarchical structure information where the three types of information are integrated into the SVM classifier in conjunction with multiple kernels.

10-00396

Hyperspectral Classification Based on Texture Feature Enhancement and Deep Belief Networks

Jiaojiao Li, Bobo Xi, Yunsong Li, Qian Du and Keyan Wang

This paper proposes a hyperspectral classification framework based on an optimal Deep Belief Networks (DBN) and a novel texture feature enhancement (TFE) to enhance classification accuracy.

Part V: Band Selection (2 papers)

10-00113

Band Subset Selection for Hyperspectral Image Classification

Chunyan Yu, Meiping Song and Chein-I Chang

This paper develops a new approach to band subset selection (BSS) for hyperspectral image classification (HSIC) which selects multiple bands simultaneously as a band subset, referred to as simultaneous multiple band selection (SMMBS), rather than one band at a time sequentially, referred to as sequential multiple band selection (SQMBS), as most traditional band selection methods do.

10-00367

Progressive Sample Processing of Band Selection for Hyperspectral Image Transmission

Keng-Hao Liu, Shih-Yu Chen, Hung-Chang Chien and Meng-Han Lu

This paper proposes an online BS method, called progressive sample processing of band selection (PSP-BS) that allows us obtain instant BS results in a progressive manner during HSI data transmission, which is carried out under band-interleaved-by-sample/pixel (BIS/BIP) format.

Part VI: Data Fusion (2 papers)

10-00373

Structure Tensor-Based Algorithm for Hyperspectral and Panchromatic Images Fusion

Jiahui Qu, Jie Lei, Yunsong Li, Wenqian Dong, Zhiyong Zeng and Dunyu Chen

This paper develops a new hyperspectral image fusion algorithm using structure tensor to fuse hyperspectral and panchromatic (PAN) images by merging spectral information of the former and spatial information of the latter.

09-001006

Integration of Absorption Feature Information from Visible to Longwave Infrared Spectral Ranges for Mineral Mapping

Veronika Kopaˇcková and Lucie Koucká

This paper proposes a fusion method to integrate two different airborne image datasets, HyMap (VIS/NIR/SWIR image data) and Airborne Hyperspectral Scanner (AHS, LWIR image data) for data analysis. It demonstrated that when the absorption feature information derived from multispectral LWIR data is integrated with the absorption feature information derived from hyperspectral VIS/NIR/SWIR data, an important improvement in terms of more complex mineral mapping is achieved.

Part VII: Applications (8 papers)

Applications (Forestry: Classification of tree species)
09-00875
Classification of Tree Species in a Diverse African Agroforestry Landscape Using Imaging Spectroscopy and Laser Scanning
Rami Piiroinen, Janne Heiskanen, Eduardo Maeda, Arto Viinikka and Petri Pellikka

This paper studies tree species classification of tree species in a diverse african agroforestry landscape using data collected by imaging spectroscopy and laser scanning at the tree crown level, with primary interest in the exotic tree species for tree species classification.

Applications (Forestry: Detection of newly grown tree leaves)
10-00096
Adaptive Window-Based Constrained Energy Minimization for Detection of Newly Grown Tree Leaves
Shih-Yu Chen, Chinsu Lin, Chia-Hui Tai and Shang-Ju Chuang

This paper applies a hyperspectral subpixel target detection technique, called Constrained Energy Minimization (CEM) along with several its variations, which is to detect of newly grown leaves via a UAV multispectral imager.

Applications (Satellite Calibration)
10-00120
Vicarious Radiometric Calibration of the Hyperspectral Imaging Microsatellites SPARK-01 and -02 over Dunhuang, China
Hao Zhang, Bing Zhang, Zhengchao Chen and Zhihua Huang

This paper develops a procedure to use dark current computation process average multiple lines of long strip imagery acquired over open oceans during nighttime and non-uniform correction process by using images acquired after the adjustment of the satellite yaw angle to 90. This technique was shown to be suitable for large-swath satellite image relative calibration.

Applications (Brain tumor detection in magnetic resonance imaging)
09-01174
A Hyperspectral Imaging Approach to White Matter Hyperintensities Detection in Brain Magnetic Resonance Images
Hsian-Min Chen, Hsin Che Wang, Jyh-Wen Chai, Chi-Chang Clayton Chen, Bai Xue, Lin Wang, Chunyan Yu, Yulei Wang, Meiping Song and Chein-I Chang

This paper applies a nonlinear band expansion (NBE) process is proposed to expand MR images to a hyperspectral image so that the well-known hyperspectral subpixel target detection, called constrained energy minimization (CEM) along with its iterative version of CEM (ICEM) for white matter hyperintensities (WMHs).

Applications (Detection of water stress in vineyards)
10-00202
Modelling Water Stress in a Shiraz Vineyard Using Hyperspectral Imaging and Machine Learning
Kyle Loggenberg, Albert Strever, Berno Greyling and Nitesh Poona

This paper applies two ensemble learners, i.e., random forest (RF) and extreme gradient boosting (XGBoost), for detection of water stress in vineyards and discriminating stressed and non-stressed Shiraz vines using terrestrial hyperspectral imaging.

Applications (Post-fire severity)
10-00389
Evaluating Endmember and Band Selection Techniques for Multiple Endmember Spectral Mixture Analysis using Post-Fire Imaging Spectroscopy
Zachary Tane, Dar Roberts, Sander Veraverbeke, Ángeles Casas, Carlos Ramirez and Susan Ustin

This paper uses imaging spectroscopy data combined with Multiple Endmember Spectral Mixture Analysis (MESMA), a form of spectral mixture analysis that accounts for endmember variability, to map fire severity of the 2013 Rim Fire. To reduce the dimensionality of the imaging spectroscopy data we used uncorrelated Stable Zone Unmixing (uSZU) for band selection.

Applications (On-Board Compression)

10-00428

A New Algorithm for the On-Board Compression of Hyperspectral Images

Raúl Guerra, Yubal Barrios, María Díaz, Lucana Santos, Sebastián López and Roberto Sarmiento

This paper develops a new transform-based lossy compression algorithm, namely Lossy Compression Algorithm for Hyperspectral Image Systems (HyperLCA), which has been shown to achieve high compression ratios with a good compression performance at a reasonable computational burden.

Applications (Hyperspectral Pansharpening)

10-00445

Hyperspectral Pansharpening Based on Intrinsic Image Decomposition and Weighted Least Squares Filter

Wenqian Dong, Song Xiao, Yunsong Li and Jiahui Qu

This paper develops a Component substitution (CS) and multiresolution analysis (MRA)-based hybrid framework based on intrinsic image decomposition and weighted least squares filter for hyperspectral pansharpening.

3. Conclusions

The Guest Editors of this Special Issue would like to thank all authors who have contributed to this volume for publishing their research findings. Our special thanks also go to anonymous reviewers for their hard working and providing their valuable and insightful comments to help the authors improve their paper presentations and quality. Last but not least, we also would like to thank the Remote Sensing editorial team for its support during the tedious process. This volume would not have completed without their help.

Funding: The research of Chein-I Chang was funded by the Fundamental Research Funds for Central Universities under Grant 3132019341. The research of Meiping Song was funded by National Nature Science Foundation of China (61601077) and Fundamental Research Funds for the Central Universities (3132017124). The research of Junping Zhang was funded by National Natural Science Foundation of China (61871150). The research of Chao-Cheng Wu was funded by MOST 108-2321-B-002-025-.

Conflicts of Interest: The Guest editors declare no conflict of interest.

References

1. Chang, C.-I. *Hyperspectral Imaging: Techniques for Spectral Detection and Classification*; Kluwer Academic/Plenum Publishers: New York, NY, USA, 2003.
2. Plaza, A.; Chang, C.-I. (Eds.) *High Performance Computing in Remote Sensing*; Chapman & Hall/CRC Press: Boca Raton, FL, USA, 2007.
3. Chang, C.-I. *Real-Time Progressive Hyperspectral Image Processing: Endmember Finding and Anomaly Detection*; Springer: Berlin/Heidelberg, Germany, 2016.
4. Chang, C.-I. *Real-Time Recursive Hyperspectral Sample and Band Processing: Algorithm Architecture and Implementation*; Springer: Berlin/Heidelberg, Germany, 2017.

5. Chang, C.-I. (Ed.) *Hyperspectral Data Exploitation: Theory and Applications*; John Wiley & Sons: Hoboken, NJ, USA, 2007.
6. Chang, C.-I. *Hyperspectral Data Processing: Algorithm Design and Analysis*; John Wiley & Sons: Hoboken, NJ, USA, 2013.

remote sensing

MDPI

Article

Nonnegative Matrix Factorization With Data-Guided Constraints For Hyperspectral Unmixing

Risheng Huang [1], Xiaorun Li [1] and Liaoying Zhao [2,*]

[1] College of Electrical Engineering, Zhejiang University, No.38, Zheda Road, Xihu District, Hangzhou 310027, China; 119@zju.edu.cn (R.H.); lxr@zju.edu.cn (X.L.)

[2] School of Computer Science, Hangzhou Dianzi University, Hangzhou 310018, China

* Correspondence: zhaoly@hdu.edu.cn

Received: 7 August 2017 ; Accepted: 18 October 2017; Published: 21 October 2017

Abstract: Hyperspectral unmixing aims to estimate a set of endmembers and corresponding abundances in pixels. Nonnegative matrix factorization (NMF) and its extensions with various constraints have been widely applied to hyperspectral unmixing. $L_{1/2}$ and L_2 regularizers can be added to NMF to enforce sparseness and evenness, respectively. In practice, a region in a hyperspectral image may possess different sparsity levels across locations. The problem remains as to how to impose constraints accordingly when the level of sparsity varies. We propose a novel nonnegative matrix factorization with data-guided constraints (DGC-NMF). The DGC-NMF imposes on the unknown abundance vector of each pixel with either an $L_{1/2}$ constraint or an L_2 constraint according to its estimated mixture level. Experiments on the synthetic data and real hyperspectral data validate the proposed algorithm.

Keywords: nonnegative matrix factorization; data-guided constraints; sparseness; evenness

1. Introduction

Hyperspectral data consists of hundreds of contiguous narrow spectral bands and has been widely used in many fields [1]. Due to the limitation of the sensor's spatial resolution, there exist mixed pixels consisting of several material signatures. To address this problem, hyperspectral unmixing (HU) has been adopted to decompose mixed pixels into endmember signatures and their corresponding proportions. According to the availablity of the prior knowledge, HU methods can be divided into three categories: unsupervised [2–5], semisupervised [6], and supervised [7] methods. We can also categorize HU methods into geometric methods and statistical methods. The pixel purity index (PPI) [8], N-FINDR [9], vertex component analysis (VCA) [10] and the simplex growing algorithm (SGA) [11] are the most famous geometric methods. The relationships among these methods are explored in [12–14]. There are also many statistical methods for hyperspectral unmixing [15–17]. Nonnegative matrix factorization (NMF) [18] is a typical statistical method [19]. It has been shown to be promising in extracting sparse and interpretable representations from a data matrix. The NMF decomposes a data matrix into two low-rank matrices with nonnegative constraint [20]. The decomposition results of NMF consist of a basis matrix and a coefficient matrix, which provide an intuitive and interpretable representation of data. As an unsupervised method, NMF is applied to hyperspectral unmixing and shows its advantages in many situations. To reduce the solution space, constraints on endmembers [21–24] and abundances [25,26] have been exploited and used in NMF. Recently, a sparseness constraint has been added to NMF to generate unique solutions and leads to better results [25,27]. The L_1 constraint is a widely-used sparseness constraint. However, L_1 regularization has the limitation that it cannot enforce further sparseness when the abundance sum-to-one constraint is used. The $L_{1/2}$ constraint is representative of $L_p(0 < p < 1)$. The solution of the $L_{1/2}$ regularizer is sparser compared with that of the L_1 regularizer. However, the $L_{1/2}$

regularizer also brings nonconvexity to the optimization problem. The nonconvex optimization problem with the $L_{1/2}$ regularizer can be solved by transforming the $L_{1/2}$ regularizer into a series of convex weighted L_1 regularizers [28]. $L_{1/2}$-NMF is a popular NMF regularization method [29]. The authors have shown that the $L_{1/2}$ regularizer can overcome the limitation of the L_1 regularizer and enforce a sufficiently sparse solution. On the contrary, L_2-NMF generates smooth results rather than sparse results [30]. In [31], piecewise smooth nonsmooth (PSnsNMF) and piecewise smooth NMF with sparseness constraints (PSNMFSC) are proposed by incorporating the piecewise smoothness of spectral data and sparseness of endmember abundances. The authors of [32] propose NMF with sparseness and smoothness constraints (NMFSSC). However, NMFSSC does not consider the sparsity level of data and just imposes sparseness and smoothness constraints simultaneously. In data-guided sparsity-regularized nonnegative matrix factorization (DgS-NMF) [33], the sudden change areas are assumed to be highly mixed and a heuristic method is proposed to employ the spatial similarity to learn the mixed level in the hyperspectral images. The pixel with a higher sparsity level corresponds to a sparser constraint (from the L_1 norm to the L_0 norm). In [34], a learning-based sparsity method is proposed to learn a guidance map from the unmixing results and impose an adaptive $l_p(0 < p \leq 1)$-constraint.

In this paper, we propose a nonnegative matrix factorization with data-guided constraints (DGC-NMF). Unlike traditional constrained NMF methods that impose the same constraint over entire data, DGC-NMF firstly evaluates the sparsity level of each pixels' abundances and then decides which kind of constraint should be imposed on the abundances of a pixel adaptively. In real hyperspectral images, the sparsity levels of the pixels' abundances are varied and the pixels do not necessarily possess spatial dependence with their neighboring pixels. To preserve the distinctive sparsity information of each pixel's abundances, the sparsity levels of pixels can be learnt via an NMF unmixing process without any sparseness constraint imposed. Therefore, each pixel's abundances could enjoy a individual constraint according to its sparsity level in our method. In evenly mixed areas, the sparseness constraint may not contribute to achieving a smooth abundance vector of a pixel. Therefore, we also introduce the L_2 constraint to reduce extreme abundance values and promote the evenness of pixels' abundance vector. Whether an $L_{1/2}$ constraint or an L_2 constraint is imposed on a pixel is learnt from its abundances' sparsity level. In this way, our method could adaptively enforce sparse or smooth abundance results in regions with different mixed levels. The experimental results of synthetic and real data demonstrate the effectiveness of DGC-NMF.

The main contributions of this paper include two aspects. Firstly, we provide a method to evaluate the sparsity level of data in different areas and obtain the sparseness map of data. The effectiveness of this method has been verified using data with various sparsity levels. Secondly, we propose a novel NMF method which makes use of the sparsity information from data and adaptively imposes constraints according to the mixed levels of pixels. We analyze the sparsity behaviors of NMF with different regularizations and indicate that NMF with fixed constraints may be not applicable for a hyperspectral image with various sparsity levels, while it has been proven that the proposed DGC-NMF is capable.

The remainder of this paper is organized as follows. Section 2 gives a brief introduction of the NMF algorithm and the NMF with the $L_{1/2}$ or the L_2 constraint. Section 3 presents the proposed DGC-NMF and provides the proof that the objective decreases along the iterates of the algorithm. Section 4 validates the effectiveness of the proposed method on synthetic data and real hyperspectral images. Finally, Section 5 concludes this paper.

2. Preliminaries

2.1. NMF

Consider the hyperspectral image data $X = [x_1, x_2, \cdots, x_N]$, where $X \in R^{L\times N}$ and N is the number of pixels. In the linear mixing model, the hyperspectral data X could be represented as:

$$X = WH + E \tag{1}$$

where $W = [\mathbf{w}_1, \mathbf{w}_2, \ldots \mathbf{w}_P] \in R^{L\times P}$ denotes the endmember matrix, $H \in R^{P\times N}$ denotes the abundances of respective endmembers, and E is a residual term. The NMF algorithm is designed to find an approximate factorization of X, such that $X \approx WH$, where $W \geq 0$ and $H \geq 0$. To quantify the quality of the approximate factorization, the Euclidean distance is commonly used to measure the distance between X and WH. The loss function of NMF based on the Euclidean distance is defined as follows:

$$f(W, H) = \frac{1}{2} \|X - WH\|_F^2 \tag{2}$$

where $||\cdot||_F$ is the Frobenius norm. The problem of NMF is globally nonconvex. The problem is convex for one of the two blocks of variables only when the other is fixed. Estimating the values of W or H is a convex optimization problem when the other is fixed. A multiplication update rule for standard NMF algorithm is presented in [18] to locally minimize the cost function in (2)

$$W = W. * (XH^T)./WHH^T \tag{3}$$

$$H = H. * (W^T X)./W^T WH \tag{4}$$

where .* and ./ denote element-wise multiplication and division, respectively.

2.2. NMF with Sparseness Constraints

2.2.1. $L_{1/2}$-NMF

Sparsity is an inherent property of hyperspectral data. To reduce the solution space and derive results with expected sparsity levels, some sparseness regularizations are added to constrain the sparseness of abundances. The L_1 regularizer is popular for generating sparse solutions. However, the L_1 regularizer may not enforce a sufficiently sparse solution while preserving the additivity constraint over the abundances since the sum-to-one constraint is a fixed L_1 norm. In [35], Qian et al. propose the $L_{1/2}$-NMF, based on the $L_{1/2}$ regularizer. The $L_{1/2}$ regularizer possesses two advantages over the L_1 regularizer. It can still enforce sparsity with the full additivity constraint imposed. Another advantage is that the $L_{1/2}$ regularizer can obtain sparser solutions than the L_1 regularizer does [36]. The model of NMF with the $L_{1/2}$ regularizer is as follows: [35]

$$f(W, H) = \frac{1}{2} \|X - WH\|_F^2 + \lambda \|H\|_{1/2} \tag{5}$$

where

$$\|H\|_{1/2} = \sum_{p,n=1}^{P,N} H_{pn}^{1/2} \tag{6}$$

and H_{pn} denotes the (p, n)-th element of H.

The objective in (5) is nonincreasing under the multiplicative update rules:

$$W = W. * (XH^T)./WHH^T \tag{7}$$

$$H = H. * (W^T X)./(W^T WH + \frac{\lambda}{2} H^{-\frac{1}{2}}) \quad (8)$$

where $H^{-\frac{1}{2}}$ denotes the reciprocal element-wise square root for each element in H.

2.2.2. L_2-NMF

Due to the needs of application, the L_2 constraint can be adopted to generate smooth results other than sparse results. For areas in hyperspectral images that are evenly mixed with signatures, we also need the L_2 regularizer to promote evenness in the abundances of pixels in these areas. In [30], Pauca et al. explore the use of L_2 regularizer in NMF algorithm. The cost function with L_2 regularization term is expressed as:

$$f(W, H) = \frac{1}{2} \|X - WH\|_F^2 + \mu \|H\|_2 \quad (9)$$

where $\|H\|_2 = \sum\limits_{p,n=1}^{P,N} H_{pn}^2$.

The objective in (9) is nonincreasing under the multiplicative update rules:

$$W = W. * (XH^T)./WHH^T \quad (10)$$

$$H = H. * (W^T X)./(W^T WH + 2\mu H) \quad (11)$$

3. Proposed NMF with Data-Guided Constraints for Hyperspectral Unmixing

3.1. *Sparsity Analysis*

The phenomena of sparsity in abundances commonly exists in hyperspectral images [31]. Sparsity is an inherent property which refers to a representative occasion where mixed pixels could be represented by a few endmember signatures. Accordingly, sparseness constraints such as the $L_{1/2}$ regularizer help to obtain unique solutions and lead to better answers in scenes with obvious sparsity. However, in hyperspectral images there exist pixels located in transition regions which are evenly mixed and own low sparsity levels. Imposing a sparseness constraint over the entire image may not contribute to the unmixing accuracy of those evenly mixed pixels. Therefore, we also adopt an L_2 regularizer to promote the evenness of pixels' abundance vectors, achieving an effect on abundances opposite to that of the $L_{1/2}$ regularizer. Through imposing the L_2 regularizer on a pixel's abundance vector, extreme abundance values are reduced and the sparseness level of abundances tends to be lower. In our method, each pixel enjoys a individual constraint related to its own sparsity level of abundance. Figure 1 represents the well known Cuprite dataset collected by an airborne visible/infrared imaging spectrometer (AVIRIS) sensor over the Cuprite mining site and the corresponding sparseness map of this scene. To evaluate the sparsity levels of pixels, the sparsity level of the nth pixel's abundances is defined as [37]

$$sparseness(\boldsymbol{H}_n) = \frac{\sqrt{P} - (\sum_1^P |H_{pn}|)/\sqrt{\sum_1^P H_{pn}^2}}{\sqrt{P} - 1} \quad (12)$$

where $\boldsymbol{H}_n$ denotes the abundance vector of nth pixel, P denotes the number of endmembers, and H_{pn} denotes the $(p, n)^{\text{th}}$ element of H. As shown in Figure 1b, some regions mainly composed by one or a few materials possess high sparsity levels, while some other regions show low sparsity levels where minerals are evenly mixed there. The estimated sparsity levels of pixels range from 0.14 to 1. For hyperspectral data consisting of regions with various sparsity levels, using a simple kind of constraint on the whole image does not meet the practical situation and may not lead to a well-defined result.

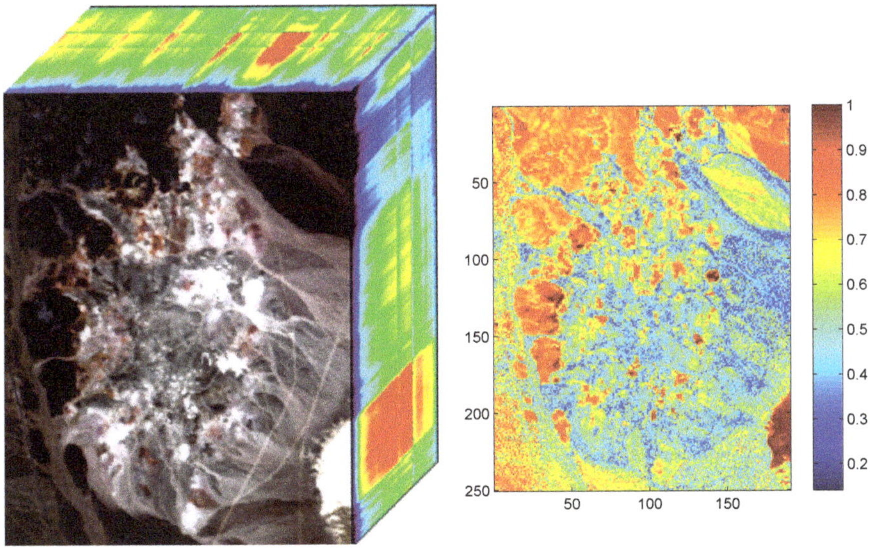

(**a**) (**b**)

Figure 1. (**a**) Airborne visible/infrared imaging spectrometer (AVIRIS) hyperspectral data of the Cuprite mining district in Nevada, USA; (**b**) Estimated sparseness map from the obtained abundance.

To solve this problem, we propose the DGC-NMF algorithm which is designed to impose constraints precisely according to the data's sparsity levels in different regions. However, the sparsity levels of abundances are previously unknown since the ground truth of abundance is not available. In the proposed DGC-NMF algorithm, we firstly carry out an unmixing process based on NMF with no constraint to derive the sparseness map of data. No sparseness constraint is employed in this unmixing process to avoid the distinctive sparsity information of a pixel's abundances being interfered with by a sparseness constraint without verification. This method of estimating sparseness maps may be a biased way. However, it is still a good choice for estimating sparsity levels since the ground truth of real hyperspectral data is not available and a small sparseness error is tolerable in our proposed DGC-NMF algorithm. To demonstrate the accuracy of the sparseness estimation, experiments are conducted to make comparison between estimated and real sparsity levels of data. Figure 2 shows that the estimated sparsity levels fit the real sparsity levels well under various sparsity levels. The estimated sparseness values could correctly reflect the general trend of the real sparsity levels. Figure 3 presents the real and the estimated sparseness map of synthetic data. The estimated map coincides with the real map well. For regions possessing high or low sparsity levels, the estimate sparseness map also shows high or low values. The estimated sparseness values can represent the real sparsity levels of pixels well. We also conduct experiments in Section 4.1 to compare DGC-NMF with the real sparseness map and DGC-NMF with the estimated sparseness map. The results also validate that it is practical to estimate sparsity levels via the unmixing result of NMF algorithm.

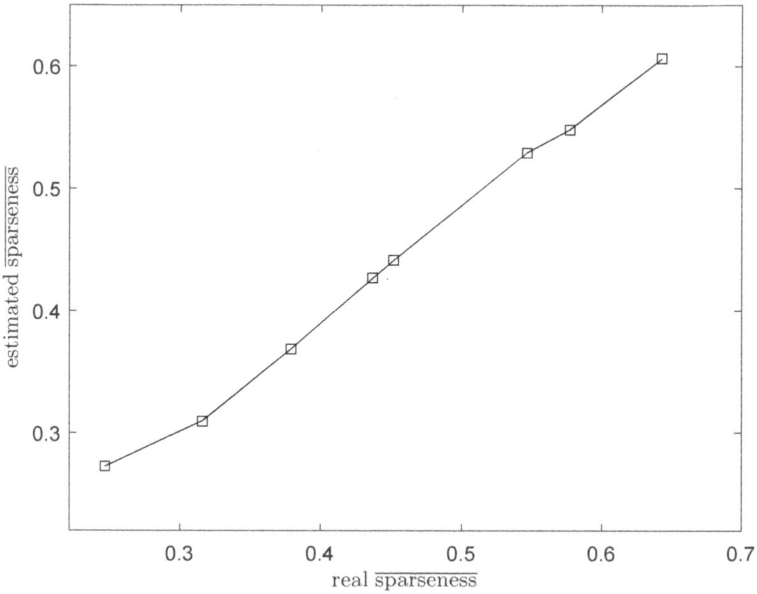

Figure 2. The real average sparsity level and the estimated average sparsity level.

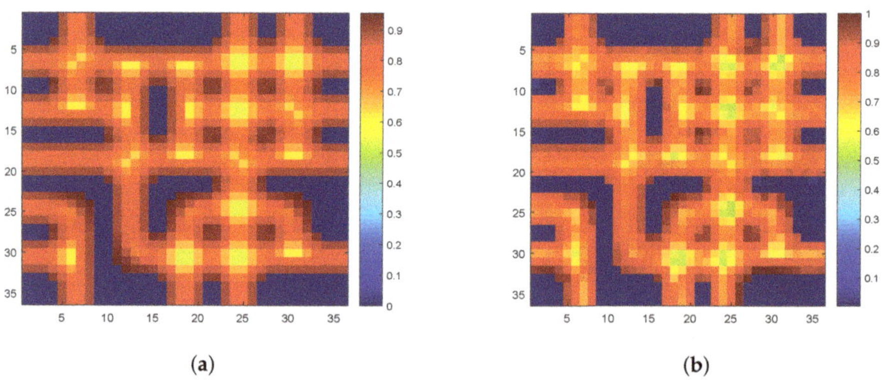

Figure 3. (**a**) The real sparseness map of synthetic data; (**b**) The estimated sparseness map of synthetic data.

3.2. DGC-NMF Algorithm

Using the sparsity information learnt in Section 3.1, the proposed NMF with data-guided constraints decides whether the $L_{1/2}$ constraint or the L_2 constraint should be assigned to a pixel. In the DGC-NMF algorithm, both the $L_{1/2}$ regularizer and the L_2 regularizer are adopted to achieve better control of sparsity in each pixel's abundances. Pixels are split into two categories according to sparsity levels. For pixels with high sparsity levels, the $L_{1/2}$ regularizer is adopted to constrain their abundance. For pixels with low sparsity levels, the L_2 regularizer is applied. The model of DGC-NMF is as follows:

$$f(W,H) = \frac{1}{2}\|X - WH\|_F^2 + \lambda\|C.*H\|_{1/2} + \mu\|D.*H\|_2 \tag{13}$$

where λ and μ are the regularization parameter, and $C = \mathbf{1}_P \mathbf{s}^T$ and $D = \mathbf{1}_P(\mathbf{1}_N - \mathbf{s})^{\mathrm{T}}$ are the indictor matrices which decide whether an $L_{1/2}$ constraint constraint or an L_2 constraint is imposed or not for each pixel. $\mathbf{s} \in \mathbb{R}_+^N$ is obtained by evaluating the sparse levels of abundances of all pixels

$$\mathbf{s}(n) = \begin{cases} 1 & sparseness(n) > \delta \\ 0 & sparseness(n) \le \delta \end{cases} \tag{14}$$

where δ is a threshold that controls which kind of constraint should be imposed. The threshold δ is decided by applying Otsu's method to maximize the separability of pixels with high sparsity level and pixels with low sparsity level [38]. Figure 4 shows the histogram of estimated sparseness values for pixels in a synthetic image and the selected value of threshold δ for this synthetic image. The sparseness histogram is obtained by counting the sparseness of estimated abundance of pixels. When the sparseness of a pixel's abundance is higher than δ, the pixel's abundance will be constrained by an $L_{1/2}$ sparsity regularization. Otherwise, the pixel's abundance will enjoy an L_2 constraint to promote evenness.

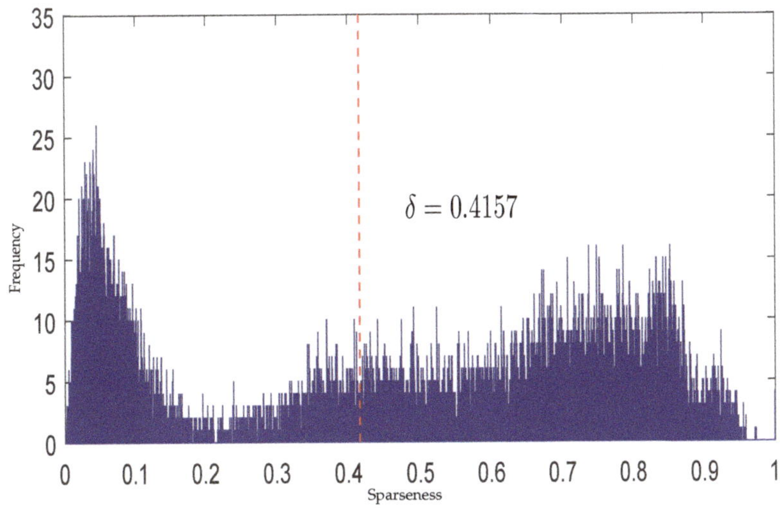

Figure 4. The histogram of estimated sparseness values for pixels in a synthetic image and the selected threshold value.

Based on the cost function in Equation (13), the update rules are derived as follows

$$W = W. * (XH^T)./WHH^T \tag{15}$$

$$H = H. * (W^T X)./(W^T WH + \frac{\lambda}{2} C. * H^{-\frac{1}{2}} + 2\mu D. * H) \tag{16}$$

The procedure of the proposed DGC-NMF is described in Algorithm 1.

Algorithm 1 DGC-NMF algorithm

Input: Hyperspectral data $X \in R^{L \times N}$; the number of endmembers P.
Initialization: Initialize endmember matrix W_1 and abundance matrix H_1 by SGA-FCLS.

repeat
  update W_1 by Equation (3).
  update H_1 by Equation (4).
until convergence
Calculate the sparseness map using H_1 by Equation (12).
Calculate the threshold δ and get the indictor matrices C and D by Equation (14)
Initialize endmember matrix W_2, abundance matrix H_2 by SGA-FCLS.
repeat
  update W_2 by Equation (15).
  update H_2 by Equation (16).
until convergence
return Endmembers matrix W_2 and abundance matrix H_2 as the final unmixing results.

The update rule for W in Equation (15) is just the same as that in [18]. The authors of [18] have proved objective (2) is nonincreasing under the update rule in Equation (3). Therefore, we only need to focus on proving objective (13) is nonincreasing under the update rule for H in Equation (16).

Theorem 1. *The objective (13) is nonincreasing under the update rule in (16).*

Since the objective function in Equation (13) is separable by columns, for each column of H, we could consider each column of H individually. For convenience, let h denote a column of H, x denotes the corresponding columns in X, and c, d denote the corresponding column in C, D, respectively. c and d are vectors with all ones or zeros. The objective function by column is expressed as follows:

$$F(h) = \frac{1}{2} \|x - Wh\|_2^2 + \lambda \|c. * h\|_{\frac{1}{2}} + \mu \|d. * h\|_2 \tag{17}$$

An auxiliary function similar to that used in the expectation-maximization algorithm is defined to prove Theorem 1 [39,40].

Definition 1. *$G(h, h')$ is an auxiliary function of $F(h)$ with*

$$G(h, h') \geq F(h), \quad G(h, h) = F(h) \tag{18}$$

satisfied

Lemma 1. *If $G(h, h')$ is an auxiliary function of $F(h)$, $F(h)$ is nonincreasing under the update*

$$h^{t+1} = \arg\min_h G(h, h^t) \tag{19}$$

Proof.

$$F(h^{t+1}) \leq G(h^{t+1}, h^t) \leq G(h^t, h^t) = F(h^t)$$

□

Following [29], we define the function G as:

$$G(h, h^t) = F(h, h^t) + (h - h^t)^T \nabla F(h^t) + \frac{1}{2}(h - h^t)^T K(h^t)(h - h^t) \tag{20}$$

where $K(h^t)$ is a diagonal matrix with diagonal k

$$k = (W^T W h^T + \frac{\lambda}{2} c. * (h^t)^{-\frac{1}{2}} + 2\mu d. * h^t)./h^t \tag{21}$$

Obviously, the second property of G defined in Definition 1 is satisfied. Writing out the Taylor expansion of $F(h)$

$$\begin{aligned} F(h) &= F(h^t) + (h - h^t)^T \nabla F(h^t) \\ &+ \frac{1}{2}(h - h^t)^T [W^T W - \frac{\lambda}{4}\mathrm{diag}(c. * (h^t)^{-\frac{3}{2}}) + 2\mu \mathrm{diag}(d)](h - h^t) \\ &+ R(\nabla^{(n \geq 3)} F(h^t)) \end{aligned}$$

where the function R denotes the Lagrange remainder term, which can be omitted.

Comparing $F(h)$ with $G(h, h^t)$ in Equation (20), we find the first property $G(h, h^t) \geq F(h)$ is satisfied when

$$0 \leq (h - h^t)^T [K(h^t) - W^T W + \frac{\lambda}{4}\mathrm{diag}(c. * (h^t)^{-\frac{3}{2}}) - 2\mu \mathrm{diag}(d)](h - h^t) \tag{22}$$

Equivalent to

$$0 \leq (h - h^t)^T [K'(h^t) + \frac{3\lambda}{4}\mathrm{diag}(c.*(h^t)^{-\frac{3}{2}})](h - h^t) \tag{23}$$

where K' is is a diagonal matrix with diagonal k'

$$k' = (W^T W h^T./h^t) - W^T W \tag{24}$$

The positive semidefiniteness of K' has been proved in [18]. Another term in Equation (23) is nonnegative since c and h both are nonnegative. Thus, Equation (22) holds due to the sum of two positive semidefinite matrices is also positive semidefinite.

It remains to select the minimum of G by taking the gradient and equating to zero

$$\nabla_s G(h, h^t) = W^T(W h^t - x) + \frac{\lambda}{2} c. * (h^t)^{-\frac{1}{2}} + 2\mu d. * h^t + K(h^t)(h - h^t) = 0 \tag{25}$$

Solving h gets h^{t+1}

$$\begin{aligned} h^{t+1} &= h^t - K(h^t)^{-1}(W^T(W h^t - x) + \frac{\lambda}{2} c. * (h^t)^{-\frac{1}{2}} + 2\mu d. * h^t) \\ &= h^t - h^t./(W^T W h^T + \frac{\lambda}{2} c. * (h^t)^{-\frac{1}{2}} + 2\mu d. * h^t) \\ &. * (W^T(W h^t - x) + \frac{\lambda}{2} c. * (h^t)^{-\frac{1}{2}} + 2\mu d. * h^t) \\ &= h^t. * (W^T x)./(W^T W h^T + \frac{\lambda}{2} c. * (h^t)^{-\frac{1}{2}} + 2\mu d. * h^t) \end{aligned} \tag{26}$$

which is the desired columnwise form of update rule in Equation (16). The proof of Theorem 1 is completed.

4. Experimental Results and Analysis

4.1. Experiments on Synthetic Data

In this section, the proposed DGC-NMF algorithm is tested on synthetic data to evaluate its performance. Three related methods, including NMF [20], $L_{1/2}$-NMF [35] and L_2-NMF [41] are used for comparison with the proposed method. The synthetic data used to test is generated following [29]. The spectral signatures are randomly selected from the United States Geological Survey (USGS) digital spectral library to simulate synthetic images [42]. The abundances are generated as follows. Firstly, a $z^2 \times z^2$ size image is divided into $z \times z$ regions. Each region is initialized with the same kind of ground material. Secondly, a $(z+1) \times (z+1)$ low-pass filter is applied to generated mixed pixels and make the abundance variation smooth. Finally, a threshold θ $(0 < \theta \leq 1)$ is used to reject pixels with high purity. The pixels with abundance larger than θ will be replaced by mixtures of all endmembers with equal abundance. θ can be used as the parameter to generate synthetic data with various sparseness levels. In addition, zero-mean white Gaussian noise is added into the synthetic data to simulate possible noise. In the experiments on synthetic data and real data, DGC-NMF and the compared NMF based algorithms are all initialized using SGA-FCLS. SGA-FCLS provides a more accurate initialization than random initialization. We also compared the unmixing performance of our proposed DGC-NMF with that of SGA-FCLS in Section 4.2.

Two criteria, spectral angle distance (SAD) and root-mean-square error (RMSE), are adopted to evaluate the unmixing performance of algorithms. They are defined as follows:

$$\mathrm{SAD}_p = \cos^{-1}\left(\frac{\mathbf{w}_i^T \hat{\mathbf{w}}_i}{\|\mathbf{w}_i\| \, \|\hat{\mathbf{w}}_i\|}\right) \tag{27}$$

$$\mathrm{RMSE}_p = \sqrt{\frac{1}{N}\sum_{n=1}^{N} (h_{pn} - \hat{h}_{pn})^2} \tag{28}$$

where $\mathbf{w}_i$ and $\hat{\mathbf{w}}_i$ are the reference endmember signatures and their estimates. Respectively, h_{pn} and $\hat{h}_{pn}$ are the reference and estimated abundances. Before calculating evaluation criteria, the estimated endmembers should firstly be reordered to match the reference endmembers. The estimated abundances should also be reordered respectively.

To present the effects of algorithms on the sparseness of unmixing results intuitively, we compare the sparseness histograms of different algorithms in Figure 5. The histogram are obtained by counting the sparseness levels of pixels' abundance estimated by different algorithms. From the histogram in Figure 5b, it can be seen that the abundance result achieved by L_2-NMF generally tends to be smoother. The histogram of L_2-NMF owns more pixels with low sparseness levels compared to other algorithms. In Figure 5c, the whole histogram of $L_{1/2}$-NMF has the tendency of a right shift, which demonstrates that $L_{1/2}$-NMF can effectively promote sparsity in the unmixing process. The sparseness of pixels' abundances will be raised when applying $L_{1/2}$-NMF. The pixels with various sparseness are not able to receive constraints accommodated to their sparsity levels in L_2-NMF and $L_{1/2}$-NMF. For the histogram of DGC-NMF in Figure 5d, the right part of the histogram has the tendency towards a right shift and the left part has a tendency towards a left shift. This validates that the DGC-NMF algorithm can impose adaptive constraints on pixels according to their sparseness of abundances. Figure 6 shows the abundance maps of NMF, $L_{1/2}$-NMF, L_2-NMF, and DGC-NMF, respectively, when applied on synthetic data. It can be seen that DGC-NMF achieves sparser abundance results than NMF and L_2-NMF in areas possessing high sparsity levels. Meanwhile, DGC-NMF obtains more accurate abundance results than $L_{1/2}$-NMF in evenly mixed areas.

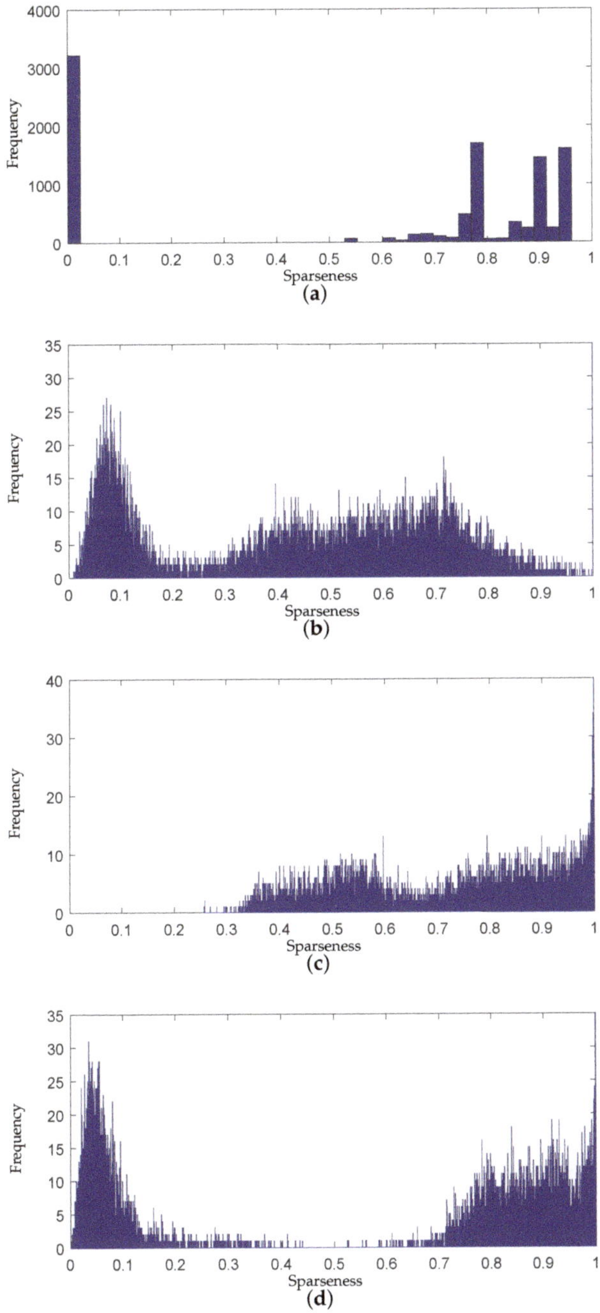

Figure 5. Comparison of sparseness histograms for true abundances different algorithms' estimated abundance. (**a**) Ground truth; (**b**) L_2-NMF; (**c**) $L_{1/2}$-NMF; (**d**) DGC-NMF. NMF: nonnegative matrix factorization; DGC-NMF: NMF with data-guided constraints.

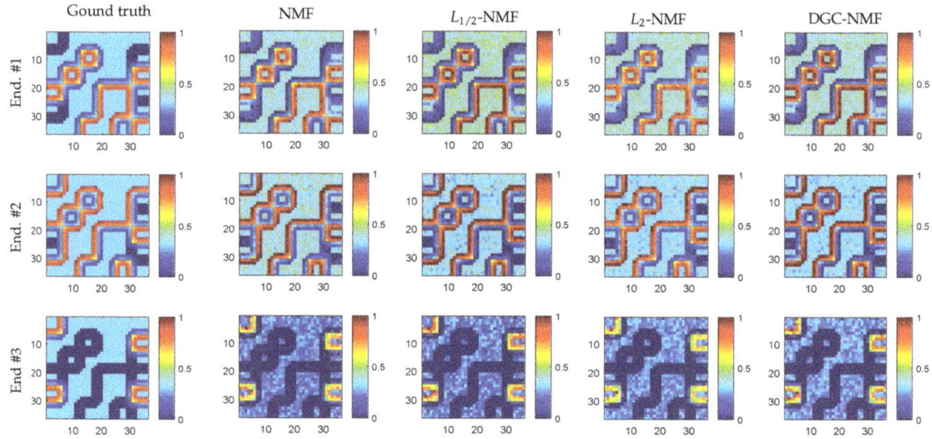

Figure 6. Abundance maps of synthetic data estimated by NMF, $L_{1/2}$-NMF, L_2-NMF, and DGC-NMF, respectively. Each row shows the corresponding abundance maps of a same endmember by different algorithms.

Due to the first unmixing process for learning sparseness maps from data, the proposed DGC-NMF is more computationally expensive than L_2-NMF and $L_{1/2}$-NMF. However, it is still in the same order of magnitude as L_2-NMF and $L_{1/2}$-NMF. Table 1 shows the running time of different algorithms on a 100×100 size synthetic image. For each algorithm, 20 independent runs are carried out and the results are averaged. All experiments are performed using a laptop PC with an Intel Core I7 CPU and 8 GB of RAM. The iteration number of the two steps in DGC-NMF is set as 200. The iteration number of comparative algorithms is also set as 200.

Table 1. Comparison of the time cost of different algorithms.

NMF	L_2-NMF	$L_{1/2}$-NMF	DGC-NMF
7.61 s	7.68 s	8.66 s	18.80 s

To further analyze the performance of algorithms, five experiments are conducted with respect to the following: (1) sparseness; (2) size of image; (3) number of endmembers; and (4) the signal-to-noise ratio (SNR). For each experiment, 20 independent runs are carried out and the results are averaged. Considering DGC-NMF has the same parameters λ and μ as L_2-NMF and $L_{1/2}$-NMF, we set λ and μ in DGC-NMF to the same values as those of $L_{1/2}$-NMF and L_2-NMF to make fair comparisons. The values of parameters λ and μ for $L_{1/2}$-NMF and L_2-NMF, respectively, are carefully determined to achieve best results as in [33]. DGC-NMF adopts the same values of parameters to validate the effectiveness.

Experiment 1: In this experiment, we investigate the performance of algorithms under various sparsity levels. Since the real abundance maps of synthetic data are available, we also make comparison between DGC-NMF with a real sparseness map and DGC-NMF with an estimated sparseness map. The algorithms are tested on synthetic data with different average sparseness levels of abundances. The size of data used here and in the following experiments is 100×100, except in Experiment 2. The endmember number K = 6 and SNR = 20 dB. Figure 7 shows that the proposed DGC-NMF performs the best at various sparseness levels. The DGC-NMF with estimated sparseness map performs quite closely with the DGC-NMF with the real sparseness map, which proves the effectiveness of the proposed method for estimating the sparsity levels of pixels' abundances. For SAD, DGC-NMF performs the best, while L_2-NMF has the poorest performance. With the sparseness level rises to 0.6,

$L_{1/2}$-NMF achieves better performance than L_2-NMF and NMF, while still being inferior to DGC-NMF. Considering RMSE, DGC-NMF also achieves the best performance under different sparseness levels. L_2-NMF achieves more accurate results than $L_{1/2}$-NMF when applied to data with relatively low sparsity levels.

Experiment 2: The algorithms are also tested on synthetic data with different sizes to validate the performance. The image size is set as $36 \times 36, 49 \times 49, ..., 144 \times 144$, respectively, with $K = 6$, and SNR = 20 dB. In this experiment and following experiments, the threshold θ is set as 0.91. Figure 8 shows that the proposed DGC-NMF achieves best results for either SAD or RMSE when applied to different sizes of images. For larger images, $L_{1/2}$-NMF and L_2-NMF may not obtain a satisfactory result since the images consist of areas with various sparsity levels and a simple constraint is not applicable. The proposed method provides a reliable way for images possessing areas with various sparsity levels and requiring adaptive constraints.

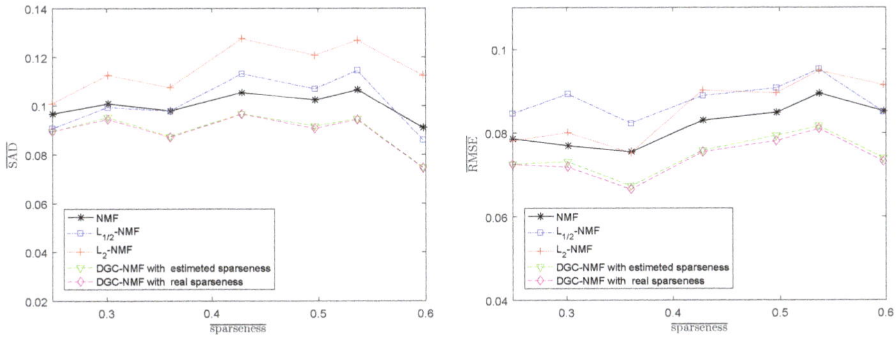

(a) (b)

Figure 7. Performance comparison of the algorithms when sparseness level of abundance varies. (**a**) spectral angle distance (SAD); (**b**) root-mean-square error (RMSE).

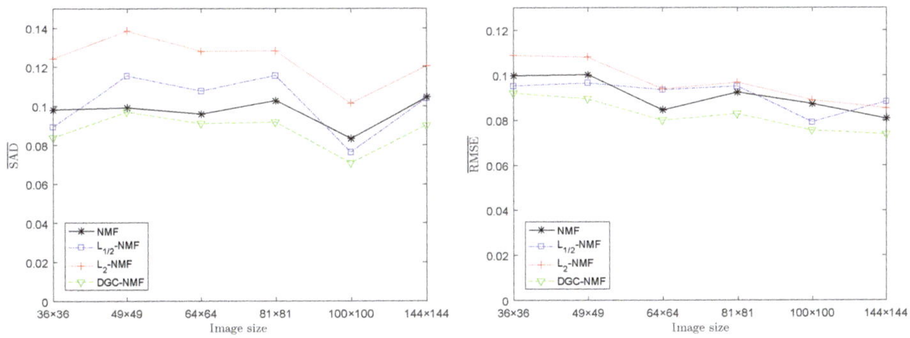

(a) (b)

Figure 8. Performance comparison of the algorithms with respect to the different sizes of images. (**a**) SAD; (**b**) RMSE.

Experiment 3: The algorithms' performance when the number of endmembers changes is presented in Figure 9a,b. The number of endmembers is set from 4 to 8 and the SNR is also set as 20 dB. Generally, DGC-NMF performs the best while L_2-NMF performs the worst when the number of endmembers varies. For SAD, DGC-NMF still gains the best results, while L_2-NMF performs the worst and $L_{1/2}$-NMF and NMF have similar performance. From Figure 9b, we can see that DGC-NMF also achieves the lowest RMSE values when applied to data with different number of endmembers.

Experiment 4: To test the robustness of the proposed method, synthetic data with different noise levels are used to examine the performance of algorithms. We change the SNR of synthetic data from 10 dB to 30 dB at the steps of 5 dB. With the increase of noise level, the performance of algorithms degrades as expected. The DGC-NMF shows the best performance as the SNR varies. For SAD, $L_{1/2}$-NMF yields better results than L_2-NMF and NMF when SNR = 20. For RMSE, NMF is better than $L_{1/2}$-NMF and L_2-NMF. It can be seen from the Figure 10 that the proposed DGC-NMF is not sensitive to noise compared to other three algorithms.

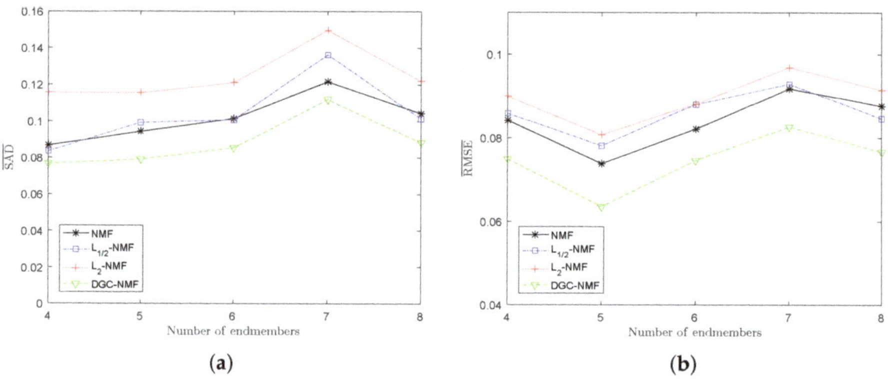

Figure 9. Performance comparison of the algorithms when the number of endmembers varies. (**a**) SAD; (**b**) RMSE.

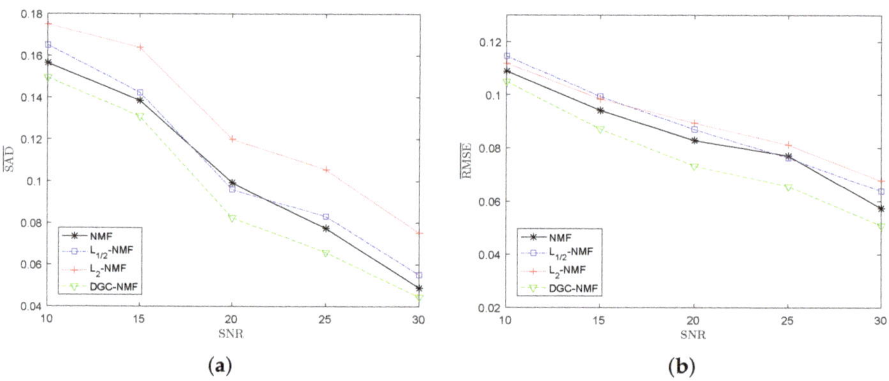

Figure 10. Performance comparison of the algorithms under various noise levels. (**a**) SAD; (**b**) RMSE.

4.2. Experiments on Real Data

In this section, we present the experimental results of the proposed method on real hyperspectral data. Two hyperspectral datasets which include regions with different sparsity levels in an urban scene and a regional mineral scene are used in the experiments. To verify the performance of the proposed method, the results of DGC-NMF are compared with NMF [20], $L_{1/2}$-NMF [35], and L_2-NMF [41]. VCA-FCLS and SGA-FCLS are also adopted to compare with the proposed method. The dimensionality reduction (DR) method adopted for SGA here is principal component analysis (PCA) [43]. The initial condition for SGA in this paper is set as starting with two endmembers with maximal segment produced by the one-dimensional two-vertex simplex with maximal distance. The experiment for VCA-FCLS is repeated 10 times. The results are averaged values and the standard deviations are taken. Since the result of SGA is consistent, there is no standard deviation reported for SGA-FCLS and NMF-based methods. The standard deviation of VCA-FCLS comes from the randomness of VCA.

The first hyperspectral scene to be used is the urban dataset collected by a Hyperspectral Digital Imagery Collection Experiment (HYDICE) sensor over an area located at Copperas Cove near Fort Hood, TX, U.S., in October 1995. The spectral and spatial resolutions are 10 nm and 2 m, respectively. After the bands with low SNR are removed from the original dataset, only 162 bands remain in the experiment (i.e., L = 162). The image is 307 × 307 pixels in size and consists of a suburban residential area as shown in Figure 11a. There are four targets of interest existing in this area: asphalt, grass, roofs, and trees. Since the ground truth of Urban dataset is not available. We use the reference abundance maps obtained from [44] to evaluate the algorithms' performance. Two criteria, spectral angle distance (SAD) and root-mean-square error (RMSE), are adopted to evaluate the accuracy of estimated endmembers and abundances, respectively.

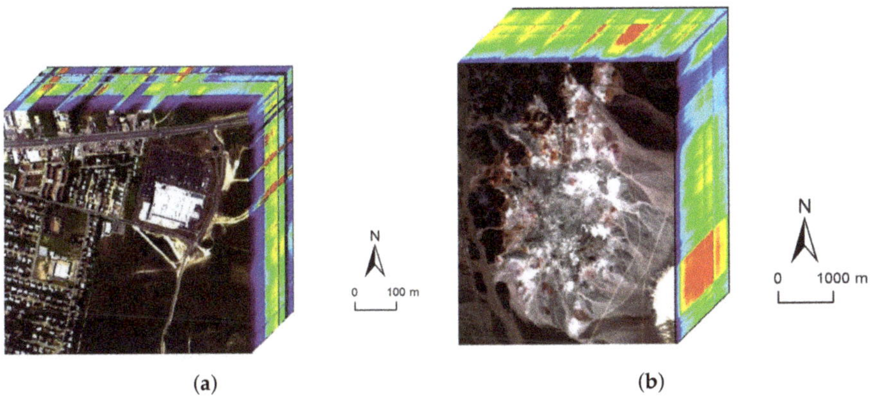

Figure 11. The two real hyperspectral data used in the experiments. (**a**) The Hyperspectral Digital Imagery Collection Experiment (HYDICE) urban dataset; (**b**) The airborne visible/infrared imaging spectrometer (AVIRIS) Cuprite dataset.

Table 2 represents the mean values and standard deviations of SAD of different methods on urban data. The rows respectively show the results of four targets of interest, i.e., 'asphalt', 'grass', 'trees' and 'roofs', along with the mean values. From the Table 2, it can be seen that the SAD results achieved by DGC-NMF are better than those of other methods in general. For target 'roofs' and the mean value, the proposed DGC-NMF achieves the best results. For 'asphalt' and 'trees', DGC-NMF achieves the second best result. The RMSE results of algorithms are illustrated in Table 3. We can also find that the DGC-NMF's results are generally better than those yielded by the other algorithms. For 'asphalt', 'grass' and the mean value, DGC-NMF achieves the best results. For 'trees' and 'roofs', DGC-NMF achieves the second best results.

In Figure 12, the endmember signatures obtained by different methods are displayed with reference to the ground truth for visual comparison. It is shown that the endmember signatures obtained by DGC-NMF are in good accordance with the ground truth. Figure 13 shows the sparseness maps of abundance results obtained by L_2-NMF, $L_{1/2}$-NMF, and DGC-NMF, respectively. It can be seen that the sparseness values in the map of L_2-NMF are low as a whole, while those of $L_{1/2}$-NMF show relatively high levels. For the proposed DGC-NMF, the sparseness values are in better accordance with the distribution of ground covers in hyperspectral data. In high sparsity level areas such as the areas composed of asphalt, DGC-NMF acts in a similar manner to $L_{1/2}$-NMF. In these areas, DGC-NMF promotes $L_{1/2}$ constraint and obtains sparser abundance results. In areas with low sparsity levels that are evenly mixed with signatures such as the areas with both trees and grass, DGC-NMF promotes L_2 constraint adaptively and obtains smoother abundance results of pixels. Therefore, the sparseness values are lower than those of $L_{1/2}$-NMF, similar to L_2-NMF. Figure 14 shows the separated abundance maps of each endmember by VCA-FCLS, SGA-FCLS, NMF, L_2-NMF, $L_{1/2}$-NMF, and DGC-NMF, respectively. As shown in the figure, all algorithms separate out the four targets successfully. Through visual comparison, we can see that L_2-NMF and $L_{1/2}$-NMF obtain smoother and sparser results than NMF, respectively. $L_{1/2}$-NMF achieves great results in high sparsity level areas, but fails to capture mixed information in evenly mixed areas. The proposed DGC-NMF achieves sparser abundance maps than L_2-NMF, and has better abundance estimation than $L_{1/2}$-NMF in transition regions. Generally, Figures 13 and 14 demonstrate that the proposed DGC-NMF could promote adaptive constraints on areas in hyperspectral images with various sparsity levels and achieve better unmixing results of abundance.

Table 2. The spectral angle distance and their standard deviations of algorithms on the urban dataset. Numbers in bold and red color represent the best results, numbers in bold and blue color represent the second-best results. FCLS: full constrained least squares.

Endmember	Spectral Angle Distance (10^{-2})					
	VCA-FCLS	**SGA-FCLS**	**NMF**	**L_2-NMF**	**$L_{1/2}$-NMF**	**DGC-NMF**
Asphalt	21.04 ± 3.64	**13.16**	32.33	23.04	96.14	**20.82**
Grass	**36.95 ± 0.28**	109.21	124.92	81.87	**36.06**	51.53
Trees	28.38 ± 7.78	**7.43**	10.19	15.93	12.70	**10.07**
Roofs	77.01 ± 0.07	**21.74**	39.54	138.98	38.94	**6.20**
Mean	40.84 ± 2.87	**37.89**	51.75	64.96	45.96	**22.16**

Table 3. RMSEs and their standard derivations of algorithms on the urban dataset.

Endmember	Root Mean Square Error (10^{-2})					
	VCA-FCLS	**SGA-FCLS**	**NMF**	**L_2-NMF**	**$L_{1/2}$-NMF**	**DGC-NMF**
Asphalt	42.42 ± 12.41	30.63	**23.68**	32.23	41.79	**20.72**
Grass	47.46 ± 1.23	47.19	**39.00**	48.24	50.00	**36.57**
Trees	26.92 ± 11.79	26.96	23.05	**19.36**	27.66	**21.23**
Roofs	18.33 ± 2.00	19.40	20.30	24.18	**8.84**	**15.08**
Mean	33.78 ± 6.86	31.05	**26.51**	31.00	32.07	**23.40**

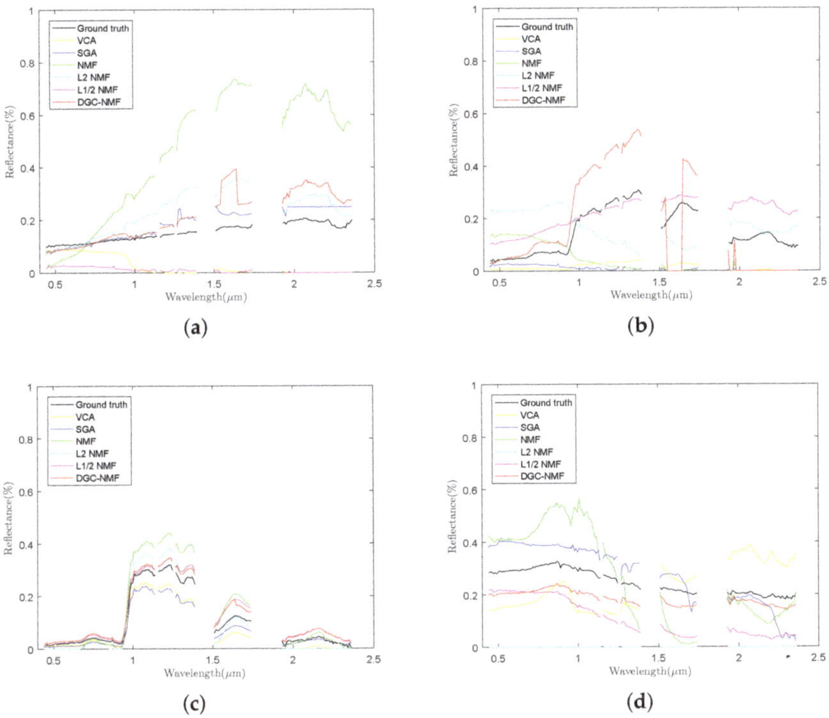

Figure 12. Comparison of endmember signatures estimated by different methods over urban data. (**a**) asphalt; (**b**) grass; (**c**) trees; (**c**) roofs. VCA: vertex component analysis; SGA: simplex growing algorithm.

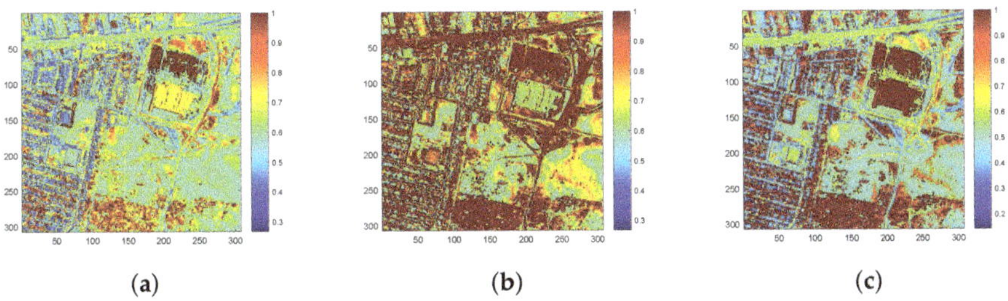

Figure 13. The sparseness maps of abundance results obtained by different algorithms. (**a**) L_2-NMF; (**b**) $L_{1/2}$-NMF; (**c**) DGC-NMF.

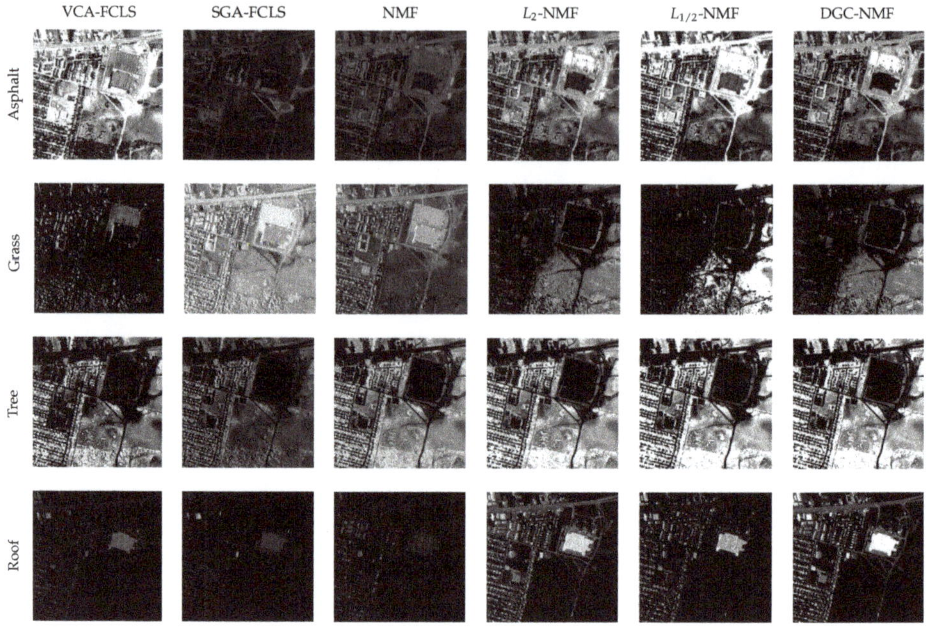

Figure 14. Abundance maps of urban data estimated by VCA-FCLS, SGA-FCLS, NMF, L_2-NMF, $L_{1/2}$-NMF, and DGC-NMF, respectively, from right column to left column. Each row shows the corresponding abundance maps of a same endmember.

To validate the performance of our proposed method on hyperspectral data with various sparsity levels, we also conduct an experiment on the Cuprite data. The well known Cuprite dataset is collected by an airborne visible/infrared imaging spectrometer (AVIRIS) sensor over Cuprite mining site, Nevada. The raw images have 224 spectral bands covering the wavelength ranging from 0.4 µm to 2.5 µm. The spatial resolution is 20 m and the spectral resolution is 10 nm. Approximate distributions of the minerals have been illustrated in many pieces of research [10,22,26]. The image used in our experiment is a 250 $\times$ 190 pixel subset of the Cuprite scene, as shown in Figure 11b. Due to the water absorption and low SNR, several bands are removed, including bands 1–2, 104–113, 148–167, and 221–224. Hence, 188 bands are used in the experiment. According to [10], there are 14 kinds of minerals existing in the scene. However, the variants of the same mineral have minor differences between each other and could be considered as the same endmember. Therefore, we set the number of endmembers in the scene to 12 [25,33]. Figure 15 presents the extracted endmembers and their corresponding abundance maps by DGC-NMF. In the figure, the extracted signatures are compared with USGS library spectra and show good accordance with them. Table 4 presents the SAD results of the proposed DGC-NMF, along with those of other methods. It shows that DGC-NMF achieves the greatest number of cases of best SAD results, outperforming NMF, $L_{1/2}$-NMF, and L_2-NMF. $L_{1/2}$-NMF obtains the most second-best SAD results of endmembers. In the terms of mean value, DGC-NMF performs the best.

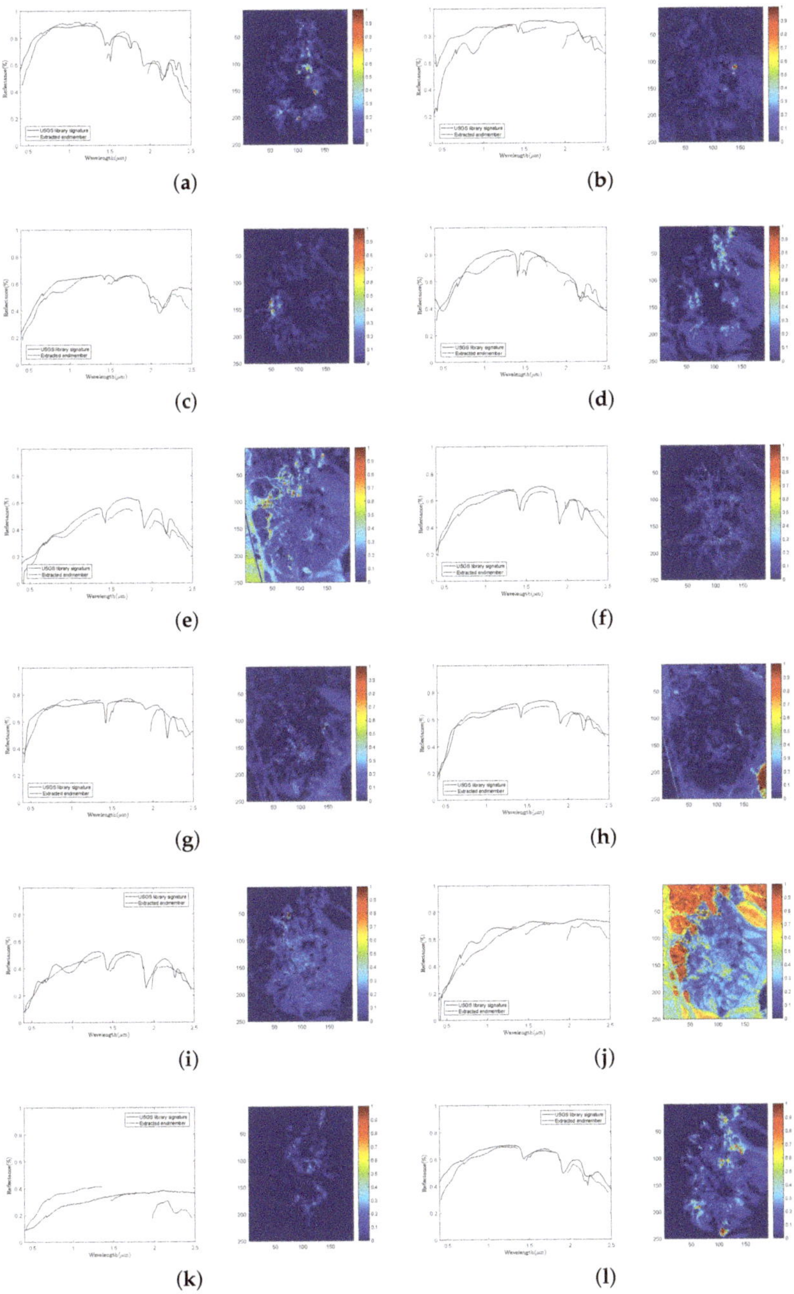

Figure 15. The extracted endmembers by DGC-NMF and their corresponding United States Geological Survey (USGS) library signatures, along with the estimated abundance maps. (**a**) alunite; (**b**) andradite; (**c**) buddingtonite; (**d**) dumortierite; (**e**) kaolinite #1; (**f**) kaolinite #2; (**g**) muscovite; (**h**) montmorillonite; (**i**) nontronite; (**j**) pyrope; (**k**) sphene; and (**l**) chalcedony.

Table 4. The spectral angle distance and their standard derivations of algorithms on the Cuprite data. Numbers in bold and red color represent the best results; numbers in bold and blue color represent the second-best results.

Endmember	Spectral Angle Distance (10^{-2})					
	VCA-FCLS	SGA-FCLS	NMF	L_2-NMF	$L_{1/2}$-NMF	DGC-NMF
Alunite	17.85 ± 9.39	11.05	**10.03**	10.18	16.01	**9.91**
Andradite	**8.21 ± 2.29**	8.44	13.12	**7.65**	12.48	12.17
Buddingtonite	9.82 ± 2.33	11.27	**6.71**	9.10	**8.24**	8.97
Dumortierite	13.36 ± 3.56	13.65	13.19	**10.37**	**6.84**	10.71
Kaolinite #1	7.68 ± 0.18	17.90	7.33	10.88	**6.88**	**6.40**
Kaolinite #2	9.82 ± 2.35	**7.00**	8.87	9.50	**8.37**	14.02
Muscovite	16.51 ± 7.09	**8.72**	**10.05**	10.47	20.31	10.42
Montmorillonite	11.07 ± 4.63	6.81	6.42	8.76	**5.89**	**5.88**
Nontronite	**7.48 ± 0.15**	13.39	12.53	10.52	10.90	**8.69**
Pyrope	9.30 ± 3.25	14.69	25.36	15.67	**6.24**	**6.12**
Sphene	**10.30 ± 5.48**	23.64	**5.58**	65.07	28.27	24.23
Chalcedony	12.31 ± 5.22	**11.66**	13.20	12.62	**12.28**	12.41
Mean	11.14 ± 3.83	12.35	**11.03**	15.07	11.89	**10.83**

5. Conclusions

In this paper, we provide a novel nonnegative matrix factorization with data-guided constraint (DGC-NMF), which is based on the data's sparsity levels in different areas. Since the sparseness of abundances is previously unknown, we provide a method to evaluate the sparsity level of each pixel's abundances. The sparseness map of data is estimated by using the obtained abundances in a NMF unmixing process with no constraint. The experiments results validate that the estimated sparseness values can represent the real sparsity levels of pixels well. Through the estimated sparseness map, sparseness constraints on pixels' abundances could be adaptively imposed and lead to better unmixing results. We have proven monotone decrease of the objective by our algorithm and illustrated the effectiveness and practicability of the algorithm by experiments on synthetic data and real hyperspectral images. For the future work, the performance of our method could be further improved by achieving a more accurate estimation of sparsity levels and by introducing more reasonable constraints imposing strategy. More methods based on mining and using the information latent in data itself would also be worthy of further study.

Acknowledgments: This work was supported by National Nature Science Foundation of China (No. 61571170, No. 61671408) and the Joint Funds of the Ministry of Education of China (No. 6141A02022314).

Author Contributions: All the authors made significant contributions to the work. Risheng Huang designed the research and analyzed the results. Xiaorun Li provided advice for the preparation and revision of the paper. Liaoying Zhao assisted in the preparation work and validation work.

Conflicts of Interest: The authors declare no conflict of interest.

References

1. Bioucas-Dias, J.M.; Plaza, A.; Dobigeon, N.; Parente, M.; Du, Q.; Gader, P.; Chanussot, J. Hyperspectral unmixing overview: Geometrical, statistical, and sparse regression-based approaches. *IEEE J. Sel. Top. Appl. Earth Obs. Remote Sens.* **2012**, 5, 354–379.

2. Altmann, Y.; Dobigeon, N.; Tourneret, J.Y. Unsupervised post-nonlinear unmixing of hyperspectral images using a Hamiltonian Monte Carlo algorithm. *IEEE Trans. Image Process.* **2014**, *23*, 2663–2675.
3. Wang, Y.; Pan, C.; Xiang, S.; Zhu, F. Robust hyperspectral unmixing with correntropy-based metric. *IEEE Trans. Image Process.* **2015**, *24*, 4027–4040.
4. Wei, Q.; Chen, M.; Tourneret, J.Y.; Godsill, S. Unsupervised nonlinear spectral unmixing based on a multilinear mixing model. *IEEE Trans. Geosci. Remote Sens.* **2017**, *55*, 4534–4544.
5. Qian, Y.; Xiong, F.; Zeng, S.; Zhou, J.; Tang, Y.Y. Matrix-vector nonnegative tensor factorization for blind unmixing of hyperspectral imagery. *IEEE Trans. Geosci. Remote Sens.* **2017**, *55*, 1776–1792.
6. Dobigeon, N.; Tourneret, J.Y.; Chang, C.I. Semi-supervised linear spectral unmixing using a hierarchical Bayesian model for hyperspectral imagery. *IEEE Trans. Signal Process.* **2008**, *56*, 2684–2695.
7. Altmann, Y.; Halimi, A.; Dobigeon, N.; Tourneret, J.Y. Supervised nonlinear spectral unmixing using a postnonlinear mixing model for hyperspectral imagery. *IEEE Trans. Image Process.* **2012**, *21*, 3017–3025.
8. Boardman, J.W. Geometric mixture analysis of imaging spectrometry data. In Proceedings of the 14th Annual International Geoscience and Remote Sensing Symposium: Surface and Atmospheric Remote Sensing: Technologies, Data Analysis and Interpretation, Pasadena, CA, USA, 8–12 August 1994; Volume 4, pp. 2369–2371.
9. Winter, M.E. N-FINDR: An algorithm for fast autonomous spectral end-member determination in hyperspectral data. In Proceedings of the SPIE's International Symposium on Optical Science, Engineering, and Instrumentation, Denver, CO, USA, 18–23 July 1999; pp. 266–275.
10. Nascimento, J.M.; Dias, J.M. Vertex component analysis: A fast algorithm to unmix hyperspectral data. *IEEE Trans. Geosci. Remote Sens.* **2005**, *43*, 898–910.
11. Chang, C.I.; Wu, C.C.; Liu, W.; Ouyang, Y.C. A new growing method for simplex-based endmember extraction algorithm. *IEEE Trans. Geosci. Remote Sens.* **2006**, *44*, 2804–2819.
12. Chang, C.I.; Wen, C.H.; Wu, C.C. Relationship exploration among PPI, ATGP and VCA via theoretical analysis. *Int. J. Comput. Sci. Eng.* **2013**, *8*, 361–367.
13. Li, H.; Chang, C.I. Linear spectral unmixing using least squares error, orthogonal projection and simplex volume for hyperspectral Images. In Proceedings of the 7th Workshop Hyperspectral Image & Signal Processing: Evolution Remote Sensing (WHISPERS), Tokyo, Japan, 2–5 June 2015; pp. 2–5.
14. Chang, C.I.; Chen, S.Y.; Li, H.C.; Chen, H.M.; Wen, C.H. Comparative study and analysis among ATGP, VCA, and SGA for finding endmembers in hyperspectral imagery. *IEEE J. Sel. Top. Appl. Earth Obs. Remote Sens.* **2016**, *9*, 4280–4306.
15. Dobigeon, N.; Tourneret, J.Y. Spectral unmixing of hyperspectral images using a hierarchical Bayesian model. In Proceedings of the IEEE International Conference on Acoustics, Speech and Signal Processing, Honolulu, HI, USA, 16–20 April 2007; Volume 3.
16. Dobigeon, N.; Moussaoui, S.; Coulon, M.; Tourneret, J.Y.; Hero, A.O. Joint Bayesian endmember extraction and linear unmixing for hyperspectral imagery. *IEEE Trans. Signal Process.* **2009**, *57*, 4355–4368.
17. Themelis, K.E.; Rontogiannis, A.A.; Koutroumbas, K.D. A novel hierarchical Bayesian approach for sparse semisupervised hyperspectral unmixing. *IEEE Trans. Signal Process.* **2012**, *60*, 585–599.
18. Lee, D.D.; Seung, H.S. Algorithms for non-negative matrix factorization. In *Advances in Neural Information Processing Systems*; MIT Press: Cambridge, MA, USA, 2001; pp. 556–562.
19. Zhu, F.; Wang, Y.; Xiang, S.; Fan, B.; Pan, C. Structured sparse method for hyperspectral unmixing. *ISPRS J. Photogramm. Remote Sens.* **2014**, *88*, 101–118.
20. Lee, D.D.; Seung, H.S. Learning the parts of objects by non-negative matrix factorization. *Nature* **1999**, *401*, 788–791.
21. Wang, N.; Du, B.; Zhang, L. An endmember dissimilarity constrained non-negative matrix factorization method for hyperspectral unmixing. *IEEE J. Sel. Top. Appl. Earth Obs. Remote Sens.* **2013**, *6*, 554–569.
22. Miao, L.; Qi, H. Endmember extraction from highly mixed data using minimum volume constrained nonnegative matrix factorization. *IEEE Trans. Geosci. Remote Sens.* **2007**, *45*, 765–777.
23. Huck, A.; Guillaume, M. Robust hyperspectral data unmixing with spatial and spectral regularized NMF. In Proceedings of the 2nd Workshop on Hyperspectral Image and Signal Processing: Evolution in Remote Sensing (WHISPERS), Reykjavik, Iceland, 14–16 June 2010; pp. 1–4.
24. Tong, L.; Zhou, J.; Qian, Y.; Bai, X.; Gao, Y. Nonnegative-matrix-factorization-based hyperspectral unmixing with partially known endmembers. *IEEE Trans. Geosci. Remote Sens.* **2016**, *54*, 6531–6544.

25. Lu, X.; Wu, H.; Yuan, Y.; Yan, P.; Li, X. Manifold regularized sparse NMF for hyperspectral unmixing. *IEEE Trans. Geosci. Remote Sens.* **2013**, *51*, 2815–2826.
26. Liu, X.; Xia, W.; Wang, B.; Zhang, L. An approach based on constrained nonnegative matrix factorization to unmix hyperspectral data. *IEEE Trans. Geosci. Remote Sens.* **2011**, *49*, 757–772.
27. Yuan, Y.; Fu, M.; Lu, X. Substance dependence constrained sparse NMF for hyperspectral unmixing. *IEEE Trans. Geosci. Remote Sens.* **2015**, *53*, 2975–2986.
28. Xu, Z.; Zhang, H.; Wang, Y.; Chang, X.; Liang, Y. $L_{1/2}$ regularization. *Sci. China Inf. Sci.* **2010**, *53*, 1159–1169.
29. Qian, Y.; Jia, S.; Zhou, J.; Robles-Kelly, A. $L_{1/2}$ Sparsity constrained nonnegative matrix factorization for hyperspectral unmixing. In Proceedings of the International Conference on Digital Image Computing: Techniques and Applications (DICTA), Sydney, Australia, 1–3 December 2010; pp. 447–453.
30. Pauca, V.P.; Piper, J.; Plemmons, R.J. Nonnegative matrix factorization for spectral data analysis. *Linear Algebra Appl.* **2006**, *416*, 29–47.
31. Jia, S.; Qian, Y. Constrained nonnegative matrix factorization for hyperspectral unmixing. *IEEE Trans. Geosci. Remote Sens.* **2009**, *47*, 161–173.
32. Wu, C.; Shen, C. Spectral unmixing using sparse and smooth nonnegative matrix factorization. In Proceedings of the International Conference on Geoinformatics, Kaifeng, China, 20–22 June 2013; pp. 1–5.
33. Zhu, F.; Wang, Y.; Fan, B.; Xiang, S.; Meng, G.; Pan, C. Spectral unmixing via data-guided sparsity. *IEEE Trans. Image Process.* **2014**, *23*, 5412–5427.
34. Zhu, F.; Wang, Y.; Fan, B.; Meng, G.; Pan, C. Effective spectral unmixing via robust representation and learning-based sparsity. *arXiv* **2014**, arXiv:1409.0685.
35. Qian, Y.; Jia, S.; Zhou, J.; Robles-Kelly, A. Hyperspectral unmixing via $L_{1/2}$ sparsity-constrained nonnegative matrix factorization. *IEEE Trans. Geosci. Remote Sens.* **2011**, *49*, 4282–4297.
36. Fan, J.; Li, R. Variable selection via nonconcave penalized likelihood and its oracle properties. *J. Am. Stat. Assoc.* **2001**, *96*, 1348–1360.
37. Hoyer, P.O. Non-negative matrix factorization with sparseness constraints. *J. Mach. Learn. Res.* **2004**, *5*, 1457–1469.
38. Otsu, N. A threshold selection method from gray-level histograms. *IEEE Trans. Syst. Man Cybern.* **1979**, *9*, 62–66.
39. Dempster, A.P.; Laird, N.M.; Rubin, D.B. Maximum likelihood from incomplete data via the EM algorithm. *J. R. Stat. Soc. Ser. B (Methodol.)* **1977**, 1–38.
40. Saul, L.; Pereira, F. Aggregate and mixed-order Markov models for statistical language processing. *arXiv* **1997**, arXiv:cmp-lg/9706007.
41. Berry, M.W.; Browne, M.; Langville, A.N.; Pauca, V.P.; Plemmons, R.J. Algorithms and applications for approximate nonnegative matrix factorization. *Comput. Stat. Data Anal.* **2007**, *52*, 155–173.
42. Clark, R.N.; Swayze, G.A.; Gallagher, A.J.; King, T.V.; Calvin, W.M. *The US Geological Survey, Digital Spectral Library*, Version 1 (0.2 to 3.0 um); Technical report; Geological Survey (US): Reston, VA, USA, 1993.
43. Dunteman, G.H. *Principal Components Analysis*; Sage Publications, Inc.: Thousand Oaks, CA, USA, 1989.
44. Zhu, F. Hyperspectral Unmixing Datasets & Ground Truths. Available online: http://www.escience.cn/people/feiyunZHU/Dataset_GT.html (accessed on 5 June 2017).

remote sensing

MDPI

Article

Joint Local Abundance Sparse Unmixing for Hyperspectral Images

Mia Rizkinia [1,2,†,*] **and Masahiro Okuda** [1,†]

[1] Faculty of Environmental Engineering, The University of Kitakyushu, Kitakyushu 808-0135, Japan; okuda-m@kitakyu-u.ac.jp
[2] Faculty of Engineering, Universitas Indonesia, Depok, Jawa Barat 16424, Indonesia
* Correspondence: w5dca002@eng.kitakyu-u.ac.jp
† This paper is partially based on the authors' conference paper, which is presented at the 2016 Asia-Pacific Signal and Information Processing Association Annual Summit and Conference (APSIPA), Jeju, Korea, 13–16 December 2016.

Received: 20 October 2017; Accepted: 22 November 2017; Published: 27 November 2017

Abstract: Sparse unmixing is widely used for hyperspectral imagery to estimate the optimal fraction (abundance) of materials contained in mixed pixels (endmembers) of a hyperspectral scene, by considering the abundance sparsity. This abundance has a unique property, i.e., high spatial correlation in local regions. This is due to the fact that the endmembers existing in the region are highly correlated. This implies the low-rankness of the abundance in terms of the endmember. From this prior knowledge, it is expected that considering the low-rank local abundance to the sparse unmixing problem improves estimation performance. In this study, we propose an algorithm that exploits the low-rank local abundance by applying the nuclear norm to the abundance matrix for local regions of spatial and abundance domains. In our optimization problem, the local abundance regularizer is collaborated with the $L_{2,1}$ norm and the total variation for sparsity and spatial information, respectively. We conducted experiments for real and simulated hyperspectral data sets assuming with and without the presence of pure pixels. The experiments showed that our algorithm yields competitive results and performs better than the conventional algorithms.

Keywords: sparse unmixing; hyperspectral; local abundance; nuclear norm

1. Introduction

The need to extract more detailed information from remote-sensing imagery has expanded from multispectral images to hyperspectral images that enable pixel-constituent-level analysis. Hyperspectral images have better spectral resolution than multispectral images due to their large number of narrow and contiguous spectral bands [1]. The detailed information provided by sensors faces a trade-off in which the sensors capture distinct materials on the Earth's surface mixed in one pixel. This is affected by one of the following factors [2–4]. The first factor is due to the low spatial resolution of the sensors; two or more separate materials occupy the same pixel. The other factor occurs when the sensors capture some distinct substances that have merged into a homogeneous mixture on the Earth's surface. This condition leads to a compelling solution, i.e., spectral unmixing.

The procedure of spectral unmixing works by decomposing the measured hyperspectral data into a collection of spectral signatures (spectral library) and a set of corresponding fractions (abundances) that represent the proportion of each spectral signature contained in the pixels [2,5–7]. The spectral signatures that exist in the mixed pixels are called endmembers. In general, endmembers correspond to familiar macroscopic objects in a scene, such as water, metal, and vegetation, as well as constituents of intimate mixtures in microscopic scale. Hyperspectral unmixing can be reconstructed from the linear mixture model (LMM) and nonlinear mixture model [2,8–10]. With the LMM, it is assumed that the spectra of each mixed pixel are linear combinations of the endmembers contained in the pixel. Despite the fact that it holds only for macroscopic mixture conditions [8,11], it is widely used due to its computational tractability and flexibility in various applications.

With the LMM, several unmixing techniques have been introduced based on either geometry [12,13], statistics [12,14], nonnegative matrix factorization (NMF) [4,15–17], or sparse regression [12,18–21]. Although the geometry and statistical techniques are unsupervised and require only a little prior information about the data, they require an assumption that at least one pure pixel (a pixel containing only one endmember) exists for each endmember [22]. The NMF techniques do not require this assumption, however, they can obtain virtual endmembers with no physical meaning [22,23]. On the other hand, in the sparse regression techniques, additional informations are introduced as prior knowledge that are added to the objective functions in the optimization problems and called regularizers, e.g., considering the abundance sparsity [24–26], information of endmembers known to exist in the data [22], or total local spatial differences [27]. An abundance sparsity regularizer algorithm, called sparse unmixing by variable splitting and augmented Lagrangian (SUnSAL), was introduced by Iordache et al. [26]. They applied the L_1 norm (the sum of the absolute values of the matrix columns) to the abundance matrix, substituting the L_0 norm (the number of nonzero elements of the matrix) to impose the sparsity. With the algorithm known as collaborative SUnSAL (CLSUnSAL), it is assumed that the pixels of a hyperspectral scene share the same active set of endmembers [28]. This assumption does not hold when an endmember is contained in several pixels instead of all pixels in the scene. For example, when the hyperspectral scene captures a location that contains locally homogeneous regions. Zhang et al. [29] proposed a local approach of the CLSUnSAL considering the fact that endmembers tend to be distributed uniformly in local spatial regions. Qu et al. [30] adopted joint sparsity combined with the low-rank model under the bilinear mixture model (BMM). The low-rank term corresponds to the low number of linearly independent columns of a matrix. They applied a local sliding window to the abundance matrix as the neighboring pixels tend to be homogeneous and constituted from the same materials.

Iordache et al. [27] proposed a spatial regularizer algorithm called sparse unmixing with the total variation regularizer (SUnSAL-TV), which uses an unmixing technique that is more powerful than the conventional unmixing ones. Nevertheless, this semi-supervised algorithm may produce over-smoothed results and blur in the edges. The spatial information is also imposed in the sparse unmixing task in a nonlocal procedure [11]. Tang et al. [22] introduced an algorithm called sparse unmixing using a priori information (SUnSPI). The required prior knowledge is that some spectral signatures (endmembers) in the hyperspectral scene are known in advance. Despite the fact that the performance is superior compared to that of conventional unmixing algorithms, it is difficult to guarantee whether the assumption can always hold. Field investigation or prior hyperspectral-data analysis may be needed to provide such information.

In a region with high spatial similarity, e.g., local spatial region, the correlation among pixels' spectral signatures can be reflected as linear dependence among their corresponding abundance vectors. The abundance matrix that is composed of these vectors should be low rank. This low-rankness has been recently applied for hyperspectral image denoising and recovery tasks [31–33], which results in superior performances. Furthermore, the low-rankness of the data also indicates high correlation among the abundance vectors corresponding to the pixels in such regions [30]. Giampouras et al. [34] proposed ADSpLRU algorithm by exploiting the low-rankness of abundance to the sparse unmixing problem to consider the spatial correlation of the abundance. However, they considered the low-rankness in the nonlocal fashion of the abundance dimension. In practice, to consider the local low-rankness of an image, Ono et al. [35] proposed the local color nuclear norm (LCNN). However, they locally applied the nuclear norm (the sum of the matrix singular values) only to the spatial dimension of RGB images. Yang et al. [36] also imposed the low-rank constraint for coupled sparse denoising and unmixing problems. However, the use of the nuclear norm is not local, and superior performance is more dominant in the denosing task rather than the unmixing one. To the best of our knowledge, there is no sparse unmixing algorithm that takes into account the low-rankness of local spectral signatures (endmembers) in the abundance dimension, whereas the high correlation between the spectral signatures can be guaranteed by the spectral angle (SA), which is a spectral similarity assesment defined as the angle between two spectral vectors. In turn, one can observe the linearity of the data distribution in local regions in terms of spatial as well as abundance dimension. This priori may lead to a novel approach for the sparse unmixing algorithm.

In this study, we developed an algorithm, which is called joint local abundance sparse unmixing (J-LASU), in which we proposed the local abundance regularizer and implanted it to the sparse unmixing problem using the nuclear norm for 3D local regions and evaluated the effect. We used the 3D local block sliding through the three dimensions of the abundance maps and imposed the nuclear norm to promote the low-rank structure of the local abundance cube. We preserve the use of the total variation (TV) regularizer for spatial consideration. The proposed algorithm was tested on simulated data as well as real hyperspectral data and compared with other sparse unmixing algorithms, i.e., CLSUnSAL, SUnSAL-TV, and ADSpLRU. The major contribution of this study is imposing our local abundance regularizer to a hybrid of state-of-the-art unmixing techniques that take into account collaborative sparsity and spatial difference. We also applied the proposed J-LASU to several scenes with and without pure pixels.

In Section 2, we discuss the problem formulation of hyperspectral unmixing as an introduction to the problem formulation of our proposed algorithm. In Section 3, we describe the proposed J-LASU algorithm starting with convincing evidence of the proposed concept. In Section 4, we describe the experiment and analysis. In Section 5, we discuss the results and findings. Finally, we conclude the paper in Section 6.

Variables and notation: Column vectors are represented as boldface lowercase letters, e.g., $\mathbf{y}$, whereas matrices are represented as boldface uppercase letters, e.g., $\mathbf{Y}$. The following variables are frequently used in this paper:

- $\mathbf{Y}$ is the hyperspectral data,
- $\mathbf{A}$ is the spectral library,
- $\mathbf{X}$ is the abundance matrix,
- $\hat{\mathbf{X}}$ is the 3D abundance data,
- m is the number of spectral signatures,
- l is the number of spectral bands,
- n is the number of pixels in $\hat{\mathbf{X}}$,
- n_c is the number of columns in $\hat{\mathbf{X}}$,
- n_r is the number of rows in $\hat{\mathbf{X}}$,
- B is the number of all local blocks in $\hat{\mathbf{X}}$,
- N is the number of pixels in each local abundance matrix,
- $\hat{\mathbf{X}}_b$ is the b-th local block,
- $\mathbf{H}_{\hat{\mathbf{x}}_b}$ is the b-th local abundance matrix.

2. Hyperspectral Unmixing

2.1. Sparse Unmixing

Let $\mathbf{Y} \in \mathbb{R}^{l \times n}$ be the observed hyperspectral data, where l is the number of bands, and n is the number of pixels. The LMM for a hyperspectral image is based on the assumption that each pixel $\mathbf{y} \in \mathbb{R}^l$ in any given spectral band is a linear combination of m spectral signatures in the spectral library $\mathbf{A} \in \mathbb{R}^{l \times m}$, that is,

$$\mathbf{y} = \mathbf{A}\mathbf{x} + \mathbf{e} \tag{1}$$

where $\mathbf{x} \in \mathbb{R}^m$ is the abundance vector, and $\mathbf{e} \in \mathbb{R}^l$ is the vector of noise and model error.

With sparse unmixing, it is assumed that the abundance vector $\mathbf{x}$ is sparse because the number of endmembers contained in a pixel is much lower than the number of spectral signatures in the library, which implies the vector $\mathbf{x}$ contains many intensities of zero. Figure 1 illustrates the LMM and sparse unmixing. Considering the ground truth, $\mathbf{x}$ has a constraint that needs to be imposed to the sparse unmixing model, i.e., the value of $\mathbf{x}$ can never be negative which is called the abundance nonnegativity constraint (ANC). The sparse unmixing problem based on the LMM for each mixed pixel can be formulated as

$$\min_{\mathbf{x}} \quad \|\mathbf{x}\|_0 \qquad \text{s.t.} \qquad \|\mathbf{y} - \mathbf{A}\mathbf{x}\|_2 \leq \delta, \quad \mathbf{x} \geq 0 \tag{2}$$

where $\|\mathbf{x}\|_0$ denotes the number of nonzero elements in $\mathbf{x} \in \mathbb{R}^m$, and δ is the error tolerance value determined from the noise and model error. The nonconvexity of the L_0 term induces an NP-hard problem; however, it has been proven that a nonconvex optimization problem can be relaxed to a convex one by replacing L_0 with L_1 [11,37]. Thus, the problem can be written as

$$\min_{\mathbf{x}} \quad \|\mathbf{x}\|_1 \qquad \text{s.t.} \qquad \|\mathbf{y} - \mathbf{A}\mathbf{x}\|_2 \leq \delta, \quad \mathbf{x} \geq 0 \tag{3}$$

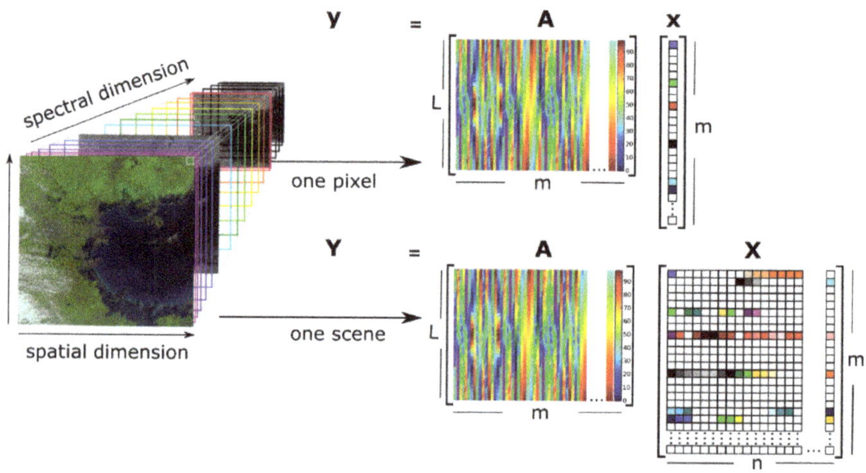

Figure 1. Illustration of hyperspectral image and sparse unmixing for pixel (**top**) and image (**bottom**).

Applying this formula to the whole image, we estimate the abundance matrix $\mathbf{X} \in \mathbb{R}^{m \times n}$ for all the pixels in the hyperspectral data $\mathbf{Y}$ using the respective Lagrangian function as

$$\min_{\mathbf{X}} \quad \frac{1}{2}\|\mathbf{A}\mathbf{X} - \mathbf{Y}\|_F^2 + \lambda\|\mathbf{X}\|_1 \qquad \text{s.t.} \qquad \mathbf{X} \geq 0 \tag{4}$$

where $\|\cdot\|_F$ denotes the Frobenius norm of a matrix, and λ is the sparsity regularizer. This problem can be solved through optimization by using alternating direction method of multipliers (ADMM).

2.2. Spatial Regularization

Despite taking into account sparsity, SUnSAL ignores spatial correlation. In SUnSAL-TV, the relationship between each pixel vector and its adjacent pixel vectors is taken into account. The regularizer is defined in [27] as

$$\text{TV}(\mathbf{X}) = \sum_{\{i,j \in \kappa\}} \|\mathbf{x}_i - \mathbf{x}_j\|_1 \tag{5}$$

which is the anisotropic TV with κ denoting the set of horizontal and vertical neighbors in $\mathbf{X}$.

Adding the TV regularizer to the problem in Equation (4) gives the optimization problem

$$\min_{\mathbf{X}} \quad \frac{1}{2}\|\mathbf{AX}-\mathbf{Y}\|_F^2 + \lambda\|\mathbf{X}\|_1 + \lambda_{TV}\mathrm{TV}(\mathbf{X}) \quad \text{s.t.} \quad \mathbf{X} \geq 0. \tag{6}$$

3. Proposed Algorithm

3.1. Local Abundance Correlation

Hyperspectral data $\mathbf{Y} \in \mathbb{R}^{l \times n}$ have linearity in their spectral [38] and spatial [30] domains. Qu et al. [30] provided prior knowledge that the high spatial correlation of the hyperspectral data, implies linearly dependent abundance vectors in the abundance matrix $\mathbf{X} \in \mathbb{R}^{m \times n}$. The high correlation also holds among the pixel members of a local region due to the spatial similarity. In a physical sense, the pixels in such regions contain the same materials, either in the same or different fractions. Hence, the abundance matrix of the region can be estimated by the low-rank property [30,34].

However, the success of sparse regression techniques is affected by the low sparsity as well as low correlation between spectral signatures in the library [27]. The former is represented by the number of endmembers existing in the scene, namely, the degree of sparsity [26]. The latter can be defined by an indicator representing the difficulty to accurately solve a linear system equation i.e., mutual coherence. The mutual coherence is defined as the largest cosine among endmembers in the library. In the hyperspectral case, the degree of sparsity is often low, but the mutual coherence is close to one. In fact, higher mutual coherence decreases the quality of the solution [28].

To overcome the high mutual coherence as well as consider the low-rank property of the abundance, we exploit the high correlation of library's spectral signatures by using our LA regularizer. In our experiment with simulated data, we confirmed the idea by observing the linearity of the data distribution in abundance domain by taking the local maximum singular value of the true abundance matrix for each local block (a block refers to the three dimensions (3D), in which the third dimension has a local coverage in the endmember direction). We found that there is one value that dominates others (the ratio is close to one) in each local block. On the other hand, the value will be less dominant as the region becomes the whole matrix (nonlocal). This implies that the linearity in abundance domain is satisfied for the abundance matrix with the local point of view. Thus, we introduce our LA regularizer using the nuclear norm for the local blocks. Instead of the image, our algorithm uses the nuclear norm to the abundance matrix that constitutes the image. Another difference is that our local block slides through all dimensions, i.e., the two spatial dimensions and the endmember direction in the abundance dimension. Figure 2 illustrates the endmember direction. The block moves within the abundance maps of the 3D abundance cube.

In addition, we guarantee high correlation by selecting endmembers from the United States Geological Survey (USGS) library to form the spectral library $\mathbf{A}$ based on the SA. The USGS library is a collection of the measured spectral signatures of hundreds of materials and used as references for material identification in hyperspectral images. We can find the most similar signatures to each endmember of the simulated data by calculating the SA, besides the mutual coherence. This parameter represents the absolute value of spectral correlation [39]. The value ranges between 0–90 degrees. The lower the SA value, the more similar the compared signature vectors are. In the simulated-data experiment, we adjusted the SA as one of our parameter settings.

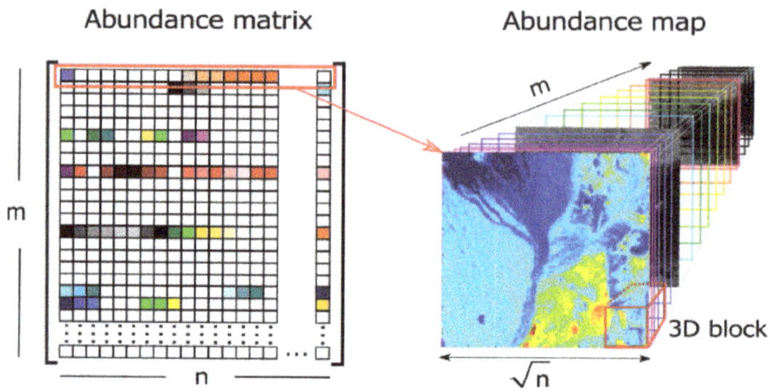

Figure 2. Illustration of endmember (m) direction in abundance dimension. 3D local block moves through pixels (n) as well as m direction of abundance maps.

3.2. Collaborative Sparsity Regularization

In practice, the abundance matrix $\mathbf{X}$ has only a few endmembers (rows) with nonzero entries. Simultaneously, all the column entries of $\mathbf{X}$ share the same active set of endmembers. In other words, $\mathbf{X}$ is sparse among the rows while dense among the columns. To implement this prior, $L_{2,1}$ norm is used instead of L_1. It takes the sum of the L_2 norm of the abundance entries to promote the collaborative sparsity of the abundance matrix.

$$\|\mathbf{X}\|_{2,1} = \sum_{i=1}^{m} \|\mathbf{x}_i\|_2 \tag{7}$$

where $\mathbf{x}_i$ represents the i-th row of $\mathbf{X}$.

3.3. Local Abundance Regularizer

First, let $\hat{\mathbf{X}} \in \mathbb{R}^{n_r \times n_c \times m}$ be the abundance data in 3D form, where m is the number of abundance matrices of the endmembers, n_c and n_r are the numbers of columns and rows, respectively, that satisfy $n = n_c \times n_r$, where n is the number of pixels in each abundance matrix. Then, for each abundance matrix $\hat{\mathbf{X}}_i \in \mathbb{R}^{n_r \times n_c}$ $(i = 1, \ldots, m)$, stacking the column on top of one another gives $\hat{\mathbf{x}}_i \in \mathbb{R}^n$, the vectorized form of the matrix.

In local regions, let $\hat{\mathbf{X}}_b \in \mathbb{R}^{n_b \times n_b \times m_b}$ denote the b-th local block, where $b = 1, \ldots, B$. The B is the number of all local blocks in $\hat{\mathbf{X}}$. Then, for each abundance of each local block $\hat{\mathbf{X}}_{j,b} \in \mathbb{R}^{n_b \times n_b}$ $(j = 1, \ldots, m_b)$, we vectorize it into $\hat{\mathbf{x}}_{j,b} \in \mathbb{R}^N$, where N is the number of pixels in each local abundance matrix that satisfies $N = n_b \times n_b$, and j is the index of local abundance matrices. Figure 3 illustrates the procedure. With this in mind, we introduce the local abundance matrix w. r. t the b-th block

$$\mathbf{H}_{\hat{\mathbf{X}}_b} = (\hat{\mathbf{x}}_{1,b}, \ldots, \hat{\mathbf{x}}_{m_b,b}) \in \mathbb{R}^{N \times m_b}. \tag{8}$$

Finally, the function of our proposed LA regularization is defined as follows

$$\|\mathbf{X}\|_{LA*} = \sum_{b=1}^{B} \|\mathbf{H}_{\hat{\mathbf{X}}_b}\|_* \tag{9}$$

where $\|\cdot\|_*$ denotes the nuclear norm, $\mathbf{X} \mapsto \sum_{i=1}^{rank(\mathbf{X})} \sigma_i(\mathbf{X})$, with σ_i denotes the i-th singular value.

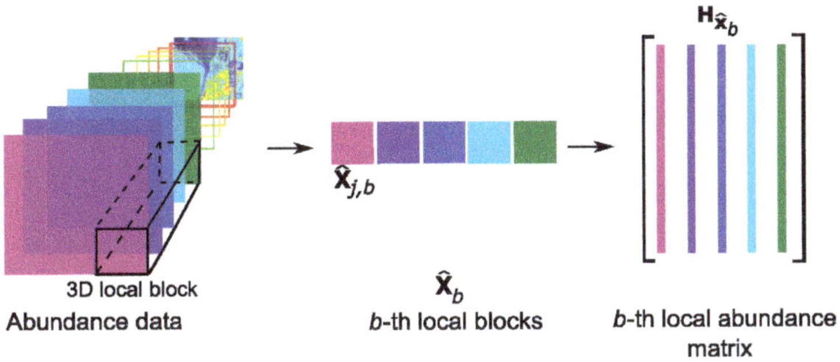

Figure 3. Process of vectorizing and arranging local abundance matrix of hyperspectral image.

3.4. J-LASU

We formulate the new problem by adding the LA term with Equation (7) and the additional TV term. Thus, the problem of the proposed J-LASU algorithm in a convex form becomes

$$\min\ g(\mathbf{X}) = \tfrac{1}{2}\|\mathbf{AX}-\mathbf{Y}\|_F^2 + \lambda\|\mathbf{X}\|_{2,1} + \gamma\|\mathbf{X}\|_{TV} + \rho\|\mathbf{X}\|_{LA*} \quad \text{s.t.} \quad \mathbf{X} \geq 0 \tag{10}$$

where λ, γ, and ρ are the regularization parameter for the collaborative sparsity, TV, and LA term, respectively. We use the anisotropic TV, which is used in SUnSAL-TV [27], defined as

$$\|\mathbf{X}\|_{TV} = \|\mathbf{DX}\|_1 \tag{11}$$

where $\mathbf{D} = [\mathbf{D}_h; \mathbf{D}_v]$, $\mathbf{D}_h : \mathbb{R}^{m\times n} \to \mathbb{R}^{m\times n}$ and $\mathbf{D}_v : \mathbb{R}^{m\times n} \to \mathbb{R}^{m\times n}$, are horizontal and vertical differential operators, respectively. The $\mathbf{D}_h\mathbf{X}$ computes the differences between the components of $\mathbf{X}$ and the corresponding right-side adjacent pixels with cyclic boundary assumption, and the same way for $\mathbf{D}_v\mathbf{X}$, which corresponds to the differences with the up-side adjacent pixels [27].

We estimate the abundance matrix $\mathbf{X}$ by solving problem in Equation (10) by using ADMM. The cost function in Equation (10) written in ADMM form becomes

$$f_1(\mathbf{X}) + f_2(\mathbf{V}) \quad \text{s.t.} \quad \mathbf{V} = \mathbf{GX} \tag{12}$$

where

$$f_1(\mathbf{X}) = \frac{1}{2}\|\mathbf{AX}-\mathbf{Y}\|_F^2 \tag{13}$$

$$f_2(\mathbf{V}) = \lambda\|\mathbf{V}_1\|_{2,1} + \gamma\|\mathbf{V}_2\|_1 + \rho\|\mathbf{V}_3\|_{LA} + \iota_{R+}(\mathbf{V}_4) \tag{14}$$

$$\mathbf{V} = \begin{bmatrix}\mathbf{V}_1\\ \mathbf{V}_2\\ \mathbf{V}_3\\ \mathbf{V}_4\end{bmatrix}, \text{ and } \mathbf{G} = \begin{bmatrix}\mathbf{I}\\ \mathbf{D}\\ \mathbf{I}\\ \mathbf{I}\end{bmatrix}. \tag{15}$$

Here, the ι_{R+} term projects the solution onto the nonnegative orthant ($\iota_{R+}(x) = 0$ if $x \geq 0$ and $\iota_{R+}(x) = +\infty$ otherwise), and $\mathbf{I}$ is an identity matrix with a proportional size. The constraint in Equation (12) satisfies the relations

$$\mathbf{V}_1 = \mathbf{X}; \quad \mathbf{V}_2 = \mathbf{DX}; \quad \mathbf{V}_3 = \mathbf{X}; \quad \mathbf{V}_4 = \mathbf{X}. \tag{16}$$

Using a positive constant μ and the Lagrange multipliers $\mathbf{B}/\mu$ corresponding to the constraint $\mathbf{V} = \mathbf{G}\mathbf{X}$, the cost function is minimized using ADMM. Then, the steps for the proposed algorithm are as follows

$$\mathbf{X}^{(k+1)} = \arg\min_{\mathbf{X}} f_1(\mathbf{X}) + \frac{\mu}{2}\|\mathbf{G}\mathbf{X} - \mathbf{V}^{(k)} - \mathbf{B}^{(k)}\|_F^2 \quad (17)$$

$$\mathbf{V}^{(k+1)} = \arg\min_{\mathbf{V}} f_2(\mathbf{V}) + \frac{\mu}{2}\|\mathbf{G}\mathbf{X}^{(k)} - \mathbf{V} - \mathbf{B}^{(k)}\|_2^2 \quad (18)$$

$$\mathbf{B}^{(k+1)} = \mathbf{B}^{(k)} - (\mathbf{G}\mathbf{X}^{(k+1)} - \mathbf{V}^{(k+1)}). \quad (19)$$

To find the solution for $\mathbf{X}$ of the augmented Lagrangian formula, we calculate the solution of Equation (17) by taking the partial derivative as follows:

$$\begin{aligned}\mathbf{X}^{(k+1)} &= \arg\min_{\mathbf{X}} \tfrac{1}{2}\|\mathbf{A}\mathbf{X} - \mathbf{Y}\|_F^2 + \tfrac{\mu}{2}\|\mathbf{G}\mathbf{X} - \mathbf{V}^{(k)} - \mathbf{B}^{(k)}\|_F^2 \\ &= \left(\mathbf{A}^T\mathbf{A} + \mu\mathbf{G}^T\mathbf{G}\right)^{-1}\left(\mathbf{A}^T\mathbf{Y} + \mu\mathbf{G}^T(\mathbf{V}^{(k)} + \mathbf{B}^{(k)})\right)\end{aligned} \quad (20)$$

The detailed steps for computing the values of variables $\mathbf{V}_1, \mathbf{V}_2, \mathbf{V}_3$, and $\mathbf{V}_4$ for each iteration are written in general form of the proximal operator (General form of the proximal operator is as follows: $\text{prox}_{\gamma h}(\bar{\mathbf{x}}) = \arg\min_{\mathbf{v}\in\mathbb{R}^N} h(\mathbf{v}) + \frac{1}{2\gamma}\|\mathbf{v} - \bar{\mathbf{x}}\|_2^2$) [40,41].

$$\begin{aligned}\mathbf{V}_1^{(k+1)} &= \text{prox}_{\frac{\lambda}{\mu}\|\cdot\|_{2,1}}(\mathbf{R}_1) \\ &= \arg\min_{\mathbf{V}_1} \lambda\|\mathbf{V}_1\|_{2,1} + \tfrac{\mu}{2}\|\mathbf{V}_1 - \mathbf{R}_1\|_2^2\end{aligned} \quad (21)$$

$$\begin{aligned}\mathbf{V}_2^{(k+1)} &= \text{prox}_{\frac{\gamma}{\mu}\|\cdot\|_1}(\mathbf{R}_2) \\ &= \arg\min_{\mathbf{V}_2} \gamma\|\mathbf{V}_2\|_1 + \tfrac{\mu}{2}\|\mathbf{V}_2 - \mathbf{R}_2\|_2^2\end{aligned} \quad (22)$$

$$\begin{aligned}\mathbf{V}_3^{(k+1)} &= \text{prox}_{\frac{\rho}{\mu}\|\cdot\|_{LA*}}(\mathbf{R}_3) \\ &= \arg\min_{\mathbf{V}_3} \rho\|\mathbf{V}_3\|_{LA*} + \tfrac{\mu}{2}\|\mathbf{V}_3 - \mathbf{R}_3\|_2^2\end{aligned} \quad (23)$$

$$\begin{aligned}\mathbf{V}_4^{(k+1)} &= \text{prox}_{\frac{1}{\mu}(\iota_{R+})}(\mathbf{R}_4) \\ &= \arg\min_{\mathbf{V}_4} \iota_{R+}(\mathbf{V}_4) + \tfrac{\mu}{2}\|\mathbf{V}_4 - \mathbf{R}_4\|_2^2\end{aligned} \quad (24)$$

where $\mathbf{R}_1 = \mathbf{X}^{(k)} - \mathbf{B}_1^{(k)}$, $\mathbf{R}_2 = \mathbf{D}\mathbf{X}^{(k)} - \mathbf{B}_2^{(k)}$, $\mathbf{R}_3 = \mathbf{X}^{(k)} - \mathbf{B}_3^{(k)}$, and $\mathbf{R}_4 = \mathbf{X}^{(k)} - \mathbf{B}_4^{(k)}$, and $\mathbf{B} = [\mathbf{B}_1; \mathbf{B}_2; \mathbf{B}_3; \mathbf{B}_4]$.

For $\mathbf{V}_1^{(k+1)}$, since the $L_{2,1}$ norm is not differentiable, the solution is obtained by the shrinkage for the group lasso as follows:

$$v_{1(i)}^{(k+1)} = \begin{cases} r_{1(i)} - \frac{\lambda}{\mu}\frac{r_{1(i)}}{\|r_{1(i)}\|_2} & \text{if } \|r_{1(i)}\|_2 > \frac{\lambda}{\mu} \\ 0 & \text{otherwise} \end{cases} \quad (25)$$

where $v_{1(i)}^{(k+1)}$ and $r_{1(i)}$ denote the i-row of $\mathbf{V}_1^{(k+1)}$ and $\mathbf{R}_1$, respectively. This operation is denoted as $group - lasso(\cdot, \tau)$, where τ is the threshold.

The TV term in Equation (22) is solved by soft-thresholding on each element of $\mathbf{V}_2^{(k+1)}$.

$$v_{2(i,j)}^{(k+1)} = \begin{cases} r_{2(i,j)} - \frac{\gamma}{\mu} & \text{if } r_{2(i,j)} > \frac{\gamma}{\mu} \\ r_{2(i,j)} + \frac{\gamma}{\mu} & \text{if } r_{2(i,j)} < -\frac{\gamma}{\mu} \\ 0 & \text{if } -\frac{\gamma}{\mu} \leq r_{2(i,j)} \leq \frac{\gamma}{\mu} \end{cases} \tag{26}$$

where $v_{2(i,j)}^{(k+1)}$ and $r_{2(i,j)}$ denote the (i,j)-element of $\mathbf{V}_2^{(k+1)}$ and $\mathbf{R}_2$, respectively. This operation is denoted as $soft(\cdot,\tau)$, where τ is the threshold.

The solution of $\mathbf{V}_3^{(k+1)}$ in Equation (23) is acquired by constructing the LA matrices, applying singular value shrinkage to each matrix, and reconstructing the output abundance matrix, which is denoted as

$$\mathbf{V}_3^{(k+1)} = shr(\mathbf{X}^{(k)} - \mathbf{B}_3^{(k)}, \tfrac{\rho}{\mu}) \tag{27}$$

where $shr(\cdot,\tau)$ denotes the singular value shrinkage ($y \mapsto \text{diag}(max\{SVD(y) - \tau, 0\})$) of the LA matrices $\mathbf{H}_{\aleph_b}$, where the singular value decomposition $SVD(\cdot)$ produces a vector containing the singular values in decreasing order and τ is the threshold.

Let $v_{4(i,j)}^{(k+1)}$ denotes the (i,j)-element of $\mathbf{V}_4^{(k+1)}$, finally, the solution of $\mathbf{V}_4^{(k+1)}$ is obtained by

$$v_{4(i,j)}^{(k+1)} = max(r_{4(i,j)}, 0) \tag{28}$$

where $r_{4(i,j)}$ denotes the (i,j)-element of $\mathbf{R}_4$.

The whole procedure of ADMM is summarized in Algorithm 1.

Algorithm 1: ADMM in pseudocode for solving problem in Equation (10)

Initialization: set $k = 0, \mathbf{V}^0 = \mathbf{0}, \mathbf{B}^0 = \mathbf{0}$, choose $\mu > 0, \lambda, \gamma, \rho$;
while *the stopping criterion is not satisfied* **do**
- $\mathbf{X}^{(k+1)} \leftarrow (\mathbf{A}^T\mathbf{A} + \mu\mathbf{G}^T\mathbf{G})^{-1}\left(\mathbf{A}^T\mathbf{Y} + \mu\mathbf{G}^T(\mathbf{V}^{(k)} + \mathbf{B}^{(k)})\right)$;
- $\mathbf{V}_1^{(k+1)} \leftarrow group - lasso(\mathbf{X}^{(k)} - \mathbf{B}_1^{(k)}, \lambda/\mu)$;
- $\mathbf{V}_2^{(k+1)} \leftarrow soft(\mathbf{D}\mathbf{X}^{(k)} - \mathbf{B}_2^{(k)}, \gamma/\mu)$;
- $\mathbf{V}_3^{(k+1)} \leftarrow shr(\mathbf{X}^{(k)} - \mathbf{B}_3^{(k)}, \rho/\mu)$;
- $\mathbf{V}_4^{(k+1)} \leftarrow max(\mathbf{X}^{(k)} - \mathbf{B}_4^{(k)}, 0)$;
- $\mathbf{B}_1^{(k+1)} \leftarrow \mathbf{B}_1^{(k)} - \mathbf{X}^{(k+1)} + \mathbf{V}_1^{(k+1)}$;
- $\mathbf{B}_2^{(k+1)} \leftarrow \mathbf{B}_2^{(k)} - \mathbf{D}\mathbf{X}^{(k+1)} + \mathbf{V}_2^{(k+1)}$;
- $\mathbf{B}_3^{(k+1)} \leftarrow \mathbf{B}_3^{(k)} - \mathbf{X}^{(k+1)} + \mathbf{V}_3^{(k+1)}$;
- $\mathbf{B}_4^{(k+1)} \leftarrow \mathbf{B}_4^{(k)} - \mathbf{X}^{(k+1)} + \mathbf{V}_4^{(k+1)}$;
- Update iteration: k $\leftarrow k + 1$;

end

4. Experiment and Analysis

We tested the proposed algorithm on several simulated data sets for three signal-to-noise ratio (SNR) levels, i.e., 10, 20, and 30 dB, and two real data sets. We evaluated the results by conducting a fair comparison with the CLSUnSAL [28] and SunSAL-TV [27]. State-of-the-art low-rank algorithm is also compared, which is sparse and low-rank unmixing by using ADMM (ADSpLRU) [34].

4.1. Simulated Data Sets

To simulate the condition of hyperspectral data with and without the presence of pure pixels, we used two types of data distribution for data generation. Both use the same library generated from 240 types of minerals selected randomly from the splib06 USGS library [42], which consists of 224 spectral bands ranging between 0.4–2.5 µm. The mutual coherence among the spectral signatures is very close to one, but we set the SA to be larger than 4.4 to make the sparse regression problem easier.

The first data set, *DS*, is a representation of the data with pure pixels and adopted from that of Iordache et al. [27] consisting of 224 bands for 75×75 pixels. The data generation follows the LMM with the abundance sum-to-one constraint imposed on each pixel. Five spectral signatures are randomly selected from the library as the endmembers and distributed spatially in the form of distinct square regions. In some pixels, the endmembers stay pure and in others they are mixed with two until there are five endmembers. In Figure 4, the red squares in each abundance map represent 100% intensity which means the pure pixel regions of each endmember. The background consists of mixed pixels with randomly fixed fractional abundance values of 0.1149, 0.0741, 0.2003, 0.2055, and 0.4051 for the five endmembers.

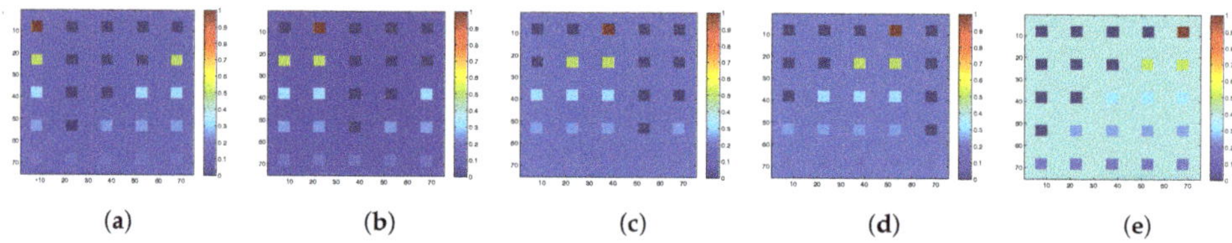

Figure 4. True abundance matrix of simulated data set 1 (*DS*). (**a**) Endmember 1; (**b**) Endmember 2; (**c**) Endmember 3; (**d**) Endmember 4; (**e**) Endmember 5.

To demonstrate the proposed algorithm under the condition without the presence of pure pixels, the distribution with a distinct spatial pattern and mixture was selected. We used the fractal database (*FR*) [39] consisting of five data sets, namely *FR1, FR2, FR3, FR4,* and *FR5*. Each is composed of 100×100 pixels with 224 spectral bands for each pixel and contains no completely pure pixels that are close to the ground-truth characteristic in which completely pure pixels are rarely found. The distribution is generated such that pixels near the edges of regions are more highly mixed than those in the center of the regions. These center pixels have a purity index between 0.95–0.99, directly proportional to the broadness of the regions. In this experiment, we set the number of endmembers to 9. Figure 5 shows *FR1, FR2, FR3, FR4,* and *FR5* represented in pseudocolor.

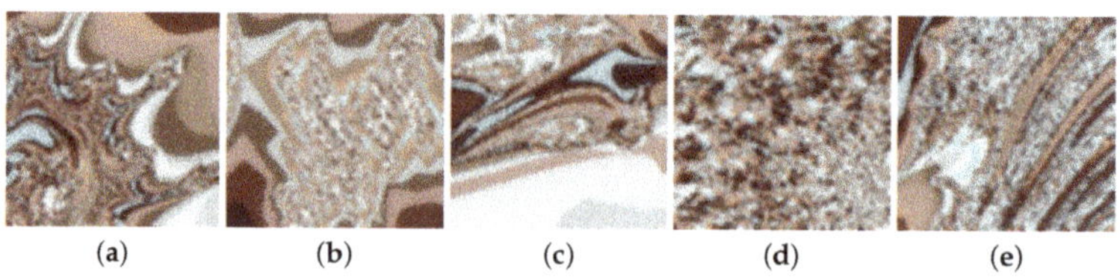

Figure 5. Fractal data sets represented in pseudocolor. (**a**) *FR1*; (**b**) *FR2*; (**c**) *FR3*; (**d**) *FR4*; (**e**) *FR5*.

4.2. Real Data Sets

For the real-data experiment, we used two real data from different sensors. The first hyperspectral scene is the widely used data set of Cuprite mining district, Nevada in 1997 [43]. We used a subscene with the size of 150×130 pixels whose area is shown in Figure 6a. The data are composed of 224 spectral bands with 3.7 m spatial resolution from the AVIRIS sensor. Prior to analysis, several bands were removed due to the low SNR; thus, remaining 188 bands. In this experiment, we used the USGS library of 498 spectral signatures as the standard spectral library for the data, with the corresponding bands removed. Figure 6b shows the USGS mineral distribution map of the Cuprite area [44]. From the figure, the area of interest contains at least three types of minerals: *alunite, chalcedony*, and *kaolinite*. The mineral map was produced using Tricorder 3.3 software in 1995, while the AVIRIS Cuprite data were collected in 1997. Hence, in our experiment, the mineral map was used only for visual qualitative evaluation, compared with the abundance maps of different sparse unmixing algorithms.

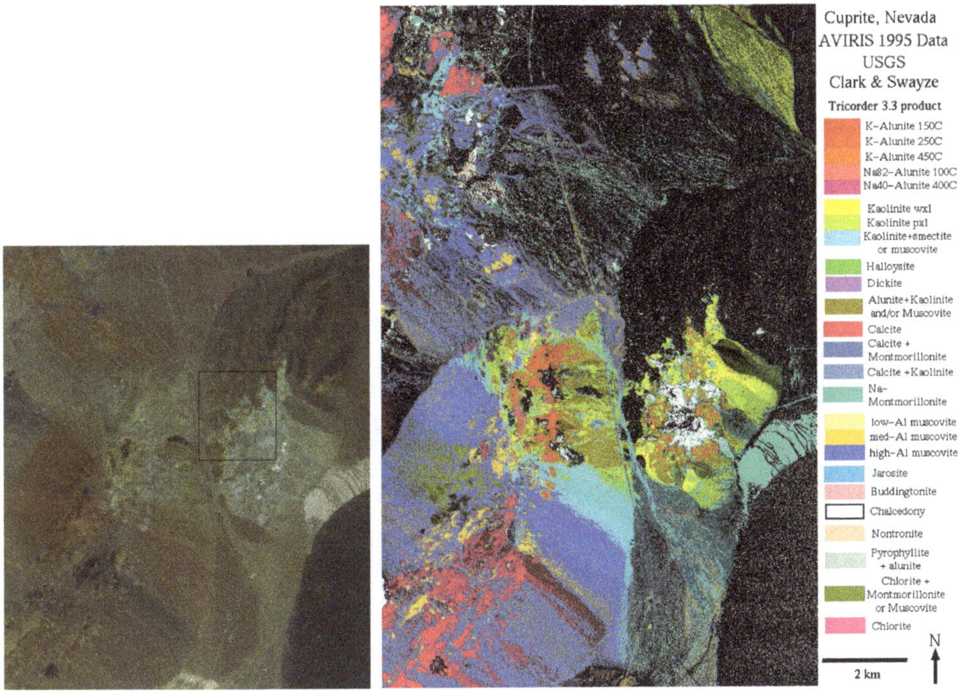

(**a**) (**b**)

Figure 6. (**a**) Cuprite data generated in pseudocolor. Black rectangle shows area of our experiment; (**b**) USGS mineral distribution map of Cuprite mining district in Nevada [44].

The second hyperspectral scene is Urban data captured by the HYDICE sensor over an area located at Copperas Cove near Fort Hood, TX, U.S., in October 1995. It consists of 307 × 307 pixels with 2 m of the pixel resolution. The wavelengths range from 0.4 to 2.5 μm divided into 210 spectral bands. After some bands with low SNRs due to dense water vapor and atmospheric effects are discarded, it remains 162 bands. We used a subscene with the size of 100 × 100 pixels. Figure 7a shows the subscene used in the experiment. The ground truth of the Urban data set is not available, however, we used the reference abundance maps obtained from [45]. The maps are achieved via the method provided in [46–48] and consist of four endmembers, i.e., *asphalt, grass, tree*, and *roof*. Figure 7b shows the spectral signatures of the four endmembers.

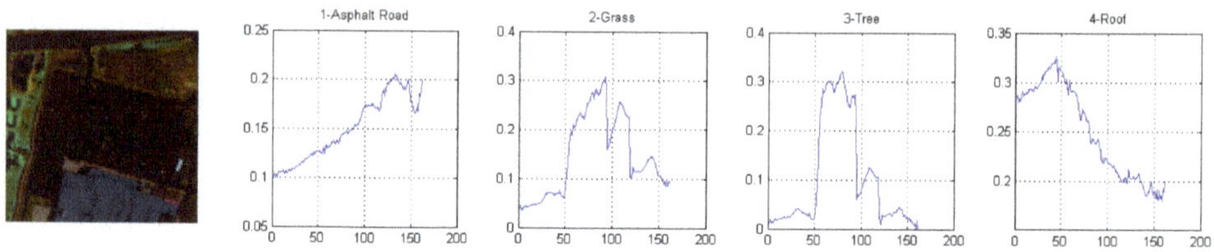

(a) (b)

Figure 7. (**a**) A subscene of Urban data used in our experiment, generated in pseudocolor; (**b**) Spectral signatures of the endmembers[48–50], x-axis and y-axis represent the band number and reflectance unit (0–1), respectively.

4.3. Parameters Setting and Evaluation Metrics

In the simulated-data experiment, to build spectral library **A**, the spectral signatures in the USGS spectral library were selected and sorted such that the SAs between the spectral signatures were not less than 4.4 degrees in increasing order. The parameter settings of J-LASU are for the collaborative sparsity (λ), TV (γ), and LA nuclear norm (ρ) regularizer. For the compared algorithms, λ_{SP} is the sparsity term for CLSUnSAL, SUnSAL-TV, and ADSpLRU [34]. For SUnSAL-TV, the TV term is controlled by λ_{TV}. The low-rank regularizer parameter is denoted as λ_{LR} for ADSpLRU. These parameters are adjusted for every data set under different SNR levels. However, we used the same parameter settings for the five fractal data sets since the characteristics of the scenes tend to be similar. Table 1 summarizes these settings. The values of λ_{LR} were the optimal ones after the experiment for some recommended values. For the LA regularization, the block size is another parameter to be set to control the coverage of adjacent pixels in the spatial and the endmember directions. After several trials in this experiment, the optimum size was found to be [5 5 5] with no overlap.

Table 1. Parameter Settings.

Data	SNR	CLSUnSAL	SUnSAL-TV		ADSpLRU		J-LASU		
		λ_{SP}	λ_{SP}	λ_{TV}	λ_{SP}	λ_{LR}	λ	γ	ρ
DS	10 dB	1×10^{1}	1×10^{-1}	1×10^{-1}	1×10^{1}	1×10^{0}	5×10^{-1}	5×10^{-1}	5×10^{-1}
	20 dB	1×10^{0}	5×10^{-4}	5×10^{-2}	1×10^{1}	1×10^{0}	2.5×10^{-1}	5×10^{-2}	3×10^{-1}
	30 dB	1×10^{0}	5×10^{-4}	1×10^{-2}	1×10^{0}	1×10^{-1}	5×10^{-2}	1×10^{-2}	8×10^{-2}
FR	10 dB	1×10^{1}	5×10^{-2}	1×10^{-1}	1×10^{1}	5×10^{0}	5×10^{-1}	1×10^{-1}	2.5×10^{-1}
	20 dB	1×10^{0}	5×10^{-3}	5×10^{-2}	1×10^{1}	3×10^{0}	2.5×10^{-1}	1×10^{-2}	1×10^{-1}
	30 dB	1×10^{0}	5×10^{-3}	2.5×10^{-2}	1×10^{1}	1×10^{-2}	5×10^{-2}	5×10^{-3}	5×10^{-2}
Cuprite	-	5×10^{-3}	5×10^{-4}	1×10^{-2}	1×10^{1}	1×10^{0}	5×10^{-2}	5×10^{-4}	1×10^{-3}
Urban	-	1×10^{-4}	1×10^{-4}	1×10^{-4}	1×10^{-2}	1×10^{-2}	1×10^{-4}	1×10^{-4}	1×10^{-4}

We evaluated the performance of the algorithms using root mean square error (RMSE) [39,51] and signal-to-reconstruction error (SRE) [26]. The RMSE measures the error between the original and reconstructed abundance matrices. The lower the RMSE, the more accurate the estimation is. The RMSE formula for the i-th endmember is defined as

$$\mathrm{RMSE}_i = \sqrt{\frac{1}{n}\sum_{h=1}^{n}(\mathbf{X}_{i,h} - \hat{\mathbf{X}}_{i,h})^2}, \tag{29}$$

where n, $\mathbf{X}$ and $\hat{\mathbf{X}}$ represent the number of pixels, true abundance matrices, and estimated abundance matrices, respectively. Then, we compute the mean value of all endmembers' RMSEs.

The SRE represents the ratio between the reconstructed abundance matrix and error, and is defined as

$$\text{SRE} = 10\log_{10}\left(\|\mathbf{X}\|_F^2 / \|\mathbf{X} - \bar{\mathbf{X}}\|_2^F\right). \tag{30}$$

For the simulated data, the original abundance matrix was generated for each data set. We compared the visual appearance among the maps of the estimated abundance matrix in addition to RMSE and SRE comparison. As for the first real data set, Cuprite, the comparison was among the estimated abundance maps of the sparse unmixing algorithms and the mineral map of each expected endmembers. For the second real data set, Urban, RMSE and SRE of each method are calculated with the ground truth abundance maps as the reference value.

4.4. Simulated-Data Experiment

Tables 2 and 3 show the RMSE and SRE values, respectively, of estimated abundances from the compared algorithms. The proposed J-LASU algorithm achieved better RMSE for all the simulated data. For the same level of SNR, J-LASU performed better than CLSUnSAL and SUnSAL-TV as well as ADSpLRU. The improvement also can be clearly seen in the *DS* data set from Figure 8. J-LASU preserved the square regions better than the others. Compared with the TV results, difference can be recognized in the small square regions in which J-LASU reconstructed the squares better. For the *FR* data sets, visually, the ADSpLRU abundance maps showed the most similar intensity with the corresponding true abundance maps. However, J-LASU is superior in preserving the gradation of intensity from edge to center of an abundance region, which is the drawback of the ADSpLRU. Compared with SUnSAL-TV, J-LASU was more accurate in determining whether an abundance is an outlier or just a low-intensity edge abundance. In addition, SUnSAL-TV produced stronger smoothing effects than J-LASU. In this case, J-LASU results are more similar with the true abundance map, which can easily be compared in the *FR2* data set.

Table 2. RMSE Comparison Result.

Data	SNR	CLSUnSAL	SUnSAL-TV	ADSpLRU	J-LASU
DS	10	0.0084	0.0078	0.0097	**0.0035**
	20	0.0102	0.0046	0.0053	**0.0013**
	30	0.0039	0.0023	0.0038	**0.0008**
FR1	10	0.0130	0.0119	0.0140	**0.0103**
	20	0.0129	0.0087	0.0107	**0.0075**
	30	0.0062	0.0068	0.0073	**0.0050**
FR2	10	0.0140	0.0119	0.0149	**0.0104**
	20	0.0138	0.0083	0.0115	**0.0076**
	30	0.0062	0.0061	0.0066	**0.0052**
FR3	10	0.0136	0.0118	0.0130	**0.0099**
	20	0.0128	0.0077	0.0107	**0.0069**
	30	0.0056	0.0058	0.0057	**0.0044**
FR4	10	0.0123	0.0120	0.0135	**0.0103**
	20	0.0126	0.0089	0.0090	**0.0074**
	30	0.0057	0.0075	0.0058	**0.0049**
FR5	10	0.0118	0.0112	0.0139	**0.0092**
	20	0.0119	0.0080	0.0106	**0.0065**
	30	0.0049	0.0062	0.0061	**0.0043**

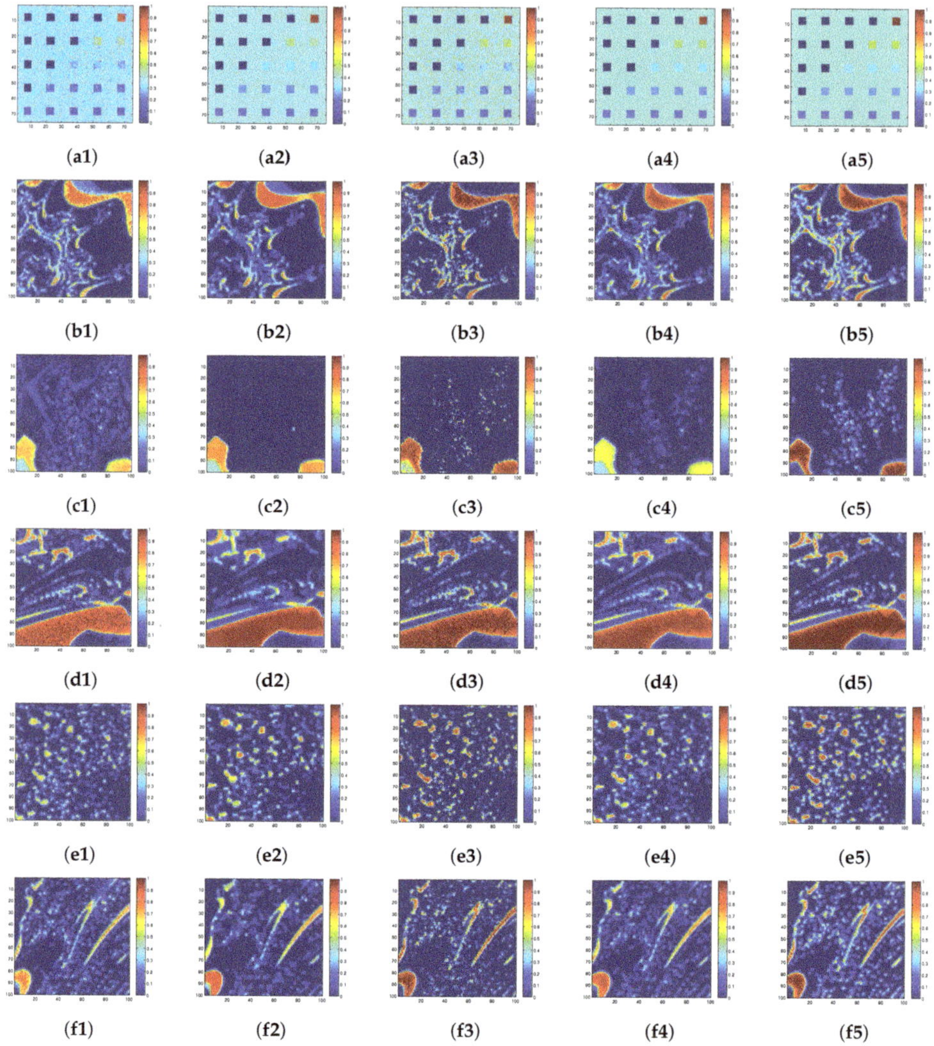

Figure 8. Estimated abundance maps for simulated data sets *DS* and *FR1–5* for SNR 30 dB (row **a–f**, respectively) using CLSUnSAL, SUnSAL-TV, ADSpLRU, and J-LASU (column **1–4**, respectively) compared to the true abundance (column **5**).

Table 3. SRE Comparison Result.

Data	SNR	CLSUnSAL	SUnSAL-TV	ADSpLRU	J-LASU
DS	10	2.5467	5.1021	0.3110	**7.2571**
	20	2.1617	6.3470	4.5515	**15.2631**
	30	6.3299	10.5770	6.1799	**20.0581**
FR1	10	0.6435	2.018	0.851	**2.3625**
	20	1.3116	3.5071	2.1257	**4.2158**
	30	4.2204	4.8625	4.0937	**6.0185**
FR2	10	0.3457	2.2395	0.2493	**2.4491**
	20	1.1915	3.8690	1.0974	**4.3822**
	30	4.4628	5.604	4.5908	**6.3273**
FR3	10	1.6928	4.0113	2.1009	**4.3074**
	20	3.1706	5.8611	2.3815	**6.5586**
	30	6.8354	6.9782	7.0605	**8.7567**
FR4	10	0.3417	1.3213	0.2092	**1.6307**
	20	1.0942	2.5735	0.3275	**3.5269**
	30	4.1734	3.263	3.5545	**5.4870**
FR5	10	1.005	2.4054	0.2591	**3.0209**
	20	1.5711	4.1026	1.228	**5.4771**
	30	6.3324	5.6279	6.0702	**7.7098**

4.5. Real-Data Experiment

The visual comparisons among the five sparse unmixing algorithms and the mineral maps for the Cuprite data can be observed in Figure 9. The images in the first column show the comparison for *alunite* abundance maps. Among the results of the compared algorithms, The proposed J-LASU produced the map that was the most similar to the mineral map, with less outliers found in the lower-left side of the map. The same superiority was also found among the *chalcedony* and *kaolinite* abundance maps in the second and third columns, respectively. Compared to SUnSAL-TV, J-LASU had less outliers or lower intensity of outliers, most of which were found on the left-side region of the maps.

It should be noted that the estimated abundance maps of any sparse unmixing algorithm are not exactly the same as the mineral maps generated from the Tricorder software in terms of intensity. The software produced the pixel-level classification maps, while the sparse unmixing algorithms executed subpixel-level classification. However, the comparison of outliers in this paper refers to the abundances that no longer exist in the mineral map. Overall, J-LASU estimated abundance maps had smooth gradation of intensity from the edge of a detected region to the center, and removed tiny regions that were found in the other algorithms' map, which seems to be the outliers in J-LASU algorithm.

For the Urban data, Figure 10 shows the ground truth and abundance maps of the four endmembers estimated by the compared algorithms. J-LASU algorithm resulted in the most similar maps to the ground truth, especially for the *asphalt* abundance map which is easier to be compared with those of the other algorithms. The quantitative comparisons also show that J-LASU yielded the best performance, with the highest SRE and lowest RMSE, as shown in Table 4. Compare to the simulated data, the Urban data experienced relatively high RMSEs for all compared algorithms. This is due to the fact that the ground truth abudance maps used for the Urban data are not achieved from a ground measurement, but from a method in which error possibly exists in term of method accuration.

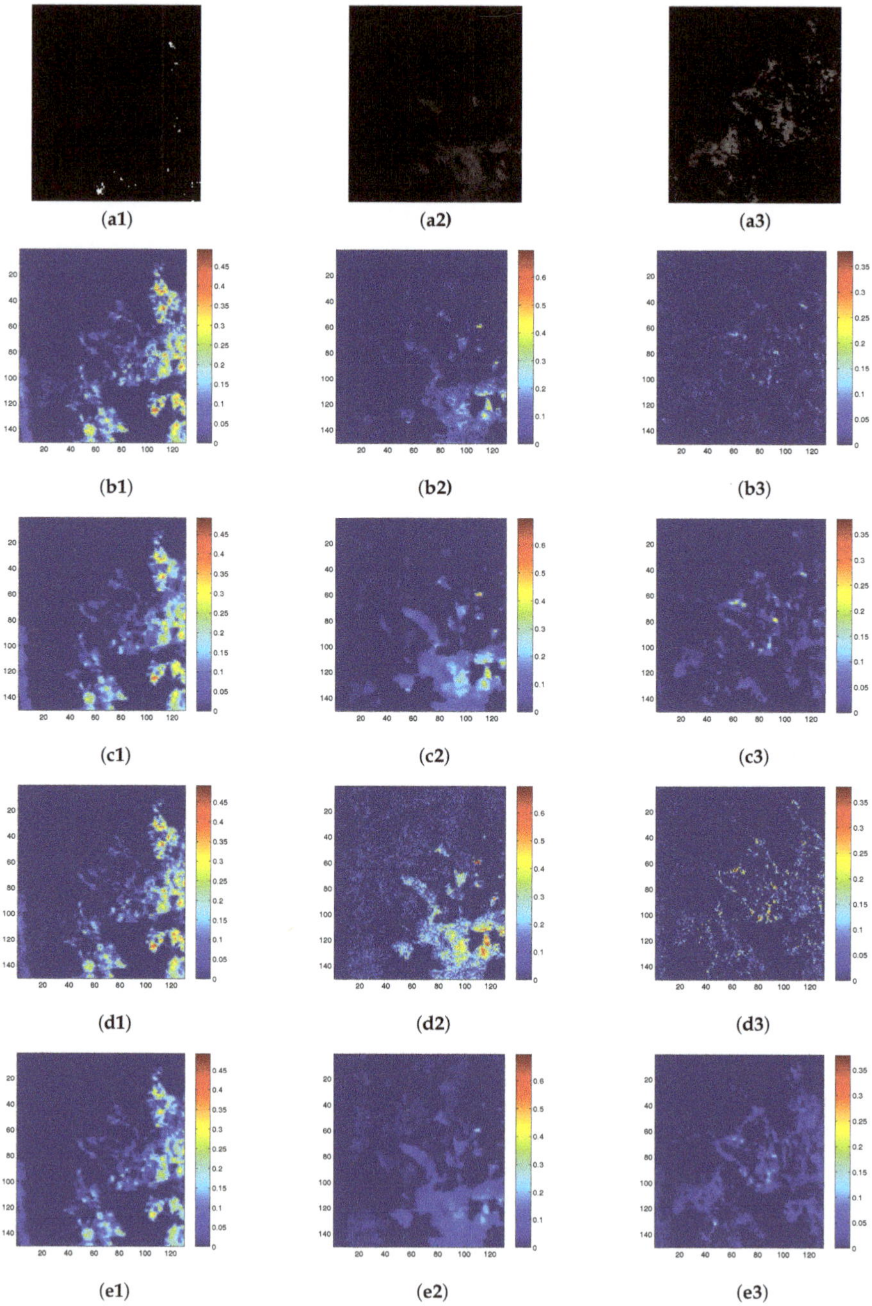

Figure 9. Estimated abundance maps of Cuprite data subscene for endmember *alunite*, *chalcedony*, and *kaolinite* (column **1–3**) using CLSUnSAL, SUnSAL-TV, ADSpLRU and J-LASU (row **b–e**). First row (**a**) shows classification maps of endmembers from USGS Tetracorder.

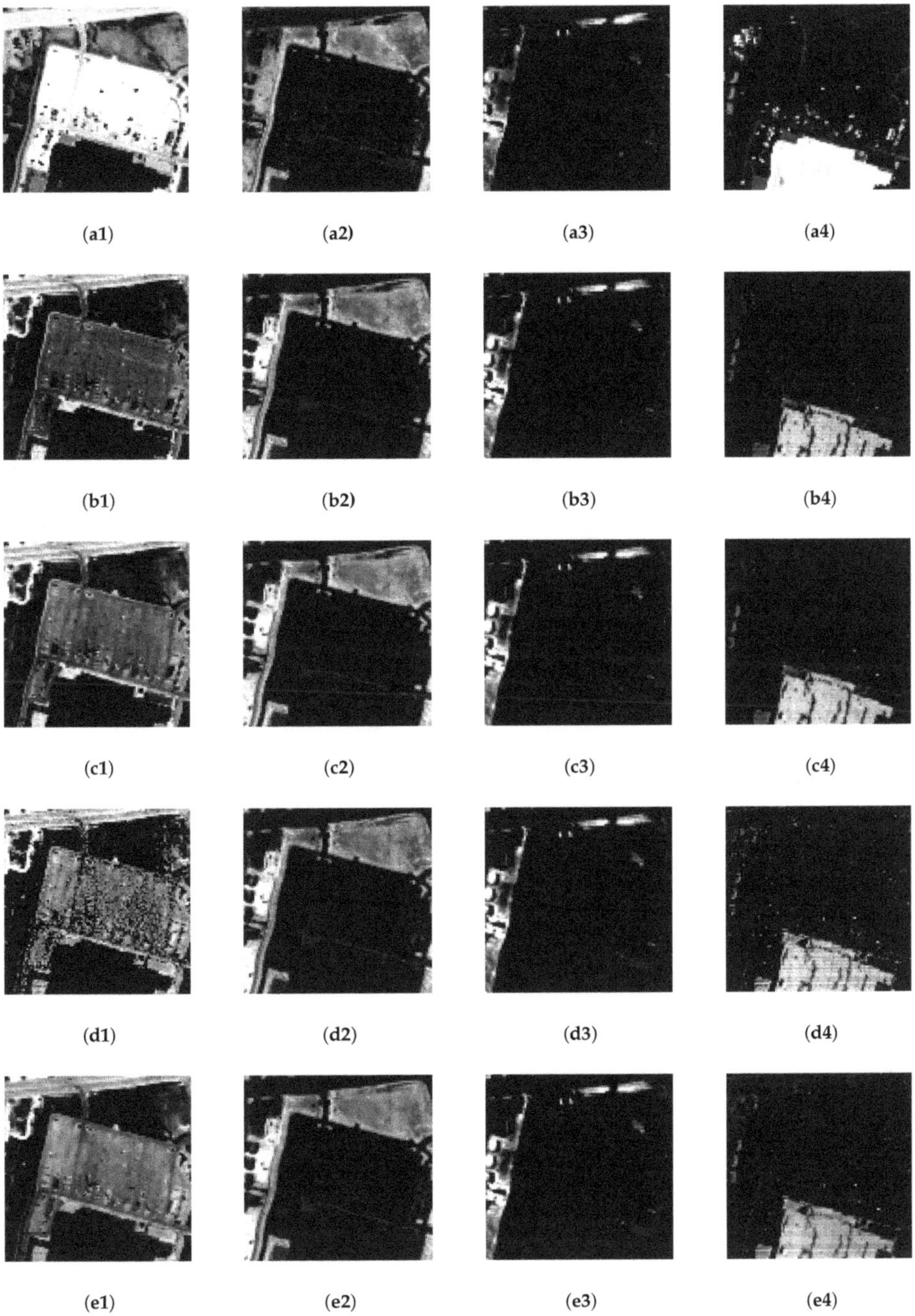

Figure 10. Estimated abundance maps of Urban data subscene for endmember *asphalt*, *grass*, *tree*, and *roof* (column **1–4**) using CLSUnSAL, SUnSAL-TV, ADSpLRU and J-LASU (row **b–e**). First row (**a**) shows the ground truth abundance maps.

Table 4. RMSE and SRE Comparison Result for Urban data.

Algorithms	CLSUnSAL	SUnSAL-TV	ADSpLRU	J-LASU
RMSE	0.2135	0.2003	0.2077	**0.1948**
SRE	4.6831	5.4738	5.0805	**5.8719**

5. Discussion

5.1. Sensitivity Test

We evaluated the performance of the proposed J-LASU algorithm when λ, γ, and ρ were not set to the optimal values. In the experiment, when a parameter was adjusted from 0 to 10, the other parameters were set to their optimal values. When the parameter increases from 0 to the optimal value, the RMSEs decrease and the SREs increase gradually. When it reaches a higher value, the results worsen. Hence, we can conclude that each parameter influences the performance of J-LASU.

To clearly evaluate the contribution of the LA regularizer, we conducted an experiment of our optimization problem with $\rho = 0$, which means no contribution of the LA regularizer. Figure 11 represent the RMSE of this condition at the three levels of SNR compared to those of J-LASU, where $\rho > 0$. For each simulated data set, other parameters were set to the optimum values. For all data, it was observed that when $\rho = 0$, the RMSE was higher than the condition when the LA regularizer was used. In other words, adding our LA regularizer with an optimal regularization parameter will contribute improvement in RMSE.

We found that improvement in visual quality corresponds to the additional low-rank regularization. Figure 12 shows visual improvement due to the abundance regularizer. The abundance maps in the figure belong to endmember 5 of the *DS* data set and endmember 7 of the *FR 2* data set. For the *FR* abundance maps, one can see that after applying our LA regularizer with an optimal ρ, the active abundances have higher intensities. The active abundances in the left-edge of the map and around the speckles clearly appear, although in lower intensities than in the true abundance map. For the *DS* data set, when ρ is set to the optimal value, the small squares are preserved better than when $\rho = 0$.

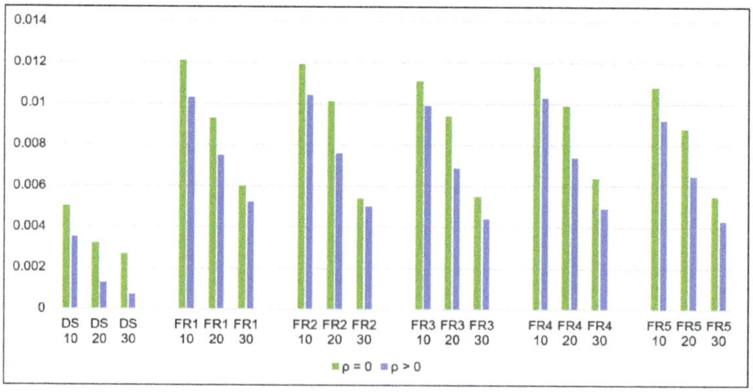

Figure 11. Effect of the LA regularizer represented by improvement in RMSE when $\rho > 0$.

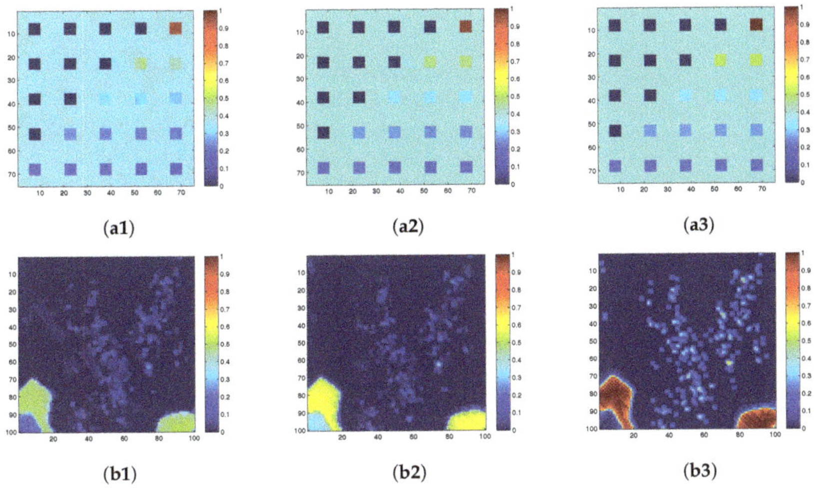

Figure 12. Effect of LA regularizer represented by improvement in RMSE when $\rho > 0$ for (**a**) *DS* data set and (**b**) *FR 2* data set. (**a1**) and (**b1**) Before, (**a2**) and (**b2**) after, (**a3**) and (**b3**) true abundance.

5.2. Effect of Block Size

The coverage of the local region affects the optimization results. In this region, the highly correlated abundance of the endmembers is taken into account by the local abundance nuclear norm. We conducted experiments to find the optimum size of the sliding block. We also observed the effect of the block size. Figure 13 shows the RMSE and SRE when the block size was adjusted in the *DS* data set. From the curves, we could determine that the radius of spatial similarity in the abundance map affects the optimum size of the sliding block. The distribution of spatial similarity in the *DS* data set, as shown in Figure 8, has a distinct pattern in which every 5×5 pixel has the same abundance value, giving the optimum block size in turn. However, the correlation does not hold for the data in which the spatial similarity is not represented in a square region, e.g., the *FR* and Cuprite data sets. In this circumstance, some trials were conducted prior to the experiment. After the trials, we found that the optimum size is 5 pixels. Hence, we selected [5 5 5] as the optimum block size for all data.

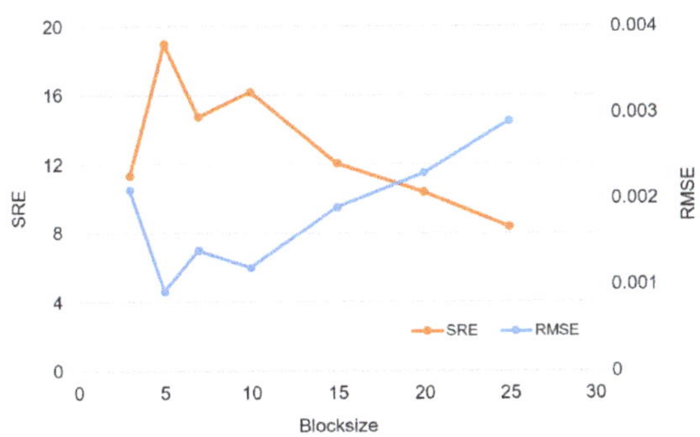

Figure 13. RMSE and SRE in relation to block size.

5.3. Computational Complexity

The running-time comparison among the algorithms is summarized in Table 5. The experiment was conducted for the *DS* simulated data, which has 75 × 75 pixels, 224 bands, and 240 spectral signatures in the library. The algorithms ran on a desktop computer with 3.50-GHz Intel Core i5 processor and 8 GB of RAM. From the table, J-LASU was the slowest due to its high computational complexity.

For the complexity analysis, recall that n, m, N, and m_b are the number of pixels, spectral signatures in the library, pixels in each LA band, and local endmembers, respectively. For each iteration of J-LASU, the computation of $\mathbf{X}$ and the SVD step in the computation of $\mathbf{V}_3$ incur the most cost. The complexity of $\mathbf{X}$ computation is due to the use of conjugate gradient solver, which costs $\mathcal{O}(m)$ per iteration. The conjugate gradient is a popular iterative technique for solving the system of linear equation $\mathbf{Ax} = \mathbf{b}$, where the matrix $\mathbf{A}$ must be symmetric possitive definite (SPD), large and sparse. The SVD step costs $\mathcal{O}(m_b^2 N)$; however, this step is repeated as many times as the number of blocks (B) due to the sliding of the local block. Since B is calculated by $mn/m_b N$, the total cost of $\mathbf{V}_3$ is $\mathcal{O}(m_b mn)$, which is more complex than the computation of $\mathbf{X}$. Hence, the overall complexity costs $\mathcal{O}(m_b mn)$.

Table 5. Comparison of running times for *DS*-data experiment.

Algorithms	CLSUnSAL	SUnSAL-TV	ADSpLRU	J-LASU
Time/iteration (s)	0.92	0.54	0.24	2.77

6. Conclusions

We proposed the local abundance regularizer algorithm for the sparse unmixing problem to improve the accuracy of abundance estimation. By imposing the term to state-of-the-art unmixing algorithms, our algorithm incorporates both spatial and abundance correlation by using the low-rankness of the abundance. We implemented the nuclear norm to the local abundance matrix, which defines the local region not only in the spatial, but also in the abundance dimension. The algorithm was run at certain SNR levels for several simulated data sets, which represent the conditions with and without pure pixels, and for two real data sets. The experimental results indicate that our proposed algorithm performs better than SUnSAL-TV and yields better results than the other state-of-the-art algorithms. Relevant future research will be concerned with exploitation of the low-rankness of abundance for overlapping local regions.

Acknowledgments: This work was supported in part by JSPS Grants-in-Aid (24560473), and MIC SCOPE (172310003). The first author acknowledges support from the Indonesia Endowment Fund for Education (LPDP), Ministry of Finance, The Republic of Indonesia.

Author Contributions: All the authors significantly contributed to different phases of this manuscript including the preparation, analysis, review and revision. The research plan and mathematical formulation were developed together. Mia Rizkinia implemented the algorithm, prepared the data, and executed the experiments. Masahiro Okuda supervised the research.

Conflicts of Interest: The authors declare no conflict of interest.

References

1. Rizkinia, M.; Okuda, M. Local abundance regularization for hyperspectral sparse unmixing. In Proceedings of the 2016 Asia-Pacific Signal and Information Processing Association Annual Summit and Conference (APSIPA), Jeju, Korea, 13–16 December 2016; pp. 1–6.
2. Keshava, N.; Mustard, J.F. Spectral unmixing. *IEEE Signal Process. Mag.* **2002**, *19*, 44–57.
3. Zhang, B.; Zhuang, L.; Gao, L.; Luo, W.; Ran, Q.; Du, Q. PSO-EM: A hyperspectral unmixing algorithm based on normal compositional model. *IEEE Trans. Geosci. Remote Sens.* **2014**, *52*, 7782–7792.

4. Song, M.; Ma, Q.; An, J.; Chang, C.I. An improved NMF algorithm based on spatial and abundance constraints. In Proceedings of the 2016 Progress in Electromagnetic Research Symposium (PIERS), Shanghai, China, 8–11 August 2016; pp. 4532–4537.
5. Shippert, P. Why use hyperspectral imagery. *Photogramm. Eng. Remote Sens.* **2004**, *70*, 377–396.
6. Landgrebe, D. Hyperspectral image data analysis. *IEEE Signal Process. Mag.* **2002**, *19*, 17–28.
7. Heiden, U.; Iwasaki, A.; Müller, A.; Schlerf, M.; Udelhoven, T.; Uto, K.; Yokoya, N.; Chanussot, J. Foreword to the special issue on hyperspectral remote sensing and imaging spectroscopy. *IEEE J. Sel. Top. Appl. Earth Obs. Remote Sens.* **2016**, *9*, 3904–3908.
8. Ahmed, A.M.; Duran, O.; Zweiri, Y.; Smith, M. Hybrid spectral unmixing: Using artificial neural networks for linear/non-linear switching. *Remote Sens.* **2017**, *9*, 775.
9. Licciardi, G.A.; Sellitto, P.; Piscini, A.; Chanussot, J. Nonlinear spectral unmixing for the characterisation of volcanic surface deposit and airborne plumes from remote sensing imagery. *Geosciences* **2017**, *7*, 46.
10. Yokoya, N.; Chanussot, J.; Iwasaki, A. Nonlinear unmixing of hyperspectral data using semi-nonnegative matrix factorization. *IEEE Trans. Geosci. Remote Sens.* **2014**, *52*, 1430–1437.
11. Zhong, Y.; Feng, R.; Zhang, L. Non-local sparse unmixing for hyperspectral remote sensing imagery. *IEEE J. Sel. Top. Appl. Earth Obs. Remote Sens.* **2014**, *7*, 1889–1909.
12. Bioucas-Dias, J.M.; Plaza, A.; Dobigeon, N.; Parente, M.; Du, Q.; Gader, P.; Chanussot, J. Hyperspectral unmixing overview: Geometrical, statistical, and sparse regression-based approaches. *IEEE J. Sel. Top. Appl. Earth Obs. Remote Sens.* **2012**, *5*, 354–379.
13. Nascimento, J.M.P.; Dias, J.M.B. Vertex component analysis: A fast algorithm to unmix hyperspectral data. *IEEE Trans. Geosci. Remote Sens.* **2005**, *43*, 898–910.
14. Berman, M.; Kiiveri, H.; Lagerstrom, R.; Ernst, A.; Dunne, R.; Huntington, J.F. ICE: A statistical approach to identifying endmembers in hyperspectral images. *IEEE Trans. Geosci. Remote Sens.* **2004**, *42*, 2085–2095.
15. Lee, D.D.; Seung, H.S. Learning the parts of objects by nonnegative matrix factorization. *Nature* **1999**, *401*, 788–791.
16. Pauca, V.P.; Piper, J.; Plemmons, R.J. Nonnegative matrix factorization for spectral data analysis. *Linear Algebra Appl.* **2006**, *416*, 29–47.
17. Liu, R.; Du, B.; Zhang, L. Hyperspectral unmixing via double abundance characteristics constraints based NMF. *Remote Sens.* **2016**, *8*, 464.
18. Iordache, M.D.; Plaza, A.J. A Sparse Regression Approach to Hyperspectral Unmixing. Ph.D. Thesis, Instituto Superior Tecnico, Lisbon, Portugal, 2011.
19. Li, C.; Ma, Y.; Mei, X.; Liu, C.; Ma, J. Hyperspectral unmixing with robust collaborative sparse regression. *Remote Sens.* **2016**, *8*, 588.
20. Ghasrodashti, E.K.; Karami, A.; Heylen, R.; Scheunders, P. Spatial resolution enhancement of hyperspectral images using spectral unmixing and bayesian sparse representation. *Remote Sens.* **2017**, *9*, 541.
21. Esmaeili Salehani, Y.; Gazor, S.; Kim, I.M.; Yousefi, S. ℓ_0-norm sparse hyperspectral unmixing using arctan smoothing. *Remote Sens.* **2016**, *8*, 187.
22. Tang, W.; Shi, Z.; Wu, Y.; Zhang, C. Sparse unmixing of hyperspectral data using spectral a priori information. *IEEE Trans. Geosci. Remote Sens.* **2015**, *53*, 770–783.
23. Chen, X.; Chen, J.; Jia, X.; Somers, B.; Wu, J.; Coppin, P. A quantitative analysis of virtual endmembers' increased impact on the collinearity effect in spectral unmixing. *Trans. Geosci. Remote Sens.* **2011**, *49*, 2945–2956.
24. Themelis, K.E.; Rontogiannis, A.A.; Koutroumbas, K. Semi-supervised hyperspectral unmixing via the weighted Lasso. In Proceedings of the IEEE International Conference on Acoustics, Speech and Signal Processing, Dallas, TX, USA, 14–19 March 2010.
25. Bioucas-Dias, J.M.; Figueiredo, M.A.T. Alternating direction algorithms for constrained sparse regression: Application to hyperspectral unmixing. In Proceedings of the 2010 2nd Workshop on Hyperspectral Image and Signal Processing: Evolution in Remote Sensing (WHISPERS), Reykjavik, Iceland, 14–16 June 2010; pp. 1–4.
26. Iordache, M.D.; Bioucas-Dias, J.M.; Plaza, A. Sparse unmixing of hyperspectral data. *IEEE Trans. Geosci. Remote Sens.* **2011**, *49*, 2014–2039.
27. Iordache, M.D.; Bioucas-Dias, J.M.; Plaza, A. Total variation spatial regularization for sparse hyperspectral unmixing. *IEEE Trans. Geosci. Remote Sens.* **2012**, *50*, 4484–4502.

28. Iordache, M.D.; Bioucas-Dias, J.M.; Plaza, A. Collaborative sparse regression for hyperspectral unmixing. *IEEE Trans. Geosci. Remote Sens.* **2014**, *52*, 341–354.
29. Zhang, S.; Li, J.; Liu, K.; Deng, C.; Liu, L.; Plaza, A. Hyperspectral unmixing based on local collaborative sparse regression. *IEEE. Trans. Geosci. Remote Sens. Lett.* **2016**, *13*, 631–635.
30. Qu, Q.; Nasrabadi, N.M.; Tran, T.D. Abundance estimation for bilinear mixture models via joint sparse and low-rank representation. *IEEE. Trans. Geosci. Remote Sens.* **2014**, *52*, 4404–4423.
31. Zhao, Y.; Yang, J. Hyperspectral image denoising via sparsity and low rank. In Proceedings of the 2013 IEEE International Geoscience and Remote Sensing Symposium (IGARSS), Melbourne, Australia, 21–26 July 2013; pp. 1091–1094.
32. Zhang, H.; He, W.; Zhang, L.; Shen, H.; Yuan, Q. Hyperspectral image restoration using low-rank matrix recovery. *IEEE. Trans. Geosci. Remote Sens.* **2014**, *52*, 4729–4743.
33. Zhao, Y.Q.; Yang, J. Hyperspectral image denoising via sparse representation and low-rank constraint. *IEEE. Trans. Geosci. Remote Sens.* **2015**, *53*, 296–308.
34. Giampouras, P.V.; Themelis, K.E.; Rontogiannis, A.A.; Koutroumbas, K.D. Simultaneously sparse and low-rank abundance matrix estimation for hyperspectral image unmixing. *IEEE. Trans. Geosci. Remote Sens.* **2016**, *54*, 4775–4789.
35. Ono, S.; Yamada, I. Color-line regularization for color artifact removal. *IEEE Trans. Comput. Imaging* **2016**, *2*, 204–217.
36. Yang, J.; Zhao, Y.Q.; Chan, J.C.W.; Kong, S.G. Coupled sparse denoising and unmixing with low-rank constraint for hyperspectral image. *IEEE. Trans. Geosci. Remote Sens.* **2016**, *54*, 1818–1833.
37. Candes, E.J.; Tao, T. Decoding by linear programming. *IEEE Trans. Inf. Theor.* **2005**, *51*, 4203–4215.
38. Rizkinia, M.; Baba, T.; Shirai, K.; Okuda, M. Local spectral component decomposition for multi-channel image denoising. *IEEE Trans. Image Process.* **2016**, *25*, 3208–3218.
39. Plaza, J.; Hendrix, E.M.T.; García, I.; Martín, G.; Plaza, A. On endmember identification in hyperspectral images without pure pixels: A comparison of algorithms. *J. Math. Imaging Vis.* **2012**, *42*, 163–175.
40. Moreau, J.J. Fonctions convexes duales et points proximaux dans un espace hilbertien. *C.R. Acad. Sci. Paris Ser. A Math.* **1962**, *255*, 2897–2899.
41. Combettes, P.L.; Pesquet, J.C. Proximal splitting methods in signal processing. In *Fixed-Point Algorithms for Inverse Problems in Science and Engineering*; Springer: New York, NY, USA, 2011; pp. 185–212.
42. USGS Digital Spectral Library 06. Available online: https://speclab.cr.usgs.gov/spectral.lib06/ (accessed on 10 June 2016).
43. AVIRIS Data. Available online: https://aviris.jpl.nasa.gov/data/free_data.html (accessed on 10 June 2016).
44. Cuprite, Nevada AVIRIS 1995 Data USGS. Available online: https://speclab.cr.usgs.gov/cuprite95.tgif.2.2um_map.gif (accessed on 11 January 2017).
45. Datasets and Ground Truths. Available online: http://www.escience.cn/people/feiyunZHU/Dataset_GT.html (accessed on 10 November 2017).
46. Jia, S.; Qian, Y. Spectral and spatial complexity-based hyperspectral unmixing. *IEEE Trans. Geosci. Remote Sens.* **2007**, *45*, 3867–3879.
47. Jia, S.; Qian, Y. Constrained nonnegative matrix factorization for hyperspectral unmixing. *IEEE Trans. Geosci. Remote Sens.* **2009**, *47*, 161–173.
48. Zhu, F.; Wang, Y.; Xiang, S.; Fan, B.; Pan, C. Structured sparse method for hyperspectral unmixing. *ISPRS J. Photogramm. Remote Sens.* **2014**, *88*, 101–118.
49. Zhu, F.; Wang, Y.; Fan, B.; Meng, G.; Xiang, S.; Pan, C. Spectral unmixing via data-guided sparsity. *IEEE Trans. Image Process.* **2014**, *23*, 5412–5427.
50. Zhu, F.; Wang, Y.; Fan, B.; Meng, G.; Pan, C. Effective spectral unmixing via robust representation and learning-based sparsity. *arXiv* **2014**, arXiv:1409.0685.
51. Zortea, M.; Plaza, A. Spatial preprocessing for endmember extraction. *IEEE Trans. Geosci. Remote Sens.* **2009**, *47*, 2679–2693.

remote sensing

MDPI

Article

Criteria Comparison for Classifying Peatland Vegetation Types Using In Situ Hyperspectral Measurements

Thierry Erudel [1,2,3,*], Sophie Fabre [3], Thomas Houet [4], Florence Mazier [2] and Xavier Briottet [3]

1. LabEx DRIIHM (Programme "Investissements D'avenir": ANR-11-LABX-0010), INEE-CNRS 3 Rue Michel-Ange, 75016 Paris, France
2. GEODE UMR 5602 CNRS, Université Toulouse Jean Jaurès, 5 Allées Antonio Machado, 31058 Toulouse CEDEX 1, France; florence.mazier@univ-tlse2.fr
3. ONERA, Optics and Associated Techniques Department, 2 Avenue Edouard Belin, 31005 Toulouse CEDEX, France; Sophie.Fabre@onera.fr (S.F.); Xavier.Briottet@onera.fr (X.B.)
4. LETG-Rennes UMR 6554 CNRS, Université Rennes 2, Place du Recteur Henri le Moal, 35043 Rennes CEDEX, France; thomas.houet@univ-rennes2.fr

* Correspondence: Thierry.Erudel@onera.fr

Academic Editors: Chein-I Chang, Meiping Song, Junping Zhang, Chao-Cheng Wu and Prasad Thenkabail
Received: 24 May 2017; Accepted: 9 July 2017; Published: 20 July 2017

Abstract: This study aims to evaluate three classes of methods to discriminate between 13 peatland vegetation types using reflectance data. These vegetation types were empirically defined according to their composition, strata and biodiversity richness. On one hand, it is assumed that the same vegetation type spectral signatures have similarities. Consequently, they can be compared to a reference spectral database. To catch those similarities, several similarities criteria (related to distances (Euclidean distance, Manhattan distance, Canberra distance) or spectral shapes (Spectral Angle Mapper) or probabilistic behaviour (Spectral Information Divergence)) and several mathematical transformations of spectral signatures enhancing absorption features (such as the first derivative or the second derivative, the normalized spectral signature, the continuum removal, the continuum removal derivative reflectance, the log transformation) were investigated. Furthermore, those similarity measures were applied on spectral ranges which characterize specific biophysical properties. On the other hand, we suppose that specific biophysical properties/components may help to discriminate between vegetation types applying supervised classification such as Random Forest (RF), Support Vector Machines (SVM), Regularized Logistic Regression (RLR), Partial Least Squares-Discriminant Analysis (PLS-DA). Biophysical components can be used in a local way considering vegetation spectral indices or in a global way considering spectral ranges and transformed spectral signatures, as explained above. RLR classifier applied on spectral vegetation indices (training size = 25%) was able to achieve 77.21% overall accuracy in discriminating peatland vegetation types. It was also able to discriminate between 83.95% vegetation types considering specific spectral range [350–1350 nm], first derivative of spectral signatures and training size = 25%. Conversely, similarity criterion was able to achieve 81.70% overall accuracy using the Canberra distance computed on the full spectral range [350–2500 nm]. The results of this study suggest that RLR classifier and similarity criteria are promising to map the different vegetation types with high ecological values despite vegetation heterogeneity and mixture.

Keywords: biodiversity; peatland; vegetation type; classification; hyperspectral; in situ measurements

1. Introduction

Peatlands represent a diverse array of wetlands that accumulate partially decomposed organic material. Whilst they may only cover a small proportion (~3%) of the Earth's land surface, these ecosystems are highly important in terms of functional and ecological values. Indeed, undisturbed, global peatland systems act as net atmospheric carbon sinks, storing approximately a third of the world's soil organic carbon [1], the vast majority of which (450–547 GtC (Gigatons of Carbon)) is held in northern peatlands (those above 45°N [2]). From an ecological perspective, these environments also provide important habitats for a number of rare plant and animal species [3].

Traditionally, species discrimination for floristic mapping needs intensive field work, including taxonomical information and the visual estimation of percentage cover for each species which are costly and time-consuming and sometimes inapplicable due to their poor accessibility [4]. Remote sensing is a technique that gathers data regularly about the earth's features. The main advantages that make remote sensing preferable to field-based methods in land cover classification, are that it has repeat coverage potential, allowing continuous monitoring, and its digital data can be easily integrated into a geographic information system (GIS) for more analysis which is less costly and less time-consuming [5,6].

Historically, aerial photography was the first remote sensing method to be employed for mapping wetland vegetation [7]. Currently, a variety of remotely sensed images are available for mapping wetland vegetation thanks to of airborne and space-borne vectors with multi-spectral sensors or hyperspectral sensors which operate within the different optical spectra [8].

Mapping and monitoring wetlands' (and even though peatland) floristic diversity is really challenging. Indeed, both temporal and spatial resolutions of remotely sensed imageries and in situ plant diversity and mixing contribute to the limitation of such techniques. Wetland plants are not as easily detectable as terrestrial plants since herbaceous wetland vegetations exhibit high spectral and spatial variabilities because of its steep environmental gradients [5,8]. Besides, the reflectance spectra of wetland vegetation canopies are often very similar and can be combined with reflectance spectra of the underlying soil, hydrologic regime and atmospheric vapour [9,10].

However, plant species have been successfully classified in estuarine [11], palustrine [12] and riparian habitats [13], as well in saltmarsh [5], in mangrove [14,15], in swamp [16] but not in peatlands, to our knowledge. Peatland mapping faces two great challenges at local and global scales due to their high environmental function (biodiversity hotspot, greenhouse gas fluxes, etc.): characterizing their internal diversity [8] and delineating their extent [17]. This study focuses on the first challenge for which only high-spectral or spatial-resolution imageries appear appropriate (see for instance [18–20]).

Plant species classification can benefit from several existing and recent techniques commonly used in remote sensing. Two main methods are applied for vegetation discrimination: the similarity measurement techniques and the supervised classification methods with sometimes application of a preliminary spectral band reduction technique. On one hand, similarity measures enable us to discriminate between similar classes from a set of spectra, extracted from images or acquired on the field. Some spectral measures, such as the Spectral Angle Mapper (SAM) are related to the difference of the spectral shape (e.g., Yagoub, H. et al. [21] identified forests of the Liege oaks from other forests, grain crops and steppes using the multispectral Advanced Very High Resolution Radiometer (AVHRR) with five bands from 580 nm to 1250 nm, 1 km spatial resolution (Overall Accuracy (OA) = 94.10%, $\kappa = 0.93$); Bahri, E.M. et al. [22] discriminated between tree species using the multispectral Advanced Spaceborne Thermal Emission and Reflection Radiometer (ASTER) sensor with 9 spectral bands from 520 nm to 2430 nm and a spatial resolution of 15 m or 30 m ($\kappa = 0.66$)). Other spectral measures, such as the Spectral Information Divergence (SID) are related to probabilistic behaviour (e.g., Sobhan, I. [23] classified different tree species at leaf and vegetation cover scales using the hyperspectral HyMap sensor: 126 spectral bands from 436 nm to 2485 nm and a spatial resolution of 4 m (OA = 91.10%, $\kappa = 0.87$)). On the other hand, the supervised classification methods may contribute as well to discriminate between (group of) spectral signatures for plant species discrimination.

The Linear Discriminant Analysis (LDA) is a method assuming that independent variables are normally distributed and which attempts to look for linear combination of variables to model the difference between the classes of the data (e.g., Clark, M.L. et al. [24] succeeded in classifying different tree species at leaf and vegetation cover scales using the HYperspectral Digital Imagery Collection Experiment (HYDICE) sensor with 210 spectral bands from 400 nm to 2500 nm, 1.6 m spatial resolution (OA = 86% using an object-based approach)). The Random Forest is an ensemble learning method based on the construction of multiple decision trees (e.g., Lawrence, R.L. et al. [25] succeeded in mapping invasive plants using the hyperspectral Probe-1 sensor: 128 bands from 450 nm to 2507 nm, 5 m spatial resolution (OA = 86% for the leafy spurge classification)). The Support Vector Machines (SVM) is a classifier that looks for the best separating hyperplane (e.g., Dalponte, M. [26] succeeded in classifying different tree species in boreal forest using HySpex VNIR-1600-instrument: 160 spectral bands ranging from 410 nm to 990 nm , with a spatial resolution of 0.4 m (OA = 79.2%); Vyas, D. et al. [27] classified successfully tropical vegetation using the Hyperion (EO-1) sensor (OA = 80%)). The Regularized Logistic Regression (RLR) is the combination of a linear model (logistic regression) and a regularization term. It is usually used for feature selection (e.g., Pant, P. et al. [28] applied it to reduce the 64 spectral bands from the hyperspectral AisaEAGLE II sensor to classify tree species in boreal forest using SVM; Pal, M. [29] applied it for reducing the 79 bands from the hyperspectral Digital Airborne Imaging Spectrometer (DAIS) sensor and the 220 bands from the hyperspectral Airborne Visible/Infrared Imaging Spectrometer (AVIRIS) sensor to classify different land covers using SVM) is investigated in this paper as a classifier.

Discriminating between and classifying plant species can be done. Firstly, using different techniques hyperspectral measurements can be made thanks to a portable spectroradiometer (FieldSpec Pro FR, Analytical Spectral Devices—ASD) which ranges on the reflective domain ([350–2500 nm] with a spectral resolution of 3 nm in Visible and Near InfraRed (VNIR) and approximatively 10 nm in the ShortWave InfraRed (SWIR)) either on laboratory [14] or immediately after the leaf was cut using the leaf clip accessory [16]. This can be an indicator of the ability of discriminating plant species using specific wavelengths or evaluating the performance of a classifier. Then, the wetlands heterogeneity mixing vegetation types can be catched still using a portable spectroradiometer: orbick, N. et al. [12] used the ASD spectroradiometer, Ground Field of View (GFOV) = 0.43 m; Schmidt, K. et al. [5] used the GER 3700 (Geophysical and Environmental Research Corporation) which ranges from 350 nm to 2509 nm) with a spectral resolution of 2 nm below 1000 nm and from 6 to 10 nm beyond 1000 nm, GFOV = 0.13 m. Secondly, with airborne imageries, hyperspectral sensors (SOC-700: 120 spectral bands between 394 and 890 nm with a 4 nm bandwidth and a spatial resolution of 0.5 m and a spatial resolution of 3 m [13]; HyMap: 128 bands in the visible and near infrared (VNIR: 0.45–1.50 µm with a 10 nm bandwidth) through the shortwave infrared (SWIR: 1.50–2.50 µm with a 15–20 nm bandwidth [11]). Thirdly, with spaceborne imageries using hyperspectral sensors (Hyperion: 242 spectral bands from 357 to 2756 nm with a spectral interval of 10 nm and a spatial resolution of 30 m [15]) or multispectral sensors (SPOT-5: 4 bands with 10 m resolution [15]) can be used to map wetlands.

This study aims at inventorying and evaluating the performance of discrimination techniques for peatland habitats based on in situ spectra. These habitats are characterized by more or less homogeneous vegetation mixing and have been chosen because of their ecological values (i.e., biodiversity). As defined by [30], mapping these habitats is therefore important to identify potential and/or effective areas with (at least) a floristic biodiversity function. For instance, we do not aim at detecting *Drosera rotundifolia* but at mapping the habitat favorable to this species (*Sphagnum* ...). Similarity measures and classifiers were applied on spectral signatures and some of their transformations (first and second derivatives, continuum removal, first derivative of continuum removal, normalized spectral signatures, log transformation). These transformations have been chosen because they enhance biophysical components which may help to distinguish plant species. These techniques were applied on different spectral ranges that either characterize specific biophysical components [31]. Classifiers were applied on spectral vegetation indices, characterizing specific biophysical components such as chlorophyll, pigments, nitrogen, cellulose, water.

This paper is organized as follows. After presenting the study site located in the Pyrenees (France) and associated data collection in Section 2 the methodology is detailed in Section 3. Then Section 4 presents and discussed the results of the different classifications that are suitable for distinguishing vegetation types. Finally, in Section 5, the conclusion summarizes the main results and some perspectives that have arisen in applying these techniques to hyperspectral imageries.

2. Material

2.1. Study Site

The study site is the Bernadouze peatbog (Latitude: 42°47′N, Longitude: 1°24′E; approximatively 2 ha), which is part of Human-Nature Observatory "Haut-Vicdessos" located in Ariège (Pyrénées, France) (Figure 1) and supported by the French CNRS and the LabEx DRIIHM. It is a long term monitored study site where hydrological, climatological, botanical, archeological, remotely sensed surveys are regularly conducted.

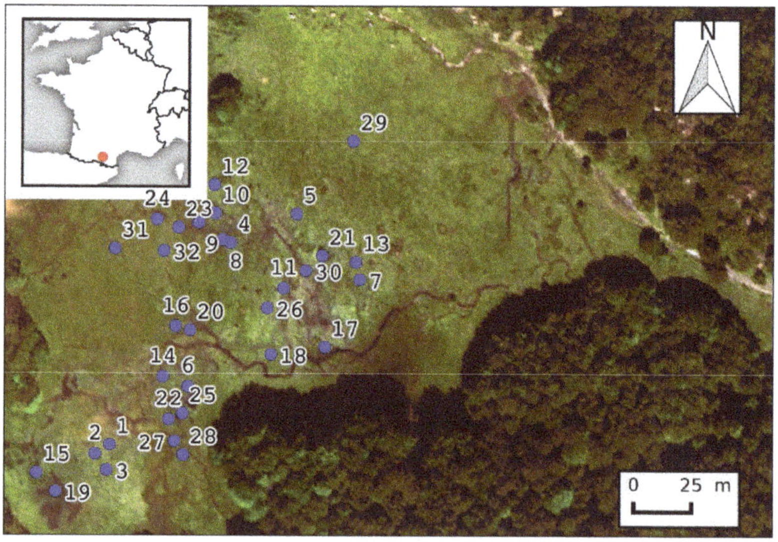

Figure 1. Location of the in situ spectroradiometer measurements—True color composite made from hyperspectral (HySpex) aerial imageries acquired on the 09/12/2014 (R = 639.98 nm, G = 549.06 nm, B = 461.79 nm).

2.2. Field Data Collection

In this study, thirteen vegetation units with ecological values and potentials (i.e., biodiversity) have been identified in the Bernadouze peatbog. These units are named hereafter "vegetation types" according to the dominant land cover type or to the potential development of interesting plant species which may have ecological values (Table 1). For each type, several locations have been surveyed to characterize their plant species composition (Table A1).

For all these 32 sample locations (Figure 1), radiances are measured at three different dates over 9 days in September 2014 (4 September 2014, 5 September 2014, 12 September 2014) under sunny and cloudless conditions between 10:00 a.m. and 1:00 p.m. and Sun's azimuth angle ranging from 106° and 160°. Data have been collected using an Analytical Spectral Device (ASD) spectroradiometer which ranges on the reflective domain (350–2500 nm) with a 3–12 nm spectral resolution depending on the spectral domain. Its spectral specifications are summarized in Table 2.

Table 1. Species names, number of measurements, number of locations and total number of spectra collected.

Vegetation Types	Code	Measurements			No. of Locations	No. of Spectra
		09/04/2014	09/05/2014	09/12/2014		
Calluna vulgaris	CAVU		2	2	2	14
Sphagnum sp.	SPHA		2	4	5	22
Eleocharis quinqueflora	ELQU	1	2	1	2	15
Pinguicula sp.	PING	1	1		1	8
Menyanthes trifoliata	METR	1	1	1	1	12
Juniperus communis	JUCO	1	2	2	2	19
Rhododendron ferrugineum	RHFR		2	2	2	14
Salix sp.	SALI	1		3	3	17
Aquatic environment a	AQ_A	3	6	7	6	53
Aquatic environment b	AQ_B		1	1	1	7
Aquatic environment c	AQ_C	1	1	1	1	12
Carex sp. homogeneous vegetation	CA_HV	2	2	3	4	26
Pinguicula sp. combined vegetation	PI_CV	1	2	1	2	15

Table 2. Analytical Spectral Device (ASD) FieldSpec Pro specifications.

	Spectral Range	Spectral Resolution	Spectral Sampling
VNIR (Visible and Near InfraRed)	0.35 µm–1.00 µm	3.00 nm at 0.70 µm	1.40 nm (0.35 µm–1.05 µm)
SWIR (Short Wave InfraRed)	1.00 µm–2.05 µm	10.00 nm at 1.40 µm 12.00 nm at 2.10 µm	2.00 nm (1.05 µm–2.50 µm)

To measure the reflectance of a sample plot (ρ) the reflectance of a white reference (ρ_{ref}) is required. This latter was obtained with a Spectralon (Labsphere, North Sutton, NH, USA) panel. Finally, after dark current correction, ρ is given by:

$$\rho = \frac{L_{\text{sam}}}{L_{\text{ref}}}\rho_{\text{ref}}, \tag{1}$$

where L_{sam} is the measured radiance from the sample plot and L_{ref} is the measured radiance from the white reference.

The sensor was positioned approximatively 1 m over the target with a 10° field of view. Consequently the ground spatial resolution is 0.18 m. The ASD was configured to collect 20 samples and automatically average in order to provide a single mean spectral measurement. Then a total of 7 to 53 field spectroradiometer measurements, i.e., spectral signatures, depending on vegetation type was taken.

2.3. Data Preprocessing

Some spectral bands (1350 nm to 1450 nm, 1810 nm to 1940 nm and 2400 nm to 2500 nm) have been removed due to a small signal-to-noise ratio resulting from strong atmospheric absorption mainly due to the presence of water vapour. More precisely, if the atmospheric transmittance value of the U.S. Standard profile was lower than 0.8 for a given wavelength, this wavelength was not taken into account in the analyse. Thus, each measured spectrum has been smoothed using a Savitzky-Golay filter [32] for reducing the noise. Figure 2 graphs the mean spectral reflectance of each vegetation type and the atmospheric transmittance. For the sake of clarity, the standard deviation of each vegetation type is not printed on Figure 2 but can be seen in Appendix B.

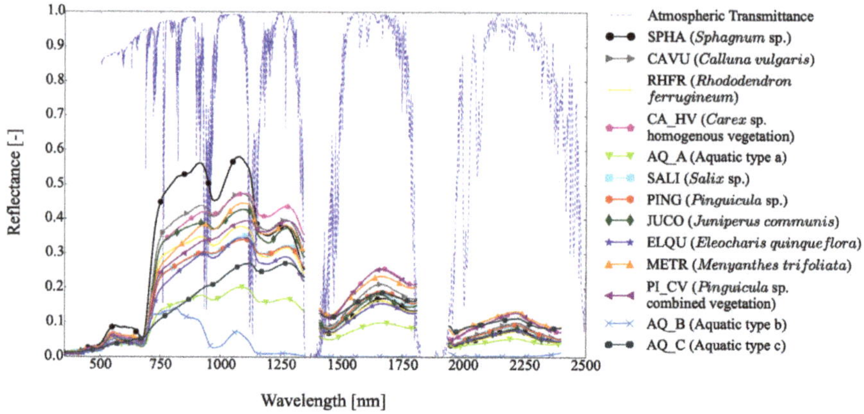

Figure 2. Mean spectral reflectances of the 13 vegetation types and the U.S. Standard atmospheric transmittance.

3. Method Description

The flowchart to evaluate the potential of hyperspectral data to discriminate between and classify wetland vegetation types is given in Figure 3. More precisely, three classes of methods have been investigated and compared:

- similarity measures calculated on spectral reflectance,
- supervised classification based on "local" information (spectral vegetation indices),
- supervised classification based on "global" information (spectral ranges).

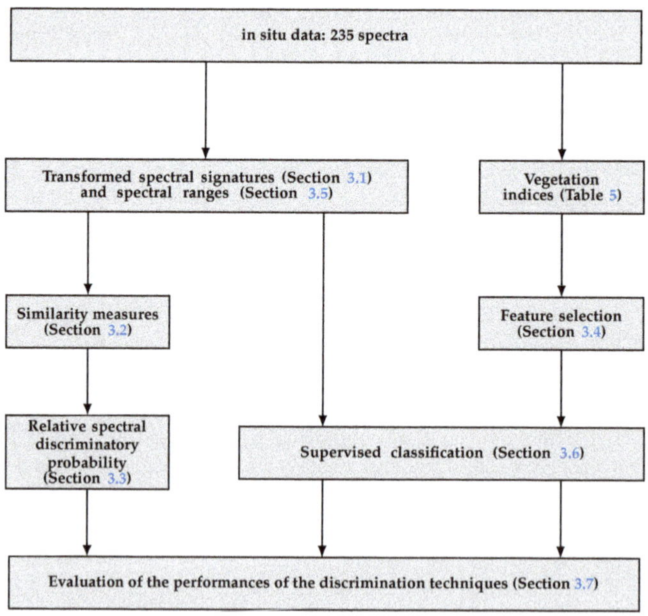

Figure 3. Flowchart showing the different methods used to classify the vegetation types.

Indeed, spectral matching can be used to discriminate between different vegetation types, because it is assumed that the spectral signatures of a given vegetation type must have similarities. To catch those similarities, several mathematical transformations—enhancing absorption features are applied on spectral signatures—(Section 3.1) and several similarity criteria—related to distances or spectral shapes or probabilistic behaviour—(Section 3.2) are investigated. Furthermore those similarity measures are applied on several spectral ranges which characterize specific biophysical properties (Section 3.5) and compared to a reference spectral database using relative spectral discriminatory probability (Section 3.3).

On the other hand as it may be difficult to have a spectral reference database, different supervised classifiers are used (Section 3.6). Besides, we assume that specific biophysical properties/components may help discriminating vegetation types. Biophysical components can be used in a local way considering spectral vegetation indices (Section 3.4.3) or in a global way considering spectral ranges and transformed spectral signatures as explained above.

To evaluate performance of similarity measures and supervised classification, the overall accuracy and F1-score are used (Section 3.7).

3.1. Transformed Spectral Signatures

As vegetation types are composed by a mix of various plant species that can be found in various vegetation types, different transformations are used (Table 3). Brightness-normalized spectral signature and second derivative are relatively insensible to variations in illumination intensity causes by changes in sun angle [33,34]. Other transformations (first derivative, second derivative, log transformation, Continuum Removal, Continuum Removed Derivative Reflectance (CRDR)) are linked to absorption features that may differ from one vegetation type to another, depending on the floristic composition.

Table 3. Transformed spectral signatures.

Transformation	Formulation	Reference
Brightness-normalized spectral signature	$\tilde{\rho}_{i,\lambda} = \frac{\rho_{i,\lambda}}{\left(\sum_{\lambda=1}^{L} \rho_{i,\lambda}^2\right)^{\frac{1}{2}}}, \forall\lambda \in [1,...,L].$	[33]
First derivative	$\left.\frac{\mathrm{d}\rho}{\mathrm{d}\lambda}\right\|_i \simeq \frac{\rho_{\lambda_j} - \rho_{\lambda_i}}{\Delta\lambda}$, where $\Delta\lambda$ is the separation between adjacent bands, $\Delta\lambda = \lambda_j - \lambda_i$ and $\lambda_j > \lambda_i$	[34]
Second derivative	$\left.\frac{\mathrm{d}^2\rho}{\mathrm{d}\lambda^2}\right\|_j = \left.\frac{\mathrm{d}}{\mathrm{d}\lambda}\left(\frac{\mathrm{d}\rho}{\mathrm{d}\lambda}\right)\right\|_j \simeq \frac{\rho_{\lambda_i} - 2\rho_{\lambda_j} + \rho_{\lambda_k}}{(\Delta\lambda)^2}$, where $\Delta\lambda = \lambda_k - \lambda_j = \lambda_j - \lambda_i, \lambda_k > \lambda_j > \lambda_i$.	[34]
log transformation or pseudo absorbance	$\tilde{\rho}_{i,\lambda} = \log\left(\frac{1}{\rho_{i,\lambda}}\right), \forall\lambda \in [1,...,L].$	[35]
Continuum Removal	$CR_\lambda = \frac{\rho_\lambda}{C_\lambda}$, where C is a convex hull fitting over the top of the spectrum to connect local spectrum maxima.	[36,37]
Continuum removal derivative reflectance	$\left.\frac{\mathrm{d}CR_\lambda}{\mathrm{d}\lambda}\right\|_i$	[38]

L is the number of wavelengths.

3.2. Similarity Measures

Let ρ_i be a spectral signature, $\rho_{i,\lambda}$ its reflectance at wavelength λ and $[1,...,L]$ its spectral range. Several criteria have been used (Table 4). Some criteria characterize the difference between reflectance levels (like the distances) and other ones are related to the difference of the spectral shape (e.g., SAM) and other ones are related to probabilistic behaviour (e.g., SID, ...). Table 4 inventories main similarity measurement techniques described in the literature.

Table 4. Similarity measures.

Similarity Measures	Formulation	Comments	Reference
Minkowski distance	$D_p(\rho_i,\rho_j) = \lVert\rho_i - \rho_j\rVert_p, = \left[\sum_{\lambda=1}^{L} \lvert\rho_{i,\lambda} - \rho_{j,\lambda}\rvert^p\right]^{1/p}.$	Spectral signatures are represented by vectors from $\mathbb{R}^L$. D_2 is the usual Euclidean distance ; D_1 is the Manhattan or City Block distance	D_2: [24,39,40] ; D_1: [41,42]
Canberra distance	$D_{\text{Canberra}}(\rho_i,\rho_j) = \sum_{\lambda=1}^{L} \frac{\lvert\rho_{i,\lambda} - \rho_{j,\lambda}\rvert}{\lvert\rho_{i,\lambda}\rvert + \lvert\rho_{j,\lambda}\rvert}.$	It is a weighted version of the Manhattan distance	[43]
Spectral Angle Mapper (SAM)	$\text{SAM}(\rho_i,\rho_j) = \cos^{-1}\left(\frac{\sum_{\lambda=1}^{L}\rho_{i,\lambda}\rho_{j,\lambda}}{\left(\sum_{\lambda=1}^{L}\rho_{i,\lambda}^2\right)^{1/2}\left(\sum_{\lambda=1}^{L}\rho_{j,\lambda}^2\right)^{1/2}}\right).$	Since the angle between two vectors is invariant with respect to the length of the vectors, this technique is relatively insensitive to illumination and albedo effects	[23,44]
Spectral Information Divergence (SID)	$SID(\rho_i,\rho_j) = D(\rho_i \Vert \rho_j) + D(\rho_j \Vert \rho_i),$ where $D(\rho_i \Vert \rho_j) = \sum_{\lambda=1}^{L} p_\lambda \log\left(\frac{p_\lambda}{q_\lambda}\right)$, where $p_\kappa = \frac{\rho_{i,\kappa}}{\sum_{\lambda=1}^{L}\rho_{i,\lambda}}, q_\kappa = \frac{\rho_{j,\kappa}}{\sum_{\lambda=1}^{L}\rho_{j,\lambda}}$	It calculates the probabilistic behaviour between spectral signatures	[45]
SAM-SID	$\text{SID-Tan}(\rho_i,\rho_j) = \text{SID}(\rho_i,\rho_j) \times \tan\left(\text{SAM}(\rho_i,\rho_j)\right),$ $\text{SID-Sin}(\rho_i,\rho_j) = \text{SID}(\rho_i,\rho_j) \times \sin\left(\text{SAM}(\rho_i,\rho_j)\right).$	It is a combination of probability and geometry spaces that improves discrimination ability	[46]
Spectral Correlation Measure (SCM)	$SCM(\rho_i,\rho_j) = \frac{L\sum_{\lambda=1}^{L}\rho_{i,\lambda}\rho_{j,\lambda} - \sum_{\lambda=1}^{L}\rho_{i,\lambda}\sum_{\lambda=1}^{L}\rho_{j,\lambda}}{\left[L\sum_{\lambda=1}^{L}\rho_{i,\lambda}^2\left(\sum_{\lambda=1}^{L}\rho_{i,\lambda}\right)^2\right]^{1/2}\left[L\sum_{\lambda=1}^{L}\rho_{j,\lambda}^2\left(\sum_{\lambda=1}^{L}\rho_{j,\lambda}\right)^2\right]^{1/2}}.$	It is calculated as the correlation coefficient of the pixel and their respective spectral signatures	[47]
Pearson Correlation Coefficient (PCC)	$PCC(\rho_i,\rho_j) = \frac{\sum_{\lambda=1}^{L}(\rho_{i,\lambda} - \mu_i)(\rho_{j,\lambda} - \mu_j)}{\left[\sum_{\lambda=1}^{L}(\rho_{i,\lambda} - \mu_i)^2\right]^{1/2}\left[\sum_{\lambda=1}^{L}(\rho_{j,\lambda} - \mu_j)^2\right]^{1/2}}$, where μ_i is the mean of ρ_i.		
Spectral Similarity Value (SSV)	$SSV(\rho_i,\rho_j) = \sqrt{D_2(\rho_i,\rho_j)^2 + \left(1 - [PCC(\rho_i,\rho_j)]^2\right)^2}.$	Low value of SSV means high similarity and *vice versa*	[48]
Spectral Correlation Angle (SCA)	$SCA(\rho_i,\rho_j) = \cos^{-1}\left(\frac{1 + PCC(\rho_i,\rho_j)}{2}\right).$	It is an improvement of SAM derivated from PCC that has been shown to be able to distinguish between positive and negative correlations and to yield better estimates in some experiments	[49,50]
Spectral Gradient Angle (SGA)	$SGA(\rho_i,\rho_j) = SAM\left(\lvert\nabla\rho_i\rvert, \lvert\nabla\rho_j\rvert\right), = \cos^{-1}\left(\frac{\sum_{\lambda=1}^{L-1}\lvert\rho_{i,\lambda+1} - \rho_{i,\lambda}\rvert\lvert\rho_{j,\lambda+1} - \rho_{j,\lambda}\rvert}{\left[\sum_{\lambda=1}^{L-1}(\rho_{i,\lambda+1} - \rho_{i,\lambda})^2\right]^{1/2}\left[\sum_{\lambda=1}^{L-1}(\rho_{j,\lambda+1} - \rho_{j,\lambda})^2\right]^{1/2}}\right).$	It is invariant to illumination conditions	[51]

3.3. Relative Spectral Discriminatory Probability

To determine if a spectral signature belongs to a class, the method proposed by [45] is used. Let $\{\rho_j\}_{j=1}^{J}$ J spectral signatures in Δ an existing spectral reference database and τ be a target signature to be identified using Δ. Let $m(\cdot,\cdot)$ be a given hyperspectral measure, the spectral discriminatory probabilities of all ρ_j in Δ with respect to τ as is defined as follows:

$$p^m_{\tau,\Delta}(i) = \frac{m(\tau,\rho_i)}{\sum_{j=1}^{J} m(\tau,\rho_j)}, \text{ for } i = 1,2,...,J, \tag{2}$$

where $\sum_{j=1}^{J} m(\tau,\rho_j)$ is a normalization constant determined by τ and Δ. The resulting probability vector is defined as

$$\mathbf{p}^m_{\tau,\Delta} = \left(p^m_{\tau,\Delta}(1), p^m_{\tau,\Delta}(2), ..., p^m_{\tau,\Delta}(J)\right)^T. \tag{3}$$

Using Equation (3), the target signature can be identified by selecting the one with the smallest spectral discriminatory probability because τ and the selected one have the minimum spectral discrimination.

Spectral Reference Database

To build the spectral reference database, spectra of mean reflectance, spectra of median reflectance and median spectra are used. Spectra of mean reflectance is defined as the mean of reflectances for each wavelength λ:

$$\overline{\rho_\lambda} = \frac{1}{N}\sum_{i=1}^{N}\rho_{i,\lambda},\ \forall\lambda \in [1,...,L], \tag{4}$$

where N is the number of spectra for a plant species. Similarly, spectra of median reflectance is defined as the median of reflectances for each wavelength λ. Median spectra is defined as the "closest" spectrum of the median reflectance considering a vegetation type. In other words, giving a spectrum of median reflectance, the spectrum that minimize the Minkowski distance between them is considered as the median spectrum (Figure 4 shows differences between the median reflectances spectrum which is an theoretic spectral signature and the different median spectra which were investigated). As distances are not equivalent considering high-dimensional data, three Minkowski distances are investigated for this study: the Euclidean distance, the Canberra distance and the City Block or Manhattan distance (which are reminded in Section 3.1).

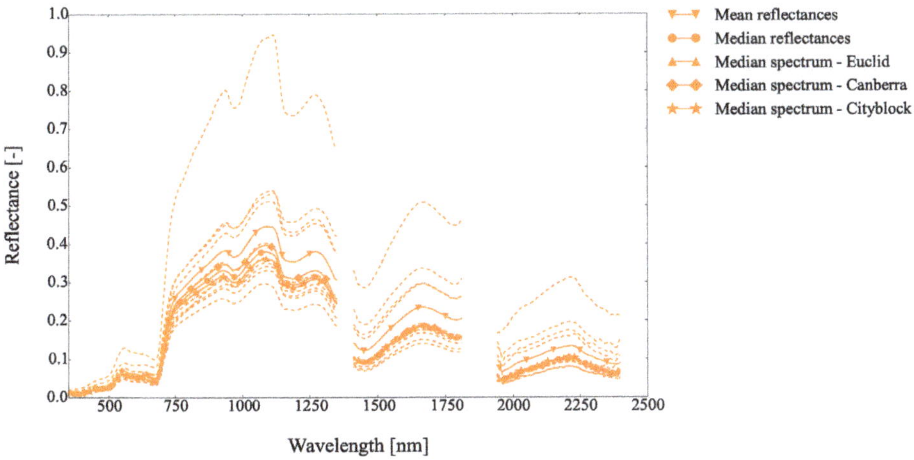

Figure 4. Median spectra, spectrum of mean reflectances, spectrum of median reflectances of *Eleocharis quinqueflora* (ELQU).

3.4. Feature Selection of Spectral Indices

3.4.1. Spectral Index Description

Spectral indices are combinations of surface reflectance (or the derivated reflectance) at two or more wavelengths or narrow spectral bands. Lots of spectral indices can be found in literature (Table 5) to characterize some biochemical components of plant species such as chlorophyll, nitrogen, lignin, cellulose, water. Although these indices have never been selected in the literature to characterize wetlands plant species, we assume that some of them can still be useful to classify them.

Table 5. Spectral vegetation indices.

Index Name	Formulation	Vegetation Properties Highlighted by the Index	String Type
Boochs	D_{703}	Chlorophyll	[52]
Boochs2	D_{720}	Chlorophyll	
CAI (Cellulose Absorption Index)	$\frac{R_{2000}+R_{2200}}{2}-R_{2100}$	Cellulose, soil litter	[53]
CARI (Chlorophyll Absorption Ratio Index)	$\frac{R_{700}\sqrt{(670a+R_{670}+b)^2}}{R_{670}\sqrt{a^2+1}}$ where $a=\frac{R_{700}-R_{550}}{150}$; $b=R_{550}-550a$	Chlorophyll	[54]
CI (Curvature Index)	$\frac{R_{675}R_{690}}{R_{683}^2}$	Chlorophyll	[55]
CCI (Canopy Chlorophyll Index)	$\frac{D_{720}}{D_{700}}$	Chlorophyll	[56]
CCCI (Canopy Chlorophyll Content Index)	$\frac{\text{NDRE}}{\text{NDVI}}$	Chlorophyll	[57]
Carter[695,420]	$\frac{R_{695}}{R_{420}}$	Stress	[58]
Carter[695,760]	$\frac{R_{695}}{R_{760}}$	Stress	
Carter[605,760]	$\frac{R_{605}}{R_{760}}$	Stress	
Carter[710,760]	$\frac{R_{710}}{R_{760}}$	Stress	
Carter[695,670]	$\frac{R_{695}}{R_{670}}$	Stress	
Carter2	R_{550}		
CaCoI[515,550] (Carotenoid Concentration Index)	$\frac{1}{R_{515}}-\frac{1}{R_{550}}$	Carotenoid	[59,60]
CaCoI[515,700]	$\frac{1}{R_{515}}-\frac{1}{R_{700}}$	Carotenoid	
CaCoI2[770,510,700]	$R_{770}\left(\frac{1}{R_{510}}-\frac{1}{R_{700}}\right)$	Carotenoid	[59,60]
CaCoI2[770,510,550]	$R_{770}\left(\frac{1}{R_{510}}-\frac{1}{R_{550}}\right)$	Carotenoid	
Datt[850]	$\frac{R_{850}-R_{710}}{R_{850}-R_{680}}$	Chlorophyll	[61]
Datt[780]	$\frac{R_{780}-R_{710}}{R_{780}-R_{680}}$	Chlorophyll	[61]
Datt2[850,710]	$\frac{R_{850}}{R_{710}}$	Chlorophyll	
Datt2[672,550]	$\frac{R_{672}}{R_{550}}$	Chlorophyll	
Datt_prime	$\frac{D_{754}}{D_{704}}$	Chlorophyll	
Datt3[672]	$\frac{R_{672}}{R_{550}R_{708}}$	Chlorophyll	[62]
Datt3[860]	$\frac{R_{860}}{R_{550}R_{708}}$	Chlorophyll	[62]
DCI	$\frac{D_{723}}{D_{703}}$		[63]
DCNI (Double-peak Canopy Nitrogen Index)	$\frac{R_{720}-R_{700}}{(R_{700}-R_{670})(R_{720}-R_{670}+0.03)}$	Nitrogen	[64]
DD (Double Difference Index)	$(R_{749}-R_{720})-(R_{701}-R_{672})$	Chlorophyll	[65]
DDn (new Double Difference Index)	$2\left(R_{710}-R_{(710-50)}-R_{(710+50)}\right)$	Chlorophyll	[66]
DPI (Double Peak Index)	$\frac{D_{688}D_{710}}{D_{967}^2}$	Chlorophyll	[55]
dG	$\max_{i\in[500,580]} D_i$	Chlorophyll, stress	
dRE	$\max_{i\in[680,750]} D_i$	Chlorophyll, stress	[67]
D[730,706]	$\frac{D_{730}}{D_{706}}$	Chlorophyll	[55]
D[705,722]	$\frac{D_{705}}{D_{722}}$		
EVI (Enhanced Vegetation Index)	$2.5\frac{R_{800}-R_{670}}{R_{800}-6R_{670}-7.5R_{475}+1}$	Chlorophyll	[68]

Table 5. *Cont.*

Index Name	Formulation	Vegetation Properties Highlighted by the Index	String Type
EGFR (Edge-Green First derivative Ratio)	$\frac{dRE}{dG}$	Chlorophyll, nitrogen	[69]
EGFN (Edge-Green first Derivative Normalized difference)	$\frac{dRE - dG}{dRE + dG}$	Chlorophyll, nitrogen	
GEMI (Global Environment Monitoring Index)	$\eta(1 - 0.25\eta) - \frac{R_{660} - 0.25}{1 - R_{660}}$, where $\eta = 2\frac{R_{830}^2 - R_{660}^2 + 1.5R_{830} + 0.5R_{660}}{R_{830} + R_{660} + 0.5}$		[70]
GI (Greeness Index)	$\frac{R_{554}}{R_{677}}$	Chlorophyll	[71]
Gitelson	$\frac{1}{R_{700}}$	Chlorophyll	[72]
Gitelson2	$\frac{R_{750} - R_{800}}{R_{965} - R_{740}} - 1$	Chlorophyll	[59]
GMI (Gitelson and Merzlyak Index)	$\frac{R_{750}}{R_{550}}$	Chlorophyll	[73]
Green NDVI	$\frac{R_{800} - R_{550}}{R_{800} + R_{550}}$	Chlorophyll	[74]
Maccioni	$\frac{R_{780} - R_{710}}{R_{780} - R_{680}}$	Chlorophyll	[75]
MARI (Modified Anthocyanin Reflectance Index)	$R_{800}\left(\frac{1}{R_{550}} - \frac{1}{R_{700}}\right)$	Anthocyanin	[76,77]
MCARI[700,670] (Modified Chlorophyll Absorption Index)	$\left((R_{700} - R_{670}) - 0.2(R_{700} - R_{550})\right)\frac{R_{700}}{R_{670}}$	Chlorophyll, Leaf Area Index	[78]
MCARI[750,705]	$\left((R_{750} - R_{705}) - 0.2(R_{750} - R_{550})\right)\frac{R_{750}}{R_{705}}$	Chlorophyll	[79]
MCARI[700,670]/OSAVI[800,670]	$\frac{\left((R_{700} - R_{670}) - 0.2(R_{700} - R_{550})\right)\frac{R_{700}}{R_{670}}}{(1 + 0.16)\frac{R_{800} - R_{670}}{R_{800} + R_{670} + 0.16}}$	Chlorophyll	[80]
MCARI[750,705]/OSAVI[750,705]	$\frac{\left(R_{750} - R_{705}) - 0.2(R_{750} - R_{550})\right)\frac{R_{750}}{R_{705}}}{(1 + 0.16)\frac{R_{750} - R_{705}}{R_{750} + R_{705} + 0.16}}$	Chlorophyll	[79]
MCARI[750,705]/MTVI2[750]	$\frac{MCARI[750,705]}{MTVI2[750]}$	Nitrogen	[81]
MNDVI[800,680] (Modified NDVI)	$\frac{R_{800} - R_{680}}{R_{800} + R_{680} - 2R_{445}}$	Chlorophyll	[82]
MNDVI[750,705]	$\frac{R_{750} - R_{705}}{R_{750} + R_{705} - 2R_{445}}$	Chlorophyll	
MSAVI (Modified Soil Adjusted Vegetation Index)	$0.5\left(2R_{800} + 1 - \sqrt{(2R_{800} + 1)^2 - 8(R_{800} - R_{670})}\right)$	Chlorophyll	[83]
MSI (Moisture Stress Index)	$\frac{R_{1599}}{R_{819}}$	Water stress	[84]
MSR[800,680] (modified Simple Ratio)	$\frac{R_{800} - R_{445}}{R_{680} - R_{445}}$	Chlorophyll	[82]
MSR[750,705]	$\frac{R_{750} - R_{445}}{R_{705} - R_{445}}$	Chlorophyll	
MSR2	$\frac{R_{750}}{R_{705}} - \frac{1}{\sqrt{\frac{R_{750}}{R_{705}} + 1}}$	Chlorophyll, Leaf Area Index	[85]
MTCI (MERIS [1] Terrestrial Chlorophyll Index)	$\frac{R_{754} - R_{709}}{R_{709} - R_{681}}$	Chlorophyll	[86]
MTVI[800] (Modified Triangular Vegetation Index)	$1.5\left(1.2(R_{800} - R_{550}) - 2.5(R_{670} - R_{550})\right)$	Leaf Area Index	[87]
MTVI[750]	$1.5\left(1.2(R_{750} - R_{550}) - 2.5(R_{670} - R_{550})\right)$	Leaf Area Index	[87]
MTVI2 [800]	$\frac{1.5\left(1.2(R_{800} - R_{550}) - 2.5(R_{670} - R_{550})\right)}{\sqrt{(2R_{800} + 1)^2 - (6R_{800} - 5\sqrt{R_{670}}) - 0.5}}$	Leaf Area Index	[87]
MTVI2 [750]	$\frac{1.5\left(1.2(R_{750} - R_{550}) - 2.5(R_{670} - R_{550})\right)}{\sqrt{(2R_{750} + 1)^2 - (6R_{750} - 5\sqrt{R_{670}}) - 0.5}}$		[87]
NDII (Normalized Difference Infrared Index)	$\frac{R_{850} - R_{1650}}{R_{850} + R_{1650}}$	Water status	[88]
	$\frac{R_{819} - R_{1649}}{R_{819} + R_{1649}}$		

Table 5. *Cont.*

Index Name	Formulation	Vegetation Properties Highlighted by the Index	String Type
NDLI (Normalized Difference Lignin Index)	$\frac{\log\left(\frac{1}{R_{1754}}\right)-\log\left(\frac{1}{R_{1680}}\right)}{\log\left(\frac{1}{R_{1754}}\right)+\log\left(\frac{1}{R_{1680}}\right)}$	Lignin	[35]
NDNI (Normalized Difference Nitrogen Index)	$\frac{\log\left(\frac{1}{R_{1510}}\right)-\log\left(\frac{1}{R_{1680}}\right)}{\log\left(\frac{1}{R_{1510}}\right)+\log\left(\frac{1}{R_{1680}}\right)}$	Nitrogen	[35]
NDRE (Normalized Difference Red Edge)	$\frac{R_{830}-R_{red}}{R_{830}+R_{red}}$, with $R_{red}=0.5(R_{670}+R_{780})$		[57]
NDVI[800,670] (Normalised Difference Vegetation Index)	$\frac{R_{800}-R_{670}}{R_{800}+R_{670}}$	Chlorophyll, Leaf Area Index	[89]
NDVI[750,705]	$\frac{R_{750}-R_{705}}{R_{750}+R_{705}}$	Chlorophyll	[73]
NDVI[682,553]	$\frac{R_{682}-R_{553}}{R_{682}+R_{553}}$	Chlorophyll	[90]
NDVI[573,440]	$\frac{R_{573}-R_{440}}{R_{573}+R_{440}}$	Nitrogen	[91]
NDWI[860,1240] (Normalized Difference Water Index)	$\frac{R_{860}-R_{1240}}{R_{860}+R_{1240}}$		
NDWI[860,1640]	$\frac{R_{860}-R_{1640}}{R_{860}+R_{1640}}$	Water status	[92]
NDWI[860,2130]	$\frac{R_{860}-R_{2130}}{R_{860}+R_{2130}}$		
NDWI[1100,1450]	$\frac{R_{1100}-R_{1450}}{R_{1100}+R_{1450}}$	Water stress	[93]
NDWI[1280,1450]	$\frac{R_{1280}-R_{1450}}{R_{1280}+R_{1450}}$	Water stress	[93]
NPCI (Normalised Pigment Chlorophyll Index)	$\frac{R_{680}-R_{430}}{R_{680}+R_{430}}$	(Total pigments)/chlorophyll	[94]
VI_opt (Vegetation Index optimal)	$(1+0.45)\frac{R_{800}^2+1}{R_{670}+0.45}$	Nitrogen	[95]
OSAVI[800,670] (Optimised Soil-Adjust Vegetation Index)	$(1+0.16)\frac{R_{800}-R_{670}}{R_{800}+R_{670}+0.16}$	Chlorophyll	[96]
OSAVI[750,705]	$(1+0.16)\frac{R_{750}-R_{705}}{R_{750}+R_{705}+0.16}$	Chlorophyll	[79]
PRI (Photochemical Reflectance Index)	$\frac{R_{531}-R_{570}}{R_{531}+R_{570}}$	Stress	[97]
RDVI (Renormalised Difference Vegetation Index)	$\frac{R_{800}-R_{670}}{\sqrt{R_{800}+R_{670}}}$	Chlorophyll, Leaf Area Index	[98]
REIP (Red-Edge Inflexion Point)	$\left\{ i \,\middle\vert\, \max_{i\in[\![680,750]\!]} D_i \right\}$	Chlorophyll, Leaf Area Index	[67,99,100]
REMI (Red-Edge Model Index)	$\frac{R_{750}}{R_{720}}-1$	Chlorophyll	[101]
REP_LE (Red-Edge Position Linear Extrapolation)	$-\frac{c_1-c_2}{m_1-m_2}$ where m_1 and c_1 represent the slope and the intercept of the far-red line and m_2 and c_2 represent the slope and the intercept of the NIR line	Nitrogen, chlorophyll	[102]
REP_LI (Red-Edge Position Linear Interpolation)	$700+40\frac{0.5(R_{800}+R_{780})}{R_{740}-R_{700}}$	Chlorophyll	[103]
RVI[810,660] (Ratio Vegetation Index)	$\frac{R_{810}}{R_{660}}$	Nitrogen	[104]
RVI[810,560]	$\frac{R_{810}}{R_{560}}$	Nitrogen	[105]
RVI[800,670]	$\frac{R_{800}}{R_{670}}$		
SIPI (Structure Insensitive Pigment Index)	$\frac{R_{800}-R_{445}}{R_{800}-R_{680}}$	Pigments/chlorophyll, stress	[106]
SPVI (Spectral Polygon Vegetation Index)	$0.4\left[3.7(R_{800}-R_{670})-1.2\sqrt{(R_{530}-R_{670})^2}\right]$	Chlorophyll × Leaf Area Index	[107]
SR[800,680] (Simple Ratio Index)	$\frac{R_{800}}{R_{680}}$	Chlorophyll	[108]
SR[750,700]	$\frac{R_{750}}{R_{700}}$		[73]
SR[752,690]	$\frac{R_{752}}{R_{690}}$		
SR[750,550]	$\frac{R_{750}}{R_{550}}$		

Table 5. *Cont.*

Index Name	Formulation	Vegetation Properties Highlighted by the Index	String Type
SR[700,670]	$\frac{R_{700}}{R_{670}}$	Chlorophyll	[109]
SR[675,700]	$\frac{R_{675}}{R_{700}}$	Chlorophyll	[110]
SR[750,710]	$\frac{R_{750}}{R_{710}}$	Chlorophyll	[111]
SR[440,690]	$\frac{R_{440}}{R_{690}}$	Stress	[112]
SRPI (Simple Ratio Pigment Index)	$\frac{R_{430}}{R_{680}}$	(Total pigments)/chlorophyll, stress	[106]
Sum_Dr[625,795]	$\sum_{i=625}^{795} D_i$	Chlorophyll	[113]
Sum_Dr[680,780]	$\sum_{i=680}^{780} D_i$	Chlorophyll, Leaf Area Index	[67]
TCARI[700,670] (Transformed Chlorophyll Absorption Ratio Index)	$3\left(R_{700} - R_{670} - 0.2(R_{700} - R_{550})\frac{R_{700}}{R_{670}}\right)$	Chlorophyll	[80]
TCARI[750,705]	$3\left(R_{750} - R_{705} - 0.2(R_{750} - R_{550})\frac{R_{750}}{R_{705}}\right)$	Chlorophyll	[79]
TCARI[700,670]/OSAVI[800,670]	$\frac{TCARI}{OSAVI}$	Chlorophyll	[80]
TCARI[750,705]/OSAVI[750,705]	$\frac{TCARI2}{OSAVI2}$	Chlorophyll	[79]
TVI (Triangular Vegetation Index)	$0.5(120(R_{750} - R_{550}) - 200(R_{670} - R_{550}))$	Leaf Area Index, Canopy chlorophyll density	[114]
Vogelmann	$\frac{R_{740}}{R_{720}}$	Chlorophyll	[115]
Vogelmann2	$\frac{R_{734} - R_{747}}{R_{715} + R_{726}}$	Chlorophyll	
Vogelmann3	$\frac{D_{715}}{D_{705}}$	Chlorophyll	
Maximum first derivatives of 8 different regions whithin the spectra	A_1D: 495–550 nm B_1D: 550–650 nm C_1D: 680–780 nm D_1D: 970–1090 nm E_1D: 1110–1205 nm F_1D: 1205–1285 nm H_1D: 1455–1640 nm J_1D: 1925–2200 nm	Pigments absorption, w., c., s., l absorption ; refer to Table 2 in [116] for a full description.	[116]
Corresponding spectral positions of the maximum first derivatives	A_WP: 495–550 nm B_WP: 550–650 nm C_WP: 680–780 nm D_WP: 970–1090 nm E_WP: 1110–1205 nm F_WP: 1205–1285 nm H_WP: 1455–1640 nm J_WP: 1925–2200 nm	Pigments absorption, w., c., s., l. absorption ; refer to Table 2 in [116] for a full description.	[116]
WI (Water Index)	$\frac{R_{900}}{R_{970}}$	Water status	[117]
WI[1100,1450]	$\frac{R_{1100}}{R_{1450}}$	Water stress	[93]
WI[1280,1450]	$\frac{R_{1280}}{R_{1450}}$	Water stress	[93]
WI2	$\frac{1}{R_{1450}}$	Water stress	[93]

R_x represents reflectance at wavelength x nm. D_x represents the derivative of the reflectance spectrum at wavelength x nm. w., c., s., l = water, cellulose, starch, lignin.

3.4.2. Classical Feature Selection Method—The Kruskal-Wallis H-Test

As some spectra per vegetation types were quite small (8 spectra for *Pinguicula* sp. (PING), 7 spectra for Aquatic type b (AQ_B)), usual ANOVA [118] test or Mann-Whitney U-test [119] can not be used. That is the reason why Kruskal-Wallis H-test [120], a non-parametric test is proposed. Moreover this test is adapted to not independent data and not normally distributed data. The H-test is used to test the hypothesis that there was no significant difference between the median spectral index value between pairs of plant species.

The null hypothesis for $N = 13$ vegetation types and $I = 129$ spectral vegetation indices per reflectance measurements is:

$$H_0 : \eta_n(i) = \eta_{n+1}(i), \tag{5}$$

where η_n is the median spectral index value for vegetation type number $n = 0, ..., N$, and $i = 1, ..., I$ the spectral index. The maximum frequency for this study is $\binom{13}{2} = \frac{13 \times (13-1)}{2} = 78$. The hypothesis was therefore tested 78 times for all possible combinations of the 13 plant species at the adjusted Bonferroni significance level of $\alpha = \frac{0.05}{78} = 6.410^{-4}$.

3.4.3. Principle of the Applied Feature Selection Method

In order to discriminate between the 78 pairs of vegetation types, the Hellinger distance, which is introduced further, is computed for each vegetation spectral index (Table 5). Then indices are ordered by frequency discrimination. A first subset of indices is composed of ones that can discriminate between pairs of vegetation types and that are not redundant. If there is no discrimination between a pair of vegetation types, the Hellinger distance is computed for a pair of vegetation indices composed of the single most discriminating one and the other ones ordered by frequency distribution amongst previous selected. Then, a second subset of pairs of indices is composed by ordering those pairs of indices by frequency discrimination. To stop the process, a maximum number of subsets is then defined. In our case, the maximum subset consists of not more than three indices. Indeed, the longer the tuple length is, the more difficult it is to explained why such combinations of indices or such biophysical components combination can discriminate between such pairs. Finally, selected vegetation indices come from each subset and single spectral vegetation indices or spectral index combinations are retained.

For a better understanding of the feature selection method, an example is given. We consider four vegetation types named: V_1, V_2, V_3, V_4 and 5 spectral vegetation indices named: I_1, I_2, I_3, I_4, I_5. We suppose that no single spectral vegetation index can discriminate between neither V_1 and V_3 nor V_2 and V_4 nor V_3 and V_4. But different single indices can separate V_1 from V_2, V_1 from V_4 and V_2 from V_3. This is summarized in the following table:

	V_2	V_3	V_4
ine V_1	I_1, I_3	$\emptyset$	I_2, I_3
V_2	-	I_2, I_3	$\emptyset$
V_3	-	-	$\emptyset$

We obtain the first subset $S_1 = \{I_1, I_2, I_3\}$. To discriminate between V_1 and V_3, V_2 and V_4, and V_3 and V_4, we are looking among the following combinations: $\{I_3 - I_2\}, \{I_3 - I_1\}, \{I_3 - I_4\}, \{I_3 - I_5\}$ because indices are ordered by frequency discrimination: $[I_3, I_2, I_1, I_4, I_5]$. We suppose that $\{I_3 - I_1\}$ can discriminate between V_1 and V_3, and V_2 and V_4 but there is still no index that can discriminate between V_3 and V_4. For the latter case, possible combinations are looking among $\{I_3 - I_1 - I_2\}, \{I_3 - I_1 - I_4\}, \{I_3 - I_1 - I_5\}$. Whatever a combination of spectral vegetation indices can be found to discriminate between those plant species or not, the process will stop in our case.

3.4.4. The Bhattacharyya Coefficient and the Hellinger Distance

For two arbitrary discrete probability distributions $\mathbf{p}$ and $\mathbf{q}$, the amount of overlap between those distributions can be measured using the Bhattacharyya coefficient:

$$C(\mathbf{p}, \mathbf{q}) = \sum_{i=1}^{n} \sqrt{p_i q_i}, \tag{6}$$

where n is the partition number. To measure the similarity between two statistical distributions in remote sensing the Hellinger distance (also known as the Matusita distance) is commonly used. It is defined as:

$$H(\mathbf{p},\mathbf{q}) = \sqrt{\frac{1}{2}\sum_{i=1}^{n}\left(\sqrt{p_i}-\sqrt{q_i}\right)^2}, \quad (7)$$

$$= \sqrt{1 - C(\mathbf{p},\mathbf{q})}. \quad (8)$$

The Hellinger distance defined in Equation (8) has upper bound equal to 1, indicating the total separability of the class pairs characterized by their distribution. As a general rule adapted from [121],

- if $H(\mathbf{p},\mathbf{q}) \geq 0.95$ then the classes can be separated,
- if $0.85 \leq H(\mathbf{p},\mathbf{q}) < 0.95$ the separation is fairly good,
- if $H(\mathbf{p},\mathbf{q}) < 0.85$ the separation is poor.

3.5. Spectral Ranges

The transformed spectral signatures defined in Section 3.2 and the spectral ranges adapted from [31] (Table 6) were investigated:

- visible: 350 nm–750 nm,
- near infrared: 750 nm–1350 nm,
- shortwave infrared a: 1410 nm–1810 nm,
- shortwave infrared b: 1940 nm–2400 nm.

The shortwave infrared domain is split into two parts. The near infrared and the shortwave infrared are not continuous because of atmospheric water absorption.

Table 6. The spectral reflectances of green vegetation on the four regions of electromagnetic spectrum from [31].

Wavelength Range [nm]	Description	Spectral Reflectance of Vegetation	References
400–700	Visible	Low reflectance and transmittance due to chlorophyll and biologically active pigments (such as carotene) absorptions	[122,123]
680–750	Red-edge	The reflectance is strongly correlated with plant biochemical and biophysical parameters	[124,125]
700–1300	Near infrared	High reflectance and transmittance, very low absorption resulting from photon scattering at the air-cell interfaces within the leaf spongy mesophyll	[126,127]
1300–2500	Shortwave infrared	Lower reflectance than other spectral regions due to strong water absorption and minor absorption of biochemical contents such as lignin and carbon constituents	[126,128]

3.6. Supervised Classification

All the classifications are performed using Python scikit-learn package [129].

3.6.1. Random Forest (RF)

RF is an ensemble classifier that uses a set of Classification And Regression Trees (CARTs) to make a prediction [130]. The trees are created by drawing a subset of training samples through replacement (a bagging approach). In standard classification trees, each node is split using the best split among all variables. In RF, each node is split using the best predictor, among a user-defined number of features (*Mtry* that is usually set to the square root of the number of input variables [131]). By growing the forest up to a user-defined number of trees (*Ntree* that is usually set to 500 but different values such as 100, 1000 or 5000 have been investigated [131]), the algorithm creates trees that have high variance and low bias. The final classification decision is taken by averaging (using the arithmetic mean) the class assignment probabilities calculated by all produced trees.

For this study, $Mtry = 500$ and $Ntree \in [500, 1000, 2000, 5000]$.

3.6.2. Support Vector Machines (SVM)

SVM is a supervised non-parametric statistical learning technique therefore there is no assumption on the distribution of the data [132]. The main idea of SVM classification is to construct a hyperplane as a decision surface in a way that the margin of separation between two classes is maximized. To do this, the original feature space is mapped into a space with a higher dimensionality, where classes can be modelled to be linearly separable. This transformation is implicitly performed by applying kernel functions to the original data. The learning of the classifier is performed using a constrained optimization process that is associated with a complex cost function. For problems that involve identification of multiple classes, adjustments are made to the simple SVM binary classifier to operate as a multi-class classifier using methods such as one-against-all, one-against-others.

For this study, two kernels are retained: a linear kernel (SVM linear) and a Gaussian kernel (SVM RBF).

3.6.3. Regularized Logistic Regression (RLR)

RLR is a linear model based on logistic regression with an additional regularization term. This classifier has been successfully used with high dimensional data (gene selection in cancer classification [133], feature selection in remote sensing [28,29,134]).

For this study, the ℓ_1-norm and ℓ_2-norm regularization term are investigated.

3.6.4. Partial Least Squares-Discriminant analysis (PLS-DA)

PLS-DA is based upon the classical partial least square regression method for constructing predictive models [135]. The goal of PLS regression is to provide dimension reduction in an application where the response variable is related to the predictor variables. In the case of PLS-DA, the response variable (i.e., vegetation types) is binary and expresses class membership [136,137]. This classifier has been successfully used with high dimensional data (gene selection [138], tree species discrimination [139]).

For this study the number of latent variables is fixed to the number of vegetation types-1 [138]. This method is not applied on spectral vegetation indices selected but on spectral signatures and their transformations on spectral ranges because it is commonly used when the number of features is much bigger than the number of spectra.

3.7. *Classification Accuracy Evaluation*

To evaluate the classification accuracy of supervised classifiers, a 30 fold cross-validation is used and six training samples size were investigated: 50%, 45%, 40%, 35%, 30% and 25% of all spectra.

To evaluate the classifier precision overall accuracy and F1-score are used. Overall accuracy computes number of correct spectra over all spectra, whereas F1-score is given by:

$$\text{F1-score} = 2 \cdot \frac{\text{PA} \cdot \text{UA}}{\text{PA} + \text{UA}}, \tag{9}$$

where PA (Producer's Accuracy) is the fraction of retrieved classes that are relevant whereas UA (User's Accuracy) is the fraction of relevant classes that are retrieved.

4. Results and Discussion

4.1. *Similarity Measures*

Considering all transformed spectral signatures, spectral ranges and similarity measures, only the Canberra distance on [350 nm to 2500 nm] gives an overall accuracy higher than 50 % whatever the spectral reference database (Table 7). Indeed, the Canberra distance gives the higher overall accuracy because it is sensitive to a small change when both coordinates are closed to zero [140,141].

Because of the high variability of some vegetation types (Appendix B), spectral reference database built from median spectra, that are *real* spectra, gave worse results than spectral reference database built from median and mean spectra, that are *theoretical* spectra not representative of a in situ measured vegetation type (Table 7). There is a need to collect more spectral signatures to build a consistent spectral database.

As spectral signatures can be considered as high dimensional vectors, a specific distance is needed to compare them. It is well known that Euclidean distance is not good when comparing high dimension data [142]. Table 8 shows that the Canberra distance always outperforms other distances, including SAM, which is commonly used in remote sensing, when considering the whole spectral range (1823 wavelengths).

Table 7. Overall accuracy (%) for Canberra distance on [350–2500 nm].

	Median Spectra			Median	Mean
	Canberra Dist.	City Block Dist.	Euclidean Dist.	Reflectance	Reflectance
Spectral signature	53.62	52.34	51.91	57.02	50.64
Normalized spectral signature	51.91	52.34	50.64	55.74	57.87
log transformation of spectral signature	52.34	52.34	51.49	55.74	51.91
First Derivative	70.64	**70.21**	**67.23**	74.47	71.49
Second Derivative	**71.06**	68.51	64.68	**81.70**	**77.45**
Continuum removed Reflectance	51.06	50.64	51.06	54.04	52.77
Continuum Removed Derivative Reflectance	64.68	62.98	61.28	78.30	75.32

Table 8. Overall accuracy (%) for different distances on [350–2500 nm] considering Median reflectances as spectral reference database.

	Distance			SAM
	Euclid	Manhattan	Canberra	
Spectral signature	50.21	51.06	**57.02**	41.70
First Derivative	62.98	70.64	**74.47**	59.15
Second Derivative	65.96	74.04	**81.70**	63.83
CRDR	71.06	74.47	**78.30**	69.36

Using the Canberra distance, best results (overall accuracy higher than 60 %) are given with the second derivative, first derivative and CRDR (Table 7), that are closely related to absorption features rather than reflectance magnitude [38]. Indeed, it is possible to discriminate between vegetation types thanks to their biophysical components which will be discussed in details in Section 4.2.1. Furthermore, Table 9 shows that the whole spectral range gives the best results. Although spectral ranges are related to specific biophysical components (Table 6), the whole spectral range is needed to discriminate between the 13 vegetation types because some of them are sharing same plant species (Table A1) and the spectral signatures are mixed. Worse results are obtained in [1940–2400 nm] whatever the transformed spectral signature. Table 9 show that worse results are obtained by the spectral signature whatever the spectral range. Indeed those transformations are related to absorption features as explained above, which confirms that transformed spectral signatures are more suitable to discriminate between vegetation types than spectral signatures.

Table 9. Overall classification accuracy (%) for different spectral ranges considering Median reflectances as spectral reference database and Canberra distance.

	350–750 nm	750–1350 nm	1410–1810 nm	1940–2400 nm	350–2500 nm
Spectral signature	47.23	47.66	37.87	34.47	**57.02**
First Derivative	59.15	64.68	60.43	55.74	**74.47**
Second Derivative	72.34	69.79	72.34	53.19	**81.70**
CRDR	74.47	57.87	59.57	59.57	**78.30**

Considering classification accuracy for each vegetation type, Table 10 shows that best F1-score is obtained by *Sphagnum* sp. (SPHA) ($\simeq$98%), *Juniperus communis* (JUCO) ($\simeq$97%), Aquatic type b (AQ_B) ($\simeq$93%) and *Salix* sp. (SALI) ($\simeq$92%). Except for JUCO, all of these vegetation types are well classified and their user's accuracy is higher than 85%. Indeed these vegetation types are less mixed than others: Table A1 shows that SPHA is mainly dominated by different kinds of *sphagnum*; AQ_B is dominated by *Utricularia* sp; JUCO is dominated by *Juniperus communis* and SALI is dominated by *Salix*. Only three other vegetation types have user's accuracy equal to 100%: *Rhododendron ferrugineum* (RHFR), *Calluna vulgaris* (CAVU) and Aquatic type a (AQ_A). However, only around 57% of spectral signatures are well identified for CAVU and AQ_A. This can be explained by the high variability of these sample plots. Contrary to SPHA, JUCO, AQ_B and SALI, there is not a single dominated plant species neither for CAVU nor for AQ_A (Table A1). Worse F1-score is obtained by *Pinguicula* sp. (PING) ($\simeq$54%) which is not dominated by only one plant species: this vegetation type is mainly dominated by *Eleocharis quinqueflora* (ELQU) (40%), bare ground (15%), *Molinia caerulea* ssp *caerulae* (10%) and *Tomenthypnum nitens* (10%). It can explain the difficulty to identify this vegetation type in particular rather than the low number of spectra: PING has eight spectra whereas AQ_B has seven spectra.

Table 10. Confusion matrix of the classification based on Second derivative, Canberra Distance on [350–2500 nm] with Median reflectance as reference spectral database. The producer's and user's accuracies, the overall accuracy and the F1-score are also shown.

	SPHA	CAVU	RH_FR	CA_HV	AQ_A	SALI	PING	JUQO	ELQU	METR	PI_CV	AQ_B	AQ_C	Producer's Accuracy (%)
SPHA	**22**	0	0	0	0	0	0	0	0	0	0	0	0	100.00
CAVU	0	**8**	0	2	0	0	1	0	0	0	3	0	0	57.14
RHFR	0	0	**11**	0	0	3	0	0	0	0	0	0	0	78.57
CA_HV	0	0	0	**22**	0	0	1	0	1	0	3	0	0	81.48
AQ_A	0	0	0	0	**30**	0	8	0	3	4	1	1	6	56.60
SALI	0	0	0	0	0	**17**	0	0	0	0	0	0	0	100.00
PING	0	0	0	1	0	0	**7**	0	0	0	0	0	0	87.50
JUCO	0	0	0	0	0	0	0	**18**	0	0	1	0	0	94.74
ELQU	0	0	0	1	0	0	0	0	**13**	1	0	0	0	86.67
METR	1	0	0	0	0	0	0	0	0	**11**	0	0	0	91.67
PI_CV	0	0	0	0	0	0	1	0	0	0	**14**	0	0	93.33
AQ_B	0	0	0	0	0	0	0	0	0	0	0	**7**	0	100.00
AQ_C	0	0	0	0	0	0	0	0	0	0	0	0	**12**	100.00
User's accuracy (%)	95.65	100.00	100.00	84.62	100.00	85.00	38.89	100.00	76.47	68.75	63.64	87.50	66.67	Overall accuracy: **81.70**
F1-score (%)	97.78	72.73	88.00	83.02	72.29	91.89	53.85	97.30	81.25	78.57	75.68	93.33	80.00	

4.2. Supervised Classification Based on Feature Selection of Spectral Vegetation Indices

4.2.1. Feature Selection

The Kruskal-Wallis method (Section 3.4.2, p. 13) does not show any significant index (frequency discrimination $> 75\,\%$) that allow discrimination between vegetation types (Figure 5, only the first 69 indices are drawn). The best vegetation index (NDWI [860, 2130]) only allows us to discriminate between 49 pairs of vegetation types, that may be explained by the plant species mixing within several vegetation types. The proposed method reduced the number of selected indices from 129 to 26 (Table 11). More precisely, on the first step of the method, only 17 single indices amongst 26 are needed to discriminate between 59 pairs of vegetation types amongst 78. On the second step, these single indices must be completed by 7 additional spectral vegetation indices to discriminate between 17 more pairs of vegetation types (Table 12; $\emptyset$ means either a pair of vegetation type can not be discriminated thanks to a pair of spectral vegetation indices built from single ones selected on the first step, either more than two vegetation indices are needed to discriminate between a pair of vegetation types).

On the last step, a single index is added to discriminate between two vegetation types whereas a combination of previous selected indices allows us to discriminate between another pair of vegetation type (Table 11). Finally several different—single or pair or triplet—vegetation indices allow us to discriminate between pairs of vegetation types. However, none single spectral index allows us to discriminate between all pairs of vegetation types nor the majority: e.g., the most discriminating single spectral index, the Water Index (WI), only discriminates between around 42% pairs of vegetation types (Table 11).

Table 13 shows that one single biophysical component can discriminate between most of the vegetation types except for *Carex* sp. homogeneous vegetation (CA_HV). More precisely, three kinds of vegetation types (*Sphagnum* sp. (SPHA), Aquatic type b (AQ_B) and Aquatic type c (AQ_C)) are separated thanks to a single biophysical component. However, some biophysical components are more discriminant than others according to vegetation types: e.g., the chlorophyll is more discriminant than the water content for AQ_C whereas the water content is the only discriminant biophysical component for AQ_B; the water content, the chlorophyll and water, cellulose, starch, lignin (w., c., s., l.) equally discriminate between SPHA and all other vegetation types.

Only two indices related to water content are needed to separate AQ_B from all other vegetation types: WI and NDWI[860,1240] (Table 13) because AQ_B vegetation type is mainly composed of *Utricularia* sp. and water (Table A1). The AQ_B spectral signatures are lower than the spectral reflectance values of the other vegetation types and the water absorption band at 900 nm and 970 nm are highlighted (Figure 6).

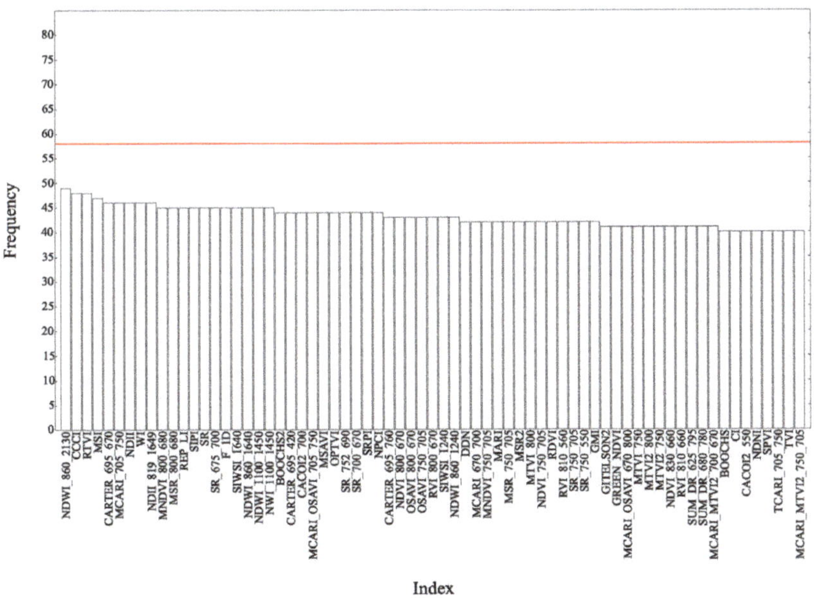

Figure 5. Frequency distribution of the Kruskal-Wallis test for the 129 spectral indices for paired species across the 13 vegetation types. The horizontal red line stands for 75 % of all 78 possible combinations of the 13 vegetation types.

Table 11. Single selected indices from the Hellinger distance and their occurrences.

Biophysical Component	Index Name	No. of All Occurrences	No. of Single Occurrences	No. of Occurrences within Pair	No. of Occurrences within Triple
Chlorophyll	CCCI *	35	24	10	1
	GMI *	33	25	8	0
	DPI **	33	16	17	0
	NDVI[750,705] **	32	25	7	0
	BOOCHS2 *	32	24	8	0
	SR[700,670] *	31	25	6	0
	OSAVI[800,670] *	31	20	8	3
	DDN **	26	18	8	0
	MNDVI[800,680] *	23	18	5	0
	GITELSON ***	13	5	5	3
Water	WI *	40	33	6	1
	MSI *	39	31	8	0
	NDWI[860,1240] *	38	31	7	0
	NDII **	38	28	9	1
	NDWI[860,2130] *	35	24	11	0
	NDWI[1100,1450] **	32	22	10	0
Stress	CARTER[695,670] *	36	26	9	1
	CARTER[695,420] **	36	16	20	0
Pigment	MARI *	75	13	62	0
	PRI *	35	9	26	0
Nitrogen	NDNI *	37	18	19	0
	MCARI/MTVI2[750,705] **	30	22	8	0
(Total pigments)/chlorophyll	NPCI *	31	18	13	0
	SRPI *	29	16	13	0
Water, cellulose, starch, lignin	F_1D *	89	27	62	2
	F_WP *	20	15	5	0

* index selected on first step. ** index selected on second step. *** index selected on third step.

Table 12. Single spectral index or pairs of spectral indices retained to discriminate between vegetation types pairs.

	CAVU	RHFR	CA_HV	AQ_A	SALI	PING
SPHA	F_WP	F_WP	WI	OSAVI[800,670]	F_WP	MSI
CAVU	-	∅	∅	F_1D	∅	GMI
RHFR	NPCI-F_1D	-	∅	∅	∅	MNDVI[800, 680]
CA_HV	MARI-WI	CARTER[695, 670]-MCARI/MTVI2[750, 705]	-	∅	∅	∅
AQ_A	-	F_1D-WI	NDNI-NDWI[1100,1450]	-	∅	∅
SALI	CARTER[695, 420]-NDII	CARTER[695, 670]-BOOCHS2	SRPI-NDVI[750,705]	F_1D-MSI	-	NPCI
PING	-	-	NDNI-WI	DDN-NDWI[860,2130]	-	-
JUCO	-	F_1D-WI	-	-	-	-
ELQU	-	-	MARI-WI	MARI-MSI	-	-
METR	-	-	CCCI-NDWI[860,1240]	-	-	DPI-F_ AD
PI_CV	-	-	-	-	-	-
AQ_B	-	-	-	-	-	-

	JUCO	ELQU	METR	PI_CV	AQ_B	AQ_C
SPHA	F_WP	CCCI	CCCI	WI	WI	OSAVI[800,670]
CAVU	MARI	CCCI	GMI	GMI	WI	SR[700, 670]
RHFR	∅	SRPI	CCCI	WI	WI	MNDVI[800, 680]
CA_HV	F_WP	∅	∅	∅	WI	CCCI
AQ_A	F_1D	∅	∅	NDNI	WI	MSI
SALI	F_WP	NPCI	NPCI	NPCI	WI	MNDVI[800,680]
PING	NDWI[860, 2130]	PRI	∅	BOOCHS2	NDWI[860,1240]	BOOCHS2
JUCO	-	F_WP	F_WP	F_WP	WI	MNDVI[800,680]
ELQU	-	-	MARI	CARTER[695,420]	NDWI[860,1240]	BOOCHS2
METR	-	-	-	∅	NDWI[860,1240]	BOOCHS2
PI_CV	-	-	PRI-WI	-	NDWI[860,1240]	OSAVI[800,670]
AQ_B	-	-	-	-	-	NDWI[860,1240]

∅ means either it is not possible to discriminate between a pair of vegetation typess thanks to a pair of spectral vegetation indices built from single ones selected on the first step, or more than two vegetation indices are needed to discriminate between a pair of vegetation types.

Table 13. Single main discriminating biophysical components for each vegetation type and their occurrences (%).

Biophysical Components	SPHA	CAVU	RHFR	CA_HV	AQ_A	SALI	PING	JUCO	ELQU	METR	PI_CV	AQ_B	AQ_C
Water	33.33	8.33	16.67	16.67	16.67	8.33	25.00	16.67	8.33	8.33	25.00	100.00	16.67
Chlorophyll	33.33	41.67	25.00	8.33	8.33	8.33	33.33	8.33	25.00	33.33	25.00	0.00	83.33
Stress	0.00	0.00	0.00	0.00	0.00	0.00	0.00	0.00	8.33	0.00	8.33	0.00	0.00
Nitrogen	0.00	0.00	0.00	0.00	8.33	0.00	0.00	0.00	0.00	0.00	8.33	0.00	0.00
Pigment	0.00	8.33	0.00	0.00	0.00	0.00	8.33	8.33	16.67	8.33	0.00	0.00	0.00
(Total pigments)/chlorophyll	0.00	0.00	8.33	0.00	0.00	33.33	8.33	0.00	16.67	8.33	8.33	0.00	0.00
W., c., s., l.	33.33	16.67	8.33	8.33	16.67	16.67	0.00	58.33	8.33	8.33	8.33	0.00	0.00
Total	100.00	75.00	58.33	33.33	50.00	66.67	75.00	91.67	83.33	66.67	83.33	100.00	100.00

W., c., s., l. = Water, cellulose, starch, lignin.

The chlorophyll is the main biophysical component (86.33 %) able to discriminate between AQ_C and all other vegetation types, except with Aquatic type a (AQ_A) and AQ_B differentiated by considering additional water indices (MSI and NDWI [860,1240]). Indeed, dry matter can be seen on spectral signatures (Figure 7): AQ_B has the lowest slope on the spectral range [705–730 nm] whereas other vegetation types (except AQ_A and AQ_B) have higher values because they still contain chlorophyll. However, as AQ_B and AQ_C have low values of Boochs2 index, it is possible to discriminate between them thanks to a water index (right side of Figure 8 shows that those vegetation types can be clearly separated; indeed, those vegetation types have different shapes and values that characterize each type).

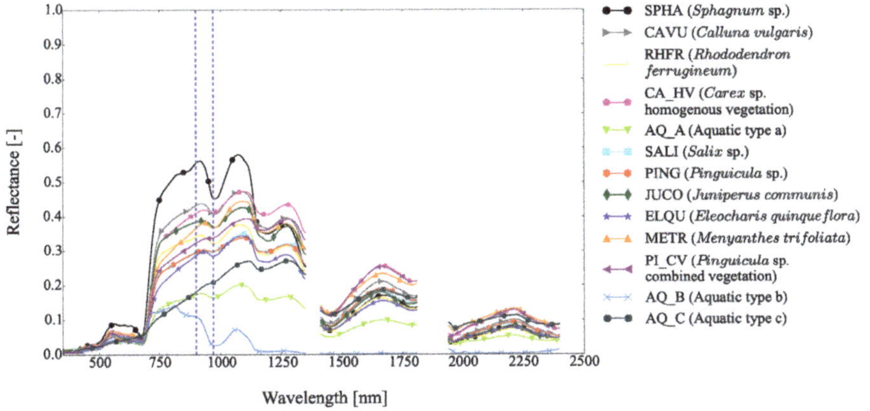

Figure 6. Mean spectral reflectance of the 13 vegetation types. Dashed lines represent the wavelengths used by Water Index (WI).

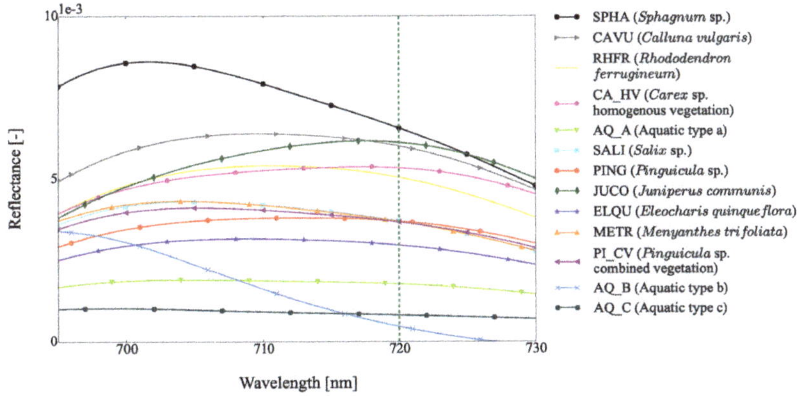

Figure 7. Mean first derivative spectral signatures of the 13 vegetation types on [695–730 nm]. The green dashed line represents the wavelength used by the Boochs2 index.

In some case, there is no single biophysical component allowing us to discriminate between vegetation types: e.g., both water content (33.33%), chlorophyll (33.33%) and w., c., s., l. (33.33%) are needed to distinguish SPHA from all other vegetation types (Table 13). More precisely, biophysical components related to water (WI, MSI) are discriminating SPHA from CA_HV, *Pinguicula* sp. (PING), *Pinguicula* sp. combined vegetation (PI_CV) and AQ_B ; biophysical components related to chlorophyll (CCCI, OSAVI [800,670]) are differentiating SPHA from AQ_A, AQ_C, *Eleocharis quinqueflora* (ELQU) and *Menyanthes trifoliata* (METR) ; biophysical components related to w., c., s., l. (F_WP) are separating

SPHA from *Calluna vulgaris* (CAVU), *Rhododendron ferrugineum* (RHFR), *Salix* sp. (SALI) and *Juniperus communis* (JUCO) (Table 13). Unlike an index related to water content (Figure 9), an index related to the chlorophyll will discriminate between SPHA and AQ_A. Indeed, the right side of Figure 9 shows that some AQ_A plant species can not be distinguished from SPHA because it is a dry moss and the left side of Figure 9 shows that SPHA and non discerned AQ_A have the same spectral signature shape. The right side of Figure 10 shows that these two vegetation species can clearly be separated despite the class variability of AQ_A. A complex biophysical component such as F_WP will differentiate SPHA from CAVU (left side of Figure 11) shows that different spectral shapes between those vegetation types can be exploited on the [1220–1280 nm] domain. The right side of Figure 10 shows that the wavelengths corresponding to the maximum of the first derivatives can clearly discern these two vegetation types even if these vegetation types can be mixed.

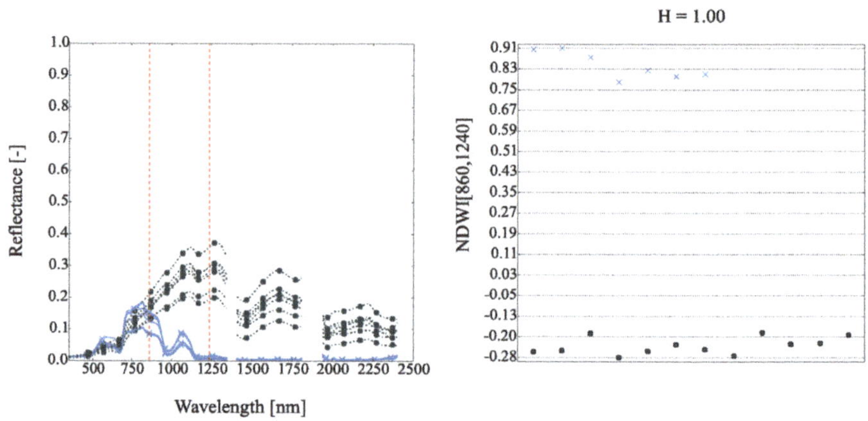

Figure 8. (**Left**) spectral signatures of AQ_B (blue) and AQ_C (dark slate gray). Red dashed lines are the wavelengths used by the Normalized Difference Water Index (NDWI) [860,1240] index; (**Right**) NDWI [860,1240] values for each vegetation type, H is the Hellinger distance.

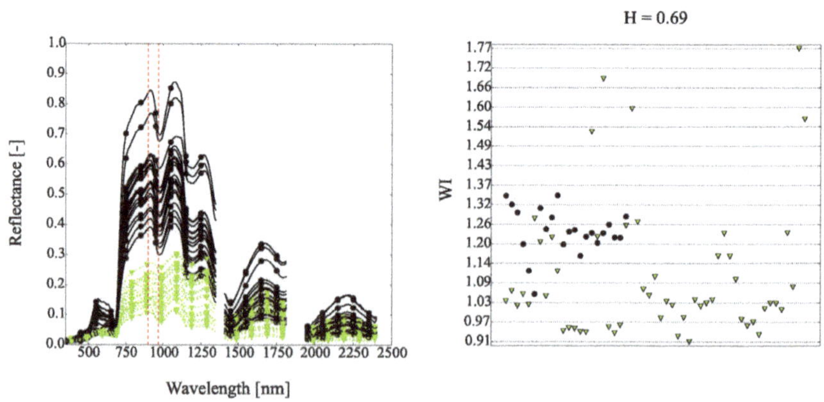

Figure 9. (**Left**) spectral signatures of Sphagnum sp. **(SPHA)!** (black) and AQ_A (green). Red dashed lines are WI wavelengths; (**Right**) WI values for each vegetation type, H is the Hellinger distance.

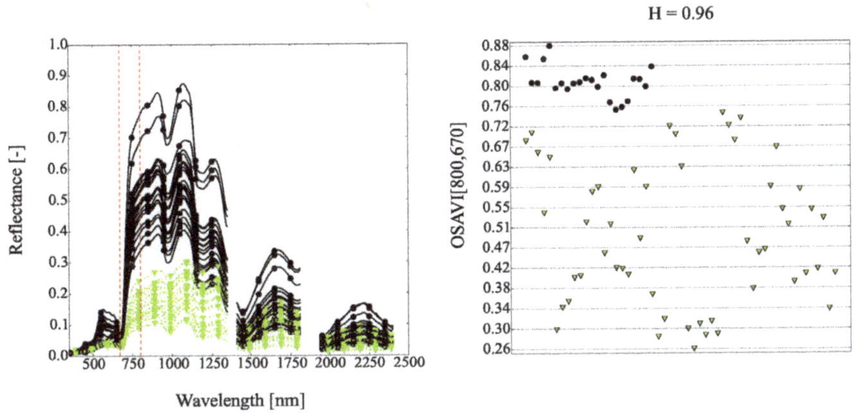

Figure 10. (**Left**) spectral signatures of SPHA (black) and AQ_A (green). Red dashed lines are Optimised Soil-Adjust Vegetation Index (OSAVI) [800,670] wavelengths; (**Right**) OSAVI [800,670] values for each vegetation type, H is the Hellinger distance.

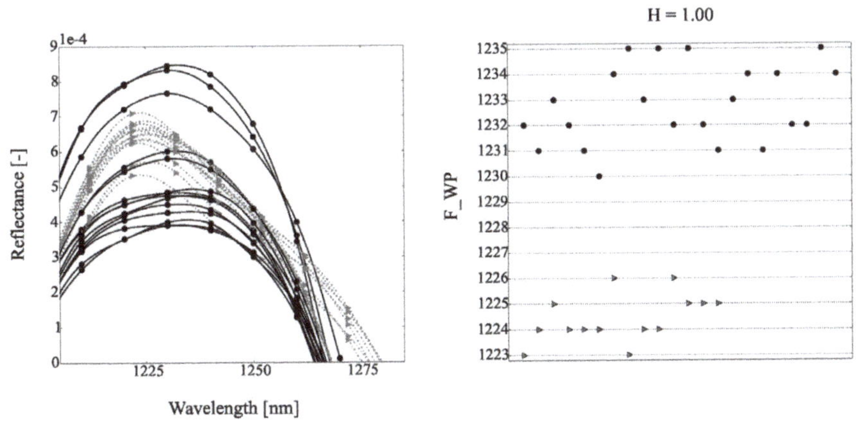

Figure 11. (**Left**) spectral signatures of SPHA (black) and Calluna vulgaris (CAVU) (gray); (**Right**) F_WP values for each vegetation type, H is the Hellinger distance.

In most cases, a single biophysical component is sufficient to class a vegetation type from the others (except for CA_HV), but a pair of biophysical components is needed to discriminate more specifically between some vegetation types (Table 12), apart from some particular cases where a pair of biophysical components is needed CA_HV (Figure 12). Indeed, CAVU and SALI are differentiated with the stress index (CARTER [695, 420]) and the water index (NDII).

Among the 78 combinations of pair of vegetation types, only two require three indices to be separated: CA_HV vs. PING and AQ_A vs. METR. Indeed, because of its within class variability (Table A1), only 33.33% of a single biophysical component can discriminate between CA_HV and all other vegetation types (Table 13). Besides, as mentioned in Section 4.1, none of the main plant species of PING represents more than 50% of this vegetation type. The advent of a third index only improves significantly their discrimination (Figure 13).

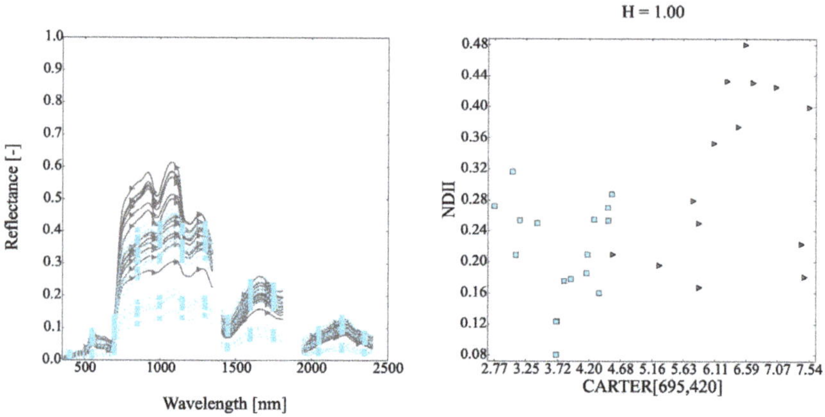

Figure 12. (**Left**) spectral signatures of CAVU (gray) and Salix sp. (SALI) (cyan); (**Right**) map of CARTER[695,420] and Normalized Difference Infrared Index (NDII) values for each vegetation type, H is the Hellinger distance.

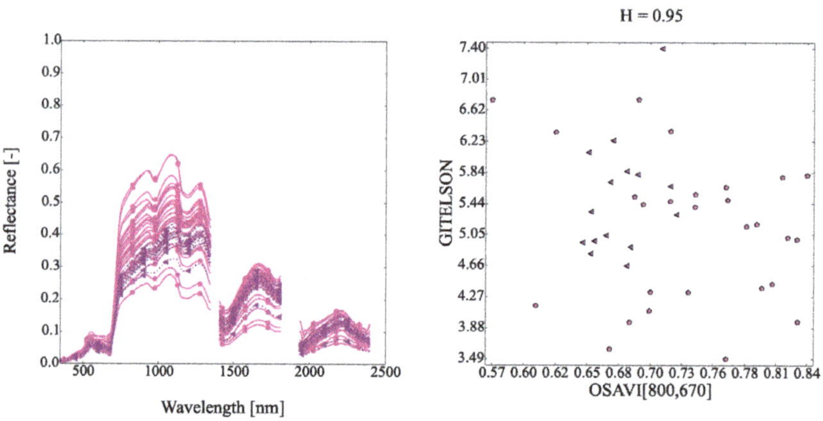

Figure 13. (**Left**) spectral signatures of CA_HV (pink) and PI_CV (magenta); (**Right**) map of Optimised Soil-Adjust Vegetation Index (OSAVI) [800,670] and GITELSON values for each vegetation type, H is the Hellinger distance value.

4.2.2. Supervised Classification

The 26 indices selected with the Hellinger distance enables overall classification accuracy scores ranging from 72.90% to 85.20% depending on the training size, whereas when considering all indices overall accuracy, scores range from 66.70% to 82.80% (Table 14). Moreover, these selected indices are robust because no significant difference between classifiers score (except for RF) regardless of the training size is noted (Figure 14). As expected, the worst results are given by the Kruskal-Wallis method (to compare performance of the two features selection methods, 26 first indices given by Kruskal-Wallis method have been selected).

Table 14. Vegetation types identification (overall accuracy (±standard deviation) in %) with indices.

Training Size	Classifier	Overall Accuracy (±Standard Deviation) (%)		
		All Indices	Kruskal-Wallis	Hellinger Distance
50%	SVM linear	79.17 (±3.51)	75.45 (±3.95)	83.31 (±3.95)
	SVM RBF	77.63 (±2.82)	75.45 (±3.65)	**83.55 (±3.65)**
	RLR-ℓ_1	**80.58 (±3.05)**	**78.37 (±3.54)**	82.84 (±3.54)
	RLR-ℓ_2	80.55 (±3.33)	78.07 (±3.48)	83.22 (±3.48)
	RF	78.71 (±3.34)	71.05 (±3.56)	81.60 (±3.56)
45%	SVM linear	78.44 (±3.09)	74.82 (±3.86)	82.46 (±3.86)
	SVM RBF	76.59 (±4.39)	74.49 (±4.53)	83.21 (±4.53)
	RLR-ℓ_1	**80.26 (±4.25)**	77.26 (±4.16)	**83.51 (±4.16)**
	RLR-ℓ_2	79.85 (±3.36)	**77.64 (±3.80)**	83.13 (±3.80)
	RF	77.26 (±4.14)	70.33 (±3.04)	80.26 (±3.04)
40%	SVM linear	76.95 (±3.59)	73.33 (±3.48)	81.89 (±3.48)
	SVM RBF	76.28 (±3.27)	73.43 (±3.84)	81.68 (±3.84)
	RLR-ℓ_1	79.69 (±3.43)	77.72 (±3.62)	**83.19 (±3.62)**
	RLR-ℓ_2	**79.74 (±2.47)**	**78.25 (±3.34)**	82.97 (±3.34)
	RF	76.86 (±3.41)	70.34 (±3.96)	80.96 (±3.96)
35%	SVM linear	76.02 (±3.35)	70.41 (±3.57)	80.02 (±3.57)
	SVM RBF	73.44 (±4.38)	71.02 (±4.17)	79.20 (±4.17)
	RLR-ℓ_1	74.98 (±2.74)	74.87 (±3.78)	80.89 (±3.78)
	RLR-ℓ_2	**77.25 (±2.80)**	**75.06 (±2.76)**	**81.04 (±2.76)**
	RF	75.32 (±3.32)	67.79 (±3.55)	79.37 (±3.55)
30%	SVM linear	73.62 (±3.84)	70.53 (±3.18)	78.34 (±3.18)
	SVM RBF	72.71 (±2.82)	69.68 (±4.33)	79.13 (±4.33)
	RLR-ℓ_1	74.08 (±4.03)	**73.66 (±3.23)**	79.25 (±3.23)
	RLR-ℓ_2	**75.74 (±3.99)**	73.39 (±3.33)	**80.36 (±3.33)**
	RF	72.53 (±2.60)	66.00 (±2.74)	77.17 (±2.74)
25%	SVM linear	71.37 (±3.18)	68.38 (±3.44)	75.91 (±3.44)
	SVM RBF	69.85 (±3.54)	67.63 (±2.67)	75.76 (±2.67)
	RLR-ℓ_1	69.42 (±4.06)	70.90 (±3.34)	76.35 (±3.34)
	RLR-ℓ_2	**73.31 (±3.34)**	**71.22 (±3.72)**	**77.21 (±3.72)**
	RF	70.79 (±2.95)	65.10 (±3.31)	75.05 (±3.31)

RLR gives better results than SVM and RF (Table 14, Figure 14) except when the size of the training set equals 50% for the Hellinger distance. That may be explained by the possible confusion between some vegetation types due to their plant species composition. Indeed, SVM aims to find the best hyperplane that can separate data, whereas RLR aims to find a probability (according to a logistic function) to separate them.

Considering RLR-$\ell 2$ some vegetation types are not easily discriminated whatever the indices. Tables 15 and 16 show that PING has the lowest F1-score (20.99 % and 33.13 % respectively) which can be explained by the mixed composition of this habitat (Appendix B) and not the low number of spectra. Indeed, AQ_B has about the same number of spectra: 7 spectra whereas 8 spectral measurements have been collected for PING. Yet it has a F1-score = 91.95 % considering all indices and F1-score = 91.66 % considering indices selected by the Hellinger distance that can be explained by its composition dominated by *Utricularia* sp.

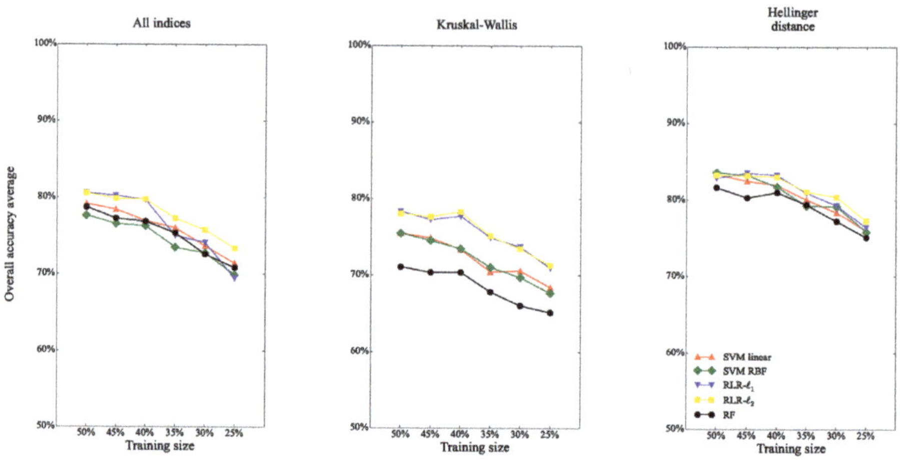

Figure 14. Vegetation types identification accuracies (overall accuracy) with indices.

Table 15. Confusion matrix of the classification based on Regularized Logistic Regression (RLR)-ℓ_2 with all indices and training size = 25%. The producer's and user's accuracies and the overall accuracy average (OAA) are also shown.

	SPHA	CAVU	RH_FR	CA_HV	AQ_A	SALI	PING	JUCO	ELQU	METR	PI_CV	AQ_B	AQ_C	Producer's Accuracy (%)
SPHA	**15.20**	0.73	0.43	0.33	0.00	0.17	0.00	0.00	0.07	0.03	0.03	0.00	0.00	89.46
CAVU	2.30	**6.20**	0.67	0.83	0.00	0.07	0.30	0.20	0.17	0.10	0.17	0.00	0.00	56.31
RHFR	1.13	0.77	**4.20**	0.00	0.07	1.67	0.70	1.57	0.50	0.17	0.23	0.00	0.00	38.15
CA_HV	0.00	0.17	0.00	**12.17**	1.03	0.00	0.53	0.07	0.57	0.57	4.90	0.00	0.00	60.82
AQ_A	0.00	0.00	0.07	0.47	**33.40**	0.20	0.83	0.00	0.80	1.60	1.00	0.17	1.47	83.48
SALI	0.00	0.30	1.00	0.13	1.33	**8.57**	0.23	0.00	0.30	0.40	0.70	0.00	0.03	65.97
PING	0.00	0.23	0.23	1.57	1.13	0.00	**1.10**	0.00	0.60	0.27	0.83	0.00	0.03	18.36
JUCO	0.07	0.00	0.10	0.00	0.13	0.00	0.10	**13.40**	0.00	0.00	0.20	0.00	0.00	95.71
ELQU	0.00	0.00	0.00	0.00	0.00	0.00	0.07	0.00	**10.93**	0.00	0.00	0.00	0.00	**99.36**
METR	0.07	0.00	0.00	1.17	1.40	0.00	0.23	0.00	0.63	**4.43**	1.03	0.00	0.03	49.28
PI_CV	0.00	0.00	0.07	1.83	0.40	0.03	0.37	0.00	0.03	0.10	**8.03**	0.00	0.13	73.07
AQ_B	0.23	0.00	0.00	0.00	0.30	0.00	0.00	0.00	0.00	0.00	0.07	**4.40**	0.00	88.00
AQ_C	0.00	0.07	0.00	0.10	0.67	0.03	0.03	0.00	0.00	0.07	0.30	0.00	**7.73**	85.89
User's accuracy (%)	80.00	73.20	62.04	65.43	83.79	79.80	24.50	87.93	74.86	57.24	45.91	**96.28**	82.06	OAA: **73.31**
F1-score (%)	84.47	63.66	47.24	63.04	83.64	72.23	20.99	91.66	85.39	52.96	56.39	**91.95**	83.93	

Focusing on shrubs, JUCO has the best performances (F1-score = 94.83%) whereas SALI and RHFR are often confounded. Table 16 shows that on average 2.53 spectra of RHFR (≃20.02%) are classified as SALI and on average 2.30 spectra of SALI (≃19.15%) are classified as RHFR. Indeed, as JUCO has a higher foliage density, the overall spatial signature is less sensitive to the ground influence and as a result JUCO spectral reflectance is close to a pure endmember (Appendix B). In the latter case, the spectral measurements are composed of soil and more affected by mixed signatures. Another pair of vegetation types is hardly discriminated: PI_CV and CA_HV. Table 16 shows that on average 4.93 spectra of CA_HV (≃25%) are classified as PI_CV which may be explained by the plant species they have in common: *Carex* (50–100% depending on the location) and *Molinia caerulea* ssp. *caerulae* (40–70%) (Appendix B).

Table 16. Confusion matrix of the classification based on RLR-ℓ_2 with indices selected by the Hellinger distance and training size = 25%. The producer's and user's accuracies and the overall accuracy average (OAA) are also shown.

	SPHA	CAVU	RH_FR	CA_HV	AQ_A	SALI	PING	JUCO	ELQU	METR	PI_CV	AQ_B	AQ_C	Producer's Accuracy (%)
SPHA	**15.40**	0.90	0.13	0.47	0.00	0.00	0.00	0.00	0.03	0.07	0.00	0.00	0.00	90.59
CAVU	0.90	**8.03**	0.67	0.47	0.00	0.03	0.70	0.00	0.03	0.03	0.13	0.00	0.00	73.07
RHFR	0.47	0.30	**6.70**	0.03	0.00	2.53	0.43	0.20	0.13	0.20	0.00	0.00	0.00	60.96
CA_HV	0.00	0.17	0.20	**11.93**	0.77	0.00	0.77	0.03	0.57	0.63	4.93	0.00	0.00	59.65
AQ_A	0.00	0.00	0.23	0.40	**33.40**	0.43	1.50	0.03	0.43	1.63	1.33	0.00	0.60	83.54
SALI	0.00	0.00	2.30	0.00	0.87	**7.77**	0.80	0.07	0.03	0.40	0.60	0.00	0.17	59.72
PING	0.00	0.27	0.17	1.67	0.37	0.00	**2.20**	0.00	0.17	0.40	0.73	0.00	0.03	36.61
JUCO	0.00	0.03	0.20	0.07	0.10	0.17	0.07	**12.93**	0.00	0.07	0.37	0.00	0.00	92.29
ELQU	0.00	0.00	0.00	0.07	0.00	0.00	0.33	0.00	**10.60**	0.00	0.00	0.00	0.00	**96.36**
METR	0.00	0.00	0.03	0.87	0.73	0.00	0.07	0.00	0.03	**6.23**	1.03	0.00	0.00	69.30
PI_CV	0.00	0.00	0.10	1.23	0.17	0.07	0.23	0.00	0.00	0.37	**8.83**	0.00	0.00	80.27
AQ_B	0.03	0.00	0.47	0.00	0.10	0.00	0.07	0.00	0.00	0.10	0.00	**4.23**	0.00	84.60
AQ_C	0.00	0.00	0.00	0.03	0.47	0.00	0.10	0.00	0.00	0.00	0.00	0.00	**8.40**	93.33
User's accuracy (%)	91.67	82.78	59.82	69.20	90.32	70.64	30.26	97.51	88.19	61.50	49.19	**100.00**	91.30	OAA: 77.21
F1-score (%)	91.12	77.62	60.39	64.07	86.80	64.72	33.13	**94.83**	92.09	65.17	61.00	91.66	92.31	

4.3. Supervised Classification According to the Spectral Ranges

Only the best results are presented, obtained with the four spectral ranges ([350–750 nm], [750–1350 nm], [350–1350 nm], [350–2500 nm]) and the spectral signature as reference and the three transformed spectral signatures (second derivative, first derivative, Continuum Removed Derivative Reflectance).

Tables 17–20 show the best results obtained with RLR-ℓ_2 on [350–1350 nm] whatever the transformed spectral signatures.

Table 17. Vegetation types identification accuracies (overall accuracy (±standard deviation) in %) on [350–750 nm].

Training Size	Classifier	Overall Accuracy (±Standard Deviation) (%)			
		Spectral Signature	Second Derivative	First Derivative	Continuum Removed Derivative Reflectance
50%	SVM linear	80.99 (±6.61)	86.94 (±5.21)	85.95 (±3.81)	**88.26 (±2.53)**
	SVM RBF	67.44 (±4.69)	78.35 (±2.74)	81.32 (±2.13)	86.94 (±3.11)
	RLR-ℓ_1	86.45 (±3.57)	86.94 (±4.10)	89.75 (±2.48)	86.94 (±1.76)
	RLR-ℓ_2	**88.10 (±3.64)**	**88.43 (±2.02)**	**90.91 (±2.86)**	87.44 (±1.84)
	RF	62.98 (±3.52)	84.79 (±4.92)	73.88 (±2.84)	86.45 (±4.07)
	PLS-DA	75.21 (±3.88)	71.90 (±4.99)	73.72 (±3.52)	75.04 (±3.28)
45%	SVM linear	81.38 (±4.80)	85.85 (±1.79)	84.62 (±1.54)	**87.69 (±1.88)**
	SVM RBF	64.15 (±2.41)	73.54 (±4.71)	76.92 (±2.06)	86.00 (±1.02)
	RLR-ℓ_1	83.85 (±4.01)	84.00 (±2.64)	85.85 (±4.63)	86.00 (±1.57)
	RLR-ℓ_2	**85.85 (±2.78)**	**86.92 (±2.01)**	**87.08 (±2.64)**	85.69 (±1.66)
	RF	59.85 (±3.35)	82.31 (±4.43)	72.46 (±3.13)	85.23 (±3.13)
	PLS-DA	75.38 (±2.18)	72.62 (±2.86)	72.15 (±1.23)	71.08 (±2.60)
40%	SVM linear	75.97 (±4.31)	83.60 (±3.23)	**84.89 (±2.69)**	**87.77 (±2.77)**
	SVM RBF	62.45 (±3.07)	73.09 (±4.50)	72.52 (±4.69)	83.45 (±2.41)
	RLR-ℓ_1	80.72 (±2.06)	82.16 (±1.47)	83.88 (±2.83)	82.73 (±1.11)
	RLR-ℓ_2	**84.46 (±3.48)**	**85.18 (±3.17)**	84.60 (±3.85)	84.32 (±1.79)
	RF	56.69 (±1.95)	80.29 (±4.50)	70.36 (±3.17)	83.74 (±2.93)
	PLS-DA	76.69 (±2.75)	72.52 (±1.79)	72.81 (±1.32)	70.22 (±1.62)
35%	SVM linear	69.74 (±7.38)	**80.52 (±5.15)**	80.00 (±3.22)	**83.77 (±2.63)**
	SVM RBF	56.23 (±3.09)	68.05 (±4.01)	68.31 (±3.86)	80.39 (±2.07)
	RLR-ℓ_1	77.92 (±4.11)	77.79 (±3.37)	80.00 (±4.78)	79.74 (±3.35)
	RLR-ℓ_2	**82.08 (±2.80)**	78.96 (±3.55)	**82.47 (±3.36)**	81.69 (±2.07)
	RF	53.25 (±3.05)	77.27 (±3.15)	67.27 (±2.12)	80.52 (±2.17)
	PLS-DA	75.45 (±3.42)	69.48 (±2.63)	70.52 (±2.12)	68.70 (±1.71)
30%	SVM linear	70.42 (±3.08)	**79.52 (±5.22)**	79.64 (±1.78)	**84.48 (±1.82)**
	SVM RBF	55.39 (±5.74)	67.03 (±4.17)	68.61 (±3.48)	80.73 (±1.50)
	RLR-ℓ_1	78.30 (±2.08)	74.91 (±7.86)	77.94 (±3.77)	78.79 (±6.37)
	RLR-ℓ_2	**80.85 (±2.98)**	77.33 (±9.20)	**81.94 (±3.42)**	81.70 (±4.01)
	RF	54.30 (±1.86)	76.97 (±4.58)	68.00 (±0.97)	79.88 (±3.33)
	PLS-DA	72.00 (±3.54)	69.09 (±4.58)	68.73 (±3.20)	68.48 (±4.85)
25%	SVM linear	65.65 (±4.57)	**74.69 (±2.46)**	74.46 (±2.33)	**80.45 (±2.49)**
	SVM RBF	52.54 (±5.26)	60.45 (±5.24)	63.28 (±4.33)	78.42 (±3.36)
	RLR-ℓ_1	75.59 (±2.49)	71.98 (±3.33)	75.25 (±4.25)	75.25 (±4.92)
	RLR-ℓ_2	**77.74 (±3.81)**	72.99 (±6.61)	**79.77 (±3.79)**	77.63 (±2.52)
	RF	52.66 (±4.40)	73.79 (±1.41)	65.42 (±1.69)	77.40 (±2.34)
	PLS-DA	71.53 (±0.92)	69.72 (±3.96)	70.40 (±2.44)	70.40 (±4.18)

Table 18. Vegetation types identification accuracies (overall accuracy (±standard deviation) in %) on [750–1350 nm].

Training Size	Classifier	Overall Accuracy (±Standard Deviation) (%) Spectral Signature	Second Derivative	First Derivative	Continuum Removed Derivative Reflectance
50%	SVM linear	83.31 (±1.10)	89.09 (±2.05)	90.91 (±1.38)	84.13 (±2.42)
	SVM RBF	57.69 (±4.03)	79.34 (±4.37)	87.60 (±2.34)	78.68 (±2.93)
	RLR-ℓ_1	**90.41 (±2.00)**	88.76 (±2.19)	89.92 (±1.42)	87.44 (±2.42)
	RLR-ℓ_2	86.28 (±3.25)	**91.07 (±1.42)**	**94.88 (±1.10)**	**90.91 (±2.45)**
	RF	53.88 (±2.05)	86.28 (±1.70)	79.83 (±1.44)	80.66 (±1.53)
	PLS-DA	77.52 (±2.30)	73.72 (±1.91)	77.69 (±2.96)	70.74 (±2.84)
45%	SVM linear	78.15 (±5.43)	84.15 (±1.86)	86.31 (±4.17)	82.77 (±3.85)
	SVM RBF	59.54 (±2.21)	72.77 (±3.82)	82.77 (±4.20)	75.85 (±2.31)
	RLR-ℓ_1	**86.46 (±3.46)**	85.38 (±3.67)	87.69 (±2.43)	82.92 (±1.78)
	RLR-ℓ_2	85.23 (±3.49)	**85.69 (±2.86)**	**90.46 (±2.46)**	**85.85 (±1.58)**
	RF	53.54 (±1.79)	80.15 (±2.73)	76.77 (±3.87)	77.54 (±2.20)
	PLS-DA	73.54 (±3.97)	70.46 (±2.31)	74.15 (±3.56)	68.15 (±3.53)
40%	SVM linear	77.70 (±5.46)	80.72 (±3.98)	83.88 (±3.82)	80.43 (±6.11)
	SVM RBF	58.85 (±2.20)	69.64 (±4.20)	80.29 (±3.04)	72.95 (±1.62)
	RLR-ℓ_1	**85.32 (±3.88)**	84.46 (±3.60)	88.06 (±3.24)	81.29 (±2.91)
	RLR-ℓ_2	82.88 (±2.25)	**86.19 (±2.38)**	**89.64 (±3.39)**	**82.73 (±3.83)**
	RF	53.24 (±2.61)	77.99 (±2.75)	74.96 (±3.29)	73.96 (±3.48)
	PLS-DA	72.09 (±1.54)	72.09 (±2.89)	74.96 (±3.07)	68.35 (±3.61)
35%	SVM linear	72.86 (±4.33)	78.44 (±4.81)	80.65 (±4.47)	75.84 (±2.83)
	SVM RBF	55.06 (±2.03)	67.14 (±4.69)	76.23 (±3.50)	66.88 (±2.87)
	RLR-ℓ_1	**80.39 (±3.71)**	79.22 (±3.60)	84.55 (±2.89)	73.90 (±3.27)
	RLR-ℓ_2	78.57 (±3.46)	**82.86 (±5.61)**	**87.27 (±3.22)**	**78.57 (±4.19)**
	RF	52.99 (±2.08)	73.64 (±2.89)	73.51 (±3.00)	69.61 (±3.14)
	PLS-DA	70.65 (±2.80)	70.52 (±2.92)	72.47 (±3.66)	66.23 (±2.82)
30%	SVM linear	74.18 (±1.70)	80.48 (±3.37)	81.58 (±2.83)	75.39 (±2.53)
	SVM RBF	55.27 (±2.93)	70.06 (±3.81)	76.24 (±4.72)	67.39 (±7.39)
	RLR-ℓ_1	**80.97 (±1.19)**	79.88 (±2.61)	84.73 (±3.05)	76.12 (±1.61)
	RLR-ℓ_2	80.00 (±3.49)	**83.88 (±3.38)**	**87.64 (±3.31)**	**78.79 (±2.06)**
	RF	52.00 (±1.69)	74.42 (±2.58)	73.21 (±2.61)	70.55 (±2.35)
	PLS-DA	72.36 (±3.69)	70.06 (±4.35)	73.45 (±3.31)	64.48 (±0.82)
25%	SVM linear	67.80 (±3.52)	75.48 (±2.59)	78.19 (±1.37)	73.11 (±0.68)
	SVM RBF	53.11 (±2.20)	60.90 (±3.90)	69.94 (±3.63)	66.78 (±2.98)
	RLR-ℓ_1	75.14 (±3.31)	77.29 (±2.93)	80.90 (±2.46)	72.77 (±1.65)
	RLR-ℓ_2	**76.84 (±2.88)**	**78.87 (±3.46)**	**83.05 (±4.55)**	**76.95 (±2.66)**
	RF	48.59 (±4.14)	71.64 (±3.87)	73.11 (±2.04)	69.83 (±2.36)
	PLS-DA	70.62 (±2.70)	69.83 (±0.68)	72.09 (±2.28)	63.95 (±3.12)

Considering wavelengths used by selected indices (Section 4.2.1), most of them use spectral bands located on [350–1350 nm] either: 50% are located in visible range and 32.35% in near-infrared range. Indeed, in this spectral range all the biophysical components discriminating the peatland vegetation types can be taken into account. That is confirmed by Figure 15 which shows that the best results are given by [350–1350 nm] considering the training size = 25% regardless the transformed spectral signatures and the the classifier, except for RF applied on the spectral signature. In this case, considering the whole spectral range improves the result by 1% compared with [350–1350 nm].

Considering RLR-ℓ_2 in [350–1350 nm], Table 21 shows that the best overall accuracies are given by first derivative, second derivative and CRDR. First and second derivatives overall accuracies are very close (difference lower than 1%). However, those transformations are sensitive to noise. However, CRDR delivered better results than spectral signatures and similar performances to the first and second derivatives (difference is lower than 4%). As mentioned in Section 4.1, those transformations are closely related to absorption features rather than reflectance magnitude [38], and are helpful to discriminate between peatland vegetation types which are clearly characterized by different biophysical components as mentioned in Section 4.2.1.

Table 19. Vegetation types identification accuracies (overall accuracy (±standard deviation) in %) in [350–1350 nm].

Training Size	Classifier	Overall Accuracy (±Standard Deviation) (%)			
		Spectral Signature	Second Derivative	First Derivative	Continuum Removed Derivative Reflectance
50 %	SVM linear	83.47 (±2.77)	93.22 (±0.96)	92.40 (±1.42)	91.57 (±2.24)
	SVM RBF	69.75 (±2.98)	55.04 (±4.10)	76.20 (±4.66)	78.02 (±1.53)
	RLR-ℓ_1	89.26 (±1.65)	92.73 (±1.69)	94.05 (±2.63)	90.41 (±1.34)
	RLR-ℓ_2	**91.07 (±3.56)**	**94.05 (±1.32)**	**96.36 (±2.00)**	**94.05 (±1.76)**
	RF	69.75 (±2.80)	90.25 (±1.91)	85.45 (±1.44)	89.26 (±2.45)
	PLS-DA	78.51 (±2.45)	80.83 (±2.05)	81.49 (±2.80)	79.17 (±2.24)
45 %	SVM linear	80.15 (±4.02)	87.38 (±2.15)	88.62 (±3.05)	91.54 (±1.61)
	SVM RBF	65.69 (±3.91)	49.38 (±3.87)	67.54 (±4.70)	72.77 (±2.31)
	RLR-ℓ_1	86.31 (±3.49)	90.46 (±1.43)	90.15 (±3.01)	88.62 (±0.58)
	RLR-ℓ_2	**90.15 (±3.35)**	**92.15 (±2.09)**	**92.77 (±1.73)**	**91.85 (±2.21)**
	RF	65.54 (±3.99)	85.85 (±3.25)	81.54 (±3.08)	86.31 (±4.28)
	PLS-DA	78.15 (±1.79)	79.85 (±3.17)	79.69 (±2.04)	76.92 (±1.54)
40 %	SVM linear	77.55 (±3.71)	86.76 (±1.62)	88.49 (±3.44)	**89.93 (±4.07)**
	SVM RBF	63.31 (±3.37)	50.79 (±3.60)	66.76 (±5.62)	69.35 (±3.24)
	RLR-ℓ_1	83.17 (±1.91)	88.06 (±1.33)	89.64 (±1.33)	85.04 (±3.26)
	RLR-ℓ_2	**87.48 (±2.79)**	**91.22 (±0.95)**	**91.80 (±1.41)**	89.64 (±1.96)
	RF	64.60 (±2.51)	84.46 (±3.17)	80.86 (±2.64)	85.32 (±4.70)
	PLS-DA	77.99 (±1.68)	80.00 (±2.00)	79.42 (±1.33)	76.40 (±1.24)
35 %	SVM linear	68.05 (±5.02)	83.90 (±3.77)	84.16 (±2.68)	85.58 (±2.74)
	SVM RBF	59.61 (±3.06)	44.03 (±3.37)	63.12 (±4.81)	64.03 (±3.69)
	RLR-ℓ_1	80.52 (±2.25)	85.71 (±2.79)	85.32 (±2.04)	80.52 (±5.08)
	RLR-ℓ_2	**84.68 (±2.83)**	**85.97 (±3.71)**	**89.09 (±1.99)**	**87.27 (±3.73)**
	RF	63.25 (±2.42)	80.26 (±3.33)	77.92 (±1.74)	82.21 (±3.35)
	PLS-DA	75.58 (±1.86)	76.36 (±2.65)	79.61 (±1.95)	75.19 (±1.04)
30 %	SVM linear	72.61 (±1.93)	84.61 (±3.22)	85.58 (±1.97)	83.76 (±4.10)
	SVM RBF	60.24 (±2.62)	42.42 (±3.36)	62.79 (±7.09)	65.21 (±3.08)
	RLR-ℓ_1	80.48 (±2.11)	82.55 (±4.01)	85.58 (±2.95)	83.03 (±4.29)
	RLR-ℓ_2	**84.12 (±4.12)**	**87.39 (±4.76)**	**89.70 (±4.22)**	**86.30 (±4.48)**
	RF	65.21 (±3.31)	79.52 (±4.22)	77.21 (±1.98)	81.58 (±3.08)
	PLS-DA	76.24 (±3.37)	76.85 (±4.99)	77.58 (±4.20)	74.79 (±3.27)
25 %	SVM linear	70.28 (±2.44)	80.90 (±2.16)	83.73 (±2.75)	82.94 (±2.59)
	SVM RBF	51.64 (±1.54)	39.89 (±1.91)	52.54 (±2.84)	61.58 (±2.34)
	RLR-ℓ_1	77.40 (±1.96)	**82.15 (±3.64)**	**83.95 (±1.70)**	79.66 (±2.02)
	RLR-ℓ_2	**81.47 (±1.10)**	80.79 (±4.42)	83.16 (±6.33)	**83.84 (±3.17)**
	RF	62.03 (±3.86)	76.16 (±3.20)	76.84 (±1.86)	80.45 (±3.67)
	PLS-DA	75.93 (±2.74)	74.58 (±2.88)	78.76 (±2.28)	72.66 (±2.49)

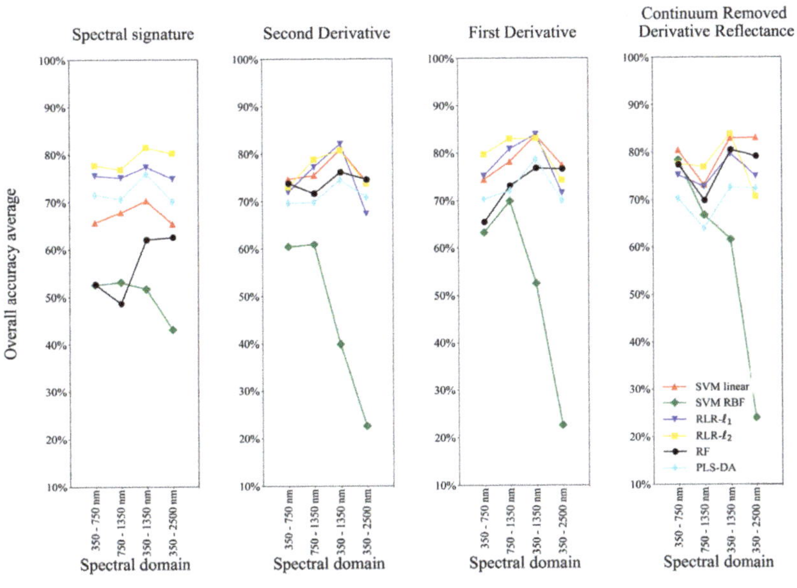

Figure 15. Vegetation type identification accuracies with the training size = 25%.

Table 20. Vegetation types identification accuracies (overall accuracy (±standard deviation) in %) on [350–2500 nm].

Training Size	Classifier	Overall Accuracy (±Standard Deviation) (%)			
		Spectral Signature	Second Derivative	First Derivative	Continuum Removed Derivative Reflectance
50%	SVM linear	83.47 (±2.34)	85.29 (±4.10)	**87.44 (±1.21)**	**91.90 (±1.76)**
	SVM RBF	61.98 (±4.31)	19.34 (±5.95)	22.81 (±0.40)	25.12 (±0.84)
	RLR-ℓ_1	91.07 (±2.30)	82.31 (±3.16)	83.80 (±3.07)	88.26 (±1.60)
	RLR-ℓ_2	**91.57 (±1.42)**	81.49 (±2.37)	82.81 (±2.05)	84.79 (±2.37)
	RF	71.24 (±2.63)	**89.92 (±1.98)**	84.96 (±2.42)	90.58 (±0.40)
	PLS-DA	75.04 (±2.05)	78.35 (±4.91)	75.70 (±2.98)	79.83 (±0.84)
45%	SVM linear	79.08 (±1.32)	79.38 (±1.57)	**82.31 (±1.61)**	**90.62 (±1.78)**
	SVM RBF	55.38 (±6.10)	22.31 (±0.00)	22.46 (±0.31)	24.15 (±1.58)
	RLR-ℓ_1	85.23 (±2.25)	79.69 (±2.86)	81.08 (±2.56)	84.77 (±2.89)
	RLR-ℓ_2	**86.00 (±2.73)**	79.23 (±2.33)	79.54 (±2.36)	77.69 (±3.61)
	RF	69.08 (±4.42)	**85.08 (±2.46)**	80.92 (±1.32)	87.69 (±2.96)
	PLS-DA	73.08 (±3.34)	75.23 (±4.31)	72.00 (±3.29)	77.69 (±1.88)
40%	SVM linear	76.12 (±0.84)	79.42 (±0.86)	**82.30 (±2.35)**	**88.06 (±1.68)**
	SVM RBF	53.24 (±3.61)	23.02 (±0.00)	23.45 (±0.58)	25.18 (±1.02)
	RLR-ℓ_1	83.88 (±3.69)	79.28 (±1.79)	79.86 (±3.83)	82.59 (±3.98)
	RLR-ℓ_2	**84.75 (±2.86)**	81.01 (±3.11)	79.57 (±2.35)	79.28 (±3.57)
	RF	65.90 (±3.48)	**84.17 (±3.34)**	79.28 (±2.67)	86.04 (±2.60)
	PLS-DA	73.67 (±1.85)	74.39 (±2.07)	71.94 (±3.75)	76.55 (±4.31)
35%	SVM linear	69.74 (±1.13)	77.27 (±1.09)	**79.87 (±1.79)**	**84.42 (±4.35)**
	SVM RBF	49.87 (±3.64)	20.00 (±5.45)	20.13 (±5.53)	22.21 (±4.69)
	RLR-ℓ_1	82.47 (±3.74)	74.42 (±2.38)	76.23 (±2.04)	78.05 (±1.26)
	RLR-ℓ_2	**83.64 (±3.19)**	77.27 (±2.87)	77.14 (±1.99)	74.94 (±2.80)
	RF	64.03 (±3.01)	**79.35 (±2.83)**	77.27 (±1.23)	82.47 (±2.82)
	PLS-DA	71.95 (±2.19)	72.34 (±2.27)	70.65 (±3.57)	74.42 (±3.20)
30%	SVM linear	69.94 (±3.90)	77.33 (±1.82)	**79.64 (±2.59)**	**84.36 (±5.88)**
	SVM RBF	48.85 (±4.05)	22.42 (±0.00)	22.42 (±0.00)	24.12 (±0.89)
	RLR-ℓ_1	79.39 (±2.24)	71.27 (±3.29)	76.36 (±3.27)	78.06 (±5.44)
	RLR-ℓ_2	**83.27 (±3.48)**	75.88 (±4.64)	75.52 (±3.03)	75.15 (±4.11)
	RF	65.21 (±3.83)	**78.06 (±2.22)**	77.21 (±2.67)	80.00 (±4.25)
	PLS-DA	70.18 (±2.80)	71.27 (±3.61)	68.85 (±4.67)	73.45 (±2.58)
25%	SVM linear	65.31 (±4.24)	74.24 (±1.54)	**77.51 (±1.49)**	**83.05 (±3.29)**
	SVM RBF	43.05 (±1.31)	22.60 (±0.00)	22.60 (±0.00)	24.07 (±0.58)
	RLR-ℓ_1	74.92 (±1.70)	67.46 (±3.44)	71.64 (±2.35)	75.03 (±5.27)
	RLR-ℓ_2	**80.23 (±0.80)**	73.79 (±3.57)	74.35 (±2.19)	70.73 (±1.84)
	RF	62.49 (±4.15)	**74.58 (±2.14)**	76.61 (±2.22)	79.10 (±2.95)
	PLS-DA	70.17 (±1.40)	70.96 (±4.00)	70.06 (±3.24)	72.43 (±2.64)

Table 21. Vegetation types identification accuracies (overall accuracy (±standard deviation) in %) on [350–1350 nm] for RLR-ℓ_2.

Training Size	Overall Accuracy (±Standard Deviation) (%)					
	Spectral Signature	Second Derivative	First Derivative	Continuum Removal	Continuum Removed Derivative Reflectance	log Transformation
50%	91.07 (±3.56)	94.05 (±1.32)	**96.36 (±2.00)**	89.59 (±1.93)	94.05 (±1.76)	93.72 (±2.13)
45%	90.31 (±3.39)	92.15 (±2.09)	**92.77 (±1.73)**	87.85 (±2.59)	91.85 (±2.21)	89.69 (±4.03)
40%	87.48 (±2.79)	91.22 (±0.95)	**91.80 (±1.41)**	83.31 (±3.79)	89.64 (±1.96)	88.35 (±2.15)
35%	84.68 (±2.83)	85.97 (±3.71)	**89.09 (±1.99)**	81.56 (±3.45)	87.27 (±3.73)	86.23 (±3.45)
30%	84.24 (±4.07)	87.39 (±4.76)	**89.70 (±4.22)**	82.79 (±4.09)	86.30 (±4.48)	84.36 (±4.22)
25%	81.47 (±1.10)	80.79 (±4.42)	83.16 (±6.33)	80.45 (±2.62)	**83.84 (±3.17)**	82.15 (±2.13)

Considering RLR, ℓ_1 regularization, which controls the selection or the removal of variables, always underperforms ℓ_2-regularization, which handles collinear variables [16]. Because of mixed plant species, it is difficult to remove variables that are not involved in the classification of all the vegetation types. Although SVM and RF are popular classifiers in remote sensing community, they are outclassed by RLR in [350 nm to 1350 nm] which is the spectral range where results are the best (Figure 16). Results given by SVM RBF are lower than those obtained with RLR and can be explained by the difficulty to find adapted parameters considering this high dimensionality problem. However, it is interesting to note that results from SVM linear are close to RLR ones considering first derivative, second derivative and CRDR. Further investigations should be conducted to better understand the link between those classifiers and improve the choice of the parameters. Figure 16 shows that PLS-DA is the least sensitive classifier to training size regardless transformed spectral signatures in [350–1350 nm].

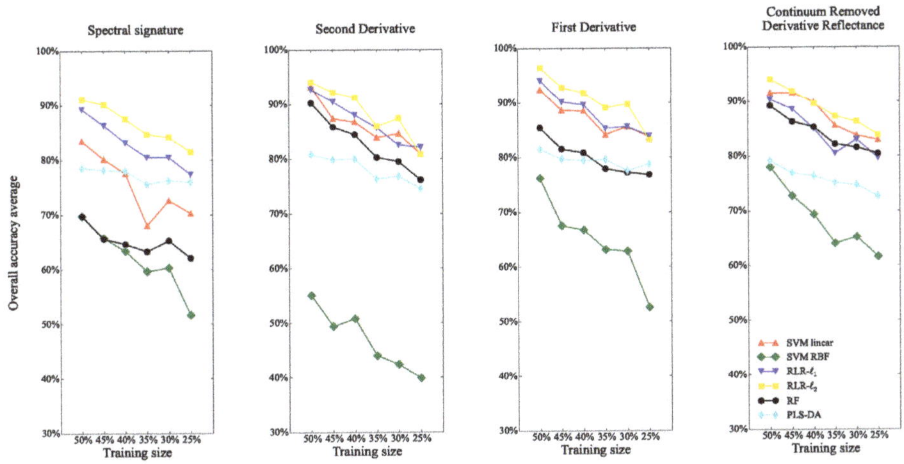

Figure 16. Vegetation type identification accuracies on [350–1350 nm].

Table 22 shows that *Pinguicula* sp. (PING) has the lowest F1-score (66.67% and 56.00% respectively) as well as for the spectral vegetation indices (Section 4.2.2). Besides, this vegetation type can hardly be discriminated from the other ones (Producer's accuracy (PA) = 53.33%) and some *Pinguicula* sp. combined vegetation (PI_CV) spectra are classified as PING). However, it should be kept in mind that PING has a small number of spectra. Considering Aquatic type b (AQ_B) which has about the same number of spectra (7 spectra against 8 for PING), User's Accuracy (UA) = 60.98% and some Aquatic type a (AQ_A) spectra are predicted as AQ_B ones. These poor UA results compared to one obtained by spectral vegetation indices can not be explained by the spectral domain. Indeed, the best spectra vegetation index (NDWI[860,1240]) that discriminate between AQ_A and AQ_B has both wavelengths in [350–1350 nm]. However, this result may be qualified by PA. Indeed, on [350–1350 nm] domain, UA = 100.00% whereas UA = 84.60% for spectral vegetation indices. Nevertheless, using a continuous spectral domain can lead to worse results for other vegetation types such as *Sphagnum* sp. (SPHA), *Calluna vulgaris* (CAVU), AQ_A: F1-score is always better considering the same classifier (RLR-ℓ_2) applied on spectral vegetation indices selected by the Hellinger distance (SPHA: 91.12% vs. 82.80%; CAVU: 77.62% vs. 71.43%; AQ_A: 86.80% vs. 82.81 %). Considering SPHA, if PA = 90.59 % for spectral vegetation indices or for [350–1350 nm], the latter predicts more SPHA than observed (UA = 76.24%) and is more confused with CAVU. This can be explained by plot 7 which is mainly composed of *Calluna vulgaris* (20%), *Carex rostrata* (25%), *Molinia caerulea* ssp. *caerulae* (20%) and *Sphagnum palustre* (20%) (Appendix B).

In our case, reducing feature space by selecting most discriminant wavelengths (using PCA or MNF) has not been implemented, whereas it can be an interesting track to explore to see if it improves results for RLR-ℓ_2. *Juniperus communis* (JUCO), *Eleocharis quinqueflora* (ELQU) and Aquatic type c (AQ_C) have about the same F1-score considering spectral vegetation indices or [350–1350 nm]: less than 2% difference. However, they have better PA on the continuous spectral range (PA = 100.00% for JUCO; 95.56% for AQ_C) which means that this spectral range contains discriminant wavelengths able to catch characteristics of those vegetation types.

Rhododendron ferrugineum (RHFR), *Carex* sp. homogeneous vegetation (CA_HV), *Salix* sp. (SALI) and *Menyanthes trifoliata* (METR) have better results considering [350–1350 nm]. This can be explained by the fact that the spectral vegetation indices used have not been built for that kind of vegetation type. Further investigations can be undertaken to find specific indices that can discriminate between those vegetation types and other ones.

Table 22. Confusion matrix of the RLR-ℓ_2 classification using Continuum Removed Derivative Reflectance (CRDR) on [350–1350 nm] (training size = 25%). The producer's and user's accuracies, the overall accuracy and the F1-score are also shown.

	SPHA	CAVU	RH_FR	CA_HV	AQ_A	SALI	PING	JUQO	ELQU	METR	PI_CV	AQ_B	AQ_C	Producer's Accuracy (%)
SPHA	**15.40**	1.40	0.00	0.20	0.00	0.00	0.00	0.00	0.00	0.00	0.00	0.00	0.00	90.59
CAVU	3.20	**7.00**	0.00	0.00	0.00	0.00	0.20	0.00	0.20	0.00	0.40	0.00	0.00	63.64
RHFR	1.40	0.00	**8.20**	0.20	0.00	0.40	0.20	0.60	0.00	0.00	0.00	0.00	0.00	74.55
CA_HV	0.00	0.00	0.00	**16.00**	1.40	0.20	0.00	0.20	0.00	0.20	2.00	0.00	0.00	80.00
AQ_A	0.20	0.00	0.00	1.80	**31.80**	0.00	0.00	0.20	0.40	1.40	0.20	3.20	0.80	79.50
SALI	0.00	0.00	0.20	0.40	0.20	**11.80**	0.00	0.40	0.00	0.00	0.00	0.00	0.00	90.77
PING	0.00	0.20	0.00	0.40	0.40	0.00	**3.20**	0.00	0.40	0.00	1.40	0.00	0.00	53.33
JUCO	0.00	0.00	0.00	0.00	0.00	0.00	0.00	**14.00**	0.00	0.00	0.00	0.00	0.00	**100.00**
ELQU	0.00	0.00	0.00	0.00	0.00	0.00	0.00	0.00	**11.00**	0.00	0.00	0.00	0.00	**100.00**
METR	0.00	0.00	0.00	0.00	2.60	0.00	0.00	0.00	0.60	**5.80**	0.00	0.00	0.00	64.44
PI_CV	0.00	0.00	0.00	0.40	0.00	0.00	0.00	0.00	0.00	0.00	**10.60**	0.00	0.00	96.36
AQ_B	0.00	0.00	0.00	0.00	0.00	0.00	0.00	0.00	0.00	0.00	0.00	**5.00**	0.00	**100.00**
AQ_C	0.00	0.00	0.00	0.00	0.40	0.00	0.00	0.00	0.00	0.00	0.00	0.00	**8.60**	95.56
User's accuracy (%)	76.24	81.40	**97.62**	82.47	86.41	95.16	88.89	90.91	87.30	78.38	72.60	60.98	91.49	OAA: **83.84**
F1-score (%)	82.80	71.43	84.54	81.22	82.81	92.91	66.67	**95.24**	93.22	70.73	82.81	75.76	93.48	

5. Conclusions and Perspectives

This study aimed at inventorying and evaluating the performance of discrimination techniques for peatland habitats based on in situ hyperspectral measurements with a high spectral resolution and high signal-to-noise ratio. To evaluate the potential of hyperspectral data to separate and classify those habitats, three classes of methods were investigated and compared:

- similarity measures calculated on spectral reflectance,
- supervised classification based on "local" information (spectral vegetation indices),
- supervised classification based on "global" information (spectral ranges).

This study demonstrated that is it possible to discriminate between peatland vegetation types by using the Canberra distance on the whole spectral range [350–2500 nm]. This distance is sensitive to a small change when both coordinates approach zero which is the case of reflectance especially in the visible ranges and in the SWIR (Figure 2). Further investigations should be conducted to see if combinations of spectral range can improve overall accuracy or if the lack of spectral signatures in the reference database (which is a weakness of this method) may explain why the whole spectral range is needed to compare spectra in that case. Besides, it is of importance to collect more spectral signatures from peatland vegetation types to build a spectral reference database of peatland vegetation types that can catch more spectral variability.

Although there are no spectral vegetation indices built to discriminate between peatland vegetation types, this study showed that some indices could be selected using the Hellinger distance. Although those indices have not been built to discriminate between peatland vegetation types, they were able to classify them because they focus on biochemical properties such as chlorophyll, nitrogen, water stress, etc. Further investigations have to be done to see the impact of spectral bandwidth around the wavelength of selected indices instead of working with one particular wavelength. For instance, there are lots of indices that catch the same biochemical property but wavelengths of interest change because they focus on specific plant species (e.g., for the chlorophyll, SR [700,670] is built for field corn, whereas SR [675,700] is built for soy beans leaves; contrary to SR [675,700], SR [700,670] has been selected with the Hellinger distance).

Contrary to similarity measures which had the best results considering the whole spectral range, supervised classification on specific spectral range as defined by [31] achieved the best overall accuracy considering [350–1350 nm] domain. This is in agreement with the spectral vegetation indices: only 4 indices (NDWI [860, 1240], NDWI [860, 2130], NDWI [1110, 1450], MSI) over the 26 selected have a discriminant wavelength which is not in this spectral range. More precisely, the discriminant wavelength is located in the SWIR and all concerned vegetation indices are linked to the water status. Further investigations should be conducted on the extraction or the reduction of features of this spectral range to understand why this domain sometimes gave worse results than spectral vegetation indices depending on the vegetation type.

Among the three methods, the best results are obtained considering a specific spectral domain [350–1350 nm] with RLR regardless of the transformed spectral signatures and the size of the training size (overall accuracy ranges from 81.47% to 96.36%). However, it should be of interest to apply feature reduction methods usually applied on remote sensing (such as PCA or MNF) to see it results are improved or specific spectral wavelength can be selected.

To our knowledge, although not popular in remote sensing for classifying (but already used for feature selection), the RLR classifier achieves the best overall classification accuracy when applied to the spectral vegetation indices selected by the Hellinger distance (77.21%) on the [350–1350 nm] domain (83.84%) considering training size = 25%.

Furthermore, this study showed that CRDR gave encouraging results event if it is slightly below those obtained by the first derivative and the second derivative considering RLR classifier.

Considering the habitats, some vegetation types were more easily separated. For instance, JUCO had the best F1-score with the spectral vegetation indices selected by the Hellinger distance (94.83%) or on the [350–1350 nm] (95.24%) with RLR and the training size = 25%. In some cases, this specific spectral domain gave better results (F1-score = 92.21% whereas with spectral vegetation indices F1-score = 64.72% for SALI) while in other case, the spectral vegetation indices gave better results (F1-score = 91.12% whereas F1-score = 82.80% for SPHA). As mentioned earlier, reducing feature space needs to be investigated to see if a particular feature space exists that can discriminate between and classify all vegetation types or if we need to consider either spectral vegetation indices or a specific spectral domain depending on the vegetation type to classify.

Although all the results strongly depended on the current dataset, this study illustrated promising methods for classifying peatland vegetation types using in situ hyperspectral measurements. The next step concerns the application or adaptation of those methods to airborne hyperspectral imageries with high spatial resolution acquired on September 2014 (simultaneously with in situ measurements). With the objective of evaluating the benefits of airborne or spaceborne sensors with a lower spectral resolution a lower signal-to-noise ratio, these conclusions may change. For that purpose, some indices (involving wavelengths lower than 480 nm) will not be used because of the camera spectral range sensitivity and some transformed spectral signatures such as second derivative will also not be used because of signal-to-noise ratio. Similarly, the first derivative transformation is very sensitive to the noise coming from the instrument but also from the atmosphere correction and this can degrade its performance.

Additional imageries acquired in October 2012 and July 2013 would allow us to test these methods with spectral signatures extracted from the ancillary dataset. Multi-temporal analysis could also be conducted to discriminate between vegetation types thanks to the phenological changes. This step would be of interest to evaluate the robustness of spectral measurements, spectral vegetation indices and classifiers selected previously from in situ hyperspectral measurements to airborne data.

Acknowledgments: The authors would like to thank Rosa Oltra-Carrió and Olivier Vaudelin for their help with field measurements and acknowledge the LabEx DRIIHM and the Observatoire Hommes-Milieux (OHM-CNRS) Haut-Vicdessos for funding and supporting the study.

Author Contributions: Thierry Erudel conducted the analyses and wrote most of the manuscript. Florence Mazier helped with floristic survey data. Thomas Houet, Sophie Fabre and Xavier Briottet helped with the field measurements and contributed as supervisors. All authors contributed to the preparation of the manuscript.

Conflicts of Interest: The authors declare no conflict of interests.

Appendix A. Composition of Vegetation Types

Table A1. Presence (+) and actual cover percentage of plant species collected on Bernadouze peatbog (Ariège, France) by Florence MAZIER & Nicolas DE MUNIK (2014/09/04 & 2014/09/11).

Plant Species/Plots	**1**	**2**	**3**	**4**	**5**	**6**	**7**	**8**	**9**	**10**	**11**	**12**	**13**	**14**	**15**	**16**
Code	SPHA	SPHA	SPHA	SPHA	SPHA	CAVU	CAVU	ELQU	ELQU	PING	METR	JUCO	JUCO	RHFR	RHFR	SALI
Alchemilla glabra																
Anthoxanthum odoratum	2	2	2	1	+											
Apiaceae																
Bare ground							1	5	4	15						
Briza media								2		+						
Calluna vulgaris		2		5	15	70	25			+						
Caltha palustris																5
Campyllium stellatum								35								
Cardamine pratensis				+							+					
Carex demissa																
Carex echinata	5	2	2	+	2				+	+	5					
Carex flava								+		+						
Carex nigra	5	2	2		2				10	5						
Carex panicea				+				+		+	5			1		
Carex paniculata																
Carex rostrata																5
Carex sp.						2	25									
Circaea lutetiana													4			
Cirsium palustre													2			
Dactylorhiza masculata	2				+		+									
Drepanocladus revolvens											30					
Drosera rotundifolia				+	+	1					+					
Dryopteraceae													+			
Eleocharis quinqueflora								60	40	40						
Epikeros pyrenaeus				+		+								+		
Equisetum sp.	1	+	+				+									5
Eriophorum angustifolium				5		10										
Festuca rubra				3												
Galium palustre																
Galium saxatile											1		2			
Gentiana ciliata										+						
Hylocomium brevirostre																
Hypnum cupressiforme													2			
Juncus alpinus										+						
Juncus bulbosus																
Juncus sp.																
Juniperus communis						5						95	80			
Lathyrus montanus						5						+				
Leotodon hispidus																
Lotus sp.				+	2											
Luzula sp.																
Lychnis floscuculi											4					

Table A1. *Cont.*

Plant Species/Plots	**1**	**2**	**3**	**4**	**5**	**6**	**7**	**8**	**9**	**10**	**11**	**12**	**13**	**14**	**15**	**16**
Code	SPHA	SPHA	SPHA	SPHA	SPHA	CAVU	CAVU	ELQU	ELQU	PING	METR	JUCO	JUCO	RHFR	RHFR	SALI
Mentha arvensis																
Menyanthes trifoliata											10					
Moliniacaerulea ssp. *caerulae*	15	25	30	15	20	10	20	15	5	10	30		5			
Narthecium ossifragum							2									
Parnassia palustris					1	4	+	1		2	3					
Pedicularis sylvatica				1	+											
Pilosella lactucella					+					1						
Pinguicula sp.									1							
Pinguicula vulgaris								+		5						
Plagiomnium elatum																
Plantago lanceolata																
Polytrichum sp.		2														
Potentilla erecta	5	5	5	5	10	5	6			5		2	+			
Potentilla sp.																
Prunella vulgaris					+					2						
Ranunculus acris													+			
Rhododendron ferrugineum																
Salix atrocinerea																
Scorpidium sp.																
Selaginella selaginoides									+	1						
Sphagnum capillifolium	10	5	5	70		25										
Sphagnum palustre	90	75	65	10	80	20	20				8					
Sphagum papillosum		15	25													
Sphagunm cuspidatum																
Succisa pratensis							+									
Tofieldia calyculata										+						
Tomenthypnum nitens							3		30	10						
Trichophorum cespitosum	+															
Trifolium arvense																
Trifolium pratense					1						1					
Utricularia sp.																
Vaccinium myrtillus										+		3				
Vicia sepium																
Viola palustris	2															
Viola sp.					+								5			
Water																
Plant Species/Plots	**17**	**18**	**19**	**20**	**21**	**22**	**23**	**24**	**25**	**26**	**27**	**28**	**29**	**30**	**31**	
Code	SALI	SALI	AQ_A	AQ_A	AQ_A	AQ_A	AQ_A	AQ_A	AQ_B	AQ_C	CA_HV	CA_HV	CA_HV	CA_HV	PI_CV	
Alchemilla glabra											2	+	3			
Anthoxanthum odoratum																
Apiaceae													+			
Bare ground										40						
Briza media											5	5				
Calluna vulgaris																
Caltha palustris	10			2						1						
Campyllium stellatum																
Cardamine pratensis																
Carex demissa																
Carex echinata			1									2	2			

Table A1. *Cont.*

Plant Species/Plots	**17**	**18**	**19**	**20**	**21**	**22**	**23**	**24**	**25**	**26**	**27**	**28**	**29**	**30**	**31**
Code	SALI	SALI	AQ_A	AQ_A	AQ_A	AQ_A	AQ_A	AQ_A	AQ_B	AQ_C	CA_HV	CA_HV	CA_HV	CA_HV	PI_CV
Carex flava															
Carex nigra															
Carex panicea															
Carex paniculata				50										100	
Carex rostrata			35	70			40	10							
Carex sp.										2			60		50
Circaea lutetiana															
Cirsium palustre														5	
Dactylorhiza masculata											+				
Drepanocladus revolvens							+								
Drosera rotundifolia															
Dryopteraceae															
Eleocharis quinqueflora						70									
Epikeros pyrenaeus															
Equisetum sp.		5		1	30		+	+		+	+				+
Eriophorum angustifolium															
Festuca rubra											1	+			10
Galium palustre				+	2										
Galium saxatile				+						+	1				
Gentiana ciliata															
Hylocomium brevirostre	+														
Hypnum cupressiforme															
Juncus alpinus															
Juncus bulbosus											1				
Juncus sp.			+												
Juniperus communis															
Lathyrus montanus													+		
Leotodon hispidus															
Lotus sp.															
Luzula sp.													+		
Lychnis floscuculi	+									1					
Mentha arvensis				+										2	
Menyanthes trifoliata	5	10		10				4							
Moliniacaerulea ssp. *caerulae*			5			5				4	60	70	40		50
Narthecium ossifragum															+
Parnassia palustris	+									2	2	2		+	1
Pedicularis sylvatica															+
Pilosella lactucella											1				
Pinguicula sp.															
Pinguicula vulgaris															
Plagiomnium elatum															
Plantago lanceolata											+	2	+		+
Polytrichum sp.															
Potentilla erecta										3	2		1		2
Potentilla sp.	+														
Prunella vulgaris											4	5	1		1
Ranunculus acris											1	+	2		+
Rhododendron ferrugineum															
Salix atrocinerea	90	100													
Scorpidium sp.											4			25	

Table A1. *Cont.*

Plant Species/Plots	17	18	19	20	21	22	23	24	25	26	27	28	29	30	31
Code	SALI	SALI	AQ_A	AQ_A	AQ_A	AQ_A	AQ_A	AQ_A	AQ_B	AQ_C	CA_HV	CA_HV	CA_HV	CA_HV	PI_CV
Selaginella selaginoides			1			1									
Sphagnum capillifolium															
Sphagnum palustre															
Sphagum papillosum															
Sphagunm cuspidatum			25												
Succisa pratensis													4		
Tofieldia calyculata															
Tomenthypnum nitens													1		
Trichophorum cespitosum															
Trifolium arvense										1					
Trifolium pratense											4	5	2		1
Utricularia sp.						5			80						
Vaccinium myrtillus															
Vicia sepium															
Viola palustris															
Viola sp.	+							1	+						
Water			50	25	70	30	60	90	20						

Appendix B. Data from Vegetation Types

Appendix B.1. Sphagnum sp. (SPHA)

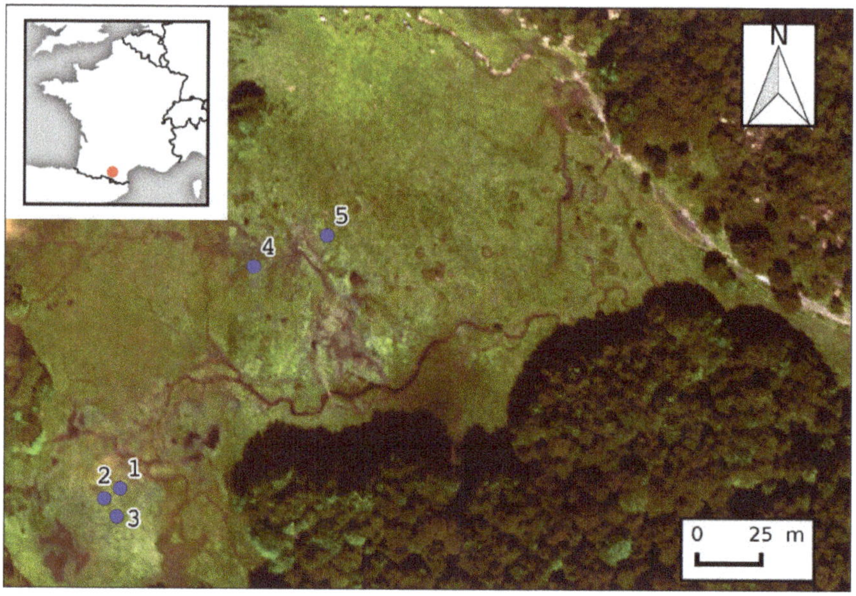

Figure A1. Location of the in situ spectroradiometer measurements for the plots of *Sphagnum* sp. (SPHA).

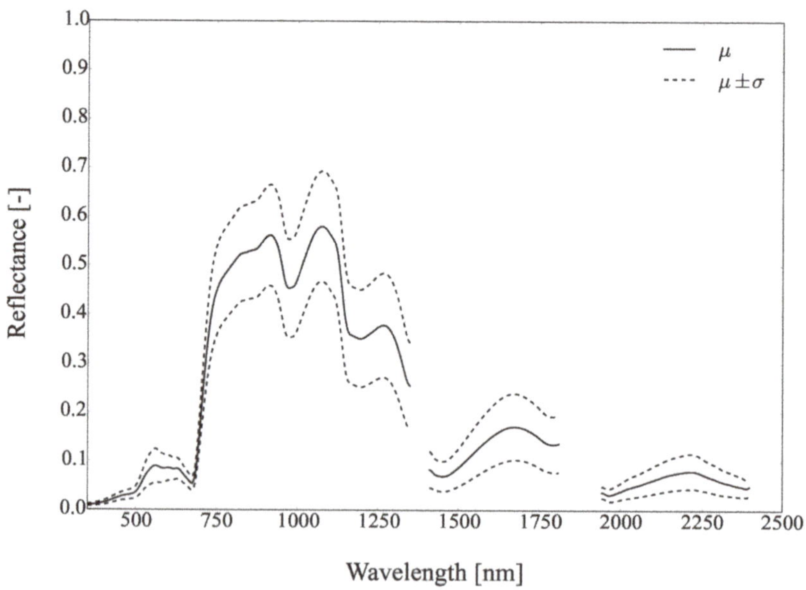

Figure A2. Mean reflectance (μ) and standard deviation (σ) of *Sphagnum* sp. (SPHA).

Table A2. Pictures, plots, geographic coordinates and number of spectra of *Sphagnum* sp. (SPHA).

Picture	Plot	Longitude (DD)	Latitude (DD)	Altitude (m)	No. of Spectra
	1	1.423156	42.802105	1343.715	4
	2	1.423080	42.802068	1344.046	4
	3	1.423143	42.802005	1344.004	4
	4	1.423771	42.802907	1344.747	7
	5	1.424118	42.803025	1346.327	3

Appendix B.2. Calluna vulgaris (CAVU)

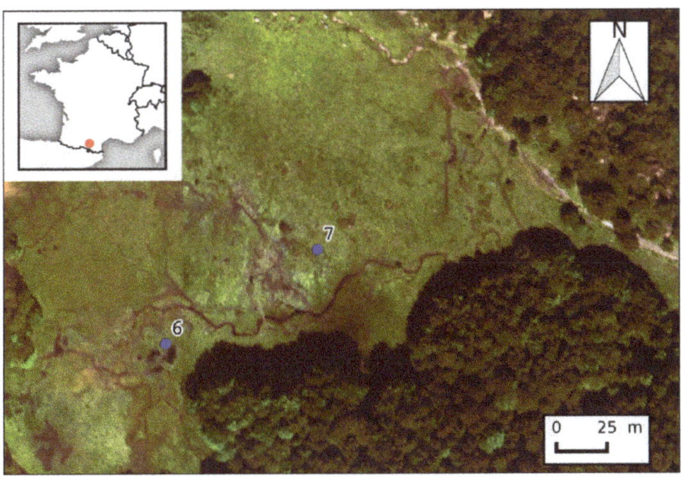

Figure A3. Location of the in situ spectroradiometer measurements for the plots of *Calluna vulgaris* (CAVU).

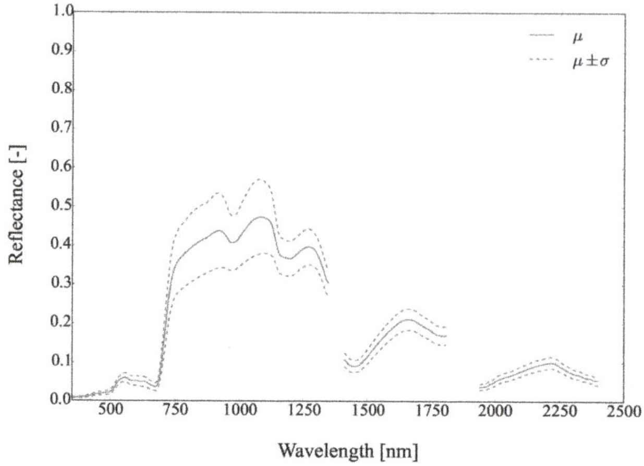

Figure A4. Mean reflectance (μ) and standard deviation (σ) of *Calluna vulgaris* (CAVU).

Table A3. Pictures, plots, geographic coordinates and number of spectra of *Calluna vulgaris* (CAVU).

Picture	Plot	Longitude (DD)	Latitude (DD)	Altitude (m)	No. of Spectra
	6	1.423564	42.80234	1343.762	7
	7	1.42446	42.802773	1343.636	7

Appendix B.3. Eleocharis quinqueflora (ELQU)

Table A4. Pictures, plots, geographic coordinates and number of spectra of *Eleocharis quinqueflora* (ELQU).

Picture	Plot	Longitude (DD)	Latitude (DD)	Altitude (m)	No. of Spectra
	8	1.423728	42.802918	1344.617	3
	9	1.423602	42.802983	1344.650	12

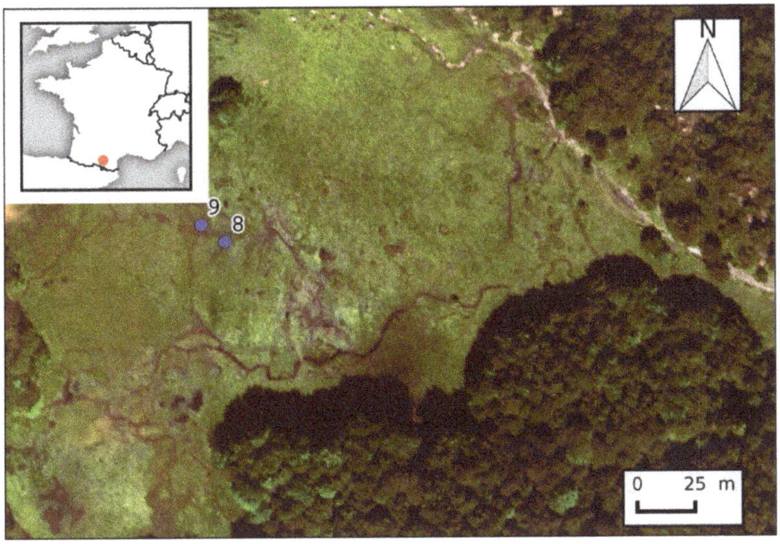

Figure A5. Location of the in situ spectroradiometer measurements for the plots of *Eleocharis quinqueflora* (ELQU).

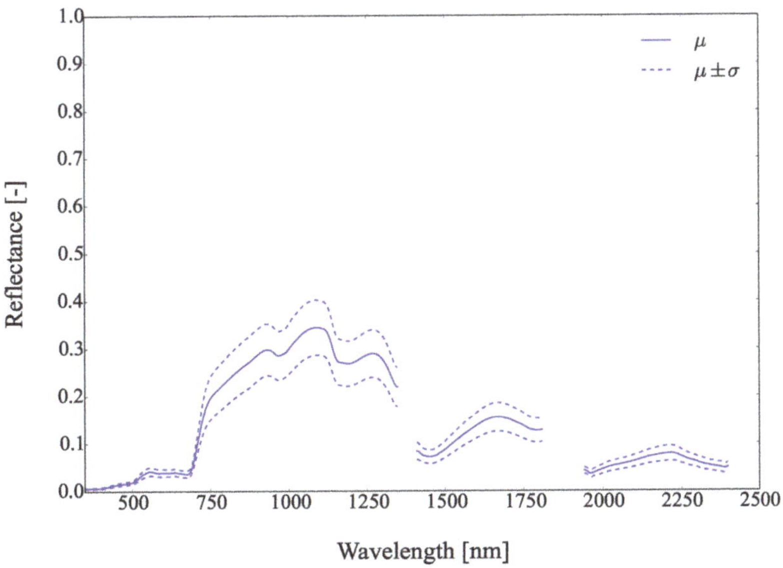

Figure A6. Mean reflectance (μ) and standard deviation (σ) of *Eleocharis quinqueflora* (ELQU).

Appendix B.4. Pinguicula sp. (PING)

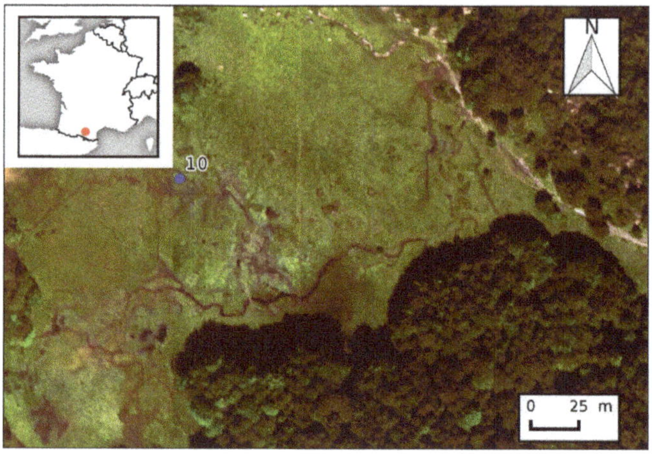

Figure A7. Location of the in situ spectroradiometer measurements for the plots of *Pinguicula* sp. (PING).

Table A5. Pictures, plots, geographic coordinates and number of spectra of *Pinguicula* sp. (PING).

Picture	Plot	Longitude (DD)	Latitude (DD)	Altitude (m)	No. of Spectra
	10	1.423687	42.803021	1345.138	8

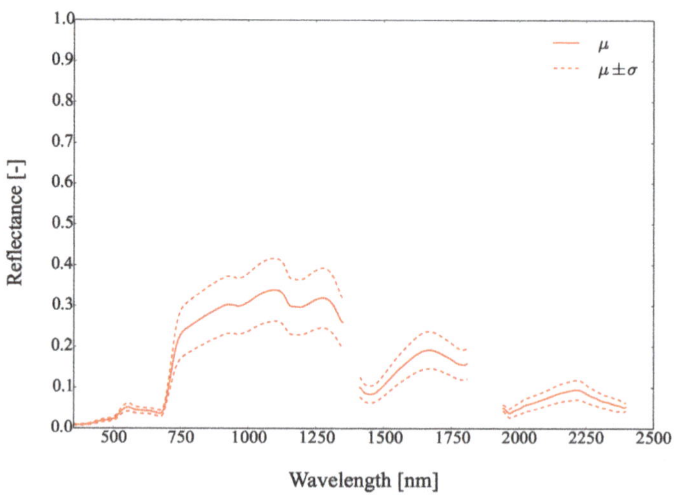

Figure A8. Mean reflectance (μ) and standard deviation (σ) of *Pinguicula* sp. (PING).

Appendix B.5. Menyanthes trifoliata (METR)

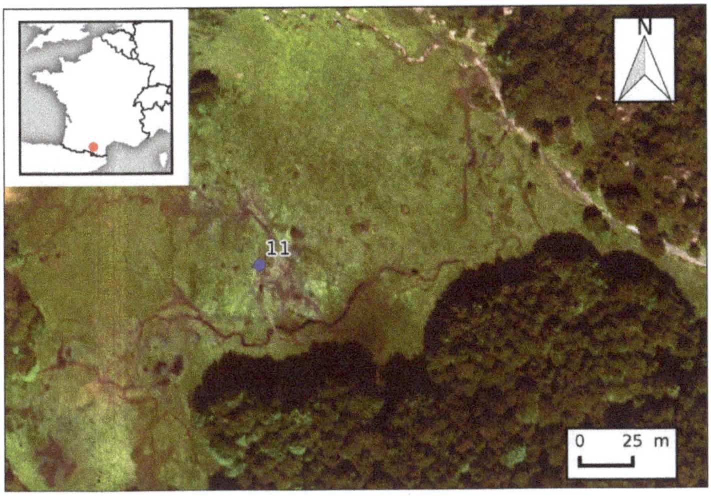

Figure A9. Location of the in situ spectroradiometer measurements for the plots of *Menyanthes trifoliata* (METR).

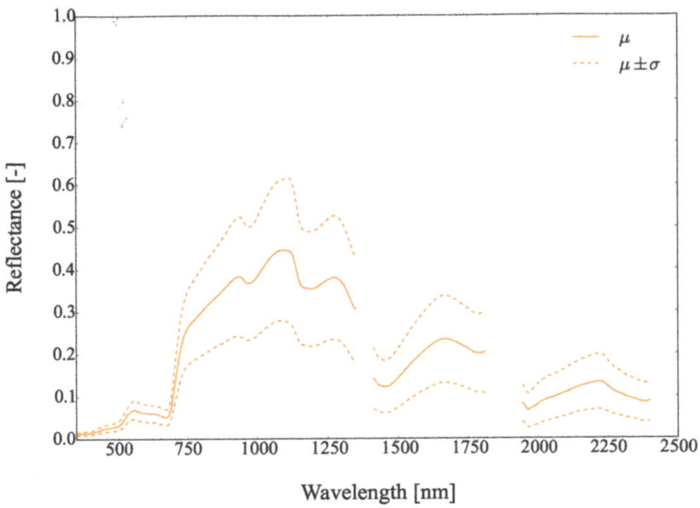

Figure A10. Mean reflectance (μ) and standard deviation (σ) of *Menyanthes trifoliata* (METR).

Table A6. Pictures, plots, geographic coordinates and number of spectra of *Menyanthes trifoliata* (METR).

Picture	Plot	Longitude (DD)	Latitude (DD)	Altitude (m)	No. of Spectra
	11	1.424057	42.802733	1343.781	12

Appendix B.6. Juniperus communis (JUCO)

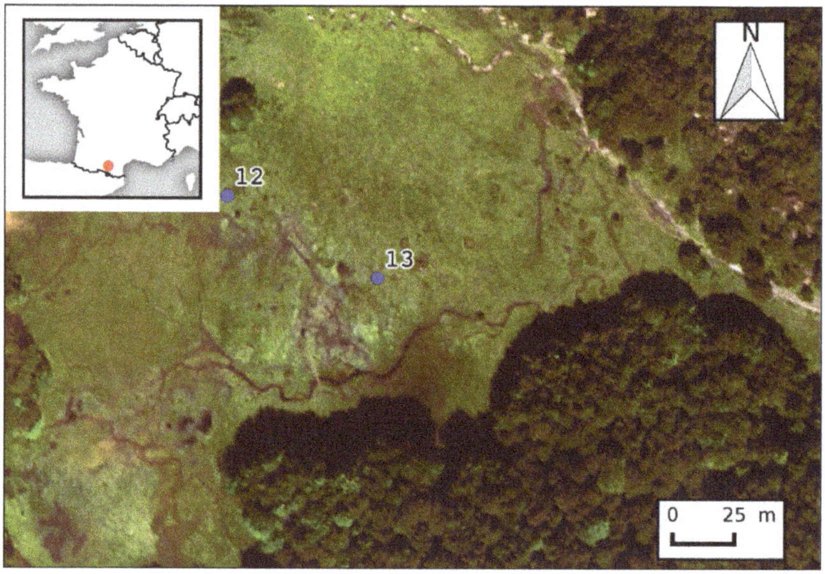

Figure A11. Location of the in situ spectroradiometer measurements for the plots of *Juniperus communis* (JUCO).

Table A7. Pictures, plots, geographic coordinates and number of spectra of *Juniperus communis* (JUCO).

Picture	Plot	Longitude (DD)	Latitude (DD)	Altitude (m)	No. of Spectra
	12	1.42368	42.803132	1345.667	12
	13	1.424437	42.802841	1344.217	7

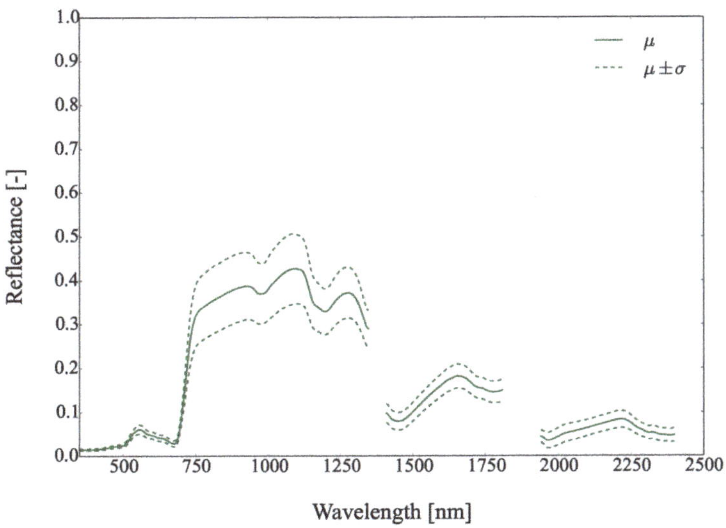

Figure A12. Mean reflectance (μ) and standard deviation (σ) of *Juniperus communis* (JUCO).

Appendix B.7. Rhododendron ferrugineum (RHFR)

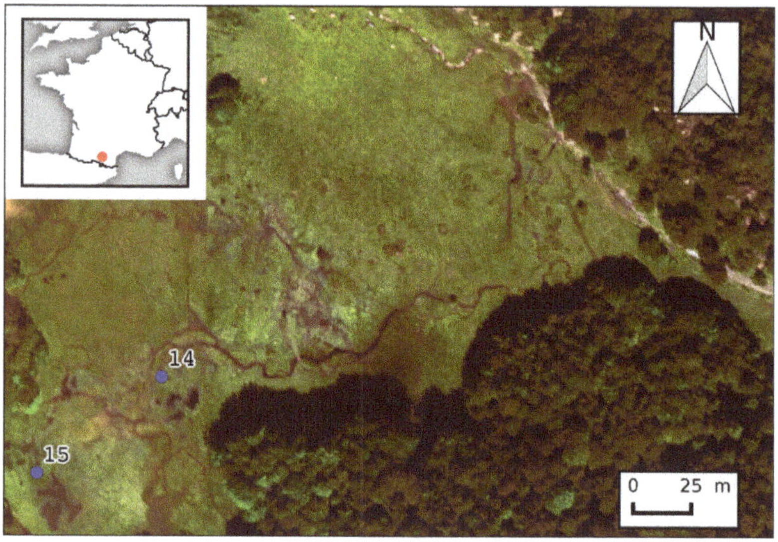

Figure A13. Location of the in situ spectroradiometer measurements for the plots of *Rhododendron ferrugineum* (RHFR).

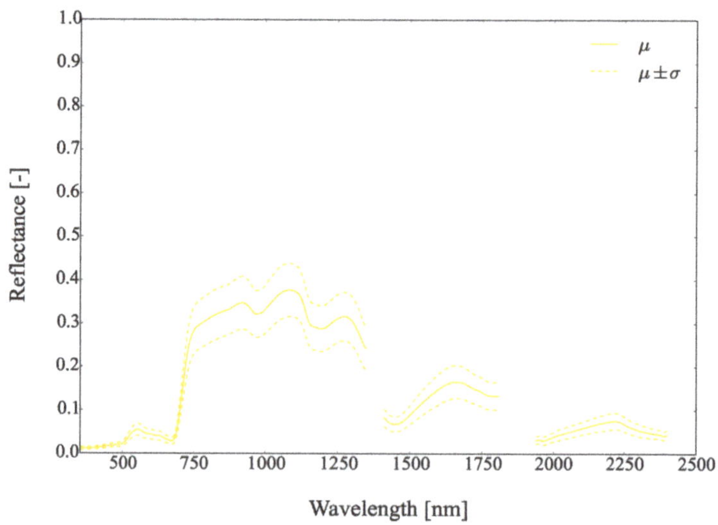

Figure A14. Mean reflectance (μ) and standard deviation (σ) of *Rhododendron ferrugineum* (RHFR).

Table A8. Pictures, plots, geographic coordinates and number of spectra of *Rhododendron ferrugineum* (RHFR).

Picture	Plot	Longitude (DD)	Latitude (DD)	Altitude (m)	No. of Spectra
	14	1.423429	42.802376	1343.301	7
	15	1.422769	42.801989	1344.606	7

Appendix B.8. Salix sp. (SALI)

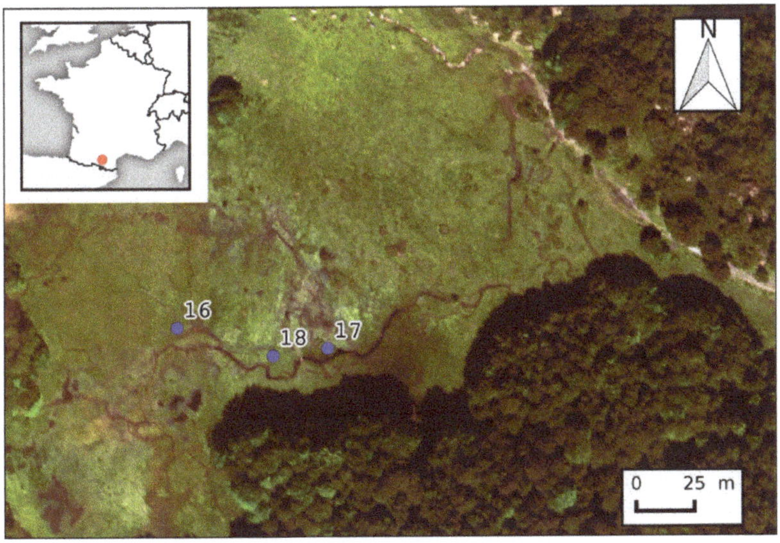

Figure A15. Location of the in situ spectroradiometer measurements for the plots of *Salix* sp. (SALI).

Table A9. Pictures, plots, geographic coordinates and number of spectra of *Salix* sp. (SALI).

Picture	Plot	Longitude (DD)	Latitude (DD)	Altitude (m)	No. of Spectra
	16	1.423492	42.802575	1343.198	9
	17	1.424283	42.802505	1343.082	4
	18	1.423997	42.802472	1343.025	4

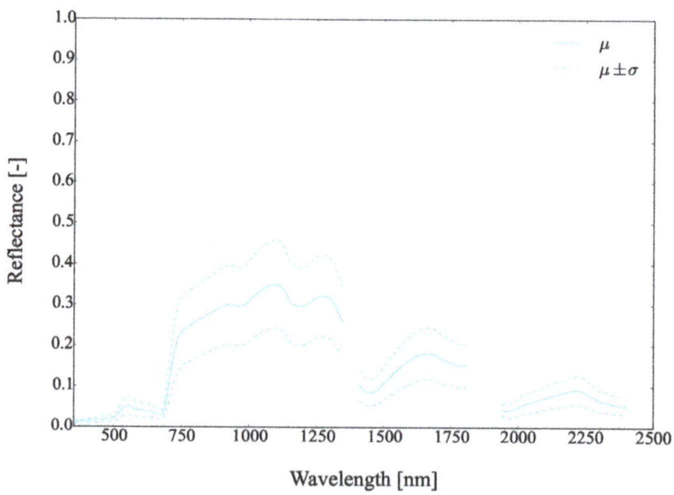

Figure A16. Mean reflectance (μ) and standard deviation (σ) of *Salix* sp. (SALI).

Appendix B.9. Aquatic Type a (AQ_A)

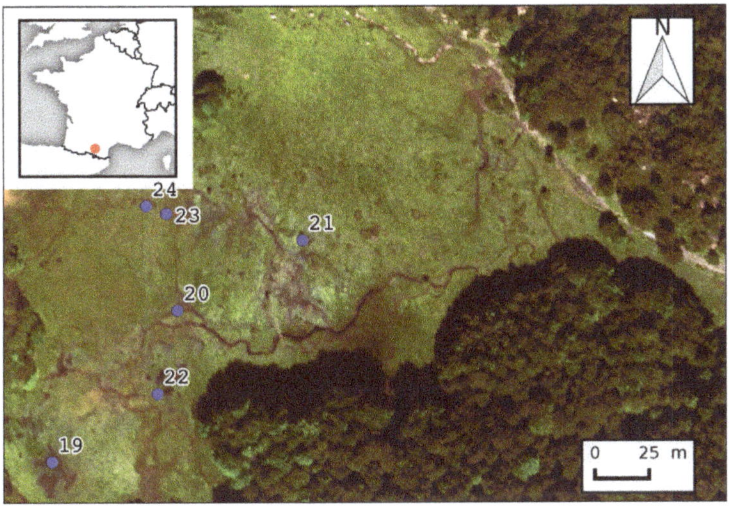

Figure A17. Location of the in situ spectroradiometer measurements for the plots of Aquatic type a (AQ_A).

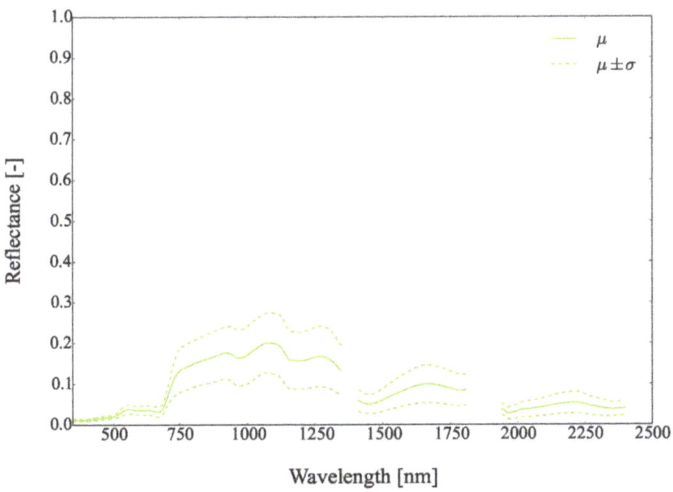

Figure A18. Mean reflectance (μ) and standard deviation (σ) of Aquatic type a (AQ_A).

Table A10. Pictures, plots, geographic coordinates and number of spectra of Aquatic type a (AQ_A).

Picture	Plot	Longitude (DD)	Latitude (DD)	Altitude (m)	No. of Spectra
	19	1.422872	42.801917	1344.375	7
	20	1.423569	42.80256	1343.070	12
	21	1.424258	42.802863	1344.285	6
	22	1.423466	42.80221	1343.305	4
	23	1.423495	42.802963	1344.493	12
	24	1.42338	42.802993	1344.632	12

Appendix B.10. Aquatic Type b (AQ_B)

Table A11. Pictures, plots, geographic coordinates and number of spectra of Aquatic type b (AQ_B).

Picture	Plot	Longitude (DD)	Latitude (DD)	Altitude (m)	No. of Spectra
	25	1.423539	42.802234	1343.04	7

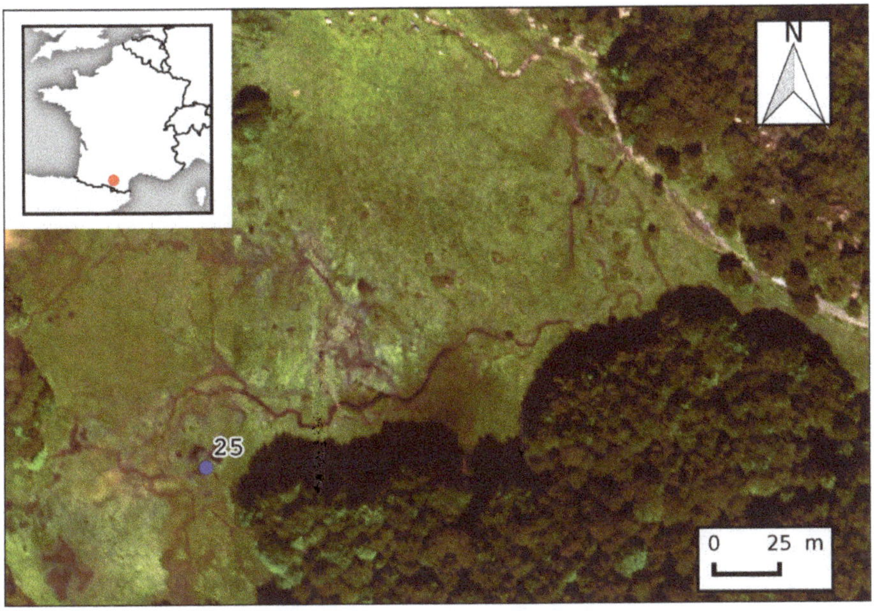

Figure A19. Location of the in situ spectroradiometer measurements for the plots of Aquatic type b (AQ_B).

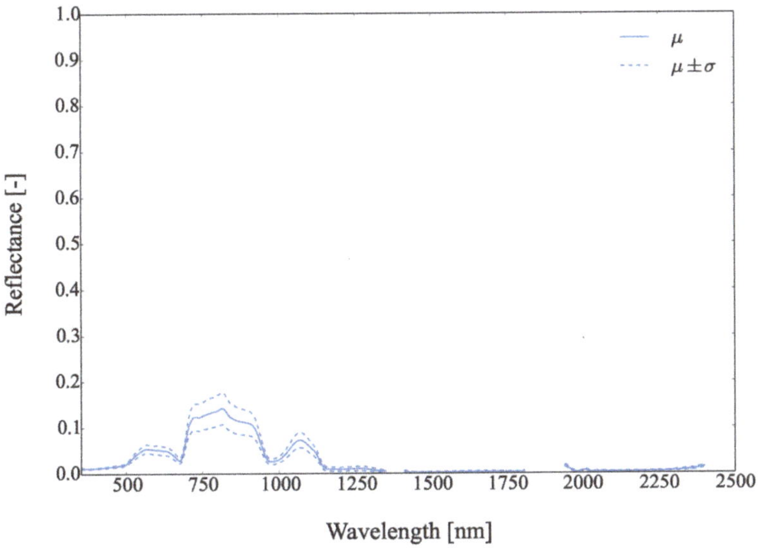

Figure A20. Mean reflectance (μ) and standard deviation (σ) of Aquatic type b (AQ_B).

Appendix B.11. Aquatic Type c (AQ_C)

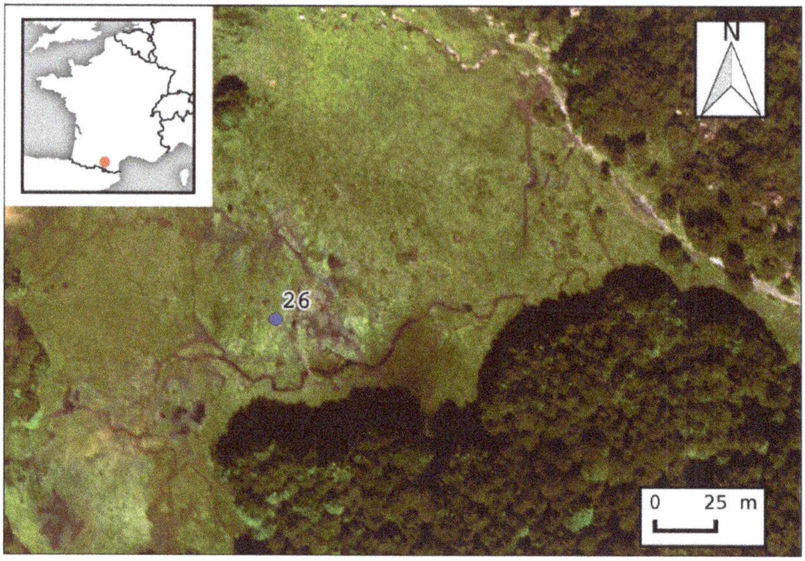

Figure A21. Location of the in situ spectroradiometer measurements for the plots of Aquatic type c (AQ_C).

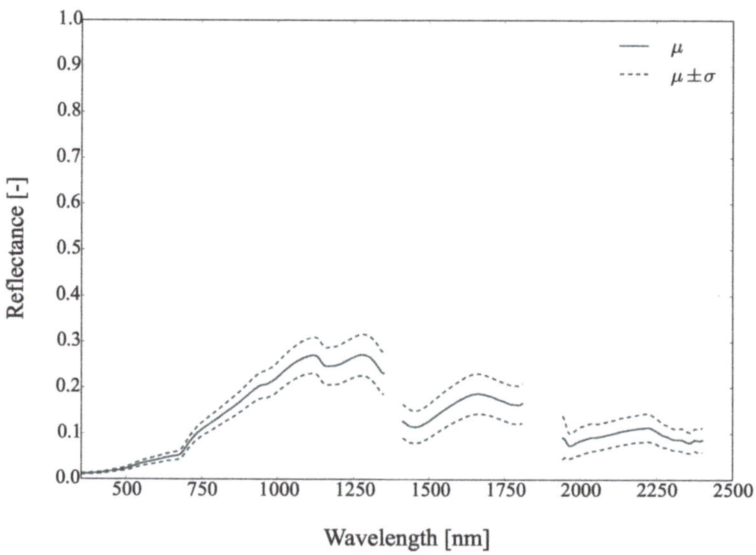

Figure A22. Mean reflectance (μ) and standard deviation (σ) of Aquatic type c (AQ_C).

Table A12. Pictures, plots, geographic coordinates and number of spectra of Aquatic type c (AQ_C).

Picture	Plot	Longitude (DD)	Latitude (DD)	Altitude (m)	No. of Spectra
	26	1.423972	42.802653	1343.362	12

Appendix B.12. Carex sp. Homogeneous Vegetation (CA_HV)

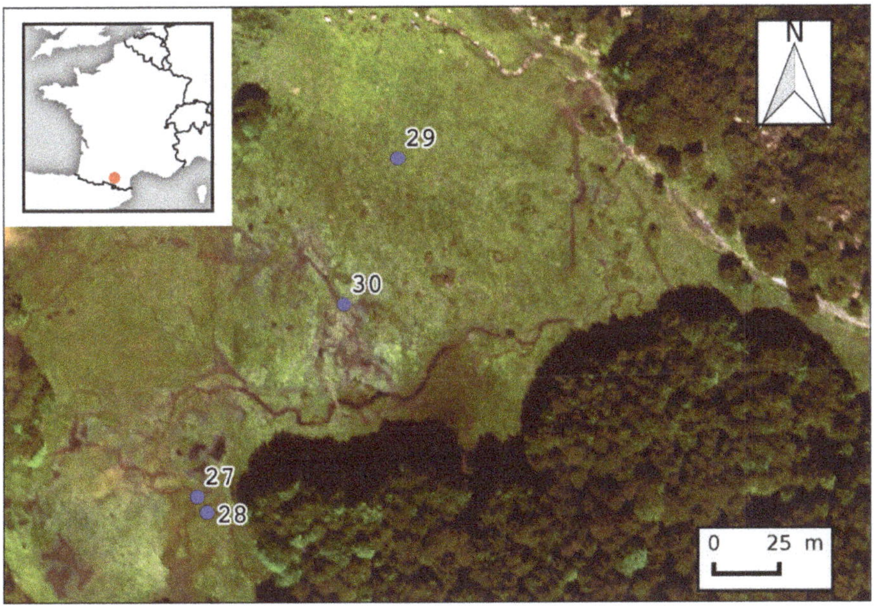

Figure A23. Location of the in situ spectroradiometer measurements for the plots of *Carex* sp. homogeneous vegetation (CA_HV).

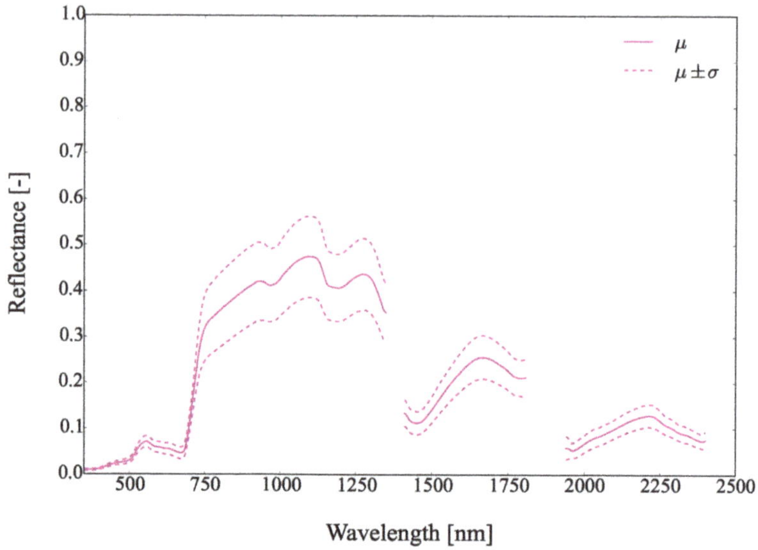

Figure A24. Mean reflectance (μ) and standard deviation (σ) of *Carex* sp. homogeneous vegetation (CA_HV).

Table A13. Pictures, plots, geographic coordinates and number of spectra of *Carex* sp. homogeneous vegetation (CA_HV).

Picture	Plot	Longitude (DD)	Latitude (DD)	Altitude (m)	No. of Spectra
	27	1.423499	42.802124	1343.533	4
	28	1.423547	42.802071	1344.568	4
	29	1.42441	42.803316	1351.678	9
	30	1.424173	42.802804	1344.481	10

Appendix B.13. Pinguicula sp. Combined Vegetation (PI_CV)

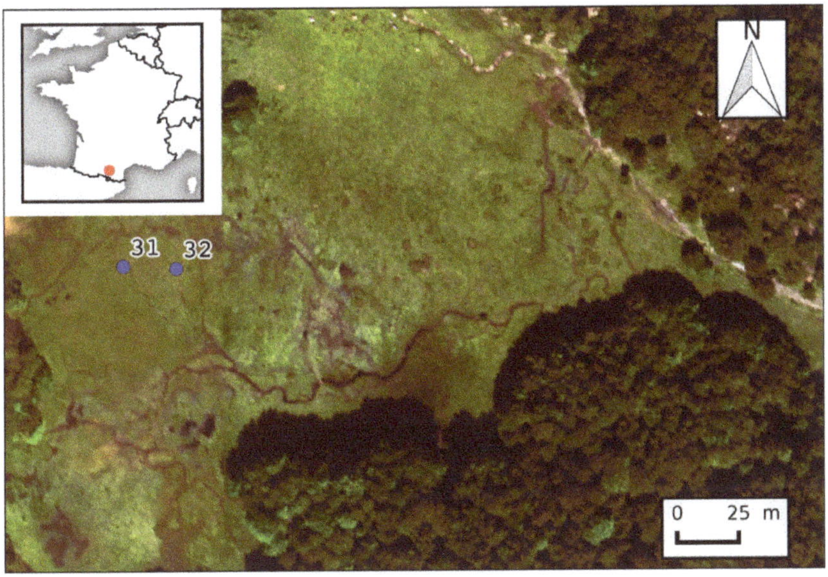

Figure A25. Location of the in situ spectroradiometer measurements for the plots of *Pinguicula* sp. combined vegetation (PI_CV).

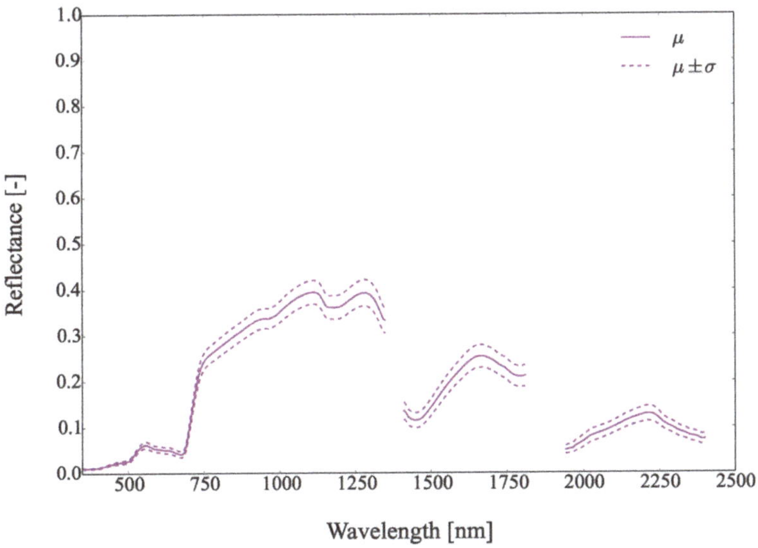

Figure A26. Mean reflectance (μ) and standard deviation (σ) of *Pinguicula* sp. combined vegetation (PI_CV).

Table A14. Pictures, plots, geographic coordinates and number of spectra of *Pinguicula* sp. combined vegetation (PI_CV).

Picture	Plot	Longitude (DD)	Latitude (DD)	Altitude (m)	No. of Spectra
	31	1.42316	42.802875	1344.344	12
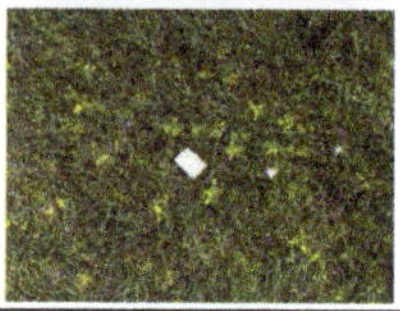	32	1.423421	42.80287	1344.247	3

References

1. Gorham, E. Northern peatlands: Role in the carbon cycle and probable responses to climatic warming. *Ecol. Appl.* **1991**, *1*, 182–195.
2. Yu, Z.; Loisel, J.; Brosseau, D.P.; Beilman, D.W.; Hunt, S.J. Global peatland dynamics since the Last Glacial Maximum. *Geophys. Res. Lett.* **2010**, *37*, doi:10.1029/2010GL043584.
3. Rydin, H.; Jeglum, J.K. *The Biology of Peatlands*, 2nd ed.; Oxford University Press: Oxford, UK, 2013.
4. Kent, M.; Coker, P. *Vegetation Description and Analysis: A Practical Approach*; Belhaven Press: Totnes, UK, 1992.
5. Schmidt, K.; Skidmore, A. Spectral discrimination of vegetation types in a coastal wetland. *Remote Sens. Environ.* **2003**, *85*, 92–108.
6. Adam, E.; Mutanga, O. Spectral discrimination of papyrus vegetation (*Cyperus papyrus* L.) in swamp wetlands using field spectrometry. *ISPRS J. Photogramm. Remote Sens.* **2009**, *64*, 612–620.
7. Seher, J.S.; Tueller, P.T. Color aerial photos for marshland. *Photogramm. Eng.* **1973**, *9*, 489–499.
8. Adam, E.; Mutanga, O.; Rugege, D. Multispectral and hyperspectral remote sensing for identification and mapping of wetland vegetation: A review. *Wetl. Ecol. Manag.* **2010**, *18*, 281–296.
9. Guyot, G. Optical properties of vegetation canopies. In *Applications of Remote Sensing in Agriculture*; Butterworths: Sevenoaks, UK, 1990; pp. 19–43.
10. Yuan, L.; Zhang, L. Identification of the spectral characteristics of submerged plant *Vallisneria spiralis*. *Acta Ecol. Sin.* **2006**, *26*, 1005–1010.
11. Hestir, E.L.; Khanna, S.; Andrew, M.E.; Santos, M.J.; Viers, J.H.; Greenberg, J.A.; Rajapakse, S.S.; Ustin, S.L. Identification of invasive vegetation using hyperspectral remote sensing in the California Delta ecosystem. *Remote Sens. Environ.* **2008**, *112*, 4034–4047.
12. Torbick, N.; Becker, B.; Qi, J.; Lusch, D. Characterizing field-level hyperspectral measurements for identifying wetland invasive plant species. In *Invasive Species: Detection, Impact and Control*; Nova Science Publishers: Hauppauge, NY, USA, 2009; pp. 97–115.
13. Hamada, Y.; Stow, D.A.; Coulter, L.L.; Jafolla, J.C.; Hendricks, L.W. Detecting Tamarisk species (*Tamarix* spp.) in riparian habitats of Southern California using high spatial resolution hyperspectral imagery. *Remote Sens. Environ.* **2007**, *109*, 237–248.
14. Vaiphasa, C.; Skidmore, A.K.; de Boer, W.F.; Vaiphasa, T. A hyperspectral band selector for plant species discrimination. *ISPRS J. Photogramm. Remote Sens.* **2007**, *62*, 225–235.
15. Jia, M.; Zhang, Y.; Wang, Z.; Song, K.; Ren, C. Mapping the distribution of mangrove species in the Core Zone of Mai Po Marshes Nature Reserve, Hong Kong, using hyperspectral data and high-resolution data. *Int. J. Appl. Earth Obs. Geoinf.* **2014**, *33*, 226–231.
16. Prospere, K.; McLaren, K.; Wilson, B. Plant species discrimination in a tropical wetland using in situ hyperspectral data. *Remote Sens.* **2014**, *6*, 8494–8523.

17. Krankina, O.; Pflugmacher, D.; Friedl, M.; Cohen, W.; Nelson, P.; Baccini, A. Meeting the challenge of mapping peatlands with remotely sensed data. *Biogeosciences* **2008**, *5*, 1809–1820.
18. Hubert-Moy, L.; Clément, B.; Lennon, M.; Houet, T.; Lefeuvre, E. Etude de zones humides de fond de vallées à partir d'images hyperspectrales CASI: Application à un bassin versant de la région de Pleine-Fougères (Bretagne, France). *Photo-Interprétation* **2003**, *39*, 33–43.
19. Thomas, V.; Treitz, P.; Jelinski, D.; Miller, J.; Lafleur, P.; McCaughey, J.H. Image classification of a northern peatland complex using spectral and plant community data. *Remote Sens. Environ.* **2003**, *84*, 83–99.
20. Knoth, C.; Klein, B.; Prinz, T.; Kleinebecker, T. Unmanned aerial vehicles as innovative remote sensing platforms for high-resolution infrared imagery to support restoration monitoring in cut-over bogs. *Appl. Veg. Sci.* **2013**, *16*, 509–517.
21. Yagoub, H.; Belbachir, A.H.; Benabadji, N. Detection and mapping vegetation cover based on the Spectral Angle Mapper algorithm using NOAA AVHRR data. *Adv. Space Res.* **2014**, *53*, 1686–1693.
22. Bahri, E.M.; Haboudane, D.; Bannari, A.; Bonn, F.; Chillasse, L. Essai de cartographie des espèces forestières dominantes dans le moyen atlas (Maroc) à l'aide des données Aster. *Revue Télédétection* **2007**, *7*, 283–301.
23. Sobhan, I. Species Discrimination from a Hyperspectral Perspective. Ph.D. Thesis, International Institute for Geo-Information Science & Earth Observation, University of Twente, Enschede, The Netherlands, 2007.
24. Clark, M.L.; Roberts, D.A.; Clark, D.B. Hyperspectral discrimination of tropical rain forest tree species at leaf to crown scales. *Remote Sens. Environ.* **2005**, *96*, 375–398.
25. Lawrence, R.L.; Wood, S.D.; Sheley, R.L. Mapping invasive plants using hyperspectral imagery and Breiman Cutler classifications (RandomForest). *Remote Sens. Environ.* **2006**, *100*, 356–362.
26. Dalponte, M.; Ørka, H.O.; Gobakken, T.; Gianelle, D.; Næsset, E. Tree species classification in boreal forests with hyperspectral data. *IEEE Trans. Geosci. Remote Sens.* **2013**, *51*, 2632–2645.
27. Vyas, D.; Krishnayya, N.; Manjuňath, K.; Ray, S.; Panigrahy, S. Evaluation of classifiers for processing Hyperion (EO-1) data of tropical vegetation. *Int. J. Appl. Earth Obs. Geoinf.* **2011**, *13*, 228–235.
28. Pant, P.; Heikkinen, V.; Korpela, I.; Hauta-Kasari, M.; Tokola, T. Logistic regression-based spectral band selection for tree species classification: Effects of spatial scale and balance in training samples. *IEEE Geosci. Remote Sens. Lett.* **2014**, *11*, 1604–1608.
29. Pal, M. Multinomial logistic regression-based feature selection for hyperspectral data. *Int. J. Appl. Earth Obs. Geoinf.* **2012**, *14*, 214–220.
30. Merot, P.; Hubert-Moy, L.; Gascuel-Odoux, C.; Clement, B.; Durand, P.; Baudry, J.; Thenail, C. A Method for Improving the Management of Controversial Wetland. *Environ. Manag.* **2006**, *37*, 258–270.
31. Asner, G.P. Biophysical and biochemical sources of variability in canopy reflectance. *Remote Sens. Environ.* **1998**, *64*, 234–253.
32. Savitzky, A.; Golay, M.J. Smoothing and differentiation of data by simplified least squares procedures. *Anal. Chem.* **1964**, *36*, 1627–1639.
33. Feilhauer, H.; Asner, G.P.; Martin, R.E.; Schmidtlein, S. Brightness-normalized partial least squares regression for hyperspectral data. *J. Quant. Spectrosc. Radiat. Transf.* **2010**, *111*, 1947–1957.
34. Tsai, F.; Philpot, W. Derivative analysis of hyperspectral data. *Remote Sens. Environ.* **1998**, *66*, 41–51.
35. Serrano, L.; Peñuelas, J.; Ustin, S.L. Remote sensing of nitrogen and lignin in Mediterranean vegetation from AVIRIS data: Decomposing biochemical from structural signals. *Remote Sens. Environ.* **2002**, *81*, 355–364.
36. Clark, R.N.; Roush, T.L. Reflectance spectroscopy: Quantitative analysis techniques for remote sensing applications. *J. Geophys. Res. Solid Earth* **1984**, *89*, 6329–6340.
37. Kokaly, R.F.; Clark, R.N. Spectroscopic determination of leaf biochemistry using band-depth analysis of absorption features and stepwise multiple linear regression. *Remote Sens. Environ.* **1999**, *67*, 267–287.
38. Mutanga, O.; Skidmore, A.K.; Prins, H. Predicting in situ pasture quality in the Kruger National Park, South Africa, using continuum-removed absorption features. *Remote Sens. Environ.* **2004**, *89*, 393–408.
39. Hu, B.; Lévesque, J.; Ardouin, J.P. Vegetation Species Identification Using Hyperspectral Imagery. In Proceedings of the IEEE International Geoscience and Remote Sensing Symposium, Boston, MA, USA, 6–11 July 2008; Volume 2, pp. II299–II302.
40. Ghiyamat, A.; Shafri, H.Z.M.; Mahdiraji, G.A.; Shariff, A.R.M.; Mansor, S. Hyperspectral discrimination of tree species with different classifications using single-and multiple-endmember. *Int. J. Appl. Earth Obs. Geoinf.* **2013**, *23*, 177–191.

41. Chang, C.I.; Ren, H. An experiment-based quantitative and comparative analysis of target detection and image classification algorithms for hyperspectral imagery. *IEEE Trans. Geosci. Remote Sens.* **2000**, *38*, 1044–1063.
42. Chauhan, H.; Mohan, B.K. Effectiveness of spectral similarity measures to develop precise crop spectra for hyperspectral data analysis. *ISPRS Ann. Photogramm. Remote Sens. Spat. Inf. Sci.* **2014**, *2*, 83–90.
43. Lance, G.N.; Williams, W.T. Computer programs for hierarchical polythetic classification ("similarity analyses"). *Comput. J.* **1966**, *9*, 60–64.
44. Kruse, F.; Lefkoff, A.; Boardman, J.; Heidebrecht, K.; Shapiro, A.; Barloon, P.; Goetz, A. The spectral image processing system (SIPS)—Interactive visualization and analysis of imaging spectrometer data. *Remote Sens. Environ.* **1993**, *44*, 145–163.
45. Chang, C.I. An information-theoretic approach to spectral variability, similarity, and discrimination for hyperspectral image analysis. *IEEE Trans. Inf. Theory* **2000**, *46*, 1927–1932.
46. Du, Y.; Chang, C.I.; Ren, H.; Chang, C.C.; Jensen, J.O.; D'Amico, F.M. New hyperspectral discrimination measure for spectral characterization. *Opt. Eng.* **2004**, *43*, 1777–1786.
47. Van der Meer, F.; Bakker, W. Cross correlogram spectral matching: application to surface mineralogical mapping by using AVIRIS data from Cuprite, Nevada. *Remote Sens. Environ.* **1997**, *61*, 371–382.
48. Farifteh, J.; Van Der Meer, F.; Carranza, E. Similarity measures for spectral discrimination of salt-affected soils. *Int. J. Remote Sens.* **2007**, *28*, 5273–5293.
49. De Carvalho, O.A., Jr.; Meneses, P.R. Spectral correlation mapper (SCM): An improvement on the spectral angle mapper (SAM). In *Summaries of the Ninth JPL Airborne Earth Science Workshop; Jet Propulsion Laboratory, National Aeronautics and Space Administration*; JPL Publication: Pasadena, CA, USA, 2000; Volume 9.
50. Robila, S. An analysis of spectral metrics for hyperspectral image processing. In Proceedings of the IEEE International Geoscience and Remote Sensing Symposium, Anchorage, AK, USA, 20–24 September 2004; IEEE: Anchorage, AK, USA, 2004; Volume 5, pp. 3233–3236.
51. Angelopoulou, E.; Lee, S.W.; Bajcsy, R. Spectral Gradient: A Material Descriptor Invariant to Geometry and Incident Illumination. In Proceedings of the Seventh International Conference on Computer Vision, Kerkyra, Greece, 23–27 September 1999; IEEE Computer Society Press: Kerkyra, Greece, 1999; Volume 2, pp. 861–867.
52. Boochs, F.; Kupfer, G.; Dockter, K.; Kühbauch, W. Shape of the red edge as vitality indicator for plants. *Remote Sens.* **1990**, *11*, 1741–1753.
53. Nagler, P.; Daughtry, C.; Goward, S. Plant litter and soil reflectance. *Remote Sens. Environ.* **2000**, *71*, 207–215.
54. Kim, M.S.; Daughtry, C.S.; Chappelle, E.; McMurtrey, J.; Walthall, C.L. The use of high spectral resolution bands for estimating absorbed photosynthetically active radiation (A par). In Proceedings of the 6th International Symposium on Physical Measurements and Signatures in Remote Sensing, D'Isere, France, 17–21 January 1994.
55. Zarco-Tejada, P.J.; Pushnik, J.; Dobrowski, S.; Ustin, S. Steady-state chlorophyll a fluorescence detection from canopy derivative reflectance and double-peak red-edge effects. *Remote Sens. Environ.* **2003**, *84*, 283–294.
56. Sims, D.A.; Luo, H.; Hastings, S.; Oechel, W.C.; Rahman, A.F.; Gamon, J.A. Parallel adjustments in vegetation greenness and ecosystem CO_2 exchange in response to drought in a Southern California chaparral ecosystem. *Remote Sens. Environ.* **2006**, *103*, 289–303.
57. Barnes, E.; Clarke, T.; Richards, S.; Colaizzi, P.; Haberland, J.; Kostrzewski, M.; Waller, P.; Choi, C.; Riley, E.; Thompson, T.; et al. Coincident detection of crop water stress, nitrogen status and canopy density using ground-based multispectral data. In Proceedings of the 5th International Conference on Precision Agriculture, Bloomington, MN, USA, 16–19 July 2000; American Society of Agronomy, Crop Science Society of America, Soil Science Society of America: Bloomington, MN, USA, 2000; pp. 1–15.
58. Carter, G.A.; Miller, R.L. Early detection of plant stress by digital imaging within narrow stress-sensitive wavebands. *Remote Sens. Environ.* **1994**, *50*, 295–302.
59. Gitelson, A.A.; Gritz, Y.; Merzlyak, M.N. Relationships between leaf chlorophyll content and spectral reflectance and algorithms for non-destructive chlorophyll assessment in higher plant leaves. *J. Plant Physiol.* **2003**, *160*, 271–282.
60. Gitelson, A.A.; Keydan, G.P.; Merzlyak, M.N. Three-band model for noninvasive estimation of chlorophyll, carotenoids, and anthocyanin contents in higher plant leaves. *Geophys. Res. Lett.* **2006**, *33*, L11402.
61. Datt, B. A new reflectance index for remote sensing of chlorophyll content in higher plants: Tests using Eucalyptus leaves. *J. Plant Physiol.* **1999**, *154*, 30–36.

62. Datt, B. Remote sensing of chlorophyll a, chlorophyll b, chlorophyll a+b, and total carotenoid content in eucalyptus leaves. *Remote Sens. Environ.* **1998**, *66*, 111–121.
63. Zarco-Tejada, P.J.; Miller, J.R.; Mohammed, G.; Noland, T.; Sampson, P. Vegetation Stress Detection through Chlorophyll+ Estimation and Fluorescence Effects on Hyperspectral Imagery. *J. Environ. Qual.* **2002**, *31*, 1433–1441.
64. Chen, P.; Haboudane, D.; Tremblay, N.; Wang, J.; Vigneault, P.; Li, B. New spectral indicator assessing the efficiency of crop nitrogen treatment in corn and wheat. *Remote Sens. Environ.* **2010**, *114*, 1987–1997.
65. Le Maire, G.; François, C.; Dufrêne, E. Towards universal broad leaf chlorophyll indices using PROSPECT simulated database and hyperspectral reflectance measurements. *Remote Sens. Environ.* **2004**, *89*, 1–28.
66. Le Maire, G.; François, C.; Soudani, K.; Berveiller, D.; Pontailler, J.Y.; Bréda, N.; Genet, H.; Davi, H.; Dufrêne, E. Calibration and validation of hyperspectral indices for the estimation of broadleaved forest leaf chlorophyll content, leaf mass per area, leaf area index and leaf canopy biomass. *Remote Sens. Environ.* **2008**, *112*, 3846–3864.
67. Filella, I.; Peñuelas, J. The red edge position and shape as indicators of plant chlorophyll content, biomass and hydric status. *Int. J. Remote Sens.* **1994**, *15*, 1459–1470.
68. Huete, A.; Liu, H.; Batchily, K.; Van Leeuwen, W. A comparison of vegetation indices over a global set of TM images for EOS-MODIS. *Remote Sens. Environ.* **1997**, *59*, 440–451.
69. Peñuelas, J.; Gamon, J.; Fredeen, A.; Merino, J.; Field, C. Reflectance indices associated with physiological changes in nitrogen-and water-limited sunflower leaves. *Remote Sens. Environ.* **1994**, *48*, 135–146.
70. Pinty, B.; Verstraete, M. GEMI: A non-linear index to monitor global vegetation from satellites. *Vegetatio* **1992**, *101*, 15–20.
71. Smith, R.; Adams, J.; Stephens, D.; Hick, P. Forecasting wheat yield in a Mediterranean-type environment from the NOAA satellite. *Crop Pasture Sci.* **1995**, *46*, 113–125.
72. Gitelson, A.A.; Buschmann, C.; Lichtenthaler, H.K. The chlorophyll fluorescence ratio F 735/F 700 as an accurate measure of the chlorophyll content in plants. *Remote Sens. Environ.* **1999**, *69*, 296–302.
73. Gitelson, A.A.; Merzlyak, M.N. Remote estimation of chlorophyll content in higher plant leaves. *Int. J. Remote Sens.* **1997**, *18*, 2691–2697.
74. Gitelson, A.A.; Kaufman, Y.J.; Merzlyak, M.N. Use of a green channel in remote sensing of global vegetation from EOS-MODIS. *Remote Sens. Environ.* **1996**, *58*, 289–298.
75. Maccioni, A.; Agati, G.; Mazzinghi, P. New vegetation indices for remote measurement of chlorophylls based on leaf directional reflectance spectra. *J. Photochem. Photobiol. B Biol.* **2001**, *61*, 52–61.
76. Gitelson, A.A.; Merzlyak, M.N.; Chivkunova, O.B. Optical Properties and Nondestructive Estimation of Anthocyanin Content in Plant Leaves. *Photochem. Photobiol.* **2001**, *74*, 38–45.
77. Gitelson, A.A.; Chivkunova, O.B.; Merzlyak, M.N. Nondestructive estimation of anthocyanins and chlorophylls in anthocyanic leaves. *Am. J. Bot.* **2009**, *96*, 1861–1868.
78. Daughtry, C.; Walthall, C.; Kim, M.; De Colstoun, E.B.; McMurtrey, J. Estimating corn leaf chlorophyll concentration from leaf and canopy reflectance. *Remote Sens. Environ.* **2000**, *74*, 229–239.
79. Wu, C.; Niu, Z.; Tang, Q.; Huang, W. Estimating chlorophyll content from hyperspectral vegetation indices: Modeling and validation. *Agric. For. Meteorol.* **2008**, *148*, 1230–1241.
80. Haboudane, D.; Miller, J.R.; Tremblay, N.; Zarco-Tejada, P.J.; Dextraze, L. Integrated narrow-band vegetation indices for prediction of crop chlorophyll content for application to precision agriculture. *Remote Sens. Environ.* **2002**, *81*, 416–426.
81. Eitel, J.; Long, D.; Gessler, P.; Smith, A. Using in-situ measurements to evaluate the new RapidEye^{TM} satellite series for prediction of wheat nitrogen status. *Int. J. Remote Sens.* **2007**, *28*, 4183–4190.
82. Sims, D.A.; Gamon, J.A. Relationships between leaf pigment content and spectral reflectance across a wide range of species, leaf structures and developmental stages. *Remote Sens. Environ.* **2002**, *81*, 337–354.
83. Qi, J.; Chehbouni, A.; Huete, A.; Kerr, Y.; Sorooshian, S. A modified soil adjusted vegetation index. *Remote Sens. Environ.* **1994**, *48*, 119–126.
84. Hunt, E.R.; Rock, B.N. Detection of changes in leaf water content using near-and middle-infrared reflectances. *Remote Sens. Environ.* **1989**, *30*, 43–54.
85. Chen, J.M. Evaluation of vegetation indices and a modified simple ratio for boreal applications. *Can. J. Remote Sens.* **1996**, *22*, 229–242.
86. Dash, J.; Curran, P. The MERIS terrestrial chlorophyll index. *Int. J. Remote Sens.* **2004**, *25*, 5403–5413.

87. Haboudane, D.; Miller, J.R.; Pattey, E.; Zarco-Tejada, P.J.; Strachan, I.B. Hyperspectral vegetation indices and novel algorithms for predicting green LAI of crop canopies: Modeling and validation in the context of precision agriculture. *Remote Sens. Environ.* **2004**, *90*, 337–352.
88. Hardisky, M.A.; Klemas, V.; Smart, R.M. The influence of soil salinity, growth form, and leaf moisture on the spectral radiance of *Spartina alterniflora* canopies. *Photogramm. Eng. Remote Sens.* **1983**, *49*, 77–83.
89. Tucker, C.J. Red and photographic infrared linear combinations for monitoring vegetation. *Remote Sens. Environ.* **1979**, *8*, 127–150.
90. Gandia, S.; Fernández, G.; García, J.; Moreno, J. Retrieval of vegetation biophysical variables from CHRIS/PROBA data in the SPARC campaign. In Proceedings of the 2nd ESA CHRIS/Proba Workshop, Frascati, Italy, 28–30 April 2004; Volume 578, pp. 40–48.
91. Hansen, P.; Schjoerring, J. Reflectance measurement of canopy biomass and nitrogen status in wheat crops using normalized difference vegetation indices and partial least squares regression. *Remote Sens. Environ.* **2003**, *86*, 542–553.
92. Gao, B.C. NDWI—A normalized difference water index for remote sensing of vegetation liquid water from space. *Remote Sens. Environ.* **1996**, *58*, 257–266.
93. Uto, K.; Kosugi, Y. Hyperspectral manipulation for the water stress evaluation of plants. *Contemp. Mater.* **2012**, *1*, 18–25.
94. Peñuelas, J.; Gamon, J.A.; Griffin, K.L.; Field, C.B. Assessing community type, plant biomass, pigment composition, and photosynthetic efficiency of aquatic vegetation from spectral reflectance. *Remote Sens. Environ.* **1993**, *46*, 110–118.
95. Reyniers, M.; Walvoort, D.J.; De Baardemaaker, J. A linear model to predict with a multi-spectral radiometer the amount of nitrogen in winter wheat. *Int. J. Remote Sens.* **2006**, *27*, 4159–4179.
96. Rondeaux, G.; Steven, M.; Baret, F. Optimization of soil-adjusted vegetation indices. *Remote Sens. Environ.* **1996**, *55*, 95–107.
97. Gamon, J.; Peñuelas, J.; Field, C.B. A narrow-waveband spectral index that tracks diurnal changes in photosynthetic efficiency. *Remote Sens. Environ.* **1992**, *41*, 35–44.
98. Roujean, J.L.; Breon, F.M. Estimating PAR absorbed by vegetation from bidirectional reflectance measurements. *Remote Sens. Environ.* **1995**, *51*, 375–384.
99. Horler, D.; Dockray, M.; Barber, J. The red edge of plant leaf reflectance. *Int. J. Remote Sens.* **1983**, *4*, 273–288.
100. Horler, D.; Dockray, M.; Barber, J.; Barringer, A. Red edge measurements for remotely sensing plant chlorophyll content. *Adv. Space Res.* **1983**, *3*, 273–277.
101. Gitelson, A.A.; Vina, A.; Ciganda, V.; Rundquist, D.C.; Arkebauer, T.J. Remote estimation of canopy chlorophyll content in crops. *Geophys. Res. Lett.* **2005**, *32*, L08403.
102. Cho, M.A.; Skidmore, A.K. A new technique for extracting the red edge position from hyperspectral data: The linear extrapolation method. *Remote Sens. Environ.* **2006**, *101*, 181–193.
103. Guyot, G.; Baret, F. Utilisation de la haute résolution spectrale pour suivre l'état des couverts végétaux. In *Signatures Spectrales d'objets en télédétection. 4 ème Colloque International*; Agence Spatiale Européenne: Aussois, France, 1988; Volume 287, pp. 279–286.
104. Zhu, Y.; Yao, X.; Tian, Y.; Liu, X.; Cao, W. Analysis of common canopy vegetation indices for indicating leaf nitrogen accumulations in wheat and rice. *Int. J. Appl. Earth Obs. Geoinf.* **2008**, *10*, 1–10.
105. Xue, L.; Cao, W.; Luo, W.; Dai, T.; Zhu, Y. Monitoring leaf nitrogen status in rice with canopy spectral reflectance. *Agron. J.* **2004**, *96*, 135–142.
106. Peñuelas, J.; Filella, I.; Lloret, P.; Muñoz, F.; Vilajeliu, M. Reflectance assessment of mite effects on apple trees. *Int. J. Remote Sens.* **1995**, *16*, 2727–2733.
107. Vincini, M.; Frazzi, E.; D'Alessio, P. Angular dependence of maize and sugar beet VIs from directional CHRIS/Proba data. In Proceedings of the 4th ESA CHRIS/Proba Workshop, ESRIN, Frascati, Italy, 19–21 September 2006.
108. Jordan, C.F. Derivation of leaf-area index from quality of light on the forest floor. *Ecology* **1969**, *50*, 663–666.
109. McMurtrey, J.; Chappelle, E.; Kim, M.; Meisinger, J.; Corp, L. Distinguishing nitrogen fertilization levels in field corn (*Zea mays* L.) with actively induced fluorescence and passive reflectance measurements. *Remote Sens. Environ.* **1994**, *47*, 36–44.

110. Chappelle, E.W.; Kim, M.S.; McMurtrey, J.E. Ratio analysis of reflectance spectra (RARS): An algorithm for the remote estimation of the concentrations of chlorophyll a, chlorophyll b, and carotenoids in soybean leaves. *Remote Sens. Environ.* **1992**, *39*, 239–247.
111. Zarco-Tejada, P.J.; Miller, J.R. Land cover mapping at BOREAS using red edge spectral parameters from CASI imagery. *J. Geophys. Res. Atmos. (1984–2012)* **1999**, *104*, 27921–27933.
112. Lichtenthaler, H.; Lang, M.; Sowinska, M.; Heisel, F.; Miehe, J. Detection of vegetation stress via a new high resolution fluorescence imaging system. *J. Plant Physiol.* **1996**, *148*, 599–612.
113. Elvidge, C.D.; Chen, Z. Comparison of broad-band and narrow-band red and near-infrared vegetation indices. *Remote Sens. Environ.* **1995**, *54*, 38–48.
114. Broge, N.H.; Leblanc, E. Comparing prediction power and stability of broadband and hyperspectral vegetation indices for estimation of green leaf area index and canopy chlorophyll density. *Remote Sens. Environ.* **2001**, *76*, 156–172.
115. Vogelmann, J.; Rock, B.; Moss, D. Red edge spectral measurements from sugar maple leaves. *Int. J. Remote Sens.* **1993**, *14*, 1563–1575.
116. Pu, R.; Foschi, L.; Gong, P. Spectral feature analysis for assessment of water status and health level in coast live oak (*Quercus agrifolia*) leaves. *Int. J. Remote Sens.* **2004**, *25*, 4267–4286.
117. Peñuelas, J.; Pinol, J.; Ogaya, R.; Filella, I. Estimation of plant water concentration by the reflectance water index WI (R900/R970). *Int. J. Remote Sens.* **1997**, *18*, 2869–2875.
118. Fisher, R.A. *Statistical Methods for Research Workers*; Genesis Publishing Pvt Ltd.: Guildford, UK, 1925.
119. Mann, H.B.; Whitney, D.R. On a test of whether one of two random variables is stochastically larger than the other. *Ann. Math. Stat.* **1947**, 50–60.
120. Kruskal, W.H.; Wallis, W.A. Use of ranks in one-criterion variance analysis. *J. Am. Stat. Assoc.* **1952**, *47*, 583–621.
121. Jensen, J.R. *Introductory Digital Image Processing: A Remote Sensing Perspective*, 2nd ed.; Prentice-Hall: Upper Saddle River, NJ, USA, 1996.
122. Salisbury, F.B.; Ross, C.W. *Plant Physiology*; Wadworth: Belmont, CA, USA, 1992.
123. Gausman, H.W. Visible light reflectance, transmittance, and absorptance of differently pigmented cotton leaves. *Remote Sens. Environ.* **1983**, *13*, 233–238.
124. Clevers, J. The use of imaging spectrometry for agricultural applications. *ISPRS J. Photogramm. Remote Sens.* **1999**, *54*, 299–304.
125. Mutanga, O.; Skidmore, A.K. Red edge shift and biochemical content in grass canopies. *ISPRS J. Photogramm. Remote Sens.* **2007**, *62*, 34–42.
126. Woolley, J.T. Reflectance and transmittance of light by leaves. *Plant Physiol.* **1971**, *47*, 656–662.
127. Boyer, M.; Miller, J.; Belanger, M.; Hare, E.; Wu, J. Senescence and spectral reflectance in leaves of northern pin oak (*Quercus palustris* Muenchh.). *Remote Sens. Environ.* **1988**, *25*, 71–87.
128. Fourty, T.; Baret, F.; Jacquemoud, S.; Schmuck, G.; Verdebout, J. Leaf optical properties with explicit description of its biochemical composition: Direct and inverse problems. *Remote Sens. Environ.* **1996**, *56*, 104–117.
129. Pedregosa, F.; Varoquaux, G.; Gramfort, A.; Michel, V.; Thirion, B.; Grisel, O.; Blondel, M.; Prettenhofer, P.; Weiss, R.; Dubourg, V.; et al. Scikit-learn: Machine Learning in Python. *J. Mach. Learn. Res.* **2011**, *12*, 2825–2830.
130. Breiman, L. Random forests. *Mach. Learn.* **2001**, *45*, 5–32.
131. Belgiu, M.; Drăguţ, L. Random forest in remote sensing: A review of applications and future directions. *ISPRS J. Photogramm. Remote Sens.* **2016**, *114*, 24–31.
132. Vapnik, V.N. *Statistical Learning Theory*; Wiley: New York, NY, USA, 1998; Volume 1.
133. Cawley, G.C.; Talbot, N.L. Gene selection in cancer classification using sparse logistic regression with Bayesian regularization. *Bioinformatics* **2006**, *22*, 2348–2355.
134. Dumont, J.; Hirvonen, T.; Heikkinen, V.; Mistretta, M.; Granlund, L.; Himanen, K.; Fauch, L.; Porali, I.; Hiltunen, J.; Keski-Saari, S.; et al. Thermal and hyperspectral imaging for Norway spruce (*Picea abies*) seeds screening. *Comput. Electron. Agric.* **2015**, *116*, 118–124.
135. Wold, S.; Sjöström, M.; Eriksson, L. PLS-regression: A basic tool of chemometrics. *Chemom. Intell. Lab. Syst.* **2001**, *58*, 109–130.
136. Barker, M.; Rayens, W. Partial least squares for discrimination. *Journal of Chemometrics* **2003**, *17*, 166–173.

137. Castillo, R.; Otto, M.; Freer, J.; Valenzuela, S. Multivariate strategies for classification of Eucalyptus globulus genotypes using carbohydrates content and NIR spectra for evaluation of their cold resistance. *J. Chemom.* **2008**, *22*, 268–280.
138. Lê Cao, K.A.; Boitard, S.; Besse, P. Sparse PLS discriminant analysis: Biologically relevant feature selection and graphical displays for multiclass problems. *BMC Bioinform.* **2011**, *12*, 253.
139. Peerbhay, K.Y.; Mutanga, O.; Ismail, R. Commercial tree species discrimination using airborne AISA Eagle hyperspectral imagery and partial least squares discriminant analysis (PLS-DA) in KwaZulu–Natal, South Africa. *ISPRS J. Photogramm. Remote Sens.* **2013**, *79*, 19–28.
140. Apolloni, B.; Pedrycz, W.; Bassis, S.; Malchiodi, D. *The Puzzle of Granular Computing*; Studies in Computational Intelligence; Springer: Berlin/Heidelberg, Germany, 2008.
141. Szmidt, E. *Distances and Similarities in Intuitionistic Fuzzy Sets*; Studies in Fuzziness and Soft Computing; Springer International Publishing: Gewerbestrasse, Switzerland, 2013.
142. Aggarwal, C.C.; Hinneburg, A.; Keim, D.A. On the surprising behavior of distance metrics in high dimensional space. In *International Conference on Database Theory*; Springer: Houten, Switzerland, 2001; pp. 420–434.

remote sensing

MDPI

Article

Reducing the Effect of the Endmembers' Spectral Variability by Selecting the Optimal Spectral Bands

Omid Ghaffari *, Mohammad Javad Valadan Zoej and Mehdi Mokhtarzade

Department of Photogrammetry and Remote Sensing, Faculty of Geodesy & Geomatics Engineering, K. N. Toosi University of Technology, Tehran 19967-15433, Iran; valadanzouj@kntu.ac.ir (M.J.V.Z.); m_mokhtarzade@kntu.ac.ir (M.M.)

* Correspondence: ghaffari@znu.ac.ir; Tel.: +98-912-4262218

Academic Editors: Chein-I Chang, Meiping Song, Junping Zhang, Chao-Cheng Wu, Qi Wang and Prasad S. Thenkabail

Received: 6 June 2017; Accepted: 23 August 2017; Published: 25 August 2017

Abstract: Variable environmental conditions cause different spectral responses of scene endmembers. Ignoring these variations affects the accuracy of fractional abundances obtained from linear spectral unmixing. On the other hand, the correlation between the bands of hyperspectral data is not considered by conventional methods developed for dealing with spectral variability. In this paper, a novel approach is proposed to simultaneously mitigate spectral variability and reduce correlation among different endmembers in hyperspectral datasets. The idea of the proposed method is to utilize the angular discrepancy of bands in the Prototype Space (PS), which is constructed using the endmembers of the image. Using the concepts of PS, in which each band is treated as a space point, we proposed a method to identify independent bands according to their angles. The proposed method comprised two main steps. In the first step, which aims to alleviate the spectral variability issue, image bands are prioritized based on their standard deviations computed over some sets of endmembers. Independent bands are then recognized in the prototype space, employing the angles between the prioritized bands. Finally, the unmixing process is done using the selected bands. In addition, the paper presents a technique to form a spectral library of endmembers' variability (sets of endmembers). The proposed method extracts endmembers sets directly from the image data via a modified version of unsupervised spatial–spectral preprocessing. The performance of the proposed method was evaluated by five simulated images and three real hyperspectral datasets. The experiments show that the proposed method—using both groups of spectral variability reduction methods and independent band selection methods—produces better results compared to the conventional methods of each group. The improvement in the performance of the proposed method is observed in terms of more appropriate bands being selected and more accurate fractional abundance values being estimated.

Keywords: hyperspectral unmixing; endmember extraction; band selection; spectral variability; prototype space

1. Introduction

In the past decade, numerous methods have been introduced for unmixing hyperspectral imagery [1,2]. Spectral mixture analysis (SMA) is one of most commonly-used methods, and is used in different applications. Basically, the spectra of mixed pixels are modeled using linear or non-linear mixture models. The spectral signature of each pixel is converted to a set of fractional abundances of its constituent spectra (endmembers) by these models [3]. The answer to this question of which one (linear or non-linear models) is superior for unmixing the hyperspectral data is not clear, and depends on the type of the mixture of objects and their applications. However, the acceptable accuracy and the simplicity of linear mixture models entice more researchers to employ them [4].

If the multiple scattering among the endmembers is negligible and the mixture could be supposed macroscopic, a linear mixture model (LMM) can be written as Equation (1):

$$\mathbf{y}(n) = \sum_{i=1}^{p} \alpha_i(n)\mathbf{m}_i + \mathbf{v}(n) = \mathbf{M}\boldsymbol{\alpha}(n) + \mathbf{v}(n), \quad (1)$$

where $\mathbf{y}(n) = [y_1(n), y_2(n), \ldots, y_B(n)]^T \in \mathbb{R}^B$ is the vector of observations; B is the number of bands; $n = 1, \cdots, N$ is the index of pixels in the image; $\mathbf{m}_i \in \mathbb{R}^B$, $i = 1, \ldots, p$ is the spectral signature of endmembers; p is the number of endmembers; $\alpha_i(n)$ is the abundance of the ith endmember in the nth pixel; $\mathbf{M} = \begin{bmatrix} \mathbf{m}_1, & \cdots & , \mathbf{m}_p \end{bmatrix} \in \mathbb{R}^{B \times p}$ is the coefficient matrix of endmembers; $\boldsymbol{\alpha}(n) = [\alpha_1(n), \cdots, \alpha_p(n)]^T \in \mathbb{R}^p$ is the vector of abundance values in the nth pixel; and $\mathbf{v}(n) \in \mathbb{R}^B$ represents noise.

The accuracy of the unmixing process highly depends on the completeness and goodness of the selected endmembers. Therefore, many endmember extraction algorithms have been developed in recent years [3,5]. The accuracy of the fractional abundances obtained from SMA is affected by the residual spectral error caused by inaccurate atmospheric correction, an insufficient signal-to-noise ratio (SNR), and the noise caused by neglecting the non-linear effect of inputs. However, the most important source of error in SMA is due to ignoring the spectral variability (SV) of endmembers caused by variable illumination and environmental, atmospheric, and temporal conditions [4]. These algorithms generally model the entire image using a constant spectral feature for each endmember. In fact, this is a simplification, because in many cases the spectrum of endmembers could change in different spatial and temporal conditions.

Generally, two types of SV can be distinguished among the samples from different classes: (1) the variability within the endmembers of a specific class (intra-class variability); and (2) the spectral similarity between the endmembers of different classes (inter-class variability) [4]. By increasing the intra-class variability, the accuracy of sub-pixel fraction estimation decreases linearly [6]. On the other hand, in some applications where the separation of similar phenomena is of interest, the spectral similarity among the different endmembers (e.g., crops and weeds in agricultural fields or spectral similarity among minerals) makes it difficult to separate these classes. The estimation of the fractional abundances using the linear mixture model could be achieved by different methods, such as least squares and sparse regression with different constraints [3,7]. In the least squares-based spectral unmixing problem, the spectral similarity among the endmembers results in a high correlation between the columns of the coefficients matrix (M) in Equation (1). Consequently, the rank deficiency of the coefficients matrix leads to an unstable solution for the least squares problem and decreasing the accuracy of the estimation of the fractional abundances. Despite the serious effects in the LMMs and the destruction of the reliability of the results of spectral unmixing (SU), this issue is typically ignored [8].

According to [4], the efforts to decrease the effect of the SV can be classified into five general categories: (1) the use of multiple endmembers for each component in an iterative mixture analysis procedure; (2) the spectral weighting of bands; (3) the spectral transformations; (4) the use of radiative transfer models in a mixture analysis; and (5) the selection of a subset of stable spectral features. In addition, to significantly improve the accuracy of the estimation of fractional abundances, the last strategy effectively reduces the computational cost.

The non-orthogonality of the endmembers appears when a linear correlation exists between two endmembers or a multi-collinearity exists among some endmembers. By increasing the correlation among the endmembers, the LMM tends to be instable and extremely sensitive to the small variations of the input spectrum and noises. According to [8], the approaches to deal with the problem could be categorized as: (1) excluding the correlated endmember; (2) de-correlating the endmembers using the spectral transformations; (3) using iterative approaches to select the independent endmembers; and (4) the regularization of the SU equations. Regarding the redundancy of bands in the hyperspectral images, it is not unexpected to identify a subset of bands that decreases the correlation of endmembers. To deal with the problem, the correlation of endmembers can be evaluated using singular value decomposition (SVD) and the condition number of the coefficient matrix of the endmembers in the unmixing procedure.

This paper presents a novel and effective approach for managing the SV and decreasing spectral correlation among the endmembers based on the selection of the optimal bands in the Prototype Space (PS) [9]. The proposed method consists of two main steps. Based on the spectral behavior of the endmembers' set, the image bands are firstly prioritized in such a way that they have the least sensitivity to the SV of the endmembers. Then, the optimal band selection is done based on this prioritization. Since the spectral correlation among the image bands is not considered in this process, in the second step the independent bands are selected using their angles in the PS. In this way, the spectral correlation among the endmembers is reduced as well. Besides, collecting a spectral library from the SV of endmembers is an expensive and time-consuming process. Therefore, these sets were directly extracted from the image in this paper.

The remaining parts of the paper are organized as follows: the theoretical background and the previous algorithms are introduced in Section 2; the proposed method is explained in Section 3; the experimental results and further discussion are provided in Section 4; and concluding remarks are found in Section 5.

2. Theoretical Background

The effect of the SV of endmembers in the SU and the approaches developed to deal with it have been taken into consideration in [4,10,11]. Furthermore, the problems caused by the spectral correlation among the endmembers and its adverse effects on the reliability of the results of the SU have been investigated in [8]. Given the centrality of the optimal band selection to enhancing the stability of the elected set against the spectral variation and also to decrease the spectral correlation among the endmembers in the proposed method, the algorithms that have dealt with the problem by the band selection approach are reviewed in this section.

2.1. Feature Selection Algorithms to Decrease the Spectral Variability Effect

The precise selection of bands that are stable against the SV (e.g., those bands that minimize the intra-class variance and maximize the inter-class dispersion) plays a significant role in the accuracy improvement of the estimation of fractional abundances. Previous studies on the variability of the optical properties of leaf, litter, and soil in semi-arid and arid areas had illustrated that the SWIR2 region from 2050–2500 nm is the least dependent on variations in structural and biochemical attributes [12]. Therefore, this region was selected as a stable spectral region for the aforementioned materials to be used in the SMA for developing the AutoSWIR algorithm in [12]. However, since the position and the spectral region of bands, as well as the number of stable spectral regions depend on the spatial, spectral, and temporal complexity as well as the mixture of endmembers that are present in the scene [4], this algorithm was not extendable for different ecosystems.

In [13], a more applicable spectral feature selection algorithm entitled the stable zone unmixing (SZU) was introduced. In this algorithm, the sensitive wavelengths to the SV are evaluated using the instability index (ISI). Then, a protocol was introduced to enhance the spectral subset selection by accounting for a tradeoff between the number of wavelengths used in the analysis (i.e., information) and the ISI (i.e., spectral variability).

The redundancy problem of the hyperspectral images and high correlation between their bands was not taken into account in either of the methods (AutoSWIR and SZU). A greater potential in computation efficiency and fraction estimate accuracy could only be provided if the independent bands were employed in the LMM [4].

2.2. Feature Selection Algorithms to Decrease the Spectral Correlation of Endmembers

The most common way to deal with the correlation of endmembers is to eliminate the collinear endmembers [8], which causes two disadvantages: (1) If two endmembers are highly correlated, which one should be excluded? (2) When an endmember is eliminated, it may contain some useful information that could result in destabilizing the LMM. Another solution is to combine the two correlated classes to form a new endmember. However, this will inevitably result in missed classes or similar problems to exclude the endmembers. There are a number of transformations (e.g., principal component analysis (PCA) [14] and the maximum noise fraction (MNF) [15]) which can be employed to reduce the correlation of endmembers by de-correlating the band-to-band correlation. The major drawback of these methods is encountering endmembers whose spectral response has no physical meaning [8].

Another obvious solution that comes to mind is to use a number of subsets of the spectral region. Regarding the high spectral resolution of the hyperspectral images, the spectral region could be decomposed into the visible, near-infrared, and shortwave infrared sections. Then, only those regions that contain the absorption features of the objects of interest could be employed in the appropriate applications. Of course, different regions of the electromagnetic wave spectrum have been theoretically considered by researchers in order to examine different phenomena (i.e., using the shortwave infrared region for analyzing minerals). However, upon decreasing the spectral region, the correlation among endmembers increases [8].

Recently, several endmember-extraction-based methods were applied to feature selection methods, which elect the distinctive spectral signatures. Some of these methods (e.g., geometrical feature selection (G-FS) [16] and linear prediction (LP) [17]) operate in the pixel space. The dimensionality of such a space is equal to the number of participating image pixels. Some others, such as prototype feature selection (PFS) and maximum tangent discrimination (MTD) [18] extract informative bands in an unsupervised manner via geometrical interpretation of distinctive bands in the Prototype Space (PS) [9]. In this way, the axes of the PS are defined based on the spectrum of the extracted endmembers or the clusters' center of the existing classes in the scene. Therefore, the dimensionality of such a space is equal to the number of constructive components of the scene, and each band is represented as a point or vector in this space. Using these methods, a subset of independent bands that are distributed throughout the spectral region of the hyperspectral sensor could be provided by eliminating the dependent bands in a meaningful manner.

2.2.1. Quantitative Evaluation of the Endmembers' Correlation

In order to quantitatively estimate the correlation of endmembers constructing the coefficient matrix (M), two measures can be employed [8]: (1) deriving a measure from the coefficient matrix based on its singular values, and (2) extracting a measure from the correlation of the endmembers' spectra. The SVD of the endmembers matrix (M) to extract singular values (i.e., the square root of eigenvalues) is $M = U \sum V^T$, where U and V are both square, unitary, and orthonormal matrices, and $\sum$ is a diagonal matrix with the singular values of M. The ratio of the largest singular value to the smallest one is called the condition number of the matrix. The ideal value of one for the condition number indicates that the matrix is fully orthogonal. By increasing the condition number of the endmember matrix, the correlation of endmembers increases, and in an extreme case, will cause the singularity of the matrix. If the correlation of endmembers exceeds 0.6, the condition number increases exponentially [8]. In this way, the condition number is a good measure to evaluate the endmembers' correlation.

The correlation matrix of the endmembers partially shows the collinearity of each pair of endmembers. The average value of the upper or the lower triangle elements of the matrix provides a measure that indicates the average correlation of the endmembers constructing the coefficient matrix, and could be used as a measure of the overall correlation of endmembers.

3. The Proposed ISI-PS Method

This section explains the proposed ISI-PS method, which is an incorporation of the methods to decrease the SV effect and to select the independent bands of the hyperspectral images. These two factors directly play a positive role in enhancing the accuracy and the reliability of the fractional abundances computed. The proposed method has been developed by assuming the existence of some pure pixels in the image for each class.

In order to explain the relation of the different parts of the proposed method, its flowchart (Figure 1) and pseudo-code are presented, and detailed information is provided in the following sections. By assuming the existence of pure pixels in the hyperspectral images, a set of pure spectra is firstly provided for each class, which indicates the class spectral variability. These sets are employed to create the endmember of each class and statistical analysis in each band. In this way, the extraction of the information of the endmembers in the proposed method has been compiled based on the geometrical endmember extraction algorithms and by assuming the existence of pure pixels in the image. By means of intra-class and inter-class analysis using measures (e.g., ISI) that consider the SV, the persistent bands against the variability are prioritized. Thereafter, the PS is established based on the spectra of the pure representative of each class, and then bands are represented in this space. Finally, the correlation among the prioritized bands is measured by computing their angles in the PS, and then those that have a similar behavior are eliminated. Thus, the remaining bands are independent and persistent against the variability, which are employed to do the SU and estimate the fractional abundance of each endmember.

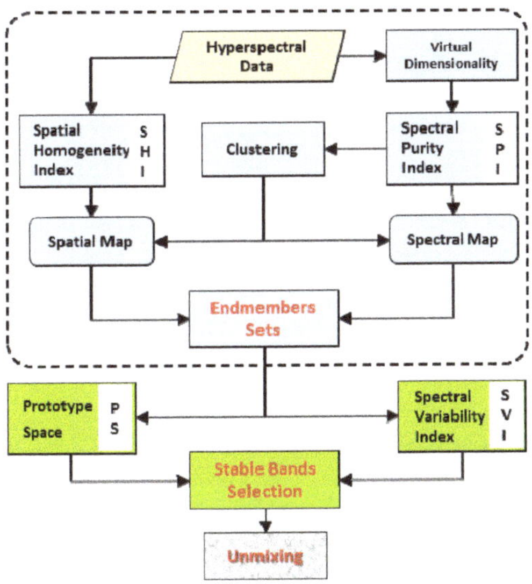

Figure 1. Flowchart of the proposed method.

Pseudocode of the proposed method:

(1) Estimating the number of classes of the image and establishing a spectral library of the SV of endmembers (i.e., the sets of endmembers).

(2) Prioritizing the persistent bands against the SV using the SV index and some training data.

$$\Omega = \{B_i\}_{i=1}^{L} \quad \text{where} \quad B_1 \geq B_2 \geq \cdots B_L \ ,$$

where Ω is the set of prioritized persistent bands (B) and ith is the prioritizing index of bands.

(3) Selecting the most different bands in the PS using the distance of the bands from the main diagonal of the space, $\Omega_0 = \{B_{PS}\}$.
(4) Considering band B_i and computing its angle from all members of set Ω_{i-1} in the PS.
(5) If the angle of band B_i from all of the previously selected bands was greater than a predefined threshold (T), then $\Omega_i = \Omega_{i-1} \cup \{B_i\}$; otherwise, $\Omega_i = \Omega_{i-1}$, and band B_i is eliminated.
(6) Unmixing the hyperspectral image using the selected bands.

3.1. Establishing a Set of Spectral Variabilities for Each Endmember

The LMM is widely used to model the spectral composition of a spectrum. However, many reasons lead to the SV of endmembers, such as the change of environmental illumination, as well as atmospheric and temporal conditions [10]. The methods for dealing with the endmembers' variability could be categorized into two classes [10]: (1) endmembers as sets, and (2) endmembers as statistical distributions. Usually, methods of the first category require a spectral library of the SV of endmembers to deal with the variability phenomena. Collecting a spectral library of the endmembers' variabilities is an expensive and time-consuming process. Therefore, automatic extraction of endmembers' sets from an image is greatly beneficial. In this regard, automated endmember bundles (AEB) [19] has established endmembers sets by executing standard endmember extraction algorithms such as N-FINDR [20], orthogonal subspace projection (OSP) [21], unsupervised fully constrained least squares (UFCLS) [22], iterative error analysis (IEA) [23] and vertex component analysis (VCA) [24] and clustering the resulted endmembers from different methods. Recently, the authors in [25] showed that VCA is essentially the same as simplex growing algorithm (SGA) [26] as long as their initial conditions are the same. So, other conventional endmember extraction algorithms such as SGA can be used in AEB.

Spectral features of those spectra that are located in each endmember set indicate the representative of that endmember. Obviously, it is due to the SV that these features are not exactly similar. Therefore, these representatives should have the general condition of an endmember, including: (1) they should lie next to the vertex of point clouds in the feature space; (2) they should situate in homogenous regions in the spatial domain; and (3) pixels of each class should have a similar spectral behavior. Recently, a module entitled spatial–spectral preprocessing (SSPP) was presented in [27], which could be used as a preprocessing function prior to endmember identification and SU. This module firstly computes the spatial homogeneity index for the pixels of the image, which is used to determine the homogenous regions of the image. Simultaneously, unsupervised clustering is employed to identify the spectral classes. Finally, by fusion of this spatial and spectral information, a subset of pixels that are spatially homogenous and spectrally pure is identified in each class, which could be used as the input of the endmember extraction algorithms.

In this way, the applied procedure to determine the endmembers has been shown in the dashed box of Figure 2 and runs as follows. Firstly, the virtual dimensionality (VD) of the image (p) is determined via Hysime [28]. Thereafter, the dimension is reduced using a PCA or MNF transformation into (p-1) bands. In this reduced feature space, endmembers are found via the well-known pixel purity index (PPI) [29] technique with a threshold value equal to zero. Then, the obtained results from the PPI are clustered into p clusters. Instead of the common K-means clustering (which is suitable for classes with spherical distributions), the Fast Density Peak Detection (FDPC) method [30] is applied since herein the classes are seen to have a variety of different distributions. Since the endmembers are found, image homogeneous regions are also detected by a Gaussian filter, and according to Equations (6) and (7). The previously found endmembers are prioritized according to both their spectral purity and homogeneity indices. Next, the spatial and spectral maps are generated based on the first 20% of the purest pixels, as well as 30% of the most homogeneous ones. The ultimate representative endmembers of each class are finally selected from the overlap of these two maps.

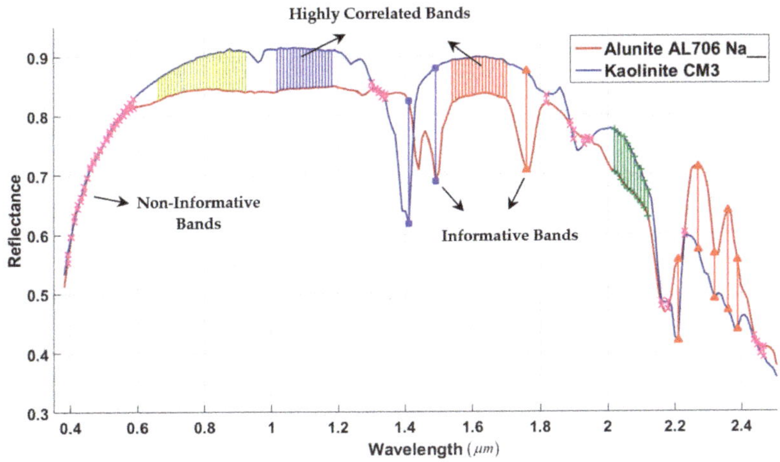

Figure 2. The spectrum of alunite and kaolinite minerals and the highly-correlated regions among the adjacent bands.

3.2. Reducing the Spectral Variability Effect by Selecting the Optimal Bands

In contrast with the conventional SMA methods which use the overall spectral region (i.e., all bands), SZU [13] has been designed based on the selection of persistent bands against the SV phenomena using ISI. The numerical value of this index is computed using Equation (2), which has been developed based on the Fisher separability function. This index is defined for each band based on the ratio of the intra-class dispersion (i.e., the total standard deviation of endmembers for each class) to the inter-class variability of endmembers (i.e., the average distance among the center of classes). The value of one indicates that the intra-class and inter-class variations are similar, and the smaller this value, the better the situation for that band to separate classes.

$$ISI_\lambda = \frac{\Delta_{within,\lambda}}{\Delta_{between,\lambda}} = \frac{2}{p(p-1)}\sum_{i=1}^{p-1}\sum_{j=i+1}^{p}\frac{1.96(\sigma_{i,\lambda}+\sigma_{j,\lambda})}{|\overline{m}_{i,\lambda}-\overline{m}_{j,\lambda}|}, \quad (2)$$

where p is the number of endmembers and $\sigma_{i,\lambda}$ and $\overline{m}_{i,\lambda}$ are the standard deviation and the average of class i in the band λ, respectively. The image's bands could be prioritized with respect to the SV of their constituent spectra using this index.

3.3. Reducing the Correlation of Endmembers by Selecting the Independent Bands in the Prototype Space

Linear correlation of two or more endmembers always exists in the SU of hyperspectral images. However, little attention has been paid to this [8]. One of the main objectives of this paper is to reduce these correlations without the elimination of the dependent endmembers. This is because—as was mentioned—it would be of interest to separate the different species in some applications. On the other hand, the correlation of the spectrum of endmembers' sets for adjacent bands has not been considered in the band prioritizing process to reduce the effect of the SV. These two issues are closely related to each other. Therefore, by eliminating those bands for which the endmembers' sets spectrum is similar, the less-correlated spectral features of these endmembers could be achieved. The spectrum of alunite and kaolinite minerals from the USGS spectral library is represented by the spectral response of the AVIRIS sensor in Figure 2. As can be seen, the bands in the blue, red, and yellow regions are redundant. Besides, these regions may be correlated with each other.

In order to estimate the correlation of bands, some methods have been proposed based on the divergence and correlation functions on the histogram of the image's bands [31,32]. However, since the goal is to improve the condition of the coefficient matrix to reduce the endmembers' correlation, in this paper, the angle of bands in the prototype space has been used as a measure of the correlation of endmembers' sets in those two bands. In other words, by establishing the PS using the endmembers and representing the bands in this space, the dependent bands could be identified. The advantage of this method is that the correlation of bands is evaluated dealing with the endmembers' sets, because the axes of the PS are the endmembers. Therefore, if the combination of the endmembers is changed in the scene, the proposed method will select a new subset of bands that also have the minimum correlation.

Bands are categorized into three classes in the PS: (1) informative bands: the lager the distance of bands from the main diagonal of the space, the better the bands can separate the image' classes; (2) correlated bands: the bands that have a similar spectral response of endmembers are gathered close together in this space—this concept is beyond the correlation of the adjacent bands in the hyperspectral images, because it would have occurred for those bands that are not adjacent; (3) non-informative bands: those bands that are located close to the main diagonal of the space, which have the exact same response for different classes [9].

The spectrum (i.e., bands) of alunite and kaolinite minerals (Figure 2) are illustrated in the PS, which was constructed based on these two endmembers in Figure 3. In this figure, examples of highly correlated bands (using blue, red, and yellow colors, which are shown with magnification in the right view), informative bands (the two bands that are shown with blue squares and that have the largest distances from the main diagonal of the space, and those bands that are illustrated with red triangles in which the two spectra have an appropriate distance from each other) and uninformative bands (those bands that are shown with magenta color and that are located close to the main diagonal of the space) are illustrated.

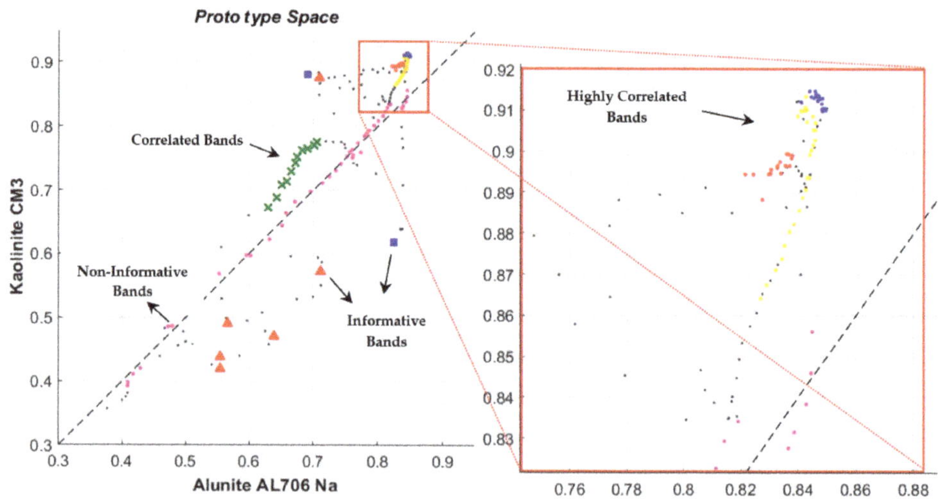

Figure 3. The Prototype Space (PS) constructed using the two endmembers alunite and kaolinite.

It is obvious that the angles of the correlated bands are close to each other, even if these regions with a similar spectrum are not adjacent. Therefore, the correlation of bands in dealing with the endmembers could be understood by extracting these angles.

3.4. Determining the Threshold Value to Identify the Independent Bands

A threshold value should be defined in order to identify and eliminate the correlated bands, and the decision to preserve or eliminate each band is made by the comparison of the angle between that band and the previously selected bands with the pre-defined threshold value. The angle between bands is computed in the prototype space. If this angle is less that the pre-defined threshold value, it means that the band evaluated is similar to a band in the set of previously selected bands. Otherwise, the band evaluated is added to the set of bands.

In order to determine the threshold value, in this paper, the independent bands are extracted by defining a range (e.g., from 0.5 to five degrees) and a step (e.g., 0.15 degrees) for the variation of this threshold. Then, an image is reconstructed from the selected bands. The root mean square error (RMSE) of discrepancies between the estimated fractional abundances from this image and the ground truth map is computed. Finally, the threshold value that leads to the minimum RMSE is selected as the optimal threshold value. However, the obtained bands using this threshold value are selected as the optimal bands.

The ground truth map is not available in most applications of spectral unmixing. In these cases, the map of the extracted pure pixels from the image in Section 3.1 of the proposed method could be used as the ground truth map. In other words, similar to the case that the ground truth map is available, the proper threshold value is determined by evaluating the RMSE obtained from the estimation of the fractional abundances at the position of the image's pure pixels.

3.5. Spectral Unmixing

Finally, in order to evaluate the fractional abundances, a valid and unique method is needed for comparing the performance of the selected bands using the different methods studied. In this study, the least squares method has been employed to solve the inverse problem. According to [33], which studied the different methods of the estimation of the fractional abundances, the fully constrained least squares (FCLS) is introduced as a proper method in this regard. In order to accurately estimate the fractional abundances of the endmembers, two constraints—namely, the abundance sum-to-one constraint (ASC) and the abundance non-negativity constraint (ANC)—have been applied to the linear mixture model. In order to use the FCLS method, the constituents of the imaging scene should be fully known. This issue is considered using the information obtained from the ground truth map or the endmembers extracted from the image in the supervised and unsupervised manners, respectively.

4. Results and Discussion

In this section, the performance of the proposed ISI-PS algorithm is evaluated using the both simulated and real hyperspectral datasets. In the first subsection, several simulated hyperspectral images were employed, which have been produced using some spectra of the USGS spectral library and different scenarios. The constituent spectra of imagery and their fractional abundances were exactly known for these datasets. Then, the effect of selecting the optimal bands using the proposed ISI-PS algorithm to reduce the SV and endmembers' correlation was evaluated on the AVIRIS hyperspectral images. Finally, the results of the proposed method were compared with the results of the SZU method [13], which only dealt with the SV, and the results of the MTD method [18], which only tried to select the independent bands in an unsupervised manner.

4.1. Simulated and Real Datasets Used

4.1.1. Simulated Dataset

In this research, several hyperspectral images with a dimension of 100 × 100 pixels and various spatial patterns were simulated using some spectra from the USGS spectral library. These spectra were selected from different combinations of minerals of hydro-thermal alteration zones in geological applications (Table 1).

Table 1. The constituent spectra of the simulated images.

	Dataset 1	Dataset 2	Dataset 3	Dataset 4	Dataset 5
Mineral	Alunite HS295.3B Calcite HS48.3B Epidote HS328.3B Kaolinite CM3 Montmorillonite CM20	Alunite HS295.3B Dickite NMNH106242 Halloysite CM13 Kaolinite CM3 Montmorillonite CM20	Alunite HS295.3B Halloysite NMNH106 Kaolinite KGa-1 (wxyl) Montmorillonite CM20 Muscovite GDS116	Alunite HS295.3B Calcite HS48.3B Chlorite HS179.3B Epidote HS328.3B Hematite GDS27 Kaolinite CM3 Montmorillonite CM20	Calcite WS272, CO2004 Halloysite NMNH106, KLH503 Kaolinite CM5, CM3 Montmorillonite CM27, CM26 Muscovite GDS116, HS24.3
Abundance Map Pattern	Spherical Gaussian Fields	Exponential Gaussian Fields	Rational Gaussian Fields	Mattern Gaussian Fields	Mattern Gaussian Fields

Several conditions were considered to simulate these images so that the resulting imagery reflected the real conditions as much as possible. In order to reconstruct the spatial patterns that appear in nature, the functions of the HYDRA software package [34] have been used, in which the two functions Legendre and Gaussian were employed to generate the fractional abundances. The Gaussian function could be performed using four different modes to generate the spatial patterns, and an example of each mode is illustrated in Figure 4.

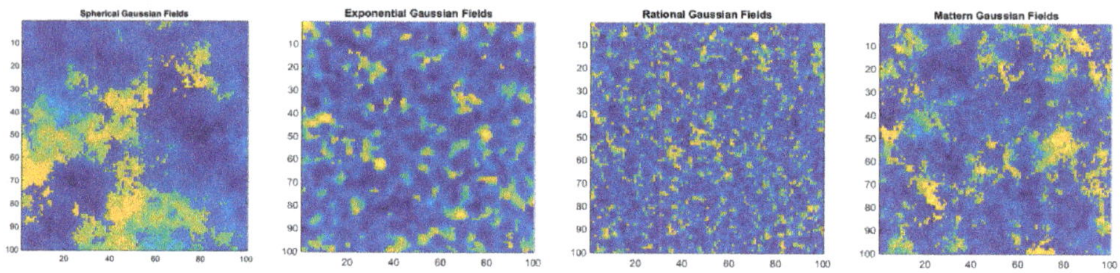

(**a**) (**b**) (**c**) (**d**)

Figure 4. An example of the fractional abundances that were generated using: (**a**) Spherical Gaussian fields; (**b**) Exponential Gaussian fields; (**c**) Rational Gaussian fields; and (**d**) Mattern Gaussian fields functions.

In this study, the effects of the endmembers' variability and the illumination fluctuation due to the topography on the spectrum of objects have been modeled according to [35]. The SV was simply characterized by spectral shape invariance [36]. In other words, while the spectral shapes of the endmembers were fairly consistent, their amplitudes varied considerably over the scene. Accordingly, the spectral variability of the *i*th endmember in each pixel can be modeled as Equation (3).

$$\mathbf{m}_i = \psi_i \mathbf{m}_i^\circ + \boldsymbol{\eta}_i, \tag{3}$$

where $\mathbf{m}_i^\circ$ is an endmember that is selected from the spectral library and $\psi_i \geq 0$ is a random factor that affects the spectral amplitude of this endmember equally in all bands. ψ_i is the variable for each endmember in each class. This factor was generated by defining the range of the spectral variations (i.e., $(1 \pm sd)$) and using the normal distribution. In the experiments, a standard deviation (sd) of 0.05 was adopted for the spectral variations. $\boldsymbol{\eta}_i$ is a random noise with a zero-mean, which was considered to model those variations that are not modeled by ψ_i.

By substituting Equation (3) in the linear mixture model (Equation 1), an equation was achieved to model the spectral variability for each pixel (Equation 4).

$$\mathbf{y} = \mathbf{M}\,\boldsymbol{\psi}\,\boldsymbol{\alpha} + \sum_{i=1}^{p} \alpha_i \boldsymbol{\eta}_i + \mathbf{n}, \tag{4}$$

where $\boldsymbol{\psi} \equiv diag(\psi_1, \psi_2, \ldots, \psi_p)$ is a $p \times p$ diagonal matrix.

In order to apply these factors, a threshold was considered for the standard deviation of variations (sd). Then, the factors were randomly generated with a normal distribution in the range of $1 \pm sd$, which were multiplied by the amplitude of the original spectra. These factors were different for each pixel and properly model the effect of SV. The range of (0.05–0.15) was employed in experiments as the standard deviation of spectral variabilities.

The effect of the illumination fluctuation due to the topography was similar for all bands, and could therefore be considered as additive noise [35]. In order to model this effect, the spectral features matrix was considered to be fixed, and the fraction of each pixel was multiplied by a factor (γ). These factors were randomly selected with a normal distribution in each pixel in the range of (0.95–1.05). In order to simulate the effects of the instrumental noises, Gaussian noise with the zero-mean and different ratios was added to the simulated scenes. Therefore,

$$\mathbf{y} = \mathbf{M}\,\gamma\,\boldsymbol{\psi}\,\boldsymbol{\alpha} + \gamma \textstyle\sum_{i=1}^{p} \alpha_i \boldsymbol{\eta}_i + \mathbf{n} = \mathbf{M}\,\boldsymbol{\alpha} + \boldsymbol{\nu}, \tag{5}$$

Equation (5) is the model that was considered in the experiments in this study, and still was linear. The illumination fluctuation, the SV, and the instrumental noise of images could be modeled using this equation. The flowchart of generating the simulated images is illustrated in Figure 5.

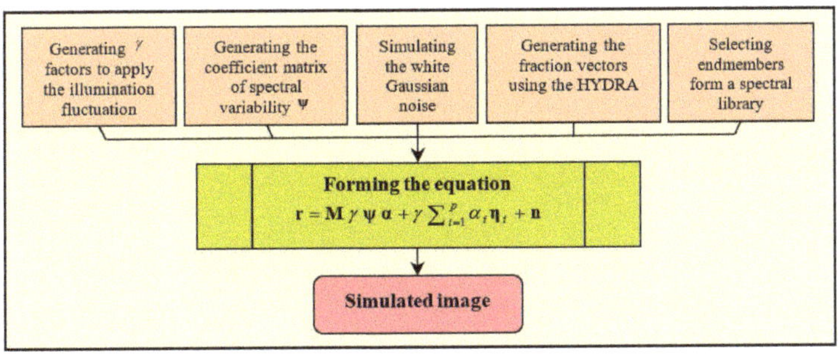

Figure 5. Flowchart of generating the simulated images.

4.1.2. LTRAS Dataset

The Russell Ranch Sustainable Agriculture Facility is a unique 300-acre facility near the UC Davis campus dedicated to investigating irrigated and dry-land agriculture in a Mediterranean climate. The goal of this facility—which is known as Long-Term Research in Agricultural Sustainability (LTRAS)—was used to investigate the impact of external factors such as crop rotation, farming systems (conventional, organic, and mixed), and the inputs of water, nitrogen, carbon, and other elements on agricultural sustainability [37].

Currently, the Century Experiment contains ten systems, which are two-year rotations and include corn/tomato, wheat/tomato, wheat/fallow, and wheat/legume rotations. Additionally, a perennial native grass system and a six-year alfalfa–corn–tomato rotation were initiated in 2012. The arrangement of each farm in this facility is illustrated in Figure 6a [38], along with the type of its irrigation and fertilization.

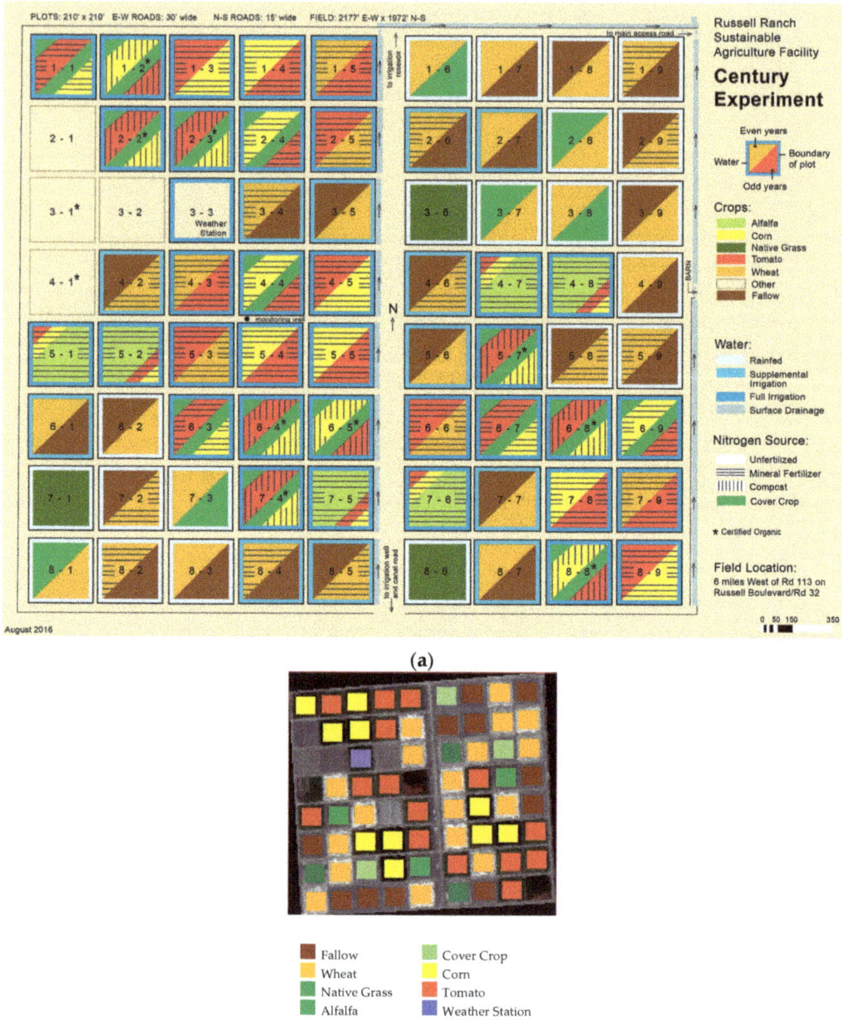

Figure 6. Long-Term Research in Agricultural Sustainability (LTRAS) dataset: (**a**) Planting schedule [38]; (**b**) Ground-truth map of the LTRAS farms.

The hyperspectral data used in this study were captured by the AVIRIS sensor on 3 August 2013, from low altitude with a 3.2-m ground pixel size. The region of LTRAS comprises 199 lines by 217 samples, and was extracted from the original image that had been radiometrically and geometrically corrected. Its true and pseudo-color composites are illustrated in Figure 7. The dimension of each farm is approximately 64 × 62 m, for which regarding the date of imaging and the planting schedule in Figure 6a, its classes have been extracted as Figure 6b. Plots 4-5 and 8-9—which had been planted with corn—were excluded from the ground truth map due to harvesting; as were Plots 5-4, which had not been planted with tomato as per the schedule. Because of the imaging date, the plots of cover crop had no covers, and were therefore considered as bare earth.

(**a**) (**b**) (**c**)

Figure 7. (**a**) True color composite; (**b**) False color composite; and (**c**) Color composite of the first three components of the maximum noise fraction (MNF) transformation.

The impacts of different irrigation systems and fertilization, as well as the effect of crop rotation were considered in this facility. Therefore, this dataset was proper to evaluate the SV phenomena in vegetables. The SV of pixels for three classes of wheat, tomato, and corn are illustrated in Figure 8.

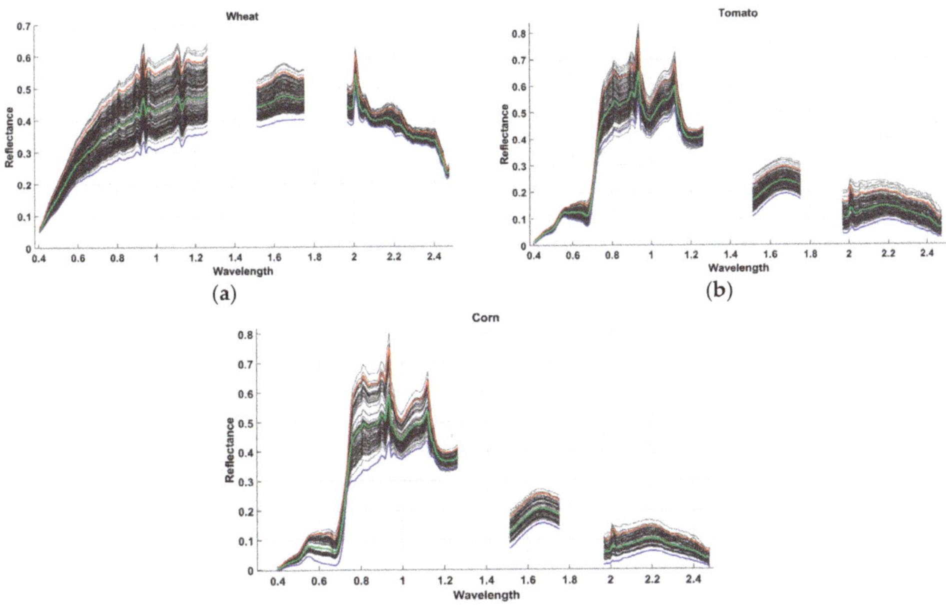

(**a**) (**b**) (**c**)

Figure 8. The spectral variability of pixels for three classes of: (**a**) Wheat; (**b**) Tomato; and (**c**) Corn.

4.1.3. Salinas Dataset

This dataset was collected by the AVIRIS sensor on 9 October 1998, over Salinas Valley, Southern California, and is available as the at-sensor radiance unit. The scene comprises 512 lines by 217 samples, with 160 spectral bands (after discarding noisy and water absorption bands) in the wavelength range of 0.4–2.5 microns. Its nominal spectral and radiometric resolutions were 10 nanometers and 16 bits, respectively. This image was captured from low altitude with a 3.7-m ground pixel size. Its false color composite is illustrated in Figure 9a.

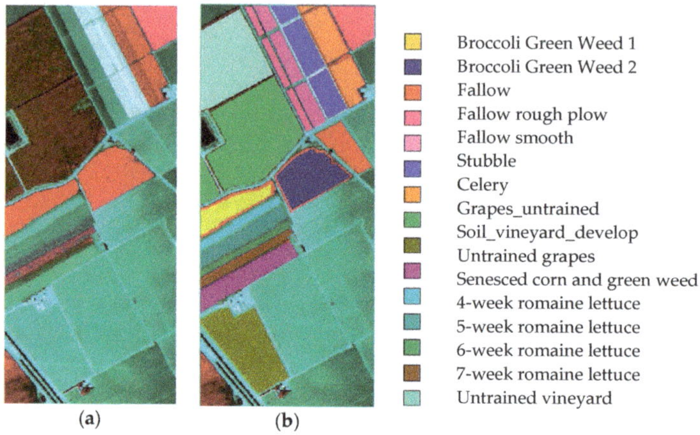

Figure 9. Salinas dataset: (**a**) False color composite of the AVIRIS image; (**b**) Ground truth map.

This dataset was collected from an agricultural region, and its ground truth map has been gathered into 15 classes, as illustrated in Figure 9b. It includes vegetables, bare soils, and vineyard fields with sub-categories as follows. The sub-categories of broccoli and green weeds were distinguished, with one having smaller and fewer weeds, while two had taller and more weeds, with both categories mostly covering the soil. The romaine lettuce sub-classes have been defined based on the planting week and their growth rates, which have different covers on the soil.

The soil was categorized into three sub-classes: the fallow rough plow class had recently been turned with larger clumps and appeared to have more moisture, while the fallow class was plowed soil with smaller clumps, and the fallow smooth class had even smaller clumps. The stubble class comprised bare soil and straw, and could also be considered a sub-class of the soil group. In the vineyard group, the untrained vineyard and the untrained grapes sub-classes were actually similar to each other. In the untrained vineyard sub-class, vine had been grown on wooden and plastic posts, and their canopies had nearly covered the soil. The situation of the selected classes at the time of imaging is shown in Figure 10.

Figure 10. Photograph of the selected classes in the region of the imaging.

4.1.4. Indiana Indian Pines dataset

The third dataset used in the experiments was collected by the AVIRIS sensor over the Indian Pines Test Site in Northwestern Indiana in 1992. This image has a size of 145 × 145 pixels, and was acquired over a mixed agricultural/forest area, early in the growing season. The spatial resolution is approximately 20 m, and the radiometric resolution is 10 bits. The image comprised 220 spectral channels in the wavelength range from 0.4–2.5 micrometers, nominal spectral resolution of 10 nanometers. Bands 1–2, 100–114, 147–167, and 216–220 were removed from the dataset due to the noise and the water absorption phenomena, leaving a total of 177 radiance channels to be used in the experiments. This scene contained two-thirds agriculture and one-third forest or other natural perennial vegetation. For illustrative purposes, Figure 11a shows a false color composition of the AVIRIS Indian Pines scene, while Figure 11b shows the ground truth map available for the scene, with 16 classes. In our experiments, we considered a real situation in which most of the similar classes were included in the evaluations. Hence, 12 classes with an adequate number of labeled samples were selected for the experiments.

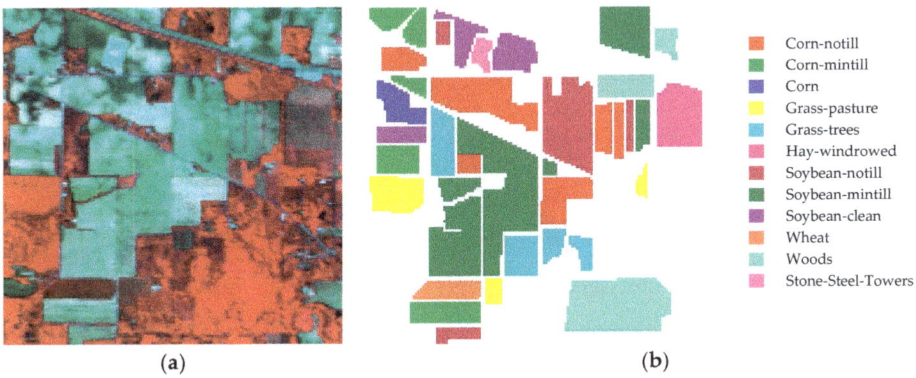

Figure 11. Indiana Indian Pines dataset: (**a**) False color composite of the AVIRIS image; (**b**) Ground truth map.

4.2. Experiments on the Simulated Dataset

The proposed ISI-PS algorithm was firstly performed on the simulated dataset, which was generated using the elements of Table 1. This data contained five datasets with different numbers and types of endmembers. In order to generate the fractional abundance for each dataset, a different pattern was employed according to Table 1. The main objective of this experiment was to evaluate the performance of the proposed method to deal with the endmembers' SV and decreasing the correlation of endmembers by selecting the optimal bands to generate accurate fractional abundances. Besides, the quality of endmembers' sets—which had been extracted in an unsupervised manner according to Figure 1—were compared with a spectral library that existed for the ground truth maps.

Several scenarios have been designed in order to evaluate the performance of the proposed method. In addition to the variety of spectral features and spatial patterns, different signal-to-noise ratios (i.e., 20:1, 25:1, and 30:1) have been employed to generate the simulated images. Equation (5) was used directly to simulate the first four datasets. In the case of the last dataset, in addition to Equation (5), two spectra from different species of one material were used to complicate the SV condition (Figure 12).

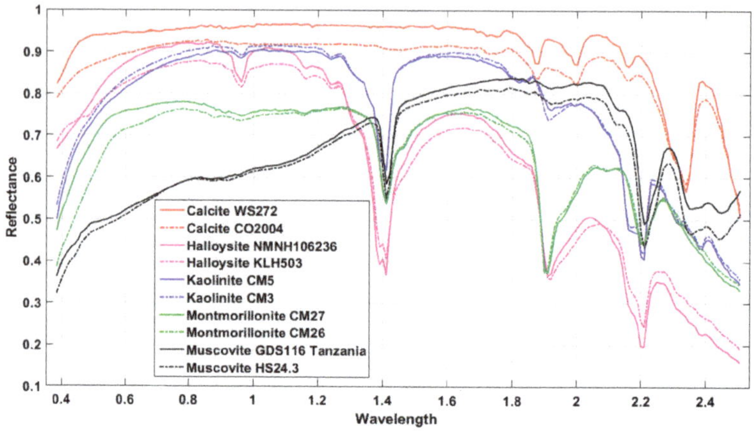

Figure 12. Endmembers used to simulate the last dataset.

As was mentioned in the fifth step of the proposed method, a threshold value (T) was employed to identify the independent bands in the PS. In other words, if the angle of the candidate band from all previously selected bands was greater than the threshold value, this band had a distinct behavior dealing with other selected bands.

The value of this threshold (T) was affected by the number, spectral similarity, and variety of endmembers in the scene. However, if a precise ground truth map were available, the root mean square error (RMSE) resulted from the comparison of this map, and the estimated fractional abundance could be used to properly evaluate this threshold value. In this regard, by increasing the threshold value in the range of 0.5–5 degrees with an increment of 0.25 degrees and evaluating the precision of the resulted fractional abundances, the threshold that led to the minimum error was selected. The results of different scenarios are provided in Table 2, along with the number of selected bands in each method.

If there were no in situ information for establishing a spectral library of the endmembers' variability, this information could be directly extracted from the image according to the early stages of the proposed algorithm. In this process—which was developed by applying some revisions to the SSPP algorithm—the virtual dimensionality (VD) of data was firstly estimated using the signal subspace identification algorithms (e.g., Hysime [28]) to be used for spectral dimension reduction by employing PCA or MNF transformations. Thereafter, the data used were also reduced in the spatial domain using spectral purity indices such as the PPI, and consequently, the number of candidate pure pixels was decreased. Using the mean spectrum of spectral clusters as the indicator of those classes without eliminating impure pixels led to the mean spectrum being affected by these pixels. In this case, the spectral changes among the clusters' pixels were not only due to the SV phenomena. By clustering those pixels that were probably pure, besides the tending of the class' mean towards the purity, the separability of classes would be properly shown.

Some other factors that challenged the performance of endmember extraction algorithms were sensitivity to noise, unusual pixels resulting from inaccurate atmospheric correction, and the image's hot spots. Therefore, extracting the homogeneous area of the image as a possible location for pure pixels could help to reduce the impact of these annoying phenomena. However, this could destroy the information of anomaly classes.

In order to identify the homogenous regions of the image used, a Gaussian filter with the functional form of $g(i',j') = \frac{1}{2\pi\sigma^2}e^{-\frac{i'^2+j'^2}{2\sigma^2}}$ was firstly applied to each band of the hyperspectral image (Equation (6)). In this filter, the parameter σ controls the amount of spatial smoothing.

$$\mathbf{X}_b^F(i,j) = \sum_{i'=-a}^{a} \sum_{j'=-a}^{a} g(i',j') \cdot x_b(i+i',j+j') \tag{6}$$

where $\mathbf{X}_b^F(i,j)$ is the value of the pixel (*i,j*) in the band b of the filtered image. The value of $a = (w-1)/2$ was determined regarding the filter dimension $g_{w\times w}$, and showed the spatial index of the filter. In order to generate the spatial homogeneity index (SHI), the RMSE of the discrepancies of the original image $\mathbf{X}$ and the filtered imaged $\mathbf{X}^F$ was computed using Equation (7) [27].

$$RMSE\left[\mathbf{X}(i,j),\mathbf{X}^F(i,j)\right] = \left(\frac{1}{B-1}\sum_{b=1}^{B}\left(\mathbf{X}_b(i,j) - \mathbf{X}_b^F(i,j)\right)^2\right)^{\frac{1}{2}} \tag{7}$$

In this way, a layer was obtained where the value of the pixel (*i,j*) indicates the homogeneity of that region of the image. The lower the pixel value, the more homogeneous that region will be.

Finally, pixels that were located in each cluster were sorted based on the spectral purity index, and the purest ones that were located in homogenous areas were selected as the indicators of those clusters. In the experiments, 20 percent of the pixels of each cluster and 30 percent of the homogenous pixels with the best scores were employed. The results obtained from the subsections of this process are illustrated in Figure 13 for the first dataset.

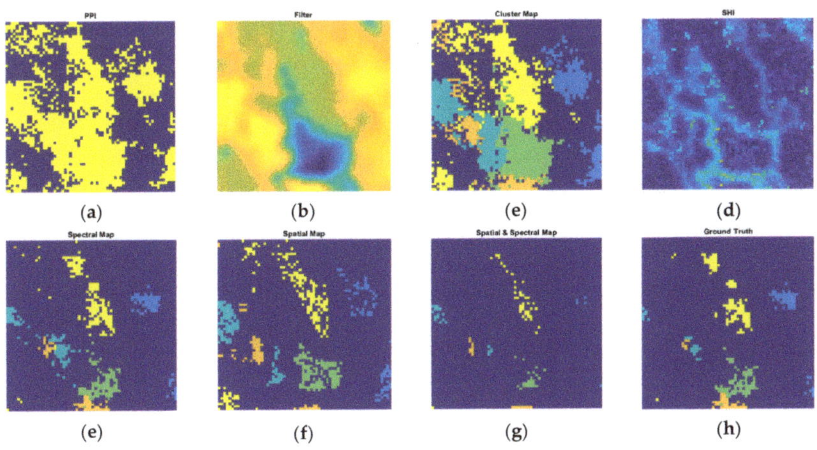

Figure 13. The results obtained from the subsections of the process of extraction of the spectral library of each class from the image: (**a**) Output of the pixel purity index (PPI) for greater than zero pixels; (**b**) The result of band filtering; (**c**) The clustering of PPI's output; (**d**) Map of the homogeneity scores of pixels; (**e**) 20 percent of pixels of each cluster with the maximum value of PPI; (**f**) 30 percent of pixels of each cluster with the best homogeneity; (**g**) Fusion of maps (**e**,**f**) to extract the final pure pixels; (**h**) Ground truth map of the image's endmembers.

The results obtained from the SU of the five simulated datasets are provided in Table 3. It is worth mentioning that the endmembers' sets were directly extracted from the image in these experiments. For comparison purposes, the data used in these experiments were quite similar to those that were used in the previous supervised experiments.

Table 2. The accuracy assessment results of the selected bands by maximum tangent discrimination (MTD), stable zone unmixing (SZU), and the proposed instability index (ISI)-prototype space (PS) algorithms, in comparison with using all bands to estimate the fractional abundances through a supervised spectral unmixing on the simulated datasets.

Data Set			Feature Selection Method													Full Dimensionality			
			ISI-PS				SZU				MTD								
Abundance Map	#S	SNR	#F	Cond	Corr	RMSE	#F	Cond	Corr	RMSE	#F	T	Cond	Corr	RMSE	#F	Cond	Corr	RMSE
Spherical	5	20:1	63	21.01	**0.39**	0.071	55	28.87	0.42	0.083	29	4.00	**20.97**	0.40	**0.071**	224	30.40	0.41	0.085
Exponential	5	25:1	43	30.09	0.70	0.105	24	117.39	0.90	0.153	27	4.00	**28.99**	**0.68**	**0.103**	224	48.72	0.87	0.148
Rational	5	30:1	68	40.98	**0.51**	0.090	67	55.82	0.58	0.101	34	3.25	**36.88**	0.55	**0.083**	224	70.58	0.53	0.112
Mattern	7	25:1	61	39.54	**0.19**	0.063	54	58.88	0.27	0.070	38	3.75	**38.04**	0.22	**0.061**	224	52.50	0.21	0.068
Mattern	10	30:1	65	49.71	0.58	0.181	67	61.08	0.64	0.183	14	4.75	**34.84**	**0.57**	**0.180**	224	81.72	0.61	0.187

#S is the number the image's components; #F is the number of selected bands; SNR is the signal-to-noise ratio; Cond is the Condition number; Corr is the mean Correlation of endmembers; T is the angle threshold and bold number is the best result.

Table 3. The accuracy assessment results of the selected bands by MTD, SZU, and the proposed ISI-PS algorithms, in comparison with using all bands to estimate the fractional abundances through an unsupervised spectral unmixing on to the simulated datasets.

Data Set			Feature Selection Method													Full Dimensionality			
			ISI-PS				SZU				MTD								
Abundance Map	#S	SNR	#F	Cond	Corr	RMSE	#F	Cond	Corr	RMSE	#F	T	Cond	Corr	RMSE	#F	Cond	Corr	RMSE
Spherical	5	20:1	61	20.77	**0.41**	**0.047**	138	25.81	0.42	0.050	30	4.00	**19.95**	0.44	**0.047**	224	32.41	0.44	0.053
Exponential	5	25:1	42	**28.09**	0.67	0.057	124	38.88	0.82	0.062	24	4.25	28.12	**0.65**	**0.055**	224	45.31	0.86	0.067
Rational	5	30:1	70	44.58	0.61	0.071	109	62.38	**0.58**	0.074	23	4.00	**36.43**	0.60	**0.068**	224	76.40	0.62	0.075
Mattern	7	25:1	61	36.41	**0.19**	**0.043**	109	45.14	0.25	0.044	40	3.75	**33.31**	0.23	**0.043**	224	46.38	0.21	0.044
Mattern	10	30:1	55	43.32	0.55	0.175	152	70.40	0.59	0.175	12	3.00	**34.58**	**0.53**	**0.174**	224	74.47	0.60	0.176

#S is the number the image's components; #F is the number of selected bands; SNR is the signal-to-noise ratio; Cond is the Condition number; Corr is the mean Correlation of endmembers; T is the angle threshold and bold number is the best result.

According to Tables 2 and 3, the proposed ISI-PS algorithm always provided proper accuracy in the estimation of fractional abundances of endmembers by selecting the optimal bands. Moreover, the results obtained from the supervised and unsupervised experiments had good agreement with each other. Therefore, the spectral library of endmembers' variability was correctly extracted. The accuracies of the fractional abundances obtained from the MTD and ISI-PS methods were compatible. However, by comparing the position of selected bands in these two methods, it is obvious that in the MTD algorithm, the SV of endmembers was neglected, and the most separable bands were selected only by considering the spectral feature of each class. When the SV of the endmembers' sets was not to the extent that the spectrum of classes highly conflicted in the overlapping regions, the results of the two methods were close to each other. However, if the SV disrupted the separability of classes, the proposed ISI-PS method led to more accurate results by selecting the bands with the minimum spectral conflicts.

4.3. Experiments on Real Datasets

In order to evaluate the effect of SV on the results of unmixing, the ground truth map of the data used was needed, as well as the spectral library of the intra-class variations of each endmember. Providing the sub-pixel fractional abundance of the image's components was practically impossible, and collecting a spectral library from the variation of each component was an expensive and time-consuming process. However, in the case of the LTRAS dataset, for which only one crop was planted in each farm, the fraction of the components could be supposed to be 100 percent in the related plot; and observed spectral variations among the pixels of each plot could be seen as their SV due to the factors mentioned in Section 4.1.2.

However, it is worth mentioning that the change of the ratio of plants and background soil in each farm will lead to the change of the received spectra. This type of change was not considered as spectral variability. In this study, regarding the rather limited area and homogeneity of the farms, it was assumed that the mixture of materials was accrued with a constant ratio.

The extraction of the statistics of each class was performed using five percent of its pixels, which were selected randomly. This subset was used as the training dataset, and the remaining part was employed as the test dataset. Regarding the intra-class variations, the spectral features of the training pixels were employed for establishing a spectral library for each class. The mean spectrum of these sets was then used as the spectral indicator of the related classes.

As previously mentioned, the correlation of bands was not considered in the prioritization process of the SZU method. The selected bands using this method are illustrated by green bars in Figure 14. As can be seen, several bands were selected in the vicinity of each other in the event that the behavior of the endmembers was close together in these bands. In other words, if this redundancy could be reduced by a meaningful selection of the optimal bands in each region, more accurate and computationally-effective results could be achieved.

In the proposed ISI-PS algorithm, the angle of bands was used in the PS to deal with the bands' correlation. The selection of a proper threshold (T) to eliminate the correlated bands was affected by the number, spectral similarity, and variety of endmembers in the scene, as well as the number of resulting bands. If a ground truth map were available, the proper angle to eliminate the redundant bands could be estimated using the estimation accuracy of the ground truth map. In this experiment, regarding the homogeneity of farms, a ground truth map was generated using the map shown in Figure 6b. In this regard, the fraction of each endmember in the related class was considered as one, and for other classes, the fractions were considered to be zero. The fraction of classes in each pixel was then estimated involving the selected bands in each method, and their box plots are illustrated in Figure 15. In this case, the closer the resulting fraction to one and the less the standard deviation of fractions, the better the performance of the algorithm for dealing with the intra-class variations. It is worth mentioning that the most SV had occurred in the corn, wheat, and tomato classes. As can be seen, in comparison with the MTD method, the proposed ISI-PS algorithm tried to reduce the SV in these classes by decreasing the median and the standard deviation of the estimated fractions. However, in the other classes, the performances of the studied methods were close to each other.

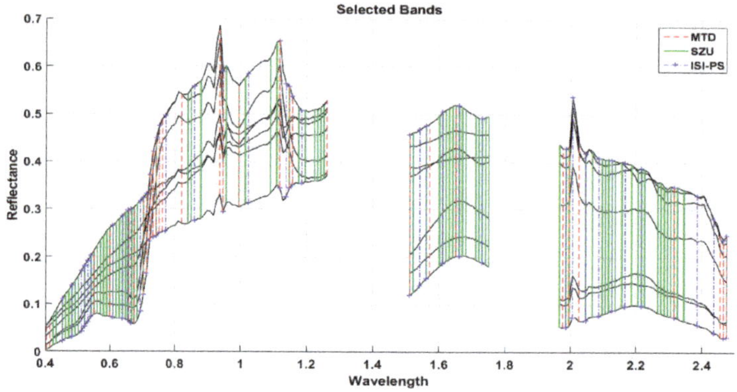

Figure 14. Representation of the selected bands using the MTD, SZU, and the proposed ISI-PS methods.

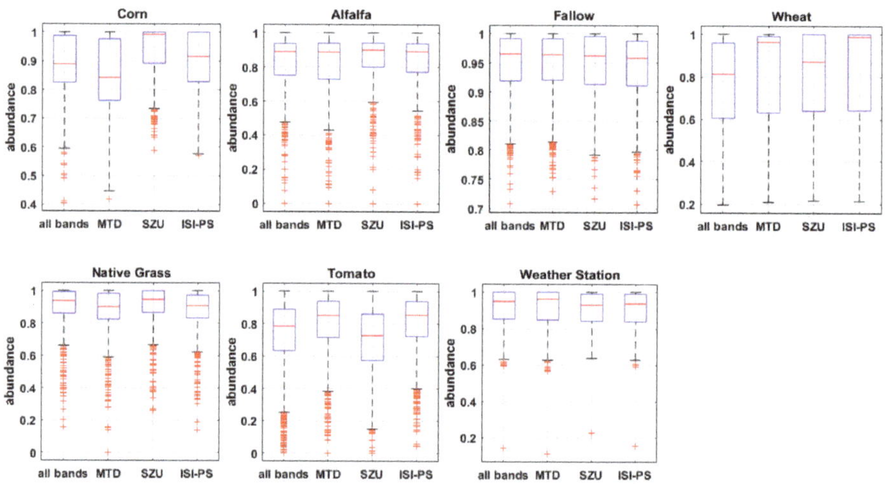

Figure 15. The box plot of the resulting fractions for each class using the selected bands by the studied method, using: (1) all bands; (2) MTD bands; (3) SZU bands; and (4) ISI-PS bands.

In this experiment, the threshold value of the correlated bands was considered as 1.25 degrees. Figure 16a shows the RMSE of the fractional abundance estimation of the LTRAS classes using the bands obtained from the MTD, the SZU, and the proposed ISI-PS algorithms, for threshold values of 0.25–5 degrees by an increment of 0.2 degrees. Index 6 was equivalent to the threshold value of 1.25 degrees, and caused the selection of 46 bands from the original 170 bands. The selected bands using the MTD, the SZU, and the ISI-PS algorithms are illustrated in Figure 14 using red, green, and dashed blue lines, respectively. It is worth mentioning that the proposed ISI-PS algorithm provided the most accurate fractions.

The singularity of the coefficient matrix, which was generated using the selected bands, was evaluated using: (1) the condition number of the endmembers matrix and (2) the average correlation of the endmembers' correlation matrix. By increasing the number of selected bands using each method, these two measures are illustrated in Figure 17, as well as for the full dimension of the data (i.e., all bands).

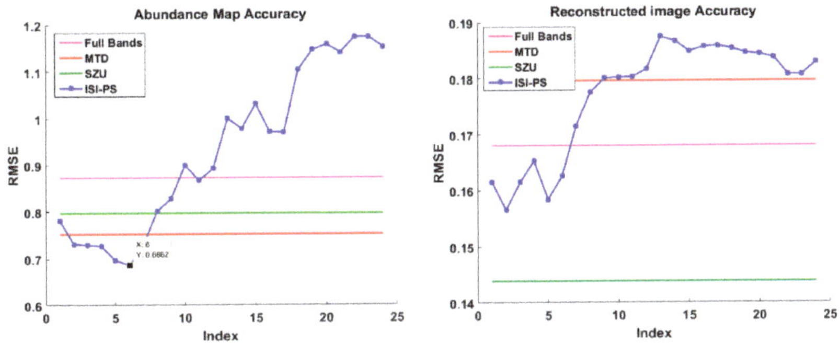

(a) (b)

Figure 16. The RMSE of estimating: (**a**) The fractions; (**b**) The LTRAS image by means of the selected bands using the MTD, the SZU, and the ISI-PS algorithms for different threshold values.

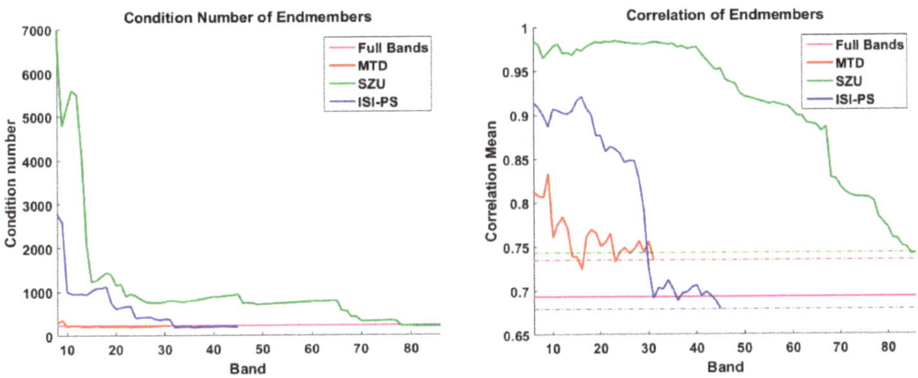

(a) (b)

Figure 17. Plot of: (**a**) The condition number of the endmembers matrix; (**b**) The average correlation of the endmembers' correlation matrix for the different band selection methods by increasing the number of the selected bands.

Finally, to evaluate the role of the selected bands on the results, the original image was reconstructed using the same set of endmembers and fractional abundances obtained from each method. In other words, by considering the accuracy of endmembers and applying the LMM, each pixel (y) of the original image could be approximated using $\hat{y} = \sum_{i=1}^{p} \alpha_i \mathbf{m}_i$, where p is the number of endmembers, α_i is the estimated fractions for each endmember, and $\mathbf{m}_i$ is the ith endmember. Accordingly, the original and the reconstructed images using the LMM could be considered as $I = (y_k)_{k=1}^{N}$ and $R = (\hat{y}_k)_{k=1}^{N}$, respectively, where N is the number of pixels. The reconstruction accuracy could be estimated using Equation (8). The average reconstruction error of each method is illustrated in Figure 16b, and the obtained results from the SU process are provided in Table 4.

$$RMSE\ (I,R) = \left(\frac{1}{N-1} \cdot \sum_{k=1}^{N} [y_k - \hat{y}_k]^2 \right)^{\frac{1}{2}}, \tag{8}$$

Table 4. The accuracies obtained from the selected bands using the MTD, the SZU, the proposed ISI-PS methods, and the full dimension real hyperspectral datasets to estimate the fractional abundances.

Data Set		Feature Selection Method										Full Dimensionality		
		MTD			SZU			ISI-PS						
Name	#S	#F	Cond Corr	Abun RMSE IMG RMSE	#F	Cond Corr	Abun RMSE IMG RMSE	#F	Cond Corr	Abun RMSE IMG RMSE	T	#F	Cond Corr	Abun RMSE IMG RMSE
LTRAS (supervised)	7	31	221.41 0.73	0.753 0.1730	83	214.41 0.74	0.797 **0.1343**	46	**203.10** **0.68**	**0.686** 0.1570	1.25	170	230.08 0.69	0.873 0.1619
LTRAS (unsupervised)	9	20	**327.48** 0.76	0.573 0.1360	81	385.38 0.72	0.556 0.1435	61	421.11 **0.67**	**0.500** **0.1309**	0.85	170	400.32 0.71	0.631 0.1393
Salinas (supervised)	15	31	12,996.30 0.89	0.860 0.0410	45	15,150.40 0.86	0.845 0.0395	43	12,911.04 **0.84**	**0.843** 0.0398	0.55	160	**11,332.69** 0.88	0.856 **0.0382**
Salinas (unsupervised)	15	29	9480.33 0.87	0.421 0.0343	52	7554.07 0.84	0.416 0.0327	49	6444.09 **0.80**	**0.395** 0.0315	0.40	160	**5896.17** 0.87	0.398 **0.0290**
Indiana (supervised)	12	29	10,563.01 0.94	2.245 0.1647	30	11,765.44 0.95	2.130 0.1680	38	9858.71 **0.92**	**2.086** 0.1655	0.35	166	**6723.01** 0.94	2.219 **0.1456**
Indiana (unsupervised)	11	20	2533.14 0.92	1.349 0.1091	29	3237.09 0.92	1.262 0.1008	28	**1459.26** **0.90**	**1.240** **0.0970**	0.50	166	1656.70 0.93	1.261 0.0977

#S is the number the image's components; #F is the number of selected bands; Cond is the Condition number; Corr is the mean Correlation of endmembers; Abun RMSE is the accuracy of fractional abundances; IMG RMSE is the accuracy of the image reconstruction; T is the angle threshold and bold number is the best result.

In the unsupervised manner, the dimension of the signal subspace was firstly estimated as nine using the HySime algorithm. Then, according to the proposed method, a spectral library was established from the spectral variability of each class, regardless of the ground truth map. In order to evaluate the spectral similarity of the endmembers obtained with the spectrum of each class from the ground truth map, the average spectra of the two sets was compared with the spectral information divergence (SID) [39] similarity measure. The results obtained are provided in Table 5. The lower the value of SID, the more similar the two spectra will be. According to the results, the corn class was split into three sub-classes. However, the spectrum of the other classes was estimated properly.

The estimated fractional abundances using the 61 selected bands by the proposed ISI-PS algorithm (Table 4) and the FCLS method are illustrated in Figure 18. As can been seen, the corn class was split into three sub-classes, as in Figure 18a,h,i. The fractional abundances of the bare soil and the wheat classes (i.e., Figure 18c,d, respectively) were partially overlaid, which was due the harvesting of wheat and the appearance of the background soil of the farms. It is worth mentioning that this issue had mostly occurred in farms with a lesser vegetation density due to the type of irrigation and fertilization. This could be obviously understood by the comparison of the obtained results and Figure 7.

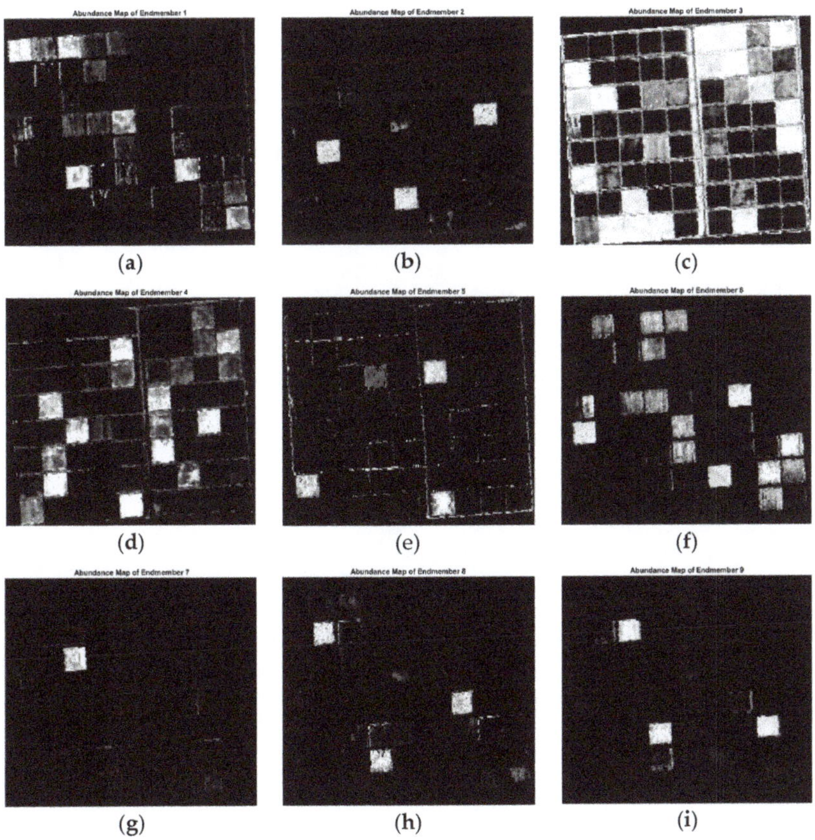

Figure 18. The estimated fractions using the selected bands by the ISI-PS method in an unsupervised manner for: (**a**) Corn; (**b**) Alfalfa; (**c**) Bare soil; (**d**) Wheat; (**e**) Native grass; (**f**) Tomato; (**g**) Weather station; (**h**) Corn; and (**i**) Corn.

Table 5. Similarity values among the extracted endmembers and the reference spectral signatures of each endmembers by spectral information divergence (SID).

SID		Extracted Endmembers								
		1	2	3	4	5	6	7	8	9
Reference Endmembers	Corn	**0.0059**	0.3061	0.7046	0.4647	0.5015	0.0193	0.0276	**0.0036**	**0.0364**
	Alfalfa	0.1727	**0.0000**	0.0879	0.0303	0.0248	0.2226	0.1535	0.3621	0.5499
	Fallow	0.4828	0.0841	**0.0001**	0.0400	0.0223	0.5478	0.4532	0.7745	1.0368
	Wheat	0.3406	0.0421	0.0231	**0.0031**	0.0112	0.3921	0.3287	0.6076	0.8386
	Native Grass	0.3090	0.0215	0.0271	0.0126	**0.0003**	0.3709	0.2861	0.5548	0.7830
	Tomato	0.0062	0.1768	0.4826	0.2778	0.3245	**0.0028**	0.0155	0.0538	0.1236
	W.S.	0.0240	0.1667	0.4879	0.2987	0.3176	0.0199	**0.0003**	0.0383	0.1133

Bold number is the minimum SID in each column.

All of the pre-mentioned process for the LTRAS dataset was performed on the Salinas and Indiana Indian Pines datasets as well (Table 4).

In this table, in order to evaluate the accuracy of the proposed method when no in situ data were available, the results have been provided for the supervised and the unsupervised manners. In the unsupervised approach, the endmember sets were extracted from the image without any prior knowledge. In this regard, using the position of the pure pixels obtained, a fractional map with a 100 percent abundance was first generated for each endmember to compute the threshold value. Then, the fractional abundances of the endmembers were estimated using the extracted endmembers and the different bands obtained from different threshold values. Finally, the threshold value that led to the minimum RMSE of the estimation of the fractional abundances of endmembers was selected as the optimal threshold value.

In order to compare the results obtained from the supervised and the unsupervised approaches, the estimated fractional maps using the FCLS method over the Salinas dataset are illustrated in Figures 19 and 20. As can be seen, the results of the unsupervised approach were compatible with the supervised approach, and the proposed method was able to separate the similar spectral classes. However, due to the similar spectral behaviors in the endmember extraction step, the two classes grapes_untrained and vineyard_untrained were considered as unique classes.

In this section, a comparison is made between the computational times of different methods and is reported in Table 6. All of the methods were executed on a PC with an i7 5820k CPU and 32 GB of RAM.

Table 6. Computational times of unmixing using full bands and reduced bands from different band selection methods.

Dataset	CPU Time (s)			
	Feature Selection Methods			Full Bands
	MTD	SZU	ISI-PS	
LTRAS	10.94	11.34	78.72	10.92
Salinas	107.43	107.71	620.90	117.73
Indiana Pines	14.93	14.92	91.13	15.66

As can be seen, ISI-PS showed disadvantages from the computational time point of view. This was mainly due to the exhaustive search, which was applied to locate an optimum value for threshold T (see Section 3.4). We had chosen a rather vast domain of T for the sake of a richer evaluation. However, the processing time of ISI-PS can be highly improved by a more exact estimation of the search domain for T. In addition, more advanced search strategies—instead of the exhaustive search applied herein—can be of great help to mitigate the computational costs of ISI-PS, which is suggested for further study.

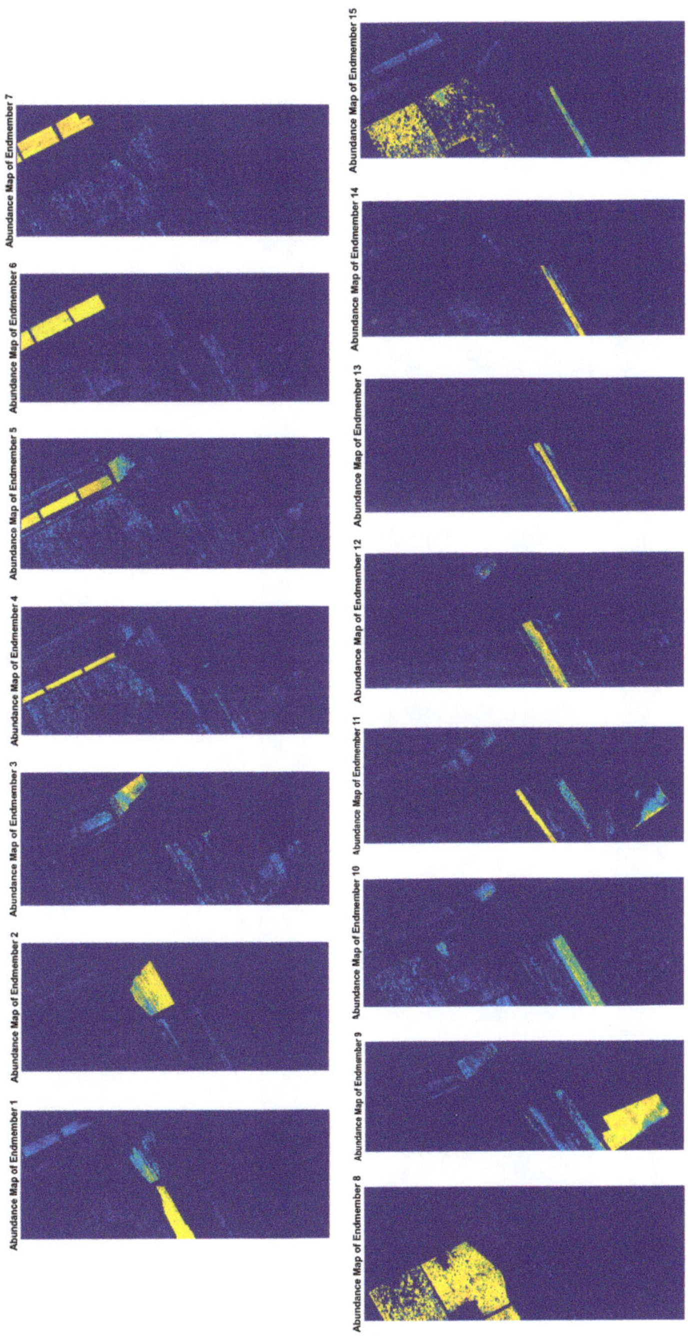

Figure 19. The fractional abundances estimated from the Salinas area using the fully constrained least squares (FCLS) method and the selected bands by the proposed ISI-PS algorithm in the supervised approach.

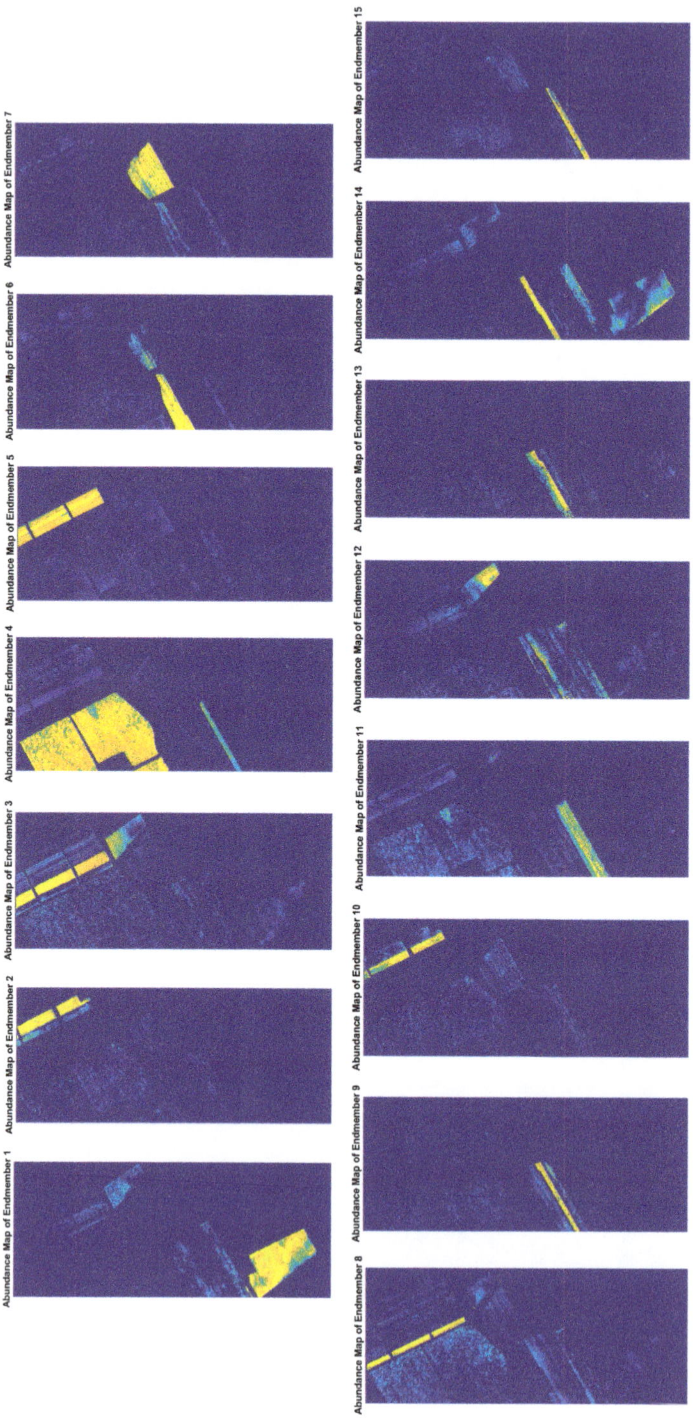

Figure 20. The fractional abundances estimated from the Salinas area using the FCLS method and the selected bands by the proposed ISI-PS algorithm in the unsupervised approach.

5. Conclusions

Band selection has been always a challenge in the processing of high-dimensional hyperspectral data. In this paper, a novel method was presented to select a subset of bands that led especially to improving the results of spectral unmixing. The proposed method—named ISI-PS—integrates two measures of band selection. Firstly, it is aimed at managing the spectral variability. To do so, the bands were prioritized in a way so as to have the least inter-class variability while at the same time achieving the highest possible between-class separation. On the other hand, the second phase takes into account the bands' dependency and makes an effort to detect and remove highly correlated bands. This phase was performed in the Prototype Space, which was formed by image endmembers. In the Prototype Space—in which the bands were treated as the space points—bands' dependencies were examined via their inter-angles.

As mentioned above, the second phase of the proposed method required the knowledge of image endmembers, which is itself a challenge in hyperspectral image processing. In this paper, as with the other contribution, an unsupervised automatic technique was proposed that can effectively extract the endmembers from the image itself and that needed no more input knowledge.

The proposed method was examined and validated on a variety of simulated and real datasets. To do so, the selected bands were used in the spectral unmixing, and the RMSE of the obtained fractional abundances was considered as the accuracy measure. The obtained results were all compatible with the in-situ observations and confirmed the effectiveness of the proposed method. In addition, the performance of the proposed method was compared with the SZU and the MTD algorithms, which proved the superiority of the proposed method.

Author Contributions: All of the authors listed contributed equally to the work presented in this paper.

Conflicts of Interest: The authors declare no conflict of interest.

References

1. Li, C.; Ma, Y.; Mei, X.; Liu, C.; Ma, J. Hyperspectral unmixing with robust collaborative sparse regression. *Remote Sens.* **2016**, *8*, 588. [CrossRef]
2. Liu, R.; Du, B.; Zhang, L. Hyperspectral unmixing via double abundance characteristics constraints based nmf. *Remote Sens.* **2016**, *8*, 464. [CrossRef]
3. Bioucas-Dias, J.M.; Plaza, A.; Dobigeon, N.; Parente, M.; Du, Q.; Gader, P.; Chanussot, J. Hyperspectral unmixing overview: Geometrical, statistical, and sparse regression-based approaches. *IEEE J. Sel. Top. Appl. Earth Obs. Remote Sens.* **2012**, *5*, 354–379. [CrossRef]
4. Somers, B.; Asner, G.P.; Tits, L.; Coppin, P. Endmember variability in spectral mixture analysis: A review. *Remote Sens. Environ.* **2011**, *115*, 1603–1616. [CrossRef]
5. Xu, M.; Zhang, L.; Du, B.; Zhang, L.; Fan, Y.; Song, D. A mutation operator accelerated quantum-behaved particle swarm optimization algorithm for hyperspectral endmember extraction. *Remote Sens.* **2017**, *9*, 197. [CrossRef]
6. Settle, J. On the effect of variable endmember spectra in the linear mixture model. *IEEE Trans. Geosci. Remote Sens.* **2006**, *44*, 389–396. [CrossRef]
7. Iordache, M.D.; Bioucas-Dias, J.M.; Plaza, A. Sparse unmixing of hyperspectral data. *IEEE Trans. Geosci. Remote Sens.* **2011**, *49*, 2014–2039. [CrossRef]
8. Van der Meer, F.D.; Jia, X. Collinearity and orthogonality of endmembers in linear spectral unmixing. *Int. J. Appl. Earth Obs. Geoinf.* **2012**, *18*, 491–503. [CrossRef]
9. Mojaradi, B.; Abrishami-Moghaddam, H.; Zoej, M.J.V.; Duin, R.P.W. Dimensionality reduction of hyperspectral data via spectral feature extraction. *IEEE Trans. Geosci. Remote Sens.* **2009**, *47*, 2091–2105. [CrossRef]
10. Zare, A.; Ho, K. Endmember variability in hyperspectral analysis: Addressing spectral variability during spectral unmixing. *IEEE Signal Proc. Mag.* **2014**, *31*, 95–104. [CrossRef]
11. Xu, X.; Tong, X.; Plaza, A.; Zhong, Y.; Xie, H.; Zhang, L. Joint sparse sub-pixel mapping model with endmember variability for remotely sensed imagery. *Remote Sens.* **2016**, *9*, 15. [CrossRef]

12. Asner, G.P.; Lobell, D.B. A biogeophysical approach for automated swir unmixing of soils and vegetation. *Remote Sens. Environ.* **2000**, *74*, 99–112. [CrossRef]
13. Somers, B.; Delalieux, S.; Verstraeten, W.; Van Aardt, J.; Albrigo, G.; Coppin, P. An automated waveband selection technique for optimized hyperspectral mixture analysis. *Int. J. Remote Sens.* **2010**, *31*, 5549–5568. [CrossRef]
14. Richards, J.A. *Remote Sensing Digital Image Analysis: An Introduction*; Springer: Berlin, Germany, 2012.
15. Green, A.A.; Berman, M.; Switzer, P.; Craig, M.D. A transformation for ordering multispectral data in terms of image quality with implications for noise removal. *IEEE Trans. Geosci. Remote Sens.* **1988**, *26*, 65–74. [CrossRef]
16. Wang, L.; Jia, X.; Zhang, Y. A novel geometry-based feature-selection technique for hyperspectral imagery. *IEEE Geosci. Remote Sens. Lett.* **2007**, *4*, 171–175. [CrossRef]
17. Du, Q.; Yang, H. Similarity-based unsupervised band selection for hyperspectral image analysis. *IEEE Geosci. Remote Sens. Lett.* **2008**, *5*, 564–568. [CrossRef]
18. Asl, M.G.; Mobasheri, M.R.; Mojaradi, B. Unsupervised feature selection using geometrical measures in prototype space for hyperspectral imagery. *IEEE Trans. Geosci. Remote Sens.* **2014**, *52*, 3774–3787.
19. Somers, B.; Zortea, M.; Plaza, A.; Asner, G.P. Automated extraction of image-based endmember bundles for improved spectral unmixing. *IEEE J. Sel. Top. Appl. Earth Obs. Remote Sens.* **2012**, *5*, 396–408. [CrossRef]
20. Winter, M.E. N-findr: An Algorithm for Fast Autonomous Spectral End-Member Determination in Hyperspectral Data. In Proceedings of the SPIE's International Symposium on Optical Science, Engineering, and Instrumentation, Denver, CO, USA, 27 October 1999.
21. Harsanyi, J.C.; Chang, C.-I. Hyperspectral image classification and dimensionality reduction: An orthogonal subspace projection approach. *IEEE Trans. Geosci. Remote Sens.* **1994**, *32*, 779–785. [CrossRef]
22. Chang, C.I. *Hyperspectral Imaging: Techniques for Spectral Detection and Classification*; Springer: New York, NY, USA, 2003; Volume 1.
23. Neville, R.; Staenz, K.; Szeredi, T.; Lefebvre, J.; Hauff, P. Automatic Endmember Extraction from Hyperspectral Data for Mineral Exploration. In Proceedings of the 21st Canadian Symposium on Remote Sens, Ottawa, ON, Canada, 21–24 June 1999.
24. Nascimento, J.M.P.; Dias, J.M.B. Vertex component analysis: A fast algorithm to unmix hyperspectral data. *IEEE Trans. Geosci. Remote Sens.* **2005**, *43*, 898–910. [CrossRef]
25. Chang, C.-I.; Chen, S.-Y.; Li, H.-C.; Chen, H.-M.; Wen, C.-H. Comparative study and analysis among atgp, vca, and sga for finding endmembers in hyperspectral imagery. *IEEE J. Sel. Top. Appl. Earth Obs. Remote Sens.* **2016**, *9*, 4280–4306. [CrossRef]
26. Chang, C.I.; Wu, C.C.; Liu, W.; Ouyang, Y.C. A new growing method for simplex-based endmember extraction algorithm. *IEEE Trans. Geosci. Remote Sens.* **2006**, *44*, 2804–2819. [CrossRef]
27. Martin, G.; Plaza, A. Spatial-spectral preprocessing prior to endmember identification and unmixing of remotely sensed hyperspectral data. *IEEE J. Sel. Top. Appl. Earth Obs. Remote Sens.* **2012**, *5*, 380–395. [CrossRef]
28. Bioucas-Dias, J.M.; Nascimento, J.M. Hyperspectral subspace identification. *IEEE Trans. Geosci. Remote Sens.* **2008**, *46*, 2435–2445. [CrossRef]
29. Boardman, J.W.; Kruse, F.A.; Green, R.O. Mapping Target Signatures via Partial Unmixing of Aviris dData. In Proceedings of the Fifth Annual JPL Airborne Earth Science Workshop, Pasadena, CA, USA, 23–26 January 1995.
30. Rodriguez, A.; Laio, A. Clustering by fast search and find of density peaks. *Science* **2014**, *344*, 1492–1496. [CrossRef] [PubMed]
31. Chang, C.-I.; Wang, S. Constrained band selection for hyperspectral imagery. *IEEE Trans. Geosci. Remote Sens.* **2006**, *44*, 1575–1585. [CrossRef]
32. Chang, C.-I.; Du, Q.; Sun, T.-L.; Althouse, M.L. A joint band prioritization and band-decorrelation approach to band selection for hyperspectral image classification. *IEEE Trans. Geosci. Remote Sens.* **1999**, *37*, 2631–2641. [CrossRef]
33. Heinz, D.C.; Chang, C.-I. Fully constrained least squares linear spectral mixture analysis method for material quantification in hyperspectral imagery. *IEEE Trans. Geosci. Remote Sens.* **2001**, *39*, 529–545. [CrossRef]
34. Hyperspectral Imagery Synthesis Tools for Matlab. Available online: http://www.ehu.es/ccwintco/index.php/Hyperspectral_Imagery_Synthesis_tools_for_MATLAB (accessed on 21 March 2017).

35. Nascimento, J.M.P. *Unsupervised Hyperspectral Unmixing*; Universidade Técnica de Lisboa: Lisbon, Portugal, 2006.
36. Shaw, G.A.; Burke, H.-H.K. Spectral imaging for remote sensing. *Linc. Lab. J.* **2003**, *14*, 3–28.
37. Russell Ranch Sustainable Agriculture Facility. Available online: http://asi.ucdavis.edu/programs/rr (accessed on 10 March 2017).
38. Photos and Maps—Agricultural Sustainability Institute—Uc Davis. Available online: http://asi.ucdavis.edu/programs/rr/photos-and-maps (accessed on 15 March 2017).
39. Chang, C.-I. An information-theoretic approach to spectral variability, similarity, and discrimination for hyperspectral image analysis. *IEEE Trans. Inf. Theory* **2000**, *46*, 1927–1932. [CrossRef]

remote sensing

MDPI

Article

Recursive Local Summation of RX Detection for Hyperspectral Image Using Sliding Windows

Liaoying Zhao [1], Weijun Lin [1], Yulei Wang [2,*] and Xiaorun Li [3]

1 Institute of Computer Application Technology, Hangzhou Dianzi University, Hangzhou 310018, China; zhaoly@hdu.edu.cn (L.Z.); jun05768@foxmail.com (W.L.)
2 Center for Hyperspectral Imaging in Remote Sensing (CHIRS), Information and Technology College, Dalian Maritime University, Dalian 116026, China
3 College of Electrical Engineering, Zhejiang University, Hangzhou 310027, China; lxr@zju.edu.cn
* Correspondence: wangyulei@dlmu.edu.cn

Received: 29 November 2017; Accepted: 9 January 2018; Published: 13 January 2018

Abstract: Anomaly detection has received considerable interest for hyperspectral data exploitation due to its high spectral resolution. Fast processing and good detection performance are practically significant in real world problems. Aiming at these requirements, this paper develops a recursive local summation RX anomaly detection approach by virtue of sliding windows. This paper develops a recursive local summation RX anomaly detection approach by virtue of sliding windows. A causal sample covariance/correlation matrix is derived for local window background. As for the real-time sliding windows, the *Woodbury* identity is used in recursive update equations, which could avoid the calculation of historical information and thus speed up the processing. Furthermore, a background suppression algorithm is also proposed in this paper, which removes the current under test pixel from the recursively update processing. Experiments are implemented on a real hyperspectral image. The experiment results demonstrate that the proposed anomaly detector outperforms the traditional real-time local background detector and has a significant speed-up effect on calculation time compared with the traditional detectors.

Keywords: hyperspectral imagery; recursive anomaly detection; local summation RX detector (LS-RXD); sliding window

1. Introduction

Attributed to the high spectral resolution, hyperspectral images are now capable of uncovering many subtle signal sources that cannot be known by prior knowledge or be visually inspected by image analysts [1,2]. Signal sources appear as anomalies in the data, such as unexpected presence, low probability of occurrence, small sample population whose signature is spectrally distinct from spectral signatures of its surrounding data samples. As a result, anomaly detection has received considerable interest in hyperspectral imaging in the last twenty years [3–6].

The RX detector developed by Reed and Yu [3] is acknowledged to be the most widely used anomaly detector. The classic RX algorithm is based on the global sample covariance matrix K, and is referred to as K-RXD. Since then, many RX-like anomaly detectors have been proposed [7–13]. Of particular interest are RXD using global sample correlation matrix R (R-RXD) [7,8], and RXD based on local background covariance matrix (L-RXD) [9]. The L-RXD uses not only spectral information but also spatial information to bring benefit for detection performance [10]. However, it may fail to obtain the best detection performance due to the penuriousness and unicity of local background distribution in every local window. A local summation anomaly detection (LSAD) is proposed in [13] by combining multiple local neighboring distributions of the pixel under test to get better performance. LSAD can be considered as a local summation RXD (LS-RXD) using subspace feature projection for the stable local covariance estimation.

The hyperspectral remote sensing has developed rapidly in recent years, but as the satellite relocation cycle becomes shorter, some new problems come out. For instance, the massive data has brought some challenges to the data transmission and storage. Moreover, for the anomaly detection problem, the anomalies such as moving targets may show up for a short time and disappear quickly. In this case, timely detection is necessary. However, data transmission is quite time-consuming, to achieve timely detection, developing the recursive anomaly detection algorithms is important and necessary. Recently, several real-time anomaly detection methods [14–19] have been proposed. Specifically, real-time causal process of K-RXD and R-RXD detector (called as RT-CK-RXD, RT-CR-RXD) were developed in [14]. The real-time R-RXD and constrained energy minimization (CEM) are optimized and integrated in a dual-mode parallel Field-Programmable Gate Array (FPGA) based hardware platform in [16]. Unlike the RT-CR-RXD, in the FPGA-based implementations, each pixel under test is located in the middle region of the background, which can improve the performance of target detection. The computational performance of real-time causal linewise progressive anomaly detection (RCLPAD) based on Cholesky decomposition along with linear system solving were developed in [17]. An advanced anomaly detector using causal sliding array windows to capture local autocorrelation matrix statistics in the sense of causality was developed (CSA-RXD) [18], by virtue of causal sliding windows, a causal sample correlation matrix can be derived for causal anomaly detection. Recursive update equations are also derived and thus speed up real-time processing. A real-time L-RXD using the local casual square window is proposed in [19]. However, the method proposed in [19] still needs to calculate the inverse of a matrix to detect each pixel. Compared with sliding array window, setting a sliding square window usually contains much more spectral-spatial integration information. This paper addresses this issue and further develops the recursive processing for LS-RXD based on sliding square window. The contribution of this work is based on two points: a recursive version of LS-RXD according to a causal relation from the *Woodbury* identity, which reduce the runtime; and a background suppression algorithm integrated with the recursive procedure, which improves the detection accuracy.

The rest of the paper is organized as follows. In Section 2, several related RX anomaly detectors are briefly covered. Section 3 provides the design of recursive sliding window detector. Section 4 demonstrates the experiments of the proposed algorithm compared with some traditional anomaly detection algorithms. Finally, Section 5 draws our conclusions.

2. Related Anomaly Detectors

In this section, we provide a short overview of K-RXD, L-RXD and LS-RXD.

Assume that $\{\boldsymbol{r}_i\}_{i=1}^{N}$ is a set of data sample vectors, and $\boldsymbol{r}_i = (r_{i1}, r_{i2}, .., r_{iL})^T$ is the ith data sample vector, where L is the total number of spectral bands.

2.1. K-RXD

The K-RXD, denoted by $\delta_{K-RXD}(\boldsymbol{r})$, is specified as follows:

$$\delta_{K-RXD}(\boldsymbol{r}) = (\boldsymbol{r}-\boldsymbol{\mu})^T \boldsymbol{K}^{-1} (\boldsymbol{r}-\boldsymbol{\mu}) \quad (1)$$

where $\boldsymbol{\mu} = (1/N)\sum_{i=1}^{N} \boldsymbol{r}_i$ is the global sample mean and $\boldsymbol{K} = (1/N)\sum_{i=1}^{N}(\boldsymbol{r}_i-\boldsymbol{\mu})(\boldsymbol{r}_i-\boldsymbol{\mu})^T$ is the sample data covariance. The form of δ_{K-RXD} in (1) is actually the well know Mahalanobis distance between the data sample being detected and global sample mean. It should be pointed out that the model assumes the data arise from two normal probability density functions with the same covariance matrix but different means.

2.2. L-RXD

Local anomaly detection is very important since the global RX anomaly detector fails to work when the anomalies are relatively small or only distinct from the local surroundings, but buried in the global background. The most widely used local anomaly detection algorithm is derived from the commonly used RXD, named as local-RX detector (L-RXD). The L-RXD, denoted by $\delta_{L-RXD}(\boldsymbol{r})$, is specified by:

$$\delta_{L-RXD}(\boldsymbol{r}) = (\boldsymbol{r} - \boldsymbol{\mu}_W)^T \Sigma_W{}^{-1} (\boldsymbol{r} - \boldsymbol{\mu}_W) \tag{2}$$

where $\boldsymbol{\mu}_W$ is the local sample mean of a square window of size $\omega \times \omega$ pixels, centered at pixel $\boldsymbol{r}$ and Σ_W is the background data sample covariance matrix of the local window $\boldsymbol{W}$.

For L-RXD, a window of the selected size should be chosen firstly. The window size should not be too large or too small to obtain considerable background estimation.

2.3. LS-RXD

The traditional L-RXD exploits only one sliding window to estimate the neighborhood background statistic for the pixel under test. It is difficult to detect a multi-pixels anomaly target by L-RXD if the local distributions of some windows are mostly occupied by anomaly pixels because the background statistic will be contaminated seriously by anomaly pixels. In order to solve this problem, a local summation RX detector is proposed in [13]. Figure 1 takes 3×3 size multiple local windows to demonstrate the implementation of the local summation strategy.

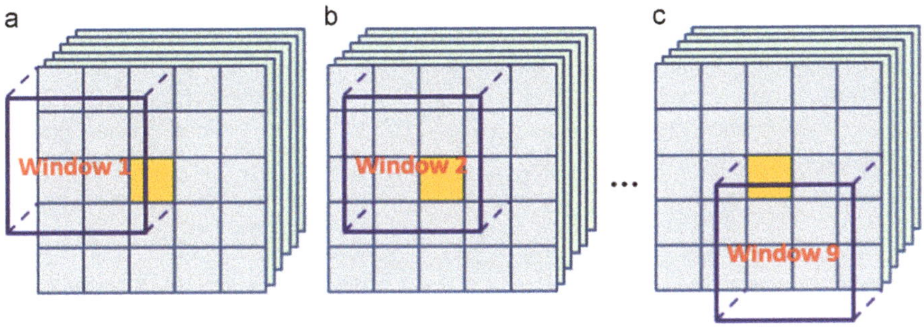

Figure 1. Multiple local window filters [13] (**a**) Window 1. (**b**) Window 2 and (**c**) Window 9.

As illustrated in Figure 1, nine local windows will be taken for the pixel under test, represented by a yellow pixel if the local window is chosen to be 3×3 size. For an $\omega \times \omega$ size local window, the sliding filter contains $\omega \times \omega$ local windows for summation. The summation detector result for the pixel under test $\boldsymbol{r}$ is specified by

$$\delta_{LS-RXD}(\boldsymbol{r}) = \sum_{i=1}^{\omega \times \omega} (\boldsymbol{r} - \boldsymbol{\mu}_{W_i})^T \Sigma_{W_i}{}^{-1} (\boldsymbol{r} - \boldsymbol{\mu}_{W_i}) \tag{3}$$

where $\boldsymbol{W}_i$ is the local pixel samples dataset from window i, μ_{W_i} and Σ_{W_i} are the mean vector and covariance matrix of $\boldsymbol{W}_i$, respectively.

Suppose that the pixel samples dataset in the local window is denoted as $\boldsymbol{W} = \{\boldsymbol{r}_{p_{ij}}\}$, where $i = 1, 2, .., \omega$, $j = 1, 2, .., \omega$ and p_{ij} is the global location of $\boldsymbol{r}_{p_{ij}}$ in the whole data set $\{\boldsymbol{r}_i\}_{i=1}^N$. As a matter of fact, the LS-RXD specified by (3) can be implemented by recursively updating the detection result of each pixel in $\boldsymbol{W}$ as the window is sliding, that is

$$\delta^{t+1}_{LS-RXD}(\boldsymbol{r}_{p_{ij}}) = \delta^{t}_{LS-RXD}(\boldsymbol{r}_{p_{ij}}) + (\boldsymbol{r}_{p_{ij}} - \boldsymbol{\mu}_W)^T \Sigma_W{}^{-1} (\boldsymbol{r}_{p_{ij}} - \boldsymbol{\mu}_W) \tag{4}$$

where μ_W and Σ_W are the mean vector and covariance matrix of W, δ^t_{LS-RXD} and δ^{t+1}_{LS-RXD} are the t, $t+1$ times updated detection result, respectively.

In doing so, the only difference between L-RXD and LS-RXD is that as the local window is sliding; only the detection result of the centered pixel in the local window is calculated by L-RXD, while the detection results of all $\omega \times \omega$ pixels in the local window are updated by LS-RXD.

It is worth noting that the local summation RX detector in [13] is called as LSAD for short. Subspace feature projection is used in LSAD to approximately calculate the $\Sigma_{W_i^{-1}}$ in Equation 3 to enable LSAD with robust background feature statistics. However, it is difficult to realize a timely process due to the subspace feature projection in practice. Band selection onboard before data transmission is feasible to avert the singularity of an inversed local covariance. Therefore, we only focus on the recursive process of L-RXD and LS-RXD in the following.

3. Recursive LS-RXD

In the aforementioned local summation detection algorithms, a new local covariance matrix inversion is repeatedly calculated as the local window slides. The key issue of the recursive process of L-RXD and LS-RXD is how to perform a recursive computation for every independent covariance matrix inversion.

In what follows, we describe how to calculate the covariance matrix inversion of a casual sliding array window recursively.

3.1. The Covariance Matrix Inversion of Causal Sliding Array Window

Figure 2 shows the causal sliding array window at $\boldsymbol{r}_{n-1}$ depicted by dotted lines and the causal sliding array window at $\boldsymbol{r}_n$ depicted by dashed lines, where a is the array window size. The farthest pixel $\boldsymbol{r}_{n-a}$ from $\boldsymbol{r}_n$ in the causal sliding array window at $\boldsymbol{r}_n$ is removed from the causal sliding array window at $\boldsymbol{r}_n$, while the most recent data sample vector $\boldsymbol{r}_n$ is added to the causal sliding array window at $\boldsymbol{r}_{n+1}$.

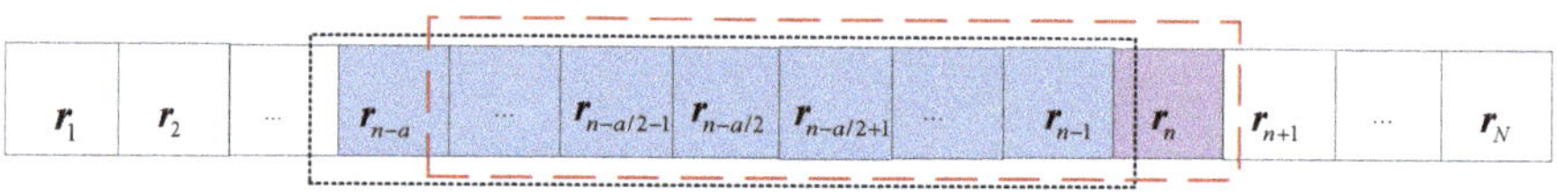

Figure 2. Casual sliding array window at $\boldsymbol{r}_n$ with width specified by a.

Defining $\boldsymbol{R}_a(n) = (1/a)\sum_{\boldsymbol{r}_i \in W} \boldsymbol{r}_i \boldsymbol{r}_i^T$, where $W = \{\boldsymbol{r}_i\}_{i=n-a+1}^{n}$. $\boldsymbol{R}_a(n)$ is called the "causal" sample auto correction matrix correlation matrix, and is formed by data sample vectors in the causal sliding array window. Then $\boldsymbol{R}_a(n)$ can further be expressed as

$$\boldsymbol{R}_a(n) = [(\boldsymbol{R}_a(n-1) - \frac{\boldsymbol{r}_{n-a}\boldsymbol{r}_{n-a}^T}{a}) + \frac{\boldsymbol{r}_n\boldsymbol{r}_n^T}{a}] \tag{5}$$

By repeatedly use of the following Woodbury matrix identity [20] twice:

$$[A + uv^T]^{-1} = A^{-1} - \frac{[A^{-1}u][v^T A^{-1}]}{1 + v^T A^{-1} u} \tag{6}$$

the inverse of $\boldsymbol{R}_a(n)$ can be updated recursively via $\boldsymbol{R}_a^{-1}(n)$ by virtue of (7) and (8) [18]

$$\boldsymbol{R}_a^{-1}(n) = (\boldsymbol{R}_a(n-1) - \frac{\boldsymbol{r}_{n-a}\boldsymbol{r}_{n-a}^T}{a})^{-1} - \frac{[(\boldsymbol{R}_a(n-1) - \frac{\boldsymbol{r}_{n-a}\boldsymbol{r}_{n-a}^T}{a})^{-1}\frac{\boldsymbol{r}_n}{\sqrt{a}}][\frac{\boldsymbol{r}_n^T}{\sqrt{a}}(\boldsymbol{R}_a(n-1) - \frac{\boldsymbol{r}_{n-a}\boldsymbol{r}_{n-a}^T}{a})^{-1}]}{1 + \frac{\boldsymbol{r}_n^T}{\sqrt{a}}(\boldsymbol{R}_a(n-1) - \frac{\boldsymbol{r}_{n-a}\boldsymbol{r}_{n-a}^T}{a})^{-1}\frac{\boldsymbol{r}_n}{\sqrt{a}}} \tag{7}$$

$$(\boldsymbol{R}_a(n-1) - \frac{\boldsymbol{r}_{n-a}\boldsymbol{r}_{n-a}^T}{a})^{-1} = \boldsymbol{R}_a{}^{-1}(n-1) + \frac{[\boldsymbol{R}_a{}^{-1}(n-1)\frac{\boldsymbol{r}_{n-a}}{\sqrt{a}}][\frac{\boldsymbol{r}_{n-a}^T}{\sqrt{a}}\boldsymbol{R}_a{}^{-1}(n-1)]}{1 - \frac{\boldsymbol{r}_{n-a}^T}{\sqrt{a}}\boldsymbol{R}_a{}^{-1}(n-1)\frac{\boldsymbol{r}_{n-a}}{\sqrt{a}}} \tag{8}$$

The "causal" covariance matrix formed by all the data sample vectors in the sliding array window can be specified by

$$\boldsymbol{K}_a(n) = \boldsymbol{R}_a(n) - \boldsymbol{\mu}_a(n)\boldsymbol{\mu}_a^T(n) \tag{9}$$

where

$$\boldsymbol{\mu}_a(n) = \boldsymbol{\mu}_a(n-1) + (1/a)(\boldsymbol{r}_n - \boldsymbol{r}_{n-a}) \tag{10}$$

is the "causal" sample mean of sliding array window. Using *Woodbury* matrix identity again, by letting $\boldsymbol{A} = \boldsymbol{R}_a(n)$, $\boldsymbol{u} = -\boldsymbol{\mu}_a(n)$, $\boldsymbol{v} = \boldsymbol{\mu}_a(n)$, then $\boldsymbol{K}_a^{-1}(n)$ can be further expressed as

$$\boldsymbol{K}_a^{-1}(n) = \boldsymbol{R}_a^{-1}(n) + \frac{[\boldsymbol{R}_a^{-1}(n)\boldsymbol{\mu}_a(n)][\boldsymbol{\mu}_a^T(n)\boldsymbol{R}_a^{-1}(n)]}{1 - \boldsymbol{\mu}_a^T(n)\boldsymbol{R}_a^{-1}(n)\boldsymbol{\mu}_a(n)} \tag{11}$$

By virtue of (7), (8), (10) and (11), $\boldsymbol{K}_a^{-1}(n)$ can be updated recursively by $\boldsymbol{R}_a^{-1}(n-1)$ and $\boldsymbol{\mu}_a(n-1)$, via deleting the pixel $\boldsymbol{r}_{n-a}$ and adding the current pixel $\boldsymbol{r}_n$.

3.2. Recursive Processing of the Covariance Matrix Inversion of Sliding Window

Figure 3 illustrates two continually sliding windows with size of $\omega \times \omega$ depicted by black dashed lines and orange dashed lines, respectively, where $\boldsymbol{r}_{p_{\omega\omega}+1}$ denotes the most recent received data sample vector. The sample data vectors update process in sliding windows can be implemented in ω steps by removing one pixel and adding one pixel each step. Suppose that $p_{\omega\omega} = n-1$, the inverses of correlation matrices of the local window at $\boldsymbol{r}_{p_{\omega\omega}}$ and $\boldsymbol{r}_{p_{\omega\omega}+1}$ are denoted as $\boldsymbol{R}_{\omega^2}^{-1}(n-1)$ and $\boldsymbol{R}_{\omega^2}^{-1}(n)$ respectively, and inverses of the covariance matrices of the local window at $\boldsymbol{r}_{p_{\omega\omega}}$ and $\boldsymbol{r}_{p_{\omega\omega}+1}$ are denoted as $\boldsymbol{K}_{\omega^2}^{-1}(n-1)$ and $\boldsymbol{K}_{\omega^2}^{-1}(n)$, respectively. In analogy with (7), (8), (10), and (11), $\boldsymbol{K}_{\omega^2}^{-1}(n)$ can be updated recursively as follows.

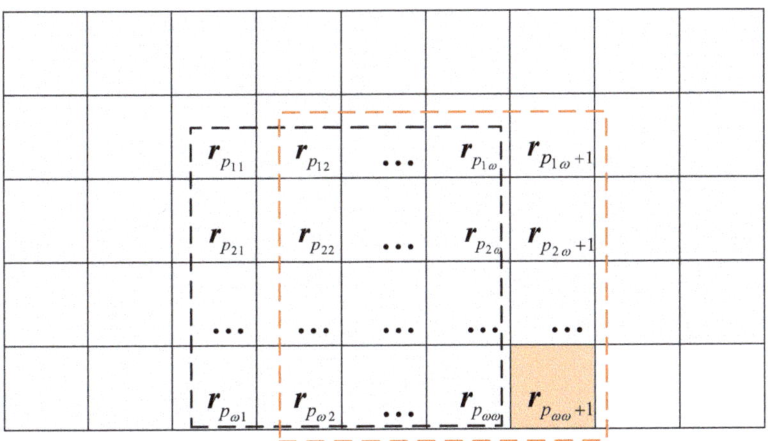

Figure 3. Sliding window with size of $\omega \times \omega$.

For $i = 1$ to ω, do

$$(\boldsymbol{R}_{\omega^2}(n-1) - \frac{\boldsymbol{r}_{p_{i1}}\boldsymbol{r}_{p_{i1}}^T}{\omega^2})^{-1} = \boldsymbol{R}_{\omega^2}^{-1}(n-1) + \frac{[\boldsymbol{R}_{\omega^2}^{-1}(n-1)\frac{\boldsymbol{r}_{p_{i1}}}{\omega}][\frac{\boldsymbol{r}_{p_{i1}}^T}{\omega}\boldsymbol{R}_{\omega^2}^{-1}(n-1)]}{1 - \frac{\boldsymbol{r}_{p_{i1}}^T}{\omega}\boldsymbol{R}_{\omega^2}^{-1}(n-1)\frac{\boldsymbol{r}_{p_{i1}}}{\omega}} \tag{12}$$

$$\boldsymbol{R}_{\omega^2}^{-1}(n) = (\boldsymbol{R}_{\omega^2}(n-1) - \frac{\boldsymbol{r}_{p_{i1}}\boldsymbol{r}_{p_{i1}}^T}{\omega^2})^{-1} - \frac{[(\boldsymbol{R}_{\omega^2}(n-1) - \frac{\boldsymbol{r}_{p_{i1}}\boldsymbol{r}_{p_{i1}}^T}{\omega^2})^{-1}\frac{\boldsymbol{r}_{p_{i\omega}}}{\omega}][\frac{\boldsymbol{r}_{p_{i\omega}}^T}{\omega}(\boldsymbol{R}_{\omega^2}(n-1) - \frac{\boldsymbol{r}_{p_{i1}}\boldsymbol{r}_{p_{i1}}^T}{\omega^2})^{-1}]}{1 + \frac{\boldsymbol{r}_{p_{i\omega}}^T}{\omega}(\boldsymbol{R}_{\omega^2}(n-1) - \frac{\boldsymbol{r}_{p_{i1}}\boldsymbol{r}_{p_{i1}}^T}{\omega^2})^{-1}\frac{\boldsymbol{r}_{p_{i\omega}}}{\omega}} \tag{13}$$

Meanwhile, update $\boldsymbol{R}_{\omega^2}^{-1}(n-1) = \boldsymbol{R}_{\omega^2}^{-1}(n)$ after each iteration.
Then

$$\boldsymbol{\mu}_{\omega^2}^{-1}(n) = \boldsymbol{\mu}_{\omega^2}^{-1}(n-1) + (1/\omega^2)(\textstyle\sum_{i=1}^{\omega}(\boldsymbol{r}_{p_{i\omega}} - \boldsymbol{r}_{p_{i1}})) \tag{14}$$

$$\boldsymbol{K}_{\omega^2}^{-1}(n) = \boldsymbol{R}_{\omega^2}^{-1}(n) + \frac{[\boldsymbol{R}_{\omega^2}^{-1}(n)\boldsymbol{\mu}_{\omega^2}(n)][\boldsymbol{\mu}_{\omega^2}^T(n)\boldsymbol{R}_{\omega^2}^{-1}(n)]}{1 - \boldsymbol{\mu}_{\omega^2}^T(n)\boldsymbol{R}_{\omega^2}^{-1}(n)\boldsymbol{\mu}_{\omega^2}(n)} \tag{15}$$

3.3. Recursive Processing of LS-RXD

Except for the recursive processing of the covariance matrix inversion of the sliding window, some other issues should also be considered.

The first issue is the edge expansion. To ensure that there is no absence of detection on the edge of an image, the edge expansion is usually operated as a preprocessing for a local window detector. Due to the low probability of anomaly targets appearance in hyperspectral images, enplaned layers can be randomly chosen from the whole image [13]. With this consideration in mind, take the window with size of 3×3 as an example. We design the sliding window strategy, depicted in Figure 4, where the yellow, blue and purple grids, respectively, denote the latest pixel received, the processed data and the pixels to be processed. As Figure 4 shows, when the sliding window meets the right board of the hyperspectral image, the next several sliding windows are across the border by moving down one line and adding new data one by one. The last sliding window moves to the right-bottom until the last sample data $\boldsymbol{r}_{Row \times Col}$ is received. This design enables the recursive processing of LS-RXD more conveniently.

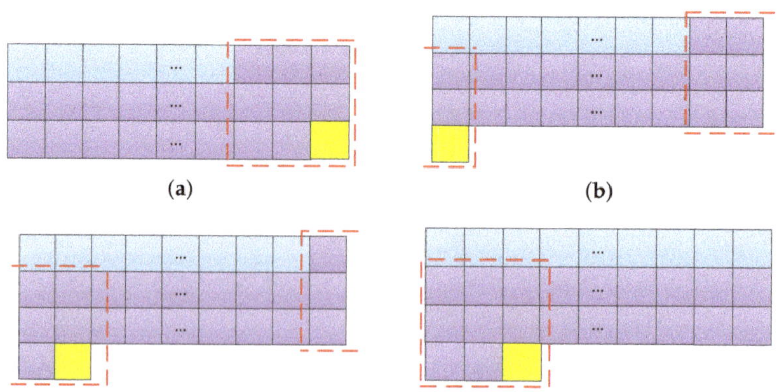

Figure 4. Sliding window strategy for recursive local summation RXD (R-LS-RXD): (**a**) $No.(Col-2)$ window; (**b**) $No.(Col-1)$ window; (**c**) $No.Col$ window; (**d**) $No.(Col+1)$ window.

The second issue is how to keep track of which data sample vector should be removed and which data sample vector should be added as a matrix window moves on. Let $\boldsymbol{W} = \{\boldsymbol{r}_{p_{ij}}\}_{i,j=1}^{\omega}$ denote the local sliding window in an image with size $Row \times Col$, where p_{ij} is the global location of $\boldsymbol{r}_{p_{ij}}$ in the whole data set $\{\boldsymbol{r}_i\}_{i=1}^{N}$. For the first local window, p_{ij} can be expressed as $p_{ij} = x_{(i-1)\times Col+j}$. Using the strategy of Figure 4, it is very easy to update the global location of pixels in the follow-up window as $p_{ij} = p_{ij+1}$ successively.

After the aforementioned issues are solved, the recursive LS-RXD, called as R-LS-RXD can be obtained by

$$\delta_{R-LS-RXD}(\boldsymbol{r}_{p_{ij}}) = \delta_{R-LS-RXD}(\boldsymbol{r}_{p_{ij}}) + (\boldsymbol{r}_{p_{ij}} - \boldsymbol{\mu}_{\omega^2}(n))^T \boldsymbol{K}_{\omega^2}^{-1}(n)(\boldsymbol{r}_{p_{ij}} - \boldsymbol{\mu}_{\omega^2}(n)) \tag{16}$$

Three comments are worthwhile:

1. It is important to note that, using the strategy of Figure 4, the updating counts of the detection value for the pixel located in several top and bottom lines of the image are less than ω. This will result in the whole detection result being inconsistent. To solve this problem, the updating number of each pixel is counted, which is denoted as $N_{p_{ij}}$, and finally the detection result is obtained as $\delta_{R-LS-RXD}(\boldsymbol{r}_{p_{ij}})/N_{p_{ij}}$
2. To avoid the singularity problem of calculating the inverse of the sample correlation and covariance matrix used by anomaly detectors, $\omega \times \omega$ must at least equal to or greater than the total number of spectral bands [13,18].
3. The whole design procedure is also suitable for recursive L-RXD which is not included here.

3.4. Background Suppression of Sliding Windows

This section mainly discusses the background suppression sliding window furthermore. It is not convenient to set the current under test pixel to conclude in the local window background with other data samples, because it will reduce the separation between background information and anomaly information separation while the current under test pixel is anomaly [21]. In order to suppress the background information and improve the detection performance, we need to remove the current under test pixel ($\boldsymbol{r}_k$) from the recursive update processing.

Assume that $\boldsymbol{R}_k$ is the correlation matrix removed $\boldsymbol{r}_k$, and $\boldsymbol{R}_k$ is specified by

$$\begin{aligned} \boldsymbol{R}_k &= \tfrac{1}{n-1}\textstyle\sum_{i=1,i\neq k}^{n} \boldsymbol{r}_i\boldsymbol{r}_i^T = \tfrac{1}{n-1}(\textstyle\sum_{i=1}^{n} \boldsymbol{r}_i\boldsymbol{r}_i^T - \boldsymbol{r}_k\boldsymbol{r}_k^T) \\ &= \tfrac{n}{n-1}\tfrac{1}{n}\textstyle\sum_{i=1}^{n} \boldsymbol{r}_i\boldsymbol{r}_i^T - \tfrac{1}{n-1}\boldsymbol{r}_k\boldsymbol{r}_k^T = \tfrac{n}{n-1}\boldsymbol{R}_n - \tfrac{1}{n-1}\boldsymbol{r}_k\boldsymbol{r}_k^T \end{aligned} \tag{17}$$

Once using *Woodbury* matrix identity, letting $A = \frac{n}{n-1}\boldsymbol{R}_n$, $u = \frac{-1}{n-1}\boldsymbol{r}_k$, $v = \boldsymbol{r}_k$, then

$$\begin{aligned} \boldsymbol{R}_k^{-1} &= (\tfrac{n}{n-1}\boldsymbol{R}_n - \tfrac{1}{n-1}\boldsymbol{r}_k\boldsymbol{r}_k^T)^{-1} \\ &= \tfrac{n-1}{n}\boldsymbol{R}_n^{-1} + \tfrac{[\frac{n-1}{n}\boldsymbol{R}_n^{-1}\frac{1}{n-1}\boldsymbol{r}_k][\boldsymbol{r}_k^T\frac{n-1}{n}\boldsymbol{R}_n^{-1}]}{1-\boldsymbol{r}_k^T\frac{n-1}{n}\boldsymbol{R}_n^{-1}\frac{1}{n-1}\boldsymbol{r}_k} \end{aligned} \tag{18}$$

Assume that $\boldsymbol{\mu}_k$ is the mean vector of background sample data removed $\boldsymbol{r}_k$, the inverse of covariance matrix could be specified by

$$\boldsymbol{K}_{BS}^{-1}(n) = \boldsymbol{R}_k^{-1}(n) + \frac{[\boldsymbol{R}_k^{-1}(n)\boldsymbol{\mu}_k(n)][\boldsymbol{\mu}_k^T(n)\boldsymbol{R}_k^{-1}(n)]}{1 - \boldsymbol{\mu}_k^T(n)\boldsymbol{R}_k^{-1}(n)\boldsymbol{\mu}_k(n)} \tag{19}$$

As a result, the background suppression recursive R-BS-LS-RXD can be specified by

$$\delta_{R-BS-LS-RXD}(\boldsymbol{r}_k) = (\boldsymbol{r}_k - \boldsymbol{\mu}_k(n))^T \boldsymbol{K}_{BS}^{-1}(n)(\boldsymbol{r}_k - \boldsymbol{\mu}_k(n)) \tag{20}$$

3.5. Computational Complexity Analysis

This section provides a detailed analysis on the computational complexity of calculating recursive update Equations (12)–(15).

The advantage of using causal sliding windows over local windows is the use of recursive Equations (12) and (13), where the *Woodbury* identity is implemented twice, instead of recalculating each time as long as a new data sample vector comes in. Table 1 shows the computation complexity of matrix algebra. Based on the information in Table 1, the matrix inversion computation complexity is higher than the matrix multiplication computation.

Table 1. Computation Complexity of Matrix Algebra.

Operation	Input	Output	Algorithm	Complexity
Matrix multiplication	Matrix a size $m \times n$; Matrix b size $n \times p$;	Matrix size $m \times p$	Schoolbook matrix multiplication	$O(mnp)$
Matrix inversion	Matrix size $m \times m$	Matrix size $m \times m$	Gauss-Jordan elimination	$O(n^3)$
			Strassen algorithm	$O(n^{2.807})$
			Coppersmith-Winograd algorithm	$O(n^{2.376})$
			Williams algorithm	$O(n^{2.373})$

The usage frequency of the *Woodbury* matrix identity is determined by the size of the sliding window. Local background information is updated by calculating ω times of Equations (12) and (13), regardless of the number of pixels in the local background. Such a significant benefit arises from the recursive specialty in (12) and (13). Hence, the computational complexity of processing a single local window specified by its window size $\omega \times \omega$ requires ω times calculations of matrix multiplication. In addition, it only needs to calculate the inverse of the covariance matrix once.

Table 2 tabulates the number of floating operations (flops) required for LS-RXD and R-LS-RXD, which update $K_a^{-1}(n)$ in different method, where the bands number is specified by L, local window size is specified by $a = \omega \times \omega$, and the pixels number to be processed is specified by N. These parameters determine the number of flops in the algorithm. Figure 5 plots the number of floating operations required for every algorithm versus L, a and N. The configurations of parameters are shown in Table 3.

Table 2. Computational Complexity for local summation RXD (LS-RXD) and recursive local summation RXD (R-LS-RXD).

Algorithm		LS-RXD			R-LS-RXD			
Operator	Initialization	Input r_n			Input r_n			
		μ_a	K	K^{-1}	μ_a	Equation(7)	Equation(8)	K^{-1}
flops	$L^3 + 2a(L^2 + L)$	$(a+1)L$	$2a(L^2 + L)$	L^3	$3L$	$\omega(6L^2 + 5L)$	$\omega(6L^2 + 5L)$	$6L^2 + L$
sum		$NL^3 + 2aNL^2 + 3a(N+1)L$			$L^3 + (2a + 12N\omega + 6N)L^2 + (2a + 4N + 10N\omega)L$			

Table 3. Configuration of The Parameters.

Figure 5	Parameters		
	L	ω	N
(a)	10:5:200	15	10,000
(b)	10	15:2:99	10,000
(c)	10	15	1000:1000:10,000

As shown in Figure 5, the comparison of different anomaly detectors depends on the specific configuration of the parameters. Generally speaking, R-LS-RXD is faster than LS-RXD.

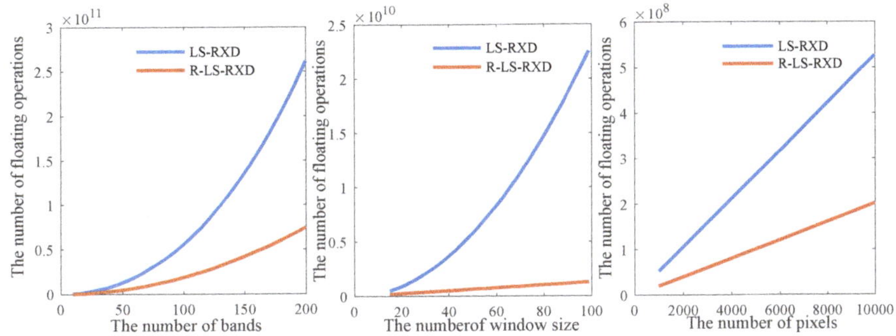

(**a**) (**b**) (**c**)

Figure 5. Numbers of floating operations in various of (**a**) bands; (**b**) ω size; (**c**) processed pixels.

4. Results and Discussion

To demonstrate the performance of anomaly detection using recursive local summation RXD, two real hyperspectral image scenes were conducted for experiments. The first image data set is the Airborne Visible/Infrared Imaging Spectrometer (AVIRIS) image scene of Sandiego airport area which is located in California. A sub-image with size 100 × 100 along with its ground truth are shown in Figure 6a,b, respectively. It was acquired through 224 spectral bands with a spectral coverage from 0.4 to 2.5 μm where the spatial resolution is 3 m and spectral resolution is 20 nm. After removing low signal-to-noise ratio (SNR) and water absorption bands, a total of 126 spectral bands were used for experiments.

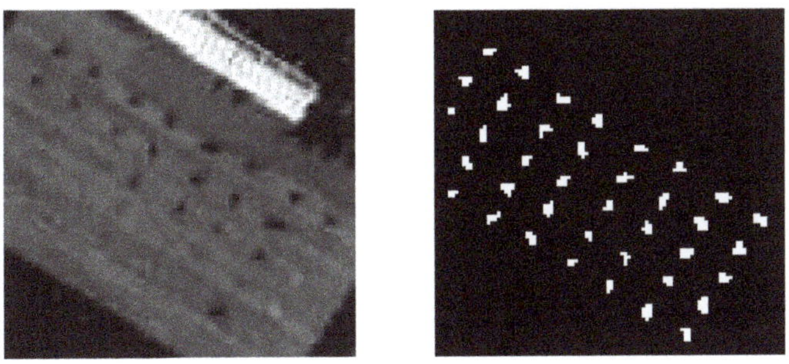

(**a**) (**b**)

Figure 6. Sandiego hyperspectral image (**a**) 30th band scene; (**b**) ground truth.

The second image data set is the Hyperspectral Digital Imagery Collection Experiment (HYDICE) image scene shown in Figure 7a which was collected in August 1995 from a flight altitude of 10,000 ft with the ground sampling distance approximately 1.56 m. This scene has been studied extensively by many reports such as [2,14]. It has a total of 169 bands which were used for the experiments with low signal/high noise bands: bands 1–3 and bands 202–210; and water vapor absorption bands: bands 101–112 and bands 137–153, removed. There are 15 panels with three different sizes of 3 m × 3 m, 2 m × 2 m and 1 m × 1 m. Figure 7b shows the precise spatial locations of these 15 panels, where red pixels (R pixels) are the panel center pixels and the pixels in yellow (Y pixels) are panel boundary pixels mixed with the background (BKG) . As a result, there are a total of 19 R panel pixels. In particular,

R panel pixels are denoted by p_{ij} with rows indexed by $i = 1, \cdots, 5$ and columns indexed by $j = 1, 2, 3$ except that the panels in the 1st column with the 2nd , 3rd , 4th, 5th rows which are two-pixel panels, denoted by $p_{211}, p_{221}, p_{311}, p_{312}, p_{411}, p_{412}, p_{511}, p_{521}$. The 1.56 m-spatial resolution of the image scene suggests that most of the 15 panels are one pixel in size.

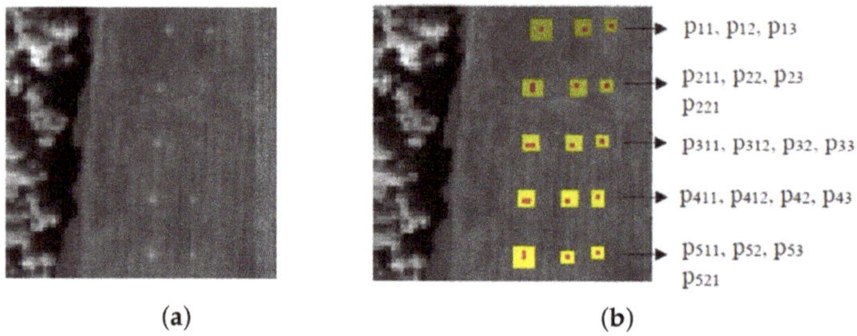

Figure 7. (**a**) A Hyperspectral Digital Imagery Collection Experiment (HYDICE) panel scene which contains 15 panels; (**b**) Ground truth map of spatial locations of the 15 panels.

In order to quantitatively evaluate detection performance, receiver operating characteristic (ROC) curves are used to compare the different algorithms. Based on the provided ground truth, we can perform an analysis via ROC curves of the false alarm ratio (Pf) versus the detection ratio (Pd) by taking all the possible thresholds (τ). We can further calculate the area under the ROC curve (AUC) for a quantitative performance analysis. The algorithm with a larger AUC value is regarded as a better performance.

Traditional ROC curves is a 2D plot represented by values of Pf and Pd. Furthermore, we can plot another 2D ROC curve of Pf and τ, which provides crucial information of progressive background suppression as the threshold τ varies. when it comes to the interpretation of anomaly detection by visual inspection with no availability of ground truth or AUC values with similar performance.

Three experiments are conducted with the purpose of: (1) evaluating the influence of window size on the detection performance of R-LS-RXD; (2) comparing the detection performance of different algorithms; and (3) comparing computing times of different algorithms, respectively.

4.1. Optimum Size of Sliding Window

Band selection is very practical in anomaly detection [22,23]; nine bands are selected by signal-to-noise ratio estimation and maximal information (SNRE-MI) [23] in the experiment to obtain better result. To investigate the influence of window size on detection performance of R-LS-RXD, two hyperspectral images of different sensors (AVIRIS and HYDICE, respectively) in the previous section are used for experiments, the size of sliding window varies from 5×5 up to 17×17 with steps of two pixels side width. Figure 8a–g and Figure 9a–g show their detection abundance fractional maps with their detected abundance fractions in gray scale of AVIRIS and HYDICE hyperspectral image, respectively. According to the experiment, the detection result is poor with a window size of 5×5, where the background and anomaly are difficult to separate for both sensors. Additionally, the performance begins to improve as the window size increases. When the window size is greater than or equal to 11×11, detection performances are similar by visual inspection as shown in Figures 8 and 9. Figures 8h and 9h show the results of global background K-RXD detector for comparison.

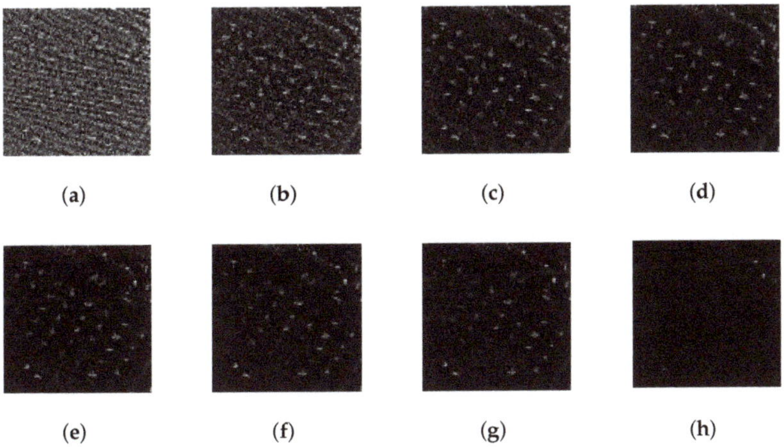

Figure 8. Detection abundance fractional maps of AVIRIS by recursive local summation RXD (R-LS-RXD) with different sliding window size: (**a**) ω =5; (**b**) ω =7; (**c**) ω =9; (**d**) ω =11; (**e**) ω =13; (**f**) ω =15; (**g**) ω =17; (**h**) global.

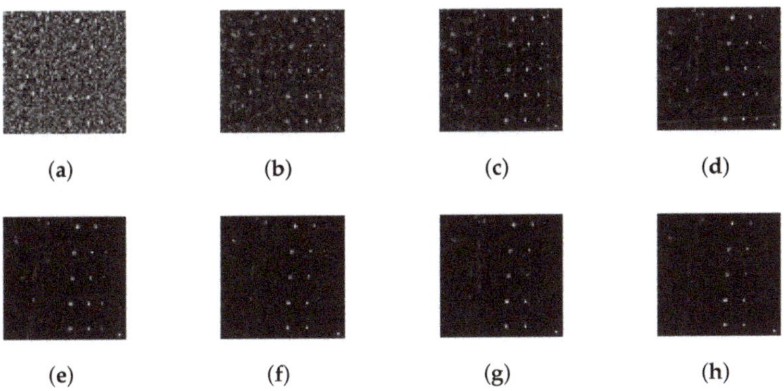

Figure 9. Detection abundance fractional maps of HYDICE by R-LS-RXD with different sliding window size: (**a**) ω =5; (**b**) ω =7; (**c**) ω =9; (**d**) ω =11; (**e**) ω =13; (**f**) ω =15; (**g**) ω =17; (**h**) global.

In order for a further quantitative evaluation of detection performance with different window sizes, the ROC curves are implemented. To simplify our study, ROC curves for HYDICE data are not given, since the results are similar for both data sources. Figure 10 shows the ROC curves for AVIRIS data with different window sizes, with a traditional (Pd, Pf) ROC in (a) and a (Pf, τ) curve analysis in (b), respectively. Additionally, the AUC values, denoted by A_z, are calculated for each (Pd, Pf) curves and (Pf, τ) curves. In general, the higher the value of $A_z(Pd, Pf)$ and the lower the value of $A_z(Pf, \tau)$, the better the detection performance is. Results are tableted in Table 4, with the best results highlighted.

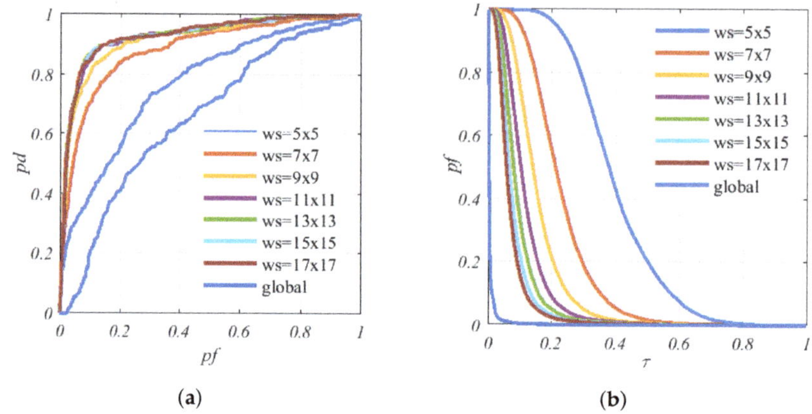

Figure 10. Receiver operating characteristic (ROC) curves analysis for AVIRIS data with different window size: (**a**) curve of (Pd, Pf); (**b**) curve of (Pf, τ)

Based on the result of Figure 10 and Table 4, as the window size goes up, the values of A_z (Pd, Pf) are increased while the values of A_z (Pf, τ) are decreased. The detector reaches the best detection power in size 13×13, and the best background suppression performs the best with size 17×17. However, the trend of A_z (Pf, τ) decreasing obviously slows down when the window size increases from 13×13 to 15×15. When it comes to the global size background, the value of A_z (Pd, Pf) decreases to an untrustworthy value and is difficult to be distinguished by visual inspection in Figures 8h and 9h.

The conclusions for the experiment are as follows. As with the size of window increases, the sliding window RXD window RXD detector obtains better detection performances. However, 13×13 is the optimum size for a Sandiego hyperspectral image. As an alternative interpretation, although a larger window size results in better background suppression, the detection performance is much more important in the detector evaluation.

Table 4. Area under the ROC curve (AUC) values of (Pd, Pf) and (Pf, τ) with different window sizes

Sensor	Window-Size	5×5	7×7	9×9	11×11	13×13	15×15	17×17	*Global*
AVIRIS	$A_z(Pd, Pf)$	0.7679	0.8813	0.9141	0.9206	0.9286	0.9281	0.9275	0.6548
	$A_z(Pf, \tau)$	0.3461	0.2059	0.1389	0.1086	0.0873	0.0752	0.0648	0.0080
HYDICE	$A_z(Pd, Pf)$	0.9895	0.9973	0.9985	0.9988	0.9988	0.9986	0.9982	0.9878
	$A_z(Pf, \tau)$	0.2636	0.1576	0.1178	0.0902	0.0713	0.0575	0.0468	0.0121

4.2. Performance Evaluation for Different Algorithms

In this section, we compare the detection performance of the LS-RXD, causal sliding array window (CSA-RXD) [18], proposed R-LS-RXD and R-BS-LS-RXD. In order to obtain the best detection results, the sliding window is implemented with size of 13×13 for both AVIRIS and HYDICE hyperspectral images.

Detection results of the four detectors using AVIRIS and HYDICE data are shown in Figures 11 and 12, respectively. The first line shows the gray scale results with detected abundance fractions, and the second line demonstrates the binary detection maps separated in an appropriate threshold, which was calculated by Otsu algorithm[24].

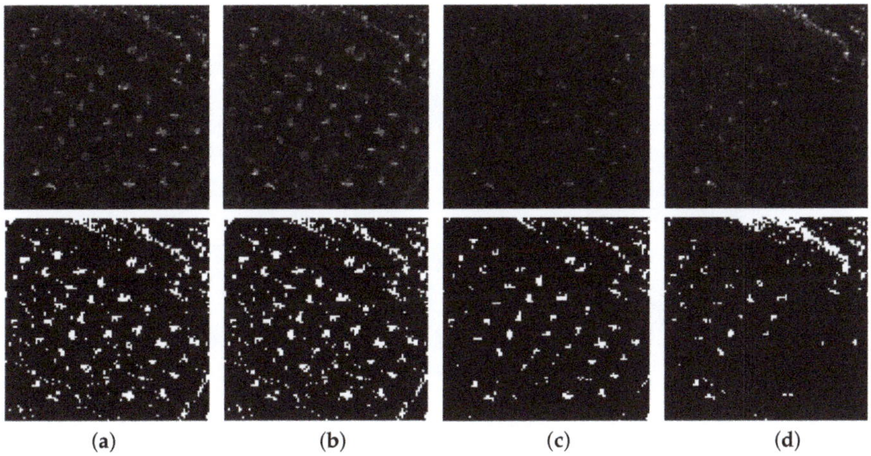

Figure 11. Detection results of AVIRIS data for different algorithms: (**a**) LS-RXD; (**b**) R-LS-RXD; (**c**) Recursive background Suppression local summation RXD (R-BS-LS-RXD); (**d**) Causal sliding array window (CSA-RXD).

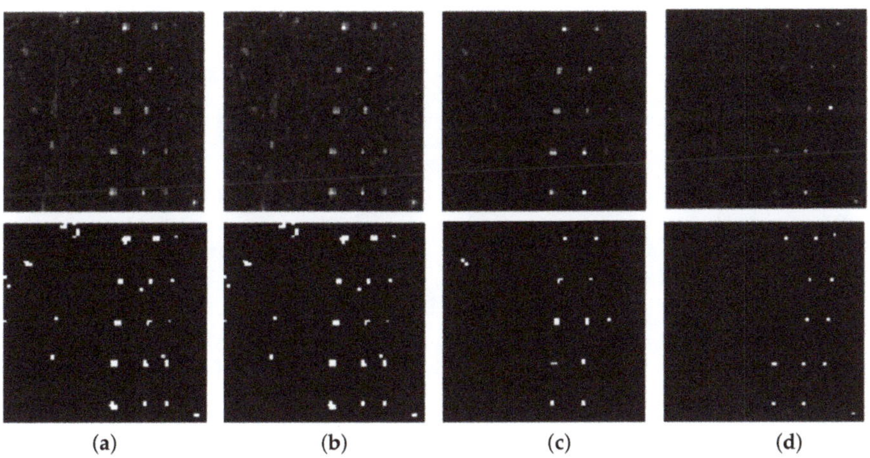

Figure 12. Detection results of HYDICE data for different algorithms: (**a**) LS-RXD; (**b**) R-LS-RXD; (**c**) R-BS-LS-RXD; (**d**) CSA-RXD.

Both hyperspectral images of different sensors came to the same conclusions as follows, showing the adaptation of proposed algorithms for different sensors. It can be found obviously from the detection results that CSA-RXD, which merely take partial advantage of spectral–spatial integration information, omit number targets by visual inspection as shown in Figures 11d and 12d. On the contrary, other anomaly detectors, which are implemented with spectral–spatial integrated information can acquire excellent detection performance. The maximum detection of ground target shows in Figures 6b and 7b can be detected by LS-RXD, R-LS-RXD and R-BS-LS-RXD by visual inspection in Figures 11a–c and 12a–c. As also shown in the figure, R-BS-LS-RXD gets better background suppression compared with R-LS-RXD and LS-RXD. This indicates that R-BS-LS-RXD can not only correctly detect anomaly target pixels as R-LS-RXD performs, but also acquires excellent background suppression as CSA-RXD performs.

Similarly, to simplify our study, a quantitative evaluation with traditional ROC curves, (Pf, τ) curves, is demonstrated in Figure 13. AUC values are listed in Table 5, only for AVIRIS Sandiego data, since the results are similar for both data sources.

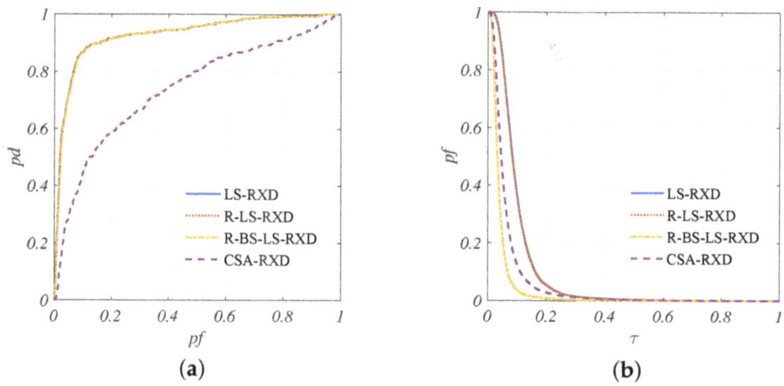

Figure 13. ROC analysis with different detectors: (**a**) curve of (Pd, Pf); (**b**) curve of (Pf, τ).

Table 5. AUC values with different detectors.

Algorithm	LS-RXD	R-LS-RXD	R-BS-LS-RXD	CSA-RXD
A_z (Pd, Pf)	0.9286	0.9286	0.9270	0.7401
A_z (Pf, τ)	0.0873	0.0873	0.0364	0.0597

It is interesting to note that the ROC curves of LS-RXD, R-LS-RXD and R-BS-LS-RXD are overlapped completely. This indicates that these algorithms get similar detection power from the traditional ROC curve analysis. Meanwhile, R-BS-LS-RXD gets a better performance in background suppression as the (Pf, τ) curve shows. It is clearly shown that anomaly detectors with spectral–spatial integration have better performance, where the ROC curves of LS-RXD, R-LS-RXD and R-BS-LS-RXD are much closer to the upper left corner than CSA-RXD.

AUC values tablet in Table 5 prove that the proposed R-LS-RXD and R-BS-LS-RXD get a similar detection performance with LS-RXD. In addition, $A_z(Pd, Pf)$ of LS-RXD, R-LS-RXD and R-BS-LS-RXD is greater than CSA-RXD. By contrast, R-BS-LS-RXD produced lowest value of $A_z(Pf, \tau)$. In general, R-BS-LS-RXD can suppress the background information and improve the detection performance.

4.3. Computing Time Comparison for Different Algorithms

In order to verify the computing effectiveness of recursive LS-RXD, we design a comprehensive comparative analysis on the computer processing time (CPT) of R-LS-RXD and LS-RXD. The computer environments used for the experiments are 64-bit operating systems with Intel i5-4590, a central processing unit (CPU) of 3.3 GHz, and 8 GB of random access memory (RAM). In order to remove the pulse error caused by the computer itself, the following data on complexity analyses are averaged after five experiments. Table 6 tablets the computing time of algorithms with different window sizes in San Diego hyperspectral image.

Table 6. Computing Time (seconds).

Windowsize	7×7	9×9	11×11	13×13	15×15	17×17
R-LS-RXD	1.366	1.648	1.951	2.247	2.504	2.764
LS-RXD	4.248	4.327	4.407	4.542	4.925	5.618
Speedup	3.110	2.627	2.259	2.022	1.967	2.033

Based on the results in Table 6, the computing time of R-LS-RXD is significantly less than LS-RXD in every window size. In addition, the acceleration is particularly noticeable when the window is in a small-scale. In the experiment, the speedup ratio is up to three when the window size is chosen as 7×7. As the window size grows up, the speedup ratio remains, at least, two.

To further evaluate computational complexity, Figure 14 plots the computing time versus the number of processed pixels for both R-LS-RXD and LS-RXD on the Sandiego hyperspectral image. Each algorithm was run and executed five times to produce an average computing time. As we can see, R-LS-RXD requires less time than LS-RXD does due to the fact that the former implements a recursive process, while the latter implements a nonrecursive process. As also shown in the figure, the computing time increases linearly as new pixels are added.

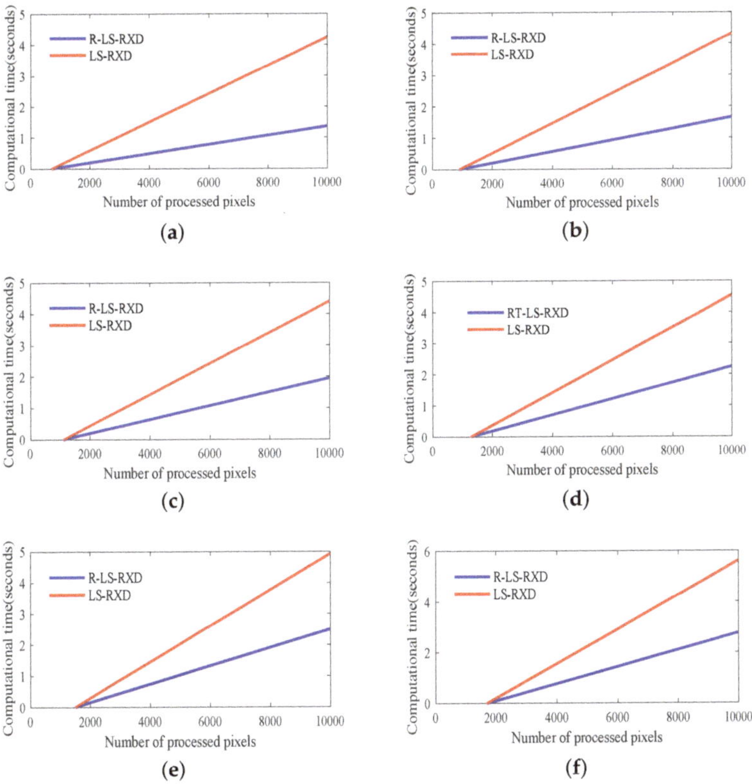

Figure 14. Plots of computing time versus number of processed pixels. (**a**) ω =7; (**b**) ω =9; (**c**) ω =11; (**d**) ω =13; (**e**) ω =15; (**f**) ω =17.

5. Conclusions

This paper proposes a recursive local summation RX algorithm for hyperspectral anomaly detection based on sliding window processing. In order for a fast implementation of a sliding window detector, a recursive update equation for the inversion of local background covariance matrices is developed. In addition, a background suppression R-BS-LS-RXD detector is also proposed in this paper, which removes the current under test pixel from the recursively update processing. This method exploits a local summation strategy in a sliding window, which could sum multiple correlated local background statistics to suppress the major background. The real hyperspectral image experiments have proven that the R-LS-RXD and LS-RXD obtain similar detection performances, which can be competitive with that of CSA-RXD based on sliding array window background. To investigate the computational complexity issue, a comprehensive comparative analysis on the CPT of running recursive updating sliding window detector and un-recursive updating method is conducted in theory and experiments. The result shows R-LS-RXD has a significant acceleration effect for calculation. Our future work mainly focuses on deriving real-time progressive processing of anomaly detection for hyperspectral imagery that was acquired by other data formats.

Acknowledgments: This work was supported by the National Nature Science Foundation of China (No. 61571170), the Joint Funds of the Ministry of Education of China No. 6141A02022314), Shanghai Aerospace Science and Technology Innovation Fund (No.SAST2015033), Fundamental Research Funds for Central Universities under Grants (No. 3132017080) and the Open Research Fund of Key Laboratory of Spectral Imaging Technology, Chinese Academy of Sciences(LSIT201707D).

Author Contributions: All the authors made significant contributions to the work. Liaoying Zhao and Weijun Lin designed the research and analyzed the results. Yulei Wang provided advice for the preparation and revision of the paper. Xiaorun Li assisted in the preparation work and validation work.

Conflicts of Interest: The authors declare no conflict of interest.

References

1. Goetz, A.F. Three Decades of Hyperspectral Remote Sensing of The Earth: A Personal View. *Remote Sens. Environ.* **2009**, *113*, S5–S16.
2. Chang, C.I. *Hyperspectral Data Processing: Algorithm Design and Analysis*; Wiley-Interscience: Hoboken, NJ, USA, 2013; pp. 441–442.
3. Reed, I.S.; Yu, X. Adaptive Multiple-band CFAR Detection of An Optical Pattern with Unknown Spectral Distribution. *IEEE Trans. Acoust. Speech Sign. Proc.* **1990**, *38*, 1760–1770.
4. Nasrabadi, N.M. Hyperspectral Target Detection: An Overview of Current and Future Challenges. *IEEE Sign. Proc. Mag.* **2014**, *31*, 34–44.
5. Manolakis, D.; Truslow, E.; Pieper, M.; Cooley, T.; Brueggeman, M. Detection Algorithms in Hyperspectral Imaging Systems: An Overview of Practical Algorithms. *IEEE Sign. Proc. Mag.* **2014**, *31*, 24–33.
6. Li, W.; Du, Q. Collaborative Representation for Hyperspectral Anomaly Detection. *IEEE Trans. Geosci. Remote Sens.* **2015**, *53*, 1463–1474.
7. Chang, C.I.; Chiang, S.S. Anomaly Detection and Classification for Hyperspectral Imagery. *IEEE Trans. Geosci. Remote Sens.* **2002**, *40*, 1314–1325.
8. Chang, C.; Hsueh, M. Characterization of Anomaly Detection in Hyperspectral Imagery. *Sensor Rev.* **2006**, *26*, 137–146.
9. Molero, J.M.; Garzón, E.M.; García, I.; Plaza, A. Analysis and Optimizations of Global and Local Versions of the RX Algorithm for Anomaly Detection in Hyperspectral Data. *IEEE J. Sel. Top. Appl. Earth Obs. Remote Sens.* **2013**, *6*, 801–814.
10. Liu, W.M.; Chang, C.I. Multiple-Window Anomaly Detection for Hyperspectral Imagery. *IEEE J. Sel. Top. Appl. Earth Obs. Remote Sens.* **2013**, *6*, 644–658.
11. Guo, Q.; Zhang, B.; Ran, Q.; Gao, L.; Li, J.; Plaza, A. Weighted-RXD and Linear Filter-Based RXD: Improving Background Statistics Estimation for Anomaly Detection in Hyperspectral Imagery. *IEEE J. Sel. Top. Appl. Earth Obs. Remote Sens.* **2014**, *7*, 2351–2366.
12. REN, X.D.; LEI, W.H. Kernel Anomaly Detection Method in Hyperspectral Imagery Based on the Spectral Discrimination Method. *Acta Photonica Sin.* **2016**, *45*, 330003.
13. Du, B.; Zhao, R.; Zhang, L.P.; Zhang, L.F. A Spectral-spatial Based Local Summation Anomaly Detection Method for Hyperspectral Images. *Sign. Proc.* **2016**, *124*, 115–131.
14. Chen, S.Y.; Wang, Y.L.; Wu, C.C.; Liu, C.; Chang, C.I. Real-time Causal Processing of Anomaly Detection for Hyperspectral Imagery. *IEEE Trans. Aerosp. Electron. Syst.* **2014**, *50*, 1511–1534.
15. Zhao, C.H.; Wang, Y.L.; Qi, B.; Wang, J. Global and Local Real-Time Anomaly Detectors for Hyperspectral Remote Sensing Imagery. *Remote Sens.* **2015**, *7*, 3966–3985.

16. Yang, B.; Yang, M.; Plaza, A.; Gao, L.; Zhang, B. Dual-Mode FPGA Implementation of Target and Anomaly Detection Algorithms for Real-Time Hyperspectral Imaging. *IEEE J. Sel. Top. Appl. Earth Obs. Remote Sens.* **2015**, *8*, 2950–2961.
17. Zhang, L.; Peng, B.; Zhang, F.; Wang, L.; Zhang, H.; Zhang, P.; Tong, Q. Fast Real-Time Causal Linewise Progressive Hyperspectral Anomaly Detection via Cholesky Decomposition. *IEEE J. Sel. Top. Appl. Earth Obs. Remote Sens.* **2017**, *10*, 4614–4629.
18. Chang, C.I.; Wang, Y.; Chen, S.Y. Anomaly Detection Using Causal Sliding Windows. *IEEE J. Sel. Top. Appl. Earth Obs. Remote Sens.* **2015**, *8*, 3260–3270.
19. Zhao, C.H.; Deng, W.W.; Yao, X.F. Hyperspectral Real-Time Anomaly Target Detection Based on Progressive Line Processing. *Acta Opt. Sin.* **2017**, *37*, 012800201–012800212.
20. Kailath, T. *Linear Systems*; Prentice-Hall: Upper Saddle River, NJ, USA, 1980; Volume 26, pp. 1–28.
21. Wang, Y.L.; Chen, S.Y.; Liu, C.H.; Chang, C.N. Background Suppression Issues in Anomaly Detection for Hyperspectral Imagery. In Proceedings of the Satellite Data Compression, Communications, and Processing X (SPIE), Baltimore, ML, USA, 8–9 May 2014; p. 912413.
22. Wang, L.; Chang, C.I.; Lee, L.C.; Wang, Y.; Xue, B.; Song, M.; Yu, C.; Li, S. Band Subset Selection for Anomaly Detection in Hyperspectral Imagery. *IEEE Trans. Geosci. Remote Sens.* **2017**, *55*, 4887–4898.
23. Wang, J.; Wang, L.; Cui, J.; Li, X. Band Selection based on Signal-to-noise Ratio Estimation and Hyperspectral Anomaly Detection. *Remote Sens. Technol. Appl.* **2015**, *30*, 292–297.
24. Otsu, N. A Threshold Selection Method from Gray-Level Histograms. *IEEE Trans. Syst. Man Cybern.* **1979**, *9*, 62–66.

remote sensing

MDPI

Article

A Sliding Window-Based Joint Sparse Representation (SWJSR) Method for Hyperspectral Anomaly Detection

Seyyed Reza Soofbaf [1], Mahmod Reza Sahebi [1,*] and Barat Mojaradi [2]

1 Faculty of Geodesy and Geomatics, K. N. Toosi University of Technology, Tehran 19667-15433, Iran; srsoofbaf@mail.kntu.ac.ir

2 Department of Geomatics Engineering, School of Civil Engineering, Iran University of Science and Technology, Tehran 16846-13114, Iran; mojaradi@iust.ac.ir

* Correspondence: sahebi@kntu.ac.ir; Tel.: +98-218-877-0218

Received: 8 December 2017; Accepted: 3 March 2018; Published: 6 March 2018

Abstract: In this paper, a new sliding window-based joint sparse representation (SWJSR) anomaly detector for hyperspectral data is proposed. The main contribution of this paper is to improve the judgments about the probability of anomaly presence in signals using the integration of information gathered during transition of sliding window for each pixel. In this method, each pixel experiences different spatial positions with respect to the spatial neighbors through the transition of this sliding window. In each position, an optimized local background dictionary is formed using a K-Singular Value Decomposition (K-SVD) algorithm and the recovery error of sparse estimation for each pixel is calculated using a simultaneous orthogonal matching pursuit algorithm (SOMP). Thus, the votes of each signal in terms of the anomaly presence in each spatial neighborhood are calculated and the variance of these recovery errors is considered as the detection criterion. The experimental results of the proposed SWJSR method on both synthetic and real datasets proved its higher performance compared to the Global RX (GRX), Local RX (LRX), Collaborative Representation Detector (CRD), Background Joint Sparse Representation (BJSR), Causal RX Detector (CR-RXD, CK-RXD), and Sliding Local RX(SLRX) detectors with an average efficiency improvement of about 7.5%, 14.25%, 8.2%, 8.25%, 6.45%, 6.5%, and 3.6%, respectively, in comparison to the mentioned algorithms.

Keywords: hyperspectral; anomaly detection; sparse coding; KSVD; sliding window

1. Introduction

Today, hyperspectral imaging has become a powerful tool in the field of remote sensing applications. It provides valuable data acquired from hundreds of narrow spectral bands across the reflective electromagnetic spectrum to distinguish different materials based on their unique spectral responses [1]. Target detection and classification could be considered as the most important information extraction approaches in hyperspectral data interpretations [2–4]. The target detection algorithms could be utilized in supervised and unsupervised categories [4]. In the former case, the spectral signatures of the targets are used in detection algorithms whereas, in the latter, no prior knowledge is available regarding the spectral characteristics of targets, and just the detection of the spectral anomalies would be on the agenda [5]. In fact, anomaly detection algorithms could be considered as an unsupervised classification with two classes (anomaly and background) [6]. Thus, the anomalies are unknown targets that are significantly different from their neighbor samples and their probabilities of occurrence are low. Detection of these differences would be independent of the spectral signature of the targets and thus, their effective parameters, including the environmental and atmospheric conditions [7]. Remote sensing application, such as search and rescue [8], detection of military vehicles and objects [9], detection of rare minerals in geology, recognition of vegetation stress [10], toxic wastes in environmental monitoring, and tumors in medical imaging could be considered as spectral anomalies that can be detected via hyperspectral anomaly detection algorithms.

All of the developed methods in the field of anomaly detection could be classified into two broad categories. Local and global methods include the first category in this area. In the global methods, the judgment criterion of each pixel in terms of anomaly presence is the generation of indicators that use all the signals recorded in the hyperspectral image [11]. In local methods, only the spatial neighbors of each signal are used for this purpose. When considering the compliance or non-compliance of hyperspectral data to the normal distribution assumption in the feature space leads to another categorization of anomaly detection algorithms. Parametric methods, such as considering the covariance/correlation matrix, assume that the background data follow a normal distribution. In contrast, methods that are based on linear un-mixing or sparse representations do not make any assumption on the statistical distribution of the hyperspectral data.

The Reed-Xiaoli (RX) method [12] is known as a traditional benchmark of hyperspectral anomaly detection algorithms. The idea of this traditional algorithm has been used as the basis of development of other similar methods which have been used in both local and global strategies, such as normalized RX, modified RX, causal RX [13,14], weighted RX [15], RX-UTD, and Adaptive Causal Anomaly Detector (ACAD) [16]. The main assumption of these algorithms is that the hyperspectral data follows the multivariate normal distribution. Thus, it is assumed that the anomalous signals would be placed in a larger Mahalanobis distance compared to the centroid of the data. Although it seems reasonable in the homogeneous regions, it is not, however, convenient to represent the background signals when the data do not follow the Gaussian distribution. In this regard, some modified version of RX, such as the Kernel-RX algorithm [17] was proposed to overcome the flaws of the mentioned RX assumption for the background. This method attempts to increase the tendency of the data in the feature space to the Gaussian distribution by mapping the signals into a higher dimensional space using non-linear kernels. When considering the use of the covariance/correlation matrix of the sampled data in RX-based methods, these methods are categorized as parametric algorithms.

Another developed algorithm to detect anomalies in hyperspectral data is the Dual Window-based Eigen Separation Transform (DWEST) algorithm [18]. Based on the linear transformation of EST, this algorithm has been designed to maximize the separation between two classes in the low-dimensional subspaces by using local windows [19]. The Nested Spatial Window-based Target Detector (NSWTD) algorithm is also another anomaly detection algorithm [20]. In this algorithm, similar to DWEST, the nested spatial windows with a pre-defined size are used as inner, middle, and outer windows. The evaluation criterion of the spectral features differences of these windows is also Orthogonal Projection Divergence (OPD). Liu and co-workers extends the concept of DWEST to propose a new approach, called multiple-window anomaly detection (MWAD), using multiple windows to perform anomaly detection adaptively. This method is able to detect anomalies of various sizes using multiple windows so that local spectral variations can be characterized and extracted by different window sizes [21]. Chang and co-workers proposed an anomaly detection method using causal sliding windows, which has the real-time capability. They suggested three types of causal windows, using causal sliding square matrix windows, causal sliding rectangular matrix windows, and causal sliding array windows. In this method a causal sample covariance/correlation matrix can be derived for causal anomaly detection. In the case of using covariance matrix and correlation matrix, they are called CK_RXD and CR-RXD, respectively. They also proposed a recursive update equation to speed up the real-time processing [22]. Moreover, Li and co-workers introduced another method, named the CRD algorithm [23]. The main assumption in this method is the possibility of precise background estimation using the neighboring pixels. Thus, it is not true for the anomaly signals and a high residual occurs. Therefore, in this method the l_2-norm of residuals of the estimated signals have been considered as an anomaly detection map. In other words, this detector locally estimates the backgrounds using a dynamic dual-window structure, and, subsequently, estimating the error vector of the signals located at the center of the window is considered as the criterion of probability of anomaly presence for each signal. The idea of background signals recovery using bases of the background subspace and utilizing these bases to recover the anomalous signals is considered as the most important innovative aspect of this detector. Yuan and co-workers proposed a novel method for fast and accurate hyperspectral anomaly detection, which is called 2DCAD [24]. In this method a high-order two-dimensional (2-D) crossing approach is proposed to find the regions of rapid change in the spectrum, which runs without any a priori assumption. This method has a low-complexity discrimination framework which can be implemented by a series of filtering operators with linear time cost. Also it has the ability to detect the true pixel-level for real-time application. Also, Yuan and co-workers proposed a graph-based method for anomaly detection without any assumptions of background distribution statistics [25]. In this method, after the construction of a vertex- and edge-weighted graph, a pixel selection process is utilized to locate the anomalies. The philosophy behind this method is that the anomalies tend to be picked out more easily than the background pixels in the constructed graph. Because an anomaly pixel generally deviates from the background, and its distinctiveness makes its connections with other background pixels vulnerable. This method has good robustness to noise and adaptability to window sizes, which makes it more applicable in the real situations.

Recently, another method by applying sparse representation theory has been introduced and successfully accepted as a strong tool for anomaly/target detection [26]. The main objective of these techniques are the recovery of the majority of high-dimensional signals via a low-dimensional subspace through a dictionary of normalized signals (atoms). In the process of sparse estimation of each signal, a limited number of atoms of a dictionary are active and a majority of coefficients related to dictionary atoms are zero [27]. In other words, signals are recovered via a linear mixing of atoms in the dictionary through the sparse coefficients.

When considering the sparse representation techniques, targets and anomalies could be detected using two different approaches. In the target detection approach, the creation of a dictionary containing background and target spectra are the main steps of sparse representation. In other words, a proper background modeling would result in the efficient presence estimation of spectral targets [5]. In this regard, Chen and co-workers [28] defined a dictionary including interested targets using their spectral signatures, at the same time another dictionary was including the local background signals. Subsequently, these two dictionaries are used to make a decision on a pixel being a target or a background. This decision can be made through sparse estimation of each pixel using two target and background dictionaries while considering the recovery error differences. Furthermore, Du and co-workers [29] presented a target detection algorithm through integration of statistical methods and sparse representation by the Hybrid Sparsity and Statistic Detector (HSSD) algorithm. The primary assumption in this method is that the pixel of interest follows the Gaussian normal distribution with the same covariance and different variance in two statistical hypotheses of being or not being a target. To achieve the efficient detection, the probable target pixels are removed from the dictionary related to the background through utilizing the SAM algorithm based on the initial target spectral signatures. Then, in an iterative process, the sparse estimation is performed by the Orthogonal Matching Pursuit (OMP) method in two stages: (1) the dictionary including only the background data; and, (2) the integrated dictionary of target and background data. Finally, the recovery error difference of the pixel in these two stages in comparison to a pre-determined threshold will yield the decision as to whether the pixel is a target or a background.

In anomaly detection methods, considering no prior knowledge about the spectral targets, the plan is to build a dictionary of atoms that can exclusively model the background elements [30]. In other words, having a dictionary that is composed of bases denoting the background subspace enables the precise recovery of background signals. Additionally, the presence of anomaly signals, assuming their deviation from the background subspace, will not have a precise estimation by the background dictionary. The main idea of anomaly detection methods based on sparse representation of signals is focused on evaluating recovery errors of signals by a dictionary that describes the background subspace. The effort of removing atoms that describe the anomaly in the background dictionary can be considered as one of the essential actions in this procedure [31]. In this field, Yuan and co-workers [32] presented a new method for anomaly detection in hyperspectral images by introducing a spatial-spectral evaluation index, which is called the Local Sparsity Divergence (LSD), where the estimation of sparse matrix elements is locally performed by the determination of the search window dimensions. Lee et al. [33] also suggested Background Joint Sparse Representation (BJSR) for anomaly detection by estimating the background locally using a limited number of subspaces extracted from the hyperspectral data through sparse coding. Zhao and co-workers [34] presented the Sparsity Score Estimation Anomaly Detector (SSEAD) algorithm for the same reason. In this method, an index is used to detect anomalies based on the frequency of the participating atoms in the dictionary learning process to estimate the background atoms. In this way, through an iterative process, the estimation of the background is optimized. Moreover, through optimization of the weighting of the forming atoms of the background, each pixel in the hyperspectral data is scored and the decision is made for being an anomaly or background. Zhang and co-workers [35] also introduced the LLTSA-SSBJSR method as an extension to the BJSR method. This method first uses the spectral space to identify anomalies, and then spatial analysis is performed on the dimensionally reduced data by the LLTSA method. Also Ma and co-workers proposed a novel anomaly detection method based on sparse dictionary learning with capped norm constraint using the sliding dual window, which is named SDLCN [36]. In this method, a number of patches with same size from the entire image are randomly selected and stacked as training data to construct the background dictionary. After that the capped l1 norm based loss function is used to suppress the effects of anomalies in the training set, which will learn a better dictionary resistant to anomalies. After learning an optimized background dictionary, through computing the sparse representation coefficient matrix, the reconstruction errors are calculated, which can be regarded as the corresponding anomaly probability values.

By focusing on local anomaly detection algorithms, in all of these methods, the assumption of the spatial symmetry of background elements is considered to judge a signal. Thus, each pixel of the hyperspectral image is tested only once in terms of anomaly presence. In such situations, if the anomalous pixels are near the edges of the image, the probability of false detection will be increased and the background signals might be considered as anomalies. Due to the lack of prior knowledge about the spatial distribution of similar signals in a geographic area, providing a voting-based approach in the definition of a diverse neighborhood could be a good solution. Accordingly, in this research, creating diversity in the definition of spatial neighborhoods of spectral signals, as well as voting-based judgment in different situations, of the spatial distribution are proposed as two approaches to confront this challenge. In other words, the most important aspect of this study is to improve the judgments about the probability of anomaly presence in signals by diversifying the definition of spatial neighborhood of their surrounding area. Since, by designing an optimized local dictionary, which is based on a sliding window with a new structure, the votes of each signal in terms of anomaly presence in each spatial neighborhood are calculated with the aim of achieving better judgment.

2. Dictionary Learning and Joint Sparse Coding

In the sparse coding techniques, the b-dimensional signals ($[s]_{b\times 1}$) are mapped to a low-dimensional subspace through a dictionary of atoms [37]. When considering $[D]_{b\times n} = [\vec{d_1}, \vec{d_2}, \ldots, \vec{d_n}]$ as a dictionary of unit length atoms ($[\vec{d_i}]_{b\times 1}$, $i = 1, 2, \ldots, n$ and $\|\vec{d_i}\|_2 = 1$) where $b << n$, the aim of the sparse estimation of a signal is to find the sparse vector $[\alpha]_{n\times 1}$ through solving an under-determined system of equations presented in Equation (1) [38]:

$$[s]_{b\times 1} = [D]_{b\times n} \times [\alpha]_{n\times 1},\ \hat{a} = \text{argmin}\|\alpha\|_0 \text{s.t.}\|r = s - D\alpha\| < \varepsilon \tag{1}$$

where $\|.\|_0$ indicates the L_0-norm, which is equivalent to the number of non-zero elements of $\hat{a}$. Since there is no explicit method to solve this equation systems, greedy algorithms [39] are used as the general approach to estimate $\hat{a}$. OMP [40] and Simultaneous Orthogonal Matching Pursuit (SOMP) [41] techniques are two common approaches for greedy and sparse estimation of signals using dictionaries. In the OMP algorithm, the sparse vector is estimated individually based on a signal, while in the SOMP algorithm, it is estimated simultaneously based on several signals. Both of these algorithms try to find atoms that describe signals iteratively to satisfy the conditions mentioned in Equation (1) [42]. In these two techniques, in each iteration, an atom from the dictionary, which has the minimum spectral angle in the estimation error of a signal/signals is added as a new atom to a set of previously-selected atoms (activate atoms). In this trend, the similarity of the residual vector/vectors related to estimated signals using previous spanned subspace is considered as the criterion for choosing new atoms. In other words, when considering R as the vector/matrix of the estimation error obtained from previously activated atoms (Equation (2)), in each iteration the atom which maximizes $\|[d_i]^T \times [R]\|_2$ $(i = 1, 2, \ldots, n)$ would be added as the new active atom to the set of previously-activated atoms in dictionary (D):

$$[R] = [S] - D \times [A] \tag{2}$$

Here, when the OMP technique is used the S is exclusively a vector including a single signal $S = [s]_{b\times 1}$ and also when SOMP technique is used a matrix including all of the signals $(S = [\vec{s_1}, \vec{s_2}, \ldots, \vec{s_t}])$ that tend to be estimated simultaneously. In the same way, it is obvious that the dimension of A will be $[\alpha]_{n\times 1}$ in the OMP technique and $[\vec{\alpha_1}, \vec{\alpha_2}, \ldots, \vec{\alpha_t}]$ in the SOMP technique, as they share the same zero rows. Notably in the first iteration R is considered S $(S = R)$ to select the first atom.

In sparse coding procedures, by creating a redundant dictionary from probable endmembers in the feature space, sparse recovery of signals is performed. Effective performance of a dictionary depends on the correct orientation of its atoms in the feature space, and also the lack of these bases in input data through the absence of end-members in the imaging process is possible. The direct use of sampled signals or learning of dictionary atoms are two main approaches of dictionary generation. In the first approach, if all sampled signals are chosen, the sparse estimation of each signal is merely converted to a minimum distance classification process and the L_0-norm of the sparse estimation vector $(\hat{a})$ of each signal will be one. Choosing a part of the sampling signals faces probable problems, such as (1) the occurrence of a minimum distance classification phenomena ($\|\hat{a}\|_0 = 1$) for chosen signals, and (2) the probability of the impossibility of the signal subspace spanning using dictionary atoms.

In the anomaly detection applications using sparse coding methods, having a dictionary where their atoms are capable of spanning the formed space by background signals is critical. In other words, because the sparse estimating error of signals by the background dictionary is considered to be a measure of being or not being an anomaly for each signal, correct extraction of the background bases subspace and their presence in the dictionary is necessary. Due to the limitations of using randomly selected signals in the formation of the background dictionary (according to the designed structure of the proposed anomaly detection algorithm) learning of dictionary atoms to match with bases that can correctly recover the space of the background signals is used in this research.

The K-SVD technique [43], as one of the dictionary learning methods, by choosing a percentage of sampled signals, randomly creates the initial dictionary and during the iterative process converges its atoms to the spanning bases of the subspace of all input signals. In each iteration of the K-SVD algorithm after the sparse estimation of all signals with the OMP technique, the effect of loss of each atom in the estimation error vector of the signals is affected by that atom is evaluated. The main idea of this technique is to correct the base of the specified atom toward the dominating base of the estimating error vectors of signals. To this aim, specific vectors corresponding to the maximum singular value that is obtained from singular value decomposition of residual matrices is chosen as the substituted base of the specified atom. This iterative procedure is continued to stabilize the base of all atoms of the dictionary.

3. Methodology

In general, as mentioned before, there are two main strategies of global and local in the field of anomaly detection for hyperspectral images where all of the so-far developed methods can be placed in one of these categories. Our proposed method is among the local strategies that evaluate the probability of anomalies' existence using various spatial neighbor conditions around each pixel of interest (*PoI*). This is performed through a transition of a sliding window with a pre-defined size around the *PoI*. In other words, when considering the location of each *PoI* in the input image, e.g., Figure 1, the *PoI* experiences different spatial positions with respect to the spatial neighbors through the transition of the sliding window. In each position, the *PoI* is investigated for the presence and absence of the anomaly. Finally, the fusion of the results obtained for each *PoI* during its presence in the sliding window will generate the anomaly detection criteria.

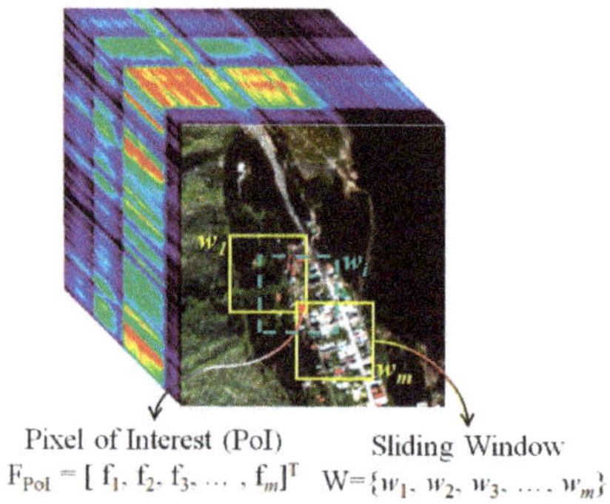

Figure 1. Structure of moving the sliding window around the *PoI*.

According to the Figure 1, m is the number of elements of sliding window. Thus, each *PoI* has the possibility of placement in m different positions with respect to the sliding window where, in each location, the consequent f_i, $I = 1, 2, \ldots, m$ index will be calculated, as described in Equation (3). In other words, through a complete transition of the sliding window on each *PoI*, m different positions of the *PoI* would occur in the sliding window (w_i, $I = 1, 2, 3, \ldots, m$). Finally, for each *PoI*, the feature vector composed of m members (F_{PoI}) will be calculated where the variance of its elements is used as the index of the anomaly detector. The main reason of proposing such an idea is the spatial asymmetry assumption of background elements around the probable spectral anomalies. Prior to this idea, all local anomaly detection methods have assumed the spatial symmetry of background elements around pixels of interest in the detection process. Furthermore, each pixel of the image is evaluated in terms of the presence of the anomaly only one time in its symmetric neighboring region. Accordingly, the proposed solution includes two main contributions: (1) the ability of judgment for each *PoI* with the variety of the spatial neighborhood; and, (2) the capability of the synergy of the obtained knowledge through the transition of the sliding window for each *PoI*.

The presentation strategy of the proposed algorithm is focused on the estimation of the f_i indices for each *PoI* in the w_i's location of the sliding window. It is obvious that generalization of this process to other w_i positions of each *PoI* will yield the generation of its m-dimensional F_{POI} vector.

Simultaneous sparse representation of all pixels occurring in each w_i using a unique set of local background atoms from a learned dictionary is the main idea of the proposed method in the field of anomaly index generation. Through this process, it is expected that simultaneous estimation of all the available signals in w_i leads to imposing the selection of the descriptive background atoms. Consequently, the increase in the l_2-norm of the estimated residuals for each signal could be interpreted as the level of anomaly. In this procedure, using the traditional RXD anomaly, the randomly selected signals used in dictionary learning are initially refined by exclusion of highly-probable anomaly signals. Figure 2 illustrates the process of f_i estimation in a w_i occurrence.

Here, a spatial subset of a hyperspectral image that coincided with the position of w_i is depicted where the *PoI* inside the w_i is shown as a red pixel (parts a, b, and c). Obviously the transition of w_i will change the location of the *PoI* in it.

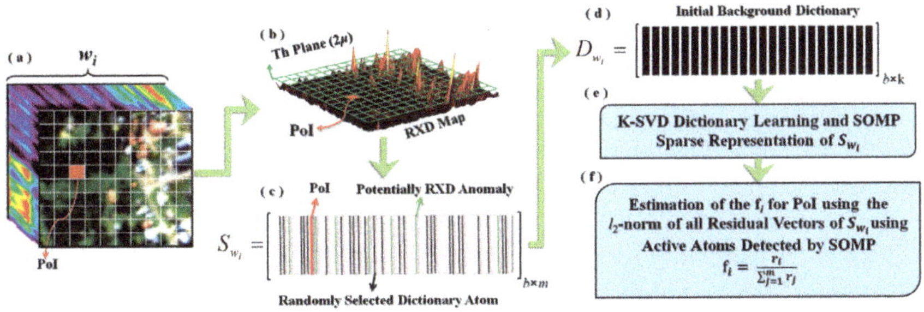

Figure 2. The process of the proposed anomaly detection algorithm.

According to Figure 2, and as the first step, using the traditional RXD, the potentially anomalous signals were removed by applying a proper threshold (*Th*-Plane). This threshold is set to twice of the average (2μ) of RXD map. In continue, the signals having higher values than *Th*-Plane were omitted from randomly selection process of dictionary learning (parts c and d). The aim of this process is to perform initial refinement of dictionary atoms. Thus, the initial dictionary atoms (D_{wi}) would be more descriptive to model the background signals.

In Figure 2, the matrix $[S_{wi}]_{b\times m}$ (b is the number of hyperspectral image bands) contains the constructive signals of wi where the green columns are indications of possible anomaly signals that are detected using RXD. The black columns are randomly-chosen candidates as initial atoms in the process of dictionary learning (D_{wi}). The number of signals used in dictionary learning (K) is equivalent to the 'q' percentage of the S_{wi} signals in the case of the omitted probably-anomalous signals. As a result, the matrix D_{wi} would be constructed from matrix S_{wi} as the initial dictionary in the dictionary learning process (part d).

To optimize the initial atoms of D_{wi}, the K-SVD method [43] has been utilized where the OMP algorithm [41] is used for the sparse coding process of each signal (part e). In this method, the direction of each dictionary atom is updated in an iterative process. This method is composed of two main steps. In the first step, sparse coding of all input signals (S_{wi}) is performed, and, in the second step, for each selected atom, a new direction is calculated using the signals coded by the selected atom. This new direction is estimated through the Singular Value Decomposition (SVD) method of a matrix composed of columnar vectors, indicating the residual of the affected signals. In the other words, the estimated residual vector of the signals is calculated by only the signals affected by that selected atom. In this process, while the selected atom is absent, through elimination of the effect of the selected atom, the residual vector of the estimated signals will be calculated. Finally, the b-dimensional eigenvector corresponds to the largest singular value will be chosen as the substitute direction of each selected atom. As can be seen in [43], dictionary learning is an iterative procedure, which includes two steps (1) sparse coding of the S_{wi} signals, and (2) optimization of the direction of D_{wi} atoms.

After the D_{wi} was learned via K-SVD, finding a common subspace for all S_{wi} signals is performed through the SOMP algorithm (part e). In this algorithm, the sparse coding of a set of signals is simultaneously carried out. This means that all of the S_{wi} signals will be simultaneously estimated through the same subspace spanned by the atoms in the learned dictionary with the minimum dimension. Furthermore, the minimization of the l_2-norm of estimated residuals should be satisfied. The aim of this process is choosing the background descriptive atoms to reconstruct the S_{wi} signals.

As discussed before, it is expected that, during this process, the background signals are estimated to be more precise than the rare and anomalous signals. When considering this expectation, after the estimation of the S_{wi} residual vectors using SOMP, their l_2-norm for all w_i signals would be calculated as r_j, $j = 1, 2, \ldots, m$. To continue, the index f_i for the *PoI* will be estimated by normalizing the r_i corresponding the location of *PoI* in w_i through the Equation (3):

$$f_i = \frac{r_i}{\sum\limits_{j=1}^{m} r_j}, \quad i = 1, 2, \ldots, m \tag{3}$$

Through the transition of w_i ($i = 1, 2, \ldots, m$) on *PoI*, a m-dimensional vector F_{PoI} will be generated. Finally, its variance would be selected as the criterion of anomaly detector after the 3σ statistical test (Equation (4). In this equation, m^* is the number of f_i for each *PoI* after the blunder separation procedure by the 3σ test. The well-known 3σ test is a standard statistical test to remove blunders from the random variable sets (f_i). In this procedure, by assuming the normal distribution of the random variables, using mean (μ) and the standard deviation (σ) of the f_i elements, those samples that are located within the range of $\mu - 3\sigma < f_i < \mu + 3\sigma$ are known as inliers and the other samples outside this interval are considered to be outliers. Finally, the outlier samples are removed from the data set and the anomaly index (Equation (4)) is calculated for the remaining samples [44]:

$$AD_{PoI} = \frac{1}{m^*}\sum_{i=1}^{m^*}\left(f_i - \sum_{i=1}^{m^*}\frac{f_i}{m^*}\right)^2 \tag{4}$$

The following pseudo-code represents the general process of the proposed algorithm (Algorithm 1).

Algorithm 1. SWJSR Anomaly Detector Algorithm.

Input: HyperCube
Size of w_i (m)
Output: Anomaly Map

```
FOR All HyperCube Signals
    PoI = ith Signal of HyperCube
  Find all w_i around the PoI (W_PoI = {w_1, w_2, ... , w_m})
    FOR all W_PoI members
      S_wi = vector matrix of all signals in w_i
      Remove signals with the high anomaly potential from S_wi using RX algorithm
      D_wi = randomly selected q% of remained Signals in S_wi
      TD_wi = Trained D_wi Using K-SVD Algorithm
      seS_wi = simultaneously estimated S_wi using TD_wi via SOMP algorithm
      rS_wi = [S_wi-seS_wi]_(b×m)
      [r]_(m,1) <- l_2-norm of column vectors of rS_wi
      Estimation of f_i for PoI using Equation (3)
    END
  F_PoI = [f_1, f_2, ... , f_m]
  {F_PoI* = [f_1, f_2, ... , f_m*], m* ≤ m} <- F_PoI after blunder removal using the 3σ statistical test
  Anomaly Map(PoI) = Variance of F_PoI*
END
```

4. Datasets and Pre-Processing

Three real and two synthetic datasets were used in this research. The first real dataset contains an urban and forestry region of Cook city in Minnesota, USA acquired by a Hymap hyperspectral sensor with 126 spectral bands ranging from 370–512 μm in 2006. This data is freely available to the public through the Rochester Institute of Technology (RIT) and includes several targets with known spatial and spectral characteristics. This data is considered as a reference for the evaluation of target and anomaly detection methods. Figure 3 shows this reference data and the location of spectral targets. The details of the spectral targets and their behavior can be studied in [45].

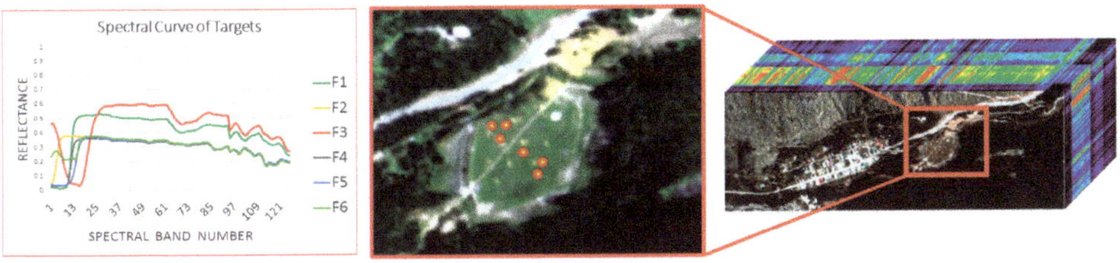

(a) (b) (c)

Figure 3. Rochester Institute of Technology (RIT) real dataset: (**a**) the spectral curve of targets; (**b**) the location of targets in selected subset; and, (**c**) original data.

A subset of 80 × 100 pixels from the first real data containing six spectral targets was selected for analysis in this study. The second real data is from an airport zone of San Diego that has been collected by the AVIRIS sensor with 224 spectral bands ranging from 350–2510 nm. This has been converted to a 100 × 100 × 189 hypercube after removal of the water absorption and noisy bands. In this data, three spectral targets (airplanes) with extents of more than a couple of pixels exist and were used to apply the anomaly detection algorithms. Figure 4 displays the original data, the selected subset, and the spectral curve of the targets.

(a) (b) (c)

Figure 4. San Diego real dataset: (**a**) the spectral curve of targets; (**b**) the location of targets in selected subset; and, (**c**) the original data.

The third real data has been acquired from a region in Viareggio city in Italy collected by the SIM.GA airborne sensor [46]. Although this original dataset has 512 spectral bands, ranging from 388–994 nm, after removal of the low Signal to Noise Ratio (SNR) bands, by applying a spectral resampling process with a 4 nm interval, it has been converted to a 100 × 100 × 123 hypercube. In the area of interest on this image, five spectral targets with extents of more than a few pixels were chosen to apply anomaly detection algorithms. Figure 5 shows the original data, the area of interest, and the spectral curve of targets.

On the other hand, in the majority of previous works, the efficiency of the developed methods for target and anomaly detection has also been evaluated using synthetic data. In this research, two synthetic datasets were also created when considering two different strategies. As the first strategy, some sub-pixel targets were implanted in a region near to the location of the original targets from the Rochester Institute of Technology (RIT) data. Figure 6 shows the implantation targets and their spectral curves. Thus, this way, the number of spectral targets will be increased and a higher number of probable spectral anomalies should be identified. According to Figure 6, seven spectral targets with 50–80% of similarity to the original spectrum were linearly added to the hyperspectral image and a total of 13 potential anomaly pixels were constructed. Then, with the aim of simulating PSF effects, a Gaussian weighted averaging process using a 3 × 3 window around the location of the implanted target was applied.

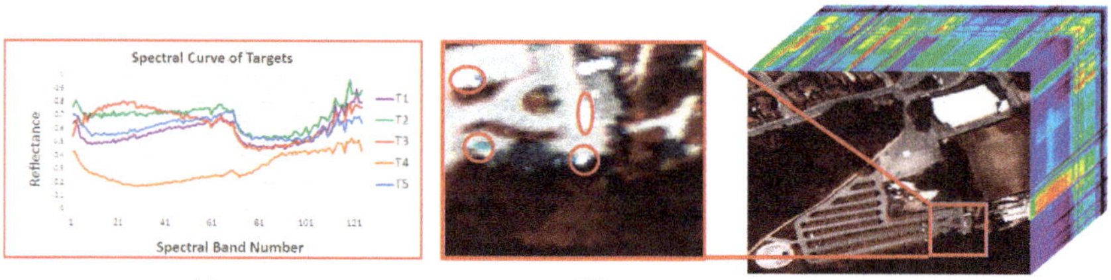

(a) (b) (c)

Figure 5. Viareggio real dataset: (**a**) the spectral curve of targets; (**b**) The location of targets in selected subset; and, (**c**) the original data.

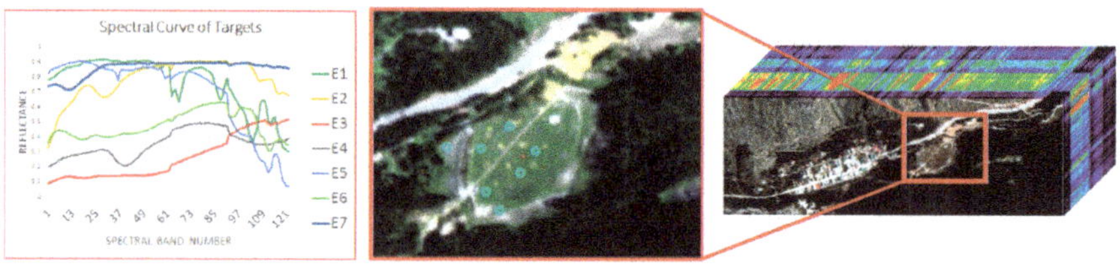

(a) (b) (c)

Figure 6. Implanted RIT dataset: (**a**) the spectral curve of implanted targets; (**b**) the location of implanted targets in selected subset; and, (**c**) the original data.

As the second strategy of synthetic data generation, spectral destruction of original signals in the real RIT dataset was performed. In this strategy, a variation between ±5 to ±20% with respect to the original signals was applied to a randomly selected number of spectral bands (ranging from 5–10% of the total image bands) for six candidate pixels and a total of 12 potential anomaly pixels were constructed. The location of candidate pixels in this strategy were also locally chosen similar to the first strategy in the relatively homogeneous regions. The position of the destructed signals, a sample of the destructed spectral curve, and its related original data are displayed in Figure 7.

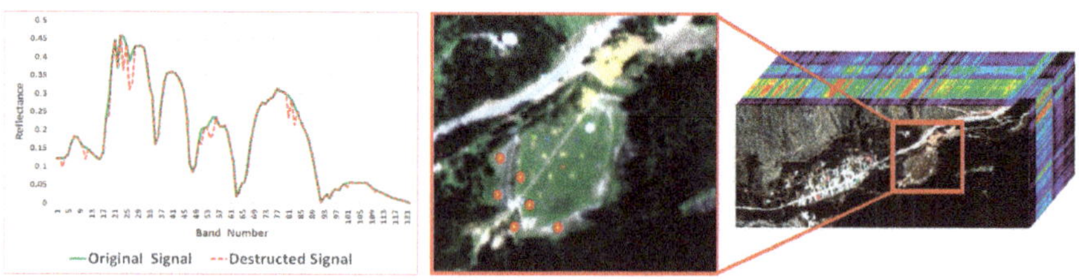

(a) (b) (c)

Figure 7. Destructed RIT dataset: (**a**) the spectral curve of destructed targets; (**b**) the location of destructed targets in the selected subset; and, (**c**) the original data.

5. Results and Discussion

As mentioned before, to evaluate the results and efficiency of the proposed algorithm, five types of different data consisting of three real and two synthetic datasets were used.

In this study, the functionality of the proposed method was assessed by performing the three-dimensional (3D)-ROC analysis [10,14] (Figure 8), evaluating the area under curves (Figure 9 and Table 1), background suppression criteria (Figure 10 and Table 1) and the generation of a target-background separation diagram [5] (Figure 11).

The traditional ROC curve is obtained by plotting of the false alarm rate (P_{FA} (versus the correct probability of detection (P_D (for different thresholds through Equation (5):

$$P_D = \frac{N_{Signal\ detected}}{N_t},\ P_{FA} = \frac{N_{False\ Alarm}}{N} \tag{5}$$

Since the output of anomaly detection algorithms is an image with two anomaly and background classes, by calculating the ratio of the number of correctly-detected anomaly pixels ($N_{Signal\ detected}$) to the total anomaly pixels (N_t)for each threshold, the probability of correct detection is calculated. Additionally, with the calculation of the ratio of number of background pixels wrongly placed in the anomaly class ($N_{False\ Alarm}$) to the all pixels of the image (N), the probability of wrong detection (known as the false alarm rate) for each threshold will be obtained.

Recently, the 3-D ROC analysis with some advantages respect to 2-D one was developed to evaluate anomaly detection algorithms [44]. In this case, varying the value of threshold (Th) enables the users to observe progressive changes in P_D and P_{FA} independently. A 3-D ROC curve can be generated when considering P_D, P_{FA}, and threshold (Th) as three components of a 3D point in the Cartesian coordinate system. In other words, it is a three-dimensional curve of (P_D, P_{FA}, Th), in which three different 2-D ROC curves could be also generated from each aspect. The 2-D ROC that was obtained from (P_D, P_{FA}) is the traditional one and the 2-D ROC obtained from (P_{FA}, Th) or (P_D, Th) are the new ones.

The 2-D ROC of (P_D, Th) could be represented as the progressive detection power versus the changes of threshold and the 2-D ROC of (P_{FA}, Th) provides important information of progressive background suppression as the threshold varies, especially in the case of visual interpretation with no availability of ground truth data.

Having obtained the detection maps of each method in different situations, the plot of this 3D curve for 5000 numbers of different thresholds with the minimum and maximum limit of the map of detection was performed and the area under curves were considered as a scale of the evaluation of the efficiency.

The separability diagram is also one of the indices of efficiency evaluation of two-class classification algorithms that shows the statistical separation of anomaly and background data. This diagram is generated with the help of the ground truth map and shows the range of the recorded values in the anomaly and background locations in the detection map. The level of separation or a presence of overlap among the domains of anomaly and background values indicate the level of success of the anomaly detection algorithm. In plotting this diagram, the following steps are considered: (1) generating anomaly detection maps for all of the compared methods; (2) normalization of detection maps considering the minimum and maximum of all anomaly detection maps simultaneously; (3) identification of anomaly and background signals through a ground truth reference map; and, (4) estimation of the minimum and maximum anomaly and background values for each detection map in two ways:

(A) without removing any of the signals that lead to the drawing of the bars in the graphs; and,
(B) removing 10% of the minimum and maximum values of background and anomaly signals and mapping down the dropped domain into colored boxes.

To compare the results obtained from the proposed algorithm, seven other anomaly detection algorithms were also implemented. The traditional Global and Local RX algorithms [12], Causal R-RX, and K-RX [22], as well as the recently-developed CRD [23] and BJSR [33] algorithms in the field of anomaly detection were chosen for this evaluations. Except the Global RX which has not any setting parameters, other six algorithms have several setting parameters. Generally, default setting parameters proposed by the developers were used in our comparisons. Window size is the only setting parameter in Local RX method. Generally, the best result obtained from windows of 11 × 11, 13 × 13 and 15 × 15 pixels were used in the comparisons. In Causal R-RX and K-RX, the window width of sliding array is the main setting parameter which set to CW = 900, according the best result obtained in [22]. The CRD setting are inner and outer windows as well as the regularization parameters. Here, a 7 × 7 inner and a 15 × 15 outer window size as well as 10^{-6} were set as the regularization parameter [23]. In continue, the BJSR setting parameters are background and guard window size, search window and level of sparsity. Window sizes (background and guard) were selected based on the optimum setting reported in [33]. So, a 17 × 17 background window, 5 × 5 guard window, and 19 × 19 search window were the spatial setting and the sparsity level was set to 3 in SOMP method [33]. However, among the detection algorithms used in this paper, the Local RX algorithm (LRX) was easily adapted to the proposed sliding window. Thus, a new version of Local RX called the Sliding-window Local RX anomaly detector (SLRX) is also used in the evaluations. In this version, the generation of F_{POI} for each *PoI* is based on the Mahalanobis distance calculated by the Local RX of samples in the w_i windows. Again, and similar to the method of SWJSR, the variance of F_{PoI} has been used as the measure of this detector for each *PoI*.

Table 1 presents the results of AUC index (P_D, P_{FA}, and P_{FA}, *Th*) for the abovementioned algorithms and the best results by the proposed algorithm (SWJSR) implemented on the five types of real and synthetic data (implanted and destructed data).

Table 1. Average improvement of efficiency (AUC) of the GRX, local RX algorithm (LRX), CRD, BJSR, CR-RXD, CK-RXD, sliding-window Local RX anomaly detector (SLRX), and SWJSR for all the datasets. (The bold one is higher in case of AUC(P_D, P_{FA}) and it is lower in case of AUC(P_{FA}, *Th*)).

Dataset \ Algorithm		GRX	LRX	CRD	BJSR	CR-RXD	CK-RXD	SLRX	SWJSR
Real RIT	AUC(P_D, P_{FA})	0.8619	0.7014	0.7935	0.8179	0.8372	0.8334	0.9563	**0.9668**
	AUC(P_{FA}, *Th*)	0.0357	0.1524	0.0487	**0.0082**	0.0355	0.0363	0.1988	0.0137
Real San Diego	AUC(P_D, P_{FA})	0.9471	0.8984	0.7078	0.9050	0.8822	0.8832	0.9165	**0.9638**
	AUC(P_{FA}, *Th*)	0.0417	0.0585	0.0544	0.0133	0.0038	0.0037	0.0039	**0.0017**
Real Viareggio	AUC(P_D, P_{FA})	0.8774	0.8433	0.8521	0.9128	0.8732	0.8773	0.8849	**0.9242**
	AUC(P_{FA}, *Th*)	0.1189	0.0590	0.0755	0.0227	0.0482	0.0440	0.0159	**0.0022**
Implanted RIT	AUC(P_D, P_{FA})	0.7425	0.7404	0.8433	0.7834	0.8792	0.8783	0.8598	**0.8922**
	AUC(P_{FA}, *Th*)	0.0323	0.0887	0.0486	0.0179	0.0354	0.0362	0.0782	**0.0084**
Destructed RIT	AUC(P_D, P_{FA})	0.9264	0.8326	0.8961	0.8968	0.9340	0.9320	0.9294	**0.9818**
	AUC(P_{FA}, *Th*)	0.0075	0.0426	0.0431	0.0180	**0.0010**	0.0011	0.0021	0.0015

When considering the traditional AUC (P_D, P_{FA}) that is provided in Table 1, in the similar conditions the higher efficiency of the proposed algorithm is observed. Thus, an average 2% improvement when compared to the best results obtained from the other methods is noticeable. On the other hand, the AUC (P_{FA}, *Th*) values of the proposed method are rather lower than the other algorithms. It should be noted that the AUC of P_{FA} vs. *Th* represents the level of background suppression and their lower values indicate the better performance of the algorithms [47]. In order to compare the anomaly detection algorithms, the 3D-ROC curves, 2-D ROCs of (P_D, P_{FA}), 2-D ROCs of (P_{FA}, *Th*), and the target-background separation diagrams that are related to each dataset (Table 1) are shown in the following figures (Figures 8–11).

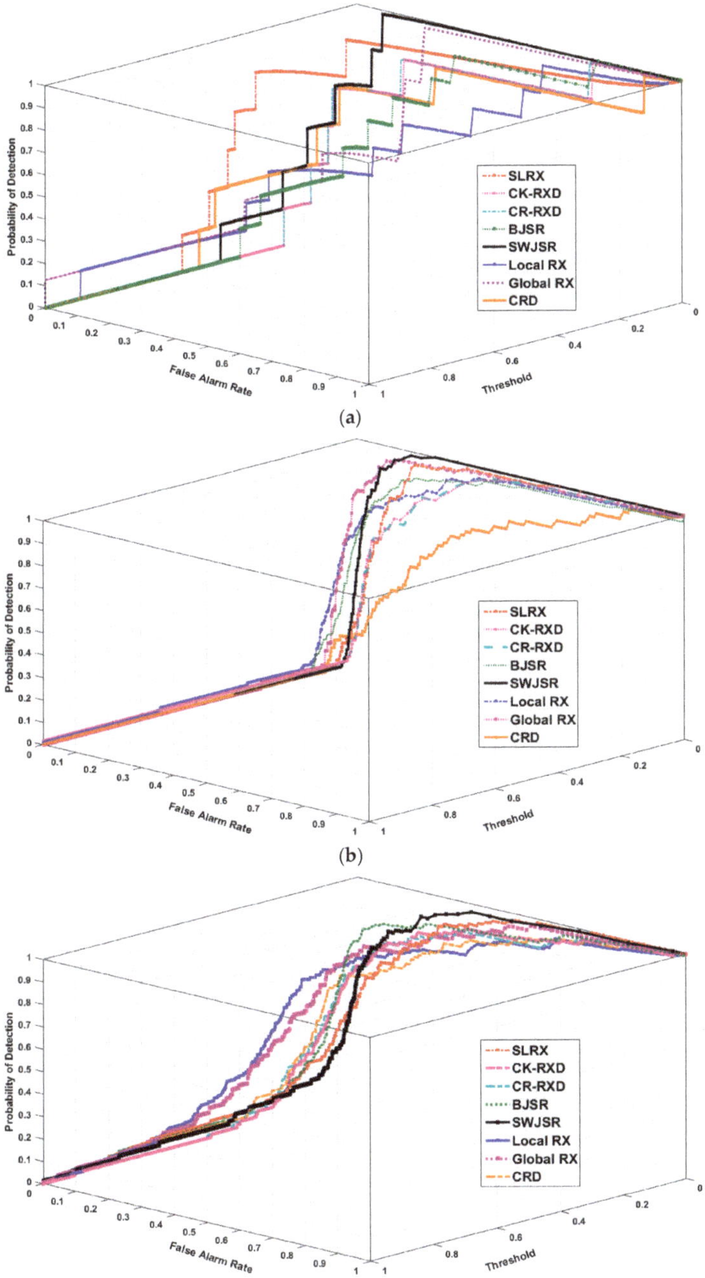

Figure 8. *Cont.*

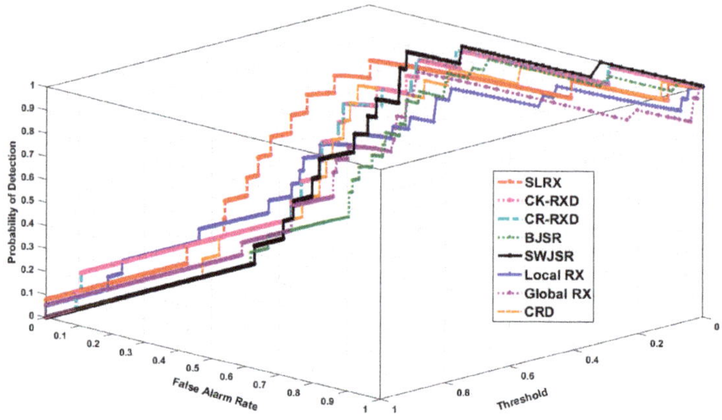

(**d**)

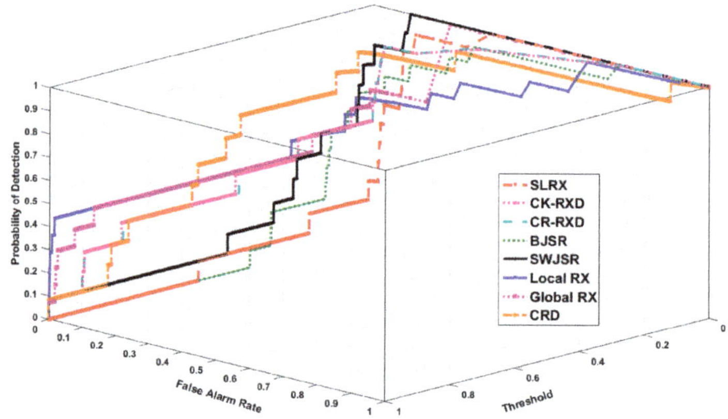

(**e**)

Figure 8. 3D ROC curves of the anomaly detection algorithms: (**a**) Real RIT dataset; (**b**) Real San Diego dataset; (**c**) Real Viareggio dataset; (**d**) Implanted RIT dataset; and, (**e**) Destructed RIT dataset.

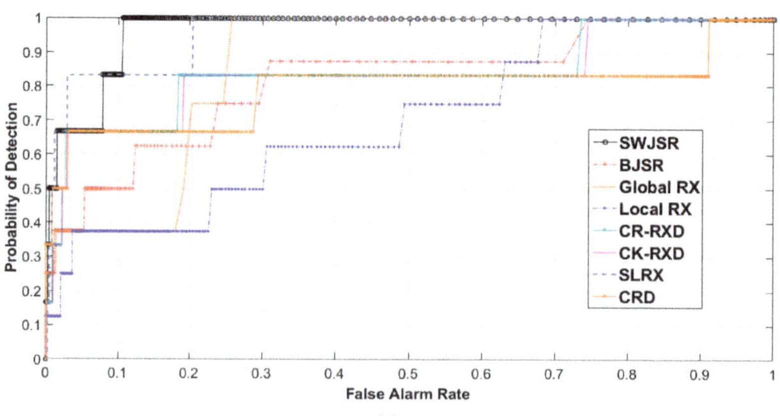

(**a**)

Figure 9. *Cont.*

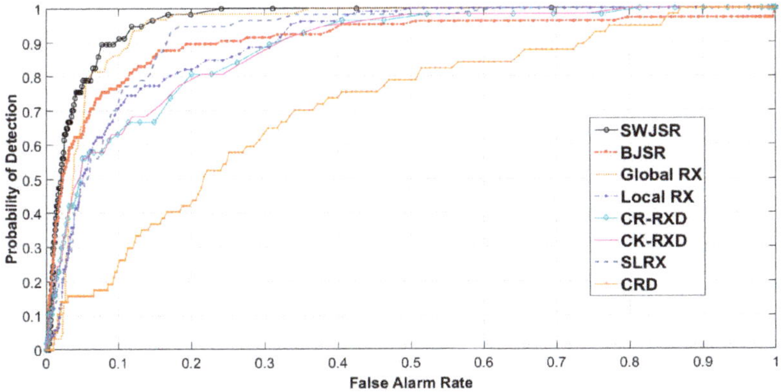

(**b**)

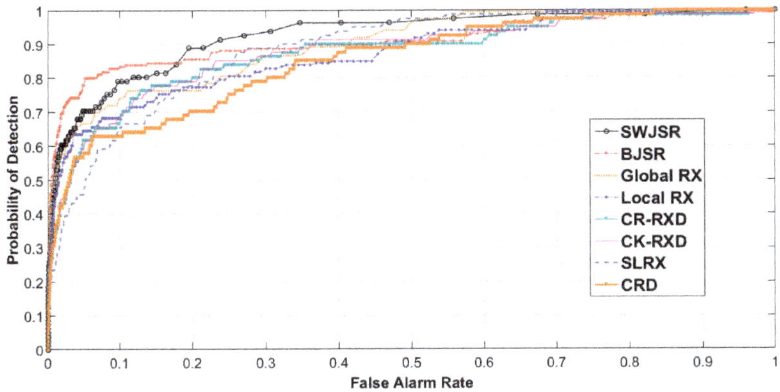

(**c**)

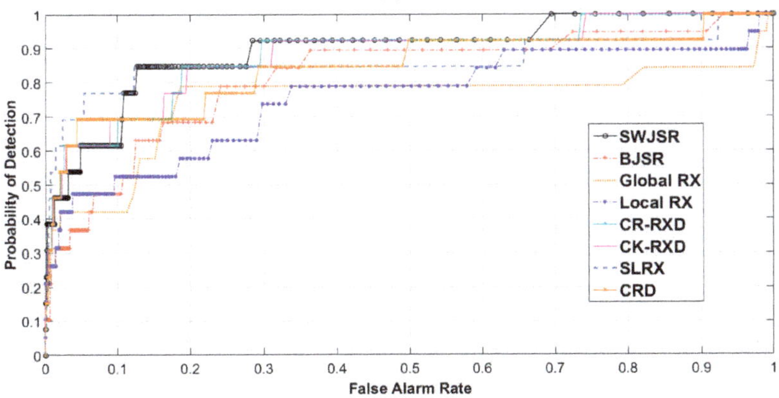

(**d**)

Figure 9. *Cont.*

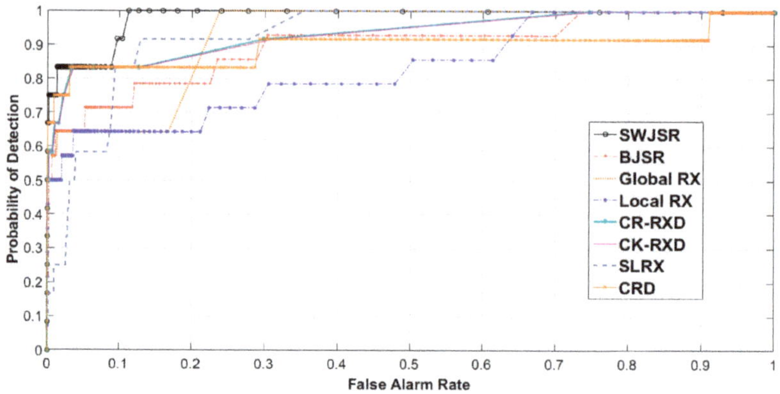

(**e**)

Figure 9. 2-D ROC (P_D, P_F) of the anomaly detection algorithms: (**a**) Real RIT dataset; (**b**) Real San Diego dataset; (**c**) Real Viareggio dataset; (**d**) Implanted RIT dataset; and, (**e**) Destructed RIT dataset.

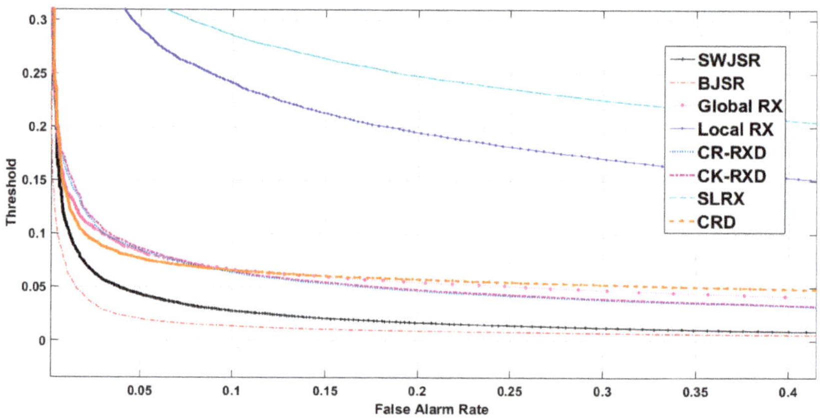

(**a**)

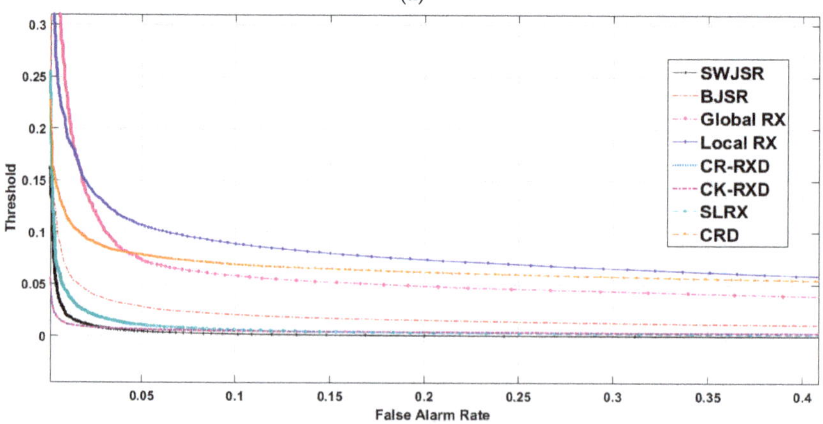

(**b**)

Figure 10. *Cont.*

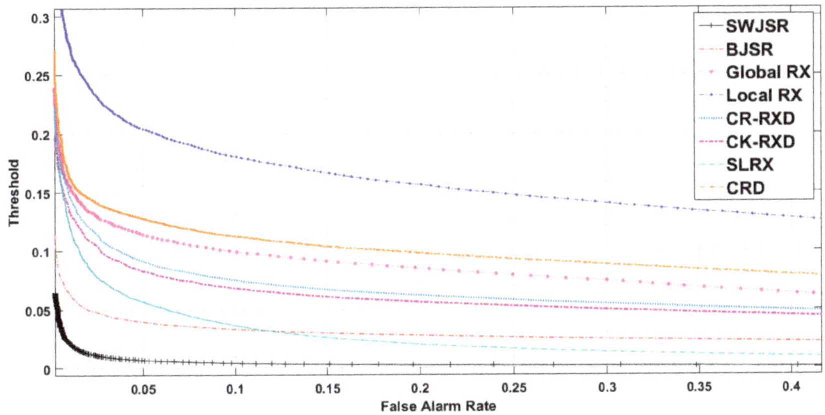

(c)

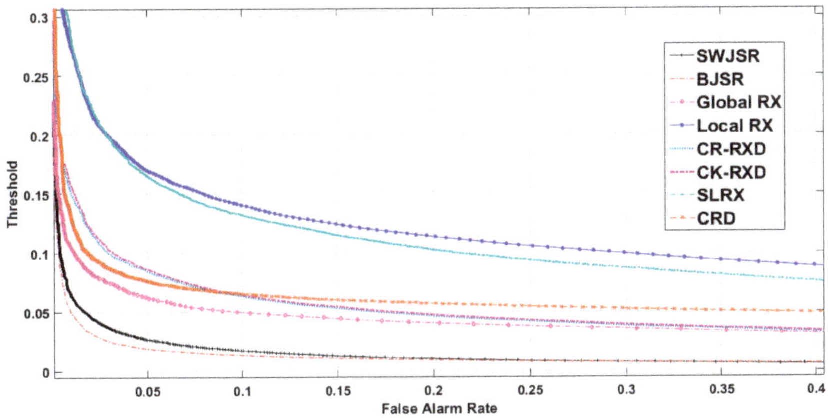

(d)

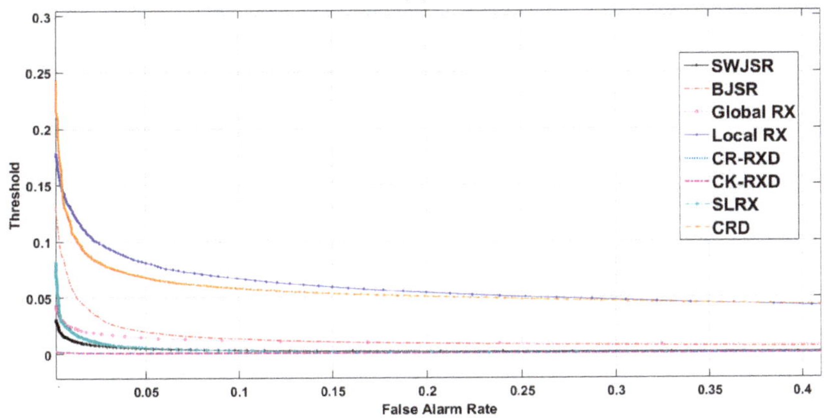

(e)

Figure 10. 2-D ROC (P_F,Th) of the anomaly detection algorithms: (**a**) real RIT dataset; (**b**) real San Diego dataset; (**c**) real Viareggio dataset; (**d**) implanted RIT dataset; and, (**e**) destructed RIT dataset.

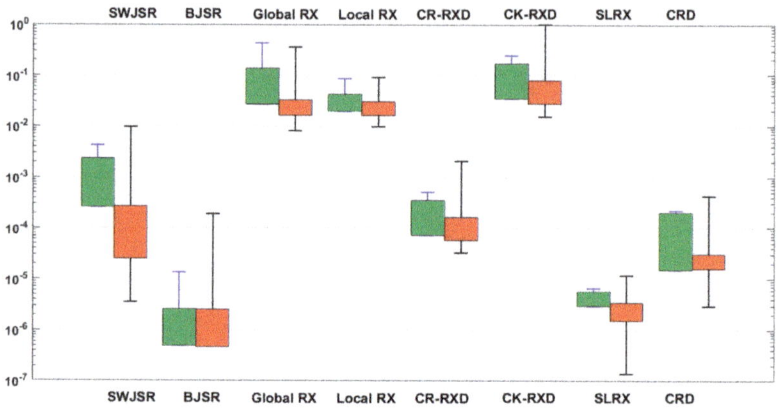

(a)

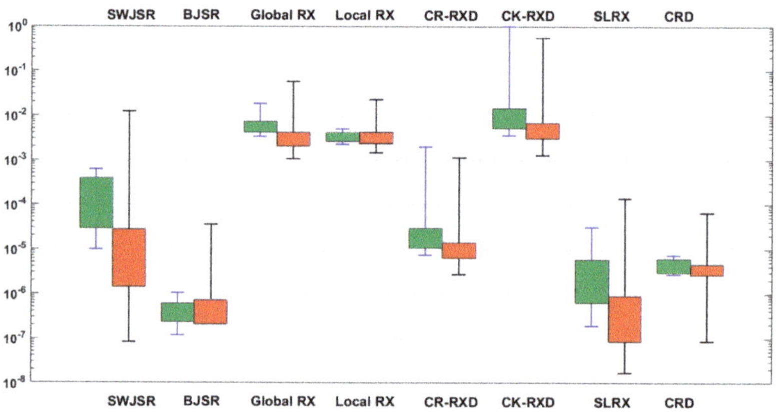

(b)

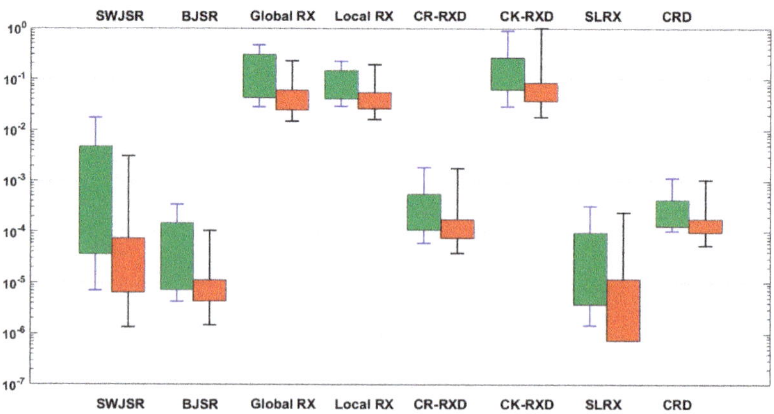

(c)

Figure 11. *Cont.*

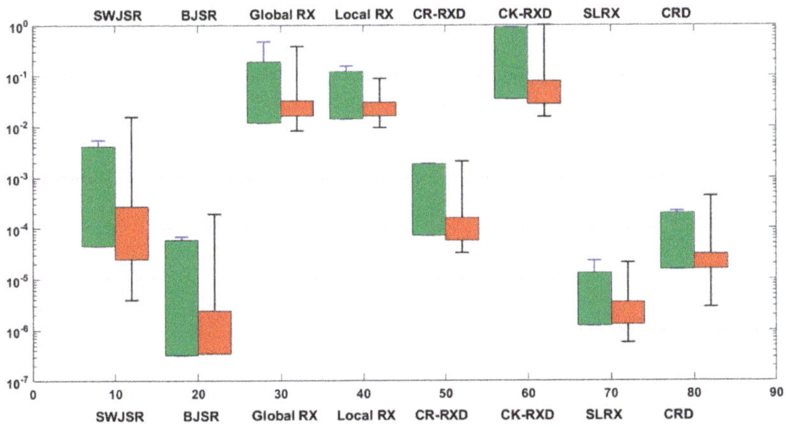

(d)

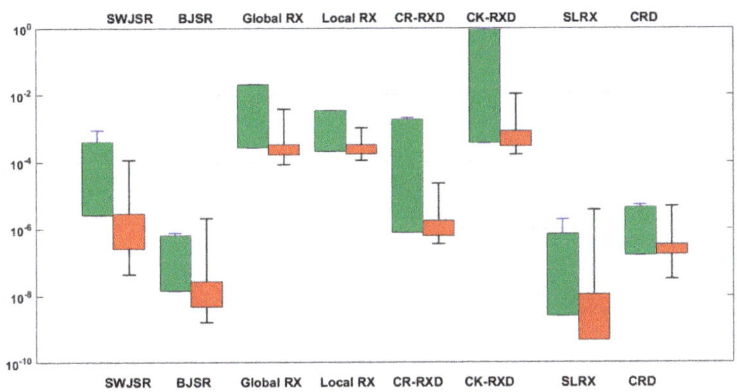

(e)

Figure 11. Target-background separation diagram of the anomaly detection algorithms (the green box shows the target and the red box shows background statistics): (**a**) real RIT dataset; (**b**) real San Diego dataset; (**c**) real Viareggio dataset; (**d**) implanted RIT dataset; and, (**e**) destructed RIT dataset.

Again, higher efficiency of the SWJSR algorithm can be seen from the formation mechanism of 3-D ROC curves and separability diagrams. These diagrams, except for the implanted synthetic data, also reveal the desirable separation between the anomaly and background elements in the proposed method, which is a verification of better functionality when compared to the other methods.

In the case of synthetic data, all of the compared methods yield similar results. In the case of ROC (P_D, P_F) curves, mainly for all of the applied examinations, the relevant curve of the SWJSR is closer to the upper-left of the diagram and this factor has yielded the increase of the AUC (P_D, P_F) value.

Figure 12 shows the obtained detection maps by the GRX, LRX, CRD, BJSR, CR-RXD, CK-RXD, SLRX, and the proposed algorithms (SWJSR) for the reference ground truth map and for all of the used data in this research.

According to the best obtained results from the suggested SWJSR algorithm, the sensitivity analysis of this algorithm with respect to its tuning parameters was also implemented. These parameters, including: (1) the dimensions of the sliding window; and, (2) the level of sparsity, were used in simultaneous reconstruction of the background signals by SOMP algorithm. As the first investigation, the results of changing the sliding window dimensions in the AUC index for all of the used data are presented in Table 2.

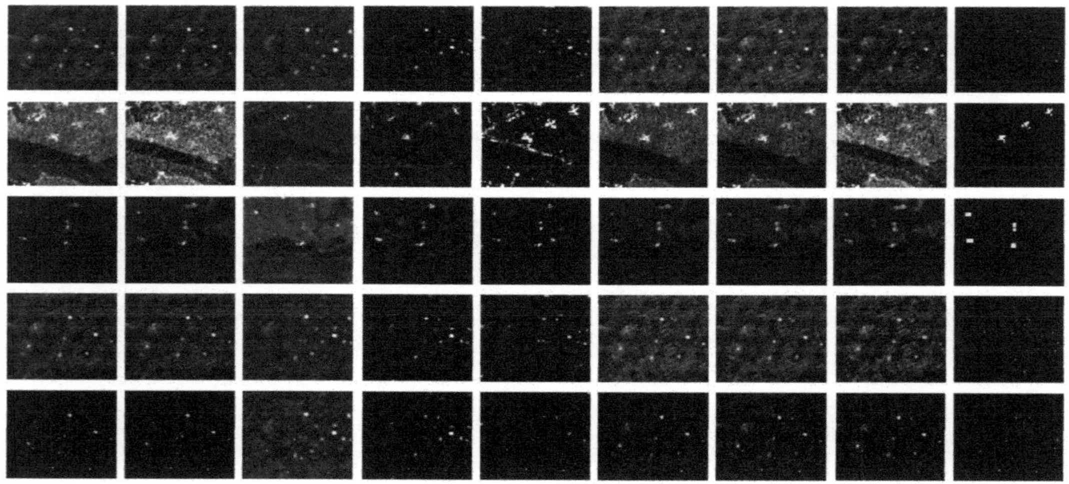

Figure 12. Detection maps of compared algorithms for all datasets: (**a**) GRX detection map; (**b**) LRX detection map; (**c**) CRD detection map; (**d**) Background Joint Sparse Representation (BJSR) detection map; (**e**) SWJSR detection map; (**f**) CR-RXD detection map; (**g**) CK-RXD detection map. (**h**) SLRX detection map; and, (**i**) Reference anomaly map.

Table 2. Effect of size of sliding window in the Sliding Window-Based Joint Sparse Representation (SWJSR) detector for all datasets. (The bold one is the higher).

Dataset \ Size of Window		SW = 8	SW = 9	SW = 10	SW = 11	SW = 12	SW = 13	SW = 14	SW = 15	SW = 17	SW = 20	Average	Standard Deviation
Real RIT	AUC	0.9216	0.8937	**0.9668**	0.9603	0.9459	0.9273	0.9182	0.8839	0.878	0.8435	0.9139	0.0391
Real San Diego	AUC	0.8743	0.8912	0.9125	0.9420	0.9470	0.9570	0.9548	0.9638	0.9614	**0.9690**	0.937	0.0330
Real Viareggio	AUC	0.8736	0.8749	0.8775	0.8780	0.8832	0.8739	0.9099	0.9127	**0.9242**	0.911	0.8919	0.0199
Implanted RIT	AUC	0.8849	0.8740	**0.8922**	0.8829	0.8641	0.8625	0.8390	0.8376	0.8189	0.7996	0.8556	0.0307
Destructed RIT	AUC	0.9639	0.9475	**0.9818**	0.9784	0.9767	0.9629	0.9604	0.9501	0.9369	0.9140	0.9573	0.0198

As shown in Table 2, the obtained results depend on the correct definition of the dimensions of the sliding window. In this regard, according to part b of Figure 2, using the traditional RXD, to remove potentially anomalous signals by applying a proper threshold (*Th*-Plane), the covariance matrix that was estimated from a small number of data samples, could involve rank-deficient (non-invertible) matrices. To overcome numerical instabilities in case of smaller sliding windows, the pseudo inverse based on the Moore-Penrose method was used. To this aim when the number of samples (sliding window elements) was less than the number of spectral bands, the "pinv" function was used as an alternative of common inversion function (inv) in MATLAB. In the case of spectral anomalies with large spatial extension, it is necessary to use large sliding windows to achieve better results. For example, since the multi-pixel anomaly regions for the San Diego airport zone and Viareggio city in Italy are more extended than one-pixel anomalies in the RIT data, the optimum sliding window is also larger. The same rule also applies to the San Diego data compared with the Viareggio data, and the larger sliding window dimensions are convenient. Thus, keeping this rule in mind and while considering the spatial resolution of the sensor, obtaining primary knowledge about the extension of probable anomalies could be effective in tuning the sliding window size, reaching reliable results faster. This knowledge is less important when dealing with anomalies in the range of one pixel or less.

One of the most important tuning parameters of the suggested algorithm is a determination of the level of simultaneous sparse estimation of the background elements, which is called level of sparsity. In other words, the maximum number of atoms used from the learned dictionary for simultaneous estimation of all signals located in the sliding window is another tuning parameter. It is obvious that the value of this parameter depends on the variety of the occurrence of endmembers in the window. Since this value is indicative of the maximum use of the dictionary atoms, it is obvious that a lower, or the same, number of dictionary atoms that are proportional to this tuning parameter are selected in the recovery of all sliding window positions. Assigning a low number for this parameter yields an incomplete modeling of the background, and, when considering a higher number than necessary, results in the possibility of cooperation of unrelated atoms in decreasing the recovery residuals during the anomaly occurrence. Accordingly, the optimum value of this parameter was selected in a way that provided a balance between the two mentioned boundaries of the consequences of the incorrect selection of this parameter.

In Table 3, by assigning the identified optimum value for each dataset to the dimension of the sliding window, the effect of changing the level of sparsity in the AUC index has also been studied for all the datasets.

As observed from the results of Table 3, an optimized selection of this parameter has a significant role in the efficiency of the proposed algorithm. Indeed, considering the variety of the input data, choosing values of 5, 6, or 7 for this parameter will mainly yield desirable results, although in the ranges close to the optimum value this parameter did not reveal a significant change in results. Incorrect determination of this will considerably influence the results.

Since the proposed method involves considerably high processing when compared to other methods, it could not be compared from the computational cost and running time point of view. For example, the running time for RIT data in MATLAB software using a computer having an Intel Core i7 2.6 GHz processor and 16 GB of RAM under the Windows 10 64-bit operating system was 129 s, which is longer than the other methods. Nevertheless, the average running time of the proposed method in comparison with other methods are tabulated in Table 4. These times are the average value of running times of all datasets in each anomaly detection algorithm.

Table 3. Effect of the level of sparsity in the SWJSR detector for all datasets. (The bold one is the higher).

Dataset \ Level of Sparsity		2	3	4	5	6	7	8	Average	Standard Deviation
Real RIT	AUC	0.6352	0.8226	0.8993	0.9434	**0.9673**	0.9668	0.9607	0.8850	0.1219
Real San Diego	AUC	0.9196	0.9303	0.9499	0.9615	**0.9638**	0.9602	0.9620	0.9496	0.0177
Real Viareggio	AUC	0.8467	0.9126	0.9265	**0.9307**	0.9127	0.9115	0.9060	0.9066	0.0279
Implanted RIT	AUC	0.5796	0.7382	0.8414	0.8668	0.8922	**0.9115**	0.9103	0.82	0.1217
Destructed RIT	AUC	0.7500	0.8947	0.9442	0.9733	**0.9818**	0.9802	0.9811	0.9293	0.0852

Table 4. Average running time of the compared algorithms using all datasets.

Algorithm	GRX	LRX	CRD	BJSR	CR-RXD	CK-RXD	SLRX	SWJSR
Running Time (s)	2.05	22.74	33.28	51.48	8.27	9.62	102.51	193.14

Finally, it seems that utilizing and developing parallel processing systems will increase the speed of running the proposed algorithm that is the focus of future studies of the authors.

6. Conclusions

Since the development of anomaly detection algorithms for hyperspectral images includes a large number of applications, many researchers are motivated to develop efficient methods in this area. In this paper, a new method based on simultaneous sparse representation of local background signals using a sliding window was proposed to detect spectral anomalies. In this method, all of the signals located in the sliding window are voted through examining the estimated error of each signal to determine if there is any anomaly or not. As the precision of recovery for each pixel of the hyperspectral image is evaluated several times during the transition of the sliding window, this potential provides better conditions for evaluation of each signal from being an anomaly or background. The learned dictionary in each position of the sliding window is affected by the signals that are located in that window, and, practically, each pixel is being recovered many times with the help of a set of different background dictionaries.

The results of implementation of the proposed SWJSR method in five used datasets in this research proved its higher functionality when compared to the GRX, LRX, CRD, BJSR, CR-RXD, CK-RXD, and SLRX detectors. According to the obtained AUC, the results show the average improvement of efficiency (AUC) of about 7.5%, 14.25%, 8.2%, 8.25%, 6.45%, 6.5%, and 3.6%, respectively, in comparison to the mentioned algorithms. The implementation of this idea and its success showed that development of voting algorithms and the combination of the results could be considered as an effective approach to detect anomalies in hyperspectral signals. This idea could also be utilized in other hyperspectral image processing algorithms to evaluate the results by comparing prior methods. The results of SLRX, which show the average improvement of efficiency (AUC) of about 10% in comparison with traditional local RX, confirms this idea.

Automatic tuning of the proposed SWJSR algorithm parameters and developing parallel processing techniques to improve the running time of this algorithm are the focus of future research of the authors. Moreover, detecting spatial anomalies by the proposed approach and using spatial-spectral features in this field include other interested future works of the authors.

Acknowledgments: The authors would like to thank the Center for Imaging Science, Rochester Institute of Technology for the "HyMap" data set and also the Remote Sensing & Image Processing Group of University of Pisa for the "Viareggio 2013 Trial" data set used in our experiments to evaluate the proposed anomaly detection algorithm.

Author Contributions: All the authors listed contributed equally to the work presented in this paper.

Conflicts of Interest: The authors declare no conflict of interest.

References

1. Landgrebe, D. *On Information Extraction Principles for Hyperspectral Data*; School of Electrical and Computer Engineering, Purdue University: West Lafayette, IN, USA, 1997; pp. 168–173.
2. Shaw, G.; Manolakis, D. Signal Processing for Hyperspectral Image Exploitation. *IEEE Signal Process. Mag.* **2002**, *19*, 12–16. [CrossRef]
3. Chang, C.-I. *Hyperspectral Imaging: Techniques for Spectral Detection and Classification*; Kluwer Academic/ Plenum Publishers: New York, NY, USA, 2003; ISBN 0-306-47483-2.
4. Matteoli, S.; Diani, M.; Corsini, G. A tutorial overview of anomaly detection in hyperspectral images. *IEEE Aerosp. Electron. Syst. Mag.* **2010**, *25*, 5–28. [CrossRef]

5. Nasrabadi, N.M. Hyperspectral Target Detection: An overview of current and future challenges. *IEEE Signal Process. Mag.* **2014**, *31*, 34–44. [CrossRef]
6. Sodemann, A.A.; Ross, M.P.; Borghetti, B.J. A Review of Anomaly Detection in Automated Surveillance. *IEEE Trans. Syst. Man Cybern. Part C Appl. Rev.* **2012**, *42*, 1257–1272. [CrossRef]
7. Acito, N.; Diani, M.; Corsini, G. Gaussian mixture model based approach to anomaly detection in multi/hyperspectral images. In Proceedings of the SPIE Remote Sensing, Bruges, Belgium, 19–22 September 2005; Volume 5982, pp. 209–217.
8. Eismann, M.T.; Stocker, A.D.; Nasrabadi, N.M. Automated hyperspectral cueing for civilian search and rescue. *Proc. IEEE* **2009**, *97*, 1031–1055. [CrossRef]
9. Yuen, P.W.T.; Bishop, G. Hyperspectral Algorithm Development for Military Applications: A Multiple Fusion Approach. In Proceedings of the 3rd EMRS DTC Technical Conference, Edinburgh, UK, 13–14 July 2006.
10. Chang, C.-I. *Hyperspectral Data Processing: Algorithm Design and Analysis*; Wiley: Hoboken, NJ, USA, 2013.
11. Zhao, C.; Wang, Y.; Qi, B.; Wang, J. Global and local real-time anomaly detectors for hyperspectral remote sensing imagery. *Remote Sens.* **2015**, *7*, 3966–3985. [CrossRef]
12. Reed, I.S.; Yu, X. Adaptive multiple-band CFAR detection of an optical pattern with unknown spectral distribution. *IEEE Trans. Acoust. Speech Signal Process.* **1990**, *38*, 1760–1770. [CrossRef]
13. Chang, C.-I.; Chiang, S.-S. Anomaly Detection and Classification for Hyperspectral Imagery. *IEEE Trans. Geosci. Remote Sens.* **2002**, *40*, 1314–1325. [CrossRef]
14. Chang, C.-I. Multiple-parameter receiver operating characteristic analysis for signal detection and classification. *IEEE Sens. J.* **2010**, *10*, 423–442. [CrossRef]
15. Ren, H.; Chen, C.-W.; Chen, H.-T. Weighted anomaly detection for hyperspectral remotely sensed images. *Proc. SPIE* **2005**, *5995*, 63–68.
16. Chang, C.-I.; Hsueh, M. Characterization of anomaly detection in hyperspectral imagery. *Sens. Rev.* **2006**, *26*, 137–146. [CrossRef]
17. Kwon, H.; Nasrabadi, N.M. Kernel RX-Algorithm: A Nonlinear Anomaly Detector for Hyperspectral Imagery. *IEEE Trans. Geosci. Remote Sens.* **2005**, *43*, 388–397. [CrossRef]
18. Kwon, H.; Der, S.Z.; Nasrabadi, N.M. Dual-window-based anomaly detection for hyperspectral imagery. *Proc. SPIE* **2003**, *5094*, 148–158.
19. Kwon, H.; Der, S.Z.; Nasrabadi, N.M. Projection-Based Adaptive Anomaly Detection for Hyperspectral Imagery. In Proceedings of the 2003 International Conference on Image Processing (ICIP 2003), Barcelona, Spain, 14–17 September 2003; Volume 1.
20. Liu, W.; Chang, C.I. A nested spatial window-based approach to target detection for hyperspectral imagery. In Proceedings of the IEEE International Geoscience and Remote Sensing Symposium (IGARSS), Anchorage, AK, USA, 20–24 September 2004; Volume 1, pp. 20–24.
21. Liu, W.; Chang, C.-I. Multiple window anomaly detection for hyperspectral imagery. *IEEE J. Sel. Top. Appl. Earth Obs. Remote Sens.* **2013**, *6*, 644–658. [CrossRef]
22. Chang, C.-I.; Wang, Y.; Chen, S.Y. Anomaly detection using causal sliding windows. *IEEE J. Sel. Top. Appl. Earth Obs. Remote Sens.* **2015**, *8*, 3260–3270. [CrossRef]
23. Li, W.; Du, Q. Collaborative Representation for Hyperspectral Anomaly Detection. *IEEE Trans. Geosci. Remote Sens.* **2015**, *53*, 1463–1474. [CrossRef]
24. Yuan, Y.; Wang, Q.; Zhu, G. Fast Hyperspectral Anomaly Detection via High-Order 2D Crossing Filter. *IEEE Trans. Geosci. Remote Sens.* **2015**, *53*, 620–630. [CrossRef]
25. Yuan, Y.; Ma, D.; Wang, Q. Hyperspectral Anomaly Detection by Graph Pixel Selection. *IEEE Trans. Cybern.* **2016**, *46*, 3123–3134. [CrossRef] [PubMed]
26. Bruckstein, A.M.; Donoho, D.L.; Elad, A. From sparse solutions of systems of equations to sparse modeling of signals and images. *SIAM Rev.* **2009**, *51*, 34–81. [CrossRef]
27. Willett, R.M.; Duarte, M.F.; Davenport, M.A.; Baraniuk, R.G. Sparsity and Structure in Hyperspectral Imaging. *IEEE Signal Process. Mag.* **2014**, *31*, 116–126. [CrossRef]
28. Chen, Y.; Nasrabadi, N.M.; Tran, T.D. Simultaneous joint sparsity model for target detection in hyperspectral imagery. *IEEE Geosci. Remote Sens. Lett.* **2011**, *8*, 676–680. [CrossRef]
29. Du, B.; Zhang, Y.; Zhang, L.; Tao, D. Beyond the Sparsity-Based Target Detector: A Hybrid Sparsity and Statistics-Based Detector for Hyperspectral Images. *IEEE Trans. Image Process.* **2016**, *25*, 5345–5357. [CrossRef] [PubMed]

30. Zhang, Y.; Du, B.; Zhang, L. A sparse representation-based binary hypothesis model for target detection in hyperspectral images. *IEEE Trans. Geosci. Remote Sens.* **2015**, *53*, 1346–1354. [CrossRef]
31. Li, F.; Zhang, Y.; Zhang, L.; Zhang, X.; Jiang, D. Hyperspectral anomaly detection using background learning and structured sparse representation. In Proceedings of the IEEE International Geoscience and Remote Sensing Symposium (IGARSS), Beijing, China, 10–15 July 2016.
32. Yuan, Z.; Sun, H.; Ji, K.; Li, Z.; Zou, H. Local sparsity divergence for hyperspectral anomaly detection. *IEEE Geosci. Remote Sens. Lett.* **2014**, *11*, 1697–1701. [CrossRef]
33. Li, J.; Zhang, H.; Zhang, L.; Ma, L. Hyperspectral anomaly detection by the use of background joint sparse representation. *IEEE J. Sel. Top. Appl. Earth Obs. Remote Sens.* **2015**, *8*, 2523–2533. [CrossRef]
34. Zhao, R.; Du, B.; Zhang, L. Hyperspectral Anomaly Detection via a Sparsity Score Estimation Framework. *IEEE Trans. Geosci. Remote Sens.* **2017**, *55*, 3208–3222. [CrossRef]
35. Zhang, L.; Zhao, C. Hyperspectral anomaly detection based on spectral-spatial background joint sparse representation. *Eur. J. Remote Sens.* **2017**, *50*, 362–376. [CrossRef]
36. Ma, D.; Yuan, Y.; Wang, Q. A Sparse Dictionary learning method for Hyperspectral Anomaly Detection with Capped Norm. In Proceeding of the IEEE International Geoscience and Remote Sensing Symposium (IGARSS), Fort Worth, TX, USA, 23–28 July 2017; pp. 648–651.
37. Chen, Y.; Nasrabadi, N.M.; Tran, T.D. Sparse representation for target detection in hyperspectral imagery. *IEEE J. Sel. Top. Signal Process.* **2011**, *5*, 629–640. [CrossRef]
38. Tropp, J.A.; Wright, S.J. Computational methods for sparse solution of linear inverse problems. *Proceed. IEEE* **2010**, *98*, 948–958. [CrossRef]
39. Cotter, S.F.; Rao, B.D.; Engan, K.; Kreutz-Delgado, K. Sparse solutions to linear inverse problems with multiple measurement vectors. *IEEE Trans. Signal Process.* **2005**, *53*, 2477–2488. [CrossRef]
40. Tropp, J.; Gilbert, A. Signal recovery from random measurements via orthogonal matching pursuit. *IEEE Trans. Inf. Theory* **2007**, *53*, 4655–4666. [CrossRef]
41. Tropp, J.A.; Gilbert, A.C.; Strauss, M.J. Algorithms for simultaneous sparse approximation. Part I: Greedy pursuit. *Signal Process.* **2006**, *86*, 572–588. [CrossRef]
42. Aravind, N.V.; Abhinandan, K.; Acharya, V.V.; Sumam, D.S. Comparison of OMP and SOMP in the reconstruction of compressively sensed hyperspectral images. In Proceedings of the International Conference on Communications and Signal Processing (ICCSP), Calicut, India, 10–12 February 2011; pp. 188–192.
43. Aharon, M.; Elad, M.; Bruckstein, A. K-SVD: An algorithm for designing over-complete dictionaries for sparse representation. *IEEE Trans. Signal Process.* **2006**, *54*, 4311–4322. [CrossRef]
44. Mikhail, E.M.; Ackermann, F. *Observations and Least Squares*; IEP: New York, NY, USA, 1976.
45. Target Detection Blind Test. Available online: http://dirsapps.cis.rit.edu/blindtest/ (accessed on 26 February 2018).
46. Acito, N.; Matteoli, S.; Rossi, A.; Diani, M.; Corsini, G. Hyperspectral Airborne "Viareggio 2013 Trial" Data Collection for Detection Algorithm Assessment. *IEEE J. Sel. Top. Appl. Earth Obs. Remote Sens.* **2016**, *9*, 2365–2376. [CrossRef]
47. Chang, C.-I. *Real-Time Progressive Hyperspectral Image Processing: Endmember Finding and Anomaly Detection*; Springer: Berlin, Germany, 2016; Part II, Chapters 5; pp. 14–18.

remote sensing

MDPI

Article

A Deep Pipelined Implementation of Hyperspectral Target Detection Algorithm on FPGA Using HLS

Jie Lei [1,*], Yunsong Li [1,*], Dongsheng Zhao [1], Jing Xie [1,*], Chein-I Chang [2], Lingyun Wu [1], Xuepeng Li [1], Jintao Zhang [1] and Wenguang Li [1]

1 State Key Laboratory of Integrated Services Networks, School of Telecommunications Engineering, Xidian University, Xi'an 710071, China; dongshengzhao@stu.xidian.edu.cn (D.Z.); lywu@stu.xidian.edu.cn (L.W.); xuepengli@stu.xidian.edu.cn (X.L.); jtzhang_1@stu.xidian.edu.cn (J.Z.); wgli@stu.xidian.edu.cn (W.L.)

2 Department of Computer Science and Electrical Engineering, University of Maryland, Baltimore County, 1000 Hilltop Circle, Baltimore, MD 21250, USA; cchang@umbc.edu

* Correspondence: jielei@mail.xidian.edu.cn (J.L.); ysli@mail.xidian.edu.cn (Y.L.); jingxie@stu.xidian.edu.cn (J.X.); Tel.: +86-029-8820-3116 (J.L.)

Received: 2 January 2018; Accepted: 22 March 2018; Published: 25 March 2018

Abstract: Real-time target detection for hyperspectral images (HSI) has received considerable interest in recent years. However, owing to enormous data volume provided by HSI, detection algorithms are generally computationally complex, thus developing rapid processing techniques for target detection has encountered several challenging issues. It seems that using a deep pipelined structure can improve the detection speed, and implementing on field programmable gate arrays (FPGAs) can also achieve concurrent operations rather than run streams of sequential instruction. This paper presents a deep pipelined background statistics (DPBS) approach to optimizing and implementing a well-known subpixel target detection algorithm, called constrained energy minimization (CEM) on FPGA by using high-level synthesis (HLS). This approach offers significant benefits in terms of increasing data throughput and improving design efficiency. To overcome a drawback of HLS on implementing a task-level pipelined circuit that includes a feedback data path, a script based circuit design method is further developed to make connections between some of the modules created by HLS. Experimental results show that the proposed method can detect targets on a real-hyperspectral data set (HyMap Data) only in 0.15 s without compromising detection accuracy.

Keywords: hyperspectral image; deep pipelined background statistics; constrained energy minimization; high-level synthesis; real-time processing

1. Introduction

Hyperspectral remote sensing imaging acquires three-dimensional (3D) data including two spatial dimensions with space information of pixels and one spectral dimension with high-dimensional reflectance vectors [1]. The rich spectral information provided by HSI is very useful and has been widely used in a range of various applications such as ecology [2], agriculture [3], environmental [4] and geology [5], where target detection plays a crucial role [6–9]. There are many algorithms have been developed for target detection in HSI [1], such as matched filter (MF) [10], spectral angle mapper (SAM) [11], constrained energy minimization (CEM) [12], target-constrained interference-minimized filter (TCIMF) [13], adaptive coherence estimator (ACE) [14], matched subspace detector (MSD) [15], orthogonal subspace projection (OSP) [16], and sparsity-based target detector (STD) [17]. Among them, CEM along with its variants have been widely used for hyperspectral target detection. The effectiveness of CEM has been shown successfully in many applications such as reconnaissance, rescue, search and on-orbit processing [6]. For such applications, the high-speed data processing of HSI is generally required for finding targets on a timely basis. However, as a trade-off the volume of data in HSI has also become unmanageable with increasing spectral and spatial resolutions. In this case, CEM needs a significantly large number of complex matrix computations. The algorithm could be implemented on one of the widely used platforms like CPUs, GPUs [18], and FPGAs [19]. Among them, FPGAs have significant advantages in supporting parallel computation and customizable deep pipeline operation with the cost of low power consumption.

Currently, great progress has been made in implementing target detection algorithms on FPGAs. For example, Chang described a new FPGA design by using the Coordinate Rotation Digital Computer (CORDIC) algorithm to solve the matrix inversion problem of the classical CEM [20]. This method of computing the inverse of a large matrix is unable to support fast target detection. Yang utilized Streaming Background Statistics (SBS) structure with an idea of continuously updating the inverse of the correlation matrix on FPGA [21]. Despite that a pixel-by-pixel processing design is realized using fewer hardware resources, its data processing speed is not high. Recently, Gonzalez C. et al. proposed an FPGA implementation of the automatic target-generation process based on an orthogonal subspace projector (ATGP-OSP) using the pseudoinverse operation [22], where Gauss-Jordan elimination method was selected for computing the inverse of a small square matrix whose size is no more than 32×32. Unfortunately, the Gauss-Jordan elimination method consumes too much logic resources for solving the large matrix inversion problem required by the CEM.

Although FPGAs gain much attention, it is still not widely deployed for accelerating many algorithms that require high computational complexity such as CEM. The main reason is that the conventional development methods of FPGAs, which are based on register transfer level (RTL) hardware description, are much more difficult than that of CPUs or GPUs. It commonly requires great efforts in achieving highly efficient results on FPGAs. Furthermore, a design method based on RTL for FPGAs lacks portability and flexibility compared to those based on C/C++ for CPUs or GPUs. To close this gap, FPGA vendors and developers have begun to take advantage of high-level synthesis (HLS) to work on FPGA applications. HLS is able to convert high abstraction languages such as C, C++ and SystemC into VHDL/Verilog hardware description language (HDL) for RTL-level circuit design. According to the user-defined constraints and C code style, the efficiency of the converted RTL designs are quite different. Until now, studies on the acceleration of hyperspectral data processing algorithms with HLS are already available. Santos proposed a novel adaptive and predictive algorithm for lossy hyperspectral image compression algorithm [23] and Lossy Compression for Exomars (LCE) algorithm [24] described in Vivado HLS. Domingo R. et al. proposed a hyperspectral image spatial-spectral classifier accelerator using Intel FPGA SDK for OpenCL [25]. What's more, HLS is popularly utilized for hardware acceleration of deep learning algorithms like convolution neural networks (CNN) [26,27]. However, no research work has been reported for implementing CEM on FPGA using HLS.

In this work, a DPBS-CEM algorithm is developed to be implemented on FPGA using HLS for real-time hyperspectral target detection. Like SBS-CEM, the inverse matrix is gradually updated according to a Sherman-Morrison formula [28]. Different from using sliding windows, DPBS-CEM takes advantage of cumulative windows instead to greatly reduce the number of calculations. As for the issue of removing data dependency in updating inverse matrices, separate memories are proposed to store the results of the successive inverse matrices, which make sure the operations on adjacent pixels can be processed independently. As a consequence, a deep pipelined implementation of DPBS-CEM can be further developed, which has an extraordinary performance improvement in terms of data throughput. Experimental results demonstrate that the proposed algorithm has the capability of operating at a high-speed rate of more than 200 MHz on FPGA. Setting the same clock frequency, the algorithm can also achieve a significant speed-up of near 7.3$\times$ than SBS-CEM [21] with no compromise for detection accuracy.

The contributions of this paper can be summarized as follows.

- A novel deep pipelined architecture is proposed to accelerate the proposed DPBS-CEM algorithm on FPGA using HLS. It outperforms the previous work designed with RTL in terms of data throughput performance.
- A solution is derived to remove the data dependency existing in SBS-CEM for updating the inverse matrices, by allowing four adjacent pixels to update their own individual inverse matrices that are stored in four different memories.
- The proposed structure can be simply rebuilt to support diverse HSI implementations with different spatial resolution and number of spectral bands through several parameters modified under HLS. Most importantly, the framework can support various operation modes including split/non-split data and local/global detection. It is easily adapted to match multiple rates of hyperspectral imagery.
- Last but not least, alternative solutions to the problems of feedback and high fanout are also provided.

The remainder of this paper is organized as follows. Section 2 briefly discusses CEM and SBS-CEM used for target detection. Section 3 describes the principle of DPBS-CEM in great detail. The FPGA implementation of DPBS-CEM is presented in Section 4. Section 5 conducts a detailed performance analysis via extensive experiments. Finally, conclusions along with some remarks were drawn in Section 6.

2. Related Algorithms

In this section, the principles of the classical CEM and SBS-CEM algorithms are described. Besides, the problems of implementing these two algorithms in practical applications are also analysed.

2.1. CEM Algorithm

2.1.1. Principle of the CEM Algorithm

Let $\mathbf{X} \in \mathbf{R}^{W \times H \times L}$ denote a HSI with $W \times H$ pixels (row of $\mathbf{X}$) and L spectral bands (column of $\mathbf{X}$). We may interpret $\mathbf{X}$ either as a collection of L 2D images (or bands) of size N ($N = W \times H$), or as a collection of $W \times H$ spectral vectors of size L. The entire data matrix $\mathbf{X} = [\mathbf{x}_1, \mathbf{x}_2, \mathbf{x}_3, ..., \mathbf{x}_N]$, where $\mathbf{x}_i$ is the ith sample pixel vector $\mathbf{x}_i = (x_{i1}, x_{i2}, ..., x_{iL})^{\mathrm{T}}$ for $1 \leq i \leq N$ and the signature $\mathbf{d} = (d_1, d_2, ..., d_L)^{\mathrm{T}}$ of target is known. The basic purpose of CEM is to design a linear finite impulse response (FIR) filter with L filter coefficients denoted by an L-dimensional vector $\mathbf{w} = (w_1, w_2, ..., w_L)^{\mathrm{T}}$ that minimizes the energy of the the output y_i $(1 \leq i \leq N)$ with the following constraint.

$$s.t.\mathbf{d}^{\mathrm{T}}\mathbf{w} = 1 \tag{1}$$

$$\min_{\mathbf{w}} \left(\mathbf{w}^{\mathrm{T}}\mathbf{R}\mathbf{w}\right) \tag{2}$$

where $\mathbf{R} = \frac{1}{N}\left[\sum_{i=1}^{N} \mathbf{x}_i\mathbf{x}_i^{\mathrm{T}}\right]$ is the global correlation matrix of $\mathbf{X}$. The weighting vector $\mathbf{w}$ solved for Equation (1) and Equation (2) is given by

$$\mathbf{w} = \frac{\mathbf{R}^{-1}\mathbf{d}}{\mathbf{d}^{\mathrm{T}}\mathbf{R}^{-1}\mathbf{d}} \tag{3}$$

which yields the CEM described by

$$y_i = \delta_{CEM}(\mathbf{x}_i) = (\mathbf{w}_{CEM})^{\mathrm{T}}\mathbf{x}_i \tag{4}$$

2.1.2. Problem Analysis

The classical CEM algorithm is a global subpixel target detector, which uses all the pixels in HSI to calculate the correlation matrix. After the correlation matrix is obtained, the process of calculating the inverse matrix is executed through many complicated steps via QR decomposition. This is a typical large matrix inversion problem, which may be the main cause of a significant latency up to obtaining the final results.

2.2. SBS-CEM Algorithm

To accelerate the task of target detection using CEM, some researchers choose to calculate local detection by using a partial set of pixel vectors instead of all data sample vectors [6,29,30]. For instance, Yang [21] proposed an FPGA-based implementation of SBS-CEM by using a new matrix inversion method to perform the correlation operation and the inversion operation simultaneously.

2.2.1. Principle of the SBS-CEM Algorithm

Unlike the classical CEM algorithm, SBS-CEM takes the inverse of the correlation matrix of the K-group pixel vectors to replace the entire pixel vectors. More specifically, SBS-CEM can be described as follows.

$$\mathbf{R}_n = (1/K)\left[\sum_{i=n-K}^{n-1} \mathbf{x}_i\mathbf{x}_i^{\mathrm{T}}\right] \tag{5}$$

$$\mathbf{S}_n = \left[\sum_{i=n-K}^{n-1} \mathbf{x}_i\mathbf{x}_i^{\mathrm{T}}\right] \tag{6}$$

Now, $\mathbf{S}_n^{-1}$ is the inverse of the correlation matrix of the K-group pixel vectors. The Sherman-Morrison formula is used to derive the following two formulas.

$$\mathbf{C}^{-1} = \left(\mathbf{S}_n + \mathbf{x}_n\mathbf{x}_n^{\mathrm{T}}\right)^{-1} = \mathbf{S}_n^{-1} - \frac{\mathbf{S}_n^{-1}\mathbf{x}_n\mathbf{x}_n^{\mathrm{T}}\mathbf{S}_n^{-1}}{\mathbf{x}_n^{\mathrm{T}}\mathbf{S}_n^{-1}\mathbf{x}_n + 1} \tag{7}$$

$$\mathbf{S}_{n+1}^{-1} = \left(\mathbf{C} - \mathbf{x}_{n-K}\mathbf{x}_{n-K}^{\mathrm{T}}\right)^{-1} = \mathbf{C}^{-1} - \frac{\mathbf{C}^{-1}\mathbf{x}_{n-K}\mathbf{x}_{n-K}^{\mathrm{T}}\mathbf{C}^{-1}}{\mathbf{x}_{n-K}^{\mathrm{T}}\mathbf{C}^{-1}\mathbf{x}_{n-K} - 1} \tag{8}$$

Based on this streaming framework, the inverse matrix can be updated by using Equations (7) and (8). When applying the Sherman-Morrison formula, the initial value of $\mathbf{S}_0^{-1}$ should be set. Let $\mathbf{S}_0^{-1} = \beta \cdot \mathbf{I}$; then $\mathbf{S}_{K+1}$ can be expressed as:

$$\mathbf{S}_{K+1} = (1/\beta) \cdot \mathbf{I} + \mathbf{x}_1\mathbf{x}_1^{\mathrm{T}} + \mathbf{x}_2\mathbf{x}_2^{\mathrm{T}} + \cdots + \mathbf{x}_K\mathbf{x}_K^{\mathrm{T}} \tag{9}$$

Among them, the matrix $(1/\beta) \cdot \mathbf{I}$ does not affect the performance of the detector. On the contrary, it makes the detection results be more stable [31]. The detection equation of the SBS-CEM algorithm is then derived as:

$$SBS - CEM(\mathbf{x}) = \frac{K\left(\mathbf{x}^{\mathrm{T}}\mathbf{S}^{-1}\mathbf{d}\right)}{K\left(\mathbf{d}^{\mathrm{T}}\mathbf{S}^{-1}\mathbf{d}\right)} = \frac{\mathbf{x}^{\mathrm{T}}\mathbf{S}^{-1}\mathbf{d}}{\mathbf{d}^{\mathrm{T}}\mathbf{S}^{-1}\mathbf{d}} \tag{10}$$

Since the pixel to be detected is located in the middle of the window, SBS-CEM can also be expressed as:

$$SBS - CEM\left(\mathbf{x}_{n-K/2}\right) = \frac{\mathbf{x}_{n-K/2}^{\mathrm{T}}\mathbf{S}_n^{-1}\mathbf{d}}{\mathbf{d}^{\mathrm{T}}\mathbf{S}_n^{-1}\mathbf{d}} \tag{11}$$

2.2.2. Problem Analysis

Sliding window problem. Compared to the classical CEM algorithm, SBS-CEM does not need the full image data sample vectors to compute the correlation matrix. Instead, a local region of the image defined by a sliding window is utilized to capture the local statistics. The size of the sliding window is fixed and set to L^2 (square of the number of spectral bands) in the SBS-CEM algorithm. The fixed size of the sliding window requires the compute-intensive task of calculating the Sherman-Morrison formula to be performed twice (as shown in Equations (7) and (8)) for each update of the inverse matrix. However, the extra calculation of the Equation (8) does not provide appreciable improvements of the target detection accuracy according to our experimental results.

Data dependency problem. The problem of data dependency exists in the process of updating the inverse matrix where the calculation for $\mathbf{S}_{n+1}^{-1}$ cannot be started until the $\mathbf{S}_n^{-1}$ is available. Unless $\mathbf{S}_n^{-1}$ is ready, it is not possible to compute $\mathbf{S}_{n+1}^{-1}$. SBS-CEM divides the process of updating the inverse matrix into several stages to reduce its complexity. Unfortunately, under such circumstance, several stages' time consumption has to spend waiting for each inverse matrix updating. This computation overhead would be the major bottleneck of the SBS-CEM's data throughput performance.

3. Algorithm Optimization

3.1. Principle of Algorithm Optimization

To solve the problems described above for SBS-CEM, an optimized algorithm is proposed in this section. The two main improvements are proposed to deal with the use of sliding windows and to remove data dependency.

Non-sliding window. We choose not to use sliding windows to update calculations of the inverse matrix, which is quite different from the SBS-CEM algorithm. With no requirement for moving out the oldest pixel, the Equation (8) can be removed and thus a large number of calculations can be therefore reduced. When a new pixel vector $\mathbf{x}_n$ is loaded into the window, we can obtain the output value $\mathbf{S}_{n+1}^{-1}$ by Equation (12).

$$\mathbf{S}_{n+1}^{-1} = \left(\mathbf{S}_n + \mathbf{x}_n\mathbf{x}_n^{\mathrm{T}}\right)^{-1} = \mathbf{S}_n^{-1} - \frac{\mathbf{S}_n^{-1}\mathbf{x}_n\mathbf{x}_n^{\mathrm{T}}\mathbf{S}_n^{-1}}{\mathbf{x}_n^{\mathrm{T}}\mathbf{S}_n^{-1}\mathbf{x}_n + 1} \tag{12}$$

Data segmentation for deep pipeline. As mentioned above, the SBS-CEM algorithm runs calculations of matrix inversions in serial. Since data dependency exists between $\mathbf{S}_{n+1}^{-1}$ and $\mathbf{S}_n^{-1}$, there is a great increase in processing time. To solve this problem, we need to complete the computation of Equation (12) in four stages and apply pipeline optimization for achieving pipeline acceleration. However, updating the inverse matrix between adjacent pixels is not independent, which prevents the use of the optimization strategy of deep pipeline. If we want to achieve a deep pipelined design, we have to make sure there is no feedback or iterations among the stages. In this case, we solve the data dependency by means of data segmentation. As a result, the current input pixel can be processed directly with no need of waiting for the previous pixel to be completed. By making the inverse calculations between neighbouring pixels independent, we are able to carry out a deep pipelined architecture, which can achieve 8× speed-up compared to SBS-CEM in theory.

Table 1, derived from the evaluation of hardware calculation, shows that the number of computations for each stage is different, but the number of clock cycles consumed by each stage is approximately equal after being parallelized. Where $\mathbf{x}_n$ $\left(\mathbf{P} = \mathbf{x}_n^{\mathrm{T}}\right)$ represents a column of $\mathbf{X}$, T and Q are scalars. $\mathbf{S}_n^{-1}$ is denoted by $\mathbf{U}$, which is an L-dimensional matrix. In addition, the detail procedure of DPBS-CEM algorithm is shown as Algorithm 1.

Table 1. Four stages of the inverse matrix update.

Stage Number	Formula	Flop $(\times : \pm)$	Parallelism	Clock Cycles
1	$\mathbf{h} = \mathbf{U}\mathbf{p}^{\mathrm{T}}$	$(L^2 : L^2)$	L	L
2	$T = \mathbf{p}\mathbf{h}$	$(L : L)$	1	L
3	$\mathbf{F} = \mathbf{h}\mathbf{h}^{\mathrm{T}}$ $Q = \frac{1}{T+1}$	$(L^2 : 0)$ $(0 : 1)$	L	L
4	$\mathbf{S}_{n+1}^{-1} = \mathbf{U} - \mathbf{F}Q$	$(L^2 : L^2)$	L	L

Algorithm 1 The deep pipelined background statistics (DPBS) target detection CEM algorithm

Input: Initialize the following parameters.
(1) HSI data size: $W \times H \times L = N \times L$;
(2) the value of β;
(3) the desired signature $\mathbf{d}$;
(4) the number of inverse matrices: $M = 4$;
(5) bn indicates the index of number;
(6) K indicates the number of pixel vectors collected before starting target detection;
Output: the final target detection results.
define an initial inverse matrix S_0^{-1}: $S_0^{-1} = \beta \cdot \mathbf{I}$
data segmentation:
for $i = 1\,; i \leq N + K\,; i{+}{+}$ **do**
    $bn = i \,\%\, M$
    calculate the inverse matrix:
    if $i \leq N$ **then**

$$\left(\mathbf{S}^{-1}\right)^{\text{bn}} = \left(\mathbf{S}^{\text{bn}} + \mathbf{x}_i\mathbf{x}_i^{\text{T}}\right)^{-1} = \left(\mathbf{S}^{-1}\right)^{\text{bn}} - \frac{\left(\mathbf{S}^{-1}\right)^{\text{bn}}\mathbf{x}_i\mathbf{x}_i^{\text{T}}\left(\mathbf{S}^{-1}\right)^{\text{bn}}}{\mathbf{x}_i^{\text{T}}\left(\mathbf{S}^{-1}\right)^{\text{bn}}\mathbf{x}_i + 1}$$

    endif
    calculate the target detection results:
    if $i \geq K$ **then**

$$DPBS - CEM\left(\mathbf{x}_{i-K}\right) = \frac{\mathbf{x}_{i-K}^{\text{T}}\left(\mathbf{S}^{-1}\right)^{\text{bn}}\mathbf{d}}{\mathbf{d}^{\text{T}}\left(\mathbf{S}^{-1}\right)^{\text{bn}}\mathbf{d}}$$

    endif
endfor

3.2. Design Challenges

Feedback. There is a feedback problem in updating the inverse matrix. In fact, the inverse matrix updated in the fourth stage has to be transmitted back to the first stage as an input operand for the next updating. All of the stages are described by individual C/C++ functions. To substantially accelerate the process of updating the inverse matrix, we have to apply the data flow optimization directly to these functions so that the HLS tool can be guided to implement a task-level pipelining. Unfortunately, the HLS tool will not take place if it detects a feedback among the functions. As a result, the task-level pipelining cannot be achieved only using HLS directly.

Fanout. Due to the use of a large number of bands, there are some high fanout cases where some registers need to drive lots of loads like multipliers, which result in longer path delay and lower clock frequency. For example, in the fourth stage as described in Table 1, the scalar Q needs to be multiplied by L elements of a column in the matrix $\mathbf{F}$ simultaneously after parallel computation applied. It means that the element of the scalar Q has a high fanout to drive as much as L slave modules. It is simple to solve the high fanout problem by means of duplicating registers when designing with RTL, but it is not easy with HLS.

4. FPGA Implementation

In this section, an overall hardware structure of DPBS-CEM is given in Section 4.1. Section 4.2 describes the internal architecture of the inverse matrix updater in detail along with its workflow of deep pipeline. The difficulties in developing the hardware framework of DPBS-CEM using the HLS tool and their solutions are discussed in Section 4.3. Section 4.4 briefly introduces a few particular features of the proposed FPGA implementation of DPBS-CEM.

4.1. Overall Hardware Architecture of DPBS-CEM

As shown in Figure 1, the framework of DPBS-CEM mainly consists of three components including an off-chip memory, a processor core, and a scheduler. The off-chip memory (DDR3 SDRAM) is utilized to cache the hyperspectral image pixels. The processor core is responsible for the data processing of

DPBS-CEM, which involves three modules: the first module is an inverse matrix updater, dedicated to update the inverse matrix in five stages; the second module is a spectral pixel filter, applied to filter pixels in four stages; and the last module is a storage component, utilized to cache the inverse matrix. Finally, the third component is scheduler which is designed to schedule the two modules of inverse matrix updater and spectral pixel filter.

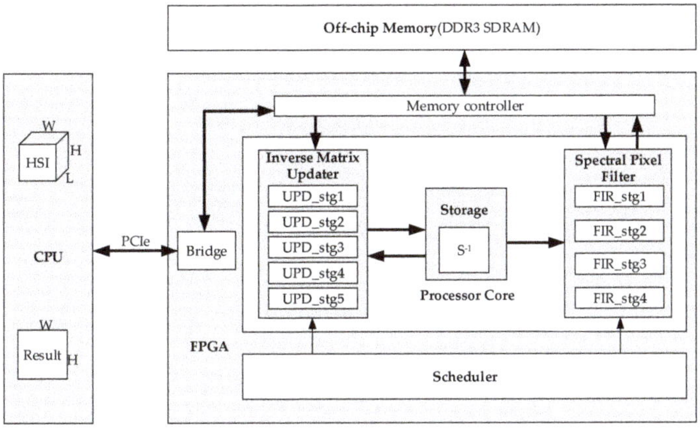

Figure 1. Overall hardware structure of DPBS-CEM.

4.2. Update Process of Inverse Matrix

4.2.1. Internal Architecture

As described in Figure 2, the inverse matrix updater contains five processing stages for updating the inverse matrices and four memory buffers for independently caching inverse matrices associated with four successive pixels. The four individual memories are allocated for solving the problem of data dependency described in Section 2.2.2. The specific calculations of each stage, the data flow, and the access mode of inverse matrices are clearly displayed in Figure 2. In addition, we arrange five blocks (Block A in Figure 3a, Block B in Figure 4a, Block C in Figure 5a, Block D in Figure 6a, and Block E in Figure 7a) to realize the last four processing stages in Figure 2, and we also provide pieces of C/C++ code written in HLS for these blocks on the right side of the Figures. In the Appendix A, the features of the #pragma used in these pieces of code are explained in Table A1.

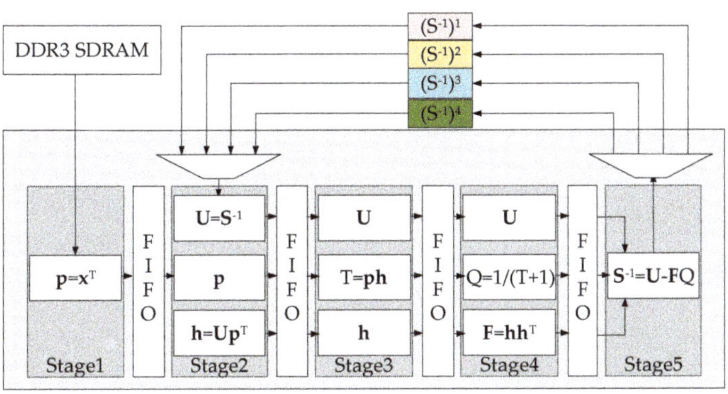

Figure 2. Block diagram of inverse matrix updater.

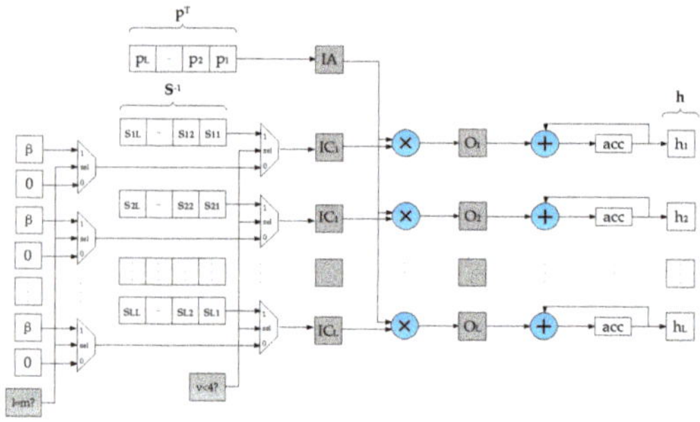

(**a**)

```
void Block_A(datatype_mid h[L],datatype_in pt[L],
        datatype_mid S_1_i[L][L][4],int v){
#pragma HLS INTERFACE ap_none port=h
#pragma HLS ARRAY_PARTITION variable=h complete dim=1
#pragma HLS INTERFACE ap_none port=pt
#pragma HLS ARRAY_PARTITION variable=pt complete dim=1
#pragma HLS INTERFACE ap_memory port=S_1_i
#pragma HLS ARRAY_PARTITION variable=S_1_i complete dim=1
    datatype_mid S_1_col[L]={0};
#pragma HLS ARRAY_PARTITION variable=S_1_col complete dim=1
    datatype_mid h_mid1[L]={0};
#pragma HLS ARRAY_PARTITION variable=h_mid1 complete dim=1
    datatype_mid h_mid2[L]={0};
#pragma HLS ARRAY_PARTITION variable=h_mid2 complete dim=1
    for(unsigned char l=0;l<L;l++)
#pragma HLS UNROLL
        h_mid1[l]=0;
    for(unsigned char m=0;m<L;m++){
#pragma HLS PIPELINE II=1
        for(unsigned char l=0;l<L;l++){
            S_1_col[l]=S_1_i[l][m][v%4];
            if (v<4){
                S_1_col[l]=0;
                if(l==m)
                    S_1_col[l]=P;
            }
            h_mid2[l]=pt[l]*S_1_col[l];
            h_mid1[l]+=h_mid2[l];
        }
    }
    for(unsigned char l=0;l<L;l++)
#pragma HLS UNROLL
        h[l]=h_mid1[l];
}
```

(**b**)

Figure 3. (**a**) Hardware structure and (**b**) C/C++ code in HLS of Block A. (v represents the pixel number, l represents the row number of the matrix $\mathbf{S}^{-1}$, and m represents the column number of the matrix $\mathbf{S}^{-1}$).

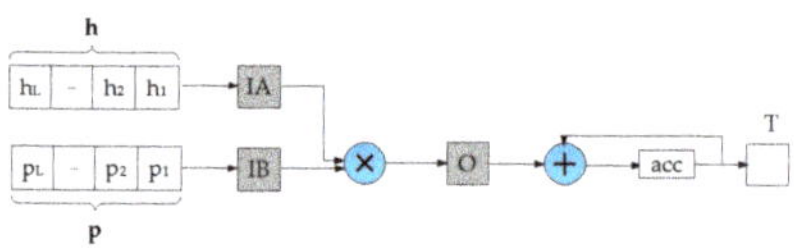

(**a**)

```
void Block_B(datatype_in h[L],datatype_mid p[L],
        datatype_mid T){
#pragma HLS INTERFACE ap_none port=h
#pragma HLS ARRAY_PARTITION variable=h complete dim=1
#pragma HLS INTERFACE ap_none port=p
#pragma HLS ARRAY_PARTITION variable=p complete dim=1
#pragma HLS INTERFACE ap_none port=T
    datatype_mid T_mid1=0;
    datatype_mid T_mid2=0;
    T_mid1=0;
    for(unsigned char m=0;m<L;m++){
#pragma HLS PIPELINE II=1
        T_mid2=h[m]*p[m];
        T_mid1+=T_mid2;
    }
    T=T_mid1;
}
```

(**b**)

Figure 4. (**a**) Hardware structure and (**b**) C/C++ code in HLS of Block B.

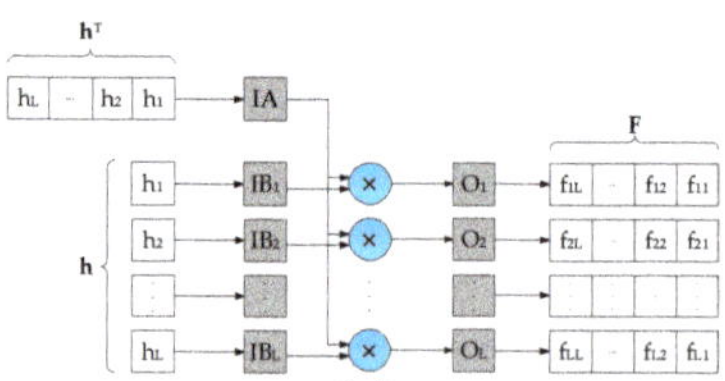

(**a**)

```
void Block_C(datatype_mid h[L],
        datatype_mid F[L][L]){
#pragma HLS INTERFACE ap_none port=h
#pragma HLS ARRAY_PARTITION variable=h complete dim=1
#pragma HLS INTERFACE ap_fifo port=F
#pragma HLS ARRAY_PARTITION variable=F complete dim=1
    for(unsigned char m=0;m<L;m++)
#pragma HLS PIPELINE II=1
        for(unsigned char l=0;l<L;l++)
            F[l][m]=h[m]*h[l];
}
```

(**b**)

Figure 5. (**a**) Hardware structure and (**b**) C/C++ code in HLS of Block C.

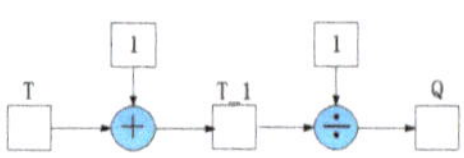

(**a**)

```
void Block_D(datatype_mid T,
        datatype_mid Q){
#pragma HLS INTERFACE ap_none port=T
#pragma HLS INTERFACE ap_none port=Q
    datatype_mid T_1=0;
    T_1=T+(datatype_mid)1;
    Q=(datatype_mid)1/T_1;
}
```

(**b**)

Figure 6. (**a**) Hardware structure and (**b**) C/C++ code in HLS of Block D.

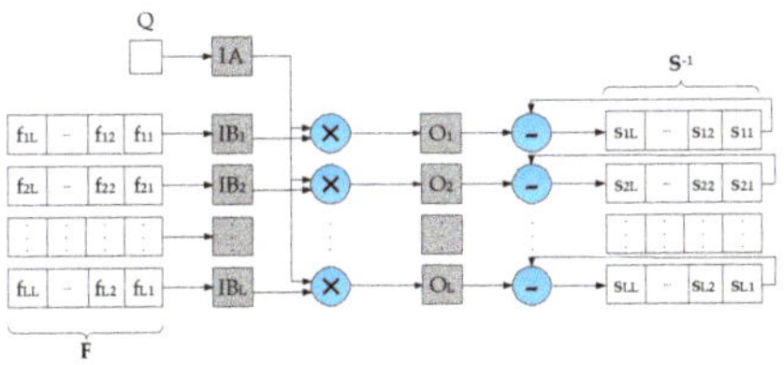

(**a**)

```
void Block_E(datatype_mid F[L][L],
        datatype_mid U[L][L],datatype_mid Q,
        datatype_mid S_1_o[L][L][4],int v){
#pragma HLS INTERFACE ap_fifo port=F
#pragma HLS ARRAY_PARTITION variable=F complete dim=1
#pragma HLS INTERFACE ap_fifo port=U
#pragma HLS ARRAY_PARTITION variable=U complete dim=1
#pragma HLS INTERFACE ap_memory port=S_1_o
#pragma HLS ARRAY_PARTITION variable=S_1_o complete dim=1
    datatype_mid S_1_col_mid1[L]={0};
#pragma HLS ARRAY_PARTITION variable=S_1_col_mid1 complete dim=1
    datatype_mid S_1_col_mid2[L]={0};
#pragma HLS ARRAY_PARTITION variable=S_1_col_mid2 complete dim=1
    for(unsigned char m=0;m<L;m++){
#pragma HLS PIPELINE II=1
        for(unsigned char l=0;l<L;l++){
            S_1_col_mid1[l]=Q*F[l][m];
            S_1_col_mid2[l]=U[l][m]-S_1_col_mid1[l];
            S_1_o[l][m][v%4]=S_1_col_mid2[l];
        }
    }
}
```

(**b**)

Figure 7. (**a**) Hardware structure and (**b**) C/C++ code in HLS of Block E.

In what follows, the complete updating process of the inverse matrices by using these blocks can be summarized in five stages as depicted in Figure 2.

Stage1 All elements of vector $\mathbf{x}_n^T$ read from the DDR3 SDRAM are loaded sequentially and passed on to the next stage.

Stage2 According to the index of the current pixel, we read a corresponding matrix $\mathbf{S}^{-1}$ from the storage module. When dealing with the first four pixels of an image, we need to overwrite the matrix $\mathbf{S}^{-1}$ with initialized matrix $\beta \cdot \mathbf{I}$. Then, we take matrix $\mathbf{U}$ and vector $\mathbf{p}^T$ as input operands into the Block A to calculate product $\mathbf{h}$. Subsequently, $\mathbf{p}$, $\mathbf{h}$, and $\mathbf{U}$ are passed to the next stage.

Stage3 We count T by applying the Block B, then transmit $\mathbf{U}$, T, and $\mathbf{h}$ to the next stage.

Stage4 The Block C is utilized to work out the product $\mathbf{F}$ of two vectors. We calculate Q by employing the Block D. Then $\mathbf{U}$, $\mathbf{F}$, and Q are delivered to the next stage.

Stage5 We figure out the new matrix $\mathbf{S}^{-1}$ through utilizing the Block E and write it to the corresponding location of the storage module according to the current pixel.

Besides, it is worth noting that the following design optimization strategies play an important role in improving the performance of the FPGA implementation.

(1) In the process of updating the inverse matrices, we allocate a single divider and execute it once for each inverse matrix updating. Thanks to such operation, a lot of logic resources and computation time consumed by the divider can be saved.

(2) There are three types of data that need to be cached between two stages, the scalar data, the vector data, and the matrix data. In order to attain the capability of parallel computation, the matrix is cached in L first in first out (FIFO) memories (In HLS, we use the STREAM directive to map these sorts of data into FIFOs). While the elements of a vector are realized as registers. In addition, L simple dual port RAMs (simple DPRAMs) are deployed to implement the storage module.

(3) The data type of input data is 16 bits signed fixed-point (15 bits fractional part), while the data type of intermediate data and detection results are not easy to assign. Due to the precision of intermediate data and detection results have a significant impact not only on the detection accuracy but also on the resource consumption, we performed some experiments to explore the relationship between the data precision and the detection accuracy. The experimental results demonstrate that the detection accuracy goes up with the increase of the bit-width of the fractional part. To better balance the trade-off between the detection accuracy and the resource consumption, we use different data types in different stages. As shown in Figure 2, the variable T and Q are defined as 38 bits signed fixed-point type (14 bits integer part, 23 bits fractional part). The elements of the matrix F are 32 bits signed fixed-point type (14 bits integer part, 17 bits fractional part).

All the other intermediate data are represented as 32 bits signed fixed-point type (7 bits integer part, 24 bits fractional part). T and Q have the significant impact on the detection accuracy. Therefore, they are assigned high data precision up to 38 bits. The elements of the matrix F are obtained by the accumulation operations, and more bits should be assigned to the integer part for avoiding data overflow. Though T and Q have larger bit-width up to 38 bits, it almost does not increase the logic resource consumption compared with the data type of 32 bits signed fixed-point. The reason is that only one single accumulation adder is allocated to compute T while one single adder and one single divider are placed to calculate Q. It is worthwhile to highlight that these data types can be defined and modified by HLS *ap_fixed* type easily.

4.2.2. Deep Pipeline

As shown in Figure 8, a full pipeline for updating inverse matrices is comprised of Task1, Task2, Task3, Task4, and Task5. These five tasks correspond to Stage1, Stage2, Stage3, Stage4, and Stage5 in Figure 2 respectively.

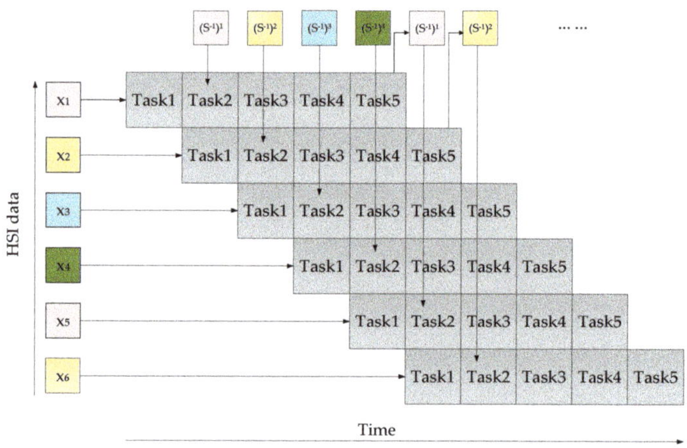

Figure 8. Timing diagram of the process of updating inverse matrix.

(1) For the purpose of reducing logic resources without compromising accuracy, we implement a high-precision division with the price of long latency. It takes near 30 clock cycles to output the division result. If the division operation is assigned to Task3, the running time of Task3 will increase a lot. As a result, Task3 will turn out to be a bottleneck in the pipeline. Therefore, we assign the division operation to Task4. Note that, the division and multiplication operations in Task4 are carried out simultaneously.

(2) For each task, it does not start until all input data are ready and all output FIFOs are not full. It can be simply realized in HLS by writing C/C++ code as shown in Figure 9. To make the pipeline run efficiently, these FIFOs, which are dedicated to bridging two adjacent tasks, are designed a little bit larger. In this work, the depth of FIFO for vector is 2, while the depth of FIFO for the matrix is $L \times 2$. Besides, the depth of simple DPRAM for the storage module is $L \times 2 \times 4$.

(3) With regard to the execution time of each task, it is consistent with $L + 12$ times of the system clock period. Among them, the input time of an L-dimensional vector is L clock cycles, the delay time of the multiplier is one clock cycle, and the remaining 11 clock cycles are used to control input/output of the task. Especially, because two extra clock cycles are required for overwriting the matrix $\mathbf{S}^{-1}$ with the initialized matrix $\beta \cdot \mathbf{I}$, the total execution time of Task2 is $L + 14$ clock cycles.

```
/*************************************************************************/
//parameter:task_cntrlfpre,task_cntrl2nxt
//implementation:The parameter task_cntrlfpre actually represents
//          a FIFO, and when the previous task is completed, the value
//          of end_flag is written to the FIFO. The value is 1 when
//          processing the last pixel of an image, 0 otherwise. The
//          parameter task_cntrl2nxt is similar to task_cntrlfpre.
/*************************************************************************/
void task(stream<bool>& task_cntrlfpre,stream<bool>& task_cntrl2nxt){
    bool end_flag;
    while(1){
        while(1)
    if((!task_cntrlfpre.empty())&&(!task_cntrl2nxt.full())) break;
        /*the calculations of this task,thus omitted*/

        ...
        task_cntrlfpre.read_nb(end_flag);
        task_cntrl2nxt.write_nb(end_flag);
        if(end_flag==1) break;
    }
}
```

Figure 9. Sample code used for implementing the data flow control of a task in HLS (The omitted lines of code are the specific computations of each stage described in Section 4.2.1).

4.3. Difficulties with Using HLS

4.3.1. Feedback

The function of task-level pipelining is available in HLS by applying DATAFLOW directive. However, one of the major difficulties with HLS is that HLS does not support to generate a task-level pipelined structure if data dependency (feedback) exists. Unfortunately, there is a feedback in the process of updating the inverse matrix as explained in Section 3.2. To solve this problem, we exploit a design method with a hybrid of RTL and HLS. As shown in Figure 10, HLS is applied to create the two complex modules, inverse matrix updater and spectral pixel filter. A small piece of RTL code is written to complete the scheduler whose function is quite simple. Verilog's generate statement is used to circularly instantiate all of the simple DPRAMs allocated in the storage module. Moreover, a TCL script for automatically connecting the above-mentioned modules is employed. When using HLS to realize the inverse matrix updater module, we define separate interface variables representing the input and output inverse matrices respectively. This separation strategy allows HLS to understand there is no data feedback in accessing the inverse matrix. In fact, the input and output inverse matrices are pointed to the same memory location in the storage module.

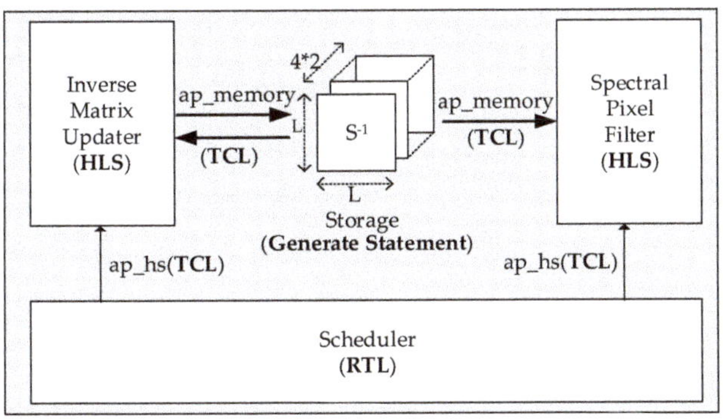

Figure 10. Diagram of development with multiple tools.

4.3.2. High Fanout

Register duplication is one of the most common ways to solve high fanout violation. It can be applied to relieve the fanout challenge as described in Section 3.2. However, the difficulty is how to make HLS replicate registers automatically since there is no inherent support of such feature in HLS. To solve this problem, we modify part of C/C++ code in HLS to split the high fanout task into two or more identical subtasks, which allows HLS to generate duplicated circuits for reducing the fanout. With this optimization, our FPGA implementation is able to work at a rate of speed higher than 200 MHz.

4.4. Specific Features

4.4.1. Scalability and Portability

Parallel computation and memory units are placed in the stages of the core architecture of DPBS-CEM to accelerate the related operations of matrix multiplication. The number of the parallel units is equal to the value L. By modifying the value L, we can easily scale the core framework of DPBS-CEM with HLS to support different HSIs with different number of bands. Parameter customized design method with HLS greatly improves the scalability of the system. Simultaneously, the framework does not rely on any specific underlying physical devices of FPGA and vendor-provided IP cores. Thus it can be easily ported to other types of FPGAs.

4.4.2. Flexibility

The flexibility feature is referred to as multiple work modes supported by the proposed DPBS-CEM. The default work mode is high-speed, at which the pipeline is fully operating. Besides, DPBS-CEM is also allowed to be configured working at low-speed mode. Then it can produce global detection results with no need to split image data into four parts for applying deep pipeline. Furthermore, through altering the control of the pipeline between the two processes of inverse matrix updater and spectral pixel filter, DPBS-CEM can output detection results while part of the image pixels are obtained.

5. Experimental Results and Analysis

A Virtex7 FPGA board (Alpha-Data ADM-PCIE-7V3) is chosen as our development platform, which provides more logic resources than the Kintex-7 board used in [21]. Besides the FPGA implementation of DPBS-CEM, the simulation versions were also implemented using the MATLAB and C++ languages. The code of MATLAB and C++ are executed on Windows 7 operating system equipped with the Intel Core (TM) quad CPU @3.2 GHz and 4 GB main memory. We compare the performance of DPBS-CEM with SBS-CEM [21] under the same condition of FPGA implementation. The rest of this section is organized as follows. Section 5.1 describes two hyperspectral data sets used in the experiment. Section 5.2 shows the detection accuracy of the DPBS-CEM algorithm evaluated on both of the hyperspectral data sets. Section 5.3 gives a comparison of the processing time of the DPBS-CEM algorithm in MATLAB, C++ and FPGA. Finally, compared to the FPGA implementation of SBS-CEM [21], we analyze the advantages of the FPGA implementation of DPBS-CEM in terms of logic resources utilization and data processing speed.

5.1. Hyperspectral Image Data Set

5.1.1. TE1 Image

As shown in Figure 11a, 25 panels created with five United States Geological Survey (USGS, Reston, VA, USA) reflectance hyperspectral signatures: alunite (A), buddingtonite (B), calcite (C), kaolinite (K), and muscovite (M). Each row of the five panels in Figure 11b is simulated by the same mineral signature and each column of five panels has the same size [32,33]. Among 25 panels are: five 4×4-pure pixel panels, $px^i_{4\times4}$ for $i = 1, \ldots, 5$ in the first column; five 2×2-pure pixel panels, $px^i_{2\times2}$ for

$i = 1, \ldots, 5$ in the second column; five 2×2-mixed pixel panels, $\left\{px^i_{3,jk}\right\}^{2,2}_{j=1,k=1}$ for $i = 1, \ldots, 5$ in the third column; five subpixel panels, $px^i_{4,11}$ for $i = 1, \ldots, 5$ in the fourth column; and five subpixel panels, $px^i_{5,11}$ for $i = 1, \ldots, 5$ in the fifth column. Table 2 tabulates the mixing details of mineral composition in the 20 panels in the third column, while subpixel panels in the fourth and fifth columns are simulated with their abundance fractions tabulated in Table 3, where the background (BKG) is simulated by the sample mean of the real cuprite image scene in USGS [33]. The Synthetic image TE1 is 200×200 pixels, 189 bands from 0.4 um to 2.5 um.

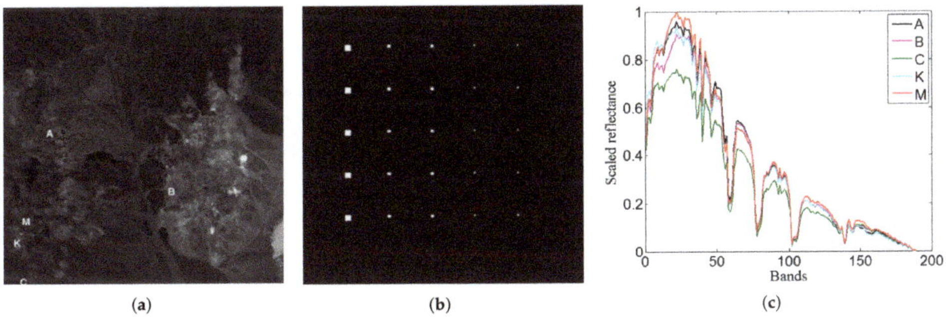

Figure 11. (**a**) Cuprite Airborne Visible/Infra Red Imaging Spectrometer (AVIRIS) image scene with spatial positions of five pure pixels corresponding to minerals: alunite (A), buddingtonite (B), calcite (C), kaolinite (K) and muscovite (M); (**b**) Synthetic image simulated by Scenario TE1; (**c**) Five reflectance USGS ground-truth mineral spectra.

Table 2. Simulated 20 mixed panel pixels in the third column.

Row1	$px^1_{3,11}$ = 0.5A + 0.5B	$px^1_{3,12}$ = 0.5A + 0.5C	$px^1_{3,21}$ = 0.5A + 0.5K	$px^1_{3,22}$ = 0.5A + 0.5M
Row2	$px^2_{3,11}$ = 0.5B + 0.5A	$px^2_{3,12}$ = 0.5B + 0.5C	$px^2_{3,21}$ = 0.5B + 0.5K	$px^2_{3,22}$ = 0.5B + 0.5M
Row3	$px^3_{3,11}$ = 0.5C + 0.5A	$px^3_{3,12}$ = 0.5B + 0.5C	$px^3_{3,21}$ = 0.5C + 0.5K	$px^3_{3,22}$ = 0.5C + 0.5M
Row4	$px^4_{3,11}$ = 0.5K + 0.5A	$px^4_{3,12}$ = 0.5K + 0.5B	$px^4_{3,21}$ = 0.5K + 0.5C	$px^4_{3,22}$ = 0.5K + 0.5M
Row5	$px^5_{3,11}$ = 0.5M + 0.5A	$px^5_{3,12}$ = 0.5M + 0.5B	$px^5_{3,21}$ = 0.5M + 0.5C	$px^5_{3,22}$ = 0.5M + 0.5K

Table 3. Abundance fractions of subpixel panels in the fourth and fifth columns.

Row	Fourth Column	Fifth Column
1	$px^1_{4,11}$ = 0.5A + 0.5BKG	$px^1_{5,11}$ = 0.25A + 0.75BKG
2	$px^2_{4,11}$ = 0.5B + 0.5BKG	$px^2_{5,11}$ = 0.25B + 0.75BKG
3	$px^3_{4,11}$ = 0.5C + 0.5BKG	$px^3_{5,11}$ = 0.25B + 0.75BKG
4	$px^4_{4,11}$ = 0.5K + 0.5BKG	$px^4_{5,11}$ = 0.25K + 0.75BKG
5	$px^5_{4,11}$ = 0.5M + 0.5BKG	$px^5_{5,11}$ = 0.25M + 0.75BKG

5.1.2. HyMap Reflectance Image

The hyperspectral data set is provided by the Digital Imaging and Remote Sensing Group, Center for Imaging Science, Rochester Institute of Technology [34]. Figure 12 shows the HyMap reflection map of Cook City, Montana, USA with a resolution of 280×800 and a total of 126 bands distributed between 0.4 and 2.4 um. There is a grass area and four real panels of fabric in the data set as shown in Table 4, where the area of interest is highlighted with a red circle.

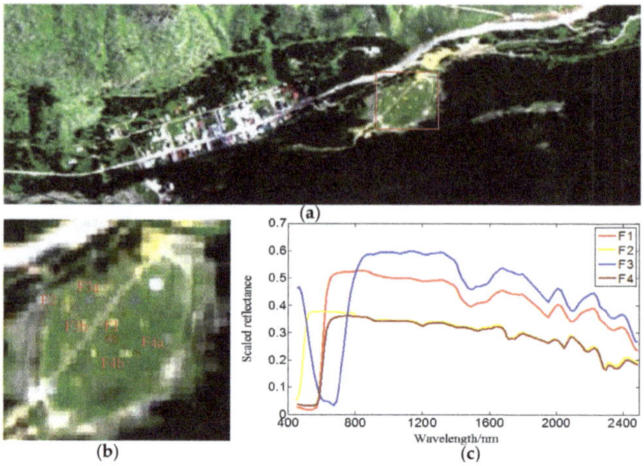

Figure 12. (**a**) HyMap reflectance image of Cook City in Montana, USA, and locations of the real targets; (**b**) Enlarged figure of red box area; (**c**) Spectral signatures of four targets.

Table 4. The characteristics of targets in the real scene of HyMap.

Name	F1	F2	F3a	F3b	F4a	F4b
Size (m)	3×3	3×3	2×2	1×1	2×2	1×1
Fabric type	Red cotton	Yellow nylon	Blue cotton	Blue cotton	Red nylon	Red nylon

5.2. Analysis of Target Detection Accuracy

In this part, we evaluate the detection accuracy of the FPGA implementation of DPBS-CEM by using the simulation/real HSI data sets described above. CEM and SBS-CEM are evaluated as well for comparison. The detection accuracy can be evaluated via Receiver Operating Characteristics (ROC) [35]. However, the ROC curves of different algorithms may be too close to determine which algorithm has better performance. Therefore, in this paper, we choose another way commonly used in medical diagnosis to calculate the area under a ROC curve, referred to as the area under the curve (AUC) [36]. The AUC values corresponding to the detection results can further quantify the differences in the accuracy of the algorithms. The higher the AUC, the better the detection accuracy.

5.2.1. Detection Accuracy of TE1

Figure 13 shows five detection maps produced by DPBS-CEM using the five-panel signatures A, B, C, K, and M in Figure 11c as the desired target signatures. The two-dimensional (2-D) results of real-time detection of target A illustrated in Figure 14. The experimental results show that all the AUC values of five desired targets detected by DPBS-CEM are one, indicating that the detection results are extremely satisfactory.

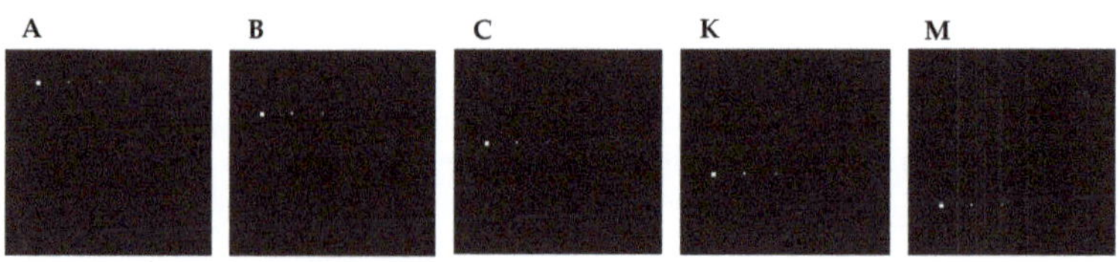

Figure 13. Detection maps of DPBS-CEM using A, B, C, K and M as desired target signature.

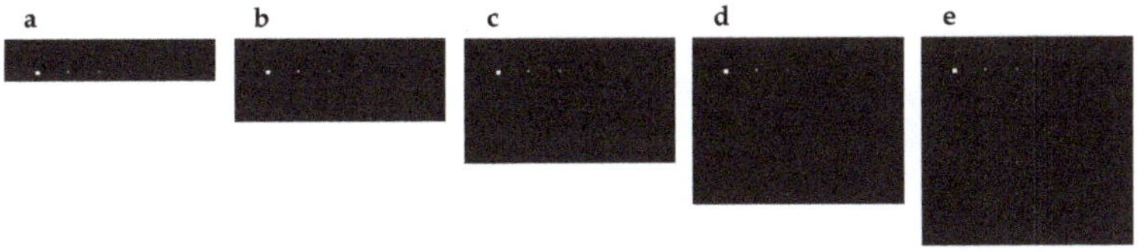

Figure 14. Real-time detection results with A used as desired target signature.

5.2.2. Detection Accuracy of HyMap

In order to further measure the performance of DPBS-CEM, we also focus on the detection results of HyMap data set. Figure 15 shows the results of the target F4 obtained by Global-CEM, SBS-CEM, and DPBS-CEM, respectively. For a more accurate representation of the detection results, we have an enlarged target region of interest, as shown in red boxes of target F4 and Figure 16 of target F1, F2, and F3. As we expected, in comparison to the target detection accuracy of SBS-CEM, DPBS-CEM has the same or even better performance. This conclusion is further verified by the AUC values in Table 5.

Table 5. AUC obtained by different algorithms for the targets.

	F1	F2	F3	F4
Global-CEM	0.9107	1	0.9067	0.9987
SBS-CEM [18]	0.9783	1	0.9862	0.9972
DPBS-CEM	0.9997	0.9999	0.9992	0.9994

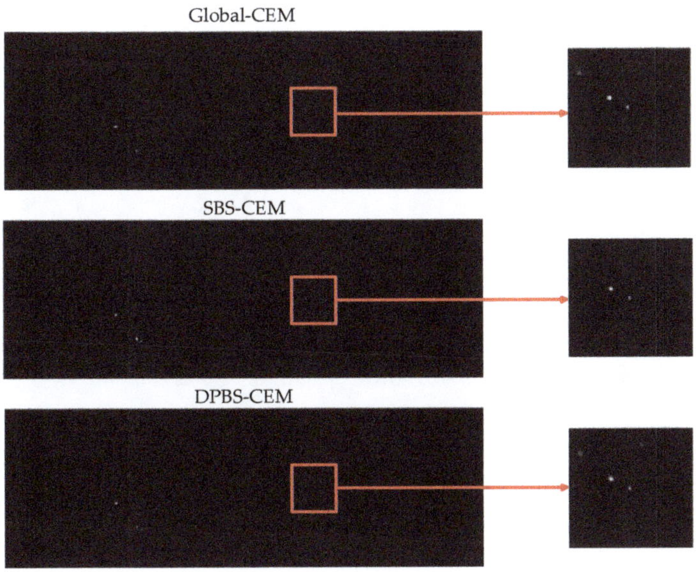

Figure 15. Detection results for target F4 obtained by different algorithms.

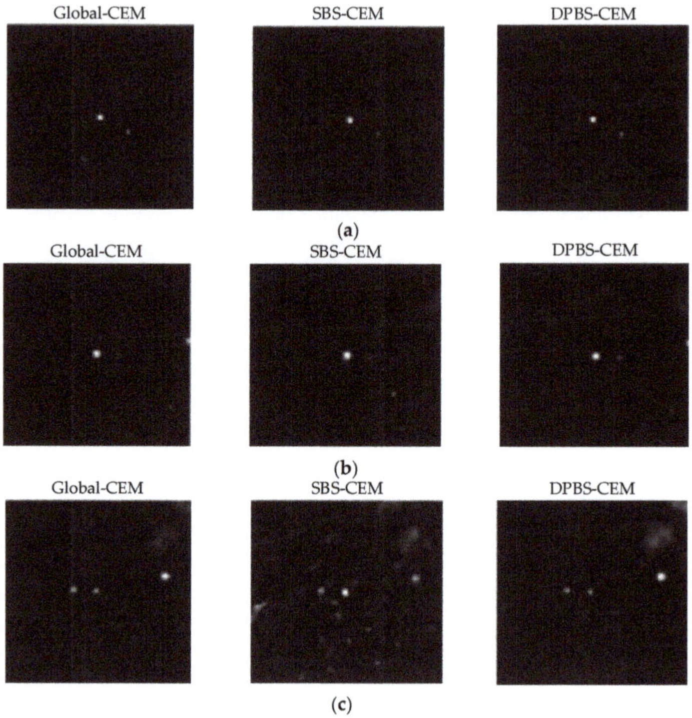

Figure 16. Detection results obtained by different algorithms for targets: (**a**) F1; (**b**) F2; and (**c**) F3.

5.3. Cross-Platform Performance Comparison

From the previous section, we can see that the proposed DPBS-CEM is very close to SBS-CEM [18] in detection accuracy, some detection results of DPBS-CEM are even superior to the latter one. Table 6 shows the processing time comparison of the proposed DPBS-CEM on different platforms (such as MATLAB, C++, and FPGA). The version of MATLAB used here is R2014a. The C++ environment directly uses the software simulation environment of Vivado HLS 2017.3. The clock frequency of FPGA is set at 200 MHz. As shown in Table 6, the processing time of DPBS-CEM implemented on FPGA has achieved significant improvements compared to MATLAB and C++ implementations. On the other hand, the processing time of our software versions is also superior to that of SBS-CEM software implementations [21] since the proposed DPBS-CEM algorithm is less computationally expensive than the SBS-CEM algorithm.

Table 6. Processing time measured for DPBS-CEM methods in MATLAB, C++, and FPGA implementations.

Platform	MATLAB (s)	C++ (s)	FPGA (s)
HyMap	60.7378	58.135	0.1568

5.4. Performance Comparison between DPBS-CEM and SBS-CEM

The FPGA design of DPBS-CEM is implemented on a Virtex7 XC7VX690T FPGA. This FPGA contains 108,300 slices, 433,200 six-input LUTs, 1470 BRAMs, and 3600 DSPs. To facilitate the performance comparison between DPBS-CEM and SBS-CEM, we selected HyMap, the same hyperspectral data source used by SBS-CEM, as our input HSI. Next, we compare the FPGA implementations of SBS-CEM and DPBS-CEM from two aspects of logic resources utilization and data processing speed.

Table 7 shows the resource utilization corresponding to SBS-CEM and DPBS-CEM. The right-hand side lists the unit's ratios and average ratios of DPBS-CEM and SBS-CEM. As Table 7 illustrates, the average resource utilization of DPBS-CEM is 5.21 times more than that of the SBS-CEM algorithm, which is caused by the deep pipelined structure. As aforementioned in Section 4.2.1, the intermediate data precision has a dramatic impact on the detection accuracy. Table 8 shows the relationship between the detection accuracy represented by AUC and the intermediate data precision. In Table 8, we set the data precision as fixed-point type with total of 32, 34, 36, 38, 40, and 42 bits, and identical 14 bits integer part. The experimental results demonstrate that the detection accuracy goes up sharply with the increase of data precision from 32 to 38 while keeps the same from 38 to 42. Due to the same bit-width of the integer part, it can be concluded that the bit-width of the fractional part mainly determines the detection accuracy. According to the experimental results, the bit-width of the fractional part should be more than 23.

The performance of DPBS-CEM has been greatly improved compared with SBS-CEM. Table 9 shows the number of clock cycles occupied by SBS-CEM and DPBS-CEM and the ratio between them. At the same clock frequency of 200 MHz, the number of clock cycles consumed by SBS-CEM is nearly 7.3 times more than that of DPBS-CEM. In other words, when processing the same image, the data processing speed of DPBS-CEM is 7.3 times faster than that of SBS-CEM. It is worthwhile to mention that our work is conducted by mainly using HLS.

Table 7. Comparison of resource utilization for the FPGA implementations of SBS-CEM and DPBS-CEM.

	SBS-CEM Units (G)	DPBS-CEM Units (Z)	Ratio $\left(\frac{Z}{G}\right)$
Number of DSP48Es	265	1396	5.27
Number of Block RAM	120	379	3.16
Number of Slices	12,088	58,167	4.81
Number of Flip Flops	28,245	217,958	7.72
Number of LUTs	21,730	111,073	5.11
Average Ratio	–	–	5.21

Table 8. Corresponding AUC values with different intermediate data accuracy of algorithm (we set F1 in HyMap image as the desired target).

Precision (Bit)	32	34	36	38	40	48
AUC	0.2530	0.4779	0.5463	0.9997	0.9997	0.9997

Table 9. Comparison of data processing speed for the FPGA implementations of SBS-CEM and DPBS-CEM.

	SBS-CEM	DPBS-CEM	Speedup
Frequency (MHz)	200	200	7.3×
Number of clock periods	229,607,996	31,360,557	

6. Discussion

CEM is an effective algorithm for subpixel target detection in hyperspectral imagery. The classical CEM needs to solve a large matrix inversion problem. SBS-CEM takes the Sherman-Morrison formula to update the inverse matrix for each pixel, which can avoid the complex calculation of large matrix inversion. However, SBS-CEM still uses sliding windows and has data dependency problems, which prevents its further performance improvement on target detection in terms of processing speed.

To solve these problems, we proposed an optimized algorithm called DPBS-CEM. It follows the same way that is used to update the inverse matrix gradually according to the Sherman-Morrison formula [28] but uses cumulative windows instead of sliding windows to reduce the number of calculations. Pixel data splitting and separating inverse matrix memories are utilized to remove the data dependency existing in the process of updating the inverse matrix. Moreover, we provide an FPGA implementation of the proposed DPBS-CEM whose deep pipelined architecture can be realized by using HLS.

According to the experimental results presented in this paper, the target detection accuracy of the proposed DPBS-CEM algorithm on two data sets are nearly the same. Compared to SBS-CEM, it has the same or even better detection accuracy. Regarding the processing speed performance, DPBS-CEM gained about 7.3 times speedup than that of SBS-CEM. It is worth noting that the proposed architecture of DPBS-CEM can also gain benefits in terms of scalability, portability, and flexibility with the help of HLS. This is particularly suitable for the real-time hyperspectral target detection applications on satellite.

7. Conclusions

In this paper, an optimized algorithm , referred to as DPBS-CEM for hyperspectral target detection, is proposed. A deep pipelined architecture of DPBS-CEM on FPGA is developed by using HLS as well. The experimental results show that the proposed FPGA implementation of DPBS-CEM has an extraordinary performance improvement in terms of data throughput without compromising for detection accuracy. Under the same test conditions, the detection speed of our proposed DPBS-CEM is about 7.3 times faster than that of SBS-CEM.

Acknowledgments: This work was partially supported by the National Natural Science Foundation of China (Nos. 61571345, 91538101, 61501346, 61502367, and 61701360) and the 111 project (B08038).

Author Contributions: Jie Lei and Yunsong Li conceived and designed the experiments; Dongsheng Zhao and Jing Xie performed the experiments; Jie Lei and Jing Xie analyzed the result data; Chein-I Chang provided suggestions on algorithm optimization and paper revision; Lingyun Wu, Xuepeng Li, Jintao Zhang, and Wenguang Li contributed reagents/materials/analysis tools; Jie Lei wrote the paper.

Conflicts of Interest: The authors declare no conflict of interest.

Abbreviations

The following abbreviations are used in this manuscript:

HSI	Hyperspectral image
FPGA	Field programmable gate array
CEM	Constrained energy minimization
HLS	High-level synthesis

Appendix A

Table A1. Vivado HLS optimization pragmas.

Directive	Description
#pragma HLS INTERFACE	Specifies how RTL ports are created from the function description.
#pragma HLS PIPELINE	Reduces the initiation interval by allowing the concurrent execution of operations within a loop or function.
#pragma HLS ARRAY_PARTITION	Partitions large arrays into multiple smaller arrays or into individual registers, to improve access to data and remove block RAM bottlenecks.
#pragma HLS UNROLL	Unroll for-loops to create multiple independent operations rather than a single collection of operations.
#pragma HLS DATAFLOW	Enable task level pipelining, allowing functions and loops to execute concurrently. Used to minimize interval.

References

1. Chang, C.I. *Hyperspectral Imaging: Spectral Techniques for Detection and Classification*; Kluwer Academic Publishers: Norwell, MA, USA, 2003.
2. Ryan, J.P.; Davis, C.O.; Tufillaro, N.B.; Kudela, R.M.; Gao, B.C. Application of the hyperspectral imager for the coastal ocean to phytoplankton ecology studies in Monterey Bay, CA, USA. *Remote Sens.* **2014**, *6*, 1007–1025.
3. Dale, L.M.; Thewis, A.; Boudry, C.; Rotar, I.; Dardenne, P.; Baeten, V.; Pierna, J.A.F. Hyperspectral imaging applications in agriculture and agro-food product quality and safety control: A review. *Appl. Spectrosc. Rev.* **2013**, *48*, 142–159.
4. Zhang, B.; Wu, D.; Zhang, L.; Jiao, Q.; Li, Q. Application of hyperspectral remote sensing for environment monitoring in mining areas. *Environ. Earth Sci.* **2012**, *65*, 649–658.
5. Cloutis, E.A. Review article hyperspectral geological remote sensing: Evaluation of analytical techniques. *Int. J. Remote Sens.* **1996**, *17*, 2215–2242.
6. Chang, C.I. Real-Time recursive hyperspectral sample and band processing: Algorithm architecture and implementation. In *Real-Time Recursive Hyperspectral Sample and Band Processing*, 1st ed.; Springer: Berlin, Germany, 2017; pp. 123–156, ISBN 978-3-319-45170-1.
7. Wang, Y.; Huang, S.; Liu, D.; Wang, H. A target detection method for hyperspectral imagery based on two-time detection. *J. Indian Soc. Remote Sens.* **2017**, *45*, 239–246.
8. Zou, Z.; Shi, Z. Hierarchical suppression method for hyperspectral target detection. *IEEE Trans. Geosci. Remote Sens.* **2016**, *54*, 330–342.
9. He, C.; Zhao, Y.; Tian, J.; Shi, P.; Huang, Q. Improving change vector analysis by cross-correlogram spectral matching for accurate detection of land-cover conversion. *Int. J. Remote Sens.* **2013**, *34*, 1127–1145.
10. Chaudhuri, S.; Chatterjee, S.; Katz, N.; Nelson, M.; Goldbaum, M. Detection of blood vessels in retinal images using two-dimensional matched filters. *IEEE Trans. Med. Imaging* **1989**, *8*, 263–269.
11. Yuhas, R.H.; Goetz, A.F.; Boardman, J.W. Discrimination among semi-arid landscape endmembers using the spectral angle mapper (SAM) algorithm. In *JPL, Summaries of the Third Annual JPL Airborne Geoscience Workshop*; NASA: Washington, DC, USA, 1992; pp.147–149.
12. Du, Q.; Ren, H.; Chang, C.I. A comparative study for orthogonal subspace projection and constrained energy minimization. *IEEE Trans. Geosci. Remote Sens.* **2003**, *41*, 1525–1529.
13. Ren, H.; Chang, C.I. A target-constrained interference-minimized filter for subpixel target detection in hyperspectral imagery. In Proceedings of the Geoscience and Remote Sensing Symposium, Honolulu, HI, USA, 24–28 July 2000; pp. 1545–1547.
14. Manolakis, D.; Marden, D.; Shaw, G.A. Hyperspectral image processing for automatic target detection applications. *J. Lincoln Lab.* **2003**, *14*, 79–116.
15. Scharf, L.L.; Friedlander, B. Matched subspace detectors. *IEEE Trans. Signal Process.* **1994**, *42*, 2146–2157.
16. Harsanyi, J.C.; Chang, C.I. Hyperspectral image classification and dimensionality reduction: An orthogonal subspace projection approach. *IEEE Trans. Geosci. Remote Sens.* **1994**, *32*, 779–785.
17. Chen, Y.; Nasrabadi, N.M.; Tran, T.D. Sparse representation for target detection in hyperspectral imagery. *IEEE J. Sel. Top. Signal Process.* **2011**, *5*, 629–640.
18. Mittal, S.; Vetter, J.S. A survey of CPU-GPU heterogeneous computing techniques. *ACM Comput. Surv.* **2015**, *47*, 69.
19. Plaza, A.; Chang, C.I. Clusters Versus FPGA for parallel processing of hyperspectral imagery. *Int. J. High Perform. Comput. Appl.* **2008**, *22*, 366–385.
20. Wang, J.; Chang, C.; Cao, M. FPGA design for constrained energy minimization. *Proc. SPIE* **2004**, 262–273, doi:10.1117/12.518559.
21. Yang, B.; Yang, M.; Plaza, A.; Gao, L.; Zhang, B. Dual-mode FPGA implementation of target and anomaly detection algorithms for real-time hyperspectral imaging. *IEEE J.-STARS* **2015**, *8*, 2950–2961.
22. Gonzalez, C.; Bernabe, S.; Mozos, D.; Plaza, A. FPGA implementation of an algorithm for automatically detecting targets in remotely sensed hyperspectral images. *IEEE J. Sel. Top. Appl. Earth Obs. Remote Sens.* **2016**, *9*, 4334–4343.
23. Santos, L.; López, J.F.; Sarmiento, R.; Vitulli, R. FPGA implementation of a lossy compression algorithm for hyperspectral images with a high-level synthesis tool. In Proceedings of the 2013 NASA/ESA Conference on Adaptive Hardware and Systems, Torino, Italy, 24–27 June 2013; pp. 107–114.

24. García, A.; Santos, L.; López, S.; Callicó, G.M.; Lopez, J.F.; Sarmiento, R. Efficient lossy compression implementations of hyperspectral images: Tools, hardware platforms, and comparisons. In *Satellite Data Compression, Communications, and Processing X*; International Society for Optics and Photonics: Bellingham, WA, USA, 2014.
25. Domingo, R.; Salvador, R.; Fabelo, H. High-level design using Intel FPGA OpenCL: A hyperspectral imaging spatial-spectral classifier. In Proceedings of the 2017 IEEE 12th International Symposium on Reconfigurable Communication-Centric Systems-on-Chip (ReCoSoC), Madrid, Spain, 12–14 July 2017; pp. 1–8.
26. Del Sozzo, E.; Solazzo, A.; Miele, A.; Santambrogio, M.D. On the automation of high level synthesis of convolutional neural networks. In Proceedings of the 2016 IEEE International Symposium on Parallel and Distributed Processing, Chicago, IL, USA, 23–27 May 2016; pp. 217–224.
27. Guan, Y.; Liang, H.; Xu, N.; Wang, W.; Shi, S.; Chen, X. FP-DNN: An automated framework for mapping deep neural networks onto FPGAs with RTL-HLS hybrid templates. In Proceedings of the 2017 IEEE International Symposium on Field-Programmable Custom Computing Machines, Napa, CA, USA, 30 April–2 May 2017; pp. 152–159.
28. Hager, W.W. Updating the inverse of a matrix. *Siam Rev.* **1989**, *31*, 221–239.
29. Chang, C.I.; Li, H.C.; Song, M.; Liu, C.; Zhang, L. Real-time constrained energy minimization for sub pixel detection. *IEEE J.-STARS* **2015**, *8*, 2545–2559.
30. Chang, C.I. Real-Time Recursive Hyperspectral Sample Processing for Active Target Detection: Constrained Energy Minimization. In *Real-Time Recursive Hyperspectral Sample and Band Processing*, 1st ed.; Springer: Berlin, Germany, 2017; pp. 123–156, ISBN 978-3-319-45170-1.
31. Nasrabadi, N.M. Regularized spectral matched filter for target recognition in hyperspectral imagery. *IEEE Signal. Proc. Lett.* **2008**, *15*, 317–320.
32. Wang, J.; Chang, C.I. Applications of independent component analysis in endmember extraction and abundance quantification for hyperspectral imagery. *IEEE Trans. Geosci. Remote Sens.* **2006**, *44*, 2601–2616.
33. Chang, C.I. Design of Synthetic Image Experiments. In *Hyperspectral Data Processing: Algorithm Design and Analysis*, 1st ed.; John Wiley & Sons: Hoboken, NJ, USA, 2013; pp. 103–113, ISBN 978-0-471-69056-6.
34. Snyder, D.; Kerekes, J.; Fairweather, I.; Crabtree, R.; Shive, J.; Hager, S. Development of a web-based application to evaluate target finding algorithms. In Proceedings of the Geoscience and Remote Sensing Symposium, Boston, MA, USA, 6–11 July 2008; pp. II-915–II-918.
35. Parker, D.R.; Gustafson, S.C.; Ross, T.D. Receiver operating characteristic and confidence error metrics for assessing the performance of automatic target recognition systems. *Opt. Eng.* **2005**, *44*, 097202.
36. Metz, C.E. Basic principles of ROC analysis. *Semin. Nucl. Med.* **1978**, *8*, 283–298.

remote sensing

MDPI

Article

Multiscale Union Regions Adaptive Sparse Representation for Hyperspectral Image Classification

Fei Tong [1,2], Hengjian Tong [1,*], Junjun Jiang [1] and Yun Zhang [2]

1 School of Computer Science, China University of Geosciences, Lumo Road 388, Wuhan 430074, China; tfcug93@163.com (F.T.); junjun0595@163.com (J.J.)

2 Department of Geodesy and Geomatics Engineering, University of New Brunswick, 15 Dineen Drive, Fredericton, NB E3B 5A3, Canada; yunzhang@unb.ca

* Correspondence: thj26@cug.edu.cn; Tel.: +86-027-6788-3716

Received: 13 July 2017; Accepted: 21 August 2017; Published: 23 August 2017

Abstract: Sparse Representation has been widely applied to classification of hyperspectral images (HSIs). Besides spectral information, the spatial context in HSIs also plays an important role in the classification. The recently published Multiscale Adaptive Sparse Representation (MASR) classifier has shown good performance in exploiting spatial information for HSI classification. But the spatial information is exploited by multiscale patches with fixed sizes of square windows. The patch can include all nearest neighbor pixels but these neighbor pixels may contain some noise pixels. Then another research proposed a Multiscale Superpixel-Based Sparse Representation (MSSR) classifier. Shape-adaptive superpixels can provide more accurate representation than patches. But it is difficult to select scales for superpixels. Therefore, inspired by the merits and demerits of multiscale patches and superpixels, we propose a novel algorithm called Multiscale Union Regions Adaptive Sparse Representation (MURASR). The union region, which is the overlap of patch and superpixel, can make full use of the advantages of both and overcome the weaknesses of each one. Experiments on several HSI datasets demonstrate that the proposed MURASR is superior to MASR and union region is better than the patch in the sparse representation.

Keywords: classification; hyperspectral image (HSI); multiscale union regions adaptive sparse representation (MURASR); multiscale spatial information

1. Introduction

Hyperspectral images have been widely applied to remote sensing image applications, such as land cover classification [1], target detection [2], anomaly detection [3], spectral unmixing [4] and others. Each pixel in HSI has hundreds of narrow contiguous bands, spanning from visible to infrared spectrum [5], which makes it possible to detect and distinguish various objects with higher accuracy [6]. However, increasing the number of spectral bands or features of an HSI pixel does not always help to increase the classification accuracy. Therefore, how to make full use of the information in HSIs is a problem in practical applications.

Many algorithms have been developed for the classification of HSIs. Among these, there are some well-known pixelwise classifiers, such as the support vector machine (SVM) [7–9], support vector conditional random classifier [10], multinomial logistic regression [11], neural network [12] and adaptive artificial immune network [13]. These pixelwise classifiers can make full use of the spectral information of HSIs, but the classification results are often noisy because the spatial information is not considered.

Therefore, some recent researches incorporated the spatial information in HSI classification to enhance the classification performance. The basic way to use spatial information is to assume that the pixels within a local region usually represent the same material and have similar spectral characteristics [1]. Various researches [14–25] have been done based on this assumption. Besides these researches, Sparse representation (SR), which is based on the observation that spectral pixels of a particular class should lie in a low-dimensional subspace spanned by dictionary atoms (training pixels) from the same class, is also employed. In [26], a Joint Sparse Representation Classification (JSRC) method has been proposed to incorporate spectral information and spatial information. The spatial information is expressed by a fixed-size local square window centered with the test pixel. Then all pixels in the window are simultaneously joint represented by a few common atoms in the specified dictionary. The JSRC can achieve a good performance but the optimal size of the window cannot be determined easily. In [27], a stepwise Markov random field (MRF) optimization was proposed to exploit spatial information based on the result of multitask joint sparse representation. In [28], MASR was proposed to release the difficulty in choosing region size. Instead of choosing a single scale, this method extends the spatial information to several scales to take advantage of correlations among multiple region scales for HSI classification. But the multiscale regions used in MASR refer to multiscale patches which may contain noise pixels. Better than patch region, shape-adaptive superpixel can provide more accurate spatial information. In [29], the superpixel was introduced to replace the patch region. Then a shape-adaptive local smooth region was generated for each test pixel by a shape-adaptive algorithm in [30]. The latest research proposed a Multiscale Superpixel-Based Sparse Representation [31]. In this research, multiscale superpixels were generated and then each scale was represented by JSRC. Finally, a fusion result was gotten from multiscale results by majority voting. But the selection of scales for superpixels is still a problem. Although it uses multiscale to release the difficulty of selecting segmentation scale, it still needs a fundamental number of superpixels determined empirically.

In fact, patch and superpixel both have their own advantages and shortages. The patch can include all nearest neighbors but it also may contain noise pixels. Shape-adaptive superpixel can exploit more accurate spatial information but there are still some mixed superpixels when the scale is not optimal. In a mixed superpixel, there must be wrong representation because all pixels in the superpixel share the same representation. Inspired by merits and demerits of patch and superpixel, we propose to use a union region to replace the patch and superpixel. Union region refers to the overlap of patch and superpixel. Compared with patch, union region includes more similar pixels for the test pixel aiming at decreasing the effect of noise pixels. Compared with superpixel, union region provides more direct neighbors for the test pixel to enhance the representation of pixels located in the wrong superpixel. In addition, the required superpixels for generating union regions don't need empirical scale. The scales are determined by the size of the image and the corresponding patch sizes. By replacing patch in MASR with union region, we get a new algorithm called Multiscale Union Regions Adaptive Sparse Representation (MURASR). MURASR also adopts a probability majority voting method to optimize the classification result generated from the sparse representation. Experiment results show that the union region based algorithms always perform better than patch region based algorithms and the proposed MURASR outperforms other algorithms in terms of quantitative metrics and visual quality on the classification maps.

The rest parts of the paper are organized as follows. The JSRC and MASR are briefly introduced in Section 2. The details of proposed MURASR method are described in Section 3. The experimental results and discussions are presented in Section 4. Finally, Section 5 summarizes the paper and future works are suggested. The outline of the MURASR is illustrated in Figure 1.

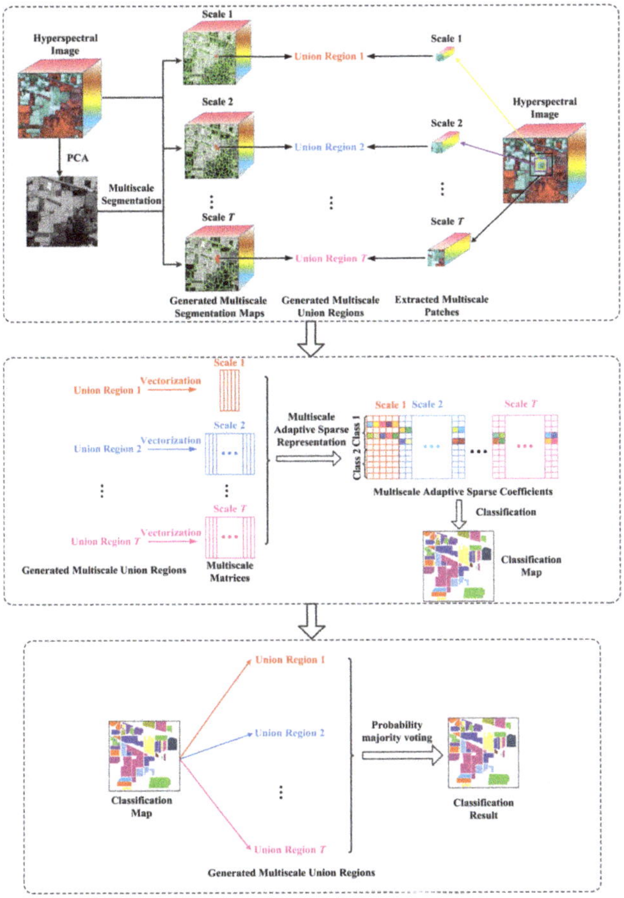

Figure 1. Outline of the proposed MURASR framework.

2. Background

2.1. JSRC

The sparse representation classification (SRC) framework was first proposed for face recognition [32]. Then Chen et al. extended the SRC to pixelwise HSI classification, which relied on the observation that spectral pixels of a particular class should lie in a low-dimensional subspace spanned by dictionary atoms (training pixels) from the same class. But spatial information is not considered by Pixelwise Sparse Representation. Therefore, based on the observation that neighboring pixels belonging to the same class usually are strongly correlated with each other, JSRC is introduced to capture such spatial correlations by assuming that neighboring pixels within a region of fixed size can be jointly represented by a few common atoms from a structural dictionary. Concretely, let $\mathbf{y} \in \mathbb{R}^{M \times 1}$ be a pixel with M denoting the number of spectral bands and $\mathbf{D} = [\mathbf{D}_1, \cdots, \mathbf{D}_c, \cdots, \mathbf{D}_C] \in \mathbb{R}^{M \times N}$ be a structure dictionary, where $\mathbf{D}_c \in \mathbb{R}^{M \times N_c}$, $c = 1, \cdots, C$ is the cth class subdictionary whose columns (atoms) are extracted from the training pixels; C is the number of classes; N_c is the number of atoms in subdictionary $\mathbf{D}_c$; and $N = \sum_{c=1}^{C} N_c$ is the total number of atoms in $\mathbf{D}$. Specifically, the size of a region surrounding the test pixel $\mathbf{y}_1$ is denoted by $W \times W$, and pixels within such a region can be denoted by a matrix $\mathbf{Y} = [\mathbf{y}_1, \mathbf{y}_2, \cdots, \mathbf{y}_{W\times W}]$. The matrix can be compactly represented as:

$$\begin{aligned}\mathbf{Y} &= [\mathbf{y}_1, \mathbf{y}_2, \cdots, \mathbf{y}_{W\times W}] = [\mathbf{D}\mathbf{A}_1, \mathbf{D}\mathbf{A}_2, \cdots, \mathbf{D}\mathbf{A}_{W\times W}] \\ &= \mathbf{D}[\mathbf{A}_1, \mathbf{A}_2, \cdots, \mathbf{A}_{W\times W}] = \mathbf{D}\mathbf{A}\end{aligned} \tag{1}$$

where $\mathbf{A} = [\mathbf{A}_1, \mathbf{A}_2, \cdots, \mathbf{A}_{W\times W}]$ is the sparse coefficients matrix corresponding to $\mathbf{Y}$. Since the indexes of the selected atoms in $\mathbf{D}$ are determined by the positions of nonzero coefficients in $[\mathbf{A}_1, \mathbf{A}_2, \cdots, \mathbf{A}_{W\times W}]$, the neighboring pixels $[\mathbf{y}_1, \mathbf{y}_2, \cdots, \mathbf{y}_{W\times W}]$ can be represented by a small set of common atoms by enforcing a few nonzero rows on the sparse coefficients matrix $\mathbf{A}$. Then, matrix $\mathbf{A}$ can be obtained by solving the following optimization problem:

$$\hat{\mathbf{A}} = \arg\min_{\mathbf{A}} \|\mathbf{Y} - \mathbf{D}\mathbf{A}\|_F \quad \text{subject to } \|\mathbf{A}\|_{\text{row},0} \leqslant K \tag{2}$$

where $\mathbf{A}_{\text{row},0}$ denotes the joint sparse norm, which is used to select a number of the most representative nonzero rows in $\mathbf{A}$, and $\|\cdot\|_F$ is the Frobenius norm. A variant of the OMP algorithm called the simultaneous OMP (SOMP) [33,34], can be used to efficiently obtain an approximate solution. After $\hat{\mathbf{A}}$ is recovered, the label of test pixel $\mathbf{y}_1$ can be decided by the minimal total error:

$$\hat{c} = \arg\min_{c} \|\mathbf{Y} - \mathbf{D}_c\hat{\mathbf{A}}_c\|_F, \qquad c = 1, \cdots, C \tag{3}$$

where $\hat{\mathbf{A}}_c$ denotes the rows in $\hat{\mathbf{A}}$ associated with the cth class.

2.2. MASR

Compared with pixelwise SRC model, the JSRC can achieve more accurate classification results because of incorporating spatial information of local regions. However, the region size (or the *region scale*) has great influence on the classification performance. It is of great importance to determine an optimal region scale for the JSRC.

Then Fang et al. proposed the MASR to release the difficulty of choosing region scale. The MASR effectively exploits spatial information at multiple scales via an adaptive sparse strategy. Not only does the adaptive sparse strategy restrict pixels from different scales to be represented by training atoms from a particular class but also allow the selected atoms for these pixels to be varied, thus providing an improved representation. Given one test pixel $\mathbf{y}_1$ in HSI, its T neighboring regions are selected via different predefined scales. Neighboring regions are defined by multiscale patches centered with test pixel. Then a multiscale matrix $\mathbf{Y}^{\text{mp}} = [\mathbf{Y}_1, \cdots, \mathbf{Y}_t, \cdots, \mathbf{Y}_T]$ can be constructed by pixels within the selected regions, where the $\mathbf{Y}_t$ includes pixels from the tth scale region. Since spatial structures and characteristics for different scales of regions are distinct, the generated multiscale matrix $\mathbf{Y}^{\text{mp}}$ for the test pixel $\mathbf{y}_1$ should provide complementary yet correlated information, which can be utilized to classify $\mathbf{y}_1$ more accurately.

In MASR, an adaptive sparse strategy is adopted to utilize the correlated information among multiscales and achieve a flexible selection process for atoms. An important part of the adaptive strategy is the adoption of a collection of adaptive sets. Each adaptive set is denoted as the indexes of a set of nonzero scalar coefficients, which belong to the same class in the multiscale sparse matrix $\mathbf{A}^{\text{mp}}$. By combining the adaptive set with the $\ell_{\text{row},0}$ norm, a new adaptive norm $\ell_{\text{adaptive},0}$ is created on $\mathbf{A}^{\text{mp}}$, which can be used to select a small number of adaptive sets from $\mathbf{A}^{\text{mp}}$. Then, $\mathbf{A}^{\text{mp}}$ matrix can be recovered by applying the adaptive norm as follows:

$$\begin{aligned}\hat{\mathbf{A}}^{\text{mp}} &= \arg\min_{\mathbf{A}^{\text{mp}}} \|\mathbf{Y}^{\text{mp}} - \mathbf{D}\mathbf{A}^{\text{mp}}\|_F \\ &\text{subject to } \|\mathbf{A}^{\text{mp}}\|_{\text{adaptive},0} \leqslant K\end{aligned} \tag{4}$$

After recovering the multiscale sparse representation matrix $\hat{\mathbf{A}}^{\mathrm{mp}}$, a single decision can be made on the test pixel $\mathbf{y}_1$ based on the lowest total representation error:

$$\hat{c} = \arg\min_{c} \|\mathbf{Y}^{\mathrm{mp}} - \mathbf{D}_c \hat{\mathbf{A}}_c^{\mathrm{mp}}\|_F, c = 1, \cdots, C \tag{5}$$

where $\hat{\mathbf{A}}_c^{\mathrm{mp}}$ represents rows in $\hat{\mathbf{A}}^{\mathrm{mp}}$ corresponding to the cth class.

3. Multiscale Union Regions Adaptive Sparse Representation

The aforementioned MASR shows good performance for HSI classification. But the MASR utilizes multiscale patches to exploit spatial information. In a patch, maybe most of the pixels are different from the test pixel, such as a pixel on the edge of a building. The classification may be misled by those noise pixels from other classes which are similar to the atoms in the dictionary, thus providing an incorrect classification for test pixel. In computer vision, superpixels have been studied to provide an efficient representation, which can facilitate visual recognition [35–37]. Each superpixel is a perceptually meaningful region, whose shape and size can be adaptively changed according to different spatial structures. But how to find an optimal scale for superpixels is still a challenge. Without optimal scale, some mixed superpixels will be generated. Based on the fact that patch and superpixel may include pixels from different classes, a multiscale union regions adaptive sparse representation model is proposed to decrease the influence of noise pixels for the test pixel. The union region is the overlap of the patch and corresponding superpixel with the same scale (see Figure 2). For a test pixel, if the patch includes some noise pixels, the superpixel can provide more similar pixels to reduce the impact of noise pixels. In the same way, if the test pixel is located in the wrong superpixel which has seldom pixels similar to test pixel, the patch can provide more similar pixels to enhance the right representation.

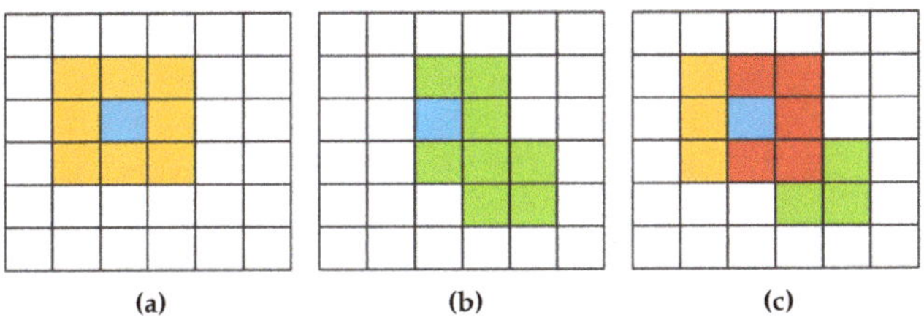

Figure 2. Three kinds of spatial regions: (**a**) fixed-size patch; (**b**) adaptive size superpixel; and (**c**) union of patch and superpixel. The blue pixel represents test pixel, orange pixels are neighbors defined by patch, green pixels are neighbors defined by superpixel and red pixels are overlap of neighbors defined by patch and superpixel.

3.1. Generation of Multiscale Union Regions

Before generating multiscale union regions, we should get multiscale superpixels. There are various researches focusing on the segmentation [36–39]. In this paper, an oversegmentation algorithm called ERS [37] is applied to generate 2-D superpixel maps on the base images because of its high efficiency. Unlike the single-band gray or three-band color image, the HSI usually has hundreds of spectral bands. To improve the computational efficiency, PCA [40] is first used to reduce the spectral bands of the HSI. Since the important information of the HSI exists in the principle components (e.g., first three principle components), they are used as the base images. In this paper, only the first principle component is chosen as the base image. Instead of choosing scales for superpixels empirically, we calculate scales of superpixels based on corresponding patch sizes. Assuming that $\mathbf{PS}_t$ refers to the patch size of tth scale and $\mathbf{N}_{total}$ is the total number of pixels in the image (note that origin image will be extended for edge pixels), the superpixels number $\mathbf{n}_t$ for tth segmentation is calculated as:

$$\mathbf{n}_t = \mathbf{N}_{total}/\mathbf{PS}_t \tag{6}$$

In this way, the average size of superpixels is equal to patch size. Then most superpixels will have similar sizes with patches. It guarantees that superpixel and patch can have similar influence on union region. What's more, with the increasing of patch size, the superpixels number decreases fast. Thus, only limited number of segmentations can be generated. According to the performance of limited number of segmentations, it will be easier for users to determine the scales number. After segmentations, T superpixels are generated for each test pixel $\mathbf{y}_1$ and these superpixels construct the corresponding multiscale matrix $\mathbf{Y}^{\mathrm{ms}} = [\mathbf{Y}_1, \cdots, \mathbf{Y}_t, \cdots, \mathbf{Y}_T]$, where the $\mathbf{Y}_t$ includes pixels from the tth superpixel. Then for a specific tth scale, the union region $\mathbf{Y}_t^{\mathrm{mu}}$ is defined as following:

$$\mathbf{Y}_t^{\mathrm{mu}} = \mathbf{Y}_t^{\mathrm{ms}} \cup \mathbf{Y}_t^{\mathrm{mp}} \tag{7}$$

3.2. Multiscale Union Regions Adaptive Sparse Representation

For a test pixel $\mathbf{y}_1$, the corresponding multiscale matrix is $\mathbf{Y}^{\mathrm{mu}} = [\mathbf{Y}_1, \cdots, \mathbf{Y}_t, \cdots, \mathbf{Y}_T]$, where $\mathbf{Y}_t$ is the union of $\mathbf{Y}_t^{\mathrm{mp}}$ and $\mathbf{Y}_t^{\mathrm{ms}}$. Then the sparse coefficients matrix $\mathbf{A}^{\mathrm{mu}}$ can be recovered by solving following problem:

$$\begin{aligned} \hat{\mathbf{A}}^{\mathrm{mu}} &= \arg\min_{\mathbf{A}^{\mathrm{mu}}} \|\mathbf{Y}^{\mathrm{mu}} - \mathbf{D}\mathbf{A}^{\mathrm{mu}}\|_F, \\ &\text{subject to } \|\mathbf{A}^{\mathrm{mu}}\|_{\mathrm{adaptive},0} \leqslant K \end{aligned} \tag{8}$$

To solve this problem, the method used in MASR is applied. At each iteration, the current residual correlation matrix is calculated firstly. Then a a new adaptive set based on the current residual correlation matrix will be selected. Once the selecting of the new adaptive set is finished, the new adaptive set will be merged with previously selected adaptive sets. Then the sparse coefficients matrix is estimated based on the merged adaptive sets. Finally, the residue is updated. The iterations will stop if the termination criterion is satisfied. After the multiscale sparse representation matrix $\hat{\mathbf{A}}^{\mathrm{mu}}$ is recovered, the final label of the test pixel $\mathbf{y}_1$ can be determined by minimal total representation error:

$$\hat{c} = \arg\min_{c} \|\mathbf{Y}^{\mathrm{mu}} - \mathbf{D}_c\hat{\mathbf{A}}_c^{\mathrm{mu}}\|_F, c = 1, \cdots, C \tag{9}$$

3.3. Probability Majority Voting

Because multiscale union regions adaptive sparse representation is a pixel-based classifier, there will be some pepper salt noise pixels in ground truth objects. Therefore, a majority voting process will be helpful to optimize the classification result. As mentioned above, for each test pixel in each scale, a union region will be generated. Then for the union region, the probabilities belonging to all classes are calculated. If a union region at ith scale contains $\mathbf{N}_i^{total}$ labeled pixels and $\mathbf{N}_i^j$ pixels classified to jth class, the probability belonging to jth class $\mathbf{P}_i^j$ is calculated as:

$$\mathbf{P}_i^j = \mathbf{N}_i^j / \mathbf{N}_i^{total} \tag{10}$$

Assuming that there are $\mathbf{k}$ classes, T scales of segmentation maps, the class label of the test pixel $\hat{j}$ can be obtained by:

$$\hat{j} = \arg\max_{j} \left(\sum_{i=1}^{T} \mathbf{P}_i^j\right), j = 1, \cdots, \mathbf{k} \tag{11}$$

4. Experimental Results and Discussion

4.1. Data Sets

To verify the effectiveness of the proposed MURASR method and superiority of the union region, experiments are conducted on the following three hyperspectral data sets: the Airborne Visible/Infrared Imaging Spectrometer (AVIRIS) Indian Pines data, the AVIRIS Salinas data, and the Reflective Optics System Imaging Spectrometer (ROSIS-03) University of Pavia data. The AVIRIS Indian Pines image has 220 data channels with the size of 145 × 145 across the spectral range from 0.2 to 2.4 μm. It was captured over the agricultural Indian Pine test site in northwestern Indiana with a spatial resolution of 20 m per pixel. Before classification, 20 water absorption bands (No. 104–108, 150–163 and 220) were discarded [41]. Figure 3a,b show the color composite of the Indian Pines image and the corresponding reference data with 16 reference classes from different types of crops.

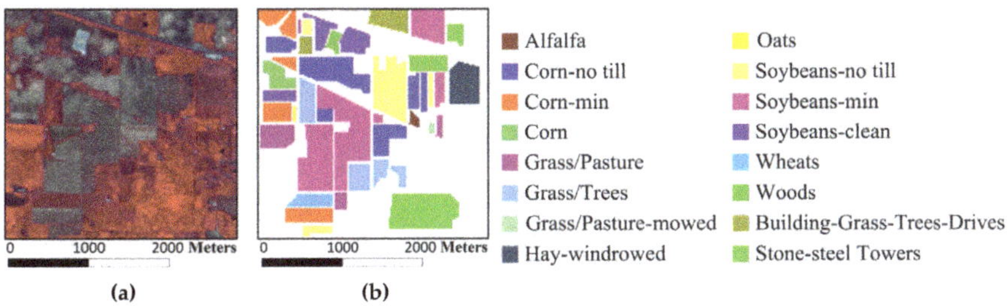

Figure 3. Indian Pines image: (**a**) three-band color composite image; (**b**) reference image.

The Salinas image was also acquired by the AVIRIS sensor over Salinas Valley, California. The image is of size 512 × 217 × 224 with a spatial resolution of 3.7 m per pixel. Similar to the Indian Pines image, 20 water absorption spectral bands (No. 108–112, 154–167 and 224) were removed and 16 different reference classes are considered for this image. Figure 4a,b show the color composite of the Salinas image and the corresponding reference data.

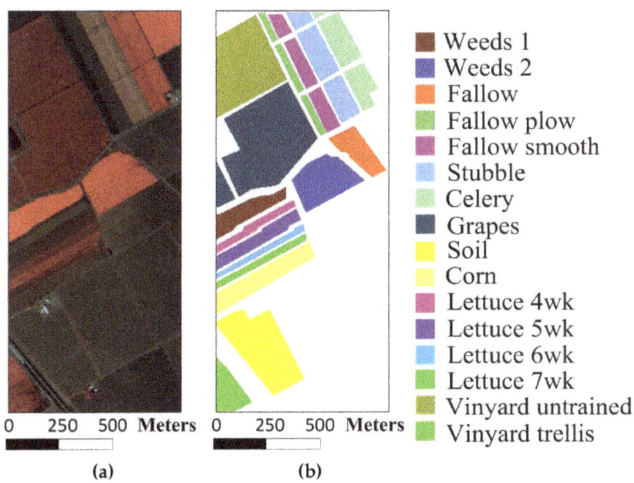

Figure 4. Salinas image: (**a**) three-band color composite image; (**b**) reference image.

The University of Pavia image, which captures an urban area surrounding the University of Pavia, Italy, was recorded by the ROSIS-03 sensor. The image is of size 610 × 340 × 115 with a spatial resolution of 1.3 m per pixel and a spectral coverage ranging from 0.43 to 0.86 μm. The 12 very noisy channels were discarded before the experiments, and nine information classes are considered for this image. Figure 5a,b show the color composite of the University of Pavia image and the corresponding reference data.

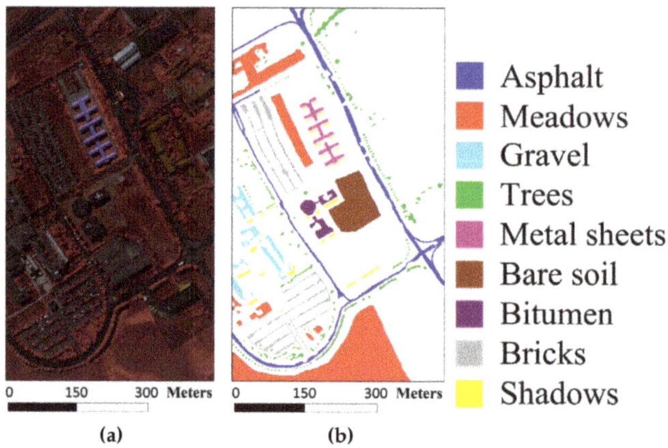

Figure 5. University of Pavia image: (**a**) three-band color composite image; (**b**) reference image.

4.2. Comparison of Experiment Results

In the experiments, all related algorithms are based on sparse representation. Except for published algorithms SRC, JSRC and MASR, JUSRC (Joint Union Sparse Representation Classification), MJSRC (Multiscale Joint Sparse Representation Classification), MJUSRC (Multiscale Joint Union Sparse Representation Classification), MURASR* and MURASR were conducted in the experiments. To verify the priority of union region further, the patch used in JSRC was replaced by JUSRC with the union region. For demonstrating the superiority of multiscale adaptive strategy, we extended the JSRC and JUSRC with a simple multiscale scheme that applied the majority voting to the results of all scales for the final decision-making. The extended algorithms are called MJSRC and MJUSRC. What's more, the MURASR* is the MURASR without probability majority voting process. The comparison between MURASR* and MURASR can show the difference of whether the probability majority voting method was used or not. The parameters for the SRC, JSRC, and JUSRC algorithms were tuned to reach the best results in these experiments. For all multiscale algorithms, seven different scales were simultaneously adopted, and the selected region scales were as follows: 3 × 3, 5 × 5, 7 × 7, 9 × 9, 11 × 11, 13 × 13, and 15 × 15. Then superpixels numbers for segmentation were calculated with Equation (6) and listed in Table 1. Other parameters in MJSRC, MJUSRC, MASR, MURASR*, and MURASR were the same as [28]. To evaluate the performance of classifiers, three objective metrics (overall accuracy (OA), average accuracy (AA) and kappa coefficient) are adopted. In addition, the McNemar's test is applied to analyse the experiment results. The McNemar's test is based on the standardized normal test statistic, as described in [42]:

$$\mathbf{Z} = \frac{h_{12} - h_{21}}{\sqrt{h_{12} + h_{21}}} \tag{12}$$

where h_{12} represents the samples correctly classified by method 1 but incorrectly classified by method 2. If $|\mathbf{Z}| > 1.96$, the accuracy between two methods can be considered statistically significant. The sign of the $\mathbf{Z}$ indicates which method is better. If $\mathbf{Z} > 0$, the method 1 is more accurate than method 2.

The Indian Pines data set was classified firstly. 10% of the labeled pixels were randomly sampled for training from each class, while the rest 90% were used to test the classifiers (see Table 2). The classification maps generated by different classifiers on the Indian Pines image are shown in Figure 6. The details of the classification results averaged by ten runs with randomly sampled training samples are tabulated in Table 3. The results of the McNemar's tests between classifiers are listed in Table 4. It is easy to find that JUSRC, MJUSRC and MURASR* perform better than JSRC, MJSRC and MASR, which demonstrates the priority of union region over patch region. In addition, the multiscale majority voting based MJSRC and MJUSRC perform worse than the multiscale adaptive strategy based MASR and MURASR* for this image. Compared with MJSRC and MJUSRC, accuracy improvements of MASR and MURASR* are more than 3%. MURASR gets a better result than MURASR* in accuracy and classification map. As can be observed from the classification maps of MURASR* and MURASR, many misclassifications in MURASR* can be eliminated efficiently by probability majority voting method. What's more, MURASR performs best among all algorithms in terms of OA and AA, and the results of the McNemar's test are statistically significant and coherent with the obtained overall accuracies.

Table 1. Number of Superpixels in Each Scale.

	1	2	3	4	5	6	7
Indian Pines	2809	1011	515	312	208	149	112
Salinas	13,500	4860	2479	1500	1004	718	540
University of Pavia	24,544	8835	4508	2727	1825	1307	981

Table 2. Sixteen reference classes in the Indian Pines image.

Class	Name	Train	Test
1	Alfalfa	5	41
2	Corn-no till	143	1285
3	Corn-min	83	747
4	Corn	24	213
5	Grass/Pasture	48	435
6	Grass/Trees	73	657
7	Grass/Pasture-mowed	3	25
8	Hay-windrowed	48	430
9	Oats	2	18
10	Soybeans-no till	97	875
11	Soybeans-min	246	2209
12	Soybeans-clean	59	534
13	Wheat	21	184
14	Woods	127	1138
15	Building-Grass-Trees-Drives	39	347
16	Stone-steel Towers	9	84
	Total	1027	9222

The second experiment was performed on the Salinas data set. To compare the classification with MASR, only 1% of the labeled pixels for each class were randomly selected for training. Then the remaining 99% labeled data were classified with the classifiers to demonstrate the superiority of the proposed MURASR (see Table 5). The classification maps for various classifiers are illustrated in Figure 7 and the average quantitative results of ten runs are tabulated in Table 6. Moreover, the results of the McNemar's tests are shown in Table 7. As can be observed, union region based algorithms JUSRC, MJUSRC and MURASR* still get more accurate results than patch region based JSRC, MJSRC and MASR in terms of OA, AA and Kappa coefficients. The classification maps of MJSRC and MJUSRC have more pepper salt noise pixels than MASR and MURASR*. Comparing classification maps of

MURASR* and MURASR, we can find that most misclassifications generated by MURASR* can be corrected by probability majority voting method. In addition, the average accuracy of MURASR is 99.70% which is very high for classification. Moreover, it should be noted that the McNemar's tests between classifiers are also statistically significant and coherent with the obtained overall accuracies.

The final experiment was conducted on the University of Pavia image. The shapes of surface objects in this image are more complex than previous two images. For each reference class, 200 train samples were randomly selected from the labeled data and the remaining pixels were used for testing the performance of various classifiers (see Table 8). The classification maps are demonstrated in Figure 8 and the detail data averaged by ten runs in term of OA, AA, and Kappa coefficients is listed in Table 9. The McNemar's tests between classifiers also were conducted on this image and the results are tabulated in Table 10. Same as previously mentioned two images, union region based classifiers also performed better than patch region based classifiers. Multiscale adaptive strategy still works better than multiscale majority voting strategy in this image. The accuracy improvement gained by probability majority voting is less than previous two images because the University of Pavia image has less large homogenous regions. And from Table 9, we can find that MASR only has more accurate result than MURASR with one class and MURASR performs best among all classifiers with 7 classes, which proves the priority of MURASR further. The results of the McNemar's tests also provide enough support for the analysis.

Compared with many presented algorithms, MASR is a time-consuming algorithm. In this paper, the proposed MURASR is designed based on the multiscale adaptive representation in MASR. Also, the generation of union regions will consume some time. Moreover, the union region has more pixels than patch region. Therefore, the MURASR is also a time-consuming algorithm and the time cost of MURASR is about twice as much as MASR. But the proposed MURASR was coded in MATLAB (R2016a, Mathworks, Portola Valley, CA, USA) and was not optimized for speed. The MURASR can be significantly sped up by changing the compiling code from MATLAB to C++ and adopting a general-purpose graphics processing unit (GPU).

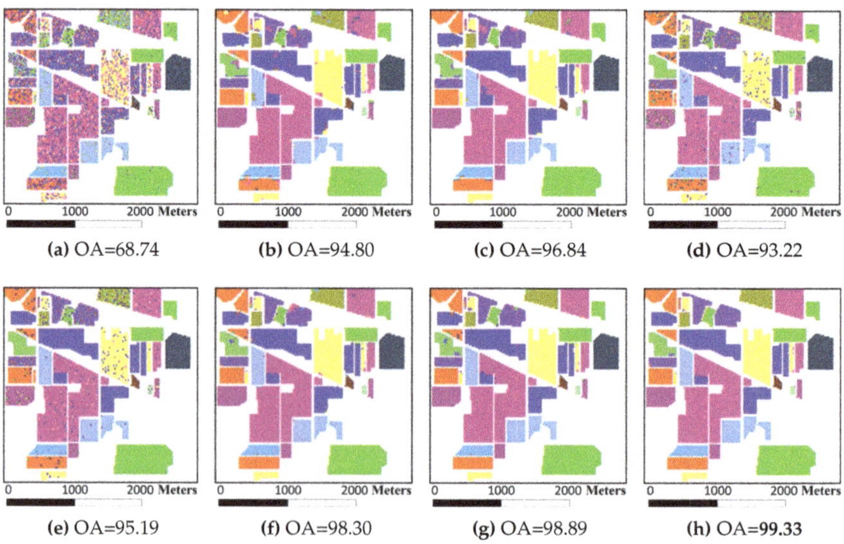

Figure 6. Classification maps for the Indian Pines image by different algorithms: (**a**) SRC-Pixel-Wise; (**b**) JSRC; (**c**) JUSRC; (**d**) MJSRC; (**e**) MJUSRC; (**f**) MASR; (**g**) MURASR*; and (**h**) MURASR.

Table 3. Classification accuracy (averaged on ten runs with randomly sampled training samples) of the Indian Pines image. The best results are highlighted in bold typeface.

Class	SRC-Pixel-Wise	JSRC	JUSRC	MJSRC	MJUSRC	MASR	MURASR*	MURASR
1	35.12	87.56	96.83	95.37	96.34	93.66	96.83	**98.54**
2	54.63	94.87	96.48	94.39	95.25	97.77	**97.93**	97.84
3	51.99	93.44	97.00	91.67	95.69	98.17	98.77	**99.54**
4	36.53	89.62	95.31	91.50	92.77	94.89	95.77	**98.78**
5	82.44	94.28	95.38	92.11	93.17	95.59	96.23	**96.51**
6	93.32	97.43	98.95	96.19	98.42	99.83	**100**	**100**
7	66.80	96.80	94.40	66.40	66.80	**98.80**	98.40	96.00
8	95.93	99.44	99.79	98.60	98.70	99.95	99.98	**100**
9	17.78	60.56	**91.11**	12.22	19.44	64.44	79.44	71.67
10	65.99	95.67	97.58	89.14	91.55	97.68	**98.23**	97.67
11	71.52	96.68	97.78	95.91	95.76	99.01	99.11	**99.85**
12	41.82	89.76	95.30	87.83	92.47	96.55	98.15	**99.25**
13	92.28	94.95	98.37	90.43	97.83	98.75	99.29	**99.89**
14	88.93	98.99	99.33	99.24	99.92	99.95	99.96	**100**
15	35.45	89.05	93.83	92.54	91.84	97.52	98.70	**99.48**
16	89.40	88.33	92.02	81.90	89.40	96.07	96.90	**98.69**
OA	68.83	95.35	97.36	93.91	95.35	98.29	98.69	**99.06**
AA	64.40	94.69	96.98	93.06	94.71	95.55	97.11	**98.93**
Kappa	0.64	0.92	0.96	0.86	0.88	0.98	**0.99**	0.97

Table 4. The McNemar's tests between classifiers (averaged on ten runs with randomly sampled training samples) of the Indian Pines image.

Method	JSRC	JUSRC	MJSRC	MJUSRC	MASR	MURASR*	MURASR
JSRC	–	−10.56	5.24	−0.25	−14.19	−15.28	−16.33
JUSRC	10.56	–	13.34	8.37	−5.15	−8.27	−10.09
MJSRC	−5.24	−13.34	–	-7.63	−17.24	−18.77	−20.26
MJUSRC	0.25	−8.37	7.63	–	−12.29	−14.84	−17.17
MASR	14.19	5.15	17.24	12.29	–	−3.51	−5.65
MURASR*	15.28	8.27	18.77	14.84	3.51	–	−3.40
MURASR	16.33	10.09	20.26	17.17	5.65	3.40	–

Table 5. Sixteen reference classes in the Salinas image.

Class	Name	Train	Test
1	Weeds_1	20	1989
2	Weeds_2	37	3689
3	Fallow	20	1956
4	Fallow plow	14	1380
5	Fallow smooth	27	2651
6	Stubble	40	3919
7	Celery	36	3543
8	Grapes	113	11,158
9	Soil	62	6141
10	Corn	33	3245
11	Lettuce 4 wk	11	1057
12	Lettuce 5 wk	19	1908
13	Lettuce 6 wk	9	907
14	Lettuce 7 wk	11	1059
15	Vinyard untrained	73	7195
16	Vinyard trellis	18	1789
	Total	543	53,586

Table 6. Classification accuracy (averaged on ten runs with randomly sampled training samples) of the Salinas image. The best results are highlighted in bold typeface.

Class	SRC-Pixel-Wise	JSRC	JUSRC	MJSRC	MJUSRC	MASR	MURASR*	MURASR
1	98.23	**100**	**100**	99.99	**100**	99.98	**100**	**100**
2	98.04	99.95	**100**	99.95	99.94	99.78	99.79	**100**
3	94.16	99.33	99.71	99.07	99.68	99.38	99.86	**100**
4	98.77	70.59	87.10	85.46	94.91	97.31	98.83	**99.88**
5	91.84	85.98	92.08	93.77	97.71	99.07	99.51	**99.52**
6	99.41	95.68	96.81	99.26	99.57	**100**	**100**	**100**
7	99.16	97.65	98.49	99.57	99.72	99.95	99.92	**100**
8	70.99	95.19	98.07	94.29	96.52	96.41	98.39	**99.61**
9	97.23	99.98	99.99	99.97	**100**	99.91	99.95	**100**
10	85.45	93.78	95.69	96.76	97.63	98.06	98.44	**99.63**
11	93.56	88.91	93.81	98.34	99.13	99.91	99.92	**100**
12	99.75	88.68	94.33	96.16	98.95	99.85	99.96	**100**
13	97.14	81.52	89.35	95.64	97.65	99.26	99.46	**99.98**
14	92.64	85.15	87.18	96.95	97.56	98.59	98.53	**99.93**
15	59.14	91.69	96.03	87.90	92.44	93.12	96.74	**98.79**
16	93.93	99.73	99.55	99.65	99.64	99.16	99.14	**99.78**
OA	85.79	94.32	96.96	95.87	97.65	97.97	98.98	**99.70**
AA	84.19	93.67	96.62	95.40	97.38	98.73	99.28	**99.66**
Kappa	0.92	0.92	0.96	0.96	0.98	0.98	0.99	**1**

Table 7. The McNemar's tests between classifiers (averaged on ten runs with randomly sampled training samples) of the Salinas image.

Method	JSRC	JUSRC	MJSRC	MJUSRC	MASR	MURASR*	MURASR
JSRC	–	−32.24	−17.31	−34.78	−37.06	−46.17	−53.08
JUSRC	32.24	–	11.43	−9.24	−11.66	−26.65	−36.90
MJSRC	17.31	−11.43	–	−26.12	−28.24	−37.55	−44.79
MJUSRC	34.78	9.24	26.12	–	−4.82	−22.67	−32.56
MASR	37.06	11.66	28.24	4.82	–	−20.12	−29.66
MURASR*	46.17	26.65	37.55	22.67	20.12	–	−18.89
MURASR	53.08	36.90	44.79	32.56	29.66	18.89	–

Table 8. Nine reference classes in the University of Pavia image.

Class	Name	Train	Test
1	Asphalt	200	6431
2	Meadows	200	18,449
3	Gravel	200	1899
4	Trees	200	2864
5	Metal sheets	200	1145
6	Bare soil	200	4829
7	Bitumen	200	1130
8	Bricks	200	3482
9	Shadows	200	747
	Total	1800	40,976

Table 9. Classification accuracy (averaged on ten runs with randomly sampled training samples) of the University of Pavia image. The best results are highlighted in bold typeface.

Class	SRC-Pixel-Wise	JSRC	JUSRC	MJSRC	MJUSRC	MASR	MURASR*	MURASR
1	62.24	86.22	93.13	86.14	94.79	89.97	96.87	**98.38**
2	80.22	96.62	97.15	97.71	98.63	98.78	99.44	**99.70**
3	69.07	98.64	99.36	99.27	99.76	99.78	99.87	**99.89**
4	91.48	91.32	90.94	95.89	94.99	**97.47**	97.13	95.66
5	99.52	97.69	99.11	99.55	99.78	**100**	**100**	**100**
6	68.84	99.54	99.71	99.53	99.93	99.87	99.93	**100**
7	86.90	97.53	99.59	99.79	99.92	**100**	**100**	**100**
8	72.67	95.89	97.92	96.70	98.68	98.76	99.72	**99.94**
9	**98.17**	66.96	75.17	84.08	86.43	92.16	94.90	95.30
OA	76.74	94.51	96.27	95.83	97.83	97.42	98.92	**99.21**
AA	69.61	92.67	95.02	94.42	97.09	96.54	98.55	**98.94**
Kappa	0.81	0.92	0.95	0.95	0.97	0.97	**0.99**	**0.99**

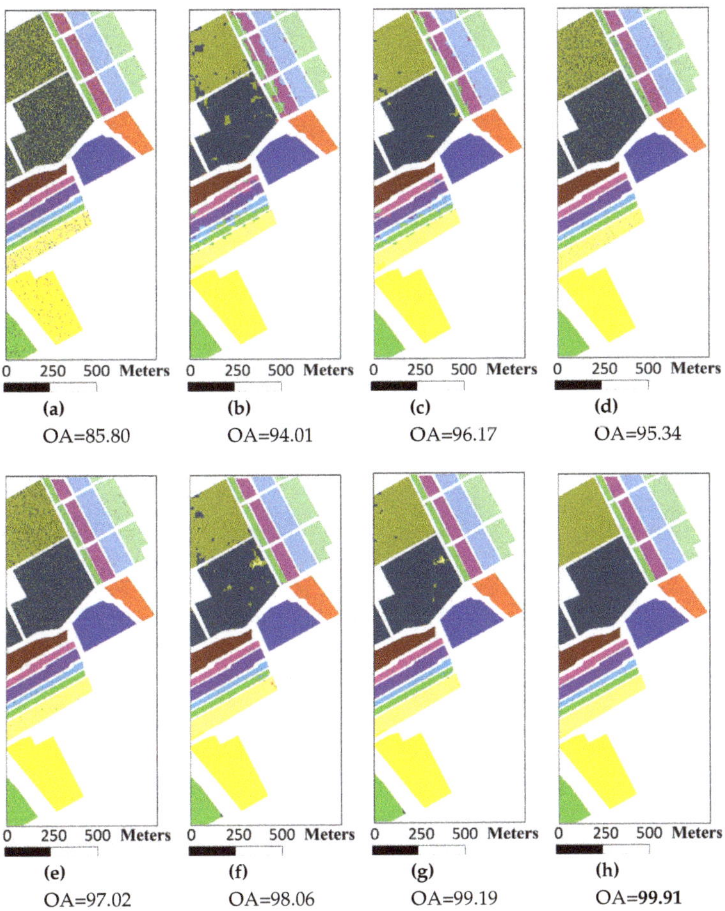

Figure 7. Classification maps for the Salinas image by different algorithms: (**a**) SRC-Pixel-Wise; (**b**) JSRC; (**c**) JUSRC; (**d**) MJSRC; (**e**) MJUSRC; (**f**) MASR; (**g**) MURASR*; and (**h**) MURASR.

Table 10. The McNemar's tests between classifiers (averaged on ten runs with randomly sampled training samples) of the University of Pavia image.

Method	JSRC	JUSRC	MJSRC	MJUSRC	MASR	MURASR*	MURASR
JSRC	–	−16.69	−13.38	−31.89	−28.46	−40.43	−42.27
JUSRC	16.69	–	4.06	−19.51	−11.12	−29.82	−32.77
MJSRC	13.38	−4.06	–	−22.57	−20.97	−32.92	−34.68
MJUSRC	31.89	19.51	22.57	–	5.09	−18.40	−21.43
MASR	28.46	11.12	20.97	−5.09	–	−21.53	−23.41
MURASR*	40.43	29.82	32.92	18.40	21.53	–	−6.93
MURASR	42.27	32.77	34.68	21.43	23.41	6.93	–

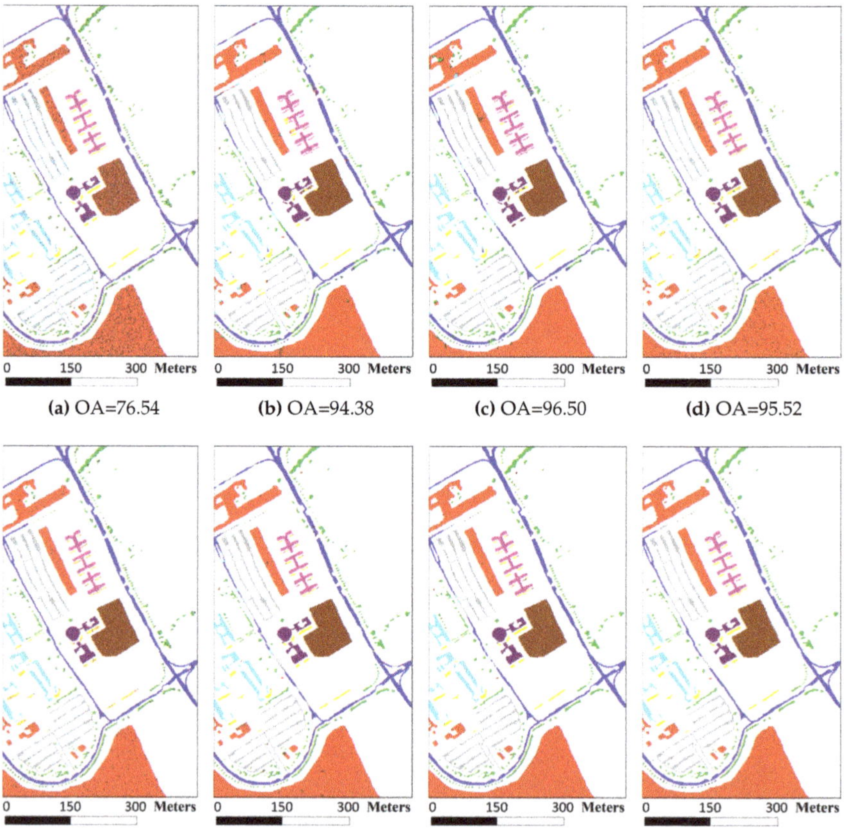

(a) OA=76.54 **(b)** OA=94.38 **(c)** OA=96.50 **(d)** OA=95.52

(e) OA=97.99 **(f)** OA=97.19 **(g)** OA=98.83 **(h)** OA=**99.05**

Figure 8. Classification maps for the University of Pavia image by different algorithms: (**a**) SRC-Pixel-Wise; (**b**) JSRC; (**c**) JUSRC; (**d**) MJSRC; (**e**) MJUSRC; (**f**) MASR; (**g**) MURASR*; and (**h**) MURASR.

4.3. Effects of Region Scales

Except for SRC-Pixel-wise, other related algorithms can be affected by different number of scales. In the previously mentioned experiments, 7 scales have been chosen to compare the performance of all algorithms. The effect of region scales for JSRC, MJSRC, and MASR has been presented in [28].

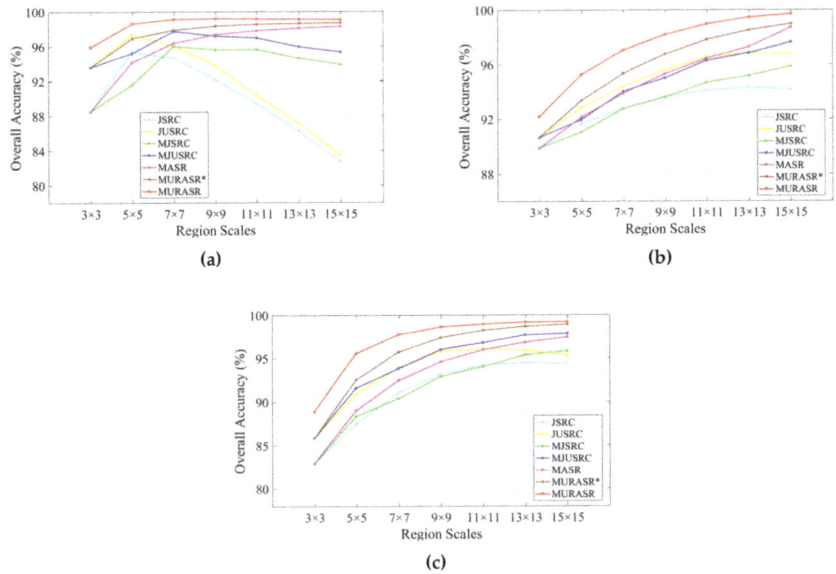

Figure 9. Effect of the region scales on single scale algorithms JSRC, JUSRC and the multiscale algorithms MJSRC, MJUSRC, MASR, MURASR* and MURASR for the: (**a**) Indian Pine image; (**b**) Salinas image; and (**c**) University of Pavia image.

From Table 1, we can find when the scale number is 7, the calculated scale for superpixels is large enough. If the scale continues increasing, there will be more mixed superpixels generated. Moreover, the classification results of MURASR on three images are encouraging when the number of scales is 7. Therefore, the effects of scales number under or equal to 7 will be analyzed in this section. It means that scales for patches range from 3 × 3 to 15 × 15. Figure 9 shows the average OA of ten runs for JSRC, JUSRC, MJSRC, MJUSRC, MASR, MURASR* and proposed MURASR. For multiscale algorithms, each scale represents the combination of the current scale and its smaller scales. It is easy to find that the union region based classifiers JUSRC, MJUSRC, and MURASR* generally outperform corresponding patch region based JSRC, MJSRC and MASR. And the probability majority voting method can optimize the classification result on each region scale. In addition, the proposed MURASR consistently outperforms other algorithms on all the region scales.

4.4. *Effects of Training Samples Number*

The number of training samples may affect the performance of the classifiers. Therefore the effects of different number of training samples on the JSRC, MJSRC, JUSRC, MJUSRC, MASR, MSPASR and proposed MURASR were examined on the three images. For the Indian Pines, the number of selected training samples for every class varies from 1% to 20% percentage. For the Salinas, the percentage is from 0.1% to 2%. For the University of Pavia, 60–500 training samples were selected for each reference class. The difference in terms of classification OA for each classifier with different number of training samples is illustrated in Figure 10. The OA is also the average of ten runs. As can be observed, the union region based classifiers JUSRC, MJUSRC and MURASR* always perform better than corresponding patch region based JSRC, MJSRC, and MASR. Comparing the result of MURASR* and MURASR, it is easy to find that the improvement obtained from probability majority voting method increases with the decreasing of training samples number. Moreover, the proposed MURASR generally outperforms other classifiers on all the training samples.

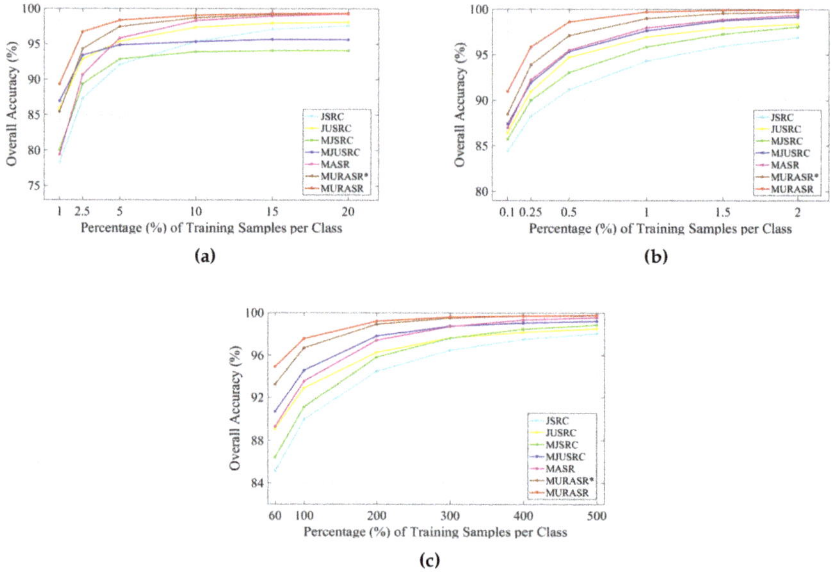

Figure 10. Effect of the number of training samples on JSRC, JUSRC, MJSRC, MJUSRC, MASR, MURASR* and MURASR for the: (**a**) Indian Pine image; (**b**) Salinas image; and (**c**) University of Pavia image.

5. Conclusions

In this paper, a novel multiscale union region adaptive sparse representation, the MURASR, which uses union region integrating patch and superpixel to exploit the spatial information, is proposed for spectral-spatial HSI classification. Unlike the patch region based MASR, the proposed MURASR extends the patch region to the union region. The union region utilizes the integration of the observation that neighboring pixels that belong to the same material usually are strongly correlated with each other and pixels in the superpixel usually belong to the same material. Before sparse representation, multiscale union regions are generated via the union operation for patch and superpixel. Then multiscale adaptive sparse representation is adopted to classify multiscale union regions and an effective probability majority voting method is applied to generate the final result. Experiments on three HSIs demonstrate that the union region based algorithms always perform better than patch region based algorithms and the proposed MURASR outperforms other algorithms in terms of quantitative metrics and visual quality for the classification maps.

As the MURASR is a pixel-based algorithm, if we replace the superpixel with a region growing up from each test pixel, the generated union region will have more accurate representation of the spatial information. Thus, the further research will generate one superpixel for each test pixel. In addition, the structure dictionary for sparse representation is constructed directly by selected training pixels. A trained structure dictionary may decrease the running time of the algorithm and provide more accurate representation for test pixels.

Acknowledgments: This research was supported by the National Natural Science Foundation of China under Grant No. 41171339 and 61501413. The authors would like to thank Ly. Fang for providing the source code of MASR and M.-Y. Liu for over-segmentation methods on their website (http://www.escience.cn/people/LeyuanFang/index.html, http://mingyuliu.net). The authors would like to thank David A. Landgrebe from Purdue University for providing the AVIRIS image of Indian Pines and Paolo Gamba from University of Pavia for providing the ROSIS data set. The authors would like to thank the National Aeronautics and Space Administration Jet Propulsion Laboratory for providing the AVIRIS image of Salinas. The authors would also like to thank the handling editor and anonymous reviewers for their valuable comments and suggestions, which significantly improved the quality of this paper.

Author Contributions: Fei Tong and Hengjian Tong proposed the model and implemented the experiments. Fei Tong wrote the manuscript. Junjun Jiang provided overall guidance of the work and edited the manuscript. Yun Zhang reviewed and edited the manuscript.

Conflicts of Interest: The authors declare no conflict of interest.

References

1. Fauvel, M.; Tarabalka, Y.; Benediktsson, J.A.; Chanussot, J.; Tilton, J.C. Advances in Spectral-Spatial Classification of Hyperspectral Images. *Proc. IEEE* **2013**, *101*, 652–675.
2. Zhang, L.; Zhang, L.; Tao, D.; Huang, X. Sparse Transfer Manifold Embedding for Hyperspectral Target Detection. *IEEE Trans. Geosci. Remote Sens.* **2014**, *52*, 1030–1043.
3. Du, B.; Zhang, L. Random-Selection-Based Anomaly Detector for Hyperspectral Imagery. *IEEE Trans. Geosci. Remote Sens.* **2011**, *49*, 1578–1589.
4. Zhong, Y.; Wang, X.; Zhao, L.; Feng, R.; Zhang, L.; Xu, Y. Blind spectral unmixing based on sparse component analysis for hyperspectral remote sensing imagery. *ISPRS J. Photogramm. Remote Sens.* **2016**, *119*, 49–63.
5. Borengasser, M.; Hungate, W.S.; Watkins, R. *Hyperspectral Remote Sensing: Principles and Applications*; CRC Press: Boca Raton, FL, USA, 2007.
6. Plaza, A.; Benediktsson, J.A.; Boardman, J.W.; Brazile, J.; Bruzzone, L.; Camps-Valls, G.; Chanussot, J.; Fauvel, M.; Gamba, P.; Gualtieri, A.; et al. Recent advances in techniques for hyperspectral image processing. *Remote Sens. Environ.* **2009**, *113* (Suppl. 1), S110–S122.
7. Melgani, F.; Bruzzone, L. Classification of hyperspectral remote sensing images with support vector machines. *IEEE Trans. Geosci. Remote Sens.* **2004**, *42*, 1778–1790.
8. Bruzzone, L.; Chi, M.; Marconcini, M. A Novel Transductive SVM for Semisupervised Classification of Remote-Sensing Images. *IEEE Trans. Geosci. Remote Sens.* **2006**, *44*, 3363–3373.
9. Chi, M.; Bruzzone, L. Semisupervised Classification of Hyperspectral Images by SVMs Optimized in the Primal. *IEEE Trans. Geosci. Remote Sens.* **2007**, *45*, 1870–1880.
10. Zhong, Y.; Lin, X.; Zhang, L. A Support Vector Conditional Random Fields Classifier With a Mahalanobis Distance Boundary Constraint for High Spatial Resolution Remote Sensing Imagery. *IEEE J. Sel. Top. Appl. Earth Obs. Remote Sens.* **2014**, *7*, 1314–1330.
11. Li, J.; Bioucas-Dias, J.M.; Plaza, A. Semisupervised Hyperspectral Image Classification Using Soft Sparse Multinomial Logistic Regression. *IEEE Geosci. Remote Sens. Lett.* **2013**, *10*, 318–322.
12. Ratle, F.; Camps-Valls, G.; Weston, J. Semisupervised Neural Networks for Efficient Hyperspectral Image Classification. *IEEE Trans. Geosci. Remote Sens.* **2010**, *48*, 2271–2282.
13. Zhong, Y.; Zhang, L. An Adaptive Artificial Immune Network for Supervised Classification of Multi-/Hyperspectral Remote Sensing Imagery. *IEEE Trans. Geosci. Remote Sens.* **2012**, *50*, 894–909.
14. Ji, R.; Gao, Y.; Hong, R.; Liu, Q.; Tao, D.; Li, X. Spectral–Spatial Constraint Hyperspectral Image Classification. *IEEE Trans. Geosci. Remote Sens.* **2014**, *52*, 1811–1824.
15. Khodadadzadeh, M.; Li, J.; Plaza, A.; Ghassemian, H.; Bioucas-Dias, J.M.; Li, X. Spectral–Spatial Classification of Hyperspectral Data Using Local and Global Probabilities for Mixed Pixel Characterization. *IEEE Trans. Geosci. Remote Sens.* **2014**, *52*, 6298–6314.
16. Ghamisi, P.; Benediktsson, J.A.; Ulfarsson, M.O. Spectral–Spatial Classification of Hyperspectral Images Based on Hidden Markov Random Fields. *IEEE Trans. Geosci. Remote Sens.* **2014**, *52*, 2565–2574.
17. Zhou, Y.; Peng, J.; Chen, C.L.P. Dimension Reduction Using Spatial and Spectral Regularized Local Discriminant Embedding for Hyperspectral Image Classification. *IEEE Trans. Geosci. Remote Sens.* **2015**, *53*, 1082–1095.

18. Falco, N.; Benediktsson, J.A.; Bruzzone, L. Spectral and spatial classification of hyperspectral images based on ICA and reduced morphological attribute profiles. *IEEE Trans. Geosci. Remote Sens.* **2015**, *53*, 6223–6240.
19. Fang, L.; Li, S.; Duan, W.; Ren, J.; Benediktsson, J.A. Classification of Hyperspectral Images by Exploiting Spectral—Spatial Information of Superpixel via Multiple Kernels. *IEEE Trans. Geosci. Remote Sens.* **2015**, *53*, 6663–6674.
20. Li, S.; Lu, T.; Fang, L.; Jia, X.; Benediktsson, J.A. Probabilistic fusion of pixel-level and superpixel-level hyperspectral image classification. *IEEE Trans. Geosci. Remote Sens.* **2016**, *54*, 7416–7430.
21. Wang, Y.; Song, H.; Zhang, Y. Spectral-spatial classification of hyperspectral images using joint bilateral filter and graph cut based model. *Remote Sens.* **2016**, *8*, 748.
22. Song, H.; Wang, Y. A spectral-spatial classification of hyperspectral images based on the algebraic multigrid method and hierarchical segmentation algorithm. *Remote Sens.* **2016**, *8*, 296.
23. Ma, L.; Ma, A.; Ju, C.; Li, X. Graph-based semi-supervised learning for spectral-spatial hyperspectral image classification. *Pattern Recogn. Lett.* **2016**, *83*, 133–142.
24. Jiang, J.; Chen, C.; Song, X.; Cai, Z. Hyperspectral image classification using set-to-set distance. In Proceedings of the ICASSP, Shanghai, China, 20–25 March 2016; pp. 3346–3350.
25. Jiang, J.; Chen, C.; Yu, Y.; Jiang, X.; Ma, J. Spatial-aware collaborative representation for hyperspectral remote sensing image classification. *IEEE Geosci. Remote Sens. Lett.* **2017**, *14*, 404–408.
26. Chen, Y.; Nasrabadi, N.M.; Tran, T.D. Hyperspectral Image Classification Using Dictionary-Based Sparse Representation. *IEEE Trans. Geosci. Remote Sens.* **2011**, *49*, 3973–3985.
27. Yuan, Y.; Lin, J.; Wang, Q. Hyperspectral image classification via multitask joint sparse representation and stepwise MRF optimization. *IEEE Trans. Cybern.* **2016**, *46*, 2966–2977.
28. Fang, L.; Li, S.; Kang, X.; Benediktsson, J.A. Spectral–Spatial Hyperspectral Image Classification via Multiscale Adaptive Sparse Representation. *IEEE Trans. Geosci. Remote Sens.* **2014**, *52*, 7738–7749.
29. Fang, L.; Li, S.; Kang, X.; Benediktsson, J.A. Spectral–Spatial Classification of Hyperspectral Images with a Superpixel-Based Discriminative Sparse Model. *IEEE Trans. Geosci. Remote Sens.* **2015**, *53*, 4186–4201.
30. Fu, W.; Li, S.; Fang, L.; Kang, X.; Benediktsson, J.A. Hyperspectral Image Classification Via Shape-Adaptive Joint Sparse Representation. *IEEE J. Sel. Top. Appl. Earth Obs. Remote Sens.* **2016**, *9*, 556–567.
31. Zhang, S.; Li, S.; Fu, W.; Fang, L. Multiscale Superpixel-Based Sparse Representation for Hyperspectral Image Classification. *Remote Sens.* **2017**, *9*, 139.
32. Wright, J.; Yang, A.Y.; Ganesh, A.; Sastry, S.S.; Ma, Y. Robust Face Recognition via Sparse Representation. *IEEE Trans. Pattern Anal. Mach. Intell.* **2009**, *31*, 210–227.
33. Leviatan, D.; Temlyakov, V.N. Simultaneous approximation by greedy algorithms. *Adv. Comput. Math.* **2006**, *25*, 73–90.
34. Tropp, J.A.; Gilbert, A.C.; Strauss, M.J. Algorithms for simultaneous sparse approximation. Part I: Greedy pursuit. *Signal Process.* **2006**, *86*, 572–588.
35. Mori, G.; Ren, X.; Efros, A.A.; Malik, J. Recovering human body configurations: combining segmentation and recognition. In Proceedings of the IEEE Conference Computer Vision and Pattern Recognition, Washington, DC, USA, 27 June–2 July 2004; Volume 2, pp. 326–333.
36. Achanta, R.; Shaji, A.; Smith, K.; Lucchi, A.; Fua, P.; Süsstrunk, S. SLIC Superpixels Compared to State-of-the-Art Superpixel Methods. *IEEE Trans. Pattern Anal. Mach. Intell.* **2012**, *34*, 2274–2282.
37. Liu, M.Y.; Tuzel, O.; Ramalingam, S.; Chellappa, R. Entropy rate superpixel segmentation. In Proceedings of the IEEE Conference Computer Vision and Pattern Recognition, Colorado Springs, CO, USA, 20–25 June 2011; pp. 2097–2104.
38. Felzenszwalb, P.F.; Huttenlocher, D.P. Efficient graph-based image segmentation. *Int. J. Comput. Vis.* **2004**, *59*, 167–181.
39. Zhong, Y.; Gao, R.; Zhang, L. Multiscale and multifeature normalized cut segmentation for high spatial resolution remote sensing imagery. *IEEE Trans. Geosci. Remote Sens.* **2016**, *54*, 6061–6075.
40. Vidal, R.; Ma, Y.; Sastry, S.S. Principal Component Analysis. In *Generalized Principal Component Analysis*; Springer: New York, NY, USA, 2016; pp. 25–62.

41. Gualtieri, J.A.; Cromp, R.F. Support vector machines for hyperspectral remote sensing classification. In Proceedings of the SPIE, Washington, DC, USA, 29 January 1999; pp. 221–232.
42. Foody, G.M. Thematic map comparison: Evaluating the statistical significance of differences in classification accuracy. *Photogramm. Eng. Remote Sens.* **2004**, *70*, 627–633.

remote sensing

MDPI

Article

Hyperspectral Image Classification Based on Semi-Supervised Rotation Forest

Xiaochen Lu *, Junping Zhang *, Tong Li and Ye Zhang

School of Electronics and Information Engineering, Harbin Institute of Technology, Harbin 150001, China; ltong@hit.edu.cn (T.L.); zhye@hit.edu.cn (Y.Z.)

* Correspondence: lxchen09@163.com (X.L.); zhangjp@hit.edu.cn (J.Z.); Tel.: +86-0451-8640-3020 (J.Z.)

Received: 20 July 2017; Accepted: 1 September 2017; Published: 6 September 2017

Abstract: Ensemble learning is widely used to combine varieties of weak learners in order to generate a relatively stronger learner by reducing either the bias or the variance of the individual learners. Rotation forest (RoF), combining feature extraction and classifier ensembles, has been successfully applied to hyperspectral (HS) image classification by promoting the diversity of base classifiers since last decade. Generally, RoF uses principal component analysis (PCA) as the rotation tool, which is commonly acknowledged as an unsupervised feature extraction method, and does not consider the discriminative information about classes. Sometimes, however, it turns out to be sub-optimal for classification tasks. Therefore, in this paper, we propose an improved RoF algorithm, in which semi-supervised local discriminant analysis is used as the feature rotation tool. The proposed algorithm, named semi-supervised rotation forest (SSRoF), aims to take advantage of both the discriminative information and local structural information provided by the limited labeled and massive unlabeled samples, thus providing better class separability for subsequent classifications. In order to promote the diversity of features, we also adjust the semi-supervised local discriminant analysis into a weighted form, which can balance the contributions of labeled and unlabeled samples. Experiments on several hyperspectral images demonstrate the effectiveness of our proposed algorithm compared with several state-of-the-art ensemble learning approaches.

Keywords: ensemble learning; hyperspectral; rotation forest; semi-supervised local discriminant analysis

1. Introduction

Hyperspectral (HS) image classification always suffers from varieties of difficulties, such as high dimensionality, limited or unbalanced training samples, spectral variability, and mixing pixels. It is well known that increasing data dimensionality and high redundancy between features might cause problems during data analysis, for example, in the context of supervised classification. A considerable amount of literature has been published with regard to overcoming these challenges, and performing hyperspectral image classification effectively [1]. Machine learning techniques such as artificial neural networks (ANNs) [2], support vector machine (SVM) [3], multinomial logistic regression [4], active learning, semi-supervised learning [5], and other methods like hyperspectral unmixing [6], object-oriented classification [7], and the multiple classifier system [8] have been popularly investigated recently as well.

Multiple classifier system (MCS), which is also sometimes named as classifier ensemble or ensemble learning (EL) in the machine learning field, is a popular strategy for improving the classification performance of hyperspectral images by combining the predictions of multiple classifiers, thereby reducing the dependence on the performance of a single classifier [8–11]. The concept of MCS, on the other hand, does not refer to a specific algorithm but to the idea of combining outputs from more than one classifier to enhance classification accuracy [12]. These outputs may result from either the same classifier of different variants or different classifiers of the same/different training samples. Previous studies have demonstrated both theoretically and experimentally that one of the main reasons for the success of ensembles is the diversity among the individual learners (namely the base classifiers) [13], because combining similar classification results would not further improve the accuracy.

MCSs have been widely applied to HS remote sensing image classification. Two approaches for constructing classifier ensembles are perceived as "classic", bagging and boosting [14,15], and afterwards numerous algorithms were successively derived from them. Bagging creates many classifiers with each base learner trained by a new bootstrapped training data set [16]. Boosting processes the data with iterative retraining, and concentrates on the difficult samples, with the goal of correctly classifying these samples in the next iteration [17,18]. Ho [19] proposed random subspace ensembles, which used random subsets of features instead of the entire feature set for each individual classifier. The rationale of the random subspace is to break down a complex high dimensional problem into several lower dimensional problems, thereby alleviating the curse of dimensionality. By integrating bagging and random subspace approaches, Breiman [20] proposed the well-known random forest (RF) algorithm [21,22]. The characteristics of RF, including reasonable computational cost, inherent support of parallelism, highly accurate predictions, and ability to handle a very large number of input variables without overfitting, make it a popular and promising classification algorithm for remote sensing data [23–25]. Generally, decision tree (DT) is used as the base classifier in ensemble learning because of its high computation efficiency, easy implementation, and sensitivity to slight changes in data. Recently, some researchers incorporated several prevalent machine learning algorithms into ensemble learning. Gurram and Kwon [26] proposed a sparse kernel-based support vector machine (SVM) ensemble algorithm that yields better performance compared with the SVM trained by cross-validation. Samat et al. [27] proposed Bagging-based and Adaboost-based extreme learning machines to overcome the drawbacks of input parameter randomness of traditional extreme learning machines. For a more detailed description about EL, refer to [28,29].

In a paper by Rodriguez and Kuncheva [30], the authors proposed a new ensemble classifier called rotation forest (RoF). By applying feature extraction (i.e., principal component analysis, PCA) to the random feature subspace, RoF greatly promotes the diversity and accuracy of the classifiers. Thereafter, several improved algorithms were proposed based on the idea of RoF, for example, Anticipative Hybrid Extreme Rotation Forest [31], rotation random forest with kernel PCA (RoRF-KPCA) [32]. Chen et al. [33] proposed to combine rotation forest with multi-scale segmentation for hyperspectral data classification, which incorporated spatial information to generate the classification maps with homogeneous regions.

A massive number of research studies show that RoF surpasses conventional RF due to the high diversity in training sample and features. Nevertheless, it is well documented in the literatures that PCA is not particularly suitable for feature extraction (FE) in classification because it does not include discriminative information in calculating the optimal rotation of the axes [30,34,35]. Although the authors explain that PCA is also valuable as a diversifying heuristic, it is expected to achieve better classification results if we try to find good class discriminative directions. Therefore, in this paper, we present an improved ensemble learning method, which uses the semi-supervised feature extraction technique instead of PCA during the "rotation" process of classical RoF approach. The proposed algorithm, named semi-supervised rotation forest (SSRoF), applies the semi-supervised local discriminant analysis (SLDA) FE method, which was proposed in our previous work [36], to fully take advantage of both the class separability and local neighbor information, with the aim of finding better rotation directions. In addition, to further enhance the diversity of features, we propose to use a weighted form of SLDA, which can balance the values of labeled samples and unlabeled samples. The main contributions of this paper are as follows: (1) an exploration of the benefit of the unlabeled samples in conventional ensemble learning methods; (2) an adjustment of the previous

SLDA technique to a weighted generalized eigenvalue problem; (3) the construction of an ensemble of classifiers, in which the weights can be randomly selected, thereby reducing the human effort for determining the optimal parameters.

The remainder of this paper is organized as follows. Section 2 describes the study data sets, and elaborates the proposed semi-supervised rotation forest algorithm. For better understanding, the SLDA feature extraction method is also briefly introduced. Section 3 reports the experiments and results. Finally, the conclusions are drawn in Section 4.

2. Materials and Methodology

In this section, we first introduce the experimental data sets, then we elaborate the proposed ensemble learning algorithm.

2.1. Study Data Sets

The experimental data sets include four HS images acquired by different sensors and resolutions. Each HS image is attached with a co-registered ground truth image.

(1) The first data set is the well-known scene taken in 1992 by the Airborne Visible Infrared Imaging Spectrometer (AVIRIS) sensor over the Indian Pines region in Northwestern Indiana. It has 144×144 pixels and 200 spectral bands with a pixel resolution of 20 m. Nine classes including different categories of crops have been labeled in the ground truth image.
(2) The second data set was collected over the University of Pavia, Italy, by the Reflective Optics System Imaging Spectrometer (ROSIS) system. It consists of 103 spectral bands after removing the noisy bands, and 610×340 pixels for each band with a pixel resolution of 1.3 m. The ground truth image contains nine classes [37,38].
(3) The third data set is a low-altitude AVIRIS HS image of a portion of the North Island of the U.S. Naval Air Station in San Diego, CA, USA. This HS image consists of 126 bands of size 400×400 pixels with a spatial resolution of 3.5 m per pixel after removing the noisy bands. The ground truth image has eight classes inside [39].
(4) The last data set is provided by the 2013 Institute of Electrical and Electronics Engineers (IEEE) Geoscience and Remote Sensing Society (GRSS) Data Fusion Contest (DFC). It was acquired by the compact airborne spectrographic imager sensor (CASI) over the University of Houston campus and neighboring urban area, and consists of 144 bands with a spatial resolution of 2.5 m. A subset of size 640×320 is used, which contains 12 classes in the corresponding ground truth image. Figure 1 shows the experimental data sets.

2.2. Weighted Semi-Supervised Local Discriminant Analysis

Semi-supervised local discriminant analysis is a semi-supervised feature extraction method that has been applied in hyperspectral image classification. It combines the supervised FE method-local Fisher discriminant analysis and unsupervised FE method-neighborhood preserving embedding, and thus attempts to discover the local discriminative information of the data while preserving the local neighbor information [36]. Compared with other typical semi-supervised FE methods, SLDA focuses more on the exploration of local information, and gives a more accurate description of the distribution of samples. For better illustration, we first briefly review the feature extraction methods.

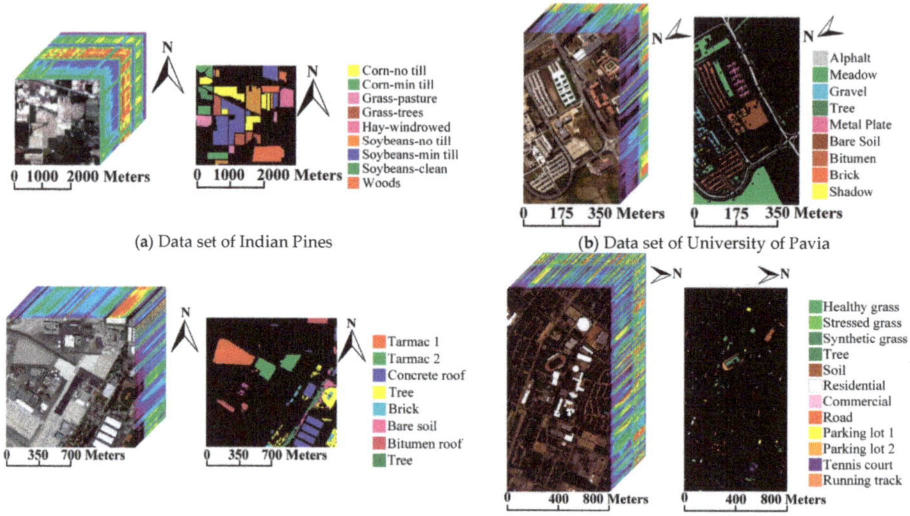

(**a**) Data set of Indian Pines (**b**) Data set of University of Pavia

(**c**) Data set of San Diego (**d**) Data set of University of Houston

Figure 1. Experimental hyperspectral and corresponding ground truth images.

Let $x_i \in \mathbb{R}^d$ be a d-dimensional sample vector, and $\boldsymbol{X} = \{x_i\}_{i=1}^{n}$ be the matrix of n samples. $\boldsymbol{Z} = \boldsymbol{T}^{\mathcal{T}}\mathbf{X}$, $(\mathbf{Z} \in \mathbb{R}^{r\times n})$ is the low-dimensional representation of the sample matrix, where $\boldsymbol{T} \in \mathbb{R}^{d\times r}$ is the transformation matrix, $\mathcal{T}$ denotes the transpose.

Many dimensionality reduction techniques developed so far involve an optimization problem of the following form [40]:

$$\boldsymbol{T} = \underset{\boldsymbol{T}}{\operatorname{argmax}} \left[\frac{\left|\boldsymbol{T}^{\mathcal{T}}\boldsymbol{S}^{b}\boldsymbol{T}\right|}{\left|\boldsymbol{T}^{\mathcal{T}}\boldsymbol{S}^{w}\boldsymbol{T}\right|} \right] \tag{1}$$

Generally speaking, $\boldsymbol{S}^b$ (and $\boldsymbol{S}^w$) corresponds to the quantity that we want to increase (and decrease), for example, between-class scatter (and within-class scatter). Equation (1) is equal to the solution of the following generalized eigenvalue problem:

$$\boldsymbol{S}^{b}\varphi = \lambda \boldsymbol{S}^{w}\varphi \tag{2}$$

where $\{\varphi_k\}_{k=1}^{d}$ is the generalized eigenvectors associated with the generalized eigenvalues $\{\lambda_k\}_{k=1}^{d}, (\lambda_1 > \lambda_2 > \ldots > \lambda_d)$. $\boldsymbol{T} = \{\varphi_k\}_{k=1}^{r}$ is composed of the first r eigenvectors corresponding to the largest eigenvalues $\{\lambda_k\}_{k=1}^{r}$. Particularly, when $\boldsymbol{S}^b$ is the total scatter matrix of all samples, and $\boldsymbol{S}^w = \boldsymbol{I}_{d\times d}$, where $\boldsymbol{I}$ denotes the identity matrix. Equation (1) turns into the PCA method.

2.2.1. Local Fisher Discriminant Analysis (LFDA)

Suppose $y_i = c\,, c \in \{1, 2, \ldots, C\}$ is the associated class labels of the sample vector x_i. C is the number of classes. n_c is the number of samples in class c, then $\sum_{c=1}^{C} n_c = n$. Let $\boldsymbol{S}^b$ and $\boldsymbol{S}^w$ be the local between-class and within-class scatter matrices, respectively, defined by [41],

$$\begin{aligned} \boldsymbol{S}^{b} &= \tfrac{1}{2}\sum_{i=1}^{n}\sum_{j=1}^{n} \boldsymbol{W}_{i,j}^{b}\left(x_i - x_j\right)\left(x_i - x_j\right)^{\mathcal{T}} \\ \boldsymbol{S}^{w} &= \tfrac{1}{2}\sum_{i=1}^{n}\sum_{j=1}^{n} \boldsymbol{W}_{i,j}^{w}\left(x_i - x_j\right)\left(x_i - x_j\right)^{\mathcal{T}} \end{aligned} \tag{3}$$

then Equation (2) turns into a local Fisher discriminant analysis problem, where W^b and W^w are $n \times n$ matrices,

$$\begin{aligned} W^b_{i,j} &= \begin{cases} A_{i,j}(1/n - 1/n_c), \; if \; y_i = y_j = c \\ 1/n, \; if \; y_i \neq y_j \end{cases} \\ W^w_{i,j} &= \begin{cases} A_{i,j}/n_c, \; if \; y_i = y_j = c \\ 0, \; if \; y_i \neq y_j \end{cases} \end{aligned} \tag{4}$$

$A_{i,j}$ is the affinity value between x_i and x_j. $A_{i,j}$ is large if the two samples are close, and vice versa. The definition of $A_{i,j}$ can be found in [42]. Note that we do not weight the values for the sample pairs in different classes. If $\forall \; i, j, \; A_{i,j} = 1$, then LFDA degenerates into the classical Fisher discriminant analysis (FDA or linear discriminant analysis, LDA) [43]. Thus, LFDA can be regarded as a localized variant of FDA, which overcomes the weakness of LDA against within-class multimodality or outliers.

2.2.2. Neighborhood Preserving Embedding (NPE)

NPE is an unsupervised feature extraction method that seeks a projection that preserves neighboring data structure in the low-dimensional feature space [44]. It can characterize the local structural information of massive unlabeled samples. The first step of NPE is also to construct an adjacency graph, and then compute the weight matrix Q by solving the following objective function,

$$\begin{aligned} &\min \sum_i \|x_i - \sum_j Q_{ij} x_j\|^2 \\ &s.t. \; \sum_j Q_{ij} = 1 \end{aligned} \tag{5}$$

In other words, for each sample, we use its K-nearest neighbors (KNN) to reconstruct it. Thus, the goal of NPE is to preserve this neighbor relationship in the projected low-dimensional space,

$$\begin{aligned} &\min \sum_i \|z_i - \sum_j Q_{ij} z_j\|^2 \\ &s.t. \; \sum_j Q_{ij} = 1 \end{aligned} \tag{6}$$

where $z_i = T^{\mathcal{T}} x_i$. Then we have

$$\min trace\left[Z(I - Q)^{\mathcal{T}}(I - Q)Z^{\mathcal{T}}\right] \tag{7}$$

By imposing the following constraint,

$$\sum_i z_i z_i^{\mathcal{T}} = I \Longrightarrow ZZ^{\mathcal{T}} = I. \tag{8}$$

the transformation matrix can be optimized by solving the following generalized eigenvalue problem,

$$XX^{\mathcal{T}} \varphi = \lambda XMX^{\mathcal{T}} \varphi \tag{9}$$

where φ denotes generalized eigenvectors, and $M = (I - Q)^{\mathcal{T}}(I - Q)$.

2.2.3. Weighted SLDA

It has been demonstrated that the performance of LFDA (and all other supervised dimensionality reduction methods) tends to degrade if only a small number of labeled samples are available [40], while PCA or NPE (and other unsupervised feature extraction (FE) methods) will generally lose the discriminative information of labeled information. Thus, combining supervised and unsupervised FE methods [45] is believed to compensate for each other's weaknesses. In this paper, we consider the combination of the aforementioned LFDA and NPE methods. As mentioned above, feature extraction techniques can be transformed into eigenvalue problems, thus, a possible way to combine LFDA and NPE is to merge the above generalized eigenvalue problems as follows [40],

$$\begin{gathered} \beta \boldsymbol{S}^b \varphi = \lambda \beta \boldsymbol{S}^w \varphi \\ (1-\beta)\boldsymbol{X}\boldsymbol{X}^T \varphi = \lambda(1-\beta)\boldsymbol{X}\boldsymbol{M}\boldsymbol{X}^T \varphi \\ \Downarrow \\ \left[\beta \boldsymbol{S}^b + (1-\beta)\boldsymbol{X}\boldsymbol{X}^T\right]\varphi \\ = \lambda\left[\beta \boldsymbol{S}^w + (1-\beta)\boldsymbol{X}\boldsymbol{M}\boldsymbol{X}^T\right]\varphi \end{gathered} \tag{10}$$

where $\beta \in [0,\ 1]$ is a trade-off parameter. Calculating the $\boldsymbol{S}^b$ and $\boldsymbol{S}^w$ of LFDA is time-consuming; an efficient implementation can be used according to [41]. Let $\boldsymbol{S}^m$ denote the local mixture scatter matrix,

$$\boldsymbol{S}^m = \boldsymbol{S}^b + \boldsymbol{S}^w = \frac{1}{2}\sum_{i=1}^{n}\sum_{j=1}^{n} \boldsymbol{W}^m_{i,j}(x_i - x_j)(x_i - x_j)^T \tag{11}$$

where

$$\boldsymbol{W}^m = \boldsymbol{W}^b + \boldsymbol{W}^w = \begin{cases} A_{i,j}/n, \ if\ y_i = y_j \\ 1/n, \ if\ y_i \neq y_j \end{cases} \tag{12}$$

Since Equation (3) can be expressed as

$$\begin{aligned} \boldsymbol{S}^w &= \sum_{i=1}^{n}\sum_{j=1}^{n} \boldsymbol{W}^w_{i,j} x_i x_i^T - \sum_{i=1}^{n}\sum_{j=1}^{n} \boldsymbol{W}^w_{i,j} x_i x_j^T \\ &= \boldsymbol{X}(\boldsymbol{D}^w - \boldsymbol{W}^w)\boldsymbol{X}^T \end{aligned} \tag{13}$$

where $\boldsymbol{D}^w$ is the n-dimensional diagonal matrix with $\boldsymbol{D}^w_{i,i} = \sum_{j=1}^{n} \boldsymbol{W}^w_{i,j}$. Similarly, $\boldsymbol{S}^m$ can be expressed as

$$\boldsymbol{S}^m = \boldsymbol{X}(\boldsymbol{D}^m - \boldsymbol{W}^m)\boldsymbol{X}^T \tag{14}$$

where $\boldsymbol{D}^m$ is the n-dimensional diagonal matrix with $\boldsymbol{D}^m_{i,i} = \sum_{j=1}^{n} \boldsymbol{W}^m_{i,j}$. Therefore, the generalized eigenvalue problem of LFDA, namely Equation (2), can be rewritten as

$$\boldsymbol{X}\boldsymbol{L}^b\boldsymbol{X}^T \varphi = \lambda \boldsymbol{X}\boldsymbol{L}^w\boldsymbol{X}^T \varphi \tag{15}$$

where $\boldsymbol{L}^w = \boldsymbol{D}^w - \boldsymbol{W}^w$, $\boldsymbol{L}^b = (\boldsymbol{D}^m - \boldsymbol{W}^m) - (\boldsymbol{D}^w - \boldsymbol{W}^w)$, from which we can see that the eigenvalue problem of LFDA has a similar form with NPE, i.e., Equation (9).

Suppose the training sample vectors are arranged by $\boldsymbol{X} = \left[\boldsymbol{X}^L,\ \boldsymbol{X}^U\right]$, where $\boldsymbol{X}^L = \left\{x_i^L\right\}_{i=1}^{n_l}$ denotes the labeled samples, and $\boldsymbol{X}^U = \left\{x_i^U\right\}_{i=1}^{n_u}$ denotes the unlabeled samples, where $n = n_l + n_u$ is the total number of available samples. We can define the following matrices

$$\begin{gathered} \boldsymbol{P}_1 = \begin{bmatrix} \boldsymbol{L}^b & \boldsymbol{0}_{n_l \times n_u} \\ \boldsymbol{0}_{n_u \times n_l} & \boldsymbol{0}_{n_u \times n_l} \end{bmatrix}, \boldsymbol{P}_2 = \begin{bmatrix} \boldsymbol{L}^w & \boldsymbol{0}_{n_l \times n_u} \\ \boldsymbol{0}_{n_u \times n_l} & \boldsymbol{0}_{n_u \times n_u} \end{bmatrix} \\ \boldsymbol{P}_3 = \begin{bmatrix} \boldsymbol{0}_{n_l \times n_l} & \boldsymbol{0}_{n_l \times n_u} \\ \boldsymbol{0}_{n_u \times n_l} & I_{n_u \times n_u} \end{bmatrix}, \boldsymbol{P}_4 = \begin{bmatrix} \boldsymbol{0}_{n_l \times n_l} & \boldsymbol{0}_{n_l \times n_u} \\ \boldsymbol{0}_{n_u \times n_l} & \boldsymbol{M} \end{bmatrix} \end{gathered} \tag{16}$$

Therefore, the weighted SLDA is equal to the solution of the following generalized eigenvalue problem

$$
\begin{array}{c}
\beta \boldsymbol{X}\boldsymbol{P_1}\boldsymbol{X}^T\varphi = \lambda\ \beta \boldsymbol{X}\boldsymbol{P_2}\boldsymbol{X}^T\varphi \\
(1-\beta)\boldsymbol{X}\boldsymbol{P_3}\boldsymbol{X}^T\varphi = \lambda\ (1-\beta)\boldsymbol{X}\boldsymbol{P_4}\boldsymbol{X}^T\varphi \\
\Downarrow \\
\boldsymbol{S}^{rb} = \boldsymbol{X}[\beta\boldsymbol{P_1} + (1-\beta)\boldsymbol{P_3}]\boldsymbol{X}^T \\
\boldsymbol{S}^{rw} = \boldsymbol{X}[\beta\boldsymbol{P_2} + (1-\beta)\boldsymbol{P_4}]\boldsymbol{X}^T \\
\Downarrow \\
\boldsymbol{S}^{rb}\varphi = \lambda\boldsymbol{S}^{rw}\varphi
\end{array}
\tag{17}
$$

and β is the trade-off parameter. In general, $0 < \beta < 1$ inherits the characteristics of both LFDA and NPE, and thus makes full use of both the class discriminative and local neighbor spatial information. In practice, searching for the optimal β is time-consuming and sometimes impractical if there are insufficient labeled samples available for validation. Several research studies suggest that ensemble learning methods can be employed to avoid the huge effort of searching for the optimal parameters [46,47]. On the other hand, different parameters also lead to diversity among features or classifiers, which benefits the generalization performance of the ensembles. Hence, we present an EL method based on the idea of RoF and the weighted SLDA algorithm.

2.3. Proposed Semi-Supervised Rotation Forest

Rotation forest was developed from conventional random forest to building independent decision trees on different sets of features. It consists of splitting the feature set into several random disjoint subsets, running PCA separately on each subset, and reassembling the extracted features [30,48]. By applying different splits of the features, diverse classifiers are obtained. The main steps of RoF are briefly presented as follows:

1. The original feature set is divided randomly into K disjoint subsets with each subset containing M features;
2. Use the bootstrap approach to select a subset of the training samples for each feature subset (typically 75% of the total training samples);
3. Run PCA on each feature subset and store the transformation coefficients;
4. Reorder the coefficients to match the original features, rotate the samples using the obtained coefficients (i.e., feature extraction);
5. Perform DT on the rotated training and testing samples;
6. The process is repeated L times to obtain multiple classifiers, followed by a majority voting rule to integrate the classification results.

By substituting SLDA for the PCA method, we propose the following SSRoF ensemble algorithm.

Apart from the different FE methods between Algorithm 1 and RoF, we use the different weights (β) to balance the discriminative information and structure information, thereby enhancing the diversity of features. Although the computation of the eigenvector matrix is repeated ten times (corresponding to different β) for each feature subset, it can be noticed that since the within-class and between-class scatter matrices are invariant for different weights, the computation cost is greatly reduced. Of course, the discrete values of β can be set by different steps; we recommend the values above by considering both the diversity and computation time.

Algorithm 1: Procedures of SSRoF

Input: Training samples $X^L = \{x_i^L\}_{i=1}^{n_l}$, testing samples $X^T = \{x_i^T\}_{i=1}^{n_l}$, unlabeled samples $X^U = \{x_i^U\}_{i=1}^{n_u}$, ensemble classifiers L, number of feature subsets K, ensemble $\mathcal{L} = \varnothing$
Output: Class labels of X^T
For $i = 1 : L$
1. Randomly split the features into K subsets;
For $j = 1 : K$
2. Randomly select a subset of samples from X^L and X^U, respectively, (typically 75% of samples) using bootstrap approach;
3. Perform the weighted SLDA algorithm by the subset of X^L and X^U to obtain the pairs of between-class and within-class scatter matrices in Equation (17);
For $\beta = 0.1 : 0.1 : 1$
4. Obtain the eigenvector matrix $T_{j,\beta}$ by solving Equation (17);
End for
End for
For $\beta = 0.1 : 0.1 : 1$
5. Construct the transformation matrix $T_\beta = \left[T_{1,\beta}, T_{2,\beta}, \ldots, T_{K,\beta}\right]$ by merging the eigenvector matrices, and rearrange the columns of T_β to match the order of original features;
6. Build DT sub-classifier using $T_\beta^T X^L$;
7. Perform classification for $T_\beta^T X^T$ by using the sub-classifier;
End for
End for
8. Use a majority voting rule for the $L \times 10$ sub-classifiers to compute the confidence of X^T and assign a class label for each testing sample;

---

3. Experimental Results and Discussion

In this section, we report the experiments on the four groups of hyperspectral images. First, the presented method is compared with several other EL algorithms to show the advantages. Then, we also introduce the performance evaluation of our method under different parameters.

3.1. Experimental Setup

In order to demonstrate the advantages of the proposed algorithm, we conducted the experiments under different numbers of training samples, and compared with several state-of-the-art ensemble learning methods, namely random forest (RF), semi-supervised feature extraction combined RF ensemble method (SSFE-RF) [22], rotation forest (RoF) [30], and rotation random forest-KPCA (RoRF-KPCA) [32]. For better comparison, the SLDA method was also used as a preprocessing step that combined with the original RoF method (we refer to it as SLDA-RoF). Finally, the LFDA and NPE methods were also used as rotation means like RoF method.

The numbers of trees were all set to $L = 10$, and the classification and regression tree (CART) was adopted as the base classifier. The numbers of features in each subset were all set to $M = 10$ for SSFE-RF, RoF, RoF-LFDA, RoF-NPE, and SSRoF. For RoRF-KPCA, Xia et al. [32] suggest that a small number of features per subset will increase the classification performance, as such, we set $M = 5$. For RF, the number of features considered at each node was set as the square root of the used feature number. The numbers of extracted features were set equal to M for RoF, RoRF-KPCA, RoF-LFDA, RoF-NPE, and SSRoF. For SLDA, the number of extracted features was set to half of the original features, and other parameters were set to the same as RoF. For RoRF-KPCA, it is quite difficult to select the optimal kernel parameters. Xia et al. [32] declares that parameter tuning is needed, but different kernel functions (linear, radial basis function, and Polynomial) provide very similar results, making this choice not critical in this context. Considering the performance enhancement and the computation cost, in our experiments, we use the polynomial kernels with the degree equals to two.

The performance is evaluated by the overall accuracy (OA), and Kappa coefficient. In all cases, we conduct ten independent Monte Carlo runs with respect to the labeled training set from the ground truth images. And the results are the average values of the 10 runs. The numbers of available samples are listed in Table 1.

Table 1. Number of available samples in each data set.

Indian Pines		University of Pavia		San Diego		University of Houston	
Class	**Samples**	**Class**	**Samples**	**Class**	**Samples**	**Class**	**Samples**
corn-no till	1434	asphalt	6304	tarmac1	7044	healthy grass	449
corn-min till	834	meadow	18146	tramac2	4721	stressed grass	454
grass-pasture	234	gravel	1815	concrete roof	5771	synthetic grass	505
grass-trees	497	tree	2912	tree	4851	tree	293
hay-windrowed	747	metal plate	1113	brick	873	soil	688
soybeans-no till	489	bare soil	4572	bare soil	1748	residential	26
soybeans-min till	968	bitumen	981	bitumen roof	2454	commercial	463
soybeans-clean	2468	brick	3364	tree	2135	road	112
woods	1294	shadow	795			parking lot 1	427
						parking lot 2	247
						tennis court	473
						running track	367

3.2. Performance Evaluation

The comparison of different EL algorithms is presented here. We randomly selected 1%, 2%, and 5% samples of each class as training samples for the first three data sets, and 5%, 10%, and 20% for the last data set. The remaining samples were used for testing purposes. Table 2 lists the classification results of the four algorithms under different numbers of samples. The upper line in each cell denotes the overall accuracies, and the lower line is the Kappa values. For clarity, the best results are shown in different colors.

From the table, it can be seen obviously that all the other methods yielded much higher accuracies than the conventional RF method. SSFE-RF achieved higher accuracies than RF due to the increment in the number of classifiers and the semi-supervised feature extraction method. Particularly, it had splendid performance on the San Diego data set. Moreover, except for the SLDA-RoF, all of the other RoF-based approaches also surpassed the RF-based methods in most cases, which demonstrates the promotion of diversity owing to the random feature extraction. RoRF-KPCA yielded similar results with RoF, although it considers the nonlinear characteristics of hyperspectral data, and would have constructed reliable rotation matrices to generate high precision classification results. A probable reason may be the selection of sub-optimal parameters for kernel functions. However, as we have mentioned, searching for the optimal parameters remains problematic, and RoRF-KPCA is not sensitive to the changes of the kernel function. A smaller value of M may also affect the classification accuracy, although a smaller M means a larger K, which leads to a higher computational complexity due to the construction of the kernel matrix. Regardless of the computation time, it can be expected that RoRF-KPCA can surpass RoF to some extent. It can also be seen that RoF-LFDA and RoF-NPE also produced similar results as RoF. RoF-LFDA sometimes performed better than RoF and RoF-NPE when more samples were available, since it only uses the discriminative information of the labeled samples. In fact, no matter which simple rotation method was used in RoF, it seems that the results were very close to each other on the whole. However, the SLDA combined RoF method has relatively lower accuracies compared with other RoF-based method, although it has been demonstrated to perform well for other conventional classifiers [36] (e.g., MLC, SVM). Thus, it seems to be not suitable for rotation forest algorithms.

By contrast, the proposed SSRoF outperformed the others clearly in most cases from both OA and Kappa values, especially on the Indian and Pavia data sets (4.35% and 1.45% higher than RoF for the Indian and Pavia data sets on average, respectively). Although the conventional RF and RoF-based algorithms performed well on the last data set, the proposed algorithm still showed slight superiority.

The main reason why the proposed SSRoF method surpasses RoF-LFDA and RoF-NPE is that SSRoF uses a weighted form to better explore the discriminative information and structure information of the available samples, thus greatly promoting the diversity of features.

Table 2. The overall accuracies (%) and Kappa coefficients of different algorithms.

		RF	SSFE-RF	RoF	RoRF-KPCA	SLDA-RoF	RoF-LFDA	RoF-NPE	SSRoF
Indian	1%	58.35 0.5018	66.87 0.5995	71.48 0.6587	70.54 0.6491	63.88 0.5660	66.17 0.5943	69.39 0.6337	74.38 0.6918
	2%	64.55 0.5746	74.89 0.6971	75.80 0.7117	77.11 0.7272	70.22 0.6437	76.72 0.7214	76.45 0.7179	80.83 0.7710
	5%	70.79 0.6502	81.04 0.7728	82.97 0.7971	82.96 0.7971	77.58 0.7330	83.01 0.7978	82.66 0.7936	86.84 0.8429
Pavia	1%	79.65 0.7143	84.93 0.7879	87.13 0.8223	87.02 0.8205	81.20 0.7373	87.09 0.8214	86.67 0.8152	88.98 0.8484
	2%	82.38 0.7538	87.27 0.8220	89.54 0.8559	89.39 0.8537	84.34 0.7840	90.15 0.8645	89.61 0.8571	91.60 0.8846
	5%	85.82 0.8029	90.26 0.8648	92.28 0.8943	92.10 0.8919	86.82 0.8186	92.52 0.8978	91.77 0.8871	93.67 0.9137
San Diego	1%	86.08 0.8333	96.07 0.9529	95.28 0.9435	94.19 0.9305	93.25 0.9192	95.20 0.9426	95.55 0.9467	95.99 0.9520
	2%	90.10 0.8814	96.78 0.9615	96.40 0.9569	95.88 0.9507	94.86 0.9385	96.50 0.9582	96.56 0.9589	97.02 0.9644
	5%	93.10 0.9175	97.69 0.9724	97.64 0.9717	97.09 0.9652	96.40 0.9569	97.62 0.9716	97.61 0.9715	98.02 0.9764
Houston	5%	91.32 0.9034	95.97 0.9551	96.06 0.9561	96.08 0.9564	93.73 0.9302	96.06 0.9561	96.33 0.9591	97.43 0.9714
	10%	94.40 0.9376	96.59 0.9620	97.08 0.9676	97.60 0.9733	94.98 0.9441	96.96 0.9662	97.33 0.9703	98.09 0.9787
	20%	96.31 0.9590	98.03 0.9780	98.18 0.9798	98.42 0.9824	96.54 0.9615	98.22 0.9802	97.77 0.9752	98.60 0.9845

RF: random forest; SSFE-RF: semi-supervised feature extraction combined random forest; RoF: rotation forest; RoRF-KPCA: rotation random forest with kernel principal component analysis; SLDA-RoF: RoF with semi-supervised local discriminant analysis pre-processing; RoF with local Fisher discriminant analysis; RoF-NPE: RoF with neighborhood preserving embedding; SSRoF: semi-supervised rotation forest.

Particularly, aside from the number of ensembles L and the number of features per subset (M), the proposed approach needs fewer additional parameters, which makes the approach much easier to implement.

3.3. Impact of Parameters

In this sub-section, we will discuss the impact of two basic parameters, i.e., the number of ensembles (L), and the number of features in each subset (M). For brevity, we simply show the results performed on the data sets of Indian Pines and University of Pavia by setting different number of trees, i.e., $L = 2, 5, 10, 20$, and 30. Likewise, the experiments are conducted under different numbers of training samples. The results are shown in Table 3. In order to give an intuitive evaluation, OAs and Kappa values are shown in different colors.

From Table 3 we can see that, obviously, with the increment of ensemble number, the overall accuracy and Kappa coefficient grow continuously, for instance, from nearly 67% to 75% under 1% samples for the Indian Pines data set, which demonstrates the benefit of EL. An interesting factor is that when the number of trees increases to 10, the classification accuracy grows slower and tends to reach convergence. This makes our approach more promising, since we can use less ensembles to achieve a relatively stable result, thereby reducing the computational burden.

Table 3. The classification results of SSRoF under different number of ensembles (L). OA: overall accuracy.

		$L = 2$		$L = 5$		$L = 10$		$L = 20$		$L = 30$	
		OA (%)	Kappa	OA (%)	Kappa	OA (%)	Kappa	OA (%)	Kappa	OA (%)	Kappa
Indian	1%	71.01	0.6516	74.16	0.6887	74.69	0.6955	74.65	0.6944	74.96	0.6978
	2%	77.91	0.7359	79.56	0.7545	80.03	0.7600	80.54	0.7660	80.95	0.7710
	5%	83.55	0.8039	85.63	0.8285	86.62	0.8403	86.87	0.8432	86.97	0.8443
	10%	86.51	0.8392	88.44	0.8622	88.87	0.8672	89.24	0.8716	89.26	0.8718
	20%	88.91	0.8682	90.71	0.8894	91.25	0.8958	91.67	0.9008	91.74	0.9016
Pavia	1%	87.71	0.8308	88.79	0.8456	89.13	0.8504	89.38	0.8538	89.45	0.8548
	2%	89.74	0.8592	91.20	0.8794	91.35	0.8814	91.65	0.8856	91.75	0.8869
	5%	92.13	0.8924	93.36	0.9094	93.70	0.9141	93.78	0.9151	93.86	0.9163
	10%	93.07	0.9053	94.10	0.9195	94.46	0.9245	94.59	0.9263	94.59	0.9262
	20%	94.48	0.9250	95.15	0.9341	95.31	0.9363	95.45	0.9382	95.46	0.9383

To investigate the impact of the number of features in each subset, we also performed tests on the Indian Pines data set regarding different feature divisions. For better comparison, the same process was also applied on RoF algorithm, and the results are shown in Figure 2, where the blue color denotes the OAs, and the magenta color denotes the Kappa values. The solid lines denote the RoF method, while the dot dash lines represent the SSRoF method. The figure indicates that when the number of features involved in each subset increases, i.e., the number of feature subsets (K) decreases, the classification results tend to degenerate for both RoF and SSRoF. In fact, this is also consistent with the conclusions of [32], and that is why we selected a small number of M for the RoRF-KPCA method. Although when the training set increased, this problem seemed to be alleviated in a manner (for instance, in Figure 2e, 91.48% for $M = 5$ and 90.94% for $M = 30$ (SSRoF), when 20% of training samples were used), a small value of M is usually preferred. However, on the other hand, a smaller M means a larger K, which means the rotation process will be executed more times, and this will lead to a huge computational cost. Apart from the above analysis, we can also see that the proposed approach seemed to be more stable than RoF with the increment in the number of features per subset.

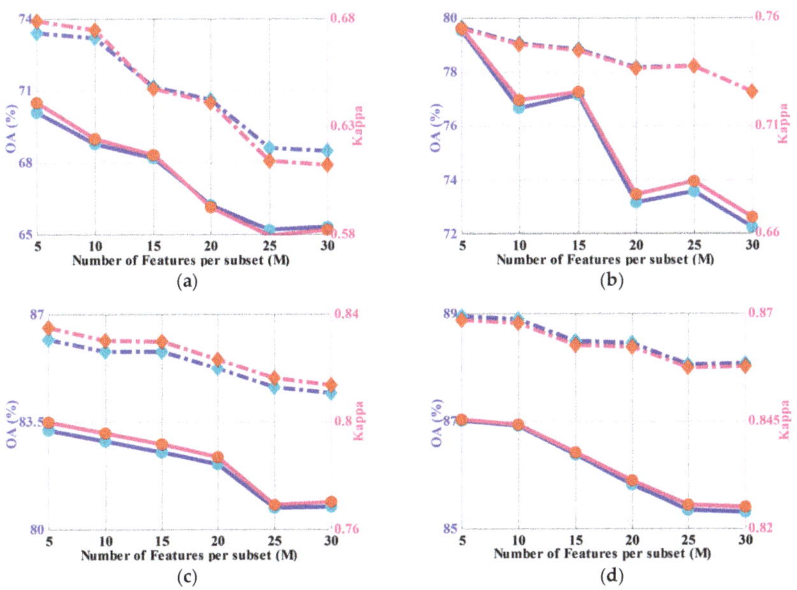

Figure 2. *Cont.*

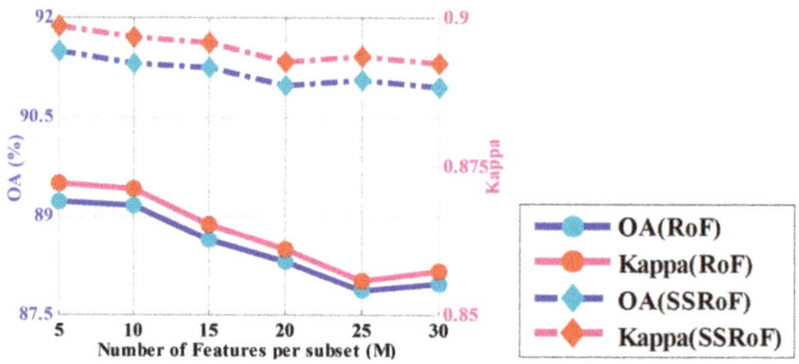

(e)

Figure 2. Impact of the number of features in each subset (M) under different numbers of training samples (1%, 2%, 5%, 10%, and 20% from (**a**–**e**), respectively).

4. Conclusions

Since existing rotation forest-based techniques fail to take account of the discriminative information of training samples during feature extraction, this paper proposed a semi-supervised rotation forest that uses the weighted semi-supervised local discriminant analysis method to jointly utilize the class discriminative information and local structural information provided by the labeled and unlabeled samples, respectively. The proposed algorithm aims to find the projection directions that provide better class separability, thus enhancing the performance of existing rotation forest algorithms. Furthermore, the proposed algorithm does not need additional parameters compared with the classical rotation forest method, which makes it easy to implement. Experiments have shown that the proposed algorithm outperforms several typical ensemble learning methods. Our future work will aim to reduce the computational time and assemble some other state-of-the-art machine learning algorithms.

Acknowledgments: This work was supported by the National Natural Science Foundation of China under Grant 61271348 and 61471148, and in part by the Foundation of Harbin Excellent Scholar under Grant 2015RAXXJ048. The authors would like to thank D. Landgrebe of Purdue University, West Lafayetter, Indiana, for providing the AVIRIS Indian Pines data set; P. Gamba of the University of Pavia, Italy, for providing the Pavia University data sets; and the Hyperspectral Image Analysis group and the NSF-NCALM at the University of Houston for providing the Houston University data set.

Author Contributions: Xiaochen Lu and Junping Zhang conceived and designed the experiments; Xiaochen Lu and Tong Li performed the experiments; all authors analyzed the data and reviewed the study; Xiaochen Lu wrote the paper.

Conflicts of Interest: The authors declare no conflict of interest.

References

1. Ballanti, L.; Blesius, L.; Hines, E.; Kruse, B. Tree species classification using hyperspectral imagery: A comparison of two classifiers. *Remote Sens.* **2016**, *8*, 445. [CrossRef]
2. Sami ul Haq, Q.; Tao, L.; Yang, S. Neural network based Adaboosting approach for hyperspectral data classication. In Proceedings of the 2011 International Conference on Computer Science and Network Technology (ICCSNT), Harbin, China, 24–26 December 2011.
3. Melgani, F.; Bruzzone, L. Classification of hyperspectral remote sensing images with support vector machines. *IEEE Trans. Geosci. Remote Sens.* **2004**, *42*, 1778–1790. [CrossRef]
4. Li, J.; Bioucas-Dias, J.; Plaza, A. Semisupervised hyperspectral image segmentation using multinomial logistic regression with active learning. *IEEE Trans. Geosci. Remote Sens.* **2010**, *11*, 4085–4098. [CrossRef]
5. Persello, C.; Bruzzone, L. Active and semisupervised learning for the classification of remote sensing images. *IEEE Trans. Geosci. Remote Sens.* **2014**, *52*, 6937–6956. [CrossRef]
6. Villa, A.; Chanussot, J.; Benediktsson, J.A.; Jutten, C. Spectral unmixing for the classification of hyperspectral images at a finer spatial resolution. *IEEE J. Sel. Top. Signal Process.* **2011**, *5*, 521–533. [CrossRef]

7. Golipour, M.; Ghassemian, H.; Mirzapour, F. Integrating hierarchical segmentation maps with MRF prior for classification of hyperspectral images in a Bayesian framework. *IEEE Trans. Geosci. Remote Sens.* **2016**, *54*, 805–816. [CrossRef]
8. Licciardi, G.; Pacifici, F.; Tuia, D.; Prasad, S.; West, T.; Giacco, F.; Thiel, C.; Inglada, J.; Christophe, E.; Chanussot, J.; et al. Decision fusion for the classification of hyperspectral data: Outcome of the 2008 GRSS data fusion contest. *IEEE Trans. Geosci. Remote Sens.* **2009**, *47*, 3857–3865. [CrossRef]
9. Wozniak, M.; Graña, M.; Corchado, E. A survey of multiple classifier systems as hybrid systems. *Inf. Fusion* **2014**, *16*, 3–17. [CrossRef]
10. Krawczyk, B.; Minku, L.L.; Gama, J.; Stefanowski, J.; Wozniak, M. Ensemble learning for data stream analysis: A survey. *Inf. Fusion* **2017**, *37*, 132–156. [CrossRef]
11. Santos, A.B.; Araújo, A.A.; Menotti, D. Combining multiple classification methods for hyperspectral data interpretation. *IEEE J. Sel. Top. Appl. Earth Obs. Remote Sens.* **2013**, *6*, 1450–1459. [CrossRef]
12. Waske, B.; Linden, S.V.D.; Benediktsson, J.A.; Rabe, A.; Hostert, P. Sensitivity of support vector machines to random feature selection in classification of hyperspectral data. *IEEE Trans. Geosci. Remote Sens.* **2010**, 2880–2889. [CrossRef]
13. Wang, S.; Yao, X. Relationships between diversity of classification ensembles and single-class performance measures. *IEEE Trans. Knowl. Data Eng.* **2013**, *25*, 206–219. [CrossRef]
14. Galar, M.; Fernandez, A.; Barrenechea, E.; Bustince, H.; Herrera, F. A review on ensembles for the class imbalance problem: Bagging-, Boosting-, and hybrid-based approaches. *IEEE Trans. Syst. Man Cybern.—Part C: Appl. Rev.* **2012**, *42*, 463–484. [CrossRef]
15. Friedman, J.H. Greedy function approximation: A gradient boosting machine. *Ann. Stat.* **2001**, *29*, 1189–1232. [CrossRef]
16. Breiman, L. Bagging predictors. *Mach. Learn.* **1996**, *24*, 123–140. [CrossRef]
17. Schapire, R.E. The strength of weak learn ability. *Mach. Learn.* **1990**, *5*, 197–227. [CrossRef]
18. Schapire, R.E.; Singer, Y. Improved boosting algorithms using confidence-rated predictions. *Mach. Learn.* **1999**, *37*. [CrossRef]
19. Ho, T.K. The random subspace method for constructing decision forests. *IEEE Trans. Pattern. Anal. Mach. Intell.* **1998**, *20*, 832–844. [CrossRef]
20. Breiman, L. Random forests. *Mach. Learn.* **2001**, *45*, 5–32. [CrossRef]
21. Ham, J.; Chen, Y.; Crawford, M.M.; Ghosh, J. Investigation of the random forest framework for classification of hyperspectral data. *IEEE Trans. Geosci. Remote Sens.* **2005**, *43*, 492–501. [CrossRef]
22. Xia, J.; Liao, W.; Chanussot, J.; Du, P.; Song, G.; Philips, W. Improving random forest with ensemble of features and semisupervised feature extraction. *IEEE Geosci. Remote Sens. Lett.* **2015**, *12*, 1471–1475. [CrossRef]
23. Chan, J.C.; Paelinckx, D. Evaluation of random forest and Adaboost tree-based ensemble classification and spectral band selection for ecotope mapping using airborne hyperspectral imagery. *Remote Sens. Environ.* **2008**, *112*, 2999–3011. [CrossRef]
24. Rodriguez-Galianoa, V.F.; Ghimireb, B.; Roganb, J.; Chica-Olmoa, M.; Rigol-Sanchezc, J.P. An assessment of the effectiveness of a random forest classifier for land-cover classification. *ISPRS J. Photogramm. Remote Sens.* **2012**, *67*, 93–104. [CrossRef]
25. Merentitis, A.; Debes, C.; Heremans, R. Ensemble learning in hyperspectral image classification: Toward selecting a favorable bias-variance tradeoff. *IEEE J. Sel. Top. Appl. Earth Obs. Remote Sens.* **2014**, *7*, 1089–1102. [CrossRef]
26. Gurram, P.; Kwon, H. Sparse kernel-based ensemble learning with fully optimized kernel parameters for hyperspectral classification problems. *IEEE Trans. Geosci. Remote Sens.* **2013**, *51*, 787–802. [CrossRef]
27. Samat, A.; Du, P.; Liu, S.; Li, J.; Cheng, L. E2LMs: Ensemble extreme learning machines for hyperspectral image classification. *IEEE J. Sel. Top. Appl. Earth Obs. Remote Sens.* **2014**, *7*, 1060–1069. [CrossRef]
28. Guo, H.; Li, Y.; Jennifer, S.; Gu, M.; Huang, Y.; Gong, B. Learning from class-imbalanced data: Review of methods and applications. *Expert Syst. Appl.* **2017**, *73*, 220–239. [CrossRef]
29. Merentitis, A.; Debes, C. Many hands make light work-on ensemble learning techniques for data fusion in remote sensing. *IEEE Geosci. Remote Sens. Mag.* **2015**, *3*, 86–99. [CrossRef]
30. Rodriguez, J.J.; Kuncheva, L.I. Rotation forest: A new classifier ensemble method. *IEEE Trans. Pattern Anal. Mach. Intell.* **2006**, *28*, 1619–1630. [CrossRef] [PubMed]

31. Ayerdi, B.; Romay, M.G. Hyperspectral image analysis by spectral-spatial processing and anticipative hybrid extreme rotation forest classification. *IEEE Trans. Geosci. Remote Sens.* **2016**, *54*, 2627–2639. [CrossRef]
32. Xia, J.; Falco, N.; Benediktsson, J.A.; Du, P.; Chanussot, J. Hyperspectral image classification with rotation random forest via KPCA. *IEEE J. Sel. Top. Appl. Earth Obs. Remote Sens.* **2017**, *10*, 1601–1609. [CrossRef]
33. Chen, J.; Xia, J.; Du, P.; Chanussot, J. Combining rotation forest and multiscale segmentation for the classification of hyperspectral data. *IEEE J. Sel. Top. Appl. Earth Obs. Remote Sens.* **2016**, *9*, 4060–4072. [CrossRef]
34. Rahulamathavan, Y.; Phan, R.C.-W.; Chambers, J.A.; Parish, J.D. Facial expression recognition in the encrypted domain based on local Fisher discriminant analysis. *IEEE Trans. Affect. Comput.* **2013**, *4*, 83–92. [CrossRef]
35. Belhumeur, P.N.; Hespanha, J.P.; Kriegman, D.J. Eigenfaces vs. Fisherfaces: Recognition using class specific linear projection. *IEEE Trans. Pattern Anal. Mach. Intell.* **1997**, *19*, 711–720. [CrossRef]
36. Lu, X.; Zhang, J.; Li, T.; Zhang, G. Synergetic classification of long-wave infrared hyperspectral and visible images. *IEEE J. Sel. Top. Appl. Earth Obs. Remote Sens.* **2015**, *8*, 3546–3557. [CrossRef]
37. Sun, B.; Kang, X.; Li, S.; Benediktsson, J.A. Random-walker-based collaborative learning for hyperspectral image classification. *IEEE Trans. Geosci. Remote Sens.* **2017**, *55*, 212–222. [CrossRef]
38. Kang, X.; Li, S.; Benediktsson, J.A. Feature extraction of hyperspectral images with image fusion and recursive filtering. *IEEE Trans. Geosci. Remote Sens.* **2014**, *52*, 3742–3752. [CrossRef]
39. Kang, X.; Zhang, X.; Li, S.; Li, K.; Li, J.; Benediktsson, J.A. Hperspectral anomaly detection with attribute and edge-preserving filters. *IEEE Trans. Geosci. Remote Sens.* **2017**, 1–12. [CrossRef]
40. Sugiyama, M.; Idé, T.; Nakajima, S.; Sese, J. Semi-supervised local Fisher discriminant analysis for dimensionality reduction. *Mach. Learn.* **2010**, *78*, 35–61. [CrossRef]
41. Sugiyama, M. Dimensionality reduction of multimodal labeled data by local Fisher discriminant Analysis. *J. Mach. Learn. Res.* **2007**, *8*, 1027–1061. [CrossRef]
42. He, X.; Niyogi, P. Locality preserving projections. In *Advances in Neural Information Processing Systems*; MIT Press: Cambridge, MA, USA, 2004; pp. 153–160.
43. Martinez, M.; Kak, A.C. PCA versus LDA. *IEEE Trans. Pattern Anal. Mach. Intell.* **2001**, *23*, 228–233. [CrossRef]
44. He, X.; Cai, D.; Yan, S.; Zhang, H. Neighborhood preserving embedding. In Proceedings of the Tenth IEEE International Conference on Computer Vision, Beijing, China, 17–20 October 2005; pp. 1208–1213.
45. Liao, W.; Pi, Y. Feature extraction for hyperspectral images based on semi-supervised local discriminant analysis. In Proceedings of the 2011 Joint Urban Remote Sensing Event (JURSE), Munich, Germany, 11–13 April 2011; pp. 401–404.
46. Bao, R.; Xia, J.; Mura, M.D.; Du, P.; Chanussot, J.; Ren, J. Combining morphological attribute profiles via an ensemble method for hyperspectral image classification. *IEEE Geosci. Remote Sens. Lett.* **2016**, *13*, 359–363. [CrossRef]
47. Xia, J.; Mura, M.D.; Chanussot, J.; Du, P.; He, X. Random subspace ensembles for hyperspectral image classification with extended morphological attribute profiles. *IEEE Trans. Geosci. Remote Sens.* **2015**, *53*, 4768–4785. [CrossRef]
48. Xia, J.; Du, P.; He, X.; Chanussot, J. Hyperspectral remote sensing image classification based on rotation forest. *IEEE Geosci. Remote Sens. Lett.* **2014**, *11*, 239–243. [CrossRef]

remote sensing

MDPI

Article

Semi-Supervised Classification of Hyperspectral Images Based on Extended Label Propagation and Rolling Guidance Filtering

Binge Cui [1,*], Xiaoyun Xie [1], Siyuan Hao [2], Jiandi Cui [1] and Yan Lu [1]

1 The College of Computer Science and Engineering, Shandong University of Science and Technology, Qingdao 266590, Shandong Province, China; xiaoyun_xie@sdust.edu.cn (X.X.); jiandi_cui@sdust.edu.cn (J.C.); luyan@sdust.edu.cn (Y.L.)

2 The School of Information and Control Engineering, Qingdao University of Technology, No.777 Jialingjiang Road, Huangdao District, Qingdao 266520, Shandong Province, China; haosiyuan@qut.edu.cn

* Correspondence: cuibinge@sdust.edu.cn; Tel.: +86-139-6968-0498

Received: 20 December 2017; Accepted: 23 March 2018; Published: 25 March 2018

Abstract: Semi-supervised classification methods result in higher performance for hyperspectral images, because they can utilize the relationship between unlabeled samples and labeled samples to obtain pseudo-labeled samples. However, how generating an effective training sample set is a major challenge for semi-supervised methods, In this paper, we propose a novel semi-supervised classification method based on extended label propagation (ELP) and a rolling guidance filter (RGF) called ELP-RGF, in which ELP is a new two-step process to make full use of unlabeled samples. The first step is to implement the graph-based label propagation algorithm to propagate the label information from labeled samples to the neighboring unlabeled samples. This is then followed by the second step, which uses superpixel propagation to assign the same labels to all pixels within the superpixels that are generated by the image segmentation method, so that some labels wrongly labeled by the above step can be modified. As a result, so obtained pseudo-labeled samples could be used to improve the performance of the classifier. Subsequently, an effective feature extraction method, i.e., RGF is further used to remove the noise and the small texture structures to optimize the features of the initial hyperspectral image. Finally, these produced initial labeled samples and high-confidence pseudo-labeled samples are used as a training set for support vector machine (SVM). The experimental results show that the proposed method can produce better classification performance for three widely-used real hyperspectral datasets, particularly when the number of training samples is relatively small.

Keywords: spectral-spatial classification; label propagation; superpixel; semi-supervised learning; rolling guidance filtering (RGF); graph; hyperspectral image

1. Introduction

Hyperspectral images have been widely used for many applications, such as classification [1], spectral unmixing [2], target detection [3], environmental monitoring [4] and anomaly detection [5]. Among these applications, classification is one of the most crucial branches. There are more than 100 spectral bands that provide detailed information to discriminate the object in a hyperspectral image [6]. However, the high dimensions of hyperspectral images require a more complicated model, while such a complicated model also requires more training samples to support it. Thus, the imbalance between the number of training samples and the high dimensions may cause the well-known "Hughes" phenomenon [7]. The existence of the "Hughes" phenomenon poses restrictions on performance improvement for the hyperspectral image classification.

In hyperspectral image classification, traditional spectral-based classification methods are widely used, such as support vector machines (SVM) [8], the back-propagation neural network (BP) [9], random forest (RF) [10] and the 1D deep convolutional neural network (1D CNN) [11]. However, all of these methods are sensitive to the quality and number of training samples; thus, the classification performance is limited when a small amount of training samples is provided. In order to further improve the classification performance, the rich spatial-contextual information is used in pixel-wise classification methods [12–15]. For instance, Pan et al. [16] introduced the hierarchical guidance filtering to extract the different spatial contextual information at different filter scales in hyperspectral images. In [17], a new network that utilizes the spectral and spatial information simultaneously was proposed to achieve more accurate classification results.

During the past few decades, the semi-supervised learning methods have shown excellent performance in hyperspectral image classification [18,19]. One goal of the semi-supervised learning method is to select the most useful unlabeled samples and to determine the label information of these new selected samples. Generally, semi-supervised learning can be classified into the generative model [20], the co-training model [21], the graph-based method [18–22], etc. All of those methods are based on an assumption that similar samples have the same labels. Hence, graph-based semi-supervised methods have attracted increasing attention in hyperspectral image classification [18,22]. For example, in [23], Wang et al. proposed a novel graph-based semi-supervised learning approach based on a linear neighborhood model to propagate the labels from the labeled samples to the whole dataset using these linear neighborhoods with sufficient smoothness. In [18], the wealth of unlabeled samples is exploited through a graph-based methodology to handle the special characteristics of hyperspectral images. The label propagation algorithm (LP) [24,25] is a widely-used method in graph-based semi-supervised learning [26,27], as in [28]; unlabeled data information is effectively exploited by combing the Gaussian random field model and harmonic function. Wang et al. [24] proposed an approach based on spatial-spectral label propagation for the semi-supervised classification method, in which labels were propagated from labeled samples to unlabeled samples with the spatial-spectral graph to update the training set. However, there are three main difficulties of the aforementioned graph-based semi-supervised classification methods: (1) how to significantly generate the pseudo-labeled samples with a high quality; (2) how to expand the propagation scope of the samples as much as possible; (3) how to modify the labels that wrongly propagate to other classes.

Recently, the superpixel technique [29] has been an effective way to introduce the spatial information for hyperspectral image classification [16,30,31]. Each superpixel is a homogeneous region, whose size and shape are adaptive. The commonly-used superpixel segmentation methods include the SLIC method [32], normalized cut method [33], regional growth method [34], etc. Moreover, superpixel-based classification methods [35,36] have shown a good robustness in the result of hyperspectral image classification. Motivated by the idea of a superpixel, we design a novel superpixel-based label propagation framework, extended label propagation (ELP), which uses a two-step propagation process to significantly extend the number of pseudo-labeled samples. In ELP, the spatial-spectral weighted graph is first constructed with the labeled samples and unlabeled samples from the spatial neighbors of the labeled samples to propagate the class labels to unlabeled samples. Second, the multi-scale segmentation algorithm [37] is used to generate superpixels, and then, superpixel propagation is introduced to assign the same label to all pixels within a superpixel. Finally, a threshold is defined; when the confidence of pseudo-labeled samples is higher than the defined threshold, they will be selected to enrich the training sample set. Note that the second step of the ELP method, i.e., extended label propagation with superpixel segmentation, is the innovation of the proposed method, because it can generate a large number of high-confidence pseudo-labeled samples.

In this paper, the motivations include three aspects. First, we would like to extend the number of high-confidence pseudo-labeled samples based on a two-step propagation process. Second, rolling guidance filtering is used to optimize the feature of the initial hyperspectral image. In the optimized image, the noise and small texture are removed, while the strong structure of the image is preserved, enhancing the discrimination within and between classes. Third, we want to modify the labels that wrongly propagate by the label propagation algorithm. The proposed ELP-RGF can effectively improve the classification performance with less training samples. The contributions of the proposed method consist of:

(1) We propose a novel extended label propagation component that is based on the label propagation algorithm. The second step of ELP, that is superpixel propagation, is the most innovative of the proposed method, because it not only expands the scope of the label propagation, but also generates a large number of high-confidence pseudo-labeled samples. Therefore, it has a good performance for hyperspectral image classification.

(2) In the step of superpixel propagation, the labels of pixels within the superpixel are obtained by a majority vote with the labeled samples belonging to that superpixel. Therefore, some pseudo-labeled samples with wrong labels that are obtained by the first step of the ELP method can be modified. Furthermore, we can show that the variation of ELP-RGF is much more stable compared to the result in [38] and [24].

(3) Optimized image features with the rolling guidance filter (RGF) [39] can eliminate the noise of the initial image. The filtered image is treated as an input to the SVM method to help improve the result of the final classification.

The remainder of this paper is organized as follows. The related work is described in Section 2. The proposed method is introduced in Section 3. The discussion is provided in Section 5. Finally, conclusions are given in Section 6.

2. Related Work

2.1. Superpixel Segmentation

The multi-scale segmentation algorithm [37] is an image segmentation method, in which the segmentation results are called patches. The essence of segmentation is to segment the image into many non-overlapping sub-regions. These patches or sub-regions are what we called superpixels. In this paper, a multi-scale segmentation algorithm is used to generate superpixels. This method uses the bottom-up region-growth strategy to group pixels with similar spectral values into the same superpixel. The key of the method is that the heterogeneity of the grouped region under the constraint term is minimal. The multi-scale segmentation method consists of three main steps:

(1) We define a termination condition T, also called the scale parameter, to control whether a regional merger is stopped. If T is smaller, the number of regions will be greater, and each region will have fewer pixels, and vice versa.

(2) Calculation of the spectral heterogeneity $h1$ and the spatial heterogeneity $h2$:

$$h_1 = \sum_{i=1}^{n} \omega_i \sigma_i \tag{1}$$

$$h_2 = \omega_u u + (1 - w_u) v \tag{2}$$

where σ_i is the standard deviation of the i-th band spectral values in the region. w_i refer to that the weight of the i-th band, and n is the band number. v and u represent the compactness and smoothness of the region, and w_u is the weight of the smoothness.

(3) The regional heterogeneity f can be obtained by combining $h1$ and $h2$:

$$f = \omega h_1 + (1 - \omega) h_2 \tag{3}$$

Here, ω is the weight of the spectral heterogeneity, and its value ranges from 0–1.

(4) Observation of the heterogeneity of regions f: if $f < T$, the region with the smallest heterogeneity will be merged with the adjacent regions.

(5) Operation of Step 4 until there are no regions that need to be merged.

2.2. Spatial-Spectral Graph-Based Label Propagation

The label propagation algorithm [24,40] is a graph-based classification method, in which the class labels are assigned to unlabeled samples by building a graph to propagate the labels. This algorithm models the input image $X = \{x_1, x_2, ..., x_n\} \in R^{d \times N}$ as a weighted graph $G = (V, E)$, in which the vertices $v \in V$ correspond to the pixels and the edges $e \in E \subseteq V \times V$ correspond to the links that connect two adjacent pixels. The label propagation algorithm consists of the following steps.

(1) A set of labeled pixels V_M is provided, where each pixel $v_i \in V_M$ has been assigned a label c, and the label set is $c \in L = \{1, ..., K\}$.

(2) The unlabeled sample set V_U of the neighbors of the labeled samples and the labeled sample set V_M are considered the nodes in the weight graph. Then, the weight matrices of spectral graph W_{ij}^{w} and W_{ij}^{s} are calculated as follows:

$$W_{ij}^{w} = e^{-(\frac{||v_i - v_j||^2}{2\varepsilon^2})} \quad if \quad x_j \in NB_k^{w}(x_i) \tag{4}$$

$$W_{ij}^{s} = e^{-(\frac{||v_i - v_j||^2}{2\varepsilon^2})} \quad if \quad x_j \in NB_d^{s}(x_i) \tag{5}$$

where $NB_k^{w}(x_i)$ is is a set of k nearest neighbors of x_i obtained by the spectral Euclidean distance and ε is a free parameter. $NB_k^{s}(x_i)$ is a set of the spatial neighbors of x_i in a spatial neighborhood system, the width of which is d.

(3) Construction of the graph W_{ij} as follow:

$$W = \mu W^{w} + (1 - \mu) W^{s} \tag{6}$$

where μ measures the weight of the spatial and spectral graph.

(4) According to the weight matrix, the propagation probability of the i-th node to the j-th node in the graph is calculated. The formula is as follows:

$$H_{ij} = \frac{W_{ij}}{\sum\limits_{k=1}^{n} W_{ik}} \tag{7}$$

(5) The labeled matrix A and probability distribution matrix P are initialized.

$$M_{ij} = \begin{cases} 1 & c_i = k, i \leq m \\ 0 & c_i \neq k, i \leq m \\ 1/K & m < i \leq n \end{cases} \tag{8}$$

$$P_{ij} = M_{ij} \qquad 1 \leq i \leq n, 1 \leq k \leq K \tag{9}$$

where the value of the labeled matrix is a probability value of each initialized node. If node i is a labeled sample, then the probability that the i-th node belongs to the k-th class is one, while the probability of belonging to the other classes is zero. If the node is an unlabeled sample, then the probability that it belongs to each class is initialized as $1/K$.

(6) Propagation process: According to the label propagation probability **P**, each node adds the weighted label information transmitted from adjacent nodes and updates the probability distribution Pto show that the nodes belong to each class. The updated formula is as follows:

$$P_{ij} = \sum_{k=1}^{n} H_{ik} P_{kj} \qquad 1 \le i \le n, 1 \le j \le K \tag{10}$$

(7) After the probability of propagation **P** is obtained, the label is assigned to the unlabeled samples based on the maximum probability.

$$c_i = \arg\max_{j \le K} P_{ij} \qquad 1 \le i \le n \tag{11}$$

All the nodes in the graph update the probability distribution based on the probability distribution of adjacent nodes. The label propagation algorithm is iteratively executed until the probability distribution of the nodes converges, then the class with the highest propagation probability is selected as the class label for the node. The propagation procedure is shown in Figure 1, and the light gray and the dark gray nodes are labeled samples from different classes, while hollow nodes represent the unlabeled samples. The values on the arrows are the propagation probabilities from the labeled samples to the unlabeled samples.

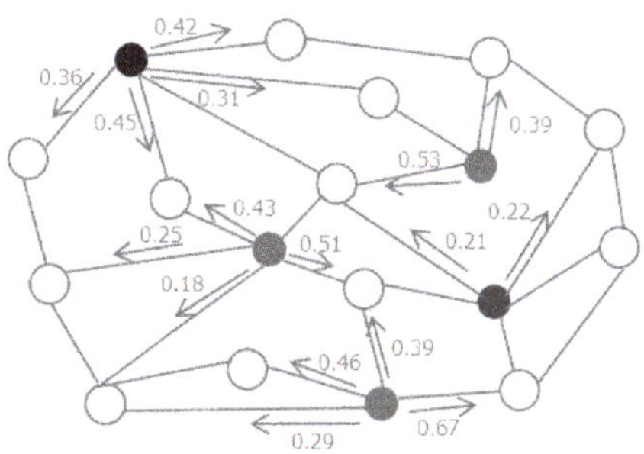

Figure 1. Procedure of label propagation.

2.3. Rolling Guidance Filtering

Filtering is an important step that removes weak edges while preserving strong ones when performing classification. In order to capture the different objects and structures in an image, the rolling guidance filtering is used to remove small-scale structures and preserve the original appearance of the large-scale structure. Therefore, the results processed by RGF are considered as the input feature of the SVM classifier, which can improve the classification accuracy. The rolling guidance filtering [39] contains two steps:

(1) Small structure removal:

In this section, a Gaussian filter is applied to blur the image, and the output is expressed as:

$$J^0(p) = \frac{\sum_{q \in R(p)} exp(-\frac{|p-q|^2}{2\sigma_s^2}).I(q)}{\sum_{q \in R(p)} exp(-\frac{|p-q|^2}{2\sigma_s^2})} \tag{12}$$

where I is the input image, p and q index the pixel coordinates in the image, $R(q)$ is a neighborhood pixel set for p and σ_s is the square of the Gaussian filter of variance σ_s^2. This means that when the scale of the image structure is smaller than σ_s, the structure will be completely removed.

(2) Large-scale edge recovery:

Large-scale edge recovery can be implemented in two steps. In the first step, the image processed by a Gaussian filter is treated as a guidance image (J^0), and then the joint bilateral filter is applied to guidance image J^0 and the initial image (I) to obtain output image J^1. In the second step, the guidance image is continuously updated by feeding the output of the previous iteration as the input to the next iteration. When the large-scale edges are recovered, the iteration of the guidance image can terminate. This procedure can be described as follows:

$$J^{t+1}(p) = \frac{1}{K_p} \sum_{q \in N(p)} \exp\left(\frac{-||p-q||^2}{2\sigma_s^2} - \frac{||J^t(p) - J^t(q)||^2}{2\sigma_r^2}\right) I(q) \tag{13}$$

$$K_p = \sum_{q \in N(p)} \exp\left(\frac{-||p-q||^2}{2\sigma_s^2} - \frac{||J^t(p) - J^t(q)||^2}{2\sigma_r^2}\right) \tag{14}$$

where Equation (11) is used for normalization and σ_s and σ_r control the spatial and range weights, respectively. t is the iteration number, and J^{t+1} is the result of the t-th iteration.

By the above two steps, RGF can perform well on the hyperspectral images. Thus, rolling guidance filtering is used to extract the information and features of the initial images, and the filtered image $\tilde{I}$ is expressed as follows:

$$\tilde{I} = RGF(I) \tag{15}$$

where RGF is the rolling-guidance filtering operator and I is the initial input image.

3. Proposed Method

Figure 2 shows the schematic diagram of the proposed semi-supervised classification method based on extended label propagation and rolling guidance filtering for the hyperspectral image, which consists of the following steps: First, the extended label propagation method is used to obtain an effective set of pseudo-labeled samples. This step is a two-step process. The first step is that the neighboring unlabeled samples from initial labeled samples are assigned labels by using the graph-based spatial-spectral label propagation method. The second step is that all pixels within the superpixel to which the labeled samples belong are assigned the same labels to achieve further label propagation. Then, pseudo-labeled samples with confidence less than the constant threshold will not be added into the training sample set. Then, rolling guidance filtering is used to optimize the feature of the original image, and the filtered result is used to extend the feature vector that is an input to the SVM. Finally, the initial labeled samples and pseudo-labeled samples are merged in training by SVM.

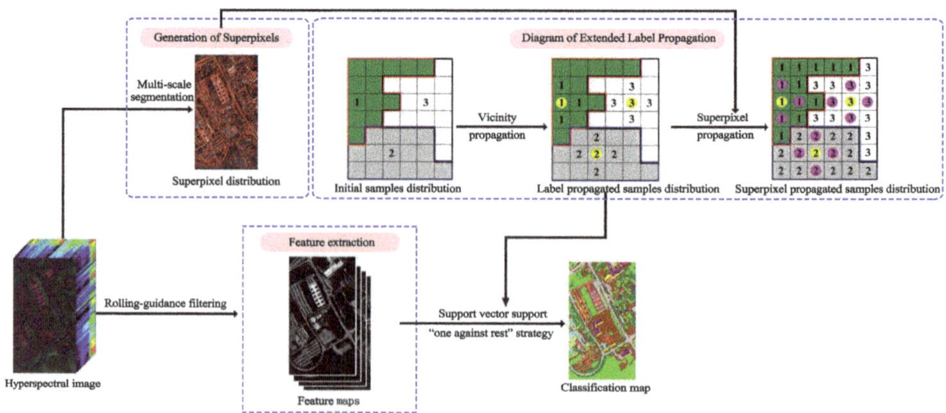

Figure 2. Schematic of the proposed semi-supervised classification method of hyperspectral images based on extended label propagation and rolling guidance filtering.

The proposed semi-supervised classification method based on extended label propagation and rolling guidance filtering (ELP-RGF) method can be shown by Algorithm 1:

Algorithm 1: proposed ELP-RGF method

Input: the dataset X, the initial labeled training sample set V_L, the weight μ, the width of spatial neighborhood system d, the segmentation scale S, the unlabeled samples V_U

1. Superpixels segmentation:
Obtain $Y = (Y_1, ..., Y_M)$, where Y_i is the i-th superpixel, based on the multi-scale segmentation algorithm for X.
2. Extended label propagation method:
Obtain the pseudo-labeled training sample set $\tilde{V}_{new}$.
(1) Label propagation:
Selection of the unlabeled training set V_U from the neighbors of the labeled samples.
Construction of the weighted graph G and weighted matrix W_{ij} by Equations (4)–(6).
Calculation of the probability matrix P by Equations (7)–(10).
Prediction of the labels of V_U by Equation (11) and generation of the pseudo-labeled sample set V_{new}.
(2) Superpixel propagation:
Observation of the labels of labeled samples belonging to superpixel Y_i, and then, the majority vote method is used to assign the labels for all pixels within Y_i.
3. Rolling guidance filtering:
Extraction of the spectral features of initial image X, and the filtered image $\tilde{I}$ is obtained by Equations (10)–(12).
4. SVM classification:
V_L and $\tilde{V}_{new}$ are merged as the final training sample set, and then, train SVM to obtain the prediction of labels of the testing set. The input feature vector to the SVM is the filtered image by the rolling guidance filtering.

Note that we perform the SVM to obtain the final classification result, because it has a good performance for the non-linear problem [41]. The goal of SVM is to find an optimal decision hyperplane that can maximize the distance between the two nearest samples on the two sides of the plane for classification. In this paper, the "one against rest" strategy [42] is adopted to achieve the multi-classification.

4. Experiment

In this section, the experimental results are performed on three real hyperspectral datasets to evaluate the performance of the proposed ELP-RGF method.

4.1. Datasets Description

In our experiments, three hyperspectral image datasets including the Indian Pines image, the University of Pavia image and the Kennedy Space Center image are utilized to evaluate the performance of the ELP-RGF.

(1) Indian Pines dataset: This image was acquired by the Airborne Visible/Infrared Imaging Spectrometer (AVIRIS) sensor, which captured an Indian Pines unlabeled agricultural site of northwestern Indiana and contains 220 × 145 × 145 bands. Twenty water absorption bands (Nos. 104–108, 150–163 and 220) were removed before hyperspectral image classification. The spatial resolution of the Indian Pines image is 20 m per pixel, and the spectral coverage ranges from 0.4–2.5 μm. Figure 3 shows a color composite and the corresponding ground-truth data of the Indian Pines image.

(2) University of Pavia dataset: This image capturing the University of Pavia, Italy, was recorded by the Reflective Optics System Imaging Spectrometer (ROSIS). This image contains 115 bands and a size 610 × 340 with a spatial resolution of 1.3 m per pixel and a spectral coverage ranging from 0.43–0.86 μm. Using a standard preprocessing approach before hyperspectral image classification, 12 noisy channels were removed. Nine classes of interest are considered for this image. Figure 4 shows the color composite and the corresponding ground-truth data of the University of Pavia image.

(3) Kennedy Space Center dataset: The Kennedy Space Center (KSC) image was captured by the National Aeronautics and Space Administration (NASA) Airborne Visible/Infrared Imaging Spectrometer instrument at a spatial resolution of 18 m per pixel. The KSC image contains 224 bands with a spatial size of 512 × 614, and the water absorption and low signal-to-noise ratio (SNR) bands

were discarded before the classification. Figure 5 shows the KSC image and the corresponding ground-truth data.

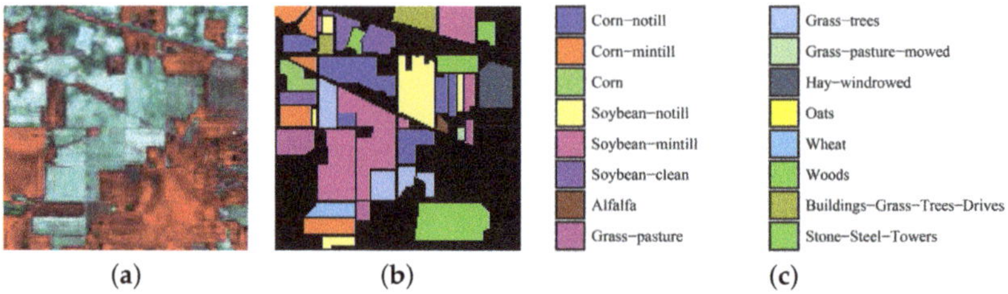

Figure 3. Indian Pines dataset. (**a**) False-color composite; (**b**,**c**) Ground-truth data.

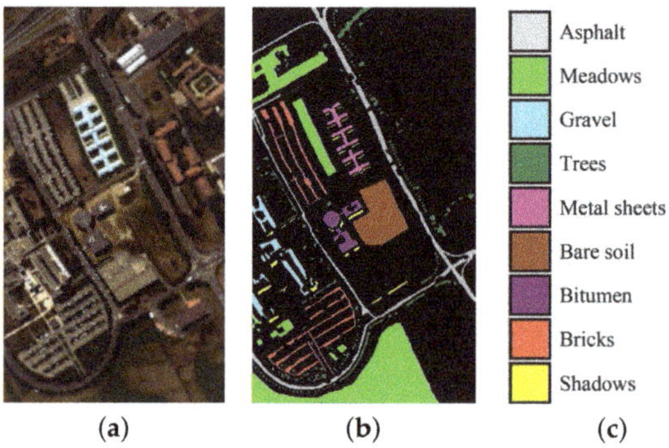

Figure 4. University of Pavia image. (**a**) False-color composite; (**b**,**c**) Ground-truth data.

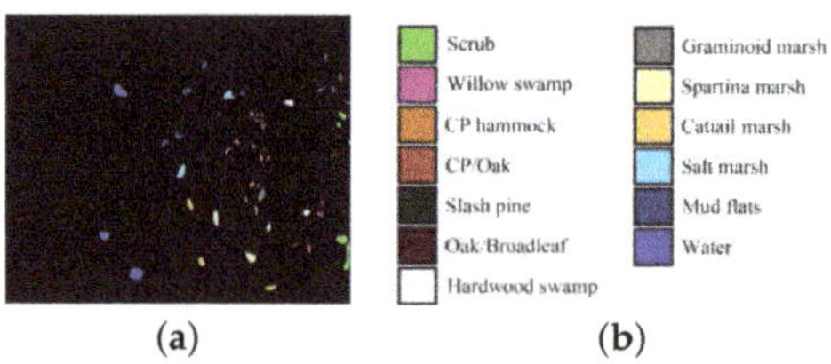

Figure 5. (**a**,**b**) Ground truth data of the Kennedy Space Center images.

4.2. Parameter Analysis of the Proposed Method

In the experiments, the original images were segmented by the multi-scale segmentation method (MSS) [37]. In this section, we fix the shape parameters $v = 0.1$ and the smoothness parameter to $u1 = 0.5$ in MSS. For the proposed method, there are three hyperparameters that have to be adjusted, namely weight parameter u, segmentation scale S and the width of spatial neighborhood d. The three hyperparameters were selected using the cross-validation strategy. Figure 6 shows the classification results obtained by the ELP-RGF method with different weight parameters μ and segmentation scales S. From Figure 6, we can see that the result of Figure 6a–b is visually more satisfactory than that of Figure 6c–d. If the process of label propagation entirely relied on the spatial graph, that is, u = 1 is applied, the result of ELP-RGF is poor, as Figure 6d shows. Therefore, $\mu = 0.001$ to $\mu = 0.01$

is considered the most optimal weight parameter range. We can see that the classification result of Figure 6f–g is better than Figure 6e,h, especially for the landscape of "Soybeans-min till" and "Hay-windrowed". In addition, Figure 7 shows the OA curves of ELP-RGF in the different u and dto illustrate the spatial weight parameter playing an important role in the process of label propagation. Furthermore, Figure 8 shows that the classification accuracies and computing time of the proposed method are significantly affected by *S*. When the two factors of the classification accuracy and the computing time are taken into full consideration (see Figure 8) and observing the selected parameters obtained by cross-validation, we can know that the optimal parameter range is 4–6.

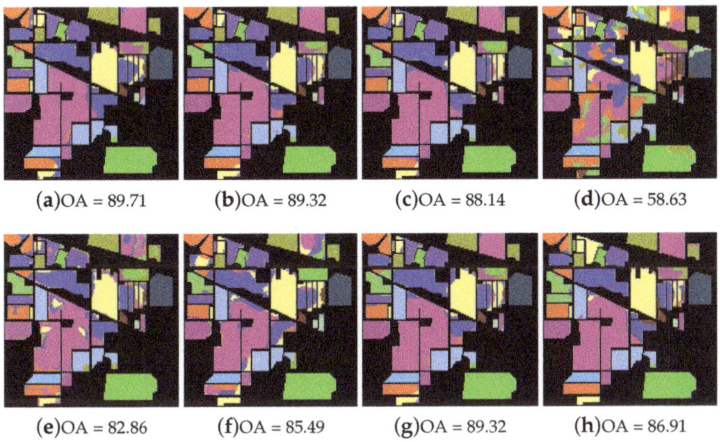

Figure 6. The analysis for the hyperparameters μ and S for the Indian Pines image. In the first row, S is fixed as five. (**a–d**) respectively show the classification results obtained by the extended label propagation (ELP)-RGF method with (**a**) u = 0.001, (**b**) u = 0.01, (**c**) u = 0.1 and (**d**) u = 1. In the second row, μ is fixed as 0.01; (**e–h**) respectively show the classification maps obtained by ELP-RGF method with (**e**) $S = 1$, (**f**) $S = 3$, (**g**) $S = 5$ and (**h**) $S = 9$.

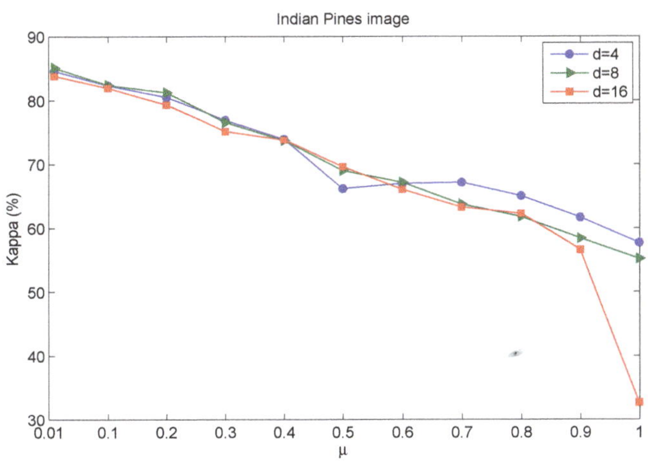

Figure 7. Influence of μ and d on the Kappa coefficient of ELP-RGF for the Indian Pines dataset.

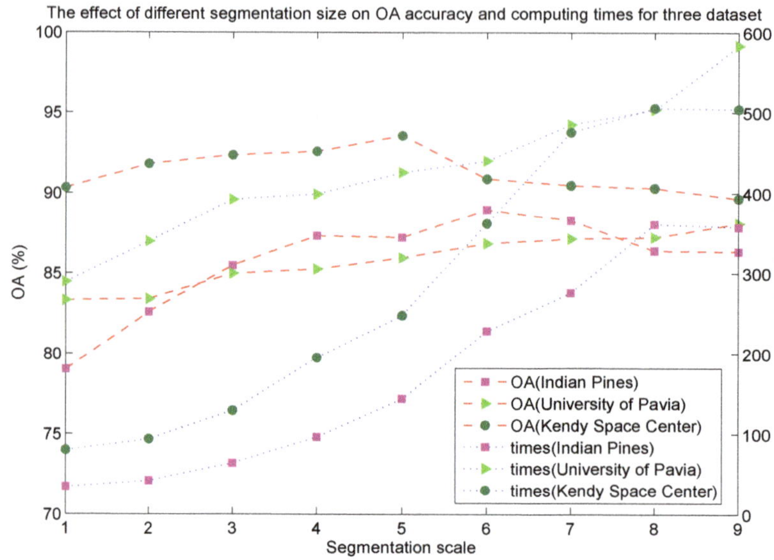

Figure 8. The effect of the different segmentation size on OA accuracy and computing times for three datasets.

4.3. Comparison with Other Classification Methods

In this section, the proposed ELP-RGF method is compared with several hyperspectral image classification methods, i.e., the typical SVM method, more advanced extended random walkers (ERW) [43] and semi-supervised methods (the Laplacian support vector machine (LapSVM) [38] and spectral-spatial label propagation (SSLP-SVM) [24]). In addition, a post-processing-based edge-preserving filtering (EPF) [44] and the rolling guidance filtering method (RGF) [39] are also used as the comparison methods. The parameter settings for the EPF, ERW and SSLP-SVM methods are given in the corresponding papers. The evaluation indexes in Tables 1–4 are given in the form of the mean $\pm$ standard deviation.

For the Indian Pines dataset, Table 1 shows the OA, AA and Kappa coefficient of different methods with the 5/10/15 training numbers per class (represented as s). From Table 1 we can see that the OA accuracy and Kappa coefficient of the proposed ELP-RGF method are better than other methods when the number of training samples is relatively small. In particular, when the $s = 5$, the OA accuracy of the proposed ELP-RGF method increases 14.29% and 36.63% compared to that of the SSLP-SVM method and the LapSVM method. The Kappa coefficient of ELP-RGF method is 13.52% higher than SSLP-SVM when $s = 10$, which fully shows the superiority of the two-step method proposed in this paper. We can see that the performance of the proposed ELP-RGF is always superior to that of the ERW method. As the number of training samples increases, the accuracy of the increase rate has decreased, however, there is still a large gap compared with other methods. Figure 9 shows the classification maps obtained by different methods. It can be seen that the classification map of the proposed method has less noise, and the boundary region in the classification map is also much clearer.

Table 1. Comparison of classification accuracies (in percentage) provided by different methods using different training samples per class (Indian Pines image). EPF, edge-preserving filtering; ERW, extended random walkers; LapSVM, Laplacian SVM; SSLP, spectral-spatial label propagation.

Methods	Metrics	Training Samples per Class (*s*)		
		s = 5	*s* = 10	*s* = 15
SVM	OA	45.31 ± 5.19	57.58 ± 2.98	63.56 ± 2.61
	AA	47.41 ± 3.71	55.19 ± 2.27	59.84 ± 2.21
	Kappa	39.21 ± 5.43	52.52 ± 3.17	59.11 ± 2.82
EPF	OA	57.97 ± 5.93	69.89 ± 3.45	77.68 ± 3.08
	AA	61.35 ± 6.40	70.07 ± 3.47	79.58 ± 2.70
	Kappa	52.91 ± 6.50	66.12 ± 3.77	74.80 ± 3.45
RGF	OA	56.14 ± 5.33	70.49 ± 4.69	78.91 ± 1.39
	AA	61.18 ± 4.89	68.49 ± 5.03	74.52 ± 2.77
	Kappa	51.13 ± 5.82	66.89 ± 5.08	76.14 ± 1.55
ERW	OA	72.30 ± 4.38	84.87 ± 4.41	90.02 ± 1.30
	AA	83.32 ± 2.41	**91.48 ± 2.30**	94.39 ± 0.94
	Kappa	68.96 ± 4.69	82.95 ± 4.86	88.69 ± 1.46
LapSVM	OA	42.50 ± 0.27	58.58 ± 0.46	58.42 ± 0.32
	AA	50.96 ± 1.36	61.54 ± 0.26	62.20 ± 0.86
	Kappa	36.89 ± 0.54	53.01 ± 0.32	53.82 ± 0.36
SSLP-SVM	OA	64.84 ± 1.43	76.05 ± 0.73	80.79 ± 1.44
	AA	65.96 ± 2.38	78.07 ± 0.64	82.08 ± 0.98
	Kappa	60.63 ± 1.49	73.17 ± 0.80	78.38 ± 1.60
ELP-RGF	OA	**79.13 ± 1.80**	**89.14 ± 1.06**	**94.31 ± 0.75**
	AA	**77.65 ± 2.94**	88.98 ± 1.37	**94.45 ± 1.54**
	Kappa	**76.37 ± 2.03**	**87.68 ± 1.20**	**93.53 ± 0.85**

For the University of Pavia image, we randomly selected 5, 10 and 15 samples from each class as the training samples. Table 2 shows the OA, AA and Kappa coefficient of the different methods with different s. According to Table 2, the proposed ELP-RGF, SSLP-SVM, LapSVM and ERW can produce greater classification accuracy than the SVM at the same s. However, the degrees of the improvement of ERW, SSLP-SVM and LapSVM are smaller compared with ELP-RGF. For example, the OA of ERW and SSLP-SVM increased by 1.16% and 10.24%. The experimental result indicated that the proposed ELP-RGF outperforms the compared methods. The results show that the OA of the proposed method is 96.02%, which is 10.25% higher than that of the SSLP-SVM and 0.91% higher than that of the ERW when $s = 15$. The OA accuracy and kappa coefficients of the ELP-RGF method are always the highest, which demonstrates that the ELP-RGF is the most accurate classifier among these methods. We can see that compared with SVM, SSLP-SVM and ERW, the OA accuracy and Kappa coefficients of the ELP method are more competitive. Figure 10 shows the classification maps of different methods when $s = 15$. The figure shows the effectiveness of the proposed method. The proposed method presents more accurate classification results for the class of MetalSheetsand Gravel, and its classification result is better than those of other methods.

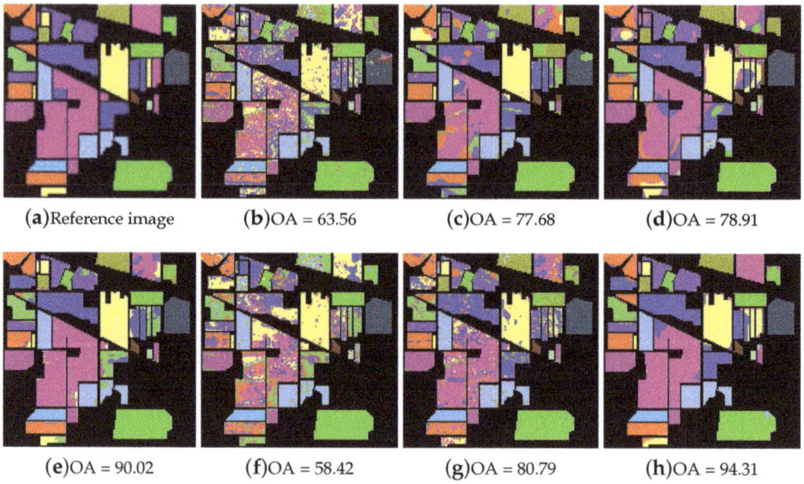

Figure 9. Classification maps of different methods for the Indian Pines image. (**a**) Reference image, (**b**) SVM method; (**c**) EPF method; (**d**) RGF method; (**e**) ERW method; (**f**) LapSVM method; (**g**) SSLP-SVM method and (**h**) ELP-RGF method.

Table 2. Overall accuracy of the various methods for the University of Pavia image (average of 10 runs with thestandard deviation; the bold values indicate the greatest accuracy among the methods in each case).

Method	$s = 5$	$s = 10$	$s = 15$
SVM	61.54 ± 5.15	67.70 ± 4.72	69.62 ± 3.35
EPF	58.98 ± 8.58	71.07 ± 8.02	80.86 ± 6.37
RGF	55.85 ± 7.22	74.82 ± 4.49	83.02 ± 4.87
ERW	80.70 ± 6.45	90.28 ± 3.71	92.57 ± 4.36
LapSVM	62.23 ± 2.03	63.03 ± 0.22	67.65 ± 0.43
SSLP-SVM	67.15 ± 2.45	82.15 ± 0.71	83.49 ± 1.30
ELP-RGF	**82.39 ± 1.42**	**91.54 ± 1.54**	**93.73 ± 1.37**

For the Kennedy Space Center dataset, we evaluated the classification accuracies of different methods using 39 training samples collected from each class. Table 3 shows the OAs, AAs, Kappa and individual classification accuracies obtained for the various methods. From Table 3, it is demonstrated that the OA, AA and Kappa accuracy of the proposed method are the highest in all comparative methods. Most of individual accuracies are significantly higher than other methods. For the class of Willowswamp, the accuracies of the proposed method and SSLP-SVM are 99.75% and 77.38%; thus, the accuracy gain is 22.37%. For the class of Oak/Broadleaf, the proposed method can produce 59.21% and 57.11% OA improvements compared with SSLP-SVM and SVM. Table 4 provides the OA accuracies of the various methods. Observing the values in Table 4, we can see that the classification accuracy is proportional to s. Moreover, the performance of the proposed method is not only higher than the other semi-supervised classification methods, but also can improve more than that of ERW. Figure 11 shows that the proposed ELP-RGF method can achieve better classification performance and produce little noise compared with other methods.

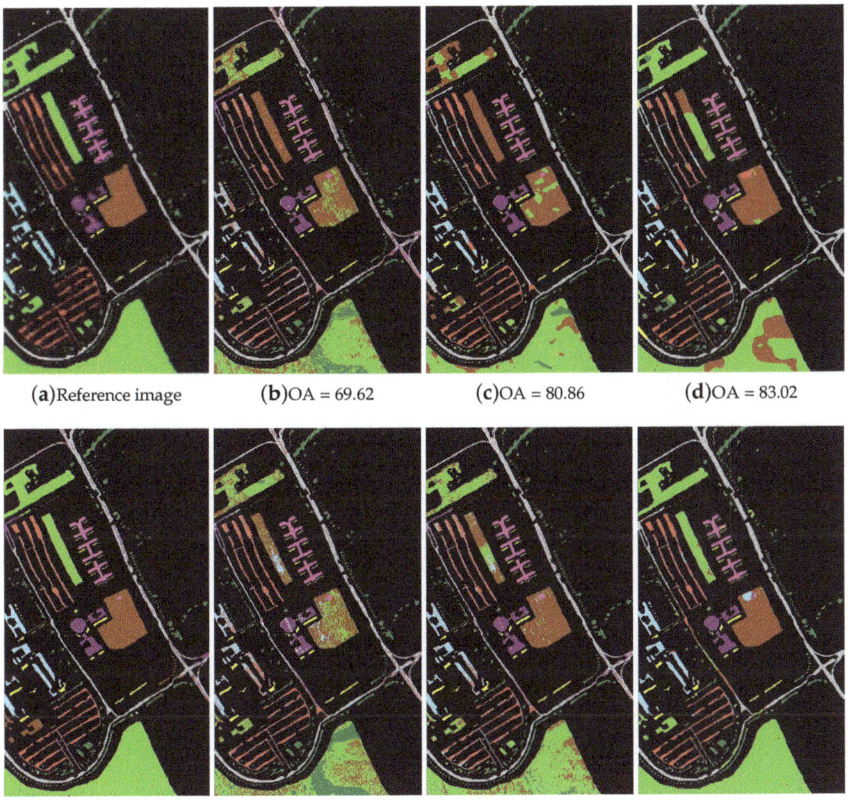

(**a**)Reference image (**b**)OA = 69.62 (**c**)OA = 80.86 (**d**)OA = 83.02

(**e**)OA = 92.57 (**f**)OA = 67.65 (**g**)OA = 83.49 (**h**)OA = 93.73

Figure 10. Classification maps of different methods for the University of Pavia image (**a**) Reference image; (**b**) SVM method; (**c**) EPF method; (**d**) RGF method; (**e**) ERW method; (**f**) LapSVM method; (**g**) SSLP-SVM method and (**h**) ELP-RGF method.

Table 3. Individual class accuracies, OA, AA and Kappa coefficient (in percentage) for the Kennedy Space Center images.

Class	Training	Test	Accuracy of Classification						
			SVM	EPF	RGF	ERW	LapSVM	SSLP-SVM	ELP-RGF
Scrub	3	758	92.25 ± 6.15	90.55 ± 2.25	96.78 ± 6.92	88.68 ± 17.99	87.17 ± 6.13	87.19 ± 6.27	**100**
Willow swamp	3	240	72.87 ± 9.38	86.74 ± 1.47	81.32 ± 14.78	68.56 ± 20.13	95.63 ± 0.69	77.38 ± 4.59	**99.75 ± 0.30**
CP hammock	3	253	70.65 ± 9.02	87.00 ± 2.36	75.16 ± 19.67	62.70 ± 23.84	70.90 ± 2.85	85.87 ± 6.56	**93.06 ± 5.65**
CP/Oak	3	249	35.41 ± 9.23	54.50 ± 2.26	37.26 ± 18.71	77.10 ± 22.45	**83.97 ± 10.32**	51.97 ± 16.92	75.49 ± 22.72
Slash pine	3	158	43.04 ± 10.74	59.64 ± 3.01	46.26 ± 39.56	**88.29 ± 8.81**	79.08 ± 1.60	41.13 ± 8.29	55.95 ± 4.09
Oak/Broadleaf	3	226	38.53 ± 18.23	62.16 ± 3.04	58.85 ± 43.78	94.48 ± 15.44	89.62 ± 3.27	36.43 ± 9.46	**95.64 ± 3.89**
Hardwood swamp	3	102	52.88 ± 17.18	77.62 ± 1.65	80.03 ± 20.80	**100**	96.34 ± 1.18	72.06 ± 10.95	98.79 ± 1.15
Graminoid marsh	3	428	43.90 ± 16.09	66.50 ± 2.03	67.77 ± 37.25	76.09 ± 21.61	93.34 ± 1.26	76.47 ± 11.11	**99.10 ± 0.74**
Spartina marsh	3	517	75.42 ± 10.30	81.25 ± 2.12	85.19 ± 12.89	78.86 ± 19.25	98.12 ± 0.67	89.52 ± 3.21	**97.21 ± 3.90**
Cattail marsh	3	401	59.72 ± 28.96	72.09 ± 2.73	61.39 ± 43.02	72.13 ± 23.05	**92.90 ± 9.77**	75.53 ± 13.87	84.79 ± 17.69
Salt marsh	3	416	89.04 ± 23.01	90.09 ± 2.42	89.94 ± 24.64	85.56 ± 22.82	94.92 ± 4.07	84.47 ± 16.34	**99.95 ± 0.10**
Mud flats	3	500	67.62 ± 16.57	78.70 ± 2.39	89.86 ± 14.49	73.77 ± 26.25	**94.22 ± 2.25**	68.16 ± 7.82	94.21 ± 4.38
Water	3	924	98.58 ± 2.35	98.90 ± 0.59	83.33 ± 40.82	93.36 ± 16.74	99.08 ± 0.72	99.15 ± 0.78	**100**
OA			65.45 ± 8.12	76.13 ± 0.96	66.81 ± 7.87	82.21 ± 4.36	91.25 ± 1.18	75.52 ± 1.24	**93.21 ± 2.44**
AA			64.61 ± 4.10	77.36 ± 0.68	73.32 ± 4.56	81.51 ± 3.59	90.41 ± 0.81	72.72 ± 1.18	**91.84 ± 2.23**
Kappa			61.76 ± 8.76	73.52 ± 1.02	63.52 ± 8.44	80.25 ± 4.79	90.25 ± 1.32	72.88 ± 1.35	**92.45 ± 2.71**

Table 4. Overall accuracy of the various methods for the Kennedy Space Center image (average of 10 runs with thestandard deviation; the bold values indicate the greatest accuracy among the method in each case).

Method	$s = 5$	$s = 10$	$s = 15$
SVM	74.05 ± 3.65	83.12 ± 1.83	85.96 ± 1.31
EPF	85.48 ± 4.26	92.66 ± 2.82	96.24 ± 1.87
RGF	87.05 ± 4.32	95.30 ± 1.76	97.42 ± 1.51
ERW	88.29 ± 3.19	96.85 ± 1.37	97.93 ± 0.94
LapSVM	61.40 ± 0.12	71.94 ± 0.10	74.09 ± 0.30
SSLP-SVM	82.01 ± 2.93	90.61 ± 0.90	93.15 ± 0.53
ELP-RGF	**94.12 ± 0.65**	**99.05 ± 0.24**	**99.38 ± 0.12**

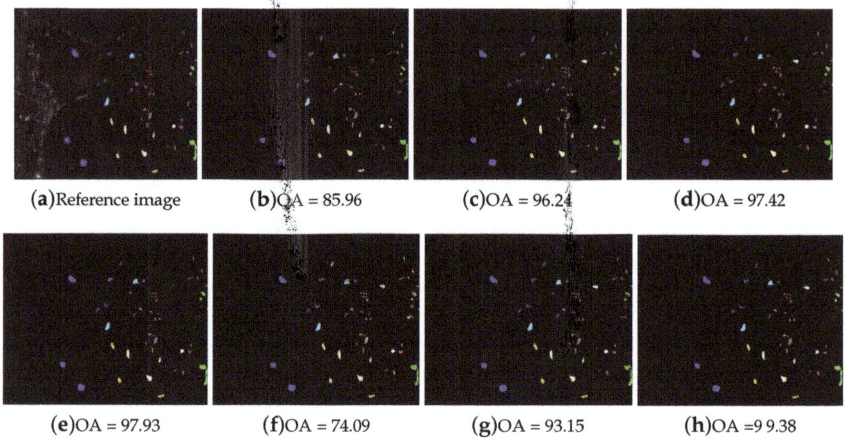

Figure 11. Classification maps of different methods for the Kennedy Space Center image. (**a**) Reference image; (**b**) SVM method; (**c**) EPF method; (**d**) RGF method; (**e**) ERW method; (**f**) LapSVM method; (**g**) SSLP-SVM method; (**h**) ELP-RGF method.

Table 5 lists the number of samples generated by the two semi-supervised methods (i.e., SSLP-SVM and ELP-RGF) under three different datasets and the correct rate of these new labeled samples. We can see that the total number of labeled samples generated by the ELP-RGF method is almost 7–22-times more than that generated by the SSLP-SVM method for three datasets. Although the correct rate of the SSLP-SVM method is slightly higher, the ELP-RGF method is also competitive. More importantly, the proposed ELP-RGF can produce more labeled samples.

Table 6 illustrates the effect of the superpixel by comparing with the RGF method, the combination of label propagation and RGF and the combination of superpixel propagation and RGF. The table shows that the superpixel propagation plays a major role in the proposed method. For example, for the Indian Pines image, the OA accuracy of the proposed ELP-RGF is 79.13%, while the accuracy obtained by SP-RGF and RGF is 75.62% and 56.14%, respectively. For the Kennedy Space Center image, the OA accuracy of the SP-RGF method is 4.64% higher than that of LP-RGF. As Table 6 shows, the accuracy of the SL-RGF is more than that of the LP-RGF method when RGF is used in those methods. While LP-RGF is higher than SP-RGF, the gap is small. Thus, the process of superpixel propagation is very useful to help improve the classification result.

Table 5. Correct rate of samples generated by the two semi-supervised methods of three datasets.

Methods	Data Set	Initial Samples	Increased Samples	Incorrect Labeled Samples	Correct Rate
SSLP-SVM	Indian Pines	108	349	1	99.71%
	University of Pavia	90	1028	4	99.61%
	Kennedy Space Center	39	115	0	100%
ELP-RGF	Indian Pines	108	2998	25	99.17%
	University of Pavia	90	7440	9	99.88%
	Kennedy Space Center	39	2558	25	99.02%

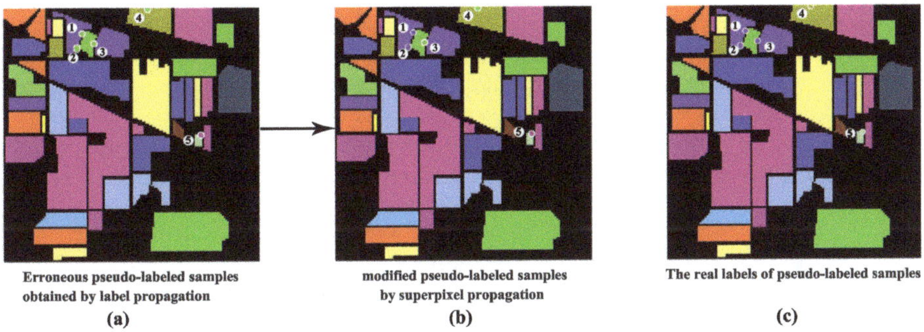

Figure 12. The process of modifying the wrongly-labeled samples. (**a**) The five wrongly-labeled pseudo-labeled samples are provided; (**b**) shows that the first and second labels of pseudo-labeled have been modified by superpixel propagation; (**c**) shows the real labels of the provided wrongly-labeled pseudo-labeled samples.

Table 6. Overall accuracy of the various combined methods involved in the proposed method for three datasets.

Data Set	RGF	LP-RGF	SP-RGF	ELP-RGF
Indian Pines	56.14	67.63	75.62	79.13
University of Pavia	55.85	75.22	74.25	82.39
Kennedy Space Center	87.05	89.33	93.97	94.12

5. Discussion

In this paper, the proposed ELP-RGF method is used to increase the number of training samples and optimize the features of the initial hyperspectral image. Previous label propagation-based works, such as SSLP-SVM, only increased a small number of training samples, which are neighboring the labeled samples, and the computational expense is large. If the scope of propagation is beyond neighbors, the computing time will rapidly increase. Furthermore, in the process of label propagation, some wrongly-labeled samples may be introduced to train the model, resulting in misclassification. In our ELP-RGF method, a two-step label prorogation process called ELP is proposed, which first utilized the spatial-spectral label propagation to propagate the label information from labeled samples to the neighboring unlabeled samples. Then, superpixel propagation is used to expand the scope of propagation to the entire superpixel to increase the huge number of training samples, and it is less time consuming compared to the propagation beyond neighbors. Compared with other semi-supervised classification methods, ELP has two obvious advantages: on the one hand, it can generate a large number of pseudo-labeled samples for model training; on the other hand, it can ensure the 'effectiveness' of the increased pseudo-labeled samples; here, 'effectiveness' means that almost all of the labels of the pseudo-labeled samples are correct, which was shown in Table 6. Moreover, as shown in Figure 12, the wrongly-labeled samples in the first step of the ELP method can be modified by the superpixel propagation. Thus, the proposed ELP-RGF method shows a better classification performance than other comparative methods. However, the greatest limitation of the proposed method is that the classification result is over-reliant on the segmentation scale. As shown in Figure 8, the difference in classification results with different segmentation scales is larger.

6. Conclusions

In this paper, a novel semi-supervised classification method of hyperspectral images based on extended label propagation and rolling guidance filtering is proposed. The first advantage of this method is that the number of pseudo-labeled training samples is significantly increased. The second advantage is that the diversity of training samples is improved to enhance the generalization of the proposed method. The third advantage is that the spatial information is fully considered using graphs and superpixels. The experimental results on three different hyperspectral datasets demonstrate that the proposed ELP-RGF method offers an excellent performance in terms of both visual quality and quantitative evaluation indexes. In particular, when the number of training samples is relatively small, the improvement is more obvious.

Acknowledgments: This work was co-supported by the National Natural Science Foundation of China (NSFC) (41406200, 61701272) and Shandong Province Natural Science Foundation of China (ZR2014DQ030, ZR2017PF004).

Author Contributions: Binge Cui conceived of the idea of this paper. Xiaoyun Xie designed the experiments and drafted the paper. Binge Cui, Siyuan Hao, Jiandi Cui and Yan Lu revised the paper.

Conflicts of Interest: The authors declare no conflict of interest.

References

1. Zhong, Z.; Fan, B.; Duan, J.; Wang, L.; Ding, K.; Xiang, S. Discriminant Tensor Spectral–Spatial Feature Extraction for Hyperspectral Image Classification. *IEEE Geosci. Remote Sens. Lett.* **2017**, *12*, 1028–1032.
2. Xu, X.; Shi, Z. Multi-objective based spectral unmixing for hyperspectral images. *ISPRS J. Photogramm. Remote Sens.* **2017**, *124*, 54–69.
3. Zhang, L.; Zhang, L.; Tao, D.; Huang, X. Sparse Transfer Manifold Embedding for Hyperspectral Target Detection. *IEEE Geosci. Remote Sens. Lett.* **2013**, *52*, 1030–1043.
4. Pan, B.; Shi, Z.; An, Z.; Jiang, Z.; Ma, Y. Sparse Transfer Manifold Embedding for Hyperspectral Target Detection. *IEEE J. Sel. Top. Appl. Earth Obs. Remote Sens.* **2017**, *99*, 1–13.
5. Kang, X.; Zhang, X.; Li, S.; Li, K.; Li, J.; Benediktsson, J.A. Hyperspectral Anomaly Detection with Attribute and Edge-Preserving Filters. *IEEE Trans. Geosci. Remote Sens.* **2017**, *99*, 1–12.
6. Chang, C.I. *Hyperspectral Imaging: Techniques for Spectral Detection and Classification*; Plenum Publishing Co.: New York, NY, USA, 2003.
7. Hughes, G. On the mean accuracy of statistical pattern recognizers. *Inf. Theory IEEE Trans.* **1968**, *14*, 55–63.
8. Gotsis, P.K.; Chamis, C.C.; Minnetyan, L. Classification of hyperspectral remote sensing images with support vector machines. *IEEE Trans. Geosci. Remote Sens.* **2004**, *42*, 1778–1790.
9. Benediktsson, J.A.; Swain, P.H.; Ersoy, O.K. Neural Network Approaches versus Statistical Methods in Classification of Multisource Remote Sensing Data. *Geosci. Remote Sens. Symp.* **1989**, *28*, 540–552.
10. Xia, J.; Du, P.; He, X.; Chanussot, J. Hyperspectral Remote Sensing Image Classification Based on Rotation Forest. *IEEE Geosci. Remote Sens. Lett.* **2013**, *11*, 239–243.
11. Chen, Y.; Jiang, H.; Li, C.; Jia, X.; Ghamisi, P. Deep Feature Extraction and Classification of Hyperspectral Images Based on Convolutional Neural Networks. *IEEE Trans. Geosci. Remote Sens.* **2016**, *54*, 6232–6251.
12. Fauvel, M.; Tarabalka, Y.; Benediktsson, J.A.; Chanussot, J.; Tilton, J.C. Advances in Spectral-Spatial Classification of Hyperspectral Images. *Proc. IEEE.* **2013**, *101*, 652–675.
13. Pan, B.; Shi, Z.; Xu, X. R-VCANet: A New Deep-Learning-Based Hyperspectral Image Classification Method. *IEEE J. Sel. Top. Appl. Earth Obs. Remote Sens.* **2017**, *99*, 1–12.
14. Kang, X.; Xiang, X.; Li, S.; Benediktsson, J.A. PCA-Based Edge-Preserving Features for Hyperspectral Image Classification. *IEEE Trans. Geosci. Remote Sens.* **2017**, *99*, 1–12.

15. Pan, B.; Shi, Z.; Xu, X. MugNet: Deep learning for hyperspectral image classification using limited samples. *ISPRS J. Photogramm. Remote Sens.* **2017**, doi:10.1016/j.isprsjprs.2017.11.003.
16. Pan, B.; Shi, Z.; Xu, X. Hierarchical Guidance Filtering-Based Ensemble Classification for Hyperspectral Images. *IEEE Trans. Geosci. Remote Sens.* **2017**, *99*, 1–13.
17. Pan, B.; Shi, Z.; Zhang, N.; Xie, S. Hyperspectral Image Classification Based on Nonlinear Spectral–Spatial Network. *IEEE Geosci. Remote Sens. Lett.* **2016**, *99*, 1–5.
18. Camps-Valls, G.; Marsheva, T.V.B.; Zhou, D. Semi-Supervised Graph-Based Hyperspectral Image Classification. *IEEE Trans. Geosci. Remote Sens.* **2007**, *45*, 3044–3054.
19. Shao, Y.; Sang, N.; Gao, C. Probabilistic Class Structure Regularized Sparse Representation Graph for Semi-Supervised Hyperspectral Image Classification. *Pat. Recognit.* **2017**, *63*, 102–114.
20. Krishnapuram, B.; Carin, L.; Figueiredo, M.A.T.; Member, S. Sparse multinomial logistic regression: Fast algorithms and generalization bounds. *IEEE Trans. Pat. Anal. Mach. Intell.* **2005**, *27*, 957–968.
21. Ando, R.K.; Zhang, T. Two-view feature generation model for semi-supervised learning. In Proceedings of the International Conference on Machine Learning, Corvalis, OR, USA, 20–24 June 2007.
22. Meng, J.; Jung, C. Semi-Supervised Bi-Dictionary Learning for Image Classification With Smooth Representation-Based Label Propagation. *IEEE Transa. Multimedia.* **2016**, *18*, 458–473.
23. Wang, F.; Zhang, C. Label Propagation through Linear Neighborhoods. *IEEE Trans. Knowl. Data Eng.* **2008**, *20*, 55–67.
24. Wang, L.; Hao, S.; Wang, Q.; Wang, Y. Semi-supervised classification for hyperspectral imagery based on spatial-spectral Label Propagation. *Isprs J. Photogramm. Remote Sens.* **2014**, *97*, 123–137.
25. Zhu, X.; Ghahramani, Z.; Mit, T.J. Semi-Supervised Learning with Graphs. In Proceedings of the International Joint Conference on Natural Language Processing, Carnegie Mellon University, Pittsburgh, PA, USA, January 2005; pp. 2465–2472.
26. Cheng, H.; Liu, Z. Sparsity induced similarity measure for label propagation. In Proceedings of the IEEE International Conference on Computer Vision, Kyoto, Japan, 29 September–2 October 2009.
27. Karasuyama, M.; Mamitsuka, H. Multiple Graph Label Propagation by Sparse Integration. *IEEE Trans. Neural Netw. Learn. Syst.* **2013**, *24*, 1999–2012.
28. Zhu, X.; Ghahramani, Z. Semi-Supervised Learning Using Gaussian Fields and Harmonic Functions. In Proceedings of the Twentieth International Conference on International Conference on Machine Learning, Washington, DC, USA, 21–24 August 2003.
29. Moore, A.P.; Prince, S.J.D.; Warrell, J.; Mohammed, U. Superpixel lattices. In Proceedings of the Computer Vision and Pattern Recognition, Anchorage, AK, USA, 23–28 June 2008.
30. Wang, L.; Zhang, J.; Liu, P.; Choo, K.K.R.; Huang, F. Spectral-spatial multi-feature-based deep learning for hyperspectral remote sensing image classification. *Soft Comput. A Fusion Found. Methodol. Appl.* **2017**, *21*, 213–221.
31. Zhang, S.; Li, S.; Fu, W.; Fang, L. Multiscale Superpixel-Based Sparse Representation for Hyperspectral Image Classification. *Remote Sens.* **2017**, *9*, 139.
32. Achanta, R.; Shaji, A.; Smith, K.; Lucchi, A.; Fua, P.; Süsstrunk, S. *SLIC Superpixels*; EPFL: Lausanne, Switzerland, 2010.
33. De Carvalho, M.A.G.; da Costa, A.L.; Ferreira, A.C.B.; Junior, R.M.C. Image Segmentation Using Watershed and Normalized Cut. In Proceedings of the SIBIGRAPI, Gramado, Brazil, 30 August–3 September 2009.
34. Ugarriza, L.G.; Saber, E.; Vantaram, S.R.; Amuso, V.; Shaw, M.; Bhaskar, R. Automatic image segmentation by dynamic region growth and multiresolution merging. *IEEE Trans. Image Process. Publ. IEEE Signal Process. Soc.* **2009**, *18*, 2275–2288.
35. Jin, X.; Gu, Y. Superpixel-Based Intrinsic Image Decomposition of Hyperspectral Images. *IEEE Trans. Geosci. Remote Sens.* **2017**, *99*, 1–11.
36. Jia, S.; Deng, B.; Zhu, J.; Jia, X.; Li, Q. Superpixel-Based Multitask Learning Framework for Hyperspectral Image Classification. *IEEE Trans. Geosci. Remote Sens.* **2017**, *99*, 1–14.
37. Zhang, D.; Yang, Y.; Song, K. Research on a Multi-Scale Segmentation Algorithm Based on High Resolution Satellite Remote Sensing Image. *International Conference on Intelligent Control and Computer Application*. 2016.
38. Belkin, M.; Niyogi, P.; Sindhwani, V. Manifold regularization: A geometric framework for learning from labeled and unlabeled examples. *Mach. Learn. Res.* **2006**, *7*, 2399–2434.

39. Zhang, Q.; Shen, X.; Xu, L. Rolling Guidance Filter. In Proceedings of the European Conference on Computer Vision, Zurich, Switzerland, 6–12 September 2014.
40. Rohban, M.H.; Rabiee, H.R. Supervised neighborhood graph construction for semi-supervised classification. *Pat. Recognit.* **2012**, *4*, 1363–1372.
41. Vapnik, V.N. The Nature of Statistical Learning Theory. *IEEE Trans. Neural Netw.* **1997**, *8*, 1564.
42. Wang, L.; Zhang, Y.; Gu, Y. The research of simplification of structure of multiclass classifier of support vector machine. *Image Graph.* **2005**, *5*, 571–572.
43. Kang, X.; Li, S.; Fang, L.; Li, M.; Benediktsson, J.A. Extended Random Walker-Based Classification of Hyperspectral Images. *IEEE Trans. Geosci. Remote Sens.* **2014**, *53*, 144–153.
44. Kang, X.; Li, S.; Benediktsson, J.A. Spectral–Spatial Hyperspectral Image Classification With Edge-Preserving Filtering. *IEEE Trans. Geosci. Remote Sens.* **2014**, *52*, 2666–2677.

remote sensing

MDPI

Article

Hashing Based Hierarchical Feature Representation for Hyperspectral Imagery Classification

Bin Pan [1,2,3], Zhenwei Shi [1,2,3,*], Xia Xu [1,2,3] and Yi Yang [4]

1 Image Processing Center, School of Astronautics, Beihang University, Beijing 100191, China; panbin@buaa.edu.cn (B.P.); xuxia@buaa.edu.cn (X.X.)
2 Beijing Key Laboratory of Digital Media, Beihang University, Beijing 100191, China
3 State Key Laboratory of Virtual Reality Technology and Systems, School of Astronautics, Beihang University, Beijing 100191, China
4 Mathematics Department, Mellon College of Science, Carnegie Mellon University, Pittsburgh, PA 15213, USA; yiy2@andrew.cmu.edu
* Correspondence: shizhenwei@buaa.edu.cn

Received: 24 September 2017; Accepted: 24 October 2017; Published: 27 October 2017

Abstract: Integrating spectral and spatial information is proved effective in improving the accuracy of hyperspectral imagery classification. In recent studies, two kinds of approaches are widely investigated: (1) developing a multiple feature fusion (MFF) strategy; and (2) designing a powerful spectral-spatial feature extraction (FE) algorithm. In this paper, we combine the advantages of MFF and FE, and propose an ensemble based feature representation method for hyperspectral imagery classification, which aims at generating a hierarchical feature representation for the original hyperspectral data. The proposed method is composed of three cascaded layers: firstly, multiple features, including local, global and spectral, are extracted from the hyperspectral data. Next, a new hashing based feature representation method is proposed and conducted on the features obtained in the first layer. Finally, a simple but efficient extreme learning machine classifier is employed to get the classification results. To some extent, the proposed method is a combination of MFF and FE: instead of feature fusion or single feature extraction, we use an ensemble strategy to provide a hierarchical feature representation for the hyperspectral data. In the experiments, we select two popular and one challenging hyperspectral data sets for evaluation, and six recently proposed methods are compared. The proposed method achieves respectively 89.55%, 99.36% and 77.90% overall accuracies in the three data sets with 20 training samples per class. The results prove that the performance of the proposed method is superior to some MFF and FE based ones.

Keywords: hashing ensemble; hierarchical feature; hyperspectral classification

1. Introduction

Hyperspectral sensors can provide images with hundreds of continuous spectral bands, which has attracted a number of applications such as environmental monitoring and mineral prospecting [1–4]. Among many surveys about hyperspectral imagery (HSI) analysis, land cover accurate classification is an important research topic. Supervised spectral classifiers are popular in the early research, including multinomial logistic regression [5], support vector machines (SVMs) [6–8] and sparse representation classifier [9].

During the last decade, a lot of endeavors have been devoted to extracting more representative features from original HSI data. It is widely recognized that joint spectral and spatial information can significantly improve the performance of HSI classification methods. Markov random field (MRF) is a powerful tool that is able to model the spatial relationship around pixels. In [10,11], MRF was combined with subspace multinomial logistic regression and Gaussian mixture model, respectively. In [12], MRF was used as a postprocessing to refine the classification maps obtained by SVM. Morphological profile (MP) is another powerful tool to utilize the spatial contextual information. In [13], Benediktsson et al. improved the original MP, and proposed the extended morphological profile (EMP) method for HSI classification. Motivated by the promising performance of EMP, two improved methods, extended attribute profile and extended multi-attribute profile, were proposed in [14].

Because a single kind of feature may not describe the integrated characteristics of HSI data, multiple feature fusion (MFF) were proposed and used to improve the performance of HSI classification models. MFF based methods can be roughly divided into four classes [15]: multiple kernel learning, band selection, subspace feature extraction and ensemble based methods. Li et al. constructed a series of generalized composite kernels where no weight parameters were required [16]. Gu et al. employed multiple kernel learning to combine different spectral-spatial features [17–19]. Band selection methods try to find the most discriminative hyperspectral channels while preserving their physical meanings. In [20], discriminative sparse multimodal learning based method was proposed for multiple feature selection. In [21], spectral and spatial information are utilized simultaneously to select the representative bands. Different from band selection, subspace methods refer to transforming the original multiple features to a new low-dimension sub-feature space. Zhang et al. introduced a patch alignment and a modified stochastic neighbor embedding based methods for feature fusion [22,23]. In [24], a low-rank representation based feature extraction method was proposed for HSI classification, where locally spatial similarity and spectral space structure were combined. In [15], Zhong et al. conducted dimension reduction on multiple features by hashing methods. Ensemble learning is another typical feature fusion method. Ensemble learning methods aim at achieving better generalization capacity by integrating different features or individual learners [25]. SVM [26,27] and random forest [28–30] based HSI classification methods were proposed in recent studies. In [31], Chen et al. improved the classification accuracy by stacked generalization of magnitude and shape feature spaces. In [32], Pan et al. combined spatial relationships in different scales via a weighted voting strategy. In addition, feature fusion methods using different data sources have also been investigated [33,34].

Recently, deep learning based methods have attracted great interest in HSI classification, e.g., [35–40]. The basic idea of these methods is to extract the "deep" feature from the original HSI data, thus hierarchical network models are designed. This idea is promising and encouraging. In some natural scene image classification tasks, deep learning methods have achieved even better results than human level [41]. In [35], the deep learning method was firstly used in HSI classification, where a stacked autoencoder was adopted. Subsequently, deep belief networks [42], convolutional neural networks [39,43,44] and recurrent neural networks [38] were investigated. In order to improve the computational efficiency, some simplified deep learning models were developed [36,37]. Most of these methods have also considered the spatial relationship via 3D networks or neighborhood information. However, the performance of deep learning methods is heavily dependant on abundant training samples that are difficult to acquire from HSI data. Compared with traditional methods, deep learning methods usually require more labeled samples. For example, in [35,36], about half of all the labeled pixels were used for training. Although deep features could really improve the classification accuracy, more research is required on finding a new way out of deep learning to extract hierarchical features.

Inspired by the ideas of MFF and deep learning, in this paper, we propose a novel hashing based hierarchical feature (H^2F) extraction method for HSI classification. The motivations of H^2F come from two aspects: (1) low-level features such as spectral variations, local texture and global texture information, should be combined to produce a comprehensive feature set. This feature set could serve as inputs of the next layer; (2) based on the obtained feature set, a further feature extraction process should be followed, so as to generate a hierarchical feature. This hierarchical feature should present better performance than every single feature. Different from traditional MFF based methods, H^2F is not a simple combination or voting of multiple features. Instead, H^2F attempts to construct more representative feature descriptor from the already extracted feature set.

Based on the two motivations, we propose a cascaded feature extraction framework with two major processes: the generation of spectral-spatial feature set and hashing based hierarchical feature extraction. In the first process, we construct a feature set which is composed of spectral variations, local and global textures. In this paper, we use rolling guidance filtering (RGF) [45], local binary pattern (LBP) [46] and global Gabor filtering [47] to form the multiple features. Although many recent works have demonstrated that there is information redundant in some popular HSI data sets [48–50], it may be not appropriate to conclude that information redundant exists in all the HSI data. Therefore, different from traditional feature fusion based methods, in this paper, we do not conduct dimension reduction so as to better preserve the distinctive classification information. All these features are collected to a feature set. In the second process, we design a hashing histograms based feature extraction strategy to give a more representative description for the HSI data. To avoid complex computation, the feature set is separated into several groups. The hashing histogram features in all the groups are concatenated as the final feature expression. It is worth noting that H^2F is actually an ensemble based method, rather than deep learning based.

At last, an extreme learning machine (ELM) classifier [51] is used to determine the label of each pixel. The most important reason of using ELM is to improve the computing speed. Usually, feature fusion methods will generate relatively high-dimensional features, and this is more apparent in H^2F since dimension reduction is not adopted. ELM has a simple structure, and it can be trained very fast because of its random weights generation in inputs and least squares solution in outputs. Furthermore, some research has proven that ELM is effective for HSI classification [46,52,53]. We compare the effectiveness and efficiency of ELM and several other classifiers in the experiments' part.

The major contribution of this paper is that a hashing based hierarchical feature ensemble method is developed. The ensemble strategy proposed in H^2F could provide a new way to utilize multiple features.

The reminder of this paper is organized as follows. In Section 2, we give a detailed description about the proposed method. In Section 3, experiments and discussion on two popular and one challenging data sets are provided. We conclude this paper in Section 4.

2. H^2F Based Classification

The proposed H^2F based HSI classification method can be divided into three steps: (1) multiple features extraction; (2) hashing based hierarchical feature representation and (3) ELM based classification. The flowchart of the proposed method is shown in Figure 1.

2.1. *Multiple Features Extraction*

Research has demonstrated that spectral-spatial joint information could significantly contribute to the performance of HSI classification methods. However, it is hard to judge which feature extraction approach performs best. Actually, each single feature has its unique emphasis. In this paper, we select three disparate features that reflect different characteristics of HSI data to construct a feature set, namely, RGF (for spectra), LBP (for local texture) and Gabor (for global texture). It is worth noting that each feature will generate one or several sub-feature sets. Take Gabor feature for example. Suppose that four wavelengths and four orientations are used. Then, for each pixel, there will be 16 sub-features. If we set eight as the number of features in a sub-feature set, two groups of sub-feature sets could be obtained. The following hierarchical feature representation operation is conducted on these subsets. Using the whole feature set directly for hierarchical feature representation is not appropriate because different types of features are heterogeneous.

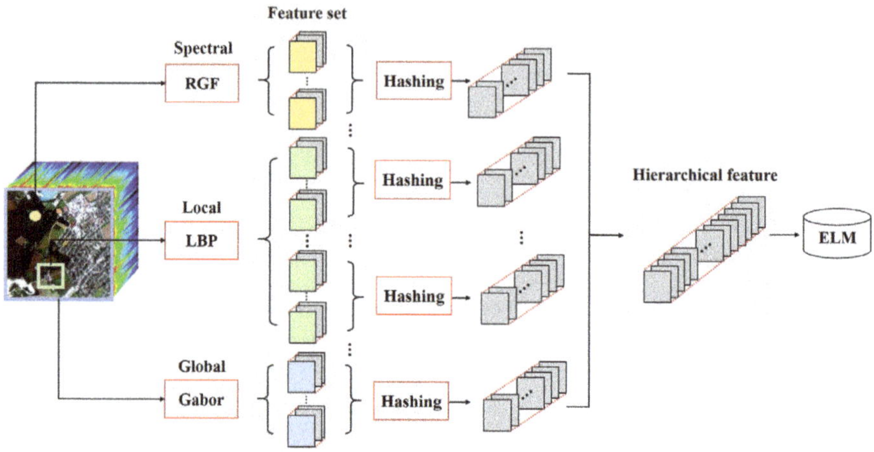

Figure 1. The flowchart of the H^2F based method.

2.1.1. RGF

Although the raw pixel spectral vectors could directly be used for training and classification, they do not perform well. Moreover, since we need sub-feature sets from spectral features, we must extend the pixels spectra to a group of features. Motivated by the effectiveness of RGF and its improvement in HSI classification [37], in this paper, we use RGF to obtain the sub-feature set using spectral information.

Let $\mathbf{Q}^p$ denote filtering result for the pth band of an hyperspectral image, we conduct guided filtering [54] by

$$\mathbf{Q}_i^p = a_k^p \mathbf{G}_i + b_k^p, \forall i \in \omega_k, \tag{1}$$

where $\mathbf{G}$ is a guidance image, i is one of a pixel in $\mathbf{G}$, ω_k is a window around pixel i, k is one of a pixel in ω_k, and a_k^p and b_k^p are coefficients to be estimated. Usually, $\mathbf{G}$ is the first principal component of HSI data. Please note that $\mathbf{G}$ only works as the guidance image, and it will not reduce the dimensionality of the filtered results. Then, minimize the following energy function:

$$E(a_k^p, b_k^p) = \sum_{i \in \omega_k} ((a_k^p \mathbf{G}_i + b_k^p - \mathbf{I}_i^p)^2 + \epsilon a_k^{p2}), \tag{2}$$

where $\mathbf{I}$ is the input HSI data, and ϵ is a hyper-parameter. Equation (2) can be solved directly by linear ridge regression [55]:

$$\begin{aligned} a_k^p &= \frac{\frac{1}{|w|} \sum_{i \in \omega_k} \mathbf{I}_i^p \mathbf{G}_i - \mu_k \bar{\mathbf{I}}_k^p}{\sigma_k{}^2 + \epsilon}, \\ b_k^p &= \mathbf{I}_k^p - a_k^p \mu_k^p, \end{aligned} \tag{3}$$

where μ_k and σ_k denote the mean value and standard variance of $\mathbf{G}$ in ω_k, $\bar{\mathbf{I}}_k^p$ is the mean value of $\mathbf{I}$ in ω_k, and $|\omega|$ is the number of pixels in ω_k.

Equation (1) is the optimization problem in guidance filtering, and a and b are the values need to be optimized. Equation (2) is the optimization object function, and Equation (3) is the solution. Rolling operation refers to replace $\mathbf{I}$ by $\mathbf{Q}$ and conduct Equations (1) and (2) repeatedly. In each rolling, we can obtain a new HSI data. Therefore, using RGF we are able to generate a series of features based on the original spectral vectors. Because RGF mainly reflects the spectral characteristics of HSI data, these features can be considered as spectral sub-feature sets.

2.1.2. LBP

In HSI data, the spatial contextual information could be described by the local texture around each pixel. LBP feature is a popular texture operator that has been investigated in [46]. The LBP map for $\mathbf{I}^p$ can be obtained by

$$\mathbf{L}^p(i) = \sum_{k=1}^{|\omega_k|-1} U(\mathbf{I}_k^p - \mathbf{I}_i^p)2^i, \quad (4)$$

where $|\omega_k|$ is the number of pixels in the window ω_k, and $U(\cdot)$ is a Heaviside step function with 1 for positive entries and 0, otherwise. In the LBP map, we can get a vector for each ω_k by counting its histogram. This vector is the new feature representation for pixel i. In this paper, uniform LBP is used. If using 8-neighbor for uniform LBP, 59 bins will be obtained totally, i.e., there are 59 sub-feature sets available based on LBP.

2.1.3. Gabor Filters

Besides local texture features, recent literature has reported that global spatial features of HSI data will also contribute to the classification accuracy, e.g., Gabor filter [47,56,57]. Suppose (x, y) is a pixel coordinate at $\mathbf{I}^p$, then the output of an Gabor filter can be expressed by

$$\mathbf{GB}(x, y) = \exp(-\frac{x'^2 + \gamma^2 y'^2}{2\sigma^2})\exp(\mathrm{j}(2\pi\frac{x'}{\delta} + \psi)), \quad (5)$$

where

$$x' = x\cos\theta + y'\sin\theta, y' = -x\sin\theta + y'\cos\theta. \quad (6)$$

γ, ψ and σ are hyper-parameters in Gabor filter, δ is the wavelength of the sinusoidal function, and θ represents the orientation of the filter. Selecting different δ and θ, the original HSI data can be transformed into many sub-feature sets.

Based on the RGF, LBP and Gabor filters, we can construct a new feature set containing many subsets. Note that the dimensionality of features in the obtained set is the same with that of the original HSI data. Traditional feature fusion based methods usually directly stack these features, or use a weighted voting strategy. In this paper, we try to extract hierarchical features from HSI data, and this feature set is used as the input of the next hierarchy.

2.2. Hashing Based Hierarchical Feature Representation

The major motivation of the proposed hashing based method is that extracting very sparse features by increasing the feature dimensionality. For the obtained feature set, we first divide it into several subsets with the same number of features. Suppose N is the number of features in a single subset, and L is the dimensionality of features. Then, the illustration of hierarchical feature representation for this subset can be exhibited by Figure 2. Generally, the hashing based hierarchical feature representation method for a subset mainly includes three steps.

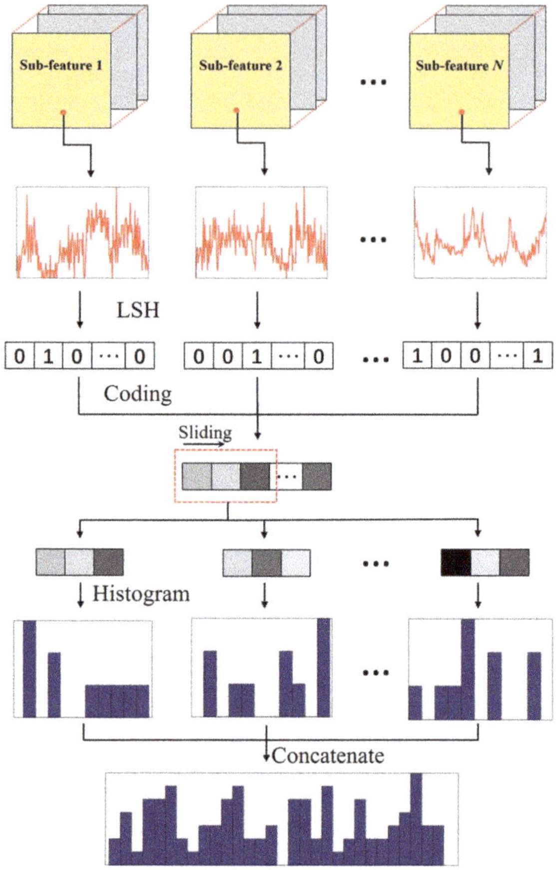

Figure 2. An illustration for the hashing based hierarchical feature representation. This figure only presents the process in one pixel and a single sub-feature set.

2.2.1. Step 1

For pixel i in position (x, y), we can obtain N features. Let $\mathbf{s}_i^n \in \mathbb{R}^{(L\times 1)}$ denote the nth sub-feature, we conduct locality-sensitive hashing (LSH) on $\mathbf{s}_i^n$, i.e.,

$$\mathbf{h}_i^n = \text{sign}(\mathbf{D}\mathbf{s}_i^n), \tag{7}$$

where $\mathbf{D} \in \mathbb{R}^{L\times L}$ is a random matrix with zero-mean normal distribution, and $\mathbf{h}_i^n$ is a binary vector. Integrating all the N vectors, we get $\mathbf{S}_i = [\mathbf{h}_i^1, \mathbf{h}_i^2, \cdots, \mathbf{h}_i^n] \in \mathbb{R}^{(L\times N)}$.

2.2.2. Step 2

Coding $\mathbf{S}_i$ by

$$\hat{\mathbf{s}}_i(j) = \sum_{\ell=1}^{N} 2^{\ell-1}\mathbf{S}_i(j, \ell), \tag{8}$$

and $\hat{\mathbf{s}}_i \in \mathbb{R}^{(L\times 1)}$ is the coding output for pixel i. Based on Equation (8), the binary results are converted to decimal vectors. Equation (8) can also indicate that the grouping strategy in H^2F is necessary because the range of the coding results is $[0, 2^N - 1]$.

2.2.3. Step 3

Use a sliding window with size $w \times 1$ to scan $\hat{\mathbf{s}}_i$, and then collect all the patches. Calculate the histogram features in all the patches (with 2^N bins), and concatenate them into a single vector $\mathbf{f}_i^m$, where $m = [1, 2, \cdots, M]$ denotes the mth sub-feature set, M is the number of sub-feature set, and $\mathbf{f}_i^m \in \mathbb{R}^{2^N(L-m+1)\times 1}$ if the step size of sliding window is set as 1. At last, the final hierarchical feature for pixel i is determined by

$$\mathbf{f}_i = [\mathbf{f}_i^1, \mathbf{f}_i^2, \cdots, \mathbf{f}_i^M] \in \mathbb{R}^{2^N M(L-m+1)\times 1}. \tag{9}$$

Obviously, increasing the step size of sliding window could reduce the dimensionality of the obtained features. Usually, 50% overlapping between patches is appropriate. Size of sliding window also has some influence on the results. Theoretically, smaller w could enhance the sparsity of the obtained features, but lead to very high dimensionality. In order to balance the sparsity and computational cost of computer memory, we set the window size as 7×1.

H^2F could be considered as a hierarchical representation for the original HSI data. According to our empirical experience, we do not recommend dimension reduction on H^2F because it may lead to loss of distinctive information. Instead, to reduce the computational cost and avoid overfitting, we use a very simple classifier, ELM, to determine the final classification results.

2.3. ELM Based Classification

ELM [51] is a simple neural network with only three layers (input, hidden and output), which performs well in small-scale data sets. ELM has two leadings characteristics: (1) the input and hidden layers are connected randomly; and (2) the weights between hidden and output layers are learned by a least squares algorithm.

Let $\mathbf{F} = [\mathbf{f}_1, \mathbf{f}_2, \cdots, \mathbf{f}_{nt}] \in \mathbb{R}^{d\times nt}$ denote the training samples matrix, d is the dimension and nt is the number of training samples. In ELM, the weights between input and hidden layers are obtained randomly, denoted by $\mathbf{W} \in \mathbb{R}^{(nh\times d)}$, where nh is the number of nodes in the hidden layer. Then, the objective function of ELM can be described by

$$\mathbf{B}\mathrm{g}(\mathbf{W}_t \cdot \mathbf{f}_i + \mathbf{b}_t) = \mathbf{Y}_i, \tag{10}$$

where $\mathbf{B} \in \mathbb{R}^{(C\times nh)}$ is the weights matrix connecting hidden and output layer, $\mathbf{b} \in \mathbb{R}^{nh\times 1}$ is the bias vector in the hidden layer, $\mathrm{g}(\cdot)$ is an activation function such as sigmoid function, C is the number of classes, and $\mathbf{Y} \in \mathbb{R}^{C\times nt}$ is the label matrix for all the training samples. Note that $\mathrm{g}(\mathbf{W}_t \cdot \mathbf{f}_i + \mathbf{b}_t)$ is the output of the hidden layer for sample $\mathbf{f}_i$. Because $\mathbf{W}$ and $\mathbf{b}$ are randomly assigned, the outputs of the hidden layer have been determined. Then, Equation (10) is actually equal to the following expression:

$$\mathbf{H} \cdot \mathbf{B} = \mathbf{Y}, \tag{11}$$

where $\mathbf{H}$ is the outputs of the hidden layer. Obviously, Equation (11) can be solved by a simple least squares method, i.e., $\mathbf{B} = \mathbf{H} + \mathbf{Y}$.

In H^2F, the final features are classified by ELM. Because the random matrix generation needs little time, the major computational cost lies in Equation (11). As long as we restrict the number of hidden nodes, the training operation could be very fast. In Algorithm 1, we provide a pseudocode for the H^2F based HSI classification method.

Algorithm 1 The H^2F based HSI classification method

Input:HSI data, ground truth
Initialize: training set, testing set
Multiple Features Extraction
1. RGF features based on Equations (1) and (2)
2. LBP features based on Equations (4)
3. Gabor features based on Equations (5) and (6)
4. Feature set generation

Hashing based Hierarchical Features
5. Separate the feature set into uniform subsets
6. For 1: Number of subsets
   Hierarchical feature extraction by Equations (7) and (8)
   End for
7. Final features generation by Equation (9)

ELM based Classification
8. Train ELM by Equations (10) and (11)
9. Classification by ELM

Output: Classification results

3. Experiments and Discussion

3.1. Experimental Setups

In this section, experimental analysis about the H^2F based classification method are provided. H^2F is compared with six recently proposed methods, i.e., Gabor + ELM (GE) [47], LBP + Gabor + ELM (LGE) [46], RGF + Network (RVCANet) [37], RGF + Ensemble (HiFi) [32], and another two methods, edge-preserving filtering (EPF) [58] and intrinsic image (IIDF) [59] based methods. Among these methods, GE and LGE directly concatenate multiple features without further operations. Thus, they could be regarded as the baselines. RVCANet also tries to extract deep features from HSI data, but it adopts a deep network manner. HiFi is a multiple feature fusion method where the results are obtained by weighted voting. These methods have similar motivation as the proposed method, so we use them for comparison. All the methods are compared on two popular (Indian Pines, Kennedy Space Center (KSC) (Available online: http://www.ehu.eus/ccwintco/index.php?title=Hyperspectral_Remote_Sensing_Scenes)) and one challenging (GRSS_DFC_2014 [60,61]) data set. We run the above methods 50 times with randomly selected train and test samples, and the average accuracies and the corresponding standard deviations are reported. Overall accuracy (OA), average accuracy (AA) and kappa coefficient (κ) are selected for evaluation [62]. For the three data sets, 20 pixels per class are used for training, and the rests for testing. Some classes (especially in Indian Pines data set) have a total of nearly 20 samples. In this case, we directly use half of them for training and the others for testing. In H^2F, we construct nine sub-feature sets (six for LBP, two for Gabor and one for RGF) with nine features per set, totally 81 features. Rolling times of RGF is set as 1–9 with $\epsilon = 1$, wavelength δ in Gabor is 16, orientation number is 18, and window size in LBP is 3×3. Under this setting, Indian Pines/KSC/GRSS_DFC_2014 could be represented by 225792/198144/92160 dimensional features, but only 1% around are non-zero. The hyper-parameters in ELM (linear kernel) is determined by five-fold cross validation. The regularization coefficient is chosen from a set {1, 10, 100, 1000}, and the hidden neuron number is chosen from $\{100, 200, \cdots, 2000\}$. According to the results of cross validation, 1000 and 100 are appropriate for the above two parameters.

3.2. Data Sets

- Indian Pines: This data is widely used in HSI classification, which was gathered by airborne visible/infrared imaging spectrometer (AVIRIS) in Northwestern Indiana. It covers the wavelengths ranges from 0.4 to 2.5 μm with 20 m spatial resolution. In total, 145×145 pixels are included and 10,249 of them are labeled. The labeled pixels are classified into 16 classes. There are 200 bands available after removing the water absorption channels. A false color composite image (R-G-B=band 36-17-11) and the corresponding ground truth are shown in Figure 3a,b.

- KSC: It is acquired by AVIRIS over the Kennedy Space Center, Florida, on March, 1996. It has 18 m spatial resolution with 512 × 614 pixels size and 10 nm spectral resolution with center wavelengths from 400 to 2500 nm. In addition, 176 bands could be used for analysis after removing water absorption and low SNR bands. There are 5211 labeled pixels available that are divided into 16 classes. A false color composite image (R-G-B=band 28-9-10) and the corresponding ground truth are shown in Figure 3c,d.
- GRSS_DFC_2014: This is a challenging HSI data set covering an urban area near Thetford Mines in Québec, Canada, and it is used in the 2014 IEEE GRSS Data Fusion Contest. It was acquired by an airborne long-wave infrared hyperspectral imager with 84 channels ranging between 7.8 to 11.5 μm wavelengths. The size of this data set is 795 × 564 pixels, and the spatial resolution is about 1 m. In total, 22,532 labeled pixels and a ground truth with seven land cover classes are provided. Some research has indicated that this data set is more challenging for HSI classification [61]. A false color composite image (R-G-B=band 30-45-66) and the corresponding ground truth are shown in Figure 3e,f.

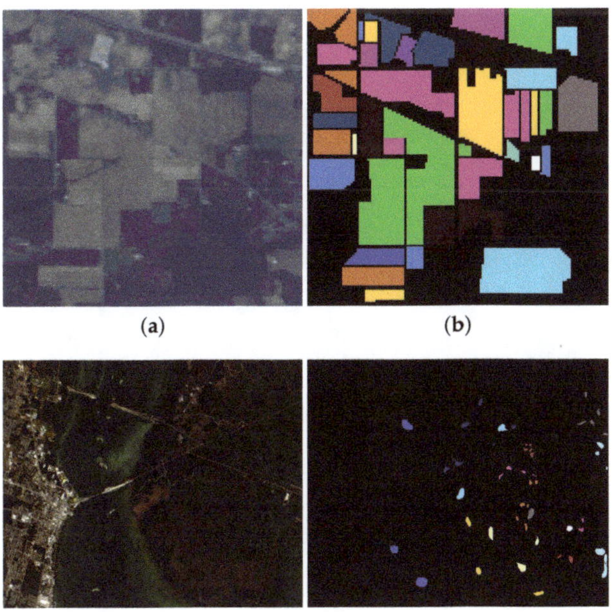

(a) (b)

(c) (d)

Figure 3. *Cont.*

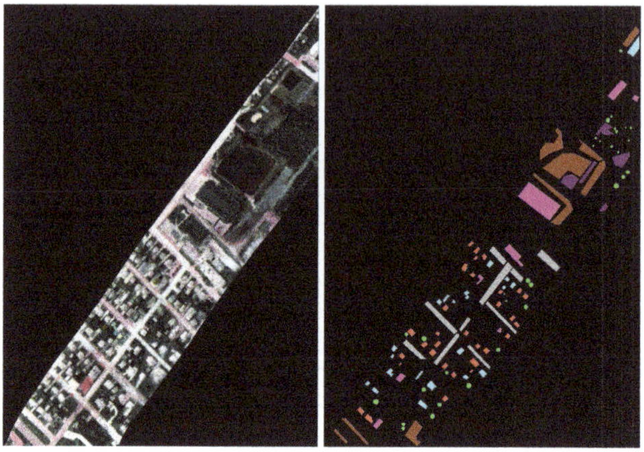

(e) (f)

Figure 3. False color composite images of (**a**) Indian Pines; (**c**) KSC and (**e**) GRSS_DFC_2014 data sets and the ground truths (**b**,**d**,**f**). Each color corresponds to a certain class.

3.3. *Classification Results*

Classification results by all the compared methods are shown in Figures 4–6 and Tables 1–3. Since the H^2F is a fusion approach of some spectral and spatial features, we especially chose the methods that use single or combine two features for comparison [32,37,46,47], so as to validate the effectiveness of the proposed fusion strategy.

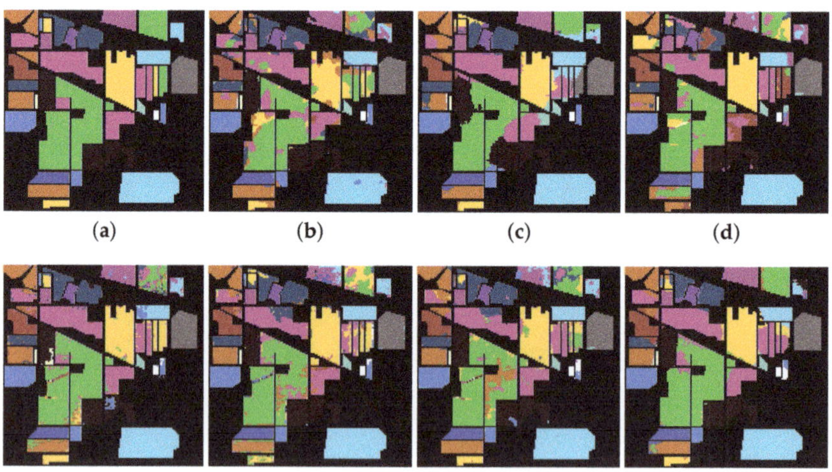

Figure 4. Classification maps by compared methods for Indian Pines data set. (**a**) The ground truth (**b**) GE (**c**) LGE (**d**) EPF (**e**) IIDF (**f**) RVCANet (**g**) HiFi (**h**) H^2F.

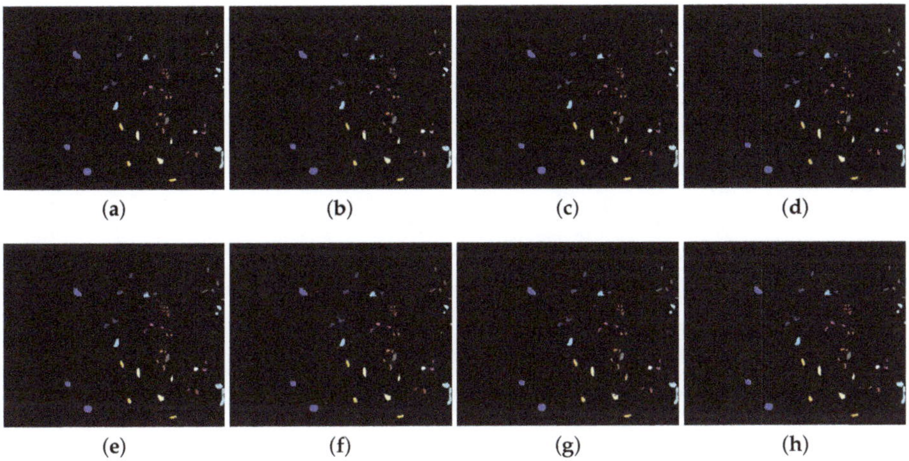

(a) (b) (c) (d)

(e) (f) (g) (h)

Figure 5. Classification maps by compared methods for KSC data set. (**a**) the ground truth; (**b**) GE; (**c**) LGE; (**d**) EPF; (**e**) IIDF; (**f**) RVCANet; (**g**) HiFi; (**h**) H^2F.

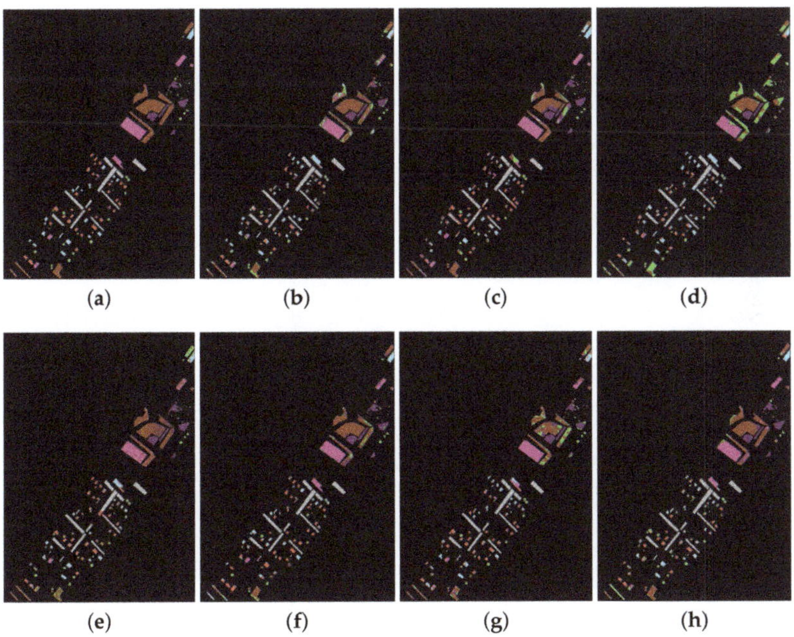

(a) (b) (c) (d)

(e) (f) (g) (h)

Figure 6. Classification maps by compared methods for GRSS_DFC_2014 data set. (**a**) the ground truth; (**b**) GE; (**c**) LGE; (**d**) EPF; (**e**) IIDF; (**f**) RVCANet; (**g**) HiFi; (**h**) H^2F.

Table 1. Classification accuracies of different methods on Indian Pines data set (%).

Class	Samples	Methods						
	Train/Test	GE	LGE	EPF	IIDF	RCANet	HiFi	H^2F
C1	20/26	99.42 ± 0.55	99.92 ± 0.54	98.84 ± 1.78	87.22 ± 14.9	99.00 ± 1.70	99.46 ± 1.35	**100.0 ± 0.00**
C2	20/1408	70.45 ± 7.42	80.89 ± 5.30	56.53 ± 11.1	80.45 ± 6.04	63.94 ± 6.85	81.91 ± 5.58	**81.88 ± 5.29**
C3	20/810	74.25 ± 7.71	85.61 ± 7.03	67.27 ± 10.5	75.89 ± 6.86	79.91 ± 7.05	**91.49 ± 4.52**	87.00 ± 5.86
C4	20/217	95.10 ± 4.59	**99.40 ± 1.14**	96.56 ± 4.60	66.03 ± 10.8	98.59 ± 2.22	96.78 ± 3.84	99.21 ± 1.30
C5	20/463	87.51 ± 5.18	92.13 ± 5.39	91.09 ± 4.56	93.49 ± 4.30	**93.60 ± 3.02**	90.06 ± 3.88	90.53 ± 4.21
C6	20/710	92.35 ± 4.19	94.99 ± 3.72	96.97 ± 3.93	97.67 ± 2.11	**98.36 ± 1.10**	97.92 ± 1.80	97.21 ± 2.11
C7	14/14	100.0 ± 0.00	100.0 ± 0.00	96.85 ± 3.58	54.08 ± 20.6	100.0 ± 0.00	96.75 ± 5.54	**100.0 ± 0.00**
C8	20/458	98.56 ± 2.30	99.83 ± 0.52	96.65 ± 5.47	99.91 ± 0.14	98.76 ± 0.63	99.39 ± 0.92	**99.98 ± 0.10**
C9	10/10	99.59 ± 0.35	100.0 ± 0.00	99.80 ± 1.41	44.83 ± 19.7	100.0 ± 0.00	100.0 ± 0.00	**100.0 ± 0.00**
C10	20/952	73.53 ± 8.16	86.55 ± 5.72	83.09 ± 7.85	73.57 ± 8.49	87.43 ± 3.79	88.16 ± 6.63	**88.97 ± 4.47**
C11	20/2435	69.93 ± 8.38	79.21 ± 5.37	69.55 ± 9.23	**92.37 ± 3.52**	72.01 ± 6.49	79.82 ± 5.86	83.97 ± 5.30
C12	20/573	81.23 ± 7.01	85.11 ± 5.95	73.26 ± 10.1	79.13 ± 6.94	90.49 ± 4.08	**93.31 ± 3.18**	87.61 ± 5.72
C13	20/185	98.76 ± 1.28	99.58 ± 1.16	99.39 ± 0.32	99.54 ± 1.53	99.49 ± 0.31	99.41 ± 0.29	**99.84 ± 0.31**
C14	20/1245	87.16 ± 5.11	96.47 ± 3.63	88.51 ± 7.76	**99.06 ± 1.04**	94.24 ± 3.49	96.96 ± 2.79	96.70 ± 3.25
C15	20/366	90.80 ± 6.09	98.21 ± 2.83	81.44 ± 10.6	84.73 ± 11.1	90.65 ± 4.05	95.23 ± 2.72	**98.46 ± 4.23**
C16	20/73	98.65 ± 2.02	98.30 ± 2.45	96.93 ± 5.68	94.62 ± 6.86	99.06 ± 1.90	99.07 ± 0.65	**99.75 ± 0.53**
OA		79.38 ± 1.82	87.61 ± 1.48	77.54 ± 3.10	85.89 ± 1.88	83.06 ± 2.32	89.06 ± 1.70	**89.55 ± 1.31**
AA		88.58 ± 1.03	93.51 ± 0.69	87.04 ± 1.88	82.66 ± 2.22	91.56 ± 0.89	94.11 ± 0.77	**94.44 ± 0.73**
κ		76.60 ± 2.03	85.99 ± 1.64	74.66 ± 3.41	84.02 ± 2.11	80.87 ± 2.56	87.51 ± 1.90	**88.15 ± 1.49**

Table 2. Classification accuracies of different methods on the KSC data set (%).

Class	Samples	Methods						
	Train/Test	GE	LGE	EPF	IIDF	RCANet	HiFi	H^2F
C1	20/741	93.54 ± 2.63	98.84 ± 1.87	99.43 ± 1.13	99.86 ± 0.22	97.80 ± 1.65	98.60 ± 1.06	**99.97 ± 0.11**
C2	20/223	68.04 ± 6.48	95.69 ± 5.98	89.65 ± 6.84	94.80 ± 5.55	95.74 ± 4.30	92.26 ± 5.02	**97.04 ± 5.21**
C3	20/236	84.24 ± 7.53	99.43 ± 1.70	97.39 ± 1.93	99.42 ± 0.91	98.33 ± 1.39	96.42 ± 3.31	**99.88 ± 0.56**
C4	20/232	75.14 ± 6.79	**98.49 ± 3.04**	93.84 ± 6.78	96.42 ± 3.19	94.17 ± 4.21	93.20 ± 3.50	97.10 ± 4.67
C5	20/141	99.06 ± 1.78	**99.91 ± 0.43**	86.45 ± 8.63	97.68 ± 3.11	95.57 ± 5.24	89.78 ± 6.69	99.58 ± 2.90
C6	20/209	93.25 ± 6.46	100.0 ± 0.00	97.96 ± 3.08	93.77 ± 4.61	94.71 ± 3.30	93.62 ± 7.67	**100.0 ± 0.00**
C7	20/85	98.49 ± 2.77	100.0 ± 0.00	99.97 ± 0.16	99.93 ± 0.49	100.0 ± 0.00	95.38 ± 7.14	**100.0 ± 0.00**
C8	20/411	78.00 ± 7.02	96.25 ± 5.49	**98.54 ± 4.29**	97.58 ± 4.47	98.27 ± 2.35	95.74 ± 4.47	96.43 ± 4.88
C9	20/500	94.05 ± 5.03	99.28 ± 3.42	99.21 ± 2.49	99.78 ± 0.15	98.33 ± 4.33	97.58 ± 1.51	**99.79 ± 0.73**
C10	20/384	91.85 ± 6.02	**100.0 ± 0.00**	98.81 ± 1.01	93.83 ± 7.04	98.66 ± 1.47	99.14 ± 1.07	99.79 ± 1.29
C11	20/399	89.73 ± 5.27	100.0 ± 0.00	99.30 ± 1.59	98.60 ± 1.37	99.51 ± 0.83	97.97 ± 3.18	**100.0 ± 0.00**
C12	20/483	91.61 ± 4.36	97.57 ± 5.27	96.28 ± 2.91	94.18 ± 4.34	97.97 ± 3.67	98.40 ± 1.46	**99.80 ± 0.76**
C13	20/907	95.09 ± 3.07	100.0 ± 0.00	99.92 ± 0.15	99.95 ± 0.30	100.0 ± 0.00	99.71 ± 0.40	**100.0 ± 0.00**
OA		89.74 ± 1.28	98.91 ± 0.64	97.84 ± 0.90	97.63 ± 0.56	98.12 ± 0.77	97.09 ± 0.84	**99.36 ± 0.54**
AA		88.62 ± 1.21	98.88 ± 0.65	96.67 ± 1.33	97.37 ± 0.69	97.62 ± 0.88	95.99 ± 1.19	**99.18 ± 0.71**
κ		88.56 ± 1.42	98.78 ± 0.72	97.58 ± 1.01	97.36 ± 0.62	97.91 ± 0.85	96.75 ± 0.93	**99.28 ± 0.61**

Table 3. Classification accuracies of different methods on the GRSS_DFC_2014 data set (%).

Class	Samples	Methods						
	Train/Test	GE	LGE	EPF	IIDF	RCANet	HiFi	H^2F
C1	20/4423	91.56 ± 3.65	96.83 ± 3.24	95.86 ± 4.44	96.39 ± 1.91	93.42 ± 3.03	96.96 ± 2.14	**98.47 ± 0.96**
C2	20/1073	**68.37 ± 6.93**	41.29 ± 7.82	53.71 ± 17.8	37.33 ± 8.82	66.34 ± 5.88	64.94 ± 5.78	65.74 ± 12.1
C3	20/1834	62.72 ± 9.78	53.89 ± 8.80	49.13 ± 17.7	53.38 ± 8.32	61.57 ± 7.37	**68.13 ± 7.49**	63.59 ± 7.07
C4	20/2106	**67.21 ± 6.62**	61.79 ± 5.72	58.63 ± 18.3	60.91 ± 6.65	68.45 ± 7.39	62.62 ± 5.83	62.23 ± 10.1
C5	20/3868	59.75 ± 6.53	73.31 ± 7.77	55.67 ± 16.7	70.54 ± 6.91	69.45 ± 6.38	76.05 ± 4.34	**80.84 ± 3.79**
C6	20/7337	66.00 ± 8.46	92.37 ± 2.42	50.78 ± 13.7	**93.37 ± 2.43**	68.64 ± 7.35	67.40 ± 6.17	70.69 ± 8.22
C7	20/1751	77.08 ± 6.94	81.13 ± 9.59	58.76 ± 12.1	83.01 ± 9.17	90.58 ± 5.31	84.49 ± 5.96	**90.86 ± 4.47**
OA		70.79 ± 2.47	76.91 ± 2.79	61.90 ± 6.56	75.37 ± 3.00	74.68 ± 2.69	75.57 ± 4.34	**77.90 ± 2.51**
AA		70.38 ± 1.42	71.52 ± 2.02	60.36 ± 4.97	70.70 ± 2.55	74.06 ± 2.03	74.37 ± 6.17	**76.06 ± 1.88**
κ		64.69 ± 2.63	71.81 ± 3.17	54.76 ± 7.12	70.07 ± 3.41	69.15 ± 3.09	70.39 ± 5.95	**73.09 ± 2.77**

3.3.1. Results on Indian Pines Data Set

Experiments on this data set appear in nearly all the HSI classification works. Maybe, it is because this data set is a little more difficult for classification than some other popular ones such as Salinas or KSC, especially when training samples number is limited. It can be seen from Table 1 that the seven compared methods present various performance with only 20 training samples per class. H^2F slightly outperforms HiFi, and achieves 5–10% advantages over other methods. It is worth noting that, in some classes with a large number of testing samples (such as classes 2, 3 and 11), all of the methods present plunges. This is because such few training samples cannot fully represent the data distribution in these classes. On the other hand, we can see from Figure 4 that the spatial consistency is roughly preserved by every one of the methods. Since all of these methods have utilized joint spatial-spectral features, Figure 4 demonstrates that spatial information is really beneficial to HSI classification.

3.3.2. Results on KSC Data Set

It is observed in Figure 5 and Table 2 that results in this data set are much better. Although only 20 samples per class are used for training, H^2F presents above 99% accuracies, which achieves about 0.4% advantage. Additionally, we find that H^2F reports more than 96% accuracy in each class. Among all of the 13 classes, H^2F performs better in nine of them. However, we must recognize that, since most methods have achieved better than 97% OA in this data set, it is not safe to conclude which one is the best. Therefore, experiments on more challenging data sets are of vital importance.

3.3.3. Results on GRSS_DFC_2014 Data Set

Apparently, this data set is more difficult for classification. Although there are still 20 samples per class used for training, accuracies by all the methods present an obvious decline, as shown in Figure 6 and Table 3. The reason may be that the imaging quality in long-wave infrared channels is relatively lower. However, H^2F still outperforms other methods by about 2%. Comparison with LGE is especially more meaningful because H^2F could be regarded as an improvement of LGE, where we extract the hierarchical features rather than a simple fusion. From Tables 1–3, we can find that H^2F is slightly better than LGE in all of the three data sets. These results may indicate that the hierarchical strategy in H^2F is effective.

3.4. *Analysis and Discussion*

Figure 7 shows the box plots of OAs by different methods. The box plot is a simple summary for the data distribution. In this paper, we have conducted all the methods 50 times, and the results in each running are displayed by box plots. In a box plot, the red line in the box denotes the median. The top and bottom of a box are the 75th and 25th percentiles, respectively. Data outside the box are mild and extreme outliers. Because LGE, HiFi and H^2F present the closest accuracies in the three data sets, we only show the box plots by these methods in Figure 7, and take OA for example. We can see that the boxes of H^2F are higher than the others in all the three data sets, and the advantage is more apparent in GRSS_DFC_2014. Moreover, we use a paired *t*-test to further validate that the improvements by H^2F are statistically significant, which is defined as follows:

$$\frac{(\overline{a_1} - \overline{a_2})\sqrt{n_1 + n_2 - 2}}{\sqrt{(\frac{1}{n_1} + \frac{1}{n_2})(n_1 s_1^2 + n_2 s_2^2)}} > t_{1-\alpha}[n_1 + n_2 - 2], \tag{12}$$

where a_1 and a_2 are the OA of H^2F and a compared method, s_1 and s_2 are the corresponding standard deviations, n_1 and n_2 are the repetition running times, which is set as 50 here, and $t_{1-\alpha}$ is the αth best quantile of the Student's law. Results indicate that the improvements by H^2F is statistically significant in all of the three data sets (at level 90%).

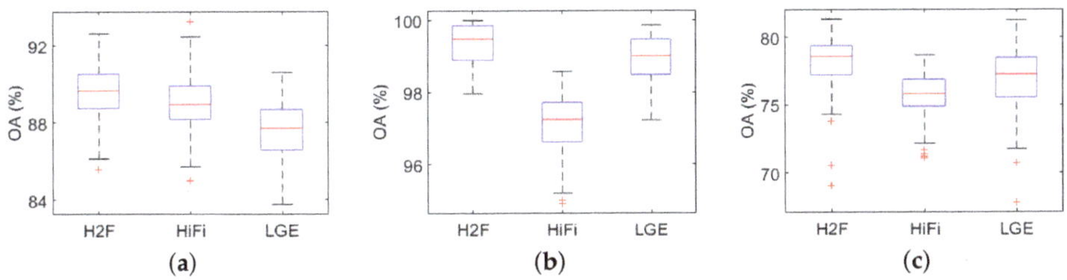

Figure 7. Box plots of different methods on (**a**) Indian Pines; (**b**) KSC and (**c**) GRSS_DFC_2014 data sets.

It is worth noting that it is not necessary to tune the parameters in each single feature such as Gabor and RGF. H^2F needs to ensemble many groups of features, and setting different parameters is a natural step to generate various sub-features. Therefore, the most important parameters in H^2F are the number of sub-feature sets M and the number of features N in each subset. In Figure 8, we provide an analysis for M and N. The results are interesting. We find that, although M and N have drastic changes, the OAs vary little in Indian Pines and KSC data sets. However, in Figure 8c, results demonstrate that more features will contribute to better accuracy. The reason may be that, in the former two data sets, the multiple features have already included some redundancy information. In other words, it is not necessary to extract too many features in Indian Pines and KSC data sets. However, it is not suitable for GRSS_DFC_2014 data set, where further increasing the multiple features would continue improving the classification accuracies. Because GRSS_DFC_2014 is long wave infrared data set, its quality is much lower than that of the other two. It is not appropriate to infer that GRSS_DFC_2014 also has information redundancy. In this case, integrating more features may further enhance the ability of feature representation in GRSS_DFC_2014. Results in Table 3 could also support this opinion. Overall, the most important point we try to emphasize in Figure 8 is that information redundancy does not exist in all of the HSI data. For some popular data sets such as Indian Pines and KSC, maybe information redundancy really exists. However, not all the HSI data includes redundancy information. It is not safe to conclude that dimension reduction could bring competitive or even better classification accuracies. In addition, this is just why we try to extract hierarchical features.

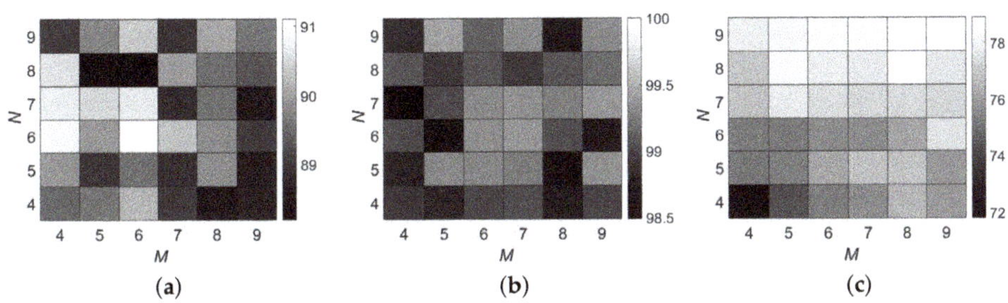

Figure 8. The influence of parameters on OA (%) in H^2F. Results on (**a**) Indian Pines; (**b**) KSC and (**c**) GRSS_DFC_2014 data sets. M is the number of sub-feature sets, and N is the number of features in each subsets.

Since the deep features extracted by H^2F are usually of high dimension, some popular classifiers such as SVM are time-consuming. In Table 4, we compare the training and testing time by ELM and SVM. To be fair and avoid parameter tuning, linear kernel is adopted by both of them. Another advantage of using linear kernel is that it could reduce the computational complexity. Furthermore, the OAs by ELM and SVM are also reported. Note that the running time in Table 4 is only composed of the classifiers' training and testing process, not including the feature extraction process. We can see from Table 4 that ELM presents slightly better performance than SVM with lower computational consumption. Because the selection of classifier is not the emphasis in H^2F, we choose ELM according to the results in Table 4.

Table 4. The OA (%)/running time (s) by ELM and SVM.

	Indian Pines	KSC	GRSS_DFC_2014
ELM	89.55/4.07	99.36/1.55	77.90/2.64
SVM	88.98/128.3	99.21/45.9	77.75/45.1

Finally, we give an evaluation for the influence of training samples number in Figure 9. Classes with totally 20 around samples are ignored because they have little influence on OA. Similar to Figure 7, HiFi and LGE are used for comparison. As is expected, the accuracy improves with the increase of training samples number. H^2F outperforms the others in most cases. In particular, we note that the gaps are more apparent when training samples are limited. This results may indicate that H^2F could provide more representative feature expression for the original HSI data.

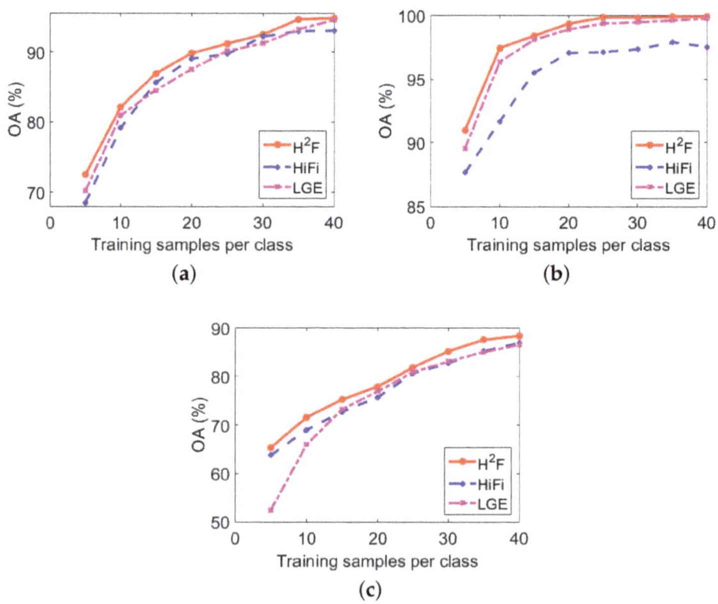

Figure 9. Influence of training samples number on (**a**) Indian Pines; (**b**) KSC and (**c**) GRSS_DFC_2014 data sets.

4. Conclusions

In this paper, we proposed a hierarchical feature extraction method for HSI classification. The proposed method is inspired by the promising performance of multiple features fusion. We hold the opinion that further utilization for the multiple features will contribute to the classification accuracy, and this idea is similar to that of deep learning methods. Therefore, instead of data dimension reduction or direct ensembling, in H^2F, we propose a hierarchical feature extraction strategy based on hashing, which attempts to explore the deep distinctive information among the original HSI data.

Spectral as well as local and global spatial features are firstly extracted, and these low-level features are further represented in a very sparse manner.

We compare H^2F with some ensemble based or deep learning based methods in the experimental part. Although the advantages are not apparent, a paired t-test has confirmed that our improvements are statistically significant. In particular, the idea of extracting hierarchical information from basic features may work as an inspiration for the further research.

In our future works, we will focus on improving the computational efficiency of the hierarchical feature extraction process. Meanwhile, the relationships among different features should also be investigated.

Acknowledgments: The authors would like to thank Telops Inc. (Québec, Canada) for acquiring and providing the data used in this study, the IEEE GRSS Image Analysis and Data Fusion Technical Committee and Michal Shimoni (Signal and Image Centre, Royal Military Academy, Belgium) for organizing the 2014 Data Fusion Contest, the Centre de Recherche Public Gabriel Lippmann (CRPGL, Luxembourg) and Martin Schlerf (CRPGL) for their contribution of the Hyper-Cam LWIR sensor, and Michaela De Martino (University of Genoa, Italy) for her contribution to data preparation. The work was supported by the National Natural Science Foundation of China under the Grant 61671037, and the Beijing Natural Science Foundation under the Grant 4152031, and the Excellence Foundation of BUAA for PhD Students under Grant 2017057.

Author Contributions: Bin Pan and Zhenwei Shi designed the algorithm; Xia Xu designed and performed the experiments; Yi Yang contributed to the English expression of this paper; Bin Pan wrote the paper.

Conflicts of Interest: The authors declare no conflict of interest.

References

1. Fu, Y.; Zhao, C.; Wang, J.; Jia, X.; Yang, G.; Song, X.; Feng, H. An Improved Combination of Spectral and Spatial Features for Vegetation Classification in Hyperspectral Images. *Remote Sens.* **2017**, *9*, 261.
2. Pan, B.; Shi, Z.; An, Z.; Jiang, Z.; Ma, Y. A Novel Spectral-Unmixing-Based Green Algae Area Estimation Method for GOCI Data. *IEEE J. Sel. Top. Appl. Earth Obs. Remote Sens.* **2017**, *10*, 437–449.
3. Kang, X.; Zhang, X.; Li, S.; Li, K.; Li, J.; Benediktsson, J.A. Hyperspectral Anomaly Detection With Attribute and Edge-Preserving Filters. *IEEE Trans. Geosci. Remote Sens.* **2017**, *55*, 5600–5611.
4. Xu, X.; Shi, Z. Multi-objective based spectral unmixing for hyperspectral images. *ISPRS J. Photogramm. Remote Sens.* **2017**, *124*, 54–69.
5. Zhong, P.; Zhang, P.; Wang, R. Dynamic Learning of SMLR for Feature Selection and Classification of Hyperspectral Data. *IEEE Geosci. Remote Sens. Lett.* **2008**, *5*, 280–284.
6. Gomez-Chova, L.; Camps-Valls, G.; Munoz-Mari, J.; Calpe, J. Semisupervised Image Classification with Laplacian Support Vector Machines. *IEEE Geosci. Remote Sens. Lett.* **2008**, *5*, 336–340.
7. Pal, M.; Foody, G.M. Feature Selection for Classification of Hyperspectral Data by SVM. *IEEE Trans. Geosci. Remote Sens.* **2010**, *48*, 2297–2307.
8. Mountrakis, G.; Im, J.; Ogole, C. Support vector machines in remote sensing: A review. *ISPRS J. Photogramm. Remote Sens.* **2011**, *66*, 247–259.
9. Castrodad, A.; Xing, Z.; Greer, J.B.; Bosch, E.; Carin, L.; Sapiro, G. Learning discriminative sparse representations for modeling, source separation, and mapping of hyperspectral imagery. *IEEE Trans. Geosci. Remote Sens.* **2011**, *49*, 4263–4281.
10. Li, J.; Bioucas-Dias, J.M.; Plaza, A. Spectral-spatial hyperspectral image segmentation using subspace multinomial logistic regression and Markov random fields. *IEEE Trans. Geosci. Remote Sens.* **2012**, *50*, 809–823.
11. Li, W.; Prasad, S.; Fowler, J.E. Hyperspectral image classification using Gaussian mixture models and Markov random fields. *IEEE Geosci. Remote Sens. Lett.* **2014**, *11*, 153–157.
12. Tarabalka, Y.; Fauvel, M.; Chanussot, J.; Benediktsson, J.A. SVM-and MRF-based method for accurate classification of hyperspectral images. *IEEE Geosci. Remote Sens. Lett.* **2010**, *7*, 736–740.
13. Benediktsson, J.A.; Palmason, J.A.; Sveinsson, J.R. Classification of hyperspectral data from urban areas based on extended morphological profiles. *IEEE Trans. Geosci. Remote Sens.* **2005**, *43*, 480–491.
14. Dalla Mura, M.; Atli Benediktsson, J.; Waske, B.; Bruzzone, L. Extended profiles with morphological attribute filters for the analysis of hyperspectral data. *Int. J. Remote Sens.* **2010**, *31*, 5975–5991.

15. Zhong, Z.; Fan, B.; Ding, K.; Li, H.; Xiang, S.; Pan, C. Efficient Multiple Feature Fusion With Hashing for Hyperspectral Imagery Classification: A Comparative Study. *IEEE Trans. Geosci. Remote Sens.* **2016**, *54*, 4461–4478.
16. Li, J.; Marpu, P.R.; Plaza, A.; Bioucas-Dias, J.M.; Benediktsson, J.A. Generalized composite kernel framework for hyperspectral image classification. *IEEE Trans. Geosci. Remote Sens.* **2013**, *51*, 4816–4829.
17. Gu, Y.; Liu, T.; Jia, X.; Benediktsson, J.A.; Chanussot, J. Nonlinear multiple kernel learning with multiple-structure-element extended morphological profiles for hyperspectral image classification. *IEEE Trans. Geosci. Remote Sens.* **2016**, *54*, 3235–3247.
18. Liu, T.; Gu, Y.; Jia, X.; Benediktsson, J.A.; Chanussot, J. Class-Specific Sparse Multiple Kernel Learning for Spectral–Spatial Hyperspectral Image Classification. *IEEE Trans. Geosci. Remote Sens.* **2016**, *54*, 7351–7365.
19. Wang, Q.; Gu, Y.; Tuia, D. Discriminative multiple kernel learning for hyperspectral image classification. *IEEE Trans. Geosci. Remote Sens.* **2016**, *54*, 3912–3927.
20. Zhang, Q.; Tian, Y.; Yang, Y.; Pan, C. Automatic Spatial–Spectral Feature Selection for Hyperspectral Image via Discriminative Sparse Multimodal Learning. *IEEE Trans. Geosci. Remote Sens.* **2015**, *53*, 261–279.
21. Zhang, L.; Zhang, Q.; Du, B.; Huang, X.; Tang, Y.Y.; Tao, D. Simultaneous Spectral-Spatial Feature Selection and Extraction for Hyperspectral Images. *IEEE Trans. Cybern.* **2017**, *PP*, 1–13, doi:10.1109/TCYB.2016.2605044.
22. Zhang, L.; Zhang, L.; Tao, D.; Huang, X. On Combining Multiple Features for Hyperspectral Remote Sensing Image Classification. *IEEE Trans. Geosci. Remote Sens.* **2012**, *50*, 879–893.
23. Zhang, L.; Zhang, L.; Tao, D.; Huang, X. A modified stochastic neighbor embedding for multi-feature dimension reduction of remote sensing images. *ISPRS J. Photogramm. Remote Sens.* **2013**, *83*, 30–39.
24. Wang, M.; Yu, J.; Niu, L.; Sun, W. Feature Extraction for Hyperspectral Images Using Low-Rank Representation With Neighborhood Preserving Regularization. *IEEE Geosci. Remote Sens. Lett.* **2017**, *14*, 836–840.
25. Zhou, Z.H. *Ensemble Methods: Foundations and Algorithms*; CRC Press: Boca Raton, FL, USA, 2012.
26. Pal, M. Ensemble of support vector machines for land cover classification. *Int. J. Remote Sens.* **2008**, *29*, 3043–3049.
27. Huang, X.; Zhang, L. An SVM ensemble approach combining spectral, structural, and semantic features for the classification of high-resolution remotely sensed imagery. *IEEE Trans. Geosci. Remote Sens.* **2013**, *51*, 257–272.
28. Liu, Z.; Tang, B.; He, X.; Qiu, Q.; Liu, F. Class-Specific Random Forest With Cross-Correlation Constraints for Spectral–Spatial Hyperspectral Image Classification. *IEEE Geosci. Remote Sens. Lett.* **2017**, *14*, 257–261.
29. Xia, J.; Bombrun, L.; Adalı, T.; Berthoumieu, Y.; Germain, C. Spectral–spatial classification of hyperspectral images using ica and edge-preserving filter via an ensemble strategy. *IEEE Trans. Geosci. Remote Sens.* **2016**, *54*, 4971–4982.
30. Xia, J.; Falco, N.; Benediktsson, J.A.; Du, P.; Chanussot, J. Hyperspectral Image Classification With Rotation Random Forest Via KPCA. *IEEE J. Sel. Top. Appl. Earth Obs. Remote Sens.* **2017**, *10*, 1601–1609.
31. Chen, J.; Wang, C.; Wang, R. Using stacked generalization to combine SVMs in magnitude and shape feature spaces for classification of hyperspectral data. *IEEE Trans. Geosci. Remote Sens.* **2009**, *47*, 2193–2205.
32. Pan, B.; Shi, Z.; Xu, X. Hierarchical Guidance Filtering-Based Ensemble Classification for Hyperspectral Images. *IEEE Trans. Geosci. Remote Sens.* **2017**, *55*, 4177–4189.
33. Debes, C.; Merentitis, A.; Heremans, R.; Hahn, J.; Frangiadakis, N.; Kasteren, T.V.; Liao, W.; Bellens, R.; Pižurica, A.; Gautama, S. Hyperspectral and LiDAR Data Fusion: Outcome of the 2013 GRSS Data Fusion Contest. *IEEE J. Sel. Top. Appl. Earth Obs. Remote Sens.* **2014**, *7*, 2405–2418.
34. Liao, W.; Pižurica, A.; Bellens, R.; Gautama, S.; Philips, W. Generalized Graph-Based Fusion of Hyperspectral and LiDAR Data Using Morphological Features. *IEEE Geosci. Remote Sens. Lett.* **2015**, *12*, 552–556.
35. Chen, Y.; Lin, Z.; Zhao, X.; Wang, G.; Gu, Y. Deep learning-based classification of hyperspectral data. *IEEE J. Sel. Top. Appl. Earth Obs. Remote Sens.* **2014**, *7*, 2094–2107.
36. Pan, B.; Shi, Z.; Zhang, N.; Xie, S. Hyperspectral Image Classification Based on Nonlinear Spectral–Spatial Network. *IEEE Geosci. Remote Sens. Lett.* **2016**, *13*, 1782–1786.
37. Pan, B.; Shi, Z.; Xu, X. R-VCANet: A new deep-learning-based hyperspectral image classification method. *IEEE J. Sel. Top. Appl. Earth Obs. Remote Sens.* **2017**, *10*, 1975–1986.
38. Wu, H.; Prasad, S. Convolutional Recurrent Neural Networks for Hyperspectral Data Classification. *Remote Sens.* **2017**, *9*, 298.
39. Ding, C.; Li, Y.; Xia, Y.; Wei, W.; Zhang, L.; Zhang, Y. Convolutional Neural Networks Based Hyperspectral Image Classification Method with Adaptive Kernels. *Remote Sens.* **2017**, *9*, doi:10.3390/rs9060618.

40. Li, W.; Wu, G.; Zhang, F.; Du, Q. Hyperspectral image classification using deep pixel-pair features. *IEEE Trans. Geosci. Remote Sens.* **2017**, *55*, 844–853.
41. He, K.; Zhang, X.; Ren, S.; Sun, J. Delving deep into rectifiers: Surpassing human-level performance on ImageNet classification. In Proceedings of the IEEE International Conference on Computer Vision, Santiago, Chile, 7–13 December 2015; pp. 1026–1034.
42. Zhong, P.; Gong, Z.; Li, S.; Schonlieb, C.B. Learning to Diversify Deep Belief Networks for Hyperspectral Image Classification. *IEEE Trans. Geosci. Remote Sens.* **2017**, *55*, 3516–3530.
43. Liang, H.; Li, Q. Hyperspectral Imagery Classification Using Sparse Representations of Convolutional Neural Network Features. *Remote Sens.* **2016**, *8*, 99.
44. Li, Y.; Zhang, H.; Shen, Q. Spectral–Spatial Classification of Hyperspectral Imagery with 3D Convolutional Neural Network. *Remote Sens.* **2017**, *9*, 67.
45. Zhang, Q.; Shen, X.; Xu, L.; Jia, J. Rolling Guidance Filter. In Proceedings of the European Conference on Computer Vision, Zurich, Switzerland, 6–12 September 2014; pp. 815–830.
46. Li, W.; Chen, C.; Su, H.; Du, Q. Local Binary Patterns and Extreme Learning Machine for Hyperspectral Imagery Classification. *IEEE Trans. Geosci. Remote Sens.* **2015**, *53*, 3681–3693.
47. Li, W.; Du, Q. Gabor-Filtering-Based Nearest Regularized Subspace for Hyperspectral Image Classification. *IEEE J. Sel. Top. Appl. Earth Obs. Remote Sens.* **2014**, *7*, 1012–1022.
48. Xu, X.; Shi, Z.; Pan, B. A New Unsupervised Hyperspectral Band Selection Method Based on Multiobjective Optimization. *IEEE Geosci. Remote Sens. Lett.* **2017**, *PP*, 1–5, doi:10.1109/LGRS.2017.2753237.
49. Cavallaro, G.; Falco, N.; Mura, M.D.; Benediktsson, J.A. Automatic Attribute Profiles. *IEEE Trans. Image Process.* **2017**, *26*, 1859–1872.
50. Kang, X.; Xiang, X.; Li, S.; Benediktsson, J.A. PCA-Based Edge-Preserving Features for Hyperspectral Image Classification. *IEEE Trans. Geosci. Remote Sens.* **2017**, *PP*, 1–12, doi:10.1109/TGRS.2017.2743102.
51. Huang, G.B.; Zhu, Q.Y.; Siew, C.K. Extreme learning machine: Theory and applications. *Neurocomputing* **2006**, *70*, 489–501.
52. Samat, A.; Du, P.; Liu, S.; Li, J.; Cheng, L. E^2LMs : Ensemble Extreme Learning Machines for Hyperspectral Image Classification. *IEEE J. Sel. Top. Appl. Earth Obs. Remote Sens.* **2014**, *7*, 1060–1069.
53. Su, H.; Cai, Y.; Du, Q. Firefly-Algorithm-Inspired Framework With Band Selection and Extreme Learning Machine for Hyperspectral Image Classification. *IEEE J. Sel. Top. Appl. Earth Obs. Remote Sens.* **2017**, *10*, 309–320.
54. He, K.; Sun, J.; Tang, X. Guided image filtering. *IEEE Trans. Pattern Anal. Mach. Intell.* **2013**, *35*, 1397–1409.
55. Friedman, J.; Hastie, T.; Tibshirani, R. *The Elements of Statistical Learning*; Springer Series in Statistics; Springer: Berlin, Germany, 2001; Volume 1.
56. Bau, T.C.; Sarkar, S.; Healey, G. Hyperspectral region classification using a three-dimensional Gabor filterbank. *IEEE Trans. Geosci. Remote Sens.* **2010**, *48*, 3457–3464.
57. Shen, L.; Jia, S. Three-dimensional Gabor wavelets for pixel-based hyperspectral imagery classification. *IEEE Trans. Geosci. Remote Sens.* **2011**, *49*, 5039–5046.
58. Kang, X.; Li, S.; Benediktsson, J.A. Spectral–Spatial Hyperspectral Image Classification with Edge-Preserving Filtering. *IEEE Trans. Geosci. Remote Sens.* **2014**, *52*, 2666–2677.
59. Kang, X.; Li, S.; Fang, L.; Benediktsson, J.A. Intrinsic Image Decomposition for Feature Extraction of Hyperspectral Images. *IEEE Trans. Geosci. Remote Sens.* **2015**, *53*, 2241–2253.
60. 2014 IEEE GRSS Data Fusion Contest. Available online: http://www.grss-ieee.org/community/technical-committees/data-fusion/ (accessed on 25 October 2017).
61. Liao, W.; Huang, X.; Coillie, F.V.; Gautama, S.; Pižurica, A.; Philips, W.; Liu, H.; Zhu, T.; Shimoni, M.; Moser, G. Processing of Multiresolution Thermal Hyperspectral and Digital Color Data: Outcome of the 2014 IEEE GRSS Data Fusion Contest. *IEEE J. Sel. Top. Appl. Earth Obs. Remote Sens.* **2015**, *8*, 2984–2996.
62. Sun, B.; Kang, X.; Li, S.; Benediktsson, J.A. Random-Walker-Based Collaborative Learning for Hyperspectral Image Classification. *IEEE Trans. Geosci. Remote Sens.* **2017**, *55*, 212–222.

remote sensing

MDPI

Article

Classification of Hyperspectral Images by SVM Using a Composite Kernel by Employing Spectral, Spatial and Hierarchical Structure Information

Yi Wang * and Hexiang Duan

Institute of Geophysics and Geomatics, China University of Geosciences, Wuhan 430074, China; dhx12cug@163.com

* Correspondence: cug.yi.wang@gmail.com; Tel.: +86-27-6788-3251

Received: 2 February 2018; Accepted: 27 February 2018; Published: 12 March 2018

Abstract: In this paper, we introduce a novel classification framework for hyperspectral images (HSIs) by jointly employing spectral, spatial, and hierarchical structure information. In this framework, the three types of information are integrated into the SVM classifier in a way of multiple kernels. Specifically, the spectral kernel is constructed through each pixel's vector value in the original HSI, and the spatial kernel is modeled by using the extended morphological profile method due to its simplicity and effectiveness. To accurately characterize hierarchical structure features, the techniques of Fish-Markov selector (FMS), marker-based hierarchical segmentation (MHSEG) and algebraic multigrid (AMG) are combined. First, the FMS algorithm is used on the original HSI for feature selection to produce its spectral subset. Then, the multigrid structure of this subset is constructed using the AMG method. Subsequently, the MHSEG algorithm is exploited to obtain a hierarchy consist of a series of segmentation maps. Finally, the hierarchical structure information is represented by using these segmentation maps. The main contributions of this work is to present an effective composite kernel for HSI classification by utilizing spatial structure information in multiple scales. Experiments were conducted on two hyperspectral remote sensing images to validate that the proposed framework can achieve better classification results than several popular kernel-based classification methods in terms of both qualitative and quantitative analysis. Specifically, the proposed classification framework can achieve 13.46–15.61% in average higher than the standard SVM classifier under different training sets in the terms of overall accuracy.

Keywords: hyperspectral images (HSIs); classification; SVM; composite kernel; algebraic multigrid methods

1. Introduction

With the rapid development of hyperspectral sensors, the present hyperspectral images (HSIs) contain rich spectral and spatial information. Therefore, different objects can be accurately recognized from HSIs using various classification algorithms for different applications, such as geological survey [1], mineral mapping [2], fine agricultural research [3,4], environmental monitoring [5], etc.

HSI classification is one of the most popular problems in the field of remote sensing and has aroused much concern, but faces the following challenges [6–8]: First, it is very difficult to acquire sufficient labeled samples. Second, information redundancy and Hughes phenomenon are inevitable due to the high dimensional features represented by hundreds of spectral bands. Finally, HSIs are often corrupted with different types of noise and dominated by mixed pixels. To solve these problems, many researchers recurred to pixel-wise methods to classify each pixel in HSIs to a certain class using its spectral information individually [9–14]. Among them, the SVM [10,15] and multinomial logistic regression (MLR) [16–18] are the two most commonly used techniques. However, these methods often result in much "salt-and-pepper" noise in classification maps, without considering spatial neighborhoods, and the classification performance cannot be further improved.

This difficulty has been greatly conquered at the appearance of spectral-spatial classification methods [19]. Generally, these methods can be divided into three categories. In the first category, spatial information is integrated with spectral information by using composite kernels [20–23]. There are many methods for spatial feature extraction, such as mean filtering [20], area filtering [24], Gabor filtering [25], gray-level co-occurrence matrix [26], edge-preserving filtering (EPF) [27], and extended morphological profiles (EMPs) [28]. In the second category, the integration of the spectral information and the spatial information is first performed by image segmentation algorithms, such as mean-shift [29], watershed [30], hierarchical segmentation [31,32], minimum spanning forest [33], graph cut [34,35], and superpixel [36] approaches. Then, the final classification map is produced by combining the pixel-wise classification map and the unsupervised segmentation map by employing a majority voting algorithm. In the third category, the two types of information are jointly included in the classification process using Markov random field (MRF) models. By applying the maximum a posteriori (MAP) decision rule, HSI classification can be effectively solved by minimizing a MAP-MRF energy function. The ensemble method of SVM and the MRF-based model is a regular scheme [37–43].

Kernel-based classification methods have been very popular for HSI classification because they can effectively deal with the intractable issues of curse of dimensionality, limited labeled samples, and noise corruption. The SVM algorithm using a single (e.g., linear, polynomial, or Gaussian radial basis function (RBF)) kernel has been widely used for image classification. To perform HSI classification, several SVM techniques using the spectral-spatial kernel were presented. For instance, Camps-Valls et al. [20] formulated a general framework of multiple kernels by exploring both the spectral and spatial information, and the spatial information is defined using basic statistical measures within a fixed-size window in the image. The selection of a suitable window size is a challenging problem because spatial structures extracted from such a region cannot be accurately represented. To solve this problem, the adaptive neighborhood system based on morphological filtering and area filtering has been considered. On the one hand, Fauvel et al. [44] applied feature extraction on the original HSI and its EMPs, respectively, and performed the SVM classification using the RBF kernel with spectral-spatial stacked vectors. Li et al. [45] developed a MLR framework using generalized composite kernels (GCK), where the spatial information is represented by using EMPs as well. The obtained spatial information by such methods is highly dependent to the Structuring Element (SE) of morphological operators. On the other hand, Fauvel et al. [22] proposed an improved SVM by using a customized spectral-spatial kernel where the spatial information is modeled as the median value on the adaptive neighbourhood of each pixel defined using morphological area filtering. The result that is achieved by such a method is very sensitive to the predefined number of areas. Recently, the superpixel-based techniques have been applied to HSI classification by Shutao Li's research group. Fang et al. [46] presented an effective SVM classifier characterized with a superpixel-based composite kernel, where the three types of the spectral, intra-superpixel and inter-superpixel information are combined, and the superpixel map is obtained using the entropy rate superpixel (ERS) algorithm. Meanwhile, texture features are crucial for object classification of HSIs. Later, we introduced an alternative SVM classifier featured with a spectral-texture kernel [23], where the textual information is modeled for each superpixel with its local spectral histogram. The number of superpixels is a data-dependent and greatly influences classification results. Lu et al. [47] developed an effective HSI classification framework by integrating the multiple feature-induced kernels into a SVM classifier, where subpixel, pixel and superpixel features are combined. More recently, Peng et al. [48] improved the spectral-spatial composite kernel by embedding label information with an ideal regularized technique. The information that is extracted from the label domain cannot describe spatial structures well.

In this paper, we develop a novel SVM classification framework with the spectral, spatial, and hierarchical kernels (SVM-SSHK), in which the spectral, spatial, and hierarchical structure information in HSIs are integrated into the SVM classifier in a way of multiple kernels. Specifically, the spectral kernel is constructed through each pixel's vector value in the original HSI, and the spatial kernel is modeled by using the EMP method due to its simplicity and effectiveness. To accurately characterize hierarchical structure features, the techniques of Fish-Markov selector (FMS), marker-based hierarchical segmentation (MHSEG) and algebraic multigrid (AMG) are combined. First, the FMS algorithm is used on the original image for feature selection to produce its spectral subset. Then, the multigrid structure of this subset is constructed using the AMG method. Subsequently, the MHSEG algorithm is exploited to obtain a hierarchy consisting of a series of segmentation maps. Finally, the hierarchical structure information is modeled by using these segmentation maps. The main contributions of this work is to present an effective composite kernel framework for HSI classification by utilizing spatial structure information in multiple scales. The previously mentioned kernel-based approaches cannot simultaneously capture salient and fine structures in the image with a predefined number of regions. However, the proposed framework can obtain a hierarchical representation of spatial structure information in HSIs. Furthermore, this hierarchical structure is only dependent to the original HSI, without considering the problem of choosing a neighborhood system or the size of a region (e.g., an area or a superpixel).

The remainder of the paper is organized as follows. In Section 2, some related techniques are reviewed. In Section 3, the proposed classification framework that is characterized with a spectral-spatial-hierarchical kernel is introduced. In Section 4, experimental results are reported in comparing to popular HSI classification methods and some issues are discussed. The last section presents some concluding remarks and the future work.

2. Related Techniques

Let $\mathbf{x}$ represent an HSI which contains B-band vectors with $\mathbf{x} \equiv \{x_1, x_2, \ldots, x_N\} \in \mathbb{R}^B$, $\mathbf{y} \equiv \{y_1, y_2, \ldots, y_N\} \in L^N$ the final classification result with $y_k \equiv \{L_1, L_2, \ldots, L_T\}$ $(k = 1, 2, \ldots, N)$, $\{(x_k, y_k)\}_{k=1}^{n}$ training samples, n_i the number of the training samples of L_i $(i = 1, 2, \ldots, Z)$.

2.1. Spatial Information with EMPs

In the spectral-spatial classification method, the first step is to extract some featured bands to model spectral information from hyperspectral images (HSIs) by dimensionality reduction, which is used to minimize redundant information and to improve computational efficiency. To this end, the most popular approaches have been used, such as principal component analysis (PCA) [28], independent component analysis (ICA) [49], Kernel PCA [50], decision boundary feature extraction, nonparametric weighted feature extraction, and Bhattacharyya distance feature selection [51]. In this work, the widely used PCA transform was used to produce the EMP. First, almost all of the spectral information in HSIs can be represented by using the first three or four principal components (PCs). Second, object boundaries in the HSIs can be better preserved in the resultant PCs [27]. Finally, it was recorded that the EMP was first constructed using PCA [7].

The main idea of EMP is to reconstruct the spatial information through morphological (opening/closing) operators, while preserving the boundaries of the image. Let k and n be the total number of the selected principal components (PCs) and the morphological operators, respectively, ψ and η the opening and closing operations, and I a gray-level image, we can build the morphological profile (MP) for each PC, as follows:

$$MP(I) = \{\psi_T(I), \ldots, \psi_2(I), \psi_1(I), I, \eta_1(I), \eta_2(I), \ldots, \eta_T(I)\} \tag{1}$$

For each PC, the MP is a $(2n + 1)$-band image. Then, the MPs are stacked to obtain the EMP as follows:

$$EMP_k = \{MP(PC_1), MP(PC_2), \ldots, MP(PC_k)\}. \tag{2}$$

where EMP is a stacked vector with the dimensionality of $k(2n+1)$ and includes both the spectral and spatial information of the HSIs. In fact, we can extract EMPs for all of the spectral bands or for some selective bands in the HSIs without PCA, which causes the following limitations. First, some redundancy can be observed in the $B(2n+1)$-band image, where B is the number of spectral bands for the HSI, which may decrease the classification accuracies. Second, the classification process should be fit for such high-dimensional data with much more computational cost.

2.2. Band Selection with FMS

In many research fields, it is necessary for supervised classification to perform feature selection. Given a set of test samples, the selected features are used to assign a class label for each sample. Feature selection and subspace methods are widely used for dimensionality reduction [52–55]. For instance, Cheng et al. [56] presented the FMS algorithm for the feature selection of high-dimensional data, whose basic idea is to find the optimal subset of features to maximize intra-class separability and minimize inter-class variations in a higher dimensional kernel space. By employing some spectral kernel functions, such as the polynomial kernel, the feature selection problem can be solved efficiently using MRF optimization techniques. In the original space, denote the within-, between- (or inter-) class, and total scatter matrices by S_w, S_b, and S_t:

$$S_w = \frac{1}{N}\sum_{j=1}^{Z}\sum_{i=1}^{n_j}\left(x_i^{(j)} - m_j\right)\left(x_i^{(j)} - m_j\right)^T \tag{3}$$

$$S_b = \frac{1}{N}\sum_{j=1}^{Z} n_j(m_j - m)(m_j - m)^T \tag{4}$$

$$S_t = \frac{1}{N}\sum_{i=1}^{n} n_j(x_i - m)(x_i - m)^T = S_w + S_b \tag{5}$$

where $x_i^{(j)}$ is the ith training sample in class L_j, and m_j and m represent the sample means for class L_j and the whole training set, respectively. The scatter matrices are denoted by $\widetilde{S}_w$, $\widetilde{S}_b$, and $\widetilde{S}_t$ in the kernel space, whose traces in algebra can be calculated as follows:

$$\mathrm{Tr}(\widetilde{S}_w) = \frac{1}{n}\mathrm{Tr}(K) - \sum_{i=1}^{Z}\frac{1}{n_i}\mathrm{Sum}\left(K^{(i)}\right) \tag{6}$$

$$\mathrm{Tr}(\widetilde{S}_b) = \frac{1}{n}\sum_{i=1}^{Z}\frac{1}{n_i}\mathrm{Sum}\left(K^{(i)}\right) - \frac{1}{n^2}\mathrm{Sum}(K) \tag{7}$$

$$\mathrm{Tr}(\widetilde{S}_t) = \frac{1}{\mathrm{n}}\mathrm{Tr}(K) - \frac{1}{n^2}\mathrm{Sum}(K) \tag{8}$$

where $\mathrm{Tr}(\cdot)$ and $\mathrm{Sum}(\cdot)$ are the summation and trace operators, respectively, and K and $K^{(i)}$ are two matrices with size of $n \times n$ and $n_i \times n_i$, respectively, and have the following forms:

$$\{K\}_{k,l} = k(x_k, x_l),\ k,l \in \{1,2,\ldots,n\} \tag{9}$$

$$\left\{K^{(i)}\right\}_{u,v} = k\left(x_u^{(i)}, x_v^{(i)}\right),\ u,v \in \{1,2,\ldots,n_i\},\ i = 1,2,\ldots,Z \tag{10}$$

The feature selector is represented by $\alpha = [\alpha_1, \alpha_2, \ldots, \alpha_B]^Z \in \{0, 1\}^B$, where "1" indicates that the kth feature is selected or "0" not selected. The selected features from the vector x are defined, as follows:

$$x(\alpha) = x \odot \alpha \tag{11}$$

where $\odot$ is the Hadamard product. Substituting (9) and (10) with (11), K and $K^{(i)}$ can be expressed as functions of α:

$$\{K(\alpha)\}_{k,l} = k(x_k \odot \alpha, x_l \odot \alpha) \tag{12}$$

$$\left\{K^{(i)}(\alpha)\right\}_{u,v} = k\left(x_u^{(i)} \odot \alpha, x_v^{(i)} \odot \alpha\right) \tag{13}$$

In such way, the previously mentioned scatter matrices can be defined as functions of α as $\mathrm{Tr}\left(\widetilde{S}_w(\alpha)\right)$, $\mathrm{Tr}\left(\widetilde{S}_b(\alpha)\right)$ and $\mathrm{Tr}\left(\widetilde{S}_t(\alpha)\right)$.

The aim of feature selection is to maximize the class separations for the most discriminative capability of the variables. According to the spirit of Fisher, the following optimization function can be obtained:

$$\underset{\alpha \in \{0,1\}^B}{\operatorname{argmax}} \left\{ \mathrm{Tr}\left(\widetilde{S}_b\right)(\alpha) - \lambda \mathrm{Tr}\left(\widetilde{S}_t\right)(\alpha) \right\} \tag{14}$$

where λ is a parameter to balance the two items.

Actually, Equation (14) is a special case of the Markov problem without pairwise interaction term and can result in the optimal solution. In this work, we can compute a coefficient for each band of HSIs to demonstrate its significance by using the FMS algorithm. The higher the coefficient, the more significant the corresponding band. In this way, we can obtain the most relevant spectral bands.

2.3. Hierarchical Representation of HSIs

To construct a scale-space representation of a HSI $\mathbf{u} = [u_1, u_2, \ldots, u_N]$, a vector-valued anisotropic diffusion PDE can be used [57,58]:

$$\frac{\partial u_i}{\partial t} = \mathrm{div}(g(\theta(\nabla \mathbf{u}_\sigma))\nabla u_i), i = 1, 2, \ldots, N \tag{15}$$

where $\mathbf{u}_\sigma$ is obtained by convolving $\mathbf{u}$ with a Gaussian kernel of standard deviation σ, and $g(\cdot)$ is the diffusivity of $|\nabla \mathbf{u}_\sigma|$. Recently, AMG has been used for multiscale representation of HSIs due to the advantage that AMG is capable of constructing a hierarchical representation of the problem from a fine grid to a coarse grid and the linear system is suitable to be effectively solved in the coarsest grid [59]. In this work, the proposed framework exploits all of the vertices in the multigrid structure as the markers.

According to the work of [60], we can construct a "pyramid" multigrid structure of HSIs as shown in Figure 1, where each grid, $s = 1, 2, \ldots, S$, can be described by a weighted graph (V^s, E^s) in which V^s and E^s are the set of vertices and edges, respectively, and the weight g_{ij} of $(i,j) \subset E^s$ expresses the similarity between the pixels of u_i^s and u_j^s in V^s. Initially, the first graph (V^0, E^0) is built from the original HSI, where V^0 denotes the set of vertices, whose size is the same as the HSI, while E^0 represents the set of edges connecting each vertex to its four-neighborhoods with weights. In our method, the initial weights g_{ij}^0 of $(i,j) \subset E^0$ are computed by using the diffusivity of the anisotropic diffusion partial differential equation:

$$g(\theta) = \begin{cases} 1, & \theta = 0 \\ 1 - e^{-\frac{3.31488}{(\theta/\beta)^8}}, & \theta > 0 \end{cases} \tag{16}$$

where θ is an indicator of the image edge strength using the Euclidean distance (ED) or the spectral angle mapper (SAM) between two pixel vectors, β denotes a gradient threshold.

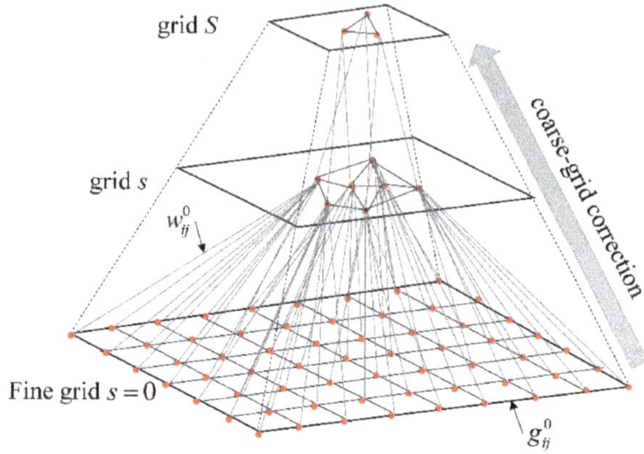

Figure 1. The multigrid structure of hyperspectral images (HSIs).

The main steps of building the multigrid structure is summarized as follows [60].

Step 1: To consecutively select a new set of V^{l+1} from V^l. To build the AMG multigrid structure, the authors of [60] introduced a mass m_i for each vertex, which is a measure for the number of pixels that are assigned to a given vertex selected to the next grid and can be initialized as $m_i^0 = 1$. The first vertex of V^{l+1} is selected as the vertex in V^l with the greatest mass. The rest vertices in V^l are sorted in decreasing order of mass. Then, a new vertex is iteratively selected if this satisfies the condition as follows [10]:

$$\frac{\sum\limits_{j \in V^{l+1}} g_{ij}^l}{\sum\limits_{(i,j) \in E^l} g_{ij}^l} \leq v \Rightarrow V^{l+1} = V^{l+1} \cup \{i\}, \text{ for each } i \in V^l \backslash V^{l+1} \tag{17}$$

where v is a threshold value with $0 \leq v \leq 1$, and $V^l \backslash V^{l+1}$ indicates the set difference between V^l and V^{l+1}. In the multigrid structure, each vertex has a mass value and the masses in the $(l+1)$th grid are calculated as follows:

$$\forall i \in V^{l+1} : m_i^{l+1} = m_i^l + \sum_{j \in V^{l+1}} w_{ij}^l \tag{18}$$

where m_i^l and m_i^{l+1} are the masses in the lth and $(l+1)$th grids, respectively, and w_{ij}^l weights how much vertex $i \in V^l \backslash V^{l+1}$ depends on the vertex $j \in V^{l+1}$:

$$\forall i \in V^l \backslash V^{l+1}, j \in V^{l+1} : w_{ij}^l = w_{ji}^l = \frac{g_{ij}^l}{\sum\limits_{k \in V^{l+1}} g_{ik}^l} \tag{19}$$

Step 2: To connect the vertices in V^{l+1} to obtain E^{l+1}. The matrix of diffusivities are obtained using the Garlekin operator $\mathbf{G}^{l+1} = I_f^c \mathbf{G}^l I_c^f$ [61], where I_f^c and I_c^f denote the restriction and interpolation operators:

$$\left[I_f^c\right]_{ij} = \frac{w_{ij}^l}{1 + \sum\limits_{j \in V^l \backslash V^{l+1}} w_{ij}^l} \tag{20}$$

$$\left[I_c^f\right]_{ij} = w_{ij}^l \tag{21}$$

According to the Garlekin operator, the weight in the $(l+1)$th grid is computed, as follows:

$$g_{ij}^{l+1} = \frac{1}{1 + \sum\limits_{j \in V^l \setminus V^{l+1}} w_{ij}^l} \sum\limits_{p,q \in V^l} w_{ip}^l g_{pq}^l w_{qj}^l \tag{22}$$

E^{l+1} are obtained by connecting the vertices, as follows:

$$E^{l+1} = \left\{(i,j) \ : \ i,j \in V^{l+1} \wedge g_{ij}^l > 0\right\} \tag{23}$$

To iteratively perform the previously two steps, a S-level multigrid structure of HSIs is constructed. The markers correspond to pixels of the smoothed image $\mathbf{u}^t$ determined by the position of the vertices in the coarse grid. The smoothed spectra can be considered as the average of spectrally similar and spatially adjacent pixels, which can decrease noise and improve the representation of the different objects in the HSI. In this work, the vertices in each grid level are used for the subsequent region growing algorithm.

2.4. AMG-MHSEG Algorithm

As described in [32], we presented a AMG-MHSEG classification framework of HSIs. The advantages of this framework are summarized as follows. First, the marker selection is performed using a AMG-derived approach, which is more effective than the classification-derived methods proposed by Tarabalka et al. [31]. The selection of markers in the classification-derived methods depend highly on the performance of the pixel-wise classifiers. Moreover, the selected markers may be greatly different due to randomly selection of training samples. The previously mentioned difficulties always cause uncertainty in the classification maps. However, the markers selected by the AMG-derived approach are only determined by structure features of HSIs. Second, the combination of the multigrid representation approach of HSIs and the MHSEG algorithm can provide the multiscale segmentation maps. The main steps of the AMG-MHSEG algorithm are introduced in Algorithm 1.

Algorithm 1: AMG-MHSEG

Input: An original hyperspectral image $\mathbf{u}$ and the coarsest grid level S.
Output: Segmentation maps

1. Input a hyperspectral image and construct an undirected graph as the finest grid.
2. AMG Relaxation
   - At the finest grid level, perform a Gauss-Seidel relaxation to solve $(\mathbf{I} - \tau\mathbf{G}^0)\mathbf{X}^0 = \mathbf{u}$ with an initial guess image $\mathbf{u}$ and compute the error $\mathbf{X}^0 = (\mathbf{I} - \tau\mathbf{G}^0)\mathbf{X}^0 - \mathbf{u}$.
   - At the coarser grid level l $(0 < l \leq S)$, perform a Gauss-Seidel relaxation to solve the residual equation $\left(\mathbf{I} - \tau\mathbf{G}^l\right)\mathbf{X}^l = \mathbf{F}^l$ with an initial guess $\mathbf{0}$, and compute the error $\mathbf{X}^l = \mathbf{F}^l - \left(\mathbf{I} - \tau\mathbf{G}^l\right)\mathbf{X}^l$ and then the residual $\mathbf{F}^l = \left(\mathbf{I} - \tau\mathbf{G}^l\right)\mathbf{X}^l$.
3. AMG Coarse-Grid Correction: Select the set of vertices in $\mathbf{V}^l$ for $\mathbf{V}^{l+1}$ to obtain $\mathbf{F}^{l+1}$ for the coarser grid level $l+1$.
4. Compute $\mathbf{G}^{l+1}$ and connect the nodes in $\mathbf{V}^{l+1}$ to obtain $\mathbf{E}^{l+1}$.
5. If $l \leq S$, go to step 2; otherwise, go to the next step.
6. Initialize the vertices in grid l as markers for the subsequent HSEG algorithm by assigning each vertex a non-zero marker label and each pixel as a separate region.
7. Perform the M-HSEG algorithm by using the markers obtained from grid l of the hyperspectral image:
   (a) Calculate the dissimilarity criterion (DC) values between all pairs of spatially adjacent regions. It should be noted that we only calculate the DC value between a markered pixel and a non-markered pixel and merge the pair of adjacent pixels that has the smallest DC value.
   (b) Merge the pair of adjacent pixels that has the smallest DC value.
   (c) Stop when there is no more merging, which means that the DC value is NaN.
8. Obtain the resultant segmentation maps for the subsequent classification.

3. The Proposed Classification Framework

In this section, the classical SVM classifier with the spectral-spatial kernel is first described. Then, the integration of the spectral, spatial and hierarchical structure information into a composite kernel framework is presented in our methodology. Figure 2 illustrates the schematic diagram of the SVM-SSHK method.

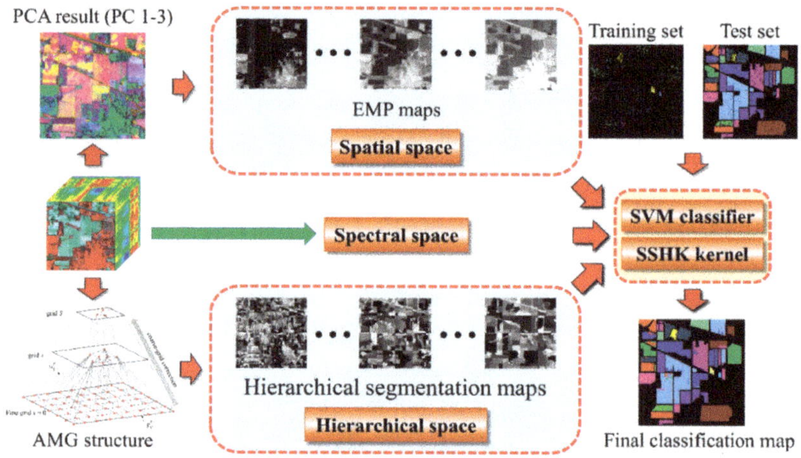

Figure 2. Schematic diagram of the SVM-SSHK method.

3.1. Spectral-Spatial Kernel

Let us consider an HSI that contains B-band vectors of $\mathbf{x}^{SPE} \equiv \{x_1^{SPE}, x_2^{SPE}, \ldots, x_N^{SPE}\} \in \mathbb{R}^{B\times N}$, its EMP $\mathbf{x}^{SPA} \equiv \{x_1^{SPA}, x_2^{SPA}, \ldots, x_N^{SPA}\} \in \mathbb{R}^{k(2n+1)\times N}$ and the hierarchical segmentation map $\mathbf{x}^{HIE} \equiv \{x_1^{HIE}, x_2^{HIE}, \ldots, x_N^{HIE}\} \in \mathbb{R}^{S\times N}$. The supervised SVM classifier is widely used for statistical classification and regression analysis due to its characteristics of geometrical margin maximization and empirical error minimization [62]. Because the HSIs are not linearly separable, the pixels are mapped from $\mathbf{x}^{SPE}$ to a kernel Hilbert space by using a mapping function $\phi(\mathbf{x}^{SPE})$ to construct the hyperplane. Then, the decision function can be defined as follows:

$$f(x) = \sum_{i=1}^{q} \alpha_i y_i K(x, x_i) + b \quad (24)$$

where $\alpha = [\alpha_1, \alpha_2, \ldots, \alpha_q]$ is a set of coefficients associated with z_q, b is the bias of the decision function f, and $K(x_i, x_j) = f(x_i)' f(x_j)$. For HSI classification using SVM, the Gaussian RBF kernel is the most widely employed as a spectral kernel, measuring the similarity between two pixels. The typical spectral kernel can be defined, as follows:

$$K^{SPE}\left(x_i^{SPE}, x_j^{SPE}\right) = \exp\left(-\gamma \|x_i^{SPE} - x_j^{SPE}\|^2\right) \quad (25)$$

where σ is the width of the RBF kernel. Similarly, the spatial kernel can be constructed using the RBF kernel. Specifically, for two vectors x_i^{SPA} and x_j^{SPA}, the spatial kernel is defined as follows:

$$K^{SPA}\left(x_i^{SPA}, x_j^{SPA}\right) = \exp\left(-\gamma \|x_i^{SPA} - x_j^{SPA}\|^2\right) \quad (26)$$

As stated in [20,63], if k_1 and k_2 are two kernels, then $\mu_1 k_1 + \mu_2 k_2$ is a new kernel with $\mu_1, \mu_2 \geq 0$. According to this property, Camps-Valls et al. [20] formulated a SVM classifier with a spectral-spatial kernel for HSI classification, and this composite kernel is shown as follows:

$$K^{SPE-SPA}(x_i, x_j) = \mu K^{SPE}\left(x_i^{SPE}, x_j^{SPE}\right) + (1-\mu)K^{SPA}\left(x_i^{SPA}, x_j^{SPA}\right) \tag{27}$$

where μ is a weight to balance the spectral kernel and the spatial one. The authors in [20] performed the spatial feature extraction for each pixel by computing the mean and variance within a fixed-size window. The SVM classifier with the composite kernel (27) can effectively combine the spectral and spatial information and achieve better results than that using the spectral kernel individually. However, the spatial structure information may not be well represented for classification within such a predefined region.

3.2. The SVM-SSHK Method

In this work, we propose an effective SVM classifier that is characterized with three kernels, which are computed on the pixels from the original, feature and hierarchical spaces to extract the spectral, spatial and hierarchical structure features, respectively. In the proposed framework, the spectral features are extracted directly through each pixel's vector value in the original HSI, and the spatial feature extraction in the proposed framework is performed using the EMP method due to its simplicity and effectiveness. As addressed in the previous question, the spectral information in HSIs can be represented by the limited PCs. It means that the spatial information of the HSI can be projected into a lower dimensional space after the PCA transform. To construct the EMP of the HSI, we can first define the MP for each PC instead of each spectral band, and then stacked the MPs of all the PCs to produce a final EMP. Specifically, the PCA transform is first applied to the original HSI for feature extraction. Then, the first three PCs are used as a feature image to obtain the EMP, where each pixel is a stacked vector, according to Equation (2).

To remedy the shortcomings of the spatial feature extraction, the hierarchical structure information can be used to as a supplement to the spatial features. Based on our previous study [32], the hierarchical structure information is helpful to improve HSI classification accuracies. As proposed in [32], the AMG method is very effective to model the spatial structure information because the multigrid structure can be used as the hierarchical representation of HSIs. To construct the hierarchical kernel, the FMS algorithm is applied to the original HSI for feature selection to obtain its spectral subset. Then, the multigrid representation of this subset is built using the AMG-based method. Next, the AMG-MHSEG algorithm is performed on each grid to obtain the corresponding segmentation map. Finally, these maps are combined to produce a stacked vector for each pixel and its value is featured with the cluster labels in different grids. The proposed hierarchical kernel is introduced, as follows:

$$K^{HIE}\left(x_i^{HIE}, x_j^{HIE}\right) = \exp\left(-\gamma \|x_i^{HIE} - x_j^{HIE}\|^2\right) \tag{28}$$

To exploit the spectral, spatial, and hierarchical structure information for HSI classification, composite kernels are considered for combining information. In this work, we present a weighted summation kernel, as follows:

$$\begin{aligned} K^{SPE-SPA-HIE}(x_i, x_j) = {} & \mu^{SPE} K^{SPE}\left(x_i^{SPE}, x_j^{SPE}\right) + \mu^{SPA} K^{SPA}\left(x_i^{SPA}, x_j^{SPA}\right) \\ & + \mu^{HIE} K^{HIE}\left(x_i^{HIE}, x_j^{HIE}\right) \end{aligned} \tag{29}$$

where μ^{SPE}, μ^{SPA} and μ^{HIE} are weights to indicate the contribution of each feature information involved in HSI classification under the condition of $\mu^{SPE} + \mu^{SPA} + \mu^{HIE} = 1$. For clarity, the SVM-SSHK method is introduced in Algorithm 2.

Algorithm 2: SVM-SSHK

Input: An original hyperspectral image **u**, the available training samples, required number of segmentation maps S, the time step size τ, Gaussian scale σ, the gradient threshold β, the critical threshold υ and the number of morphological operators n.

Step 1: Initialize S, τ, σ, υ and n.
Step 2: Obtain the first three PCs of **u**;
Step 3: Construct the EMP by computing the MPs for all the PCs in **Step 2** as described in **Section** 2.1.
Step 4: Perform the FMS algorithm on **u** for feature selection to produce its spectral subset **u1** with the most relevant spectral bands as described in **Section** 2.2.
Step 5: For $i = 1, 2, \ldots, S$
(a) Construct the ith grid of **u1** using the procedures described in **Section** 2.3.
(b) Select all the vertices in the ith grid as makers for the HSEG algorithm and initialize each vertex with a non-zero marker label.
(c) Obtain the ith segmentation map by using the MHSEG algorithm described in **Algorithm 1**.
End
Step 6: Normalize **u**, the EMP and the S-scale HSEG maps to [0,1].
Step 7: Construct the spectral, spatial and hierarchical kernels as described in **Section** 3.2.
Step 8: Apply the SVM classifier with the proposed SSHK kernel in (29) to classify **u** using the training samples by choosing the optimal C and γ.
Step 9: Obtain the final classification map.

4. Experiments

4.1. Image Description

The effectiveness of the SVM-SSHK method was validated using two hyperspectral remote sensing images of the AVIRIS Indian Pines (IP) and the ROSIS-03 University of Pavia (UP). The 145×145 IP image was obtained over northwestern Indiana, USA, and its ground truth data (GTD) includes 16 agricultural objects, and the 610×340 UP image was acquired over an urban area in Pavia, Italy, and its GTD has been produced with nine classes available. In this work, two spectral subsets of the IP and UP images with 185 bands and 103 bands, respectively, are used in our experiments, because those discarded bands locate in the absorption spectrum of water or are too noisy. Figure 3 shows the RGB color composite of the two images and GTD. Note that the class of *background* in the two HSIs was removed from further consideration in the following experiments.

4.2. Experimental Settings

To evaluate the performance of the SVM-SSHK method, seven state-of-the-art kernel-based classification methods were selected for comparison, including SVM, EMP [28], EPF [27], SVM using a composite kernel (SVM-CK) [20], MLR-GCK [45], the two superpixel-based classifiers using spectral-spatial kernel (SC-SSK) [22], and multiple kernels (SC-MK) [46]. The overall accuracy (OA), average accuracy (AA), and kappa coefficient (κ) were used for quantitative evaluation. Before demonstrating the experimental results, a brief description on the parameter settings and related issues are provided. To fix the optimal parameter settings for each method, we tuned these parameters in a certain range based on the original references to obtain the best classification performance, which can be comparative to the classification results from these original references for the IP and UP images with the same number of training samples. The parameter settings for each method are provided as follows:

(1) The SVM algorithm with the RBF kernel was exploited by all of the methods, except for MLR-GCK, and the optimal C and γ for each method were obtained by five-fold cross validation ranging from 2^{-5} to 2^{15} and 2^{-15} to 2^{5}, respectively.

(2) For EMP, the first three PCs were used for building the MPs, which were computed using a flat disk-shaped SE with radius from 1 to 15 with the step size of 2.
(3) For EPF, the first PC was used as a guidance image, a local 5 × 5 window was used for the joint bilateral filter, two Gaussian scales were fixed as $\delta_s = 3$ and $\delta_r = 0.2$.
(4) For SVM-CK, the weight was fixed as $\mu = 0.4$, and a local 5 × 5 window was used for each pixel to compute the mean and variance.
(5) For MLR-GCK, the spectral and spatial variances were fixed as $\sigma_{SPE} = 1.5$ and $\sigma_{SPA} = 2$, respectively, and $\lambda = 10^{-5}$.
(6) For SC-SSK, the two parameters were fixed as $\mu = 0.8$ and $\sigma = 0.8$. The number of superpixels was fixed as 200 and 3500 for the IP and UP images, respectively.
(7) For SC-MK, the three weights were fixed as $\mu_{Spec} = 0.2$, $\mu_{IntraS} = 0.4$ and $\mu_{\mathrm{InterS}} = 0.4$, respectively, and the number of superpixels was fixed as 200.

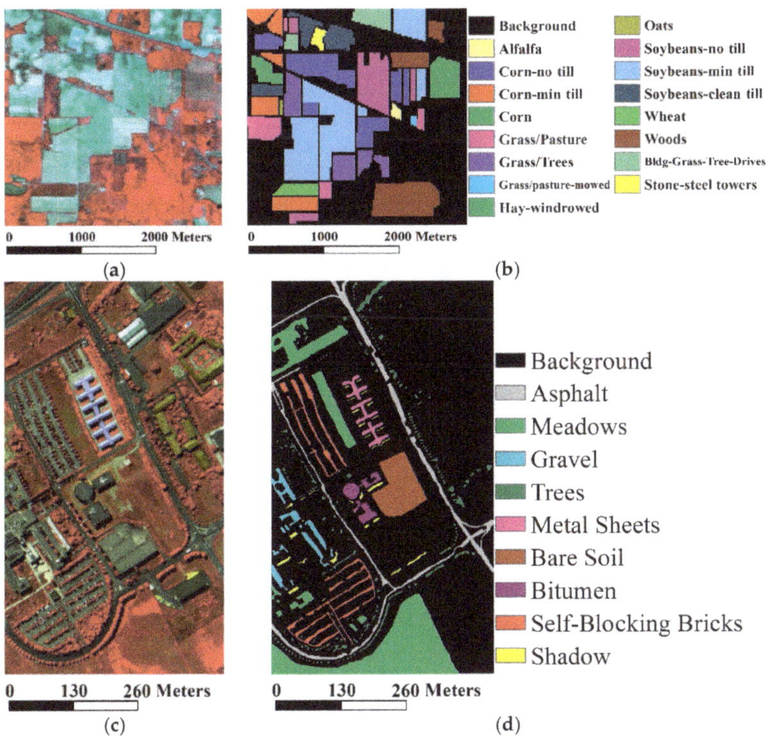

Figure 3. Hyperspectral images and the corresponding ground truth data (GTD). (**a**) A false color composite image (bands 47, 23, and 13) of the Indian Pines (IP) image and (**b**) its GTD; (**c**) a false color composite image (bands 103, 56, and 31) of the University of Pavia (UP) image and (**d**) its GTD.

In our experiments, we randomly divided the GTD for training and test and followed the scheme in [46] by setting training samples M ranging from 15 to 40 with a step size of 5 for each class and the rest for test. For some minority classes in the IP image, the labeled samples were divided into the equal training and test samples when the total of the labeled samples is less than M. Table 1 demonstrates that the percentage of the total samples (pixels) that were used for training and test for the two HSIs under different values of M. The classification experiments using each training set were repeated 10 times for reliable evaluation of the results.

Table 1. The percentage of the total pixels used as training and test for the IP and UP images under different values of M.

Class	M											
	15		20		25		30		35		40	
	Training	Test	Training	Test	Training	Test	Training	Test	Training	Test	Training	Test
The IP image												
Alfalfa	32.61%	67.39%	43.48%	56.52%	54.35%	45.65%	65.22%	34.78%	76.09%	23.91%	86.96%	13.04%
Corn-no till	1.05%	98.95%	1.40%	98.60%	1.75%	98.25%	2.10%	97.90%	2.45%	97.55%	2.80%	97.20%
Corn-min till	1.81%	98.19%	2.41%	97.59%	3.01%	96.99	3.62%	96.38%	4.22%	95.78%	4.82%	95.18%
Corn	6.33%	93.67%	8.44%	91.56%	10.55%	89.45%	12.66%	87.34%	14.77%	85.23%	16.88%	83.12%
Grass-pasture	3.11%	96.89%	4.14%	95.86%	5.18%	94.82%	6.21%	93.79%	7.25%	92.75%	8.28%	91.72%
Grass-trees	2.05%	97.95%	2.74%	97.26%	3.42%	96.58%	4.11%	95.89%	4.79%	95.21%	5.48%	94.52%
Grass-pasture-mowed	53.57%	46.43%	71.43%	28.57%	89.29%	10.71%	50.00%	50.00%	50.00%	50.00%	50.00%	50.00%
Hay-windrowed	3.14%	96.86%	4.18%	95.82%	5.23%	94.77%	6.28%	93.72%	7.32%	92.68%	8.37%	91.63%
Oats	75%	25%	50.00%	50.00%	50.00%	50.00%	50.00%	50.00%	50.00%	50.00%	50.00%	50.00%
Soybean-no till	1.54%	98.46%	2.06%	97.94%	2.57%	97.43%	3.09%	96.91%	3.60%	96.4%	4.12%	95.88%
Soybean-min till	0.61%	99.39%	0.81%	99.19%	1.02%	98.98%	1.22%	98.78%	1.43%	98.57%	1.63%	98.37%
Soybean-clean	2.53%	97.47%	3.37%	96.63%	4.22%	95.78%	5.06%	94.94%	5.90%	94.1%	6.75%	93.25%
Wheat	7.32%	92.68%	9.76%	90.24%	12.20%	87.80%	14.63%	85.37%	17.07%	82.93%	19.51%	80.49%
Woods	1.19%	98.81%	1.58%	98.42%	1.98%	98.02%	2.37%	97.63%	2.77%	97.23%	3.16%	96.84%
Buildings-Grass-Trees-Drives	3.89%	96.11%	5.18%	94.82%	6.48%	93.52%	7.77%	92.23%	9.07%	90.93%	10.36%	89.64%
Stone-Steel-Towers	16.13%	83.87%	21.51%	78.49%	26.88%	73.12%	3.23%	96.77%	37.63%	62.37%	43.01%	56.99%
The UP Image												
Asphalt	0.23%	99.77%	0.30%	99.7%	0.38%	99.62%	0.45%	99.55%	0.53%	99.47%	0.60%	99.4%
Meadows	0.08%	99.92%	0.11%	99.89%	0.13%	99.87%	0.16%	99.84%	0.19%	99.81%	0.21%	99.79%
Gravel	0.71%	99.29%	0.95%	99.05%	0.12%	99.88%	1.43%	98.57%	1.67%	98.33%	1.91%	98.09%
Trees	0.49%	99.51%	0.65%	99.35%	0.82%	99.18%	0.98%	99.02%	1.14%	98.86%	1.31%	98.69%
Metal Sheets	1.12%	98.88%	1.49%	98.51%	1.86%	98.14%	2.23%	97.77%	2.60%	97.4%	2.97%	97.03%
Bare Soil	0.30%	99.7%	0.40%	99.6%	0.50%	99.5%	0.60%	99.4%	0.70%	99.3%	0.80%	99.2%
Bitumen	1.13%	98.87%	1.50%	98.5%	1.88%	98.12%	2.26%	97.74%	2.63%	97.37%	3.01%	96.99%
Self-Blocking Bricks	0.41%	99.59%	0.54%	99.46%	0.68%	99.32%	0.81%	99.19%	0.95%	99.05%	1.09%	98.91%
Shadow	1.58%	98.42%	2.11%	97.89%	2.64%	97.36%	3.17%	96.83%	3.70%	96.3%	4.22%	95.78%

4.3. Experimental results

4.3.1. The IP Image

In the first experiment, we reported the classification results in the case of $M = 40$ in Table 2 to show the contribution of each kernel in the proposed method with $\mu^{SPE} = 0.3$, $\mu^{SPA} = 0.1$ and $\mu^{HIE} = 0.6$ in $K^{SPE-SPA-HIE}$, $\tau = 1$, $\sigma = 0.1$, $\upsilon = 0.3$, $\beta = 0.01$ and $S = 11$ were used by the AMG-MHSEG algorithm, and the PC 1-3 and $n = 8$ were used for the constructions of the EMP. For the IP image, the most relevant 30 spectral bands were selected by the FMS algorithm. Table 1 shows that the hierarchical structure information can further increase discriminative capability of the SVM classifier. Specifically, SVM with $K^{SPE-HIE}$ can increases the OA, AA, and κ by 10.31%~15.77%, 6.21%~9.81%, and 11.7%~17.82%, respectively, when compared to SVM with K^{SPE}. Furthermore, SVM with $K^{SPE-SPA-HIE}$ can improve the OA, AA, and κ over the others in this table by 0.61%~13.65%, 0.25%~8.26%, and 0.69%~15.45% in average, respectively. The improvement of $K^{SPE-SPA-HIE}$ over the other kernels in Table 1 demonstrates that the combination of the spectral, spatial, and hierarchical kernels can generate better classification results than using a single or double kernels in terms of OA, AA, and κ. Finally, the SVM classifier with $K^{SPE-SPA-HIE}$ can achieve the highest CAs for 12 of 16 classes above 90%.

Table 2. Classification Results [Mean Accuracy (%) ± Standard Deviation] by the SVM Classifier with the Spectral, Spatial and Hierarchical Kernels for the IP Image. The best accuracies are indicated in bold in each raw.

Class	Kernels Used in the SVM Classifier					
	K^{SPE}	K^{SPA}	K^{HIE}	$K^{SPE-SPA}$	$K^{SPE-HIE}$	$K^{SPE-SPA-HIE}$
Alfalfa	**98.33 ± 5.00**	**98.33 ± 5.00**	**98.33 ± 5.00**	**98.33 ± 5.00**	**98.33 ± 5.00**	**98.33 ± 5.00**
Corn-no till	76.40 ± 3.96	82.50 ± 3.26	85.44 ± 5.02	87.07 ± 1.84	**88.90 ± 4.59**	87.37 ± 5.48
Corn-min till	76.01 ± 3.37	90.82 ± 3.01	94.07 ± 1.27	91.56 ± 2.14	**94.89 ± 1.43**	94.02 ± 1.86
Corn	89.55 ± 3.98	92.60 ± 3.74	93.06 ± 4.68	93.06 ± 3.63	95.00 ± 3.29	**95.25 ± 3.23**
Grass-pasture	92.32 ± 2.85	91.98 ± 2.49	90.65 ± 2.85	93.07 ± 2.62	93.79 ± 3.04	**94.02 ± 2.70**
Grass-trees	94.42 ± 1.28	97.50 ± 1.84	91.69 ± 3.91	97.84 ± 1.46	96.31 ± 2.19	**98.04 ± 1.48**
Grass-pasture-mowed	95.71 ± 4.74	95.00 ± 4.57	95.00 ± 3.27	98.57 ± 2.86	**99.29 ± 2.14**	**99.29 ± 2.14**
Hay-windrowed	97.66 ± 0.67	99.63 ± 0.15	98.30 ± 2.08	99.66 ± 0.15	99.52 ± 0.86	**99.79 ± 0.22**
Oats	99.00 ± 3.00	97.89 ± 4.23	97.78 ± 6.67	99.00 ± 3.00	**100 ± 0**	**100 ± 0**
Soybean-no till	79.00 ± 7.18	83.89 ± 3.82	92.53 ± 4.57	86.39 ± 4.58	93.62 ± 3.66	**93.91 ± 3.87**
Soybean-min till	66.59 ± 5.06	85.34 ± 5.54	88.01 ± 4.73	84.30 ± 4.82	88.16 ± 3.14	**90.55 ± 3.84**
Soybean-clean	85.04 ± 5.23	85.23 ± 4.82	95.43 ± 2.43	90.91 ± 4.07	95.54 ± 1.80	**95.57 ± 1.80**
Wheat	**99.15 ± 0.49**	98.78 ± 0.72	95.12 ± 3.01	98.90 ± 0.65	99.09 ± 0.49	98.96 ± 0.55
Woods	90.38 ± 2.90	98.42 ± 2.27	92.41 ± 3.68	98.11 ± 2.00	98.03 ± 2.05	**98.89 ± 0.76**
Buildings-Grass-Trees-Drives	71.73 ± 4.09	97.80 ± 1.76	94.35 ± 2.78	97.71 ± 1.40	97.39 ± 1.43	**97.94 ± 1.67**
Stone-Steel-Towers	96.77 ± 1.48	98.48 ± 0.76	97.34 ± 1.75	**98.67 ± 0.87**	98.48 ± 0.76	98.29 ± 1.02
OA	80.08 ± 1.41	89.64 ± 1.60	90.93 ± 1.86	90.72 ± 1.53	93.12 ± 1.32	**93.73 ± 1.36**
AA	88.01 ± 0.94	93.39 ± 0.93	93.72 ± 1.35	94.57 ± 0.70	96.02 ± 0.86	**96.27 ± 0.74**
κ	77.38 ± 1.56	88.16 ± 1.80	89.65 ± 2.10	89.40 ± 1.72	92.14 ± 1.50	**92.83 ± 1.54**

In the second experiment, we applied each classification method to the IP image under different training sets. Table 3 lists the classification results and the last row of this table records the average rank for each method. All of the accuracies of the same row in this table are ranked in descending order and average rank is defined as the mean of the rankings for the same column. We can observe from Table 3 that using composite or multiple kernels in the SVM classifier can well combine the spectral and spatial information and provide higher results in all of the cases than the single feature-stacked kernel methods, including SVM and EMP, except for EPF, which can obtain a lower average rank of 4.94 than that of SVM-CK. The average rank values of SVM-CK and MLR-GCK are 5.72 and 4, respectively, and the superpixel-based methods of SC-SSK and SC-MK are better than these two methods and achieve similar performances with 2.5 and 2.67, respectively, in terms of the average rank. SVM-SSHK can outperform the other methods in terms of OA, AA, and κ in the case of different training samples and its average rank reaches 1.33.

Table 3. Classification Results [Mean Accuracy (%) ± Standard Deviation] by Different Methods using Different Number of Training Samples for the IP Image. The best accuracies are indicated in bold in each raw.

M		Methods							
		SVM	**EMP**	**EPF**	**SVM-CK**	**MLR-GCK**	**SC-SSK**	**SC-MK**	**SVM-SSHK**
15	OA	70.85 ± 1.72 (8)	74.64 ± 3.03 (7)	82.40 ± 2.46 (3)	76.11 ± 2.42 (6)	82.17 ± 1.41 (4)	83.29 ± 2.08 (2)	81.83 ± 2.39 (5)	83.31 ± 1.79 (1)
	AA	81.41 ± 1.32 (8)	85.18 ± 1.35 (7)	90.89 ± 0.91 (1)	85.25 ± 1.42 (6)	89.13 ± 0.73 (5)	89.76 ± 1.39 (4)	90.68 ± 1.16 (2)	90.30 ± 1.31 (3)
	κ	67.22 ± 1.90 (8)	71.42 ± 3.25 (7)	80.13 ± 2.70 (3)	73.05 ± 2.65 (6)	79.86 ± 1.53 (4)	81.08 ± 2.33 (2)	79.50 ± 2.66 (5)	81.14 ± 1.97 (1)
20	OA	73.64 ± 1.11 (8)	81.36 ± 1.86 (7)	85.45 ± 1.37 (5)	81.57 ± 1.53 (6)	85.70 ± 0.75 (4)	87.17 ± 2.25 (2)	86.90 ± 1.55 (3)	87.69 ± 1.39 (1)
	AA	83.31 ± 1.61 (8)	88.78 ± 1.25 (7)	91.86 ± 2.25 (4)	89.27 ± 1.31 (6)	91.12 ± 0.74 (5)	91.97 ± 1.60 (3)	93.32 ± 0.88 (1)	92.69 ± 1.21 (2)
	κ	70.33 ± 1.23 (8)	78.86 ± 2.07 (7)	83.53 ± 1.54 (5)	79.14 ± 1.73 (6)	83.76 ± 0.84 (4)	85.44 ± 2.51 (2)	85.15 ± 1.74 (3)	86.01 ± 1.55 (1)
25	OA	76.25 ± 1.41 (8)	84.51 ± 1.29 (7)	87.57 ± 2.49 (5)	85.68 ± 1.87 (6)	88.53 ± 1.01 (3)	89.46 ± 1.19 (2)	88.39 ± 1.44 (4)	89.87 ± 1.18 (1)
	AA	84.66 ± 1.54 (8)	90.72 ± 1.34 (7)	92.74 ± 2.01 (5)	91.63 ± 0.98 (6)	92.79 ± 0.76 (4)	92.89 ± 1.26 (3)	94.26 ± 0.86 (1)	93.87 ± 0.62 (2)
	κ	73.18 ± 1.54 (8)	82.39 ± 1.44 (7)	85.89 ± 2.79 (5)	83.76 ± 2.08 (6)	86.94 ± 1.13 (3)	88.00 ± 1.33 (2)	86.83 ± 1.62 (4)	88.48 ± 1.33 (1)
30	OA	77.04 ± 1.13 (8)	87.24 ± 1.32 (6)	86.92 ± 2.05 (7)	87.89 ± 1.25 (5)	89.40 ± 0.80 (4)	91.70 ± 1.39 (2)	90.41 ± 1.44 (3)	92.28 ± 0.83 (1)
	AA	85.74 ± 0.89 (8)	91.88 ± 0.98 (6)	90.65 ± 1.80 (7)	93.01 ± 0.52 (5)	93.25 ± 0.44 (4)	94.21 ± 0.89 (3)	94.94 ± 1.06 (1)	94.90 ± 0.80 (2)
	κ	74.05 ± 1.26 (8)	85.45 ± 1.47 (6)	85.11 ± 2.32 (7)	86.22 ± 1.40 (5)	87.92 ± 0.91 (4)	90.52 ± 1.57 (2)	89.09 ± 1.62 (3)	91.18 ± 0.93 (1)
35	OA	79.08 ± 1.29 (8)	87.87 ± 1.64 (7)	90.22 ± 1.58 (5)	89.11 ± 1.52 (6)	90.55 ± 0.67 (4)	92.63 ± 0.82 (3)	92.75 ± 1.21 (2)	92.83 ± 1.31 (1)
	AA	86.76 ± 0.91 (8)	92.51 ± 0.53 (7)	92.71 ± 1.31 (6)	94.00 ± 0.55 (5)	94.16 ± 0.54 (4)	94.84 ± 0.73 (3)	96.07 ± 0.79 (1)	95.84 ± 0.76 (2)
	κ	76.28 ± 1.41 (8)	86.16 ± 1.84 (7)	88.83 ± 1.78 (5)	87.60 ± 1.70 (6)	89.20 ± 0.75 (4)	91.58 ± 0.94 (3)	91.72 ± 1.36 (2)	91.82 ± 1.49 (1)
40	OA	79.49 ± 1.47 (8)	89.32 ± 1.27 (7)	90.50 ± 2.09 (5)	89.46 ± 1.29 (6)	91.51 ± 0.76 (4)	93.21 ± 0.95 (2)	92.77 ± 1.63 (3)	93.73 ± 1.36 (1)
	AA	87.53 ± 0.95 (8)	93.38 ± 0.47 (7)	93.80 ± 0.97 (6)	94.29 ± 0.53 (5)	94.96 ± 0.51 (4)	95.48 ± 0.69 (3)	96.24 ± 0.65 (2)	96.27 ± 0.74 (1)
	κ	76.72 ± 1.63 (8)	87.78 ± 1.44 (7)	89.15 ± 2.36 (5)	88.00 ± 1.46 (6)	90.29 ± 0.86 (4)	92.22 ± 1.08 (2)	91.74 ± 1.84 (3)	92.83 ± 1.54 (1)
Average Rank		8	6.83	4.94	5.72	4	2.5	2.67	1.33

Figure 4 illustrates some classification maps by the different methods with 40 training samples per class, corresponding to Table 3 with $M = 40$. The noise in the SVM classification maps in Figure 4a was obviously visible and can be greatly removed by the other kernel methods, which validated that the spatial information is significant for improving the classification results. However, the noise effect was still observed in two classes of *Soybeans-no till* and *Soybeans-min till* in the EMP and MLR-GCK results. The classification maps can be improved by removing the noise in the two previously mentioned classes by SVM-CK and SC-SSK. Nevertheless, the edges of the image were corrupted with the noise by EPF and SVM-CK due to using a fixed-size window for feature extraction. The adaptive neighborhood system of SC-SSK can solve the problem of SVM-CK, but cannot completely remove the noise effect. The SC-MK and SVM-SSHK classification maps were comparable and much better than the others and less noise and classification errors were seen in the SVM-SSHK result by comparison.

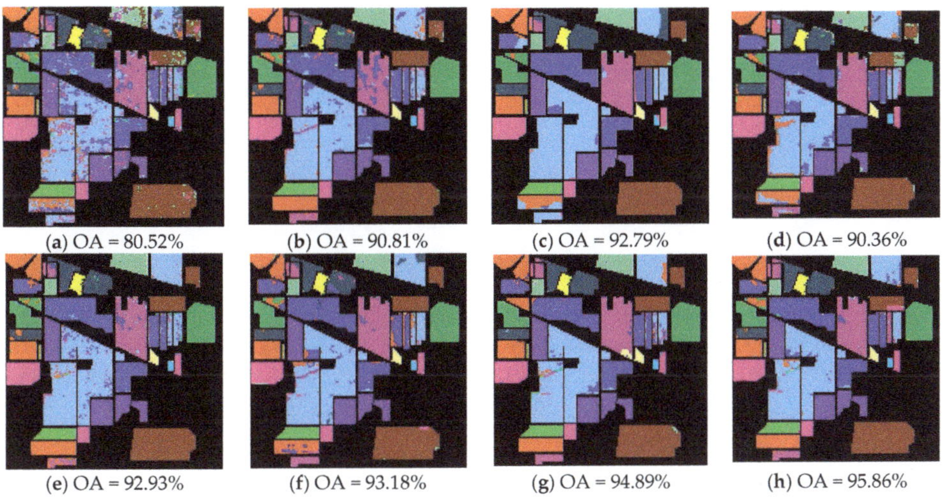

Figure 4. Classification results of the IP image. (**a**) SVM; (**b**) EMP; (**c**) EPF; (**d**) SVM-CK; (**e**) MLR-GCK; (**f**) SC-SSK; (**g**) SC-MK; and, (**h**) SVM-SSHK.

4.3.2. The UP Image

Similarly, the classification results in the case of $M = 40$ are recorded in Table 4 to evaluate the contribution of each kernel in the SVM-SSHK method, $\mu^{SPE} = 0.2$, $\mu^{SPA} = 0.6$, and $\mu^{HIE} = 0.2$ in $K^{SPE-SPA-HIE}$, $\tau = 1$, $\sigma = 0.1$, $v = 0.2$, $\beta = 0.01$, and $S = 13$ were used by the AMG-MHSEG algorithm, and the PC 1-3 and $n = 8$ were used for the constructions of the EMP. For the UP image, the most relevant 30 spectral bands were selected by the FMS algorithm. It can be observed from Table 4 that SVM with $K^{SPE-HIE}$ can increases the OA, AA and κ by 6.51%~13.33%, 5.73%~9.13%, and 8.42%~16.66%, respectively, when compared to SVM with K^{SPE}. Furthermore, SVM with $K^{SPE-SPA-HIE}$ can improve the OA, AA, and κ over the others in this table by 3.16%~16.56%, 1.47%~11.52%, and 4.12%~21.08% in average, respectively. In addition, SVM with $K^{SPE-SPA-HIE}$ is capable of obtaining the highest CAs for all of the classes above 96% for the UP image, except for the class of *Self-Blocking Bricks*.

Table 4. Classification Results [Mean Accuracy (%) ± Standard Deviation] by the SVM Classifier with the Spectral, Spatial and Hierarchical Kernels for the UP Image. The best accuracies are indicated in bold in each raw.

Class	Kernels Used in the SVM Classifier					
	K^{SPE}	K^{SPA}	K^{HIE}	$K^{SPE-SPA}$	$K^{SPE-HIE}$	$K^{SPE-SPA-HIE}$
Asphalt	82.58 ± 3.97	98.42 ± 0.48	88.83 ± 5.70	98.48 ± 0.34	94.27 ± 2.25	**98.75 ± 0.67**
Meadows	81.00 ± 3.71	97.37 ± 2.66	71.67 ± 6.58	97.42 ± 1.97	93.73 ± 2.74	**98.14 ± 0.89**
Gravel	75.90 ± 2.48	95.79 ± 1.38	57.29 ± 5.77	95.95 ± 1.36	87.91 ± 3.03	**96.20 ± 1.18**
Trees	77.53 ± 4.02	90.60 ± 3.24	57.50 ± 3.75	91.38 ± 2.71	87.25 ± 2.45	**96.62 ± 1.16**
Metal Sheets	79.34 ± 4.40	91.44 ± 2.19	82.13 ± 5.29	92.07 ± 2.98	97.08 ± 1.36	**98.46 ± 0.82**
Bare Soil	99.75 ± 0.20	99.49 ± 0.42	76.55 ± 4.17	99.66 ± 0.34	99.67 ± 0.26	**99.89 ± 0.12**
Bitumen	99.56 ± 0.20	99.32 ± 0.74	93.90 ± 1.56	99.63 ± 0.20	99.07 ± 0.98	**99.66 ± 0.20**
Self-Blocking Bricks	94.10 ± 2.78	**98.50 ± 1.19**	52.91 ± 3.78	98.43 ± 1.11	92.92 ± 2.66	98.25 ± 0.96
Shadow	91.46 ± 2.11	98.69 ± 0.58	90.71 ± 4.69	98.64 ± 0.68	96.18 ± 1.99	**98.90 ± 0.49**
OA	80.79 ± 1.92	93.73 ± 1.47	65.31 ± 1.95	94.19 ± 1.21	90.71 ± 1.49	**97.35 ± 0.52**
AA	86.80 ± 0.86	96.62 ± 0.49	74.61 ± 1.50	96.85 ± 0.50	94.23 ± 0.84	**98.32 ± 0.31**
κ	75.42 ± 2.25	91.80 ± 1.88	57.28 ± 2.15	92.38 ± 1.55	87.96 ± 1.87	**96.50 ± 0.68**

Next, we applied each classification method to the UP image under different training sets and the classification result of each method is listed in Table 5. In this table, the average rank of SVM is lowest with 8, which is the same as in Table 3. EMP, EPF and SVM-CK performed HSI classification with similar average rank values of 5.38, 5.94, and 5.56, achieving the fifth, sixth, and seventh positions in this table, respectively. The remaining methods using composite or multiple kernels can obtain higher average rank values than the previously mentioned methods. For instance, the average rank values of SC-SSK, MLR-GCK, and SC-MK are 4.72, 3.28, and 2.11, respectively. The proposed SVM-SSHK method can achieve the best classification accuracies in all cases of training samples in terms of OA, AA, and κ. The improvement of the SSHK over the other composite or multiple kernels indicates that the introduction of the hierarchical structure information for classification can further improve discriminative capability of the kernel methods.

Table 5. Classification Results [Mean Accuracy (%) ± Standard Deviation] by Different Methods using Different Number of Training Samples for the UP Image. The best accuracies are indicated in bold in each raw.

M		Methods							
		SVM	EMP	EPF	SVM-CK	MLR-GCK	SC-SSK	SC-MK	SVM-SSHK
15	OA	75.66 ± 4.01 (8)	84.95 ± 3.20 (7)	85.59 ± 3.95 (6)	86.88 ± 3.00 (5)	87.93 ± 2.32 (4)	88.19 ± 1.69 (3)	88.84 ± 1.99 (2)	90.61 ± 2.50 (1)
	AA	81.15 ± 2.00 (8)	92.33 ± 1.36 (3)	89.78 ± 1.84 (6)	87.91 ± 1.70 (7)	92.58 ± 0.93 (2)	91.74 ± 1.13 (5)	92.20 ± 1.87 (4)	93.45 ± 1.66 (1)
	κ	68.94 ± 4.66 (8)	80.82 ± 3.88 (7)	81.41 ± 4.82 (6)	82.93 ± 3.77 (5)	84.46 ± 2.82 (4)	84.70 ± 2.10 (3)	85.54 ± 2.47 (2)	87.94 ± 3.23 (1)
20	OA	77.62 ± 3.78 (8)	87.62 ± 2.90 (6)	87.35 ± 4.11 (7)	88.89 ± 1.97 (5)	90.33 ± 3.03 (3)	90.23 ± 1.05 (4)	91.69 ± 2.29 (2)	92.26 ± 2.25 (1)
	AA	83.14 ± 0.94 (8)	94.30 ± 1.31 (3)	91.26 ± 1.50 (6)	89.82 ± 1.56 (7)	92.58 ± 0.93 (5)	93.66 ± 0.83 (4)	94.71 ± 1.35 (2)	95.80 ± 0.88 (1)
	κ	71.49 ± 4.20 (8)	84.15 ± 3.52 (6)	83.73 ± 4.89 (7)	85.47 ± 2.52 (4)	84.46 ± 2.82 (5)	87.30 ± 1.30 (3)	89.21 ± 2.89 (2)	89.97 ± 2.84 (1)
25	OA	79.42 ± 3.67 (8)	90.70 ± 1.99 (6)	89.06 ± 4.15 (7)	91.06 ± 1.87 (4)	93.03 ± 1.15 (3)	90.82 ± 1.41 (5)	94.02 ± 1.68 (2)	95.03 ± 1.46 (1)
	AA	84.66 ± 1.49 (8)	95.03 ± 1.24 (4)	92.92 ± 1.83 (6)	92.10 ± 1.01 (7)	96.01 ± 0.65 (3)	94.17 ± 0.88 (5)	96.16 ± 0.80 (2)	96.53 ± 0.89 (1)
	κ	73.72 ± 4.25 (8)	87.91 ± 2.54 (6)	85.91 ± 5.05 (7)	88.30 ± 2.39 (4)	90.90 ± 1.46 (3)	88.04 ± 1.80 (5)	92.15 ± 2.19 (2)	93.47 ± 1.88 (1)
30	OA	82.25 ± 1.61 (8)	90.93 ± 1.48 (7)	91.91 ± 2.51 (5)	91.62 ± 1.68 (6)	93.04 ± 1.23 (3)	92.18 ± 1.43 (4)	94.46 ± 1.33 (2)	95.98 ± 1.19 (1)
	AA	86.07 ± 1.06 (8)	95.28 ± 0.63 (4)	93.75 ± 1.51 (6)	92.05 ± 1.24 (7)	95.93 ± 0.94 (3)	94.48 ± 0.61 (5)	96.59 ± 0.58 (2)	97.25 ± 0.44 (1)
	κ	77.09 ± 2.00 (8)	88.21 ± 1.85 (7)	89.44 ± 3.23 (5)	89.00 ± 2.16 (6)	90.91 ± 1.58 (3)	89.76 ± 1.78 (4)	92.75 ± 1.70 (2)	94.66 ± 1.44 (1)
35	OA	82.38 ± 1.17 (8)	92.45 ± 1.28 (6)	91.93 ± 1.56 (7)	93.07 ± 0.86 (4)	94.70 ± 1.35 (3)	92.99 ± 0.67 (5)	95.94 ± 0.87 (2)	96.43 ± 1.00 (1)
	AA	86.84 ± 0.94 (8)	96.30 ± 0.66 (4)	94.42 ± 1.04 (6)	93.33 ± 0.41 (7)	96.97 ± 0.42 (3)	95.36 ± 0.73 (5)	97.39 ± 0.42 (2)	97.85 ± 0.40 (1)
	κ	77.32 ± 1.42 (8)	90.16 ± 1.63 (6)	89.49 ± 1.98 (7)	90.88 ± 1.08 (4)	93.05 ± 1.72 (3)	90.81 ± 0.87 (5)	94.66 ± 1.14 (2)	95.31 ± 1.28 (1)
40	OA	83.46 ± 1.43 (8)	93.55 ± 1.91 (6)	93.61 ± 2.06 (4)	93.55 ± 1.12 (5)	95.14 ± 0.90 (3)	93.05 ± 1.07 (7)	96.27 ± 1.11 (2)	97.35 ± 0.52 (1)
	AA	87.26 ± 1.00 (8)	96.46 ± 1.02 (4)	95.52 ± 0.72 (5)	93.51 ± 0.71 (7)	97.12 ± 0.32 (3)	95.30 ± 0.38 (6)	97.68 ± 0.81 (2)	98.32 ± 0.31 (1)
	κ	78.61 ± 1.73 (8)	91.56 ± 2.47 (5)	91.64 ± 2.61 (4)	91.49 ± 1.44 (6)	93.60 ± 1.16 (3)	90.89 ± 1.34 (7)	95.09 ± 1.44 (2)	96.50 ± 0.68 (1)
Average Rank		8	5.38	5.94	5.56	3.28	4.72	2.11	1

Figure 5 shows the classification results corresponding to Table 5 with $M = 40$. From this figure, we can see that the SVM classification map was corrupted with much noise. Some pixels that belonging to Meadows are incorrectly assigned with a Bare Soil label in the EMP classification map. This problem can be partially resolved by SVM-CK and MLR-GCK to generate better classification results in Figure 5d,e, respectively. The EPF and SC-SSK classification maps became smoother, but several misclassified areas were produced in the middle and bottom of the image. SC-MK improved the SC-SSK classification map by greatly correcting such areas and caused classification errors in other parts of the image as well. For instance, two large areas of two classes of *Asphalt* and *Meadows* in the GTD were labelled to *Bare Soil* and *Self-Blocking Bricks* in the upper-left and right of the image, respectively. SVM-SSHK can better discriminate all of the objects, though very few pixels had false class labels.

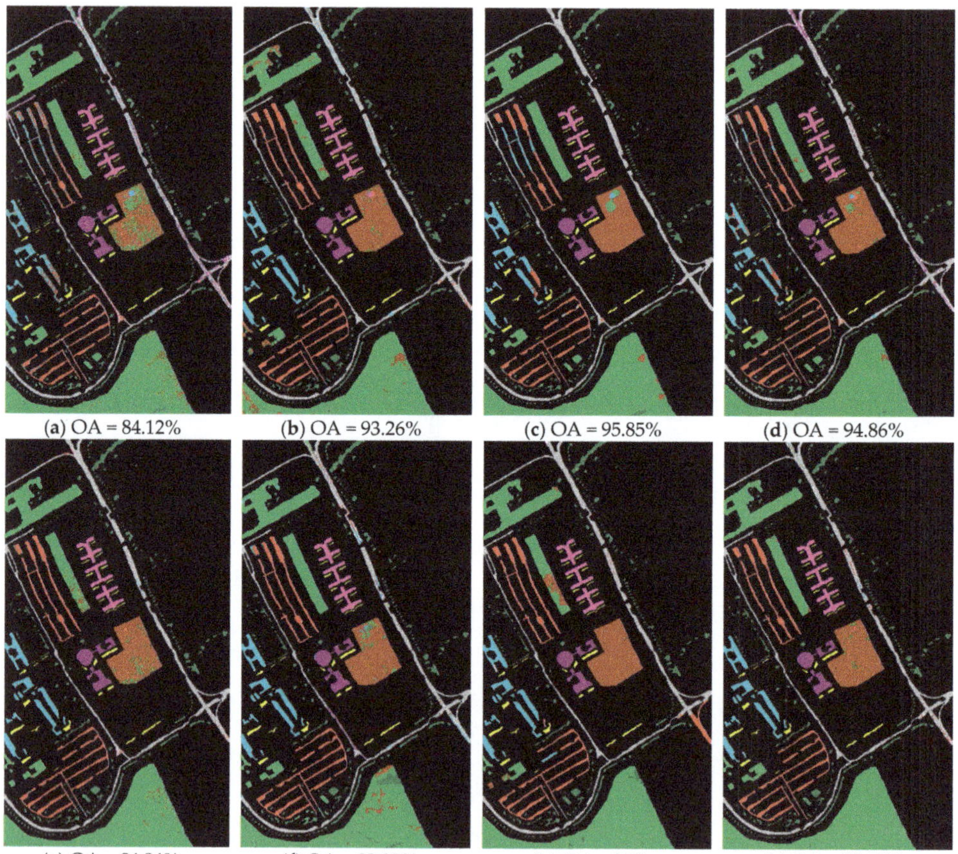

(**a**) OA = 84.12% (**b**) OA = 93.26% (**c**) OA = 95.85% (**d**) OA = 94.86%

(**e**) OA = 94.94% (**f**) OA = 92.33% (**g**) OA = 95% (**h**) OA = 98.10%

Figure 5. Classification results of the UP image. (**a**) SVM; (**b**) EMP; (**c**) EPF; (**d**) SVM-CK; (**e**) MLR-GCK; (**f**) SC-SSK; (**g**) SC-MK; and, (**h**) SVM-SSHK.

5. Discussion

As mentioned in Sections 2 and 3, some parameters should be fixed in the SVM-SSHK method. All of our experiments on HSIs, including those that are not mentioned here, confirmed that the number of the morphological operators n and the selected PCs play an important role for the construction of the EMP, and the critical threshold v greatly influences the hierarchical information extraction.

Furthermore, the weights in the spectral-spatial-hierarchical kernel make a significant impact on the classification performance of the proposed method. In this section, the impact of all the previously mentioned parameters is further analyzed to better understand the application of SVM-SSHK method for HSI classification.

5.1. Impact of n

To exploit the spatial kernel in the proposed framework, the number of the opening/closing operators (n) should be appropriately selected. In this subsection, the impact of n on the performance of the SVM-SSHK method is firstly analyzed. Experiments were performed on the IP and UP images in the case of $M = 40$ and the parameter settings were fixed the same to the previous experiments in Sections 4.3.1 and 4.3.2. Table 6 lists the classification accuracies by the proposed framework under different values of n. In this table, the highest classification accuracies for the IP image can be obtained when $n = 8$, and the OA, AA and κ for the UP image were very stable when $n \geq 8$ around 98.1%, 98.6%, and 97.5%, respectively, and the highest OA, AA, and κ were achieved when $n = 16$, 14, and 12, respectively. A large value of n means that more number of MPs should be computed for spatial information extraction. To ensure computational efficiency of the SVM-SSHK method, we fixed this parameter as $n = 8$ for both the IP and UP images.

Table 6. Classification accuracy (%) by the SVM-SSHK method under different values of n for the IP and UP images. The best accuracies are indicated in bold in each column.

n	Classification Accuracy					
	IP			UP		
	OA	AA	κ	OA	AA	κ
2	94.64	96.52	93.86	94.86	95.79	93.21
4	94.96	96.58	94.22	96.15	97.08	94.9
6	94.93	96.56	94.19	96.85	97.91	95.84
8	**95.86**	**97.12**	**95.25**	98.10	98.73	97.49
10	93.07	95.64	92.07	98.15	98.68	97.55
12	93.11	95.67	92.12	98.20	98.71	**97.65**
14	93.05	95.63	92.04	98.19	**98.75**	97.60
16				**98.21**	98.70	97.63
18				98.09	98.54	97.47
20				98.08	97.75	97.01

5.2. Impact of Different Number of PCs

To present further inspections with respect to the most appropriate number of PCs, three combinations were analyzed for spatial information extraction. Experiments were performed on the two HSIs in the case of $M = 40$ and the parameter settings were fixed the same to the previous experiments in Sections 4.3.1 and 4.3.2. Table 7 lists the classification accuracies by the proposed framework under different number of PCs. In this table, as the number of PCs is increased, which means that more spatial information can be exploited for constructing the EMP of the HSI, the improved classification accuracies can be obtained. For instance, the SVM-SSHK method using the first three PCs can increase the OA for the IP image by 0.84% and 0.88%, and for the UP image by 0.43% and 4.74%, than using PC 1 + PC 2 and PC 1, respectively. For conciseness and efficiency, the first three PCs were exploited for spatial information extraction.

Table 7. Classification accuracy (%) by the SVM-SSHK method under different number of PCs for the IP and UP images.

Option	Classification Accuracy					
	IP			UP		
	OA	AA	κ	OA	AA	κ
PC 1	94.98	96.53	94.25	93.36	96.28	91.35
PC 1 + PC 2	95.02	96.54	94.28	97.67	98.42	96.91
PC 1 + PC 2 + PC 3	95.86	97.12	95.25	98.1	98.73	97.49

5.3. Impact of v

To figure out the impact of v, experiments were performed on the IP and UP images in the case of $M = 40$ and the parameter settings were fixed the same to the previous experiments in Sections 4.3.1 and 4.3.2. Table 8 provides the classification accuracies by the proposed framework under different values of v. As v is increased from 0.05 to 0.1 for the two HSIs, the variation of the classification accuracies is very similar. Specifically, the highest OA, AA, and κ of 95.86%, 97.12%, and 95.25% for the IP image and 98.1%, 98.73%, and 97.49% for the UP image were achieved when $v = 0.3$ and $v = 0.2$, respectively. To ensure that the SVM-SSHK is capable of achieving the optimal results, the parameter settings were used in the previous experiments for comparison.

Table 8. Classification accuracy (%) by the SVM-SSHK method under different values of v for the IP and UP images. The best accuracies are indicated in bold in each column.

v	Classification Accuracy					
	IP			UP		
	OA	AA	κ	OA	AA	κ
0.05	95.19	96.58	94.49	96.98	97.20	96.00
0.1	94.58	96.2	93.79	97.23	98.26	96.34
0.2	93.49	95.87	92.53	**98.10**	**98.73**	**97.49**
0.3	**95.86**	**97.12**	**95.25**	96.08	97.32	94.80
0.4	94.82	96.13	94.05	91.91	95.52	89.43
0.5	90.88	94.81	89.57	95.26	97.40	93.76

5.4. Impact of Weights

In the SVM-SSK method, the weights in $K^{SPE-SPA-HIE}$ critically determine the classification performance, since their values indicate the contribution of spectral, spatial, and hierarchical structure information for classification. An appropriate combination of their values may obtain better results. To obtain the interaction effect of μ^{SPE}, μ^{SPA}, and μ^{HIE}, we can perform a four-dimensional (4-D) analysis to evaluate the influence of these three weights on our method's performance. Based on the constraint of $\mu^{SPE} + \mu^{SPA} + \mu^{HIE} = 1$, we converted this 4-D analysis to a problem of analyzing different combinations of μ^{SPE} and μ^{SPA} in terms of classification accuracies. Figure 6 illustrates the three-dimensional (3-D) plot of the classification accuracies with the change of μ^{SPE} and μ^{SPA} from 0 to 1 with a step size of 0.1. Several conclusions can be observed from this figure.

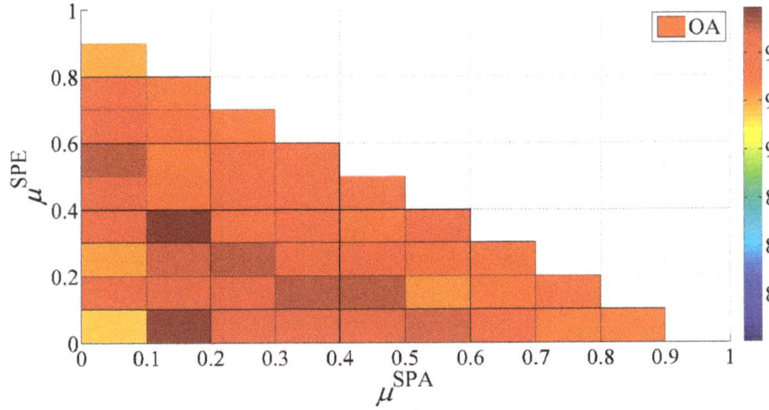

(a)

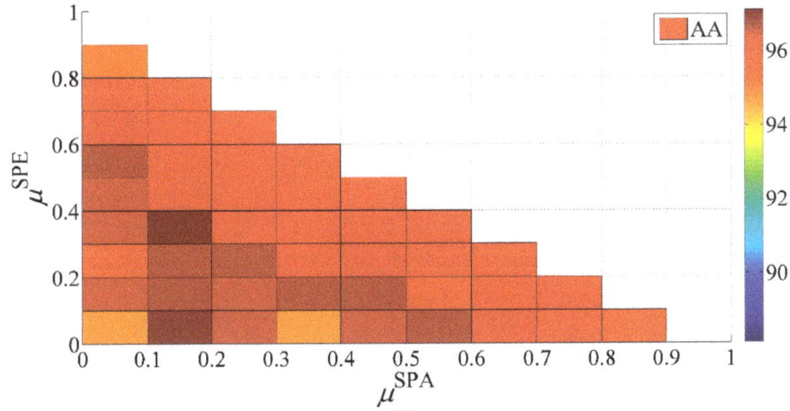

(b)

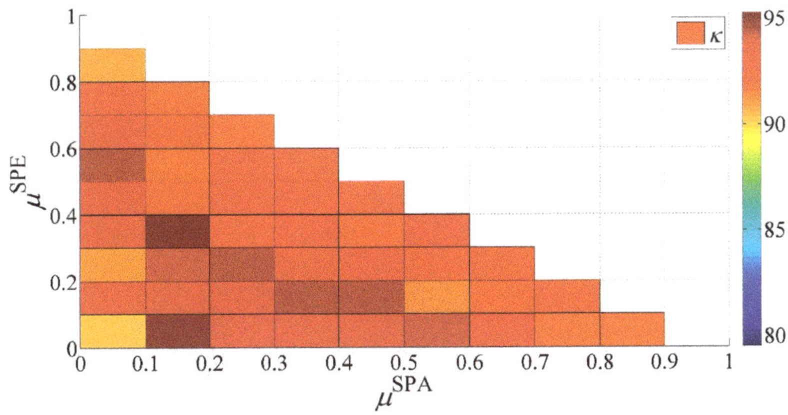

(c)

Figure 6. *Cont.*

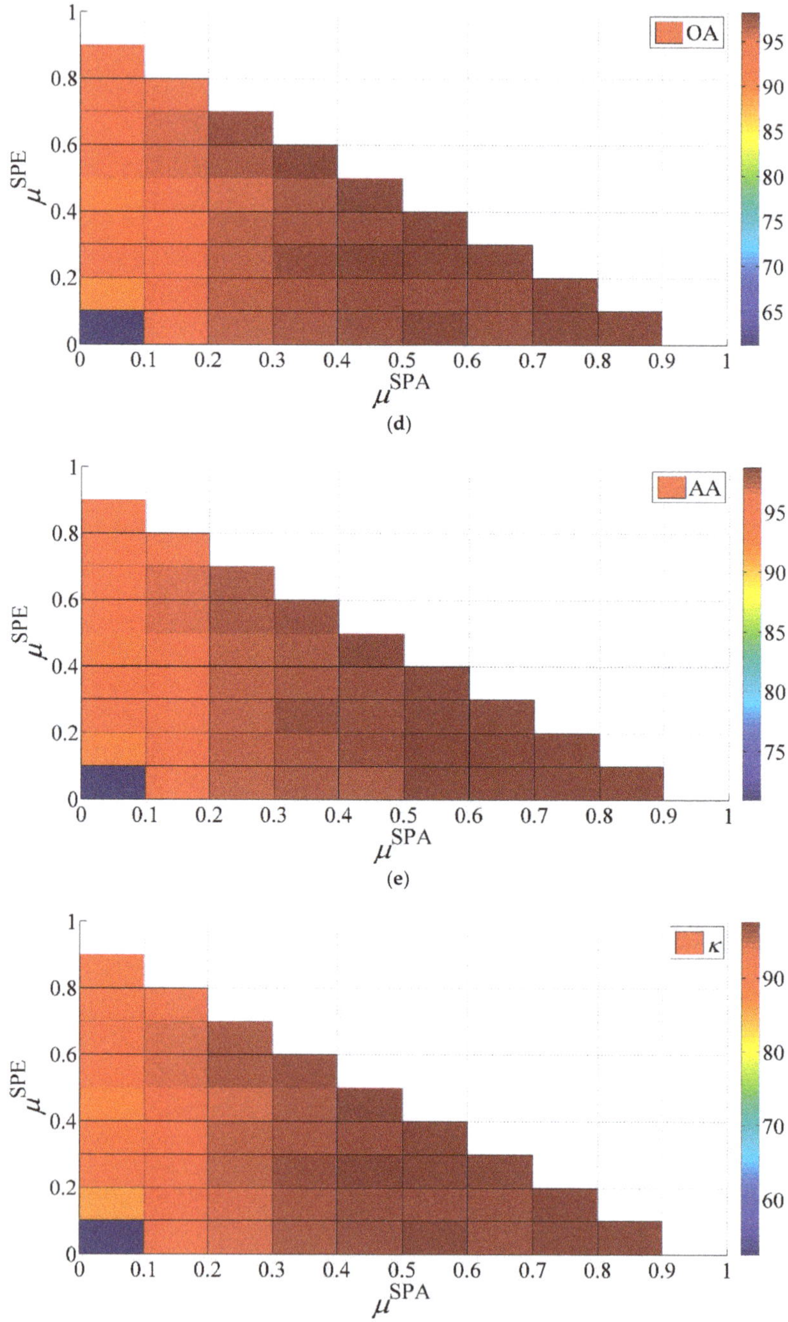

Figure 6. Impact of μ^{SPE} and μ^{SPA} using the two images on SVM-SSHK's performance. (**a**) overall accuracy (OA); (**b**) average accuracy (AA); and, (**c**) kappa coefficient (κ) for the IP image; (**d**) OA; (**e**) AA and (**f**) κ for the UP image.

First, for the IP image, if $\mu^{SPE} = 0$, it means that the proposed framework includes only two kernels of K^{SPA} and K^{HIE}. In such case, we can obtain the OA, AA and κ with 91.01%~95.59%, 93.53%~96.91%, and 89.69%~94.95%, respectively; if $\mu^{SPA} = 0$, it indicates that the proposed framework includes only two kernels of K^{SPE} and K^{HIE}. In such a case, we can obtain the OA, AA and κ with 82.06%~95.1%, 88.12%~96.65%, and 79.51%~94.38%, respectively. Specifically, the OA, AA and κ of 91.52%, 94.7%, and 90.31% were achieved when $\mu^{SPE} = \mu^{SPA} = 0$. For the UP image, the OA, AA, and κ can be obtained with 61.28%~98.05%, 70.91%~98.68%, and 52.53%~97.42% when $\mu^{SPE} = 0$, respectively, and 61.28%~92.86%, 70.91%~95.27%, and 52.53%~90.65% when $\mu^{SPA} = 0$, respectively. Specifically, the very poor OA, AA, and κ of 61.28%, 70.91%, and 52.53% were achieved when $\mu^{SPE} = \mu^{SPA} = 0$, respectively.

Second, the appropriate selection of μ^{SPE}, μ^{SPA}, and μ^{HIE} can result in the best classification accuracies. For instance, the highest OA, AA, and κ for the IP and UP images can reach to 95.86%, 97.12%, and 95.25% under the condition of $\mu^{SPE} = 0.3$, $\mu^{SPA} = 0.1$, and $\mu^{HIE} = 0.6$, and to 98.14%, 98.75%, and 97.53% under the condition of $\mu^{SPE} = 0.1$, $\mu^{SPA} = 0.7$, and $\mu^{HIE} = 0.2$, respectively. Compared to Tables 2 and 4, it can be confirmed again that the combination of the spectral, spatial and hierarchical kernels is really essential to produce better classification accuracies than using a single or double kernels in the SVM classifier.

Finally, the SVM-SSHK method can demonstrate very stable classification performance in most cases of different parameter settings on μ^{SPE} and μ^{SPA}. According to Figure 6, there are 66 combinations of the two weights in total. For the IP image, the SVM-SSHK method can obtain the OA, AA and κ higher than 92%, 94%, and 90% for 53 of 66 (80.30%) different parameters settings, respectively. For the UP image, the proposed method is capable of achieving the OA, AA, and κ higher than 95%, 95%, and 90% for 40 of 66 (60.60%) different parameters settings, respectively.

6. Conclusions

In this paper, we present an effective classification framework by integrating the spectral, spatial, and hierarchical structure information into the SVM classifier in a way of multiple kernels. In this framework, the spectral kernel is constructed using directly the original HSI, the spatial kernel is modeled using the EMP method and the hierarchical kernel is introduced by combining the techniques of FMS and AMG-MHSEG. The main advantage of the proposed framework is to utilize spatial structure information in multiple scales for HSI classification. Experimental results on two benchmark HSIs confirmed the following conclusions: (1) The combination of the spectral, spatial and hierarchical kernels in the SVM-SSHK method can generate better classification results than using any single or double of these three kernels; (2) The SVM-SSHK method can achieve the most accurate classification results under different training sets, when compared to the popular kernel-based classification methods. Specifically, SVM-SSHK can be 0.02–15.24% and 0.08–15.61% higher than the other methods in average in the terms of OA for the IP and UP images, respectively; (3) SVM-SSHK can demonstrate stable classification performance in most cases of different parameter settings on the weights of the three kernels. In conclusion, the SVM-SSHK method is very promising for the improvement of classification of hyperspectral images. In the future, advanced studies will be performed by exploring more efficient SVMs with multiple kernels.

Acknowledgments: This work was supported by the National Natural Science Foundation of China (61271408). The authors would like to thank D. Landgrebe from Purdue University for providing the AVIRIS image of Indian Pines and the Gamba from University of Pavia for providing the ROSIS data set.

Author Contributions: Y.W. and H.D. implemented all the proposed classification method and conducted the experiments. H.D. finished the first draft, Y.W. supervised the research and contributed to the editing and review of the manuscript.

Conflicts of Interest: The authors declare no conflict of interest.

References

1. Cloutis, E.A. Review article hyperspectral geological remote sensing: Evaluation of analytical techniques. *Int. J. Remote Sens.* **1996**, *17*, 2215–2242. [CrossRef]
2. Kruse, F.A.; Boardman, J.W.; Huntington, J.F. Comparison of airborne hyperspectral data and eo-1 hyperion for mineral mapping. *IEEE Trans. Geosci. Remote Sens.* **2003**, *41*, 1388–1400. [CrossRef]
3. Cochrane, M.A. Using vegetation reflectance variability for species level classification of hyperspectral data. *Int. J. Remote Sens.* **2000**, *21*, 2075–2087. [CrossRef]
4. Strachan, I.B.; Pattey, E.; Boisvert, J.B. Impact of nitrogen and environmental conditions on corn as detected by hyperspectral reflectance. *Remote Sens. Environ.* **2002**, *80*, 213–224. [CrossRef]
5. Dahan, M. Compressive fluorescence microscopy for biological and hyperspectral imaging. *Proceed. Natl. Acad. Sci. USA* **2012**, *109*, 1679–1687.
6. Hughes, G. On the mean accuracy of statistical pattern recognizers. *Inf. Theory IEEE Trans.* **1968**, *14*, 55–63. [CrossRef]
7. Fauvel, M.; Tarabalka, Y.; Benediktsson, J.A.; Chanussot, J.; Tilton, J.C. Advances in spectral-spatial classification of hyperspectral images. *Proc. IEEE* **2013**, *101*, 652–675. [CrossRef]
8. Camps-Valls, G.; Tuia, D.; Bruzzone, L.; Benediktsson, J.A. Advances in hyperspectral image classification: Earth monitoring with statistical learning methods. *IEEE Signal Process. Mag.* **2014**, *31*, 45–54. [CrossRef]
9. Yang, H. A back-propagation neural network for mineralogical mapping from aviris data. *Int. J. Remote Sens.* **1999**, *20*, 97–110. [CrossRef]
10. Melgani, F.; Bruzzone, L. Classification of hyperspectral remote sensing images with support vector machines. *IEEE Trans. Geosci. Remote Sens.* **2004**, *42*, 1778–1790. [CrossRef]
11. Ham, J.; Chen, Y.; Crawford, M.M.; Ghosh, J. Investigation of the random forest framework for classification of hyperspectral data. *IEEE Trans. Geosci. Remote Sens.* **2005**, *43*, 492–501. [CrossRef]
12. Chen, C.H.; Peter Ho, P.-G. Statistical pattern recognition in remote sensing. *Pattern Recognit.* **2008**, *41*, 2731–2741. [CrossRef]
13. Camps-Valls, G.; Bruzzone, L. *Kernel Methods for Remote Sensing Data Analysis*; Wiley: New York, NY, USA, 2009.
14. Ratle, F.; Camps-Valls, G.; Weston, J. Semisupervised neural networks for efficient hyperspectral image classification. *IEEE Trans. Geosci. Remote Sens.* **2010**, *48*, 2271–2282. [CrossRef]
15. Cortes, C.; Vapnik, V. Support vector machine. *Mach. Learn.* **1995**, *20*, 273–297. [CrossRef]
16. Böhning, D. Multinomial logistic regression algorithm. *Ann. Inst. Stat. Math.* **1992**, *44*, 197–200. [CrossRef]
17. Li, J.; Bioucas-Dias, J.M.; Plaza, A. Semisupervised hyperspectral image segmentation using multinomial logistic regression with active learning. *IEEE Trans. Geosci. Remote Sens.* **2010**, *48*, 4085–4098. [CrossRef]
18. Li, J.; Bioucas-Dias, J.M.; Plaza, A. Spectral-spatial hyperspectral image segmentation using subspace multinomial logistic regression and markov random fields. *IEEE Trans. Geosci. Remote Sens.* **2012**, *50*, 809–823. [CrossRef]
19. Plaza, A.; Benediktsson, J.A.; Boardman, J.W.; Brazile, J.; Bruzzone, L.; Camps-Valls, G.; Chanussot, J.; Fauvel, M.; Gamba, P.; Gualtieri, A. Recent advances in techniques for hyperspectral image processing. *Remote Sens. Environ.* **2009**, *113*, S110–S122. [CrossRef]
20. Camps-Valls, G.; Gomez-Chova, L.; Munoz-Mari, J.; Vila-Frances, J.; Calpe-Maravilla, J. Composite kernels for hyperspectral image classification. *IEEE Geosci. Remote Sens. Lett.* **2006**, *3*, 93–97. [CrossRef]
21. Camps-Valls, G.; Shervashidze, N.; Borgwardt, K.M. Spatio-spectral remote sensing image classification with graph kernels. *IEEE Geosci. Remote Sens. Lett.* **2010**, *7*, 741–745. [CrossRef]
22. Mathieu, F.; Jocelyn, C.; Atli, B.J. A spatial-spectral kernel-based approach for the classification of remote-sensing images. *Pattern Recognit.* **2012**, *45*, 381–392.
23. Wang, Y.; Zhang, Y.; Song, H. A spectral-texture kernel-based classification method for hyperspectral images. *Remote Sens.* **2016**, *8*, 919. [CrossRef]
24. Fauvel, M.; Chanussot, J.; Benediktsson, J.A. A spatial-spectral kernel-based approach for the classification of remote-sensing images. *Pattern Recognit.* **2012**, *45*, 381–392. [CrossRef]
25. Shen, L.; Zhu, Z.; Jia, S.; Zhu, J.; Sun, Y. Discriminative gabor feature selection for hyperspectral image classification. *IEEE Geosci. Remote Sens. Lett.* **2013**, *10*, 29–33. [CrossRef]
26. Huang, X.; Zhang, L. A comparative study of spatial approaches for urban mapping using hyperspectral rosis images over pavia city, northern italy. *Int. J. Remote Sens.* **2009**, *30*, 3205–3221. [CrossRef]

27. Kang, X.; Li, S.; Benediktsson, J.A. Spectral-spatial hyperspectral image classification with edge-preserving filtering. *IEEE Trans. Geosci. Remote Sens.* **2014**, *52*, 2666–2677. [CrossRef]
28. Benediktsson, J.A.; Palmason, J.A.; Sveinsson, J.R. Classification of hyperspectral data from urban areas based on extended morphological profiles. *IEEE Trans. Geosci. Remote Sens.* **2005**, *43*, 480–491. [CrossRef]
29. Huang, X.; Zhang, L. An adaptive mean-shift analysis approach for object extraction and classification from urban hyperspectral imagery. *IEEE Trans. Geosci. Remote Sens.* **2008**, *46*, 4173–4185. [CrossRef]
30. Tarabalka, Y.; Chanussot, J.; Benediktsson, J.A. Segmentation and classification of hyperspectral images using watershed transformation. *Pattern Recognit.* **2010**, *43*, 2367–2379. [CrossRef]
31. Tarabalka, Y.; Tilton, J.C.; Benediktsson, J.A.; Chanussot, J. A marker-based approach for the automated selection of a single segmentation from a hierarchical set of image segmentations. *IEEE J. Sel. Top. Appl. Earth Obs. Remote Sens.* **2012**, *5*, 262–272. [CrossRef]
32. Song, H.; Wang, Y. A spectral-spatial classification of hyperspectral images based on the algebraic multigrid method and hierarchical segmentation algorithm. *Remote Sens.* **2016**, *8*, 296. [CrossRef]
33. Tarabalka, Y.; Chanussot, J.; Benediktsson, J.A. Segmentation and classification of hyperspectral images using minimum spanning forest grown from automatically selected markers. *IEEE Trans. Syst. Man Cybern.* **2010**, *40*, 1267–1279. [CrossRef] [PubMed]
34. Tarabalka, Y.; Rana, A. Graph-Cut-Based Model for Spectral-Spatial Classification of Hyperspectral Images. In Proceedings of the IEEE Geoscience and Remote Sensing Symposium, Quebec City, QC, Canada, 13–18 July 2014; pp. 3418–3421.
35. Wang, Y.; Song, H.; Zhang, Y. Spectral-spatial classification of hyperspectral images using joint bilateral filter and graph cut based model. *Remote Sens.* **2016**, *8*, 748. [CrossRef]
36. Fang, L.; Li, S.; Kang, X.; Benediktsson, J.A. Spectral-spatial classification of hyperspectral images with a superpixel-based discriminative sparse model. *IEEE Trans. Geosci. Remote Sens.* **2015**, *53*, 4186–4201. [CrossRef]
37. Farag, A.A.; Mohamed, R.M.; El-Baz, A. A unified framework for map estimation in remote sensing image segmentation. *IEEE Trans. Geosci. Remote Sens.* **2005**, *43*, 1617–1634. [CrossRef]
38. Tarabalka, Y.; Fauvel, M.; Chanussot, J.; Benediktsson, J.A. Svm-and mrf-based method for accurate classification of hyperspectral images. *IEEE Geosci. Remote Sens. Lett.* **2010**, *7*, 736–740. [CrossRef]
39. Zhang, B.; Li, S.; Jia, X.; Gao, L.; Peng, M. Adaptive markov random field approach for classification of hyperspectral imagery. *IEEE Geosci. Remote Sens. Lett.* **2011**, *8*, 973–977. [CrossRef]
40. Moser, G.; Serpico, S.B. Combining support vector machines and markov random fields in an integrated framework for contextual image classification. *IEEE Trans. Geosci. Remote Sens.* **2013**, *51*, 2734–2752. [CrossRef]
41. Bai, J.; Xiang, S.; Pan, C. A graph-based classification method for hyperspectral images. *IEEE Trans. Geosci. Remote Sens.* **2013**, *51*, 803–817. [CrossRef]
42. Ghamisi, P.; Benediktsson, J.A.; Ulfarsson, M.O. Spectral-spatial classification of hyperspectral images based on hidden markov random fields. *IEEE Trans. Geosci. Remote Sens.* **2014**, *52*, 2565–2574. [CrossRef]
43. Golipour, M.; Ghassemian, H.; Mirzapour, F. Integrating hierarchical segmentation maps with mrf prior for classification of hyperspectral images in a bayesian framework. *IEEE Trans. Geosci. Remote Sens.* **2016**, *54*, 805–816. [CrossRef]
44. Fauvel, M.; Benediktsson, J.A.; Chanussot, J.; Sveinsson, J.R. Spectral and spatial classification of hyperspectral data using svms and morphological profiles. *IEEE Trans. Geosci. Remote Sens.* **2008**, *46*, 3804–3814. [CrossRef]
45. Li, J.; Marpu, P.R.; Plaza, A.; Bioucas-Dias, J.M.; Benediktsson, J.A. Generalized composite kernel framework for hyperspectral image classification. *IEEE Trans. Geosci. Remote Sens.* **2013**, *51*, 4816–4829. [CrossRef]
46. Fang, L.; Li, S.; Duan, W.; Ren, J.; Benediktsson, J.A. Classification of hyperspectral images by exploiting spectral-spatial information of superpixel via multiple kernels. *IEEE Trans. Geosci. Remote Sens.* **2015**, *53*, 6663–6674. [CrossRef]
47. Lu, T.; Li, S.; Fang, L.; Jia, X.; Benediktsson, J.A. From subpixel to superpixel: A novel fusion framework for hyperspectral image classification. *IEEE Trans. Geosci. Remote Sens.* **2017**, *55*, 4398–4411. [CrossRef]
48. Peng, J.; Chen, H.; Zhou, Y.; Li, L. Ideal regularized composite kernel for hyperspectral image classification. *IEEE J. Sel. Top. Appl. Earth Obs. Remote Sens.* **2017**, *10*, 1563–1574. [CrossRef]

49. Palmason, J.A.; Benediktsson, J.A.; Sveinsson, J.R.; Chanussot, J. Classification of Hyperspectral Data from Urban Areas Using Morphological Preprocessing and Independent Component Analysis. In Proceedings of the 2005 IEEE International Geoscience and Remote Sensing Symposium, Seoul, Korea, 25–29 July 2005; p. 4.
50. Fauvel, M.; Chanussot, J.; Benediktsson, J.A. Kernel principal component analysis for the classification of hyperspectral remote sensing data over urban areas. *EURASIP J. Adv. Signal Process.* **2009**, *2009*, 783194. [CrossRef]
51. Castaings, T.; Waske, B.; Atli Benediktsson, J.; Chanussot, J. On the influence of feature reduction for the classification of hyperspectral images based on the extended morphological profile. *Int. J. Remote Sens.* **2010**, *31*, 5921–5939. [CrossRef]
52. Pal, M.; Foody, G.M. Feature selection for classification of hyperspectral data by svm. *IEEE Trans. Geosci. Remote Sens.* **2010**, *48*, 2297–2307. [CrossRef]
53. Tuia, D.; Camps-Valls, G.; Matasci, G.; Kanevski, M. Learning relevant image features with multiple-kernel classification. *IEEE Trans. Geosci. Remote Sens.* **2010**, *48*, 3780–3791. [CrossRef]
54. Jia, X.; Kuo, B.C.; Crawford, M.M. Feature mining for hyperspectral image classification. *Proc. IEEE* **2013**, *101*, 676–697. [CrossRef]
55. Taşkın, G.; Kaya, H.; Bruzzone, L. Feature selection based on high dimensional model representation for hyperspectral images. *IEEE Trans. Image Process.* **2017**, *26*, 2918–2928. [CrossRef] [PubMed]
56. Cheng, Q.; Zhou, H.; Cheng, J. The fisher-markov selector: Fast selecting maximally separable feature subset for multiclass classification with applications to high-dimensional data. *IEEE Trans. Pattern Anal. Mach. Intell.* **2011**, *33*, 1217–1233. [CrossRef] [PubMed]
57. Perona, P.; Malik, J. Scale-space and edge detection using anisotropic diffusion. *IEEE Trans. Pattern Anal. Mach. Intell.* **1990**, *12*, 629–639. [CrossRef]
58. Weickert, J.; Romeny, B.M.T.H.; Viergever, M.A. Efficient and reliable schemes for nonlinear diffusion filtering. *IEEE Trans. Geosci. Remote Sens.* **1998**, *7*, 398–410. [CrossRef] [PubMed]
59. Falgout, R.D. An introduction to algebraic multigrid. *Comput. Sci. Eng.* **2006**, *8*, 24–33. [CrossRef]
60. Duarte-Carvajalino, J.M.; Sapiro, G.; Vélez-Reyes, M.; Castillo, P.E. Multiscale representation and segmentation of hyperspectral imagery using geometric partial differential equations and algebraic multigrid methods. *IEEE Trans. Geosci. Remote Sens.* **2008**, *46*, 2418–2434. [CrossRef]
61. Briggs, W.L.; Henson, V.E.; McCormick, S.F. *A Multigrid Tutorial*, 2nd ed.; SIAM: Philadelphia, PA, USA, 2000; pp. 7–48.
62. Cristianini, N.; Scholkopf, B. Support vector machines and kernel methods: The new generation of learning machines. *Ai Mag.* **2002**, *23*, 31.
63. Schölkopf, B.; Smola, A.J. *Learning with Kernels. Support Vector Machines, Regularization, Optimization, and Beyond*; MIT Press: Cambridge, MA, USA, 2002.

remote sensing

MDPI

Article

Hyperspectral Classification Based on Texture Feature Enhancement and Deep Belief Networks

Jiaojiao Li [1,*], Bobo Xi [1], Yunsong Li [1], Qian Du [2] and Keyan Wang [1]

1 The State Key Lab. of Integrated Service Networks, School of Telecommunications Engineering, Xidian University, Xi'an 710071, China; xibobo1301@foxmail.com (B.X.); ysli@mail.xidian.edu.cn (Y.L.); kywang@mail.xidian.edu.cn (K.W.)

2 The Department of Electronic and Computer Engineering, Mississippi State University, Starkville, MS 39762, USA; du@ece.msstate.edu

* Correspondence: jjli@xidian.edu.cn

Received: 2 January 2018; Accepted: 2 March 2018; Published: 4 March 2018

Abstract: With success of Deep Belief Networks (DBNs) in computer vision, DBN has attracted great attention in hyperspectral classification. Many deep learning based algorithms have been focused on deep feature extraction for classification improvement. Multi-features, such as texture feature, are widely utilized in classification process to enhance classification accuracy greatly. In this paper, a novel hyperspectral classification framework based on an optimal DBN and a novel texture feature enhancement (TFE) is proposed. Through band grouping, sample band selection and guided filtering, the texture features of hyperspectral data are improved. After TFE, the optimal DBN is employed on the hyperspectral reconstructed data for feature extraction and classification. Experimental results demonstrate that the proposed classification framework outperforms some state-of-the-art classification algorithms, and it can achieve outstanding hyperspectral classification performance. Furthermore, our proposed TFE method can play a significant role in improving classification accuracy.

Keywords: deep belief networks; deep learning; texture feature enhancement; hyperspectral classification; band grouping

---

1. Introduction

Hyperspectral imagery with hundreds of narrow spectral channels provides wealthy spectral information. With very high spectral resolution, hyperspectral data has been of great interest in many practical applications, such as in agriculture, environment, surveillance, medicine [1–4] etc. Hyperspectral classification is a key technique employed in aforementioned applications. A majority of classification methods have been promoted in the last several decades to distinguish physical objects and classify each pixel into a unique land-cover label, such as maximum likelihood [5], minimum distance [6], K-nearest neighbors [7,8], random forests [9], Bayesian models [10,11], neural networks, etc., and their improvements [12–15]. Among these supervised classifiers, one of the most important classifiers is kernel-based support vector machine (SVM), which can also be considered as a kind of neural network. It can achieve superior hyperspectral classification accuracy via building an optimal hyperplane to best separate training samples.

In addition, sparse representation based on an over-complete signal dictionary has gained great attention in the literature. Sparse representation-based classification (SRC) [16–18] and collaborative representation classification (CRC) [19,20] are proposed from a different aspect: they do not adopt the traditional training–testing fashion. Such classification methods do not need any prior knowledge about probability density distribution of the data. To further enhance the performance of SRC and CRC, Du and Li [21] utilized a diagonal weight matrix to adaptively adjust the regularization parameter.

To address the issues of Hughes phenomenon in hyperspectral classification, majority of feature extraction and selection algorithms are utilized to delete redundant features from the original data. To further improve performance of hyperspectral classification, multi-features are extracted and employed for classification. For instance, Kang et al. combined spectral and spatial features through a guided filter to process pixel-wise classification map in each class [22]. Several studies [23–25] focused on integrating spatial and spectral information in hyperspectral imagery. In addition, texture features are considered to assist hyperspectral classification [26], and modeling of hyperspectral image textures is significant for classification and material identification.

Recent research has highlighted deep learning with deep neural networks, which can learn high-level features hierarchically. They have demonstrated their potential in image classification, which also motivated successful applications of deep models on hyperspectral image classification. The classic deep learning method is convolutional neural networks (CNN), which plays a dominant role in visual-based issues. The local receptive fields of CNN can extract spatial-related features at high levels. Fukushima [27] introduced the motivations of CNNs. Ciresan and Lee et al. [28,29] depicted the invariants of CNNs. Chen et al. proposed 2-D CNN and 3-D CNN [30] to capture deep abstract and robust features, yielding superior hyperspectral classification performance. Although CNNs are typical supervised models, a massive training dataset is needed to trigger their powers. Unfortunately, a limited number of labeled samples are usually given in hyperspectral imagery. Deep belief networks (DBNs) [31] and stacked autoencoders (SAEs) [32] are also very promising deep learning methods for hyperspectral classification with limited training samples.

In this paper, we mainly investigate the DBN for its suitability and practicality to hyperspectral classification. A novel hyperspectral classification framework is proposed based on an optimum DBN. To acquire desirable performance, we also promote an advanced algorithm to enhance the texture features of hyperspectral imagery. The main contributions of this paper are summarized below.

1. We first promote a band group method to separate the bands of hyperspectral data into different band groups. Multi-texture features are used to select a sample band in each band group.
2. We propose a novel algorithm to enhance the texture features of hyperspectral data. We advocate the use of guided filter to complete the procedure of texture feature enhancement (TFE).
3. An optimal DBN structure is proposed with consideration of learning and deep features extraction. The learned features are exploited in Softmax to address the classification problem. Furthermore, with enhanced texture features, accurate classification maps can be generated by considering spatial information.

The rest of the paper is organized into four sections. Section 2 is a brief description of related work. In Section 3, we detail our proposed DBN model. Datasets and parameters setting are demonstrated in Section 4. Experimental results and discussions are depicted in Section 5. Section 6 draws the conclusion of this paper.

2. The Related Work

A deep belief network (DBN) is a model that is first pre-trained in an unsupervised way, and then the available labeled training samples are used to fine-tune the pre-trained model through optimizing a cost function defined over the labels of training samples and their predictions.

The original DBN, published in Science [33], uses a generative model in the pre-training procedure, and uses back-propagation in the fine-tuning stage. This is very useful when the number of training samples is limited, such as in the case of hyperspectral remote sensing. DBN can be efficiently trained in an unsupervised, layer-by-layer manner where the layers are typically made of restricted Boltzmann machines (RBM). Thus, to explain the structure and theory of the DBN, we first describe its main component, the RBM.

2.1. Restricted Boltzmann Machines (RBM)

An RBM generally uses unsupervised learning, which can be interpreted as stochastic neural networks. It was originally developed to form a distributed representation. It is a two layer-wise network, which is composed of visible and hidden units. Learning RBM only allows the full connection between visible and hidden units, and does not allow connection between two visible units or connections between two hidden units. With the given visible units, hidden units can be obtained via mapping of visible units. The activations of each neuron in hidden layers are independent. Meanwhile, with the given hidden units, visible units have the same effects. A typical RBM structure is depicted in Figure 1.

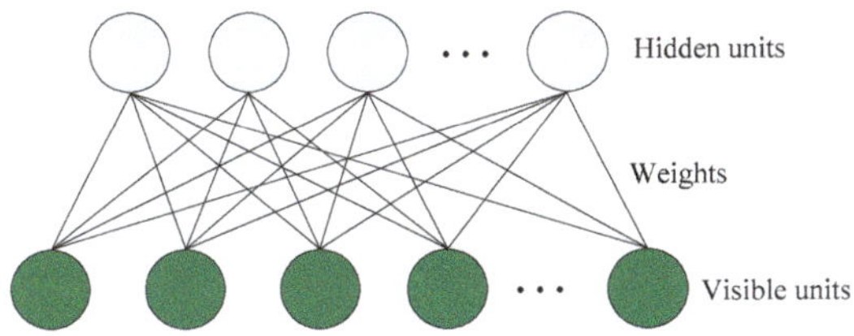

Figure 1. Architecture of Restricted Boltzmann Machines.

The visible units can be represented as $\mathbf{h}$, and the hidden units can be expressed as $\mathbf{v}$. The RBM model is a kind of energy-based models in which the joint distribution of the layers can be expressed as Boltzmann distribution. Energy-based probabilistic models define a probability distribution through an energy function as:

$$p(\mathbf{v},\mathbf{h}|\theta) = \frac{\exp(-E(\mathbf{v},\mathbf{h}|\theta))}{Z(\theta)}, \tag{1}$$

where the normalization constant $Z(\theta)$ is called the partition function by analogy with physical systems:

$$Z(\theta) = \sum_{\mathbf{v}}\sum_{\mathbf{h}} E(\mathbf{v},\mathbf{h};\theta) \tag{2}$$

A joint configuration of the units has an energy given by:

$$\begin{aligned} E(\mathbf{v},\mathbf{h};\theta) &= -\textstyle\sum_{i=1}^{n} a_i v_i - \sum_{j=1}^{m} b_j h_j - \sum_{i=1}^{n}\sum_{j=1}^{m} v_i w_{ij} h_j \\ &= -\mathbf{a}^T\mathbf{v} - \mathbf{b}^T\mathbf{h} - \mathbf{v}^T\mathbf{w}\mathbf{h} \end{aligned}, \tag{3}$$

where $\theta = \{a_i, b_j, w_{ij}\}$; w_{ij} represents the weight connecting the visible unit i and the hidden unit j; a_i and b_j denote the bias terms of visible and hidden layers, respectively; n and m are the total visible and hidden unit numbers; and v_i and h_j represent the states of visible unit i and hidden unit j.

Due to the specific structure of RBMs, visible and hidden units are conditionally independent, as given by:

$$\begin{aligned} P(v_i = 1|\mathbf{h},\theta) &= \sigma(a_i + \textstyle\sum_i w_{ij} v_i) \\ P(h_j = 1|\mathbf{v},\theta) &= \sigma(b_j + \textstyle\sum_j w_{ij} h_j) \end{aligned}, \tag{4}$$

where $\sigma(\bullet)$ is the logistic function defined as

$$\sigma(x) = \frac{1}{1+\exp(-x)} \tag{5}$$

Overall, an RBM has five parameters: $\mathbf{h}$, $\mathbf{v}$, $\mathbf{w}$, $\mathbf{a}$ and $\mathbf{b}$, where $\mathbf{w}$, $\mathbf{a}$ and $\mathbf{b}$ are achieved via learning, $\mathbf{v}$ is input, and $\mathbf{h}$ is output. $\mathbf{w}$, $\mathbf{a}$ and $\mathbf{b}$ can be learned and updated via the contrastive divergence (CD) method as

$$w_{ij} \leftarrow w_{ij} + \lambda(P(h_j|v_i)v_i - P(h_j^r|v_i^r)v_i^r) \tag{6}$$

$$a_i \leftarrow a_i + \lambda(v_i - v_i^r) \tag{7}$$

$$b_j \leftarrow b_j + \lambda(h_j - h_j^r) \tag{8}$$

where λ denotes the learning rate, $P(h_j^r|v_i^r)$ represents the reconstructed probability distribution, and v_i^r and h_j^r are the reconstruction of visible and hidden unit, respectively. Once the states of hidden units are chosen, the visible units can be reconstructed via the hidden units sampled via Gibbs method. Then, the states of hidden units are updated through the visible units, so that the hidden units demonstrate the features of reconstruction. The distribution of visible units approximates the distribution of the real data. The learning ability of an RBM depends on whether the hidden units contain enough information of the input data.

2.2. Deep Belief Learning

The learning ability of a single hidden layer is limited. To capture the comprehensive information of data, the hidden units of the RBM can be feed as the input (visible units) of another RBM. This kind of layer-by-layer learning structure trained in a greedy manner forms so-called Deep Belief Networks. In this way, DBN can extract deep features of image data. The structure of three-layer DBN is depicted in Figure 2.

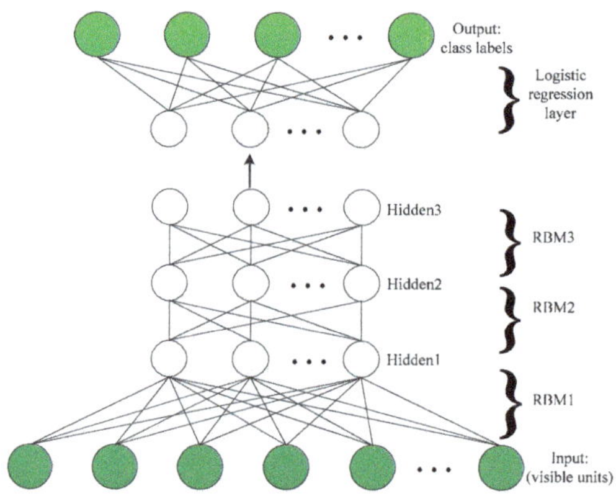

Figure 2. An illustration of three-layer DBN with logistic regression.

The process of training of DBN consists of two parts: pre-training and fine-tuning. The pre-training is an unsupervised training stage that initializes the model in such a way to enhance the efficiency of supervised training. The fine-tuning process can be realized as supervised training stage, which adjusts the classifier's prediction to match the ground truth of the data.

3. The Proposed Framework

To extract more powerful and invariant features, we propose a novel DBN hyperspectral classification algorithm based on TFE. DBN is composed of several layers of latent factors, which can be deemed as neurons of neural networks. However, the limited training samples in the real hyperspectral image classification task usually lead to many "dead" (never responding) or "potential over-tolerant" (always responding) latent factors (neurons) in the trained DBN. Our proposed framework mainly consists of three steps: band grouping and sample band selection, TFE, and DBN-based classification.

3.1. Band Grouping and Sample Band Selection

Compared to multispectral imagery, hyperspectral imagery with hundreds of spectral bands has relatively narrow bandwidths. The correlation between spectral bands needs to be considered.

In our framework, we calculated all the pair wise correlation coefficient of bands, and then utilized the correlations between adjacent bands. The spectral correlation coefficients in different datasets are depicted in Figure 3.

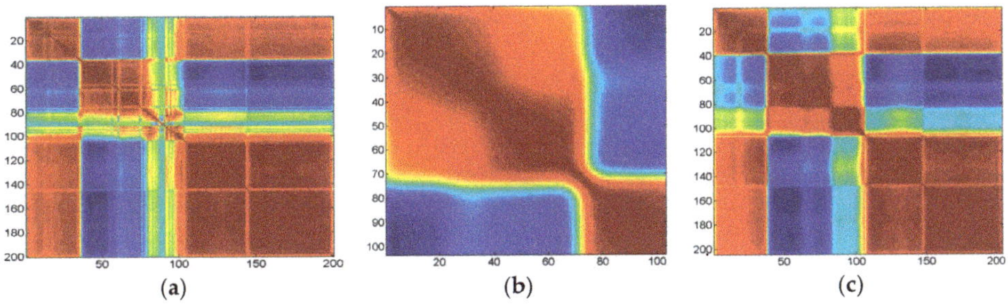

Figure 3. The maps of correlation coefficients of spectral bands in different datasets: (**a**) Indian Pines; (**b**) University of Pavia; and (**c**) Salinas.

We can obtain the correlation coefficient between adjacent bands as:

$$\rho_{i,j} = corr(\mathrm{B}_i, \mathrm{B}_j) = \mathrm{cov}(\mathrm{B}_i, \mathrm{B}_j) / \sqrt{\mathrm{var}(\mathrm{B}_i)\mathrm{var}(\mathrm{B}_j)} \tag{9}$$

where cov is covariance and var means variance. B_i and B_j represent the i-th and j-th band channels, respectively. $i = 1, 2, \ldots, L-1$. Here, L denotes the number of bands of the hyperspectral dataset. Based on Equation (9), the correlation coefficients between adjacent bands in different datasets are calculated, as shown in Figure 4. We can see that the highest correlation coefficient in Indian Pines is 0.9997, and the lowest correlation coefficient is 0.0686. The spectral bands of university of Pavia have strong correlations overall, where the highest correlation coefficient is 0.9998, and the lowest correlation coefficient is 0.9294. The highest correlation coefficient in Salinas is 0.9999, and the lowest correlation coefficient is 0.5856.

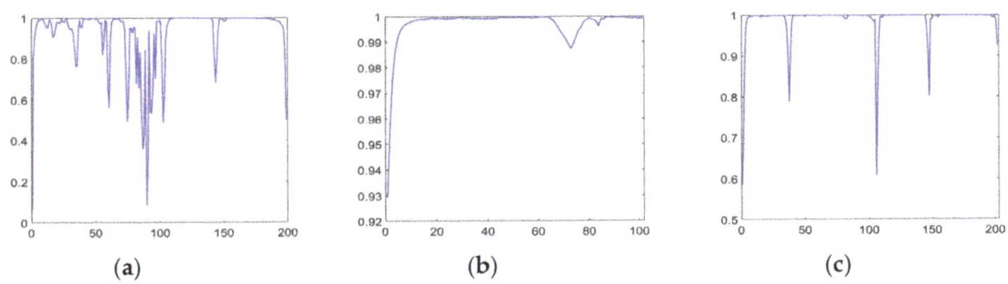

Figure 4. The correlation coefficients of adjacent spectral bands in different datasets: (**a**) Indian Pines; (**b**) University of Pavia; and (**c**) Salinas.

Here, we design an algorithm for grouping bands rationally.

Firstly, calculate the average correlation coefficients of the adjacent bands, denoted as $\overline{\mathrm{C}}$, which is utilized as the threshold in the following steps. It can be calculated through:

$$\overline{\mathrm{C}} = \frac{1}{L-1} \sum_{i=1}^{L-1} \rho_{i,j} \tag{10}$$

where $j = i + 1$. If the correlation coefficients of adjacent bands are greater than $\overline{\mathrm{C}}$, these two bands are considered to have strong correlation.

Second, search local minimum values from the correlation coefficients between the adjacent bands, denoted as $\boldsymbol{\rho}^{\min}$, where $\boldsymbol{\rho}^{\min} = \{\rho_{i,j} | \rho_{i,j} \leq \rho_{i+1,j+1} \,||\, \rho_{i,j} \leq \rho_{i-1,j-1}\}$. All the elements in $\boldsymbol{\rho}^{\min}$ are compared with $\overline{C}$. If the inequality $\{\rho_{i,j} \in \boldsymbol{\rho}^{\min}\} < \overline{C}$ is satisfied, it indicates that the correlation between the i-th band and the j-th band is lower than the average correlation value, and the correlation between these two bands is considered to be weak. Then, the corresponding index group $\{i, j\}$ is recorded and added to the set $\boldsymbol{\rho}^{Loc}$.

Third, band grouping depends on the stored index pairs in $\boldsymbol{\rho}^{Loc}$. For instance, with regard to index pair $\{i, j\}$, the i-th band is set as the end band of the former band group and the j-th band is set as the first band of the next band group. Thus, based on the aforementioned rules, all the bands are divided into different band groups $\{\mathbf{G}_1, \mathbf{G}_2, \cdots, \mathbf{G}_K\}$.

After dividing all the bands of hyperspectral dataset into different band groups, a sample band with the strongest and clearest texture features is searched and selected from each group.

To extract texture features, the gray level co-occurrence matrix (GLCM) has been employed successfully. GLCM [34] is defined as a matrix of frequencies which can extract second order statistics from a hyperspectral image. The distribution in the matrix depends on the angular and distance relationship between pixels. After the GLCM is created, it can be used to compute various features. We choose the five most commonly used features in Table 1 to select a sample band from each band group. The texture feature score of each band can be calculated by Equation (11):

$$T = \sum_{i=1}^{5} F_i \tag{11}$$

Table 1. Feature calculated from the normalized co-occurrence matrix $P(i,j)$.

No.	Feature	Formula
F_1	Energy	$\sum_i \sum_j P^2(i,j)$
F_2	Entropy	$\sum_i \sum_j P(i,j) \log P(i,j)$
F_3	Contrast	$\sum_i \sum_j (i-j)^2 P(i,j)$
F_4	Mean	$\frac{1}{m*n} \sum_i \sum_j \lvert i-j \rvert P(i,j)$
F_5	Homogeneity	$\sum_i \sum_j \frac{P(i,j)}{1+\lvert i-j \rvert}$

The sample band in each band group can be selected through:

$$\mathbf{g}_k = \underset{\mathrm{B}_{l^k}}{\operatorname{argmax}} \left\{ T_{\mathrm{B}_{l^k}} \middle| \mathrm{B}_{l^k} \in \mathbf{G}_k \right\}, \tag{12}$$

where $\mathbf{G}_k$ represents the k-th band group of the dataset, $l^k \in \{1, 2, \cdots, N^k\}$, N^k is the number of bands in the k-th band group, and B_{l^k} represent the l^k-th band in the k-th band group. Finally, the sample band set are comprised of $\{\mathbf{g}_1, \mathbf{g}_2, \cdots, \mathbf{g}_k\}$.

3.2. Texture Feature Enhancement

As an effective edge-preserving filter, guided filter (GF) was proposed by He in 2012. It can enhance the detail of an image. Texture feature is a kind of important spatial characteristics and also has long history in image processing. In this paper, we utilize the GF in each band group to enhance the texture features of the image.

The general guided image filtering was designed for gray-scale images or color images. It is very easy to extend to multi-channel image. Firstly, the guidance image in our proposed framework is multi-channel image, denoted as $\mathbf{I}^M$, which is comprised of the copies of the band with the strongest texture features in each band group. We assume q^M is a linear transform of $\mathbf{I}^M$ in a window ω_k centered at the pixel k, and the multi-channel guided filter model can be expressed as

$$q_i^M = \left(\mathbf{a}_k^M\right)^T \mathbf{I}_i^M + b_k^M, \forall i \in \omega_k \tag{13}$$

where $\mathbf{I}_i^M$ is a $C \times 1$ vector, and C is the channel number of the input image, $\mathbf{a}_k^M$ is a $C \times 1$ coefficient vector, and q_i^M and b_k^M are scalars. The guided filter for multi-channel guidance image becomes

$$\begin{aligned} \mathbf{a}_k^M &= (\Sigma_k + \varepsilon \mathbf{U})^{-1} \left(\frac{1}{|\omega|} \sum_{i \in \omega_k} \mathbf{I}_i^M p_i^M - \mu_k \overline{p_k^M}\right) \\ b_k^M &= \overline{p_k^M} - \left(\mathbf{a}_k^M\right)^T \mu_k \\ q_i^M &= \left(\overline{\mathbf{a}_i^M}\right)^T \mathbf{I}_i^M + \overline{b_i^M} \end{aligned}, \tag{14}$$

where $\sum_k$ is the $C \times C$ covariance matrix of $\mathbf{I}^M$ in ω_k, $\mathbf{U}$ is an $C \times C$ identity matrix, p^M denotes a filtering input image which is given beforehand according to the application, μ_k is the mean of $\mathbf{I}^M$ in ω_k, $\overline{p_k^M}$ is the mean of p^M in ω_k, and $|\omega|$ represents the number of pixels in ω_k.

Then, the extending guided image filtering for multi-channel images will be applied to each band group. For instance, each channel of the guidance image I^M in Equation (14) for the k-th band group $\mathbf{G}_k$ is the copy of the sample band $\mathbf{g}_k$ selected previously.

After guided filtering for all groups is completed, the output bands are restored to a hyperspectral image cube according to the band number. Finally, the reconstructed image data with enhanced texture features are obtained through the aforementioned steps. Figure 5 demonstrates the procedure of band grouping and TFE. We can see that, after sample bands with strongest textures are obtained, the reconstructed image data with enhanced texture feature can be achieved through the GF process.

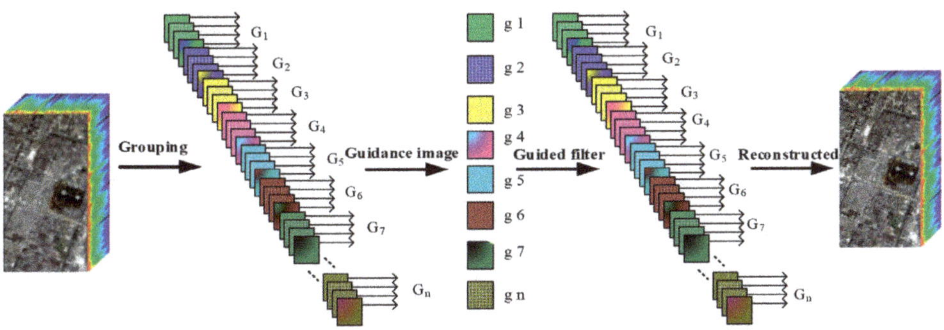

Figure 5. The procedure of band grouping and texture features enhancement.

3.3. DBN Classification Model

In this section, a DBN-based framework for hyperspectral classification with feature enhanced data is developed.

Spectral information is the most significant and direct feature, and can be directly utilized for classification. Architectures of existing methods, such as SVM and KNN, can extract spectral features but not deep enough. Therefore, only a deep architecture can make full use of the texture enhanced hyperspectral image characteristics. However, as the training samples are limited, the overfitting problem often occurs if the network is too deep, so we advocate a novel DBN framework, which has only two hidden layers (Figure 6).

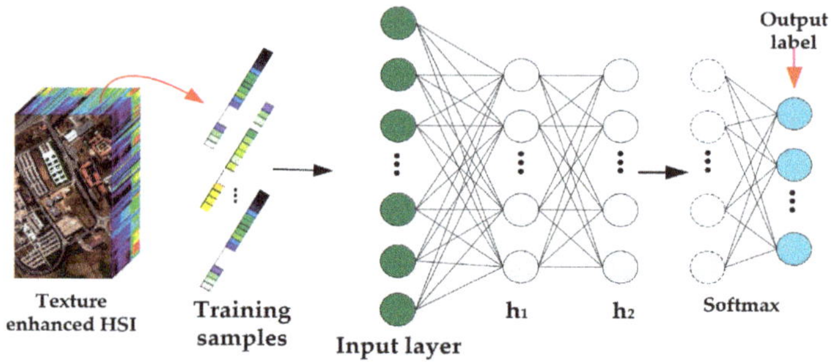

Figure 6. Our proposed DBN network for classification.

The input data consist of training samples that are one-dimensional (1-D) vectors, and each pixel of a training sample is collected from the texture enhanced HSI data. For ease of description, the first hidden layer is denoted as $\mathbf{h}_1$ and the second $\mathbf{h}_2$. The first layer is learned for extracting features from the input data, and the learned features are preserved in $\mathbf{h}_1$. Then, to pursue refined and abstract features, using the features contained in $\mathbf{h}_1$ as the visible data of the second layer, $\mathbf{h}_2$ keeps the refined features. This procedure is generally called recursive greedy learning for pre-training a DBN.

In practice, learning each layer is often performed through the n-step CD, and the weights are updated using Equations (6)–(8).

To fine-tune the DBN and accomplish classification, a Softmax layer is added to the end of the network.

Now, let $\mathbf{X} = \{\mathbf{x}_1, \mathbf{x}_2, \ldots, \mathbf{x}_K\}$ be a set of training samples and $\mathbf{Y} = \{y_1, y_2, \ldots, y_K\}$ be the corresponding labels, where $\mathbf{x}_k = [x_{k1}, x_{k2}, \ldots, x_{kL}]^T$ is the spectral signature of the k-th sample with L bands. Utilizing the maximum likelihood method, the objective function can be written as

$$C(\boldsymbol{\theta}) = -\sum_{k=1}^{K} \log(P(y_k|\mathbf{x}_k), \boldsymbol{\theta}) = -\sum_{k=1}^{K} \log(S_{y_k}(\mathbf{x}_k, \boldsymbol{\theta})) \tag{15}$$

where K is the number of training samples, $(P(y_k|\mathbf{x}_k), \boldsymbol{\theta})$ means the distribution of y_k when given $\mathbf{x}_k$ with the parameters $\boldsymbol{\theta}$ of the Softmax layer, and $S_{y_k}(\mathbf{x}_k, \boldsymbol{\theta})$ denotes the output of the Softmax layer of the k-th training sample, that is

$$S_{y_k}(\mathbf{x}_k, \boldsymbol{\theta}) = \frac{\exp\{-\sum_{m=1}^{M} \delta(y_k = m)\boldsymbol{\theta}_m^T \mathbf{h}^{H_L}\}}{\sum_{n=1}^{M} \exp\{-\boldsymbol{\theta}_n^T \mathbf{h}^{H_L}\}}, \tag{16}$$

where H_L is the number of the hidden layers, which is set to 2 in our proposed framework, and M is the number of the classes. $\boldsymbol{\theta}_m$ and $\boldsymbol{\theta}_n$ are the parameter vectors for the m-th and n-th unit of the softmax layer, respectively. $\mathbf{h}^{H_L}$ is the output of the H_L-th hidden layer, which is calculated via the input data, the weights and bias from the first layer to the H_L-th hidden layer. To optimize the objective function, the stochastic gradient descent (SGD) algorithm is used. Finally, the label of each testing pixel is determined via the weights and biases from aforementioned steps.

4. Experiments

4.1. Datasets

In this section, three typical hyperspectral datasets, namely Indian Pines, University of Pavia and Salinas, are employed to compare the proposed DBN classification method with other state-of-the-art methods. In these experiments, we randomly select 300 labeled pixels per class for training, of which 20 samples are utilized for validation. The remaining pixels of labeled data are used for testing. Furthermore, each pixel is uniformly scaled to the range of −1 to 1.

The first experiment is Indian Pines dataset, which was gathered by Airborne Visible/Infrared Imaging Spectrometer (AVIRIS) sensor in northwestern Indiana. There are 220 spectral channels in 0.4 to 2.45 μm region with spatial resolution of 20 m. It consists of 145 × 145 pixels with 200 bands after removing 20 noisy and water absorption bands. Here, we employ 8 large classes in this experiment. The numbers of training and testing samples are listed in Table 2.

Table 2. Number of training and testing samples used in the Indian Pines dataset.

No.	Classes	Training	Testing
1	Corn-notill	300	1160
2	Corn-mintill	300	534
3	Grass-pasture	300	197
4	Hay-windrowed	300	189
5	Soybean-notill	300	668
6	Soybean-mintill	300	2168
7	Soybean-clean	300	314
8	Woods	300	994
	Total	2400	6224

The second dataset with 610 × 340 pixels is the University of Pavia, which was acquired by the Reflective Optics System Imaging Spectrometer (ROSIS) during a flight campaign over Pavia, northern Italy. The ROSIS sensor cover 115 spectral bands from 0.43 to 0.86 μm and the geometric resolution is 1.3 m. Each pixel has 103 bands after discarding bad bands. There are 9 ground-truth classes with the number of labeled samples shown in Table 3.

Table 3. Number of training and testing samples used in the Pavia University dataset.

No.	Classes	Training	Testing
1	Asphalt	300	6331
2	Meadows	300	18,349
3	Gravel	300	1799
4	Trees	300	2764
5	Painted metal sheets	300	1045
6	Bare Soil	300	4729
7	Bitumen	300	1030
8	Self-Blocking Bricks	300	3382
9	Shadows	300	647
	Total	2700	40,076

The third experiment is on Salinas dataset, which was also collected by the AVIRIS sensor, capturing an area over Salinas Valley, California, with a spatial resolution of 3.7 m. The area comprises 512 × 217 pixels with 204 bands after removing noisy and water absorption bands. It mainly contains vegetables, bare soils, and vineyard fields. There are 16 different ground-truth classes, and the numbers of training and testing samples are listed in Table 4.

Our experiments are implemented using Matlab 2015b which is manufactured by Mathworks in Massachusetts, US. The CPU we employed is Intel Core i5-3470. The basic frequency is 3.200 GHz. The operation system is Win7 with 64 bits.

Table 4. Number of training and testing samples used in the Salinas dataset.

No.	Classes	Training	Testing
1	Brocoli_green_weeds_1	300	1709
2	Brocoli_green_weeds_2	300	3426
3	Fallow	300	1676
4	Fallow_rough_plow	300	1094
5	Fallow_smooth	300	2378
6	Stubble	300	3659
7	Celery	300	3279
8	Grapes_untrained	300	10,971
9	Soil_vinyard_develop	300	5903
10	Corn_senesced_green_weeds	300	2978
11	Lettuce_romaine_4wk	300	768
12	Lettuce_romaine_5wk	300	1627
13	Lettuce_romaine_6wk	300	616
14	Lettuce_romaine_7wk	300	770
15	Vinyard_untrained	300	6968
16	Vinyard_vertical_trellis	300	1507
	Total	4800	49,329

4.2. Parameters Tuning and Analysis

In our proposed framework, we have several parameters that need to be adjusted: the number of hidden units, the learning rate, the max epoch and the number of hidden layers. In this section, some tuning experimental results are listed for selecting proper values. Both the number of hidden layers and the number of hidden units in hidden layers play an important role in classification performance. A suitable number of hidden layers and neurons can make full use of texture enhanced hyperspectral data without over-training, and can support a fitting mapping from original hyperspectral data to hyperspectral features. In the training process of DBN, the learning rate controls the pace of learning. It implies that a too large learning rate will lead an unstable output of training, and a too small learning rate will lead a longer training process. Therefore, an appropriate learning rate can expedite our training procedure with satisfactory performance.

In Figure 7, we can see that our proposed framework achieves best classification accuracy with 200 hidden neurons in each hidden layer. It demonstrates that 200 is a suitable number of hidden neurons. Figure 8 depicts the relationship between accuracies and the learning rates. It can be seen that the values of learning rate from 0.15 to 0.2 can obtain better performance. Therefore, we select 0.15 for the first RBM, and 0.2 for the second RBM. To determine the max epoch, we set the range of max epoch from 50 to 500. Figure 9 demonstrates that, when max epoch reaches 300, our proposed framework can achieve best classification performance. Consequently, the max epoch is set to 300. Table 5 lists the accuracies achieved with different numbers of hidden layers in DBN. When employing two hidden layers, the classification performance of DBN can achieve superior results. Thus, in our proposed framework, we set the number of hidden layers to 2.

In our paper, we utilize Graycomatrix function in Matlab to calculate the GLCM. The parameters used in experiments are "NumLevels" and "Offset", and they are set to 8 and $[0, 3; -3, 3; -3, 0; -3, -3]$, respectively.

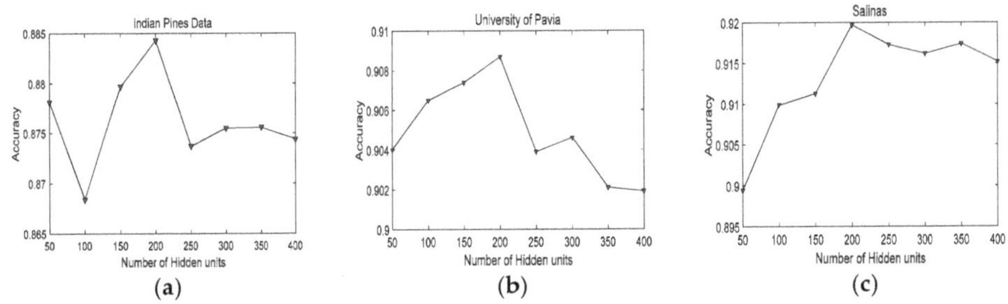

Figure 7. The relationship between accuracies and the number of hidden units in different datasets: (**a**) Indian Pines; (**b**) University of Pavia; and (**c**) Salinas.

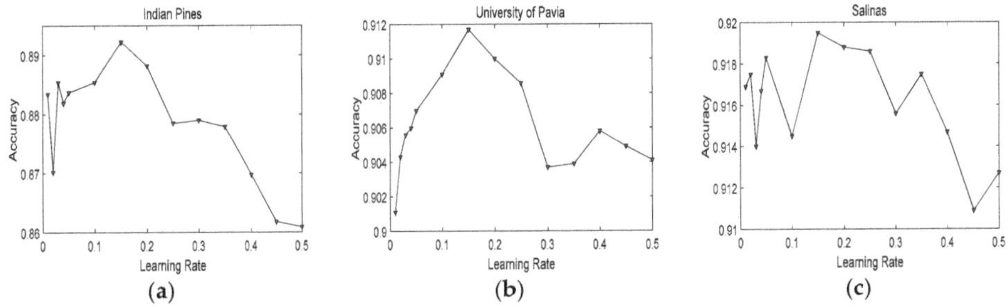

Figure 8. The relationship between accuracies and the learning rates in different datasets: (**a**) Indian Pines; (**b**) University of Pavia; and (**c**) Salinas.

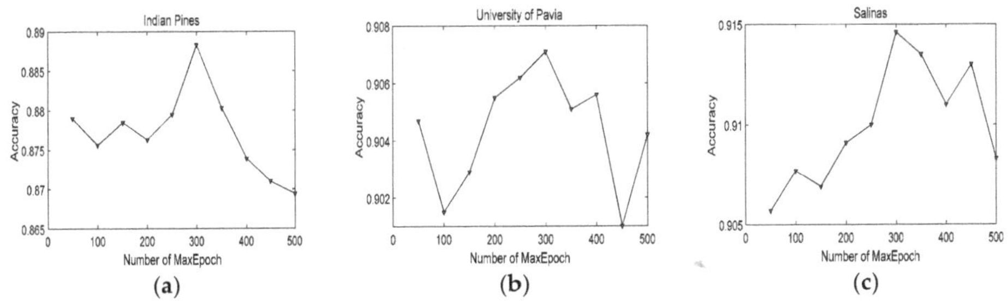

Figure 9. The relationship between accuracies and the numbers of Max epoch in different datasets: (**a**) Indian Pines; (**b**) University of Pavia; and (**c**) Salinas.

Table 5. The accuracies obtained via different numbers of hidden layers in DBN.

Datasets	1 Layer	2 Layers	3 Layers	4 Layers
Indian Pines	0.8919	**0.8948**	0.8892	0.8432
University of Pavia	0.9090	**0.9123**	0.9065	0.8994
Salinas	0.9123	**0.9228**	0.9104	0.9064

4.3. Evaluation Criteria

The evaluation criteria used in our paper are overall accuracy (OA), average accuracy (AA), precision, and Kappa. Especially, OA, Precision and Kappa are highlighted for assessment of the proposed framework.

Figure 10 demonstrates a p-class confusion matrix. Based on Figure 10, AA and precision can be derived as [35]

$$\mathrm{P}_{\mathrm{AA}} = \frac{1}{p}\left(\sum_{i=1}^{p} \frac{n_{ii}}{\sum_{j=1}^{p} n_{ji}}\right) \tag{17}$$

$$\mathrm{P}_{precision} = \frac{1}{p}\left(\sum_{i=1}^{p} \frac{n_{ii}}{\sum_{j=1}^{p} n_{ij}}\right) \tag{18}$$

where p is the number of classes. N is the total number of the hyperspctral image data samples and $N = \sum_{i=1}^{p} n_i$. n_{ii} is the number of hyperspectral image samples in the i-th class to be classified into the i-th class, and n_{ji} is the number of hyperspectral image samples in the i-th class to be classified into the j-th class.

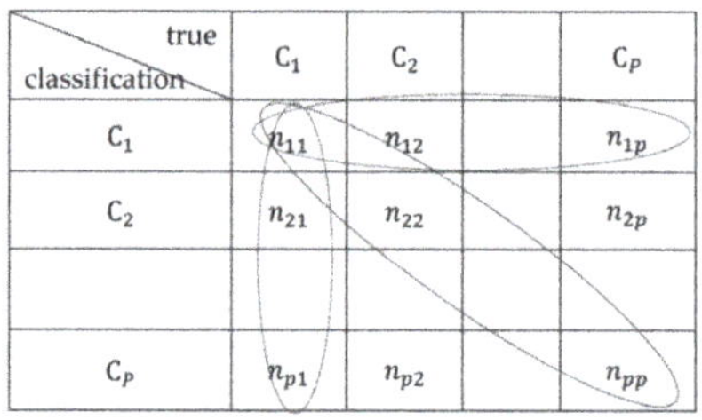

Figure 10. *P*-class confusion matrix.

We also take the nonparametric McNemar's test based on the standardized normal test statistic to evaluate the statistical significance in the improvement of OA with different hyperspectral classification algorithms. The McNemar's test statistic for two different algorithms noted as Algorithm 1 and Algorithm 2 can be calculated as [36]:

$$z = (f_{12} - f_{21}) / \sqrt{f_{12} + f_{21}}, \tag{19}$$

where f_{12} denotes the number of samples misclassified using Algorithm 2 but not Algorithm 1, and f_{21} means the number of samples misclassified using Algorithm 1 but not Algorithm 2. $|z|$ is the absolute value of z. For 5% level of significance, the $|z|$ value is 1.96. If a $|z|$ value is greater than this quantity, the two classification algorithms have significant discrepancy.

5. Experimental Results and Discussion

In this section, the proposed TFE and the novel classification framework will be evaluated and the relevant results will be summarized and discussed in detail.

5.1. Compared Methods and Band Groups

To analyze and evaluate our proposed algorithm, which combines the TFE and the optimal DBN efficiently, existing algorithm, such as SVM with Radial Basis Function kernel (SVM-RBF), the Radical Basis Function neural network (RBFNN) and CNN, are employed for comparison purpose. Besides, we also compare with a state-of-the-art spectral–spatial algorithm called EPF-G-c [22]. All these algorithms are widely used with excellent performance in hyperspectral image classification tasks, especially EPF-G-c. In addition, for evaluating our proposed texture feature enhancement (TFE) algorithm, we also applied TFE algorithm on the traditional SVM-RBF and RBFNN. All experiments are repeated 10 times with the average classification results demonstrated for comparison.

According to our proposed band grouping solution, the bands of Indian Pines can be divided into 41 groups: 1, 2, 3, 4–17, 18, 19–33, 34, 35, 36, 37–56, 57, 58–60, 61, 62, 63–74, 75, 76, 77–82, 83, 84, 85, 86,

87, 88, 89, 90, 91, 92–93, 94, 95, 96–97, 98–102, 103, 104, 105, 106–143, 144, 145, 146–198, 199 and 200. The bands of University of Pavia can be divided into 19 groups: 1, 2, 3, 4, 5, 6, 7, 8–68, 69, 70, 71, 72, 73, 74, 75, 76, 77, 78–84 and 85–103. The bands of Salinas can be divided into 21 groups: 1, 2, 3, 4, 5–35, 36, 37, 38, 39, 40, 41–104, 105–106, 107, 108, 109–146, 147, 148, 149–201, 202, 203 and 204. All these band groups are employed in the TFE algorithm.

5.2. Discussion on Effectiveness of the Proposed TFE

Figure 11 demonstrates the reconstructions of border and inner pixels of four classes after TFE in Indian Pines dataset. The first image of each row depicts the locations of border and inner pixels. The reconstruction and reconstructed error of the border pixel are demonstrated in the second image of each row. Meanwhile, the reconstruction and reconstructed error of the inner pixel are demonstrated in the third image.

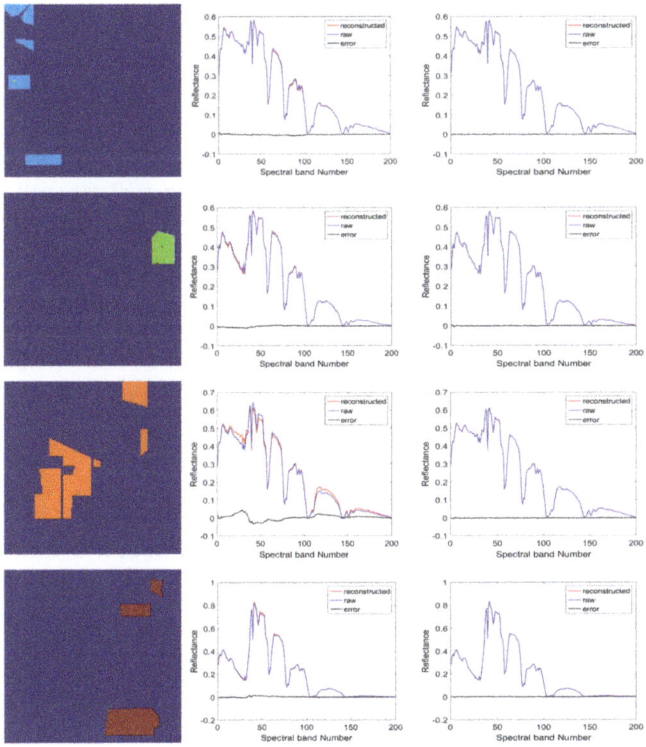

Figure 11. The reconstructions of the border-pixels and inner-pixels of different classes in Indian Pines. First row is the reconstruction information of Class 2, second row is the reconstruction information of Class 4, third row is the reconstruction information of Class 6 and last row is the reconstruction information of Class 8.

In hyperspectral classification, some spectra of the hyperspectral image are distorted through imaging noise or low spatial resolution, especially border-pixels, therefore the difficulty of hyperspectral classification primarily focuses on the correct classification of the border pixels. In Figure 11, it can be seen that, by utilizing TFE, the reconstructed border pixels become different from the original border pixels, and the reconstructed inner pixels are nearly the same as the original inner pixels, which implies that TFE plays an important role for border pixels. TFE can make border pixels distinct with its characteristics and more similar to their original spectra. Hence, the texture feature of the hyperspectral image become more obvious and clear. Consequently, the pixels that are difficult to distinguish can be recognized more easily than before with clearer texture feature. In other words, TFE has a positive effect for enhancing hyperspectral classification performance.

5.3. Discussion on Classification Results and Statistical Test

Table 6 provides the classification performance on Indian Pines achieved by different classification algorithms: SVM, RBFNN, optimal DBN (O_DBN), SVM combined with TFE (SVM_TFE), RBFNN combined with TFE (RBFNN_TFE), CNN, EFP-G-c and our proposed framework. O_DBN denotes the optimal DBN we proposed but without TFE. The SVM_TFE and RBFNN_TFE are two algorithms combined with the TFE method. The classification accuracy of each class is also listed in this table. In Table 6, we can see that our proposed framework can obtain the superior performance compared with other algorithms. Meanwhile, the optimal DBN has the best classification accuracy compared to the other algorithms without TFE, such as SVM and RBFNN. Although EFP-G-c is an outstanding spectral–spatial hyperspectral classification algorithm, our proposed framework utilizing TFE still has slightly better classification accuracy. Besides, SVM_TFE and RBFNN_TFE outperform SVM and RBFNN, respectively. The OA of SVM_TFE is 5.06% greater than SVM, and the OA of RBFNN_TFE is 8.97% higher than RBFNN. Compared with O_DBN, the OA obtained via our proposed framework improved by 8.08% and the Kappa increased by 9.98%. All these facts indicate the successful effects of TFE and demonstrates that our proposed framework and TFE have good influence on Indian Pines in hyperspectral classification.

Table 6. Classification accuracy of different algorithms on Indian Pines.

Class	SVM	RBFNN	O_DBN	SVM_TFE	RBFNN_TFE	CNN	EPF-G-c	Our Proposed
1	0.8578	0.8672	0.8562	0.9069	0.9638	0.9107	0.9757	0.9690
2	0.9251	0.9288	0.9532	0.9625	0.9944	0.7783	0.9736	0.9888
3	0.9391	0.9543	0.9594	0.9594	0.9949	0.8462	0.9314	0.9594
4	0.9841	1	0.9947	1	1	0.9793	0.9793	1
5	0.9162	0.9237	0.9172	0.9506	0.9910	0.7842	0.9268	0.9880
6	0.8054	0.7975	0.8189	0.8962	0.9553	0.9348	0.9855	0.9613
7	0.9363	0.9459	0.9490	0.9522	0.9809	0.8442	0.9873	0.9682
8	0.9940	0.9950	0.9909	1	1	0.9929	0.9881	1
OA	0.8837	0.8854	0.8948	0.9343	0.9751	0.8983	0.9754	**0.9756**
AA	0.9197	0.9265	0.9270	0.9535	0.9850	0.8838	0.9685	**0.9793**
Kappa	0.8559	0.8582	0.8617	0.9180	0.9688	0.8736	0.9692	**0.9694**

Table 7 lists the classification precision achieved via these different classification algorithms. In Table 7, we can see that the precision of our proposed algorithm outperforms SVM, RBFNN, O_DBN, SVM_TFE, RBFNN_TEF, CNN and EPF-G-c. In addition, the methods associated with TFE have better classification precision than without TFE.

Table 7. Classification precision of different algorithms on Indian Pines.

Class	SVM	RBFNN	O_DBN	SVM_TFE	RBFNN_TFE	CNN	EPF-G-c	Our Proposed
1	0.8585	0.8643	0.8563	0.9132	0.9646	0.8440	0.9474	0.9571
2	0.7577	0.7631	0.7496	0.8877	0.9620	0.9270	0.9606	0.9661
3	0.9113	0.9353	0.8400	0.8873	0.9849	0.9492	0.9645	0.9692
4	0.9688	0.9844	0.9495	0.9895	1	1.0000	1	1
5	0.7917	0.7434	0.8037	0.8675	0.9272	0.9087	0.965	0.9396
6	0.9307	0.9341	0.9417	0.9643	0.9862	0.8538	0.9686	0.9836
7	0.7861	0.8710	0.8466	0.8617	0.9716	0.9490	0.9936	0.9882
8	0.9930	0.9940	0.9970	0.9990	1	0.9909	1	1
Precision	0.8747	0.8862	0.8731	0.9213	0.9746	0.9278	0.9750	**0.9755**

Tables 8 and 10 present the classification accuracy acquired via different algorithms for University of Pavia and Salinas datasets. Meanwhile, Tables 9 and 11 also list the precisions obtained through our proposed model and other classification algorithms on different datasets. It is obvious in Tables 8 and 10 that our proposed framework has better performance than other classification methods. Especially, we can see that all algorithms that integrate TFE outperform those without TFE. By employing the TFE, the performance of SVM increased by 5.78% in University of Pavia and 1.75% in Salinas, while the performance of RBFNN improved by 6.8% in University of Pavia and 1.55% in Salinas. The OA achieved by the proposed framework is 6.55% higher than the OA achieved via optimal DBN in University of Pavia and 3.94% larger than the OA achieved via optimal DBN in Salinas. Furthermore, the proposed classification framework has better performance than CNN and EPF-G-c. As for kappa coefficients, we can see that our proposed framework has better consistency. The possible reason is the ability of our proposed framework, as a deep network, to extract high-level features of data is stronger than the RBFN and the SVM, as shallow networks, thus the description ability of our proposed framework is more stable. In Tables 9 and 11, the precisions obtained through our proposed model on different datasets are better than precisions achieved via other algorithms. Furthermore, our proposed TFE has a positive effect on classification accuracy.

Table 8. Classification accuracy of different algorithms on University of Pavia.

Class	SVM	RBFNN	O_DBN	SVM_TFE	RBFNN_TFE	CNN	EPF-G-c	Our Proposed
1	0.7466	0.7733	0.8650	0.8534	0.9029	0.9758	0.9579	0.9458
2	0.8442	0.8980	0.9281	0.9058	0.9601	0.9832	0.9993	0.9728
3	0.8533	0.8377	0.8410	0.8922	0.9305	0.7795	0.9511	0.9550
4	0.9801	0.9602	0.9765	0.9772	0.9787	0.9096	0.9677	0.9881
5	0.9990	0.9990	0.9990	0.9981	0.9971	0.9830	0.9372	0.9990
6	0.9108	0.9492	0.9125	0.9558	0.9903	0.8153	0.9263	0.9873
7	0.9456	0.9583	0.8990	0.9544	0.9932	0.6680	0.9885	0.9893
8	0.8430	0.8628	0.8613	0.9101	0.9571	0.8562	0.9421	0.9438
9	1	1	0.9985	1	1	0.9985	0.9895	1.0000
OA	0.8555	0.8888	0.9123	0.9133	0.9568	0.9211	0.9671	**0.9696**
AA	0.9025	0.9154	0.9201	0.9385	0.9678	0.8855	0.9622	**0.9757**
Kappa	0.8103	0.8525	0.8824	0.8845	0.9418	0.8943	0.9590	**0.9590**

Table 9. Classification precision of different algorithms on University of Pavia.

Class	SVM	RBFNN	O_DBN	SVM_TFE	RBFNN_TFE	CNN	EPF-G-c	Our Proposed
1	0.9795	0.9798	0.9675	0.9836	0.9877	0.8531	0.9822	0.9837
2	0.9720	0.9841	0.9763	0.9869	0.9980	0.9341	0.9756	0.9978
3	0.6657	0.6905	0.7568	0.7803	0.8876	0.8766	0.9711	0.9261
4	0.7657	0.8906	0.8207	0.9122	0.9808	0.9678	0.9642	0.9437
5	0.9831	0.9981	0.9849	0.9943	1	0.9981	0.9900	0.9877
6	0.6714	0.7388	0.7735	0.7456	0.8639	0.9484	0.9450	0.9189
7	0.5084	0.5583	0.6515	0.7567	0.8575	0.9592	0.9157	0.9586
8	0.8312	0.8028	0.8645	0.8394	0.8772	0.8752	0.9864	0.9117
9	1	1	0.9985	0.9985	1	0.9985	0.8779	0.9969
Precision	0.8197	0.8492	0.8660	0.8886	0.9392	0.9345	0.9565	**0.9583**

Table 10. Classification accuracy of different algorithms on Salinas Dataset.

Class	SVM	RBFNN	O_DBN	SVM_TFE	RBFNN_TFE	CNN	EPF-G-c	Our Proposed
1	0.9965	0.9971	0.9947	0.9982	0.9988	1.0000	1.0000	0.9947
2	0.9947	0.9947	1	0.9956	0.9950	0.9933	0.9994	0.9962
3	0.9976	0.9988	0.9976	0.9976	0.9982	0.9589	0.9994	0.9976
4	0.9963	0.9963	0.9963	0.9954	0.9954	0.9838	0.9973	0.9973
5	0.9886	0.9849	0.9811	0.9874	0.9899	0.9898	0.9992	0.9853
6	0.9981	0.9986	0.9978	0.9981	0.9981	0.9995	0.9984	0.9973
7	0.9970	0.9963	0.9957	0.9960	0.9966	0.9988	0.9989	0.9963
8	0.8606	0.8567	0.8315	0.8761	0.8893	0.8379	0.8690	0.9085
9	0.9934	0.9985	0.9939	0.9942	0.9966	0.9896	0.9911	0.9949
10	0.9661	0.9758	0.9426	0.9698	0.9775	0.8848	0.9715	0.9614
11	0.9987	0.9961	0.9961	0.9987	0.9987	0.8919	1	1
12	0.9994	1	1	0.9994	1	0.9685	0.9992	0.9994
13	0.9968	0.9951	0.9984	0.9951	0.9951	0.9534	0.9987	0.9968
14	0.9792	0.9857	0.9948	0.9857	0.9805	0.9159	0.9978	0.9948
15	0.6972	0.7336	0.7646	0.7941	0.7916	0.7673	0.8856	0.9127
16	0.9920	0.9900	0.9854	0.9920	0.9914	0.9695	1	0.9887
OA	0.9212	0.9266	0.9228	0.9387	0.9421	0.9155	0.9543	**0.9622**
AA	0.9658	0.9687	0.9669	0.9733	0.9746	0.9439	0.9816	**0.9826**
Kappa	0.9114	0.9175	0.9133	0.9312	0.9350	0.9051	0.9486	**0.9575**

Table 11. Classification precision of different algorithms on Salinas Dataset.

Class	SVM	RBFNN	O_DBN	SVM_TFE	RBFNN_TFE	CNN	EPF-G-c	Our Proposed
1	0.9988	0.9994	0.9971	0.9994	1	0.9801	1	1
2	0.9985	0.9985	0.9980	0.9994	0.9994	0.9947	0.9995	0.9991
3	0.9744	0.9721	0.9489	0.9830	0.9824	0.9976	0.9782	0.9682
4	0.9909	0.9864	0.9847	0.9918	0.9900	0.9973	0.9991	0.9900
5	0.9941	0.9970	0.9978	0.9920	0.9895	0.9315	0.9987	0.9924
6	0.9995	0.9997	0.9884	0.9992	0.9995	0.9978	0.9997	0.9940
7	0.9966	1	1	0.9951	1	0.9957	0.9991	0.9973
8	0.8209	0.8372	0.8592	0.8729	0.8726	0.8952	0.9162	0.9415
9	0.9956	0.9916	0.9898	0.9931	0.9927	0.9810	0.9475	0.9926
10	0.9517	0.9735	0.8699	0.9534	0.9674	0.9325	0.9627	0.9487
11	0.9808	0.9922	0.8242	0.9935	0.9948	0.9831	0.9994	0.9785
12	0.9909	0.9897	0.9748	0.9933	0.9921	1.0000	0.9987	0.9933
13	0.9777	0.9919	0.9935	0.9871	0.9839	0.9968	0.9920	0.9731
14	0.8737	0.9245	0.8235	0.9256	0.8945	0.9506	0.9359	0.8899
15	0.7803	0.7747	0.7344	0.8187	0.8341	0.5128	0.7777	0.8559
16	0.9701	0.9920	0.9861	0.9658	0.9953	0.9854	0.9946	0.9900
Precision	0.9559	0.9638	0.9356	0.9665	0.9680	0.9458	0.9687	**0.9690**

Figures 12–14 demonstrate the classification maps obtained in Indian Pines, University of Pavia and Salinas, respectively. Clearly, the classification maps shown in Figures 12–14 achieved by our proposed framework are the smoothest and clearest. The classification accuracy of border pixels in these datasets is improved greatly and the boundaries of different classes are more distinct. Compared to other classification algorithms, the results of our proposed framework are better because they contain less salt-and-pepper noise.

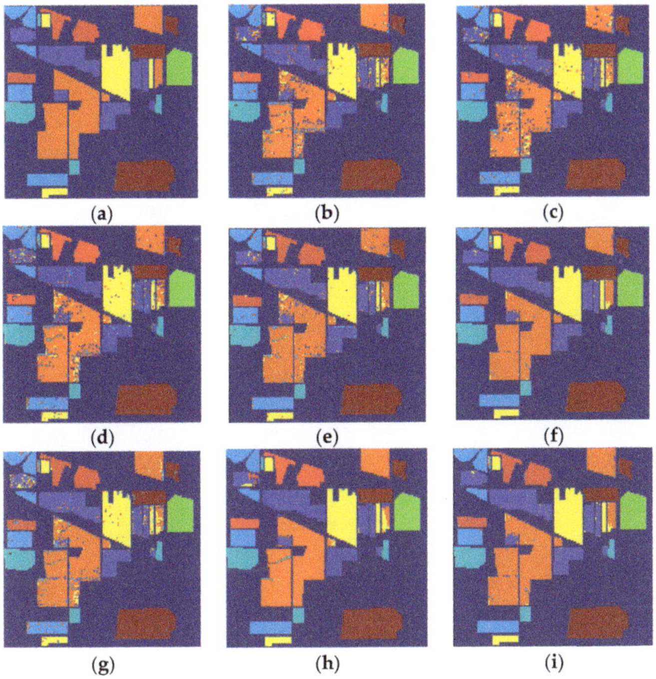

Figure 12. The classification maps obtained via different algorithms in Indian Pines: (**a**) Ground truth; (**b**) SVM; (**c**) RBFNN; (**d**) O_DBN; (**e**) SVM_TFE; (**f**) RBFNN_TFE; (**g**) CNN; (**h**) EFP-G-c; and (**i**) the proposed framework.

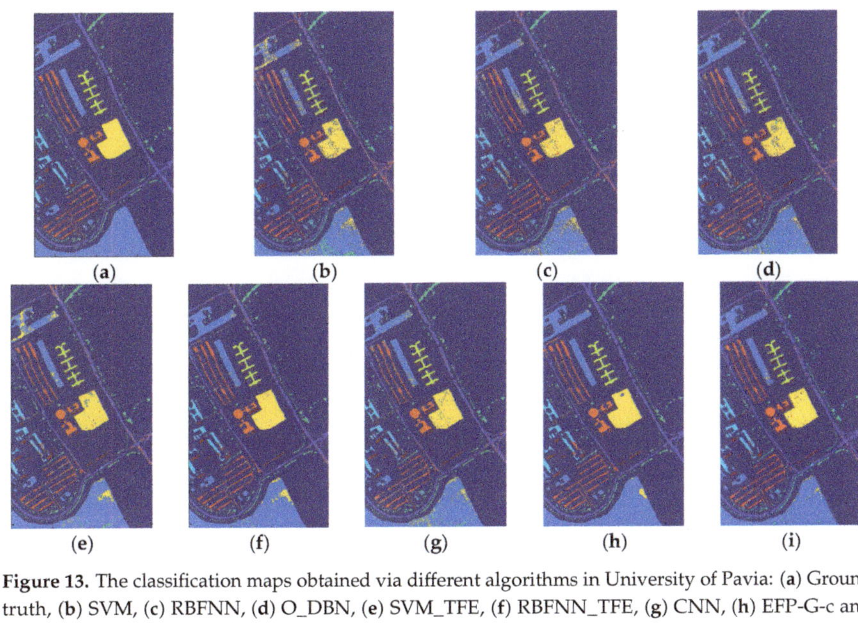

Figure 13. The classification maps obtained via different algorithms in University of Pavia: (**a**) Ground truth, (**b**) SVM, (**c**) RBFNN, (**d**) O_DBN, (**e**) SVM_TFE, (**f**) RBFNN_TFE, (**g**) CNN, (**h**) EFP-G-c and (**i**) the proposed framework.

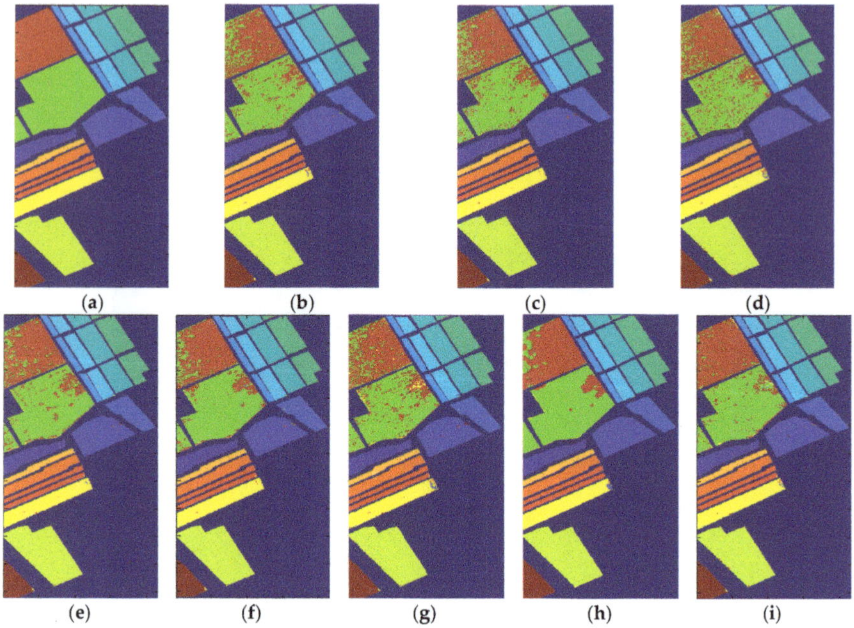

Figure 14. The classification maps obtained via different algorithms in Salinas Dataset: (**a**) Ground truth, (**b**) SVM, (**c**) RBFNN, (**d**) O_DBN, (**e**) SVM_TFE, (**f**) RBFNN_TFE, (**g**) CNN, (**h**) EFP-G-c and (**i**) the proposed framework.

Table 12 presents the average $|z|$ values achieved from Indian Pines, Pavia University and Salinas of the proposed classification framework as well as other classification algorithms. A "yes" here denotes the two classification algorithms in McNemar's test have significant performance discrepancy. Obviously, the proposed classification framework is statistically different from its counterparts with 5% significance level.

Table 12. ($|z|$ values/Siginificant?) in the McNemar's Test.

Algorithms	Indian Pines	Pavia University	Salinas
SVM	31.16/Yes	68.33/Yes	41.19/Yes
RBFNN	31.34/Yes	69.27/Yes	41.39/Yes
O_DBN	2.78/Yes	3.74/Yes	3.32/Yes
SVM_TFE	31.95/Yes	73.29/Yes	41.21/Yes
RBFNN_TFE	32.82/Yes	74.84/Yes	42.49/Yes
CNN	3.50/Yes	3.00/Yes	4.49/Yes
EPF_G_c	32.16/Yes	75.13/Yes	41.21/Yes

Note: 5% significance level is selected.

6. Conclusions

In this paper, we investigate a novel hyperspectral classification framework based on an optimal DBN algorithm. In our proposed framework, we develop a new TFE algorithm that employs multi-texture features and band grouping method. The resulting classification framework can offer better classification accuracy than other classic algorithms. To further test our proposed TFE algorithm, a series of experiments based on the combination of the state-of-the-art algorithms and the TFE algorithm are applied on the three classic hyperspectral datasets. Experimental results demonstrate that the algorithms with TFE outperform those without TFE, which implies that our proposed TFE can play an important role in improving hyperspectral classification performance. We believe that the proposed hyperspectral classification framework based on the optimal DBN and TFE is more suitable to process hyperspectral data in practical applications when training samples are limited.

Acknowledgments: This work was partially supported by the National Nature Science Foundation of China (Nos. 61571345, 91538101, 61501346, 61502367 and 61701360) and the 111 project (B08038). It was also partially supported by the Fundamental Research Funds for the Central Universities JB170109, the Natural Science Basic Research Plan in Shaanxi Province of China (No. 2016JQ6023) and General Financial Grant from the China Postdoctoral Science Foundation (No. 2017M623124).

Author Contributions: L.J. and L.Y.S. conceived and designed the study; L.J. performed the experiments; X.B. analyzed the data; L.J. and X.B. wrote the paper; and W.K.Y. and D.Q. reviewed and edited the manuscript. All authors read and approved the manuscript.

Conflicts of Interest: The authors declare no conflict of interest. The funding sponsors had no role in the design of the study; in the collection, analyses, or interpretation of data; in the writing of the manuscript, and in the decision to publish the results.

References

1. Adão, T.; Hruška, J.; Pádua, L.; Bessa, J.; Peres, E.; Morais, R.; Sousa, J.J. Hyperspectral Imaging: A Review on UAV-Based Sensors, Data Processing and Applications for Agriculture and Forestry. *Remote Sens. Agric. Veg.* **2017**, *9*, 1110. [CrossRef]
2. Yokoya, N.; Chan, J.C.W.; Segl, K. Potential of Resolution-Enhanced Hyperspectral Data for Mineral Mapping Using Simulated EnMAP and Sentinel-2 Images. *Remote Sens.* **2016**, *8*, 172. [CrossRef]
3. Merentitis, A.; Debes, C.; Heremans, R. Ensemble Learning in Hyperspectral Image Classification: Toward Selecting a Favorable Bias-Variance Tradeoff. *IEEE J. STARS* **2014**, *7*, 1089–1102. [CrossRef]
4. He, J.; He, Y.; Zhang, C. Determination and Visualization of Peimine and Peiminine Content in Fritillaria thunbergii Bulbi Treated by Sulfur Fumigation Using Hyperspectral Imaging with Chemometrics. *Molecules* **2017**, *22*, 1402. [CrossRef] [PubMed]
5. Richards, J.A.; Jia, X. Using Suitable Neighbors to Augment the Training Set in Hyperspectral Maximum Likelihood Classification. *IEEE Geosci. Remote Sens. Lett.* **2008**, *5*, 774–777. [CrossRef]
6. Leonenko, G.; Los, S.O.; North, P.R.J. Statistical Distances and Their Applications to Biophysical Parameter Estimation: Information Measures, M-Estimates, and Minimum Contrast Methods. *Remote Sens.* **2013**, *5*, 1355–1388. [CrossRef]
7. Zhang, J.; Mani, I. KNN Approach to Unbalanced Data Distributions: A Case Study Involving Information Extraction. In Proceedings of the ICML 2003 Learning Imbalanced Datasets, Washington, DC, USA, 21–24 August 2003.
8. Mathew, J.; Luo, M.; Pang, C.K.; Chan, H.L. Kernel-based SMOTE for SVM classification of imbalanced datasets. In Proceedings of the IECON 2015 41st Annual Conference of the IEEE Industrial Electronics Society, Yokohama, Japan, 9–12 November 2015; pp. 001127–001132.
9. Immitzer, M.; Atzberger, C.; Koukal, T. Tree Species Classification with Random Forest Using Very High Spatial Resolution 8-Band WorldView-2 Satellite Data. *Remote Sens.* **2012**, *4*, 2661–2693. [CrossRef]
10. Dobigeon, N.; Tourneret, J.Y.; Chang, C.I. Semi-Supervised Linear Spectral Unmixing Using a Hierarchical Bayesian Model for Hyperspectral Imagery. *IEEE Trans. Signal Process.* **2008**, *56*, 2684–2695. [CrossRef]
11. Zhang, L.; Wei, W.; Zhang, Y.; Li, F.; Yan, H. Structured sparse BAYESIAN hyperspectral compressive sensing using spectral unmixing. In Proceedings of the 2014 6th Workshop on Hyperspectral Image and Signal Processing: Evolution in Remote Sensing (WHISPERS), Lausanne, Switzerland, 24–27 June 2014.
12. Yu, H.; Gao, L.; Li, J.; Li, S.S.; Zhang, B.; Benediktsson, J.A. Spectral-Spatial Hyperspectral Image Classification Using Subspace-Based Support Vector Machines and Adaptive Markov Random Fields. *Remote Sens.* **2016**, *8*, 355. [CrossRef]
13. Chen, H.M.; Wang, H.C.; Chai, J.W.; Chen, C.C.C.; Xue, B.; Wang, L.; Yu, C.; Wang, Y.; Song, M.; Chang, C.I. A Hyperspectral Imaging Approach to White Matter Hyperintensities Detection in Brain Magnetic Resonance Images. *Remote Sens.* **2017**, *9*, 1174. [CrossRef]

14. Kayabol, K. Bayesian Gaussian mixture model for spatial-spectral classification of hyperspectral images. In Proceedings of the 2015 23rd European Signal Processing Conference (EUSIPCO), Nice, France, 31 August–4 September 2015; pp. 1805–1809.
15. Ramo, R.; Chuvieco, E. Developing a Random Forest Algorithm for MODIS Global Burned Area Classification. *Remote Sens.* **2017**, *9*, 1193. [CrossRef]
16. Starck, J.; Elad, M.; Donoho, D. Image decomposition via the combination of sparse representation and a variational approach. *IEEE Trans. Image Process.* **2005**, *14*, 1570–1582. [CrossRef] [PubMed]
17. Wright, J.; Yang, A.Y.; Ganesh, A.; Sastry, S.S.; Ma, Y. Robust face recognition via sparse representation. *IEEE Trans. Pattern Anal. Mach. Intell.* **2009**, *31*, 210–226. [CrossRef] [PubMed]
18. Zhang, L.; Zhou, W.D.; Chang, P.C.; Yan, Z.; Wang, T.; Li, F.Z. Kernel Sparse Representation-Based Classifier. *IEEE Trans. Signal Process.* **2012**, *60*, 1684–1695. [CrossRef]
19. Chen, Y.; Nasrabadi, N.M.; Tran, T.D. Hyperspectral image classification using dictionary-based sparse representation. *IEEE Trans. Geosci. Remote Sens.* **2011**, *49*, 3973–3985. [CrossRef]
20. Chen, Y.; Nasrabadi, N.M.; Tran, T.D. Hyperspectral image classification via kernel sparse representation. *IEEE Trans. Geosci. Remote Sens.* **2013**, *51*, 217–231. [CrossRef]
21. Li, W.; Tramel, E.W.; Prasad, S.; Fowler, J.E. Nearest regularized subspace for hyperspectral classification. *IEEE Trans. Geosci. Remote Sens.* **2014**, *52*, 477–489. [CrossRef]
22. Kang, X.; Li, S.; Benediktsson, J.A. Spectral–Spatial Hyperspectral Image Classification with Edge-Preserving Filtering. *IEEE Trans. Geosci. Remote Sens.* **2014**, *52*, 2666–2677. [CrossRef]
23. Chen, C.; Li, W.; Su, H.; Liu, K. Spectral-Spatial Classification of Hyperspectral Image Based on Kernel Extreme Learning Machine. *Remote Sens.* **2014**, *6*, 5795–5814. [CrossRef]
24. Wang, T.; Zhang, H.; Lin, H.; Fang, C. Textural–Spectral Feature-Based Species Classification of Mangroves in Mai Po Nature Reserve from Worldview-3 Imagery. *Remote Sens.* **2016**, *8*, 24. [CrossRef]
25. Zhong, Y.; Jia, T.; Zhao, J.; Wang, X.; Jin, S. Spatial-Spectral-Emissivity Land-Cover Classification Fusing Visible and Thermal Infrared Hyperspectral Imagery. *Remote Sens.* **2017**, *9*, 910. [CrossRef]
26. Peng, B.; Li, W.; Xie, X.; Du, Q.; Liu, K. Weighted-Fusion-Based Representation Classifiers for Hyperspectral Imagery. *Remote Sens.* **2015**, *7*, 14806–14826. [CrossRef]
27. Fukushima, K. Neocognitron: A hierarchical neural network capable of visual pattern recognition. *Neural Netw.* **1988**, *1*, 119–130. [CrossRef]
28. Ciresan, D.C.; Meier, U.; Masci, J.; Gambardella, L.M.; Schmidhuber, J. Flexible, high performance convolutional neural networks for image classification. In Proceedings of the Joint Conference Artificial Intelligence (IJCAI'11), Barcelona, Catalonia, Spain, 16–22 July 2011; pp. 1237–1242.
29. Lee, H.; Kwon, H. Going Deeper with Contextual CNN for Hyperspectral Image Classification. *IEEE Trans. Image Process.* **2017**, *26*, 4843–4855. [CrossRef] [PubMed]
30. Chen, Y.; Jiang, H.; Li, C.; Jia, X.; Ghamisi, P. Deep Feature Extraction and Classification of Hyperspectral Images Based on Convolutional Neural Networks. *IEEE Trans. Geosci. Remote Sens.* **2016**, *54*, 6232–6251. [CrossRef]
31. Chen, Y.; Zhao, X.; Jia, X. Spectral–Spatial Classification of Hyperspectral Data Based on Deep Belief Network. *IEEE J. STARS* **2015**, *8*, 2381–2392. [CrossRef]
32. Özdemir, A.O.B.; Gedik, B.E.; Çetin, C.Y.Y. Hyperspectral classification using stacked autoencoders with deep learning. In Proceedings of the 2014 6th Workshop on Hyperspectral Image and Signal Processing: Evolution in Remote Sensing (WHISPERS), Lausanne, Switzerland, 24–27 June 2014.
33. Hinton, G.E.; Slakhutdinov, R.R. Reducing the Dimensionality of Data with Neural Networks. *Science* **2006**, *313*, 504–507. [CrossRef] [PubMed]
34. Ohanian, P.P.; Dubes, R.C. Performance evaluation for four classes of textural features. *Pattern Recognit.* **1992**, *25*, 819–833. [CrossRef]
35. Xue, B.; Yu, C.; Wang, Y.; Song, M.; Li, S.; Wang, L. A subpixel target detection approach to hyperspectral image classification. *IEEE Trans. Geosci. Remote Sens.* **2017**, *55*, 5093–5114. [CrossRef]
36. Foody, G.M. Thematic map comparison: Evaluating the statistical significance of differences in classification accuracy. *Photogramm. Eng. Remote Sens.* **2004**, *70*, 627–633. [CrossRef]

remote sensing

MDPI

Article

Band Subset Selection for Hyperspectral Image Classification

Chunyan Yu [1], Meiping Song [1,2,*] and Chein-I Chang [1,3,4,5]

1 Center for Hyperspectral Imaging in Remote Sensing (CHIRS), Information and Technology College, Dalian Maritime University, Dalian 116026, China; yucy@dlmu.edu.cn (C.Y.); cchang@umbc.edu (C.-I.C.)
2 State Key Laboratory of Integrated Services Networks, School of Telecommunications Engineering, Xidian University, Xi'an 710071, China
3 Department of Computer Science and Information Engineering, National Yunlin University of Science and Technology, Douliu 64002, Taiwan
4 Department of Computer Science and Information Management, Providence University, Taichung 02912, Taiwan
5 Department of Computer Science and Electrical Engineering, University of Maryland, Baltimore County, Baltimore, MD 21250, USA
* Correspondence: smping@dlmu.edu.cn or smping@163.com; Tel.: +86-131-9016-4770

Received: 24 November 2017; Accepted: 10 January 2018; Published: 15 January 2018

Abstract: This paper develops a new approach to band subset selection (BSS) for hyperspectral image classification (HSIC) which selects multiple bands simultaneously as a band subset, referred to as simultaneous multiple band selection (SMMBS), rather than one band at a time sequentially, referred to as sequential multiple band selection (SQMBS), as most traditional band selection methods do. In doing so, a criterion is particularly developed for BSS that can be used for HSIC. It is a linearly constrained minimum variance (LCMV) derived from adaptive beamforming in array signal processing which can be used to model misclassification errors as the minimum variance. To avoid an exhaustive search for all possible band subsets, two numerical algorithms, referred to as sequential (SQ) and successive (SC) algorithms are also developed for LCMV-based SMMBS, called SQ LCMV-BSS and SC LCMV-BSS. Experimental results demonstrate that LCMV-based BSS has advantages over SQMBS.

Keywords: band selection (BS); band subset selection (BSS); hyperspectral image classification; linearly constrained minimum variance (LCMV); Otsu's method; successive LCMV-BSS (SC LCMV-BSS); sequential LCMV-BSS (SQ LCMV-BSS)

1. Introduction

Hyperspectral image classification has received considerable interest in recent years [1–23]. Its band selection (BS) issue has been also studied extensively [24–57]. In general, there are two approaches to BS. One is to select bands one at a time, sequentially; this is referred to as sequential multiple band selection (SQMBS). In this case, a criterion that can be used to select bands, according to priorities ranked by the criterion, is usually required. Such a criterion is referred to as a band prioritization (BP) criterion, and it can be designed according to two perspectives. One type of BP criterion is based on data characteristics or statistics such as variance, signal-to-noise ratio (SNR), entropy, and information divergence (ID) to calculate a priority score for each of the individual bands in order to rank them [25]. As a result, such BP-based SQMBS is generally unsupervised and is not adaptive to any particular application. In other words, the same selected bands are also applied to all different applications. The other type of BP criterion is supervised and is adaptive to a particular application, such as classification [26–57], target detection [49,50], endmember extraction [51], spectral unmixing [52], etc. Unfortunately, one of major problems with BP-derived BS methods is how to deal with band correlation. Since hyperspectral imagery has very high interband correlation, the fact that a band has a high priority to be selected implies that its adjacent bands also have high priorities to be selected. To avoid this dilemma, band decorrelation may be required to remove redundant bands from a group of selected bands. However, this also comes with two issues, i.e., how to select a band correlation criterion to measure the correlation between two bands, and how to determine the threshold for two bands that are sufficiently decorrelated.

As an alternative to BP-based SQMBS methods, another approach, referred to as simultaneous multiple band selection (SMMBS), is to select multiple bands simultaneously as a band subset. This approach does not have issues in prioritizing bands or decorrelating bands that are encountered in SQMBS. However, the price paid for these advantages is how to develop an effective search strategy to find an optimal band subset, since it generally requires an exhaustive search, which is practically infeasible. To address this issue, several works have been recently proposed, such as band clustering [58–60], particle swarm optimization (PSO) in [35], firefly algorithm (FA) in [36], multitask sparsity pursuit (MTSP) [38], multigraph determinantal point process (MDPP) [43], dominant set extraction BS (DSEBS) in [40], etc. Of particular interest is a new concept of band subset selection (BSS) to address this issue which is quite different from the aforementioned SMMBS methods in the sense of the search strategy to be used for finding an optimal set of multiple bands. It considers a selected band as a desired endmember. Accordingly, finding an optimal set of endmembers from all data sample vectors can be translated to selecting an optimal band subset simultaneously from all bands. With this interpretation, two sequential algorithms designed to realize an N-finder algorithm (N-FINDR) [61] numerically, called sequential N-FINDR (SQ N-FINDR) and successive N-FINDR (SC N-FINDR) [62–65] can be redesigned to find desired band subsets, called SQ BSS and SC BSS algorithms. These two SQ BSS and SC BSS algorithms were recently developed for SMMBS in applications of anomaly detection [66] and spectral unmixing and classification [67,68]. This paper further extends BSS to hyperspectral image classification and has several different aspects not found in [66–68]. First and foremost is the criterion used for BSS, which is the minimum variance resulting from a linearly constrained finite impulse response filter arising in adaptive beamforming in array signal processing [69–72]. This linearly constrained minimum variance (LCMV)-based BSS interprets signal sources as class signature vectors and linearly constrains the class signature vectors, finding an optimal band subset for classification. It is very different from constrained energy minimization (CEM)-based BS [26], which constrains a single selected band, and also from constrained multiple band selection (CMBS) [68], which extends CEM-BS by constraining multiple bands as band subsets, not as class signature vectors as LCMV-BSS does. Secondly, two new SQ BSS and SC BSS algorithms are developed for LCMV-BSS, specifically for classification, referred to as SQ LCMV-BSS and SC LCMV-BSS. Thirdly, the classifier used to evaluate BS performance is also an LCMV classifier which is particularly designed to best utilize the bands selected by LCMV-BSS. Fourthly, despite the fact that LCMV-BSS may not exhaust all possible band combinations, to the authors' best knowledge, LCMV-BSS is probably the only BSS algorithm to search band subsets among all possible band combinations numerically compared to other SMMBS algorithms such as PSO, FA, MTSP, MDPP, DSEBS which are indeed designed to run only a very small selected set of band subsets. Finally, and most importantly, the proposed LVMV-BSS is very easy to implement because there are no parameters that need to be tuned, as many BS methods have. This is a tremendous advantage since such parameters must be adaptive to various applications.

2. LCMV Criterion for BSS

Suppose that there are M classes of interest and each class is specified by a class signature vector, denoted by $\mathbf{d}_1$, $\mathbf{d}_2$, $\cdots$, $\mathbf{d}_M$. We can now form a class signature matrix, denoted by $\mathbf{D} = [\mathbf{d}_1\mathbf{d}_2\cdots\mathbf{d}_M]$. The goal is to design an FIR linear filter with L filter coefficients $\{w_1, w_2, \cdots, w_L\}$, denoted by an L-dimensional vector $\mathbf{w} = (w_1, w_2, \cdots, w_L)^T$ that minimizes the filter output energy subject to the following constraint:

$$\mathbf{D}^T\mathbf{w} = \mathbf{c} \text{ where } \mathbf{d}_j^T\mathbf{w} = \sum_{l=1}^{L} w_l t_{jl} \text{ for } 1 \leq j \leq M \tag{1}$$

where $\mathbf{c} = (c_1, c_2, \cdots, c_k)^T$ is a constraint vector. Using (1), we derive the following linearly constrained optimization problem:

$$\min_{\mathbf{w}}\left\{\mathbf{w}^T\mathbf{R}\mathbf{w}\right\} \text{ subject to} \mathbf{D}^T\mathbf{w} = \mathbf{c} \tag{2}$$

where $\mathbf{R} = (1/N)\sum_{i=1}^{N}\mathbf{r}_i\mathbf{r}_i^T$ is the autocorrelation sample matrix of the image. The solution to (2) is called the LCMV-based classifier and can be obtained in [69,71,72] by

$$\delta^{\mathrm{LCMV}}(\mathbf{r}) = \left(\mathbf{w}^{\mathrm{LCMV}}\right)^T\mathbf{r} \tag{3}$$

with

$$\mathbf{w}^{\mathrm{LCMV}} = \mathbf{R}^{-1}\mathbf{D}\left(\mathbf{D}^T\mathbf{R}^{-1}\mathbf{D}\right)^{-1}\mathbf{c}. \tag{4}$$

Substituting (3) into (4) yields

$$\begin{aligned} &\left(\mathbf{w}^{\mathrm{LCMV}}\right)^T\mathbf{R}^{-1}\mathbf{w}^{\mathrm{LCMV}} \\ &= \left[\mathbf{R}^{-1}\mathbf{D}\left(\mathbf{D}^T\mathbf{R}^{-1}\mathbf{D}\right)^{-1}\mathbf{c}\right]^T\mathbf{R}^{-1}\left[\mathbf{R}^{-1}\mathbf{D}\left(\mathbf{D}^T\mathbf{R}^{-1}\mathbf{D}\right)^{-1}\mathbf{c}\right] \\ &= \mathbf{c}^T\left(\mathbf{D}^T\mathbf{R}^{-1}\mathbf{D}\right)^{-1}\mathbf{D}^T\mathbf{R}^{-1}\mathbf{D}\left(\mathbf{D}^T\mathbf{R}^{-1}\mathbf{D}\right)^{-1}\mathbf{c} = \mathbf{c}^T\left(\mathbf{D}^T\mathbf{R}^{-1}\mathbf{D}\right)^{-1}\mathbf{c} \end{aligned}. \tag{5}$$

According to [70], (5) is the minimum variance weighted by $\mathbf{R}^{-1}$. As a matter of fact, (5) can be also viewed as the minimal $\mathbf{R}^{-1}$-weighted least squares error (LSE) caused by misclassification errors from operating δ^{LCMV} on the entire image cube. For those who would like to learn more about LCMV, its details can be found in [69–71].

3. Band Subset Selection

A BS problem is generally described as follows. Assume that $J(.)$ is a generic objective function of $\boldsymbol{\Omega}_{\mathrm{BS}}$ for the BS to be optimized where $\boldsymbol{\Omega}_{\mathrm{BS}}$ is a band subset selected from a full band set $\boldsymbol{\Omega}$. For a given number n_{BS} of selected bands, a BS method is to find an optimal band subset $\boldsymbol{\Omega}_{\mathrm{BS}}^*$ with $|\boldsymbol{\Omega}_{\mathrm{BS}}| = n_{\mathrm{BS}}$ which satisfies the following optimization problem:

$$\boldsymbol{\Omega}_{\mathrm{BS}}^* = \arg\left\{\max/\min_{\boldsymbol{\Omega}_{\mathrm{BS}}\subset\boldsymbol{\Omega},\ |\boldsymbol{\Omega}_{\mathrm{BS}}|=n_{\mathrm{BS}}} J(\boldsymbol{\Omega}_{\mathrm{BS}})\right\}. \tag{6}$$

Depending upon how the objective function $J(\boldsymbol{\Omega}_{\mathrm{BS}})$ is designed, the optimization in (6) can be performed by either maximization or minimization over all possible band subsets $\boldsymbol{\Omega}_{\mathrm{BS}}$ contained in $\boldsymbol{\Omega}$ with $|\boldsymbol{\Omega}_{\mathrm{BS}}| = n_{\mathrm{BS}}$.

Since solving (6) requires exhausting all possible n_{BS}-band combinations to find an optimal band subset, $\boldsymbol{\Omega}_{\mathrm{BS}}^*$, it is practically impossible to do so. Accordingly, many approaches have been investigated by designing various criteria or features to define $J(\boldsymbol{\Omega}_{\mathrm{BS}})$ and solve (6). One traditional approach is to design a BP criterion to rank all bands from which BS can be carried out by selecting bands according to their calculated priorities by a particular BP criterion. Such an approach generally results in an SQMBS method which selects multiple bands one at a time sequentially. As noted in the introduction, one major issue arising from this approach is how to deal with redundant bands caused by band correlation. As an alternative, another BP-derived SQMBS method is to specify a particular application such as minimum estimated abundance covariance (MEAC) for classification [34], which can generate feature vectors for BP and then takes advantage of the sequential forward floating search (SFFS) and sequential backward floating search (SBFS) developed in [73] to derive forward and backward BS methods. However, the band correlation issue still remains.

In contrast to SQMBS, many recent efforts have been directed to SMMBS, which selects multiple bands simultaneously at the same time. Associated with SMMBS are also two main issues needed to be addressed. One is determining the number n_{BS} of bands to be selected, which is also an issue in SQMBS. Generally, n_{BS} can be determined by either trial-and-error or the virtual dimensionality (VD) developed in [69,74]. The other is a more critical issue, which is to how to find appropriate n_{BS} bands. Suppose that $n_{BS} = p$ is the number of bands needed to be selected, $\mathbf{\Omega}_p = \left\{\mathbf{B}_{l_1}, \mathbf{B}_{l_2}, \ldots, \mathbf{B}_{l_p}\right\}$ is a p-band band subset selected from a full band set $\mathbf{\Omega} = \{\mathbf{B}_1, \mathbf{B}_2, \cdots, \mathbf{B}_L\}$ where L is the total number of bands, and $\mathbf{B}_{l_j}$ is the selected jth band. In order to find an optimal band subset $\mathbf{\Omega}_p^*$, we must run through all possible $\begin{pmatrix} L \\ p \end{pmatrix} = \frac{L!}{p!(L-p)!}$ p-combinations among L bands. Practically, this is infeasible if L is large such as in hyperspectral imagery. In this case, developing an effective search strategy for finding an optimal set of multiple bands that does not exist in SQMBS is a great challenge to SMMBS.

A simple SMMBS approach is to group or combine bands into clusters, each of which produces a representative band for BS using certain band measure criteria [58–60]. In particular, the concept in [58] is similar to Fisher's ratio, using mutual information as a band prioritization criterion for clustering. Most interestingly, a band group-wise method was developed [38], which used band combinations by compressive sensing and a multitask sparsity pursuit (MTSP)-based criterion to select band combinations based on linear sparse representation via an evolution-based algorithm-derived search strategy. Another SMMBS approach is to narrow the search range by specifying particular parameters to limit a small number of band subsets as candidate optimal sets, then follow an optimization algorithm such as PSO [35] or FA [36] to find an optimal band subset from the selected candidate set of band subsets.

Most recently, two other promising approaches have been reported. One is to use graph-based representations with each path used to specify a particular band subset. For example, Yuan et al. [43] proposed a graph-based SMMBS method, called multigraph determinantal point process (MDPP), which makes use of multiple graphs to discover a structure and diverse band subset from a graph where each node represents a band and the edges are specified by similarity between bands. Accordingly, a path represents a possible band subset. Then, a search algorithm called mixture determinantal point process (Mix-DPP) was further developed to find a diverse subset that can be a potential optimal band combination. The other is DSEBS, which exploits structure information via a set of local spatial–spectral filters and uses a graph-based clustering search strategy derived from dominant set extraction to find a potential optimal band subset [40].

In addition to the above-mentioned approaches there is also a new approach, called BSS, which considers the problem of multiple band selection as an endmember finding problem. If a desired selected band is interpreted as an endmember and the full band set as the entire data set, then a band subset can be interpreted as a set of endmembers. Consequently, finding an optimal set of n_{BS} bands can be carried out in a similar way to finding an optimal set of n_{BS} endmembers. This BSS-based approach has recently proved to be very promising and has great potential in various applications such as anomaly detection in [65], spectral unmixing in [66], and target detection in [67]. This paper presents another new application of BSS to hyperspectral image classification with LCMV used as a criterion particularly designed for classification.

4. LCMV-BSS Algorithms

Now, if we replace the full band set $\boldsymbol{\Omega}$ in $\mathbf{R}^{-1}$ of (5) with a selected band subset $\boldsymbol{\Omega}_{\mathrm{BS}}$, then (5)

$$\mathrm{MV}(\boldsymbol{\Omega}_{\mathrm{BS}}) = \mathbf{c}^T \left(\mathbf{D}_{\boldsymbol{\Omega}_{\mathrm{BS}}}^T \mathbf{R}_{\boldsymbol{\Omega}_{\mathrm{BS}}}^{-1} \mathbf{D}_{\boldsymbol{\Omega}_{\mathrm{BS}}} \right)^{-1} \mathbf{c} \tag{7}$$

which is the minimum variance weighted by $\mathrm{R}_{\boldsymbol{\Omega}_{\mathrm{BS}}}^{-1}$ resulting from the LCMV filter using a partial band subset specified by $\boldsymbol{\Omega}_{\mathrm{BS}}$. There is another interpretation of (7) which can be also considered as the least $\mathrm{R}_{\boldsymbol{\Omega}_{\mathrm{BS}}}^{-1}$-weighted square error. It should be noted that the constraint vector $\mathbf{c}$ is specifically designed to take care of M class signatures, $\mathbf{d}_1, \mathbf{d}_2, \cdots, \mathbf{d}_M$, not bands. Accordingly, $\mathbf{c}$ has nothing to do with the selected band subset $\boldsymbol{\Omega}_{\mathrm{BS}}$ and, thus, it remains a constant in (7) for any selected band subset $\boldsymbol{\Omega}_{\mathrm{BS}}$.

Using the $\mathrm{MV}(\boldsymbol{\Omega}_{\mathrm{BS}})$ in (7), a criterion can be designed to find an optimal band subset $\boldsymbol{\Omega}_{\mathrm{BS}}^*$ which solves

$$\boldsymbol{\Omega}_{\mathrm{BS}}^* = \arg\{\min_{\boldsymbol{\Omega}_{\mathrm{BS}} \subset \boldsymbol{\Omega}} \mathrm{MV}(\boldsymbol{\Omega}_{\mathrm{BS}})\}. \tag{8}$$

By virtue of (8), two types of algorithms from SQ N-FINDR and SC N-FINDR, called the sequential LCMV-BSS (SQ LCMV-BSS) algorithm and the successive LCMV-BSS (SC LCMV-BSS) algorithm, can be further developed as follows.

4.1. SQ LCMV-BSS

The idea of SQ LCMV-BSS is to use two loops to iterate band subsets $\boldsymbol{\Omega}_{\mathrm{BS}}$ in an outer loop and compute $\mathrm{MV}(\boldsymbol{\Omega}_{\mathrm{BS}})$ in (7) in an inner loop. Depending upon how $\mathrm{MV}(\boldsymbol{\Omega}_{\mathrm{BS}})$ is computed in the inner loop, two versions can be developed. The first one is called SQ LCMV-BSS-1, and finds the minimum variance $\mathrm{MV}(\boldsymbol{\Omega}_{\mathrm{BS}}^{(j)})$ currently being iterated for $1 \leq j \leq n_{\mathrm{BS}}$ in the inner loop compared to the minimum variance $\mathrm{MV}(\boldsymbol{\Omega}_{\mathrm{BS}}^{(l)})$ obtained at the lth iteration in the outer loop. A detailed step-by-step implementation is described below.

Algorithm 1 SQ LCMV-BSS-1

Step 1: Initial conditions

(i) $n_{\mathrm{BS}} = p$, which is the number of selected multiple bands determined by VD.

(ii) Let $\boldsymbol{\Omega}_p^{(0)} = \left\{\mathbf{B}_1^{(0)}, \mathbf{B}_2^{(0)}, \cdots, \mathbf{B}_p^{(0)}\right\}$ with $\mathbf{B}_1^{(0)} = \mathbf{B}_1, \mathbf{B}_2^{(0)} = \mathbf{B}_2, \cdots, \mathbf{B}_p^{(0)} = \mathbf{B}_p$ uniformly selected from the band set $\boldsymbol{\Omega}$.

(iii) Calculate

$$\mathrm{MV}(\boldsymbol{\Omega}_p^{(0)}) = \mathbf{c}^T \left(\mathbf{D}_{\boldsymbol{\Omega}_p^{(0)}}^T \mathbf{R}_{\boldsymbol{\Omega}_p^{(0)}}^{-1} \mathbf{D}_{\boldsymbol{\Omega}_p^{(0)}} \right)^{-1} \mathbf{c}.$$

Step 2: Outer loop

For $l = 1, \cdots, L$ do

Step 3: Inner loop

Compute $\mathrm{MV}(\boldsymbol{\Omega}_p^{(l)})$

For $j = 1, \cdots, p$ do

Find an index j^* by

$$j^* = \arg\left\{ \left(\min_{1 \leq j \leq p} \mathrm{MV}(\boldsymbol{\Omega}_p^{(j)})\right) < \mathrm{MV}(\boldsymbol{\Omega}_p^{(l)})\right\}$$

with

$$\mathrm{MV}(\boldsymbol{\Omega}_p^{(j)}) = \mathbf{c}^T \left(\mathbf{D}_{\boldsymbol{\Omega}_p^{(j)}}^T \mathbf{R}_{\boldsymbol{\Omega}_p^{(j)}}^{-1} \mathbf{D}_{\boldsymbol{\Omega}_p^{(j)}} \right)^{-1} \mathbf{c}$$

which specifies the band to be replaced by the lth band $\mathbf{B}_l$. Such a band is now denoted by $\mathbf{B}_j^{(l+1)}$. A new set of bands is then produced by letting $\mathbf{B}_{j^*}^{(l+1)} = \mathbf{B}_l$ and $\mathbf{B}_j^{(l+1)} = \mathbf{B}_j^{(l)}$ for $j \neq j^*$

A second version of SQ LCMV-BSS, referred to as SQ LCMV-BSS-2, always finds the minimum variance $\text{MV}(\boldsymbol{\Omega}_{\text{BS}}^{(j)})$ currently being iterated for $1 \leq j \leq n_{\text{BS}}$ at each iteration in the inner loop; its detailed step-by-step implementation is summarized as follows.

Algorithm 2 SQ LCMV-BSS-2

Step 1: Initial conditions

(i) $n_{\text{BS}} = p$, which is the number of selected multiple bands determined by VD.

(ii) Let $\boldsymbol{\Omega}_p^{(0)} = \left\{\mathbf{B}_1^{(0)}, \mathbf{B}_2^{(0)}, \cdots, \mathbf{B}_p^{(0)}\right\}$ with $\mathbf{B}_1^{(0)} = \mathbf{B}_1, \mathbf{B}_2^{(0)} = \mathbf{B}_2, \cdots, \mathbf{B}_p^{(0)} = \mathbf{B}_p$ uniformly selected from the band set $\boldsymbol{\Omega}$.

(iii) Calculate

$$\text{MV}(\boldsymbol{\Omega}_p^{(0)}) = \mathbf{c}^T \left(\mathbf{D}_{\boldsymbol{\Omega}_p^{(0)}}^T \mathbf{R}_{\boldsymbol{\Omega}_p^{(0)}}^{-1} \mathbf{D}_{\boldsymbol{\Omega}_p^{(0)}}\right)^{-1} \mathbf{c}.$$

Step 2: Outer loop

For $l = 1, \cdots, L$ do

Step 3: Inner loop

For $j = 1, \cdots, p$ do

Find an index j^* by

$$j^* = \arg\left\{\min_{1 \leq j \leq p} \text{MV}(\boldsymbol{\Omega}_p^{(j)})\right\}$$

with

$$\text{MV}(\boldsymbol{\Omega}_p^{(j)}) = \mathbf{c}^T \left(\mathbf{D}_{\boldsymbol{\Omega}_p^{(j)}}^T \mathbf{R}_{\boldsymbol{\Omega}_p^{(j)}}^{-1} \mathbf{D}_{\boldsymbol{\Omega}_p^{(j)}}\right)^{-1} \mathbf{c}$$

which specifies the band to be replaced by the lth band $\mathbf{B}_l$. Such a band is now denoted by $\mathbf{B}_j^{(l+1)}$. A new set of bands is then produced by letting $\mathbf{B}_{j^*}^{(l+1)} = \mathbf{B}_l$ and $\mathbf{B}_j^{(l+1)} = \mathbf{B}_j^{(l)}$ for $j \neq j^*$

4.2. SC LCMV-BSS

A second type of LCMV-BSS algorithm is SC LCMV-BSS, which reverses the two loops implemented in SQ LCMV-BSS by iterating the computation of $\text{MV}(\boldsymbol{\Omega}_{\text{BS}})$ in (7) in an outer loop, while iterating band subsets n_{BS} in an inner loop. Its detailed step-step implementation is provided in the following.

Algorithm 3 SC LCMV-BSS

Step 1: Initial conditions

(i) $n_{\text{BS}} = p$, which is the number of selected multiple bands determined by VD.

(ii) Let $\boldsymbol{\Omega}_p^{(0)} = \left\{\mathbf{B}_1^{(0)}, \mathbf{B}_2^{(0)}, \cdots, \mathbf{B}_p^{(0)}\right\}$ with $\mathbf{B}_1^{(0)} = \mathbf{B}_1, \mathbf{B}_2^{(0)} = \mathbf{B}_2, \cdots, \mathbf{B}_p^{(0)} = \mathbf{B}_p$ uniformly selected from the band set $\boldsymbol{\Omega}$.

(iii) Calculate

$$\text{MV}(\boldsymbol{\Omega}_p^{(0)}) = \mathbf{c}^T \left(\mathbf{D}_{\boldsymbol{\Omega}_p^{(0)}}^T \mathbf{R}_{\boldsymbol{\Omega}_p^{(0)}}^{-1} \mathbf{D}_{\boldsymbol{\Omega}_p^{(0)}}\right)^{-1} \mathbf{c}$$

Step 2: Outer loop

For $j = 1, \cdots, p$ do

Step 3: Inner loop

For $l = 1, \cdots, L$ do

Find

$$\mathbf{B}_j^{(*)} = \arg\left\{\min_{\mathbf{B}_l \in \tilde{\boldsymbol{\Omega}}} \text{MV}(\tilde{\boldsymbol{\Omega}}_p^l)\right\}$$

where $\tilde{\boldsymbol{\Omega}} = \boldsymbol{\Omega} - \left\{\mathbf{B}_1^{(*)}, \cdots, \mathbf{B}_{j-1}^{(*)}, \mathbf{B}_{j+1}^{(p)}, \cdots, \mathbf{B}_p^{(p)}\right\}$, $\tilde{\boldsymbol{\Omega}}_p^l = \left\{\mathbf{B}_1^{(*)}, \cdots, \mathbf{B}_{j-1}^{(*)}, \mathbf{B}_l, \mathbf{B}_{j+1}^{(p)}, \cdots, \mathbf{B}_p^{(p)}\right\}$.

Step 4: Output the final band subset, $\left\{\mathbf{B}_1^{(*)}, \mathbf{B}_2^{(*)}, \cdots, \mathbf{B}_p^{(*)}\right\}$.

---

5. Real Image Experiments

Three popular real hyperspectral images, Purdue University's Indiana Indian Pines, Salinas, and University of Pavia, available at http://www.ehu.eus/ccwintco/index.php?title=Hyperspectral_Remote_Sensing_Scenes, were used in experiments. The detailed data descriptions and matlab data files can be also found on this website.

5.1. Purdue Indiana Indian Pines Scene

The first image scene used for experiments is an airborne visible/infrared imaging spectrometer (AVIRIS) hyperspectral data set from the Purdue Indiana Indian Pines test site shown in Figure 1a, with its ground truth of 17 class maps in Figure 1b. It has a size of 145 × 145 pixel vectors, taken from an area of mixed agriculture and forestry in Northwestern Indiana, USA with details of band and wavelength given in the caption. The data set is available at website https://purr.purdue.edu/publications/1947/serve/1?el=1. It was recorded in June 1992 with 220 bands which include water absorption bands (bands 104–108 and 150–163, 220).

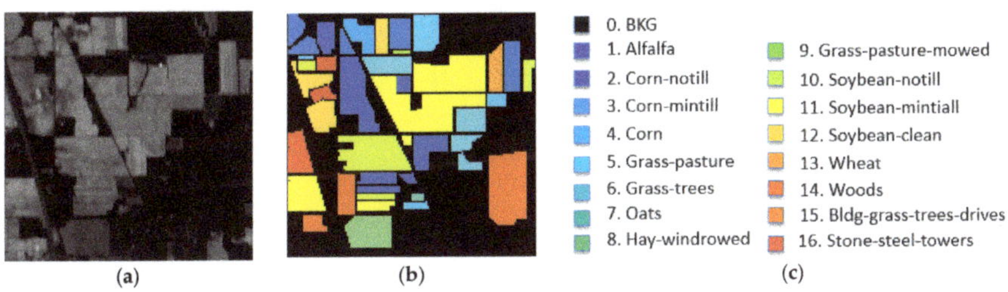

Figure 1. Purdue's Indiana Indian Pines scene with 16 classes. (**a**) Band 186, (**b**) ground truth map, (**c**) ground truth class labels.

5.2. Salinas

A second set of AVIRIS data used for experiments was the Salinas scene shown in Figure 2a, which was captured by the AVIRIS sensor over Salinas Valley, California, with a spatial resolution of 3.7 m per pixel and spectral resolution of 10 nm. It has a size of 512 × 217 × 224. Figure 2b,c show the color composite of the Salinas image along with the corresponding ground truth class labels.

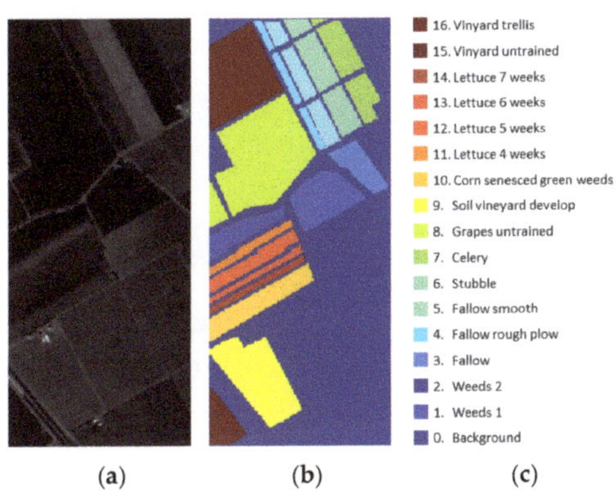

Figure 2. Ground truth of Salinas scene with 16 classes. (**a**) Band 126, (**b**) color ground-truth image, (**c**) class labels.

5.3. ROSIS Data

The last hyperspectral image data used for experiments was the University of Pavia image shown in Figure 3, which is an urban area surrounding the University of Pavia, Italy. It was recorded using the ROSIS-03 satellite sensor. It is of size 610 × 340 × 115 with a spatial resolution of 1.3 m per pixel and spectral coverage ranging from 0.43 to 0.86 μm with spectral resolution of 4 nm (the 12 most noisy channels were removed before experiments). Nine classes of interest, plus a background (BKG) class (class 0), were considered for this image.

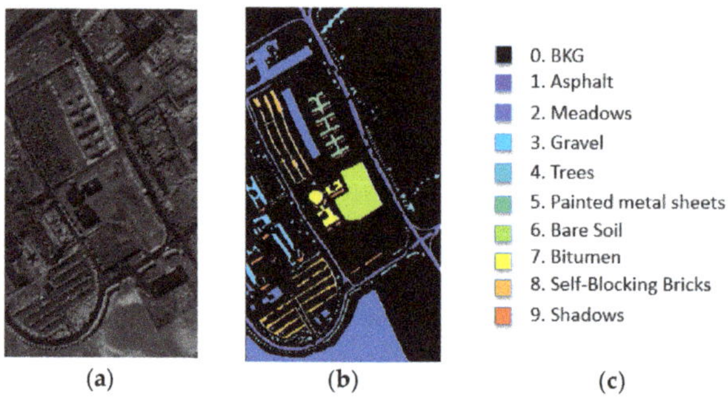

Figure 3. Ground truth of University of Pavia scene with nine classes. (**a**) Band 95, (**b**) color ground truth image, (**c**) class labels.

In the following experiments, four types of BS methods were tested for a comparative study and analysis.

1. Uniform band selection (UBS): According to our extensive experiments, UBS is a reasonably good BS method which is also reported in the literature. It does not require any prior knowledge or BS criterion. It is the simplest BS method.
2. MEAC: This uses the minimum covariance derived from the estimated abundance matrix, which is similar to the minimum variance in (5). In addition, it can also represent the category of SQMBS methods.
3. MDPP and DSEBS: Both represent the category of SMMBS methods. They make use of graph representations to specify band groups. Most importantly, these two methods were compared with CEM/LCMV-based methods in [26] and both are also based on the LCMV formulation specified by (2).
4. LCMV-BSS developed in this paper: This represents the category of BSS methods using the LCMV formulation in (2).

As noted in the introduction and in Section 3, although PSO, FA, and MTSP are also SMMBS methods, they are not compared in this paper for the following reasons. One is that their design rationale is completely different from that of LCMV-BSS. Secondly, the initial candidate sets from which their search algorithms find an optimal band subset are random and are also too small. So, their results are not representative and also are not reproducible. Thirdly, the details of their used parameters were not specified and provided in their papers. Therefore, it is very difficult to implement their algorithms for fair comparisons.

Table 1 tabulates the number n_{BS} of selected bands estimated for three scenes using Harsanyi-Farrand-Chang (HFC) method/noise whitened HFC (NWHFC method developed for VD in [69,74,75] where n_{BS} was determined to be n_{BS} = 18 for Purdue's data, 21 for Salinas and 14 for University of Pavia with a false alarm probability of 10^{-4}.

Table 1. n_{BS} estimated by HySime and HFC/NWHFC.

	$P_F = 10^{-1}$	$P_F = 10^{-2}$	$P_F = 10^{-3}$	$P_F = 10^{-4}$	$P_F = 10^{-5}$
Purdue	73/21	49/19	35/18	27/18	25/17
Salinas	32/33	28/24	25/21	21/21	20/20
Univ. of Pavia	25/34	21/27	16/17	14/14	13/12

Table 2 lists the bands selected by seven BS methods—uniform BS (UBS), minimum estimated abundance covariance (MEAC), multigraph determinantal point process (MDPP), dominant set extraction BS (DSEBS), SQ LCMV-BSS-1, SQ LCMV-BSS-2, and SC LCMV-BSS—for the three scenes; $n_{BS} = 18$ for Purdue's Indian Pines, $n_{BS} = 21$ for Salinas, and $n_{BS} = 14$ for University of Pavia.

Table 2. Bands selected by UBS, SQ LCMV-BSS-1, SQ LCMV-BSS-2, SC LCMV-BSS.

Data	Methods	Selected Bands
Purdue Indian Pines (18 bands)	UBS	1, 14, 27, 40, 53, 66, 79, 92, 105, 118, 131, 144, 157, 170, 183, 196, 209, 220
	MEAC	159, 3, 92, 96, 82, 36, 39, 55, 41, 1, 2, 33, 206, 38, 163, 17, 204, 9
	MDPP	10, 39, 59, 75, 79, 85, 92, 130, 140, 146, 147, 149, 150, 152, 160, 164, 175, 193
	DSEBS	42, 129, 97, 131, 174, 16, 176, 177, 172, 43, 192, 193, 98, 171, 99, 132, 40, 33
	SQ LCMV-BSS-1	39, 164, 29, 155, 108, 66, 79, 8, 105, 42, 44, 17, 156, 150, 3, 43, 213, 41
	SQ LCMV-BSS-2	38, 109, 29, 52, 163, 66, 158, 8, 164, 219, 43, 78, 157, 220, 3, 49, 218, 2
	SC LCMV-BSS	54, 156, 42, 159, 53, 41, 79, 91, 105, 57, 51, 43, 157, 48, 107, 160, 115, 163
Salinas (21 bands)	UBS	1, 12, 23, 34, 45, 56, 67, 78, 89, 100, 111, 122, 133, 144, 155, 166, 177, 188, 199, 210, 224
	MEAC	107, 148, 203, 149, 5, 8, 105, 3, 28, 12, 18, 10, 44, 36, 25, 17, 51, 32, 110, 68, 58
	MDPP	1, 8, 11, 22, 27, 28, 50, 57, 58, 65, 90, 99, 105, 119, 123, 134, 142, 157, 175, 191, 204
	DSEBS	99, 101, 16, 119, 177, 112, 44, 46, 120, 47, 131, 175, 196, 121, 17, 102, 174, 180, 187, 135, 42
	SQ LCMV-BSS-1	7, 50, 23, 48, 45, 73, 65, 15, 40, 19, 80, 122, 38, 41, 42, 46, 78, 47, 200, 37, 2
	SQ LCMV-BSS-2	7, 42, 56, 28, 45, 58, 67, 15, 41, 19, 50, 122, 38, 34, 36, 47, 224, 46, 183, 37, 172
	SC LCMV-BSS	18, 39, 41, 31, 45, 44, 67, 78, 90, 101, 40, 91, 42, 141, 46, 48, 102, 185, 47, 86, 50
Univ. of Pavia (14 bands)	UBS	1, 9, 17, 25, 33, 41, 49, 57, 65, 73, 81, 89, 97, 103
	MEAC	1, 23, 24, 40, 42, 58, 56, 59, 48, 31, 47, 83, 25, 54
	MDPP	2, 23, 44, 46, 50, 62, 66, 73, 89, 91, 92, 93, 96, 102
	DSEBS	86, 102, 64, 20, 21, 63, 65, 6, 19, 22, 7, 66, 95, 67
	SQ LCMV-BSS-1	1, 4, 55, 16, 95, 83, 84, 93, 39, 77, 91, 102, 92, 103
	SQ LCMV-BSS-2	1, 4, 38, 76, 85, 55, 84, 102, 16, 83, 93, 89, 92, 103
	SC LCMV-BSS	1, 4, 84, 16, 38, 102, 85, 92, 83, 72, 95, 91, 96, 103

In order to perform HSIC, choosing an appropriate classifier is crucial. Recently, Yu et al. [76] developed a new classifier, called the iterative multiclass constrained background suppression classifier (IMCBSC), and further demonstrated that IMCBSC performed well in both overal accuarcy rate (P_{OA}) and precision rate (P_R) Since IMCBSC was also derived from LCMV and implemented by LCMV in an iterative manner, the iterative linearly constrained minimum variance (ILCMV) is used in this paper instead of IMCBSC to reflect its idea arising from LCMV and its iterative nature in algorithm implementation. Most importantly, ILCMV was adopted for two main reasons. One is because of the work in [76], which showed that ILCMV could perform at least comparably in P_{OA} but significantly better than the work in [12]. The other is that ILCMV is indeed derived from the LCMV criterion specified by (2). So, it is natural to use ILCMV to perform classification.

Two remarks on the implementation of ILCMV are noteworthy.

1. Unlike most supervised classifiers used for HSIC which require training samples, ILCMV only needs the knowledge of the class signatures **D**, which can be obtained by either prior knowledge or class sample means. Specifically, the class signatures in **D** are not necessarily real data samples.
2. Also, unlike most supervised classifiers used for HSIC which require test and training data samples from the same class, the test samples for ILCMV can be selected from any arbitrary class including the BKG class, and are not necessarily limited to the same class trained by the training samples. This is a crucial difference between ILCMV and existing hyperspectral image classification algorithms reported in the literature. For more details, we refer to [23,76].

Figure 4c–i, Figures 5c–i and 6c–i show classification maps produced by ILCMV, using bands selected in Table 2 by seven BS methods—UBS, MEAC, MDPP, DSEBS, SQ LCMV-BSS-1, SQ LCMV-BSS-2, and SC LCMV-BSS, respectively—where the ground truth map and classification map produced by the full bands are also included in (a) and (b), respectively, for comparison.

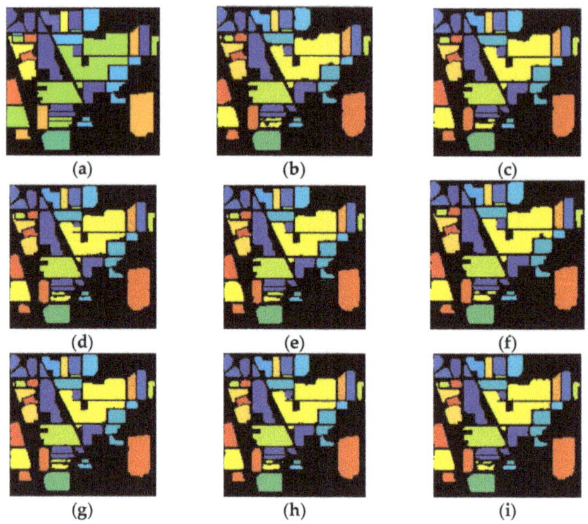

Figure 4. Classification maps produced by iterative LCMV (ILCMV) for Purdue's data using bands selected in Table 2. (**a**) Ground truth, (**b**) Full bands, (**c**) UBS, (**d**) MEAC, (**e**) MDPP, (**f**) DSEBS, (**g**) SQ LCMV-BSS-1, (**h**) SQ LCMV-BSS-2, (**i**) SC LCMV-BSS.

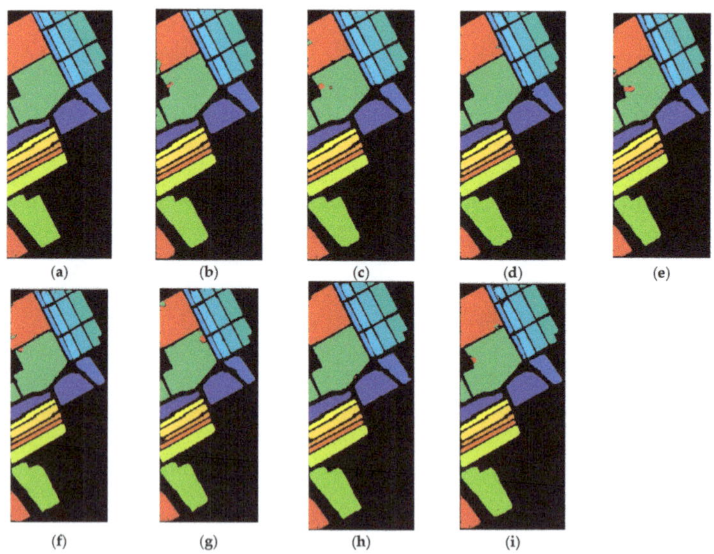

Figure 5. Classification maps produced by ILCMV for Salinas using bands selected in Table 2. (**a**) Ground truth, (**b**) Full bands, (**c**) UBS, (**d**) MEAC, (**e**) MDPP, (**f**) DSEBS, (**g**) SQ LCMV-BSS-1, (**h**) SQ LCMV-BSS-2, (**i**) SC LCMV-BSS.

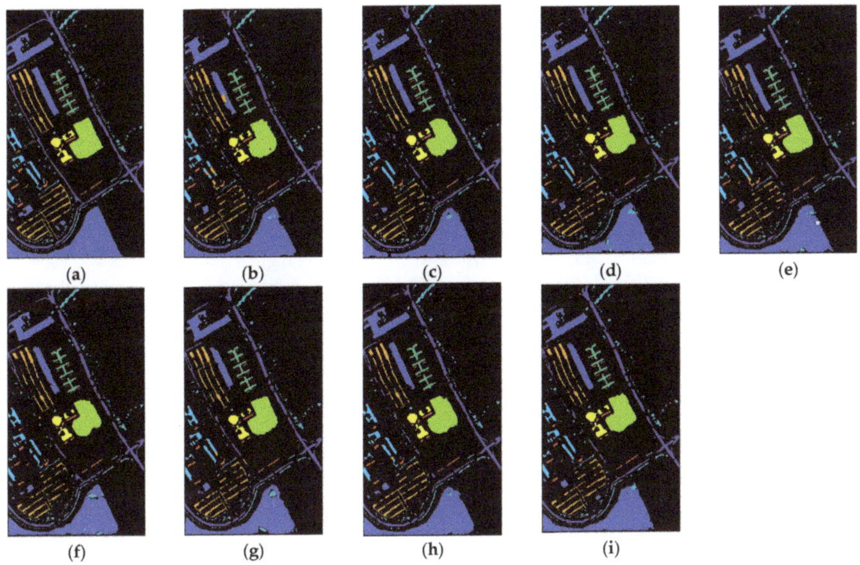

Figure 6. Classification maps produced by ILCMV for Pavia using bands selected in Table 2. (**a**) Ground truth, (**b**) Full bands, (**c**) UBS, (**d**) MEAC, (**e**) MDPP, (**f**) DSEBS, (**g**) SQ LCMV-BSS-1, (**h**) SQ LCMV-BSS-2, (**i**) SC LCMV-BSS.

Apparently, it is difficult to see any appreciable difference among all the classification results in Figures 4–6 by visual inspection. In this case, to better evaluate each BS method, conducting a quantitative analysis is necessary. It has been shown in [23,76] that using overall accuracy (OA), P_{OA} may not be sufficient to evaluate the effectiveness of classification performance. To address this issue, two additional measures, called precision rate, P_R, and detection rate, P_D (also known as recall rate), developed in [23,76] were introduced for HSIC where P_R and P_D have been widely used in pattern recognition such as medical imaging, handwritten character recognition, and biometric recognition. The definitions and details of P_{OA}, P_R, and P_D can be found in [23,76].

Tables 3–5 show P_D, P_{OA}, and P_R calculated by the ILCMV classification results in Figures 4–6 using the bands selected in Table 2 for Purdue's data, Salinas, and University of Pavia, respectively, where the best results with highest rates are shown in boldface. Here, we would like to point out a crucial fact used in the experiments, as noted in the second remark described above, where the P_D, P_{OA}, and P_R were calculated by including the background (BKG) for classification because LCMV is particularly designed to take care of the BKG issue in classification, as shown in [76]. This is quite different from many reports which calculate P_{OA} excluding BKG from classification, such as [12].

Since P_D varies with each class, it is difficult to evaluate the overall classification performance. So, our analysis is conducted based on P_{OA} and P_R. As we can see from the tables, SQ LCMV-BSS-2 and SC LCMV-BSS outperformed all the other five BS methods in terms of P_{OA} and P_R for Salinas and University of Pavia scenes, but were slightly worse than MDPP in P_{OA} and DSEBS in P_R. Interestingly, both MDPP and DSEBS produced the best results in terms of P_{OA} and P_R respectively for the Purdue data. As also noted in Tables 3–5, the P_{OA} and P_R using full bands were generally not as good as those produced by most of the test BS methods, but also worse than that produced by UBS. These experiments showed that hyperspectral image classification can benefit greatly from the judicious selection of bands with appropriately determined n_{BS}.

Table 3. P_D, P_{OA}, and P_R calculated from the classification results in Figure 4 for Purdue's data.

Class	Full Bands		UBS		MEAC		MDPP		DSEBS		SQ LCMV-BSS-1		SQ LCMV-BSS-2		SC LCMV-BSS	
	P_D	P_R	P_D	P_R	P_D	P_R	P_D	P_R	P_D	P_R	P_D	P_R	P_D	P_R	P_D	P_R
1	95.65	100	95.65	100	93.48	100	95.65	100	95.65	100	95.65	100	97.83	100	100	100
2	96.01	100	97.13	99.57	93.07	99.63	96.08	100	96.99	100	95.59	99.71	93.78	99.85	94.89	99.85
3	96.99	99.88	96.51	100	96.27	100	97.35	100	97.23	99.88	96.39	100	95.67	100	94.10	100
4	98.73	100	98.73	100	98.31	100	99.58	100	98.31	100	97.89	100	98.31	100	98.31	100
5	89.44	100	90.68	100	91.51	100	92.34	100	93.58	100	91.93	100	92.34	100	92.96	100
6	97.12	100	97.67	100	97.40	99.58	96.71	100	97.12	100	96.44	100	97.95	100	95.75	100
7	100	100	100	100	100	100	100	100	100	100	100	100	100	100	100	100
8	98.78	100	98.54	100	99.16	100	97.49	100	97.91	100	97.91	100	99.16	100	98.95	100
9	100	100	100	100	90.00	100	100	100	100	100	100	90.91	100	95.24	100	100
10	93.93	99.78	91.98	100	93.31	100	94.65	99.78	93.00	100	94.24	100	91.98	99.58	91.98	100
11	94.70	99.87	96.13	99.96	94.55	98.22	95.48	99.87	95.85	100	95.48	99.96	96.17	100	95.93	99.49
12	95.45	100	94.94	100	96.29	100	96.80	100	97.30	100	95.95	100	95.11	100	96.63	100
13	98.54	100	98.54	100	99.02	100	97.56	100	96.59	100	97.56	100	98.54	100	98.54	100
14	93.52	100	94.15	100	94.78	100	94.70	100	94.55	100	95.89	100	96.05	100	96.13	100
15	90.67	100	95.60	100	92.49	100	96.89	100	93.52	100	94.82	100	94.56	100	96.11	100
16	98.92	98.92	98.92	98.92	98.92	100	98.92	98.92	98.92	100	98.92	100	97.85	100	95.70	97.80
P_{OA}	95.09		95.69		94.91		**95.89**		95.88		95.67		95.48		95.46	
P_R	97.61		97.90		97.52		**98.00**		97.99		97.89		97.80		97.79	

Table 4. P_D, P_{OA}, and P_R calculated from the classification results in Figure 5 for Salinas.

Class	Full Bands		UBS		MEAC		MDPP		DSEBS		SQ LCMV-BSS-1		SQ LCMV-BSS-2		SC LCMV-BSS	
	P_D	P_R	P_D	P_R	P_D	P_R	P_D	P_R	P_D	P_R	P_D	P_R	P_D	P_R	P_D	P_R
1	95.52	100	97.16	100	97.71	100	97.76	100	97.16	100	96.37	100	97.01	100	96.91	100
2	98.42	100	98.85	100	98.44	100	97.99	100	99.17	100	98.79	100	98.36	100	98.71	100
3	93.78	99.70	95.50	100	94.03	100	93.98	100	95.65	100	90.44	100	95.14	100	95.95	100
4	95.62	100	94.69	98.80	94.33	97.84	97.49	98.76	94.74	99.62	96.56	98.39	95.91	99.11	92.04	94.83
5	96.90	100	96.45	100	95.19	99.88	95.22	100	96.90	99.85	95.87	100	95.94	100	90.78	99.79
6	98.79	100	98.59	100	98.56	100	98.79	100	98.56	100	97.95	100	98.91	100	97.75	100
7	98.63	100	98.21	100	98.18	100	97.99	100	97.65	100	98.35	100	98.44	100	98.32	100
8	96.69	98.26	95.81	99.39	97.40	99.84	95.23	99.74	96.11	99.38	95.84	99.06	97.47	100	96.61	99.42
9	95.87	100	95.60	100	94.74	100	95.29	100	95.73	100	94.79	100	94.89	100	95.44	100
10	96.67	100	96.37	100	96.34	100	96.46	100	97.25	100	95.73	100	96.58	100	96.77	100
11	97.75	100	97.85	100	91.10	100	97.75	100	98.31	100	95.79	100	97.38	100	97.66	100
12	97.15	100	96.16	100	95.54	100	97.46	100	97.66	100	96.32	100	95.39	100	95.43	100
13	96.51	100	96.94	99.44	93.35	99.88	96.40	100	95.63	100	87.77	100	97.38	99.78	94.00	98.97
14	95.89	100	98.14	100	97.66	99.90	97.01	100	98.04	100	97.76	99.05	97.20	100	96.93	99.81
15	94.00	98.66	95.27	98.09	96.52	100	95.42	96.70	95.25	98.84	95.42	97.73	96.27	99.86	95.84	98.60
16	93.30	100	96.07	100	93.86	100	95.07	100	95.02	100	95.68	100	95.13	100	95.41	100
P_{OA}	96.37		96.49		96.45		96.25		96.63		95.93		**96.81**		96.21	
P_R	98.23		98.29		98.27		98.17		98.36		98.02		**98.45**		98.15	

Table 5. P_D, P_{OA}, and P_R calculated from the classification results in Figure 6 for University of Pavia.

Class	Full Bands		UBS		MEAC		MDPP		DSEBS		SQ LCMV-BSS-1		SQ LCMV-BSS-2		SC LCMV-BSS	
	P_D	P_R	P_D	P_R	P_D	P_R	P_D	P_R	P_D	P_R	P_D	P_R	P_D	P_R	P_D	P_R
1	86.42	99.90	87.67	99.45	86.44	99.76	87.97	99.68	87.71	99.74	84.44	99.63	88.05	99.44	88.67	99.77
2	73.34	99.99	84.38	99.95	83.33	99.89	82.14	99.96	84.63	99.92	84.14	99.89	85.21	99.98	86.76	99.95
3	79.85	96.30	78.90	100	76.66	100	76.17	99.02	79.22	100	76.49	100	74.71	100	78.56	99.95
4	98.81	96.65	97.84	95.16	98.88	87.91	96.95	91.85	97.99	88.71	98.14	88.96	97.77	95.30	97.70	93.11
5	91.49	100	89.93	100	91.33	100	93.32	100	87.11	100	93.50	100	90.57	100	90.77	100
6	89.10	99.98	91.35	100	82.78	100	87.53	100	87.44	100	86.19	100	90.00	100	91.13	100
7	81.10	100	83.32	100	76.26	100	75.64	100	76.34	100	82.84	100	82.92	100	82.46	100
8	78.46	85.20	79.09	97.37	79.51	97.20	79.83	95.71	79.30	97.44	77.09	97.16	77.09	98.45	79.30	98.96
9	77.24	99.87	75.86	99.47	76.32	100	74.01	98.44	76.17	100	80.21	99.46	78.22	99.86	77.08	99.87
P_{OA}	84.32		85.19		83.85		84.33		84.25		84.45		85.41		**85.92**	
P_R	96.76		96.93		96.64		96.76		96.75		96.78		**96.96**		96.95	

Table 6 tabulates the computing times in seconds for each of six BS methods in a computer environment with a 1.6 GHz Intel Core i5 with OS X EI Capitan and 4 GB 1600 MHz DDR3; the software used to run experiments was Matlab_R2014b. Obviously, the best time was achieved by DSEBS, followed by SC LCMV-BSS and SQ LCMV-BSS. The worst time was achieved by MDPP for the Purdue data and MEAC for Salinas and University of Pavia.

Table 6. Computing time in seconds required by six test BS methods: MEAC, MDPP, DSEBS, SQ LCMV-BSS-1, SQ LCMV-BSS-2, SC LCMV-BSS.

	MEAC	MDPP	DSEBS	SQ LCMV-BSS-1	SQ LCMV-BSS-2	SC LCMV-BSS
Purdue	13.70	41.14	0.58	7.00	7.10	6.93
Salinas	83.64	44.66	5.27	43.43	46.63	43.55
University of Pavia	44.53	29.22	4.62	16.67	17.52	16.84

As noted above, a classifier can also have a significant impact on BS, especially when BKG is included for consideration. A recent work [12] developed four edge preserving filtering (EPF)-based techniques—EPF-B-c, EPF-G-c, EPF-B-g, and EPF-G-g for HSIC—and also conducted a comprehensive comparative analysis to show that their methods indeed performed better than most recently developed spectral–spatial techniques. Therefore, in what follows, we conducted experiments to evaluate the performance of ILCMV in comparison with these four EPF-based techniques with BKG particularly included for classification. To see this, we also implemented these four EPF-based techniques with "B" and "G" used to specify bilateral filter and guided filter, respectively, and "g" and "c" indicate that the first principal component and color composite of the three principal components are used as reference images [12].

Tables 7–15 tabulate the results in terms of P_{OA} and P_R rates produced by the four EFP-based methods and ILCMV, all of which included BKG for classification and also used the bands selected in Table 2 to implement the three image scenes. Data for the Purdue image is shown in Tables 7–9 using bands selected by SQ LCMV-BSS-1, SQ LCMV-BSS-2, and SC LCMV-BSS; data for Salinas is shown in Tables 10–12 using bands selected by SQ LCMV-BSS-1, SQ LCMV-BSS-2, and SC LCMV-BSS; and data for University of Pavia is shown in Tables 13–15 using bands selected by SQ LCMV-BSS-1, SQ LCMV-BSS-2, and SC LCMV-BSS. In addition, their computing times in seconds are included in the tables for comparison.

Table 7. P_{OA} and P_R calculated by the classification results using the bands selected by SQ LCMV-BSS-1 for the Purdue data.

Class	EPF-B-g with Full Bands	EPF-B-c with Full Bands	EPF-G-g with Full Bands	EPF-G-c with Full Bands	ILCMV with Full Bands	EPF-B-g -BS	EPF-B-c -BS	EPF-G-g -BS	EPF-G-c -BS	ILCMV -BS
1	100	100	97.83	100	95.65	100	100	100	100	95.65
2	76.47	87.82	80.60	82.98	96.01	74.79	90.76	71.99	79.90	95.59
3	93.49	83.98	79.64	65.42	96.99	63.37	75.66	84.10	72.77	96.39
4	99.16	100	100	96.20	98.73	100	100	100	97.89	97.89
5	93.79	94.00	97.10	94.82	89.44	89.86	97.52	96.07	94.41	91.93
6	100	99.59	99.59	99.45	97.12	98.22	99.86	99.73	99.32	96.44
7	92.86	92.86	96.43	96.43	100	92.86	64.29	92.86	89.29	100
8	100	100	100	100	98.78	100	100	100	100	97.91
9	80.00	65.00	100	100	100	65.00	95.00	65.00	10.00	100
10	90.53	91.46	87.14	93.00	93.93	75.31	73.97	84.88	76.54	94.24
11	90.67	92.67	86.27	88.88	94.70	78.98	70.79	64.73	80.94	95.48
12	98.31	96.46	93.93	92.07	95.45	47.55	42.50	79.76	57.67	95.95
13	99.02	99.51	99.51	99.51	98.54	100	100	100	100	97.56
14	97.71	97.00	97.87	98.26	93.52	94.86	94.86	87.67	92.57	95.89
15	100	100	99.74	82.90	90.67	95.85	96.37	95.85	97.41	94.82
16	97.85	100	100	100	98.92	98.92	100	98.92	97.85	98.92
P_{OA}	92.27	93.45	90.32	89.79	95.09	81.62	82.94	81.17	84.15	95.67
P_R	44.98	44.56	44.03	43.77	97.61	39.79	40.43	39.86	41.02	97.89
Time(s)	196.58	200.84	194.09	200.87	25.37	31.27	36.77	31.14	36.16	37.25

Table 8. P_{OA} and P_R calculated by the classification results using full bands and the bands selected by SQ LCMV-BSS-2 for the Purdue data.

Class	EPF-B-g with Full Bands	EPF-B-c with Full Bands	EPF-G-g with Full Bands	EPF-G-c with Full Bands	ILCMV with Full Bands	EPF-B-g-BS	EPF-B-c-BS	EPF-G-g-BS	EPF-G-c-BS	ILCMV-BS
1	100	100	100	100	95.65	100	100	100	100	97.83
2	86.83	82.49	86.62	84.52	96.01	85.78	81.86	64.22	66.60	93.78
3	90.84	85.90	82.53	80.72	96.99	81.93	74.58	63.98	64.82	95.67
4	99.16	99.16	100	99.58	98.73	99.16	100	99.58	100	98.31
5	95.03	92.96	97.72	92.34	89.44	93.17	95.03	91.72	92.13	92.34
6	100	99.86	99.59	99.73	97.12	99.73	97.81	100	99.59	97.95
7	89.29	89.29	96.43	96.43	100	92.86	78.57	96.43	92.86	100
8	100	100	100	100	98.78	100	100	100	100	99.16
9	60.00	75.00	70.00	50.00	100	100	95.00	65.00	35.00	100
10	92.59	90.33	90.74	92.49	93.93	70.27	66.26	64.71	54.42	91.98
11	89.12	89.33	88.51	86.44	94.70	64.89	70.26	82.12	82.24	96.17
12	96.46	98.31	98.15	98.99	95.45	67.45	52.78	51.10	34.74	95.11
13	99.02	99.02	99.51	99.51	98.54	99.51	99.51	99.51	99.51	98.54
14	98.81	98.18	96.36	95.02	93.52	90.59	97.79	92.81	89.41	96.05
15	99.74	100	94.04	95.34	90.67	90.41	90.41	93.26	83.42	94.56
16	100	100	100	100	98.92	95.70	98.92	100	100	97.85
P_{OA}	93.37	92.17	92.10	90.96	95.09	81.49	81.28	80.01	77.66	95.48
P_R	45.52	44.93	44.89	44.34	97.61	39.72	39.61	39.00	37.85	97.80
Time(s)	194.14	199.37	194.13	200.36	25.37	31.16	37.93	32.56	36.51	41.58

Table 9. P_{OA} and P_R calculated by the classification results using full bands and the bands selected by SC LCMV-BSS for the Purdue data.

Class	EPF-B-g with Full Bands	EPF-B-c with Full Bands	EPF-G-g with Full Bands	EPF-G-c with Full Bands	ILCMV with Full Bands	EPF-B-g-BS	EPF-B-c-BS	EPF-G-g-BS	EPF-G-c-BS	ILCMV-BS
1	100	100	100	97.83	95.65	100	97.83	97.83	100	100
2	83.26	83.26	83.05	78.92	96.01	48.25	42.09	48.74	39.92	94.89
3	79.16	69.64	80.48	66.87	96.99	62.05	55.18	60.36	44.10	94.10
4	100	98.31	100	99.58	98.73	100	100	100	99.58	98.31
5	93.79	93.17	95.86	93.79	89.44	91.93	94.00	93.58	93.79	92.96
6	99.73	99.73	98.08	99.32	97.12	99.32	99.45	94.93	99.45	95.75
7	96.43	92.86	96.43	89.29	100	85.71	71.43	92.86	64.29	100
8	100	99.79	100	100	98.78	100	100	99.79	100	98.95
9	20.00	75.00	75.00	65.00	100	20.00	45.00	35.00	0	100
10	75.31	84.16	84.47	89.61	93.93	59.05	48.97	47.22	49.90	91.98
11	91.65	91.00	87.98	91.73	94.70	76.25	78.09	70.22	72.87	95.93
12	96.29	92.07	96.12	97.47	95.45	70.83	60.20	44.86	58.85	96.63
13	99.51	99.51	99.51	99.51	98.54	100	100	100	100	98.54
14	92.89	98.34	96.36	96.68	93.52	98.34	93.04	97.23	97.39	96.13
15	100	100	98.19	99.48	90.67	93.26	89.12	79.27	67.36	96.11
16	100	100	100	100	98.92	100	100	93.55	95.70	95.70
P_{OA}	90.06	90.42	90.56	90.37	95.09	77.37	74.12	72.31	71.25	95.46
P_R	43.90	44.08	44.15	44.05	97.61	37.72	36.13	35.25	34.73	97.79
Time(s)	187.76	203.60	195.20	201.21	25.37	32.01	38.78	31.92	38.07	42.73

Table 10. P_{OA} and P_R calculated by the classification results using full bands and the bands selected by SQ LCMV-BSS-1 in Table 2 for Salinas.

Class	EPF-B-g with Full Bands.	EPF-B-c with Full Bands	EPF-G-g with Full Bands	EPF-G-c with Full Bands	ILCMV with Full Bands	EPF-B-g-BS	EPF-B-c-BS	EPF-G-g-BS	EPF-G-c-BS	ILCMV-BS
1	100	100	100	100	95.52	100	99.75	100	100	96.37
2	100	100	100	100	98.42	99.97	100	99.97	100	98.79
3	100	100	100	100	93.78	100	100	100	100	90.44
4	100	100	100	99.93	95.62	99.93	99.86	99.93	100	96.56
5	99.37	99.25	98.69	99.25	96.90	98.92	98.95	98.92	98.95	95.87
6	100	100	100	100	98.79	99.97	99.97	99.97	99.97	97.95
7	100	99.92	100	100	98.63	99.66	99.72	99.66	99.83	98.35
8	90.92	90.40	89.26	90.61	96.69	87.11	91.94	87.11	91.70	95.84
9	99.98	100	99.97	99.97	95.87	99.58	100	99.58	99.97	94.79
10	96.95	98.08	98.60	98.51	96.67	96.19	98.57	96.19	98.60	95.73
11	99.91	99.91	99.91	100	97.75	99.91	99.91	99.91	99.81	95.79
12	100	100	100	100	97.15	100	100	100	100	96.32
13	99.89	99.56	99.02	99.89	96.51	98.47	99.13	98.47	99.56	87.77
14	99.91	99.25	99.63	100	95.89	98.97	99.35	98.97	98.04	97.76
15	89.01	85.65	85.65	87.08	94.00	90.66	82.21	90.66	85.99	95.42
16	99.83	100	99.45	99.83	93.30	99.67	100	99.67	100	95.68
P_{OA}	96.40	95.89	95.64	96.17	96.37	95.64	95.73	95.64	96.19	95.93
P_R	46.97	46.26	46.95	46.85	98.23	46.60	46.64	46.60	46.86	98.02
Time(s)	1060.77	741.84	1082.51	1134.06	167.80	75.17	78.73	75.17	104.31	134.96

Table 11. P_{OA} and P_R calculated by the classification results using full bands and the bands selected by SQ LCMV-BSS-2 in Table 2 for Salinas.

Class	EPF-B-g with Full Bands	EPF-B-c with Full Bands	EPF-G-g with Full Bands	EPF-G-c with Full Bands	ILCMV with Full Bands	EPF-B-g-BS	EPF-B-c-BS	EPF-G-g-BS	EPF-G-c-BS	ILCMV-BS
1	100	100	100	100	95.52	100	100	100	100	97.01
2	100	100	100	100	98.42	100	100	100	99.97	98.36
3	100	100	100	100	93.78	100	100	99.95	100	95.14
4	100	100	100	99.78	95.62	100	99.93	100	100	95.91
5	99.48	98.95	98.58	99.14	96.90	98.66	98.73	99.14	98.84	95.94
6	100	100	100	100	98.79	99.95	100	100	100	98.91
7	100	99.89	100	99.89	98.63	99.92	99.97	99.89	99.80	98.44
8	88.63	91.98	87.90	90.86	96.69	87.84	89.50	90.61	91.86	97.47
9	99.90	99.98	99.95	100	95.87	99.97	99.94	99.60	100	94.89
10	97.56	98.90	97.04	97.50	96.67	99.33	99.51	99.21	98.29	96.58
11	100	99.91	99.91	99.72	97.75	100	99.91	99.72	100	97.38
12	100	100	100	100	97.15	100	100	100	100	95.39
13	99.24	99.89	98.80	99.13	96.51	99.02	100	98.36	99.13	97.38
14	99.91	99.91	99.91	99.25	95.89	100	100	99.91	99.35	97.20
15	88.66	85.43	94.25	91.43	94.00	79.35	93.04	83.94	86.13	96.27
16	99.94	100	99.61	99.50	93.30	100	100	100	100	95.13
P_{OA}	95.91	96.24	96.42	96.69	96.37	94.56	96.77	95.71	96.24	96.81
P_R	46.73	46.89	46.97	47.11	98.23	46.07	47.15	46.63	46.89	98.45
Time(s)	1128.55	755.52	1050.34	722.76	167.80	73.49	101.80	71.57	98.16	159.26

Table 12. P_{OA} and P_R calculated by the classification results using full bands and the bands selected by SC LCMV-BSS in Table 2 for Salinas.

Class	EPF-B-g with Full Bands	EPF-B-c with Full Bands	EPF-G-g with Full Bands	EPF-G-c with Full Bands	ILCMV with Full Bands	EPF-B-g-BS	EPF-B-c-BS	EPF-G-g-BS	EPF-G-c-BS	ILCMV-BS
1	100	100	100	100	95.52	100	100	100	100	96.91
2	100	100	100	100	98.42	99.95	99.84	99.92	99.97	98.71
3	100	100	100	100	93.78	99.90	100	100	99.90	95.95
4	100	100	100	100	95.62	100	99.57	100	99.86	92.04
5	99.07	99.22	99.10	98.92	96.90	98.88	99.55	99.33	98.99	90.78
6	100	100	100	100	98.79	100	100	100	99.90	97.75
7	100	99.97	100	99.97	98.63	99.61	99.89	100	99.94	98.32
8	90.18	89.51	89.40	91.38	96.69	90.88	89.02	90.68	90.30	96.61
9	100	99.98	99.81	99.94	95.87	99.95	99.95	99.97	100	95.44
10	97.47	97.71	97.41	99.48	96.67	98.90	99.24	98.11	98.29	96.77
11	100	100	100	100	97.75	100	99.72	99.34	99.72	97.66
12	100	100	100	100	97.15	100	100	100	100	95.43
13	100	99.89	98.91	99.89	96.51	99.34	99.89	99.24	99.13	94.00
14	100	100	100	99.81	95.89	99.35	99.63	98.79	98.32	96.93
15	88.99	84.26	88.08	86.90	94.00	89.74	93.35	78.92	88.92	95.84
16	99.28	99.89	100	99.94	93.30	100	100	100	99.94	95.41
P_{OA}	96.25	95.52	95.95	96.34	96.37	96.50	96.70	95.02	96.26	96.21
P_R	46.89	46.53	46.75	46.94	98.23	47.02	47.11	46.29	46.90	98.15
Time(s)	1139.99	1106.95	1154.78	1089.61	167.80	73.06	94.17	69.92	91.50	147.30

Table 13. P_{OA} and P_R calculated by the classification results using full bands and the bands selected by SQ LCMV-BSS-1 in Table 2 for University of Pavia.

Class	EPF-B-g with Full Bands	EPF-B-c with Full Bands	EPF-G-g with Full Bands	EPF-G-c with Full Bands	ILCMV with Full Bands	EPF-B-g-BS	EPF-B-c-BS	EPF-G-g-BS	EPF-G-c-BS	ILCMV-BS
1	98.04	98.04	98.08	97.81	77.24	96.15	92.47	96.34	97.29	80.21
2	98.66	99.39	97.79	98.28	86.42	94.31	94.84	97.39	98.06	84.44
3	91.09	93.52	95.00	94.33	73.34	92.62	94.81	95.57	95.33	84.14
4	93.47	95.27	92.92	98.01	79.85	97.52	97.45	97.00	96.87	76.49
5	100	100	100	99.85	98.81	100	100	100	100	98.14
6	99.98	100	100	100	91.49	99.64	98.15	99.64	100	93.50
7	100	99.32	99.92	99.77	89.10	99.92	99.32	99.40	100	86.19
8	99.02	99.00	97.80	99.78	81.10	95.60	96.22	96.41	96.28	82.84
9	100	100	100	100	78.46	100	99.89	100	100	77.09
P_{OA}	98.12	98.67	97.80	98.46	84.32	95.96	97.49	97.49	98.01	84.45
P_R	20.24	20.35	20.17	20.31	96.76	19.79	19.71	20.11	20.21	96.82
Time(s)	225.93	265.79	232.05	252.50	401.08	51.16	83.46	52.90	78.32	1387.01

Table 14. P_{OA} and P_R calculated by the classification results using full bands and the bands selected by SQ LCMV-BSS-2 in Table 2 for University of Pavia.

Class	EPF-B-g with Full Bands	EPF-B-c with Full Bands	EPF-G-g with Full Bands	EPF-G-c with Full Bands	ILCMV with Full Bands	EPF-B-g-BS	EPF-B-c-BS	EPF-G-g-BS	EPF-G-c-BS	ILCMV-BS
1	94.89	97.80	97.84	98.21	77.24	95.43	95.96	88.76	92.01	78.22
2	98.58	97.63	99.25	98.01	86.42	96.43	94.88	92.59	93.13	88.05
3	94.90	93.57	92.57	95.19	73.34	92.19	92.47	95.52	93.62	85.21
4	95.63	94.97	93.02	98.83	79.85	98.69	96.02	98.37	98.47	74.71
5	100	100	100	99.85	98.81	100	100	100	100	97.77
6	100	100	100	100	91.49	100	99.52	99.72	99.94	90.57
7	100	100	99.40	99.77	89.10	100	99.92	100	100	90.00
8	96.93	98.89	98.94	98.48	81.10	93.97	94.70	95.22	93.56	82.92
9	100	100	100	100	78.46	100	100	100	100	77.09
P_{OA}	97.76	97.85	98.36	98.39	84.32	96.74	95.97	94.25	94.78	85.41
P_R	20.16	20.18	20.29	20.29	96.76	19.95	19.79	19.44	19.55	96.94
Time(s)	219.10	249.76	226.85	238.81	401.08	48.52	82.95	53.53	79.05	971.56

Table 15. P_{OA} and P_R calculated by the classification results using full bands and the bands selected by SC LCMV-BSS in Table 2 for University of Pavia.

Class	EPF-B-g with Full Bands	EPF-B-c with Full Bands	EPF-G-g with Full Bands	EPF-G-c with Full Bands	ILCMV with Full Bands	EPF-B-g-BS	EPF-B-c-BS	EPF-G-g-BS	EPF-G-c-BS	ILCMV-BS
1	99.43	97.16	97.95	97.98	77.24	95.93	92.73	98.27	97.99	77.08
2	98.19	99.15	98.80	98.79	86.42	95.44	94.31	97.00	96.23	88.67
3	99.24	95.09	94.85	93.57	73.34	93.52	91.04	92.38	94.14	86.76
4	93.93	94.65	94.61	98.27	79.85	97.52	96.74	98.27	96.96	78.56
5	100	99.85	100	99.93	98.81	100	100	100	100	97.70
6	99.96	99.34	99.68	100	91.49	99.70	98.11	99.05	100	90.77
7	100	100	99.70	99.77	89.10	99.62	99.47	99.92	99.55	91.13
8	96.85	98.13	97.77	99.51	81.10	92.88	92.61	92.99	92.97	82.46
9	100	100	100	100	78.46	100	100	100	100	79.30
P_{OA}	98.37	98.32	98.28	98.67	84.32	96.22	94.85	97.21	96.92	85.92
P_R	20.29	20.28	20.27	20.35	96.76	19.85	19.56	20.05	19.99	97.09
Time(s)	238.01	270.24	234.01	259.33	401.08	51.13	83.41	49.37	76.48	998.17

Several interesting findings can be derived from the results in Tables 7–15.

1. It is very obvious to note that BSS did improve ILCMV classification results. Such an improvement cannot be found in the four EPF-based methods, where the classification results of the four EPF-based methods using band subsets could only get worse compared with the results using full bands. This may be due to the fact that the four EPF-based methods used principal component analysis (PCA) to compress the original data in preprocessing which retains some crucial information provided by full bands.
2. The precision rates produced by the four EPF-based methods were very low as also noted in [23,76]. However, ILCMV using bands selected by LCMV-BSS consistently performed very well in both P_{OA} and P_R.
3. According to Tables 7–9, ILCMV performed slightly better than the four EPF-based methods in P_{OA} but significantly better in P_R for Purdue's data and Salinas. The scene of the University of Pavia is interesting, as shown in Tables 13–15. The four EPF-based methods performed very well in P_{OA} but did very poorly in P_R with about only 20%. Furthermore, P_{OA} produced by ILCMV may not be as good as those produced by the four EPF-based methods (about 10% less) but the P_R produced by ILCMV were around 96% which is nearly 4.8 times better than the 20% produced by the four EPF-based methods. These experiments demonstrated that the BKG issue is critical in data analysis of the University of Pavia and cannot be ignored or discarded in data processing. Unfortunately, this BKG issue has never been investigated in the past.
4. Unlike the four EPF-based methods, which performed well in P_{OA} but very poorly in P_R, ILCMV consistently performs well in both P_{OA} and P_R, and even better when it is implemented in conjunction with BSS—a case that the EPF-based methods actually failed, as shown in Tables 7–15.
5. Last but not least, BS is heavily determined by three factors: the data to be processed, the BS method selected, and the classifier used. Unfortunately, most works on BS for hyperspectral image classification have been focused on the design and development of BS methods but very little has been reported on performance evaluation of different classifiers which use the same set of bands selected by a BS method. For example, as shown in Tables 7–15, if the four EPF methods were implemented by BS, their classification results could not be improved, but those of ILCMV could.
6. It should be noted that P_D results are not included in Tables 7–15 due to two reasons. One is that the results of P_D using full bands are already available in [23,76]. The other is that EPF-based methods using partial bands did not perform better than their counterparts using full bands. So, it does not make sense to include their results in tables. Besides this, due to limited space, there is no need to include their results.

6. Conclusions

This paper developed an SMMBS method, called LCMV-BSS, which selects multiple bands as a band subset using LCMV to linearly constrain class signature vectors as a criterion to select an optimal band subset. It is completely different from existing BS methods, with the following contributions: (i) It is a BSS method particularly developed for HSIC; (ii) It is quite different from single band-constrained methods in [26] and multiple-band constrained methods in [68], by constraining multiple class signature vectors instead of multiple bands; (iii) It develops three numerical search algorithms to find optimal band subsets which are different from the graph-based approaches [40,43] used by other SMMBS methods; (iv) It is very simple to implement via (7) with no parameters needing to be tuned; (v) Most importantly, it shows that HSIC can be improved by BS provided that the number n_{BS} of selected bands and the set of n_{BS} bands are properly selected.

Acknowledgments: The work of C.Y. is supported by National Nature Science Foundation of Liaoning Province (20170540095). The work of M.S. is supported by National Nature Science Foundation of China (61601077) and State Key Laboratory of Integrated Services Networks. The work of C.-I.C. is supported by the Fundamental Research Funds for Central Universities under Grant 3132016331. The authors would like to thank Xiaoqiang Lu for helping run the MDPP method with the same parameters used in [43] and the authors of ref. [40] for providing their software to run DSEBS.

Author Contributions: C.Y. and M.S. conceived and designed the experiments; C.Y. performed the experiments; C.Y. and C.-I.C. analyzed the data; C.Y. and M.S. contributed reagents/materials/analysis tools; C.-I.C. wrote the paper.

Conflicts of Interest: All authors have declared no conflict of interest.

References

1. Melgani, F.; Bruzzone, L. Classification of hyperspectral remote sensing images with support vector machines. *IEEE Trans. Geosci. Remote Sens.* **2004**, *42*, 1778–1790. [CrossRef]
2. Benediktsson, J.A.; Palmason, J.A.; Sveinsson, J.R. Classification of hyperspectral data from urban areas based on extended morphological profiles. *IEEE Trans. Geosci. Remote Sens.* **2005**, *43*, 480–491. [CrossRef]
3. Camps-Valls, G.; Bruzzone, L. Kernel-based methods for hyperspectral image classification. *IEEE Trans. Geosci. Remote Sens.* **2005**, *43*, 1351–1362. [CrossRef]
4. Tarabalka, Y.; Benediktsson, J.A.; Chanussot, J. Spectral-spatial classification of hyperspectral imagery based on partitional clustering techniques. *IEEE Trans. Geosci. Remote Sens.* **2009**, *47*, 2973–2987. [CrossRef]
5. Tarabalka, Y.; Fauvel, M.; Chanussot, J.; Benediktsson, J.A. SVM- and MRF-based method for accurate classification of hyperspectral images. *IEEE Geosci. Remote Sens. Lett.* **2010**, *9*, 736–740. [CrossRef]
6. Li, J.; Bioucas-Dias, J.M.; Plaza, A. Semisupervised hyperspectral image segmentation using multinominal logistic regression model with active learning. *IEEE Trans. Geosci. Remote Sens.* **2010**, *48*, 4085–4098.
7. Zhang, B.; Li, S.; Jia, X.; Gao, L.; Peng, M. Adpative hyperspectral Markov random field approach for classification of hyperspectral imagery. *IEEE Geosci. Remote Sens. Lett.* **2011**, *8*, 4085–4098. [CrossRef]
8. Li, J.; Bioucas-Dias, J.M.; Plaza, A. Hyperspectral image segmentation using a new Bayesion approach with active learning. *IEEE Trans. Geosci. Remote Sens.* **2011**, *49*, 3947–3960. [CrossRef]
9. Chen, Y.; Nasabardi, N.; Tran, T.D. Hyperspectral image segmentation using dictionary-based sparse representation. *IEEE Trans. Geosci. Remote Sens.* **2011**, *49*, 3973–3985. [CrossRef]
10. Tarabalka, Y.; Fauvel, M.; Chanussot, J.; Benediktsson, J.A. Segmentation and classification of hyperspectral images using minimum spanning forest grown from automatically selected makers. *IEEE Trans. Syst. Man Cybern. Part B Cybern.* **2011**, *40*, 1267–1279. [CrossRef] [PubMed]
11. Fauvel, M.; Tarabalka, Y.; Benediktsson, J.A.; Chanussot, J.; Tilton, J. Advances in spectral-spatial classification of hyperspectral images. *Proc. IEEE* **2013**, *101*, 652–675. [CrossRef]
12. Kang, K.; Li, S.; Benediktsson, J.A. Spectral-spatial hyperspectral image classification with edge-preserving filtering. *IEEE Trans. Geosci. Remote Sens.* **2014**, *52*, 2666–2677. [CrossRef]
13. Fu, W.; Li, S.; Fang, L.; Kang, X.; Benediktsson, J.A. Hyperspectral image classification via shape-adaptive joint sparse representation. *IEEE J. Sel. Top. Appl. Earth Obs. Remote Sens.* **2016**, *9*, 556–567. [CrossRef]
14. Kang, K.; Li, S.; Fang, L.; Li, M.; Benediktsson, J.A. Extended random walker-based classification of hyperspectral images. *IEEE Trans. Geosci. Remote Sens.* **2015**, *53*, 144–153. [CrossRef]
15. Lu, T.; Li, S.; Fang, L.; Bruzzone, L.; Benediktsson, J.A. Set-to-set distance-based spectral spatial classification of hyperspectral images. *IEEE Trans. Geosci. Remote Sens.* **2016**, *54*, 7122–7134. [CrossRef]
16. Guo, X.; Huang, X.; Zhang, L.; Plaza, A.; Benedikisson, J.A. Support tensor machines for classification of hyperspectral remote sesning imagery. *IEEE Trans. Geosci. Remote Sens.* **2016**, *54*, 3248–3264. [CrossRef]
17. Ma, X.; Wang, H.; Wang, J. Semisupervised classification of hyperspectral image based on multi- decision labeling and deep feature learning. *ISPRS J. Photogramm. Remote Sens.* **2016**, *120*, 99–107. [CrossRef]
18. Camps-Valls, G.; Gomez-Chova, L.; Munoz-Mari, J.; Vila-Frances, J.; Calpe-Maravilla, J. CKs for hyperspectral image classification. *IEEE Geosci. Remote Sens. Lett.* **2006**, *3*, 93–97. [CrossRef]
19. Gurram, P.; Kwon, H. Contextual SVM using Hilbert space embedding for hyperspectral classification. *IEEE Geosci. Remote Sens. Lett.* **2013**, *10*, 1031–1035. [CrossRef]
20. Peng, J.; Zhou, Y.; Chen, C. Region-kernel-based support vector machines for hyperspectral image classification. *IEEE Trans. Geosci. Remote Sens.* **2015**, *53*, 4810–4824. [CrossRef]

21. Sun, S.; Zhong, P.; Xiao, H.; Wang, R. Active learning with Gaussian process classifier for hyperspectral image classification. *IEEE Trans. Geosci. Remote Sens.* **2015**, *53*, 1746–1760. [CrossRef]
22. Pullanagari, R.; Kereszturi, G.; Yule, I.J.; Ghamisi, P. Assessing the performance of multiple spectral-spatial features of a hyperspectral image for classification of urban land cover classes using support vector machines and artificial neural network. *J. Appl. Remote Sens.* **2017**, *11*, 026009-1–026009-21. [CrossRef]
23. Xue, B.; Yu, C.; Wang, Y.; Song, M.; Li, S.; Wang, L.; Chen, H.M.; Chang, C.-I. A subpixel target detection approach to hyperpsectral image classification. *IEEE Trans. Geosci. Remote Sens.* **2017**, *55*, 5093–5114. [CrossRef]
24. Chang, C.-I. *Hyperspectral Data Processing: Signal Processing Algorithm Design and Analysis*; Wiley: Hoboken, NJ, USA, 2013.
25. Chang, C.-I.; Du, Q.; Sun, T.S.; Althouse, M.L.G. A joint band prioritization and band decorrelation approach to band selection for hyperspectral image classification. *IEEE Trans. Geosci. Remote Sens.* **1999**, *37*, 2631–2641. [CrossRef]
26. Chang, C.-I.; Wang, S. Constrained band selection for hyperspectral imagery. *IEEE Trans. Geosci. Remote Sens.* **2006**, *44*, 1575–1585. [CrossRef]
27. Huang, R.; He, M. Band selection based on feature weighting for classification of hyperspectral data. *IEEE Trans. Geosci. Remote Sens. Lett.* **2005**, *2*, 156–159. [CrossRef]
28. Keshava, N. Distance metrics and band selection in hyperspectral processing with applications to material identification. *IEEE Trans. Geosci. Remote Sens.* **2004**, *42*, 1552–1565. [CrossRef]
29. Mausel, P.W.; Kramber, W.J.; Lee, J.K. Optimum band selection for supervised classification of multispectral data. *Photogramm. Eng. Remote Sens.* **1990**, *56*, 55–60.
30. Jia, S.; Ji, Z.; Qian, Y.-Y.; Shen, L.-L. Unsupervised band selection for hyperspectral imagery classification without manual band removal. *IEEE J. Sel. Top. Appl. Earth Obs. Remote Sens.* **2012**, *5*, 531–543. [CrossRef]
31. Stearns, S.D.; Wilson, B.E.; Peterson, J.R. Dimensionality reduction by optimal band selection for pixel classification of hyperspectral imagery. *Proc. SPIE* **1993**, *2028*, 118–127.
32. Backer, S.; Kempeneers, P.; Debruyn, W.; Scheunders, P. Band selection for hyperspectral remote sensing. *Pattern Recognit.* **2005**, *2*, 319–323.
33. Du, Q.; Yang, H. Similarity-based unsupervised band selection for hyperspectral image analysis. *IEEE Geosci. Remote Sens. Lett.* **2008**, *5*, 564–568. [CrossRef]
34. Yang, H.; Du, Q. An efficient method for supervised hyperspectral band selection. *IEEE Geosci. Remote Sens. Lett.* **2011**, *8*, 138–142. [CrossRef]
35. Su, H.; Du, Q.; Chen, G.; Du, P. Optimized hyperspectral band selection using particle swarm optimization. *IEEE J. Sel. Top. Appl. Earth Obs. Remote Sens.* **2014**, *7*, 2659–2670. [CrossRef]
36. Su, H.; Yang, B.; Du, Q. Hyperspectral band selection using improved firefly algorithm. *IEEE Geosci. Remote Sens. Lett.* **2016**, *13*, 68–72. [CrossRef]
37. Wei, W.; Du, Q.; Younan, N.H. Fast supervised hyperspectral band selection using graphics processing unit. *J. Appl. Remote Sens.* **2012**, *6*, 061504-1–061504-12.
38. Yuan, Y.; Zhu, G.; Wang, Q. Hyperspectral band selection by multitask sparisty pursuit. *IEEE Trans. Geosci. Remote Sens.* **2015**, *53*, 631–644. [CrossRef]
39. Lu, X.; Li, X.; Mou, L. Semi-supervised multitask learning for scene recognition. *IEEE Trans. Cybern.* **2015**, *45*, 1967–1976. [PubMed]
40. Zhu, G.; Huang, H.; Lei, J.; Bi, Z.; Xu, F. Unsupervised hyperspectral band selection by dominant set extraction. *IEEE Trans. Geosci. Remote Sens.* **2016**, *54*, 227–239. [CrossRef]
41. Yuan, Y.; Lin, J.; Wang, Q. Dual-clustering-based hyperspectral band selection by contextual analysis. *IEEE Trans. Geosci. Remote Sens.* **2016**, *54*, 1431–1445. [CrossRef]
42. Wang, Q.; Lin, J.; Yuan, Y. Salient band selection for hyperspectral image classification via manifold ranking. *IEEE Trans. Neural Netw. Learn. Syst.* **2016**, *27*, 1279–1289. [CrossRef] [PubMed]
43. Yuan, Y.; Zheng, X.; Lu, X. Discovering diverse subset for unsupervised hyperspectral band selection. *IEEE Trans. Image Process.* **2017**, *26*, 51–64. [CrossRef] [PubMed]
44. Lu, X.; Yuan, Y.; Zheng, X. Joint dictionary learning for multispectral change detection. *IEEE Trans. Cybern.* **2017**, *47*, 884–897. [CrossRef] [PubMed]
45. Lu, X.; Zheng, X.; Yuan, Y. Remote sensing scene classification by unsupervised representation learning. *IEEE Trans. Geosci. Remote Sens.* **2017**, *55*, 5148–5157. [CrossRef]

46. Feng, J.; Jiao, L.C.; Zhang, X.; Sun, T. Hyperspectral band selection based on trivariate mutual information and clonal selection. *IEEE Trans. Geosci. Remote Sens.* **2014**, *52*, 4092–4105. [CrossRef]
47. Feng, J.; Jiao, L.C.; Liu, F.; Sun, T.; Zhang, X. Mutual-information0-based semi-supervised hyperspectral band selection with high discrimination, high information and low redundancy. *IEEE Trans. Geosci. Remote Sens.* **2015**, *53*, 2956–2969. [CrossRef]
48. Wang, C.; Gong, M.; Zhang, M.; Chan, Y. Unsupervised hyperspectral image band selection via column subset selection. *IEEE Geosci. Remote Sens. Lett.* **2015**, *12*, 1411–1415. [CrossRef]
49. Geng, X.; Sun, K.; Ji, L. Band selection for target detection in hyperspectral imagery using sparse CEM. *Remote Sens. Lett.* **2014**, *5*, 1022–1031. [CrossRef]
50. Sun, K.; Geng, X.; Ji, L. A new sparsity-based band selection method for target detection of hyperspectral image. *IEEE Geosci. Remote Sens. Lett.* **2015**, *12*, 329–333.
51. Zare, A.; Gader, P. Hyperspectral band selection and endmember detection using sparisty promoting priors. *IEEE Geosci. Remote Sens. Lett.* **2008**, *5*, 256–260. [CrossRef]
52. Ball, J.E.; Bruce, L.E.; Younan, N.H. Hyperspectral pixel unmixing via spectral band selection and DC-insensitive singular value decomposition. *IEEE Geosci. Remote Sens. Lett.* **2007**, *4*, 382–386. [CrossRef]
53. Sun, K.; Geng, X.; Ji, L.; Lu, Y. A new band selection method for hyperspectral image based on data quality. *IEEE J. Sel. Top. Appl. Earth Obs. Remote Sens.* **2014**, *7*, 2697–2703.
54. Sun, K.; Geng, X.; Ji, L. Exemplar component analysis: A fast band selection method for hyperspectral imagery. *IEEE Geosci. Remote Sens. Lett.* **2015**, *12*, 998–1002.
55. Koonsanit, K.; Jaruskulchai, C.; Eiumnoh, A. Band selection for dimension reduction in hyper spectral image using integrated information gain and principal components analysis technique. *Int. J. Mach. Learn. Comput.* **2012**, *2*, 248–251. [CrossRef]
56. Xia, W.; Wang, B.; Zhang, L. Band selection for hyperspectral imagery: A new approach based on complex networks. *IEEE Geosci. Remote Sens. Lett.* **2013**, *10*, 1229–1233. [CrossRef]
57. Chang, C.-I.; Liu, K.-H. Progressive band selection for hyperspectral imagery. *IEEE Trans. Geosci. Remote Sens.* **2014**, *52*, 2002–2017. [CrossRef]
58. Martínez-Usó, A.; Pla, F.; Sotoca, J.M.; García-Sevilla, P. Clustering-based hyperspectral band selection using information measures. *IEEE Trans. Geosci. Remote Sens.* **2007**, *45*, 4158–4171. [CrossRef]
59. Yang, H.; Du, Q.; Sheng, Y. Semisupervised band clustering for dimensionality reduction of hyperspectral imagery. *IEEE Geosci. Remote Sens. Lett.* **2011**, *8*, 1135–1139.
60. Su, H.; Du, Q. Hyperspectral band clustering and band selection for urban land cover classification. *Geocarto Int.* **2012**, *27*, 395–411. [CrossRef]
61. Winter, M.E. N-finder: An algorithm for fast autonomous spectral endmember determination in hyperspectral data. *Proc. SPIE* **1999**, *3753*, 266–277.
62. Wu, C.-C.; Chu, S.; Chang, C.-I. Sequential N-FINDR algorithm. In Proceedings of the SPIE Conference on Imaging Spectrometry XIII, San Diego, CA, USA, 10–14 August 2008.
63. Xiong, X.; Wu, C.-C.; Chang, C.-I.; Kapalkis, K.; Chen, H.M. Fast algorithms to implement N-FINDR for hyperspectral endmember extraction. *IEEE J. Sel. Top. Appl. Earth Obs. Remote Sens.* **2011**, *4*, 545–564. [CrossRef]
64. Chang, C.-I. Maximum Simplex Volume-Based Endmember Extraction Algorithms. U.S. Patent 8,417,748 B2, 9 April 2013.
65. Chang, C.-I. *Real Time Progressive Hyperspectral Image Processing: Endmember Finding and Anomaly Detection*; Springer: New York, NY, USA, 2016.
66. Wang, L.; Chang, C.-I.; Wang, Y.; Xue, B.; Song, M.; Yu, C.; Li, S. Band subset selection for anomaly detection in hyperspectral imagery. *IEEE Trans. Geosci. Remote Sens.* **2017**, *55*, 4887–4898. [CrossRef]
67. Chang, C.-I.; Lee, L.C.; Xue, B.; Song, M.; Chen, J. Channel capacity approach to band subset selection for hyperspectral imagery. *IEEE J. Sel. Top. Appl. Earth Obs. Remote Sens.* **2017**, *10*, 4630–4644. [CrossRef]
68. Wang, L.; Li, H.C.; Xue, B.; Chang, C.-I. Constrained band subset selection for hyperspectral imagery. *IEEE Geosci. Remote Sens. Lett.* **2017**, *14*, 2032–2036. [CrossRef]
69. Chang, C.-I. *Hyperspectral Imaging: Techniques for Spectral Detection and Classification*; Kluwer Academic/Plenum Publishers: New York, NY, USA, 2003.
70. Frost, O.L., III. An algorithm for linearly constrained adaptive array processing. *Proc. IEEE* **1972**, *60*, 926–935. [CrossRef]

71. Chang, C.-I. Target signature-constrained mixed pixel classification for hyperspectral imagery. *IEEE Trans. Geosci. Remote Sens.* **2002**, *40*, 1065–1081. [CrossRef]
72. Harsanyi, J.C. *Detection and Classification of Subpixel Spectral Signatures in Hyperspectral Image Sequences*; Department of Electrical Engineering, University of Maryland: Baltimore County, MD, USA, 1993.
73. Pudil, P.; Novovicova, J.; Kittler, J. Floating search methods in feature selection. *Pattern Recognit. Lett.* **1994**, *15*, 1119–1125. [CrossRef]
74. Chang, C.-I.; Du, Q. Estimation of number of spectrally distinct signal sources in hyperspectral imagery. *IEEE Trans. Geosci. Remote Sens.* **2004**, *42*, 608–619. [CrossRef]
75. Harsanyi, J.C.; Farrand, W.; Chang, C.-I. Detection of subpixel spectral signatures in hyperspectral image sequences. In Proceedings of the American Congress on Surveying & Mapping (ACSM)/American Society of Photogrammetry & Remote Sensing (ASPRS) Annual Converntion and Exposition, Baltimore, MD, USA, 25–28 April 1994; Volume 1, pp. 236–247.
76. Yu, C.; Xue, B.; Wang, Y.; Song, M.; Wang, L.; Li, S.; Chang, C.-I. Multi-class constrained background suppression approach to hyperspectral image classification. In Proceedings of the 2017 IEEE/GRSS International Geoscience and Remote Sensing Symposium (IGARSS), Fort Worth, TX, USA, 23–28 July 2017.

remote sensing

MDPI

Article

Progressive Sample Processing of Band Selection for Hyperspectral Image Transmission

Keng-Hao Liu [1,*], Shih-Yu Chen [2], Hung-Chang Chien [1] and Meng-Han Lu [1]

1 Department of Mechanical and Electro-Mechanical Engineering, National Sun Yat-sen University, Kaohsiung 80424, Taiwan; m043020080@student.nsysu.edu.tw (H.-C.C.); m053020108@student.nsysu.edu.tw (M.-H.L.)
2 Department of Computer Science and Information Engineering, National Yunlin University of Science and Technology, Yunlin 64002, Taiwan; sychen@yuntech.edu.tw
* Correspondence: keng3@mail.nsysu.edu.tw; Tel.: +886-7-5252000-4220

Received: 18 December 2017; Accepted: 16 February 2018; Published: 26 February 2018

Abstract: Band selection (BS) is one of the important topics in hyperspectral image (HSI) processing. Many types of BS algorithms were proposed in the last decade. However, most of them were designed for off-line use. They can only be used with pre-collected data, and are sometimes ineffective for applications that require timeliness, such as disaster prevention or target detection. This paper proposes an online BS method that allows us obtain instant BS results in a progressive manner during HSI data transmission, which is carried out under band-interleaved-by-sample/pixel (BIS/BIP) format. Such a revolutionary method is called progressive sample processing of band selection (PSP-BS). In PSP-BS, BS can be done recursively pixel by pixel, so that the instantaneous BS can be achieved without waiting for all the pixels of an image. To develop a PSP-BS algorithm, we proposed PSP-OMPBS, which adopted the recursive version of a self-sparse regression BS method (OMPBS) as a native algorithm. The experiments conducted on two real hyperspectral images demonstrate that PSP-OMPBS can progressively output the BS with extremely low computing time. In addition, the convergence of BS results during transmission can be further accelerated by using a pre-defined pixel transmission sequence. Such a significant advantage not only allows BS to be done in a real-time manner for the future satellite data downlink, but also determines the BS results in advance, without waiting to receive every pixel of an image.

Keywords: band selection (BS); progressive sample processing (PSP); real-time processing

1. Introduction

Due to the use of hundreds of spectral bands, hyperspectral imaging (HSI) generally has enormous data volume and contains vast amount of information. This special characteristic results in several issues. First, the inter-band correlation of HSI is very high, and adjacent bands may contain redundant spectral information. This would lead to the well-known problem called the "curse of dimensionality" in data analysis. Second, the computational complexity of processing HSI data is very high. Third, storing HSI data usually requires large amount of disc space. Finally, transmitting HSI data requires higher bandwidths. Under such circumstances, removing partial data without significant loss of an image's spectral information is necessary. One commonly used approach is band selection (BS) [1]. BS takes advantage of such high-band correlation to remove the redundant bands, in order to achieve a wide range of applications, such as dimensionality reduction, data storage, data transmission, target detection, and classification.

Many different types of BS algorithms [2–15] have been proposed in the past two decades. Most of them make an assumption that the BS problem is an optimization problem, which maximizes or minimizes a pre-defined objective function that can measure the amount of information or inter-band redundancy contained by the currently selected bands. For instance, Keshava et al. [2] adopts common distance measures, such as Euclidean distance and a spectral angle mapper, to measure the similarity of bands. Du et al. [3] uses linear regression and orthogonal subspace projection to sequentially select a new band based on the complementary vector space. Chang et al. [4] regards a spectral band as a signal, and designed a BS method based on a constrained energy minimization algorithm. Martínez-Usó et al. [5] introduces a hierarchical clustering-based method to select bands from the band groups that were pre-clustered by a specific method, with some informational measures. In addition, there has been a lot of research that adopts the perspective of information theory. The works in [6–8] adopt mutual information as the similarity index to find the maximum information spanned by the bands. In recent years, sparse regression models were used to perform BS. For instance, Sun et al. [9–11] uses a self-sparse regression (SSR) model to select bands. In an SSR problem, finding a new basis is equivalent to selecting the most representation bands in an HSI image. Lai et al. [12] proposed a SSR-based BS method, which adopts an orthogonal matching pursuit (OMP) algorithm to sequentially find the next band in an efficient manner. For the purpose of HSI data transmission, Chang et al. [13,14] proposed a progressive BS (PBS) method for spectral unmixing. In PBS, the bands are first prioritized by a certain criterion, and are transmitted by the order of prioritization. Du et al. [15] proposed a BS-based dimensional reduction method for change detection in multi-temporal hyperspectral images. Except for the above-mentioned work, there were still various kinds of BS research published in the literature [16–25].

Since many existing HSI algorithms were designed to deal with off-line problems, they usually require more computing time for pursuing extremely high accuracy. Undoubtedly, those sophisticated methods may not be appropriate for dealing with timeliness applications. Hence, developing real-time and efficient approaches is urgently needed. Based on that point, some research started to develop on-board computing approaches [26–29], where the data compression could be carried out on satellites to reduce the demand of downlink bandwidth, focusing additionally on real-time image classification with the aid of GPU acceleration [30,31].

On the other hand, processing HSI data in a real-time progressive fashion under the data transmission held from satellite to ground station has become another attractive topic. The term "progressive processing" for HSI data originated from [32]. According to [32], the term refers to a type of signal processing method. It decomposes entire data processing into a finite number of stages, and processes data progressively stage by stage, in the sense that the results obtained by previous stages can be used to update or improve the results to be processed in subsequent stages. Recently, the concept of progressive processing was utilized to process HSI data. For example, in order to achieve real-time spectral unmixing during band transmission, a revolutionary concept called progressive band processing (PBP) [33], was developed. PBP assumes that the HSI data is transmitted by band sequential (BSQ) format, and processes the image immediately after receiving a new band. Under the PBP framework, many common topics, such as spectral unmixing, target detection, and anomaly detection could be realized [33–35]. Unlike the traditional algorithms, which can only be implemented on the collected data, PBP makes use of recursive processing, to instantly process the current data segment. More specifically, PBP methods preserve the past information obtained at previous stage, and use it to accelerate the computation at the current stage. The most attractive feature of PBP is that the computing time will not significantly increase as the growth of number of received bands. Therefore, PBP algorithms have significant potential for the uses of applications in data transmission and communication.

Based on the importance of BS, and the urgent demand of real-time algorithms, this paper proposes a real-time progressive BS method for applications related to data transmission. To our best knowledge, there were no BS methods proposed for this purpose before. In our case, we assume that HSI data is transmitted in band-interleaved-by-pixel (BIP) or band-interleaved-by-sample (BIS) format, and is processed pixel by pixel (sample by sample). According to [32], such a process operated on BIP/BIS transmission is called progressive sample processing (PSP). To sum up, we developed a novel concept called progressive sample processing of band selection (PSP-BS), which combines BS with PSP. Figure 1 illustrates the difference of traditional BS and PSP-BS.

To realize PSP-BS, we had to choose an appropriate BS algorithm as the core of PSP-BS. In this paper, a self-sparse regression BS algorithm, called orthogonal matching pursuit-based BS (OMPBS) [12], has been adopted because (1) the OMPBS can sequentially find the most information-complementary bands from the image, without any prior knowledge or any complicated optimization scheme; and (2) The algorithm is easily re-formulated to fulfill recursive processing through mathematical decomposition of the optimization equations of OMPBS. In the later sections, we first introduce OMPBS, and then derive its recursive version, called recursive OMPBS (Re-OMPBS), as the key algorithm to realize PSP-BS. In addition, we make an assumption that the pixel transmission sequence (PTS) could be defined before transmission. We propose three PTSs to make BS results converge more quickly during the PSP-BS. We call Re-OMPBS combined with PTS PSP-OMPBS.

To evaluate the effectiveness of PSP-OMPBS, two real hyperspectral datasets were used in the experiments. We conducted a comparison of computing time and the corresponding BS accuracy at every time stage in a progressive manner. Land cover classification was also adopted to evaluate the quality of the selected bands. According to the experimental results, the computational efficiency of PSP-OMPBS is stably high so that the process could be carried out in real time. The BS accuracy can be improved early on by using particular PTSs.

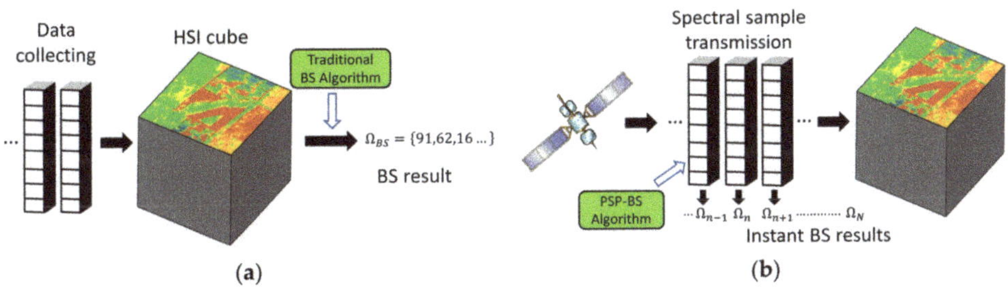

Figure 1. An illustration of the difference of traditional band selection (BS) and progressive sample processing of band selection (PSP-BS). (**a**) The traditional BS algorithm is implemented on the collected hyperspectral image (HSI) data. (**b**) The PSP-BS is implemented in data sample transmission. In PSP-BS, the real-time BS monitoring can be achieved.

2. Orthogonal Matching Pursuit-Based BS (OMPBS)

The OMPBS [12] is a sequential BS method based on a self-sparse regression (SSR) model [36]. It aims to find a set of representative bands that can represent all bands based on the minimization of reconstructed errors in the SSR model, by using an orthogonal matching pursuit (OMP) algorithm [37]. Suppose L is the number of bands and N is number of pixels in a hyperspectral image. Let $\mathbf{b}_i \in R^{N\times 1}$ presents the ith band vector. In an SSR model, both the observation matrix as well as the dictionary matrix are the set of band vectors denoted by $\mathbf{B} = [\mathbf{b}_1, \mathbf{b}_2, \dots, \mathbf{b}_L]$ and the coefficient matrix is denoted by $\mathbf{C}$. Obviously, the goal is to find the optimal coefficient matrix $\mathbf{C}$ that minimizes the reconstructed error $||\mathbf{B} - \mathbf{BC}||_F^2$. Since only k bands need to be selected, we impose the sparsity constraint $||\mathbf{C}||_{0,2} \leqq p$ where l_0/l_2 norm $||\mathbf{C}||_{0,2}$ counts the number of the non-zero rows in $\mathbf{C}$. Thus, our sparse-BS optimization problem is

$$\hat{\mathbf{C}} = \arg_{\mathbf{c}} \min||\mathbf{B} - \mathbf{BC}||_F^2 \dots \text{s.t. } ||\mathbf{C}||_{0,2} \leqq p, \quad (1)$$

where $||\mathbf{B} - \mathbf{BC}||_F^2$ is the reconstructed error, $||\mathbf{C}||_{0,2} \leqq p$ is the sparsity constraint, and p is the sparsity level. The sub-optimization problem (find $\mathbf{C}$) is solved by least square estimator $\hat{\mathbf{Q}} = \left(\mathbf{P}^T\mathbf{P}\right)^{-1}\mathbf{P}^T\mathbf{B}$, where **P** is temporary basis matrix composed by the candidate bands. The original OMPBS algorithm is composed of a doubly-nested loop. So, it is required to solve $L!/(L-p)!$ times of (1), so the overall computation is time-consuming. To improve its efficiency, we revised the original algorithm by residual perspective. The revised algorithm (Algorithm 1) is shown as follows.

Algorithm 1. OMPBS

Objective: Find p representative bands for a HSI cube
Input: $\mathbf{B}$ ($N \times L$ matrix), sparsity = p
Output: Band indices set $\Omega_p = \{b_1, b_2, \ldots, b_p\}$
Steps:
1. Initialization: Set $\Omega_0 = \varnothing$, $\mathbf{R}_0 = \mathbf{B}$, $\mathbf{P}_0 = \varnothing$, $j = 1$.
2. At j-th iteration, calculate the row norm vector $\mathbf{k} \in L \times 1$ of matrix $\mathbf{R}_j^T \mathbf{R}_j$.
3. Find the row index b_j whose value is maximum at $\mathbf{k}$. Then set $\mathbf{P}_j = \left[\mathbf{P}_{j-1}\ \mathbf{b}_{b_j}\right]$, $\Omega_j = \Omega_{j-1} \cap \left\{b_j\right\}$.
4. Solve the sub optimization problem $\hat{\mathbf{Q}}_j = \arg\min_{\mathbf{Q}_j} \left|\left|\mathbf{B} - \mathbf{P}_j \mathbf{Q}\right|\right|_F^2$ by $\mathbf{Q}_j = \left(\mathbf{P}_j^T \mathbf{P}_j\right)^{-1} \mathbf{P}_j^T \mathbf{B}$.
5. Update the residual matrix $\mathbf{R}_{j+1} = \mathbf{B} - \mathbf{P}_j \mathbf{Q}_j$.
6. If $j = p$, break; otherwise, set $j \leftarrow j + 1$ and go to Step 2.

3. Progressive Sample Processing-Based OMPBS (PSP-OMPBS)

The proposed PSP-OMPBS method is fully introduced in this section. It consists of two parts, as shown in Figure 2. The first part controls the data transmission order, in which the pixel transmission sequence will be formed. The second part is the real-time BS computing core, recursive OMPBS (Re-OMPBS). Section 3.1 introduces the derivation of Re-OMPBS, where we convert some steps in OMPBS to their recursive forms. Section 3.2 describes the condition for implementing recursive equations introduced in Section 3.1. Section 3.3 summaries the details of the Re-OMPBS algorithm. Finally, the formation of the pixel transmission sequence is introduced in Section 3.4.

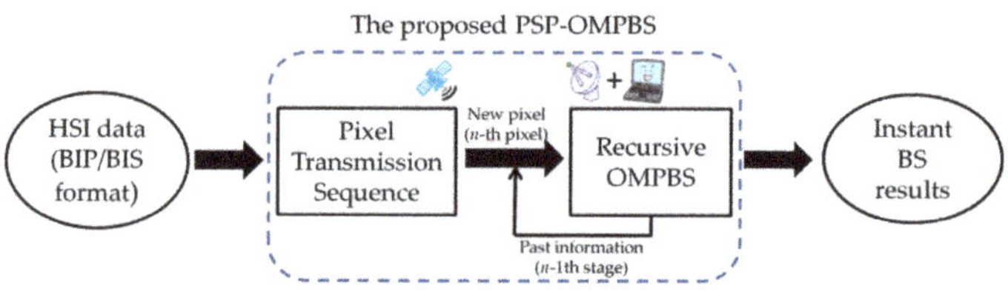

Figure 2. Conceptual flowchart of progressive sample processing-based OMPBS (PSP-OMPBS). The center part is the proposed PSP-OMPBS method which includes two blocks: formation of the pixel transmission sequence (PTS), and implementation of the recursive OMPBS (Re-OMPBS) algorithm. The transmission is carried out from the PTS block to the Re-OMPBS block.

3.1. Derivation of Recursive OMPBS (Re-OMPBS)

In this section, we aim to integrate OMPBS into PSP. According to the definition, PSP-BS can instantly acquire the current BS result during the data transmission by using two types of information: the processed information provided in previous stage (i.e., the $(n-1)$th stage), and the new information provided in current stage (i.e., nth stage). More specifically, PSP-BS can immediately update the BS result when a new pixel (the nth pixel) is received. Assume that n presents the number of currently received pixels. We would like to utilize the past information obtained in the $(n-1)$th stage to reduce the computational burden of the nth stage. In doing so, the optimization process of OMPBS must be broken down.

3.1.1. Decomposition of the Least Square Estimator

In the following derivation, we assume that j is a fixed number and do not annotate it for each term. In the OMPBS algorithm, there are two calculations that can be decomposed. The first one is the least square estimator in Step 4:

$$\mathbf{Q}(n) = \left(\mathbf{P}(n)^T \times \mathbf{P}(n)\right)^{-1} \times \mathbf{P}(n)^T \times \mathbf{B}(n). \tag{2}$$

As we know, computing Equation (2) is required for each iteration j in the inner loop of OMPBS. This computation would be time-consuming when n is large. This would be unfavorable to progressive processing. To resolve this issue, we must convert Equation (2) to its recursive version. To do that, we define two matrices:

$$\mathbf{B}(n) = \begin{bmatrix} b_{11} & b_{12} & \dots & b_{1L} \\ b_{21} & b_{22} & \dots & b_{2L} \\ \vdots & \vdots & \ddots & \vdots \\ b_{n1} & b_{n2} & \dots & b_{nL} \end{bmatrix}, \ \mathbf{P}(n) = \begin{bmatrix} p_{11} & p_{12} & \dots & p_{1p} \\ p_{21} & p_{22} & \dots & p_{2p} \\ \vdots & \vdots & \ddots & \vdots \\ p_{n1} & p_{n2} & \dots & p_{np} \end{bmatrix}. \tag{3}$$

$\mathbf{B}(n)$ represents the total pixel data after the nth pixel is received, and $\mathbf{P}(n)$ presents the data's temporary basis matrix, composed by selected band vectors. Let $\mathbf{r}_n = [b_{n1}, b_{n2}, \dots, b_{nL}]$ be the new pixel and $\widetilde{\mathbf{p}}_n = [p_{n1}, p_{n2}, \dots, p_{np}]$. Next, we define auxiliary matrix $\mathbf{H}(n)$

$$\mathbf{H}(n) = \mathbf{P}(n)^T \times \mathbf{P}(n) = \begin{bmatrix} \mathbf{P}(n-1)^T & \widetilde{\mathbf{p}}_n^T \end{bmatrix} \begin{bmatrix} \mathbf{P}(n-1) \\ \widetilde{\mathbf{p}}_n \end{bmatrix} = \mathbf{P}(n-1)^T \times \mathbf{P}(n-1) + \widetilde{\mathbf{p}}_n^T \widetilde{\mathbf{p}}_n. \tag{4}$$

Substituting (4) into (2), we get

$$\mathbf{Q}(n) = \left(\mathbf{P}(n-1)^T \times \mathbf{P}(n-1) + \widetilde{\mathbf{p}}_n^T \widetilde{\mathbf{p}}_n\right)^{-1} \times \mathbf{P}(n)^T \times \mathbf{B}(n). \tag{5}$$

Now use Woodberry's Identity [38]

$$\left(\mathbf{A} + \mathbf{c}\mathbf{c}^T\right)^{-1} = \mathbf{A}^{-1} - \frac{(\mathbf{A}^{-1}\mathbf{c})(\mathbf{c}^T\mathbf{A}^{-1})}{1 + \mathbf{c}^T\mathbf{A}^{-1}\mathbf{c}} \tag{6}$$

to decompose Equation (5). Let $\mathbf{A} = \mathbf{P}(n-1)^T\mathbf{P}(n-1), \mathbf{c} = \widetilde{\mathbf{p}}_n^T, v = \mathbf{A}^{-1}$ and $\rho = \mathbf{c}^T\mathbf{A}^{-1}\mathbf{c}$, and Equation (6) becomes

$$\mathbf{Q}(n) = \left(\mathbf{H}(n-1)^{-1} - \frac{\mathbf{v}_{n|(n-1)}\mathbf{v}_{n|(n-1)}^T}{1 + \rho_{n|(n-1)}}\right) \begin{bmatrix} \mathbf{P}(n-1) \\ \widetilde{\mathbf{p}}_n \end{bmatrix}^T \begin{bmatrix} \mathbf{B}(n-1) \\ \mathbf{r}_n \end{bmatrix} \tag{7}$$

where $\mathbf{v}_{n|(n-1)} = \left(\mathbf{P}_{n-1}^T\mathbf{P}_{n-1}\right)^{-1}\widetilde{\mathbf{p}}_n^T = \mathbf{H}(n-1)^{-1}\widetilde{\mathbf{p}}_n^T, \rho_{n|(n-1)} = \widetilde{\mathbf{p}}_n\left(\mathbf{P}_{n-1}^T\mathbf{P}_{n-1}\right)^{-1}\widetilde{\mathbf{p}}_n^T = \widetilde{\mathbf{p}}_n\mathbf{v}_{n|(n-1)}$.

We expand Equation (7), and further simplify it by $\mathbf{Q}(n-1) = \mathbf{H}(n-1)^{-1}\mathbf{P}(n-1)^T\mathbf{B}(n-1)$, to get

$$\begin{aligned} \mathbf{Q}(n) &= \mathbf{H}(n-1)^{-1}\left(\mathbf{P}(n-1)^T\mathbf{B}(n-1)\right) + \widetilde{\mathbf{p}}_n^T\mathbf{r}_n \\ &\quad - \tfrac{\mathbf{v}_{n|(n-1)}\mathbf{v}_{n|(n-1)}^T}{1+\rho_{n|(n-1)}}\left(\mathbf{P}(n-1)^T\mathbf{B}(n-1) + \widetilde{\mathbf{p}}_n^T\mathbf{r}_n\right) \\ &= \mathbf{Q}(n-1) + \mathbf{H}(n-1)^{-1}\widetilde{\mathbf{p}}_n^T\mathbf{r}_n - \tfrac{\mathbf{V}_{n|(n-1)}\mathbf{V}_{n|(n-1)}^T}{1+\rho_{n|(n-1)}}\left(\mathbf{P}(n-1)^T\mathbf{B}(n-1) + \widetilde{\mathbf{p}}_n^T\mathbf{r}_n\right). \end{aligned} \tag{8}$$

Multiply the 3rd term of Equation (8) by $\mathbf{H}(n-1)\mathbf{H}(n-1)^{-1}$, and then re-organize it. Finally we obtain

$$\begin{aligned}\mathbf{Q}(n) &= \mathbf{Q}(n-1) - \tfrac{\mathbf{v}_{n|(n-1)}\mathbf{v}_{n|(n-1)}^T}{1+\rho_{n|(n-1)}}\mathbf{H}(n-1)\mathbf{Q}(n-1) \\ &\quad + \left(\mathbf{H}(n-1)^{-1} - \tfrac{\mathbf{v}_{n|(n-1)}\mathbf{v}_{n|(n-1)}^T}{1+\rho_{n|(n-1)}}\right)\widetilde{\mathbf{p}}_n^T\mathbf{r}_n \\ &= \left(\mathbf{I}_{L\times L} - \tfrac{\mathbf{v}_{n|(n-1)}\mathbf{v}_{n|(n-1)}^T}{1+\rho_{n|(n-1)}}\mathbf{H}(n-1)\right)\mathbf{Q}(n-1) \\ &\quad + \left(\mathbf{H}(n-1)^{-1} - \tfrac{\mathbf{v}_{n|(n-1)}\mathbf{v}_{n|(n-1)}^T}{1+\rho_{n|(n-1)}}\right)\widetilde{\mathbf{p}}_n^T\mathbf{r}_n\end{aligned} \tag{9}$$

Equation (9) is the recursive version of Equation (2). In Equations (7)–(9), $\mathbf{Q}(n-1)$ and $\mathbf{H}(n-1)$ are the past information obtained at the $(n-1)$th stage; $\mathbf{v}_{n|(n-1)}$ and $\rho_{n|(n-1)}$ are the so-called innovation information [32], whose calculation is involved with both previous information and current information; and $\widetilde{\mathbf{p}}_n$ and $\mathbf{r}_n$ are provided by the new incoming pixel. Once $\mathbf{Q}(n)$ is obtained, we calculate $\mathbf{H}(n) = \mathbf{H}(n-1) + \widetilde{\mathbf{p}}_n^T\widetilde{\mathbf{p}}_n$ for the use in the next stage. It should be noted that Equation (9) can only be used when the updating condition is met. This detail will be presented in Section 3.2.

3.1.2. Decomposition of the Residual Multiplication Term

On the other hand, in OMPBS, the other calculation that can be simplified is the residual multiplication term $\mathbf{R}(n)_{j+1}^T\mathbf{R}(n)_{j+1}$ in Step 5 of OMPBS. Calculating this term would be time-consuming when n becomes large. According to OMPBS, the residual term is formed by

$$\mathbf{R}_{j+1}(n) = \mathbf{B}(n) - \mathbf{P}_j(n)\mathbf{Q}_j(n). \tag{10}$$

We expand $\mathbf{R}(n)_{j+1}^T\mathbf{R}(n)_{j+1}$ in terms of Equation (10):

$$\begin{aligned}\mathbf{R}_{j+1}(n)^T \times \mathbf{R}_{j+1}(n) &= \left(\mathbf{B}(n) - \mathbf{P}_j(n)\mathbf{Q}_j(n)\right)^T\left(\mathbf{B}(n) - \mathbf{P}_j(n)\mathbf{Q}_j(n)\right) \\ &= \mathbf{B}(n)^T\mathbf{B}(n) - \mathbf{Q}_j(n)^T\mathbf{P}_j(n)^T\mathbf{B}(n) - \mathbf{B}(n)^T\mathbf{P}_j(n)\mathbf{Q}_j(n) + \mathbf{Q}_j(n)^T\mathbf{P}_j(n)^T\mathbf{P}_j(n)\mathbf{Q}_j(n).\end{aligned} \tag{11}$$

There are four terms required to be computed in Equation (11). The first term is $\mathbf{B}(n)^T\mathbf{B}(n)$, which can be obtained by the following recursive equation

$$\mathbf{B}(n)^T\mathbf{B}(n) = \begin{bmatrix} \mathbf{B}(n-1)^T & \mathbf{r}_n^T \end{bmatrix}\begin{bmatrix} \mathbf{B}(n-1) \\ \mathbf{r}_n \end{bmatrix} = \mathbf{B}(n-1)^T\mathbf{B}(n-1) + \mathbf{r}_n^T\mathbf{r}_n. \tag{12}$$

Similarly, the second term can be represented by

$$\mathbf{Q}_j(n)^T\mathbf{P}_j(n)^T\mathbf{B}(n) = \mathbf{Q}_j(n)^T\begin{bmatrix} \mathbf{P}_j(n-1)^T & \widetilde{\mathbf{p}}_n^T \end{bmatrix}\begin{bmatrix} \mathbf{B}(n-1) \\ \mathbf{r}_n^T \end{bmatrix} = \mathbf{Q}_j(n)^T\left(\mathbf{P}_j(n-1)^T\mathbf{B}(n-1) + \widetilde{\mathbf{p}}_n^T\mathbf{r}_n\right), \tag{13}$$

where $\mathbf{Q}_j(n)^T$ is supposed to be known by Equation (9). The third term $\mathbf{B}(n)^T\mathbf{P}_j(n)\mathbf{Q}_j(n)$ is the transposition of the second term, so it is unnecessary to re-compute it. Finally, the fourth term can be calculated by

$$\mathbf{Q}_j(n)^T\mathbf{P}_j(n)^T\mathbf{P}_j(n)\mathbf{Q}_j(n) = \mathbf{Q}_j(n)^T\mathbf{H}_j(n)\mathbf{Q}_j(n) = \mathbf{Q}_j(n)^T\left(\mathbf{H}_j(n-1) + \widetilde{\mathbf{p}}_n^T\widetilde{\mathbf{p}}_n\right)\mathbf{Q}_j(n) \tag{14}$$

In conclusion, term $\mathbf{R}(n)_{j+1}^T\mathbf{R}(n)_{j+1}$ can be quickly calculated by using Equations (11)–(14).

3.2. Condition for Applying Recursive Equations

Unlike related work about PBP [33–35], where the recursive calculation can be applied to each iteration of PBP, Equation (9) of Re-OMPBS can only be implemented under particular conditions. To understand it, we assume that $\Omega_j(n)$ presents the BS result at the *j*th iteration in the *n*th stage, and the result was just obtained. Now we move to calculate $\mathbf{Q}_j(n)$ for finding the residual. If the condition $\Omega_j(n) = \Omega_j(n-1)$ holds, we can use Equation (9) to calculate $\mathbf{Q}_j(n)$. More specifically, if the selected bands' indices of the *j*th iteration of the previous stage are the same as the band indices of the *j*th iteration in the current stage, we can use the recursive equation to save computing time. This is simply because the sparse coefficient matrix $\mathbf{Q}_j(n)$ can only be "updated" when the currently-selected bases (i.e., band indices) are exactly the same as the previous ones. More specifically, the information in the horizontal dimension of temporary matrices $\mathbf{P}_j(n-1)$ and $\mathbf{P}_j(n)$ are aligned so that the information of previous stage can be shared with the current stage. In the following, we call it the "updating condition".

Satisfying the updating condition not only reduces computing time for $\mathbf{Q}_j(n)$, but also slightly accelerates Equations (12)–(14). One key of implementing Re-OMPBS is that we have to store the computing results obtained at the $(n-1)$th stage, such as terms $\{\mathbf{Q}_{j-1}(n)\}_{j=1}^{p}$, $\{\mathbf{H}_{j-1}(n)\}_{j=1}^{p}$, $\mathbf{B}(n-1)^T\mathbf{B}(n-1)$, and $\left\{\mathbf{P}(n-1)^T\mathbf{B}(n-1)\right\}_{j=1}^{p}$, for applying in Equation (9) and Equations (11)–(14), when the updating condition is held at the *n*th stage. One should note that meeting this condition is not necessary for Re-OMPBS. For those cases that do not meet the conditions, we directly use the original least square formula $\mathbf{Q}_j(n) = \left(\mathbf{P}_j(n)^T\mathbf{P}_j(n)\right)^{-1}\mathbf{P}_j(n)^T\mathbf{B}(n)$ to calculate $\mathbf{Q}_j(n)$.

One might think that the probability of reaching these conditions would be low. However, this is not the case. In fact, when the amount of received pixels *n* reaches a certain amount, a single new incoming pixel will not obviously influence the BS result. In other words, the BS result is expected to be consistent. This phenomenon will be demonstrated in our experiment section. Moreover, Re-OMPBS is regarded as the "approximate" version of OMPBS. Theoretically, there would be some tiny numerical errors accumulated in $\mathbf{Q}_j(n)$ if the updating condition has been reached continuously for a long period. In this case, an occasional lack of meeting the updating condition exactly gives the chance to reset the numerical error to zero.

3.3. Re-OMPBS Algorithm

Based on the above-mentioned content, the Re-OMPBS algorithm (Algorithm 2) is listed below.

Algorithm 2. Re-OMPBS

Objective: Find the BS of the nth stage during BIP/BIS transmission

Input:

1. Sparsity p
2. Processed information:
   - Previous BS result: $\Omega_p(n-1) = \left\{b_1^{(n-1)}, b_2^{(n-1)}, \dots, b_p^{(n-1)}\right\}$
   - Previous sparse coefficient matrices: $\left\{\mathbf{Q}_{j-1}(n)\right\}_{j=1}^{p}$
   - Previous auxiliary matrix: $\left\{\mathbf{H}_{j-1}(n)\right\}_{j=1}^{p}$
   - Other terms $\mathbf{B}(n-1)^T\mathbf{B}(n-1)$ and $\left\{\mathbf{P}(n-1)^T\mathbf{B}(n-1)\right\}_{j=1}^{p}$
3. New information: n-th pixel: $\mathbf{r}_n = [\mathrm{r}_{n1}, \mathrm{r}_{n2}, \dots, \mathrm{r}_{nL}]$

Output: Band indices $\Omega_p(n) = \left\{b_1^{(n)}, b_2^{(n)}, \dots, b_p^{(n)}\right\}$

Steps:

1. Initialization: Set $\Omega_0(n) = \emptyset$, $\mathbf{P}_0(n) = \emptyset$, j=1, $\mathbf{R}_1(n) = \mathbf{B}(n-1)^T\mathbf{B}(n-1) + \mathbf{r}_n^T\mathbf{r}_n$.
2. At the jth iteration, calculate the vector-wise norms of all rows in matrix $\mathbf{R}(n)_j^T\mathbf{R}(n)_1$, and denote it by column vector $\mathbf{k}$, $\mathbf{k} \in L \times 1$.
3. Find the row index whose value is maximum is $\mathbf{k}$, denoted by b_j. Then set $\mathbf{P}_j(n) = [\mathbf{B}_{j-1}(n) \quad \mathbf{b}_{b_j}(n)]$, $\Omega_j(n) = \Omega_{j-1}(n) \cap b_j$.
4. If $\Omega_j(n) = \Omega_j(n-1)$, using Equation (9) to calculate $\mathbf{Q}_j(n)$; otherwise, $\mathbf{Q}_j(n) = \left(\mathbf{P}_j(n)^T\mathbf{P}_j(n)\right)^{-1}\mathbf{P}_j(n)^T\mathbf{B}(n)$.
5. Update auxiliary matrix $\mathbf{H}_j(n)$ by equation $\mathbf{H}_j(n) = \mathbf{H}_j(n-1) + \tilde{\mathbf{p}}_n^T\tilde{\mathbf{p}}_n$.
6. Update the residual term $\mathbf{R}(n)_{j+1}^T\mathbf{R}(n)_{j+1}$ by Equations (11)–(14).
7. If $j = p$, break; otherwise, $j \leftarrow j+1$ and go to Step 2.

3.4. Design of Pixel Transmission Sequence (PTS)

The original pixel transmission order in BIP/BIS format is based on a horizontal line scan [35]. If we regard a hyperspectral image as a 2D matrix, the pixel transmission starts from the upper left corner. Once the first row has been transmitted, then it transmits the second row, and so on. However, using this sequence for Re-OMPBS might mean that the BS cannot converge until the almost all pixels are transmitted. Ideally, in a progressive processing, we might expect to see the final results as soon as possible. In this case, adopting an appropriate sampling technique in the spatial domain to select the representative pixels for early transmission is necessary. There are three pixel transmission sequences we considered in this paper.

3.4.1. Original Band-Interleaved-by-Sample/Pixel (BIS/BIP) Sequence

As mentioned above, the original pixel transmission sequence is the same as the storage sequence in the BIS/BIP format. Suppose the size of a hyperspectral image cube I is $\mathrm{x} \times \mathrm{y} \times L$. The number of total pixels is $N = \mathrm{x} \times \mathrm{y}$. We reshape I to an $L \times N$ matrix form in row–major order, $[\mathbf{r}_1\mathbf{r}_2 \dots \mathbf{r}_N]_{L \times N}$, where the column vectors are spectral pixels of I. The BIS/BIP transmission sequence is defined by $\{\mathbf{r}_1, \mathbf{r}_2, \dots, \mathbf{r}_N\}$. A simple illustration is shown in Figure 3a.

3.4.2. Step Sequence

The step method refers to uniformly selecting pixels in an original sequence with interval k. To do this, we uniformly segment pixel set $[\mathbf{r}_1\mathbf{r}_2 \dots \mathbf{r}_N]$ into k groups. For each transmission, one pixel is selected from each group in turn. If k = 5, the resulting transmission pixel sequence would be $\{\mathbf{r}_1, \mathbf{r}_6, \mathbf{r}_{11}, \dots \mathbf{r}_2, \mathbf{r}_7, \mathbf{r}_{12\dots}\}$, as shown in Figure 3b. As a result, the transmitted pixels are the samples uniformly picked from the pixel set. Theoretically, using such sequence would make the PSP-BS converge more quickly, because in this case fewer received pixels can represent the image better.

3.4.3. Block Sequence

Since the step method only considers image sampling in the horizontal direction, we further consider another method, which can consider both vertical and horizontal directions simultaneously, called the block method. The block method first splits the image into several $b \times b$ square blocks, as shown in Figure 3c. Then the new pixel is picked from each block in turn. In such a design, the sampling process is performed on the full spatial domain. Theoretically, it would shorten convergence time of PSP-BS, too.

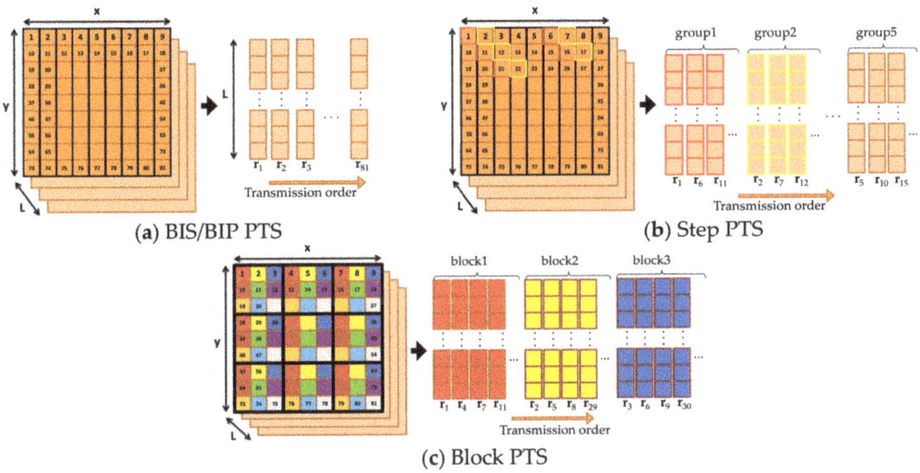

Figure 3. Illustration of the pixel transmission sequence of a 9 × 9 hyperspectral image formed by different methods. (**a**) A band-interleaved-by-sample/pixel (BIS/BIP) sequence. (**b**) Step method with $k = 5$. (**c**) The block method with $b = 3$.

In the later sections, we call the PSP-OMPBS implemented with original BIS/BIP sequence, step sequence, and block sequence PSP-OMPBS, S-PSP-OMPBS, and B-PSP-OMPBS, respectively.

Finally, we summarize the distinct features of OMPBS and the proposed PSP-OMPBS in Table 1.

Table 1. The basic properties of orthogonal matching pursuit-based BS (OMPBS) and the proposed PSP-OMPBS.

	OMPBS	**PSP-OMPBS (proposed)**
Description	A traditional BS method which is implemented on a pre-obtained image data.	A PSP-BS method. It is used in the moment of spectral sample transmission or collecting.
Core algorithm	OMP-BS [37]	Re-OMPBS (Algorithm 2)
Use of past information	No	Yes
Use of PTS	No	Yes (Original/Step/Block)
Least square estimator used in the optimization	$\mathbf{Q}_j = \left(\mathbf{P}_j^T\mathbf{P}_j\right)^{-1}\mathbf{P}_j^T\mathbf{B}$	Recursive Equation (9)
Residual term $R_j^T R_j$ calculation	Direct computation	Recursive Equations (11)–(14)
Capable for progressive processing	No, the computing time will increase with the data volume.	Yes

4. Experiments

In this section, the proposed PSP-OMPBS is tested on two publicly-available real hyperspectral images. To evaluate the performance of PSP-OMPBS, we conduct three different studies, including BS accuracy analysis, land cover classification, and computational efficiency.

4.1. Hyperspectral Dataset and Experimental Setting

The first dataset used in the experiments was a real hyperspectral image, which was collected by the reflective optics system imaging spectrometer (ROSIS) optical sensor over an urban area of the University of Pavia. The Pavia image measures 610 × 340, with very high spatial resolution of about 1.3 m per ground pixel. The original data contains 115 spectral bands, with a spectral range from 0.43 to 0.86 μm. After removing 12 noisy bands, the remaining 103 bands were used for the experiments. Figure 4a–c respectively show the image scene of the 50th band, the geometric locations of all target classes, and the corresponding spectral signatures. According to the ground truth in Figure 5b, there are 9 classes in this image scene, consisting of several urban targets, such as vegetation, soil, and roads.

The second dataset used in our experiments is another real hyperspectral image data, Purdue's Indiana Indian Pine test site, which was collected by the airborne visible-infrared imaging spectrometer (AVIRIS) system. It has been extensively studied in the literature and provides a good candidate for those who are interested in algorithm design and analysis. It has a 20 m spatial resolution and a 10 nm spectral resolution in the range of 0.4–2.5 μm with size 145 × 145 pixel vectors, taken from an area of mixed agriculture and forestry in Northwestern Indiana, U.S. It was recorded in June 1992 with 220 bands, among which bands 104–108 and 150–162 were removed, whereas the remaining 202 bands were retained. Figure 5a,b shows the image of band 20 and the ground truth map, respectively. The ground truth map contains 16 crops classes and one background class.

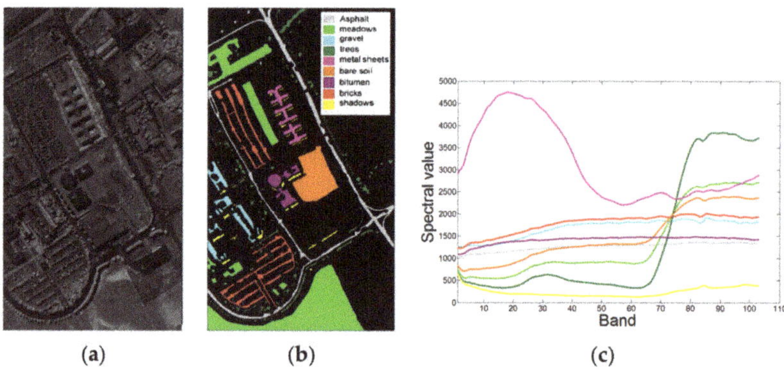

Figure 4. ROSIS image scene: University of Pavia. (**a**) Band 80. (**b**) Ground truth map for nine classes. (**c**) Spectral signatures of nine classes.

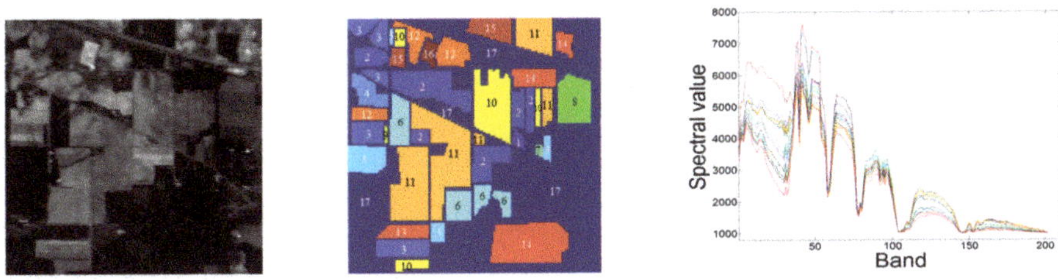

(a) (b) (c)

Figure 5. AVIRIS image scene: Purdue Indiana Pine test site. (**a**) Band 20. (**b**) Ground truth map. (**c**) Spectral signatures of 17 classes.

According to the introduction, the development of PSP-BS is not for achieving outstanding BS for a particular data analysis, such as image classification or spectral unmixing, but for the real-time BS monitoring during data transmission. In this case, the experiment is conducted based on self-comparison, instead of comparing with state-of-the-art methods. To evaluate the BS accuracy and efficiency of PSP-OMPBS, we still can adopt OMPBS as the compared method (baseline).

In the following progressive experiments, we set n = 100–207,400 for the Pavia data, and n = 100–21,025 for the Purdue data. For the p value, we set 16 for the Pavia data and 34 for the Purdue data, according to the virtual dimensionality (VD) algorithm [39] with false alarm 0.01. Three PTS methods mentioned in Section 3.4 were adopted for PSP-OMPBS methods. We empirically set k as 200 and 100 and b as 50 and 25 for the Pavia and Purdue data, respectively. Table 2 lists the parameters used in the experiment.

Table 2. The parameters used in the experiments.

Parameter	Pavia Data	Purdue Data
N (number of total pixel)	207,400	21,025
p (number of selected bands)	16	34
k (step size of step PTS)	200	100
b (block size of block PTS)	50	25

4.2. BS Accuracy Analysis

The accuracy of the instant BS result during data transmission is an index to evaluate the effectiveness of PSP-OMPBS. Theoretically, the BS would gradually converge to the final BS result (i.e., the result of OMPBS performed on the complete image cube) over time. To observe this phenomenon, we use the results of OMPBS implemented on both images, as the ground truths. For instance, Pavia's ground truth is $\Omega_{\text{OMPBS}}^{\text{Pavia}}(N) = \{91, 62, 16, 1, 34, 3, 73, 105, 5, 46, 85, 8, 83, 2, 11, 78\}$. To evaluate BS correctness, the accuracy index (ACC) is defined by

$$\text{ACC}(n) = \frac{||\Omega_{\text{PSP-OMPBS}}(n) \cap \Omega_{\text{OMPBS}}(n)||_0}{p}. \quad (15)$$

Base on Equation (15), ACC(n) indicates the ratio of the target bands in overall p-selected bands, when the number of received pixels is n. Higher ACC values mean higher BS correctness. Figure 6a–d plots the ACC curves of OMPBS, PSP-OMPBS, S-PSP-OMPBS, and B-PSP-OMPBS, all implemented on the Pavia data. Several observations can be seen:

1. Comparing Figure 6a with Figure 6b, the tendency of the ACC curves of OMPBS and PSP-OMPBS are nearly the same. This implies that the derivation of Re-OMPBS is correct. In fact, these two curves are still slightly different in some regions. Based on our study, the difference

was caused by the numerical errors produced by using the recursive equations derived from Woodbury's Identity.

2. In the OMPBS and PSP-OMPBS curves, the ACC values kept at 40~60% when n is less than 1.5×10^5. This is mainly due to the fact that these values were generated by using the original BIS/BIP PTS. The BS results could not stabilize quickly.
3. The ACC results of using step sequence are shown in Figure 6c. We find that the ACC values located between 60~80% at $n \in [2000,125,000]$. After receiving 1.25×10^5 pixels, the ACCs increased to 90%. In other words, using the PTS formed by uniform sampling drastically accelerated the speed of convergence. It suggests that using full pixels is not necessary to obtain the correct BS result using PSP-OMPBS. The nearly-correct BS result could be obtained during transmission.
4. Using a block pixel transmission sequence also accelerated the BS convergence. Figure 6d shows the corresponding ACC results of B-PSP-OMPBS. Similar conclusions can be drawn. It only requires 40,000 pixels to reach 65–70% accuracy. After receiving 60,000 pixels, the ACC grew to over 80%. The overall ACC performance is undoubtedly better than OMPBS, PSP-OMPBS, and even S-PSP-OMPBS.
5. In Figure 6a–d , it can be seen that the ACC curves remain flat in some segments. In those regions, the BS results kept the same. That is, the new incoming pixels did not change the BS results.
6. All the curves reached 100% at $n = N = 207,400$.

In the remote sensing community, the Purdue image was considered as a tough image for algorithm evaluation because of its noise and heavily-mixed properties. Figure 7a–d plots the ACC curves of OMPBS, PSP-OMPBS, S-PSP-OMPBS, and B-PSP- OMPBS, implemented on the Purdue data, respectively. Theoretically, the larger p and the special properties of the Purdue image, may lead to different consequences. Compared with the Pavia case, we have several interesting findings. Firstly, the oscillation of the four ACC curves is more notable. This is probably due to the noise property of the Purdue image. Secondly, the tendency of OMPBS and PSP-OMPBS curves are the same. This is in accordance with our expectations, because both methods are essentially the same. However, their ACC values are slightly different at some places. The PSP-OMPBS curve of the Purdue image is more unstable and lower at middle region. We think such a phenomenon is caused by the larger setting of p. The greater p is set, the more numerical errors will be accumulated while selecting later bands at each n. Thirdly, and most importantly, we found that the advantages of using sampled sequences (i.e., step and block) is not obvious in the Purdue experiment. We think the strange phenomenon is caused by the homogeneity of spectral profiles of the 16 ground classes. In this case, using any kind of PTS may obtain analogous pixel sets, so that the BS results are similar. From this point of view, in the case of the Purdue data, the OMPBS selects bands based more on spectral variation, instead of spatial/geographical variation. Despite this issue, it is still observed that S-PSP-OMPBS and B-PSP-OMPBS outperformed OMPBS a little bit with regard to ACC stability. The ACC trends of S-PSP-OMPBS and B-PSP-OMPBS did not drop after they reached over 80%, particularly at $n \in [6000,13,000]$. In conclusion, the ACC performance of PSP-OMPBS methods varies with the properties of images. Using PSP-OMPBS on the images with lower noise seemed to produce more stable ACC curves. The sampled pixel sequences were more suitable for the images with larger heterogeneity in the spatial domain. The spectral similarity of ground classes was another important factor. Finally, we can find the curve of B-PSP-OMPBS did not end at 100% when $n = N = 21,025$. This was caused by the numerical error.

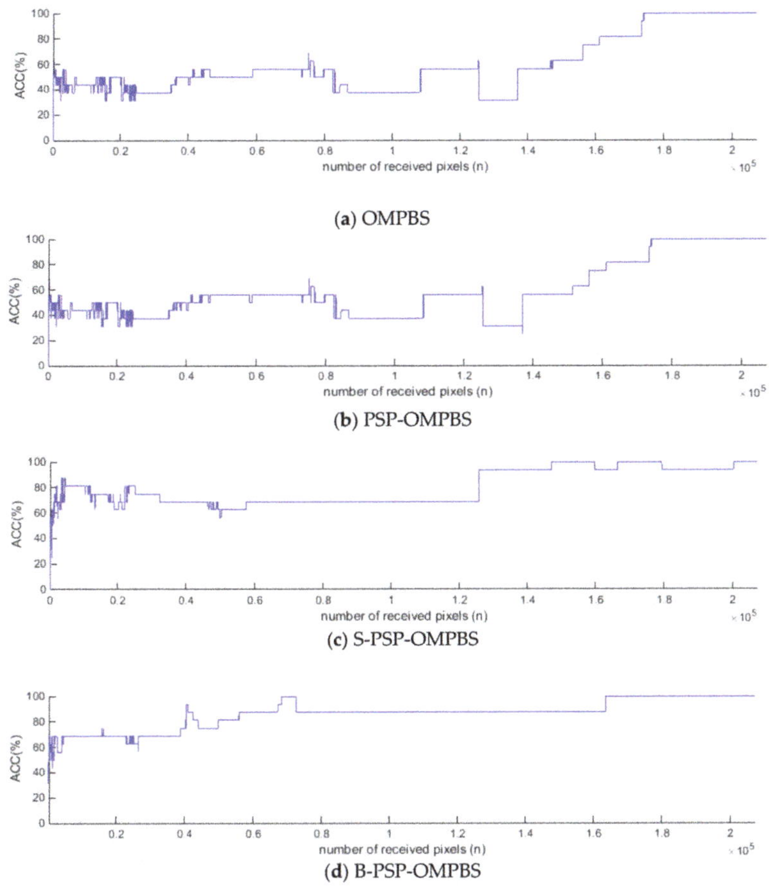

Figure 6. Accuracy index (ACC) curves of different methods implemented on ROSIS Pavia dataset: (**a**) OMPBS, (**b**) PSP-OMPBS, (**c**) step sequence PSP-OMPBS (S-PSP-OMPBS), and (**d**) block sequence PSP-OMPBS (B-PSP-OMPBS).

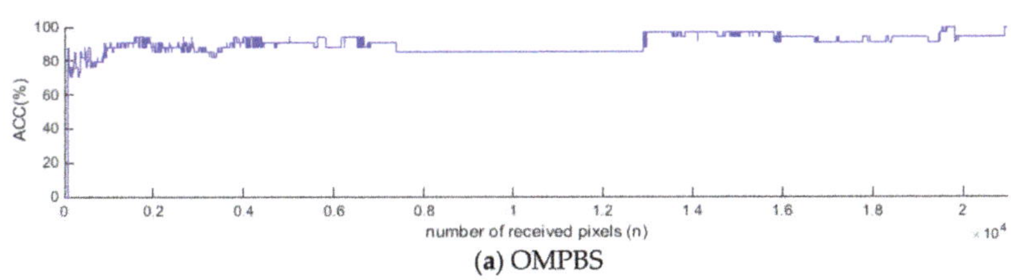

Figure 7. *Cont.*

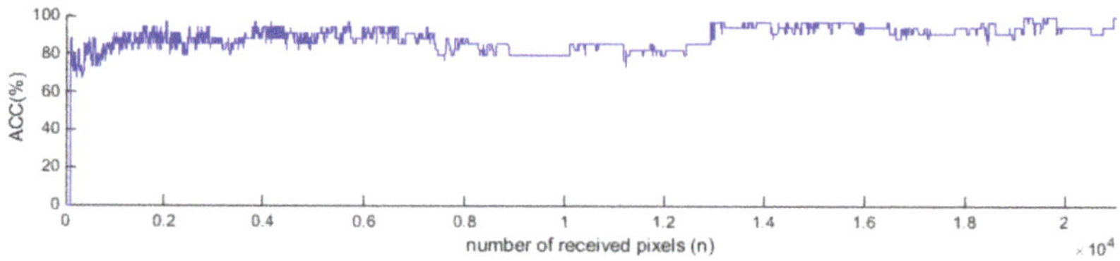

(b) PSP-OMPBS

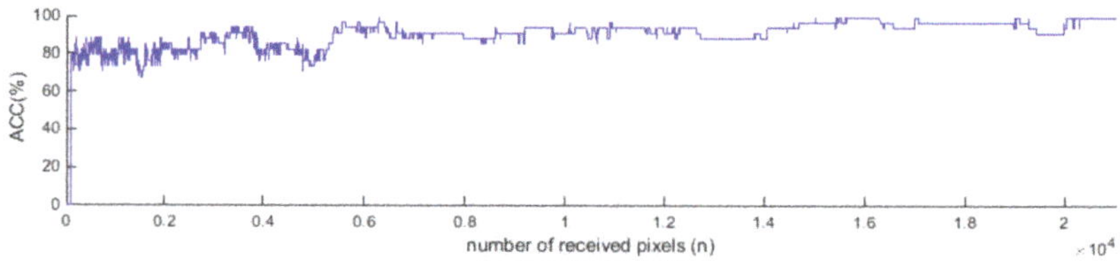

(c) S-PSP-OMPBS

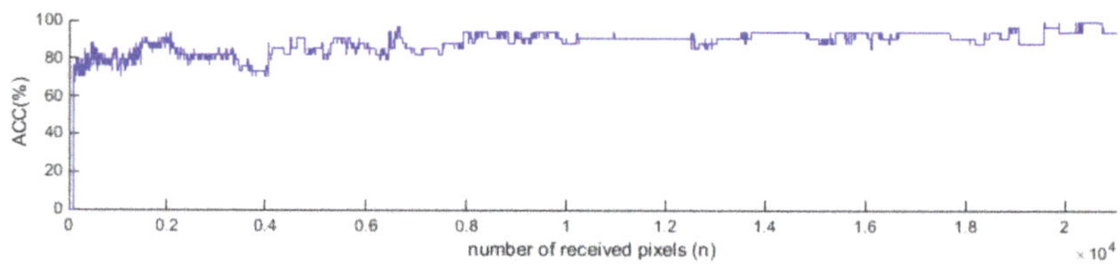

(d) B-PSP-OMPBS

Figure 7. ACC curves of different methods implemented on the AVIRIS Purdue dataset: (**a**) OMPBS, (**b**) PSP-OMPBS, (**c**) step sequence PSP-OMPBS (S-PSP-OMPBS), and (**d**) block sequence PSP-OMPBS (B-PSP-OMPBS).

Tables 3 and 4 list the corresponding BS results of the Pavia and Purdue experiments, where n is selected by 50% of N for each case. The last row shows the BS ground truth. It can be seen that the BS accuracies of PSP-OMPBSs are better than OMPBS at the 50% middle of transmission.

Table 3. List of the 16 bands, selected by four different methods, implemented on the Pavia data at n = 103,700.

Method	Selected Bands	ACC
OMPBS	86,20,63,1,103,3,45,73,6,32,83,9,4,55,12,2	37.5
PSP-OMPBS	86,20,63,1,103,3,45,73,6,32,83,9,4,55,12,2	37.5
S-PSP-OMPBS	91,62,15,1,33,3,73,103,5,46,85,7,2,83,10,22	68.75
B-PSP-OMPBS	91,62,16,1,34,3,73,103,5,47,46,48,85,8,83,2	87.5
OMPBS (full pixels)	91,62,16,1,34,3,73,103,5,46,85,8,83,2,11,78	N/A

Table 4. List of the 34 bands, selected by four different methods, implemented on the Purdue data at n = 10,513.

Method	Selected Bands	ACC
OMPBS	42,29,2,89,6,35,1,53,117,3,4,15,38,5,32,72,41,104, 7,39,34,8,37,43,9,44,97,63,45,10,33,36,46,40	85.31
PSP-OMPBS	42,29,2,89,6,35,1,53,117,3,4,15,16,38,5,32,72,41,104, 7,39,34,8,37,43,44,45,97,9,63,33,10,46,36	85.29
S-PSP-OMPBS	42,29,3,89,35,1,2,70,9,118,4,38,5,32,14,44,16,6,144, 43,39,97,34,7,45,37,63,8,41,48,52,47,50,10	88.24

Table 4. *Cont.*

Method	Selected Bands	ACC
B-PSP-OMPBS	42,29,2,89,35,6,1,70,118,3,38,4,14,5,32,43,144,44, 7,39,34,97,8,45,11,48,37,63,41,100,46,9,33,17	91.18
OMPBS (full pixels)	42,29,3,89,35,1,2,70,9,118,4,38,5,32,15,48,6,144, 41,34,39,97,7,37,8,47,45,63,44,10,43,33,46,36	N/A

4.3. Land Cover Classification

Image classification is a common procedure to evaluate the effectiveness of a BS approach. For each dataset, we implemented supervised classification using a PSP-OMPBS-selected band $\Omega_P(n)$, with a support vector machine (SVM) classifier [40]. The radial basis function (RBF) kernel was adopted for SVM with the selected parameter $[\sigma/4, \sigma/2, \sigma, 2\sigma, 4\sigma]$ where r is calculated by the average pairwise distance among training data $\sigma = E||x_i - x_j||$. The training samples were obtained by randomly selecting 10% of date samples in each class, according to the ground truth maps shown in Figures 4b and 5b. The other 90% of samples were used as the test samples.

For measuring the classification performance, average accuracy (AA), overall accuracy (OA), and Cohen's kappa coefficient were used as the performance metrics. Since PSP-OMPBS provided 207,400 and 21,025 BS results for two datasets respectively, we simply chose the BS results using 20%, 40%, 60%, 80%, and 100% of N for the experiment. Three PSP-OMPBS algorithms were considered. The classification result of using full bands was also used for the comparison.

Table 5 lists all the AA, OA, and kappa coefficient values of the SVM results performed on PSP-OMPBS selected bands for both the Pavia and Purdue datasets. We observed that using the selected bands of PSP-OMPBS preserved sufficient spectral information to achieve comparative classification performance, compared to the results of using full bands. The highest accuracies of the AA, OA, and kappa coefficient values reached were 0.907, 0.868, and 0.874, respectively, which is close to the full bands results of 0.914, 0.880, and 0.884, respectively. This simply implies that the bands selected by PSP-OMPBSs are informatively complementary.

On the other hand, in the Purdue case, the most accurate results for the AA, OA, and kappa coefficient values were 0.736, 0.600, and 0.693, respectively, and were slightly worse than the full bands results of 0.756, 0.586, and 0.716. This might be simply because only using 34 bands is not enough for the Purdue image classification. This issue can be resolved by increasing p, fine-tuning kernel parameters, or use other types of SVM classifiers that utilize spectral–spatial information to improve overall performance. Such a study is beyond the scope of this paper, and thus is not included.

To evaluate classification performance in visual assessment, Figures 8 and 9 plot the classification maps of SVM using the bands selected by PSP-OMPBS, S-PSP-OMPBS, and B-PSP-OMPBS at n = 60% of N for the Pavia and Purdue data, respectively. The maps using full bands are included for comparison. In the Pavia case, it can be seen that using PSP-OMPBS-selected bands could generate nearly the same classification maps compared to those obtained by using full bands. In the Purdue case, the quality of the PSP-OMPBS-generated maps are a little worse than the full bands map, due to the insufficient number of selected bands.

Table 5. Classification performance, in the form of the overall accuracy (OA)/average accuracy (AA)/kappa coefficient, of support vector machine (SVM) performed on the PSP-OMPBS-selected bands $\Omega_p(n)$ in different n stages.

	Pavia Data (16 Selected Bands)			Purdue Data (34 Selected Bands)		
	PSP-OMPBS	S-PSP-OMPBS	B-PSP-OMPBS	PSP-OMPBS	S-PSP-OMPBS	B-PSP-OMPBS
n = 20% of N	0.904/0.863/ 0.872	0.906/**0.868**/ **0.874**	0.891/0.845/ 0.853	0.724/0.560/ 0.678	0.729/0.592/ 0.686	0.721/0.573/ 0.677
n = 40% of N	0.904/0.861/ 0.870	0.899/0.852/ 0.864	0.899/0.853/ 0.865	**0.736**/**0.600**/ **0.693**	0.725/0.570/ 0.681	0.719/0.564/ 0.673
n = 60% of N	0.897/0.854/ 0.861	0.901/0.853/ 0.867	0.900/0.860/ 0.866	0.713/0.551/ 0.666	0.711/0.537/ 0.665	0.724/0.550/ 0.680
n = 80% of N	0.898/0.853/ 0.862	0.902/0.859/ 0.868	0.899/0.853/ 0.864	0.724/0.589/ 0.679	0.708/0.573/ 0.659	0.712/0.579/ 0.664
n = 100% of N (equivalent to OMPBS)	0.904/0.861/ 0.871	**0.907**/0.865/ **0.874**	0.901/0.856/ 0.866	0.711/0.533/ 0.663	0.728/0.549/ 0.684	0.735/0.545/ 0.680
Full bands	0.914/0.880/0.884			0.756/0.586/0.716		

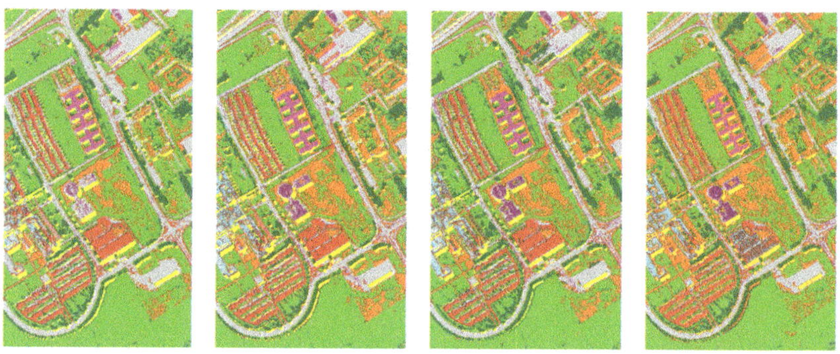

(**a**) PSP-OMPBS (**b**) S-PSP-OMPBS (**c**) B-PSP-OMPBS (**d**) Full bands

Figure 8. Examples of Pavia classification maps of SVM (RBF) performed on the 16 bands selected by (**a**) PSP-OMPBS, (**b**) S-PSP-OMPBS, (**c**) B-PSP-OMPBS, and (**d**) full 103 bands. For three PSP-OMPBSs, n is set at 50% of N. The corresponding OA values are 0.86, 0.89, 0.90, and 0.91.

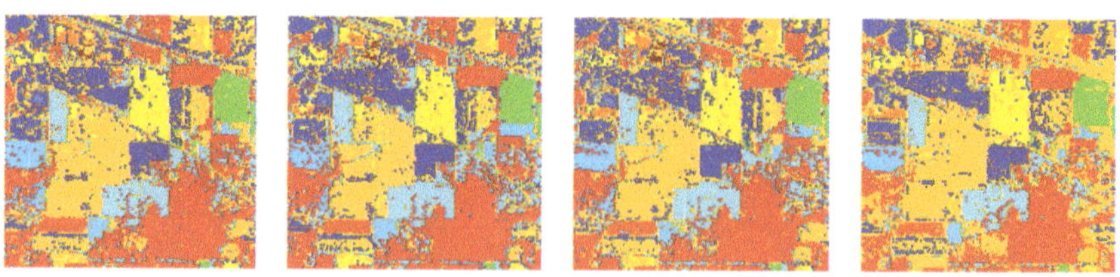

(**a**) PSP-OMPBS (**b**) S-PSP-OMPBS (**c**) B-PSP-OMPBS (**d**) Full bands

Figure 9. Examples of Purdue classification maps of SVM (RBF) performed on the 34 bands selected by (**a**) PSP-OMPBS, (**b**) S-PSP-OMPBS, (**c**) B-PSP-OMPBS, and (**d**) full 202 bands. For three PSP-OMPBSs, n is set at 50% of N. The corresponding OA values are 0.72, 0.73, 0.73, and 0.76, respectively.

To analyze the classification performance in the progressive manner, Figure 10 plots the kappa OA/AA/kappa coefficient curves of the SVM classification using uniformly selected BS results, performed by PSP-OMPBSs on the n-axis with interval 25 for the Pavia dataset. In those figures, the x-axis (n) indicates band set $\Omega_p(n)$, and the y-axis denotes the corresponding accuracy metrics of the classification. Several observations can be found. First, all the metric curves of PSP-OMPBS and S-PSP-OMPBS are not stable in the beginning $n \in [1{,}200]$ because of the poor BS quality that occurs with a low number of received pixels. With too few pixels, PSP-OMPBS could not select correct bands to fulfill the complementary spectral information. This resulted in unstable and lower classification performance. In contrast, the curves of B-PSP-OMPBS could be consistent because the transmitted pixels were uniformly sampled from the image. Second, the overall averaged performances of S-PSP-OMPBS and B-PSP-OMPBS are slightly better than OMP-BS for these three criteria. Third, when n is greater than 250, the curves of all three methods tended to be consistent. According to our extended study, all the curves will stay roughly at the same level, with extremely low deviation in the future time $n \in [600{,}N]$.

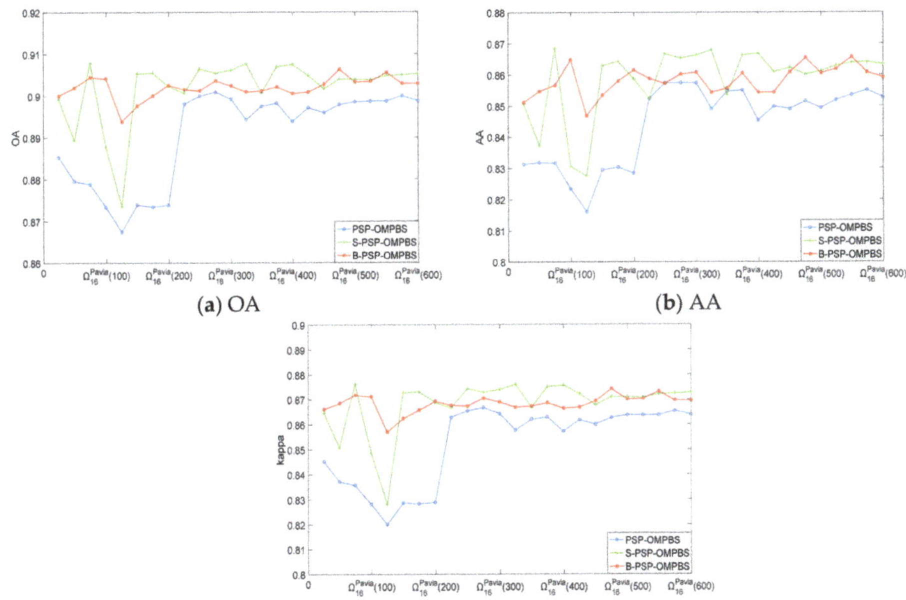

(**a**) OA (**b**) AA (**c**) kappa coefficient

Figure 10. SVM classification of using the bands selected by three PSP-OMPBS methods at n = [1,600]: (**a**) OA, (**b**) AA, and (**c**) kappa coefficient, for the ROSIS Pavia dataset.

Similarly, Figure 11 shows the OA, AA, and kappa coefficient curves for the Purdue experiment. Similarly, there are some interesting findings. First, the classification performance is low at the beginning $n \in [1{,}100]$. Second, the Purdue image seemed to require more pixels to stabilize the BS quality. We can find that the curves of PSP-OMPBS, S-PSP-OMPBS, and B-PSP-OMPBS could not converge at n $\in$ [100,600] in each figure, particular Figure 10b. This is probably because of the heavily mixed and noisy properties of the Purdue image. Each newly incoming pixel may easily disturb the result of band selection. According to our investigation, this phenomenon will last up to n = 4000. Finally, the S-PSP-OMPBS and B-PSP-OMPBS did not outperform PSP-OMPBS in the Purdue case. We think this is caused by the homogeneity of spectral profiles of the 16 ground classes. In this case, using a sampled transmission sequence could not significantly improve the collecting of spectral information.

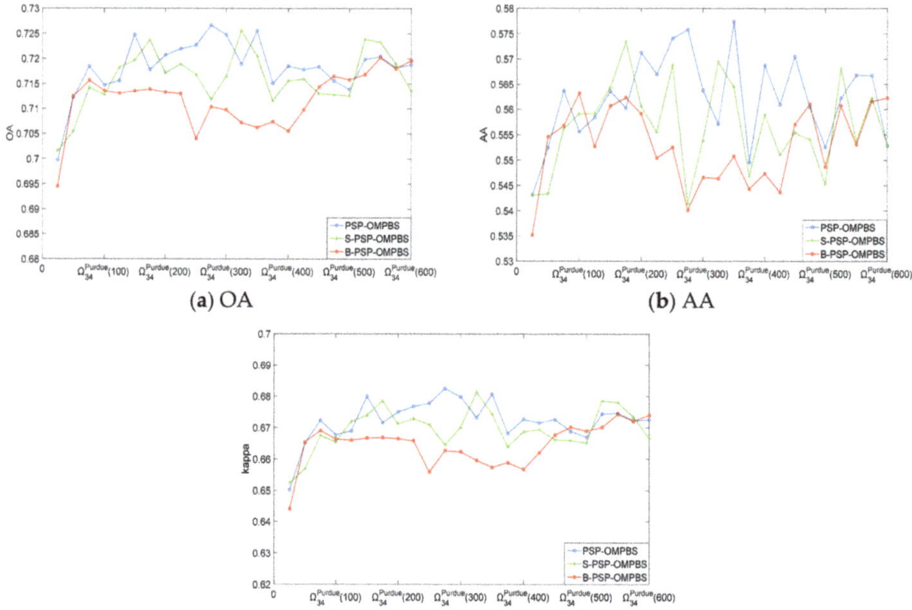

(a) OA

(b) AA

(c) kappa coefficient

Figure 11. SVM classification of using the bands selected by three PSP-OMPBS methods at n = [1,600]: (a) OA, (**b**) AA, and (**c**) kappa coefficient, for AVIRIS Purdue dataset.

4.4. Computing Time

To validate the real-time capability in BIS/BIP transmission, we first compare three PSP-OMPBSs with the OMPBS performed in a progressive manner. The number of transmitted pixels, n, was set from 1 to N for each image. The experiments were conducted on a computer with Intel i7-4790 3.6 GHz CPU, 16 GB RAM, Windows 7 and Matlab 2015. The values of computing time are reported by the average of ten random runs.

Figure 12a shows the required computing time of progressive OMPBS, PSP-OMPBS, S-PSP-OMPBS, and B-PSP-OMPBS, implemented on the Pavia dataset, where the x-axis presents the n value and the y-axis presents the corresponding computing time. It can be observed that the computing time of OMPBS significantly increases when n increases. This is due to the increase in size of $\mathbf{B}$, which results in greater computational complexity for OMPBS. The variation of computing time increases with n, too. On the contrary, by virtue of recursive processing, the PSP-BSs produced almost flat curves. The computing time is stably under 0.1 s in most n indices. This implies that PSP-BS has the potential to be run in a real-time manner during transmission. Figure 12b–d further shows the individual time curve for PSP-OMPBS, S-PSP-OMPBS, and B-PSP-OMPBS, respectively. It could be seen that the computing times of most of the n regions stick on the straight trend line. In those cases, the BS results remained the same, so that the recursive equations could be applied continuously. On the other hand, we can find some of the "peak phenomenon" shown in the curves. Those peaks occurred when the updating condition was not reached. Thus, the optimization problems were solved by non-recursive equations instead. The extra computing time is required.

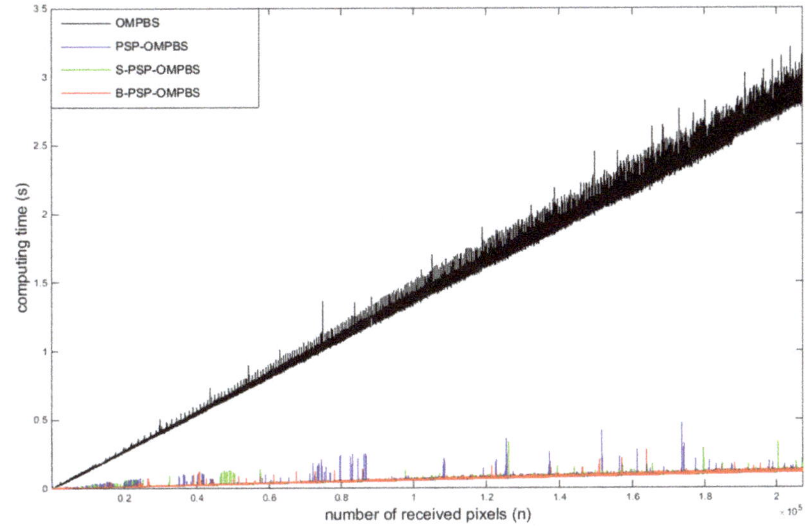

(**a**) All methods

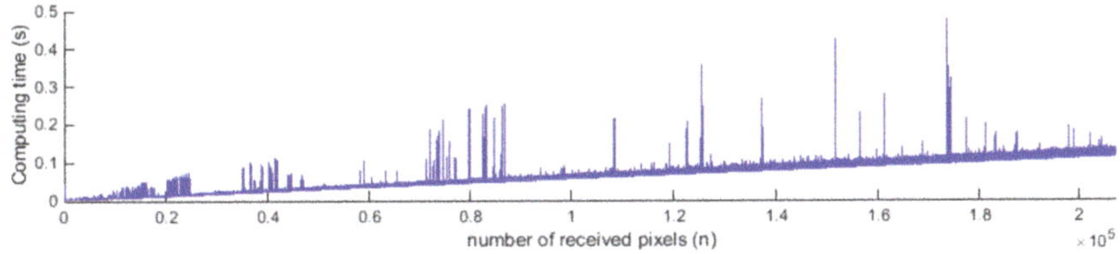

(**b**) PSP-OMPBS

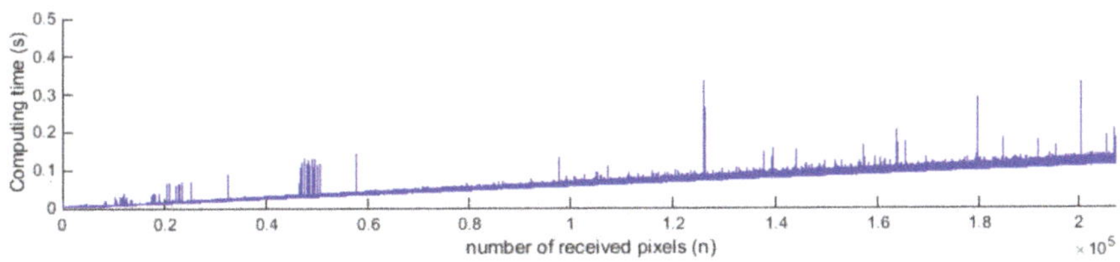

(**c**) S-PSP-OMPBS

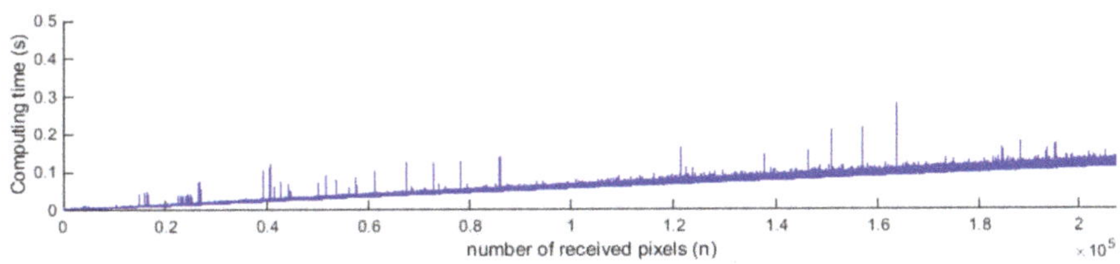

(**d**) B-PSP-OMPBS

Figure 12. Plots showing the computing time required for different progressive BS methods implemented on the ROSIS Pavia dataset: (**a**) all methods, (**b**) PSP-OMPBS, (**c**) step sequence PSP-OMPBS (S-PSP-OMPBS), and (**d**) block sequence PSP-OMPBS (B-PSP-OMPBS).

Similarly, Figure 13a shows the computing time of progressive OMPBS, PSP-OMPBS, S-PSP-OMPBS, and B-PSP-OMPBS implemented on the Purdue dataset, and Figure 13b–d further shows the individual curves of the three PSP-OMPBSs. Similarly, the three PSP-OMPBS methods required significantly less computing time than OMPBS. The required time for PSP-OMPBSs at is less than 0.07 s from $n = 1$–21,025, while OMPBS needs over 1 s when n is larger than 14,000. Again, this implies the superior computational efficiency provided by PSP-OMPBS. Comparing Figure 13 with Figure 12, we can interestingly find that the peak phenomenon occurred more frequently in the Purdue case. This is probably due to the noisy property of the Purdue image. Under the circumstances, the BS result varies easily if the new incoming pixel is a noisy pixel. This results in the instability of BS on the n-axis, and thus reduces the opportunity of using recursive equations. Fortunately, the peak phenomenon disappeared gradually after sufficient pixels were received, since in later periods of transmission, the received pixels represented the image well. In other words, the BS results tended to be more stable.

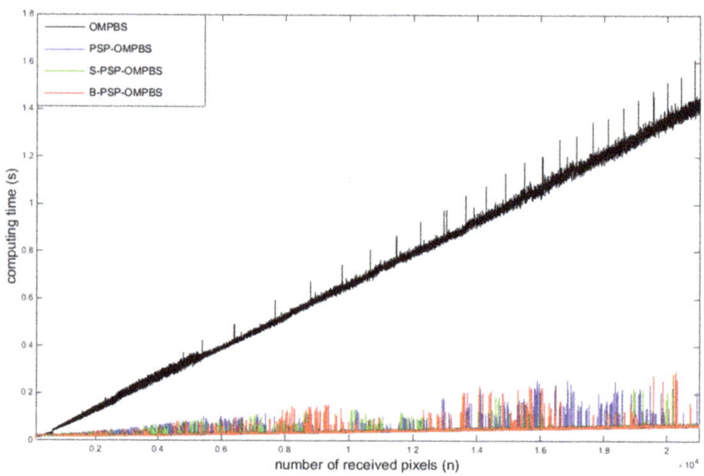

(**a**) All methods

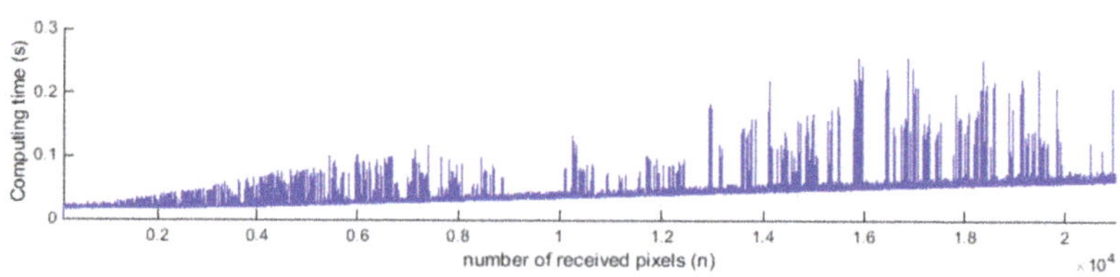

(**b**) PSP-OMPBS

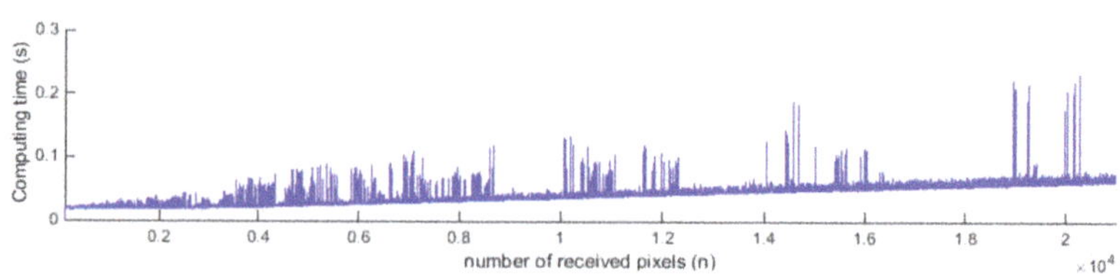

(**c**) S-PSP-OMPBS

Figure 13. *Cont.*

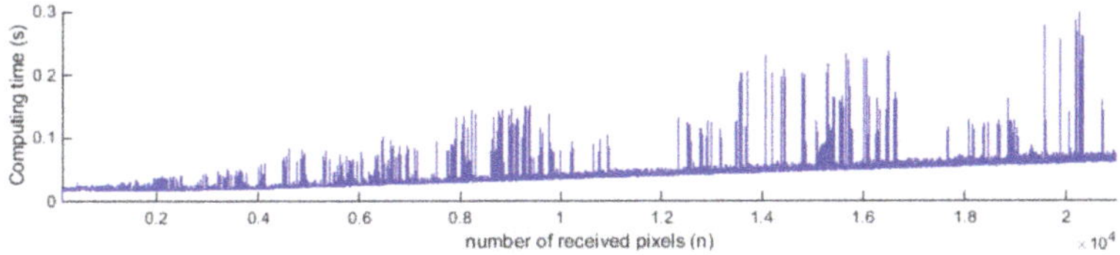

(**d**) B-PSP-OMPBS

Figure 13. Plots of computing time required for different progressive BS methods implemented on the AVIRIS Purdue dataset: (**a**) all methods, (**b**) PSP-BS OMPBS, (**c**) step sequence PSP-OMPBS (S-PSP-OMPBS), and (**d**) block sequence PSP-OMPBS (B-PSP-OMPBS).

It is worth mentioning that PSP-OMPBSs required a little more computing time compared to OMPBS at the beginning of transmission. Figure 14a,b shows the zoom-in computing time plots of the Pavia and Purdue datasets at $n = 100–1000$ and $n = 100–600$. There is a noticeable intersection between the OMPBS and PSP-BS curves at roughly $n = 300$. When n is less than 300, PSP-OMPBS requires more computing time because it needs to perform an additional logic operation (i.e., Step 4 in PSP-OMPBS algorithm), which leads to an increase of overall computing time. As n grows, the proportion of computing logic operations decreases, and this additional burden is relatively diminished.

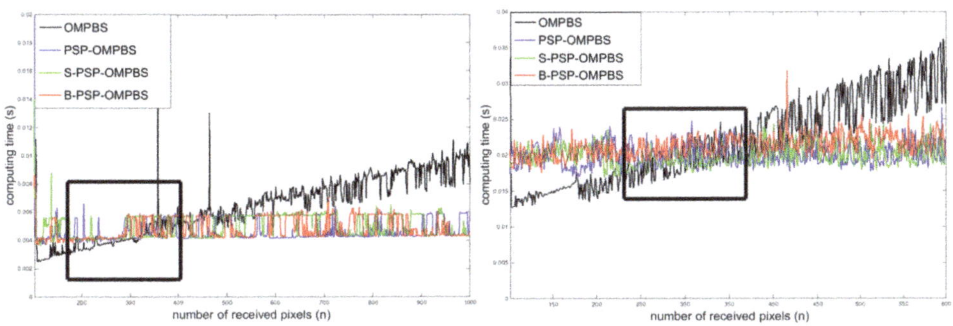

(**a**) Pavia data (**b**) Purdue data

Figure 14. Zoom-in plots of computing time at the beginning of progressive process: (**a**) the Pavia data, and (**b**) the Purdue data.

Finally, Table 6 lists the overall accumulative computing time of the progressive experiments of Figures 12 and 13. It was found that using PSP-BS could significantly reduce the amount of calculation. In addition, there was no significance difference between three PSP-OMPBS methods.

Table 6. The overall required computing time (in seconds) in the experiments.

Method	Pavia Data	Purdue Data
OMPBS	296,540	14,750
PSP-OMPBS	13,031	915
S-PSP-OMPBS	13,002	893
B-PSP-OMPBS	13,006	906

4.5. Graphical User Intervace Design

In order to analyze the relationship between the ground location of the received pixels, the produced values of quantitative indexes, and the BS results of progressive processing, a Matlab graphical user interface (GUI) was developed, as shown in Figure 15. It allows users to load different image data, input p values, and choose different PTS methods (original, step, block) with the corresponding parameters. Once all the inputs are loaded and the stage button is pressed, the PSP-OMPBS starts to do real-time BS simulation. The red square in the left will show the image scene, where the red dots present the locations of the received pixels. In the top-right corner, the green square will record the statistics, including the number of received pixels (n), the processing time, and ACC value for the current n. The time curve of a short period is drawn. In the bottom-right corner, the yellow square shows the BS results of the previous stage and the current stage, in which the red bins denote the instant results of PSP-OMPBS and the black bins denote the BS ground truth.

For an easy illustration, the experiment shown in Figure 15 was performed by B-PSP-OMPBS on the Pavia dataset. The parameters were set by p = 10 and b = 10. Over time, we can observe that the image gradually filled with red dots (transmitted sample pixels), and the ACC increased until it reached 100%. As a result, the BS can be fully monitored in the whole progressive process.

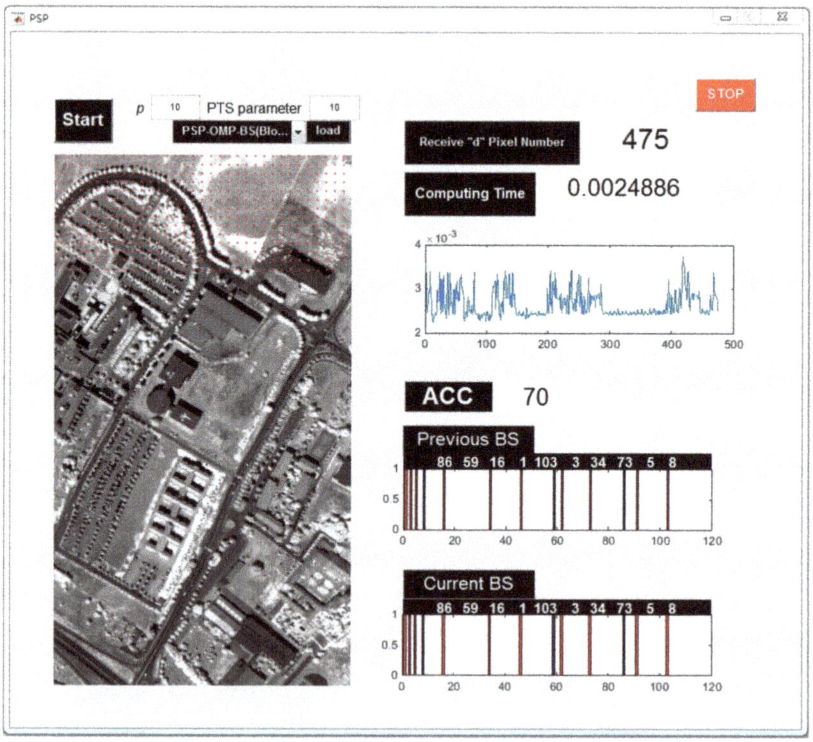

Figure 15. Graphical user interface (GUI) design for PSP-OMPBS.

5. Discussion

With the perspective of HSI application, there are several advantages of using PSP-BS. First, we can observe all the BS results in the whole transmission, where the local BS results (i.e., the BS results in a particular time segment) would not be missed. This is one of the most attractive features in progressive processing [34,35]. For instance, some "weaker bands" would appear during the transmission and disappear later on. Those bands might be very important for analyzing some specific types of ground materials, and should be considered to be selected. Besides, by virtue of progressive processing, the PSP-BS can easily be implemented on specific spatial regions (i.e., sub-image) within the image scene. Analyzing the partial data is also important. The second advantage of PSP-BS is the saving of storage volume. After PSP-BS is done, the redundant bands can be immediately removed, without waiting to apply the BS algorithm again, which saves time and storage space. The third advantage of PSP-BS is the saving of transmission bandwidth. Once the BS results converge in the early stage of transmission, we can ask the transmitter to re-transmit the rest of the image data only with the selected spectral dimensions, in order to save bandwidth. This is particularly important if the transmission bandwidth is limited, or the volume of original data cube is extremely large.

Similar to PBP, that PSP-BS obeys the principle that the time of processing one band must be less than the time of transmitting one band; in PSP-BS, the processing time at each stage must be less than the time to receive a new spectral pixel. In fact, both PBP and PSP methods suffer from the computation complexity issue when the image size or spectral dimension increases. So a balance between computation capability and transmission bandwidth is a prerequisite for the real-time process. This issue has not been emphasized in the related literature. Unlike the PBP works [33–35], the PSP-OMPBS algorithm is further involved with the iterations related to sequential search. The computing time may increase with the drastic increase of p. Thus, the computation time must be further reduced to satisfy the prerequisite. Fortunately, the calculation of PSP-OMPBS can be accelerated by parallel processing. For instance, the task of Step 2, and any terms about matrix multiplication in other steps, can be allocated to different cores of the central processing unit (CPU) to compute. The calculation of the matrix inverse can also be reduced by using a graphics processing unit (GPU). If the processing time cannot be reduced to fulfill the real-time process, we still can perform BS with larger n interval. In this case, deriving new recursive equations is necessary. We leave that to our future work.

6. Conclusions

This paper presents an instant BS method based on progressive sample processing (PSP), called PSP-OMPBS. PSP-OMPBS uses a recursive algorithm to efficiently accelerate computation. It processes BS pixel-by-pixel according to the BIS/BIP format, or by re-arranged pixel transmission sequences. Unlike traditional BS methods, which must re-implement the total received data, PSP-OMPBS can immediately obtain BS results when receiving a new pixel by referring to past information. The experiments conducted on two hyperspectral datasets show that PSP-OMPBS can instantly output the BS of each stage with very low computing time. Besides, by adopting different types of pixel transmission sequences, it is proved that using sampled sequences can significantly accelerate the BS convergence speed. Those advantages have allowed PSP-OMPBS to be applied effectively in a real-time manner during pixel transmission, and have potential for real-time BS monitoring. Our future work includes reducing the computational complexity by parallel processing, minimizing the numerical errors, considering other pixel transmission strategies, fusing different BS algorithms with PSP, and extending PSP to other topics in the field of hyperspectral image processing.

Acknowledgments: This work was supported by the Ministry of Science and Technology in Grants No.: MOST 105-2221-E-110-085 and MOST 105-2221-E-224-054.

Author Contributions: Keng-Hao Liu proposed the idea, derived the theoretical part, designed the experiments, and wrote the paper; Hung-Chang Chien and Meng-Han Lu built the Matlab codes, performed the experiments, and designed the Matlab GUI; Shih-Yu Chen reviewed the paper, inspected the theoretical part, and gave constructive advice.

Conflicts of Interest: The authors declare no conflict of interest.

References

1. Maisel, P.W.; Kramber, W.J.; Lee, J.K. Optimum band selection for supervised classification of multispectral data. *Photogramm. Eng. Remote Sens.* **1990**, *56*, 55–60.
2. Keshava, N. Distance metrics and band selection in hyperspectral processing with applications to material identification and spectral libraries. *IEEE Trans. Geosci. Remote Sens.* **2004**, *42*, 1552–1565. [CrossRef]

3. Du, Q.; Yang, H. Similarity-based unsupervised band selection for Hyperspectral Image Analysis. *IEEE Geosci. Remote Sens. Lett.* **2008**, *5*, 564–568. [CrossRef]
4. Chang, C.-I.; Wang, S. Constrained band selection for hyperspectral imagery. *IEEE Trans. Geosci. Remote Sens.* **2006**, *44*, 1575–1585. [CrossRef]
5. Martínez-Usó, A.; Pla, F.; Sotoca, J.M.; García-Sevilla, P. Clustering-Based Hyperspectral Band Selection Using Information Measures. *IEEE Trans. Geosci. Remote Sens.* **2007**, *45*, 4158–4171. [CrossRef]
6. Guo, B.; Gunn, S.; Damper, B.; Nelson, J. Adaptive band selection for hyperspectral image fusion using mutual information. In Proceedings of the 8th IEEE International Conference on Information Fusion, Philadelphia, PA, USA, 25–28 July 2005; p. 8.
7. Feng, J.; Jiao, L.C.; Zhang, X.; Sun, T. Hyperspectral Band Selection Based on Trivariate Mutual Information and Clonal Selection. *IEEE Trans. Geosci. Remote Sens.* **2014**, *52*, 4092–4105. [CrossRef]
8. Feng, J.; Jiao, L.; Liu, F.; Sun, T.; Zhang, X. Mutual-Information-Based Semi-Supervised Hyperspectral Band Selection With High Discrimination, High Information, and Low Redundancy. *IEEE Trans. Geosci. Remote Sens.* **2015**, *53*, 2956–2969. [CrossRef]
9. Sun, W.; Zhang, L.; Du, B. A Sparse Self-Representation Method for Band Selection in Hyperspectral Imagery Classification. In Proceedings of the 7th IEEE Workshop on Hyperspectral Image and Signal Processing: Evolution in Remote Sensing (Whispers), Tokyo, Japan, 2–5 June 2015; IEEE: New York, NY, USA; pp. 1–4.
10. Sun, W.; Zhang, L.; Zhang, L.; Lai, Y.M. A Dissimilarity-Weighted Sparse Self-Representation Method for Band Selection in Hyperspectral Imagery Classification. *IEEE J. Sel. Top. Appl. Earth Obs. Remote Sens.* **2016**, *9*, 4374–4388. [CrossRef]
11. Sun, W.; Jiang, M.; Li, W.; Liu, Y. A Symmetric Sparse Representation Based Band Selection Method for Hyperspectral Imagery Classification. *Remote Sens.* **2016**, *8*, 238. [CrossRef]
12. Lai, C.H.; Chen, C.S.; Chen, S.Y.; Liu, K.H. Sequential band selection method based on orthogonal matching pursuit. In Proceedings of the 8th IEEE Workshop on Hyperspectral Image and Signal Processing: Evolution in Remote Sensing (WHISPERS), Los Angeles, CA, USA, 21–24 August 2016; IEEE: New York, NY, USA; pp. 1–4.
13. Fisher, K.; Chang, C.-I. Progressive band selection for satellite hyperspectral data compression and transmission. *J. Appl. Remote Sens.* **2010**, *4*, 041770.
14. Chang, C.-I.; Liu, K.-H. Progressive Band Selection of Spectral Unmixing for Hyperspectral Imagery. *IEEE Trans. Geosci. Remote Sens.* **2014**, *52*, 2002–2017. [CrossRef]
15. Liu, S.; Du, Q.; Tong, X.; Samat, A.; Pan, H.; Ma, X. Band Selection-Based Dimensionality Reduction for Change Detection in Multi-Temporal Hyperspectral Images. *Remote Sens.* **2017**, *9*, 1008. [CrossRef]
16. Geng, X.; Sun, K.; Ji, L.; Zhao, Y. A Fast Volume-Gradient-Based Band Selection Method for Hyperspectral Image. *IEEE Trans. Geosci. Remote Sens.* **2014**, *52*, 7111–7119. [CrossRef]
17. Iordache, M.D.; Bioucas-Dias, J.M.; Plaza, A. Potential and limitations of band selection and library pruning in sparse hyperspectral unmixing. Proceedings of 2015 7th Workshop on Hyperspectral Image and Signal Processing: Evolution in Remote Sensing (WHISPERS), Tokyo, Japan, 2–5 June 2015; IEEE: New York, NY, USA; pp. 1–4.
18. Chang, Y.L.; Chang, L.; Fang, J.P. Particle swarm optimization/impurity function class overlapping scheme based on multiple attribute decision making model for hyperspectral band selection. Proceedings of 2015 IEEE International CONFERENCE Geoscience and Remote Sensing Symposium (IGARSS), Milan, Italy, 26–31 July 2015; IEEE: New York, NY, USA; pp. 441–444.
19. Huber-Lerner, M.; Hadar, O.; Rotman, S.R.; Huber-Shalem, R. Hyperspectral Band Selection for Anomaly Detection: The Role of Data Gaussianity. *IEEE J. Sel. Top. Appl. Earth Obs. Remote Sens.* **2016**, *9*, 732–743. [CrossRef]
20. Cao, X.; Xiong, T.; Jiao, L. Supervised Band Selection Using Local Spatial Information for Hyperspectral Image. *IEEE Geosci. Remote Sens. Lett.* **2016**, *13*, 329–333. [CrossRef]
21. Zhan, Y.; Hu, D.; Xing, H. Hyperspectral Band Selection Based on Deep Convolutional Neural Network and Distance Density. *IEEE Geosci. Remote Sens. Lett.* **2017**, *14*, 2365–2369. [CrossRef]
22. Wang, L.; Li, H.C.; Xue, B.; Chang, C.I. Constrained Band Subset Selection for Hyperspectral Imagery. *IEEE Geosci. Remote Sens. Lett.* **2017**, *14*, 2032–2036. [CrossRef]
23. Zhu, G.; Huang, Y.; Li, S.; Tang, J.; Liang, D. Hyperspectral Band Selection via Rank Minimization. *IEEE Geosci. Remote Sens. Lett.* **2017**, *14*, 2320–2324. [CrossRef]

24. Park, H.; Choi, J.; Park, N.; Choi, S. Sharpening the VNIR and SWIR Bands of Sentinel-2A Imagery through Modified Selected and Synthesized Band Schemes. *Remote Sens.* **2017**, *9*, 1080. [CrossRef]
25. Yang, C.; Tan, Y.; Bruzzone, L.; Lu, L. Discriminative Feature Metric Learning in the Affinity Propagation Model for Band Selection in Hyperspectral Images. *Remote Sens.* **2017**, *9*, 782. [CrossRef]
26. Hihara, H.; Moritani, K.; Inoue, M.; Hoshi, Y.; Iwasaki, A.; Takada, J.; Inada, H.; Suzuki, M.; Seki, T.; Ichikawa, S.; Tanii, J. On board Image Processing System for Hyperspectral Sensor. *Sensors* **2015**, *15*, 24926–24944. [CrossRef] [PubMed]
27. Conoscenti, M.; Coppola, R.; Magli, E. Constant SNR, Rate Control, and Entropy Coding for Predictive Lossy Hyperspectral Image Compression. *IEEE Trans. Geosci. Remote Sens.* **2016**, *54*, 7431–7441. [CrossRef]
28. Giordano, R.; Lombardi, A.; Guccione, P. Efficient clustering and on-board ROI-based compression for Hyperspectral Radar. In Proceedings of the IARIA Conference, Lisbon, Portugal, 26–30 June 2016; pp. 33–38.
29. Giordano, R.; Guccione, P. ROI-Based On-Board Compression for Hyperspectral Remote Sensing Images on GPU. *Sensors* **2017**, *17*, 1160. [CrossRef] [PubMed]
30. Quesada-Barriuso, P.; Argüello, F.; Heras, D.B. Computing Efficiently Spectral-Spatial Classification of Hyperspectral Images on Commodity GPUs. In *Recent Advances in Knowledge-based Paradigms and Applications. Advances in Intelligent Systems and Computing*; Springer: Cham, Switzerland, 2014.
31. Ma, N.; Wang, S.; Ali, S.M.; Cui, X.; Peng, Y. High Efficiency On-Board Hyperspectral Image Classification with Zynq SoC. Proceedings of 2016 7th International Conference on Mechatronics and Manufacturing (ICMM 2016), Harbin, China, 16 March 2016; Volume 45. [CrossRef]
32. Chang, C.I. Overview and Introduction. In *Real-Time Progressive Hyperspectral Image Processing*; Springer-Verlag: New York, NY, USA, 2016; pp. 1–32.
33. Chang, C.I.; Wu, C.C.; Liu, K.H.; Chen, H.M.; Chen, C.C.C.; Wen, C.H. Progressive Band Processing of Linear Spectral Unmixing for Hyperspectral Imagery. *IEEE J. Sel. Top. Appl. Earth Obs. Remote Sens.* **2015**, *8*, 2583–2597. [CrossRef]
34. Chang, C.I.; Li, Y.; Hobbs, M.C.; Schultz, R.C.; Liu, W.M. Progressive Band Processing of Anomaly Detection in Hyperspectral Imagery. *IEEE J. Sel. Top. Appl. Earth Obs. Remote Sens.* **2015**, *8*, 3558–3571. [CrossRef]
35. Chang, C.I.; Schultz, R.C.; Hobbs, M.C.; Chen, S.Y.; Wang, Y.; Liu, C. Progressive Band Processing of Constrained Energy Minimization for Subpixel Detection. *IEEE Trans. Geosci. Remote Sens.* **2015**, *53*, 1626–1637. [CrossRef]
36. Elhamifar, E.; Sapiro, G.; Vidal, R. See all by looking at a few: Sparse modeling for finding representative objects. Proceedings of IEEE Conference on Computer Vision and Pattern Recognition (CVPR), Providence, RI, USA, 16–21 June 2012; IEEE: New York, NY, USA; pp. 1600–1607.
37. Tropp, J.A.; Gilbert, A.C. Signal Recovery from Random Measurements via Orthogonal Matching Pursuit. *IEEE Trans. Inf. Theory* **2007**, *53*, 4655–4666. [CrossRef]
38. Woodbury Matrix Identity. Available online: https://en.wikipedia.org/wiki/Woodbury_matrix_identity (accessed on 10 December 2017).
39. Chang, C.-I.; Du, Q. Estimation of number of spectrally distinct signal sources in hyperspectral imagery. *IEEE Trans. Geosci. Remote Sens.* **2004**, *42*, 608–619. [CrossRef]
40. LibSVM. Available online: https://www.csie.ntu.edu.tw/~cjlin/libsvm/ (accessed on 18 January 2018).

remote sensing

MDPI

Article

Structure Tensor-Based Algorithm for Hyperspectral and Panchromatic Images Fusion

Jiahui Qu [1,2,*], Jie Lei [1,2,*], Yunsong Li [1,2,*], Wenqian Dong [2], Zhiyong Zeng [3] and Dunyu Chen [4]

1 Joint Laboratory of High Speed Multi-source Image Coding and Processing, School of Telecommunications Engineering, Xidian University, Xi'an 710071, China
2 State Key Lab. of Integrated Service Networks, School of Telecommunications Engineering, Xidian University, Xi'an 710071, China; Dongwq_xd@163.com
3 School of Mathematics and Information, Fujian Normal University, Fuzhou 350108, China; zzyong@fjnu.edu.cn
4 School of Electrical Communication, Yuan Ze University, Taoyuan City 32003, Taiwan; dychen@saturn.yzu.edu.tw
* Correspondence: qujh_xd@163.com (J.Q.); jielei@mail.xidian.edu.cn (J.L.); ysli@mail.xidian.edu.cn (Y.L.); Tel.: +86-178-0292-9680 (J.Q.); +86-139-9195-7389 (J.L.); +86-029-8820-2721 (Y.L.)

Received: 19 December 2017; Accepted: 24 February 2018; Published: 1 March 2018

Abstract: Restricted by technical and budget constraints, hyperspectral (HS) image which contains abundant spectral information generally has low spatial resolution. Fusion of hyperspectral and panchromatic (PAN) images can merge spectral information of the former and spatial information of the latter. In this paper, a new hyperspectral image fusion algorithm using structure tensor is proposed. An image enhancement approach is utilized to sharpen the spatial information of the PAN image, and the spatial details of the HS image is obtained by an adaptive weighted method. Since structure tensor represents structure and spatial information, a structure tensor is introduced to extract spatial details of the enhanced PAN image. Seeing that the HS and PAN images contain different and complementary spatial information for a same scene, a weighted fusion method is presented to integrate the extracted spatial information of the two images. To avoid artifacts at the boundaries, a guided filter is applied to the integrated spatial information image. The injection matrix is finally constructed to reduce spectral and spatial distortion, and the fused image is generated by injecting the complete spatial information. Comparative analyses validate the proposed method outperforms the state-of-art fusion methods, and provides more spatial details while preserving the spectral information.

Keywords: image fusion; hyperspectral image; panchromatic image; structure tensor; image enhancement; weighted fusion

1. Introduction

Hyperspectral (HS) remote sensing is an emerging discipline. Traditional remote sensing sensors obtain the image in a few discrete bands, and lose a large amount of useful information. A hyperspectral remote sensing sensor is capable of acquiring numerous contiguous narrow bands in a certain wavelength range [1]. As a result, the HS imagery has very high spectral resolution, and is a three-dimensional data cube, of which two spatial dimensions contain the space information, and one spectral dimension at each pixel includes the high-dimensional reflectance vectors [2,3]. Such HS image with abundant spectral information has been widely utilized in many domains, such as military surveillance [4], environmental monitoring [5], mineral exploration [6,7], and agriculture [8,9]. However, due to the constraints of technical difficulties and budget, the HS image usually has low spatial resolution. Although the high spectral resolution is crucial for identifying the materials, high spatial resolution is also important for locating the objects with high accuracy. There are various techniques to improve the spatial resolution of the HS image. Hyperspectral image fusion is one of the important spatial resolution enhancement approaches. Panchromatic (PAN) sensors can provide the PAN imagery with high spatial resolution. Fusion of an HS image and a PAN image is able to obtain a fused HS image with high spectral and spatial resolution by integrating the spectral information of the HS image and the spatial information of the PAN image.

A large number of hyperspectral image fusion methods have been proposed, and can be roughly divided into five families [10]. The first family is component substitution (CS), which first separates the spatial and spectral information of an HS image. The separated spatial component is then substituted by the PAN image, and a fused HS image can be obtained by applying the inverse transformation [11]. The CS includes algorithms such as intensity-hue-saturation (IHS) [12–14], principal component analysis (PCA) [15–17], Gram-Schmidt (GS) [18], adaptive GS (GSA) [19], Brovey transform (BT) [20], and partial replacement adaptive CS (PRACS) [21]. These CS based methods work well from a spatial aspect [19], and have fast and simple implementation [13]. However, they may suffer from serious spectral distortion, cause by the difference between the PAN image and the substituted spatial component [22]. The second family is multiresolution analysis (MRA) which aims to extract the spatial details of a PAN image through the multiscale decomposition or spatial filtering. The extracted spatial details are then injected into an HS image. Some well-known examples in the MRA family are smoothing filter based intensity modulation (SFIM) [23], decimated wavelet transform (DWT) [24], Laplacian pyramid [25], modulation transfer function (MTF) generalized Laplacian pyramid (MTF-GLP) [26], and MTF-GLP with high pass modulation (MTF-GLP-HPM) [27]. The MRA algorithms have temporal coherence [28], and good spectral preservation performance. On the negative side, these MRA algorithms have heavy computational burden and complicated implementation when compared to CS-based algorithms [28]. The CS and MRA approaches are the traditional fusion methods, and have been also extended from multispectral (MS) pansharpening to hyperspectral pansharpening.

The other three families, Bayesian methods, matrix factorization based methods, and hybrid methods, have been proposed recently. Bayesian methods transform a fusion problem into an explicit probabilistic framework, and then define a suitable prior distribution of interest to regularize the optimization model [29]. Bayesian sparsity promoted Gaussian prior (Bayesian sparse) [30], Bayesian HySure [31], and Bayesian naive Gaussian prior (Bayesian naive) [32] belong to this class of hyperspectral pansharpening. Matrix factorization based methods employ the nonnegative matrix factorization (NMF) model [33], and utilize the estimated solution of the NMF model to generate the fused HS image. The matrix factorization family contains algorithms such as nonnegative sparse coding (NNSC) [34], and constrained nonnegative matrix factorization (CNMF) [35]. Bayesian approaches and matrix factorization approaches perform well in terms of the preservation of spectral information. However, they have high computational cost. Hybrid methods combine algorithms from different families, for example, CS and MRA families, to form a new algorithm. Such obtained new algorithms generally take advantages of algorithms in both families [36]. Examples include the curvelet and ICA fusion method [37], the guided filter PCA (GFPCA) method [38], and the non-linear PCA (NLPCA) and indusion method [39].

The key to hyperspectral pansharpening is to provide more spatial information while preserving the spectral information of the original HS image. In order to accomplish this goal, this paper presents a new hyperspectral image fusion algorithm based on structure tensor. In this work, the structure tensor which describes the geometry structure and spatial details is applied to the fusion of HS and PAN images for the first time. Traditional methods extract the spatial details only from the PAN image without considering the structure information of the HS image, and thus, cause spectral distortion or deficient spatial enhancement. The proposed method considers the spatial details of the HS and PAN images simultaneously. The spatial details of the PAN image are extracted by calculating and analyzing the structure tensor and its eigenvalues. The spatial details of the HS image are synchronously generated by the adaptive weighted method. In order to consider the HS and PAN images simultaneously and obtain the complete spatial details, an appropriate weighted fusion strategy is introduced to merge the extracted spatial information from the PAN image with the spatial information obtained from the HS image. To avert artifacts at the boundaries, a guided filter which is an edge-preserving filter is applied to the obtained merged spatial information image. Consequently, we can effectively provide spatial information and accomplish sufficient spatial enhancement. In order to maintain the spectral information, an injection gains matrix is constructed. This gains matrix can also further reduce the spatial distortion by a defined tradeoff parameter. After a desired gains matrix is constructed, the fused HS image is obtained by adding spatial details to the interpolated HS image. Extensive experiments have been conducted on both simulated and real hyperspectral remote sensing datasets to verify the excellent fusion performance in spatial and spectral aspects.

The rest of this paper is organized as follows. Section 2 briefly introduces the basic theory of structure tensor. The proposed hyperspectral image fusion method is described in detail in Section 3. In Section 4, the experimental results and analysis for different datasets are presented. Conclusions are drawn in Section 5.

2. Related Work

A structure tensor can represent the structure and spatial information of images and has been shown to be an important tool in the field of image analysis [40,41]. The structure tensor has been successfully applied to many image processing problems, such as texture analysis [42], anisotropic filtering [43], and motion detection [44].

For a gray image $I(x,y)$, the change generated by a shift $(\Delta x, \Delta y)$ can be described as

$$r = \sum_{(x,y)} \mathrm{w}(x,y)[I(x+\Delta x, y+\Delta y) - I(x,y)]^2 \tag{1}$$

where $(\Delta x, \Delta y)$ includes $\{(0,1),(1,0),(1,1),(-1,1)\}$, and w is a smooth window, such as a Gaussian window [40]. Then, by using the first-order Taylor series $I(x+\Delta x, y+\Delta y) = I(x,y) + I_x\Delta x + I_y\Delta y + O(\Delta x^2, \Delta y^2)$, the change r are described as

$$r = \sum_{(x,y)} \mathrm{w}(x,y)[I_x\Delta x + I_y\Delta y + O(\Delta x^2, \Delta y^2)]^2 \tag{2}$$

where $I_x = \frac{\partial I}{\partial x}$ and $I_y = \frac{\partial I}{\partial y}$ are the horizontal and vertical components of the gradient vector. For the small shift, the change r can be simplified as

$$r = [\Delta x, \Delta y] T [\Delta x, \Delta y]^{\mathrm{T}} \tag{3}$$

where a matrix T is the structure tensor, defined as

$$T = \begin{bmatrix} \sum\limits_{(x,y)} \mathrm{w}(x,y)(I_x)^2 & \sum\limits_{(x,y)} \mathrm{w}(x,y)I_xI_y \\ \sum\limits_{(x,y)} \mathrm{w}(x,y)I_xI_y & \sum\limits_{(x,y)} \mathrm{w}(x,y)(I_y)^2 \end{bmatrix} \tag{4}$$

This structure tensor T is a semi-definite matrix and can be decomposed as

$$T = \begin{bmatrix} e_1 & e_2 \end{bmatrix} \begin{bmatrix} \mu_1 & 0 \\ 0 & \mu_2 \end{bmatrix} \begin{bmatrix} e_1 & e_2 \end{bmatrix}^{\mathrm{T}} \tag{5}$$

where μ_1 and μ_2 are the nonnegative eigenvalues, and e_1 and e_2 are the eigenvectors corresponding to the two eigenvalues. The two nonnegative eigenvalues describe the structure information of an image. When $\mu_1 \approx \mu_2 \approx 0$, the windowed image region is the flat area. If $\mu_1 > \mu_2 \approx 0$, the area belongs to the edge region. When $\mu_1 \geq \mu_2 > 0$, this indicates a corner. The trace is the sum of the eigenvalues and the determinant is the product of the eigenvalues, and a thresholding is used to classify and detect a edge or a corner [40]. For one pixel of an image, structure tensor matrix J is defined as

$$J = \begin{bmatrix} (I_x)^2 & I_x I_y \\ I_x I_y & (I_y)^2 \end{bmatrix} = \nabla I \cdot \nabla I^{\mathrm{T}} \tag{6}$$

where $\nabla I = [I_x, I_y]^{\mathrm{T}}$ is the gradient operator, and $\cdot$ is matrix product.

3. Proposed Hyperspectral Image Fusion Algorithm

Figure 1 shows a diagram of the proposed method, which consists of the following steps. First, the spatial information of an HS image is obtained by using an adaptive weighted method. Then, an image enhancement approach is applied to the PAN image to sharpen the spatial information. This is followed up by a structure tensor which is introduced to extract the spatial details of the enhanced PAN image. Subsequently, the extracted spatial information of the HS and PAN images is merged via a matching weighted fusion method, and a guided filter is performed on the merged spatial information to prevent artifacts. Finally, an injection gains matrix is constructed to avoid the spectral and spatial distortion, and a fused image is produced through injecting the integrated spatial details into each band of the interpolated HS image.

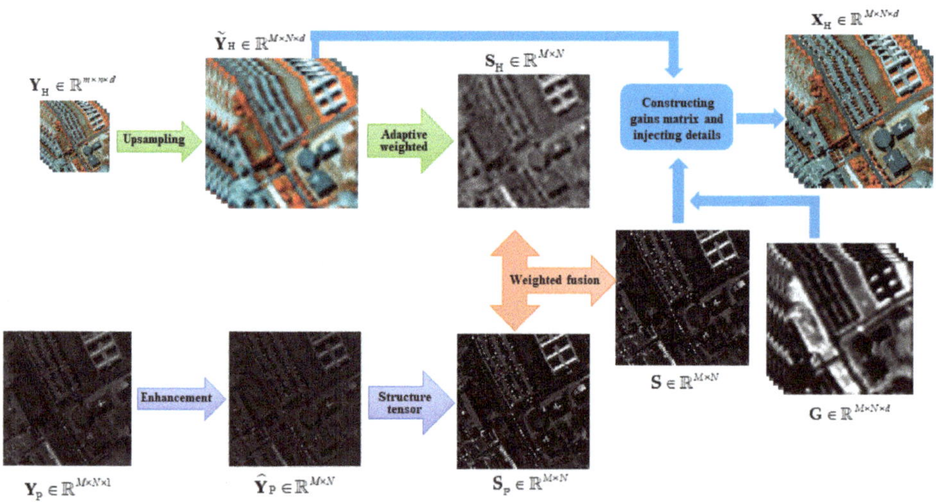

Figure 1. Diagram of the proposed hyperspectral image fusion algorithm. (m and M ($m < M$) represent the image height of the original HS and PAN images, respectively. n and N ($n < N$) represent the image width of the two images, and d represents the number of the HS image bands.).

3.1. Upsamping and Adaptive Weighted for the HS Image

For the same scene, let $\mathbf{Y}_{\mathrm{H}} \in \mathbb{R}^{m\times n\times d}$ represent the original low spatial resolution HS image, and $\mathbf{Y}_{\mathrm{P}} \in \mathbb{R}^{M\times N\times 1}$ represent the high spatial resolution PAN image. Fusing the HS and PAN images aims to obtain a fused high spatial resolution HS image $\mathbf{X}_{\mathrm{H}} \in \mathbb{R}^{M\times N\times d}$. Here, m and M ($m < M$) denote the image height of the HS and PAN images, respectively. n and N ($n < N$) denote the image width of these two images, and d denotes the number of the HS image bands. The low spatial resolution HS image is upsampled to the scale of the PAN image by the finite impulse response (FIR) filter interpolation method. The FIR filter interpolation method first performs zero interpolation on the low spatial resolution HS image, and then carries out the FIR filter processing to obtain the interpolated image.

$$\widetilde{\mathbf{Y}}_{\mathrm{H}}^{l} = \uparrow \mathbf{Y}_{\mathrm{H}}^{l} \tag{7}$$

for $l = 1, 2, \ldots, d$, where $\uparrow$ is the upsampling operation, $\widetilde{\mathbf{Y}}_{\mathrm{H}} \in \mathbb{R}^{M \times N \times d}$ is the interpolated HS image, $\widetilde{\mathbf{Y}}_{\mathrm{H}}^{l} \in \mathbb{R}^{M \times N}$ is the lth band of the interpolated HS image, and $\mathbf{Y}_{\mathrm{H}}^{l} \in \mathbb{R}^{m \times n}$ is the lth band of the original HS image.

For the purpose of extracting the spatial information of the HS image, an adaptive weighted method [19] is applied to the interpolated HS image.

$$\mathbf{S}_{\mathrm{H}} = \sum_{l=1}^{d} \omega_l \widetilde{\mathbf{Y}}_{\mathrm{H}}^{l} \tag{8}$$

where $\mathbf{S}_{\mathrm{H}} \in \mathbb{R}^{M \times N}$ is the spatial information of the HS image, and $[\omega_1, \omega_2, \ldots, \omega_d]^{\mathrm{T}}$ is the weight vector. To obtain the weights $\{\omega_l\}_{l=1,\ldots,d}$, the PAN image is first reduced to the same spatial scale of the low spatial resolution HS image. Let us denote this reduced PAN image as $\overline{\mathbf{Y}}_{\mathrm{P}}$. Then, let us assume that $\hat{\mathbf{S}}_{\mathrm{H}} = \sum_{l=1}^{d} \omega_l \mathbf{Y}_{\mathrm{H}}^{l}$. The optimal set of weights $\{\omega_l\}_{l=1,\ldots,d}$ can be calculated by linear ridge regression to minimize the mse between $\overline{\mathbf{Y}}_{\mathrm{P}}$ and $\hat{\mathbf{S}}_{\mathrm{H}}$. We obtain the closed-form solution of the weight ω_l as follows

$$\omega_l = \left((\mathbf{Y}_{\mathrm{H}}^{l})^{\mathrm{T}} (\mathbf{Y}_{\mathrm{H}}^{l}) \right)^{-1} (\mathbf{Y}_{\mathrm{H}}^{l})^{\mathrm{T}} \overline{\mathbf{Y}}_{\mathrm{P}} \tag{9}$$

for $l = 1, 2, \ldots, d$, where $()^{\mathrm{T}}$ is the transpose operation, and $()^{-1}$ is the matrix inversion. In Equation (9), $\mathbf{Y}_{\mathrm{H}}^{l} \in \mathbb{R}^{m \times n}$ and $\hat{\mathbf{Y}}_{\mathrm{P}} \in \mathbb{R}^{m \times n}$ are converted to the $mn \times 1$ dimensional form to calculate the solution.

3.2. Image Enhancement and Structure Tensor Processing for the PAN Image

To sharpen the spatial structure information of the PAN image, image enhancement processing is applied to the PAN image. The spatial filtering method is adopted to sharpen the PAN image. Compared with the Laplace algorithm, Laplacian of Gaussian (LOG) image enhancement algorithm can improve the robustness to noise and discrete points. We choose the LOG enhancement algorithm to sharpen the PAN image. The LOG algorithm first reduces noise by Gaussian convolution filtering. Subsequently, Laplace operator is utilized to enhance the spatial details. The Laplacian filtered image is finally combined with the PAN image to obtain the enhanced PAN image. This LOG enhancement procedure can be described as

$$\hat{\mathbf{Y}}_{\mathrm{P}} = \mathbf{Y}_{\mathrm{P}} + c[\mathbf{Y}_{\mathrm{P}} * f_{LOG}(x, y)] \tag{10}$$

where $\hat{\mathbf{Y}}_{\mathrm{P}} \in \mathbb{R}^{M \times N}$ denotes the enhanced PAN image, $f_{LOG}(x, y)$ denotes the kernel function of LOG operator, $*$ denotes the convolution operator, and c is a constant. If the central coefficient of the kernel $f_{LOG}(x, y)$ is negative, c is equal to -1. If the central coefficient of the kernel $f_{LOG}(x, y)$ is a positive value, c is 1. In this work, the size of the kernel is set to 15×15. The central coefficient of the kernel is a negative value, and c is -1. Based on the principle of the LOG operator, the kernel function $f_{LOG}(x, y)$ is defined as

$$f_{LOG}(x, y) = \frac{\partial^2}{\partial x^2} f_G(x, y) + \frac{\partial^2}{\partial y^2} f_G(x, y) \tag{11}$$

where $f_G(x, y)$ is the Gaussian convolution function, defined as

$$f_G(x, y) = \frac{1}{\sqrt{2\pi\sigma^2}} \exp\left(-\frac{x^2 + y^2}{2\sigma^2}\right) \tag{12}$$

where σ is the standard deviation, and σ is set to 0.43. Thus, the kernel function $f_{LOG}(x, y)$ is calculated by

$$f_{LOG}(x, y) = \frac{x^2 + y^2 - 2\sigma^2}{\sigma^4} \exp\left(-\frac{x^2 + y^2}{2\sigma^2}\right) \tag{13}$$

In order to extract the spatial details of the enhanced PAN image, the structure tensor processing is introduced. Based on Equation (6), structure tensor matrix of the enhanced PAN image at pixel i is defined by

$$\mathbf{T}_i = \begin{bmatrix} \hat{Y}_{Px,i}^{2} & \hat{Y}_{Px,i}\hat{Y}_{Py,i} \\ \hat{Y}_{Px,i}\hat{Y}_{Py,i} & \hat{Y}_{Py,i}^{2} \end{bmatrix} \tag{14}$$

where $\hat{\mathbf{Y}}_{\mathrm{P}x} = \frac{\partial \hat{\mathbf{Y}}_{\mathrm{P}}}{\partial x}$ and $\hat{\mathbf{Y}}_{\mathrm{P}y} = \frac{\partial \hat{\mathbf{Y}}_{\mathrm{P}}}{\partial y}$ are the horizontal and vertical components of the gradient vector on the enhanced PAN image, and $\mathbf{T}_i \in \mathbb{R}^{2\times 2}$ is the structure tensor matrix at pixel i on the enhanced PAN image. To include the local spatial structure information, a Gaussian kernel function is convoluted with the above structure tensor.

$$\overline{\mathbf{T}}_i = \begin{bmatrix} g_r * \hat{\mathbf{Y}}_{\mathrm{P}x,i}^{2} & g_r * \hat{\mathbf{Y}}_{\mathrm{P}x,i}\hat{\mathbf{Y}}_{\mathrm{P}y,i} \\ g_r * \hat{\mathbf{Y}}_{\mathrm{P}x,i}\hat{\mathbf{Y}}_{\mathrm{P}y,i} & g_r * \hat{\mathbf{Y}}_{\mathrm{P}x,i}^{2} \end{bmatrix} = \begin{bmatrix} P_{11} & P_{12} \\ P_{12} & P_{22} \end{bmatrix} \tag{15}$$

where g_r represents a Gaussian kernel with standard deviation r, the kernel size and the standard deviation of the Gaussian kernel are set to 1×2 and 0.5, $\overline{\mathbf{T}}_i$ represents the resulting structure tensor matrix at pixel i, and $\begin{bmatrix} P_{11} & P_{12} \\ P_{12} & P_{22} \end{bmatrix}$ simply represents the tensor $\overline{\mathbf{T}}_i$. According to the content of related work, the structure tensor $\overline{\mathbf{T}}_i$ can be decomposed as the form shown in Equation (5). Two nonnegative eigenvalues which represent the spatial structure information are calculated using the following formula

$$\xi_{1,2} = \frac{1}{2}\left[P_{11} + P_{22} \pm \sqrt{(P_{11} - P_{22})^2 + 4P_{12}^2}\right] \tag{16}$$

where ξ_1 and ξ_2 are the nonnegative eigenvalues for the structure tensor matrix shown in Equation (15). The values of the two nonnegative eigenvalues divide the structure information into three types, i.e., flat area, edge area, and corner. If ξ_1 and ξ_2 are near zero, the area of this pixel is the flat area. When $\xi_1 > \xi_2 \approx 0$ and $\xi_1 \geq \xi_2 > 0$, the area of this pixel belongs to the edge area and corner, respectively. For an image, we consider that the effective spatial information includes edge region and corner.

The trace of a matrix, denoted by R, is the sum of the eigenvalues, and is also the sum of P_{11} and P_{22}. The determinant denoted by D, is the product of the eigenvalues. We test on a large number of enhanced PAN images to study the trace and determinant at each pixel. Figure 2 shows the trace and determinant at each pixel on two of the enhanced PAN images. For a pixel, if R is near zero, the two eigenvalues are all near zero, and the area is the flat area. If R is larger than zero, at least one of the nonnegative eigenvalues is greater than zero, and the area at this pixel belongs to edge region or corner. Similarly, when D is near zero, at least one of the eigenvalues is near zero, and the area is flat area or edge region. when D is larger than zero, the two eigenvalues at this pixel are all larger than zero, and this pixel is a corner. The edge regions and corners are important spatial information. Based on the analysis and study on numerous enhanced PAN images, we suggest the following guidelines. When $R > 1 * 10^{(-5)}$, the pixel is identified as edge region or corner, and the area of this pixel is the effective spatial information. Thus, the value of this pixel should be retained. Otherwise, the area of this pixel is classified as the flat area, and the value of this pixel is not retained. This procedure of extracting the spatial details of the enhanced PAN image can be described as

$$\mathbf{S}_{\mathrm{P},i} = \begin{cases} \hat{\mathbf{Y}}_{\mathrm{P},i}, & \text{if } R > 1 * 10^{(-5)} \\ 0, & \text{otherwise} \end{cases} \tag{17}$$

where $\mathbf{S}_{\mathrm{P}} \in \mathbb{R}^{M\times N}$ is the spatial information of the enhanced PAN image, $\mathbf{S}_{\mathrm{P},i}$ is the value of the spatial information at pixel i, and $\hat{\mathbf{Y}}_{\mathrm{P},i}$ is the value of the enhanced PAN image at pixel i.

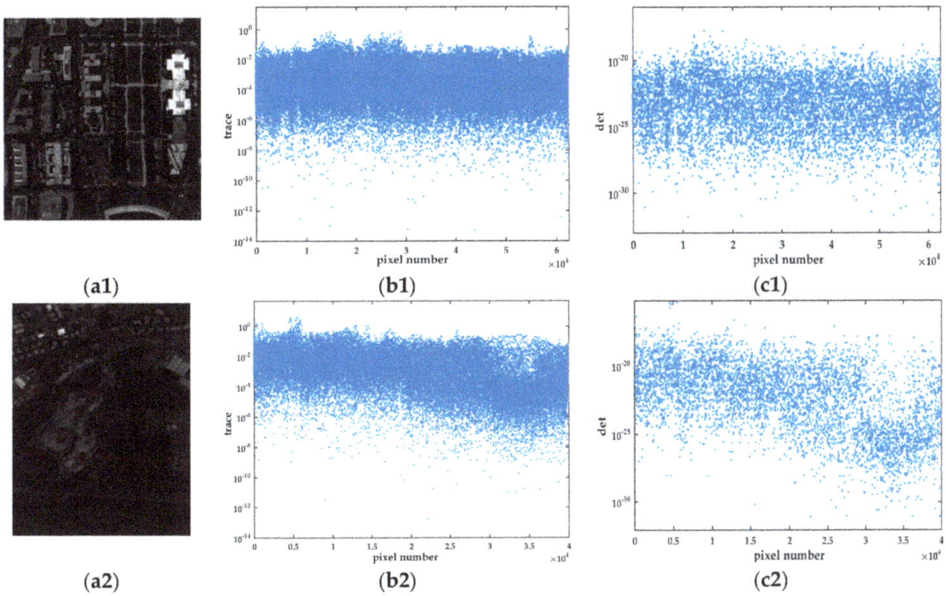

Figure 2. Trace and determinant at each pixel on two enhanced PAN images. (**a1**,**a2**) PAN image; (**b1**,**b2**) Trace of structure tensor at each pixel; (**c1**,**c2**) Determinant of structure tensor at each pixel.

Figure 3 shows the spatial information obtained by the gradient methods and the structure tensor method. Figure 3a shows a PAN image. Figure 3b shows the enhanced PAN image which is sharpened by using the LOG image enhancement algorithm. Figure 3c,d show the spatial information extracted by the horizontal gradient processing and the vertical gradient processing, respectively. According to Equation (17), the spatial information extracted by the structure tensor method is obtained and shown in Figure 3e. As shown in Figure 3, the extracted spatial information of the horizontal gradient method and the vertical gradient method only retain part of edge information of the original image. By contrast, the spatial information obtained by the structure tensor method contains most of the edge and structure information. This illustrates the structure tensor processing method in this subsection can effectively extract the spatial details of the enhanced PAN image.

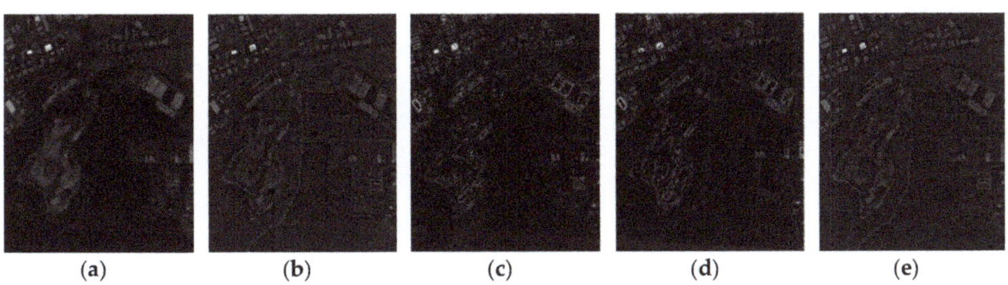

Figure 3. Spatial information of an enhanced PAN image extracted by the gradient methods and the structure tensor method. (**a**) PAN image; (**b**) Enhanced PAN image; (**c**) Horizontal gradient method; (**d**) Vertical gradient method; (**e**) Structure tensor method.

3.3. Weighted Fusion of Spatial Details

As shown in Figure 1, $\mathbf{S}_\mathrm{H}$ contains the spatial information of the HS image, and $\mathbf{S}_\mathrm{P}$ retains the spatial details of the enhanced PAN image. For a same scene, the HS image and the PAN image all include spatial information, and the spatial information of the two images is different and complementary. The PAN image has more spatial information, but may not include the details about the spatial structure of the HS image. Most conventional approaches only extract the spatial information from the PAN image, and not consider the spatial structure of the HS image. They may lead to spectral distortion or inadequate spatial enhancement. To obtain the complete spatial details and consider the spatial information of the HS and PAN images simultaneously, a weighted fusion method is presented to integrate the spatial information of the HS image with the spatial information of the PAN image.

$$\mathbf{S}_{\mathrm{F},i} = \begin{cases} \mathbf{S}_{\mathrm{H},i}, & \text{if } \mathbf{S}_{\mathrm{P},i} = 0 \\ \lambda_1 \cdot \mathbf{S}_{\mathrm{P},i} + \lambda_2 \cdot \mathbf{S}_{\mathrm{H},i}, & \text{if } \mathbf{S}_{\mathrm{P},i} \neq 0 \end{cases} \tag{18}$$

where λ_1 and λ_2 are weight coefficients, $\mathbf{S}_\mathrm{F} \in \mathbb{R}^{M \times N}$ is the complete spatial details, $\mathbf{S}_{\mathrm{F},i}$ is the value of $\mathbf{S}_\mathrm{F}$ at pixel i, $\mathbf{S}_\mathrm{P}$ and $\mathbf{S}_\mathrm{H}$ are the spatial information of the enhanced PAN image and the HS image, respectively, $\mathbf{S}_{\mathrm{P},i}$ and $\mathbf{S}_{\mathrm{H},i}$ are the values of $\mathbf{S}_\mathrm{P}$ and $\mathbf{S}_\mathrm{H}$ at pixel i. Since the PAN image contains more spatial details compared with the HS image, λ_1 and λ_2 are set to 0.9 and 0.1, respectively. Subsequently, to avoid artifacts at the boundaries, a guided filter is applied to the obtained fused spatial information image. The guided filter is an edge-preserving filter. It can smooth the input image while transferring the structure information from the guidance image to the output image [45]. The fused spatial information $\mathbf{S}_\mathrm{F}$ is served as both the guidance image and the input image. The filtered image which is the continuous and smooth result of the input image has the spatial structure information of the guidance image. Thus, the filtered output image which is continuous can avert artifacts at the boundaries, and preserves the spatial details of the fused spatial information $\mathbf{S}_\mathrm{F}$, simultaneously. According to the principle of the guided filter, the output image is a local linear transformation of the guidance image. This procedure are described as

$$\mathbf{S}_i = \bar{a}_i \mathbf{S}_{\mathrm{F},i} + \bar{b}_i = \frac{1}{|s|} \sum_{k \in v_i} a_k \mathbf{S}_{\mathrm{F},i} + \frac{1}{|s|} \sum_{k \in v_i} b_k, \forall i \in v_k \tag{19}$$

where $\mathbf{S} \in \mathbb{R}^{M \times N}$ is the output image, v_k is a local square window centered at pixel k, the local window size is set to 40, a_k and b_k are linear coefficients assumed to be constant, $\bar{a}_i$ and $\bar{b}_i$ are the average coefficients of all windows overlapping i, and $|s|$ is the number of pixels in v_k. The linear coefficients a_k and b_k are computed by minimizing the difference between the input image $\mathbf{S}_\mathrm{F}$ and the output image $\mathbf{S}$ while maintaining the linear transformation in the window v_k. The solution of a_k and b_k is obtained by calculating the linear ridge regression model.

$$a_k = \frac{\frac{1}{|s|} \sum_{i \in v_k} \mathbf{S}_{\mathrm{F},i} \mathbf{S}_{\mathrm{F},i} - \theta_k \overline{\mathbf{S}}_{\mathrm{F},k}}{\chi_k^2 + \varepsilon} \tag{20}$$

$$b_k = \overline{\mathbf{S}}_{\mathrm{F},k} - a_k \theta_k \tag{21}$$

where θ_k and χ_k^2 are the mean and variance of the guidance image $\mathbf{S}_\mathrm{F}$ in v_k, $\overline{\mathbf{S}}_{\mathrm{F},k}$ is the mean of the input image $\mathbf{S}_\mathrm{F}$ in v_k, ε is a regularization parameter, and parameter ε is set to 10^{-4}.

3.4. Constructing Gains Matrix and Injecting Spatial Details

Before including the integrated continuous spatial details into the interpolated HS image, a injection gains matrix is constructed to control the spectral and spatial distortion. To reduce the spectral distortion, the ratios between each pair of the HS bands should preserve unchanged. It is significant for maintaining the spectral information to preserve such ratios. It is depicted as

$$\mathbf{G}^l \propto \frac{\widetilde{\mathbf{Y}}_{\mathrm{H}}^l}{(1/d)\sum_{l=1}^{d} \widetilde{\mathbf{Y}}_{\mathrm{H}}^l} \tag{22}$$

for $l = 1, 2, \ldots, d$, where $\mathbf{G} \in \mathbb{R}^{M \times N \times d}$ denotes the injection gains matrix, and $\mathbf{G}^l \in \mathbb{R}^{M \times N}$ denotes the lth band of the gains matrix. For the sake of ensuring the spatial quality, we define a following tradeoff parameter to regulate the amount of the injected spatial details.

$$\mathbf{G}^l = \tau \frac{\widetilde{\mathbf{Y}}_{\mathrm{H}}^l}{(1/d)\sum_{l=1}^{d} \widetilde{\mathbf{Y}}_{\mathrm{H}}^l} \tag{23}$$

for $l = 1, 2, \ldots, d$, where τ is the defined tradeoff parameter. The influence and setting of the tradeoff parameter τ have been expounded in the experimental part. Then, the spatial details are injected into the interpolated HS image to generate the fused HS image for each band.

$$\mathbf{X}_{\mathrm{H}}^l = \widetilde{\mathbf{Y}}_{\mathrm{H}}^l + \mathbf{G}^l \cdot \mathbf{S} \tag{24}$$

where $\cdot$ is element-wise multiplication.

4. Experimental Results and Discussion

In this section, we design the experimental setup, and analyze the setting of the tradeoff parameter. To evaluate the fusion performance of the proposed method, four hyperspectral remote sensing datasets are used for experiments.

4.1. Experimental Setup

The proposed STF method is tested on four public hyperspectral datasets, which are shown in Table 1. Table 1 summarizes their characteristic.

- *Pavia University dataset*: Pavia University dataset was acquired by the Reflective Optics System Imaging (ROSIS) over Pavia, Italy. The HS image consists of 115 bands covering the spectral range 0.4–0.9 μm. The dimensions of the experimental PAN image are 250 × 250 with the spatial resolution of 1.3 m. The test HS image is of size 50 × 50 pixels with the spatial resolution of 6.5 m. For Pavia University dataset, 103 bands are applied to experimentation.
- *Moffett field dataset*: Moffett field dataset is a standard data product which has been provided by the Airborne Visible Infrared Imaging Spectrometer (AVIRIS) [29]. This dataset contains 224 bands in the spectral range of 0.4–2.5 μm. The size of the PAN and HS images that are used for experimentation are 250 × 160 and 50 × 32. The spatial resolution of the experimental PAN and HS images are 20 m and 100 m, respectively. The water absorption and noise corrupted bands are removed, and 176 bands are used for experimentation.
- *Washington DC dataset*: Washington DC dataset is an airborne hyperspectral data over the Washington DC Mall. This dataset includes 210 bands in the spectral range of 0.4–2.4 μm. Bands in the opaque atmosphere region are removed from the dataset, and 191 bands are left for experimentation. The test PAN image is of size 250 × 250 pixels, and the size of the HS image is of 50 × 50 pixels.
- *Hyperion dataset*: The EO-I spacecraft launched in 2000, and carried two primary instruments which were Advanced Land Imager (ALI) and Hyperion [29]. Hyperion instrument can provide the HS image which contains 242 bands covering the spectral range of 0.4–2.5 μm. ALI instrument is capable of providing the PAN image. For Hyperion dataset, 128 bands are applied to experimentation. The size of the test PAN image is 216 × 174 with the spatial resolution of 10 m. The experimental HS image is of size 72 × 58 pixels with the spatial resolution of 30 m.

Table 1. Characteristic of the four used datasets.

Dataset	Size	Spatial Resolution	Band Number	Spectral Range
Pavia University	PAN 250 × 250 HS 50 × 50	1.3 m 6.5 m	103	0.4–0.9 μm
Moffett field	PAN 250 × 160 HS 50 × 32	20 m 100 m	176	0.4–2.5 μm
Washington DC	PAN 250 × 250 HS 50 × 50	3 m 15 m	191	0.4–2.4 μm
Hyperion	PAN 216 × 174 HS 72 × 58	10 m 30 m	128	0.4–2.5 μm

Pavia University dataset, Moffett field dataset, and Washington DC dataset are semi-synthetic dataset. Given a reference high spatial resolution HS image, the simulated low spatial resolution HS image and the simulated PAN image are generated. The simulated PAN image is generated by averaging the bands of the visible range of the reference image. According to the Wald's protocol [46], the low spatial resolution HS image is simulated by applying a 9 × 9 Gaussian kernel blurring and downsampling to the reference HS image, and the downsampling factor is 5. Hyperion dataset is a real dataset to evaluate the capability of the proposed method in real hyperspectral remote sensing image.

The proposed method is compared with six hyperspectral pansharpening methods, namely MTF-GLP with High Pass Modulation (MTF-GLP-HPM) [27], Bayesian sparsity promoted Gaussian prior (Bayesian Sparse) [30], constrained nonnegative matrix factorization (CNMF) [35], guided filter PCA (GFPCA) [38], Brovey transform (BT) [20] and principal component analysis (PCA) [15]. MTF-GLP-HPM (abbreviated as MGH) belongs to multiresolution analysis (MRA) class. Bayesian Sparse fusion method (abbreviated as BSF) is one of the Bayesian methods. The CNMF algorithm and the GFPCA fusion approach belong to matrix factorization based methods and hybrid methods, respectively. These four methods which give the state-of-the-art fusion performance were all presented in recent years. The BT and PCA method which are the simple and classical fusion methods belong to component substitution (CS) family. These compared methods cover the recent effective works and the existing five categories which have been described in introduction section. In the experiments, the number of endmembers is set to 20 for the CNMF approach. For the GFPCA algorithm, the window size and the blur degree of the guided filter are set to 17 and 10^{-6} respectively. The pixel values of every test image are normalized to the range of 0–1.0 to reduce the amount of calculation.

To assess the capability of the proposed fusion method, several widely used evaluation indices are adopted, i.e., cross correlation (CC) [47], spectral angle mapper (SAM) [47], root mean squared error (RMSE), and erreur relative global adimensionnelle de synthse (ERGAS) [48]. CC is a spatial index and the best value is 1. SAM measures the degree of spectral similarity. The RMSE and ERGAS indices show the global quality of the fused image. The optimal value of SAM, RMSE, and ERGAS are 0. The experiments for the four datasets were all performed using MATLAB R2015b, and tested on a PC with an Intel Core i5-7300HQ CPU @ 2.50 GHz and 8 GB memory.

4.2. Tradeoff Parameter Setting

In the proposed method, the complete spatial details are finally included into the interpolated HS image. In order to reduce the spatial distortion, we define the tradeoff parameter τ to control the amount of the injected spatial details. The setting of the tradeoff parameter τ has an important impact on the spatial quality. Since the tradeoff parameter regulates the spatial distortion, the best value of τ can be chosen via the spatial index. Thus, for the sake of concluding the influence of τ, the proposed approach is tested on the Moffett field dataset and the Washington DC dataset to observe the CC values with different τ settings. Figure 4 shows the CC index values with different tradeoff parameter settings. When the tradeoff parameter τ is set to 0.1, the proposed method acquires the optimal CC values. We have also performed on numerous hyperspectral remote sensing images, and discovered that $\tau = 0.1$ also provides the largest CC values. Therefore, for the proposed method, the tradeoff parameter τ is set as 0.1.

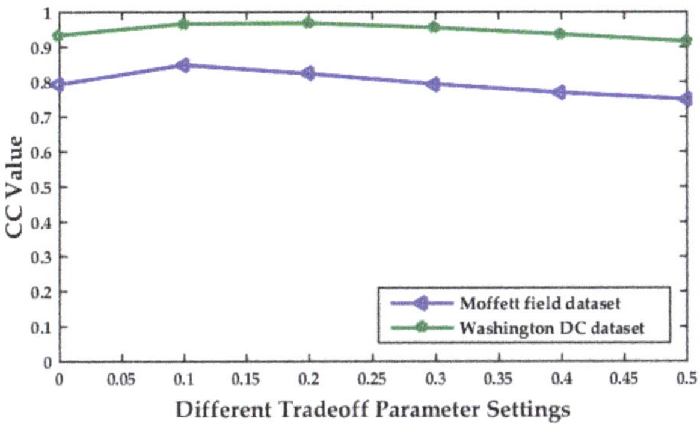

Figure 4. CC values with different tradeoff parameter settings.

4.3. *Experiments on Simulated Hyperspectral Remote Sensing Datasets*

In this part, the experiments are performed on three simulated hyperspectral remote sensing datasets to evaluate the fusion performance of the proposed method. Three datasets are Pavia University dataset, Moffett field dataset, and Washington DC dataset, respectively.

4.3.1. Pavia University Dataset

Figure 5a shows the reference high resolution HS image of Pavia University dataset. Figure 5d–j shows the fused HS images of each method for the Pavia University dataset. By comparing the fused images with the reference HS image visually, it can be observed that the GFPCA method looks blurry. This is because the GFPCA method utilizes the guided filter to transfer the spatial details from the PAN image to the HS image, but the spatial details are injected insufficiently. The BT approach provides enough spatial information, but the fused image obtained by the BT approach has spectral distortion in some areas, such as the trees and roads. Although the CNMF method has good fidelity of the spectral information, the CNMF method has deficient improvement of the spatial quality in some marginal areas, such as the edges of the trees and roofs. By contrast, we find that the PCA, BSF, MGH, and proposed STF method have the satisfactory fusion performance, and the MGH and STF methods achieve the better capability in preserving the spectral information compared with the PCA and BSF methods. In order to further compare the fusion performance, Figure 6 shows the error images (absolute values) of the competing methods for Pavia University dataset. Yellow means large differences, and blue means small differences. From Figure 6, it can be seen that the proposed SFT method shows the smallest differences between the fused HS image and the reference HS image.

Quantitative results of different fusion methods are shown in Table 2, which indicates that the proposed method achieves the best performance. The SAM, RMSE, and ERGAS values of the proposed method are the best, and the CC value of the proposed method is the second best. These results demonstrate that the proposed STF algorithm performs well in both the objective and subjective evaluations.

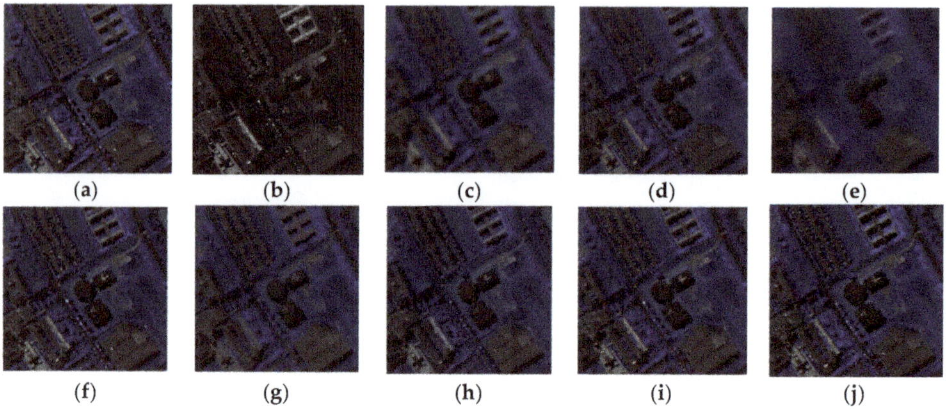

Figure 5. Fusion results obtained by each method for Pavia University dataset. (**a**) Reference HS image; (**b**) Simulated PAN image; (**c**) Interpolated HS image; (**d**) PCA; (**e**) GFPCA; (**f**) BT; (**g**) CNMF; (**h**) BSF; (**i**) MGH; (**j**) STF.

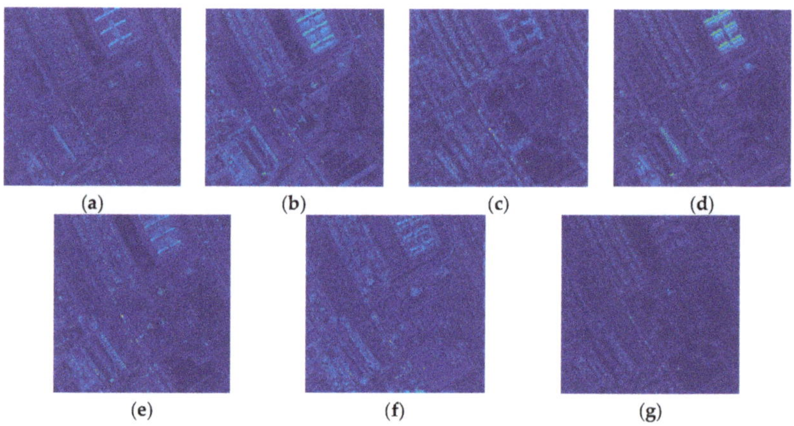

Figure 6. Error images of the competing methods for Pavia University dataset. (**a**) PCA; (**b**) GFPCA; (**c**) BT; (**d**) CNMF; (**e**) BSF; (**f**) MGH; (**g**) STF.

Table 2. Quantitative results of different fusion methods for Pavia University dataset.

Index	PCA	GFPCA	BT	CNMF	BSF	MGH	STF
CC	0.9342	0.8142	**0.9375**	0.8598	0.9059	0.8608	0.9336
SAM	7.2570	9.5526	6.6324	7.6670	8.8048	7.2589	**6.6212**
RMSE	0.0387	0.0596	0.0389	0.0493	0.0428	0.0867	**0.0386**
ERGAS	4.2443	6.8524	3.9901	5.7962	4.8990	7.7826	**3.9733**

4.3.2. Moffett Field Dataset

The fusion results obtained by each method for Moffett field dataset are displayed in Figure 7d–j. Visually, the PCA and BT methods have high fidelity in rendering the spatial details, but cause spectral distortion. This is due to the mismatching between the PAN image and the replaced spatial component. Compared with the PCA and BT approaches, the GFPCA seems to have less spectral distortion, but the spatial details are not sufficient. The fused result obtained by the CNMF method has good spectral fidelity, but the edges and spatial structures are not sharp enough, especially in the rural areas. The visual analysis shows that the BSF, MGH, and STF methods give the better fused results. The MGH, and STF algorithms are clearer, especially in the rural regions and rivers. However, the pansharpened image obtained by the MGH approach is too sharp in some areas, such as the tall buildings in urban areas. By contrast, the proposed STF method has superior performance in terms of providing the spatial information while preserving the spectral information. Table 3 reports the objective quantitative results for each method. From Table 3, we can apparently see that the proposed STF method has the largest CC value, and smallest SAM, RMSE, and ERGAS values.

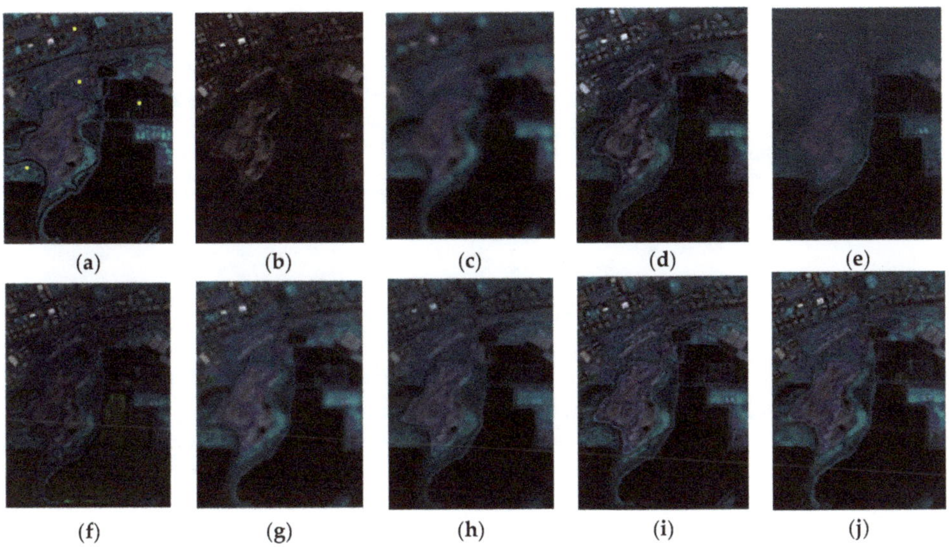

Figure 7. Fusion results obtained by each method for Moffett field dataset. (**a**) Reference HS image; (**b**) Simulated PAN image; (**c**) Interpolated HS image; (**d**) PCA; (**e**) GFPCA; (**f**) BT; (**g**) CNMF; (**h**) BSF; (**i**) MGH; (**j**) STF.

Table 3. Quantitative results of different fusion methods for Moffett field dataset.

Index	PCA	GFPCA	BT	CNMF	BSF	MGH	STF
CC	0.9046	0.9163	0.8705	0.9398	0.9558	0.9586	**0.9647**
SAM	12.0820	10.1200	8.3690	7.3153	7.9628	6.4328	**6.2690**
RMSE	0.0479	0.0444	0.0524	0.0372	0.0321	0.0489	**0.0308**
ERGAS	6.5091	6.1392	8.2161	5.1683	4.5358	6.6523	**3.9744**

The spectral reflectance curve difference values between the reference image and each fused image on one single pixel are compared to assess the spectral preservation performance. Figure 8 shows the spectral reflectance difference values on four pixels which are marked in yellow in Figure 7a. As shown in Figure 8, a gray dotted line is served as the benchmark. The closer the spectral reflectance difference values between the reference image and the fused image get to the dotted line, the more the spectral information is preserved. From Figure 8, it can be observed that the spectral reflectance difference values of the proposed method are most approximate to the dotted line (benchmark line) on the whole. These results validate the proposed method has the smallest difference when compared to other fusion methods.

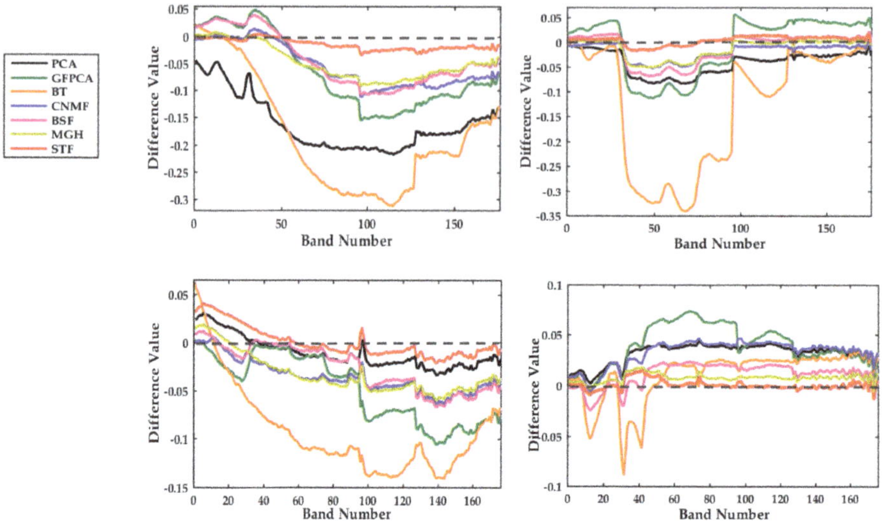

Figure 8. Spectral reflectance difference values comparison on four single pixels shown in Figure 7a.

4.3.3. Washington DC Dataset

The visual experimental results obtained by each method for the Washington DC dataset are shown in Figure 9d–j. In spite of good spatial quality, the fused images produced by the PCA and BT approaches cause spectral distortion in the roads and buildings. According to visual comparison of these results, the fused image generated by the MGH method has good fidelity of the spectral information. However, the MGH method suffers from spectral distortion in some areas, such as the roof areas. Compared with the PCA and MGH methods, the GFPCA algorithm has less spectral distortion. But the result of the GFPCA method has insufficient enhancement in the spatial aspect, and the fused image is blurry. The BSF, and STF method provide more spatial details compared to the CNMF method, since the CNMF method loses a little spatial information in the edges, such as in the roads and buildings. In contrast, the BSF, and STF method enhance more spatial information while preserving the spectral information of the original HS image.

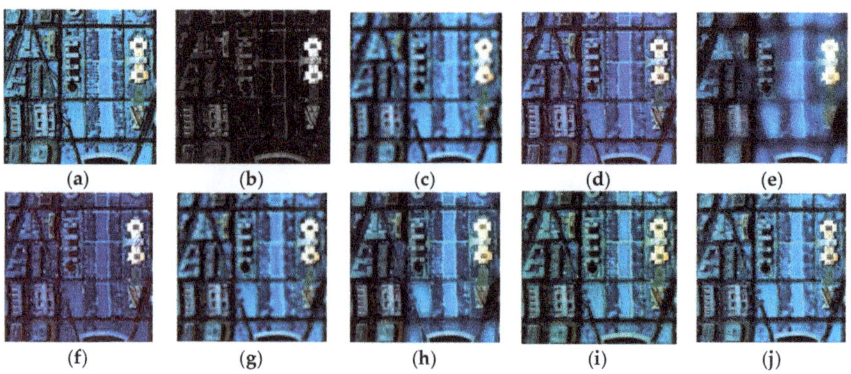

Figure 9. Fusion results obtained by each method for Washington DC dataset. (**a**) Reference HS image; (**b**) Simulated PAN image; (**c**) Interpolated HS image; (**d**) PCA; (**e**) GFPCA; (**f**) BT; (**g**) CNMF; (**h**) BSF; (**i**) MGH; (**j**) STF.

To further compare the fusion capability, the error images (absolute values) of different approaches for the Washington DC dataset are shown in Figure 10. Yellow means large differences, and blue means small differences. As shown in Figure 10, the SFT method shows the smallest differences in most regions, which testifies the preeminent fusion performance of the proposed method. The values of objective quality evaluation of each method for the Washington DC dataset are tabulated in Table 4. As shown in Table 4, for the proposed method, the CC, SAM, and RMSE values are the best, which prove once again that the proposed method is superior to the compared hyperspectral pansharpening methods.

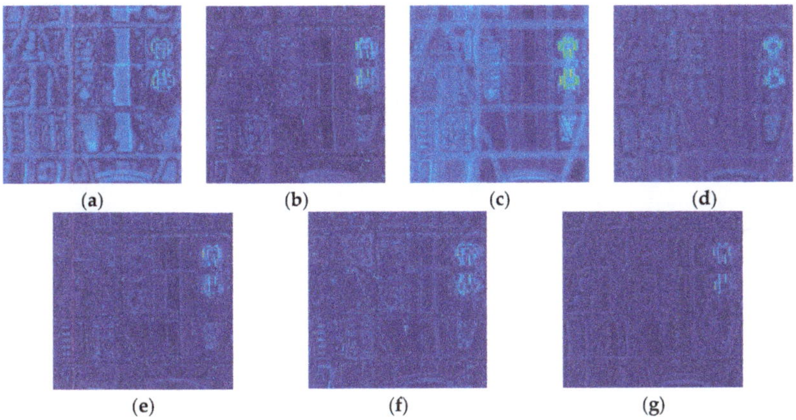

Figure 10. Error images of the competing methods for Washington DC dataset. (**a**) PCA; (**b**) GFPCA; (**c**) BT; (**d**) CNMF; (**e**) BSF; (**f**) MGH; (**g**) STF.

Table 4. Quantitative results of different fusion methods for Washington DC dataset.

Index	PCA	GFPCA	BT	CNMF	BSF	MGH	STF
CC	0.8485	0.7650	0.8157	0.7655	0.8294	0.8502	**0.8636**
SAM	7.9107	9.9500	7.9970	8.4167	10.0846	7.5508	**7.3970**
RMSE	0.0145	0.0148	0.0205	0.0148	0.0149	0.0359	**0.0140**
ERGAS	80.7202	59.4649	45.0533	**43.2606**	73.4928	95.2458	72.3328

4.4. Experiments on Real Hyperspectral Remote Sensing Datasets

In this part, the experiments are performed on the real hyperspectral remote sensing dataset to assess the fusion capability of the proposed method. The real HS dataset is the Hyperion dataset. Figure 11a,b show the low spatial resolution original HS image and the high spatial resolution PAN image. The fusion results of the competing methods are shown in Figure 11d–j. By a visual comparison of the pansharpened images, the PCA method has significant spectral distortion. For the GFPCA method, the spatial details are injected insufficiently, and the fused HS image looks fuzzy. The BSF method is better than the PCA method in preserving the spectral information, while the spatial details is a little less in the regard to some regions, such as the roads and grass. By contrast, the BT, CNMF, MGH, and STF method achieve the superior property. Since the low spatial resolution original HS image is unclear, the spectral information of the BT, CNMF, MGH, and STF method cannot accurately be compared. In the spatial aspect, the proposed STF method has the better performance, since it adds more spatial details.

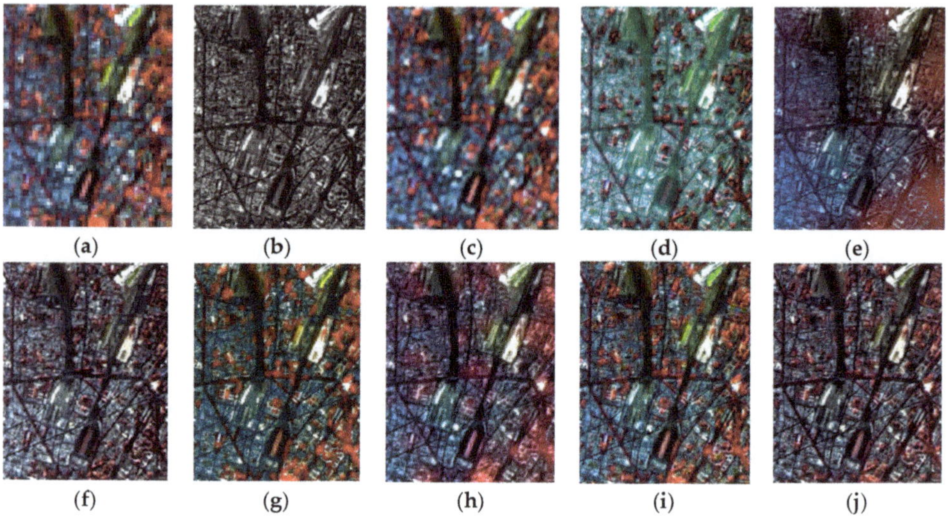

Figure 11. Fusion results obtained by each method for Hyperion dataset. (**a**) HS image; (**b**) PAN image; (**c**) Interpolated HS image; (**d**) PCA; (**e**) GFPCA; (**f**) BT; (**g**) CNMF; (**h**) BSF; (**i**) MGH; (**j**) STF.

For the real HS dataset, a reference high spatial resolution HS image is commonly not available. The original low resolution HS image can be served as the reference image. According to the Wald's protocol [45], the available original HS image is degraded to generate a degraded HS image. The available PAN image is also degraded to obtain a degraded PAN image. The degraded HS and PAN images are fused by each method to obtain the fusion results. These fusion results are compared to the original HS image to evaluate the objective fusion performance of different methods. Table 5 reports the objective fusion results for each method. From Table 5, we can apparently see that the proposed STF has the largest CC value, and smallest SAM and ERGAS values. These results demonstrate that the proposed algorithm obtains the excellent fusion performance.

Table 5. Quantitative results of different fusion methods for Hyperion dataset.

Index	PCA	GFPCA	BT	CNMF	BSF	MGH	STF
CC	0.7154	0.7309	0.7545	0.8702	0.8233	0.8661	**0.8780**
SAM	4.2361	4.8197	2.9466	3.1359	4.7309	2.7979	**2.6465**
RMSE	0.0476	0.0488	0.0775	0.0453	0.0459	**0.0389**	0.0421
ERGAS	8.9573	9.8286	9.9465	8.6040	8.9578	8.5842	**8.0167**

5. Conclusions

In this paper, a novel hyperspectral remote sensing image fusion using structure tensor approach is presented. The proposed method is believed to be the first work using the structure tensor to fuse the HS and PAN images. The PAN image is first sharpened by the LOG image enhancement method. Then, structure tensor is applied to the enhanced PAN image to extract the spatial information, while the spatial details of the HS image are obtained by an adaptive weighted method, simultaneously. To obtain the complete spatial details and accomplish spatial consistency, a suitable weighted fusion algorithm is proposed to integrate the extracted spatial details of the HS and PAN images. Experimental results from the Pavia University, Moffett field, Washington DC, and Hyperion datasets have shown that the proposed method is superior to the other fusion methods in retaining the spectral information and improving the spatial quality. In the future, we will investigate the issue of how to determine the weight coefficients λ_1 and λ_2 adaptively.

Acknowledgments: This work was supported by the National Science Foundation of China under Grants 61222101, 61272120, 61301287, 61301291 and 61350110239.

Author Contributions: Jiahui Qu and Yunsong Li devised the approach and analyzed the data. Jiahui Qu, Jie Lei, and Wenqian Dong performed the experiments. Zhiyong Zeng, and Dunyu Chen contributed materials and analysis tools. Jiahui Qu drafted the manuscript, which was revised by all authors. All authors read and approved the submitted manuscript.

Conflicts of Interest: The authors declare no conflict of interest.

References

1. Richards, J.A. *Remote Sensing Digital Image Analysis: An Introduction*; Springer: New York, NY, USA, 2013.
2. Li, Y.S.; Xie, W.Y.; Li, H.Q. Hyperspectral image reconstruction by deep convolutional neural network for classification. *Pattern Recognit.* **2017**, *63*, 371–383. [CrossRef]
3. Li, J.; Dias, J.B.; Plaza, A. Spectral–spatial hyperspectral image segmentation using subspace multinomial logistic regression and Markov random fields. *IEEE Trans. Geosci. Remote Sens.* **2012**, *50*, 809–823. [CrossRef]
4. Ertürk, A.; Iordache, M.D.; Plaza, A. Sparse unmixing with dictionary pruning for hyperspectral change detection. *IEEE J. Sel. Top. Appl. Earth Obs. Remote Sens.* **2017**, *10*, 321–330. [CrossRef]
5. Ellis, R.J.; Scott, P.W. Evaluation of hyperspectral remote sensing as a means of environmental monitoring in the St. Austell China clay (kaolin) region, Cornwall, UK. *Remote Sens. Environ.* **2004**, *93*, 118–130. [CrossRef]
6. Carrino, T.A.; Crósta, A.P.; Toledo, C.L.; Silva, A.M. Hyperspectral remote sensing applied to mineral exploration in southern Peru: A multiple data integration approach in the Chapi Chiara gold prospect. *Int. J. Appl. Earth Obs. Geoinform.* **2018**, *64*, 287–300. [CrossRef]
7. Bishop, C.; Liu, J.G.; Mason, P. Hyperspectral remote sensing for mineral exploration in Pulang, Yunnan Province, China. *Int. J. Remote Sens.* **2011**, *32*, 2409–2426. [CrossRef]
8. Mahesha, S.; Jayasa, D.S.; Paliwala, J.; Whiteb, N.D.G. Hyperspectral imaging to classify and monitor quality of agricultural materials. *J. Stored Prod. Res.* **2015**, *61*, 17–26. [CrossRef]
9. Chion, C.; Landry, J.A.; Da Costa, L. A Genetic-programming-based method for hyperspectral data information extraction: Agricultural Applications. *IEEE Trans. Geosci. Remote Sens.* **2008**, *46*, 2446–2457. [CrossRef]
10. Qu, J.H.; Li, Y.S.; Dong, W.Q. Hyperspectral pansharpening with guided filter. *IEEE Geosci. Remote Sens. Lett.* **2017**, *14*, 2152–2156. [CrossRef]
11. Liu, J.M.; Liang, S.L. Pan-sharpening using a guided filter. *Int. J. Remote Sens.* **2016**, *37*, 1777–1800. [CrossRef]
12. Carper, W.; Lillesand, T.M.; Kiefer, P.W. The use of Intensity-Hue-Saturation transformations for merging SPOT panchromatic and multispectral image data. *Photogramm. Eng. Remote Sens.* **1990**, *56*, 459–467.
13. Tu, T.M.; Su, S.C.; Shyu, H.C.; Huang, P.S. A new look at IHS-like image fusion methods. *Inf. Fusion* **2001**, *2*, 117–186. [CrossRef]
14. Anderson, J.A.; Sides, S.C.; Chavez, P.S. Comparison of three different methods to merge multiresolution and multispectral data: Landsat TM and SPOT panchromatic. *Photogramm. Eng. Remote Sens.* **1991**, *57*, 295–303.
15. Chavez, P.S.; Kwarteng, A.Y.A. Extracting spectral contrast in Landsat thematic mapper image data using selective principal component analysis. *Photogramm. Eng. Remote Sens.* **1989**, *55*, 339–348.
16. Shettigara, V. A generalized component substitution technique for spatial enhancement of multispectral images using a higher resolution data set. *Photogramm. Eng. Remote Sens.* **1992**, *58*, 561–567.
17. Shah, V.P.; Younan, N.; King, R.L. An efficient pan-sharpening method via a combined adaptive PCA approach and contourlets. *IEEE Trans. Geosci. Remote Sens.* **2008**, *56*, 1323–1335. [CrossRef]
18. Laben, C.; Brower, B. Process for Enhacing the Spatial Resolution of Multispectral Imagery Using Pan-Sharpening. U.S. Patent 6,011,875, 4 January 2000.
19. Aiazzi, B.; Baronti, S.; Selva, M. Improving component substitution pansharpening through multivariate regression of MS + Pan data. *IEEE Trans. Geosci. Remote Sens.* **2007**, *45*, 3230–3239. [CrossRef]
20. Gillespie, A.R.; Kahle, A.B.; Walker, R.E. Color enhancement of highly correlated images-II. Channel ratio and "chromaticity" transformation techniques. *Remote Sens. Environ.* **1987**, *22*, 343–365. [CrossRef]
21. Choi, J.; Yu, K.; Kim, Y. A new adaptive component-substitution based satellite image fusion by using partial replacement. *IEEE Trans. Geosci. Remote Sens.* **2011**, *49*, 295–309. [CrossRef]

22. Thomas, C.; Ranchin, T.; Wald, L.; Chanussot, J. Synthesis of multispectral images to high spatial resolution: A critical review of fusion methods based on remote sensing physics. *IEEE Trans. Geosci. Remote Sens.* **2008**, *46*, 1301–1312. [CrossRef]
23. Liu, J.G. Smoothing filter based intensity modulation: A spectral preserve image fusion technique for improving spatial details. *Int. J. Remote Sens.* **2000**, *21*, 3461–3472. [CrossRef]
24. Mallat, S. A theory for multiresolution signal decomposition: The wavelet representation. *IEEE Trans. Pattern Anal. Mach. Intell.* **1989**, 674–693. [CrossRef]
25. Burt, P.J.; Adelson, E.H. The Laplacian pyramid as a compact image code. *IEEE Trans. Commun.* **1983**, *31*, 532–540. [CrossRef]
26. Aiazzi, B.; Alparone, L.; Baronti, S.; Garzelli, A.; Selva, M. MTF-tailored multiscale fusion of high-resolution MS and pan imagery. *Photogramm. Eng. Remote Sens.* **2006**, *72*, 591–596. [CrossRef]
27. Vivone, G.; Restaino, R.; Mura, M.D.; Licciardi, G.; Chanussot, J. Contrast and error-based fusion schemes for multispectral image pansharpening. *IEEE Trans. Geosci. Remote Sens. Lett.* **2014**, *11*, 930–934. [CrossRef]
28. Aiazzi, B.; Alparone, L.; Baronti, S.; Garzelli, A.; Selva, M. 25 years of pansharpening: A critical review and new developments. In *Signal Image Processing for Remote Sensing*, 2nd ed.; Chen, C.H., Ed.; CRC Press: Boca Raton, FL, USA, 2011; Chapter 28; pp. 533–548.
29. Mookambiga, A.; Gomathi, V. Comprehensive review on fusion techniques for spatial information enhancement in hyperspectral imagery. *Multidimens. Syst. Signal Process.* **2016**, *27*, 863–889. [CrossRef]
30. Wei, Q.; Dobigeon, N.; Tourneret, J.Y. Bayesian fusion of multiband images. *IEEE J. Sel. Top. Signal Process.* **2015**, *9*, 1117–1127. [CrossRef]
31. Simoes, M.; Dias, J.B.; Almeida, L.; Chanussot, J. A convex formulation for hyperspectral image superresolution via subspace-based regularization. *IEEE Trans. Geosci. Remote Sens.* **2015**, *53*, 3373–3388. [CrossRef]
32. Wei, Q.; Dobigeon, N.; Tourneret, J.Y. Fast fusion of multiband images based on solving a sylvester equation. *IEEE Trans. Image Process.* **2015**, *24*, 4109–4121. [CrossRef] [PubMed]
33. Lee, D.D.; Seung, H.S. Learning the parts of objects by non negative matrix factorization. *Nature* **1999**, *401*, 788–791. [PubMed]
34. Hoyer, P.O. Non negative sparse coding. In Proceedings of the IEEE Workshop Neural Network Signal Processing, Martigny, Switzerland, 6 September 2002; pp. 557–565.
35. Yokoya, N.; Yairi, T.; Iwasaki, A. Coupled nonnegative matrix factorization unmixing for hyper-spectral and multispectral data fusion. *IEEE Trans. Geosci. Remote Sens.* **2012**, *50*, 528–537. [CrossRef]
36. Ghassemian, H. A review of remote sensing image fusion methods. *Inf. Fusion* **2016**, *32*, 75–89. [CrossRef]
37. Ghahremani, M.; Ghassemian, H. Remote-sensing image fusion based on Curvelets and ICA. *Int. J. Remote Sens.* **2015**, *36*, 4131–4143. [CrossRef]
38. Liao, W.; Huang, X.; Coillie, F.; Gautama, S.; Pizurica, A.; Philips, W.; Liu, H.; Zhu, T.; Shimoni, M.; Moser, G.; et al. Processing of multiresolution thermal hyperspectral and digital color data: Outcome of the 2014 IEEE GRSS data fusion contest. *IEEE J. Sel. Top. Appl. Earth Obs. Remote Sens.* **2015**, *8*, 2984–2996. [CrossRef]
39. Licciardi, G.; Khan, M.M.; Chanussot, J.; Montanvert, A.; Condat, L.; Jutten, C. Fusion of hyperspectral and panchromatic images using multiresolution analysis and nonlinear PCA band reduction. *EURASIP J. Adv. Signal Process.* **2012**, *1*, 1–17. [CrossRef]
40. Harris, C. A combined corner and edge detector. *Proc. Alvey Vis. Conf.* **1988**, *3*, 147–151.
41. Zenzo, S.D. A note on the gradient of a multi-image. *Comput. Vis. Graph. Image Process.* **1986**, *33*, 116–125. [CrossRef]
42. Köthe, U. Edge and Junction Detection with an Improved Structure Tensor. *Pattern. Recognit.* **2003**, *2781*, 25–32.
43. Li, S. An improved algorithm for anisotropic nonlinear diffusion for denoising cryo-tomograms. *J. Struct. Biol.* **2003**, *144*, 152–161.
44. Weickert, J.; Schuster, O.; Richter, S. A tensor-driven active contour model for moving object segmentation. In Proceedings of the 2001 International Conference on Image Processing, Thessaloniki, Greece, 7–10 October 2001.
45. He, K.; Sun, J.; Tang, X. Guided image filtering. *IEEE Trans. Pattern Anal. Mach. Intell.* **2013**, *35*, 1397–1409. [CrossRef] [PubMed]
46. Wald, L.; Ranchin, T.; Mangolini, M. Fusion of satellite images of different spatial resolutions: Assessing the quality of resulting images. *Photogramm. Eng. Remote Sens.* **1997**, *63*, 691–699.

47. Alparone, L.; Wald, L.; Chanussot, J.; Thomas, C.; Gamba, P.; Bruce, L. Comparison of pansharpening algorithms: Outcome of the 2006 GRS-S data-fusion contest. *IEEE Trans. Geosci. Remote Sens.* **2007**, *45*, 3012–3021. [CrossRef]
48. Zhang, L.; Zhang, L.; Tao, D.; Huang, X. On combining multiple features for hyperspectral remote sensing image classification. *IEEE Trans. Geosci. Remote Sens.* **2012**, *50*, 879–893. [CrossRef]

remote sensing

MDPI

Article

Integration of Absorption Feature Information from Visible to Longwave Infrared Spectral Ranges for Mineral Mapping

Veronika Kopačková * and Lucie Koucká

Remote Sensing Department, Czech Geological Survey, Prague 11821, Czech Republic; lucie.koucka@gmail.com
* Correspondence: veronika.kopackova@seznam.cz; Tel.: +420-257-089-481

Received: 4 September 2017; Accepted: 24 September 2017; Published: 28 September 2017

Abstract: Merging hyperspectral data from optical and thermal ranges allows a wider variety of minerals to be mapped and thus allows lithology to be mapped in a more complex way. In contrast, in most of the studies that have taken advantage of the data from the visible (VIS), near-infrared (NIR), shortwave infrared (SWIR) and longwave infrared (LWIR) spectral ranges, these different spectral ranges were analysed and interpreted separately. This limits the complexity of the final interpretation. In this study a presentation is made of how multiple absorption features, which are directly linked to the mineral composition and are present throughout the VIS, NIR, SWIR and LWIR ranges, can be automatically derived and, moreover, how these new datasets can be successfully used for mineral/lithology mapping. The biggest advantage of this approach is that it overcomes the issue of prior definition of endmembers, which is a requested routine employed in all widely used spectral mapping techniques. In this study, two different airborne image datasets were analysed, HyMap (VIS/NIR/SWIR image data) and Airborne Hyperspectral Scanner (AHS, LWIR image data). Both datasets were acquired over the Sokolov lignite open-cast mines in the Czech Republic. It is further demonstrated that even in this case, when the absorption feature information derived from multispectral LWIR data is integrated with the absorption feature information derived from hyperspectral VIS/NIR/SWIR data, an important improvement in terms of more complex mineral mapping is achieved.

Keywords: imaging spectroscopy; optical spectral region; thermal infrared spectral region; mineral mapping; data integration; HyMap; AHS; raw material; remote sensing

1. Introduction

Modern remote sensing has become a novel tool, not only for detecting and quantifying geological materials [1], but also for monitoring dynamic processes and induced changes in their physical/chemical properties [2–6]. Multispectral and superspectral imagery have been effectively used for mapping geology/minerals [7–11] as well as for monitoring mining impacts [12–19]. However, with a low number of rather broad spectral bands, these systems provide only discrete spectral information (e.g., the state-of-the-art Sentinel-2 sensor has 13 spectral bands [20]. On the other hand, data with very high spectral resolution (hundreds of narrow bands)—known in the remote sensing community as hyperspectral (HS) or imaging spectroscopy (IS) data—are nowadays capable of providing a continuous spectrum throughout the whole spectral range (0.4–13 μm). These systems are mainly available for aerial data acquisition; however, new hyperspectral satellite systems will be lunched in the near future [21–25].

The IS data can cover different spectral ranges from visible (VIS, 0.4–0.7 μm) through the near infrared (NIR, 0.7–1.0 μm) and shortwave infrared (SWIR, 1.0–2.5 μm) to even longer wavelengths of the thermal region (longwave infrared: LWIR, 8–13 μm). Within the VIS/NIR/SWIR/LWIR regions specific or combined absorptions (called absorption features from now on) can be found, caused by the electronic transition of Fe-bearing minerals (VIS/NIR region) and by the molecular vibration of specific chemical groups (e.g., OH^-, CO_3, Si–O) (SWIR and LWIR spectral regions). Considering the main mineralogical groups, the VIS/NIR parts of the electromagnetic (EMS) spectrum allow for mapping surfaces with a high concentration of Fe^{3+}-bearing minerals (e.g., hematite, goethite and jarosite) [6,26–28] and SWIR is useful in detecting carbonates, clay minerals and salts [29–31]. On the other hand, the VNIR and SWIR portions of the EMS are not optimal for detecting the main constituents of igneous rocks, quartz and feldspars due to their lack of absorption features in the optical part of the EMS. These minerals can be mapped using the thermal LWIR region [32–36].

Clearly, optical and thermal IS data, when used together, allow different varieties of minerals to be mapped and thus allow lithology mapping in a more complex way. The synergy effect of merging both ranges (optical and thermal) was demonstrated in examples of soil proximal sensing [37–39]. Considering mineral mapping, as stated by McDowell and Kruse [40], the majority of the previous work exploiting spectral IS data has focused on data from a single wavelength range, typically the VNIR, SWIR or LWIR. Few studies have taken advantage of data from the full VIS/NIR, SWIR, and LWIR spectral range, whereas the different spectral ranges were analysed and interpreted separately [41–43] and the full-range information was not actually combined into a single integrated data product. This limits the complexity of the final interpretation as spectral and spatial associations or patterns may be too complex to be seen by the naked eye and thus may remain hidden.

Recently, Kruse [41] proposed integrating the individual mapping results derived from AVIRIS (VIS/NIR/SWIR) and HyTES data (LWIR) and combining them using geologically directed logical operators. In the following study by McDowell and Kruse [40], spectral information from the individual VIS, NIR, SWIR and LWIR ranges was first analysed independently and then the resulting compositional information, in the form of image endmembers and apparent abundances, was integrated using ISODATA cluster analysis. They demonstrated that the integrated map provided additional compositional information that was not evident in the VIS, NIR, SWIR, or LWIR data alone, and concluded that their analysis allowed for more complete and accurate compositional mapping.

This study tested whether the multiple absorption features, which are directly linked to the mineral composition and are present though the VIS/NIR/SWIR and LWIR ranges, can be:

- automatically derived throughout the different spectral ranges
- integrated and, moreover, if this new dataset can be successfully used for final mineral/lithology mapping

To map multiple absorption feature parameters automatically, a toolbox was used that was developed using Interactive Data Language (IDL). The biggest advantage of such an approach is that it allows the issue of prior definition of the endmembers to be overcome; this is a requested routine used for all widely-used spectral mapping techniques (e.g., Spectral angle mapping SAM [44], Spectral feature fitting SFF e.g., [45,46] and Spectral unmixing [47,48]. Two different airborne image datasets were analysed, HyMap (HyVista Corp., Australian airborne imaging spectrometer, VIS/NIR/SWIR image data) and Airborne Hyperspectral Scanner (AHS, LWIR image data), both datasets were acquired over the Sokolov open-cast lignite mines in the Czech Republic. It is further demonstrated that even in this case, when the absorption feature information derived from multispectral LWIR data is integrated with the absorption feature information derived from hyperspectral VIS/NIR/SWIR data, it is an important contribution and improvement in terms of more complex mineral mapping.

2. Materials and Methods

2.1. Test Site

The study was performed in the Sokolov basin in the western part of the Czech Republic (Figure 1), in a region affected by long-term extensive lignite mining. The basement of the Sokolov Basin is formed of pre-Variscan and Variscan metamorphic complexes (recorded metamorphism from Devonian to Lower Carboniferous periods) of the Eger, Erzgebirge, Slavkov Forest, Thuring-Vogtland Crystalline Units and granitoids of the Karlovy Vary Pluton. The upper portions of these rocks are frequently weathered to kaolinitic residue. The basal late Eocene Staré Sedlo Formation is formed of well-sorted fluvial sandstones and conglomerates and is overlain by a volcano–sedimentary complex up to 350 m thick, which contains three lignite seams with variable sulphur (S) content. Long-term open cast mining required the removal of up to 180 m of thick overburden (Cypris clays), which was stockpiled and replaced after the lignite was extracted. At the dumps, the material consists mostly of weathered volcanic tuffs and Cypris clays, which can be characterised as well-laminated clays with a dominant kaolinite content; however, different varieties of mineralogical composition are common (e.g., the presence of montmorillonite, illite with admixtures of Ca–Mg–Fe carbonates, sulphates, sulphides, analcite, Mg–micas and bitumen [49]. Due to the presence of S in the coal, both active and abandoned lignite mines are affected by acid mine drainage (AMD) [50,51].

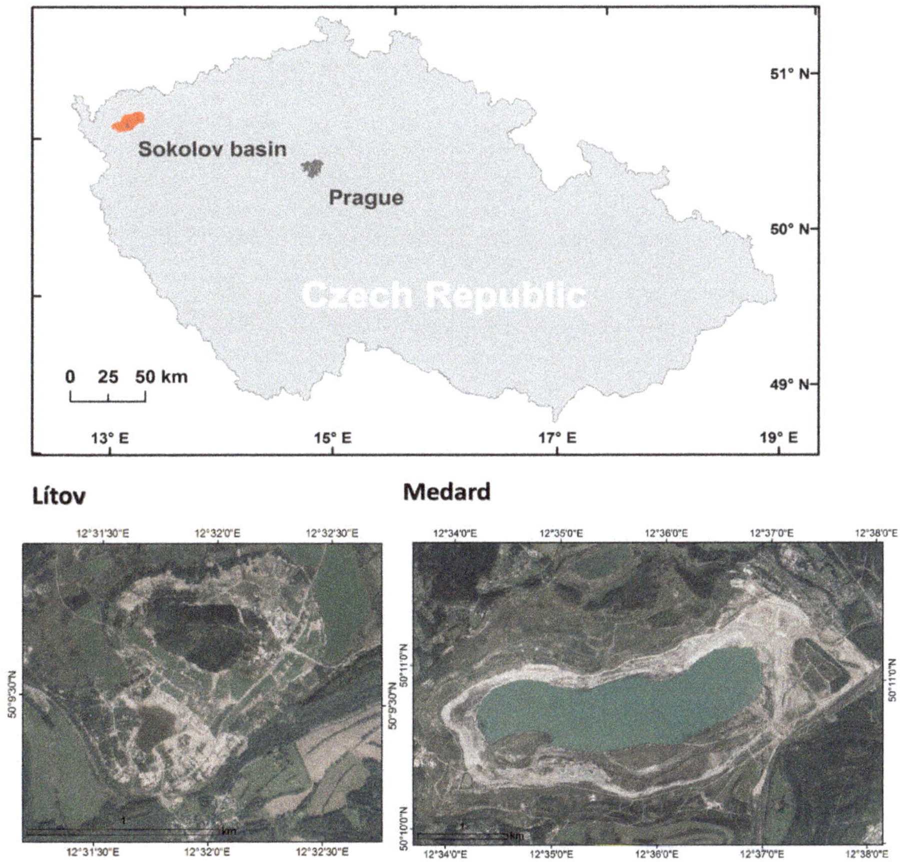

Figure 1. Geographic position of the two sites under study: Lítov dump and Medard Lake, Sokolov basin, Czech Republic.

Considering the Sokolov site, under various research projects (HypSo, EO-MINERS, DeMinTIR), numerous studies have been published demonstrating how hyperspectral imaging data can be utilised to quantitatively model the substrate pH [6,52], map mineral composition [41], estimate mine water pollution [53] and assess tree health [54–56]. This study focused on the Lítov dump and the abandoned open pit called Medard (Figure 1), as no human activities were conducted between 2010 and 2011 at these two sites, the years when the two aerial image datasets, HyMap and AHS, were acquired, thus these two sites faced no changes regarding the relief and material transport between 2010 and 2011.

2.2. Data

2.2.1. Airborne Imaging Datasets and Their Pre-Processing

Two different airborne image datasets acquired over the Sokolov basin were used in this study. The HyMap image data was acquired in 2010 (August 27) during the HyEUROPE 2010 flight campaign using the HyMap (HyVista Corp., HyVista Corporation Pty Ltd., Baulkham Hills, NSW, Australia) airborne imaging spectrometer. The HyMap sensor records image data in 126 narrow spectral bands covering the entire spectral interval between 0.450 and 2.480 μm of the spectral range with a Full Width Half Maximum (FWHM) of 15 nm and a ground field of view of 4 m. The resulting ground pixel resolution of the image datasets was 5 m. In order to successfully pre-process the hyperspectral data, supportive calibration and validation ground campaigns were organised simultaneously with the HyMap data acquisition. At the selected homogenous targets the ground measurements were acquired by an ASD FieldSpec-3 spectroradiometer to properly calibrate as well as validate the image data and to enable: (i) atmospheric correction of the airborne hyperspectral images and (ii) retrieval of surface reflectance values for further verification. The final atmospheric correction was performed in the ATCOR-4 software package [57] using the MODTRAN 4 physical model of the atmosphere [58]. A detailed description of the HyMap data preprocessing can be found in Adar et al. [59].

The second image dataset was acquired by the Airborne Hyperspectral Scanner (AHS) in collaboration with the Spanish Aerospace Institute (INTA) as a set of day (19 July 2011) and night image data (22 July 2011). The AHS is an imaging 80-band line-scanner radiometer with 63 bands in the visible-near infrared (VNIR) and shortwave infrared (SWIR) regions, seven bands in the mid-wave infrared (MWIR) region and 10 bands in the longwave infrared (LWIR) region [60]. However, due to cloud cover in the daytime image, it was only possible to use the cloud-free night-time LWIR data. These were acquired after a dry day, on a clear night with no precipitation. The flight lines were acquired at an altitude of 2 km above ground level, resulting in a 5-m pixel size (the same pixel size as the HyMap dataset). The temperature and emissivity were derived from the sensor radiance using the approach described in detail by Notesco et al. [42].

Both datasets were geo-corrected using on-board navigation information. After that both datasets were further georeferenced to the very high spatial resolution aerial orthophotos (pixel size = 0.5 m) achieving sub-pixel positional accuracy. To avoid any spatial misalignments, both datasets were resampled to a 10-m spatial resolution using nearest-neighbour resampling.

2.2.2. Ground Verification Data

Various soil/substrate samples were collected in the field in both years (2010 and 2011) and analysed with a Philips X'Pert X-ray Diffractometer (XRD) at the Czech Geological Survey to resolve their mineralogy. The X-ray powder diffraction patterns were obtained using monochromatic (CuKα) radiation and a graphite secondary monochromator. The whole-sample random patterns were collected in the angular range from 2° to 70° (2θ) with steps of 0.05° (2θ). Oriented clay-fraction specimens (fraction < 2 μm) were prepared by a conventional sedimentation method [61]. The oriented clay specimens were analysed after air-drying and after saturation for 10 h with ethylene–glycol vapour at 60 °C. Their diffraction data were acquired in the angular range of 2–50° (2θ) with steps of 0.05° (2θ). Mixed-layered minerals were identified by comparing the analysed XRD patterns of the ethylene-glycolated oriented clay fraction with the modelled XRD patterns obtained by NEWMOD code [61].

2.3. *Methods*

2.3.1. Absorption Wavelength Mapping

To map absorption features, namely absorption wavelengths and depths, new tools, which are described here for the very first time, were programmed to allow automatic detection of multiple absorption feature parameters. The tools—e.g., called QUANTools—have been created using IDL programming language (ENVI/IDL: version 5.0 and higher, [62]).

The tools can process both spectral libraries and hyperspectral image data. They consist of basic modules (GUIs) (Figure 2) allowing users to:

1. Define a spectral range within the visible (VIS), near-infrared (NIR), shortwave infrared (SWIR) or thermal (TIR) spectral regions. Different spectral ranges can be defined and analysed consequently, one after another.
2. Employ Continuum Removal (CR)—A standard method to normalise the spectrum, to a departure from the norm [63].
3. Detect bad spectral bands—A user can use a graphical interface to detect and correct bad (noisy) spectral bands.
4. Define a number of desired absorption features to be detected within a set spectral range: the user can decide whether to detect an absolute absorption (the most pronounced one) or to define a number of multiple absorption features that can be identified within a set spectral range.
5. Calculate absorption feature parameters (absorption wavelengths and depths): after correcting noisy bands, the trend of a spectral curve is analysed and saddle points—the local absorption maximum wavelengths (loc_max)—are detected and assigned to an image matrix. The detected absorptions are sorted in ascending order from shorter to longer wavelengths. Additionally, a corresponding absorption depth matrix is also calculated for each absorption feature.

The only decisions made by an operator/expert are to define the spectral region (Figure 2, step 1) and the number of desired absorption features mapped in each spectral region (Figure 2, step 4). All the other steps are done automatically when processing the datasets. A detailed explanation of the processing steps available in the toolbox is given in Section 2.3.2, below.

2.3.2. The Toolbox (QuanTools) Description

QuanTools can be used for spectral absorption band mapping using spectral libraries or hyperspectral image data (Figure 2). A user first defines the spectral range to be analysed within the visible (VIS), near-infrared (NIR), shortwave infrared (SWIR) or longwave infrared (LWIR) regions (Figure 2, step 1). Different spectral ranges can be defined and analysed after one another. Next comes continuum removal (CR, [63]), by which the continuum—that is, a convex hull of straight-line segments, is fitted over a reflectance spectrum and subsequently removed by division or rationing (Figure 2, step 2). The next step is noise detection and removal using local minima of the spectral curve defined as:

$$R = \left\{ \begin{array}{l} 1\, if\ \rho_{CR}(\lambda - 1) \geq \rho_{CR}(\lambda) \\ 0\, if\ \rho_{CR}(\lambda - 1) < \rho_{CR}(\lambda) \end{array} \right\} \tag{1}$$

$$if\ [R(\lambda - 1) = 1] \wedge [R(\lambda) = 0] \Rightarrow localminimum, \tag{2}$$

where $\rho_{CR}(\lambda)$ and $\rho_{CR}(\lambda - 1)$ are the values of the spectral curve after Continuum Removal and R is an auxiliary variable. When the local minima are found, it has to be decided whether the value represents noise or not. The decision is based on two sets of values (a defined number of surrounding bands, e.g., called a spectral window)—derived from bands placed before and after the local minimum:

$$A(\lambda) = \left\{ \begin{array}{c} \rho_{CR}(\lambda - nn); (\lambda - nn) \\ \vdots \\ \rho_{CR}(\lambda - 2); (\lambda - 2) \\ \rho_{CR}(\lambda - 1); (\lambda - 1) \end{array} \right\} \tag{3}$$

$$(\lambda) = \left\{ \begin{array}{c} \rho_{CR}(\lambda+1); (\lambda+1) \\ \rho_{CR}(\lambda+2); (\lambda+2) \\ \vdots \\ \rho_{CR}(\lambda+nn); (\lambda+nn) \end{array} \right\}, \tag{4}$$

where $A(\lambda)$ represents the set of surrounding bands before the local minimum, $B(\lambda)$ represents the set after the local minimum, and nn is the number of surrounding bands. Then the minimum values of these sets are detected as:

$$A_{min}(\lambda) = \min \rho_{CR} \in A(\lambda) \tag{5}$$

$$B_{min}(\lambda) = \min \rho_{CR} \in B(\lambda). \tag{6}$$

The minimum of the spectral window ($A_{min}(\lambda)$, $B_{min}(\lambda)$) is compared to the local minimum $\rho_{CR}(\lambda)$. Using this comparison, the noise is detected as follows:

$$[B_{min}(\lambda) < \rho_{CR}(\lambda)] \wedge [A_{min}(\lambda) > \rho_{CR}(\lambda)]: \tag{7}$$

$$if \rho_{CR} \in A(\lambda) < \rho_{CRmax} \in B(\lambda) \Rightarrow print \left\{ \begin{array}{c} Err(\lambda-1) \\ Err(\lambda-2) \\ \vdots \\ Err(\lambda-n) \end{array} \right\}; n \leq nn$$

$$if \left\{ \begin{array}{c} (\lambda+1) \\ (\lambda+2) \\ \vdots \\ (\lambda+nn) \end{array} \right\} \in B(\lambda) < \rho_{CRmax}(\lambda) \in B(\lambda) \Rightarrow print \left\{ \begin{array}{c} Err(\lambda+1) \\ Err(\lambda+2) \\ \vdots \\ Err(\lambda+n) \end{array} \right\}; n \leq nn$$

$$[A_{min}(\lambda) < \rho_{CR}(\lambda)] \wedge [B_{min}(\lambda) > \rho_{CR}(\lambda)]: \tag{8}$$

$$if \rho_{CR} \in B(\lambda) < \rho_{CRmax} \in A(\lambda) \Rightarrow print \left\{ \begin{array}{c} Err(\lambda+1) \\ Err(\lambda+2) \\ \vdots \\ Err(\lambda+n) \end{array} \right\}; n \leq nn$$

$$if \left\{ \begin{array}{c} (\lambda-1) \\ (\lambda-2) \\ \vdots \\ (\lambda-nn) \end{array} \right\} \in A(\lambda) < \rho_{CRmax}(\lambda) \in A(\lambda) \Rightarrow print \left\{ \begin{array}{c} Err(\lambda-1) \\ Err(\lambda-2) \\ \vdots \\ Err(\lambda-n) \end{array} \right\}; n \leq nn,$$

where $Err(\lambda)$ represents noise ("bad" bands).

These error values are recalculated using the values of the surrounding bands (their number is defined by a spectral window size). For sample spectral data, a graphical interphase can be used to set up a spectral window size and check the corrected spectral curves (Figure 3, step 3). This allows the user to compare the results of different settings and finally employ a correction that is optimally tuned for the data under analysis.

After a noise-cleaned image is retained, the operator decides how many absorption features are to be derived (Figure 2, step 4) and the analysis is done automatically (Figure 2, step 5) using the following processing:

$$CRdepth(\lambda) = 1 - \rho_{CR}(\lambda). \tag{9}$$

Then the trend of the spectrum is analysed in a similar way to noise detection. First the saddle points (the local absorption maxima) are detected using these conditions:

$$R_{ij} = \left\{ \begin{array}{l} 1\, if\ CRdepth(\lambda)_{ij} \geq CRdepth(\lambda-1)_{ij} \\ 0\, if\ CRdepth(\lambda)_{ij} < CRdepth(\lambda-1)_{ij} \end{array} \right\} \tag{10}$$

$$if\ [R(\lambda-1)_{ij}=1]\wedge[R(\lambda)_{ij}=0]\Rightarrow print\ Loc\_max_{ij},$$

where R is an auxiliary variable that records increase (1) or decrease (0). During this step, local absorption maximum depths and wavelengths are registered:

$$Loc_{max\,ij}=\left\{\begin{array}{c} Loc_{max1}depth(\lambda)_{ij}; Loc_{max1}(\lambda)_{ij}\\ Loc_{max2}depth(\lambda)_{ij}; Loc_{max2}(\lambda)_{ij}\\ \vdots\\ Loc_{max\,nm}depth(\lambda)_{ij}; Loc_{max\,nm}(\lambda)_{ij}\end{array}\right\}. \tag{11}$$

A set number of desired absorption features is detected as the most pronounced local absorption maxima (a set number of the most pronounced absorptions according to the absorption depths). If the desired number is larger than the number of local maxima detected, the final array is completed by zero values:

$$Loc_{max\,ij}=\left\{\begin{array}{c} Loc_{max1}depth(\lambda)_{ij}; Loc_{max1}(\lambda)_{ij}\\ Loc_{max2}depth(\lambda)_{ij}; Loc_{max2}(\lambda)_{ij}\\ \vdots\\ Loc_{max\,Tot_{nm}}depth(\lambda)_{ij}; Loc_{max\,nm}(\lambda)_{ij}\\ 0;0\\ \vdots\\ 0_{nm};0_{nm}\end{array}\right\}. \tag{12}$$

Finally, the local absorption maximum features are assigned to image matrices. For z absorption features, two matrices with z bands are created; one has the absorption wavelengths assigned, while the second one has corresponding absorption depths assigned, respectively.

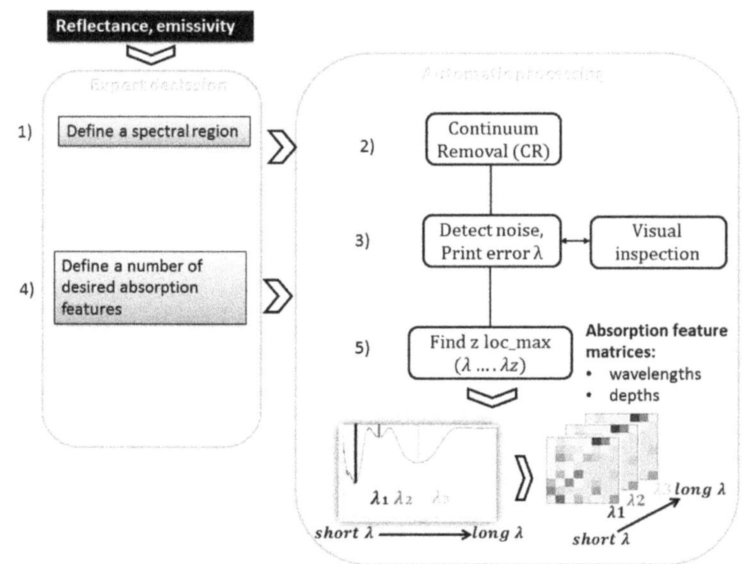

Figure 2. QUANTools: a simplified processing scheme (CR: continuum removal, λ : wavelength loc_mac: local absorption feature maximum). The detected absorption feature parameters are sorted in ascending order by wavelength (λ1, λ2 − λz).

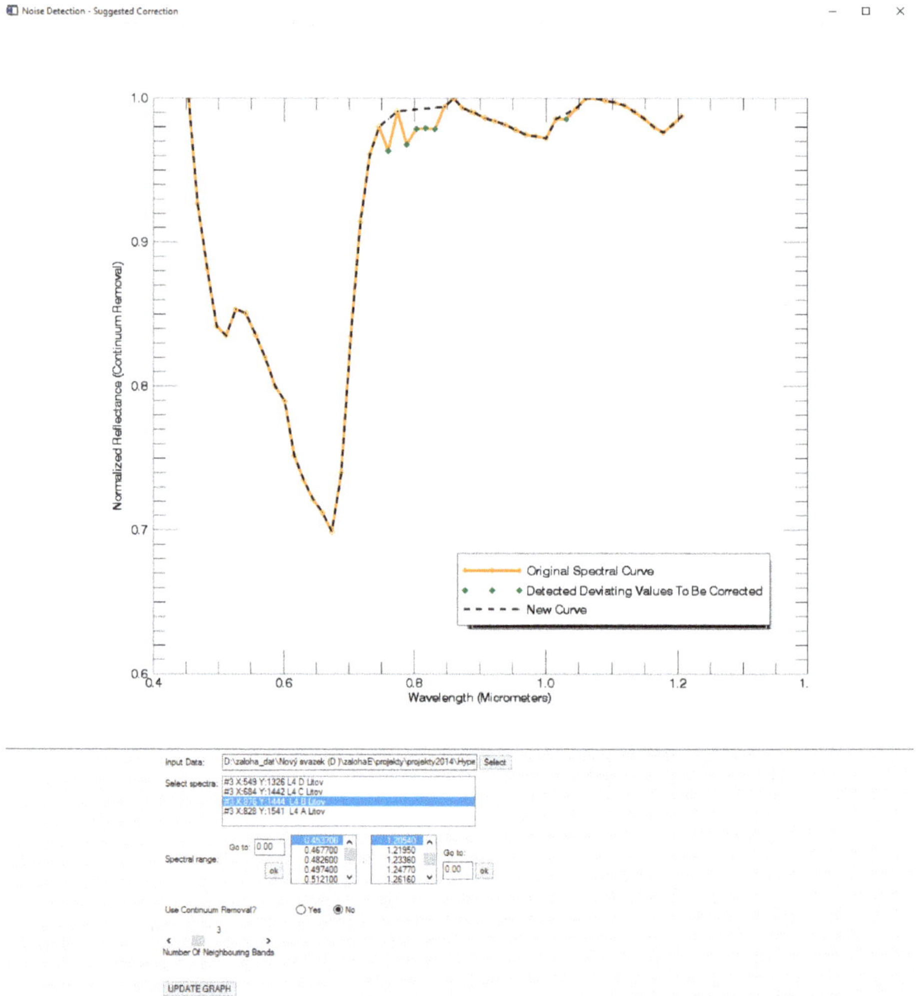

Figure 3. Bad band detection (spectral range 0.450–1.200 μm) in QuanTools where the bad bands detected are shown by the green square points and the corrected curve as a dashed line (set spectral window: three neighbouring bands).

2.3.3. Specific Setting of the QUANTools

The main aim was to test if the major absorption features present though the VIS/NIR/SWIR and LWIR ranges can be (i) derived and integrated into one raster dataset; or (ii) successfully used for a final mineral mapping. The spectral ranges from which the absorption feature parameters were derived were defined as follows: VIS/NIR (HyMap): 0.450–1.200 μm, SWIR (HyMap): 2.100–2.400 μm and LWIR (AHS): 8.500–12.500 μm. Noise detection and correction was employed to the VIS/NIR reflectance (0.450–1.200 μm) and the spectral window was set to 3 (three spectral bands before and after the local minimum were analysed). The number of absorption features to be mapped in each spectral region was set to 2. As a result, two new raster datasets were derived for each spectral range, one having assigned the two major absorption wavelengths and the second raster dataset with corresponding absorption depths.

2.3.4. Integration of Absorption Feature Information Detected In VIS/NIR/SWIR and LWIR Data and Further Classification

After the wavelength mapping was employed to different spectral ranges, it was necessary to find a way to further integrate the absorption feature mapping results (Figure 4), more specifically the absorption wavelength and depth matrices. The MNF transformation [64] was employed to compress the data variability of these image matrices (six raster absorption wavelength matrices and six raster absorption depth matrices).

The intention was to test what the advantage will be of adding LWIR data (AHS) to the further mineral classification compared to using only the VIS/NIR/SWIR data (HyMap). Therefore, the MNF transformation was employed for two different scenarios:

1. the MNF transformation was used so as to be comprised of only eight absorption wavelength/ depth matrices derived on the basis of the HyMap data (VIS/NIR: two absorption wavelength and two absorption depth matrices, SWIR: two absorption wavelength and two absorption depth matrices)
2. the MNF was employed to comprise of all 12 absorption wavelength/depth matrices derived from both HyMap and AHS datasets (in addition to eight absorption wavelength and depth matrices derived from the HyMap data, two absorption wavelength and two absorption depth matrices derived for the AHS data were added).

For each scenario, the first three MNF bands were visually analysed further to identify training areas (ROIs) representing different material/surfaces. The ROIs were defined as representative pixels of different colour clusters/regions. These were easy to identify when MNF1/MNF2/MNF3 images were displayed as RGB (Figure 5). The ROIs were then used for further supervised classification. In this case the simple non-parametric supervised (parallelepiped) classification was employed [65–67], which uses a simple decision rule to classify multispectral data. The parallelepiped classifier uses the threshold of each class signature to determine if a given pixel falls within the class or not. The thresholds specify the dimensions (in standard deviation units) of each side of a parallelepiped surrounding the mean of the class in the feature space. The best results (e.g., class separability, lowest number of unclassified pixels) were achieved when the upper and lower limits of each parallelepiped were set to ±1.5 standard deviations. In the last step, the classifications were overlaid over the original HyMap and AHS image data and the average spectrum from the original HyMap and AHS image data was computed for each class to (i) ensure that each class represents different surface material (is represented by a unique spectral signature) and (ii) be able to describe the mineral composition of each class when interpreting the spectral property.

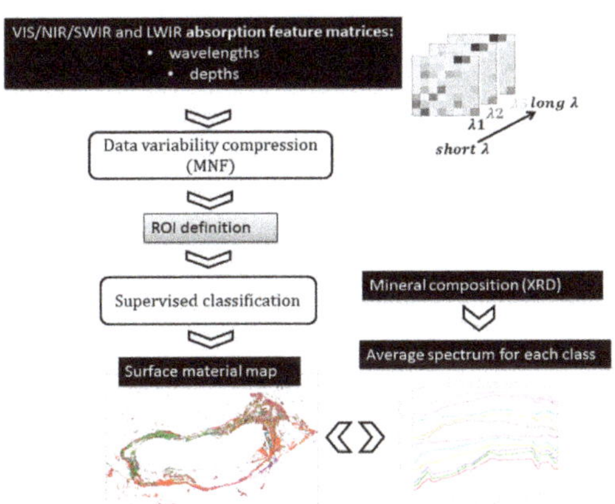

Figure 4. Scheme of the processing allowing integration of the absorption feature information detected in VIS/NIR/SWIR and LWIR data and further classification.

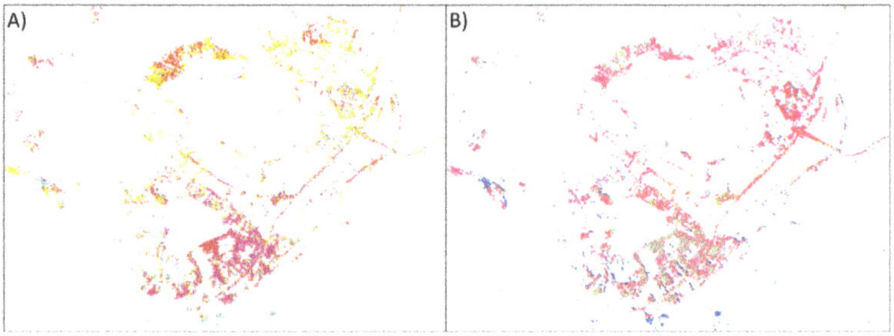

Figure 5. Lítov: MNF1/MNF2/MNF3 images displayed as RGB: (**A**) Scenario 1 (absorption feature parameters derived only from the HyMap data were used); (**B**) Scenario 2 (absorption feature parameters derived from both datasets, HyMap and AHS, were used), the MNF bands were further visually analysed to identify training areas (ROIs) representing different material/surfaces.

3. Results

3.1. Full-Range (VNIR, SWIR and LWIR) Absorption Wavelength Mapping and Further Classification

The intention was to detect the two most pronounced absorptions within the VIS/NIR, SWIR (both HyMap dataset) and LWIR (AHS dataset) regions. To validate if the wavelengths were detected correctly, the absorption wavelength positions were manually derived from 50 different image pixels of the HyMap and AHS data representing different material/surfaces, and they were compared with the absorption wavelength positions of corresponding pixels detected by QaunTools (Figure 6). Only a small deviation in the NIR range (around 0.900 µm) was detected due to the smoothing employed for the VIS/NIR spectral region prior to wavelength mapping. The wavelength positions of all the other absorption features in the SWIR and LWIR regions were detected without any shifts or errors.

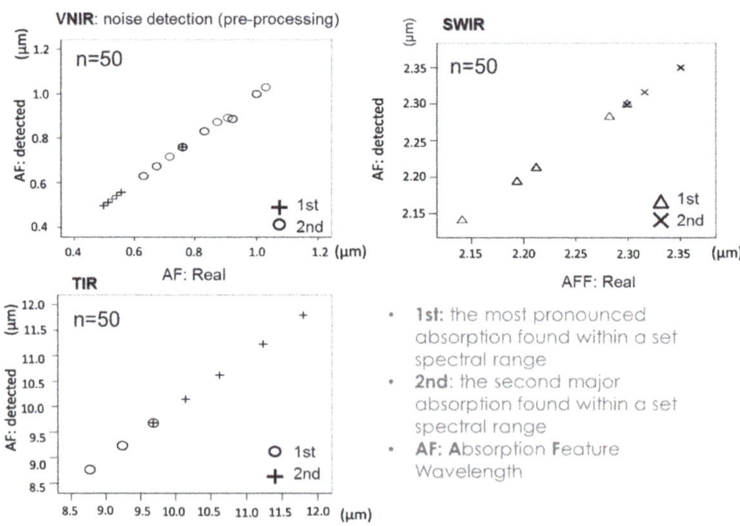

Figure 6. The manually derived absorption wavelength positions from 50 different image pixels of the the HyMap and the AHS image data representing different material/surfaces compared to the absorption wavelength positions derived automatically when using QaunTools.

As described in the Methods an MNF transformation was employed to the absorption feature mapping results, first when using those derived from the HyMap data only (Scenario 1) and secondly to those derived from both the HyMap and the AHS data (Scenario 2). For each scenario, the ROIs were constructed using RGB colour compositions of the first three MNF bands and these ROIs were consequently used to employ a supervised parallelepiped classification. As a result, eight classes, which were the same for both classifications (Scenarios 1 and 2), were mapped (Figures 7B and 8B). In the case of Scenario 2, where absorption feature parameters derived from both datasets (HyMap and AHS) were used for the consequent mineral mapping, it was possible to map the addition of two classes (Figures 7C and 8C: *Class 2 and *Class 9). Figures 9 and 10 show the average class spectrum derived from the HyMap data and the AHS data, respectively. The spectral property of these classes is further discussed and linked with the mineralogy in the following text.

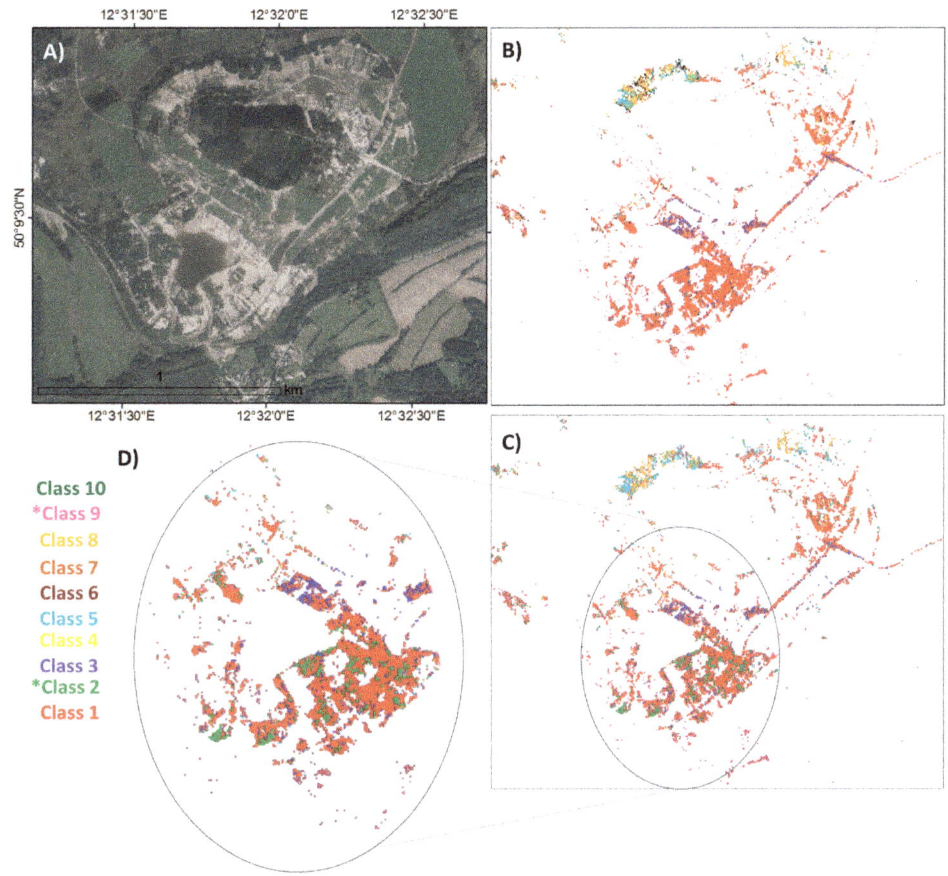

Figure 7. Lítov, mineral mapping: (**A**) orthophoto showing the Lítov dump; (**B**) classification using Scenario 1 (absorption feature parameters derived only from the HyMap data were used for the consequent mineral mapping), (**C**) classification using Scenario 2 (absorption feature parameters derived from both datasets, HyMap and AHS, were used for the consequent mineral mapping), (**D**) enlargement of the area of interest. The mapped classes correspond to the mineral classes in Table 1, *Class 2 and *Class 9: two additional classes mapped when using Scenario 2.

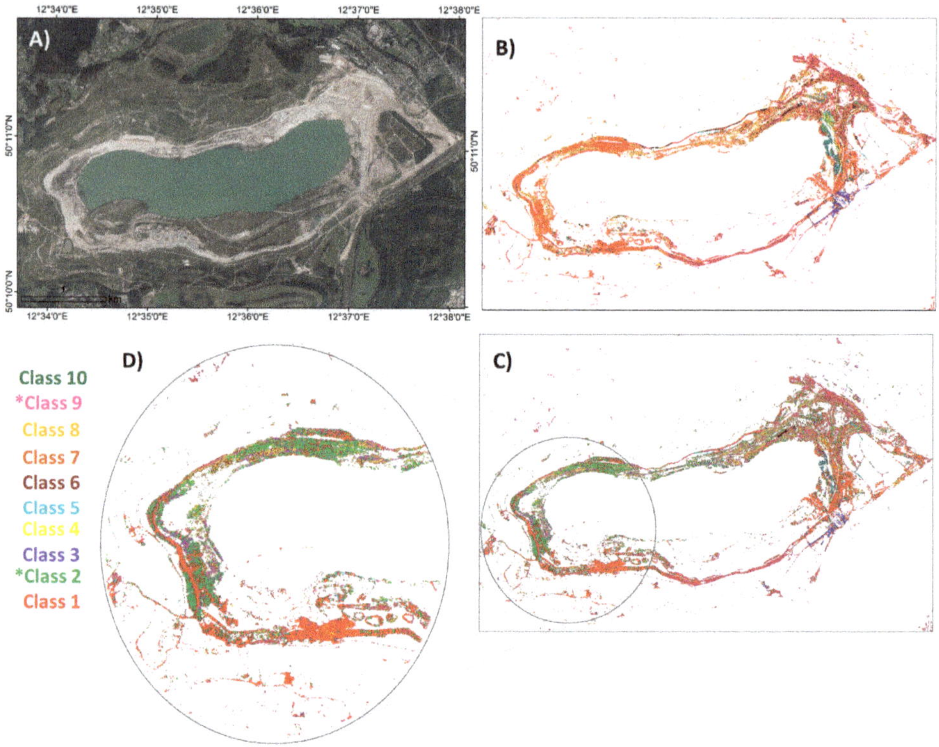

Figure 8. Medard Lake, mineral mapping: (**A**) orthophoto showing the Lítov dump, (**B**) classification using Scenario 1 (absorption feature parameters derived only from the HyMap data were used for the consequent mineral mapping), (**C**) classification using Scenario 2 (absorption feature parameters derived from both datasets, HyMap and AHS, were used for the consequent mineral mapping), (**D**) enlargement of the area of interest. The mapped classes correspond to the mineral classes in the Table 1, *Class 2 and *Class 9: two additional classes mapped when using Scenario 2.

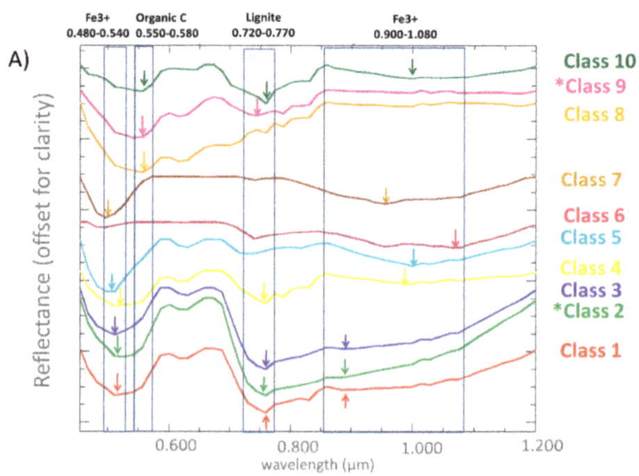

Figure 9. *Cont.*

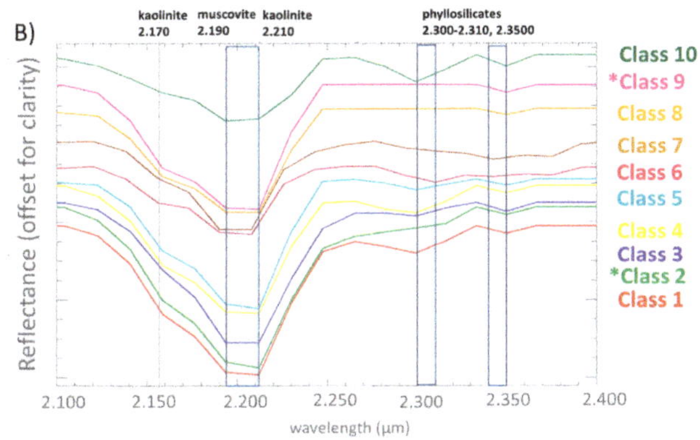

Figure 9. The average class spectrum derived from the HyMap data: (**A**) the VIS/NIR spectral range (0.450–1.200 μm); (**B**) the SWIR range (2.100–2.400 μm).

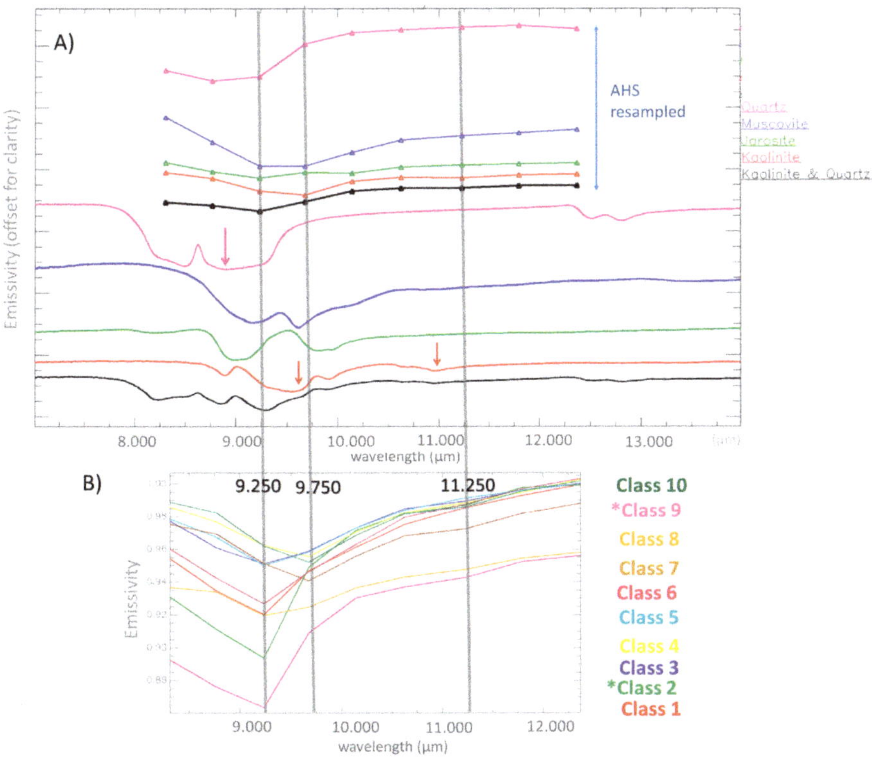

Figure 10. (**A**) Emissivity of some silicates is displayed using the Arizona University Spectral library [68], the original spectra are displayed together with the equivalent spectra resampled to the spectral resolution of the AHS data. (**B**) The average class spectrum derived from the AHS data. The mapped classes correspond to the mineral classes in Table 1, *Class 2 and *Class 9: two additional classes mapped when using Scenario 2.

3.2. Linking the Spectral and Mineral Properties

In the VIS/NIR region, the average class spectra (Figure 9A) reflect the variations in absorption features characterising diverse iron oxy-hydroxides (secondary minerals with Fe^{3+}) and the organic component (organic C and lignite). Secondary minerals with Fe^{3+} (hydroxysulfates and oxyhydroxides) exhibit absorption features around 0.500 μm and before 1.00 μm [29,69]. The absorption wavelengths and intensities of the absorption features in this region depend on the nature of the crystal field around the Fe atom and on the nature of the bonds around it, because the nature of the magnetic coupling between the Fe^{3+} ions (as influenced by the crystal field) facilitates the transition of electrons between energy states [70]. In addition, shifts in the wavelength positions can also reflect material mixing, when the Fe^{3+} secondary minerals coexist together or with different minerals and an organic component [6]. When interpreting the average class spectra (Figure 9A), it can be seen that Classes 1–3, 5 and 7 exhibit the typical absorption features of secondary Fe^{3+}-bearing minerals. The wavelength position of the first absorption shifts from 0.480 to 0.540 μm, while the second one varies between 0.900 and 1.080 nm. The shift of the second absorption indicates that Classes 1–3 represent the lithology where jarosite is present at higher amounts or coexist together with other Fe^{3+}-bearing minerals, as jarosite exhibits maximum absorption closer to 0.900 μm (shorter wavelengths). On the other hand, the oxyhydroxides (e.g., goethite and hematite) have the second absorption centred around 1.000 μm (longer wavelengths), therefore this mineral is present at high amounts in Classes 4–7 and 10. The absorptions characteristic of organic C and lignite are at 0.550–0.580 μm and 0.720–0.770 μm, respectively [39], and these two are the most visible in the spectra of Classes 4, 8, 9 and 10.

The overtones and combinations of the fundamental OH and H–O–H vibrations can be mainly observed in the SWIR (Figure 9B). In general, the OH combination bands occur due to the two Al cations in the octahedral sites near 2.210 μm; the spectra of kaolinite shows a distinct absorption doublet at 2.170 and 2.210 μm, whereas muscovite has the main absorption at the shorter wavelength (2.200 μm). The absorption around 2.300 μm characterises carbonates [29]; however, no carbonates were identified at primary, secondary or accessory abundances by the XRD analyses conducted for numerous samples collected in Sokolov. Therefore, in this case, additional smaller absorptions at the longer wavelength (around 2.300 and 2.350 μm) are characteristic of the AlFe–OH and Fe_2–OH combination bands in phyllosilicates [71]. Kaolinite dominates the spectra of Classes 1–2 and 5–6, whereas Classes 3–4 and 7–10 represent mixtures between kaolinite and muscovite as they have less pronounced absorption at 2.170 μm and flat absorption with the same absorption intensities for the wavelength 2.190 and 2.210 μm.

In the LWIR region absorption features, resulting from fundamental molecular vibration modes, show additional information about mineral constituents, such as Si-bearing minerals (mainly quartz and clay minerals). Quartz, due to the molecular vibrations of the Si–O stretching (reststrahlen bands) displays a broad emissivity doublet in the TIR between 8 and 10 μm [72]. In addition, it is possible to differentiate among diverse clay minerals (e.g., kaolinite, illite, and montmorillonite) using the LWIR range [34]. In Figure 10A the emissivity of some silicates is displayed using the Arizona University Spectral library [68], the original emissivity is displayed together with the equivalent emissivity resampled to the spectral resolution of the AHS data. It shows that the AHS data allows quartz, kaolinite or quartz and kaolinite mixtures to be detected. In the case of the AHS data (Figure 10B), quartz affects the emissivity of the bands centred at 9.25 μm (low emissivity) and 9.75 μm (high emissivity) as well as the slope of the emissivity between these two bands. The distinct absorption features detectable by the AHS bands placed at 9.75 μm and 11.2 μm characterise kaolinite. Muscovite exhibits a wide absorption between 9 and 10 μm, and in general lowers the emissivity between 10–12 μm. It can be concluded that Classes 1 and 9 represent lithologies with a dominant quartz content, Classes 1, 2, 5, 6 and 8 then represent a lithology that has both kaolinite and quartz coexisting together. On the other hand, Classes 4, 7 and 10 predominantly have kaolinite.

The classifications (Figures 7 and 8) were compared to the field documentation and XRD analysis of the samples that fall spatially within each class was carried out. Table 1 shows the mineralogy (XRD analysis) that characterises six out of 10 defined classes (Classes 1–3, 6, 7 and 10)l for Classes 4, 5, 8 and 9 there were no samples (no XRD analysis) that would fall within their spatial extent. When comparing the mineralogy determined by the XRD with the class spectral properties discussed above, there is good agreement. Kaolinite and quartz represent the dominant minerals for Classes 1–3, whereas jarosite is present together with hematite at secondary abundances, as well as muscovite. However, Class 2 has quartz dominating and was described as a quartz-rich crust developed on the tuffs; on the other hand, Class 3 represents less weathered tuffs exposed by erosion. Class 6 represents the fresh clays of the Cypris formation, where kaolinite is the dominant mineral, followed by quartz and muscovite. Class 7 represents the material of backfill overburden, which was described as quartz-rich hard pack material with a clay matrix also containing lignite fragments. Class 10 represents weathered tuffs in which quartz, muscovite and kaolinite are the dominant, minerals whereas lignite and hematite are present at secondary abundances. Although there were no XRD analyses available for Classes 4, 5, 8 and 9, a mineral description (Table 1) was added; however, it should be emphasised that it is based on the interpretation of the class spectral property (Figures 9 and 10) and in this case it was not possible to differentiate between primary and secondary/accessory abundances of different mineral phases.

Table 1. Map classes from Figures 7–10 compared to the XRD analysis and field documentation of the samples that fall spatially within each class. For Classes 4, 5, 8 and 9 there were no XRD analyses available; however, the mineral description based on the interpretation of the spectral property was added.

Class	Primary Minerals	Secondary and Accessory Minerals	Description	Name
Class 1	kaolinite (60–80%), quartz (up to 10%)	jarosite, hematite, Muscovite, lignite fragments	weathered tuffs on the surface	tuffs
Class 2	quartz > kaolinite	jarosite, hematite, Muscovite	crust developed on the surface of the tuffs	quartz-rich crust
Class 3	kaolinite > quartz	jarosite, hematite, muscovite lignite, pyrite	fresh layer of tuffs exposed by erosion	tuffs
Class 4	kaolinite, quartz, lignite, muscovite		No XRD analysis available	-
Class 5	kaolinite, quarz, hematite, muscovite		No XRD analysis available	-
Class 6	kaolinite	Quartz, muscovite	well-laminated clays with kaolinite content dominating and other admixtures	Cypris clays
Class 7	quartz (>50%), clay content (10–15%)	lignite	back-fill overburden: quartz-rich hard pack with clay matrix, lignite fragments	Back-fill overburden
Class 8	organic C, kaolinite, quartz		no XRD analysis available	soil substrate
Class 9	quartz, kaolinite, lignite, muscovite		no XRD analysis available	-
Class 10	quartz, muscovite, kaolinite	lignite, hematite	weathered Tuffs	tuffs with lignite and hematite

4. Discussion

Absorption feature parameters—wavelength position and depth—are the most essential information used in spectroscopy; to put it simply, the maximum absorption wavelength position defines what material it is, whereas absorption depth defines its relative abundance. Different absorption feature mapping/matching techniques have been used by the remote sensing community since hyperspectral data were made available. Among these the spectral feature fitting [73] or its improved version—multi-range spectral feature fitting (MRSFF)—are frequently used [74,75]. From the recently developed toolboxes integrating absorption feature matching techniques the Tetracoder [76] or EnGeoMAP 2.0 toolbox [77] should be listed. These approaches are based on the common principle that they compare and statistically assess the fit of the image spectra to the reference spectra. The reference spectra, called endmembers, are scaled to match the image spectra, and can be either laboratory or field spectral measurements or can be extracted directly from the image (image endmembers). Basically, these classifiers require a routine whereby the endmembers (reference spectra) need to be defined prior to spectral mapping. Moreover, if laboratory or field spectra are used, the successful definition of such endmembers usually requires prior knowledge of the material composition and its spatial distribution within the area of interest. If image endmembers are to be used, and an expert with a background in spectroscopy is required to perform image analysis such as pixel purity analysis (PPI, [78,79]).

In contrast, the approach used here does not require a prior definition of the endmembers or any knowledge of the site conditions; this is a clear advantage. Instead, the absorption wavelength positions and depths of the major absorption features present in different spectral ranges (VNIR/SWIR/LWIR) are extracted automatically and integrated into new raster datasets (multiple absorption feature wavelength and depth matrices). After compressing the main data variability of these new multi-band datasets, using, for instance, an MNF transformation, it is possible to assess the material spatial variability, define training areas (ROIs) and employ a supervised classification. The approach used here is thus unsupervised at the beginning, as no data or knowledge is required prior to absorption feature mapping. Throughout the automatic processing, there is a point where material variability is visualised in a spatial context; afterwards, the training areas (ROIs) for supervised classification are defined. The selection of the spectral ranges under analysis and the definition of the ROIs are the only expert-dependent parts of the analysis. It is thus recommended to divide spectral ranges into VIS/NIR, SWIR and LWIR regions where the diverse mineral groups exhibit distinct absorptions. The definition of the ROIs is rather intuitive as the expert interprets different colour clusters within the image.

Previously, techniques that utilise different types of interpolations for estimating absorption feature wavelength and depth were proposed. Van der Meer [80] proposed a simple linear interpolation technique in order to derive absorption-band position, depth and asymmetry from hyperspectral image. Such parameters were used to interpret the data in terms of the known alteration phases or to estimate heavy metal contents [81]. Rodger et al. [82] proposed a simple quadratic method (SQM) to estimate the wavelengths of absorption features in the shortwave infrared (SWIR) spectral region. The SQM method was tested using spectral data convolved to four different instrument configurations differing in sampling regimes and spectral resolutions. The SQM method was found to estimate feature wavelengths within a reasonable accuracy and to perform well even in noisy environments. Ruitenbeek et al. [83] mapped the wavelength position of the deepest absorption features between 2.100 and 2.400 μm using a second-order polynomial fitting. They concluded that mapping the wavelength position of absorption features between 2.100 and 2.400 μm provided a new method for exploratory analysis of the surface mineralogy and that it would be particularly useful in areas where field validation is sparse and imagery contains shallow spectral absorption features. Moreover, Van der Meer et al. [84] tested two approaches—the 'Wavelength Mapper' [83] and the QUANTools [62] (also used for this study)—and demonstrated for the Rodalquilar epithermal system that deriving absorption feature characteristics, such as the wavelength position and the depth, can be directly linked to mineral type and abundance, and, even more, to subtle changes in mineral chemical composition.

However, as seen from the literature, this is the first demonstration of how the multiple absorption features, and their respective parameters, can be extracted automatically from different spectral ranges (VIS/NIR/SWIR and LWIR), and, furthermore, how, when using this rather simple approach, it is possible to successfully ingrate optical (VNIR and SWIR) and thermal (LWIR) spectral information, gathered in this case by two different sensors. It is also demonstrated how further integration can lead to more complex mineral classification. This is a crucial and highly relevant issue nowadays, when, in addition to the optical sensors, the sensors acquiring the data in the LWIR range are more often available and used (e.g., TASI, AISA OWEL) and even more for the future, when a hyperspectral satellite collecting spectral band through the VIS/NIR/SWIR and LWIR will be operating in orbit (e.g., HySPIRI).

So far the different spectral ranges are more frequently analysed and interpreted separately, without combining the full range of information (VIS/NIR/SWIR/LWIR) into a single integrated data product. Too few approaches have been proposed allowing for real optical and thermal data integration; still, they require an expert decision to be made prior to spectral mapping (e.g., endmember definition). As previously explained, the approach used here does not require any prior definition of the endmembers; moreover, there is no need for any knowledge or ground data from the site under study.

In the approach used here, the level of noise present in the image datasets can be a limitation. However, in an example of HyMap VIS/NIR data, it was demonstrated that this problem can be minimised by employing spectral smoothing and that it can be tailored specifically to the level of noise present in the data. In addition, there might be other variables that may create false alarms, such as varying topography causing brightness differences across the image produced by shadow and slope variations, varying mineral particle size and soil moisture content. QUANTools process the data in such a way that at the beginning they are normalised by employing the continuum removal method, which helps to remove the effect of scattering. Then the absorption feature wavelengths are mapped and, as long as the noisy/false absorptions are eliminated, the absorption wavelength positions do not tend to change spectral locations as regards shadows or slope variations. The other absorption feature parameter—absorption depth—reflects the material quantity but is also sensitive to the sizes of mineral particles. Therefore, varying grain size is also part of a final classification. It is thus important to decide if the grain size is also a criterion for material mapping or not. If only the chemical composition of the targets is requested, then it is possible to use only absorption feature wavelength matrices in a classification. In this study, 2010 HyMap and 2011 AHS image data were used to classify two sites, Lítov and Medard, which faced no changes regarding the relief and material transport between 2010 and 2011; however, some differences regarding changes in microtopography or moisture content may have been present. Both absorption feature parameters—wavelength and depth—were used for classifications and brought results that were in good agreement with the XRD analysis, showing that, at a general level, this approach allows diverse minerals to be mapped, including ones that exhibit multiple absorption features though the VIS/NIR/SWIR and LWIR ranges. There are also other useful absorption feature parameters, such as shoulder positions, symmetry or width, that can be used for gathering the information on minor material components, featureless parameters (e.g., heavy metals) or chemical processes/changes. Mapping these additional absorption feature parameters will be a subject for future QUNATools development.

Using this approach, it was possible to integrate the absorption feature information derived from the VIS/NIR/SWIR regions (HyMap data), together with the absorption feature information derived from the LWIR region (AHS data). This integration led to a mineral classification that differentiated between the presence and abundance of diverse Fe^{3+}-bearing minerals and phyllosilicates as well as lignite and quartz contents. In addition to HyMap reflectance, the AHS emissivity data allowed a better discrimination between a quartz-dominating crust and substrates (Classes 2 and 9) from the other classes where quartz did not have such dominating abundances and was present together with other mineral phases at primary or secondary abundance (e.g., Classes 1 and 7).

5. Conclusions

In this study, it was demonstrated how the multiple absorption features, respectively their parameters such as wavelengths and depths, can be automatically extracted from different spectral ranges and further integrated into one raster dataset. The absorption feature information gathered from the different spectral-ranges of the two different sensors (HyMap and AHS) was integrated and led to a mineral classification that differentiated between the diverse Fe^{3+}-bearing minerals and phyllosilicates as well as lignite and quartz contents and overall resulted in a more complex mineral/lithology classification. This is a crucial and highly relevant issue nowadays, when, in addition to optical sensors, sensors acquiring data in the LWIR range are available and more often used (e.g., TASI, AISA OWEL); and even more for the future, when a hyperspectral satellite collecting a spectral band through the VIS/NIR/SWIR and LWIR will be operating in orbit (e.g., HySPIRI).

It can be concluded that:

- the approach used here does not require prior definition of the endmembers; moreover, there is no need for prior knowledge or data on the specific conditions
- QUANTools, the new toolbox developed, allows automatic and errorless multiple-absorption feature parameters extraction from different spectral ranges, and these parameters can be further integrated into one product, which can consequently be successfully used for mineral mapping/classification
- this multi-range spectral integration leads to more complex mineral/lithology classification
- the approach can be used to integrate the spectral information acquired by different sensors (e.g., HyMap and AHS).

Acknowledgments: Development of the toolbox (QUANTools) was supported by a grant from the Ministry of Education Youth and Sports (LH 13266). The study and manuscript writing were done under the support of Czech Science Foundation grant 17-05743S. The AHS data were acquired under the EUFAR programme (DeMinTIR project). The QUANTools toolbox is available for free (February 2016); if interested, read more at http://www.cgs-rs.g6.cz/hyperalgo.html or contact quantoolscgs@gmail.com.

Author Contributions: Veronika Kopačková designed the study, performed the analysis and wrote the paper, Lucie Koucká programmed the 'QUANTools'; both authors contributed to the interpretation of the result and further manuscript revisions.

Conflicts of Interest: The authors declare no conflict of interest.

References

1. Van der Meer, F.D.; van der Werff, H.M.; Van Ruitenbeek, F.J.; Hecker, C.A.; Bakker, W.H.; Noomen, M.F.; van der Meijde, M.; Carranza, E.J.M.; De Smeth, J.B.; Woldai, T. Multi-and hyperspectral geologic remote sensing: A review. *Int. J. Appl. Earth Obs. Geoinf.* **2012**, *14*, 112–128. [CrossRef]
2. Chabrillat, S.; Goetz, A.F.; Krosley, L.; Olsen, H.W. Use of hyperspectral images in the identification and mapping of expansive clay soils and the role of spatial resolution. *Remote Sens. Environ.* **2002**, *82*, 431–445. [CrossRef]
3. Escribano, P.; Palacios-Orueta, A.; Oyonarte, C.; Chabrillat, S. Spectral properties and sources of variability of ecosystem components in a Mediterranean semiarid environment. *J. Arid Environ.* **2010**, *74*, 1041–1051. [CrossRef]
4. Haubrock, S.N.; Chabrillat, S.; Kuhnert, M.; Hostert, P.; Kaufmann, H. Surface soil moisture quantification and validation based on hyperspectral data and field measurements. *J. Appl. Remote Sens.* **2008**, *2*, 023552. [CrossRef]
5. Kokaly, R.F.; Despain, D.G.; Clark, R.N.; Livo, K.E. Mapping vegetation in Yellowstone National Park using spectral feature analysis of AVIRIS data. *Remote Sens. Environ.* **2003**, *84*, 437–456. [CrossRef]
6. Kopačková, V. Using multiple spectral feature analysis for quantitative pH mapping in a mining environment. *Int. J. Appl. Earth Obs. Geoinf.* **2014**, *28*, 28–42. [CrossRef]
7. Cudahy, T.; Hewson, R. ASTER geological case histories: Porphyry-skarnepithermal, iron oxide Cu-Au and Broken hill Pb–Zn–Ag. In Proceedings of the Annual General Meeting of the Geological Remote Sensing Group 'ASTER Unveiled', London, UK, 6–7 December 2002.
8. Kruse, F.A.; Perry, S.L. Mineral mapping using simulated Worldview-3 short-wave-infrared imagery. *Remote Sens.* **2013**, *5*, 2688–2703. [CrossRef]
9. Van der Meer, F.D.; Van der Werff, H.M.A.; Van Ruitenbeek, F.J.A. Potential of ESA's Sentinel-2 for geological applications. *Remote Sens. Environ.* **2014**, *148*, 124–133. [CrossRef]
10. Van der Werff, H.; van der Meer, F. Sentinel-2 for mapping iron absorption feature parameters. *Remote Sens.* **2015**, *7*, 12635–12653. [CrossRef]
11. Van der Werff, H.; van der Meer, F. Sentinel-2A MSI and Landsat 8 OLI provide data continuity for geological remote sensing. *Remote Sens.* **2016**, *8*, 883. [CrossRef]
12. Mielke, C.; Boesche, N.K.; Rogass, C.; Kaufmann, H.; Gauert, C.; de Wit, M. Spaceborne mine waste mineralogy monitoring in South Africa, applications for modern push-broom missions: Hyperion/OLI and EnMAP/Sentinel-2. *Remote Sens.* **2014**, *6*, 6790–6816. [CrossRef]

13. De Morais, M.C.; Junior, P.P.M.; Paradella, W.R. Multi-scale approach using remote sensing images to characterize the iron deposit N1 influence areas in Carajás Mineral Province (Brazilian Amazon). *Environ. Earth Sci.* **2012**, *66*, 2085–2096. [CrossRef]
14. Khalifa, I.H.; Arnous, M.O. Assessment of hazardous mine waste transport in west central Sinai, using remote sensing and GIS approaches: A case study of Um Bogma area, Egypt. *Arabian J. Geosci.* **2012**, *5*, 407–420. [CrossRef]
15. Matejicek, L.; Kopackova, V. Changes in croplands as a result of large scale mining and the associated impact on food security studied using time-series Landsat images. *Remote Sens.* **2010**, *2*, 1463–1480. [CrossRef]
16. Kopačková, V.; Chevrel, S.; Bourguignon, A.; Rojík, P. Application of high altitude and ground-based spectroradiometry to mapping hazardous low-pH material derived from the Sokolov open-pit mine. *J. Maps* **2012**, *8*, 220–230. [CrossRef]
17. Fernandes, G.W.; Goulart, F.F.; Ranieri, B.D.; Coelho, M.S.; Dales, K.; Boesche, N.; Bustamante, M.; Carvalho, F.A.; Carvalho, D.C.; Dirzo, R.; et al. Deep into the mud: Ecological and socio-economic impacts of the dam breach in Mariana, Brazil. *Natureza Conservação* **2016**, *14*, 35–45. [CrossRef]
18. Rodríguez-Hernández, A.; Briones-Gallardo, R.; Razo, I.; Noyola-Medrano, C.; Lázaro, I. Processing Methodology Based on ASTER Data for Mapping Mine Waste Dumps in a Semiarid Polysulphide Mine District. *Can. J. Remote Sens.* **2016**, *42*, 643–655. [CrossRef]
19. Davies, G.E.; Calvin, W.M. Mapping acidic mine waste with seasonal airborne hyperspectral imagery at varying spatial scales. *Environ. Earth Sci.* **2017**, *76*, 432. [CrossRef]
20. GMES Sentinel-2 Mission Requirements Document. European Space Agency. Available online: http://esamultimedia.esa.int/docs/GMES/Sentinel-2_MRD.pdf (accessed on 1 September 2017).
21. Galeazzi, C.; Sacchetti, A.; Cisbani, A.; Babini, G. The PRISMA program. In Proceedings of the 2008 IEEE International Geoscience and Remote Sensing Symposium, Boston, MA, USA, 6–11 July 2008; Volume 4, p. IV-105.
22. Amato, U.; Antoniadis, A.; Carfora, M.F.; Colandrea, P.; Cuomo, V.; Franzese, M.; Pignatti, S.; Serio, C. Statistical classification for assessing PRISMA hyperspectral potential for agricultural land use. *IEEE J. Sel. Top. Appl. Earth Obs. Remote Sens.* **2013**, *6*, 615–625. [CrossRef]
23. Kaufmann, H.; Segl, K.; Guanter, L.; Hofer, S.; Foerster, K.P.; Stuffler, T.; Mueller, A.; Richter, R.; Bach, H.; Hostert, P.; et al. Environmental mapping and analysis program (EnMAP)-Recent advances and status. In Proceedings of the 2008 IEEE International Geoscience and Remote Sensing Symposium, Boston, MA, USA, 6–11 July 2008; Volume 4, p. IV-109.
24. Guanter, L.; Kaufmann, H.; Segl, K.; Foerster, S.; Rogass, C.; Chabrillat, S.; Kuester, T.; Hollstein, A.; Rossner, G.; Chlebek, C.; et al. The EnMAP spaceborne imaging spectroscopy mission for earth observation. *Remote Sens.* **2015**, *7*, 8830–8857. [CrossRef]
25. Roberts, D.A.; Quattrochi, D.A.; Hulley, G.C.; Hook, S.J.; Green, R.O. Synergies between VSWIR and TIR data for the urban environment: An evaluation of the potential for the Hyperspectral Infrared Imager (HyspIRI) Decadal Survey mission. *Remote Sens. Environ.* **2012**, *117*, 83–101. [CrossRef]
26. Hunt, G.R. Spectral signatures of particulate minerals in the visible and near infrared. *Geophysics* **1977**, *42*, 501–513. [CrossRef]
27. Richter, N.; Jarmer, T.; Chabrillat, S.; Oyonarte, C.; Hostert, P.; Kaufmann, H. Free iron oxide determination in Mediterranean soils using diffuse reflectance spectroscopy. *Soil Sci. Soc. Am. J.* **2009**, *73*, 72–81. [CrossRef]
28. Montero, I.C.; Brimhall, G.H.; Alpers, C.N.; Swayze, G.A. Characterization of waste rock associated with acid drainage at the Penn Mine, California, by ground-based visible to short-wave infrared reflectance spectroscopy assisted by digital mapping. *Chem. Geol.* **2005**, *215*, 453–472. [CrossRef]
29. Clark, R.N.; King, T.V.; Klejwa, M.; Swayze, G.A.; Vergo, N. High spectral resolution reflectance spectroscopy of minerals. *J. Geophys. Res. Solid Earth* **1990**, *95*, 12653–12680. [CrossRef]
30. Gobrecht, A.; Roger, J.M.; Bellon-Maurel, V. Major issues of diffuse reflectance NIR spectroscopy in the specific context of soil carbon content estimation: A review. *Adv. Agron.* **2014**, *123*, 145–175.
31. Sørensen, L.K.; Dalsgaard, S. Determination of clay and other soil properties by near infrared spectroscopy. *Soil Sci. Soc. Am. J.* **2005**, *69*, 159–167. [CrossRef]
32. Gillespie, A.; Rokugawa, S.; Matsunaga, T.; Cothern, J.S.; Hook, S.; Kahle, A.B. A temperature and emissivity separation algorithm for Advanced Spaceborne Thermal Emission and Reflection Radiometer (ASTER) images. *IEEE Trans. Geosci. Remote Sens.* **1998**, *36*, 1113–1126. [CrossRef]

33. Van der Meijde, M.; Knox, N.M.; Cundill, S.L.; Noomen, M.F.; Van der Werff, H.M.A.; Hecker, C. Detection of hydrocarbons in clay soils: A laboratory experiment using spectroscopy in the mid-and thermal infrared. *Int. J. Appl. Earth Obs. Geoinf.* **2013**, *23*, 384–388. [CrossRef]
34. Eisele, A.; Chabrillat, S.; Hecker, C.; Hewson, R.; Lau, I.C.; Rogass, C.; Segl, K.; Cudahy, T.J.; Udelhoven, T.; Hostert, P.; et al. Advantages using the thermal infrared (TIR) to detect and quantify semi-arid soil properties. *Remote Sens. Environ.* **2015**, *163*, 296–311. [CrossRef]
35. Yitagesu, F.A.; Van der Meer, F.; Van der Werff, H.; Hecker, C. Spectral characteristics of clay minerals in the 2.5–14μm wavelength region. *Appl. Clay Sci.* **2011**, *53*, 581–591. [CrossRef]
36. Hecker, C.; van der Meijde, M.; van der Meer, F.D. Thermal infrared spectroscopy on feldspars—Successes, limitations and their implications for remote sensing. *Earth-Sci. Rev.* **2010**, *103*, 60–70. [CrossRef]
37. Horta, A.; Malone, B.; Stockmann, U.; Minasny, B.; Bishop, T.F.A.; McBratney, A.B.; Pallasser, R.; Pozza, L.; et al. Potential of integrated field spectroscopy and spatial analysis for enhanced assessment of soil contamination: A prospective review. *Geoderma* **2015**, *241*, 180–209. [CrossRef]
38. Vohland, M.; Ludwig, M.; Thiele-Bruhn, S.; Ludwig, B. Determination of soil properties with visible to near-and mid-infrared spectroscopy: Effects of spectral variable selection. *Geoderma* **2014**, *223*, 88–96. [CrossRef]
39. Kopačková, V.; Ben-Dor, E.; Carmon, N.; Notesco, G. Modelling Diverse Soil Attributes with Visible to Longwave Infrared Spectroscopy Using PLSR Employed by an Automatic Modelling Engine. *Remote Sens.* **2017**, *9*, 134. [CrossRef]
40. McDowell, M.L.; Kruse, F.A. Enhanced Compositional Mapping through Integrated Full-Range Spectral Analysis. *Remote Sens.* **2016**, *8*, 757. [CrossRef]
41. Kruse, F.A. Integrated visible and near-infrared, shortwave infrared, and longwave infrared full-range hyperspectral data analysis for geologic mapping. *J. Appl. Remote Sens.* **2015**, *9*, 096005. [CrossRef]
42. Notesco, G.; Kopačková, V.; Rojík, P.; Schwartz, G.; Livne, I.; Dor, E.B. Mineral classification of land surface using multispectral LWIR and hyperspectral SWIR remote-sensing data. A case study over the Sokolov lignite open-pit mines, the Czech Republic. *Remote Sens.* **2014**, *6*, 7005–7025. [CrossRef]
43. Feng, J.; Rogge, D.; Rivard, B. Comparison of lithological mapping results from airborne hyperspectral VNIR-SWIR, LWIR and combined data. *Int. J. Appl. Earth Obs. Geoinf.* **2017**, in press. [CrossRef]
44. Kruse, F.A.; Lefkoff, A.B.; Boardman, J.W.; Heidebrecht, K.B.; Shapiro, A.T.; Barloon, P.J.; Goetz, A.F.H. The spectral image processing system (SIPS)—Interactive visualization and analysis of imaging spectrometer data. In Proceedings of the AIP Conference Proceedings, St. Louis, MO, USA, 10 August 1993; pp. 192–201. [CrossRef]
45. Debba, P.; Van Ruitenbeek, F.J.A.; Van Der Meer, F.D.; Carranza, E.J.M.; Stein, A. Optimal field sampling for targeting minerals using hyperspectral data. *Remote Sens. Environ.* **2005**, *99*, 373–386. [CrossRef]
46. Haest, M.; Cudahy, T.; Laukamp, C.; Gregory, S. Quantitative mineralogy from infrared spectroscopic data. I. Validation of mineral abundance and composition scripts at the rocklea channel iron deposit in Western Australia. *Econ. Geol.* **2012**, *107*, 209–228. [CrossRef]
47. Settle, J.J.; Drake, N.A. Linear mixing and the estimation of ground cover proportions. *Int. J. Remote Sens.* **1993**, *14*, 1159–1177. [CrossRef]
48. Rogge, D.M.; Rivard, B.; Zhang, J.; Feng, J. Iterative spectral unmixing for optimizing per-pixel endmember sets. *IEEE Trans. Geosci. Remote Sens.* **2006**, *44*, 3725–3736. [CrossRef]
49. Rojík, P. New stratigraphic subdivision of the Tertiary in the Sokolov Basin in Northwestern Bohemia/Nové stratigrafické clenení terciéru sokolovské pánve v sz. Cechách. *J. Geosci.* **2004**, *49*, 173.
50. Kopačková, V.; Chevrel, S.; Bourguignon, A. Spectroscopy as a tool for geochemical modeling. In Proceedings of the SPIE, Prague, Czech Republic, 19–22 September 2011. [CrossRef]
51. Kopačková, V.; Chevrel, S.; Bourguignon, A.; Rojík, P. Mapping hazardous low-pH material in mining environment: Multispectral and hyperspectral aproaches. In Proceedings of the 2012 IEEE International on Geoscience and Remote Sensing Symposium (IGARSS), Munich, Germany, 22–27 July 2012; pp. 2695–2698.
52. Götze, C.; Beyer, F.; Gläßer, C. Pioneer vegetation as an indicator of the geochemical parameters in abandoned mine sites using hyperspectral airborne data. *Environ. Earth Sci.* **2016**, *75*, 613. [CrossRef]
53. Kopačková, V.; Hladíková, L. Applying Spectral Unmixing to Determine Surface Water Parameters in a Mining Environment. *Remote Sens.* **2014**, *6*, 11204–11224. [CrossRef]

54. Mišurec, J.; Kopačková, V.; Lhotáková, Z.; Hanuš, J.; Weyermann, J.; Entcheva-Campbell, P.; Albrechtová, J. Utilization of hyperspectral image optical indices to assess the Norway spruce forest health status. *J. Appl. Remote Sens.* **2012**, *6*, 063545.
55. Kopačková, V.; Mišurec, J.; Lhotáková, Z.; Oulehle, F.; Albrechtová, J. Using multi-date high spectral resolution data to assess the physiological status of macroscopically undamaged foliage on a regional scale. *Int. J. Appl. Earth Obs. Geoinf.* **2014**, *27*, 169–186. [CrossRef]
56. Lhotáková, Z.; Brodský, L.; Kupková, L.; Kopačková, L.; Potůčková, M.; Mišurec, J.; Klement, A.; Kovářová, M.; Albrechtová, J. Detection of multiple stresses in Scots pine growing at post-mining sites using visible to near-infrared spectroscopy. *Environ. Sci. Process. Impacts* **2013**, *15*, 2004–2015.
57. Richter, R.; Schläpfer, D. Geo-atmospheric processing of airborne imaging spectrometry data. Part 2: Atmospheric/topographic correction. *Int. J. Remote Sens.* **2002**, *23*, 2631–2649. [CrossRef]
58. Adler-Golden, S.M.; Matthew, M.W.; Bernstein, L.S.; Levine, R.Y.; Berk, A.; Richtsmeier, S.C.; Acharya, P.K.; Anderson, G.P.; Felde, J.W.; Gardner, J.A.; et al. Atmospheric correction for short-wave spectral imagery based on MODTRAN 4. In Proceedings of the International Society for Optical Engineering, Denver, CO, USA, 21–22 July 1999.
59. Adar, S.; Shkolnisky, Y.; Notesco, G.; Ben-Dor, E. Using Visible Spectral Information to Predict Long-Wave Infrared Spectral Emissivity: A Case Study over the Sokolov Area of the Czech Republic with an Airborne Hyperspectral Scanner Sensor. *Remote Sens.* **2013**, *5*, 5757–5782. [CrossRef]
60. Miguel, E.D.; Jiménez, M.; Pérez, I.; Cámara, Ó.G. AHS and CASI processing for the REFLEX remote sensing campaign: Methods and results. *Acta Geophys.* **2015**, *63*, 1485–1498. [CrossRef]
61. Yuan, H.; Bish, D.L. NEWMOD+, a new version of the NEWMOD program for interpreting X-ray powder diffraction patterns from interstratified clay minerals. *Clays Clay Miner.* **2010**, *58*, 318–326. [CrossRef]
62. Kopačková, V.; Koucká, L. Mineral Mapping Based on Automatic Detection of Multiple Absorption Features. Available online: http://www.eproceedings.org/static/vol13_S1/13_S1_kopackova2.pdf (accessed on 3 September 2017).
63. Clark, R.N.; Roush, T.L. Reflectance spectroscopy: Quantitative analysis techniques for remote sensing applications. *J. Geophys. Res. Solid Earth* **1984**, *89*, 6329–6340. [CrossRef]
64. Green, A.A.; Berman, M.; Switzer, P.; Craig, M.D. A transformation for ordering multispectral data in terms of image quality with implications for noise removal. *IEEE Trans. Geosci. Remote Sens.* **1988**, *26*, 65–74. [CrossRef]
65. Richards, J.A. *Remote Sensing Digital Image Analysis*; Springer: Berlin, Germany, 1999; Volume 3.
66. Hudak, A.T.; Brockett, B.H. Mapping fire scars in a southern African savannah using Landsat imagery. *Int. J. Remote Sens.* **2004**, *25*, 3231–3243. [CrossRef]
67. Manzo, C.; Mei, A.; Salvatori, R.; Bassani, C.; Allegrini, A. Spectral modelling used to identify the aggregates index of asphalted surfaces and sensitivity analysis. *Constr. Build. Mater.* **2014**, *61*, 147–155. [CrossRef]
68. The Spectral Library Hosted by the Mars Space Flight Facility at Arizona State University. Available online: http://speclib.asu.edu (accessed on 20 May 2017).
69. Murphy, R.J.; Monteiro, S.T. Mapping the distribution of ferric iron minerals on a vertical mine face using derivative analysis of hyperspectral imagery (430–970nm). *ISPRS J. Photogramm. Remote Sens.* **2013**, *75*, 29–39. [CrossRef]
70. Sherman, D.M.; Waite, T.D. Electronic spectra of Fe^{3+} oxides and oxide hydroxides in the near IR to near UV. *Am. Mineral.* **1985**, *70*, 1262–1269.
71. Bishop, J.L.; Lane, M.D.; Dyar, M.D.; Brown, A.J. Reflectance and emission spectroscopy study of four groups of phyllosilicates: Smectites, kaolinite-serpentines, chlorites and micas. *Clay Miner.* **2008**, *43*, 35–54. [CrossRef]
72. Salisbury, J.W.; D'Aria, D.M. Emissivity of terrestrial materials in the 8–14 μm atmospheric window. *Remote Sens. Environ.* **1992**, *42*, 83–106. [CrossRef]
73. Clark, R.N.; Swayze, G.A.; Gallagher, A.; Gorelick, N.; Kruse, F.A. Mapping with imaging spectrometer data using the complete band shape least-squares algorithm simultaneously fit to multiple spectral features from multiple materials. In Proceedings of the Third Airborne Visible/Infrared Imaging Spectrometer (AVIRIS) Workshop; Jet Propulsion Laboratory: La Cañada Flintridge, CA, USA, 1991; Volume 42, pp. 2–3.
74. Pan, Z.; Huang, J.; Wang, F. Multi range spectral feature fitting for hyperspectral imagery in extracting oilseed rape planting area. *Int. J. Appl. Earth Obs. Geoinf.* **2013**, *25*, 21–29. [CrossRef]

75. Shanmugam, S.; SrinivasaPerumal, P. Spectral matching approaches in hyperspectral image processing. *Int. J. Remote Sens.* **2014**, *35*, 8217–8251. [CrossRef]
76. Clark, R.N.; Swayze, G.A.; Livo, K.E.; Kokaly, R.F.; Sutley, S.J.; Dalton, J.B.; McDougal, R.R.; Gent, C.A. Imaging spectroscopy: Earth and planetary remote sensing with the USGS Tetracorder and expert systems. *J. Geophys. Res. Planets* **2003**. [CrossRef]
77. Mielke, C.; Rogass, C.; Boesche, N.; Segl, K.; Altenberger, U. EnGeoMAP 2.0—Automated hyperspectral mineral identification for the german EnMAP space mission. *Remote Sens.* **2016**, *8*, 127. [CrossRef]
78. Boardman, J.W.; Kruse, F.A. Automated spectral analysis: A geological example using AVIRIS data, north Grapevine Mountains, Nevada. In Proceedings of the Thematic Conference on Geologic Remote Sensing, San Antonio, TX, USA, 9–12 May 1994; p. I-407.
79. Kruse, F.A.; Richardson, L.L.; Ambrosia, V.G. Techniques Developed for Geologic Analysis of Hyperspectral Data Applied to Near-Shore Hyperspectral Ocean Data. Available online: http://www.w.hgimaging.com/PDF/Kruse_erim97_marine.pdf (accessed on 3 September 2017).
80. Van Der Meer, F. Analysis of spectral absorption features in hyperspectral imagery. *Int. J. Appl. Earth Obs. Geoinf.* **2004**, *5*, 55–68. [CrossRef]
81. Choe, E.; van der Meer, F.; van Ruitenbeek, F.; van der Werff, H.; de Smeth, B.; Kim, K.W. Mapping of heavy metal pollution in stream sediments using combined geochemistry, field spectroscopy, and hyperspectral remote sensing: A case study of the Rodalquilar mining area, SE Spain. *Remote Sens. Environ.* **2008**, *112*, 3222–3233. [CrossRef]
82. Rodger, A.; Laukamp, C.; Haest, M.; Cudahy, T. A simple quadratic method of absorption feature wavelength estimation in continuum removed spectra. *Remote Sens. Environ.* **2012**, *118*, 273–283. [CrossRef]
83. Van Ruitenbeek, F.J.; Bakker, W.H.; van der Werff, H.M.; Zegers, T.E.; Oosthoek, J.H.; Omer, Z.A.; Marsh, S.H.; van der Meer, F.D. Mapping the wavelength position of deepest absorption features to explore mineral diversity in hyperspectral images. *Planet. Space Sci.* **2014**, *101*, 108–117. [CrossRef]
84. Van der Meer, F.D.; Kopackova, V.; Koucká, L.; van der Werff, H.M.; van Ruitenbeek, F.J.; Bakker, W.H. Wavelength feature mapping as a proxy to mineral chemistry for investigating geologic systems: An example from the Rodalquilar epithermal system. *Int. J. Appl. Earth Obs. Geoinf.* **2017**, in print.

remote sensing

MDPI

Article

Classification of Tree Species in a Diverse African Agroforestry Landscape Using Imaging Spectroscopy and Laser Scanning

Rami Piiroinen [1,*], Janne Heiskanen [1], Eduardo Maeda [2], Arto Viinikka [3] and Petri Pellikka [1]

1 Earth Change Observation Laboratory, Department of Geosciences and Geography, University of Helsinki, P.O. Box 64, FI-00014 Helsinki, Finland; janne.heiskanen@helsinki.fi (J.H.); petri.pellikka@helsinki.fi (P.P.)
2 Fisheries and Environmental Management Group, Department of Environmental Sciences, University of Helsinki, P.O. Box 68, FI-00014 Helsinki, Finland; eduardo.maeda@helsinki.fi
3 Finnish Environmental Institute (SYKE), Environmental Policy Centre, P.O. Box 140, FI-00251 Helsinki, Finland; arto.viinikka@ymparisto.fi
* Correspondence: rami.piiroinen@helsinki.fi; Tel.: +358-45-632-5313

Received: 14 July 2017; Accepted: 18 August 2017; Published: 23 August 2017

Abstract: Airborne imaging spectroscopy (IS) and laser scanning (ALS) have been explored widely for tree species classification during the past decades. However, African agroforestry areas, where a few exotic tree species are dominant and many native species occur less frequently, have not yet been studied. Obtaining maps of tree species would provide useful information for the characterization of agroforestry systems and detecting invasive species. Our objective was to study tree species classification in a diverse tropical landscape using IS and ALS data at the tree crown level, with primary interest in the exotic tree species. We performed multiple analyses based on different IS and ALS feature sets, identified important features using feature selection, and evaluated the impact of combining the two data sources. Given that a high number of tree species with limited sample size (499 samples for 31 species) was expected to limit the classification accuracy, we tested different approaches to group the species based on the frequency of their occurrence and Jeffries–Matusita (JM) distance. Surface reflectance at wavelengths between 400–450 nm and 750–800 nm, and height to crown width ratio, were identified as important features. Nonetheless, a selection of minimum noise fraction (MNF) transformed reflectance bands showed superior performance. Support vector machine classifier performed slightly better than the random forest classifier, but the improvement was not statistically significant for the best performing feature set. The highest F1-scores were achieved when each of the species was classified separately against a mixed group of all other species, which makes this approach suitable for invasive species detection. Our results are valuable for organizations working on biodiversity conservation and improving agroforestry practices, as we showed how the non-native *Eucalyptus* spp., *Acacia mearnsii* and *Grevillea robusta* (mean F1-scores 76%, 79% and 89%, respectively) trees can be mapped with good accuracy. We also found a group of six fruit bearing trees using JM distance, which was classified with mean F1-score of 65%. This was a useful finding, as these species could not be classified with acceptable accuracy individually, while they all share common economic and ecological importance.

Keywords: imaging spectroscopy; airborne laser scanning; minimum noise fraction; class imbalance; Africa; agroforestry; tree species

1. Introduction

Globally, more than 55% of new agricultural land in tropics was converted from forests between 1980 and 2000 [1]. In eastern Africa, the yearly increase rate of agricultural land has been 1.4% during

1990–2010, while the yearly deforestation rate increased from 0.2% during 1990–2000 to 0.4% during 2000–2010 [2]. Agroforestry systems are considered as an option for mitigating the negative impacts of this change [3,4]. In addition, selecting proper tree species is important for a productive and environmentally sustainable agroforestry system [5–7]. However, the transformation of forests and woodlands into agroforestry might decrease biodiversity as native tree species are often replaced with exotic species. In the Afromontane highlands of the Taita Hills (southeast Kenya), 66.5% of tree species observed in the croplands (agroforestry) are exotic, and were associated in a recent study with functional traits such as economic function and nitrogen fixation [8].

Remote sensing based tree species mapping has great potential to reduce costs of observing changes in the tree species composition in comparison to field based approaches that require large number of field plots [9]. In their recent review, Fassnacht et al. [9] identified that common motives for tree species classification using remote sensing include biodiversity assessment and monitoring, monitoring of invasive species, hazard and stress management, wildlife habitat mapping, sustainable forest management, and resource inventory. Airborne laser scanning (ALS) and imaging spectroscopy (IS) were the most commonly used data types in the recent studies. Most studies had been conducted in temperate forests, while tropical forests had been studied in South America and savannah systems in South Africa [9]. The review did not list any studies from a diverse agroforestry landscape in Africa with patches of shrubland and native forest, where a few exotic tree species are dominant and a high number of native species occur less frequently. This leads to imbalance in the training data used in classification [10]. Although the cost of IS and ALS based tree species mapping is low in comparison to covering the study area on foot, airborne remote sensing is more expensive than satellite-based remote sensing data. However, mapping trees on species level with satellite-based data is challenging in agroforestry landscape, where trees are often isolated on farmland and high resolution data are needed to detect the trees at the crown level.

Previous studies with a high number of tree species have shown decline in the classification accuracy with the increasing number of classes [11], and increase in accuracy with the greater number of samples per species [12]. However, collecting comprehensive field reference data for all the species in a high species diversity system is challenging. The negative impact of imbalanced or limited training data on tree species classification has been approached, for example, by standardizing class sizes using down-sampling [10,13], and by using semi-supervised approaches to increase the size of training data from unlabeled observations [14]. However, we did not find studies where class sizes were balanced using up-sampling, or where different approaches to divide the species into groups based on their spectral and structural characteristics using Jeffries–Matusita (JM) distance, were compared.

According to Fassnacht et al. [9], non-parametric support vector machine (SVM) [15] and random forest (RF) [16] are the most commonly used classifiers for tree species classification. Both classifiers have performed well in remote sensing based classifications while neither has constantly outperformed the other [17–19]. Feature extraction and/or feature selection methods are commonly used with high dimensional data to improve the classification accuracy; in particular, the minimum noise fraction (MNF) transformation [20] has performed well in the previous studies [9,18,21].

The main aim of this study was to study tree species classification in an African agroforestry landscape with high species diversity and imbalanced training set. The specific objectives of the study were to:

1. compare the performance of the different feature sets derived from IS and ALS data using SVM and RF classifiers;
2. find species or groups of species with ecological or economical function that can be detected relatively accurately; and
3. evaluate the impact of up-sampling and different approaches to group the species on the classification accuracy.

2. Material and Methods

2.1. Study Site

The study area (10 km × 10 km) was located in the elevation range of 1100–2200 m a.s.l. in the Taita Hills (3°25′ S, 38°19′ E) in southeast Kenya (Figure 1). The Taita Hills are part of the Eastern Arc Mountains (EAM). There are two rainy seasons with long rains occurring in the March–June and short rains in October–December [22]. The potential natural vegetation for the Taita Hills is moist Afromontane forest or cloud forest [23] while most of the area is transformed to agricultural use [24,25]. The hills are a biodiversity hotspot with high endemism and exceptional loss of habitat with 80 endemic woody plant species occurring in EAM [8,26]. Only 4.2 km^2 of montane forests persist in 12 forest relicts [27].

In agricultural land, the most common tree species are exotic [8]. *Eucalyptus* spp. native to Australia have been planted to produce lumber, but have been reported by local communities to cause degradation of water quality and lower the availability of water [28] (Figure 2). *Eucalyptus* plantations may even dry up rivers completely [5]. *Grevillea robusta*, also native to Australia, have been planted on farmland to produce lumber since the 1980s, with positive impacts reported by locals [28]. Other studies have shown that *Grevillea robusta* trees may improve agricultural productivity by increasing rainfall utilization, while careful consideration must be given to the distribution of trees among crops to avoid mutually detrimental effects on the tree establishment and the crop growth yield [6,7]. *Acacia mearnsii*, native to Australia, was originally brought to the area to produce vegetable tannins for the tanning of leather [29], but is presently used mainly as firewood. It is considered an invasive species that may reduce local biodiversity [30]. Another commonly found species is *Cupressus lucitanica*, native to Mexico and Central America, which has been planted as fences as they grow thick and dense needle canopy that is difficult to pass through. Most common fruit bearing tree species is *Persea Americana* and less common ones include *Eriobotrya japonica* and *Mangifera indica*.

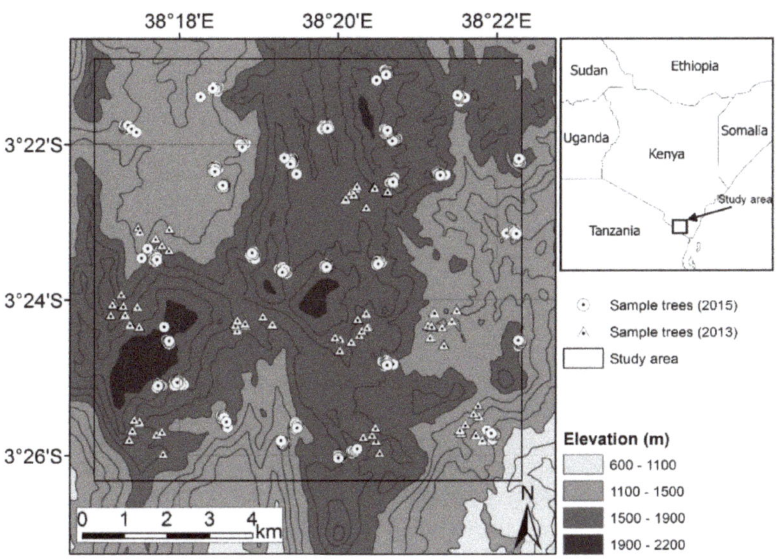

Figure 1. Location of the study area in Coast Province of Kenya and trees sampled in 2013 and 2015.

Figure 2. (**a**) Aerial view of town of Wundanyi and surrounding agroforestry landscape with forest fragments; (**b**) agroforestry landscape near Ngangao forest; and (**c**) *Grevillea robusta*; (**d**) *Eucalyptus saligna*; (**e**) *Acacia mearnsii*; (**f**) *Persea americana*; and (**g**) *Euphorbia kibwezensis* trees in a valley in the north-western corner of the study area.

2.2. Field Data

The fieldwork was conducted in two stages. The first campaign was organized between 17 January and 8 February 2013. The 100 km^2 study area was divided into 16 tiles (2.5 km × 2.5 km), which were sampled by 100 ha clusters. Each cluster had ten circular 0.1 ha study plots (17.84 m radius). Ten clusters were selected for the detailed tree sampling, and as one plot was treeless, this resulted in 99 study plots. From each study plot, every tree that had a diameter at breast height greater than 10 cm was measured. The central point of each study plot was measured with GNSS (Trimble GeoExplorer GeoXH 6000, Trimble Inc., Sunnyvale, CA, USA). Measuring tape and compass were used to measure the relative position of each tree from the plot center. To enable the differential correction of the data, a GNSS base station (Trimble Pro 6H receiver, Trimble Inc., Sunnyvale, CA, USA) was logging in a known position during the field measurements. The plots located in the closed indigenous forest were omitted, as we were primarily interested in exotic species that have been planted on agricultural or otherwise managed land. The data from 2013 contained 531 individual trees from 55 different tree species.

The second campaign was organized during 1–30 October 2015. The study area was divided into 1 km × 1 km tiles and 30 tiles were randomly selected. Each tile was further divided into rectangular 1 ha study plots and one study plot was selected within each tile, with the exception of one tile that had two study plots. In total, there were 31 study plots. Within each study plot, nine sampling points with 33.3 m intervals were established. At each sampling point, two trees were selected using the T-square plotless sampling method [8,31]. The same GNSS receiver and base station were used as in 2013 but each tree was measured directly with GNSS. A tree was defined as any woody plant taller than five meters. The data from 2015 contained 538 trees, while we excluded 98 trees that were either located under higher trees (not visible from the air) or that had GNSS positional accuracy <4 m, which left us 440 crowns. In total, there were 950 tree measurements with sufficient positional accuracy and visible canopy from 2013 and 2015 combined.

2.3. Remote Sensing Data

A flight campaign was conducted in 3–8 February 2013 during the dry season. Two sensors were used for collecting the ALS and IS data from a mean flying height of 750 m. Optech ALTM 3100 (Teledyne Optech, Vaughan, Ontario, Canada) is an oscillating mirror laser scanner capable of recording up to four echoes (returns). The sensor was operated at a pulse rate of 100 kHz and a scan rate of 36 Hz. Scan angle was ±16°. Achieved pulse density was 9.6 pulses/m^2. Mean footprint diameter was 23 cm (Figure 3). The IS data were acquired with AisaEAGLE (Spectral Imaging Ltd., Oulu, Finland) sensor, a pushbroom scanner with an instantaneous field of view of 0.648 mrad and field of view of 36.04°. The sensor was used with four times spectral binning mode that produced output images with 129 bands and a full width at half maximum of 4.5–5.0 nm in the spectral range of 400–1000 nm. The output pixel size was one meter (Figure 3).

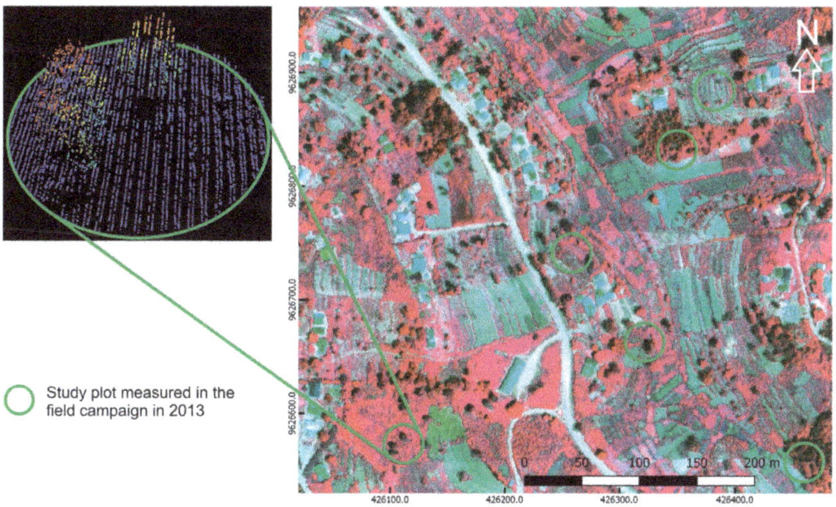

Figure 3. An example of AisaEAGLE data (color-infrared) and point cloud derived from laser scanning data. The coordinates are in UTM37S/WGS84 coordinate system.

2.4. Remote Sensing Data Preprocessing

ALS data were preprocessed by the data vendor (Topscan Gmbh, Rheine, Germany) and delivered as a georeferenced point cloud in UTM37S/WGS84 coordinate system with ellipsoidal heights. The buildings and power lines were excluded and some erroneous measurements from steep slopes were removed using TerraScan software (Terrasolid Ltd., Helsinki, Finland). A canopy height model (CHM) was created using LAStools (version 170201, rapidlasso GmbH, Gilching, Germany) software from point cloud using a pit-free method [32].

The raw IS data were radiometrically corrected and orthorectified with CaliGeoPro 2.2 (Spectral Imaging Ltd. Oulu, Finland). Atmospheric correction was applied using ATCOR-4 (ReSe Applications Schläpfer, Wil, Switzerland), [33]. After the orthorectification, it was noted that there were geometric mismatches between IS and ALS data. As the LiDAR sensor system had higher quality inertial measurement unit and the IS data had obvious distortions, we co-registered the ALS and IS data using control points collected manually from the CHM. The processed IS scanning lines were clipped so that the side overlap was minimized to reduce the distortions on the sides of the flight lines. In total,

50–100 control points were collected for each flight line and first order polynomial transformation was applied to co-register the images. After the co-registration, RMSE was 1.06 m, which was considered appropriate for the data fusion [34].

Before the classification, we filtered the spectral data by excluding pixels with NIR (836 nm) reflectance < 20% and NDVI < 0.5. Higher NDVI threshold has been used in some studies [35], but we selected 0.5 threshold as even the brightest pixels of some of the species in the study area showed lower NDVI values (e.g., *Euphorbia kibwezensis*).

2.5. Segmentation and Preparing Training Data

Tree crowns were segmented using the dalponte2016 algorithm [36] implemented in the lidR package [37] in R software (R version 3.4.0, R Foundation for Statistical Computing) [38]. The algorithm finds local maxima from rasterized CHM, designates these as tree tops, and then uses a decision tree method to grow individual crowns around the local maxima.

We matched the 950 field measured trees to 543 tree crowns (Figure 4). If multiple field measurements from the same species were detected for the same segmented crown, only one of the observations would be kept. Thus, one segmented tree crown may consist of multiple trees from the same species, while one segment was considered as one sample. If a crown contained more than one species, it would be excluded. In total 61 species were observed for 543 crowns, while 19 of the species had only 1 observation and 10 had 2–3 observations. These were excluded from the classification as done in a similar setting earlier [10]. The classifications included in total 499 crowns from 31 species (Table 1). All species belonging to *Eucalyptus* and *Syzygium* genuses were labeled as *Eucalyptus* spp. and *Syzygium* spp., respectively, as we could not identify the exact species in all instances. However, the majority of the *Eucalyptuses* were *Eucalyptus saligna*.

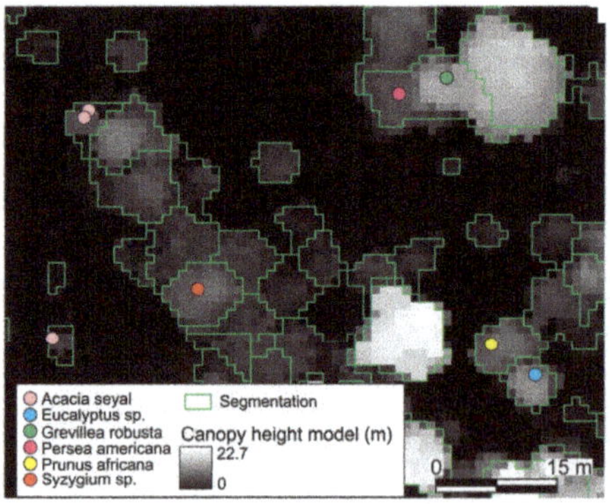

Figure 4. An example of segmented tree crowns and the field measured trees on top of the canopy height model.

Table 1. Species with more than three samples, abbreviation used in figures, type (exotic or native), the frequency of the species (crown level) and number of pixels per species.

Species	Abbreviation	Type	Crowns	Pixels
Grevillea robusta	Grerob	exotic	109	5485
Acacia mearnsii	Acamea	exotic	53	2437
Eucalyptus spp.	Eucspp	exotic	42	2577
Persea americana	Perame	exotic	42	1989
Cupressus lusitanica	Cuplus	exotic	31	1641
Euphorbia kibwezensis	Eupkib	native	31	1181
Eriobotrya japonica	Erijap	exotic	14	516

Table 1. *Cont.*

Species	Abbreviation	Type	Crowns	Pixels
Ficus thonningii	Fictho	native	14	948
Maesa lanceolata	Maelan	native	14	612
Mangifera indica	Manind	exotic	14	668
Zimmermania ovata	Zimova	native	13	674
Zimmermania commiphora	Zimcom	native	11	426
Psidium guajava	Psigua	exotic	10	459
Erythrina abyssinica	Eryaby	native	9	415
Acacia seyal	Acasey	native	8	417
Phoenix reclinata	Phorec	native	8	442
Albizia gummifera	Albgum	native	7	387
Prunus africana	Pruafr	native	7	410
Bridelia micrantha	Brimic	native	6	335
Dombeya kirkii	Domkir	native	6	138
Ficus sur	Ficsur	native	6	311
Combretum collinum	Comcol	native	5	234
Cussonia spicata	Cusspi	native	5	215
Macademia spp.	Macspp	exotic	5	122
Millettia oblata	Milobl	native	5	401
Acacia tortilis	Acator	native	4	90
Dombeya rotundifolia	Domrot	native	4	169
Ficus sycomorus	Ficsyc	native	4	228
Newtonia buchananii	Newbuc	native	4	385
Syzygium spp.	Syzspp	native	4	174
Xymalos monospora	Xymmon	native	4	123

2.6. Minimum Noise Fraction Transformation

MNF transformation [20] was applied to the atmospherically corrected reflectance data to reduce dimensionality and pack coherent information in a smaller set of features. The algorithm was applied in ENVI software (version 5.0, Research Systems Inc., Boulder, CO, USA) [39]. We determined the usefulness of the MNF components by the evaluation of their eigenvalues and through visual interpretation [40]. We selected 15 first MNF components and disregarded the rest based on their low eigenvalues. The further visual inspection confirmed that the MNF bands from 16 onwards contained mostly noise. Each MNF component has corresponding eigenvectors that can be used to interpret the weight that each original reflectance band has on the component.

2.7. Narrowband Vegetation Indices

We calculated a set of narrowband vegetation indices (NVI) that have been linked in earlier studies with vegetation structure, biochemistry and plant physiology (Table S1) [41]. Some of the indices have originally been developed for broadband data but were calculated using narrowband IS data and thus referred as NVIs.

2.8. Point cloud Features

We used lidR package [37] in R to calculate features for tree crown segments from the ALS point cloud. The average density was 9.6 pulses m^2, but there was notable variation in pulse density due to elevation variations across the study area (Figure 1). Especially in the valleys, there were not enough returns per tree crown to calculate complex features. Intensity values were not considered as they were uncalibrated and IS data were available. As ALS derived height information alone is of limited value [9], we calculated variance (var), minimum height (min), 95th percentile (P95), median absolute deviation around median (MAD_{median}) , median absolute deviation around mean (MAD_{mean}), average absolute deviation around median (AAD_{median}), average absolute deviation around mean (AAD_{mean}), quadratic mean (QM), the count of returns (count), maximum height divided by the maximum crown diameter (HD) and maximum height divided by the count of returns (HC).

2.9. Feature Selection

We used VSURF (Variable Selection Using Random Forests) package in R [42,43] to perform the feature selection. VSURF uses the RF variable importance to identify the features that are the most important for the classification task. It was developed especially for handling high dimensional data. VSURF performs three steps: (1) irrelevant features are eliminated; (2) all features related to the response are selected (interpretation step); and (3) selection is refined by eliminating redundancy in the set of features selected in the second step (prediction step). The features retained in the final step were used to test the impact of variable selection on the classification accuracy and to identify the features that were important for the classification.

2.10. Classification Methods

All classifications were realized using "caret" package in R [44]. Specifically, we used "Kernlab" and "RandomForest" packages for SVM and RF, respectively [45,46]. We used the radial basis kernel with SVM and optimized C and *sigma* values through grid search. For RF, we set *ntree* to 500 and searched *mtry* value by testing values 4, 8, 16 and 32. As our dataset was imbalanced, we selected the models that produced the highest Kappa instead of overall accuracy (OA) [44].

The class balancing was done for the training data during the cross validation using the up-sampling method from "caret" package, while test data were left intact. The up-sampling method randomly samples (with replacement) the minority classes to be the same size as the majority class (class with most samples). Classifications using up-sampling are referred to as balanced classification from here on after.

2.11. Measures of Performance

OA was calculated as the total number of correctly classified samples divided by the total number of samples. We used precision and recall, equivalent to user's and producer's accuracy [47], to evaluate the performance on the class level. We calculated also F1-score, which is harmonic mean of precision and recall as following:

$$F1 = 2 \times \left(\frac{precision\ \times recall}{precision + recall} \right) \tag{1}$$

F1-score increases with higher precision or recall and with the higher similarity between the two values. The values range between 0 and 1, while the best value is 1 and worst is 0.

2.12. Jeffries–Matusita Distance

As we expected that some of the species cannot be classified individually and grouping all the species under an arbitrarily defined limit for minimum number of samples might not be meaningful, we used JM distance to find spectrally and structurally similar subgroups of species. The multiclass JM distances were calculated using the varSel package in R [48] using the best performing feature set. First, the JM distances were calculated for all species pairs. Next, the species with the least number of samples were grouped with the species with the lowest matching JM distance. The process was repeated until each species belonged to one of the groups. Finally, the process was repeated to achieve a smaller number of groups. The process was started from the species with the fewest samples to achieve groups with enough samples for building a stable classification model.

2.13. Statistical Significance Tests

McNemar's test without continuum correction was used for testing statistical significance [18,49,50]. It is an appropriate method when the sample size is small [51]. Specifically, McNemar's test was used to assess: (1) whether there were significant differences in the OAs between the different feature sets; (2) whether there were significant differences in the OAs between SVM and RF classification results; and (3) whether the feature selection had a significant impact. McNemar's tests were calculated from leave-one-out cross validation (LOOCV) results. The limitation of McNemar's test is that it does not measure the variation resulting from the choice of training sets or internal randomness of the algorithm. As the RF results vary between iterations, we repeated the LOOCV for RF classifier 50 times and used the mode of the prediction results of each sample when McNemar's test results were classified.

2.14. Classification Trials

All the classification trials are summarized in Figure 5. First, we classified all species with more than three samples using SVM and RF classifiers with the following feature sets: (1) reflectance; (2) NVI; (3) MNF; (4) ALS; (5) reflectance+ALS; (6) NVI+ALS; and (7) MNF + ALS features. Next, we run VSURF feature selection on all feature sets and repeated the classifications. Then, we evaluated the impact of feature selection on classification accuracy and identified important features. We selected the classifier and feature set that produced the highest Kappa values for later analysis of different approaches for grouping the species. In the final step, we tested different approaches for grouping species with fewer samples and tested the impact of up-sampling on the classification results.

The statistical significance between the different grouping approaches was not evaluated as the number of classes varied. Instead, we used 3-fold cross validation that was repeated 10 times to evaluate the stability of the classifier and interpreted the precision, recall and F1-scores. We tested three different approaches to group species. First, we grouped all species with less than 20 samples together to group "other" and classified the rest of the species separately as done earlier in a similar setting [10]. Next, we classified each of the species separately against a mixed group of all other species. The last approach was to use JM distance to combine species together, while species with high F1-scores and low variability were classified individually.

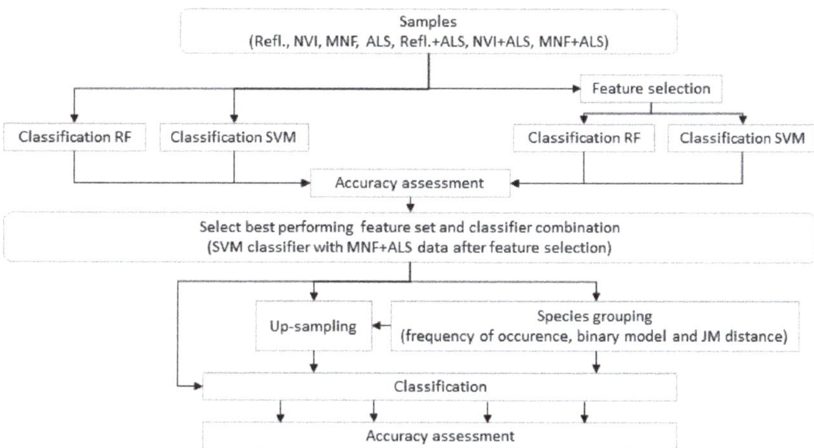

Figure 5. Workflow for the classification trials.

3. Results

3.1. Comparison of Feature Sets and Classifiers

The highest OA using both classifiers was achieved with MNF + ALS feature set (Table 2). There was a statistically significant difference ($p < 0.05$) between the SVM and RF classifications only when reflectance or NVI feature set was used (Table S2). MNF feature set outperformed reflectance,

NVI and ALS feature sets with statistical significance. For SVM classification the fusion of ALS features with MNF features improved the OA with statistical significance, compared with the classification with only MNF features (Tables S3 and S4). For RF classification there was not statistically significant improvement between these feature sets. The highest classification accuracy was achieved with SVM classifier and MNF + ALS features set, but the improvement to RF classification with the same feature set was not statistically significant. Generally, the OAs were low, as many species with fewer samples performed poorly.

Table 2. Classification results for the different feature sets using support vector machine and random forest classifiers with all of the species with more than three samples classified separately.

	Support Vector Machine		Random Forest	
Feature Set	**Accuracy**	**Kappa**	**Accuracy**	**Kappa**
Refl.	37.9	30.9	31.7	21.6
NVI	45.5	37.5	35.9	25.8
MNF	53.3	46.8	51.3	44.8
ALS	31.7	21.6	30.7	21.8
Refl.+ALS	42.9	35.7	42.9	35.6
NVI+ALS	43.5	37.2	44.3	37.5
MNF + ALS	57.1	52.1	54.1	48.2

3.2. Feature Selection

Feature selection had only small impact on OA and Kappa (Table S5), while the change in OA was statistically significant only for the NVI feature set classified with the SVM classifier. However, we could achieve the same level of accuracy with a smaller number of input features (Table 3). The important spectral regions were found at 400–450 nm, while 550–570 and 700–800 nm were also important. The most important MNF component (MNF9) had high weights around the same areas where we found important spectral bands (Table 4 and Figure 6). Most important vegetation indices were anthocyanin content index (ACI) and anthocyanin reflectance index (ARI) (Table S1 and Table 4) that were calculated from spectral bands centered at 549, 698 and 788 nm that are also seen as spikes in the MNF9 component.

Table 3. Features selected by VSURF at prediction phase for the different feature sets. The features are ordered based on their importance starting from the most important.

Feature Set	No Var.	Feature Names
Refl.	19	R406, R401, R553, R549, R414, R562, R419, R572, R769, R717, R576, R530, R526, R521, R581, R458, R688, R632, R674
NVI	8	ACI, ARI, $CI_{red\ edge}$, PRI, PSSR, mCARI, CRI1, EVI
MNF	10	MNF9, MNF1, MNF5, MNF7, MNF6, MNF4, MNF8, MNF10, MNF14, MNF2
ALS	5	HD, MAD_{median}, P95, AAD_{median}, *min*
Refl. + ALS	15	HD, *max*, MAD_{median}, MAD_{mean}, AAD_{median}, R406, R562, R558, *min*, R414, R423, R726, R731, R540, R774
NVI + ALS	17	ACI, HD, ARI, P95, *max*, MAD_{median}, CARI, AAD_{median}, AAD_{mean}, CRI2, $CI_{red\ edge}$, HC, min, PRI, PSSR, SR, VI_{green}
MNF + ALS	13	MNF9, HD, MNF5, MNF1, MNF4, MNF8, MNF7, MNF6, MNF11, MNF12, P95, MNF10, MNF14
R = reflectance, MNF = minimum noise fraction		

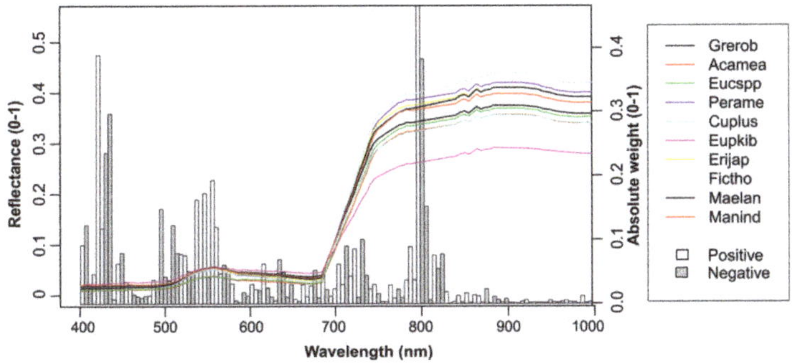

Figure 6. Contribution (weight) of different wavelengths on the most important MNF component (MNF9; vertical bars) plotted over mean spectra of 10 species with most samples. Bars represent the absolute weight and sign is indicated with color (positive, negative).

3.3. Jeffries–Matusita Distance

The spectral regions with the highest JM distances between species (nine species with most samples selected for closer inspection) were found most often near 400 nm and 550 nm (Figure 7). There were notable differences between species as, for example, *Euphorbia kibwezensis* did not have any bands with the highest distance between species around 470–740 nm, while we can see an important region around 750 nm, where the reflectance is notably lower in comparison with other species.

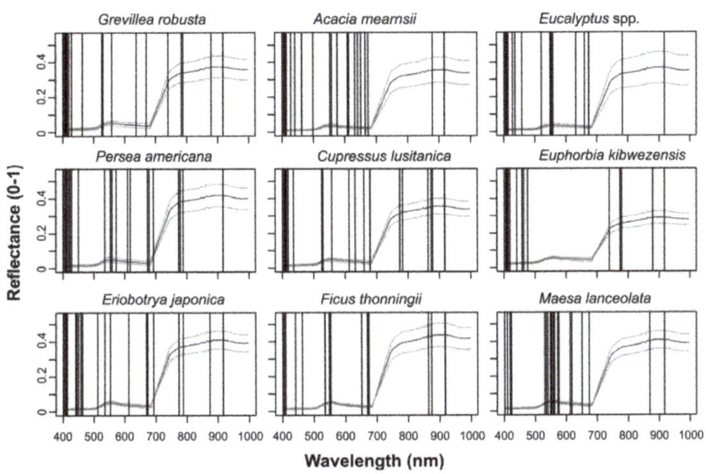

Figure 7. Reflectance (mean and standard deviation) for selected species and the wavelengths with the greatest JM distances between species (vertical lines).

3.4. Data Balancing

The mean OA (10 iterations with the best performing feature set MNF + ALS with feature selection) was 57.1% for the imbalanced and 56.0% the balanced classification (all 31 species). The mean F1-scores ranged between 0% (*Ficus sur*) and 84.9% (*Eucalyptus* spp) in the imbalanced setting (Figure 8). *Acacia mearnsii*, *Grevillea robusta*, *Eucalyptus* spp. and *Euphorbia kibwezensis* had mean F1-scores of 73.8%, 74.6%, 84.9% and 71.7%, respectively, with low variability. The species with fewer samples had high variability and lower F1-scores. However, *Erythrina abyssinica*, *Acacia tortilis* and *Ficus sycomorus* with

9, 4 and 8 samples, respectively, performed better than *Persea Americana* and *Cupressus lusitanica* with 42 and 31 samples, respectively. Up-sampling had only minor impact on the results.

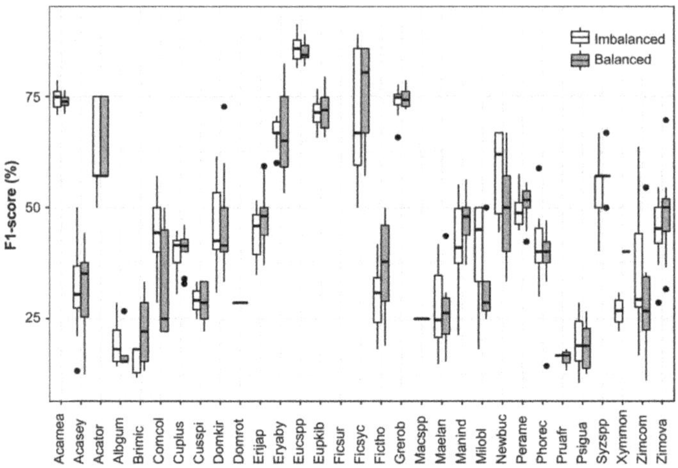

Figure 8. F1-scores for all species (more than three samples) in imbalanced and balanced (up-sampling) setting using support vector machine classifier and features selected by VSURF (MNF + ALS).

3.5. Grouping by Frequency

Combining the species with less than 20 samples increased the mean OA to 70.2% (imbalanced) and 69.2% (balanced), while mean Kappa was 62.9% and 61.3% for imbalanced and balanced classification, respectively (Figure 9). Up-sampling increased recall for *Acacia mearnsii, Cupressus lucitanica* and *Persea Americana* while precision decreased. We found a notable increase in the mean F1-score only for *Persea americana* while the F1-scores of species with more samples and higher initial classification accuracy decreased slightly.

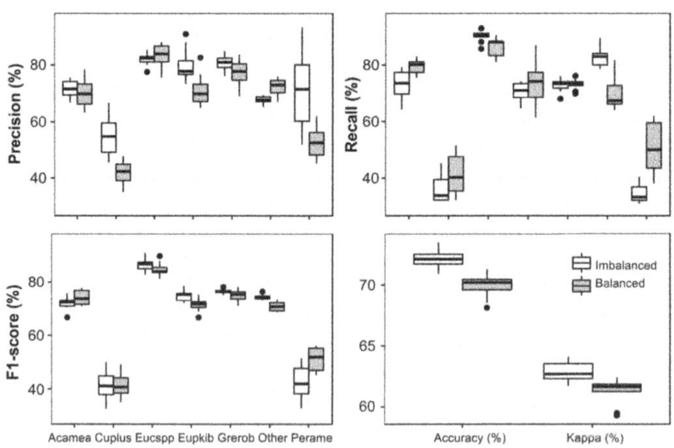

Figure 9. Precision, Recall and F1-scores for the species with more than 20 samples and Other class in balanced and imbalanced setting using support vector machine classifier and features selected by VSURF (MNF + ALS).

3.6. Single Species Classfication

The up-sampling had the biggest impact on the results when the species were classified individually against all remaining species (Figure 10). The mean recall increased and mean precision decreased for most species. The mean F1-scores increased notably for *Acacia seyal*, *Acacia tortilis* and *Ficus sycomorus* from 40.2%, 64.3% and 82.6% to 51.3%, 77.3%, and 87.2%, respectively.

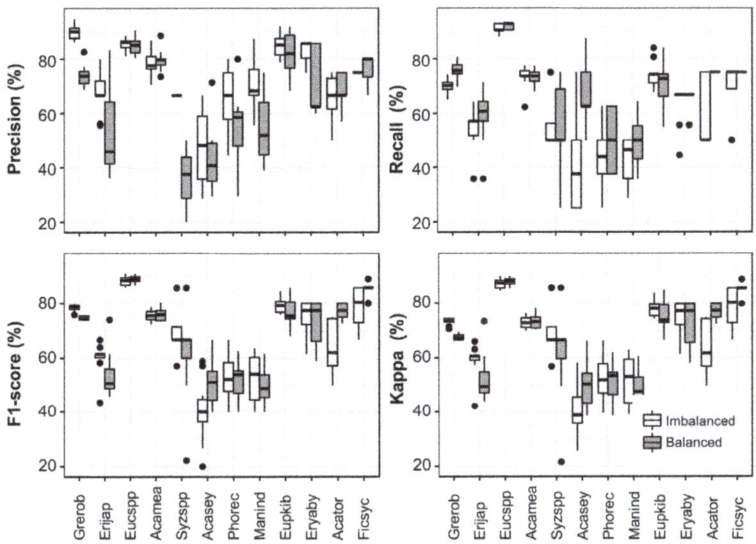

Figure 10. Classification results when each species was classified individually against mixed group of all other species (results shown for species with F1-score > 50%) in balanced and imbalanced setting (SVM classifier and MNF + ALS feature set with feature selection). The class level accuracies for the "other" class are not shown.

3.7. Grouping Species Based on Jeffries–Matusita Distance

The species with high F1-scores and low variability (*Acacia mearnsii, Grevillea robusta, Eucalyptus* spp. and *Euphorbia kibwezensis*) were classified separately, while four groups were created for the remaining species based on JM-distance (Table 4). Two of the groups were classified with F1-score > 60% while the other two had mean F1-scores around 50% (Figure 11). The mean OA was 66% and mean Kappa 61% while up-sampling had only a little impact on the overall performance. All species in Group 3 are fruit bearing trees with economic importance.

Table 4. Groups generated using JM distances and the total number of samples in each group.

Group	Species in the Group	Samples
1	Macspp, Maelan, Erijap, Domrot, Xymmon, ComCol, Zimova	59
2	Zimcom, Eryaby, Acasey, Acator, Domkir	38
3	Perame, Ficsur, Phorec, Manind, Brimic, Syzspp	80
4	Psigua, Cusspi, Cuplus, Pruafr, Ficsyc, Newbuc, Fictho, Milobl, Albgum	87

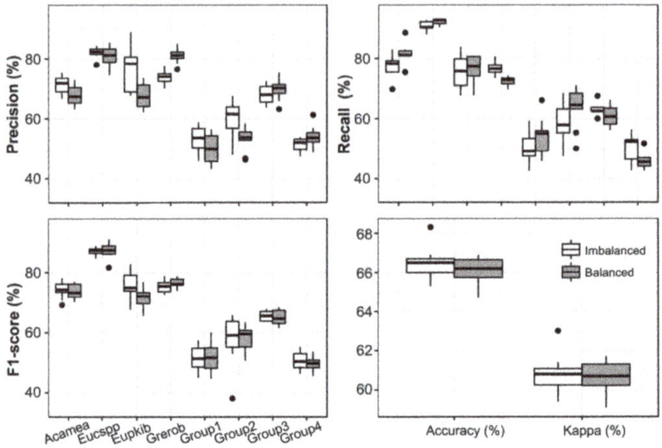

Figure 11. Classification results with JM distance based class grouping in balanced and imbalanced setting using support vector machine classifier and features selected by VSURF (MNF + ALS).

3.8. Comparison of Different Aproaches

We selected four of the species (*Acacia mearnsii*, *Grevillea robusta*, *Eucalyptus* spp. and *Euphorbia kibwezensis*) with the highest F1-scores for a closer comparison of how they were affected depending on how the remaining species were grouped (Figure 12). For the selected species, the highest mean precision and F1-scores were achieved when the species were classified individually against mixed groups of all other species. The highest recall for *Acacia mearnsii* and *Grevillea robusta*, the species with the greatest number of samples, was achieved when all 31 species were classified separately.

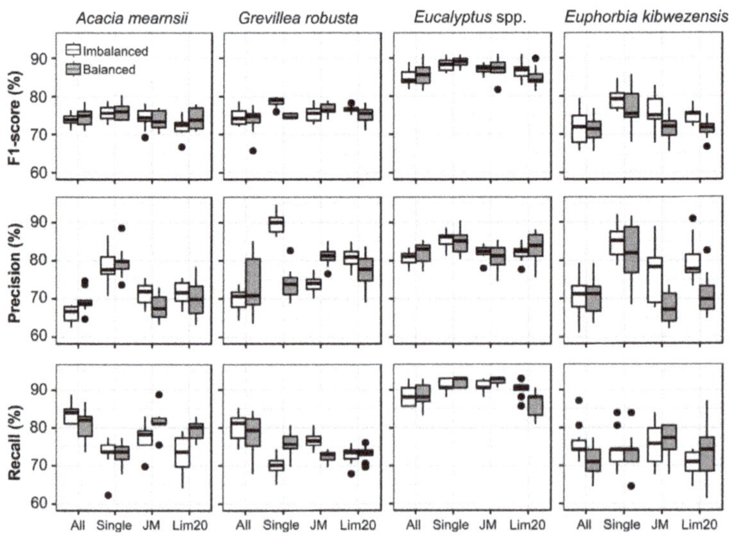

Figure 12. Comparison of precision, recall and F1-score for the selected tree species with different grouping methods and up-sampling (All = all 31 species classified individually, Single = each species is classified against mixed group of all other species, JM = JM distance used to group species, Lim20 = species with fewer than 20 samples grouped together).

4. Discussion

4.1. Impacts of Classifier, Feature Selection and Data Fusion

Both SVM and RF classifiers have been reported performing well in remote sensing based classifications, while neither is constantly outperforming the other [17–19]. In our results, there was no statistically significant difference between the two classifiers with the best performing feature set (MNF + ALS). Previous research has shown that fusing ALS data with IS data may increase the classification accuracy, while canopy height information alone is often not enough for the improved results [9,18,52]. In our results, the impact of data fusion depended on the classifier and how the IS data were used. For example, SVM classifier benefited from data fusion when ALS data were combined with MNF data, while RF classifier did not. The most important ALS derived feature was the maximum height divided by the maximum crown diameter. This is logical as, for example, *Eucalyptus* spp. and *Grevillea Robusta* are very tall trees with narrow crowns, while acacias tend to have wider crowns in relation to their height. Our ALS feature set was limited because of discrete return ALS data while full waveform ALS data could yield even bigger improvements [53].

Feature selection did not have statistically significant impact on the best performing feature set MNF + ALS. However, it was still useful, given that the same accuracies were achieved with fewer features, making the training and execution of the models faster. In addition, we used feature selection to find the features that were important for the classification procedure. In a recent review, the near-infrared wavelength regions that were important for tree species classification were found most commonly at 450–550 nm and 650–700 nm [9]. In our results, the most important regions were found at 400–450 nm, 550–570 nm and 700–800 nm. The most common important spectral region in the summary was at 650 nm, which was not important in our study. In our results, the wavelengths near 400 nm were especially important, which was selected as important wavelength area in 38% of the studies in the review [9]. The wavelength regions between 800–1000 nm were not commonly important, which was the same in our study. We did not have SWIR data available, which has important wavelengths, for example, near 1200 nm and 1450 nm, which might increase the classification accuracy.

4.2. The Impact of Up-Sampling and Grouping of Species on the Classification Results

The OA and Kappa were low when all the 31 species were classified separately. However, when species with fewer than 20 samples were combined, we reached higher OA and Kappa than in a recent study conducted in a similar landscape in Panama [10]. However, our data had fewer species with more than 20 samples. In addition, the F1-scores ranged similarly with high variability between species.

The high F1-score (with low variability) for *Eucalyptus* spp. enables map production for conservation planning. However, it needs to be considered that species with less than four samples were removed from the model (8.1% field measurements with matching tree crown). The highest F1-score for *Eucalyptus* spp. was achieved when it was classified against a mixed group of all the other species using up-sampling. However, the difference between up-sampling and imbalanced model was marginal. In addition, *Acacia mearnsii*, *Grevillea robusta* and *Euphorbia kibwezensis*, could be mapped with relatively high accuracy. *Acacia mearnsii* is highly invasive and monitoring its distribution in the long term would be valuable for conservation planning. *Euphorbia kibwezensis*, a dry land species, could possibly be used to study linkages between changes in climate and the occurrence of the species, as done earlier with *Euphorbia ingens* in South Africa [54]. Species with fewer samples, *Ficus sycomorus*, *Acacia tortilis*, *Erythrina abyssinica* and *Syzygium* spp., were also classified with relatively high mean F1-scores in the single species setting. However, the small sample size and high variability in the results make it difficult to assess how the models would perform when extended over the whole study area. *Cupressus lusitanica* was classified with poor accuracy, possibly because it is used in fences where it grows densely and achieve low maximum heights, but it is also used in plantations for lumber production, where it reaches much higher heights. *Persea americana* was classified with poor accuracy, which could be explained by spectral and structural similarities with a number of other fruit trees. As hyperspectral data also capture the phenological states of trees [55], it is possible that some of the misclassifications can be explained by the spectral variation caused by different phenology resulting from the varying local climate caused by topography.

In the previous studies with a high number of species, a mixed group of species with fewer samples has been commonly used [10,56]. However, combining all the species under a fixed limit (e.g., 20 samples) creates large and highly heterogeneous mixed class. On the other hand, spectral similarity measures, like JM distance, can be used to find spectrally and structurally similar species, which enables creating smaller and more homogeneous groups that also balance the training data. For example, Group 3 created with this approach has two exotic (*Persea Americana* and *Mangifera indica*) and three native (*Ficus sur*, *Syzygium* spp. and *Bridelia micrantha*) fruit bearing tree species. The sixth species, *Phoenix reclinata*, is a palm that produces edible fruits (dates). Thus, it is also an ecologically meaningful group as all of the species are fruit bearing. Based on our results, JM distance may help in identifying groups of species that could be classified together with acceptable accuracy. However, grouping the species using JM distance makes sense only if the created groups have a common ecological or economic function. Up-sampling did not improve OA or Kappa, but we did see improvements on the species level, usually for species with smaller sample sizes or lower initial classification accuracy.

4.3. Evaluation of the Quality of Airborne Data, Field Measurements and Segmentation

As the airborne data were acquired in 2013 and the later field campaign was conducted in 2015, it was difficult to estimate if a tree had been over five meters tall two years ago. Thus, some of the trees measured in 2015 might have been left undetected by the segmentation algorithm. In addition, some of the species (e.g., *Acacia mearnsii*) grow in dense bush-like formations, which were problematic for the segmentation algorithm, as the crowns were difficult to separate even in the visual interpretation of the CHM. For example, one of the segmented tree crowns contained six *Acacia mearnsii* field measurements. Generally, isolated trees on farmland were easier for the segmentation algorithm, but naturally growing trees with tightly knit canopy structure were challenging, which underlies the difficulty of accurate tree crown segmentation in the tropical areas with dense canopies [11,57]. Furthermore, the positional accuracy was low in some areas due to the mountainous nature of the study area and only one available GNSS base station. Removing all field measurements with positional accuracy lower than four meters helped, but still some of the field measurements might have been matched with wrong tree crown.

5. Conclusions

Our results demonstrate that, even in diverse African agroforestry landscapes with high species diversity and imbalanced training data, the classification of some species, or groups of species, is possible when proper pre-processing, feature transformation and species grouping approaches are used. The MNF transformed data combined with the ALS features was superior in performance when compared with the other feature sets, and the best results were achieved with SVM classifier. If the aim is only to map the distribution of one species at the time, we suggest combining all the other species into one mixed group, as the highest accuracies were achieved with this approach. If many species are classified at the same time, the spectral separability measures like JM distance can be used to find spectrally and structurally similar groups of species. With this approach, we found an ecologically and economically meaningful group of six fruit bearing trees that can be mapped with moderate accuracy. The up-sampling improved the F1-scores for some species with fewer samples. For example, *Acacia tortilis* with only four samples was classified with high mean F1-score after up-sampling. However, due to the small sample size, it is difficult to assess the performance of the model when predicted over the whole study area. Our results also provide important insights into the spectral and structural features that differentiate the tree species in the study area, while we found notable differences in the important spectral regions compared with previous studies. The three non-native tree species (*Eucalyptus* spp., *Grevillea robusta*, and *Acacia mearnsii*) that could be mapped with the highest accuracies account for 40.1% of the samples. Thus, it is possible to map the decrease in biodiversity indirectly by mapping changes in the distribution of these species, as the increase in their distribution could mean decrease in the number of native tree species. In addition, *Eucalyptus* spp. and *Acacia mearnsii* are highly invasive species, and mapping their distribution would be valuable for conservation planning. *Grevillea robusta* is an important agroforestry tree and mapping their distribution would provide valuable information for the characterization of agroforestry practices in the study area. Although the number of species that were classified accurately is relatively low, better results could be achieved with more representative field data.

Supplementary Materials: The following are available online at www.mdpi.com/2072-4292/9/9/875/s1, Table S1: List of narrowband vegetation indices, Table S2: McNemar's score and statistical significance of difference in overall accuracy between support vector machine and random forest classifiers for different feature sets, Table S3: McNemar's score (lower triangular part) and statistical significance of difference in overall accuracy (upper triangular part) between different feature sets using support vector machine; Table S4: McNemar's score (lower triangular part) and statistical significance of change in overall accuracy (upper triangular part) between different feature sets using random forest, Table S5: Change in overall accuracy and Kappa after feature selection.

Acknowledgments: This research was funded by Ministry for Foreign Affairs of Finland (CHIESA and BIODEV projects) and Academy of Finland (TAITAWATER project). The authors want to thank Jessica Broas and Kirsi Kivistö from University of Helsinki and Darius Kimuzi from Taita Research Station for helping to collect the field data in 2013 and 2015. Jesse Hietanen preprocessed the ALS point clouds. Elisa Schäfer, Kirsi Kivistö and Hari Adhikari helped to orthorectify the AisaEAGLE data. Pekka Hurskainen and Fabian Fassnacht offered valuable comments on the manuscript. Taita Research Station of the University of Helsinki in Kenya and its staff are warmly acknowledged for logistical support and safe accommodation during the fieldwork. Research permit NCST/RCD/17/012/33 from National Council for Science and Technology of Republic of Kenya is warmly acknowledged.

Author Contributions: R.P., J.H., E.M. and P.P. conceived and designed the experiments. R.P. performed the experiments and analyzed the data. J.H., E.M. and A.V. contributed materials and analysis tools. R.P. wrote the paper.

Conflicts of Interest: The authors declare no conflict of interest.

References

1. Gibbs, H.K.; Ruesch, A.S.; Achard, F.; Clayton, M.K.; Holmgren, P.; Ramankutty, N.; Foley, J.A. Tropical forests were the primary sources of new agricultural land in the 1980s and 1990s. *Proc. Natl. Acad. Sci. USA* **2010**, *107*, 16732–16737. [CrossRef] [PubMed]
2. Brink, A.B.; Bodart, C.; Brodsky, L.; Defourney, P.; Ernst, C.; Donney, F.; Lupi, A.; Tuckova, K. Anthropogenic pressure in East Africa—Monitoring 20 years of land cover changes by means of medium resolution satellite data. *Int. J. Appl. Earth Obs. Geoinf.* **2014**, *28*, 60–69. [CrossRef]
3. Mbow, C.; van Noordwijk, M.; Prabhu, R.; Simons, T. Knowledge gaps and research needs concerning agroforestry's contribution to Sustainable Development Goals in Africa. *Curr. Opin. Environ. Sustain.* **2014**, *6*, 162–170. [CrossRef]
4. Luedeling, E.; Kindt, R.; Huth, N.I.; Koenig, K. Agroforestry systems in a changing climate—Challenges in projecting future performance. *Curr. Opin. Environ. Sustain.* **2014**, *6*, 1–7. [CrossRef]
5. Rodriguez-Suarez, J.A.; Soto, B.; Perez, R.; Diaz-Fierros, F. Influence of Eucalyptus globulus plantation growth on water table levels and low flows in a small catchment. *J. Hydrol.* **2011**, *396*, 321–326. [CrossRef]
6. Lott, J.E.; Khan, A.A.H.; Black, C.R.; Ong, C.K. Water use in a Grevillea robusta-maize overstorey agroforestry system in semi-arid Kenya. *For. Ecol. Manag.* **2003**, *180*, 45–59. [CrossRef]
7. Omoro, L.M.A.; Nair, P.K.R. Effects of mulching with multipurpose-tree prunings on soil and water run-off under semi-arid conditions in Kenya. *Agrofor. Syst.* **1993**, *22*, 225–239. [CrossRef]
8. Thijs, K.W.; Aerts, R.; van de Moortele, P.; Aben, J.; Musila, W.; Pellikka, P.; Gulinck, H.; Muys, B. Trees in a human-modified tropical landscape: Species and trait composition and potential ecosystem services. *Landsc. Urban Plan.* **2015**, *144*, 49–58. [CrossRef]
9. Fassnacht, F.E.; Latifi, H.; Sterenczak, K.; Modzelewska, A.; Lefsky, M.; Waser, L.T.; Straub, C.; Ghosh, A. Review of studies on tree species classification from remotely sensed data. *Remote Sens. Environ.* **2016**, *186*, 64–87. [CrossRef]

10. Graves, S.J.; Asner, G.P.; Martin, R.E.; Anderson, C.B.; Colgan, M.S.; Kalantari, L.; Bohlman, S.A. Tree species abundance predictions in a tropical agricultural landscape with a supervised classification model and imbalanced data. *Remote Sens.* **2016**, *8*, 161. [CrossRef]
11. Feret, J.-B.; Asner, P.G. Tree species discrimination in tropical forests using airborne imaging spectroscopy. *IEEE Trans. Geosci. Remote Sens.* **2013**, *51*, 73–84. [CrossRef]
12. Baldeck, C.A.; Asner, G.P. Improving remote species identification through efficient training data collection. *Remote Sens.* **2014**, *6*, 2682–2698. [CrossRef]
13. Freeman, E.A.; Moisen, G.G.; Frescino, T.S. Evaluating effectiveness of down-sampling for stratified designs and unbalanced prevalence in Random Forest models of tree species distributions in Nevada. *Ecol. Model.* **2012**, *233*, 1–10. [CrossRef]
14. Dalponte, M.; Ene, L.T.; Marconcini, M.; Gobakken, T.; Næsset, E. Semi-supervised SVM for individual tree crown species classification. *ISPRS J. Photogramm. Remote Sens.* **2015**, *110*, 77–87. [CrossRef]
15. Vapnik, V. *Statistical Learning Theory*; John Wiley & Sons: New York, NY, USA, 1998.
16. Breiman, L.E.O. Random Forests. *Mach. Learn.* **2001**, *45*, 5–32. [CrossRef]
17. Belgiu, M.; Drăguţ, L. Random forest in remote sensing: A review of applications and future directions. *ISPRS J. Photogramm. Remote Sens.* **2016**, *114*, 24–31. [CrossRef]
18. Ghosh, A.; Fassnacht, F.E.; Joshi, P.K.; Kochb, B. A framework for mapping tree species combining hyperspectral and LiDAR data: Role of selected classifiers and sensor across three spatial scales. *Int. J. Appl. Earth Obs. Geoinf.* **2014**, *26*, 49–63. [CrossRef]
19. Ghosh, A.; Joshi, P.K. A comparison of selected classification algorithms for mappingbamboo patches in lower Gangetic plains using very high resolution WorldView 2 imagery. *Int. J. Appl. Earth Obs. Geoinf.* **2014**, *26*, 298–311. [CrossRef]
20. Green, A.; Berman, M.; Switzer, P.; Craig, M.D. A transformation for ordering multispectral data in terms of image quality with implications for noise removal. *IEEE Trans. Geosci. Remote Sens.* **1988**, *26*, 65–74. [CrossRef]
21. Piiroinen, R.; Heiskanen, J.; Mõttus, M.; Pellikka, P. Classification of crops across heterogeneous agricultural landscape in Kenya using AisaEAGLE imaging spectroscopy data. *Int. J. Appl. Earth Obs. Geoinf.* **2015**, *39*. [CrossRef]
22. Jaetzold, R.; Schmidt, H.; Hornetz, B.; Shisanya, C. *Farm Management Handbook of Kenya: Natural Conditions and Farm Management Information*; Ministry of Agriculture, Kenya & German Agricultural Team; German Agency for Technical Cooperation: Rossdorf, Germany, 1983; Volume 2.
23. Aerts, R.; Thijs, K.W.; Lehouck, V.; Beentje, H.; Bytebier, B.; Matthysen, E.; Gulinck, H.; Lens, L.; Muys, B. Woody plant communities of isolated Afromontane cloud forests in Taita Hills, Kenya. *Plant Ecol.* **2011**, *212*, 639–649. [CrossRef]
24. Clark, B.; Pellikka, P. Landscape analysis using multi-scale segmentation and objectoriented classification. *Recent Adv. Remote Sens.* **2009**, 323–341.
25. Pellikka, P.K.E.; Clark, B.J.F.; Gosa, A.G.; Himberg, N.; Hurskainen, P.; Maeda, E.; Mwang'ombe, J.; Omoro, L.M.A.; Siljander, M. Agricultural expansion and its consequences in the Taita Hills, Kenya. *Dev. Earth Surf. Process.* **2013**, *16*, 165–179. [CrossRef]
26. Burgess, N.D.; Butynski, T.M.; Cordeiro, N.J.; Doggart, N.H.; Fjeldså, J.; Howell, K.M.; Kilahama, F.B.; Loader, S.P.; Lovett, J.C.; Mbilinyi, B.; et al. The biological importance of the Eastern Arc Mountains of Tanzania and Kenya. *Biol. Conserv.* **2007**, *134*, 209–231. [CrossRef]
27. Pellikka, P.K.E.; Lötjönen, M.; Siljander, M.; Lens, L. Airborne remote sensing of spatiotemporal change (1955–2004) in indigenous and exotic forest cover in the Taita Hills, Kenya. *Int. J. Appl. Earth Obs. Geoinf.* **2009**, *11*, 221–232. [CrossRef]
28. Hohenthal, J.; Räsänen, M.; Owidi, E.; Andersson, B.; Minoia, P.; Pellikka, P.K.E. *Community and Institutional Perspectives on Water Management and Environmental Changes in the Taita Hills, Kenya*; University of Helsinki: Helsinki, Finland, 2015; ISBN 9789515113412.
29. Mugedo, J.Z.A.; Waterman, P.G. Sources of tannin: Alternatives to wattle (Acacia mearnsii) among indigenous Kenyan species. *Econ. Bot.* **1992**, *46*, 55–63. [CrossRef]
30. Boudiaf, I.; Le Roux, C.; Baudoin, E.; Galiana, A.; Beddiar, A.; Prin, Y.; Duponnois, R. Soil Bradyrhizobium population response to invasion of a natural Quercus suber forest by the introduced nitrogen-fixing tree Acacia mearnsii in El Kala National Park, Algeria. *Soil Biol. Biochem.* **2014**, *70*, 162–165. [CrossRef]

31. Engeman, R.M.; Sugihara, R.T.; Pank, L.F.; Dusenberry, W.E. A comparison of plotless density estimators using Monte Carlo simulation. *Ecology* **1994**, *75*, 1769–1779. [CrossRef]
32. Khosravipour, A.; Skidmore, A.K.; Isenburg, M.; Wang, T.; Hussin, Y. Generating pit-free canopy height models from airborne lidar. *Photogramm. Eng. Remote Sens.* **2014**, *80*, 863–872. [CrossRef]
33. Richter, R.; Schläpfer, D. Geo-atmospheric processing of airborne imaging spectrometry data. Part 2: Atmospheric/topographic correction. *Int. J. Remote Sens.* **2002**, *23*, 2631–2649. [CrossRef]
34. Valbuena, R. Integrating airborne laser scanning with data from global navigation satellite systems and optical sensors. In *Forestry Applications of Airborne Laser Scanning: Concepts and Case Studies*; Maltamo, M., Naesset, E., Vauhkonen, J., Eds.; Springer: Berlin, Germany, 2014; Volume 27, pp. 63–88.
35. Baldeck, C.A.; Asner, G.P.; Martin, R.E.; Anderson, C.B.; Knapp, E.; Kellner, J.R.; Wright, S.J. Operational tree species mapping in a diverse tropical forest with airborne imaging spectroscopy. *PLoS ONE* **2015**. [CrossRef] [PubMed]
36. Dalponte, M.; Coomes, D.A. Tree-centric mapping of forest carbon density from airborne laser scanning and hyperspectral data. *Methods Ecol. Evol.* **2016**, *7*, 1236–1245. [CrossRef] [PubMed]
37. Roussel, J.-R.; Auty, D. lidR: Airborne LiDAR Data Manipulation and Visualization for Forestry Applications, Version 1.2.0. Available online: https://rdrr.io/cran/lidR/ (accessed on 13 June 2017).
38. R Core Team. *R: A Language and Environment for Statistical Computing*, Version 3.4.0; R Foundation for Statistical Computing: Vienna, Austria, 2017.
39. RSI. *ENVI User's Guide*; Research Systems, Inc.: Boulder, CO, USA, 2004; pp. 1–1150.
40. Zhang, C.; Xie, Z. Combining object-based texture measures with a neural network for vegetation mapping in the Everglades from hyperspectral imagery. *Remote Sens. Environ.* **2012**, *124*, 310–320. [CrossRef]
41. Roberts, D.A.; Roth, K.L.; Perroy, R.L. Hyperspectral vegetation indices. In *Hyperspectral Remote Sensing of Vegetation*; Thenkabail, P.S., Lyon, P.S., Huete, J.G., Eds.; CRC press: Boca Raton, FL, USA, 2011; pp. 309–328.
42. Genuer, R.; Poggi, J.-M.; Tuleau-Malot, C. Variable selection using random forests. *Pattern Recognit. Lett.* **2010**, *31*, 2225–2236. [CrossRef]
43. Genuer, R.; Poggi, J.-M.; Tuleau-Malot, C. VSURF: An R package for variable selection using random forests. *R J.* **2015**, *7*, 19–33.
44. Kuhn, M. Building Predictive Models in R Using the caret Package. *J. Stat. Softw.* **2008**, *28*, 1–26. [CrossRef]
45. Karatzoglou, A.; Smola, A.; Hornik, K.; Zeileis, A. kernlab—An S4 Package for Kernel Methods in R. *J. Stat. Softw.* **2004**, *11*, 1–20. [CrossRef]
46. Liaw, A.; Wiener, M. Classification and Regression by randomForest. *R News* **2002**, *2*, 18–22. [CrossRef]
47. Congalton, R.G. A review of assessing the accuracy of classifications of remotely sensed data. *Remote Sens. Environ.* **1991**, *37*, 35–46. [CrossRef]
48. Dalponte, M.; Ørka, H.O. varSel: Sequential Forward Floating Selection Using Jeffries-Matusita Distance. Available online: https://rdrr.io/cran/varSel/ (accessed on 5 November 2016).
49. Ballanti, L.; Blesius, L.; Hines, E.; Kruse, B. Tree species classification using hyperspectral imagery: A comparison of two classifiers. *Remote Sens.* **2016**, *8*, 445. [CrossRef]
50. Foody, G.M. Thematic map comparison: evaluating the statistical significance of differences in classification accuracy. *Photogramm. Eng. Remote Sens.* **2004**, *70*, 627–633. [CrossRef]
51. Dietterich, T.G. Approximate statistical tests for comparing supervised classi cation learning algorithms. *Neural Comput.* **1998**, *10*, 1895–1923. [CrossRef] [PubMed]
52. Jones, T.G.; Coops, N.C.; Sharma, T. Assessing the utility of airborne hyperspectral and LiDAR data for species distribution mapping in the coastal Pacific Northwest, Canada. *Remote Sens. Environ.* **2010**, *114*, 2841–2852. [CrossRef]
53. Hovi, A.; Korhonen, L.; Vauhkonen, J.; Korpela, I. LiDAR waveform features for tree species classification and their sensitivity to tree- and acquisition related parameters. *Remote Sens. Environ.* **2016**, *173*, 224–237. [CrossRef]
54. Van Der Linde, J.A.; Roux, J.; Wingfield, M.J.; Six, D.L. Die-off of giant Euphorbia trees in South Africa: Symptoms and relationships to climate. *S. Afr. J. Bot.* **2012**, *83*, 172–185. [CrossRef]
55. Landmann, T.; Piiroinen, R.; Makori, D.M.; Abdel-Rahman, E.M.; Makau, S.; Pellikka, P.; Raina, S.K. Application of hyperspectral remote sensing for flower mapping in African savannas. *Remote Sens. Environ.* **2015**, *166*, 50–60. [CrossRef]

56. Colgan, M.S.; Baldeck, C.A.; Féret, J.-B.; Asner, G.P. Mapping savanna tree species at ecosystem scales using support vector machine classification and BRDF correction on airborne hyperspectral and LiDAR data. *Remote Sens.* **2012**, *4*, 3462–3480. [CrossRef]
57. Clark, M.L.; Roberts, D.A. Species-Level Differences in Hyperspectral Metrics among Tropical Rainforest Trees as Determined by a Tree-Based. *Remote Sens.* **2012**, *4*, 1820–1855. [CrossRef]

remote sensing

MDPI

Article

Adaptive Window-Based Constrained Energy Minimization for Detection of Newly Grown Tree Leaves

Shih-Yu Chen [1], Chinsu Lin [2,*], Chia-Hui Tai [1] and Shang-Ju Chuang [1]

[1] Department of Computer Science and Information Engineering, National Yunlin University of Science and Technology, Douliu 64002, Taiwan; sychen@yuntech.edu.tw (S.-Y.C.); second59420@gmail.com (C.-H.T.); m10517016@yuntech.edu.tw (S.-J.C.)
[2] Department of Forestry and Natural Resources, National Chiayi University, Chiayi 60004, Taiwan
* Correspondence: chinsu@mail.ncyu.edu.tw

Received: 23 November 2017; Accepted: 8 January 2018; Published: 12 January 2018

Abstract: Leaf maturation from initiation to senescence is a phenological event of plants that results from the influences of temperature and water availability on physiological activities during a life cycle. Detection of newly grown leaves (NGL) is therefore useful for the diagnosis of tree growth, tree stress, and even climatic change. This paper applies Constrained Energy Minimization (CEM), which is a hyperspectral target detection technique to spot grown leaves in a UAV multispectral image. According to the proportion of NGL in different regions, this paper proposes three innovative CEM based detectors: Subset CEM, Sliding Window-based CEM (SW CEM), and Adaptive Sliding Window-based CEM (AWS CEM). AWS CEM can especially adjust the window size according to the proportion of NGL around the current pixel. The results show that AWS CEM improves the accuracy of NGL detection and also reduces the false alarm rate. In addition, the results of the supervised target detection depend on the appropriate signature. In this case, we propose the Optimal Signature Generation Process (OSGP) to extract the optimal signature. The experimental results illustrate that OSGP can effectively improve the stability and the detection rate.

Keywords: hyperspectral detection; target detection; sprout detection; constrained energy minimization; iterative algorithm; adaptive window

1. Introduction

The persistence of forest ecosystem resources is the key to protecting wild coverage of reproductive trees in order to alleviate global warming or the impact of climate change. Specifically, the variance in the area of woods, the accumulation of forest biomass/carbon storage, and the improvement of healthy forests are the periodic evaluation indices of global forest resources for forest sustainability, according to Forest Resources Assessment FAO (Food and Agriculture Organization of the United Nations) [1]. Therefore, using telemetry to monitor the health level of forest ecosystems has a critical effect on the subject of global warming control. Climate change may affect phenological events such as the onset of green-up and dormancy [2]. Trees start to sprout once they sense the growing signals in early spring. After the leaf initiation stage, the newly sprouted leaflet will gradually develop and further facilitate tree growth in crown width, height, diameter, and carbon storage [3]. As a result, the newly grown leaves (NGL) can be seen as the first objects of trees in response to a change in temperature, and can therefore provide critical information for the early detection of climate changes [4]. UAV-sensed images are generally collected at low altitude; the images are supposedly free of atmospheric effects [5], and provide very high spatial resolution for applications. Taking the strengths of UAV-sensed images, NGL over a forest area can be detected via appropriate remote sensing techniques.

In forest science, remote sensing has been applied to investigate species classification [6,7], tree delineation [8,9], and biomass productivity estimation [10,11]. The detection of NGL is a new application in respect to the previous applications. The target of interest may occur under a very low probability or probably has a relatively smaller size than the background, such as the damaged portion of crown in the forest canopy or the new leaf crown in the forest canopy. In this case, the traditional spatial domain (i.e., literal)-based image processing techniques [12–15] may fail to extract these targets effectively, especially when the target size is smaller than the pixel resolution. In contrast, the technique using spectral characteristics to detect the subpixel level is one of the more feasible methods. From the angle of multispectral/hyperspectral detection, the spectral information-based target detection technique [16–18] should be able to solve these problems.

Hyperspectral subpixel detection techniques can be divided into active and passive manners. In the active methods, the detectors only use single or multiple spectral signatures of targets of interest for detection purposes; e.g., Constrained Energy Minimization (CEM) [19], Correlation Mahalanobis Distance (RMD) [7,20], Mahalanobis Distance (KMD) [17,21], Adaptive Coherence Estimator (ACE) [22], Target-Constrained Interference-Minimized Filter (TCIMF) [17,18,23], etc. The correctness of target information plays an important role. Incorrect information results in a false alarm and omission of a target detection result. Therefore, how to provide the correct target object information is a very important step. The Optimal Signature Generation Process (OSGP) that is proposed in this paper increases the accuracy of selecting a target iteratively so as to solve the previously mentioned problem.

Hyperspectral algorithms have been developed for many different target detections in recent years and are used in different areas [24–29]. Many studies have proposed CEM based algorithms [30,31] in the last decade. However, using hyperspectral algorithms to detect targets in RGB images presents several issues since spectral information is insufficient and spatial information is not used, and thus a false alarm is likely to occur where there is a similar spectral characteristic. In order to solve this dilemma, this paper proposes three innovative CEM based detectors-Subset CEM, Sliding Window-based CEM, and Adaptive Sliding Window-based CEM to establish their own autocorrelation matrix, and uses sliding window point-to-point scanning for calculation. As the sliding window passes through, the contrast between a NGL and the background can be enhanced. When compared with traditional CEM results that incur too many false alarms, our proposed methods can solve this issue and produce more stable results.

2. Materials and Methods

2.1. Constrained Energy Minimization

Traditional Constrained Energy Minimization (CEM) [16–19] of active hyperspectral target detection is the major technique that is adopted in this paper. The CEM only needs the spectral signature of one specific target of interest during target detection; it is free of the spectral signature of other targets or background. Many target detection methods have been proposed in recent years, among those, the CEM only needs the signature of one target of interest and the detection is stable. Thus, this paper selects CEM to compare local improvement methods. In the CEM algorithm, only one spectral signature (desired signature or target of interest) is given, referred to as **d**, any prior knowledge is not required, e.g., multiple targets of interest or background. In other words, users can extract the specific target of interest without any background information, implementing target detection. This is one of the major advantages of CEM. Another advantage of CEM is that as many signal sources cannot be recognized or observed with the naked eye, some materials may be detected by sensors leading to false alarm. However, the CEM transposes the correlation matrix **R** of data samples before the desired signature **d** is extracted. The sample autocorrelation matrix can be defined as $\mathbf{R} = (1/N) \sum_{i=1}^{N} \mathbf{r}_i \mathbf{r}_i^{\mathrm{T}}$, so that the background can be suppressed by **R**, and the filter is matched with signature d to enhance the capability of detecting signature, the execution is more efficient. The CEM is evolved from the LCMV proposed by Frost [32]. If there are N pixels **r** in a hyperspectral image with L band, which are $\{\mathbf{r}_1, \mathbf{r}_2, \mathbf{r}_3, \ldots, \mathbf{r}_N\}$, where $\mathbf{r}_i = (\mathbf{r}_{i1}, \mathbf{r}_{i2}, \mathbf{r}_{i3}, \ldots, \mathbf{r}_{iL})^{\mathrm{T}}$, the desired target to be looked for is represented by $\mathbf{d}$, defined as $\mathbf{d} = (\mathbf{d}_1, \mathbf{d}_2, \mathbf{d}_3, \ldots, \mathbf{d}_L)^{\mathrm{T}}$, the desired target to be looked for can be detected by finite impulse response filter (FIR) based on CEM and desired target $\mathbf{d}$. The filter coefficient is defined as $\mathbf{w} = (\mathbf{w}_1, \mathbf{w}_2, \mathbf{w}_3, \ldots, \mathbf{w}_L)^{\mathrm{T}}$, the $\mathbf{w}$ can be obtained with the minimum average energy, defined as $\mathbf{d}^{\mathrm{T}}\mathbf{w} = \mathbf{w}^{\mathrm{T}}\mathbf{d} = 1$. Therefore, if $\mathbf{y}_i$ is defined as $\mathbf{r}_i$ and imported into FIR, and then $\mathbf{y}_i$ can be expressed as

$$\mathbf{y}_i = \sum_{l=1}^{L} \mathbf{w}_l \mathbf{r}_{il} = \mathbf{w}^{\mathbf{T}}\mathbf{r}_i = \mathbf{r}_i{}^{\mathbf{T}}\mathbf{w} \tag{1}$$

The average energy is

$$(1/N)\sum_{i=1}^{N} \mathbf{y}_i{}^2 = (1/N)\sum_{i=1}^{N} \left(\mathbf{r}_i{}^{\mathbf{T}}\mathbf{w}\right)^2 = \mathbf{w}^{\mathbf{T}}\left[(1/N)\sum_{i=1}^{N} \mathbf{r}_i\mathbf{r}_i{}^{\mathbf{T}}\right]\mathbf{w} = \mathbf{w}^{\mathbf{T}}\mathbf{R}_{L\times L}\mathbf{w} \tag{2}$$

With the minimum average energy, the optimal solution of w can be obtained

$$\min_{\mathbf{w}}\left\{\mathbf{w}^{\mathbf{T}}\mathbf{R}_{L\times L}\mathbf{w}\right\} \text{ subject to } \mathbf{d}^{\mathbf{T}}\mathbf{w} = \mathbf{w}^{\mathbf{T}}\mathbf{d} = 1 \tag{3}$$

According to the theory of Harsanyi [19], the optimal solution to the weight vector of one L band is

$$\mathbf{w}^{\mathrm{CEM}} = \frac{\mathbf{R}_{L\times L}^{-1}\mathbf{d}}{\mathbf{d}^{\mathbf{T}}\mathbf{R}_{L\times L}^{-1}\mathbf{d}} \tag{4}$$

Equation (3) is substituted in Equation (2), the result of CEM is

$$\delta^{\mathrm{CEM}} = \left(\mathbf{w}^{\mathrm{CEM}}\right)^{\mathbf{T}}\mathbf{r} = \left(\mathbf{d}^{\mathbf{T}}\mathbf{R}_{L\times L}^{-1}\mathbf{d}\right)^{-1}\left(\mathbf{R}_{L\times L}^{-1}\mathbf{d}\right)^{\mathbf{T}}\mathbf{r} \tag{5}$$

2.2. Subset CEM

The first target detection algorithm using autocorrelation matrix $\mathbf{S}$ that is proposed in this paper is a novel method using subsets, known as Subset CEM. The Subset CEM splits the image into several small square images; these small images are the subsets of the original image. The CEM detection is then implemented, and the results of subsets are patched up to obtain a complete resulting image. In other words, the small image of each subset has its own autocorrelation matrix $\mathbf{S}$. For example, a 1200×1500 image is divided into nine small images; the resolution of each small image is 400×500, the autocorrelation matrix $\mathbf{S}$ is $\mathbf{S}_1$, $\mathbf{S}_2$, $\mathbf{S}_3$, … , $\mathbf{S}_9$, respectively, and the corresponding $\mathbf{S}_n$ of each pixel is substituted into CEM for detection, thus obtaining the result. Figure 1 is the schematic diagram of the autocorrelation matrix in the image. The subset image size of the local autocorrelation matrix $\mathbf{S}$ is obtained by trial and error. Normally, using five times smaller than the original size is a good first try. The image resolution used in this paper is 1000×1300, and thus the image is divided into 200×260 subset images.

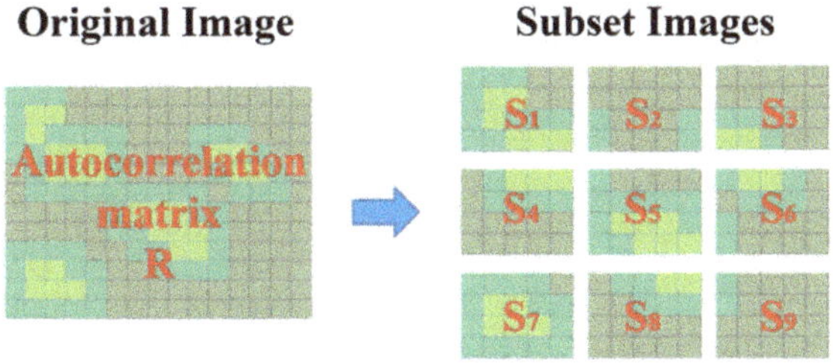

Figure 1. Schematic diagram for the autocorrelation matrix of the original image and subset images.

The results show that this method can effectively suppress the background pixels that are too similar to the desired target. Because the image is divided into the set of small images, the autocorrelation matrix of each small image changes accordingly. There is a different **S** for suppressing the background according to different images, and **S** is calculated according to the pixels in a small area; thus, the small difference between similar spectral signatures is enlarged. It is easier to judge the difference between two RGB values for suppression, so as to increase the detection rate. Figure 2 shows the detection process after the Subset CEM splits the image into subsets.

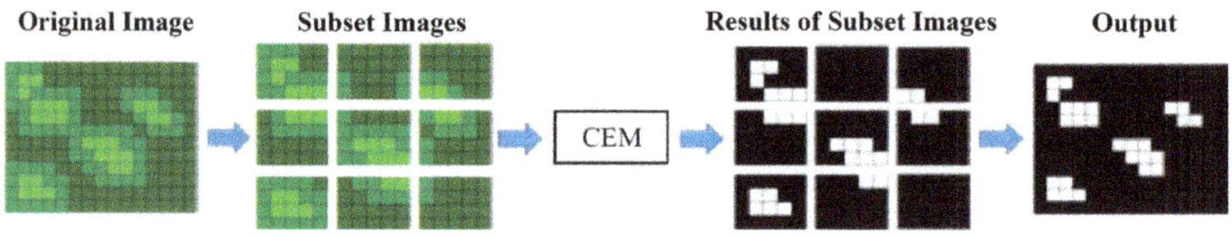

Figure 2. Schematic diagram of the division of subset images by Subset CEM.

2.3. *Sliding Window-Based CEM (SW CEM)*

Section 2.2 mentioned that the Subset CEM can effectively reduce the false alarm rate of similar background spectrums using the concept of the local autocorrelation matrix **S**. The Subset CEM divides the image into small square images, where each small image has its own autocorrelation matrix **S**. In other words, every two small images have dissimilar **S**. However, Subset CEM uses non-overlapped windows, which causes artifacts at the borders of subset images. In order to resolve such an issue, this paper proposes the pixel-by-pixel sliding window-based CEM (SW CEM) for detection. The sliding window concept is used in many studies [27,33–35]. The pixel-by-pixel CEM uses a sliding window of fixed size to obtain the RGB values around each pixel, according to different spectral characteristics around each pixel, so as to determine its autocorrelation matrix $\mathbf{S}_n$. In other words, if the Subset CEM divides the image into small square images, then the pixel-by-pixel CEM divides the image into pixels, combined with a sliding window, to acquire the pixels around the pixel to determine the autocorrelation matrix $\mathbf{S}_n$. Figure 3 shows the sliding window and direction.

r(m-2,n-2)	r(m-1,n-2)	r(m,n-2)	r(m+1,n-2)	r(m+2,n-2)	r(m+3,n-2)	r(m+4,n-2)
r(m-2,n-1)	r(m-1,n-1)	r(m,n-1)	r(m+1,n-1)	r(m+2,n-1)	r(m+3,n-1)	r(m+4,n-1)
r(m-2,n)	r(m-1,n)	r(m,n)	r(m+1,n)	r(m+2,n)	r(m+3,n)	r(m+4,n)
r(m-2,n+1)	r(m-1,n+1)	r(m,n+1)	r(m+1,n+1)	r(m+2,n+1)	r(m+3,n+1)	r(m+4,n+1)
r(m-2,n+2)	r(m-1,n+2)	r(m,n+2)	r(m+1,n+2)	r(m+2,n+2)	r(m+3,n+2)	r(m+4,n+2)
r(m-2,n+3)	r(m-1,n+3)	r(m,n+3)	r(m+1,n+3)	r(m+2,n+3)	r(m+3,n+3)	r(m+4,n+3)
r(m-2,n+4)	r(m-1,n+4)	r(m,n+4)	r(m+1,n+4)	r(m+2,n+4)	r(m+3,n+4)	r(m+4,n+4)
r(m-2,n+5)	r(m-1,n+5)	r(m,n+5)	r(m+1,n+5)	r(m+2,n+5)	r(m+3,n+5)	r(m+4,n+5)

Figure 3. Sliding window matrices of $\mathbf{r}_{(m,n)}$ and $\mathbf{r}_{(m+1,n)}$.

This means that each pixel in the image has its $\mathbf{S}_n$, and each $\mathbf{S}_n$ is independent and different. Hence, the SW CEM can be defined as:

$$\text{SW\_CEM} = \frac{\mathbf{d}^{\mathrm{T}}\mathbf{R}_{mn}^{-1}\mathbf{r}_{mn}}{\mathbf{d}^{\mathrm{T}}\mathbf{R}_{mn}^{-1}\mathbf{d}} \tag{6}$$

where, $\mathbf{r}_{mn}$ is the current pixel value, and $\mathbf{R}_{mn}$ is the autocorrelation matrix of $\mathbf{r}_{mn}$, if the size of the sliding window is $2k + 1$, as shown in Figure 4.

When the size of a sliding window is known, $\mathbf{R}_{mn}$ can be defined as:

$$\mathbf{R}_{mn} = \frac{1}{(2k+1)^2}\sum_{i=m-k}^{m+k}\sum_{j=n-k}^{n+k}\mathbf{x}_{ij}\mathbf{x}_{ij}^{\mathbf{T}} \tag{7}$$

where, $\mathbf{x}_{ij}$ represents each pixel in the sliding window, and $\frac{1}{(2k+1)^2}$ is a constant, if $\mathbf{R}_{mn}$ is simplified by $\mathbf{S}_{mn}$.

$$\mathbf{R}_{mn} = \frac{1}{(2k+1)^2}\sum_{i=m-k}^{m+k}\sum_{j=n-k}^{n+k}\mathbf{x}_{ij}\mathbf{x}_{ij}^{\mathbf{T}} = \frac{1}{(2k+1)^2}.\mathbf{S}_{\mathbf{mn}} \tag{8}$$

We substitute Equation (8) into Equation (6) to obtain:

$$\textbf{SW\_CEM} = \frac{\mathbf{d}^{\mathrm{T}}\mathbf{R}_{mn}^{-1}\mathbf{r}_{mn}}{\mathbf{d}^{\mathrm{T}}\mathbf{R}_{mn}^{-1}\mathbf{d}} = \frac{(2k+1)^2(\mathbf{d}^{\mathrm{T}}\mathbf{S}_{mn}^{-1}\mathbf{r}_{mn})}{(2k+1)^2(\mathbf{d}^{\mathrm{T}}\mathbf{S}_{mn}^{-1}\mathbf{d})} = \frac{\mathbf{d}^{\mathrm{T}}\mathbf{S}_{mn}^{-1}\mathbf{r}_{mn}}{\mathbf{d}^{\mathrm{T}}\mathbf{S}_{mn}^{-1}\mathbf{d}} \tag{9}$$

$$\mathbf{S}_{mn} = \sum_{i=m-k}^{m+k}\sum_{j=n-k}^{n+k}\mathbf{x}_{ij}\mathbf{x}_{ij}^{\mathbf{T}} \tag{10}$$

where, $\mathbf{S}_{mn}$ is the autocorrelation matrix of current pixel $\mathbf{r}_{mn}$, and the capability of suppressing the background can be enhanced for each region by $\mathbf{S}_{mn}$.

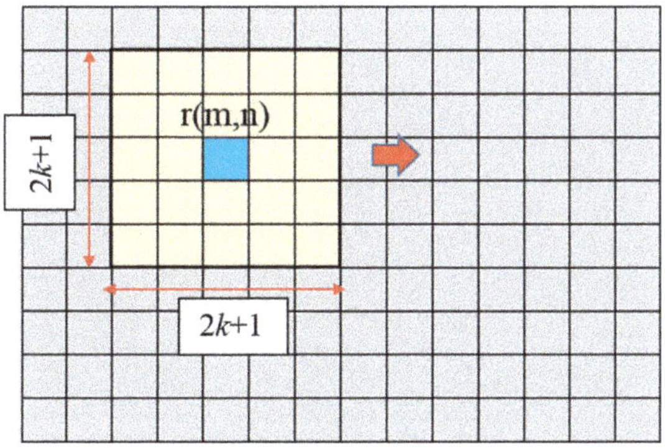

Figure 4. Schematic diagram of a sliding window.

2.4. Adaptive Sliding Window-Based CEM (ASW CEM)

The adaptive window concept has been applied to targets detection in many applications in the last decade, such as vehicles detection [36,37], adaptive filters [38], and anomaly detection [39,40]. When compared to SW CEM, which uses a fixed window size to calculate autocorrelation matrices **S**, AWS CEM determines the window size according to the spatial and spectral characteristics around each pixel, so as to suppress the background. The optimum size of the sliding window varies with the quantity of NGL around each pixel, and thus the sliding window improves based on the local CEM in this paper. The size of sliding window is determined by acquiring the proportion of sprouts around the current pixel. When the sliding window size $2K + 1$ of the current pixel is determined, the result of SW CEM target detection can be obtained. In this case, we developed Adaptive Sliding Window-based CEM (ASW CEM) to combine adaptive window concept in CEM. Figure 5 illustrates the flowchart of ASW CEM. ASW CEM can change the size of the sliding window according to the ratio of the NGL around the current pixel to enhance NGL and suppress background.

The execution of ASW CEM comprises six steps:

1. Input image
2. Decide the default size of the sliding window
3. Calculate the rate of the sprout in the sliding window
4. If the rate of the sprout meets the set condition or the window size has reached the default maximum or minimum window, then the **S** in Equation (10) is calculated according to the pixel values of the current window size. If the rate of the sprout is not met or the window size has not reached the limit value, then the size of the sliding window is changed and return to Step 3; otherwise, proceed to Step 5.
5. The **S** obtained in Step 4 is used to calculate CEM to obtain the result value of the current pixel.
6. If all pixels of the image have been calculated, then ASW CEM detection is finished; otherwise, return to Step 2.

The calculation of the rate of the sprout in the sliding window in Step 3 and the sliding window change conditions in Steps 2 and 4 are introduced below.

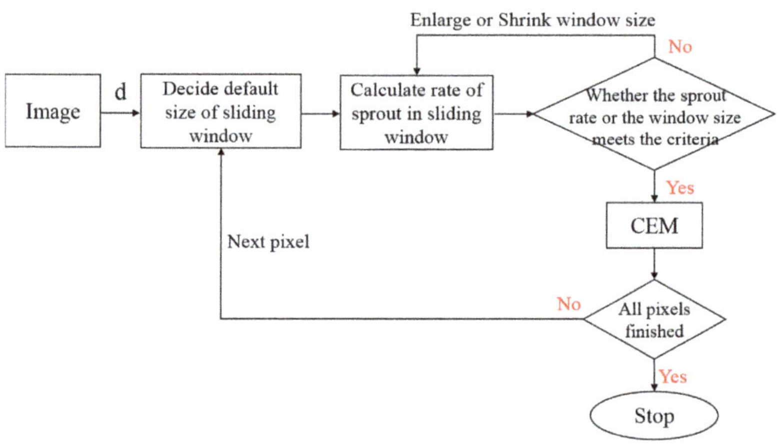

Figure 5. Flowchart of Adaptive Sliding Window-based CEM.

2.4.1. Acquire the Rate of the NGL around the Current Pixel (NGL Map)

In order to enable the sliding window of each pixel to decide whether or not to enlarge or to shrink the current window according to the proportion of peripheral NGL, this paper requires a sprout map for judgment and reading. To set up the sprout distribution map, the spectral comparison is conducted on the he Spectral Information Divergence (SID) [17,41] for experimental multispectral images before ASW CEM, so as to obtain a preliminary sprout detection result. This result contains a small false alarm, but according to this result, when deciding whether or not to enlarge or shrink the sliding window, only the relative rate of the sprout in the sliding window shall be calculated, as the actual number of NGL is not required. The SID resulting image is segmented several times by using Otsu's method [42]. The rate of the sprout in the image is preliminarily estimated at 1~2%, so as to minimize the false alarm, and the major sprout points are maintained for calculation. When the image with a preliminary estimate of the sprout is obtained, the proportion of NGL can be calculated by using the default sliding window size, and the size of the sliding window is changed according to the number of NGL. Figure 6 is the flowchart of this step.

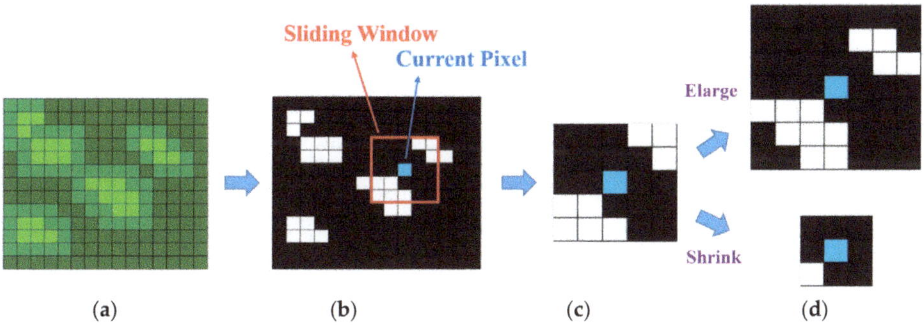

Figure 6. The acquired rate of the sprout in the window from the sprout proportion chart and resizing the window: (**a**) original image; (**b**) preliminarily estimated rate of the sprout; (**c**) calculated rate of the sprout in the sliding window; and, (**d**) changed window size.

2.4.2. Adaptive Sliding Window

To calculate the rate of the sprout around the current pixel, a sliding window of preset size needs to be made. The rate of the sprout in this window decides whether or not to enlarge or shrink the sliding window for subsequent algorithmic detection. Figure 7 is the flowchart of this step. A larger sliding window size is required in the region with a higher rate of the sprout; the optimum size of maximum window is set as m^2; on the contrary, a smaller sliding window size is required in the region with a lower rate of the sprout, and the optimum size of the smallest window is set as n^2. When the maximum window m^2 and minimum window n^2 are obtained, the default window is set as an intermediate between maximum and minimum windows, i.e., $\left(\frac{m+n}{2}\right)^2$. The advantage is that as the initial window is intermediately sized, the window is enlarged or shrunk to the limit relatively fast. Afterwards, the sliding window is enlarged or shrunk gradually, according to the rate of the sprout in the window.

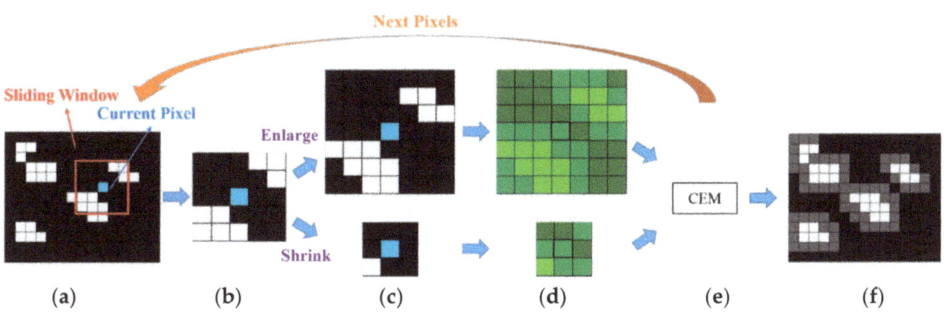

Figure 7. Flowchart of Adaptive Sliding Window CEM: (**a**) extract the newly grown leaves (NGL) around the current pixel from the preliminarily estimated sprout distribution map; (**b**) calculate the proportion of NGL in the sliding window; (**c**) enlarge or shrink the sliding window according to the proportion of NGL; (**d**) extract the pixel values from the window in relative position of the original image; (**e**) calculate the autocorrelation matrix for the pixel values in the window and substitute it in CEM for calculation; if all pixel values in the image are not finished, calculate the next pixel; and, (**f**) export the result of all pixel values.

As the distribution of NGL is not even, when the window is shrunk or enlarged, the rate of the sprout does not always increase or decrease. This method can enlarge and shrink the window. In order to avoid the non-uniform rate of the sprout leading to an infinite circulation of window enlargement or shrinkage, the initial window is used to calculate the sprout as a watershed. When the rate of the sprout in the initial window is higher than a threshold, the sliding window is enlarged gradually until the rate of the sprout in the window is lower than a threshold or the window is maximized before CEM detection. If the rate of the sprout in the initial window is lower than a threshold, then the sliding window is shrunk gradually until the rate of the sprout in the window is higher than a threshold, or the window is minimized before CEM detection. In order to adjust the window size conditionally in the stable level, this paper includes a parameter ε as initial NGL rate in the window, where ε is the proportion of the number of NGL in the window to the total number of pixels in the sliding window. When the rate of the sprout in the window is lower than ε, the window is shrunk; if the rate of the sprout in the window is higher than ε, then the window is enlarged.

2.5. Optimal Signature Generation Process (OSGP)

In order to remedy the defect in the CEM-related algorithm in that only one desired target can be selected, the Optimal Signature Generation Process (OSGP) is used herein to obtain a stable desired target by the iterative process. The idea of the iterative process is similar to K-means [43], iterative self-organizing data (ISODATA) [44] and iterative FLDA [45]. The OSGP implements Subset CEM target detection for the image iteratively. When the result of CEM is obtained, the image is segmented by using Otsu's method until the number of result pixels is 2–3%, which is the target pixel with the highest probability of the sprout. These pixels correspond to the same pixel RGB values in the original image, averaged as a new target **d′**. If the pixel value of **d′** is not similar enough to that of **d**, then **d′** is substituted in the next CEM, a new desired target is obtained. This is repeated until this and the last Spectral Angle Mapper [46] are smaller than a value θ, and then the current target **d** is exported. The threshold of spectral angle was tested continuously, and the threshold was set as 0.003. Figure 8 is the flowchart of OSGP.

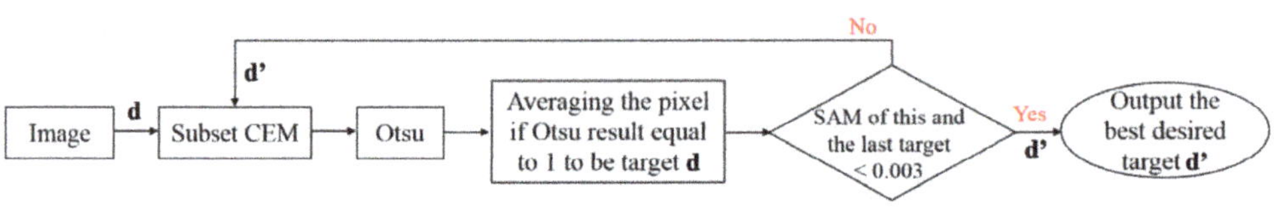

Figure 8. Flowchart of Optimal Signature Generation Process (OSGP).

It is noteworthy that CEM uses the global correlation matrix **R** to suppress the background, and it is likely that the background has similar spectral signatures. Simply iterating CEM is the same condition, and some background pixels that do not belong to the desired target will be misrecognized as the result pixels, and the averaging of them influences the new desired target. A stable target **d** is still obtained after iteration, but the misrecognized pixels may result in errors in **d**, thus reducing the detection accuracy slightly. To solve this dilemma, this paper proposes Optimal Signature Generation Process (OSGP) and replaces CEM by subset CEM. The Subset CEM splits the image into many small sets; in the global view of the full image, the small square image of a subset is the local concept. The local algorithm can effectively suppress the pixel RGB values similar to the desired target, which is substituted into the iterative algorithm. This not only reduces the misrecognized result of pixels during iteration, but also obtains a better desired target **d**, so that the precision of detection increases and the probability of a false alarm decreases.

2.6. Parameter Settings of Different Algorithms

For different parameter settings of algorithms in this paper, the results are different, and so the parameter settings of the multiple CEM algorithm are listed in this section, as shown in Table 1. The input **d** of CEM, Subset CEM, SW CEM, and ASW CEM is the randomly selected desired target. The desired target **d′** is iterated by using OSGP. The OSGP iteration stop condition is that the SAM value of two adjacent targets **d** shall be smaller than θ; if tenable, then the desired target **d′** is exported. Here, ε is the condition value of the rate of the sprout when the sliding window of ASW CEM is enlarged or shrunk. In the following experiments in this study, ε is set to 1%. However, this parameter depends on the proportion of the number of target pixel in the entire image.

Table 1. Parameter settings of different algorithms.

Algorithm	Input Desired Target	Autocorrelation Matrix R	OGSP Stopping Rule	NGL Rate
Traditional CEM	**d**/**d′**	Global **R**	θ	
Subset CEM	**d**/**d′**	Local **R** (Fixed window size)	θ	
SW CEM	**d**/**d′**	Local **R** (Fixed window size)	θ	
ASW CEM	**d**/**d′**	Local **R** (Adaptive window size)	θ	ε

In the following experiments the Subset image size is set as 200 × 260 pixel, and each small image uses the same **d** and **d′**, where each small image takes the pixels of its image size to calculate **S**. The subset image size of the local autocorrelation matrix **S** is obtained by trial and error. When each small image has calculated CEM, the results are exported and merged into the original picture size, and the merged picture is the result of Subset CEM.

The window size of SW CEM varied by different application and images. Normally, 5–6 times smaller than the entire images are the good try as the initial setting. This study sets the sliding window size as 151 × 151 pixel. It extracts the pixels of 151 × 151 pixel around each pixel to calculate **S** in Equation (10), which are substituted into CEM in order to work out the result value of the pixel. When each pixel is calculated, the output result of SW CEM is obtained.

ASW CEM uses the original image for SID measurement and gives the preliminarily estimated NGL map, and then ASW CEM employs the sliding window of preset size to calculate the rate of the sprout. When the rate of the sprout mismatches the stop condition, the window size is changed and this rate in the window is recalculated, until the window size reaches the threshold or the rate of the sprout is equivalent to ε. The pixel values in this window are used to calculate S, which is substituted into CEM to work out the result value of the pixel. When each pixel has been calculated, the output result of ASW CEM is obtained. For this experiment, the maximum size of the sliding window is set as 151 × 151, the minimum window is 31 × 31, and the initial window is 91 × 91. The initial window size can be set as the average of the maximum and minimum size.

3. Experiments

3.1. Experimental Procedure

This section introduces the experimental process of all detection algorithms used in this paper, including CEM, Subset CEM, Sliding Window-based CEM, and Adaptive Sliding Window-based CEM. In order to remedy the defect in the target detection algorithm that only one d can be selected at one time, the Optimal Signature Generation Process (OSGP) is used, and the optimum target of interest **d′** is iterated by iterative learning. The incorrect results of selection errors are thus reduced effectively. Figure 9 shows the approximate process of all the detection algorithms for this experiment.

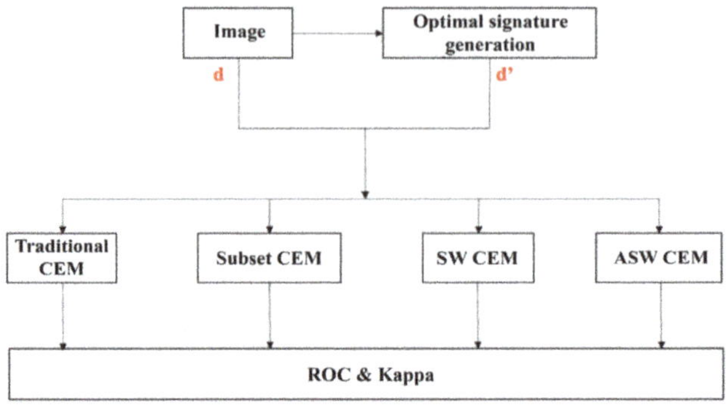

Figure 9. Flowchart of all algorithms for this experiment.

There are two methods of hyperspectral target detection for selecting the desired target. One method only selects the single target **d** for detection. The other method selects the pixel values of multiple desired targets for detection. The CEM is the first type. CEM only selects a single target for detection, and so the quality of detection result highly depends on the selected desired target.

The desired targets **d** used by all of the detection algorithms for the experiments are randomly selected from the ground truth and **d'** is iterated by using OSGP to increase the precision of detection. The full image is used to calculate autocorrelation global **R** for target detection of CEM.

3.2. Description of the Study Site

The study site is in Baihe District (23°20′N, 120°27′E) which is part of Tainan City, Taiwan. Tainan City is characterized by a tropical savanna climate. The weather is generally hot and humid. The mean annual temperature is 24.38 °C. The authors have deployed a few permanent plots over the broadleaf forest for research of forest growth [47,48] in 2008. In which, a series of ground inventory is annually conducted for stand dynamics [49].

3.2.1. UAV Data Collection

We applied the picture of a forest in the middle of Taiwan taken by a Canon PowerShot S110 camera on an eBee RTK drone flying at an altitude of 239.2 ft on 12 July 2014. This image has R, G, and B bands. Data acquisition took place under wonderful weather conditions. The ground pixel size is 6 cm, the original image resolution is 1000 × 1300 pixels, and the actual area of the full image is 60 m × 78 m. The data have been successfully used to derive forest canopy height model [50] and the desired target **d** for the target detection algorithm in this paper is the desired target that is selected randomly in the experimental image, as shown in Figure 10. The red circle is the target **d** for detection.

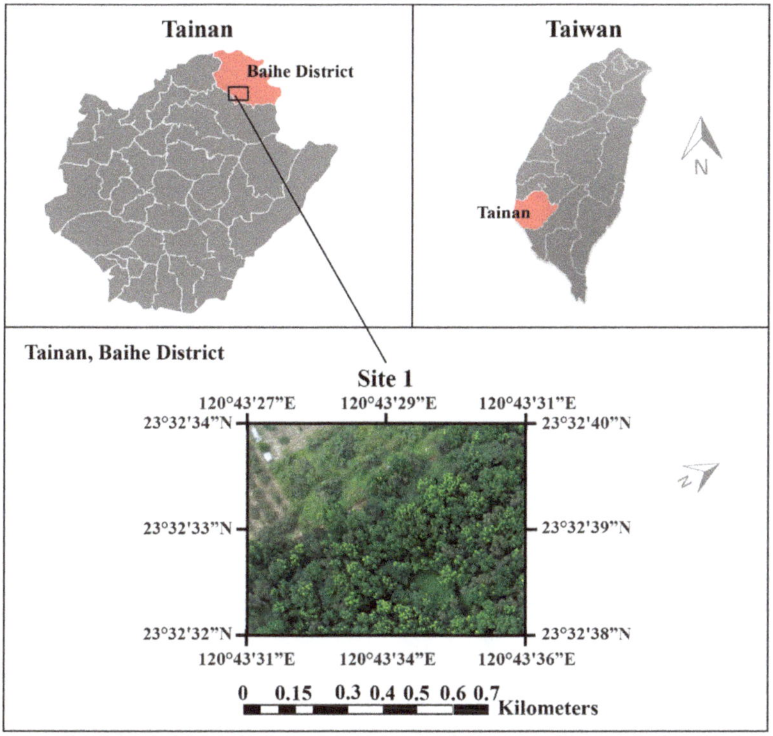

Figure 10. Study site at Baihe District, Tainan, Taiwan.

3.2.2. Ground Truth

As shown in Figure 11, the NGLs can be visually interpreted due to their appearance of being bright and light green and is aggregated over tree crowns. According to a row of several years of inventory, the ground truth of the NGL over the images were visually interpreted and also validated in situ. In order to quantify and compare the effects of different target detection methods, there must be a NGL detection map as the standard and measure, i.e., ground truth of Region 1 and Region 2, as shown in Figure 12. Table 2 tabulates the number of pixels in NGL and non-NGL for Region 1 and 2. The NGL are only about 3–4% of the entire images.

(a) (b)

Figure 11. 1000 × 1300 actual images of a forest in central Taiwan: (**a**) actual image of Region 1; and, (**b**) actual image of Region 2.

(a) (b)

Figure 12. 1000 × 1300 Groundtruth of forest in central Taiwan: (**a**) Groundtruth of Region 1 in the original picture; (**b**) Groundtruth of Region 2 in the original picture.

Table 2. Rates of Sprout and Non-Sprout in the ground truth.

Image	Sprout		Non-Sprout	
	Pixels	Rate	Pixels	Rate
Region 1	49,427	3.80%	1,250,573	96.20%
Region 2	55,140	4.24%	1,244,860	95.76%

3.3. Evaluation of Detection Results

The research methods used in this study were introduced in previous sections. In order to validate whether the three methods that are proposed herein can improve the original global CEM, two methods for evaluating the precision are used in this paper. The first method is the ROC curve [51–53], which is used to calculate the detection effect of a hyperspectral algorithm. The second method is Cohen's kappa [54], which is an evaluation method extensively used in biology to calculate the model precision. In order to perform quantitative analysis, we further calculated the area under curve (AUC) for each ROC curves and overall accuracy (ACC).

3.3.1. ROC Curve

The main concept of ROC analysis [51–53] is a binary classification model, meaning there are only two classes of output, such as positive/negative, pass/failure, animal/non-animal, etc.

For classification, a threshold must be given, and the threshold separates two classes. The probability of detection power (P_D) and false alarm probability (P_F) may differ under different thresholds. If the threshold is too high, then too many NGL will be estimated as non-NGL. If it is too low, then there will be more false alarms. To avoid this condition, P_D and P_F are calculated, respectively, by using different thresholds, and all threshold (τ) and P_D and P_F are drawn to obtain a ROC curve, as shown in Figure 13.

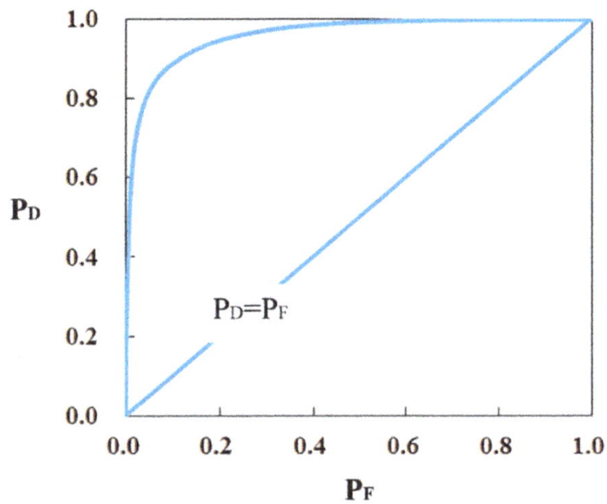

Figure 13. ROC Curve.

The optimum threshold (τ) depends on P_D and P_F to separate the NGL from the background, defined as:

$$\tau = \text{Arg Max}\ (P_D(\tau) + (1 - P_F(\tau))) \tag{11}$$

As we hope that P_D is larger the better and that P_F is smaller the better, then the optimum threshold (τ) can be obtained on this condition. According to the detection result and practical situation, an Error Matrix can be given.

Generally, the weights of P_D and $(1 - P_F(\tau)$ in Equation (11) are 0.5; as their weights are identical, it is often ignored. The weight of Equation (11) can be adjusted depending on different applications.

$$\tau = \text{Arg Max}\ (a * P_D(\tau) + b * (1 - P_F(\tau))) \tag{12}$$

3.3.2. Cohen's Kappa

Cohen's kappa coefficient is a statistical evaluation method for measuring the consistency between the two classes. In image processing, the effect of a detector is generally measured by the ROC Curve, whereas Cohen's kappa is extensively used in biology to measure the efficiency of a detector. Cohen's kappa is an algorithm using the result of binarization to evaluate and calculate consistency. It uses the error matrix identical with the ROC Curve to calculate the kappa value.

According to Table 3, Cohen's kappa can be defined as

$$K = \frac{P_o - P_e}{1 - P_e} = 1 - \frac{1 - P_o}{1 - P_e} \tag{13}$$

$$P_o = \frac{P_a + P_d}{P_a + P_b + P_c + P_d} = \frac{P_a + P_d}{N} \tag{14}$$

$$P_e = P_{Yes} + P_{No} = \frac{P_a + P_b}{N} . \frac{P_a + P_c}{N} + \frac{P_c + P_d}{N} . \frac{P_b + P_d}{N} \tag{15}$$

where P_o represents the observation consistency (observed proportionate agreement) and P_e represents the desired consistency (probability of random agreement). The K value ranges from -1 to 1. If the K value is smaller than 0, the detected result is worse than the stochastic prediction.

Table 3. Error Matrix of Cohen's kappa.

Error Matrix		**Ground Truth**		**Total**
		Sprout (p)	**Non-Sprout (n)**	
Detection	Sprout (p′)	True Positive P_a	False Positive P_b	$P_a + P_b$
	Non-Sprout (n′)	False Negative P_c	True Negative P_d	$P_c + P_d$
Total		$P_a + P_c$	$P_b + P_d$	Total Pixels N

3.4. Experimental Results

The brighter pixels in the detection maps of Figures 14 and 15 represent the higher probability of NGL and highlight the pixels of targets hit in red points, the pixels of a false alarm in blue points, the pixels of targets missing in yellow points in Regions 1. By visually inspecting the Figures, the Subset CEM and SW CEM detectors seemed to perform slightly better than traditional CEM in terms of NGL pixel detection. Figures 16 and 17 represent the higher probability of NGL and highlight the pixels of targets in Region 2. Obviously, the results of ASW CEM in Figures 15d and 17d reduce plenty of false alarm pixels.

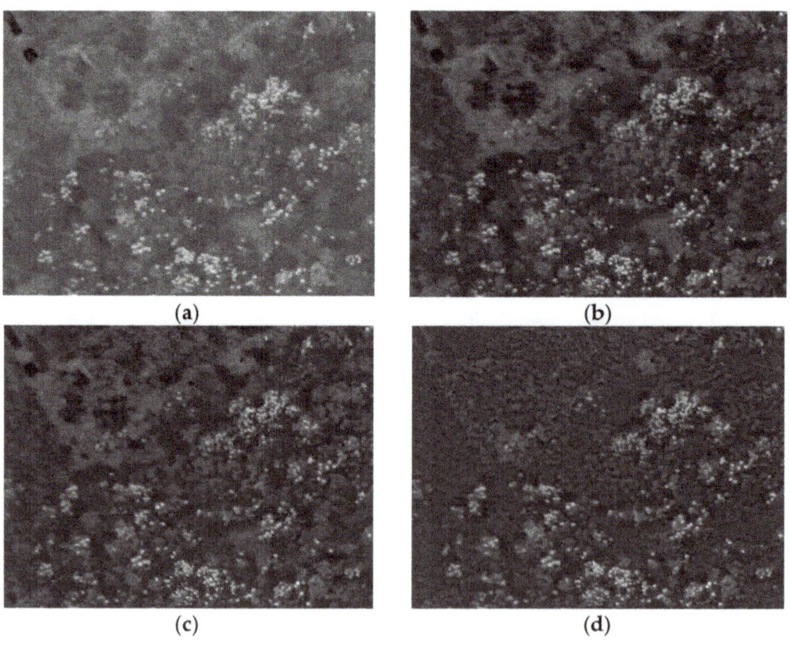

Figure 14. Detection maps of 4 algorithms in Region 1. (**a**) CEM (**b**) Subset CEM (**c**) Sliding Window-based CEM (SW CEM) (**d**) Adaptive Window-based CEM (ASW CEM).

Figure 15. Detection results highlighted with different colors of 4 algorithms in Region 1. (**a**) CEM (**b**) Subset CEM (**c**) Sliding Window-based CEM (SW CEM) (**d**) Adaptive Window-based CEM (ASW CEM).

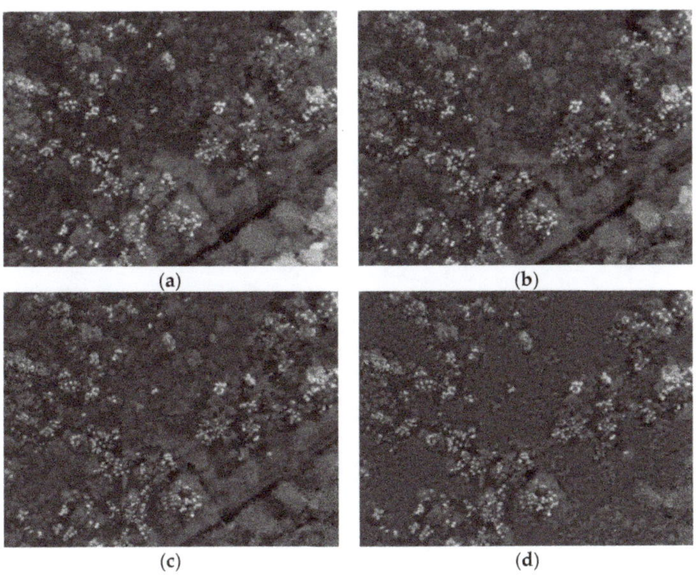

Figure 16. Detection maps of 4 algorithms in Region 2. (**a**) CEM (**b**) Subset CEM (**c**) Sliding Window-based CEM (SW CEM) (**d**) Adaptive Window-based CEM (ASW CEM).

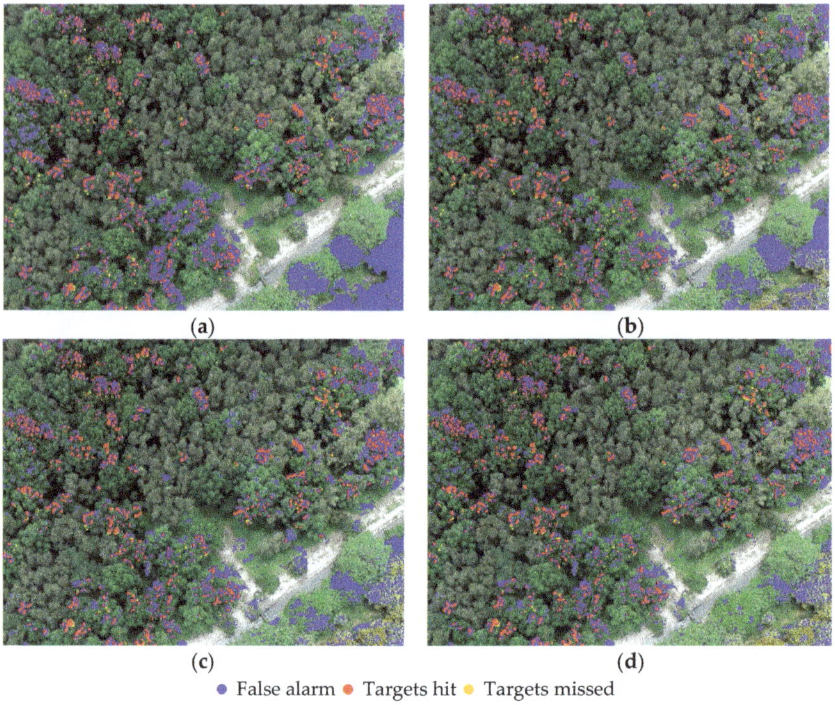

Figure 17. Detection results highlighted with different colors of 4 algorithms in Region 2. (**a**) CEM (**b**) Subset CEM (**c**) Sliding Window-based CEM (SW CEM) (**d**) Adaptive Window-based CEM (ASW CEM).

Figures 18 and 19 show the ROC curves of traditional CEM and our proposed three window based CEMs. Tables 4 and 5 show the AUC calculated, according to the ROC Curve in the experimental images of different regions and the evaluation of P_D, P_F, overall accuracy (ACC), and kappa under the optimum threshold.

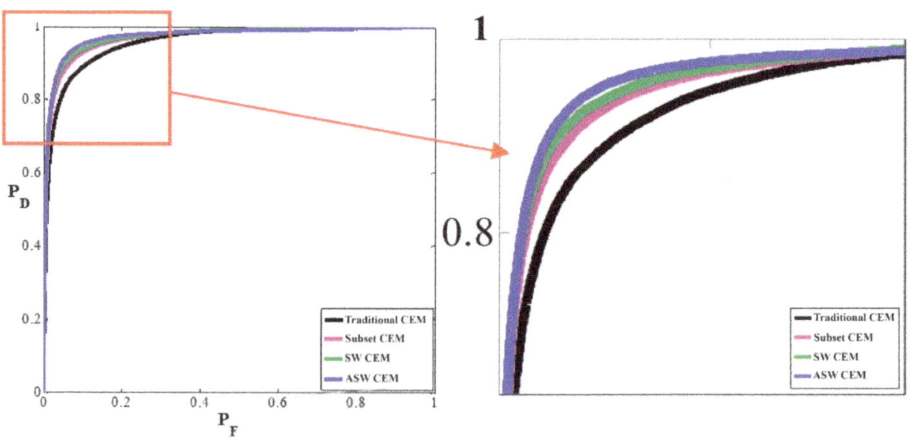

Figure 18. ROC curves of local and global CEMs on Region 1.

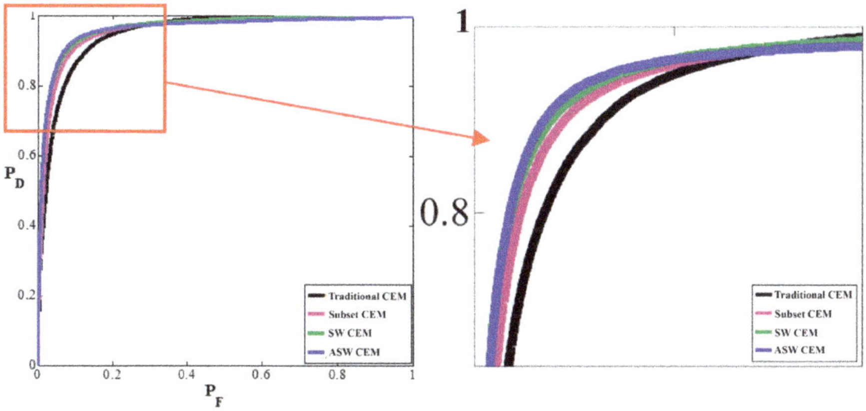

Figure 19. ROC curves of local and global CEMs on Region 2.

Table 4. Detection results of traditional CEM and our proposed CEMs in the image of Region 1.

Detection	AUC	P_D	P_F	ACC	Kappa
Traditional CEM	0.9556	0.8794	0.1114	0.8882	0.3345
Subset CEM	0.9714	0.9122	0.0789	0.9207	0.4347
SW CEM	0.9737	0.9190	0.0723	0.9274	0.4604
Adaptive-SW CEM	0.9755	0.9299	0.0676	0.9323	0.4822

Table 5. Detection results of traditional CEM and our proposed CEMs in the image of Region 2.

Detection	AUC	P_D	P_F	ACC	Kappa
Traditional CEM	0.9510	0.8952	0.1237	0.8771	0.3377
Subset CEM	0.9596	0.9099	0.0999	0.9005	0.3981
SW CEM	0.9649	0.9133	0.0885	0.9116	0.4310
Adaptive-SW CEM	0.9653	0.9212	0.0848	0.9155	0.4456

The performance of each detection method can be judged according to its ROC curve. Different target detection algorithms have different AUCs (Area under the Curve of ROC). Generally speaking, the value of AUC is 0~1, and the performance of a detection method can be judged according to the AUC value. If AUC = 1, then the detector is almost perfect. When this detector is used, there are at least two thresholds, so the result appears to be ideal. If AUC is 0.5~1, then this detector is better than a random guess. If AUC is just equal to 0.5—as shown in Figure 13, when the detection power (P_D) is equal to the false alarm probability (P_F), meaning that the result is the same as a random guess, like flipping a coin—then the probability of front and back is 1/2. If AUC < 0.5, then the result is worse than a random guess, and the resulting target and the background may be inverted. Put briefly, the larger the AUC value is, the more correct is the detection method. According to Figures 16 and 17, the three proposed CEMs have higher AUC than the traditional CEM.

Finally, according to the data of ROC, kappa, and the error matrix in Tables 4 and 5, in the three images of different resolutions of the two regions, respectively, ASW performs better than the other algorithms. The performance of TPR is slightly different from that of the other algorithms. However, in terms of the false alarm, ASW CEM can effectively reduce the detection of non NGL pixels, which thus can increase overall accuracy and the Kappa value of image detection. This result means that ASW CEM has a better detection result than the other algorithms. Figures 20 and 21 illustrate the comparison of traditional CEM and our proposed three window based CEMs in the results of AUC and Kappa.

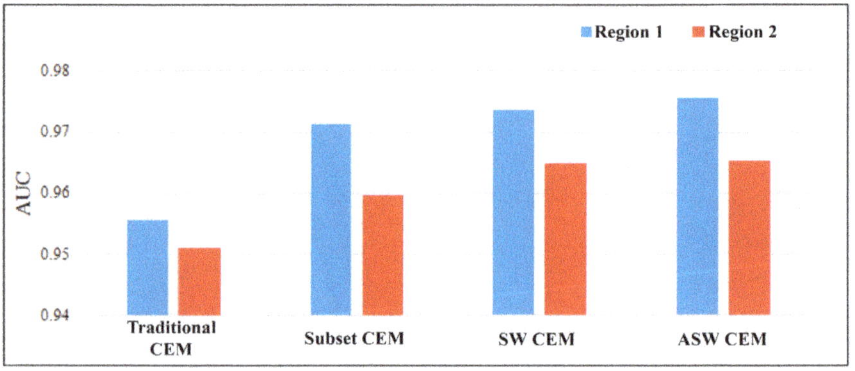

Figure 20. Area under curve (AUC) detection results of Region 1 and Region 2.

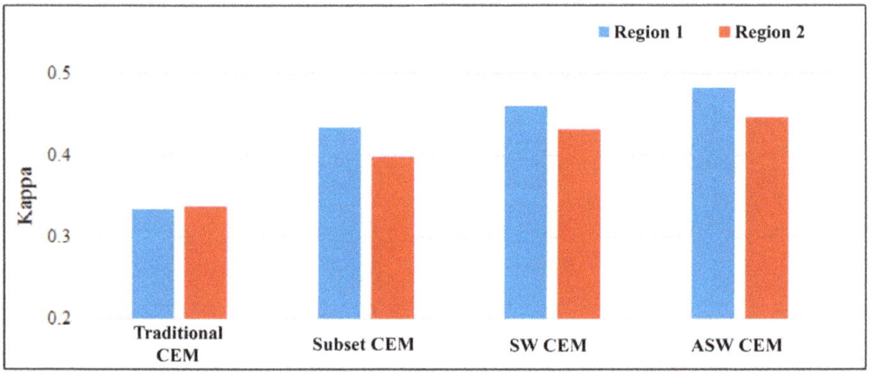

Figure 21. Kappa detection results of Region 1 and Region 2.

Figures 22 and 23 highlight parts of two regions where a false alarm is likely to occur. In those regions where the false alarm is likely to occur in the two images, it is observed that the CEM algorithm is likely to misrecognize similar RGB values as NGL. Our proposed algorithms using local autocorrelation matrix **S** in Equation (10), such as Subset CEM, SW CEM, and ASW CEM, are likely to suppress the background of the region, so as to reduce the false alarm rate. Based on the experimental results, ASW CEM performs the best effect.

Figure 22. Resulting images of the region where a false alarm is likely to occur in Region 1: (**a**) Traditional CEM; (**b**) Subset CEM; (**c**) SW CEM; (**d**) ASW CEM.

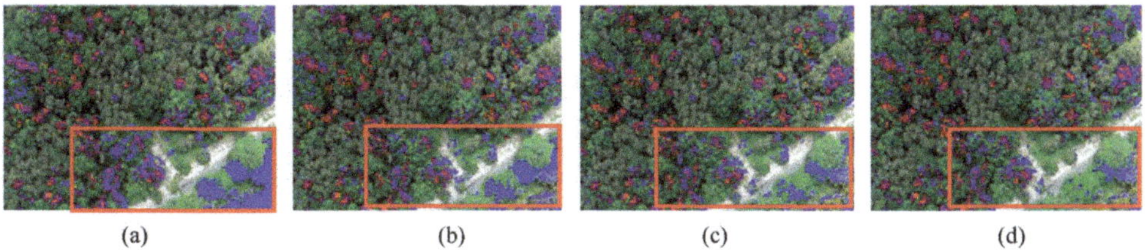

Figure 23. Resulting images of the region where a false alarm is likely to occur in Region 2: (**a**) Traditional CEM; (**b**) Subset CEM; (**c**) SW CEM; and, (**d**) ASW CEM.

In order to the validate the influence of the window size for Subset CEM and SW CEM, Tables 6 and 7 tabulate the detection results of various window sizes in Region 1. As seen, various window sizes of Subset CEM and SW CEM produce very similar results. On the other hand, the window size of Subset CEM and SW CEM are not sensitive to the final performance; thus, it is not the critical parameter for the detectors.

Table 6. Detection results of Subset CEM with various window sizes.

Subset CEM Window Size	AUC	P_D	P_F	ACC	Kappa
50 × 65	0.9688	0.9108	0.0770	0.9227	0.4424
200 × 260	0.9714	0.9122	0.0789	0.9207	0.4347
500 × 650	0.9674	0.9048	0.0923	0.9075	0.3912

Table 7. Detection results of SW CEM with various window sizes.

SW CEM Window Size	AUC	P_D	P_F	ACC	Kappa
101 × 101	0.9732	0.9179	0.0706	0.9290	0.4667
151 × 151	0.9737	0.9190	0.0723	0.9274	0.4604
301 × 301	0.9697	0.9135	0.0863	0.9137	0.4121

3.5. Computing Time

This section calculates the computing time in seconds by running CEM, Subset CEM, SW CEM, and ASW CEM on two real image scenes using MATLAB where the computer environment used for experiments was 64-bit Windows operating system with Intel i7-4710, CPU 2.5 Ghz, and 16 GB memory (RAM). In the two real image scenes, ASW CEM improves detection accuracy, the false alarm rate, and evaluation consistency better than the other algorithms, but the detection time of using local autocorrelation matrix **S** is longer than the traditional CEM, as shown in Table 8. It is noted that OSGP is not included in the computing time. The time listed in Table 8 is the execution time for each algorithm only. The time in ASW CEM also includes the time of acquiring the rate of NGL.

From computation perspective, calculating the inverse of the matrix takes most of time during computation. Since SW CEM needs to recalculate the inverse of **S** in Equation (10) in every pixel with the fixed window, in this case, it takes the longest time. AWS CEM can adjust the window size, and so the computing time is the second longest. In the best results of ASW CEM, the detection time is longer than CEM, but all of the evaluated data are enhanced significantly, meaning the ASW CEM algorithm consumes more detection time to increase accuracy, but the increment rate of result is higher. When compared to real time processing [16,55–58], the time is not a main consideration in this study. On the contrary, if the computing time is the issue, Subset CEM provides the reasonable improvement, with no computing time penalty. In this case, Subset CEM also can be applied in the some other applications.

Table 8. Computing time of different algorithms and CEM evaluation.

Detection	Time	AUC Increment Rate	P_D Increment Rate	P_F Sink Rate	ACC Increment Rate	Kappa Increment Rate
Traditional CEM	0.08	0.00%	0.00%	0.00%	0.00%	0.00%
Subset CEM	0.08	1.22%	2.38%	2.82%	2.80%	8.03%
SW CEM	326.24	1.60%	2.89%	3.72%	3.69%	10.96%
Adaptive-SW CEM	232.33	1.71%	3.83%	4.14%	4.13%	12.78%

Disregarding the minor defect of a long detection time, ASW CEM performs better in enhancement than the other algorithms, because ASW CEM can change the size of the sliding window according to the rate of the sprout around the current pixel.

4. Discussion

A variety of target detection techniques have been published during the last few decades [12–15,25,59], with several studies applying support vector machines (SVM) [60] or Fisher's linear discriminant analysis (FLDA) [60] to solve target detection problems as a binary classification problem [61–65]. These algorithms require a number of classes, and their class distribution model must be known in advance. In order to avoid any biased selection of training samples, the partition must be performed randomly. In other words, training samples must be randomly selected from a dataset to form a training sample set for cross validation. As a result, such a validation is not repeatable and cannot be re-produced. The results are inconsistent. To alleviate this dilemma, this paper proposes a novel Constrained Energy Minimization (CEM) based technique that takes advantage of spectral and spatial information and developed Optimal Signature Generation Process (OSGP) in terms of the iterative process point of view to solve the issues mentioned above. CEM only requires one desired target information for the specific target of interest, regardless of other background information, which is its major advantage. Theoretically, CEM subpixel detection is generally performed by two operations that involve background suppression and matched filter [16]. First, it performs background suppression via the inverse of $\mathbf{R}$ so as to enhance a detected target contrast against the background. Second, CEM operates a matched filter using $\mathbf{d}$ as the desired matched signature so as to increase intensity of the target of interest. Since only one target signature can be used as the $\mathbf{d}$ in Equation (5), selecting an appropriate $\mathbf{d}$ is a very crucial step for detection results. Although CEM has many applications [27,30,31], very few studies investigated the issues of selecting a desire target signature. Therefore, this paper developed the Optimal Signature Generation Process (OSGP) to resolve this issue.

When compared to the classification based approaches that require very precise prior knowledge to generate a set of training samples and features, applying OSGP on the proposed CEM based methods required only one target signature information and provided stable results even if the initial desire target information is bias or not reliable. In the iterative process of OSGP in Figure 24, the iteration results of different desired signatures $\mathbf{d}$ after different numbers of iterations give a stable AUC result, so that the originally worse desired target obtains a relatively better desired target. Figure 25 shows different $\mathbf{d}$'s have different results in the same algorithm. However, the $\mathbf{d}'$ iterated by OSGP used in CEM, Subset CEM, SW CEM, and ASW CEM can enhance the original desired target to some extent. Moreover, the results are approximately identical, meaning OSGP can determine the appropriate desired target automatically when selecting inappropriate $\mathbf{d}$ as initial, and the result is still very stable.

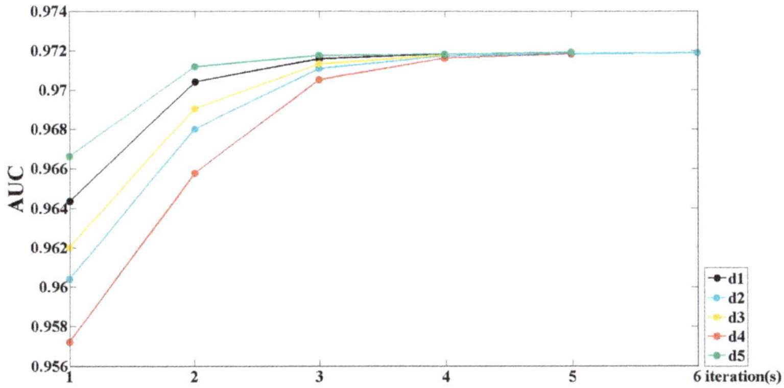

Figure 24. Iterative process of OSGP and corresponding AUC detection result.

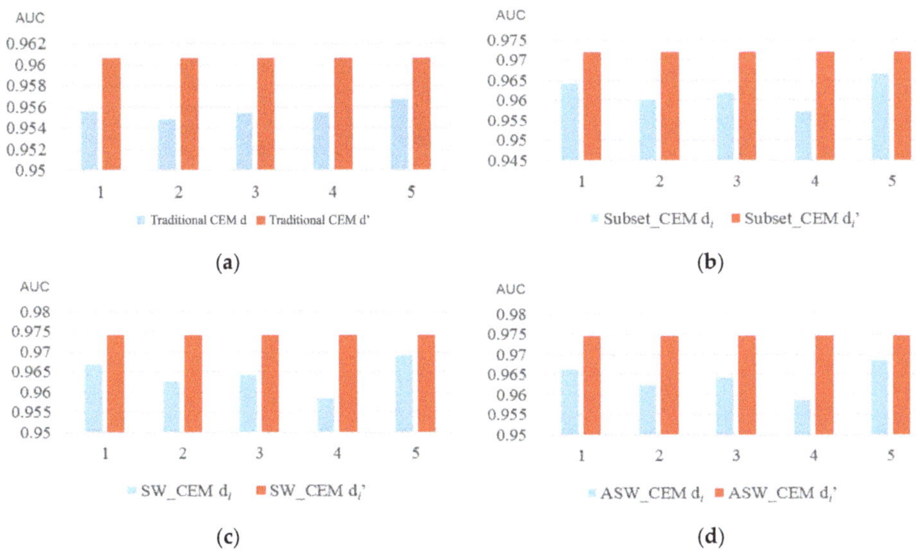

Figure 25. AUC detection results of various algorithms executing five different **d** and corresponding **d'** generated by OSGP (**a**) CEM; (**b**) Subset CEM; (**c**) SW CEM; and, (**d**) ASW CEM.

CEM technique only takes advantages of spectral information to detect target of interests. However, when spectral information is insufficient to distinguish between targets and some materials have similar spectral signature, this likely causes false alarms in the multispectral images. In this case, our proposed window-based techniques actually include spatial information into the CEM algorithms via fixed or adaptive windows to compensate for the insufficient spectral information. According to the experimental results and the resulting images in Figures 14–17, among our proposed local CEM algorithms, the Subset CEM, Sliding Window-based CEM (SW CEM) of the fixed window size, or Adaptive Window-based CEM (ASW CEM) enhances the contrast between the target and the background better than the general CEM. Because the autocorrelation matrix **R** of the CEM algorithm is different, CEM uses **R** of the full image, whereas our proposed local CEMs uses local autocorrelation matrix **S** in Equation (10) to suppress the background. According to effect of the background suppression [58], it is obvious that using local autocorrelation **S** is better than global autocorrelation **R** in this study. Figures 26 and 27 show the RGB signatures corresponding to different objects in the study site. As seen, some RGB signatures of leaves and grass are very close to NGL. In the upper left of Region 1, as the grass is too similar to the sprout shown in Figure 26, the CEM detection is likely to give a false alarm. Because the **R** that is used by CEM is generated according to the pixel value of the full image, the difference between NGL and grass is not obvious in the full image. In the entire image, the house and soil are larger than the RGB difference between grass and the sprout, and so the grass is likely to be misrecognized as NGL. On the contrary, in our proposed CEM based algorithms using **S**, because **S** is generated by pixels around the current pixel value and the proportion of soil and house is not high in a small area, the difference in RGB values between grass and NGL is enlarged, and the grass is likely to be suppressed, thus reducing the false alarm rate. In the same way, the lower right of Region 2 also easily gives a false alarm. Because the pixel values of some leaves are very similar to NGL in the region shown in Figure 27, when **R** is used to suppress the background, it is likely to be influenced by pixel values with a larger difference, and this problem can be solved by using **S**.

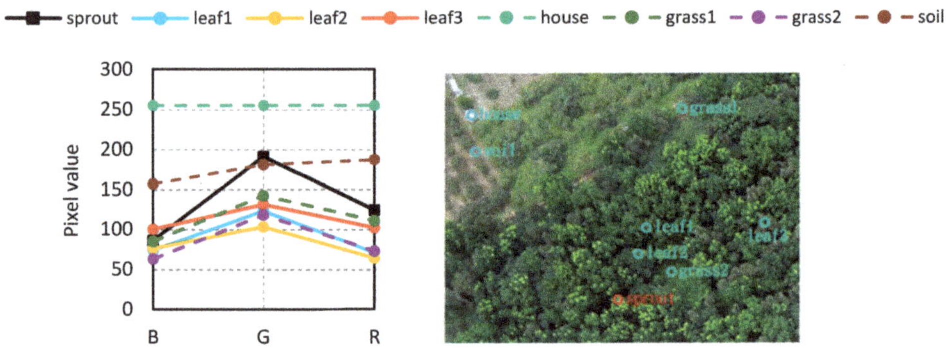

Figure 26. RGB signatures corresponding to different objects in Region 1 [4].

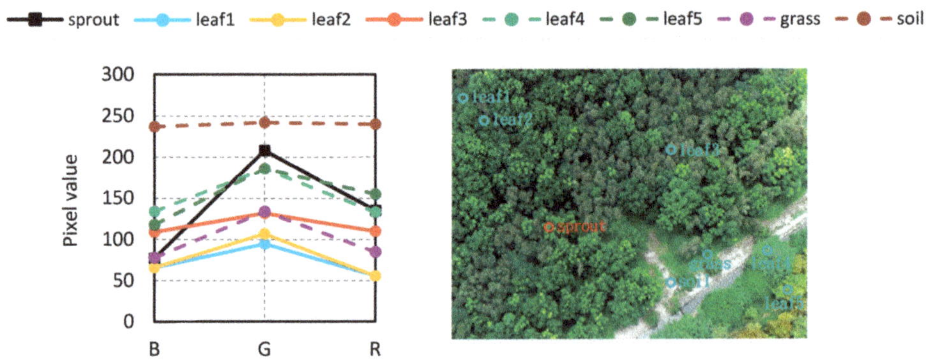

Figure 27. RGB signatures with different objects in Region 2 [4].

ASW CEM can change the size of the sliding window, according to the ratio of NGL around the current pixel. When there are too many NGL in the window, the difference between them is likely to be enlarged, and NGL that is very different from the desired target will be suppressed, leading to detection omission. Therefore, the sliding window shall be enlarged to reduce the rate of NGL and enhance the difference from the background, thus increasing the detected value. On the contrary, if the rate of NGL is too low, then the difference between backgrounds increase, and the RGB values that are similar to NGL is likely to be misrecognized. At this point, the sliding window is shrunk, the rate of NGL increases, the difference between NGL and background are more apparent, and the result value of the non-NGL is reduced for suppression. When NGL are enhanced and background is suppressed; their difference is enlarged, so as to highlight NGL.

Briefly, CEM technique was originally designed to catch (1) low probability of infrequent occurrence, (2) relatively small sample size, and (3) most importantly, the target pixel has spectrally distinct from its surrounding pixels [16]. Obviously, the NGL in RGB images shows the same features. This explains why the window-based CEM techniques can achieve satisfied results of NGL detection even only three spectral signatures is used.

5. Conclusions

Constant leaf sprouting and development can be an indication of healthy trees in beneficial environmental circumstances. This paper investigated the feasibility of NGL detection using hyperspectral detection algorithms in UAV bitmap images. Since the bitmap images only provide RGB values, using a traditional subpixel detector CEM presents false alarm issue. In order to address this issue, three window based CEMs are proposed in this paper. First, Subset CEM is developed to split the image into different small images, according to different regions. Second, the sliding window-base CEM was proposed to extract the RGB values around the current pixel to calculate autocorrelation matrix **S**. Third, this paper further proposed adaptive window-based CEM (ASW CEM), which can change the window size automatically according to NGL around the current pixel. ASW CEM extracts and calculates autocorrelation matrix **S**, increasing the contrast between NGL and the background, so as to highlight NGLs and to suppress the background. Last but not the least, in order to reduce the effect of the quality of the desired target selected for CEM, this paper designed OSGP to generate a stable desired target during iterations. The experimental results show that our proposed approaches can effectively reduce the errors resulting from a false alarm so as to obtain more appropriate desired target and stable results for newly grown tree leaves in UAV images.

Acknowledgments: The authors would like to acknowledge the support provided by projects MOST 106-2221-E-224-055 and MOST 105-2119-M-415-002 funded by the Ministry of Science and Technology, Taiwan, ROC.

Author Contributions: S.-Y.C. conceived and designed the algorithms and wrote the paper; C.L. analyzed the data, contributed data collection, reviewed the paper and organized the revision. C.-H.T. and S.-J.C. performed the experiments.

Conflicts of Interest: The authors declare no conflict of interest.

References

1. Forest Resources Assessment (FAO). *Global Forest Resources Assessment 2015—How Are the World's Forests Changing*; Food and Agricultural Organization of United Nations: Rome, Italy, 2015.
2. Lin, C.; Dugarsuren, N. Deriving the Spatiotemporal NPP Pattern in Terrestrial Ecosystems of Mongolia using MODIS Imagery. *Photogram. Eng. Remote Sens.* **2015**, *81*, 587–598. [CrossRef]
3. Lin, C.; Thomson, G.; Popescu, S.C. An IPCC-Compliant Technique for Forest Carbon Stock Assessment Using Airborne LiDAR-Derived Tree Metrics and Competition Index. *Remote Sens.* **2016**, *8*, 528. [CrossRef]
4. Lin, C.; Chen, S.-Y.; Chen, C.-C.; Tai, C.-H. Detecting Newly Grown Tree Leaves from Unmanned-Aerial-Vehicle Images using Hyperspectral Target Detection Techniques. *ISPRS J. Photogramm. Remote Sens.* **2017**. in review.
5. Lin, C.; Wu, C.C.; Tsogt, K.; Ouyang, Y.C.; Chang, C.I. Effects of Atmospheric Correction and Pansharpening on LULC Classification Accuracy using WorldView-2 Imagery. *Inf. Process. Agric.* **2015**, *2*, 25–36. [CrossRef]
6. Götze, C.; Gerstmann, H.; Gläßer, C.; Jung, A. An approach for the classification of pioneer vegetation based on species-specific phenological patterns using laboratory spectrometric measurements. *Phys. Geogr.* **2017**, *38*, 524–540. [CrossRef]

7. Burai, P.; Deák, B.; Valkó, O.; Tomor, T. Classification of Herbaceous Vegetation Using Airborne Hyperspectral Imagery. *Remote Sens.* **2015**, *7*, 2046–2066. [CrossRef]
8. Mohan, M.; Silva, C.A.; Klauberg, C.; Jat, P.; Catts, G.; Cardil, A.; Hudak, A.T.; Dia, M. Individual Tree Detection from Unmanned Aerial Vehicle (UAV) Derived Canopy Height Model in an Open Canopy Mixed Conifer Forest. *Forest* **2017**, *8*, 340. [CrossRef]
9. Wallace, L.; Lucieer, A.; Watson, C.S. Evaluating Tree Detection and Segmentation Routines on Very High Resolution UAV LiDAR Data. *IEEE Trans. Geosci. Remote Sens.* **2014**, *52*, 7619–7628. [CrossRef]
10. Dugarsuren, N.; Lin, C. Temporal variations in phenological events of forests, grasslands and desert steppe ecosystems in Mongolia: A Remote Sensing Approach. *Ann. For. Res.* **2016**, *59*, 175–190.
11. Popescu, S.C.; Zhao, K.; Neuenschwander, A.; Lin, C. Satellite lidar vs. small footprint airborne lidar: Comparing the accuracy of aboveground biomass estimates and forest structure metrics at footprint level. *Remote Sens. Environ.* **2011**, *115*, 2786–2797. [CrossRef]
12. Zeng, M.; Li, J.; Peng, Z. The design of Top-Hat morphological filter and application to infrared target detection. *Infrared Phys. Technol.* **2006**, *48*, 67–76. [CrossRef]
13. Gao, C.; Meng, D.; Yang, Y.; Wang, Y.; Zhou, X.; Hauptmann, A.G. Infrared patch-image model for small target detection in a single image. *IEEE Trans. Image Process.* **2013**, *22*, 4996–5009. [CrossRef] [PubMed]
14. Debes, C.; Zoubir, A.M.; Amin, M.G. Enhanced detection using target polarization signatures in through-the-wall radar imaging. *IEEE Trans. Geosci. Remote Sens.* **2012**, *50*, 1968–1979. [CrossRef]
15. Qi, S.; Ma, J.; Tao, C.; Yang, C.; Tian, J. A robust directional saliency-based method for infrared small-target detection under various complex backgrounds. *IEEE Trans. Geosci. Remote Sens. Lett.* **2013**, *10*, 495–499.
16. Chang, C.I. *Real-Time Progressive Hyperspectral Image Processing: Endmember Finding and Anomaly Detection*; Springer: Berlin, Germany, 2016.
17. Chang, C.-I. *Hyperspectral Data Processing: Algorithm Design and Analysis*; Wiley: Hoboken, NJ, USA, 2013.
18. Chang, C.-I. *Hyperspectral Imaging: Techniques for Spectral detection and Classification*; Kluwer Academic/Plenum Publishers: Dordrecht, The Netherlands, 2003.
19. Harsanyi, J.C. Detection and Classification of Subpixel Spectral Signatures in Hyperspectral Image Sequences. Ph.D. Thesis, Department of Electrical Engineering, University of Maryland, Baltimore, MA, USA, 1993.
20. Rees, G.; Rees, W.G. *Physical Principles of Remote Sensing*; Cambridge University Press: Cambridge, UK, 2013.
21. Mahalanobis, P.C. On the generalized distance in statistics. *Proc. Natl. Inst. Sci. India* **1936**, *2*, 49–55.
22. Kraut, S.; Scharf, L.L.; Butler, R.W. The adaptive coherence estimator: A uniformly most-powerful-invariant adaptive detection statistic. *IEEE Trans. Signal Process.* **2005**, *53*, 427–438. [CrossRef]
23. Ren, H.; Chang, C.-I. Target-constrained interference-minimized approach to subpixel target detection for hyperspectral images. *Opt. Eng.* **2000**, *39*, 3138–3145. [CrossRef]
24. Xue, B.; Yu, C.; Wang, Y.; Song, M.; Li, S.; Wang, L.; Chen, H.; Chang, C.I. A Subpixel Target Detection Approach to Hyperspectral Image Classification: Iterative Constrained Energy Minimization. *IEEE Trans. Geosci. Remote Sens.* **2017**, *55*, 5093–5114. [CrossRef]
25. Sun, H.; Sun, X.; Wang, H.; Li, Y.; Li, X. Automatic target detection in high-resolution remote sensing images using spatial sparse coding bag-of-words model. *IEEE Trans. Geosci. Remote Sens.* **2012**, *9*, 109–113. [CrossRef]
26. Chang, C.I.; Wang, Y.; Chen, S.Y. Anomaly Detection Using Causal Sliding Windows. *IEEE J. Sel. Top. Appl. Earth Obs. Remote Sens.* **2015**, *8*, 3260–3270. [CrossRef]
27. Chang, C.I.; Schultz, R.C.; Hobbs, M.C.; Chen, S.Y.; Wang, Y.; Liu, C. Progressive Band Processing of Constrained Energy Minimization for Subpixel Detection. *IEEE Trans. Geosci. Remote Sens.* **2015**, *53*, 1626–1637. [CrossRef]
28. Manolakis, D.; Marden, D.; Shaw, G.A. Hyperspectral image processing for automatic target detection applications. *Linc. Lab. J.* **2003**, *14*, 79–116.
29. Nasrabadi, N.M. Hyperspectral target detection: An overview of current and future challenges. *IEEE Signal Process. Mag.* **2014**, *31*, 34–44. [CrossRef]
30. Zou, Z.; Shi, Z. Hierarchical Suppression Method for Hyperspectral Target Detection. *IEEE Trans. Geosci. Remote Sens.* **2016**, *54*, 330–342. [CrossRef]
31. Sun, K.; Geng, X.; Ji, L. A New Sparsity-Based Band Selection Method for Target Detection of Hyperspectral Image. *IEEE Trans. Geosci. Remote Sens. Lett.* **2015**, *12*, 329–333.
32. Frost, O.L., III. An algorithm for linearly constrained adaptive array processing. *Proc. IEEE.* **1972**, *60*, 926–935. [CrossRef]

33. Weinberg, G.V. An invariant sliding window detection process. *IEEE Signal Process. Lett.* **2017**, *24*, 1093–1097. [CrossRef]
34. Castella, F.R. Sliding window detection probability. *IEEE Trans. Aerosp. Electron. Syst.* **1976**, *6*, 815–819. [CrossRef]
35. Zhang, L.; Lin, J.; Karim, R. Sliding window-based fault detection from high-dimensional data streams. *IEEE Trans. Syst. Man Cybern. Syst.* **2017**, *47*, 289–303. [CrossRef]
36. Noh, S.; Shim, D.; Jeon, M. Adaptive Sliding-Window Strategy for Vehicle Detection in Highway Environments. *IEEE Trans. Intell. Transp. Syst.* **2016**, *17*, 323–335. [CrossRef]
37. Gao, G.; Liu, L.; Zhao, L.; Shi, G.; Kuang, G. An Adaptive and Fast CFAR Algorithm Based on Automatic Censoring for Target Detection in High-Resolution SAR Images. *IEEE Trans. Geosci. Remote Sens.* **2009**, *47*, 1685–1697. [CrossRef]
38. Akkoul, S.; Ledee, R.; Leconge, R.; Harba, R. A New Adaptive Switching Median Filter. *IEEE Signal Process. Lett.* **2010**, *17*, 587–590. [CrossRef]
39. Matteoli, S.; Diani, M.; Corsini, G. Impact of signal contamination on the adaptive detection performance of local hyperspectral anomalies. *IEEE Trans. Geosci. Remote Sens.* **2014**, *52*, 1948–1968. [CrossRef]
40. Matteoli, S.; Veracini, T.; Diani, M.; Corsini, G. A locally adaptive background density estimator: An evolution for RX-based anomaly detectors. *IEEE Geosci. Remote Sens. Lett.* **2014**, *11*, 323–327. [CrossRef]
41. Chang, C.-I. An information theoretic-based approach to spectral variability, similarity and discriminability for hyperspectral image analysis. *IEEE Trans. Inf. Theory* **2000**, *46*, 1927–1932. [CrossRef]
42. Otsu, N. A threshold selection method from gray-level histograms. *IEEE Trans. Syst. Man Cybern.* **1979**, *9*, 62–66. [CrossRef]
43. Hall, D.; Ball, G. *Isodata: A Novel Method of Data Analysis and Pattern Classification*; Stanford Research Institude: Menlo Park, CA, USA, 1965.
44. MacQueen, J. Some methods for classification and analysis of multivariate observations. In *Proceedings of the 5th Berkeley Symposium on Mathematical Statistics and Probability, Berkeley, CA, USA, 21 June–18 July 1965 and 27 December 1965–7 January 1966*; University of California Press: Berkeley, CA, USA, 1967; pp. 281–296.
45. Chen, H.M.; Lin, C.; Chen, S.Y.; Wen, C.H.; Chen, C.C.C.; Ouyang, Y.C.; Chang, C.I. PPI SVM-Iterative FLDA Approach to Unsupervised Multispectral Image Classification. *IEEE J. Sel. Top. Appl. Earth Obs. Remote Sens.* **2013**, *6*, 1834–1842. [CrossRef]
46. Zhuang, H.; Deng, K.; Fan, H.; Yu, M. Strategies combining spectral angle mapper and change vector analysis to unsupervised change detection in multispectral images. *IEEE Trans. Geosci. Remote Sens.* **2016**, *13*, 681–685. [CrossRef]
47. Lin, C.; Lin, C.H. Comparison of Carbon Sequestration Potential in Agricultural and Afforestation Farming Systems. *Sci. Agric.* **2013**, *70*, 93–101. [CrossRef]
48. Lin, C.; Wang, J.J. The effect of trees spacing on the growth of trees in afforested broadleaf stands on cultivated farmland. *Q. J. Chin. For.* **2013**, *46*, 311–326.
49. Lin, C.; Tsogt, K.; Zandraabal, T. A decompositional stand structure analysis for exploring stand dynamics of multiple attributes of a mixed-species forest. *For. Ecol. Manag.* **2016**, *378*, 111–121. [CrossRef]
50. Lin, C.; Lo, K.L.; Huang, P.L. A classification method of unmanned-aerial-systems-derived point cloud for generating a canopy height model of farm forest. In Proceedings of the 2016 IEEE International Geoscience and Remote Sensing Symposium (IGARSS), Beijing, China, 10–15 July 2016; pp. 740–743.
51. Burnett, R.; Brunstrom, A.; Nilsson, A.G. *Perspectives on Multimedia: Communication, Media and Information Technology*; John Wiley & Sons: Hoboken, NJ, USA, 2005.
52. Chang, C.-I. Multiple-parameter receiver operating characteristic analysis for signal detection and classification. *IEEE Sens. J.* **2010**, *10*, 423–442. [CrossRef]
53. Swets, J.A.; Pickett, R.M. *Evaluation of Diagnostic Systems: Methods from Signal Detection Theory*; Academic Press: Cambridge, MA, USA, 1982.
54. Pontius, R.; Millones, M. Death to Kappa: Birth of quantity disagreement and allocation disagreement for accuracy assessment. *Int. J. Remote Sens.* **2011**, *32*, 4407–4429. [CrossRef]
55. Chang, C.-I.; Ren, H.; Chiang, S.S. Real-time processing algorithms for target detection and classification in hyperspectral imagery. *IEEE Trans. Geosci. Remote Sens.* **2001**, *39*, 760–768. [CrossRef]
56. Stellman, C.M.; Hazel, G.G.; Bucholtz, F.; Michalowicz, J.V.; Stocker, A.D.; Schaaf, W. Real-time hyperspectral detection and cuing. *Opt. Eng.* **2000**, *39*, 1928–1935. [CrossRef]

57. Du, Q.; Ren, H. Real-time constrained linear discriminant analysis to target detection and classification in hyperspectral imagery. *Pattern Recognit.* **2003**, *36*, 1–12. [CrossRef]
58. Chang, C.I. *Real-Time Recursive Hyperspectral Sample and Band Processing: Algorithm Architecture and Implementation*; Springer: Berlin, Germany, 2017.
59. Zhang, L.; Zhang, L.; Tao, D.; Huang, X. Sparse transfer manifold embedding for hyperspectral target detection. *IEEE Trans. Geosci. Remote Sens.* **2014**, *52*, 1030–1043. [CrossRef]
60. Vapnik, V.N. *Statistical Learning Theory*; Wiley: New York, NY, USA, 1998.
61. Wu, Y.; Yang, X.; Plaza, A.; Qiao, F.; Gao, L.; Zhang, B.; Cui, Y. Approximate Computing of Remotely Sensed Data: SVM Hyperspectral Image Classification as a Case Study. *IEEE J. Sel. Top. Appl. Earth Obs. Remote Sens.* **2016**, *9*, 5806–5818. [CrossRef]
62. Bo, C.; Lu, H.; Wang, D. Hyperspectral Image Classification via JCR and SVM Models with Decision Fusion. *IEEE Trans. Geosci. Remote Sens. Lett.* **2016**, *13*, 177–181.
63. Xue, Z.; Du, P.; Su, H. Harmonic Analysis for Hyperspectral Image Classification Integrated With PSO Optimized SVM. *IEEE J. Sel. Top. Appl. Earth Obs. Remote Sens.* **2014**, *7*, 2131–2146. [CrossRef]
64. Kang, X.; Xiang, X.; Li, S.; Benediktsson, J.A. PCA-Based Edge-Preserving Features for Hyperspectral Image Classification. *IEEE Trans. Geosci. Remote Sens.* **2017**, *55*, 7140–7151. [CrossRef]
65. Roli, F.; Fumera, G. Support vector machines for remote-sensing image classification. *Proc. SPIE* **2001**, *4170*, 160–166.

remote sensing

MDPI

Article

Vicarious Radiometric Calibration of the Hyperspectral Imaging Microsatellites SPARK-01 and -02 over Dunhuang, China

Hao Zhang [1], Bing Zhang [1,2], Zhengchao Chen [1,*] and Zhihua Huang [1,3]

1 Key Laboratory of Digital Earth Science, Institute of Remote Sensing and Digital Earth, Chinese Academy of Sciences, No. 9 Dengzhuang South Road, Beijing 100094, China; zhanghao612@radi.ac.cn (H.Z.); zb@radi.ac.cn (B.Z.); 15831606243@163.com (Z.H.)

2 University of Chinese Academy of Sciences, No. 19(A) Yuquan Road, Shijingshan District, Beijing 100049, China

3 College of Geoscience and Surveying Engineering, China University of Mining & Technology, No. 11 Xueyuan Road, Haidian District, Beijing 100083, China

* Correspondence: chenzc@radi.ac.cn; Tel.: +86-10-8217-8775; Fax: +86-10-8217-8177

Received: 8 December 2017; Accepted: 12 January 2018; Published: 17 January 2018

Abstract: Two wide-swath hyperspectral imaging microsatellites, SPARK-01 and -02, were launched on 22 December 2016. Radiometric calibration coefficients were determined for these two satellites via a calibration experiment performed from the end of February to the beginning of March 2017 at the high-altitude, homogenous Dunhuang calibration site in the Gobi Desert in China. In-situ measurements, including ground reflectance, direct transmittance, diffuse-to-global irradiance ratio, and radiosonde vertical profile, were acquired. A unique relative calibration procedure was developed using actual satellite images. This procedure included dark current computation and non-uniform correction processes. The former was computed by averaging multiple lines of long strip imagery acquired over open oceans during nighttime, while the latter was computed using images acquired after the adjustment of the satellite yaw angle to 90°. This technique was shown to be suitable for large-swath satellite image relative calibration. After relative calibration, reflectance, irradiance, and improved irradiance-based methods were used to conduct absolute radiometric calibrations in order to predict the top-of-atmosphere (TOA) radiance. The SPARK-01 and -02 satellites passed over the calibration site on 7 March and 28 February 2017, during which time fair and non-ideal weather occurred, respectively. Thus, the SPARK-01 calibration coefficient was derived using reflectance- and irradiance-based methods, while that of SPARK -02 was derived using reflectance- and improved irradiance-based methods. The sources of calibration uncertainty, which include aerosol-type assumptions, transmittance measurements, water vapor content retrieval, spectral wavelength shift and satellite image misregistration, were explored in detail for different calibration methods. Using the reflectance and irradiance-based methods, the total uncertainty for SPARK-01 was estimated to be 4.7% and 4.1%, respectively, in the <1000 nm spectral range. For SPARK-02, total uncertainties of 8.1% and of 5.9% were estimated using the reflectance- and improved irradiance-based methods, respectively. The calibration methods were also verified using MODIS images, which confirmed that the calibration accuracies were within the expected range. These in-situ measurements, analyses, and results provide a basis for in-orbit radiometric calibration of the SPARK-01 and -02 satellites. These experiments strongly support the use of diffuse-to-global ratio measurements in in-situ vicarious calibration experiments and the addition of spectrally continuous measurements for direct transmittance, which is important for hyperspectral satellite sensors.

Keywords: vicarious calibration; reflectance-based method; irradiance-based method; Dunhuang site; 90° yaw imaging

1. Introduction

At 3:22 am UTC on 22 December 2016, two wide-swath pushbroom hyperspectral imaging microsatellites, SPARK-01 and -02, which were manufactured by the Shanghai Engineering Center for Microsatellites, were successfully launched at the Jiuquan satellite launch center by the CZ-2D rocket. The spectrometers on the satellites were developed by the Academy of Opto-electronics, Chinese Academy of Sciences, less than one year previously. SPARK-01 and -02 have spectral ranges of 400–1000 nm, a swath of ~100 km, a spatial resolution of 50 m and 2048 pixels along the cross-track direction. The spectrometers use prisms to split the beam into different bands, and thus, the spectral resolution (or full width at half maximum, FWHM) varies from 1 to 10 nm. Figure 1 shows a schematic of the satellite; major satellite characteristics are described in Table 1.

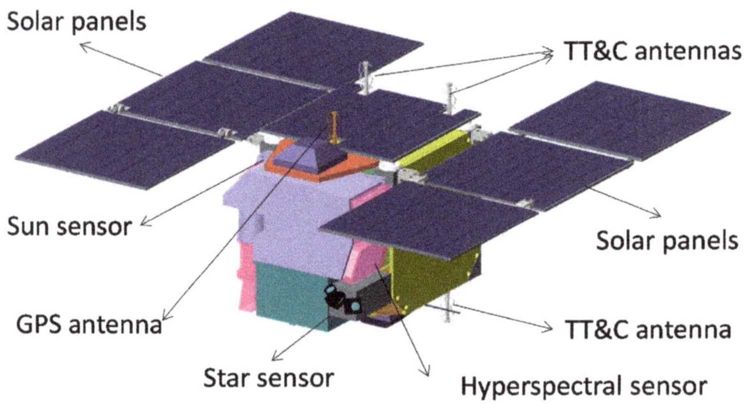

Figure 1. Diagram of the SPARK satellite.

Table 1. Main characteristics of SPARK satellite and imaging sensor.

Satellite Characteristic	Description
Spectral bands	160 bands ranging from 400 to 1000 nm (151 and 153 valid bands for SPARK-01 and -02, respectively)
Swath and spatial resolution	100 km with a resolution of 50 m at nadir viewing
Signal-to-noise ratio (SNR)	$\geq$100:1 on average with conditions: solar zenith = 45°, ground reflectance = 0.3, and visibility = 23 KM.
Revisit period	13–25 days
Observation area	2500 km × 6000 km per day
Mass	50 kg
Dimensions	300 mm × 300 mm × 450 mm
Cost	US$ 3 million per satellite
Production cycle	1 year
Lifetime	Over 1 year

The SPARK satellites are lightweight and inexpensive. They provide the advantages of fine spectral resolution and large swath. These two hyperspectral satellites can be used for applications such as environmental and disaster monitoring, target detection, and precise classification. They provide basic information to support quantitative applications, resource exploration, and business applications [1]. However, due to size, weight, and cost limitations, SPARK-01 and -02 do not have on-board calibration systems. Also, complete preflight radiometric calibrations were not performed in the laboratory due to the short manufacture time and prioritization of more urgent tasks before the satellite launch. Only the spectral calibration for each detector was conducted by a monochromator in the laboratory. The spectral response curves followed the Gauss function quite well after data processing, and thus, the central wavelengths and the full-width at half-maximum (FWHM) of the SPARK satellites were determined. The averaged central spectral wavelengths of these two satellites are slightly different (Figure 2). Moreover, the spectral smile effect is minor for SPARK-01 but is evident in SPARK-02 (Figure 3). This aspect should be considered in the data processing flow. Therefore, in-orbit vicarious calibration must be used to transform the satellite data into meaningful physical information. Previous studies used reflectance-, irradiance-, and radiance-based techniques [2,3] to successfully calibrate satellites such as the SPOT HRV [4], Landsat TM/ETM [3,5,6], Airborne Visible and Infrared Spectrometer [7], EO-1 Hyperion [8,9], and FY [10,11], MISR [12], Landsat OLI [13], CBERS-4 [14], and many other optical remote sensors [15]. The reflectance- and irradiance-based methods have been compared with cross-calibration methods to derive the calibration coefficients for the BJ-1 microsatellite [16]. The results showed the irradiance-based method to be superior to the reflectance- based method, especially under low-visibility atmosphere conditions. In reality, vicarious calibration methods have always been used in combination with the pre-launch calibration and on-board calibration to determine calibration accuracy and monitor the sensor's radiometric stability [17–19]. Apart from the vicarious calibration methods frequently applied to multispectral remote sensing satellites, some novel methods for hyperspectral sensors have also been proposed in recent years, such as the improved irradiance-based method [20] and supervised vicarious calibration method [21,22]. A distinguishing characteristic of a hyperspectral sensor is its high spectral resolution, and spectral smile effect and spectral shift may greatly affect the radiometric accuracy near the atmospheric absorption wavelength regions [7,23,24]. Due to the lack of on-board calibrator and pre-launch radiometric calibration, the in-orbit calibration of the SPARK-01 and -02 satellites was achieved via a calibration experiment performed at the dry Dunhuang site in the Gobi Desert in western China from 28 February to 10 March 2017. In-situ measurements, including both ground reflectance and atmospheric parameters, were also acquired during this calibration period. Two vicarious calibration methods (i.e., reflectance-based and irradiance-based) were used independently to predict the top-of-atmosphere (TOA) radiance (L_{TOA}) using MODTRAN® 5 software. The vicarious method results were then used to obtain the final SPARK-01 and -02 calibration coefficients.

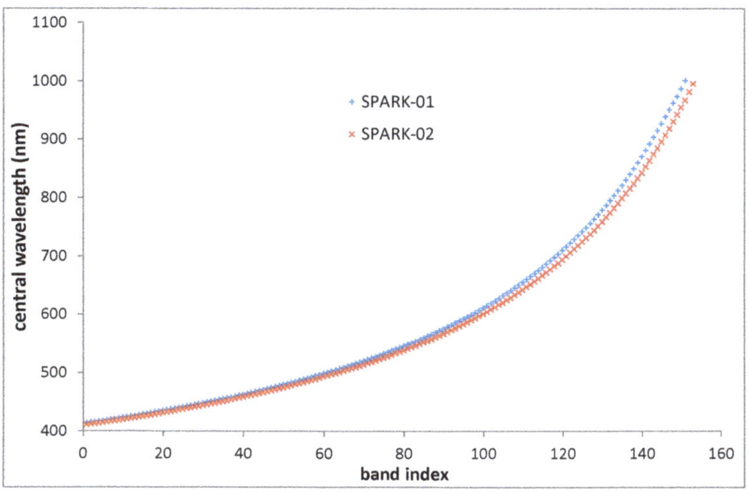

Figure 2. Central spectral wavelengths of various SPARK-01 and -02 bands.

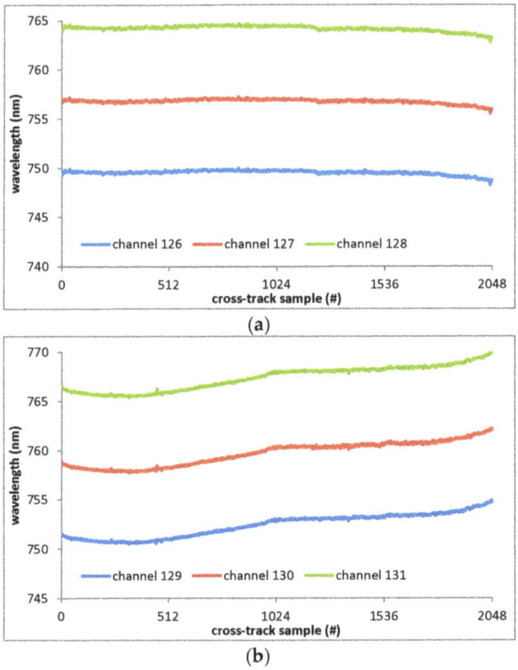

Figure 3. Cross-track central spectral wavelengths for channels centered near 760 nm for SPARK-01 (**a**) and -02 (**b**), respectively.

2. Calibration Site and Measurements

Three simultaneous measurement datasets from the Dunhuang calibration site were required for the SPARK radiometric calibrations: raw data from the SPARK satellites, surface reflectance measurements, and atmospheric measurements. Also, in order to correct for non-uniform phenomenon detection due to differing detector responses, two more observations were performed around the time when the calibration experiment occurred: a 90° yaw observation or slide slither over the bright desert region during daytime and a dark current observation over the open ocean during nighttime. Use of the 90° yaw observation is efficient for correcting the non-uniform radiometric response among different detectors. This technique has been utilized for Hyperion [9], Quickbird [25], RapidEye [26], and Landsat 8 [27]. Using this technique, all the pixels along the cross-track direction would observe nearly the same scene. Owing to the wide swath (~100 km) of the SPARK satellites, it is difficult to find a uniform ground site wider than 100 km to permit normalization of the different responses among pixels in the cross-track direction. Thus, 90° yaw observation is necessary to perform the relative radiometric calibration. The surface reflectance measurements were conducted by a spectroradiometer (FieldSpec-4, ASD Inc., Longmont, CO, USA) one hour before and after the SPARK satellite overpass. The atmospheric measurements were acquired by a CE318 sunphotometer, a Microtops II sunphotometer (Solar Light Company, Inc., Glenside, PA, USA), an irradiance sphere combined with an SVC GER1500 spectrograph and radiosonde balloons. The details are illustrated as follows.

2.1. Calibration Site

The Dunhuang calibration site (40°5′32.80″N, 94°23′35.78″E) is located on the eastern edge of the Kumutage Penniform Desert, which is in the Gobi Desert in northwestern China, about 35 km west of the city of Dunhuang, Gansu Province. The calibration area is approximately 1.2 km above sea level. The entire vicarious calibration target area (30 km × 30 km) is situated on a stabilized alluvial fan (see Figure 4). The area used for the vicarious calibration measurements for the high- and medium-spatial resolution sensors is approximately 400 m × 400 m and is located in the center of the alluvial fan; the surface is covered by cemented gravels. Several years ago, this calibration site was protected by the addition of protective fens along the edges to form a 500 m × 500 m square region.

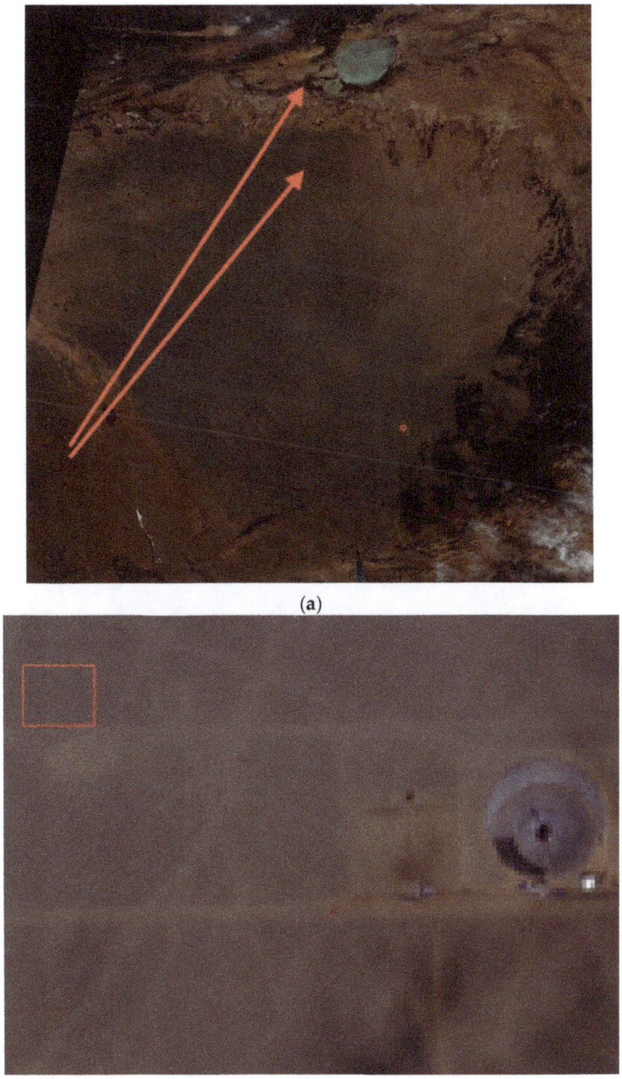

(a)

(b)

Figure 4. Dunhuang calibration site for medium-high resolution satellites, as illustrated in a Landsat 8/OLI image acquired on 2 February 2017. (**a**) Scaled subset image; (**b**) 5× magnification of a portion of the original image, where the red rectangle is the outline of the calibration site.

Figure 4 shows a Landsat/OLI image of the Dunhuang calibration site in which the surrounding fens can just be discerned. The local atmosphere is dry with low aerosol loading, which is beneficial for the calibration experiments. The atmospheric aerosol characteristics at the site are typical of a rural continental location, although some larger particles have been observed, possibly originating from sand dunes located to the northwest [10,11,28,29].

2.2. SPARK Satellite Observations

SPARK-01 and -02 data were acquired over the Dunhuang calibration site at 06:48:30 UTC on 7 March 2017 and at 06:52:32 UTC 28 February 2017, respectively. The dark current data and 90° yaw data for the relative calibration were acquired on 13 March and 11 March 2017, respectively, for the SPARK-01 satellite and on 27 February and 28 February 2017 for the SPARK-2 satellite. Figure 5 shows the SPARK-01 and -02 raw data; a number of vertical strips are evident. Clouds are evident in the SPARK-2 image over the southern and eastern areas of the calibration site. Although these atmospheric conditions are not ideal for SPARK-02 calibration, the observations over the calibration site were not affected by either clouds or shadows (Figure 5b), and, thus, the calibration results are expected to be comparable. Detailed imaging information for the calibration site and the relative radiometric calibration is listed in Tables 2 and 3, respectively.

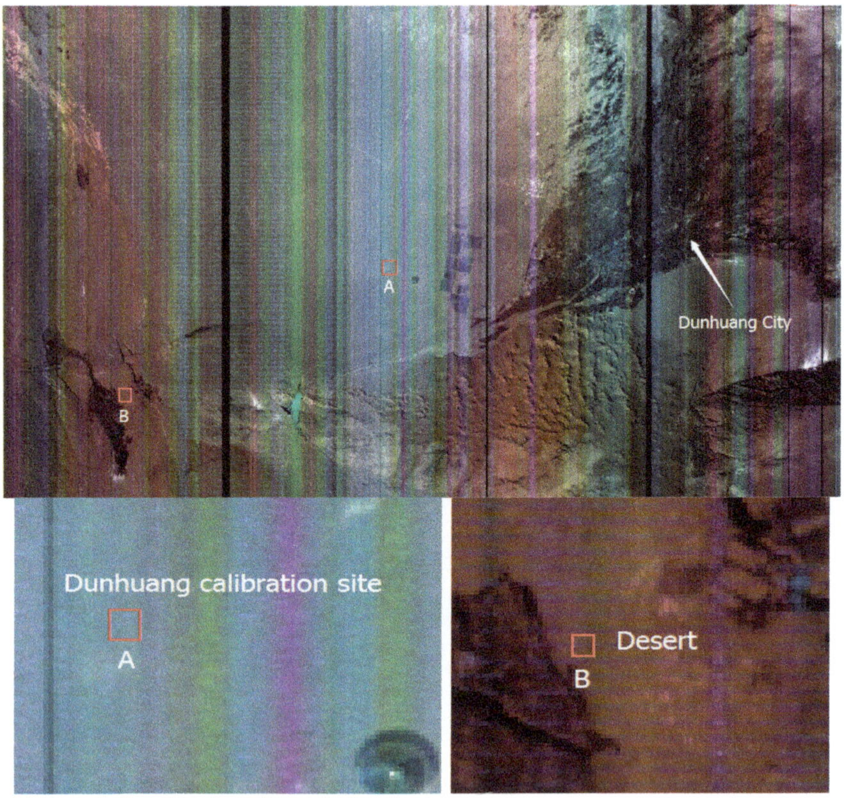

(a)

Figure 5. *Cont.*

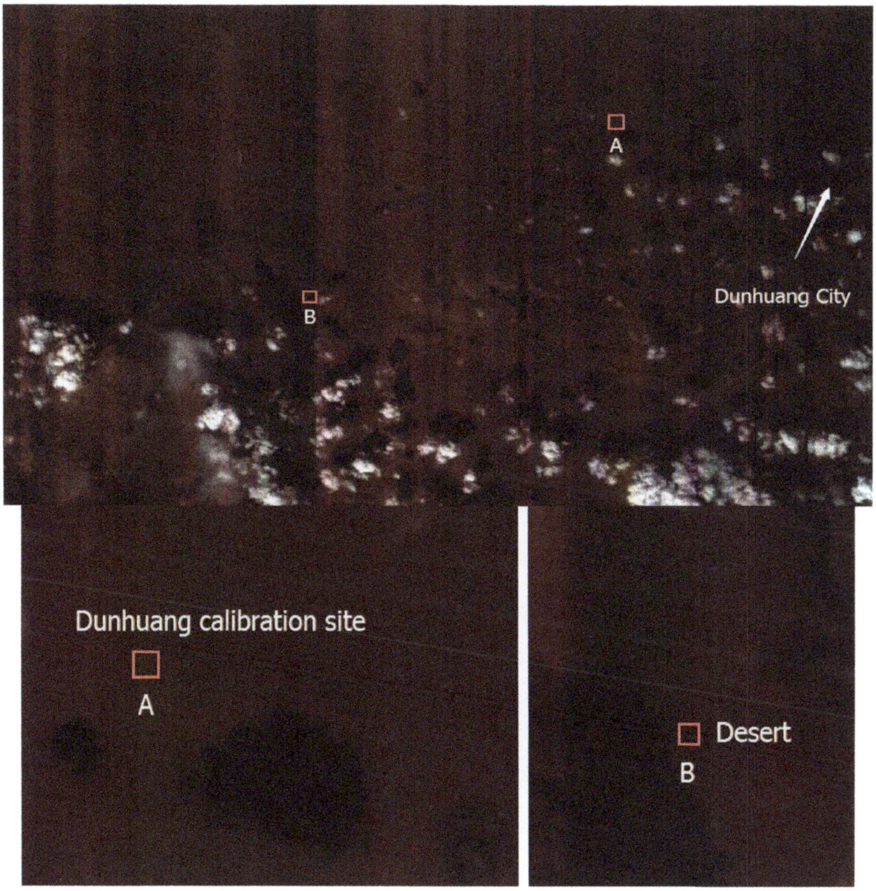

(**b**)

Figure 5. Subsets of SPARK-01 and -02 data acquired over the Dunhuang calibration site featuring pseudo color composited from the 141 (856.60 nm), 111 (648.60 nm) and 84 (550.30 nm) bands. (**a**) SPARK-01 data acquired on 7 March 2017 at 06:48:30 UTC; (**b**) SPARK-02 data acquired on 28 February 2017 at 06:52:32 UTC. These images were 180° rotated from the original raw data to maintain the northern and eastern directions on the top and the right hand, respectively.

Table 2. SPARK image acquisition information at the Dunhuang calibration site.

Date	Satellite	Pass Time (UTC)	Solar Zenith (°)	Solar Azimuth (°)	Viewing Zenith (°)	Viewing Azimuth (°)
7 March	SPARK-01	6:48	47.0579	198.5470	5.0	93.101
28 Febraury	SPARK-02	6:52	49.8249	198.4959	1.9	88.011

Table 3. SPARK image acquisition information for the relative calibration.

Date	Satellite	Imaging Time (UTC)	Lines	Location	Imaging Manner
13 March	SPARK-01	19:29	79,640	26.335°N, 160.668°E (Coral Sea)	Normal
11 March	SPARK-01	21:12	39,911	12.352°N, 29.810°E (Sudan)	90° yaw
27 Febraury	SPARK-02	12:04	42,645	60.634°N, 22.301°W (Labrador Sea)	Normal
11 Febraury	SPARK-02	20:36	22,960	13.849°N, 12.874°E (Niger)	90° yaw

2.3. Ground Reflectance Measurements

In-situ ground surface reflectance was measured over a 400 × 400 m square region one hour before and after the SPARK satellite overpass. The surface consists of cemented gravels of different colors and sizes (from mm to cm), as well as sand just beneath the gravel (Figure 6a). The measurements were taken by an ASD, Inc. spectroradiometer along a fixed route as shown in Figure 6b. Adjacent measurement points were ~40 m apart, and 10 measurements were taken around each measurement point. As a result, a total of nearly 1000 surface measurements were acquired from the Dunhuang calibration site. Thorough site measurements were repeated several times during the experiment in order to verify the stability of the surface reflectance. Measurements taken under clear atmospheric conditions were examined carefully and, after the exclusion of any erroneous measurements, averaged to produce the average reflectance of the calibration site. The ground reflectance measured on different dates during the experimental period is shown in Figure 7. The ground reflectance is relatively stable on different dates, with differences of less than 2%. The desert reflectance during the experimental period was also measured on 7 March 2017 as shown in Figure 8. These data were used to verify the radiometric calibration results.

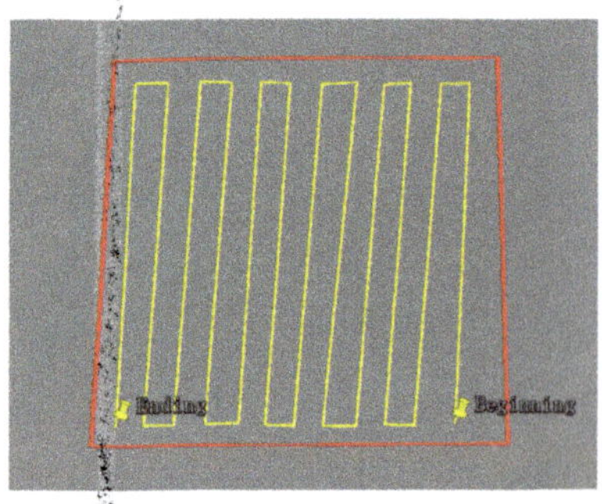

Figure 6. (**a**) Photo of the calibration site; (**b**) Schematic of the surface measurement route.

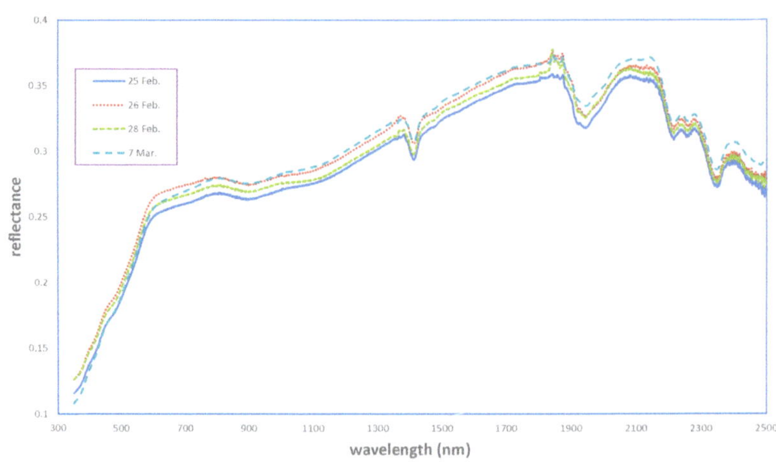

Figure 7. Ground reflectance at the Dunhuang calibration site on different dates. Measurements on 7 March and 28 February 2017 correspond to SPARK-01 and SPARK-02, respectively.

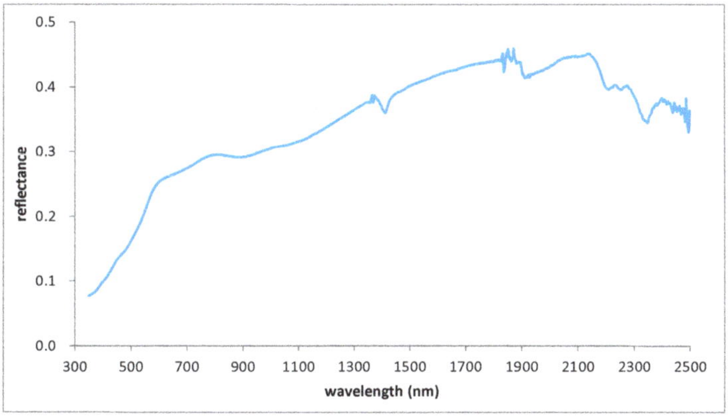

Figure 8. Ground reflectance measured on 7 March 2017 over the desert area south of the Dunhuang calibration site.

2.4. Atmospheric Data

Atmospheric measurements acquired during the experimental period include columnar atmospheric parameters (i.e., AOD and total columnar water vapor, or CWV), the diffuse-to-global irradiance ratio, and the radiosonde vertical profile. A CE318 photometer was used to measure the total AOD and CWV. A total of 5 days of valid data were acquired on 25, 26, and 28 February and 4 and 7 March 2017. The Langley calibration method [30] and a modified calibration method were used for non-water and water absorption channels, respectively, to update the calibration coefficients for the sun measurement channels. These calculations were made with the measurements acquired on 7 March 2017, as the atmosphere was stable and aerosol burden was low. Then, the AOD was calculated in each channel using Beer's law and a spectral response function [31]. We used the measured pressure from a barometer and columnar ozone and nitrogen dioxide content from the Ozone Monitoring Instrument (OMI). The CWV was retrieved using a 4-parameter method [32]. These parameters, which were retrieved within the SPARK overpass period, were averaged over 15 minutes and used as inputs in the calibration process. The 550 nm channel AOD was calculated via logarithmic interpolation of the 440 nm and 675 nm channel AODs. The 550 nm channel AOD and CWV are shown in Figure 9 for the SPARK satellite overpass dates; data influenced by clouds were excluded. Stable atmospheric conditions are indicated by the AOD and water vapor content patterns on 7 March 2017.

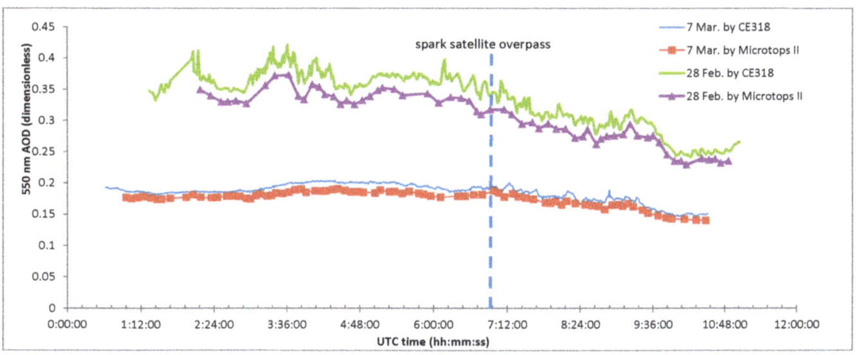

(**a**)

Figure 9. *Cont.*

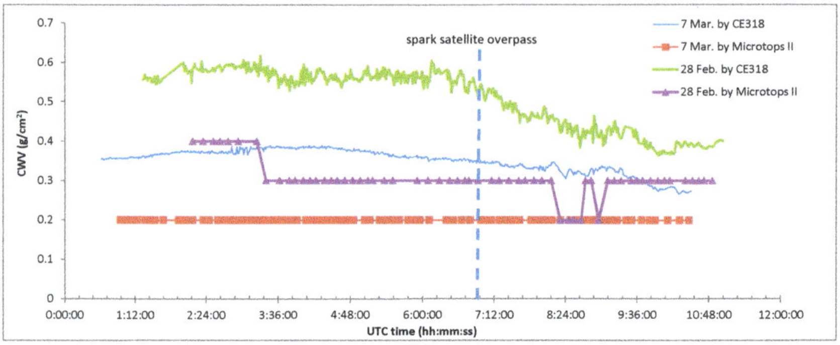

(**b**)

Figure 9. 550 nm AOD (**a**) and CWV (**b**) retrieved from CE318 measurements and Microtops II measurements on 7 March and 28 February 2017, respectively.

Measurements from a Microtops II sunphotometer were used to verify the accuracy of the CE318 observations, as shown in Figure 9. The AOD measurements are more accurate, with differences of less than 0.02 between the two instruments. However, the water vapor content differs greatly between the two instruments. We speculate that the calibration coefficients for the Microtops II 940 nm channel need to be updated. A lack of such updates would introduce additional error in water vapor retrievals. Nevertheless, the CE318 results are expected to be more reliable, as the CE318 automatic operation mode is used extensively worldwide. The 550 nm AOD and CWV were averaged over 15 min intervals within the satellite overpasses on 7 March and 28 February 2017; the average AOD and CWV values are 0.1928 and 0.3513 g/cm^2 for 7 March and 0.3476 and 0.5379 g/cm^2 for 28 February. A rural aerosol type was chosen for use in MODTRAN due to the barren Gobi Desert surroundings. Also, the angstrom exponent coefficients derived from the 440 nm to 675 nm channel AOD measurements are 0.75 and 0.3519 for the SPARK-01 and -02 overpass times, respectively. The ozone density was 299 and 305 DU on 7 March and 28 February 2017, respectively; these values were derived from NASA OMI data [33].

In addition, radiosonde balloons were released during SPARK satellite overpass periods to measure the vertical profiles of atmospheric pressure, temperature, and humidity; balloons were released at 05:49:55 and 05:05:08 UTC on 28 February and 7 March 2017, respectively. Figure 10 shows the vertical profiles measured on each date. The variations in pressure and temperature with altitude are similar between the two dates. The humidity changes little at altitudes greater than 5000 m.

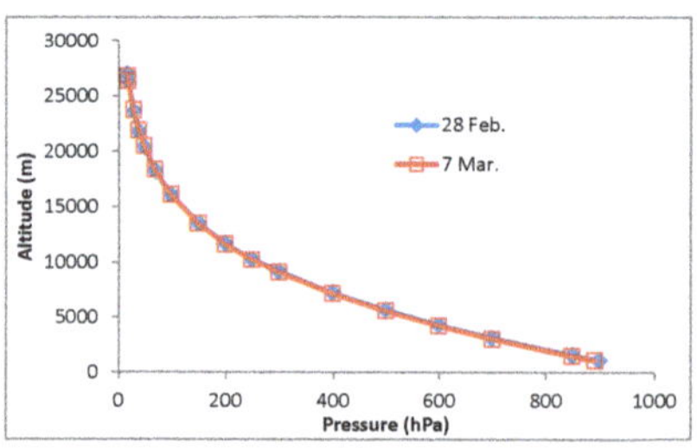

(**a**)

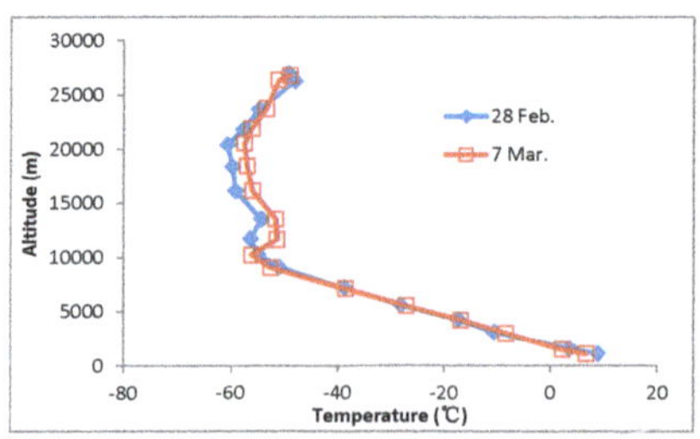

(**b**)

Figure 10. *Cont.*

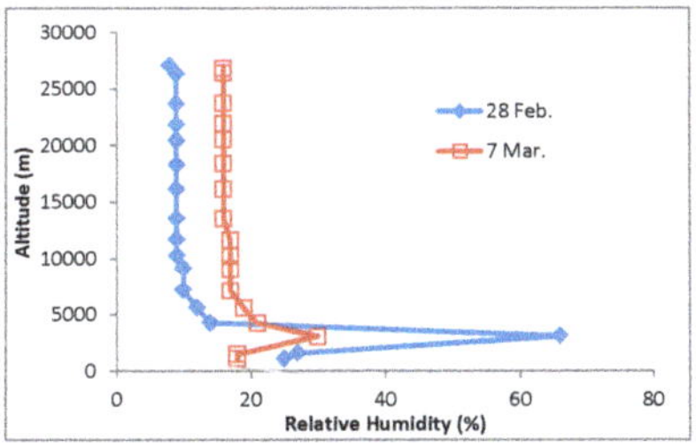

(c)

Figure 10. Vertical profiles of (**a**) pressure; (**b**) temperature; and (**c**) relative humidity measured using radiosondes released on 7 March and 28 February 2017.

To acquire the diffuse-to-global irradiance ratio data, an irradiance sphere was used with a SVC GER1500 spectrograph at the calibration site to measure the irradiance at ten minutes intervals throughout the day. Each of the measurements outlined below were repeated three consecutive times. The global solar irradiance (L1) was measured fist, followed by the sky diffuse irradiance (L2), which was assessed with a light barrier. Finally, the global solar irradiance (L3) was determined [20]. Figure 11a shows the diffuse-to-global irradiance ratios at 550 nm on the date of the SPARK satellite overpass during the Dunhuang experiment. The smooth diffuse-to-global irradiance ratio curve indicates a very stable atmosphere on 7 March 2017. Figure 11b shows diffuse-to-global irradiance ratio for the entire spectrum at the time of the SPARK satellite overpass; the lower aerosol burden on 7 March 2017 caused lower diffuse-to-global irradiance ratios in comparison to those measured on 28 February 2017. Lastly, the diffuse-to-global irradiance ratios were convolved with the spectral response functions of the corresponding SPARK satellites channels.

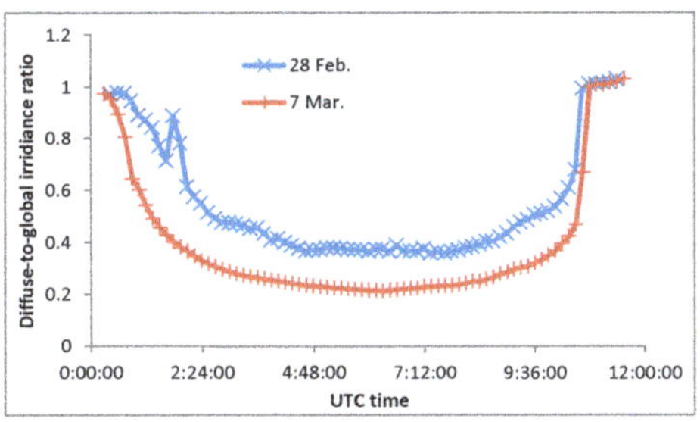

(a)

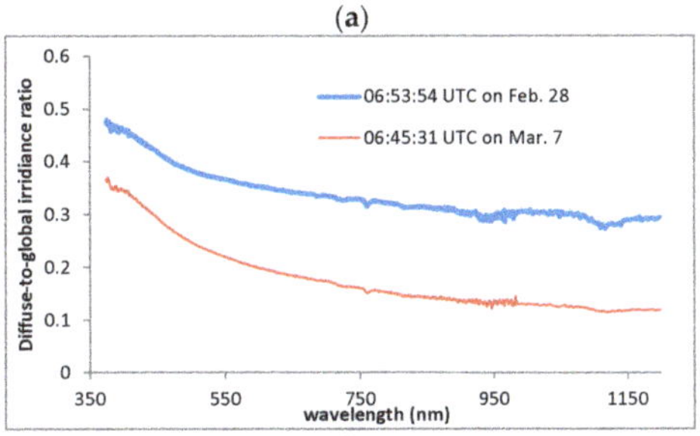

(b)

Figure 11. Diffuse-to-global irradiance ratios measured (**a**) at 550 nm on the date of the SPARK satellite overpass during the Dunhuang experiment, and (**b**) for the entire spectrum at the time of the SPARK satellite overpass.

3. Methods

Vertical striping effects were evident in the raw SPARK satellites images and were caused by several factors, such as odd-even detector processing, "smile" effects due to optical aberrations and misalignments, and signal output variations caused by electrical design. In addition, dark stripes caused by bad pixels are evident in the raw SPARK images (Figure 5). Therefore, several pre-processing steps should be conducted before the absolute radiometric calibration, including: (1) bad pixel fixing using the cross-track neighboring two pixels; (2) dark current subtraction; (3) de-smiling by cubic spline interpolation according to the pre-launch spectral calibration; and (4) relative calibration to normalize the different responses among cross-track detectors. The relative calibration procedure was applied to SPARK satellite data to correct different detector responses within a band. The dark current image and 90° yaw bright image were used to calculate the relative calibration coefficients for each detector. Then, the SPARK radiance over the calibration site was propagated to the top of the atmosphere (TOA) by using the measured ground reflectance and atmospheric parameters. The radiometric calibration coefficients were derived by dividing the predicted TOA radiance from the averaged calibration site digital number (DN) curves. Figure 12 shows the DN curves extracted from SPARK-01 and -02 data after spectral smile correction and relative radiometric calibration. These curves were derived from 6×6 pixel averaged values. In the next section, the reflectance- and irradiance-based methods are compared with each other.

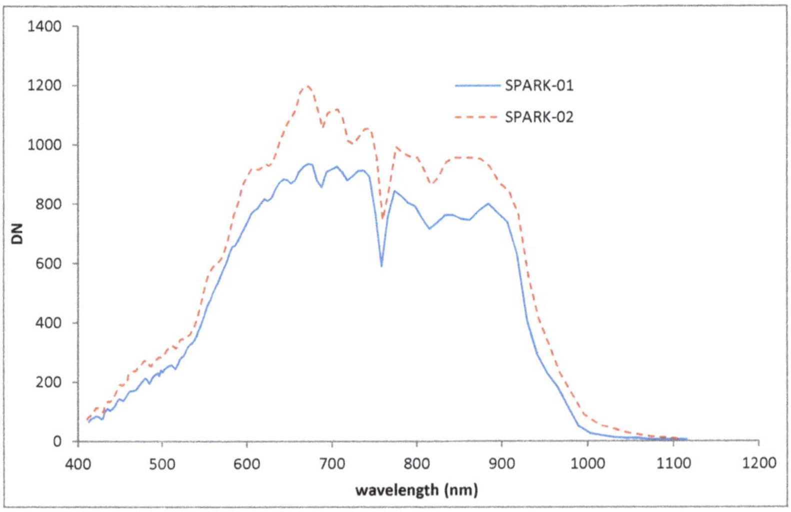

Figure 12. DN curves from SPARK-01 and -02 at the Dunhuang calibration site, averaged over 6×6 pixels.

3.1. Relative Radiometric Calibration and Spectral De-Smiling Correction

Enabled by flexible satellite controls and the wide swath (100 km), 90° yaw imaging was performed over the bright desert. This unique imaging method involves turning all detectors to observe nearly the same scene along the orbit direction. The number of rows of the SPARK-01 and -02 90° yaw images exceeds 20,000, and this number is sufficient to normalize the different responses among pixels along the cross-track direction. The average column value for each pixel was used to normalize the differing response behaviors in the given pixel. Figure 13 illustrates the normal and 90° yaw imaging methods.

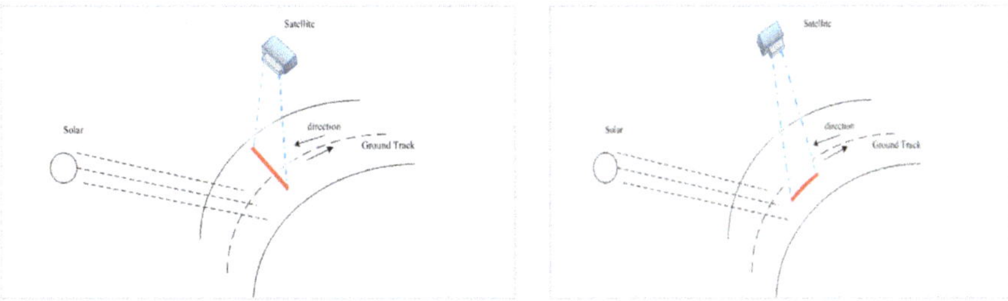

(**a**) (**b**)

Figure 13. Schematics of (**a**) normal and (**b**) 90° yaw imaging methods in the SPARK satellite.

3.1.1. Dark Current Subtraction

The dark current (DC) image was acquired over the open ocean during nighttime. The signal remaining in the image represents the DC; accurate DC values were calculated from the average along the row direction, which can be expressed as:

$$B(i,k) = \frac{1}{N}\sum_{j=1}^{N} DC(i,j,k) \tag{1a}$$

$$DN^{*}(i,j,k) = DN(i,j,k) - B(i,k) \tag{1b}$$

where i is the column index, j is the row index, k is the band index, DC is the dark current acquired over the ocean during nighttime, DN is the raw data from the SPARK image, DN^* is the SPARK data after dark current correction, and B is the dark current averaged along the row direction.

3.1.2. De-Smiling Correction

Cubic spline interpolation was used for de-smiling Hyperion data given its advantage of retaining spectral curve features [24,34]. It was also adopted to interpolate the image after dark current subtraction from the central wavelengths for each pixel provided by pre-launch spectral calibration into the average wavelength (Figure 2) of all 2048 pixels. Then, the hyperspectral cube data after dark current subtraction DN^* in the original spectral wavelength was transformed into the new cube data DN^{**} in the average wavelength, and is expressed as follows:

$$\langle DN^{**}(i,j,k)\rangle|_{i,j} = cubic\_spline(\langle DN^{*}(i,j,k)\rangle|_{i,j}, \langle\lambda(j)\rangle, \langle\overline{\lambda}\rangle) \tag{2}$$

where $<DN^*>$ and $<DN^{**}>$ represent spectral data at the spatial position (i, j) before and after de-smiling correction, respectively; $<\lambda(j)>$ is the central wavelength values for j pixel from pre-launch spectral calibration; $\langle\overline{\lambda}\rangle$ is the average wavelength of all 2048 pixels; and *cubic_spline* represents the cubic spline interpolation method.

3.1.3. Uniform Normalization

After dark current subtraction, the differing response in each detector was corrected through columnar normalization as follows:

$$DN^{***}(i,k) = \frac{1}{M}\sum_{j=1}^{M} DN^{**}(i,j,k) \tag{3a}$$

$$\overline{DN^{***}}(k) = \frac{1}{N}\sum_{i=1}^{N} DN^{***}(i,k) \tag{3b}$$

$$A(i,k) = \overline{DN^{***}}(k)/DN^{***}(i,k) \quad (3c)$$

where DN^{**} is the SPARK data averaged along the row direction after dark current correction and de-smiling correction, $\overline{DN^{***}}$ is the average of quantity DN^{**} in all the column and A is the relative radiometric correction coefficient used in the uniform normalization.

Theoretically, if the satellite flies with a strict yaw angle of 90°, the use of the same imaging path for each detector would cause a one-pixel delay between two adjacent detectors. However, the delay distance may be less than one pixel due partially to inexact yaw angle control and partially to minor differences between the ground sample distances (GSDs) along the orbit direction and across the orbit direction. This pattern of delays forms an evident line on the 90° yaw image (Figure 14a,c); the correction methods are listed in Figure 14b,d. The total delay, in pixels, is easily visually estimated from the 90° yaw image. This approximate estimation is sufficiently accurate for use because the image row number is quite large, which allows us to ignore minor errors in delay estimations. Equation (3a) can be modified to Equation (4), which applies to a similar imaging path along the orbit direction, through the addition of the delay factor as follows:

$$\begin{gathered} DN^{***}(i,k) = \frac{1}{N-D}\sum_{j=S(i)}^{N-D} DN^{**}(i,j,k) \\ S(i) = D/M \times i \text{ (for SPARK-01)} \\ S(i) = N - D/M \times (M-i) \text{ (for SPARK-02)} \end{gathered} \quad (4)$$

where N and M represent the total column number and row number for SPARK 90° yaw image, respectively, D denotes the delay lines, and S is the starting line number to perform the average operation.

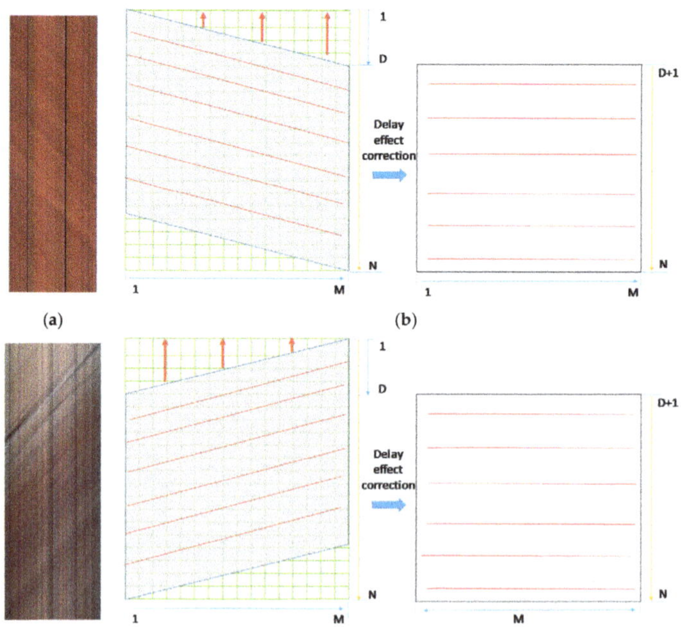

Figure 14. 90° yaw images and delay lines in different columns, including the (**a**) subset image and (**b**) delay effect correction scheme for SPARK-01 satellite images and the (**c**) subset image and (**d**) delay effect correction scheme for SPARK-02 satellite images.

3.2. Absolute Radiometric Calibrations

Reflectance-based and irradiance-based methods are widely used for absolute vicarious radiometric calibrations in in-situ experiments. Both methods require accurate measurements of spectral reflectance for the ground target, as well as spectral AOD, vertical columnar water content, and other meteorological parameters. For the reflectance-based method, atmospheric scattering and absorption are computed based on these measurements using MODTRAN 5. In principle, the reflectance-based method aerosol model is assumed based on experience, which may introduce much uncertainty as AOD increases. The irradiance-based method uses the measured data in the reflectance-based method with measurements of the diffuse-to-global spectral irradiance ratio at ground level. This additional measurement helps reduce uncertainty in the aerosol model used for the scattering calculations [35]. The principles used to calculate the TOA spectral reflectance in the reflectance- and irradiance-based methods are shown by Equations (5) and (6), respectively [20]. Both methods use Equation (7) to transform the TOA spectral reflectance into the TOA radiance.

$$\rho^*(\theta_s, \theta_v, \phi_{v-s}) = \rho_a(\theta_s, \theta_v, \phi_{v-s}) + \frac{\rho_t}{1 - \rho_t \times s} \times T(\theta_s) \times T(\theta_v) \tag{5}$$

$$\rho^*(\theta_s, \theta_v, \phi_{v-s}) = \rho_a(\theta_s, \theta_v, \phi_{v-s}) + \frac{e^{-\tau/\mu_s}}{1 - \alpha_s} \times \rho_t \times (1 - \rho_t \times s) \times \frac{e^{-\tau/\mu_v}}{1 - \alpha_v} \tag{6}$$

$$L = \rho^* \times \mu_s \times E_0 / (d^2 \times \pi) \tag{7}$$

In Equations (5)–(7), θ_s is the sun zenith angle, θ_v is the view zenith angle of the sensor, and ϕ_{v-s} is the relative azimuth angle between the view azimuth angle and the sun azimuth angle. ρ_t is the measured spectral reflectance of the ground target, and ρ_a is the reflectance that corresponds to the atmospheric path radiance (or atmospheric intrinsic reflectance). S is the atmospheric hemisphere reflectance. $T(\theta_s)$ and $T(\theta_v)$ are the total transmittance of the solar path and the view path, respectively, while ρ^* and L are the TOA spectral reflectance and the TOA radiance of the ground target, respectively. μ_s and μ_v are the values of $\cos\theta_s$ and $\cos\theta_v$, respectively, and a_s and a_v are the diffuse-to-global ratios of the sun direction and the view direction, respectively. E_0 is the TOA solar irradiance, and d is the Sun-Earth distance in astronomical units (AU).

If the atmospheric conditions are stable, a linear relationship exists between the relative optical air mass (m) (i.e., inverse of the cosine of the solar zenith ($1/\mu_s$)) and the natural logarithm of 1 minus the diffuse-to-global irradiance ratio ($\ln(1 - a_s)$) [2].

$$\ln(1 - \alpha_s) = \ln(1 - \rho s) - (1 - b)\tau m \tag{8}$$

On stable days, the fitted slope value $(1 - b)\tau$ can be used to compute the diffuse-to-global irradiance ratio for both the solar direction and the viewing direction. During the experiment, diffuse-to-global measurements were taken every 10 min throughout the day. Therefore, the α_s can be interpolated with sufficient accuracy via the use of an adjacent measurement in Equation (8). However, the α_v must be extrapolated to a zenith angle approximating zero from measurements at observation angles quite different from zero. Thus, if the atmosphere was not very stable, as on 28 February 2017, only α_s was used to replace the scattering effect in the reflectance-based method; the upward transmittance was also calculated from MODTRAN 5. Similar modifications have been used previously for UAV hyperspectral sensor vicarious calibration [20].

$$\rho^*(\theta_s, \theta_v, \phi_{v-s}) = \rho_a(\theta_s, \theta_v, \phi_{v-s}) + \frac{\rho \times e^{-\tau/\mu_s}}{1 - \alpha_s} \times T(\theta_v) \tag{9}$$

The ratio of $\ln(1 - \alpha_s)$ to relative optical air mass (m) (which had values of no more than 6) at 549.89 nm was scattered (see Figure 15) for both measurements taken early in the day on 28 February and 7 March 2017. Two clear outliers measured on 28 February were removed due to the influence of clouds. The measurements on 7 March show a nearly linear relationship, which indicates stable atmospheric conditions. However, non-linear behavior is observed on 28 February. Therefore, the irradiance-based method applied to the 7 March measurements took the form of Equation (6), while that applied to the 28 February measurements took the form of Equation (9). For comparison, Equation (6) was also applied to the 28 February measurements despite the atmospheric instability. Then, the spectral diffuse-to-global irradiance ratio was convolved with the spectral response functions to derive the band-weighted values; the diffuse-to-global irradiance ratio at the viewing direction was calculated according to Equation (8). Then, linear regression was performed for each band. The goodness-of-fit (R^2) values are shown in Figure 16. Linear relationships are evident in each band for the 7 March measurements, but rather lower linear correlations are noted for the 28 February data. The diffuse-to-global irradiance ratios for both SPARK-01 and SPARK-02 are shown in Figure 17 at their calibration site overpass times.

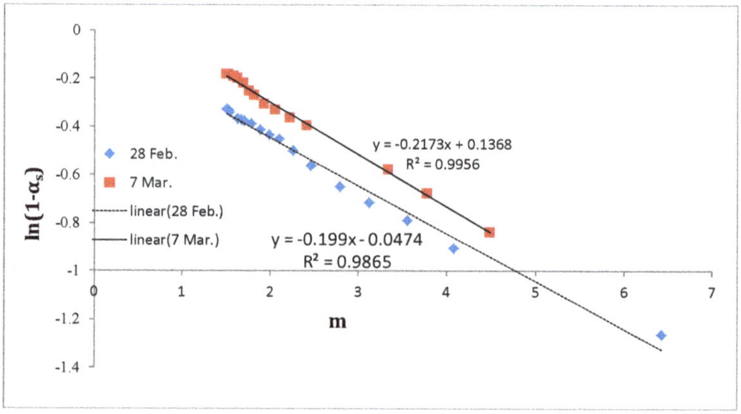

Figure 15. A scatter plot of $\ln(1-\alpha_s)$ versus m at 549.89 nm.

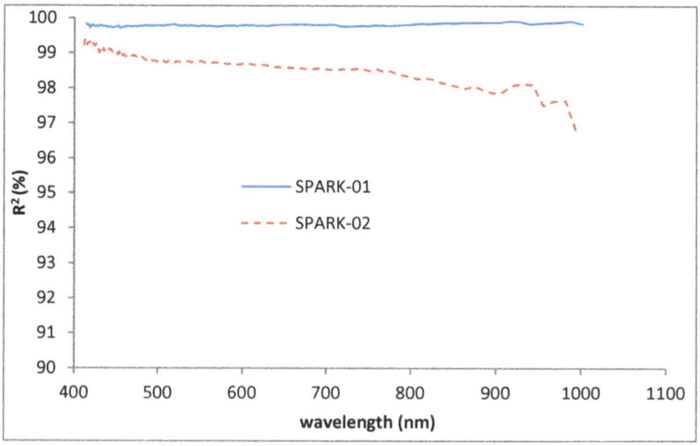

Figure 16. Goodness-of-fit statistics for diffuse-to-global irradiance ratio measurements according to Equation (8).

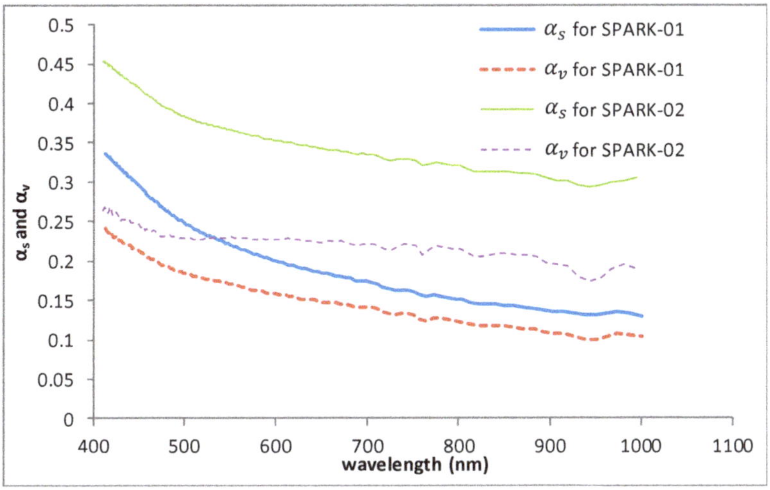

Figure 17. Diffuse-to-global irradiance ratio extrapolated and interpolated to the solar direction (α_s) and viewing direction (α_v) during the satellite overpasses.

4. Results

4.1. Relative Radiometric Calibration

The relative radiometric calibration coefficients were derived for the SPARK-01 and -02 satellites. The results in terms of blue, green, and red bands are shown in Figures 18 and 19. In total, 32 sub-regions were evident in the SPARK-01 and -02 dark current curves; this number coincides with the design, which features 32 electrical outputs. These coefficients were applied to SPARK images acquired over the calibration site in on 28 February and 7 March 2017. The non-uniformities and variations were largely eliminated after relative radiometric correction using the row-averaged curves (Figure 20).

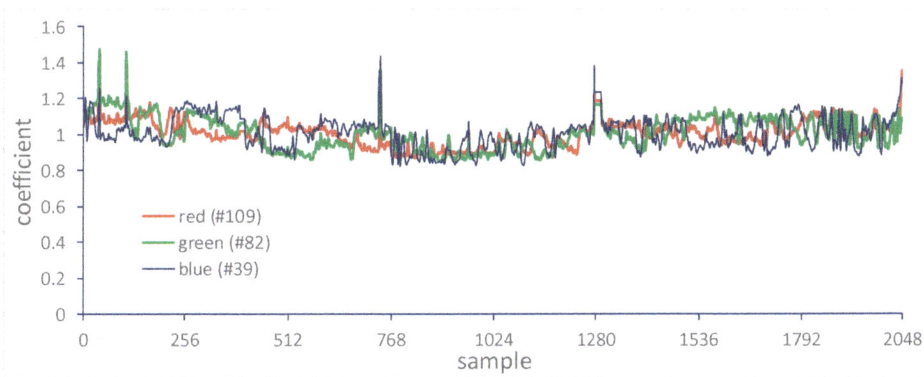

(a)

Figure 18. *Cont.*

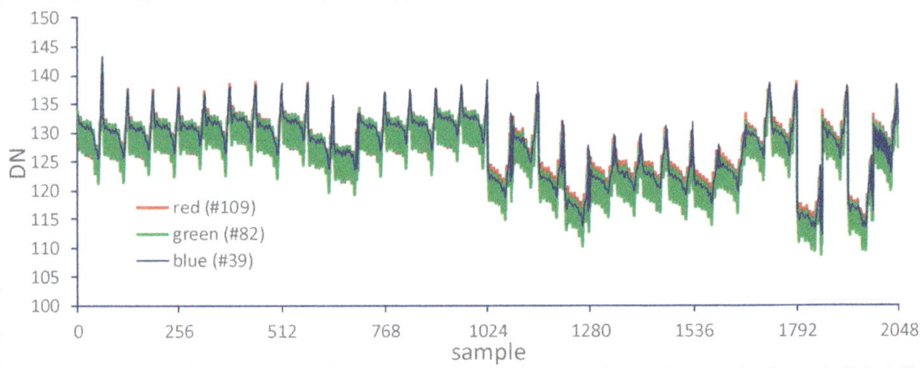

(b)

Figure 18. Relative radiometric correction coefficients for SPARK-01 satellite images at 650.4 nm, 551.5 nm, and 461.7 nm showing (**a**) gain curves (non-uniform correction coefficients) and (**b**) offset curves (dark current).

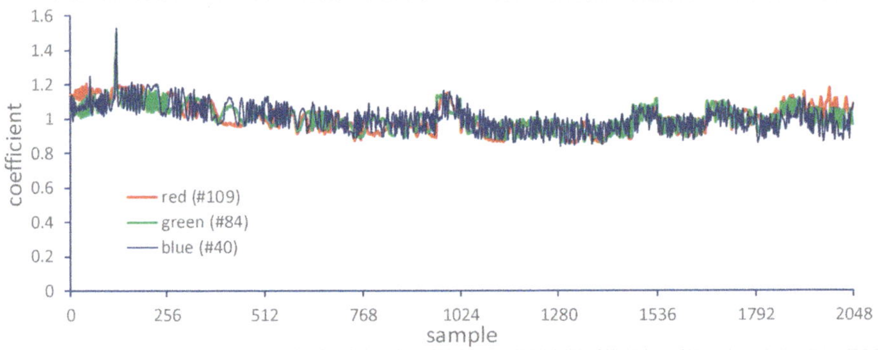

(a)

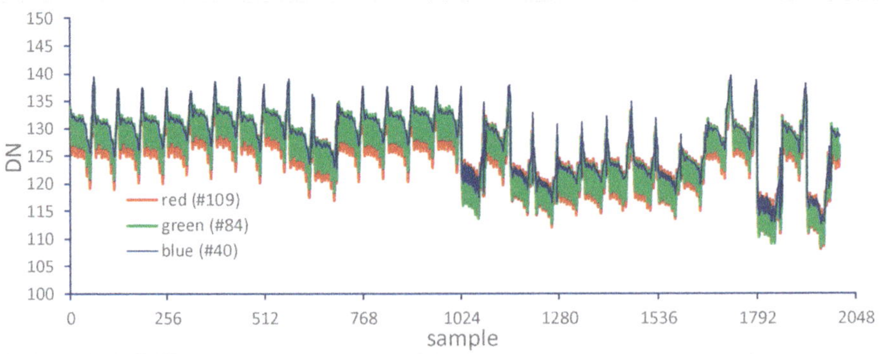

(b)

Figure 19. Relative radiometric correction coefficients for SPARK-02 satellite images at 638.0 nm, 549.5 nm, and 459.0 nm showing (**a**) gain curves (non-uniform correction coefficients) and (**b**) offset curves (dark current).

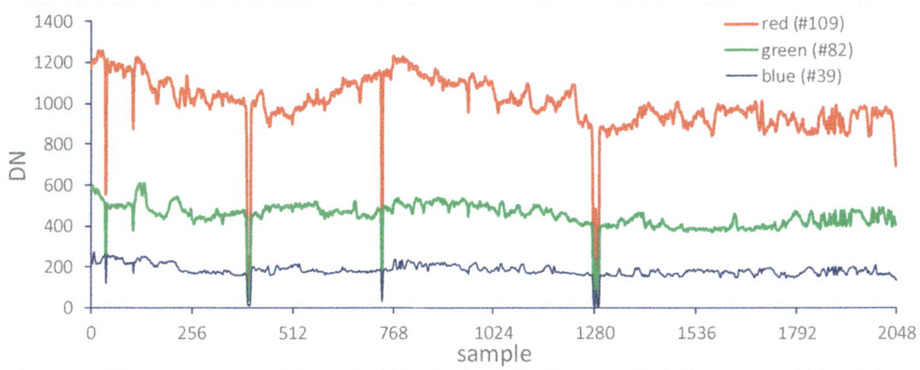

(a)

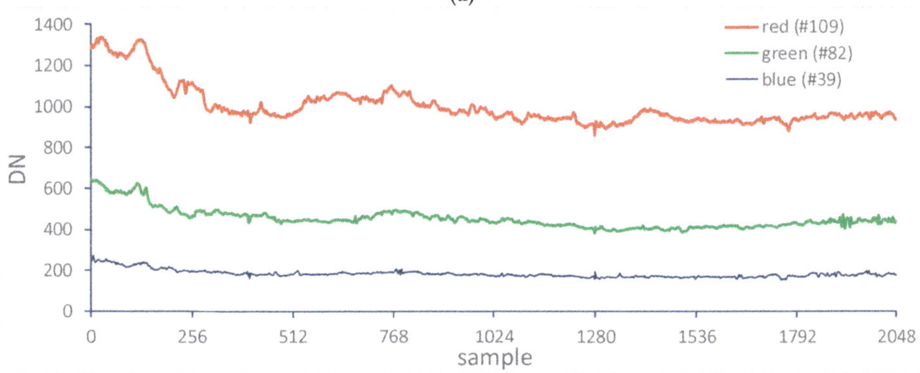

(b)

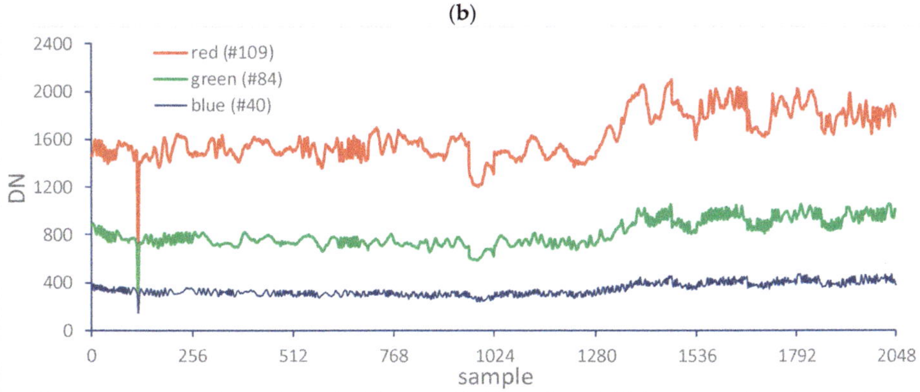

(c)

Figure 20. *Cont.*

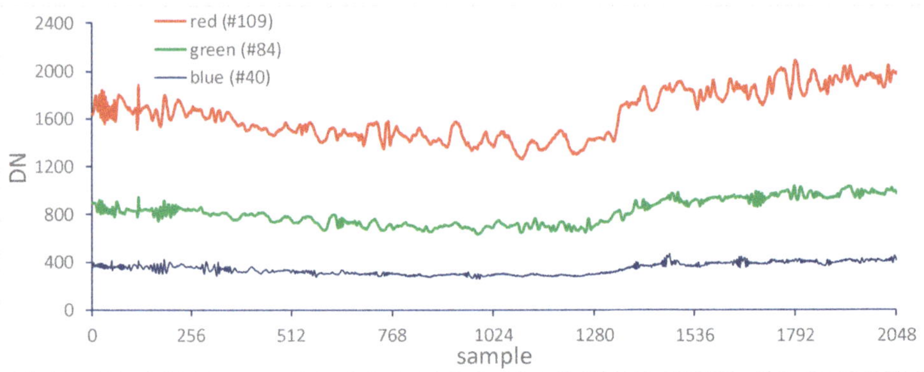

(**d**)

Figure 20. Row-averaged values at 650.4 nm, 551.5 nm, and 461.7 nm from images acquired over the calibration site on 7 March and 28 February 2017 using SPARK-01 (**a**) before and (**b**) after relative calibration, and at 638.0 nm, 549.5 nm, and 459.0 nm using SPARK-02 (**c**) before and (**d**) after relative calibration.

4.2. Absolute Radiometric Calibrations

The MODTRAN-simulated radiance calculated using both the reflectance- and irradiance-based methods is shown in Figures 21 and 22, respectively, for the SPARK-01 and -02 satellites. The absolute radiometric calibration is simple to derive by dividing the radiance from the 6 × 6 averaged DN values. The difference between the results from reflectance- and irradiance-based methods does not exceed 6% for the SPARK-01 satellite and shows evident discrepancies in spectral bands <600 nm. However, the differences between the reflectance- and irradiance-based results are greater than 9% for the SPARK-02 satellite in spectral bands <500 nm. These large discrepancies are caused partially by the relatively large AOT (AOT at 550 nm = 0.35) and partially by the unstable weather conditions on 28 February 2017. In comparison, the improved irradiance-based method, which uses only the downward diffuse-to-global irradiance ratio, derived approximately the same radiance for the SPARK-02 on 28 February 2017 as did the reflectance-based method.

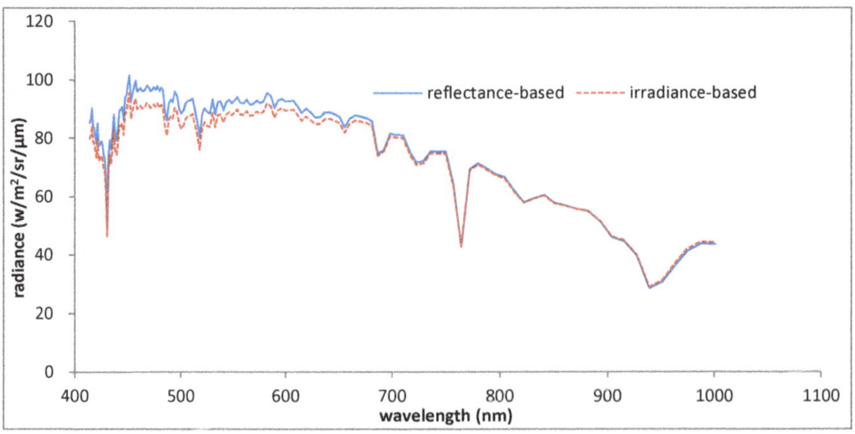

Figure 21. MODTRAN-simulated radiance for the SPARK-01 satellite calculated using both reflectance- and irradiance-based methods.

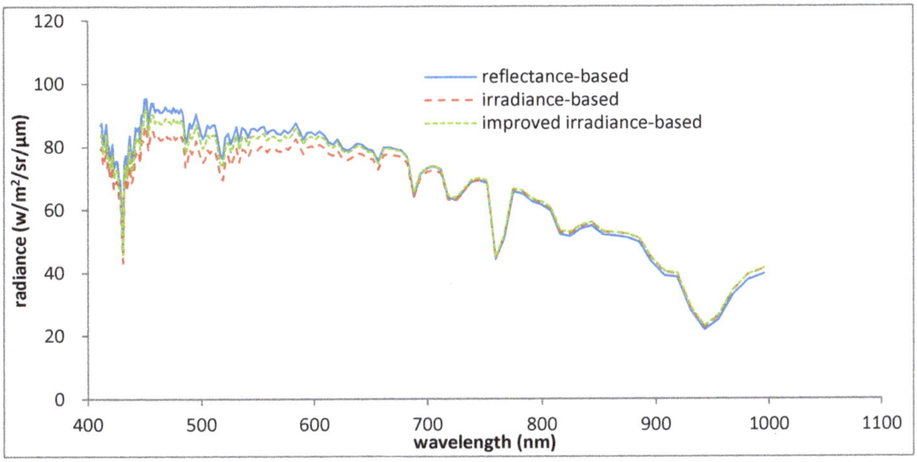

Figure 22. MODTRAN-simulated radiance for the SPARK-02 satellite calculated using reflectance-, irradiance-, and improved irradiance-based methods.

5. Discussion

Errors associated with in-situ measurements, data processing, and calibration method selection all contribute to uncertainty in satellite calibration coefficients [20]. Given that calibration uncertainty sources, such as ground reflectance measurements, inherent code accuracy, etc. have been thoroughly discussed in the literature, we focused our analysis on uncertainties caused by aerosol type assumptions, AOD measurements, water vapor content measurements, atmospheric profile measurements, and satellite image misregistration. Furthermore, the wavelength shift occurring in hyperspectral data would also impose an additional influence on the radiometric calibration accuracy, especially near the atmospheric absorption wavelengths, due to gases like oxygen, water vapor, carbon dioxide, etc.

5.1. Uncertainty Due to Aerosol Type Assumptions

The aerosol type used in the Radiative Transfer Model (RTM) introduces great uncertainty in vicarious calibrations, especially in situations in which the AOD is large. It was impractical to measure the vertical distribution of aerosol characteristics during the calibration campaign. However, the actual aerosol type in Dunhuang, which is in arid northwestern China, is generally close to the RURAL and DESERT types described in MODTRAN. To evaluate uncertainty due to aerosol type, three additional aerosol types similar and dissimilar to local conditions (i.e., urban, desert, and maritime) were chosen to replace the rural aerosol type used in the original calculations. The radiance was computed again using these three aerosol types, and the results were compared to those derived using the rural aerosol type. The angstrom exponent coefficients derived from CE318 measurements were also used as inputs for MODTRAN and were 0.75 and 0.3519 for the SPARK-01 and -02 satellites, respectively. The resulting differences in radiance can be used to evaluate uncertainty due to aerosol type (SPARK-01: Figure 23; SPARK-02: Figure 24). The average relative differences are listed in Table 4.

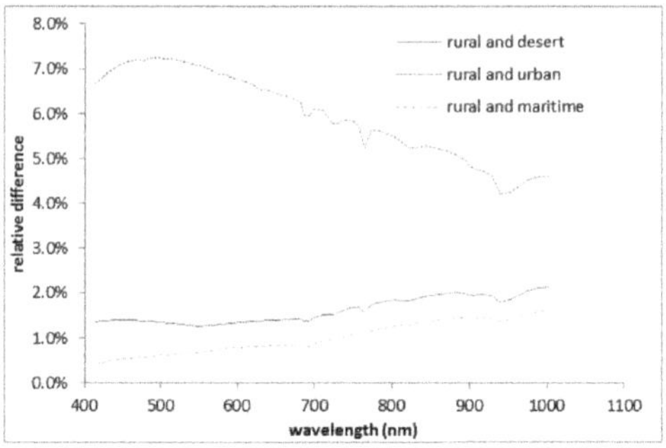

(**a**)

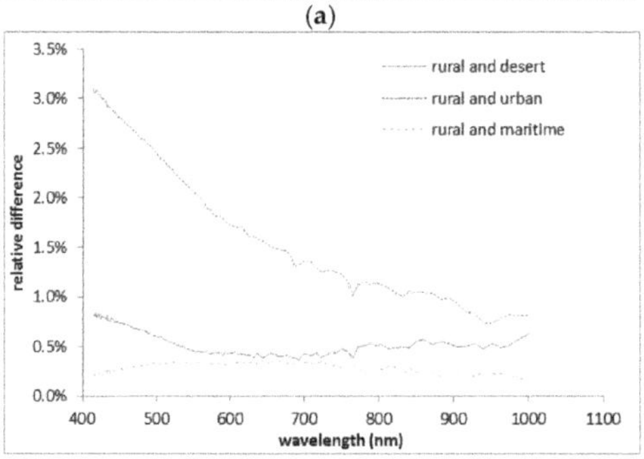

(**b**)

Figure 23. Relative differences in radiance using different aerosol types for the SPARK-01 calibration for 7 March 2017 using (**a**) the reflectance-based method and (**b**) the irradiance-based method.

Table 4. Average relative differences in radiance for the SPARK-01 and -02 satellites.

Satellite	**SPARK-01**			**SPARK-02**		
Method	**Rural and Urban**	**Rural and Maritime**	**Rural and Desert**	**Rural and Urban**	**Rural and Maritime**	**Rural and Desert**
Reflectance-based	6.58%	0.76%	1.46%	14.46%	1.77%	3.78%
Irradiance-/improved irradiance-based	2.07%	0.30%	0.58%	9.23%	0.23%	2.63%

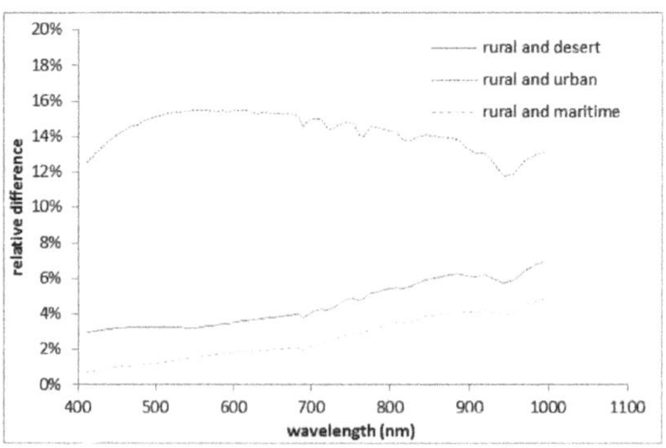

(**a**)

Figure 24. *Cont.*

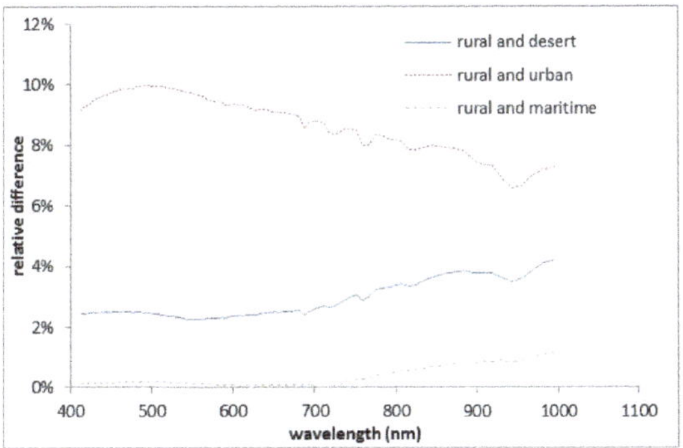

(**b**)

Figure 24. Relative differences in radiance using different aerosol types for the SPARK-02 calibration for 28 February 2017 using (**a**) the reflectance-based method and (**b**) the irradiance-based method.

Both the irradiance-based and improved irradiance-based methods use the diffuse-to-global ratio to minimize the uncertainty associated with aerosol-type assumptions. The average maximum uncertainty of the irradiance-based method used for SPARK-01 on 7 March 2017 is 2%, or less than half of that calculated for the reflectance-based method. The uncertainty in the reflectance-based method due to the aerosol-type assumption increased to 14% for the SPARK-02 satellite on 28 February 2017 because of the relatively large AOD at 550 nm (0.35). The improved irradiance-based method, despite replacing only the downward transmittance features, considerably decreased uncertainty, i.e., by 9%. In reality, the Dunhuang calibration site is surrounded by the Gobi Desert and, thus, the local aerosol is likely fall into the rural or desert types. However, the uncertainty due to aerosol type was conservatively estimated by using the setting of half the difference between the predicted TOA radiance with the urban aerosol type and that with the rural aerosol type.

5.2. Uncertainty Due to AOD Measurements

AOD is retrieved from CE318 measurements with a total uncertainty of ~0.01–0.021, which is spectrally dependent and features higher errors in the UV bands [36]; this uncertainty is validated using CE318 and Microtops II measurements in Section 2.4. Therefore, an uncertainty of ± 0.02 was added to the 550 nm AOD used for the SPARK-01 and -02 reflectance-based calibrations. For the reflectance-based method, the uncertainty was estimated by comparing the predicted TOA radiance using different AOD values in MODTRAN 5. For the irradiance-based methods (Equations (6) and (9)), the uncertainty in the predicted TOA radiance can be attributed to both errors in the directly measured transmittance $e^{-\tau/\mu_s}$ and errors in the retrieved CE318 measurements. In reality, the transmittance values consist of the CE318 direct retrievals divided by the calibration coefficient for each channel. Thus, the retrieved transmittance uncertainty is a combination of calibration uncertainty from the CE318 calibration coefficient and uncertainty due to the process of interpolating measured transmittance in a few bands into the SPARK satellite bands. The former (calibration) uncertainty is estimated to be ~0.01–0.02 (higher in the UV bands) [36], while the latter is 0.5% of the transmittance [37]. It is reasonable to set a relative uncertainty of 0.015 for the measured transmittance in the solar direction because the SPARK satellite spectral range spans from the visible to the near-infrared bands, without UV bands. The transmittance uncertainty in the view direction can be inferred from that in the solar direction by applying the cosine of the view zenith angle. Also, the uncertainty in the retrieved AOD would cause the path radiance and sphere albedo to change with different signs [2]. In order to simplify the calculation, the downward and upward direct transmittances, $e^{-\tau/\mu_s}$ and $e^{-\tau/\mu_v}$, were replaced with $T_{dir}(\theta_s)$ and $T_{dir}(\theta_v)$. The transmittance and AOD uncertainties were assumed to be independent;

thus, the error propagation equations for the TOA reflectance uncertainties using the irradiance-based (Equation (6)) and improved irradiance-based (Equation (9)) methods can be written as:

$$\begin{aligned}\Delta\rho^* = [(\Delta\rho_a - \tfrac{T_{dir}(\theta_s)}{1-\alpha_s} \times \tfrac{T_{dir}(\theta_v)}{1-\alpha_v} \times \rho^2 \times \Delta s)^2 + \\ (\tfrac{1}{1-\alpha_s} \times \rho \times (1-\rho \times s) \times \tfrac{T_{dir}(\theta_v)}{1-\alpha_v} \times (1+\tfrac{\mu_s}{\mu_v}))^2 \times ((\Delta T_{dir}(\theta_s))^2 + (0.005T_{dir}(\theta_s))^2)]^{1/2}\end{aligned} \quad (10)$$

$$\begin{aligned}\Delta\rho^* = [(\Delta\rho_a + \tfrac{\rho \times T_{dir}(\theta_s)}{1-\alpha_s} \cdot \Delta T(\theta_v))^2 + \\ (\tfrac{\rho}{1-\alpha_s} \times T(\theta_v))^2 \times ((\Delta T_{dir}(\theta_s))^2 + (0.005T_{dir}(\theta_s))^2)]^{1/2}\end{aligned} \quad (11)$$

The uncertainty estimated for SPARK-01 and -02 using the reflectance- and irradiance-based methods is shown in Figure 25. For the reflectance-based method, an AOD uncertainty of 0.02 contributes little (maximum values of 0.6% and 0.7%, respectively, for SPARK-01 and -02) to the total TOA radiance prediction uncertainty. However, the uncertainties for the irradiance- and improved irradiance-based methods appear higher than that for the reflectance-based method. The average and maximum uncertainties are 2.17% and 2.60%, respectively, for SPARK-01 and 1.20% and 1.45% for SPARK-02. The higher uncertainties for the irradiance- and improved irradiance-based methods may be attributed primarily to the direct transmittance uncertainty, which would be partially decreased by the diffuse transmittance uncertainty calculated by MODTRAN, although with the opposite sign.

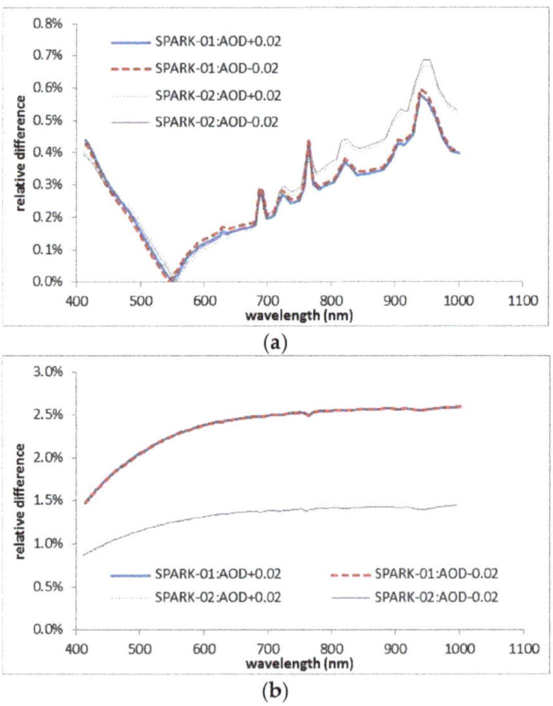

Figure 25. Calibration uncertainties caused by the AOD measurements for SPARK-01 and -02 using the (**a**) reflectance- and (**b**) irradiance- based (or improved irradiance-based) methods, respectively.

5.3. Uncertainty Due to Water Vapor Measurements

The water vapor content retrieval from the CE318 measurements is expected to have an uncertainty of 10%. Therefore, an uncertainty of ±10% was added to the water vapor content retrievals during the SPARK satellite calibration site overpass. The TOA radiance was computed again with MODTRAN, and the difference represents the calibration uncertainty caused by the water vapor measurement (Figure 26). Large uncertainties are apparent in the water vapor absorption bands near 720, 820, and 940 nm. The uncertainties for the water vapor non-absorption bands are lower than 0.2% and, thus, can be omitted. The reflectance- and irradiance-based methods show similar results (Figure 26). The highest values occur in the 940 nm band, amounting to 4.45% and 4.39% for the reflectance- and irradiance-based (or improved irradiance) methods, respectively, in SPARK-01, and 4.17% and 4.04% in SPARK-02. Due to the low water vapor content in arid areas like Dunhuang, the uncertainty caused by the water vapor measurement is relatively small.

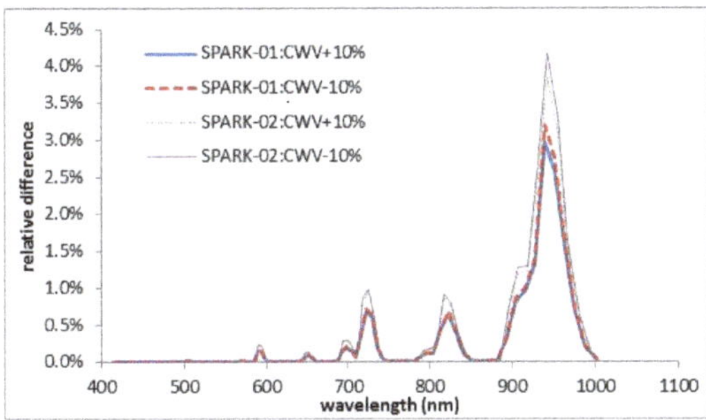

(a)

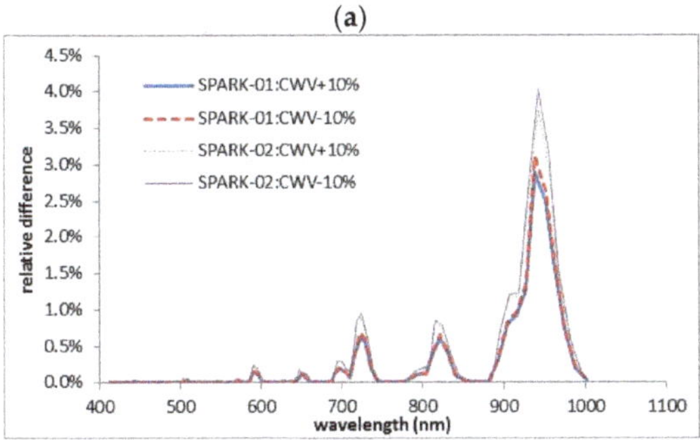

(b)

Figure 26. Calibration uncertainties caused by the water vapor content measurements for SPARK-01 and -02 using the (**a**) reflectance- and (**b**) irradiance- based (or improved irradiance-based) methods, respectively.

5.4. Uncertainty Due to Atmospheric Profile Measurements

The vertical distributions of temperature, humidity, pressure, and other atmospheric constituents also influence the TOA radiance prediction. In order to explore uncertainty due to the atmospheric profile, the measured radiosonde data used in MODTRAN 5 were replaced with three atmospheric models (i.e., the Mid-Latitude Summer, MS; Mid-Latitude Winter, MW; and 1976 US Standard Atmosphere, US models). The differences in TOA radiance predicted by the three additional atmospheric models and those by the measured radiosonde data represent the uncertainty due to atmospheric profile measurement, as shown in Figure 27. The irradiance- and improved irradiance-based methods show slightly higher uncertainties due to the atmospheric profile than does the reflectance-based method; however, their uncertainties are less than 1.3% in all bands apart from the water vapor absorption bands near 940 nm and 1135 nm. In addition, the MW model appears to be more similar to the radiosonde measurements, as evidenced by the relatively small difference in the radiances predicted using these two inputs. The MS model is likely to represent the actual conditions, considering the location and season of the calibration experiment. Therefore, the maximum differences, which were derived from replacing the radiosonde measurements with US and MW models, were applied in the calibration uncertainty calculations.

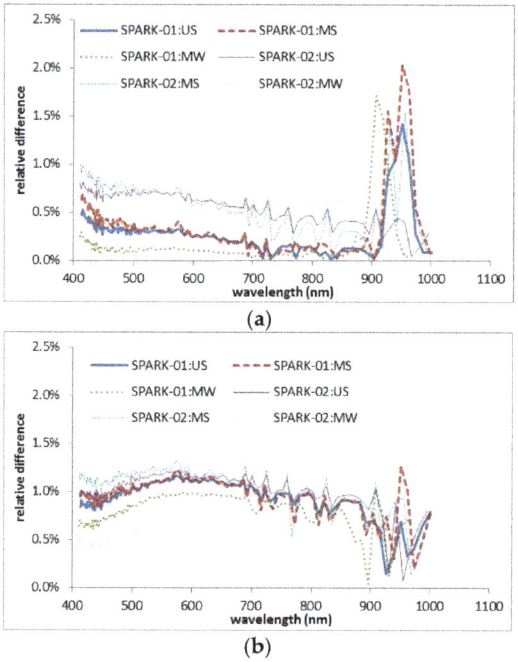

Figure 27. Calibration uncertainties caused by the vertical atmospheric profile measurements for SPARK-01 and -02 using the (**a**) reflectance- and (**b**) irradiance-based methods, respectively. "US", "MS", and "MW" refer to the relative difference in predicted radiance derived from replacing radiosonde measurements with these three atmospheric models.

5.5. Uncertainty Due to Image Misregistration Errors

To locate the calibration site, the OLI image acquired on 28 February 2017 was used to geo-rectify the SPARK satellite images. The first-order polynomial method was applied with nearest-neighbor resampling around the calibration site to retain the raw DN acquired by the sensors. A one-pixel misregistration around the calibration site is reasonable between the SPARK and OLI images. In addition, the border of the calibration site can be seen in the OLI image due to its 30 m spatial resolution and high radiometric resolution. Therefore, a misregistration of up to two pixels was assumed in the computation of the average DNs at the calibration site. The average DNs and minimum and maximum average DNs determined by shifting the 6 × 6 pixel area by up to two pixels in all directions are shown in Figure 28. The difference between the averaged DNs and the shifted averaged DNs reflect the uncertainty due to image misregistration errors. The differences caused by misregistration are <1.5% in the SPARK-01 495–955 nm spectral range and the SPARK-02 459–995 nm range. The differences are large at the ends of the spectral range due to high noise; data in these ranges are not generally used.

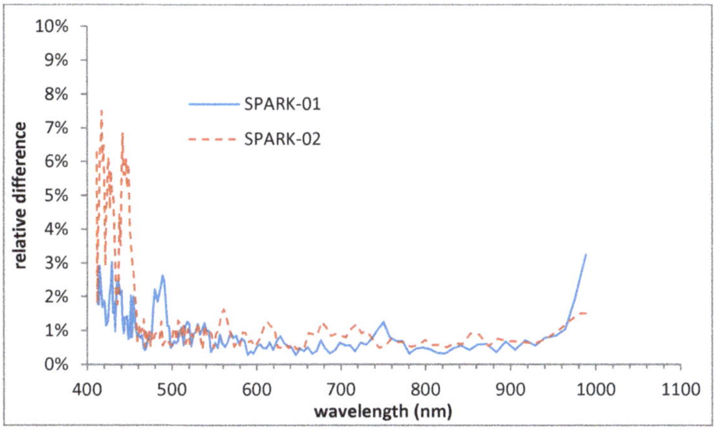

Figure 28. Averaged DNs from 6 × 6 pixel areas in the calibration site and DNs calculated by shifting the 6 × 6 pixel area by up to two pixels in all directions for SPARK-01 and -02.

5.6. Uncertainty Due to Spectral Wavelength Shift

Although the central wavelength values were measured for all the 2048 pixels of SPARK-01 and -02 in the laboratory before launch, the wavelength shift may affect the radiometric calibration result in the band near atmospheric gas absorption wavelength. The spectral shifts of Hyperion were estimated to be 0.38–1.39 nm at the 760 nm oxygen band by a spectral fitting algorithm compared with the laboratory spectral calibration [8]. The same method was also applied to TG-1 hyperspectral imager and the spectral shifts were 2–3 nm, with an uncertainty of 0.3 nm [38]. Thus, the spectral fitting algorithm was also used to estimate the spectral shifts for the SPARK satellite. The measured radiance spectrum over the desert was compared with a MODTARN 5-modeled radiance spectrum using the SPARK spectral calibration parameters to derive the spectral shift value. Figure 29a,c show the comparison near the 760 nm oxygen band in the desert (Figure 5) for SPARK-01 and -02, respectively. The MODTRAN 5 spectrum was normalized to match the SPARK-measured radiance level and the spectral wavelength was shifted in 0.1 nm increments. The optimal shifts were estimated to be −0.1 nm for both SPARK-01 and -02. Such a minor spectral shift indicated that SPARK satellites do not undergo an evident spectral shift. Figure 29b,d show the slightly minimized radiance difference after applying a −0.1 nm shift in the SPARK pre-launch laboratory spectral calibration position.

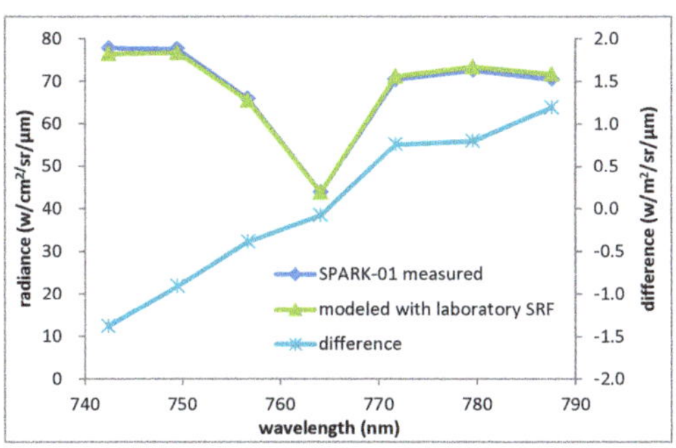

(a)

Figure 29. *Cont.*

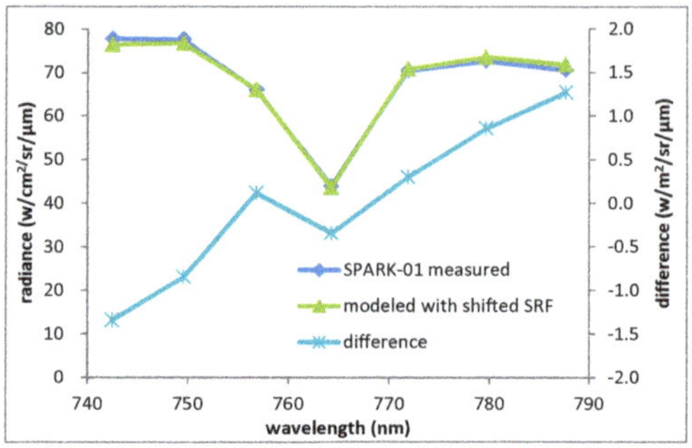

(b)

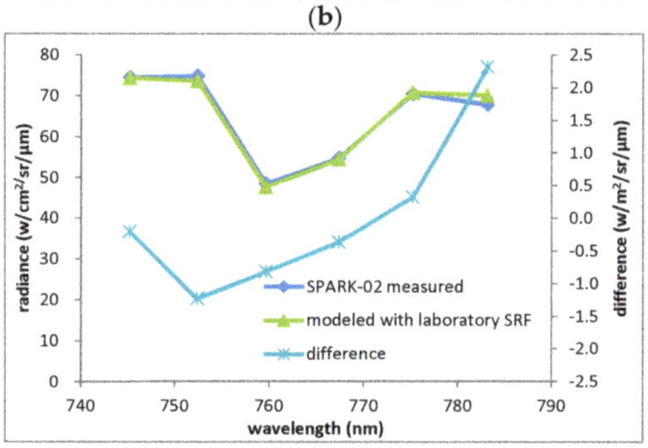

(c)

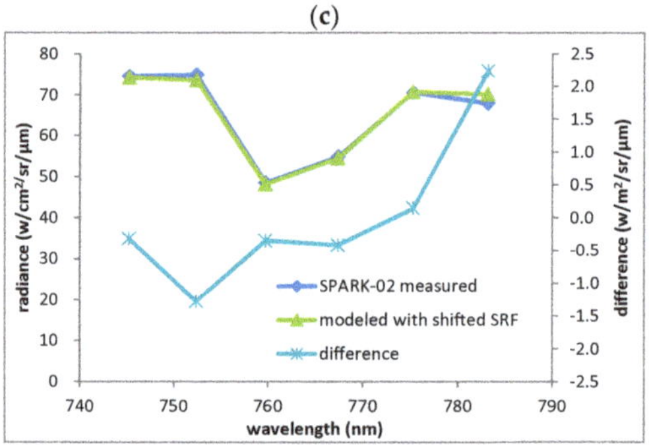

(d)

Figure 29. Spectral fit result and optimal spectral calibration result with an –0.1 nm shift at the desert in Dunhuang (Figure 5). (**a**,**c**) are for SPARK-01 before and after spectral shifting; and (**b**,**d**) are for SPARK-02 before and after spectral shifting.

The spectral shift of −0.1 nm was applied to SPARK-01 and -02 to calculate its contribution to radiometric calibration accuracy (Figure 30). As the SPARK satellite uses a prism, the absolute spectral shift for each band will be linear to its FMHW, expressed as:

$$\Delta\lambda(i) = \frac{FWHM(i)}{FWHM(i\,|_{\lambda=760\ nm})} \times \Delta \tag{12}$$

Considering the additional errors caused by the spectral fitting algorithm itself and laboratory calibration, a ±1 nm spectral shift at the 760 nm band was assumed to further estimate the influence on the radiometric calibration of SPARK satellites. The wavelength position at the 760 nm band was shifted by ±1 nm and the radiance difference was calculated for SPARK-01 and -02. As expected, the uncertainty is evident near the atmospheric gas absorption wavelengths, e.g., Fraunhofer 430 nm and 685 nm, the 760 nm and 690 nm oxygen bands, and the 720 nm, 820 nm, and 940 nm water vapor bands. The maximum value occurred for the 760 nm oxygen band. The uncertainties are less than 2% in the oxygen bands, and less than 1% in the water vapor bands if a spectral shift of −0.1 nm was ignored during SPARK satellite calibration. However, if the spectral shift was increased to 1 nm, the uncertainties would increase considerably in these atmospheric gas absorption bands (e.g., 8% by the spectral shift of +1 nm for SPARK-01 and >10% by the spectral shift of −1 nm for SPARK-02 (Figure 31). In addition, the uncertainty in the 940 nm band for SPARK-02 is higher than that of SPARK-01 due to the larger water vapor content occurring in the daytime for SPARK-02 radiometric calibration (0.35 g/cm^2 for SPARK-01 versus 0.54 g/cm^2 for SPARK-02).

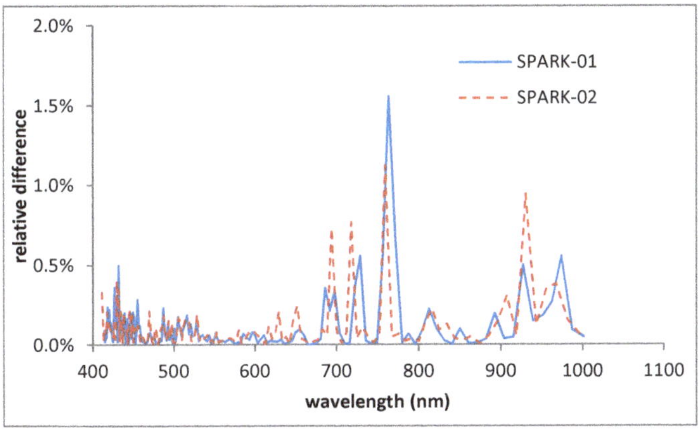

Figure 30. Calibration uncertainties caused by spectral shift of −0.1 nm for SPARK-01 and -02.

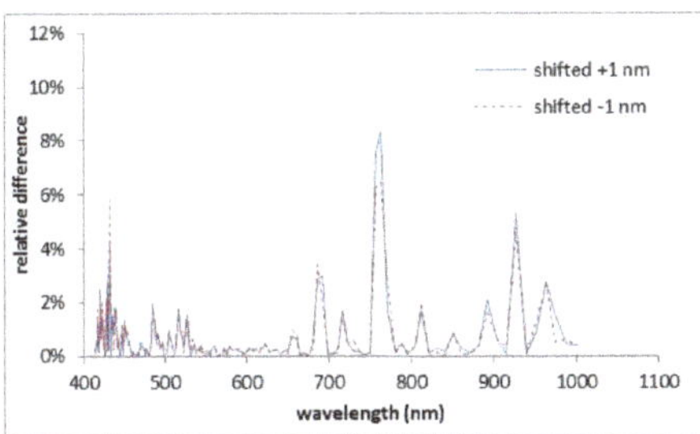

(a)

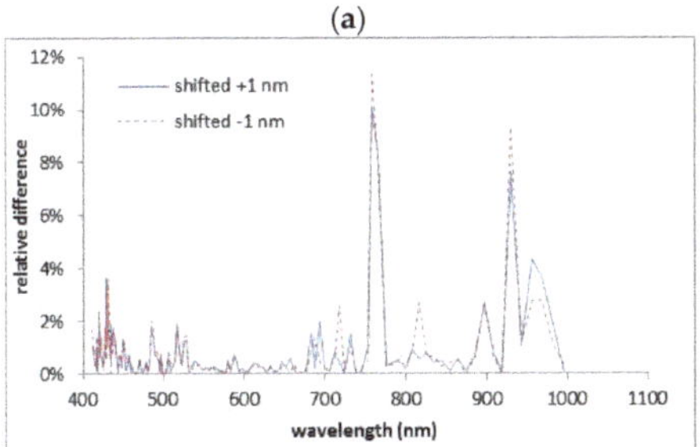

(b)

Figure 31. Calibration uncertainties caused by spectral shift of ±1 nm for SPARK-01 (**a**) and -02 (**b**).

5.7. Total Calibration Uncertainty Estimation

Calibration uncertainties caused by other sources are relatively constant and have been discussed previously [20,39,40]. The uncertainty due to ground reflectance measurements has been estimated at 2% in field experiments, and this estimation was validated measurements taken on different days during the calibration experiment. The uncertainty in measurement of diffuse-to-global irradiance ratio contributes 2.0% to the total calibration uncertainty [40]. The uncertainty due to ozone measurements with an error of 20% was estimated to be 1.3% [39]. Thus, because the ozone acquired from OMI has the uncertainty of 4%, it is reasonable to set this uncertainty to 0.6% [41]. Although the accuracy of MODTRAN 5 is much improved and comparable to that of the benchmark Line-by-Line Radiative Transfer Model (LBLRTM) [42], this uncertainty is conservatively estimated to be 1%. The uncertainty due to non-Lambertian ground characteristics was estimated at 2% for the Dunhuang calibration site [16]. In total, the overall vicarious calibration uncertainty contains uncertainties and errors caused by atmospheric characterization, surface characterization, radiative transfer calculations, and site-average DN calculations [43]. The uncertainties discussed above associated with the reflectance-based method are summarized in Table 5 for the SPARK-01 and -02 satellite calibrations; those associated with the irradiance-based method (used for SPARK-01) and the improved irradiance-based method (used for SPARK-02) are summarized in Table 6. The uncertainties associated with different methods are shown for each spectral band of both satellites in Figure 32. Total uncertainty statistics are listed for different spectral ranges in Table 7. For SPARK-01, uncertainties of 4.71 ± 0.34% and 4.11 ± 0.21% were estimated using the reflectance- and irradiance-based methods, respectively. For SPARK-02, uncertainties of 8.12 ± 0.29% and 5.86 ± 0.29% were estimated at >456 nm using the reflectance- and improved irradiance-based methods, respectively. The uncertainty is greatly increased in other spectral ranges due to high image noise. As expected, the uncertainties in both the irradiance- and improved irradiance-based methods are lower than that in the reflectance-based method, especially when the aerosol optical depth is large (e.g., in the SPARK-02 results). However, the irradiance and improved irradiance-based methods depend greatly on the accuracy of the direct transmittance measurements (in Section 5.2) and the diffuse-to-global irradiance ratios; it is therefore important to improve the accuracy of these measurements.

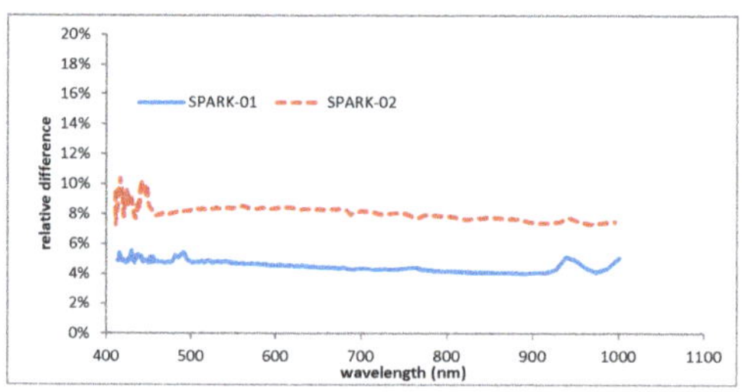

(**a**)

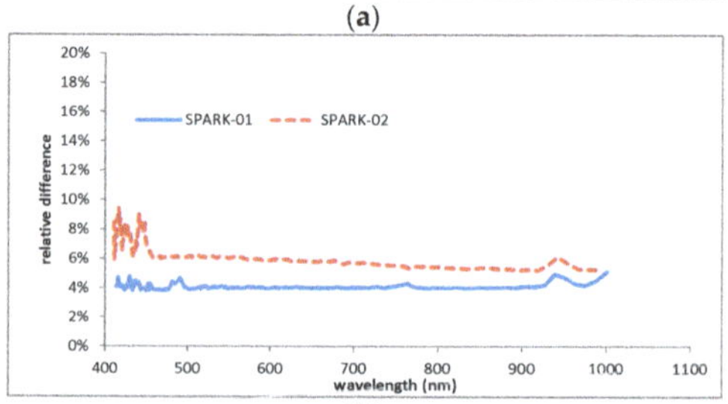

(**b**)

Figure 32. Total calibration uncertainty estimated for SPARK-01 and -02 using (**a**) the reflectance-based method and (**b**) the irradiance-based method.

Table 5. Estimated uncertainty associated with the reflectance-based method.

Source of Uncertainty	SPARK-01 Uncertainty (%)	SPARK-02 Uncertainty (%)
Ground reflectance measurement	2.0	2.0
Non-Lambertian ground properties	2.0	2.0
AOD retrieval	0–0.6	0–0.7
Water vapor retrieval	0–3.0	0–3.8
Ozone absorption computation	0.6	0.6
Assumption of aerosol type	2.1–3.6	5.9–7.8
Assumption of atmospheric model	0–1.7	0.3–0.8
Radiative transfer code accuracy	1	1
Spectral wavelength shift	0–1.6	0–1.1
Image uncertainty errors	0.3–3.3	0.4–7.5
Total uncertainty (root sum of squares)	4.0–5.5	7.3–10.4

Table 6. Estimated uncertainty associated with the irradiance-based method used for SPARK-01 and the improved irradiance-based method used for SPARK-02.

Source of Uncertainty	SPARK-01 Uncertainty (%)	SPARK-02 Uncertainty (%)
Ground reflectance measurement	2.0	2.0
AOD retrieval	1.5–2.6	0.9–1.4
Water vapor retrieval	0–2.9	0–3.8
Ozone absorption computation	0.6	0.6
Assumption of aerosol type	0.4–1.5	3.3–5.0
Assumption of atmospheric model	0.2–1.2	0.2–1.2
Radiative transfer code accuracy	/	1.0
Spectral wavelength shift	0–1.6	0–1.1
Image uncertainty errors	0.3–3.3	0.4–7.5
Ratio of diffuse-to-global irradiance measurement	2.0	2.0
Total uncertainty (root sum of squares)	3.9–5.1	5.2–9.4

Table 7. Average relative differences for various wavelength ranges.

Wavelength Range (nm)	Reflectance-Based Method (SPARK-01)	Irradiance-Based Method (SPARK-01)	Wavelength Range (nm)	Reflectance-Based Method (SPARK-02)	Improved Irradiance-Based Method (SPARK-02)
-	-	-	411–454	8.90 ± 0.75%	7.60 ± 0.90%
-	-	-	>456	8.12 ± 0.29%	5.86 ± 0.29%
All bands	4.71 ± 0.34%	4.11 ± 0.21%	All bands	8.31 ± 0.56%	6.28 ± 0.90%

5.8. Spectral Smile Effect Correction

The smile effect is a common phenomenon in the pushbroom sensor. It is mainly caused by optical aberrations and misalignments and cannot be completely avoided. Figure 33 shows relative differences of the central wavelength positions of all the 2048 pixels for SPARK-01 and -02 satellites, compared with the average central wavelength, expressed as:

$$\Delta\eta(i,j) = \frac{CWV(i,j) - \overline{CWV}(j)}{FWHM(j)} \tag{13}$$

where $\Delta\eta$ is the relative difference of the central spectral wavelength, CWV is the central spectral wavelength for each pixel, i and j denote the cross-track position and band index, respectively, and $\overline{CWV}$ is the average central spectral wavelength.

The spectral smile is more pronounced in SPARK-02 than in SPARK-01, with the maximum difference even exceeding half of the FWHM, which is nearly of the same magnitude as that in the Hyperion data [8]. The de-smiling technique is always applied to interpolate the raw data from individual spectral positions into the commonly defined spectral central wavelengths. The de-smiling processing would not affect most spectral bands to a great degree, but it may introduce artificial features near the atmospheric gas absorption wavelengths when deriving ground surface reflectance through atmospheric correction. Thus, it is strongly recommended that the spectral polishing technique be applied after atmospheric correction to remove the spectral artificial features near the gas absorption wavelengths.

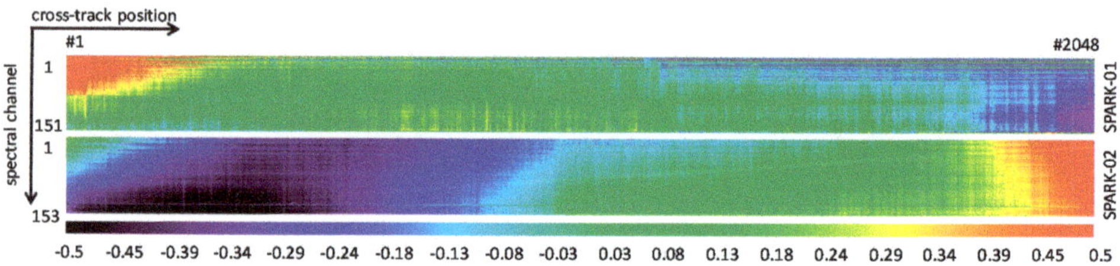

Figure 33. Central wavelength positions measured pre-launch in laboratory for SPARK-01 and -02.

5.9. Preliminary Validation

The calibration coefficients were derived from the irradiance-based method and applied to the SPARK-01 image. Then, the ground reflectance from the 7 March 2017 SPARK-01 image was calculated using measured atmospheric parameters. The retrieved desert reflectance is compared to the in-situ measured reflectance in Figure 8 (Figure 34). Retrieved values are close to the measured values; the discrepancy is within 8% in 500–1000 nm. The difference is partly attributed to the radiometric calibration and partly attributed to the slight terrain fluctuation, inhomogeneous surface and BRDF effect of the desert.

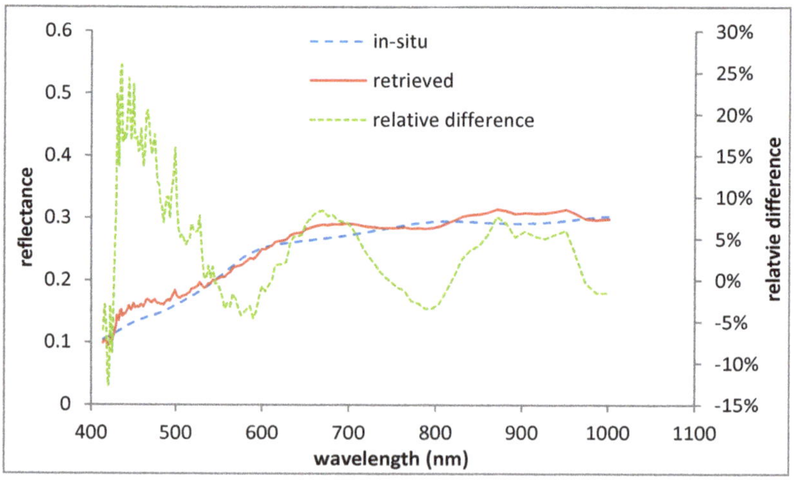

Figure 34. Comparison of in-situ desert reflectance measurements with retrieved reflectance from the SPARK-01 image acquired on 7 March 2017.

Terra MODIS images acquired on 28 February and 7 March 2017 were also used to verify the vicarious calibration methods. The processes used to predict the TOA radiance were similar to those used in the SPARK calibration. Because MODIS has an on-board calibration system, and thus, its calibration accuracy is expected within 3% [44], the MODIS image radiance was taken as a reference to calculate the relative accuracy of the TOA radiance predicted using the vicarious calibration methods (Table 8). For the MODIS image acquired on 28 February, the improved irradiance-based method appears superior to the reflectance-based method in the infrared and shortwave infrared spectral bands, but inferior in the third and fourth bands. However, the weather on 28 February was poor, and thus this comparison shows only that the improved irradiance-based method may be appropriate during non-ideal conditions. On 7 March, the atmosphere was stable and aerosol burden was low. Thus, both the reflectance- and irradiance-based methods predicted values approaching those from the MODIS image, with no more than 4% error in the first four bands. In the 1.2 and 1.6 μm bands, the difference between MODIS and the reflectance-based method is larger than that between MODIS and the irradiance-based method, which is likely due to the aerosol-type assumption. The irradiance-based method shows large difference from MODIS in the 2.1 μm bands, which may be attributed to the low signal-to-noise ratio of the instrument in this band. Comparison of TOA radiances predicted by MODIS and the vicarious calibration methods used for the SPARK satellites show that the SPARK calibration methods achieved the accuracy expected.

Table 8. Differences between the TOA radiance predicted by the vicarious methods (i.e., the Reflectance-, Irradiance-, and Improved irradiance-based methods) and that from MODIS image radiance.

MODIS Band	Central Wavelength (nm)	28 February		7 March	
		Reflectance	Improved Irradiance	Reflectance	Irradiance
1	646	−4.41%	−5.54%	1.18%	−1.85%
2	856	−10.35%	−8.41%	−3.57%	−3.95%
3	466	−1.24%	−4.84%	2.37%	−4.30%
4	554	−2.56%	−5.33%	2.39%	−2.68%
5	1242	−15.47%	−10.94%	−6.60%	−2.74%
6	1629	−13.84%	−7.48%	−5.10%	1.65%
7	2114	−9.89%	3.58%	−3.63%	7.97%

6. Conclusions

This study presents the first in-situ vicarious calibration experiments at the Dunhuang site for the SPARK-01 and -02 satellites. Reflectance-, irradiance-, and improved irradiance-based calibration methods were used on images acquired on 7 March and 28 February 2017 by these two satellites. We proposed a 90° yaw imaging technique for use in the relative calibration method; such methods are very useful for microsatellites without on-board calibration instruments, and especially for satellites with large swaths. An absolute calibration was performed using MODTRAN 5 data, and the methodological and measurement errors in the calibration results were analyzed in detail. Because the SPARK-01 image was acquired during fair weather (e.g., stable atmosphere and low AOD), the calibration uncertainties of the reflectance- and irradiance-based methods are 4.7% and 4.1%, respectively. However, the SPARK-02 image, which was acquired during poor weather, has an uncertainty of 8.12% using the reflectance-based method from 456 to 1000 nm. Under these conditions, the improved irradiance-based method was superior, producing a lower uncertainty of 5.86%. Thus, the additional diffuse-to-global ratio measurements included in the irradiance- and improved irradiance-based methods considerably decreases the calibration uncertainty, likely due to its aerosol property assumptions. The improved irradiance-based method is superior to the reflectance-based method under non-ideal atmospheric conditions as it improves the simulated downward transmittance. Although the irradiance- and improved irradiance-based methods are superior to the reflectance-based method on average, the accuracy of the diffuse-to-global ratio measurements may limit the use of these two methods. Indeed, the instrument used to measure the diffuse-to-global ratio has a lower signal-to-noise ratio in the dark blue bands (i.e., <400 nm) and shortwave infrared bands (i.e., >2.1 μm). Moreover, spectral calibration accuracy is a crucial factor to guarantee accurate radiometric calibration. A 1 nm spectral shift for a hyperspectral sensor with a 10 nm spectral resolution would cause as much as a 10% radiometric calibration error near the gas absorption wavelengths. The precise pre-launch spectral calibration in the laboratory as well as the on-orbit monitoring of spectral wavelength shifting are needed. Also, we strongly suggest combining the calibration results derived by the reflectance- and irradiance- (or improved irradiance-) based methods for optimized results. In the future, irradiance-based methods for hyperspectral satellites should be evaluated in more detail by adding spectrally continuous direct transmittance measurements. This could improve calibration accuracy in the gas absorption bands near 940 nm, 1135 nm, 820 nm, etc.

Acknowledgments: This research was supported by the National Natural Science Foundation of China (41325004, 41771397) and the National Key R&D Program of China (2016YFB0500304). The authors acknowledge the Shanghai Engineering Center for Microsatellites, Chinese Academy of Sciences (CAS), and the Academy of Opto-electronics, CAS for providing satellite data and documentation. We also thank China RS Geo-informatics Co. Ltd. for financial support for the field experiment. We highly appreciate Juanjuan Jing from the Academy of Opto-electronics, CAS for the assistance in spectral calibration data processing and in-depth discussions.

Author Contributions: Hao Zhang, Bing Zhang, and Zhengchao Chen conceived and designed the experiments. Hao Zhang and Zhihua Huang performed the experiments and processed the data. Hao Zhang and Zhengchao Chen contributed to the analysis and discussion. Hao Zhang wrote the paper.

Conflicts of Interest: The authors declare no conflict of interest.

References

1. Tong, Q.; Xue, Y.; Zhang, L. Progress in hyperspectral remote sensing science and technology in China over the past three decades. *IEEE J. Sel. Top. Appl. Earth Obs. Remote Sens.* **2013**, *7*, 70–91. [CrossRef]
2. Biggar, S.F. In-Flight Methods for Satellite Sensor Absolute Radiometric Calibration. Ph.D. Thesis, College of Optical Sciences, University of Arizona, Tucson, AZ, USA, 1990; p. 157.
3. Slater, P.N.; Biggar, S.F.; Holm, R.G.; Jackson, R.D.; Mao, Y.; Moran, M.S.; Yuan, B. Reflectance- and Radiance-Based Methods for the In-Flight Absolute Calibration of Multispectral Sensors. *Remote Sens. Environ.* **1987**, *22*, 11–37. [CrossRef]
4. Gellman, D.I.; Biggar, S.F.; Dinguirard, M.C.; Henry, P.J.; Moran, M.S.; Thome, K.J.; Slater, P.N. Review of SPOT-1 and -2 Calibrations at White Sands from Launch to the Present. *Proc. SPIE* **1993**, *1938*, 118–125.
5. Thome, K.J.; Gellman, D.I.; Parada, R.J.; Biggar, S.F.; Slater, P.N.; Moran, M.S. In-Flight Radiometric Calibration of Landsat-5 Thematic Mapper from 1984 to Present. *Proc. SPIE* **1993**, *1938*, 126–130.
6. Thome, K.J.; Helder, D.L.; Aaron, D.; Dewald, J.D. Landsat-5 TM and Landsat-7 ETM+ Absolute Radiometric Calibration Using the Reflectance-Based Method. *IEEE Trans. Geosci. Remote Sens.* **2004**, *42*, 2777–2785. [CrossRef]
7. Vane, G.; Green, R.O.; Chrien, T.G.; Enmark, H.T.; Hansen, E.G.; Porter, W.M. The Airborne Visible/Infrared Imaging Spectrometer (AVIRIS). *Remote Sens. Environ.* **1993**, *44*, 127–143. [CrossRef]
8. Green, R.O.; Pavri, B.E.; Chrien, T.G. On-Orbit Radiometric and Spectral Calibration Characteristics of EO-1 Hyperion Derived with an Under-flight of AVIRIS and in-situ Measurements at Salar de Arizaro, Argentina. *IEEE Trans. Geosci. Remote Sens.* **2003**, *41*, 1194–1203. [CrossRef]
9. Pearlman, J.S.; Barry, P.S.; Segal, C.C.; Shepanski, J.; Beiso, D.; Carmanet, S.L. Hyperion, a space-based imaging spectrometer. *IEEE Trans. Geosci. Remote Sens.* **2003**, *41*, 1160–1173. [CrossRef]
10. Hu, X.; Zhang, Y.; Qiu, K. In-Flight Radiometric Calibration for VIR Channels of FY-IC Satellite Sensor by Using Irradiance-Based Method. *J. Remote Sens.* **2003**, *7*, 458–464.
11. Hu, X.; Zhang, Y.; Liu, Z.; Zhang, G.; Huang, Y.; Qiu, K.; Wang, Y.; Zhang, L.; Zhu, X.; Rong, Z. Optical Characteristics of China Radiometric Calibration Site for Remote Sensing Satellite Sensors. *Proc. SPIE* **2001**, *4151*, 77–86.
12. Bruegge, C.J.; Diner, D.J.; Kahn, R.A.; Chrien, N.; Helmlinger, M.C.; Gaitley, B.J.; Abdou, W.A. The MISR radiometric calibration process. *Remote Sens. Environ.* **2007**, *107*, 2–11. [CrossRef]
13. Czapla-Myers, J.; McCorkel, J.; Anderson, N.; Thome, K.; Biggar, S.; Helder, D.; Aaron, D.; Leigh, L.; Mishra, N. The Ground-Based Absolute Radiometric Calibration of Landsat 8 OLI. *Remote Sens.* **2015**, *7*, 600–626. [CrossRef]
14. Pinto, C.; Ponzoni, F.; Castro, R.; Leigh, L.; Mishra, N.; Aaron, D.; Helder, D. First in-flight radiometric calibration of MUX and WFI on-board CBERS-4. *Remote Sens.* **2016**, *8*, 405. [CrossRef]

15. Mattar, C.; Hernández, J.; Santamaría-Artigas, A.; Durán-Alarcón, C.; Olivera-Guerra, L.; Inzunza, M.; Tapia, D.; Escobar-lavín, E. A first in-flight absolute calibration of the Chilean Earth Observation Satellite. *ISPRS J. Photogramm. Remote Sens.* **2014**, *92*, 16–25. [CrossRef]
16. Chen, Z.C.; Zhang, B.; Zhang, H.; Zhang, W.J. Vicarious Calibration of Beijing-1 Multispectral Imagers. *Remote Sens.* **2014**, *6*, 1432–1450. [CrossRef]
17. Slater, P.N.; Biggar, S.F.; Palmer, J.M.; Thome, K.J. Unified approach to absolute radiometric calibration in the solar-reflective range. *Remote Sens. Environ.* **2001**, *77*, 293–303. [CrossRef]
18. de Vries, C.; Danaher, T.; Denham, R.; Scarth, P.; Phinn, S. An operational radiometric calibration procedure for the Landsat sensors based on pseudo-invariant target sites. *Remote Sens. Environ.* **2007**, *107*, 414–429. [CrossRef]
19. Markham, B.L.; Helder, D.L. Forty-year calibrated record of earth-reflected radiance from Landsat: A review. *Remote Sens. Environ.* **2012**, *122*, 30–40. [CrossRef]
20. Li, H.W.; Zhang, H.; Chen, Z.C.; Zhang, B. A Method Suitable for Vicarious Calibration of a UAV Hyperspectral Remote Sensor. *IEEE J. Sel. Top. Appl. Earth Obs. Remote Sens.* **2015**, *8*, 3209–3223.
21. Brook, A.; Bendor, E. Supervised Vicarious Calibration (SVC) of Multi-Source Hyperspectral Remote-Sensing Data. *Remote Sens.* **2015**, *7*, 6196–6223. [CrossRef]
22. Brook, A.; Polinova, M.; Ben-Dor, E. Fine tuning of the SVC method for airborne hyperspectral sensors: The BRDF correction of the calibration nets targets. *Remote Sens. Environ.* **2018**, *204*, 861–871. [CrossRef]
23. Dadon, A.; Ben-Dor, E.; Karnieli, A. Use of Derivative Calculations and Minimum Noise Fraction Transform for Detecting and Correcting the Spectral Curvature Effect (Smile) in Hyperion Images. *IEEE Trans. Geosci. Remote Sens.* **2010**, *48*, 2603–2612. [CrossRef]
24. Yokoya, N.; Miyamura, N.; Iwasaki, A. Detection and correction of spectral and spatial misregistrations for hyperspectral data using phase correlation method. *Appl. Opt.* **2010**, *49*, 4568–4579. [CrossRef] [PubMed]
25. Henderson, B.G.; Krause, K.S. Relative radiometric correction of QuickBird imagery using the side-slither technique on-orbit. *Proc. SPIE* **2004**, *5542*, 426–436.
26. Anderson, C.; Naughton, D.; Brunn, A.; Thiele, M. Radiometric correction of RapidEye imagery using the on-orbit side-slither method. *Proc. SPIE* **2011**, *8180*, 1–15.
27. Gerace, A.; Schott, J.; Gartley, M.; Montanaro, M. An Analysis of the Side Slither On-Orbit Calibration Technique Using the DIRSIG Model. *Remote Sens.* **2014**, *6*, 10523–10545. [CrossRef]
28. Wu, D.; Yin, Y.; Wang, Z.; Gu, X.; Verbrugghe, M.; Guyot, G. Radiometric Characterisation of Dunhuang Satellite Calibration Test Site (China). In *Physical Measurements and Signatures in Remote Sensing*; Guyot, G., Phulpin, T., Eds.; Balkema: Rotterdam, The Netherlands, 1997; Volume 1, pp. 151–160.
29. Hu, X.; Liu, J.; Sun, L.; Rong, Z.; Li, Y.; Zhang, Y.; Zheng, Z.; Wu, R.; Zhang, L.; Gu, X. Characterization of CRCS Dunhuang Test Site and Vicarious Calibration Utilization for Fengyun (FY) Series Sensors. *Can. J. Remote Sens.* **2010**, *36*, 566–582. [CrossRef]
30. Holben, B.N.; Eck, T.F.; Slutsker, I.; Tanré, D.; Buis, G.P.; Setzer, A.; Vermote, E.; Reagan, J.A.; Kaufman, Y.J.; Nakajima, T.; et al. AERONET-A Federated Instrument Network and Data Archive for Aerosol Characterization. *Remote Sens. Environ.* **1998**, *66*, 1–16. [CrossRef]
31. Zhang, H.; Zhang, B.; Chen, D.M.; Li, J.S.; Zhao, G.N. Influence of Filter Band Function on Retrieval of Aerosol Optical Depth from Sunphotometer Data. *J. Atmos. Ocean. Technol.* **2013**, *30*, 929–941. [CrossRef]
32. Zhang, H.; Chen, Z.C.; Zhang, B.; Chen, C.Y. Improved Algorithm for Water Vapor Retrieval and Field Calibration in the Channel 940-nm of Sun-photometer. In Proceedings of the 23rd CALCON Technical Conference, Logan, UT, USA, 11–14 August 2014.
33. Seftor, C. Ozone over Your Head. Available online: https://ozoneaq.gsfc.nasa.gov/tools/ozonemap/ (accessed on 27 July 2017).
34. Datt, B.; Jupp, D.L.B. *Hyperion Data Processing Workshop, Hands-On Processing Instructions*; CSIRO Office of Space Science & Applications Earth Observation Centre: Canberra, Australia, 2004.
35. Dinguirard, M.; Slater, P.N. Calibration of Space-Multispectral Imaging Sensors: A Review. *Remote Sens. Environ.* **1999**, *68*, 194–205. [CrossRef]
36. Eck, T.F.; Holben, B.N.; Reid, J.S.; Dubovik, O.; Smirnov, A.; O'Neill, N.T.; Slutsker, I.; Kinne, S. Wavelength dependence of the optical depth of biomass burning, urban, and desert dust aerosols. *J. Geophys. Res.* **1999**, *104*, 31333–31349. [CrossRef]

37. Schiller, S.J. Technique for estimating uncertainties in top-of-atmosphere radiances derived by vicarious calibration. *Proc. SPIE* **2013**, *5151*, 502–516.
38. Zhang, Y.Q.; Zhang, W.J.; Chen, Z.C.; Zhang, B. A comparative analysis of the accuracy of alternative spectral calibration methods based on TG-1 hyperspectral imager. *J. Infrared Millim. Waves* **2015**, *34*, 360–367.
39. Biggar, S.; Slater, P.; Gellman, D. Uncertainties in the in-flight calibration of sensors with reference to measured ground sites in the 0.4–1.1 μm range. *Remote Sens. Environ.* **1994**, *48*, 245–252. [CrossRef]
40. Thome, K.J. *Validation Plan for MODIS Level 1 At-Sensor Radiance*; Remote Sensing Group of the Optical Sciences Center, University of Arizona: Tucson, AZ, USA, 1999.
41. Liu, X.; Bhartia, P.; Chance, K.; Froidevaux, L.; Spurr, R.; Kurosu, T. Validation of Ozone Monitoring Instrument (OMI) ozone profiles and stratospheric ozone columns with Microwave Limb Sounder (MLS) measurements. *Atmos. Chem. Phys.* **2010**, *10*, 2539–2549. [CrossRef]
42. Berk, A.; Anderson, G.P.; Acharya, P.K.; Shettle, E.P. *MODTRAN®5.2.1 User's Manual*; Spectral Science, Inc.: Burlington, MA, USA, 2011.
43. Thome, K.J. Absolute Radiometric Calibration of Landsat-7 ETM+ Using the Reflectance-Based Method. *Remote Sens. Environ.* **2001**, *78*, 27–38. [CrossRef]
44. Chang, T.; Xiong, X.; Angal, A.; Wu, A.; Geng, X. Aqua and Terra MODIS RSB Calibration Comparison Using BRDF Modeled Reflectance. *IEEE Trans. Geosci. Remote Sens.* **2017**, *55*, 2288–2298. [CrossRef]

remote sensing

MDPI

Article

A Hyperspectral Imaging Approach to White Matter Hyperintensities Detection in Brain Magnetic Resonance Images

Hsian-Min Chen [1,2], Hsin Che Wang [1], Jyh-Wen Chai [3,4,*], Chi-Chang Clayton Chen [3], Bai Xue [5], Lin Wang [6], Chunyan Yu [7], Yulei Wang [7,8], Meiping Song [7] and Chein-I Chang [5,7,9,10]

1 Center for Quantitative Imaging in Medicine (CQUIM), Department of Medical Research, Taichung Veterans General Hospital, Taichung 40705, Taiwan; hsmin6511@gmail.com (H.-M.C.); leononwang@gmail.com (H.C.W.)
2 Department of Biomedical Engineering, Hungkuang University, Taichung 43302, Taiwan
3 Department of Radiology, Taichung Veterans General Hospital, Taichung 40705, Taiwan; ccc@vghtc.gov.tw
4 College of Medicine, China Medical University, Taichung 40402, Taiwan
5 Remote Sensing Signal and Image Processing Laboratory, Department of Computer Science and Electrical Engineering, University of Maryland, Baltimore County, Baltimore, MD 21250, USA; baixue1@umbc.edu (B.X.); cchang@umbc.edu (C.-I.C.)
6 School of Physics and Optoelectronic Engineering, Xidian University, Xi'an 710126, China; slwang@mail.xidian.edu.cn
7 Center for Hyperspectral Imaging in Remote Sensing (CHIRS), Information and Technology College, Dalian Maritime University, Dalian 116026, China; yuchunyan1997@126.com (C.Y.); wangyulei.wyl@163.com (Y.W.); smping@163.com (M.S.)
8 State Key Laboratory of Integrated Services Networks, Xi'an 710071, China
9 Department of Computer Science and Information Engineering, National Yunlin University of Science and Technology, Yunlin 64002, Taiwan
10 Department of Computer Science and Information Management, Providence University, Taichung 43301, Taiwan
* Correspondence: hubt@vghtc.gov.tw; Tel.: +886-4-23592525 (ext. 3737)

Received: 4 September 2017; Accepted: 13 November 2017; Published: 16 November 2017

Abstract: White matter hyperintensities (WMHs) are closely related to various geriatric disorders including cerebrovascular diseases, cardiovascular diseases, dementia, and psychiatric disorders of elderly people, and can be generally detected on T2 weighted (T2W) or fluid attenuation inversion recovery (FLAIR) brain magnetic resonance (MR) images. This paper develops a new approach to detect WMH in MR brain images from a hyperspectral imaging perspective. To take advantage of hyperspectral imaging, a nonlinear band expansion (NBE) process is proposed to expand MR images to a hyperspectral image. It then redesigns the well-known hyperspectral subpixel target detection, called constrained energy minimization (CEM), as an iterative version of CEM (ICEM) for WMH detection. Its idea is to implement CEM iteratively by feeding back Gaussian filtered CEM-detection maps to capture spatial information. To show effectiveness of NBE-ICEM in WMH detection, the lesion segmentation tool (LST), which is an open source toolbox for statistical parametric mapping (SPM), is used for comparative study. For quantitative analysis, the synthetic images in BrainWeb provided by McGill University are used for experiments where our proposed NBE-ICEM performs better than LST in all cases, especially for noisy MR images. As for real images collected by Taichung Veterans General Hospital, the NBE-ICEM also shows its advantages over and superiority to LST.

Keywords: band expansion process (BEP); constrained energy minimization (CEM); correlation band expansion process (CBEP); iterative CEM (ICEM); nonlinear band expansion (NBE); Otsu's method

1. Introduction

White matter hyperintensities (WMHs) are commonly observed on T2W or FLAIR brain MR images of elderly people and related to various geriatric disorders including cerebrovascular diseases, cardiovascular diseases, dementia, and psychiatric disorders [1]. According to [2], WMHs are brain lesions that generally show up as brighter areas and can be visualized by T2W and FLAIR MRI sequences. It is also referred to as Leukoaraiosis and is often found in computed tomography (CT) or MRI of older patients. It is a marker of small-vessel vascular disease. In clinical practice, it is indicative of cognitive and emotional dysfunction, particularly in the ageing population. Its initial discovery was observed in the late 1980s by Hachinski and colleagues [2] who described WMH as patchy low attenuation in the periventricular and deep white matter. Since then, detection of WMH has received considerable interest. Although a supervised method may produce better results, it requires human intervention which is very time-consuming and also suffer the issues of intra- and inter-observer variation [3]. Accordingly, segmentation of WMH has been recently directed to semi-unsupervised and automatic methods which rely on computer assisted tools to help diagnosis to avoid human subjective interpretation. Most importantly, such computer assisted diagnosis can be further used to quantify WMH and calculate its volume [3–7]. However, it also comes with two major issues. One is that most works are based on T1 weighted (T1W), T2W and photon density (PD) or FLAIR images to produce spatial statistics to segment WMH. The other is selection of an appropriate threshold, which ultimately determines the detection results of WMH. Generally, such automatic method is not fully automatic but rather semi-unsupervised because it requires adaptively adjusting threshold values by visual inspection. This paper takes a quite different approach to designing a joint spectral–spatial method that takes advantage of spectral properties provided by MR image sequences to perform subvoxel detection in conjunction with a Gaussian spatial filter to capture spatial contextual information surrounding the WMH detected voxels.

One of the strengths of magnetic resonance imaging (MRI) is its ability in imaging structures of soft tissues. Because an MR image is collected by specifically designed image sequences such as T1W, T2W or PD, it can be considered as a multispectral image [8]. Hyperspectral imaging has recently emerged as an advanced technique in remote sensing to deal with many issues that cannot be resolved by multispectral imaging, specifically, subpixel target detection and mixed pixel classification [9]. Its applications to MRI classification have been also explored in [10–15]. However, it seems that using the concept of hyperspectral imaging techniques for WMH detection in brain MRI has not been investigated. This paper presents a new application of hyperspectral imaging in WMH detection of MR brain images.

To expand capability of multispectral imaging to hyperspectral imaging in data analysis, it suffers from insufficient band dimensionality. To resolve this dilemma, a nonlinear band expansion (NBE) process was previously proposed in [16] which used nonlinear functions to produce a new set of nonlinear band images that can be incorporated into the original images to create a new data set. As more such nonlinearly generated images are included, the resulting multispectral image has become a hyperspectral image. In this case, we can take advantage of the well-known hyperspectral subpixel target detection technique, called constrained energy minimization (CEM) [9,17–19], to detect the lesion of interest [20]. However, the nonlinearly expanded band images by NBE used in [20] can only capture spectral information nonlinearly but not spatial information. As noted above, effectively detecting the boundary of a lesion may also require spatial information due to the shape of the boundary that is closely related to spatial correlation.

This paper develops a novel NBE approach that expands NBE [20] to produce new band images that can capture not only nonlinear spectral information but also spatial information. Since CEM is a pixel-based technique and does not take into account spatial information. In order for CEM to capture spatial information an iterative version of CEM, to be called Iterative CEM (ICEM), is developed for this purpose. Its idea is to apply a Gaussian filter to a CEM-detection map so that the Gaussian-filtered CEM detection map will contain spatial information to be further fed back as a new band image to create a new image cube. The same process of operating CEM on this new data cube is repeated over again in an iterative manner via feedback loops. To terminate ICEM an automatic stopping rule is also designed, which uses Otsu's method [21] to threshold the Gaussian-filtered CEM detection map obtained at each iteration as a binary image. If the two consecutive binary images agree within an error threshold measured by Dice similarity index (DSI) [22], then ICEM is terminated and the final Otsu's thresholded binary image is the desired lesion detection map.

There are several main contributions derived from NBE-ICEM. One is using NBE to create new band images to make a multispectral image a hyperspectral image. Another is including Gaussian filters to capture spatial information. Thirdly, such Gaussian-filtered spatial information is further fed back to be included in the data cube being processed as new images to account for spatial information of detected WMH lesions. Fourthly, the spatial information included in CEM is increased via repeated feedback loop in an iterative manner. That is, the more feedbacks the more spatial information to be included in the data cube for better boundary detection. Fifthly, Otsu's method is introduced to automatically terminate the iterative process carried out by ICEM. Finally, once the ICEM is terminated, the resulting Otsu's thresholded binary image is the desired final lesion detection result.

2. Methods

2.1. Nonlinear Band Dimensionality Expansion

An early attempt to expand the original set of band images is to utilize nonlinear functions, for example, auto-correlation and cross-correlation, an idea derived from [16,20]. This type of NBE process is referred to as correlation band expansion process (CBEP). Combining these new CBEP-generated band images with the original set of band images produces a hyperspectral image with sufficient band images.

The CBEP presented in this section is an NBE process using correlation functions to generate new band images from the original set of multispectral images. Its original idea was developed in [16,20].

Correlation Band Expansion Process (CBEP)

Step 1. First-order band image: $\{\mathbf{B}_l\}_{l=1}^{L}$ = set of original band images

Step 2. Second-order correlated band images:

(i) $\{\mathbf{B}_l^2\}_{l=1}^{L}$ = set of auto-correlated band images

(ii) $\{\mathbf{B}_k\mathbf{B}_l\}_{k=1,l=1,k\neq l}^{L,L}$ = set of cross-correlated band images

Step 3. Third order correlated band images

(i) $\{\mathbf{B}_l^3\}_{l=1}^{L}$ = set of auto-correlated band images

(ii) $\{\mathbf{B}_k^2\mathbf{B}_l\}_{k=1,l=1,l\neq k}^{L,L}$ = set of two cross-correlated band images

(iii) $\{\mathbf{B}_k\mathbf{B}_l\mathbf{B}_m\}_{k=1,l=1,m=1,k\neq l\neq m}^{L,L,L}$ = set of three cross-correlated band images

Step 4. Other nonlinear correlated band images

(i) $\{\sqrt{\mathbf{B}_l}\}_{l=1}^{L}$= set of band images stretched out by the square-root.

(ii) $\{\log(\mathbf{B}_l)\}_{l=1}^{L}$ = set of band images stretched out by the logarithmic function.

It should be noted that, according to the nonlinear functions described in Steps (1)–(4), the band images generated by CBEP contain only nonlinear spectral information but not spatial information. In what follows, we develop an iterative CEM (ICEM) to address this issue where spatial information can be captured by using a Gaussian filter and feed it back to expand images currently being processed to create a new set of image data cubes.

2.2. Iterative CEM

ICEM, presented in this section, is implemented in conjunction with CBEP in an iterative manner. More specifically, it utilizes CBEP to create new band images via an NBE process. Once CBEP process is completed, a new set of image data cubes is generated for CEM to perform subpixel target detection. To obtain class spatial information, a Gaussian filter is introduced in the CEM-detected maps so that spatial contextual information of data sample vectors can be captured by a Gaussian filter. The resulting Gaussian-filtered CEM-detection abundance fractional map is fed back to create a new band incorporated into NBE to form a new hyperspectral cube which will be further used for re-processing CEM again. The same process is repeated over and over again until a stopping rule is satisfied. This repeated implementation of CEM via feedback loops in an iterative fashion is ICEM.

Specifically, at each iteration, say kth iteration, a Gaussian filter is used to blur $\left|\mathbf{B}\right|_{\text{CEM}}^{(k)}$ which is the absolute value of CEM-detection abundance fractional map, $\mathbf{B}_{\text{CEM}}^{(k)}$. This Gaussian-filtered band image, $\left|\mathbf{B}\right|_{\text{GF(CEM)}}^{(k)}$ provides spatial classification information as similar filters used in [23] and will be further fed back to $\mathbf{\Omega}_{\text{NBE}}^{(k)}$ to create a new set of hyperspectral images, $\mathbf{\Omega}_{\text{NBE}}^{(k+1)} = \mathbf{\Omega}_{\text{NBE}}^{(k)} \cup \left\{ \left|\mathbf{B}\right|_{\text{GF(CEM)}}^{(k)} \right\}$ to be used by CEM again for next iteration. The same procedure is continued. To terminate the process, an automatic stopping rule is designed. It applies Otsu's method [21] to $\left|\mathbf{B}\right|_{\text{GF(CEM)}}^{(k)}$ to produce a binary classification map, $\mathbf{B}_{\text{binary}}^{(k)}$ that will be used to calculate DSI [22]. If two consecutive DSI values are within an error threshold, ICEM will be terminated and $\left|\mathbf{B}\right|_{\text{CEM}}^{(k)}$ and $\mathbf{B}_{\text{binary}}^{(k)}$ will be the desired final real-valued WMH lesion detection map for visual inspection and binary value detection maps of WMH lesions for quantitative analysis.

In the following, we describe detailed step-by-step implementation of ICEM in great detail.

ICEM

1. Initial condition: Let $\{\mathbf{B}_l\}_{l=1}^{L}$ be the original set of band images.
2. Use an NBE process to create a new set of nonlinear band images, $\{\mathbf{B}_i^{\text{NB}}\}_{i=1}^{n_{\text{NB}}}$ where n_{NB} is the number of new band images by the NBE process.
3. Form a new set of band images, $\mathbf{\Omega}^{(0)} = \{\mathbf{B}_l\}_{l=1}^{L} \cup \{\mathbf{B}_i^{\text{NB}}\}_{i=1}^{n_{\text{NB}}}$. Let $\mathbf{d}^{(0)} = \left(d_1, \cdots, d_L, d_1^{\text{NB}}, \cdots, d_{n_{\text{NB}}}^{\text{NB}}\right)^T$ be the desired target pixels in $\mathbf{\Omega}^{(0)}$. Let δ_0^{CEM} be CEM using $\mathbf{d}^{(0)}$ and $\mathbf{R}^{(0)}$ which are obtained from $\mathbf{\Omega}^{(0)}$. Let $k = 1$.
4. At the kth iteration, update $\mathbf{d}^{(k)}$ and $\mathbf{R}^{(k)} = \sum_{i=1}^{N} \mathbf{r}_i^{(k)} \left(\mathbf{r}_i^{(k)}\right)^T$ from $\mathbf{\Omega}^{(k)}$.
5. Use new generated $\mathbf{d}^{(k)}$ and $\mathbf{R}^{(k)}$ for δ_k^{CEM} to be implemented on $\mathbf{\Omega}^{(k)}$. Let $\mathbf{B}_{\text{CEM}}^{(k)}$ be the detection abundance fractional map produced by δ_k^{CEM}.
6. Use a Gaussian filter to blur $\left|\mathbf{B}\right|_{\text{CEM}}^{(k)}$ where $\left|\mathbf{B}\right|_{\text{CEM}}^{(k)}$ is the absolute value of $\mathbf{B}_{\text{CEM}}^{(k)}$. The resulting image is denoted by Gaussian-filter $\left|\mathbf{B}\right|_{\text{GFCEM}}^{(k)}$.
7. Check if $\left|\mathbf{B}\right|_k^{\text{CEM}}$ satisfies a given stopping rule. If no, continue. Otherwise, go to Step 9.
8. Form $\mathbf{\Omega}^{(k+1)} = \mathbf{\Omega}^{(k)} \cup \left\{ \left|\mathbf{B}\right|_{\text{GFCEM}}^{(k)} \right\}$. Let $k \leftarrow k+1$ and go to Step 4.
9. $\mathbf{B}_{\text{CEM}}^{(k)}$ is the desired detection abundance fractional map and ICEM is terminated.

Figure 1 delineates how ICEM is processed as a detector where ICEM uses Gaussian filters to smooth CEM-detection abundance fractional maps and feeds back Gaussian-filtered CEM detection abundance fractional maps to provide spatial information for re-processing CEM iteratively. By gradually increasing more spatial information through feedback loops the boundaries of WMH lesions can be detected more effectively. It should also be noted that, if a particular NBE technique is used such as CBEP, then NBE-ICEM can be specified by CBEP-ICEM.

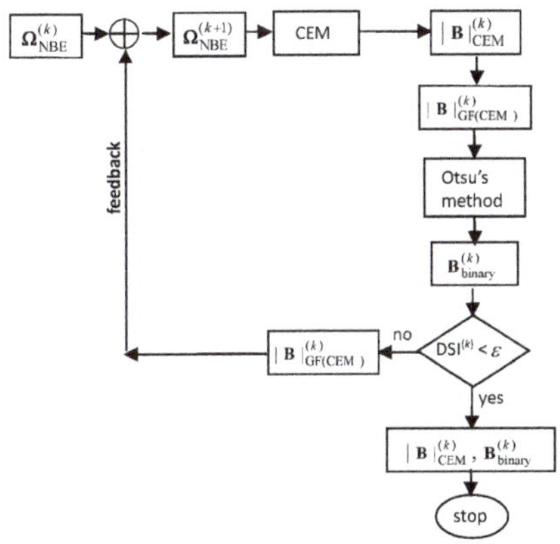

Figure 1. A diagram of the kth iteration carried out by hyperspectral image classification implementing ICEM on $\boldsymbol{\Omega}_{\text{NBE}}^{(k)}$

2.3. Stopping Rule for ICEM

To effectively terminate ICEM, DSI defined in [22] as

$$\text{DSI}^{(k)} = \frac{2|S_k \cap S_{k-1}|}{|S_k \cup S_{k-1}|} \tag{1}$$

is used a stopping criterion where $|S|$ is size of a set S, S_k and S_{k-1} are the kth thresholded binary image of the kth CEM detection abundance fractional map, $|\mathbf{B}|_k^{\text{CEM}}$ and $k-1$st thresholded binary image of the $k-1$st CEM detection abundance fractional map, $|\mathbf{B}|_{k-1}^{\text{CEM}}$. Figure 2 depicts a flow chart of a stopping rule using DSI with ε as a prescribed error threshold.

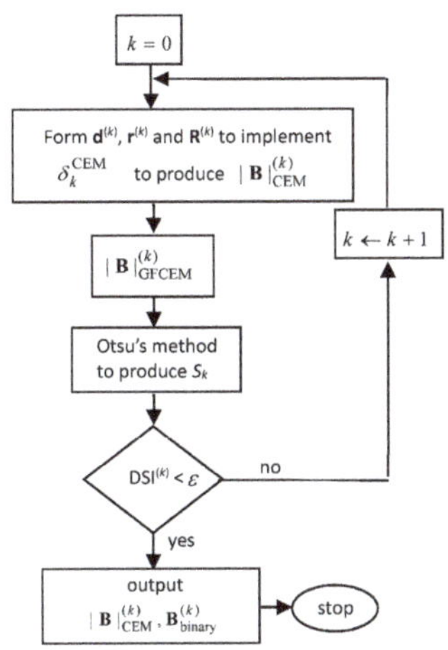

Figure 2. A flow chart of the stopping rule used for NBE-ICEM.

2.4. Algorithm for NBE-ICEM

Using Figures 1 and 2, an algorithm developed to implement ICEM in conjunction with NBE can be described as follows. Figure 3 describes a graphic flow chart of implementing NBE-ICEM.

NBE-ICEM

1. Initial conditions:
    - For each class, find its sample mean to calculate the desired signature **d** for the particular class.
    - Select the values of the parameter σ used for Gaussian filters in ICEM,
    - Prescribe an error threshold ε for DSI in Equation (1)
2. Use the NBE process described in Section 2.1 to generate a set of nonlinear band images, $\{\mathbf{B}_l^{\mathrm{NB}}\}_{l=1}^{n_{\mathrm{NB}}}$.
3. Apply ICEM described in Figure 1 to $\boldsymbol{\Omega}^{(0)} = \{\mathbf{B}_l\}_{l=1}^{L} \cup \{\mathbf{B}_l^{\mathrm{NB}}\}_{l=1}^{n_{\mathrm{NB}}}$.
4. Use DSI described in Figure 2 as a stopping rule to terminate ICEM.
5. Output $\left|\mathbf{B}|_{\mathrm{CEM}}^{(k)}\right.$, which is real-valued, and $\mathbf{B}_{\mathrm{binary}}^{(k)}$, which is binary-valued, to produce a confusion matrix for classification.

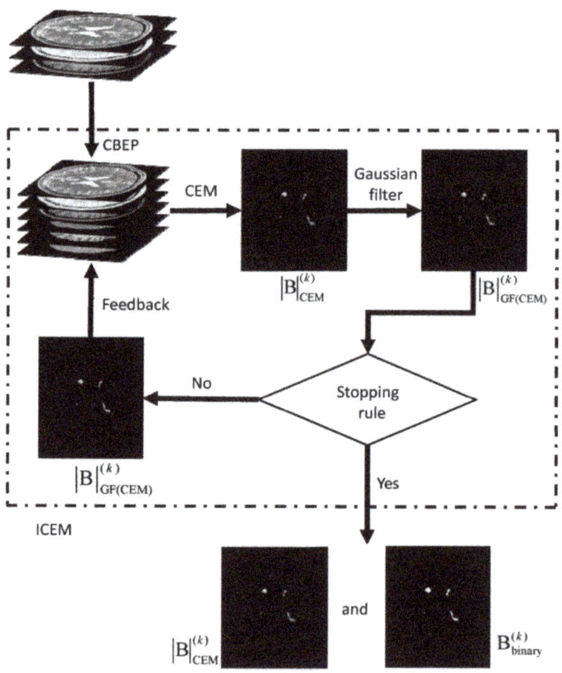

Figure 3. Graphic implementation of NBE-ICEM in Figure 1.

3. Results

3.1. Synthetic Image Experiments

To conduct an objective quantitative study, the synthetic MR brain images containing multiple sclerosis (MS) lesions obtained from the MR imaging simulator of McGill University, Montreal, Canada were used for experiments [24]. MS lesions are typically hyperintense on T2W or FLAIR sequence image. Figure 4a–c shows a slice MR brain image along with the ground truth of MS lesion shown in Figure 4d. The MR brain images are acquired by the modalities of T1W, T2W and PD with specifications provided in BrainWeb site [24]. The thickness of slice is 1 mm with size of 181 × 217 × 181. Each slice is specified by INU (intensity non-uniformity) 0% or 20%, denoted by rf0 and rf20 with six different levels of noise, 0%, 1%, 3%, 5%, 7% and 9%. The noise in the background of the simulated images is simulated by Rayleigh statistics and signal regions are simulated by Rician statistics. The "percentage (%) of noise" represents the ratio of the standard deviation of the white Gaussian noise to the signal for a reference tissue [24] in terms of %. There were 23 MR images from 91 to 113 slices for our study. To implement ICEM, we need to know the desired target signature **d**. Two ways were suggested to select training samples to calculate **d**. One is called all slices-selected training samples, which selects a small set of training samples from all MR image slices. The other is called single slice-selected training samples, which selects a small set of training samples from a particular single MR image slice that can be further used to find training samples for entire MR image slices. Its idea was derived from the extrapolation process used in volume sphering analysis (VSA) developed in [14,15]. Table 1 specifies the values of parameters used for experiments where two Gaussian filters using window sizes of 3 × 3 and 5 × 5, and two different σ = 0.1 and 0.5. The experiments were conducted for all MR image slices according to Table 1 where the results obtained by Gaussian window of 5 × 5 with σ = 0.5 are tabulated in parentheses.

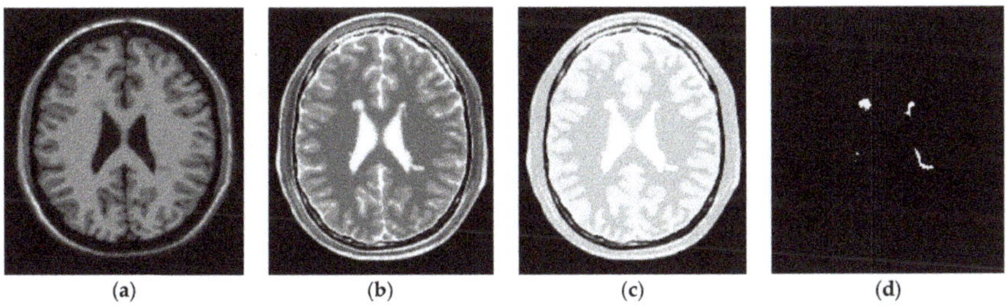

Figure 4. Three MR images containing MS lesions acquired by T1W, T2W and PD with 0% noise and 0% INU. (**a**) T1W; (**b**) T2W; (**c**) PD; (**d**) ground truth (lesions)

Table 1. Specifications of parameters used by NBE-ICEM for BrainWeb images.

Band images	T1W, T2W, PD (3 bands)
Correlation Band Expansion Process (CBEP)	3rd order correlated band images
d	found by all slices-selected or single slice-selected training samples
Gaussian window size	3 × 3(5 × 5)
σ used in Gaussian filter	0.1 with window size 3 × 3 (0.5 with window size 5 × 5)
Thresholding method	Otsu's method
error threshold (DSI)	0.80

To further evaluate the performance of the proposed NBE-ICEM, a commonly used segmentation approach, called lesion segmentation tool (LST) [25,26], was used for comparative study. It was originally developed for the segmentation of MS lesions and has also been proven to be useful for the segmentation of brain lesions. Table 2 tabulates DSI values calculated by Equation (1) averaged over 23 MR image slices 91–113 of lesion detection produced by CBEP-ICEM1, CBEP-ICEM2 and LST for six different noise levels and two INU levels where all slices-selected training samples were used to find the knowledge of **d**. The shaded DSI values in Table 2 are the best results. As we can see from the table, CBEP-ICEM1 performed better than CBEP-ICEM2 when noise level is low. However, when noise level is high, CBEP-ICEM2 performed better than CBEP-ICEM1. Nonetheless, both NBE-ICEM-based methods, i.e., CBEP-ICEM1 and CBEP-ICEM2, performed better than LST. It should be also noted that, since LST produced real-valued gray scale images, it required a threshold value to segment WMH lesions. The LST results in Table 2 were obtained by manually adjusting threshold values in order to yield the highest detection rate.

Table 2. Averaged DSI values of lesions detection by CBEP-ICEM1, CBEP-ICEM2 and LST over MR image slices 91–113.

Methods / Noise/INU Level	CBEP-ICEM1	CBEP-ICEM2	LST
n0/rf0	0.865	0.808	0.739
n1/rf0	0.886	0.864	0.749
n3/rf0	0.893	0.863	0.750
n5/rf0	0.806	0.839	0.731
n7/rf0	0.652	0.822	0.693
n9/rf0	0.579	0.801	0.714
n0/rf20	0.861	0.829	0.733
n1/rf20	0.867	0.834	0.753
n3/rf20	0.881	0.827	0.746
n5/rf20	0.814	0.831	0.732
n7/rf20	0.714	0.825	0.694
n9/rf20	0.540	0.806	0.655

Similarly, Table 3 also tabulates DSI values calculated by Equation (1) averaged over 23 MR image slices 91–113 of lesion detection produced by CBEP-ICEM1, CBEP-ICEM2 and LST for six different noise levels and two INU levels where single slice-selected training samples were used to find the knowledge of **d** and the slice 102 was chosen as the desired single slice. The selection of slice 102 is empirical as long as it includes sufficient tissue information, in which case the middle MR image slice can serve as this purpose. The same conclusions drawn from Table 2 were also valid for Table 3, even though the results in Table 3 were slightly degraded compared to the results in Table 2 because the single slice-selected training samples were used. It should be noted that the results of LST in Tables 2 and 3 were the same because LST did not allow users to select training samples. This disadvantage is further offset by a need of finding an appropriate threshold value to segment lesion out from the background.

Table 3. Averaged DSI values of lesions detection by CBEP-ICEM1, CBEP-ICEM2 and LST over MR image slices 91–113 using slice 102 to select training samples.

Methods / Noise/INU Level	CBEP-ICEM1	CBEP-ICEM2	LST
n0/rf0	0.798	0.784	0.739
n1/rf0	0.848	0.847	0.749
n3/rf0	0.871	0.858	0.750
n5/rf0	0.776	0.836	0.731
n7/rf0	0.625	0.816	0.693
n9/rf0	0.389	0.778	0.714
n0/rf20	0.844	0.834	0.733
n1/rf20	0.859	0.837	0.753
n3/rf20	0.854	0.814	0.746
n5/rf20	0.811	0.819	0.732
n7/rf20	0.710	0.804	0.694
n9/rf20	0.549	0.799	0.655

For an illustrative purpose, Figures 5–10 only show detection results of WMH lesions of the 97th MR image slice with six levels of noise and 0% INU by two versions of CBEP-ICEM, using the Gaussian window size of 3 × 3 specified by $\sigma = 0.1$ and the Gaussian window size of 5 × 5 specified by $\sigma = 0.5$, referred to as CBEP-ICEM1 and CBEP-ICEM2, respectively, where two sets of training samples selected by all slices and the single 102nd slice were used to calculate the desired target signatures **d** to implement NBE-ICEM. As we can see by visual inspection against the ground truth in Figure 4d, CBEP-ICEM1 and CBEP-ICEM2 using two sets of training samples, i.e., all slices-selected and single slice-selected training samples, produced very close results and they both also performed better lesion detection than LST did.

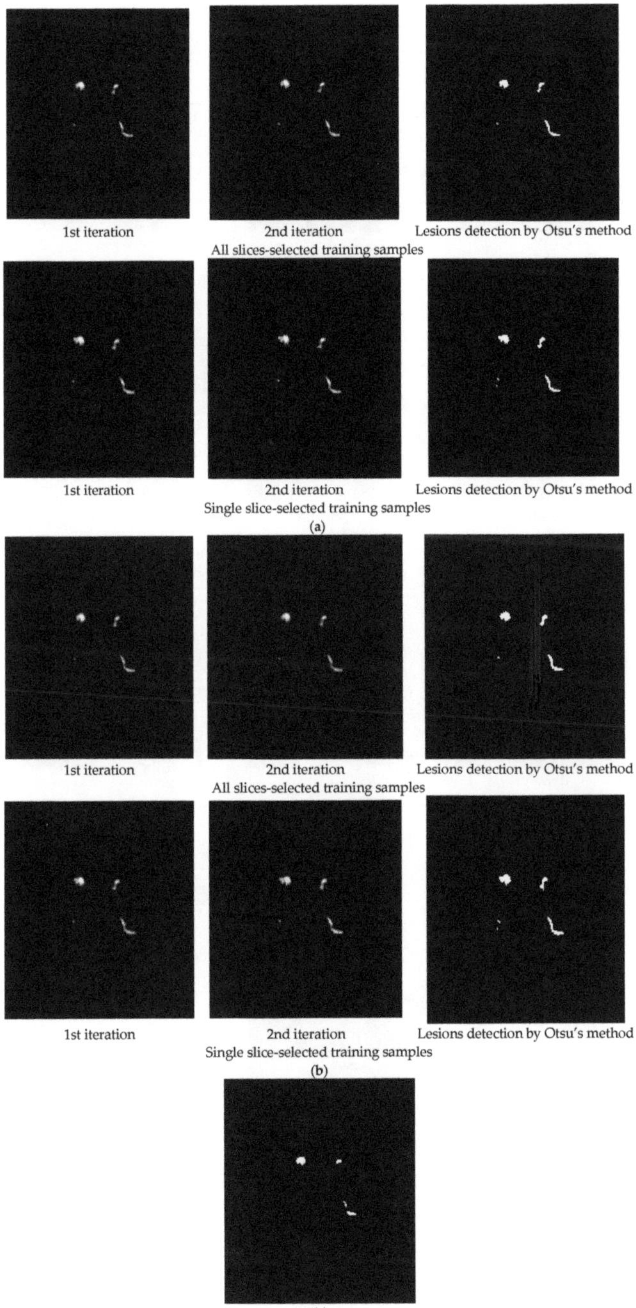

Figure 5. Lesion detection of Slice 97 with 0% noise and 0% INU by CBEP-ICEM1 and CBEP-ICEM2. (**a**) CBEP-ICEM1; (**b**) CBEP-ICEM2; (**c**) Lesion detection LST.

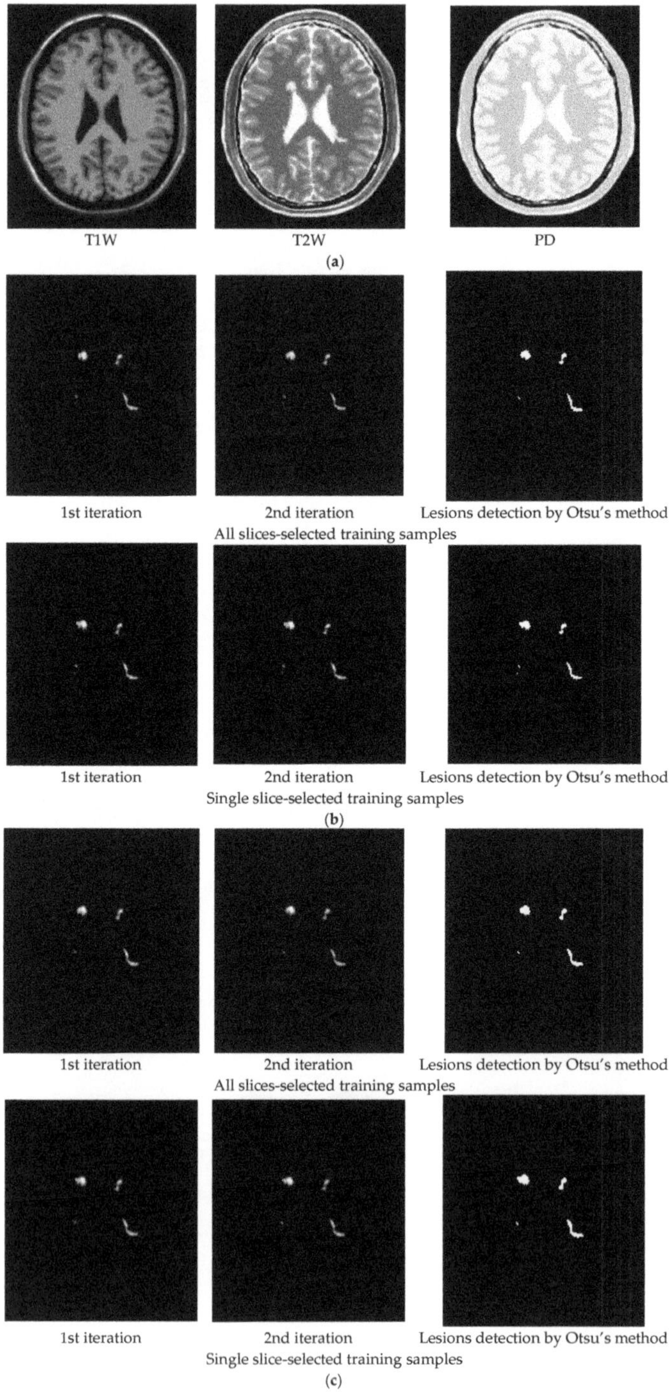

Figure 6. *Cont.*

(d)

Figure 6. Lesion detection of Slice 97 with 1% noise and 0% INU by CBEP-ICEM1 and CBEP-ICEM2. (**a**) Original 97th slice of MS MR brain images with 1% noise and 0% INU; (**b**) CBEP-ICEM1; (**c**) CBEP-ICEM2; (**d**) Lesion detection LST.

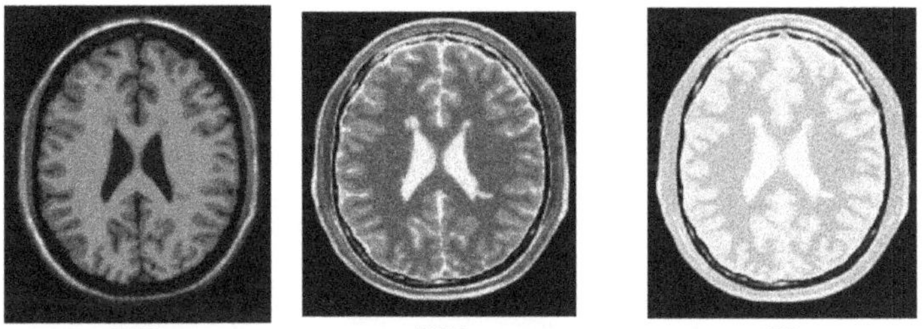

T1W T2W PD

(**a**)

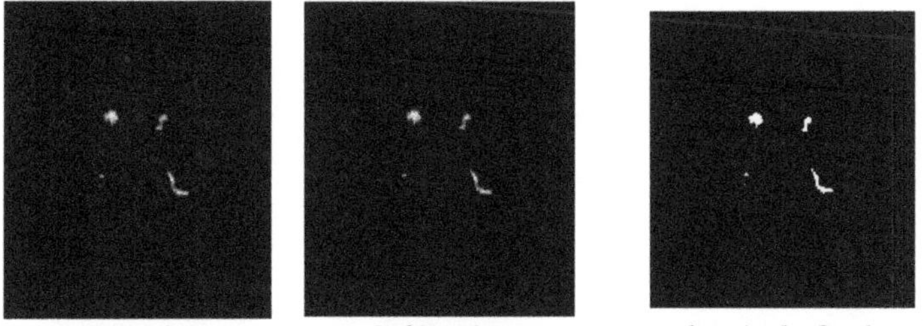

1st iteration 2nd iteration Lesions detection by Otsu's method

All slices-selected training samples

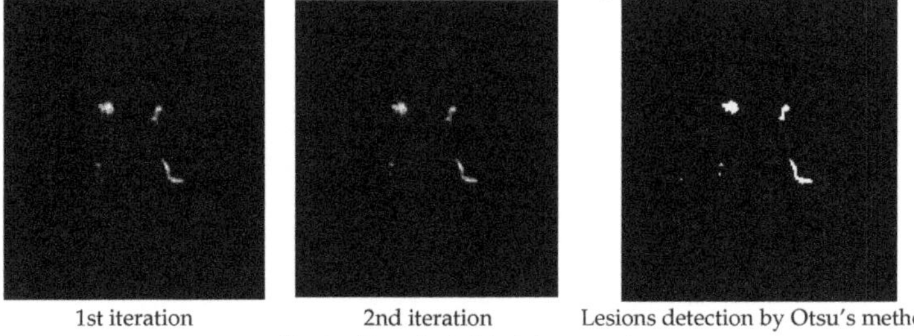

1st iteration 2nd iteration Lesions detection by Otsu's method

Single slice-selected training samples

(**b**)

Figure 7. *Cont.*

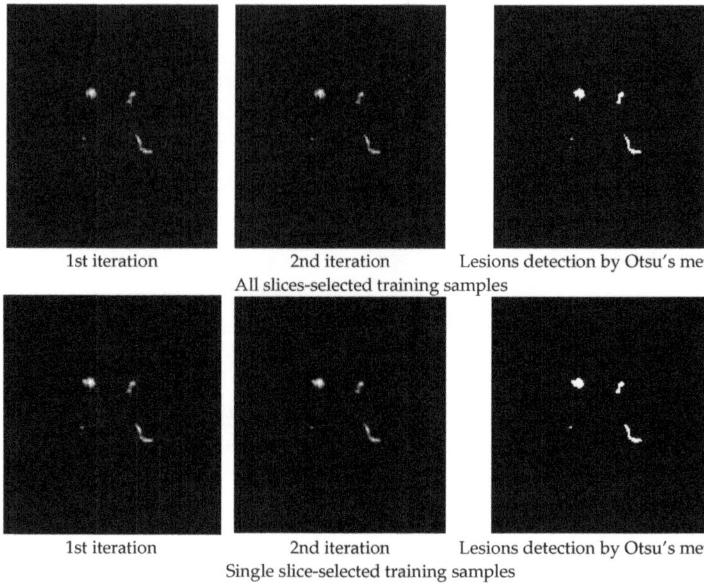

(c)

(d)

Figure 7. Lesion detection of Slice 97 with 3% noise and 0% INU by CBEP-ICEM1 and CBEP-ICEM2. (**a**) Original 97th slice of MS MR brain images with 3% noise and 0% INU; (**b**) CBEP-ICEM1; (**c**) CBEP-ICEM2; (**d**) Lesion detection LST.

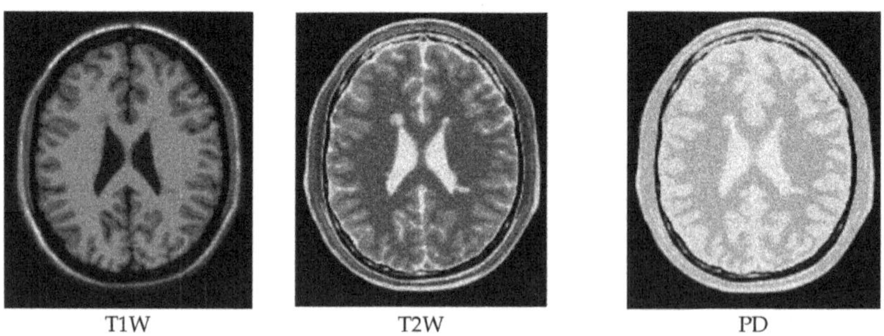

(a)

Figure 8. *Cont.*

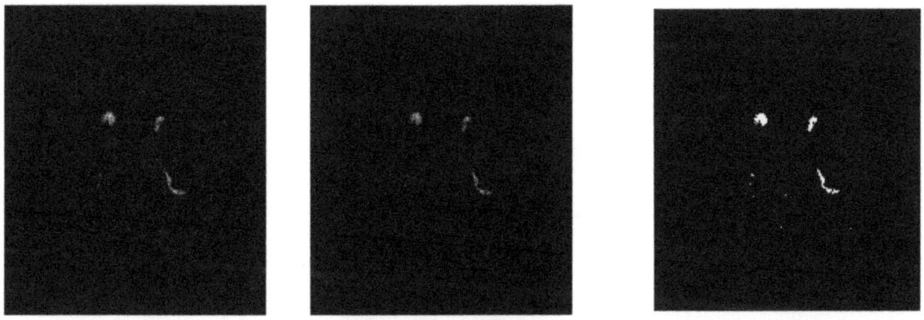

1st iteration 4th iteration (final) Lesions detection by Otsu's method

All slices-selected training samples

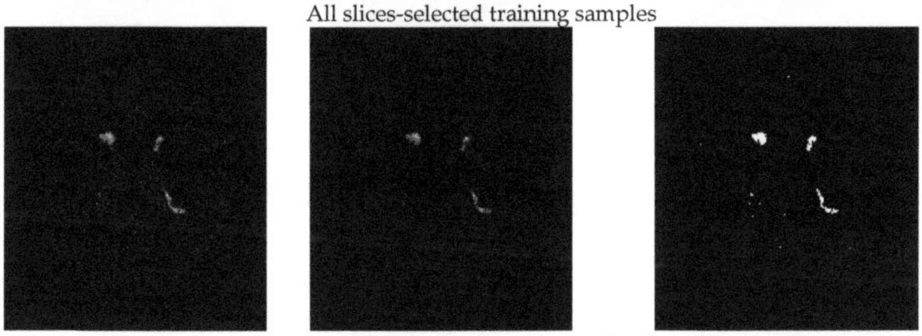

1st iteration 4th iteration (final) Lesions detection by Otsu's method

Single slice-selected training samples

(**b**)

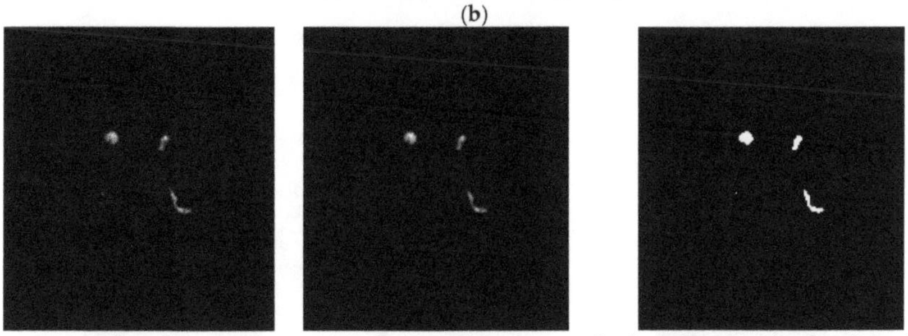

1st iteration 2nd iteration Lesions detection by Otsu's method

All slices-selected training samples

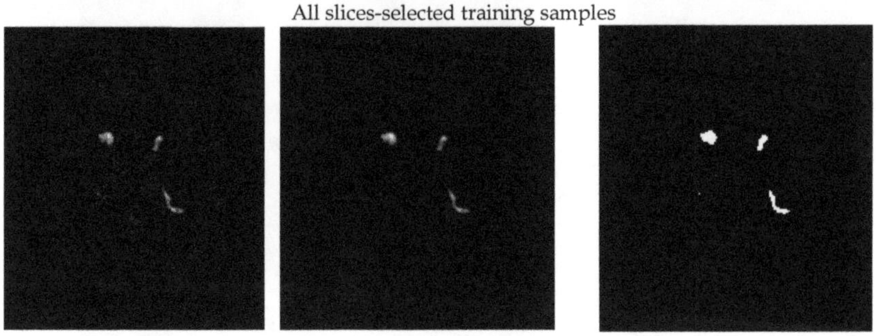

1st iteration 3rd iteration (final) Lesions detection by Otsu's method

Single slice-selected training samples

(**c**)

Figure 8. *Cont.*

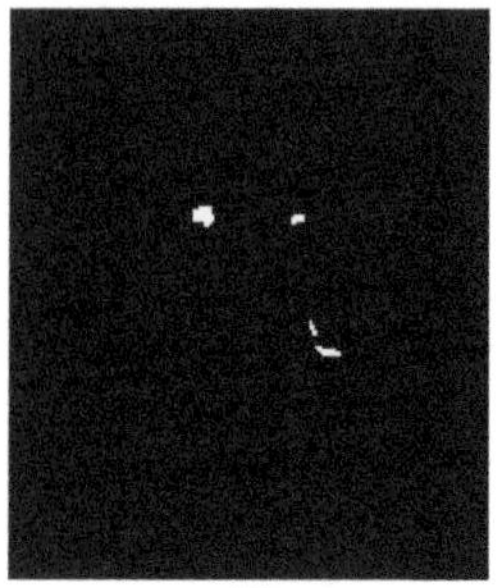

(d)

Figure 8. Lesion detection of Slice 97 with 5% noise and 0% INU by CBEP-ICEM1 and CBEP-ICEM2. (**a**) Original 97th slice of MS MR brain images with 5% noise and 0% INU; (**b**) CBEP-ICEM1; (**c**) CBEP-ICEM2; (**d**) Lesion detection LST.

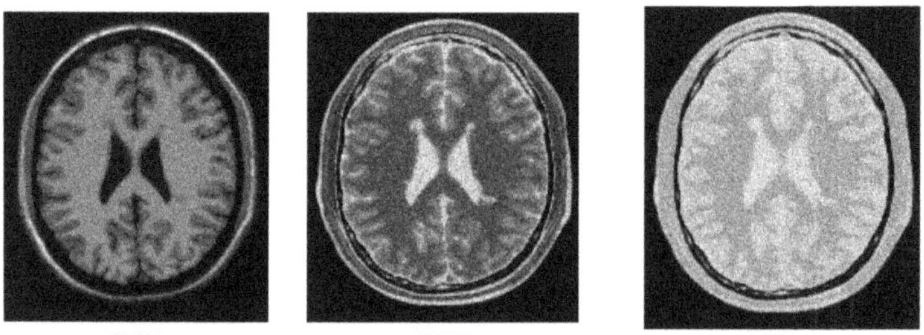

T1W T2W PD

(a)

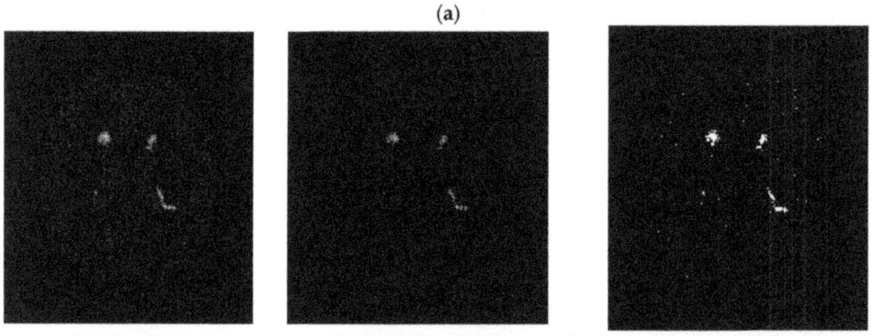

1st iteration 6th iteration (final) Lesions detection by Otsu's method

All slices-selected training samples

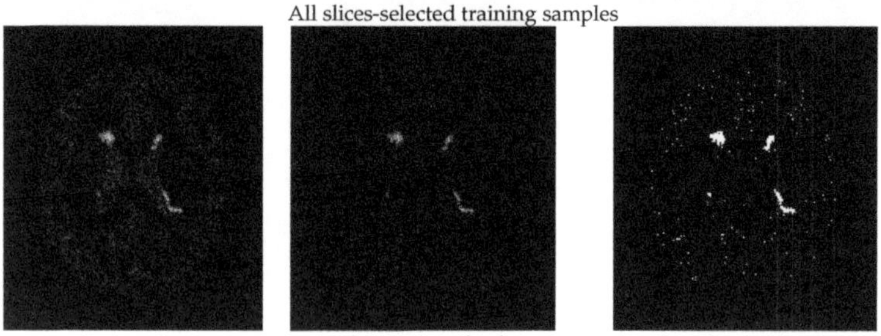

1st iteration 5th iteration (final) Lesions detection by Otsu's method

Single slice-selected training samples

(b)

Figure 9. *Cont.*

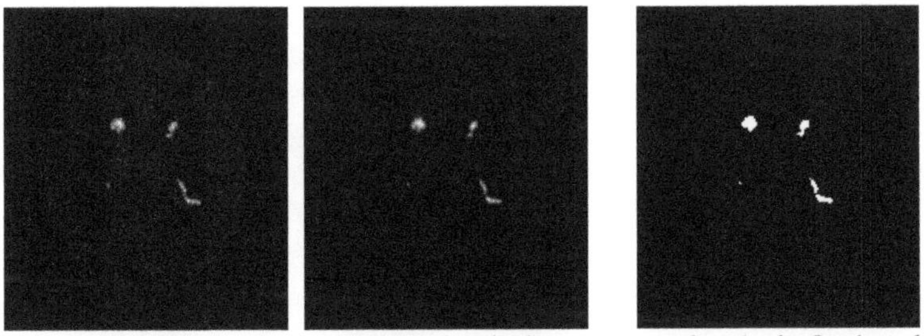

1st iteration 3rd iteration (final) Lesions detection by Otsu's method

All slices-selected training samples

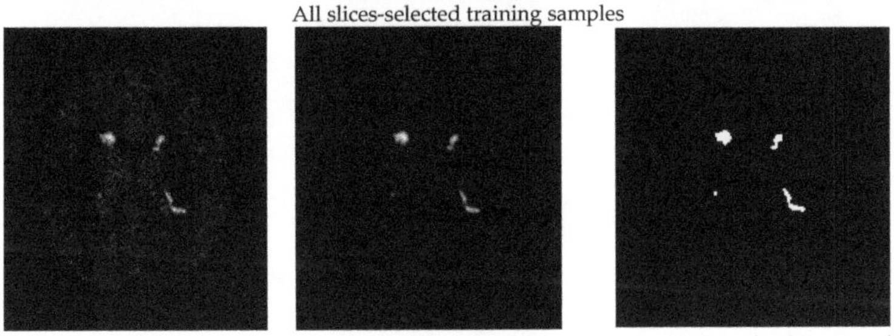

1st iteration 3rd iteration (final) Lesions detection by Otsu's method

Single slice-selected training samples

(**c**)

(**d**)

Figure 9. Lesion detection of Slice 97 with 7% noise and 0% INU by CBEP-ICEM1 and CBEP-ICEM2. (**a**) Original 97th slice of MS MR brain images with 7% noise and 0% INU; (**b**) CBEP-ICEM1; (**c**) CBEP-ICEM2; (**d**) Lesion detection LST.

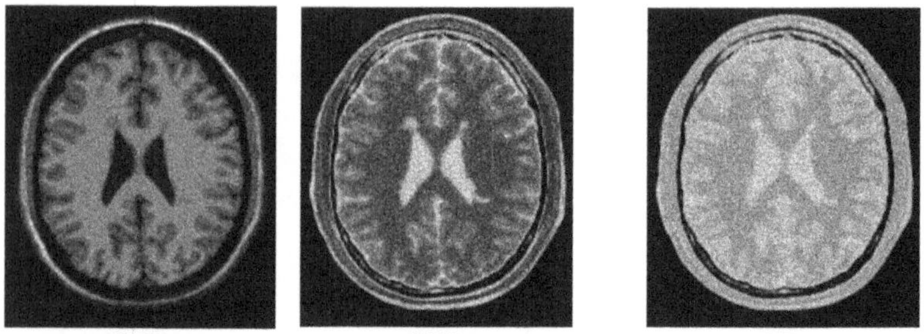

T1W T2W PD

(a)

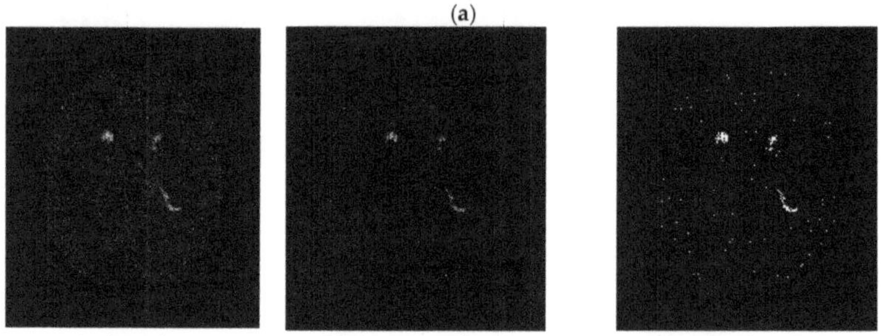

1st iteration 7th iteration (final) Lesions detection by Otsu's method

All slices-selected training samples

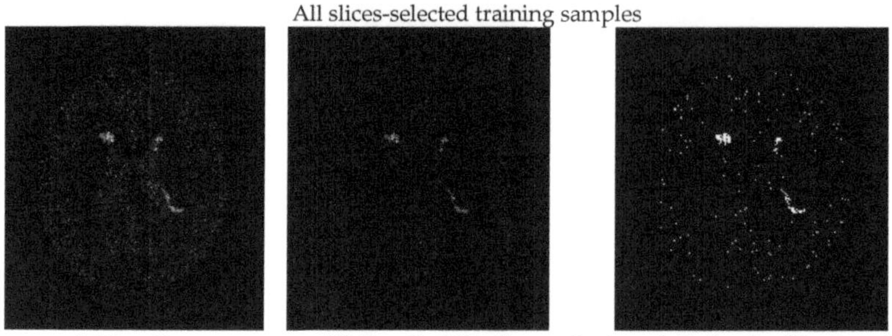

1st iteration 6th iteration (final) Lesions detection by Otsu's method

Single slice-selected training samples

(b)

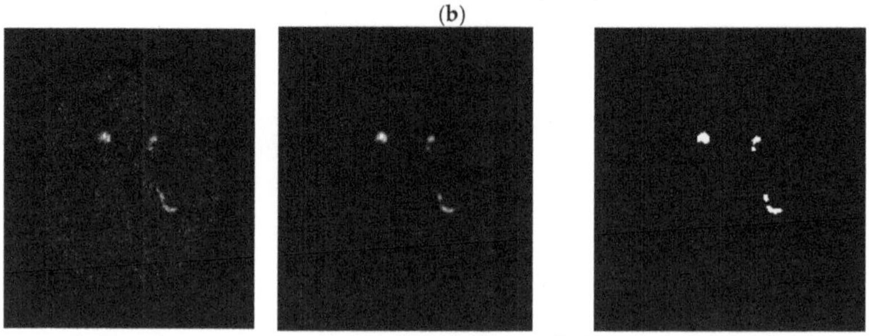

1st iteration 4th iteration (final) Lesions detection by Otsu's method

All slices-selected training samples

Figure 10. *Cont.*

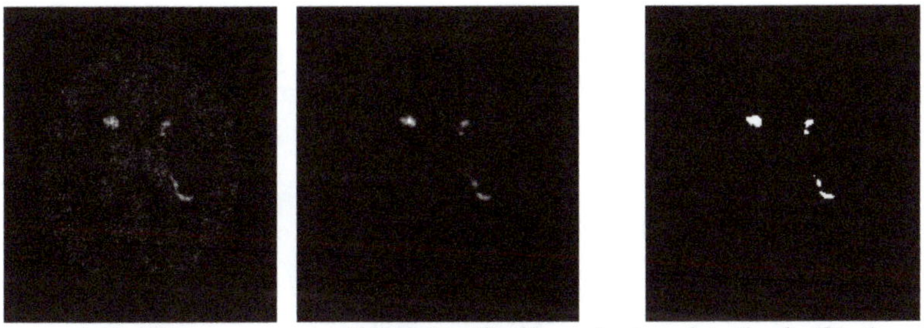

(**c**)

(**d**)

Figure 10. Lesion detection of Slice 97 with 9% noise and 0% INU by CBEP-ICEM1 and CBEP-ICEM2. (**a**) Original 97th slice of MS MR brain images with 9% noise and 0% INU; (**b**) CBEP-ICEM1; (**c**) CBEP-ICEM2; (**d**) Lesion detection LST.

Two comments are noteworthy.

1. Despite the fact that the training samples used for our proposed NBE-ICEM were selected based on 2D images such as by all slices and a single slice, these training samples were either stacked as voxels from all slices or extrapolated from a single slice as voxels by VSA. Accordingly, NBE-ICEM is actually run on 3D images as image cubes.
2. There is an issue in implementing LST. Since it is packaged as a software algorithm, there is no flexibility for users to choose parameters at their discretion. Besides, it cannot implement T1W, T2W or FLAIR images alone. Instead, it must require T1W images as reference images to segment WMHs [26]. Most importantly, it produces real valued gray level images, which require users selecting a threshold value from a range from 0.05 to 0.95 with a step size of 0.05 to detect WMHs. In [26], this threshold value was suggested between 0.25 and 0.4. However, in practical applications, the best value is generally selected manually. Thus, technically speaking, LST is not fully automatic. Specifically, when synthetic images from the BainWeb were used for experiments, it was found that using both T1W and T2W could not segment WMHs. It must use T1W and PD to detect WMHs and the threshold value must be set to around 0.2 to segment WMHs.

3.2. Real Image Experiments

Real MRI brain images were acquired at the Taichung Veterans General Hospital (TCVGH) by Siemens Magnetom Aera 1.5 Tesla (Erlangen, Germany) MR scanner with a 16-channel phase-array head coil. MR imaging protocol included T1W with 3D MPRAGE, T2W and FLAIR. Since T1W, T2W and FLAIR images used for experiments were collected by 3D high resolution sequences with each voxel of size, $1 \times 1 \times 1$ mm^3, the interpolation artifacts and partial volume do not have much effect on imaging. However, as a part of trade-off, this also requires additional 2 min for image acquisition. Other imaging parameters used for data acquisition were voxel size of $1 \times 1 \times 1$ mm^3, matrix size = $256 \times 256 \times 176$, NEX = 1. According to a clinical visual inspection criterion [27], the WMH lesions can be graded by Fazekas with three grades of Fazekas shown in Figure 11 for illustration.

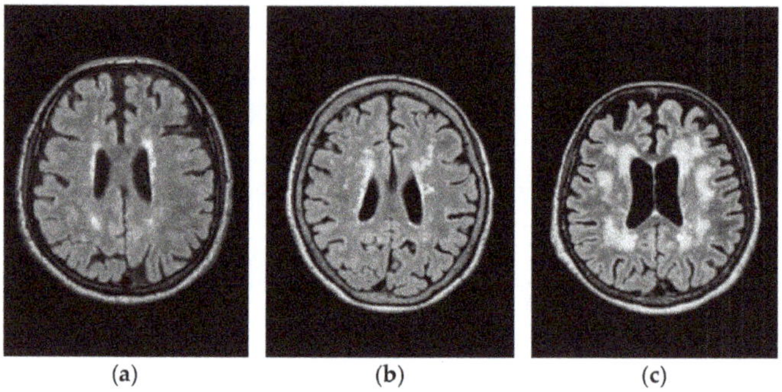

Figure 11. Lesion categorized by three grades of Fazekas shown in FLAIR images. (**a**) Fazekas grade 1; (**b**) Fazekas grade 2; (**c**) Fazekas grade 3

A total of 111 cases were collected and all the participants have been well-informed and signed their consents. In addition, the study conducted in this paper was approved by the Ethics Committee of Clinical Research, Taichung Veterans General Hospital (IRB number: CE16138A). Among all the 111 cases there are 58 cases of Fazekas grade 1, 44 cases of Fazekas grade 2 and 9 cases of Fazekas grade 3. Thus, in this study, we selected 10 cases from Fazekas grade 1, 11 cases from Fazekas grade 2, and 9 cases from Fazekas grade 3.

As demonstrated by synthetic image experiments, CBEP-ICEM2 was shown to be a better WMH detection technique. Thus, CBEP-ICEM 2 was used in real image experiments. Table 4 tabulates the values of parameters used by CBEP-ICEM2 where two sets of training samples selected by all slices and the single 90th slice were selected to calculate the desired target signature **d** to implement NBE-ICEM.

Table 4. Parameters used by CBEP-ICEM2.

Band	T1W, T2W, FLAIR (3 bands)		
CBEP	3rd order correlated band images		
d	found by all slices-selected or single slice-selected training samples		
Fazekas grade	1	2	3
Gaussian window size	5×5		
σ used in Gaussian filter	0.5 with window size 5×5		
Thresholding method	Otsu's method		
stopping threshold (DSI)	0.80		

Figures 12–14 show the WMH lesion detection results produced by CBEP-ICEM2 and LST for three Fazekas grades, respectively, where Figures 12a, 13a and 14a are original T1W, T2W and FLAIR MR images; Figures 12b, 13b and 14b are iterative WMH lesion detection images by CBEP-ICEM2 along with final WMH lesion detection by Otsu's method; and Figures 12c, 13c and 14c show comparisons between lesion detections by CBEP-ICEM2 and LST.

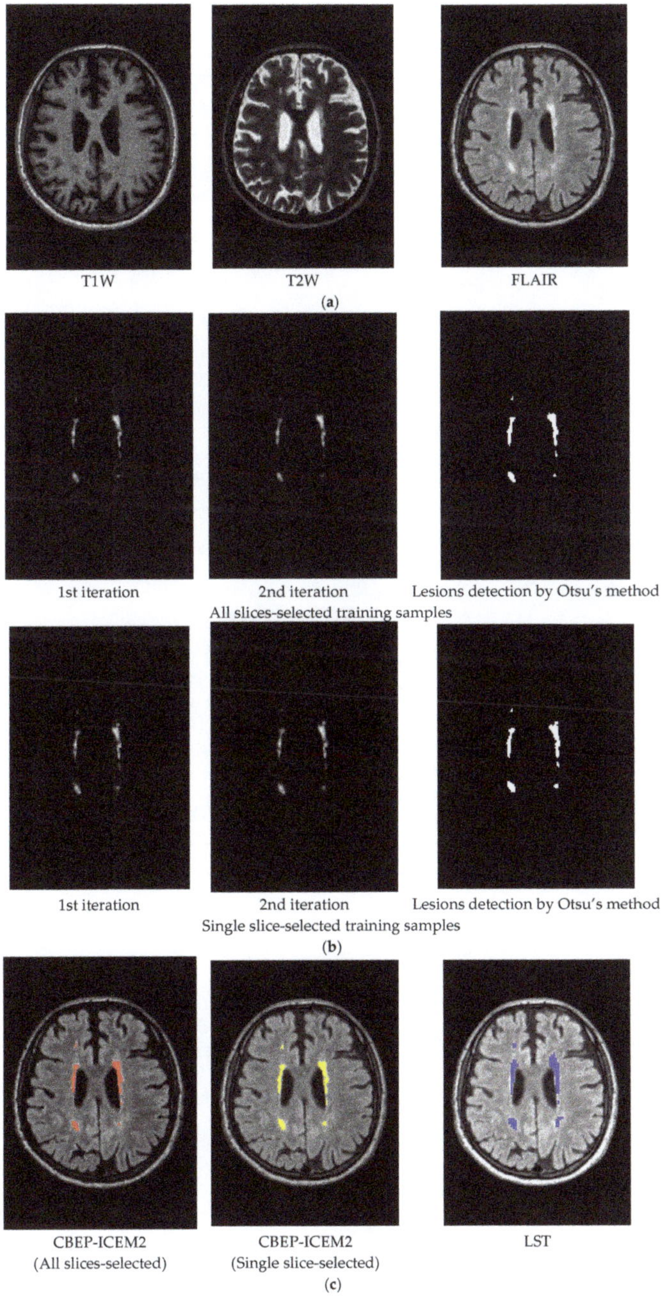

Figure 12. Lesion detection of Fazekas grade 1 by CBEP-ICEM2 and LST (**a**) Original MR images (T1W, T2W, FLAIR) with lesions of Fazekas grade 1; (**b**) CBEP-ICEM2-detected lesion of Fazekas grade 1; (**c**) Comparison between lesion detections by CBEP-ICEM2 and LST.

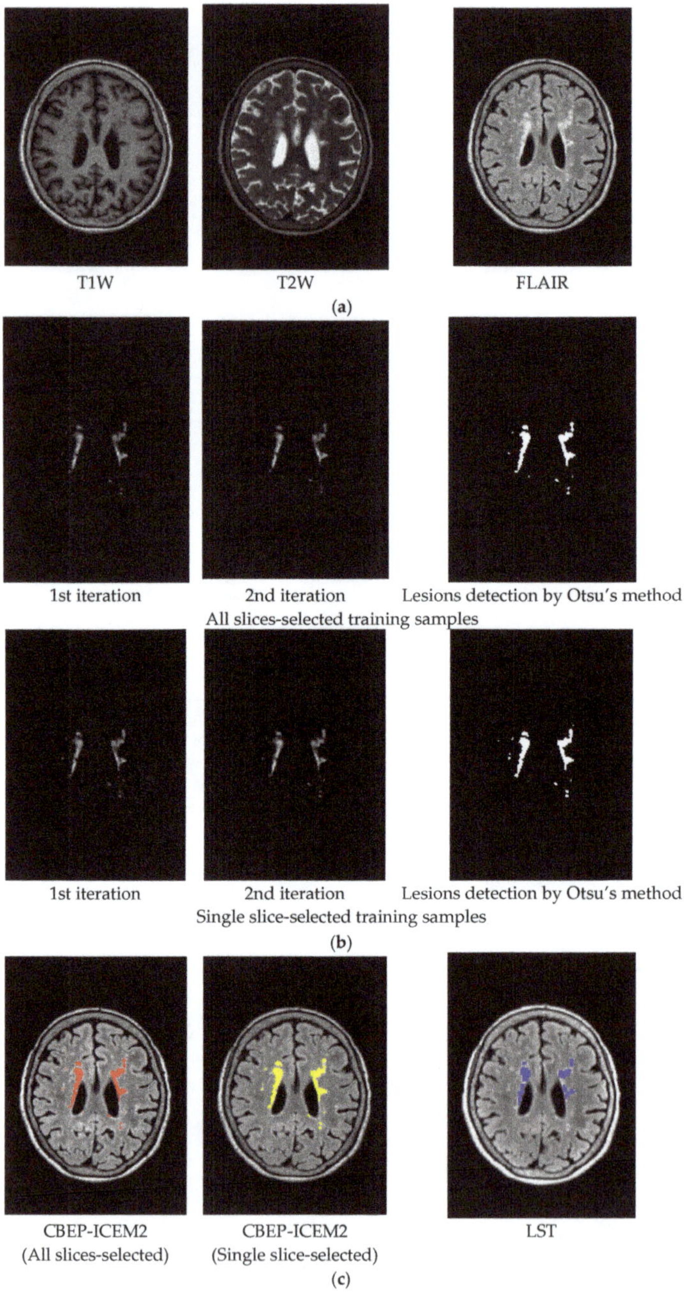

Figure 13. Lesion detection of Fazekas grade 2 By CBEP-ICEM2 and LST. (**a**) Original MR images (T1W, T2W, FLAIR) with lesions of Fazekas grade 2; (**b**) Lesion detection by CBEP-ICEM2; (**c**) Comparison between lesion detections by CBEP-ICEM2 and LST.

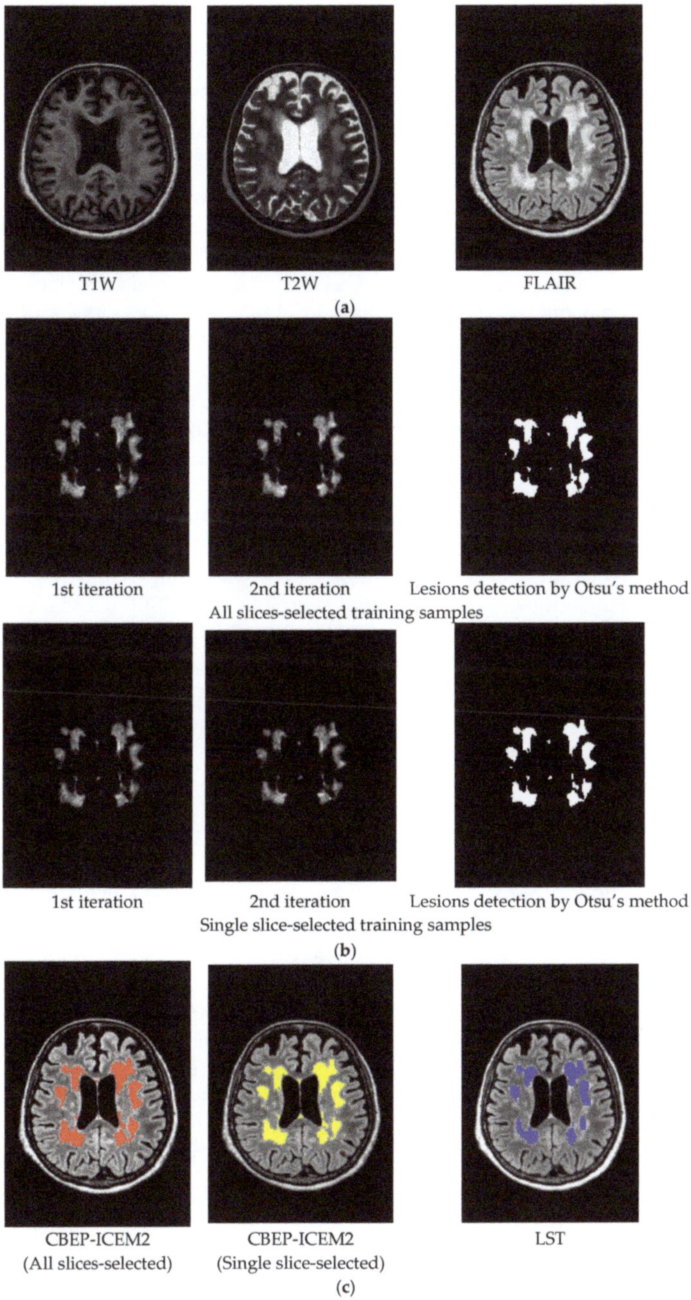

Figure 14. Lesion detection of Fazekas grade 3 By CBEP-ICEM2 and LST. (**a**) Original MR images (T1W, T2W, FLAIR) with lesions of Fazekas grade 3; (**b**) Lesion detection by CBEP-ICEM2; (**c**) Comparison between lesion detections by CBEP-ICEM2 and LST.

As demonstrated in Figures 12–14, our proposed NBE-ICEM using two sets of training samples performed very similarly and also better than LST according to clinical visual evaluation criterion, Fazekas grades [27].

4. Discussion

This paper is believed to be the first work ever reported in the literature to attempt to use a hyperspectral subpixel detection, NBE-ICEM, to detect WMHs on MRI. As demonstrated in Tables 2 and 3 and Figures 5–10 and Figures 12–14 the synthetic and real image experiments confirm significant improvements using NBE-ICEM over the LST method in WMH detection.

In comparison between CBEP-CEM1 and CBEP-CEM2, we found from Tables 2 and 3 in the synthetic image experiments that, when the noise level is low (0%, 1%, 3%), CBEP-ICEM1 using a smaller Gaussian window performed better than CBEP-ICEM2 using a larger Gaussian window. However, when the noise level is high (5%, 7%, 9%), the conclusion is reversed, i.e., CBEP-ICEM2 performed better than CBEP-ICEM1. It is also interesting to note that CBEP-ICEM1 performed very poorly when noise level reached 7% and above and even worse than LST. Tables 2 and 3 also shown that it was noise not INU that had an impact on lesion detection. On the other hand, CBEP-ICEM2 generally performed well regardless of noise level if DSI value was set to at least or above 0.8 compared to LST whose DSI values did not go beyond 0.8. The synthetic image experiments suggested that CBEP-ICEM2 was a better technique due to its robustness to noise level and ability in WMH detection.

In addition, based on the results of real image experiments from Figures 12–14, there are three interesting findings. Firstly, the number of iterations carried out by CBEP-ICEM is always two for all three Fazekas grades. Secondly, in Figure 12c, CBEP-ICEM2 and LST performed similarly but quite different in Figures 13c and 14c, where the areas of lesions detected by LST were much smaller than CBEP-ICEM2. Thirdly, the iterative images produced in Figures 12b, 13b and 14b by CBEP-ICEM2 showed that including spatial information captured by Gaussian-filtered CEM detection images did improve lesion detection, particularly edge and boundary pixels.

This paper makes several main contributions to WMH lesions detection in MR brain images. First, it develops NBE to resolve two issues arising in WMH detection, insufficient spectral dimensionality and linear non-separability problem. NBE plays a similar role that kernels play in pattern classification such as support vector machine (SVM). Second, it introduces Gaussian filters to be included in CEM to expand capability of CEM in capturing spatial information surrounding CEM-detected WMH lesions. Third, the real-valued CEM-detection abundance fractional maps provide soft decisions for visual inspection. Fourth, Otsu's method is incorporated in ICEM to produce thresholded binary maps as hard decisions that show WMH lesions detection. This resolves the main issue encountered in LST. Fifth, the feedbacks of Gaussian filtered CEM detection abundance fractional maps allow CEM to perform better detection in WMH lesions when spatial information of lesions is crucial, specifically, their boundaries. Finally, an automatic stopping rule is particularly designed to determine how much spatial information is needed for CEM to perform its best in detection of WMH lesions.

5. Conclusions

In conclusion, this paper develops a novel approach, called NBE-ICEM, for WMH lesions detection in MR brain images. It is derived from a hyperspectral imaging-based subpixel target detection method (CEM), but is rather different from CEM in two aspects. One is an introduction of NBE into CEM, which expands the original MR images by including nonlinearly correlated band images generated by NBE to make a multispectral MR image into a hyperspectral MR image since CEM is a hyperspectral imaging technique. The other is the development of an iterative version of CEM, ICEM, which can feed back spatial information captured by Gaussian filters in an iterative manner. More specifically, it applies a Gaussian filter to CEM-detection maps to produce Gaussian-filtered CEM-detected abundance fractional maps that can be further fed back iteratively to form a new set of MR image cubes which will be used a new data set to be re-processed by CEM to improve WMH lesions detection performance. ICEM can be considered as a joint spectral–spatial filter. The more iterations carried out by ICEM, the more spatial information captured. The work on NBE-ICEM presents a potential and promising technique for WMH lesions detection. It is our belief that there would be new applications of NBE-ICEM to MRI yet to explore in the future, specifically, partial volume estimation for specific tissues of interest.

Acknowledgments: The work of this paper was supported by grants from Taichung Veterans General Hospital (TCVGH-1047315C, and TCVGH-1057315C), and grants from Ministry of Science and Technology (MOST 104-2221-E-075A-002, and MOST-105-2221-E-075A-001). The work of L. Wang is supported by the Fundamental Research Funds for Central Universities under Grant JB150508 and the 111 Project (B17035). The works of Y. Wang and M. Song are supported by Fundamental Research Funds for Central Universities (3132016028 and 3132016331) and National Nature Science Foundation of China (61601077 and 61301228), respectively. The work of C.-I Chang is supported by the Fundamental Research Funds for Central Universities under Grant 3132016331.

Author Contributions: H.-M.C., J.-W.C., and C.-C.C.C. conceived and designed the experiments; L.W., C.Y., and Y.W. performed the experiments; H.C.W., M.S., and B.X. analyzed the data; C.-I.C., H.-M.C., J.-W.C., and C.-C.C.C. contributed reagents/materials/analysis tools; C.-I C. wrote the paper.

Conflicts of Interest: All authors have declared no conflict of interest.

References

1. Callisaya, M.L.; Beare, R.; Phan, T.; Blizzard, L.; Thrift, A.G.; Chen, J.; Srikanth, V.K. Progression of White Matter Hyperintensities of Presumed Vascular Origin Increases the Risk of Falls in Older People. *J. Gerontol. A Biol. Sci. Med. Sci.* **2015**, *70*, 360–366. [CrossRef] [PubMed]
2. Hachinski, V.C.; Potter, P.; Merskey, H. Leuko-araiosis: an ancient term for a new problem. *Can. J. Neurol. Sci. J. Can. Sci. Neurol.* **1986**, *13*, 533–534. [CrossRef]
3. Boutet, C.; Rouffiange-Leclair, L.; Schneider, F.; Camdessanché, J.-P.; Antoine, J.-C.; Barral, F.-G. Visual Assessment of Age-Related White Matter Hyperintensities Using FLAIR Images at 3 T: Inter- and Intra-Rater Agreement. *Neurodegener. Dis.* **2015**. [CrossRef] [PubMed]
4. Valverde, S.; Oliver, A.; Roura, E.; González-Villà, S.; Pareto, D.; Vilanova, J.C.; Ramió-Torrentà, L.; Rovira, À.; Lladó, X. Automated tissue segmentation of MR brain images in the presence of white matter lesions. *Med. Image Anal.* **2017**, *35*, 446–457. [CrossRef] [PubMed]
5. Roura, E.; Oliver, A.; Cabezas, M.; Valverde, S.; Pareto, D.; Vilanova, J.C.; Ramió-Torrentà, L.; Rovira, À.; Lladó, X. A toolbox for multiple sclerosis lesion segmentation. *Neuroradiology* **2015**, *57*, 1031–1043. [CrossRef] [PubMed]
6. Samaille, T.; Fillon, L.; Cuingnet, R.; Jouvent, E.; Chabriat, H.; Dormont, D.; Colliot, O.; Chupin, M. Contrast-Based Fully Automatic Segmentation of White Matter Hyperintensities: Method and Validation. *PLoS ONE* **2012**, *7*. [CrossRef] [PubMed]
7. Gibson, E.; Gao, F.; Black, S.E.; Lobaugh, N.J. Automatic segmentation of white matter hyperintensities in the elderly using FLAIR images at 3T. *J. Magn. Reson. Imaging* **2010**, *31*, 1311–1322. [CrossRef] [PubMed]
8. Vannier, M.W.; Butterfield, R.L.; Jordan, D.; Murphy, W.A.; Levitt, R.G.; Gado, M. Multispectral analysis of magnetic resonance images. *Radiology* **1985**, *154*, 221–224. [CrossRef] [PubMed]
9. Chang, C.-I. *Hyperspectral Imaging: Techniques for Spectral Detection and Classification*; Springer: New York, NY, USA, 2003; ISBN 978-0-306-47483-5.
10. Nakai, T.; Muraki, S.; Bagarinao, E.; Miki, Y.; Takehara, Y.; Matsuo, K.; Kato, C.; Sakahara, H.; Isoda, H. Application of independent component analysis to magnetic resonance imaging for enhancing the contrast of gray and white matter. *NeuroImage* **2004**, *21*, 251–260. [CrossRef] [PubMed]
11. Ouyang, Y.-C.; Chen, H.-M.; Chai, J.-W.; Chen, C.-C.; Chen, C.C.-C.; Poon, S.-K.; Yang, C.-W.; Lee, S.-K. Independent Component Analysis for Magnetic Resonance Image Analysis. *EURASIP J. Adv. Signal Process.* **2008**, *2008*. [CrossRef]
12. Ouyang, Y.-C.; Chen, H.-M.; Chai, J.-W.; Chen, C.C.-C.; Poon, S.-K.; Yang, C.-W.; Lee, S.-K.; Chang, C.-I. Band Expansion-Based Over-Complete Independent Component Analysis for Multispectral Processing of Magnetic Resonance Images. *IEEE Trans. Biomed. Eng.* **2008**, *55*, 1666–1677. [CrossRef] [PubMed]

13. Chai, J.-W.; Chen, C.C.-C.; Chiang, C.-M.; Ho, Y.-J.; Chen, H.-M.; Ouyang, Y.-C.; Yang, C.-W.; Lee, S.-K.; Chang, C.-I. Quantitative analysis in clinical applications of brain MRI using independent component analysis coupled with support vector machine. *J. Magn. Reson. Imaging* **2010**, *32*, 24–34. [CrossRef] [PubMed]
14. Chai, J.-W.; Chen, C.C.; Wu, Y.-Y.; Chen, H.-C.; Tsai, Y.-H.; Chen, H.-M.; Lan, T.-H.; Ouyang, Y.-C.; Lee, S.-K. Robust Volume Assessment of Brain Tissues for 3-Dimensional Fourier Transformation MRI via a Novel Multispectral Technique. *PLOS ONE* **2015**, *10*. [CrossRef] [PubMed]
15. Chiou, Y.-J.; Chen, C.C.-C.; Chen, S.-Y.; Chen, H.-M.; Chai, J.-W.; Ouyang, Y.-C.; Su, W.-C.; Yang, C.-W.; Lee, S.-K.; Chang, C.-I. Magnetic resonance brain tissue classification and volume calculation. *J. Chin. Inst. Eng.* **2015**, *38*, 1055–1066. [CrossRef]
16. Ren, H.; Chang, C.-I. A generalized orthogonal subspace projection approach to unsupervised multispectral image classification. *IEEE Trans. Geosci. Remote Sens.* **2000**, *38*, 2515–2528.
17. Harsanyi, J.C. *Detection and Classification of Subpixel Spectral Signatures in Hyperspectral Image Sequences*; Department of Electrical Engineering, University of Maryland, Baltimore County: Baltimore, MD, USA, 1993.
18. Farrand, W.H.; Harsanyi, J.C. Mapping the distribution of mine tailings in the Coeur d'Alene River Valley, Idaho, through the use of a constrained energy minimization technique. *Remote Sens. Environ.* **1997**, *59*, 64–76. [CrossRef]
19. Chang, C.-I. Target signature-constrained mixed pixel classification for hyperspectral imagery. *IEEE Trans. Geosci. Remote Sens.* **2002**, *40*, 1065–1081. [CrossRef]
20. Xue, B.; Wang, L.; Li, H.-C.; Chen, H.M.; Chang, C.-I. Lesion Detection in Magnetic Resonance Brain Images by Hyperspectral Imaging Algorithms. In *Proceedings Volume 9874, Remotely Sensed Data Compression, Communications, and Processing XII*; International Society for Optics and Photonics: Baltimore, MD, USA, 2016; Vol. 9874, p. 98740M.
21. Otsu, N. A threshold selection method from gray-level histgram. *IEEE Trans. Syst. Man Cybern.* **1979**, *9*, 62–66. [CrossRef]
22. Dice, L.R. Measures of the amount of ecologic association between species. *Ecology* **1945**, *26*, 297–302. [CrossRef]
23. Kang, X.; Li, S.; Benediktsson, J.A. Spectral-Spatial Hyperspectral Image Classification with Edge-Preserving Filtering. *IEEE Trans. Geosci. Remote Sens.* **2014**, *52*, 2666–2677. [CrossRef]
24. BrainWeb: Simulated Brain Database. Available online: http://www.bic.mni.mcgill.ca/brainweb/ (accessed on 15 November 2017).
25. LST: A Lesion Segmentation Tool for SPM. Available online: http://www.statistical-modelling.de/lst.html (accessed on 15 November 2017).
26. Schmidt, P.; Gaser, C.; Arsic, M.; Buck, D.; Förschler, A.; Berthele, A.; Hoshi, M.; Ilg, R.; Schmid, V.J.; Zimmer, C.; et al. An automated tool for detection of FLAIR-hyperintense white-matter lesions in Multiple Sclerosis. *NeuroImage* **2012**, *59*, 3774–3783. [CrossRef] [PubMed]
27. Fazekas, F.; Chawluk, J.B.; Alavi, A.; Hurtig, H.I.; Zimmerman, R.A. MR signal abnormalities at 1.5 T in Alzheimer's dementia and normal aging. *Am. J. Roentgenol.* **1987**, *149*, 351–356. [CrossRef] [PubMed]

remote sensing

MDPI

Article

Modelling Water Stress in a Shiraz Vineyard Using Hyperspectral Imaging and Machine Learning

Kyle Loggenberg [1], Albert Strever [2], Berno Greyling [2] and Nitesh Poona [1,*]

[1] Department of Geography and Environmental Studies, Stellenbosch University, Private Bag X1, Matieland 7602, South Africa; kyleloggenberg254@gmail.com

[2] Department of Viticulture and Oenology, Stellenbosch University, Private Bag X1, Matieland 7602, South Africa; aestr@sun.ac.za (A.S.); btg@sun.ac.za (B.G.)

* Correspondence: poona@sun.ac.za; Tel.: +27-21-808-9105

Received: 18 December 2017; Accepted: 25 January 2018; Published: 30 January 2018

Abstract: The detection of water stress in vineyards plays an integral role in the sustainability of high-quality grapes and prevention of devastating crop loses. Hyperspectral remote sensing technologies combined with machine learning provides a practical means for modelling vineyard water stress. In this study, we applied two ensemble learners, i.e., random forest (RF) and extreme gradient boosting (XGBoost), for discriminating stressed and non-stressed Shiraz vines using terrestrial hyperspectral imaging. Additionally, we evaluated the utility of a spectral subset of wavebands, derived using RF mean decrease accuracy (MDA) and XGBoost gain. Our results show that both ensemble learners can effectively analyse the hyperspectral data. When using all wavebands (p = 176), RF produced a test accuracy of 83.3% (KHAT (kappa analysis) = 0.67), and XGBoost a test accuracy of 80.0% (KHAT = 0.6). Using the subset of wavebands (p = 18) produced slight increases in accuracy ranging from 1.7% to 5.5% for both RF and XGBoost. We further investigated the effect of smoothing the spectral data using the Savitzky-Golay filter. The results indicated that the Savitzky-Golay filter reduced model accuracies (ranging from 0.7% to 3.3%). The results demonstrate the feasibility of terrestrial hyperspectral imagery and machine learning to create a semi-automated framework for vineyard water stress modelling.

Keywords: terrestrial hyperspectral imaging; vineyard; water stress; machine learning; tree-based ensemble

1. Introduction

Water stress in vineyards is a common phenomenon that occurs in the Western Cape of South Africa during the summer [1]. Water stress promotes stomatal closure [2], which inhibits photosynthesis and transpiration, leading to an increase in vine leaf temperature [3,4]. Reduced water availability impacts on vine health and productivity, and ultimately on grape quality [5]. Additionally, under increased climate change scenarios, greater drought periods may be experienced in the near future [6], with this strain on water resources further inhibiting the development of grapes [5]. There is consequently an imminent need for the real-time monitoring of water stress in vineyards.

Remote sensing provides a fast and cost-effective method for detecting vineyard water stress [4], and can thereby help alleviate devastating losses in crop production [7] and safeguard high-quality grape yields [8]. Several studies, for example [7,9], have modelled water stress in vineyards using spectral remote sensing techniques. Plant leaves reflect the majority of the near-infrared (NIR) spectrum, with the majority of the visible (VIS) spectrum, i.e., 400–680 nm, being absorbed by plant chlorophyll pigments [3]. Water stress changes the spectral signatures of plants due to decreased photosynthetic absorbance [3], resulting in decreased NIR reflectance [10]. This phenomenon is known as the "blue-shift", where the red-edge (680–730 nm) shifts toward the VIS end of the spectrum [11]. Therefore, the red-edge position has subsequently been used to detect water stress in plants [10].

The high spectral resolution of hyperspectral (spectroscopy) data allows for a more detailed analysis of plant properties [11], and provides a non-destructive approach for assessing vineyard water stress [12]. Consequently, the application of hyperspectral remote sensing techniques to model vineyard water stress is becoming common practice in precision viticulture [8]. For example, De Bei et al. [12] used near infrared (NIR) field spectroscopy to predict the water status of vines using leaf spectral signatures and in-field leaf water potential measurements. Similar studies were conducted by [13,14]. All three studies found that wavebands ranging between the 1000–2500 nm were ideal for detecting the water stress of vines. Alternatively, studies conducted by Zarco-Tejada et al. [7] and Pôças et al. [15] successfully demonstrated the viability of the VIS and red-edge, i.e., 400–730 nm, regions of the electromagnetic (EM) spectrum to predict water stress in vines.

Moreover, the advancement of remote sensing technology in recent years has prompted an increased availability of hyperspectral imaging (imaging spectroscopy) sensors. Hyperspectral imaging integrates spectroscopy with the advantages of digital imagery [16]. Each image provides contiguous, narrow-band (typically 10 nm) data, collected across the ultraviolet (UV), VIS, NIR, and shortwave infrared (SWIR) spectrum; typically 350–2500 nm, coupled with high spatial resolutions; typically 1 mm–2 m [16,17]. A major limitation to the application of hyperspectral data is the inherent "curse of dimensionality" [18], which gives rise to the Hughes effect [19] in a classification framework [20]. High dimensionality can result in reduced classification accuracies [21], as the number of wavebands (p) are often many times more than the number of training samples (n), i.e., $p > n$ [22]. However, using variable importance (VI) to create an optimised feature space, i.e., to create an optimal subset of input features, has been shown to be effective in reducing the effects of high dimensionality [23]. For example, Pedergnana et al. [20] exploited the RF mean decrease Gini (MDG) measure of VI to reduce the dimensionality of AVIRIS hyperspectral imagery. The study found that the subset selected based on RF VI produced an increase in accuracy of approximately 1.0%. Alternatively, Abdel-Rahman et al. [23] utilised the RF mean decrease accuracy (MDA) measure to rank the waveband importance of an AISA Eagle hyperspectral image dataset. The subset produced using MDA VI resulted in a 3.5% increase in accuracy. Contrary to this, Abdel-Rahman et al. [23] and Corcoran et al. [24] also utilised RF MDA values to create an optimal subset of features but observed a 4.0% decrease in accuracy. However, in both studies, it was concluded that RF VI could effectively be utilised to increase classification efficiency. Machine learning algorithms, such as Random Forest (RF) [25], have proven to be particularly adept at mitigating the Hughes effect (for example, see [22,26,27]). RF is an ensemble of weak decision trees used for classification and regression [22]. It uses bagging (i.e., bootstrap aggregation) and random variable selection to grow a multitude of unpruned trees from randomly selected training samples [25]. RF classification has recently gained significant recognition for its applications in precision viticulture. For example, Sandika et al. [28] used RF and digital terrestrial imagery to classify Anthracnose, Powdery Mildew, and Downy Mildew diseases within vine leaves. The study found that RF produced the highest accuracy with 82.9%, outperforming Probabilistic Neural Network (PNN), Back Propagation Neural Network (BPNN), and Support Vector Machine (SVM) models. Similar results were found by Knauer et al. [29] using RF and terrestrial hyperspectral imaging. RF produced an overall accuracy of 87% for modelling Powdery Mildew on grapes. Additionally, Knauer et al. [29] found that dimensionality reduction led to an increase in classification accuracy.

More recently, another tree-based classifier known as Extreme Gradient Boosting (XGBoost) [30], has shown considerable promise in various applications (for example, see [31–33]). XGBoost is an optimised implementation of gradient boosting [34], designed to be fast, scalable, and highly efficient [35]. Gradient boosting (or boosted trees) combines multiple pruned trees of low accuracies, or weak learners, to create a more accurate model [36]. The difference between RF and XGBoost is the way the tree ensemble is constructed. RF grows trees that are independent of one another [25], whereas XGBoost grows trees that are dependent on the feedback information provided by the previously grown tree [30]. Essentially, each tree in an XGBoost ensemble learns from previous trees and tries to reduce the error produced in subsequent iterations.

Mohite et al. [37] is the only known study to have employed XGBoost classification in precision viticulture. The study used hyperspectral data to detect pesticide residue on grapes. Four classifiers were compared, i.e., XGBoost, RF, SVM, and artificial neural network (ANN). Additionally, the study also investigated the utility of LASSO and Elastic Net feature selection. Results indicated that RF produced the most accurate classification models when using both the LASSO and Elastic Net selected wavebands.

A review of the literature indicated that no study to date has investigated the use of terrestrial hyperspectral imaging to model vineyard water stress. Furthermore, no study has utilised RF or XGBoost classification to detect leaf level water stress in the precision viticulture domain. The aim of the present study was to develop a remote sensing-machine learning framework to model water stress in a Shiraz vineyard. The specific objectives of the study are to evaluate the utility of terrestrial hyperspectral imaging to discriminate stressed and non-stressed Shiraz vines, and investigate the efficacy of the RF and XGBoost algorithms for modelling vineyard water stress.

2. Materials and Methods

2.1. Study Site

The study was conducted at the Welgevallen experimental farm in Stellenbosch (33°56′38.5″S, 18°52′06.8″E), situated in the Western Cape Province of South Africa (Figure 1). Stellenbosch has a Mediterranean climate characterised by dry summers and mild winters, with a mean annual temperature of 16.4 °C [38]. Stellenbosch receives low to moderate rainfall, mainly during the winter months (June, July, and August), with an annual average of 802 mm [38], making water scarcity a real threat to irrigated vineyards. Soil deposits in the region comprise rich potassium minerals that are favourable for vineyard growth [38]. The Welgevallen experimental farm comprises well-established grape cultivars, including Shiraz and Pinotage; Pinotage being a red cultivar unique to South Africa. Welgevallen is used by Stellenbosch University for research and training, and additionally produces high-quality grapes for commercial use.

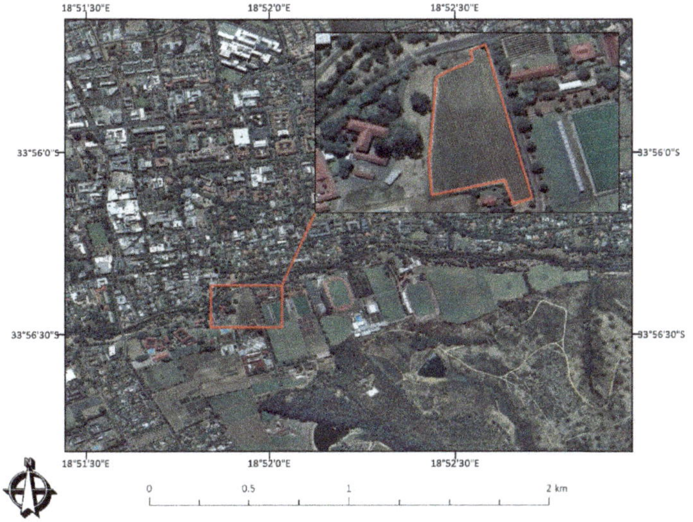

Figure 1. Location of the Welgevallen Shiraz vineyard plot used in this study. Background image provided by National Geo-Spatial Information (NGI) (2012).

2.2. Data Acquisition and Pre-Processing

To confirm the water stress status of vines, in-field stem water potential (SWP) measurements were captured using a customised pressure chamber (Figure 2) as used by [39,40]. Based on the experiments by [39,41], vines with SWP values ranging from −1.0 MPa to −1.8 MPa were classified as water-stressed, whereas vines with SWP values ≥ −0.7 MPa were classified as non-stressed. Imaging spectrometer data was subsequently acquired for a water-stressed and non-stressed Shiraz vine. Images were captured between 10:00 and 12:00, on 24 February 2017, to ensure that the side of the vine canopy being captured was fully sunlit.

Figure 2. Customised pressure chamber used to measure Stem Water Potential.

Images were captured using the SIMERA HX MkII hyperspectral sensor (SIMERA Technology Group, Somerset West South Africa). The sensor is a line scanner that captures 340 spectral wavebands across the VIS and NIR, i.e., 450–1000 nm, with a sensor bandwidth ranging from 0.9 nm to 5 nm. The sensor was mounted on a tripod (Figure 3A) to facilitate the collection of terrestrial imagery from a side-on view of the vine canopy. The sensor-tripod assembly was placed at a constant distance of one metre from the vine canopy to ensure that the full canopy of a single vine (approximately 1.4 m W × 1 m H) was captured per image (Figure 3B).

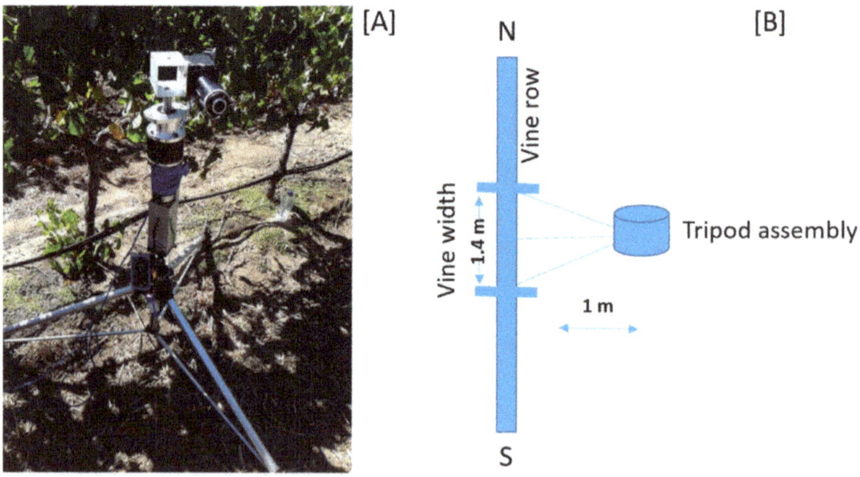

Figure 3. The hyperspectral sensor tripod assembly (**A**); and in-field setup when collecting terrestrial imagery of the vine canopy (**B**).

Due to sensor sensitivity and a deteriorating silicon chip, not all the wavebands could be utilised. Spectral subsets were, therefore, created per image. The spectral subsets consisted of 176 wavebands with a spectral range of 473–708 nm. Thereafter, raw image DN's were converted to reflectance using the empirical line correction algorithm [42]. Empirical line correction uses known field (or reference) reflectance spectra and linear regression to equate digital number (DN) values to surface reflectance by estimating correction coefficients for each waveband [42]. Following [42], a white reference panel, positioned in the vine canopy prior to image capture, was used for image correction. Image pre-processing was performed in the Environment for Visualising Images (ENVI) version 5.3.1 software. Using a 2 × 2 pixel region of interest (ROI), a total of 60 leaf spectra were extracted from each image—30 samples per class (stressed and non-stressed)—and used as the input for classification.

2.3. Spectral Smoothing

In-field spectral measurements are often subjected to noise due to variable sun illumination [43]. Therefore, it is recommended that spectral smoothing be performed in order to produce a spectral signal that represents the original spectra without the interference of noise [44]. The Savitzky-Golay filter [45] is a common smoothing technique used in hyperspectral remote sensing [43,46,47]. Savitzky-Golay is based on least-squares approximation, which determines smoothing coefficients by applying a polynomial equation of a given degree and cluster size [45]. The filter is ideal for spectroscopic data as it minimises signal noise whilst preserving the originality and shape of the input spectra. A second order polynomial filter with a filter size of 15 was applied to the spectral samples prior to classification, following the recommendations of [47]. The Savitzky-Golay filter was applied using the 'signal' package [48] in the R statistical software environment [49]. Classification models were produced for both the unsmoothed and smoothed datasets.

2.4. Classification

2.4.1. Random Forest (RF)

The RF ensemble uses a bootstrap sample, i.e., 2/3 of the original dataset (referred to as the "in-bag" sample), to train decision trees. The remaining 1/3 of the data is used to compute an internal measure of accuracy (referred to as the "out-of-bag" or OOB error) [25]. To produce the forest of decision trees, two parameters need to be set: The number of unpruned trees to grow, known as ntree; and the number of predictor variables (i.e., wavebands) selected, known as mtry [25]. Mtry variables are tested at each node to specify the best split when growing trees. These randomly selected variables produce low correlated trees that prevent over-fitting. In a classification framework, the final classification results are determined by averaging the results of all the decision trees produced. For a detailed account of RF, see [25,50]. RF was implemented using the 'randomForest' package [51] in the R statistical software environment [49]. The default values for ntree (ntree = 500) and mtry (mtry = $\sqrt{p}$) were used following [50,52].

2.4.2. Extreme Gradient Boosting (XGBoost)

XGBoost, like gradient boosting, is based on three essential elements; (i) a loss function that needs to be optimised; (ii) a multitude of weak decision trees that are used for classification; and (iii) an additive model that combines weak decision trees to produce a more accurate classification model [31]. XGBoost simultaneously optimises the loss function while constructing the additive model [30,31]. The loss function accounts for the errors in classification that were introduced by the weak decision trees [31]. For a detailed account of XGBoost, see [30]. XGBoost was implemented using the 'xgboost' package [53] in the R statistical software environment [49]. XGBoost requires the optimisation of several key parameters (Table 1). However, to facilitate a fair comparison of RF and XGBoost, the default values for all parameters were used to construct the XGBoost models, with nrounds set to 500.

Furthermore, to ensure a more robust model and prevent overfitting, a 10-fold cross validation was performed for both RF and XGBoost.

Table 1. Key parameters used for XGBoost classification.

Parameter	Description	Default Value
max_depth	controls the maximum depth of each tree (used to control over-fitting)	6
subsample	specifies the fraction of observations to be randomly sampled at each tree (adds randomness)	1
eta	the learning rate	0.3
nrounds	the number of trees to be produced (similar to ntree)	100–1000
gamma	controls the minimum loss reduction required to make a node split (used to control over-fitting)	0
min_child_weight	Specifies the minimum sum of instance weight of all the observations required in a child (used to control over-fitting)	1
colsample_bytree	Specifies the number of features to consider when searching for the best node split (adds randomness)	1

2.5. Dimensionality Reduction

Both RF and XGBoost provide an internal measure of VI. RF provides two measures of VI, namely mean decrease Gini (MDG) and mean decrease accuracy (MDA) [25]. MDG quantifies VI by measuring the sum of all decreases in the Gini index, produced by a particular variable. MDA measures the changes in OOB error, which results from comparing the OOB error of the original dataset to that of a dataset created through random permutations of variable values. In this study, MDA was utilised to compute VI following the recommendations of [22,54,55]. The MDA VI for a waveband X_j is defined by [56]:

$$VI(X_j) = \frac{1}{\text{ntree}} \sum_t (errOOB_{tj} - errOOB_t) \tag{1}$$

where $errOOB_t$ is the misclassification rate of tree t on the OOB_t bootstrap sample not used to construct tree t, and $errOOB_{tj}$ is the error of predictor t on the permuted OOB_{tj} sample.

XGBoost ranks VI based on Gain [30]. Gain measures the degree of improved accuracy brought on by the addition of a given waveband. VI is calculated for each waveband, used for node splitting at a given tree, and then averaged across all trees to produce the final VI per waveband [30]. Similar to [23,24], the top 10% (p = 18) of the ranked waveband importance as determined by RF and XGBoost was used to create a subset of important wavebands. RF and XGBoost models were produced for both the original dataset and the subset of 18 wavebands.

2.6. Accuracy Assessment

To provide an independent estimate of model accuracy, an independent test set was used to evaluate all RF and XGBoost models. Therefore, a second dataset of spectral samples (n = 60) was collected for both stressed (n = 30) and non-stressed (n = 30) vines. Both algorithms were trained using the first dataset of 60 samples and tested using the second dataset. Overall classification accuracies were computed using a confusion matrix [57]. Additionally, Kappa analysis was used to evaluate model performance. The KHAT statistic [58] provides a measure of the difference between the actual and the chance agreement in the confusion matrix:

$$\hat{K} = \frac{p_a - p_c}{1 - p_c} \tag{2}$$

where p_a describes the actual agreement and p_c describes the chance agreement. Following [20,23,59], the McNemar's test was employed to determine whether the differences in accuracies yielded by RF and XGBoost were statistically significant. Abdel-Rahman et al. [23] stated that the McNemar's test can be expressed as the following chi-squared formula:

$$v^2 = \frac{(f_{xgb} - f_{rf})^2}{f_{xgb} + f_{rf}} \tag{3}$$

where f_{xgb} denotes the number of samples misclassified by RF but correctly classified by XGBoost, and f_{rf} denotes the number of samples misclassified by XGBoost but correctly classified by RF. A v^2 value of greater than 3.84, at a 0.05 level of significance, indicates that the results of the two classifiers are significantly different [23,59].

3. Results

3.1. Spectral Smoothing Using the Savitzky-Golay Filter

Figure 4 shows the results of smoothing the spectral data using the Savitzky-Golay filter. It is evident that the Savitzky-Golay filter produced smoothed spectra without changing the shape of the original spectra. Additionally, the filter successfully preserved the original reflectance values, with the mean difference in reflectance values being less the than 0.3% with a standard deviation of 0.003 across all wavebands. All spectra (n = 120) were subsequently smoothed, and the smoothed spectra used as the input to classification.

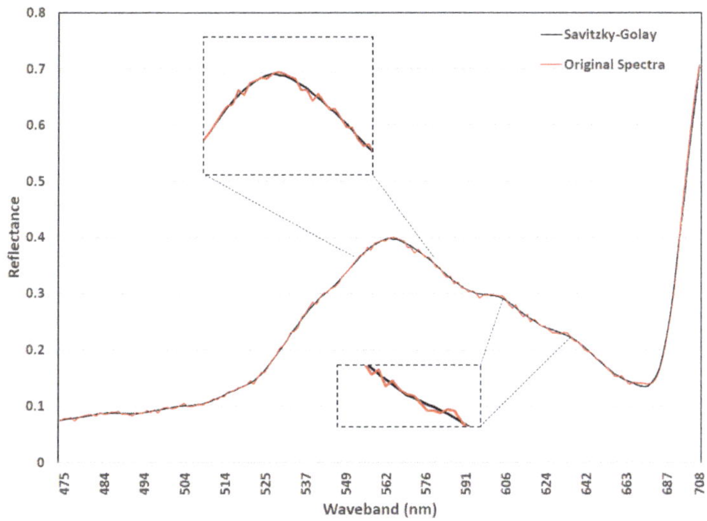

Figure 4. Spectra comparison before (red) and after (black) applying Savitzky-Golay filter.

3.2. Important Waveband Selection

The top 10% (p = 18) of importance wavebands as determined by RF MDA and XGBoost gain are shown in Figure 5A,B, respectively. The results in Table 2 show that RF selected wavebands across blue and green (473.92–585.12 nm) regions of the EM spectrum. In comparison, XGBoost selected wavebands across the VIS (473.92–646.04 nm) and red-edge (686.69–708.32 nm) regions. It is evident from Figure 5 that the location of the wavebands selected by RF and XGBoost are significantly different. We attribute the difference in waveband location to the difference in VI measures used for RF and XGBoost. Nevertheless, as illustrated in Figure 5C, there were common wavebands selected by both RF and XGBoost. The overlapping wavebands ($p = 6$) were located across blue and green (473.92–585.12 nm) regions. Consequently, those wavebands may be the most important for discriminating between stressed and non-stressed Shiraz vines.

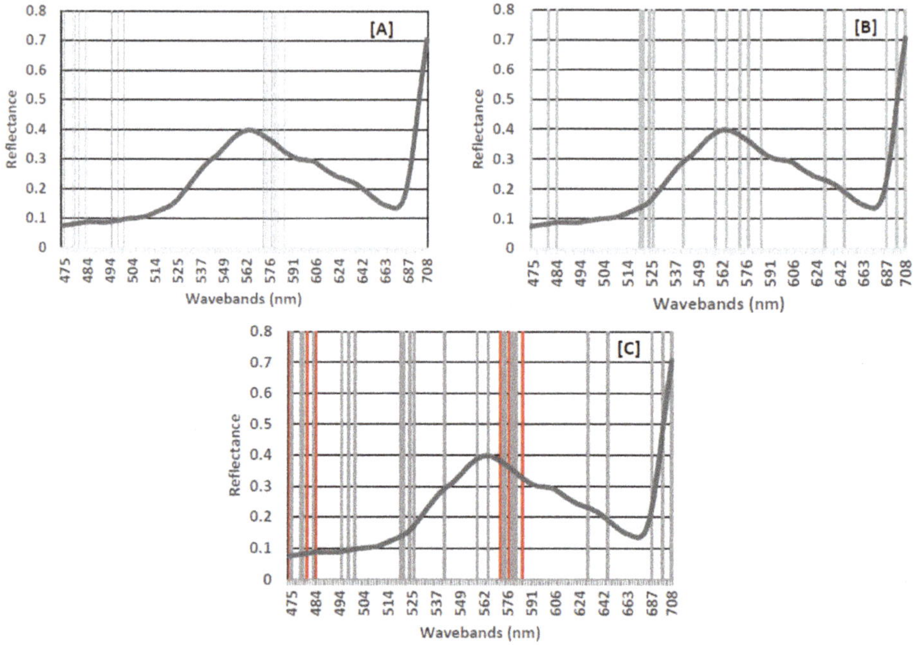

Figure 5. The importance wavebands as determined by RF (**A**); XGBoost (**B**); and overlapping (**C**). The grey bars represent the important wavebands selected by RF and XGBoost, respectively. The red bars indicate the overlapping wavebands. The mean spectral signature of a sample is shown as a reference.

Table 2. Location of the RF and XGBoost selected important wavebands in the EM spectrum.

	p	VIS (473 nm–680 nm)	p	Red-Edge (680 nm–708 nm)
RF	12	474.74, 478.09, 478.94, 483.2, 494.64, 497.36, 500.11, 573.31, 574.59, 578.48, 579.79, 581.11	0	-
XGBoost	9	520.31, 521.32, 524.36, 526.42, 541.34, 558.52, 564.56,630.23, 646.04	3	686.69, 698.39, 708.32
Overlap	6	473.92, 480.63, 484.06 572.04, 577.17, 585.12	0	-

3.3. Classification Using Random Forest and Extreme Gradient Boosting

The classification results for RF and XGBoost are shown in Table 3 Training accuracies for all models were above 80.0%, with test accuracies ranging from 77.6% to 83.3% (with KHAT values ranging from 0.60 to 0.87). Overall, the results indicate that RF outperformed XGBoost, producing the highest accuracies for all the classification models.

Table 3. Classification accuracies of both the RF and XGBoost models constructed using all the wavebands and the subset of important wavebands.

		All Wavebands (p = 176)				Important Wavebands (p = 18)			
		Train		Test		Train		Test	
		Accuracy (%)	Kappa	Accuracy (%)	Kappa	Accuracy (%)	Kappa	Accuracy (%)	Kappa
XGBoost	Unsmoothed	85.0	0.70	78.3	0.57	90.0	0.80	80.0	0.60
	Smoothed	83.3	0.67	77.6	0.53	86.7	0.73	78.3	0.57
RF	Unsmoothed	90.0	0.80	83.3	0.67	93.3	0.87	83.3	0.67
	Smoothed	90.0	0.80	81.7	0.63	91.7	0.83	81.7	0.63

Using all wavebands (p = 176), RF yielded a training accuracy of 90.0% (KHAT = 0.80) and a test accuracy of 83.3% (KHAT = 0.67). In comparison, XGBoost produced significantly lower accuracies, i.e., a training accuracy of 85.0% (KHAT = 0.7) and a test accuracy of 78.3% (KHAT = 0.57). These results indicate that the XGBoost ensemble resulted in reduced accuracies (approximately −5.0%) when using all wavebands to classify stressed and non-stressed Shiraz leaves.

Using the subset of important wavebands (p = 18) resulted in an overall improvement in classification accuracies for the RF and XGBoost. Training accuracy for RF increased by 3.3% to 93.3% (KHAT = 0.87). However, the test accuracy remained unchanged. Although XGBoost produced less accurate results, it did experience a greater increase in accuracy (5.0%), producing a training accuracy of 90.0% and a KHAT value of 0.8. The greater increase in accuracy may be attributed to the red-edge wavebands that were only present in the XGBoost subset. Moreover, the XGBoost subset also produced a slight increase (1.7%) in test accuracy (80.0%, KHAT = 0.6). We attribute the superior performance of the RF algorithm to its use of bootstrap sampling [25], which introduces model stability, and its robustness to noise [50].

Classification using the Savitzky-Golay smoothed spectra resulted in reduced accuracies overall. The decrease in accuracy ranged from 0.7–3.3% for all models. Furthermore, according to the McNemar's test results, the difference in classifier performance was not statistically significant. For all the classification models, the chi-squared values were less than 3.84 with v^2 values ranging from 0.14 to 1.29.

4. Discussion

Ensemble classifiers, like RF and XGBoost, have been widely used to address the classification challenges inherent in high dimensional data [52]. The present study evaluated the use of terrestrial hyperspectral imaging to model vineyard water stress. More specifically, we tested the utility of two tree-based ensemble classifiers, namely RF and XGBoost, to model water stress in a Shiraz vineyard. The experimental results are discussed in further detail in the following sections.

4.1. Efficacy of the Savitzky-Golay Filter

The Savitzky-Golay filter has become a popular algorithm for smoothing spectroscopic data (for example, see [43,47,60]). In this study, the filter proved adept at smoothing the hyperspectral signature without significantly altering the originality of the input data. However, the results of this study showed that the filter negatively impacted the classification accuracy, producing reduced accuracies for RF (−1.6%) and XGBoost (−3.3%). The decrease in classification accuracy may be attributed to the specific parameter values used to implement the filter. The study only meant to test the functionality of the Savitzky-Golay filter. Therefore, the filter was implemented using the hyperparameter values as recommended by [47]. Consequently, the recommended values may not be optimal for the specific dataset used in this study.

Carvalho et al. [61] utilised the Savitzky-Golay filter to smooth magnetic flux leakage (MFL) signals. Similar to our study, the authors found that using the smoothed data with an ANN classifier, resulted in reduced classification accuracies. It is, therefore, evident that careful consideration has to be taken when applying the Savitzky-Golay filter.

4.2. Classification Using All Wavebands

Both tree-based ensemble classifiers tested in our study successfully demonstrated their efficiency for analysing hyperspectral data. However, our analysis found the RF bagging ensemble to outperform the boosting-based XGBoost ensemble when using all wavebands ($p = 176$).

Published comparisons between RF and boosting classifiers, similar to XGBoost, have reported mixed results. For example, Miao et al. [62] found that RF (93.5%) and AdaBoost (95.3%) produced similar overall accuracies when classifying ecological zones using multi-temporal and multi-sensor data. This is contrary to [62,63], which reported that RF outperformed boosting ensemble classifiers when classifying RADARSAT-1 imagery. Moreover, when directly comparing RF and XGBoost, within the context of spectroscopic classification, our findings contradict the results reported by [31]. Their study reported that XGBoost (96.0%) yielded significantly better results than RF (87.0%) when classifying supernovae. However, it should be noted that their study optimised RF and XGBoost parameters. More specifically, within viticulture, the results of our study compare favourably to those reported by [37]. The authors found that RF (87.8%) produced an improved accuracy compared to XGBoost (81.6%) when using hyperspectral data in combination with feature selection. A review by [50] concluded that RF generally achieves greater accuracies compared with boosting methods when used for the classification of high dimensional data such as hyperspectral imagery.

When comparing the utility of both algorithms, a key advantage shared between them is that RF and XGBoost effectively prevent overfitting [25,30]. However, given that RF grows trees independently (i.e., parallel to one another), whereas XGBoost grows trees sequentially, it is less complex and, therefore, less computationally intensive. Furthermore, RF requires the optimisation of only two parameters [25], whereas XGBoost has various parameters that could be optimised for a given dataset [30].

4.3. Classification Using Subset of Important Wavebands

Dimensionality reduction of hyperspectral data using machine learning has been extensively researched (for example see [20,22,23]). The results of our study indicate the VI ranking provided by RF and XGBoost can successfully be used to select a subset of wavebands for classification. This was evident from the increased accuracies obtained for both RF and XGBoost.

Our results compare favourably to those reported by [20,23], who demonstrated the feasibility of VI to reduce the high dimensionality of hyperspectral data and improve the classification accuracy. We, therefore, attribute the improved classification performance to the subset of most important wavebands. Although the subset of important wavebands did not result in massive accuracy gains (accuracy increase of RF ranged from 1.7% to 3.3% and from 0.7% to 3.3% for XGBoost), it did improve classification accuracy using only 10% of the original data. The majority of important wavebands, for RF ($p = 9$) and XGBoost ($p = 10$), were located in the green region of the EM spectrum (Table 2). The selected wavebands correspond to similar wavebands reported by [7,15]. The green region (i.e., between 500–600 nm) is highly sensitive to plant chlorophyll absorption [15]. Consequently, water stress in plants is closely related to lowered chlorophyll leaf concentrations [15], which can present a possible explanation for the selection of these wavebands.

Moreover, Shimada et al. [10] reported the use of the blue (490 nm) and red wavebands (620 nm) as indicators of plant water stress, and these wavebands correspond to similar wavebands present in the XGBoost subset (484.06 nm and 630.23 nm). In this study, only three red-edge wavebands (Table 2) were selected by XGBoost with none selected by RF. These results contradict those reported by [4], which found that wavebands in the red-edge region (695–730 nm) were ideal for early water stress detection in vineyards. However, given the overlapping wavebands that occur in the blue and green regions, and the results of our study, we can conclude that the red-edge wavebands may not be important for discriminating between stressed and non-stressed Shiraz vines. The results of this study subsequently demonstrate the feasibility of VIS wavebands to model water stress in a Shiraz vineyard.

Various aspects of the current research lend themselves to be operationalised within precision viticulture. For instance, the developed remote sensing-machine learning framework can be readily applied to model vegetative water stress. Furthermore, the identification of important wavebands can potentially lead to the construction of custom multispectral sensors that are less expensive and application specific.

5. Conclusions

This study presents a novel remote sensing-machine learning framework for modelling water stress in a Shiraz vineyard using terrestrial hyperspectral imaging. Based on the results of our study, we can draw the following conclusions:

1. Both RF and XGBoost may be utilised to model water stress in a Shiraz vineyard.
2. Wavebands in the VIS region of the EM spectrum may be used to model water stress in a Shiraz vineyard.
3. It is imperative that future studies carefully consider the impact of applying the Savitzky-Golay filter for smoothing spectral data.
4. The developed framework requires further investigation to evaluate its robustness and operational capabilities.

Given the results obtained in the present study, we recommend the employment of RF, rather than XGBoost, for the classification of hyperspectral data to discriminate stressed from non-stressed Shiraz vines.

Acknowledgments: Winetech funded the research. The authors sincerely thank the SIMERA Technology Group for providing the hyperspectral sensor.

Author Contributions: Kyle Loggenberg, Nitesh Poona, and Albert Strever conceptualised the research; Kyle Loggenberg conducted the field work, carried out the main analysis, and wrote the paper; Nitesh Poona and Berno Greyling assisted in data analysis and field work; Nitesh Poona contributed to the interpretation of the results. Nitesh Poona, Albert Strever, and Berno Greyling contributed to the editing of the manuscript.

Conflicts of Interest: The authors declare no conflict of interest.

References

1. Costa, J.M.; Vaz, M.; Escalona, J.; Egipto, R.; Lopes, C.; Medrano, H.; Chaves, M.M. Modern viticulture in southern Europe: Vulnerabilities and strategies for adaptation to water scarcity. *Agric. Water Manag.* **2016**, *164*, 5–18. [CrossRef]
2. Zarco-Tejada, P.J.; González-Dugo, V.; Berni, J.A.J. Fluorescence, temperature and narrow-band indices acquired from a UAV platform for water stress detection using a micro-hyperspectral imager and a thermal camera. *Remote Sens. Environ.* **2012**, *117*, 322–337. [CrossRef]
3. Kim, Y.; Glenn, D.M.; Park, J.; Ngugi, H.K.; Lehman, B.L. Hyperspectral image analysis for water stress detection of apple trees. *Comput. Electron. Agric.* **2011**, *77*, 155–160. [CrossRef]
4. Maimaitiyiming, M.; Ghulam, A.; Bozzolo, A.; Wilkins, J.L.; Kwasniewski, M.T. Early Detection of Plant Physiological Responses to Different Levels of Water Stress Using Reflectance Spectroscopy. *Remote Sens.* **2017**, *9*, 745. [CrossRef]
5. Bota, J.; Tomás, M.; Flexas, J.; Medrano, H.; Escalona, J.M. Differences among grapevine cultivars in their stomatal behavior and water use efficiency under progressive water stress. *Agric. Water Manag.* **2016**, *164*, 91–99. [CrossRef]
6. Chirouze, J.; Boulet, G.; Jarlan, L.; Fieuzal, R.; Rodriguez, J.C.; Ezzahar, J.; Er-Raki, S.; Bigeard, G.; Merlin, O.; Garatuza-Payan, J.; et al. Intercomparison of four remote-sensing-based energy balance methods to retrieve surface evapotranspiration and water stress of irrigated fields in semi-arid climate. *Hydrol. Earth Syst. Sci.* **2014**, *18*, 1165–1188. [CrossRef]

7. Zarco-Tejada, P.J.; González-Dugo, V.; Williams, L.E.; Suárez, L.; Berni, J.A.J.; Goldhamer, D.; Fereres, E. A PRI-based water stress index combining structural and chlorophyll effects: Assessment using diurnal narrow-band airborne imagery and the CWSI thermal index. *Remote Sens. Environ.* **2013**, *138*, 38–50. [CrossRef]
8. González-Fernández, A.B.; Rodríguez-Pérez, J.R.; Marcelo, V.; Valenciano, J.B. Using field spectrometry and a plant probe accessory to determine leaf water content in commercial vineyards. *Agric. Water Manag.* **2015**, *156*, 43–50. [CrossRef]
9. Baluja, J.; Diago, M.P.; Balda, P.; Zorer, R.; Meggio, F.; Morales, F.; Tardaguila, J. Assessment of vineyard water status variability by thermal and multispectral imagery using an unmanned aerial vehicle (UAV). *Irrig. Sci.* **2012**, *30*, 511–522. [CrossRef]
10. Shimada, S.; Funatsuka, E.; Ooda, M.; Takyu, M.; Fujikawa, T.; Toyoda, H. Developing the Monitoring Method for Plant Water Stress Using Spectral Reflectance Measurement. *J. Arid Land Stud.* **2012**, *22*, 251–254.
11. Govender, M.; Dye, P.; Witokwski, E.; Ahmed, F. Review of commonly used remote sensing and ground based technologies to measure plant water stress. *Water SA* **2009**, *35*, 741–752. [CrossRef]
12. De Bei, R.; Cozzolino, D.; Sullivan, W.; Cynkar, W.; Fuentes, S.; Dambergs, R.; Pech, J.; Tyerman, S.D. Non-destructive measurement of grapevine water potential using near infrared spectroscopy. *Aust. J. Grape Wine Res.* **2011**, *17*, 62–71. [CrossRef]
13. Diago, M.P.; Bellincontro, A.; Scheidweiler, M.; Tardaguila, J.; Tittmann, S.; Stoll, M. Future opportunities of proximal near infrared spectroscopy approaches to determine the variability of vineyard water status. *Aust. J. Grape Wine Res.* **2017**, *23*, 409–414. [CrossRef]
14. Beghi, R.; Giovenzana, V.; Guidetti, R. Better water use efficiency in vineyard by using visible and near infrared spectroscopy for grapevine water status monitoring. *Chem. Eng. Trans.* **2017**, *58*, 691–696. [CrossRef]
15. Pôças, I.; Rodrigues, A.; Gonçalves, S.; Costa, P.M.; Gonçalves, I.; Pereira, L.S.; Cunha, M. Predicting grapevine water status based on hyperspectral reflectance vegetation indices. *Remote Sens.* **2015**, *7*, 16460–16479. [CrossRef]
16. Carreiro Soares, S.F.; Medeiros, E.P.; Pasquini, C.; de Lelis Morello, C.; Harrop Galvão, R.K.; Ugulino Araújo, M.C. Classification of individual cotton seeds with respect to variety using near-infrared hyperspectral imaging. *Anal. Methods* **2016**, *8*, 8498–8505. [CrossRef]
17. Mulla, D.J. Twenty five years of remote sensing in precision agriculture: Key advances and remaining knowledge gaps. *Biosyst. Eng.* **2013**, *114*, 358–371. [CrossRef]
18. Poona, N.K.; Van Niekerk, A.; Nadel, R.L.; Ismail, R. Random Forest (RF) Wrappers for Waveband Selection and Classification of Hyperspectral Data. *Appl. Spectrosc.* **2016**, *70*, 322–333. [CrossRef] [PubMed]
19. Hughes, G.F. On the mean accuracy of statistical pattern recognizers. *IEEE Trans. Inf. Theory* **1968**, *14*, 55–63. [CrossRef]
20. Pedergnana, M.; Marpu, P.R.; Mura, M.D. A Novel Technique for Optimal Feature Selection in Attribute Profiles Based on Genetic Algorithms. *IEEE Trans. Geosci. Remote Sens.* **2013**, *51*, 3514–3528. [CrossRef]
21. Tong, Q.; Xue, Y.; Zhang, L. Progress in hyperspectral remote sensing science and technology in China over the past three decades. *IEEE J. Sel. Top. Appl. Earth Obs. Remote Sens.* **2014**, *7*, 70–91. [CrossRef]
22. Poona, N.K.; Ismail, R. Using Boruta-selected spectroscopic wavebands for the asymptomatic detection of fusarium circinatum stress. *IEEE J. Sel. Top. Appl. Earth Obs. Remote Sens.* **2014**, *7*, 3764–3772. [CrossRef]
23. Abdel-Rahman, E.M.; Mutanga, O.; Adam, E.; Ismail, R. Detecting Sirex noctilio grey-attacked and lightning-struck pine trees using airborne hyperspectral data, random forest and support vector machines classifiers. *ISPRS J. Photogramm. Remote Sens.* **2014**, *88*, 48–59. [CrossRef]
24. Corcoran, J.M.; Knight, J.F.; Gallant, A.L. Influence of multi-source and multi-temporal remotely sensed and ancillary data on the accuracy of random forest classification of wetlands in northern Minnesota. *Remote Sens.* **2013**, *5*, 3212–3238. [CrossRef]
25. Breiman, L. Random forests. *Mach. Learn.* **2001**, *45*, 5–32. [CrossRef]
26. Abdel-Rahman, E.M.; Makori, D.M.; Landmann, T.; Piiroinen, R.; Gasim, S.; Pellikka, P.; Raina, S.K. The utility of AISA eagle hyperspectral data and random forest classifier for flower mapping. *Remote Sens.* **2015**, *7*, 13298–13318. [CrossRef]
27. Adam, E.; Deng, H.; Odindi, J.; Abdel-Rahman, E.M.; Mutanga, O. Detecting the Early Stage of Phaeosphaeria Leaf Spot Infestations in Maize Crop Using In Situ Hyperspectral Data and Guided Regularized Random Forest Algorithm. *J. Spectrosc.* **2017**, *2017*. [CrossRef]

28. Sandika, B.; Avil, S.; Sanat, S.; Srinivasu, P. Random forest based classification of diseases in grapes from images captured in uncontrolled environments. In Proceedings of the IEEE 13th International Conference, Signal Processing Proceedings, Chengdu, China, 6–10 November 2016; pp. 1775–1780.
29. Knauer, U.; Matros, A.; Petrovic, T.; Zanker, T.; Scott, E.S.; Seiffert, U. Improved classification accuracy of powdery mildew infection levels of wine grapes by spatial-spectral analysis of hyperspectral images. *Plant Methods* **2017**, *13*. [CrossRef] [PubMed]
30. Chen, T.; Guestrin, C. XGBoost: A Scalable Tree Boosting System. In Proceedings of the 22nd ACM Sigkdd International Conference on Knowledge Discovery and Data Mining, San Francisco, CA, USA, 13–17 August 2016; pp. 785–794.
31. Möller, A.; Ruhlmann-Kleider, V.; Leloup, C.; Neveu, J.; Palanque-Delabrouille, N.; Rich, J.; Carlberg, R.; Lidman, C.; Pritchet, C. Photometric classification of type Ia supernovae in the SuperNova Legacy Survey with supervised learning. *J. Cosmol. Astropart. Phys.* **2016**, *12*. [CrossRef]
32. Torlay, L.; Perrone-Bertolotti, M.; Thomas, E.; Baciu, M. Machine learning–XGBoost analysis of language networks to classify patients with epilepsy. *Brain Inform.* **2017**, *4*, 159–169. [CrossRef] [PubMed]
33. Fitriah, N.; Wijaya, S.K.; Fanany, M.I.; Badri, C.; Rezal, M. EEG channels reduction using PCA to increase XGBoost's accuracy for stroke detection. *AIP Conf. Proc.* **2017**, *1862*, 30128. [CrossRef]
34. Friedman, J. Greedy function approximation: A gradient boosting machine. *Ann. Stat.* **2001**, *29*, 1189–1232. [CrossRef]
35. Ren, X.; Guo, H.; Li, S.; Wang, S. A Novel Image Classification Method with CNN-XGBoost Model. In *International Workshop on Digital Watermarking*; Springer: Cham, Switzerland, 2017; pp. 378–390.
36. Friedman, J.H. Stochastic gradient boosting. *Comput. Stat. Data Anal.* **2002**, *38*, 367–378. [CrossRef]
37. Mohite, J.; Karale, Y.; Pappula, S.; Shabeer, T.P.; Sawant, S.D.; Hingmire, S. Detection of pesticide (Cyantraniliprole) residue on grapes using hyperspectral sensing. In *Sensing for Agriculture and Food Quality and Safety IX, Proceedings of the SPIE Commercial+ Scientific Sensing and Imaging Conference, Anaheim, CA, USA, 1 May 2017*; Kim, M.S., Chao, K.L., Chin, B.A., Cho, B.K., Eds.; International Society for Optics and Photonics: Bellingham, WA, USA, 2017.
38. Conradie, W.J.; Carey, V.A.; Bonnardot, V.; Saayman, D.; Van Schoor, L.H. Effect of Different Environmental Factors on the Performance of Sauvignon blanc Grapevines in the Stellenbosch/Durbanville Districts of South Africa I. Geology, Soil, Climate, Phenology and Grape Composition. *S. Afr. J. Enol. Vitic.* **2002**, *23*, 78–91.
39. Deloire, A.; Heyms, D. The leaf water potentials: Principles, method and thresholds. *Wynboer* **2011**, *265*, 119–121.
40. Choné, X.; Van Leeuwen, C.; Dubourdieu, D.; Gaudillère, J.P. Stem water potential is a sensitive indicator of grapevine water status. *Ann. Bot.* **2001**, *87*, 477–483. [CrossRef]
41. Myburgh, P.; Cornelissen, M.; Southey, T. Interpretation of Stem Water Potential Measurements. WineLand. 2016, pp. 78–80. Available online: http://www.wineland.co.za/interpretation-of-stem-water-potential-measurements/ (accessed on 26 January 2018).
42. Aasen, H.; Burkart, A.; Bolten, A.; Bareth, G. Generating 3D hyperspectral information with lightweight UAV snapshot cameras for vegetation monitoring: From camera calibration to quality assurance. *ISPRS J. Photogramm. Remote Sens.* **2015**, *108*, 245–259. [CrossRef]
43. Schmidt, K.S.; Skidmore, A.K. Smoothing vegetation spectra with wavelets. *Int. J. Remote Sens.* **2004**, *25*, 1167–1184. [CrossRef]
44. Člupek, M.; Matějka, P.; Volka, K. Noise reduction in Raman spectra: Finite impulse response filtration versus Savitzky–Golay smoothing. *J. Raman Spectrosc.* **2007**, *38*, 1174–1179. [CrossRef]
45. Savitzky, A.; Golay, M.J. Smoothing and differentiation of data by simplified least squares procedures. *Anal. Chem.* **1964**, *36*, 1627–1639. [CrossRef]
46. Liu, L.; Ji, M.; Dong, Y.; Zhang, R.; Buchroithner, M. Quantitative Retrieval of Organic Soil Properties from Visible Near-Infrared Shortwave Infrared Feature Extraction. *Remote Sens.* **2016**, *8*, 1035. [CrossRef]
47. Prasad, K.A.; Gnanappazham, L.; Selvam, V.; Ramasubramanian, R.; Kar, C.S. Developing a spectral library of mangrove species of Indian east coast using field spectroscopy. *Geocarto Int.* **2015**, *30*, 580–599. [CrossRef]
48. Ligges, U.; Short, T.; Kienzle, P.; Schnackenberg, S.; Billinghurst, S.; Borchers, H.-W.; Carezia, A.; Dupuis, P.; Eaton, J.W.; Farhi, E.; et al. Signal: Signal Processing. 2015. Available online: http://docplayer.net/24709837-Package-signal-july-30-2015.html (accessed on 26 January 2018).

49. R Development Core Team, R. *R: A Language and Environment for Statistical Computing*; R Foundation for Statistical Computing: Vienna, Austria, 2017.
50. Belgiu, M.; Drăguţ, L. Random forest in remote sensing: A review of applications and future directions. *ISPRS J. Photogramm. Remote Sens.* **2016**, *114*, 24–31. [CrossRef]
51. Liaw, A.; Wiener, M. Classification and Regression by randomForest. *R News* **2002**, *2*, 18–22. [CrossRef]
52. Poona, N.; van Niekerk, A.; Ismail, R. Investigating the utility of oblique tree-based ensembles for the classification of hyperspectral data. *Sensors* **2016**, *16*. [CrossRef] [PubMed]
53. Chen, T.; He, T.; Benesty, M.; Khotilovich, V.; Tang, Y. Xgboost: Extreme Gradient Boosting. 2017. Available online: https://cran.r-project.org/package=xgboost (accessed on 26 January 2018).
54. Immitzer, M.; Atzberger, C.; Koukal, T. Tree species classification with Random forest using very high spatial resolution 8-band WorldView-2 satellite data. *Remote Sens.* **2012**, *4*, 2661–2693. [CrossRef]
55. Belgiu, M.; Tomljenovic, I.; Lampoltshammer, T.J.; Blaschke, T.; Höfle, B. Ontology-based classification of building types detected from airborne laser scanning data. *Remote Sens.* **2014**, *6*, 1347–1366. [CrossRef]
56. Genuer, R.; Poggi, J.-M.; Tuleau-Malot, C. Variable selection using random forests. *Pattern Recognit. Lett.* **2010**, *31*, 2225–2236. [CrossRef]
57. Kohavi, R.; Provost, F. Glossary of terms. *Mach. Learn.* **1998**, *30*, 271–274.
58. Congalton, R.G.; Green, K. *Assessing the Accuracy of Remotely Sensed Data: Principles and Practices*, 2nd ed.; CRC Press: Boco Raton, FL, USA, 2008.
59. Foody, G.M. Thematic map comparison: Evaluating the statistical significance of differences in classification accuracy. *Photogramm. Eng. Remote Sens.* **2004**, *70*, 627–633. [CrossRef]
60. Gutiérrez, S.; Tardaguila, J.; Fernández-Novales, J.; Diago, M.P. Data mining and NIR spectroscopy in viticulture: Applications for plant phenotyping under field conditions. *Sensors* **2016**, *16*, 236. [CrossRef] [PubMed]
61. Carvalho, A.A.; Rebello, J.M.A.; Sagrilo, L.V.S.; Camerini, C.S.; Miranda, I.V.J. MFL signals and artificial neural networks applied to detection and classification of pipe weld defects. *NDT E Int.* **2006**, *39*, 661–667. [CrossRef]
62. Miao, X.; Heaton, J.S.; Zheng, S.; Charlet, D.A.; Liu, H. Applying tree-based ensemble algorithms to the classification of ecological zones using multi-temporal multi-source remote-sensing data. *Int. J. Remote Sens.* **2012**, *33*, 1823–1849. [CrossRef]
63. Xu, L.; Li, J.; Brenning, A. A comparative study of different classification techniques for marine oil spill identification using RADARSAT-1 imagery. *Remote Sens. Environ.* **2014**, *141*, 14–23. [CrossRef]

remote sensing

MDPI

Article

Evaluating Endmember and Band Selection Techniques for Multiple Endmember Spectral Mixture Analysis using Post-Fire Imaging Spectroscopy

Zachary Tane [1,2,*], Dar Roberts [1], Sander Veraverbeke [3,4], Ángeles Casas [5], Carlos Ramirez [2] and Susan Ustin [6]

1 Department of Geography, University of California Santa Barbara, Santa Barbara, CA 93106, USA; dar@geog.ucsb.edu
2 United States Department of Agriculture, Forest Service, Pacific Southwest Region, Remote Sensing Lab, McClellan, CA 95652, USA; carlosramirez@fs.fed.us
3 Faculty of Science, Vrije Universiteit Amsterdam, 1081 HV Amsterdam, The Netherlands; s.s.n.veraverbeke@vu.nl
4 Department of Earth System Science, University of California Irvine, Irvine, CA 92697, USA
5 Independent Researcher; angelescasasp@gmail.com
6 Center for Spatial Technologies and Remote Sensing (CSTARS), Department of Land, Air, and Water Resources, University of California Davis, Davis, CA 95616, USA; slustin@ucdavis.edu
* Correspondence: ztane@fs.fed.us

Received: 21 December 2017; Accepted: 27 February 2018; Published: 2 March 2018

Abstract: Fire impacts many vegetated ecosystems across the world. The severity of a fire is major component in determining post-fire effects, including soil erosion, trace gas emissions, and the trajectory of recovery. In this study, we used imaging spectroscopy data combined with Multiple Endmember Spectral Mixture Analysis (MESMA), a form of spectral mixture analysis that accounts for endmember variability, to map fire severity of the 2013 Rim Fire. We evaluated four endmember selection approaches: Iterative Endmember Selection (IES), count-based within endmember class (In-CoB), Endmember Average Root Mean Squared Error (EAR), and Minimum Average Spectral Angle (MASA). To reduce the dimensionality of the imaging spectroscopy data we used uncorrelated Stable Zone Unmixing (uSZU). Fractional cover maps derived from MESMA were validated using two approaches: (1) manual interpretation of fine spatial resolution WorldView-2 imagery; and (2) ground plots measuring the Geo Composite Burn Index (GeoCBI) and the percentage of co-dominant and dominant trees with green, brown, and black needles. Comparison to reference data demonstrated fairly high correlation for green vegetation and char fractions (r^2 values as high as 0.741 for the MESMA ash fractions compared to classified WorldView-2 imagery and as high as 0.841 for green vegetation fractions). The combination of uSZU band selection and In-CoB endmember selection had the best trade-off between accuracy and computational efficiency. This study demonstrated that detailed fire severity retrievals based on imaging spectroscopy can be optimized using techniques that would be viable also in a satellite-based imaging spectrometer.

Keywords: spectral mixture analysis; fire severity; AVIRIS

1. Introduction

Fire behavior, size, and severity are changing in the western United States [1–3]. To fully comprehend these changes, techniques to reliably map fire effects over large areas are required. The most common means of assessing fire severity with remote sensing is using the Normalized Burn

Ratio (NBR) and its derivatives [4,5]. NBR based techniques are popular because of their simplicity, ease of use, and relatively good performance when compared to field measurements [6,7]. The NBR, however, is sensitive to variations in soil brightness [8], vegetation type [9], and vegetation density [10]. These limitations constrain the use of NBR-based indices for consistent fire severity assessments over large areas within heterogeneous ecosystems [10]. Variation in fire severity can effectively be broken down into detectable differences in the relative abundance of char, green vegetation, dead vegetation, and bare soil; remotely sensed fire severity assessments are therefore essentially based on mixtures composed of these four constituents. Under this paradigm, Spectral Mixture Analysis (SMA), in which reflectance is assumed to a be a linear combination of components or endmembers at a subpixel level [11,12], represents a viable alternative to NBR-based analysis, potentially overcoming NBR's sensitivity to different cover type and soil brightness variation [13,14]. SMA has been used previously to characterize tree mortality and soil char cover [10,15–18].

Lentile et al. [19] defines fire severity as the immediate ecosystem impacts of fire; burn severity, in contrast, is defined as the combined shorter and longer term ecosystem impacts and response to fire. Key and Benson [4] also separates first-order effects, which are related to the fire only, and second-order effects, which are related to other environmental (e.g., wind and rain) and vegetative processes. Veraverbeke et al. [20] followed these conventions and used the term fire severity when images were acquired soon after the fire (e.g., within the first month) and second-order effects can safely be neglected.

In SMA, an endmember class is one of the categories the image will be divided into, and an endmember is the individual training spectra that are constituents of each endmember class. In simple SMA, only one endmember represents each endmember class across the image. Multiple Endmember SMA (MESMA), in contrast, allows the endmember representing each endmember class to vary on a per-pixel basis [21]. This approach accounts for the variability that may exist within each endmember class and further allows for consistency and accuracy across ecosystems where there could be considerable spectral variability within an endmember class [21,22].

SMA is particularly well suited for imaging spectrometry data, as the large number of bands provide additional information for cover type discrimination in critical, but spectrally narrow regions; for example, the red edge for green vegetation discrimination [23]. The Airborne Visible/Infrared Imaging Spectrometer (AVIRIS), samples spectra from 350 nm to 2500 nm at a nominal sampling interval of 10 nm [24]. There is currently no spaceborne imaging spectrometer with a similar signal-to-noise ratio to AVIRIS and global coverage with frequent return intervals, which currently impedes the use of imaging spectroscopy for fire severity comparisons at regional scales and over longer time periods. However, there are several proposed spaceborne imaging spectrometers, including the Environmental Mapping and Analysis Program (EnMAP, [25]), the PRecursore IperSpettrale della Missione Applicativa (PRISMA, [26]), and the Hyperspectral Infrared Imager (HyspIRI, [27]), that would make using imaging spectroscopy to monitor fire severity possible at regional to global scales within the decade.

A drawback of MESMA, particularly if three or more endmember classes are allowed to be modeled within a single pixel, is that the number of endmember combinations tested can be large. A variety of approaches have been developed that attempt to select the minimum number of spectra in a library needed to represent within endmember class variation and therefore eliminate redundant spectra. Many techniques focus on automated or semi-automated endmember selection, which identifies pure pixels through the extremes of image data [28] or the construction of synthetic endmembers based on image data [29]. Other techniques first create a large spectral library from various sources and then reduce its size to achieve a library that is both parsimonious and captures each endmember class's variability [30–32]. Several means exist to evaluate which endmembers to keep. Generally, criteria evaluated focus on either which endmembers best represent their endmember class [33–35], or which endmembers best model the library as a whole [31]. All techniques have slightly difference balances between capturing spectral variability and creating efficient libraries. While there are several studies reviewing endmember extraction techniques [36–38], the comparison of endmember selection techniques for MESMA is rare [32], and to our knowledge, there has been no evaluation of various endmember selections techniques for cover fraction identification (MESMA with multiple non-shade endmembers allowed per pixel).

Individual bands in imaging spectroscopy tend to be highly correlated, and the inclusion of all bands in image analysis techniques increases computational times and can decrease accuracy [36,39]. Data reduction techniques are therefore often applied imaging spectroscopy data sets prior to SMA [36]; two of the more common are Principal Component Analysis (PCA; [40]) and maximum noise fraction (MNF; [41]). PCA, MNF, and similar techniques reduce data dimensionality based on the spectral properties of the image; however, they do not necessarily do so in a manner that improves separability between endmember classes [42]. Asner and Lobell [43] proposed a data reduction technique designed specifically to improve the accuracy of SMA when applied to plant cover with AutoSWIR. AutoSWIR uses a priori optical properties of leaf, litter, and soil material to select critical spectral regions to use for spectral unmixing. Somers et al. [44] proposed a technique similar to AutoSWIR in stable zone unmixing (SZU), which uses variability within and between endmember classes to select spectral regions to use in SMA. Since SZU is based solely on the input spectral library, it has the built-in assumption that the endmembers in a spectral library are representative of the variability of the spectra in the image to be analyzed; however, it has the advantage of selecting spectral regions specific to the problem. SZU has been demonstrated to be effective in invasive species monitoring, soil type classification, and oil spill detection and tracking [44–46]. Neither autoSWIR nor SZU specifically addresses the highly correlated nature of adjacent bands. Somers and Asner [47] proposed a further refinement on SZU, uncorrelated SZU (uSZU), which selects bands that capture the maximum variability within and between endmember classes while eliminating highly correlated bands. uSZU was shown to have improved cover abundance estimate accuracy and performance times compared to SZU [47].

In this study, we test the ability of imaging spectroscopy based MESMA to derive cover fractions that correspond with field measurements. Additionally, we test the specific effects of four different endmember selection techniques and one band reduction technique on MESMA's accuracy and processing time in the context of deriving indicators of fire severity of a large wildfire using imaging spectroscopy.

2. Methodology

In order to test the effectiveness of derived cover fractions at modeling the actual mixed composition of pixels, four post-fire AVIRIS flight lines were processed using MESMA. We used four endmember selection techniques and tested each technique with both the full AVIRIS spectra and a reduced spectral subset determined from a band selection algorithm. The result was eight separate spectral libraries. Each of these eight libraries was used to perform a separate run of the MESMA algorithm (Figure 1). In order to evaluate MESMA derived fractions correspondence with fire severity, two separate validation data sets were used. First, the various cover fractions generated by the eight MESMA runs were compared through linear regression against randomly selected and manually classified plots on WorldView-2 scenes. Using MESMA derived fractions from a spectral library that performed well, the relationship of the cover fraction with two field measurements, Geo Composite Burn Index (GeoCBI), and the percent cover of green, brown, and black trees over an area, was evaluated.

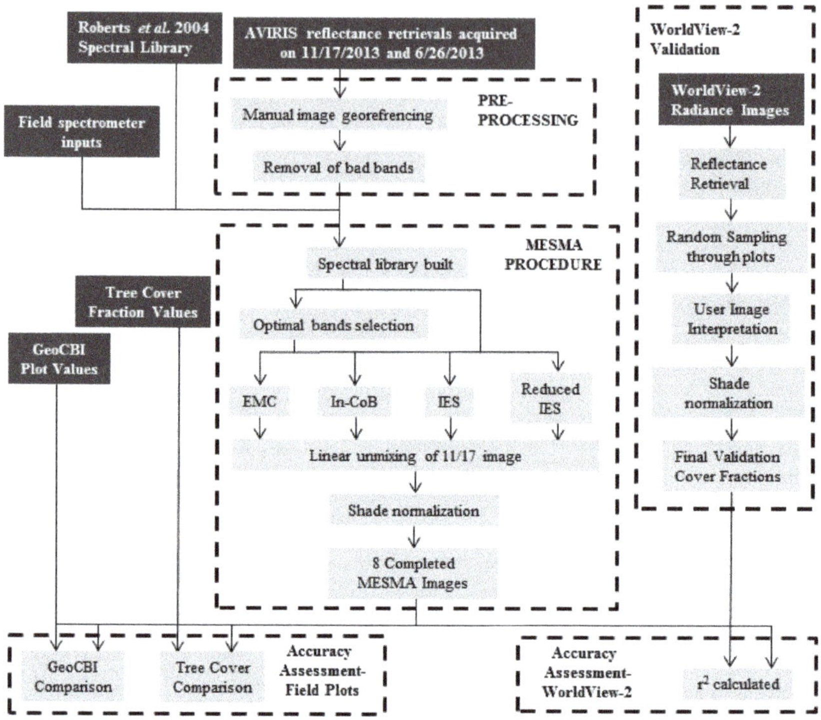

Figure 1. Flow chart of methods. Endmembers collected via Airborne Visible/Infrared Imaging Spectrometer (AVIRIS), field spectrometry, and an existing spectral library are used to perform Multiple Endmember Spectral Mixture Analysis (MESMA). Before MESMA, four different means of determining the optimal endmembers from a large spectral library were used: a technique that, for any given endmember within an endmember class, evaluates the count of the other endmembers modeled under an error threshold (In-CoB); a technique that uses the endmember average root mean square error (EAR), minimum average spectral angle (MASA), and In-Cob (EMC); a technique which uses forward-selection to iteratively add or remove spectra until an optimal state is reached (IES); and a technique that post-processes the IES library to produce a more parsimonious result (Reduced IES). Each endmember selection technique is tested with a full AVIRIS spectrum and with a spectral subset. The fractions are tested for goodness of linear fit with the WorldView-2 data via coefficient of determination value (r^2) and also compared with field data.

2.1. Study Area

The Rim Fire occurred in California's Sierra Nevada, starting in the Stanislaus National Forest and burning into Yosemite National Park (Figure 2). The Rim Fire started on 17 August, 2013 and was contained on 24 October 2013. According to the official fire perimeter statistics from the state of California (http://frap.fire.ca.gov/) the Rim Fire burned 104,131 ha and was the largest fire ever recorded in the Sierra Nevada and the fourth largest in California [3]. Several vegetation types were affected by the fire: coniferous forests dominated by Ponderosa pine (*Pinus ponderosa*), Lodgepole pine (*Pinus contorta*), red fir (*Abies magnifica*), and mixed conifer stands; deciduous and evergreen hardwood forests made up of various lower montane species including blue oak (*Quercus douglasii*), valley oak (*Quercus lobata*), interior live oak (*Quercus wislizeni*), and California buckeye (*Aesculus Californica*); annual grass; and shrubs, primarily composed of many common upper and lower montane Sierra Nevada chaparral species such as manzanita (*Arctostaphylos* spp.) and California lilac (*Ceanothus* spp.).

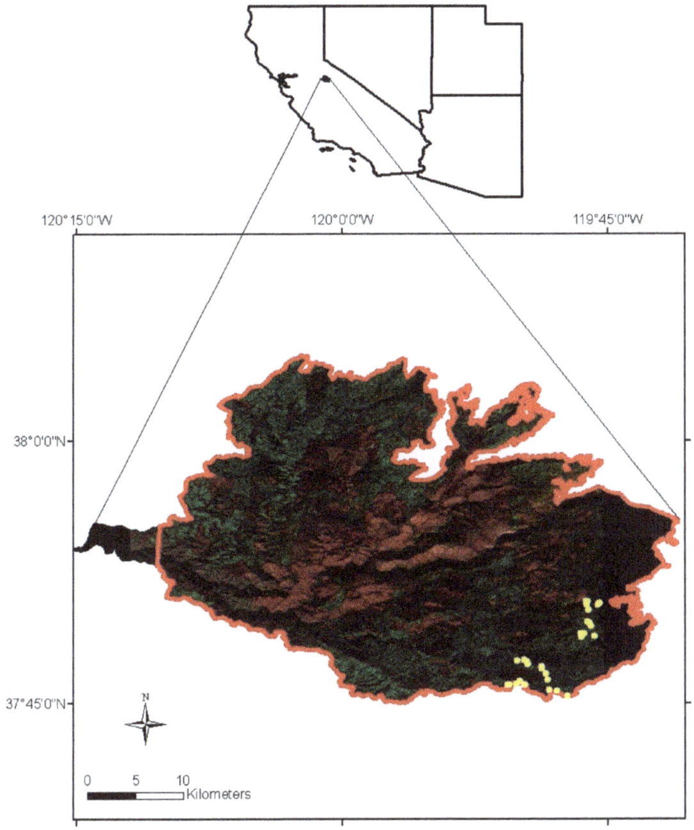

Figure 2. Airborne Visible/Infrared Imaging Spectrometer (AVIRIS) false color composite of the Rim Fire from 17 November 2013 imagery. The composite used bands centers at 2217 nm (red), 832 nm (green), and 657 nm (blue). Areas of no data within the Rim Fire are portrayed as black. Areas included in the mosaic of 28 September 2013 WorldView-2 imagery are outlined in red. The location of the field plots used in this analysis are portrayed with yellow dots.

2.2. AVIRIS Imagery and MESMA

2.2.1. Preprocessing

Level 2 ortho-reflectance AVIRIS flight lines were acquired from the Jet Propulsion Lab (JPL) ftp site (https://aviris.jpl.nasa.gov/data/AV_HyspIRI_Prep_Data.html). The flight lines used in this analysis were f131117t01p00r07, f131117t01p00r08, f131117t01p00r09, and f131117t01p00r10, all acquired on 17 November 2013. Before retrieval, the images had been orthorectified. Surface reflectance was retrieved using ATmospheric REMoval program (ATREM; [48]). Images were further manually georeferenced using United States Geological Survey (USGS) 2004 1m digital orthophoto quarter quad (DOQQ) maps (http://nationalmap.gov/viewer.html) as base maps. Images were warped using a first-degree polynomial with nearest neighbor resampling. Images were resampled to a uniform 14.8 m on a side pixel size. The flight lines cover approximately 93% of the Rim Fire burn scar. Wavelengths 365–405 nm, 1325–1420 nm, 1810–2040 nm, and 2450–2500 nm were known to either be associated with atmospheric water absorption features or found to contain significant atmospheric artifacts and a low signal-to-noise ratio from visual inspection; bands within these wavelengths were not considered in further processing, leaving 177 AVIRIS bands for use in the analysis. In addition, flight lines over the same area from 26 June 2013 were used as sources for the spectral library used for the MESMA process. The flight lines used for this purpose were f130626t01p00r13, f130626t01p00r14, and f130626t01p00r15. These flight lines were processed using procedures already described.

2.2.2. Spectral Library

Five different endmember classes were used to unmix the Rim Fire images: green vegetation (GV); non-photosynthetic vegetation (NPV), which includes dead needles, forest litter, bark senesced grass and other non-photosynthetically active plant material; soil, which includes bare soil and rock; char, which includes charred plant material and ash; and shade. Typical sources for endmembers include either spectra derived from the field, laboratory, or spectra derived from the imagery itself using pure pixels [49].

In this study, a combination of sources were used (Table 1). Field samples of charcoal, non-photosynthetic vegetation and substrate were collected and analyzed at the JPL spectroscopy lab [50]. Measurements in the 400–2500 nm spectral domain were obtained with an Analytic Spectral Devices (ASD) Full Range Spectrometer under artificial lighting conditions with an ASD Pro lamp. Reflectance was calibrated using a white Spectralon panel. We also used spectra collected from AVIRIS imagery, acquired on 26 June 2013 and 17 November 2013. To extract these spectra, polygons of areas composed solely of one of the four endmember classes (not including shade) were identified in the images. The spectra of all pixels within each polygon were extracted and labeled by endmember class. The spectral profile of each pixel was then manually reviewed to assure that it was consistent with the endmember class it was labeled as. To capture spatial scales that were not otherwise available in our analysis, an existing spectral library created from Wind River Research Forest (in Southern Washington state) was used [51]. This library included collections of leaves (branches) and tree bark, measured with an ASD in the field. Tree crowns measured in situ via a canopy crane were also included in this library. Although the spectra in the Wind River library were acquired in a different geographic location, many of the species and genera are common for both locations. Although Roberts et al. [51] also included stand level spectra measured with AVIRIS, these spectra were not included in our study.

Table 1. The number and origin of endmember used for the source spectral library. Analytic Spectral Devices is abbreviated as (ASD) and denotes field collected samples spectrally analyzed at a lab. Images acquired through Airborne Visible/Infrared Imaging Spectrometer are abbreviated as (AVIRIS). Spectra developed during a study in Wind River Research Forest [51] are denoted as Wind River. Green vegetation is abbreviated as (GV) and non-photosynthetic vegetation as (NPV).

	Char	**GV**	**NPV**	**Soil**	**Total**
17 November AVIRIS	457	308	0	358	1227
26 July AVIRIS	0	1739	245	510	2634
ASD	21	0	3	46	70
Wind River	0	498	129	139	766
Total	478	2545	377	1053	4453

2.2.3. Band Selection

Once the spectral library was finalized, the uSZU band selection technique [47] was performed to identify bands that optimize endmember class discrimination. uSZU is an automated band selection technique which attempts to select bands which maximize variation between endmember classes in a spectral library while minimizing the correlation between selected bands. uSZU assigns each band a Stability Index (SI) value based on the interclass variability divided by the intraclass variability for each class [47]:

$$SI_i = \frac{\Delta inter, i}{\Delta intra, i} = \frac{\left|\sum_{j=1}^{k} R_{mean,j,i}\right|}{1.96 \times \sum_{j=1}^{k} \sigma_{j,i}} \tag{1}$$

where k is the total number endmembers, $R_{mean,j,i}$ is the mean reflectance for endmember class j at wavelength i, and $\sigma_{j,i}$ is the standard deviation of class j for wavelength i. The band with the highest SI value is selected. Then a spectral correlation value (*Corr*) is calculated:

$$Corr(X,Y) = \frac{cov(X,Y)}{\sigma_x \sigma_y} \tag{2}$$

where $cov(X,Y)$ is the covariance between the selected waveband, X and each remaining waveband, Y; and σ is the standard deviation. All bands with *corr* values above a predetermined threshold c are then discarded. The process is repeated, with the threshold for correlation needed to discard a band decreasing by the value i with each iteration. For this study, values of c = 0.99 and I = 0.001 were used, both of these values were tested in Somers and Asner [47] and found to give acceptable accuracy. MESMA that was based on uSZU band reduced spectral libraries will be noted by including the term "uSZU" in the name.

2.2.4. Endmember Selection

Several approaches have been developed for determining the relative value of individual endmembers for representing their endmember class. One method for doing this is a count-based (CoB) approach, where each endmember is selected iteratively, and, using simple SMA, tested to see how many endmembers it can successfully model within the spectral library, using a predetermined Root Mean Square Error (*RMSE*) threshold to define success [52]. When this approach is applied to an endmember within its own endmember class, it is called (In-CoB); a desirable endmember will have a high In-CoB number (indicating a large number of other endmembers within the endmember class are derivative). This technique can also be applied against all the endmember classes which the endmember does not belong (Out-CoB), in this case, a desirable endmember will have a low Out-CoB number, indicating that this endmember would minimize confusion with other endmember classes.

Another approach for selecting endmembers is Endmember Average Root Mean Square Error (*EAR*), which evaluates each endmember's ability to model all other within class endmembers based on a summed RMSE [33]. EAR is calculated using the following formula:

$$EAR_{A_i} = \frac{\sum_{j=1}^{n} RMSE_{A_i A_j}}{n-1} \tag{3}$$

where A is an endmember class, A_i is the selected single endmember, and A_j are each of the other endmembers within the endmember class, and n is the total number of spectra in class A. A smaller EAR value is more desirable. Another approach, Minimum Average Spectral Angle (MASA, [34]), is similar to EAR but evaluates the summed spectral angle [53] instead of the RMSE fit. Both EAR and MASA evaluate only within an endmember class and do not evaluate interclass confusion.

Two techniques based on EAR, MASA, and CoB were used in this study. The first technique used a combination of EAR, MASA, and CoB to select three endmembers, it will therefore be abbreviated to "EMC". The combined EMC technique selected three endmembers for each endmember class: an endmember which minimizes the EAR value, an endmember which minimizes the MASA value, and an endmember which maximizes the In-CoB value. If multiple endmembers had the same In-CoB value, the endmember with the smallest Out-CoB value was selected. If the same endmember is selected through multiple EMC criteria (for example, the spectra that minimizes MASA also minimizes EAR), then fewer than three endmembers were used for that endmember class. The In-CoB [35] technique was also used in this study. For this technique, the In-CoB value of each endmember in an endmember class was evaluated, and any endmember with a unique In-CoB value was selected. If multiple endmembers have the same In-CoB value, the endmember with the minimum EAR value was selected. Another approach to endmember selection is Iterative Endmember Selection (IES) [31], which first picks two endmembers that maximize the performance of two-endmember SMA classifying the entire spectral library, as determined by using Cohen's kappa [54]. The algorithm then iteratively adds and removes endmembers in order to maximize kappa until the smallest spectral library that maximizes kappa is developed. While the other endmember selection approaches focus on within-class variability, IES explicitly considers confusion of endmembers between classes. In this study, the parameter of 0.025 was used as an RMSE constraint for the two-endmember SMA needed to classify the spectral library [32]. Spectral libraries processed in this way will be referred to as "IES" for the remainder of this study.

IES generally results in relatively large spectral libraries that can make the resultant MESMA analysis computationally expensive. Roberts et al. [55] proposed a method for reducing the size of IES generated spectral libraries through an iterative process, wherein the brightest endmember in each endmember class is initially chosen and all endmembers are modeled as a mixture of that endmember and shade are subsequently removed from the spectral library. Following the first pass, a new set of bright endmembers are selected from the reduced library, targeting endmembers that are most spectrally distinct from the first set. The process is repeated iteratively until either there are no more endmembers within a class in the reduced library, or new endmember selections fail to further reduce the library. The goal of this process is to identify the smallest set of endmembers for each class that also fully characterizes the spectral diversity of that class. The final result is a spectral library that retains the high accuracy of IES, while significantly reducing the number of endmember combinations. Spectral libraries processed in this way will be referred to as "Reduced IES".

2.2.5. MESMA

Four endmember MESMA (a maximum mixture of three endmember classes and shade could be modeled for any given pixel) was performed on the post-fire images using the resultant eight spectral libraries. In order for a model with more endmembers to be chosen over a model with fewer endmembers, the model with more endmembers must improve the model's RMSE for the pixel by a minimum threshold of 0.007 [35]. The endmember and shade fraction were constrained to be physically reasonable (no fractions below 0 or above 1 were allowed). The shade fraction was further constrained to be no higher than 0.8. A maximum allowed RMSE of 0.025 was also used to constrain the MESMA model [21].

2.3. *Validation*

2.3.1. WorldView-2 Imagery

High spatial resolution imagery provides a validation source for cover fractions over large portions of the fire area. WorldView-2 is a commercially operated very high spatial resolution, multispectral satellite imager. Twelve WorldView-2 images from 28 September 2013 that together captured about 97% of the Rim Fire burn scar (Figure 2) were purchased from Satellite Imaging Corp. The images had a pixel resolution of 2 m. After acquisition, images were converted into reflectance using Atmospheric and Topographic CORrection (ATCOR) as implemented in the PCI Geomatica software package (PCI Geomatics, Geomatica version 2014) using a LiDAR data set acquired by the National Center for Airborne Laser Mapping (NCALM-University of Houston).

In order to compare the AVIRIS-derived MESMA analysis and the WorldView-2 images, 120 square samples were randomly selected. The samples were 119.2 m on a side, or equivalent to an eight by eight square of AVIRIS pixels, this was chosen to reduce problems of differences in georeferencing between the images, as well as reduce the effect of pixels which are only partially within the sample area in the AVIRIS imagery. Areas within the sample squares were manually interpreted and digitized from the WorldView-2 imagery into five cover classes: GV, NPV, char, soil, and shadow (see Figure 3). The spectral resolution of WorldView-2 typically allowed for easy discrimination of GV. The NPV, char, and soil classes were not always easy separable spectrally, but contextual clues such as shape and texture were often helpful in separating these classes. There were likely some pixels that were misclassified, but overall this was considered a minor source of error.

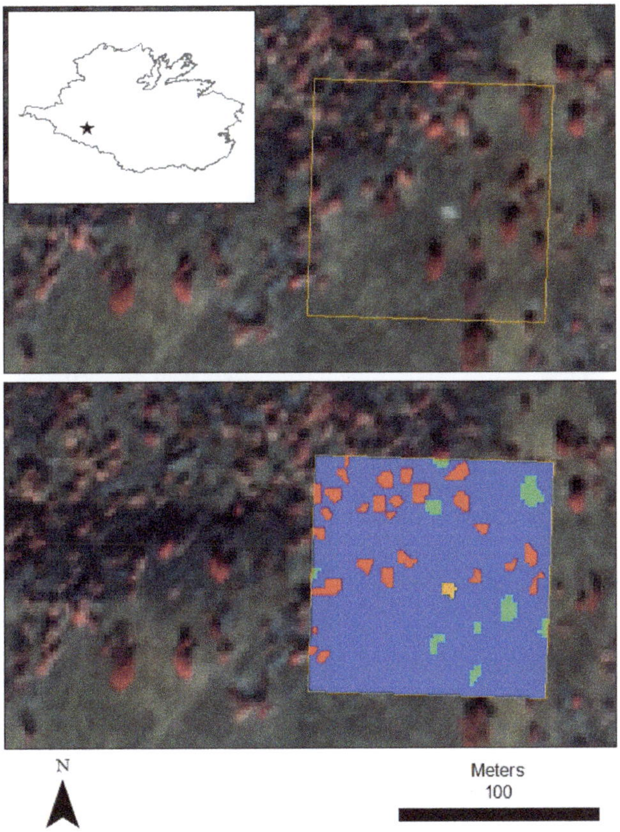

Figure 3. An example of a randomly selected manually classified WorldView-2 plot, the false color infrared WorldView-2 imagery uses band centers at 830 nm for red, 660 nm for green, and 545 nm for blue. The upper image displays just the WorldView-2 imagery and the sampling box, the lower box has the classified area. In the lower image blue represent char dominated pixels, green represents green vegetation dominated pixels, red represents non-photosynthetic vegetation dominated pixels, orange represents soil dominated pixels. An overview of the plot location is featured in the top left.

For both the eight MESMA unmixing results and the manually interpreted high-resolution imagery, the average cover for each cover class was calculated for each point, then the fractions were shade normalized, in which the shade cover fraction was removed by dividing each endmember by the total percent cover of all non-shade endmembers [56]. This allowed the relative abundance of the non-shade endmembers classes to be compared by partly correcting for the effects of differential solar geometry between the two images [13]. The quality of a linear fit for the relationship between each of the eight MESMA analysis and the manually identified imagery based on validation plots was assessed.

2.3.2. Field Plots

Thirty-four field plots of fire severity were collected in October 2013. The plots were located in the southeastern part of the fire and were, due to permitting restrictions, all taken within Yosemite National Park. Five severity was assessed in 30 by 30 m square shaped plots. The plot's center coordinates were recorded with a handheld Trimble GeoExplorer 6000 series GeoXH GPS device (these unit average approximately 1-m x and y error, see Figure 2 for plot locations). The plots were located no less than 200 m apart and were taken in areas of relatively homogeneous fuel type and fire severity. Four of the 34 original plots were removed from the analysis since they were located outside of AVIRIS images of 17 November 2013.

GeoCBI [57], a modified version of Composite Burn Index (CBI) [4], was measured in the field plots. GeoCBI divides the ecosystem into five different strata: (i) substrates; (ii) herbs, low shrubs and trees less than 1 m; (iii) tall shrubs and trees of 1 to 5 m; (iv) intermediate trees of 5 to 20 m; and (v) tall trees taller than 20 m. In the field form, 20 different factors can be rated (e.g., soil and rock cover/color change, percent leaf area index change, char height) but only those factors present and able to be reliably evaluated for that strata are considered. Ratings are given on a continuous scale between zero and three and the results are averaged per stratum. These stratum averages are weighted according to their fractional cover within each plot to compute a final GeoCBI rating. GeoCBI scores range from zero to three, with three representing the highest severity. In addition to GeoCBI, every dominant and co-dominant tree within each plot was identified. Trees were then assigned to three possible classes based on their needle color: green, brown, or black.

To compare the 30 by 30 m GeoCBI plots with the 14.8 m pixel size, a weighted plot level averaging of pixels was performed. Weights were assigned based on each pixel's percentage of area within a plot. A weighted average of all pixels within an individual plot was then calculated, and compared with the plot level data.

3. Results

3.1. uSZU Band Selection

After uSZU of the 177 original bands, 20 were retained in the analysis. Selected bands ranged across the spectrum with a preference for the red edge (five bands were selected between 670 and 735 nm). uSZU also appears to have selected several bands that are at the edges of wavelengths removed due to atmospheric noise, suggesting the sensitivity of the technique to spectral artifacts. Figure 4 shows the final selection of bands using the uSZU algorithm.

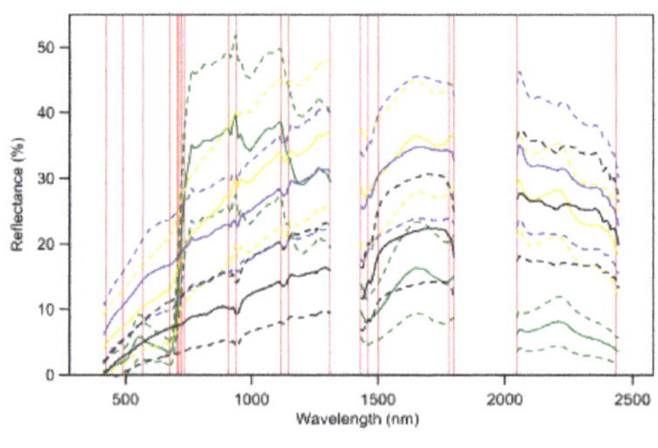

Figure 4. Spectral profile of the average spectra in each endmember class (solid line), plus or minus one standard deviation of the spectral average (dotted line). Green vegetation is represented by dark green, non-photosynthetic vegetation is represented by yellow, soil is represented by blue, and char is represented by black. Spectral bands selected by the uncorrelated Stable Zone Unmixing (uSZU) algorithm are represented by red vertical lines.

3.2. Endmember Selection and Processing Times

The number of endmembers selected by each technique varied considerably (Table 2). EMC has an inherent limit to the number of endmembers that may be selected; in EMC no more than the number of endmember classes multiplied by three endmembers are kept (the maximum possible number of endmembers in an EMC reduced library is therefore 12 in this study). uSZU generally had the effect of increasing the number of endmembers selected for all techniques except EMC. The increase was generally moderate, resulting in a 10–25% increase in the number of endmembers. Despite the larger number of endmembers selected, the processing time for the MESMA algorithm was dramatically reduced for all band reduction techniques, with time reductions ranging between 30% and 55%.

Table 2. The number of endmembers and modeling times of various combinations of endmember and band selection techniques. Modeling time was assessed on a 640 × 530 pixel Airborne Visible/Infrared Imaging Spectrometer (AVIRIS) scene within the Rim Fire. Models were run using a computer with an Intel Xeon central processing unit E7-4850v2 at 2.3 GHz with 48 cores and 512 gigabytes of random-access memory. Models were run in a manner to minimize run time with computer resources. Uncorrelated Stable Zone Unmixing (uSZU) represents a feature reduction technique. EMC, In-COB, and IES are all endmember reduction technique for Multiple Endmember Spectral Mixture Analysis (MESMA). Total models represents the number of models need to run four endmember MESMA.

Class	EMC	uSZU EMC	In-CoB	uSZU In-CoB	IES	uSZU IES	Reduced IES	uSZU Reduced IES
Char	2	3	5	5	10	10	5	7
GV	2	2	14	17	31	42	25	40
NPV	3	2	11	11	36	40	32	37
Soil	2	2	15	16	55	53	35	38
Total Models	83	83	5729	7071	115,543	156,821	45,327	92,415
Processing Time	0.98	0.44	14.88	8.70	151.85	66.76	78.36	54.45

3.3. Unmixed Images and Overall Model Comparison

The eight endmember libraries produced broadly similar patterns (Figure 5). The EMC method without band selection stands out as modeling more NPV than the others, notably modeling NPV in areas that other approaches modeled as GV. Unmixed images generated from libraries without uSZU band selection appear to have slightly more NPV than their uSZU counterparts, but otherwise give similar results.

While the models were qualitatively similar, there were quantifiable differences between them (Table 3). The number of pixels successfully modeled, defined by the number of pixels where any endmember combination tested by the MESMA algorithm resulted in an RMSE below the threshold of 0.025, ranged from 79.1% of the fire area (uSZU In-CoB) to 93.0% of the fire area (IES). Soil was the most variable endmember class between the models, with one scene containing as few as 7.5% of pixels with any modeled soil cover (uSZU EMC) to 41.8% of pixels (IES). Char and GV were the most commonly modeled endmember classes with between 36.9% and 53.1% of pixels containing char and between 35.3% and 54.1% of pixels containing GV.

Table 3. Percent of total pixels within the Rim Fire boundary that were successfully modeled and the percent of pixels containing each endmember class for various endmember selection techniques. In total, approximately 7,850,000 pixels within the Rim Fire boundary were considered. A single pixel was allowed to contain up to three endmember classes (not including shade).

	EMC	uSZU EMC	In-CoB	uSZU In-CoB	IES	uSZU IES	Reduced IES	uSZU Reduced IES
Modeled	87.0%	81.7%	83.2%	79.1%	93.0%	92.0%	85.7%	86.4%
Char	36.9%	53.1%	47.3%	36.9%	33.4%	42.9%	41.6%	34.4%
GV	54.1%	52.9%	43.1%	39.9%	35.3%	39.4%	36.8%	38.1%
NPV	38.8%	21.9%	27.7%	22.0%	23.5%	21.2%	30.0%	21.65
Soil	7.9%	7.5%	11.5%	26.2%	41.8%	28.6%	17.5%	28.4%

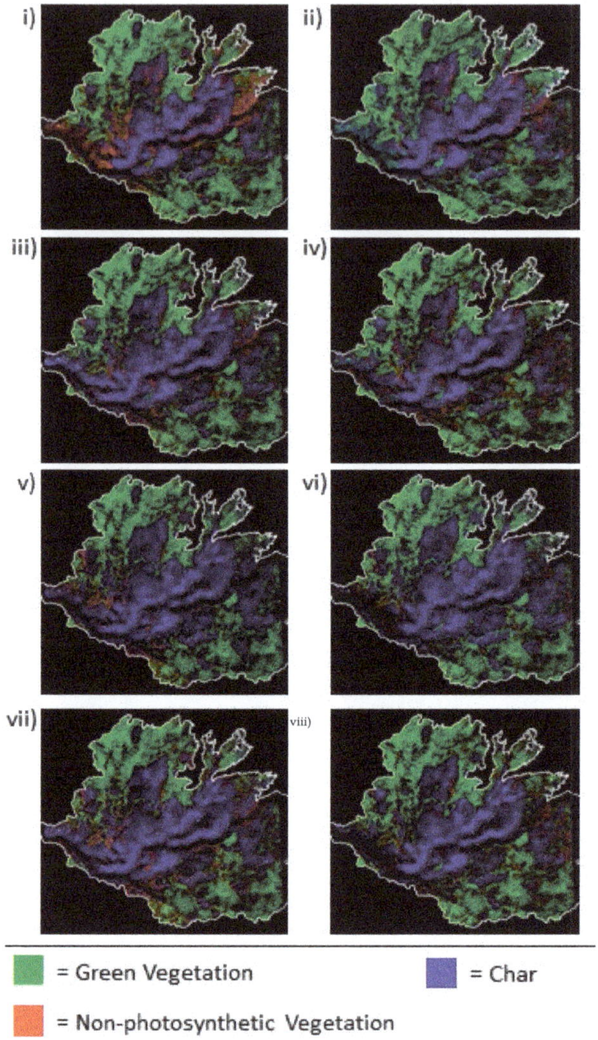

Figure 5. MESMA cover fraction images within the Rim Fire burn scar, soil cover is not shown in the image series. The boundary of the Rim Fire is represented by a white line. Black indicates areas that were not classified or only had soil cover. Endmember selection and band selection techniques are as follows: (**i**) EMC; (**ii**) uSZU EMC; (**iii**) In-CoB; (**iv**) uSZU In-CoB; (**v**) IES; (**vi**) uSZU IES; (**vii**) Reduced IES, (**viii**) uSZU Reduced IES.

3.4. Endmember Sources in Model Selection and the Image

Most spectral libraries selected endmembers from all sources (Table 4). In the initial spectral library, a majority of spectra were derived from AVIRIS imagery, however, a disproportionally small amount of spectra from this source were selected by most endmember reduction techniques. This suggests that, by the metrics used by the endmember reduction techniques, most image derived spectra were redundant and were therefore eliminated. In contrast, spectra created in Wind River [51] were disproportionally retained by most endmember selection techniques. This indicates spectra from that source were generally distinct from other spectra included in the initial spectral library.

Table 4. The source of endmembers for each endmember reduction technique.

Spectra Source	EMC	uSZU EMC	In-CoB	uSZU In-CoB	IES	uSZU IES	Reduced IES	uSZU Reduced IES
AVIRIS	9	9	15	18	41	45	20	30
ASD	0	0	5	4	12	14	9	12
Wind River	0	0	25	27	79	86	68	80

When spectral libraries were used to model cover fractions across the Rim Fire, spectra generated from the AVIRIS images were disproportionally selected, given their abundance in the reduced spectral libraries (Table 5). All spectral sources were used to some degree. It should be noted that not all endmember classes modeled in this study were collected for each spectral source and there were large differences in the number of spectra collected for each source.

Table 5. Percentage of pixels modeled by source for each endmember reduction technique. A single pixel can be modeled by multiple different sources, resulting in totals that exceed 100%.

Spectra Source	EMC	uSZU EMC	In-CoB	uSZU In-CoB	IES	uSZU IES	Reduced IES	uSZU Reduced IES
Not Modeled	13%	18.3%	16.8%	20.9%	7%	8%	14.3%	13.6%
AVIRIS	87%	81.7%	42.1%	53.2%	61.8%	55.3%	48.6%	32.5%
ASD	0	0	15.7%	6.2%	5.4%	10.3%	6.9%	26.4%
Wind River	0	0	45.8%	38%	47.1%	44.4%	47.6%	41.6%

3.5. Validation

3.5.1. WorldView-2 Based Validation

The classified WorldView-2 imagery and the AVIRIS-based MESMA cover fractions were compared using the 120 randomly selected polygons across the two image types (Table 6). The best linear correlations between WorldView-2 fractions and MESMA were observed for the GV cover, which displayed a near 1:1 relationship and had generally high r^2 values regardless of the endmember selection technique used. Correlation coefficients for char were generally fairly high (between 0.59 and 0.741), however, all models showed a tendency to under-model char, as demonstrated by linear models consistently having an intercept above 0.2. Correlations were generally poor for NPV and soil, with r^2 values often below 0.2. Larger endmember libraries did not always translate to higher fractional cover accuracy; for example, the highest r^2 for the char fraction was found for the uSZU EMC library (0.741).

Table 6. The coefficient of determination (r^2), intercept, and slope value of a linear fit between WorldView-2 estimated cover fractions as the dependent variable and AVIRIS derived MESMA cover fraction for different endmember and band selection techniques as the independent variable. Green vegetation is abbreviated as GV, non-photosynthetic vegetation as NPV.

		EMC	uSZU EMC	In-CoB	uSZU In-CoB	IES	uSZU IES	Reduced IES	uSZU Reduced IES
Char	r^2	0.605	0.741	0.727	0.642	0.620	0.538	0.687	0.594
	Intercept	0.355	0.303	0.212	0.202	0.321	0.223	0.255	0.310
	Slope	0.721	0.755	0.841	0.835	0.765	0.783	0.771	0.750
GV	r^2	0.750	0.770	0.836	0.871	0.848	0.846	0.853	0.861
	Intercept	−0.054	−0.023	−0.026	−0.065	0.020	0.000	-0.010	−0.019
	Slope	0.653	0.769	0.708	0.675	0.895	0.812	0.804	0.748
NPV	r^2	0.086	0.099	0.164	0.249	0.209	0.237	0.165	0.245
	Intercept	0.110	0.103	0.109	0.126	0.099	0.107	0.099	0.094
	Slope	0.240	0.521	0.426	0.375	0.968	0.943	0.377	0.699
Soil	r^2	0.273	0.261	0.088	0.042	0.014	0.049	0.075	0.057
	Intercept	0.013	0.015	0.015	0.013	0.019	0.016	0.011	0.012
	Slope	0.641	0.102	0.222	0.965	0.041	0.088	0.2	0.103

Plots of the linear regression for the uSZU IES MESMA run and the WorldView-2 classifications are provided below (Figure 6). The plots are broadly representative of the trend of most MESMA versus WorldView-2 relationships. There is general support for the linear relationship of the GV identification by MESMA and the WorldView-2, with a high r^2 value and little systematic error. Based on the manual interpretation of the WorldView-2 imagery, MESMA appears to be systematically modeling a lower fraction of NPV and char cover, and modeling a higher fraction of soil cover.

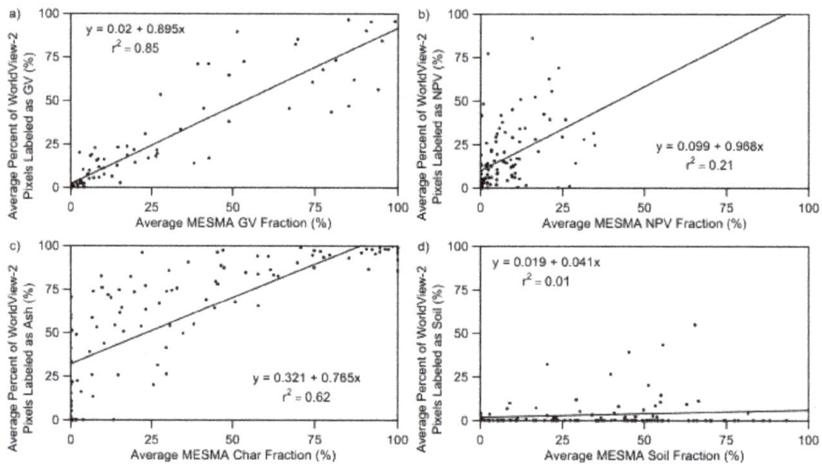

Figure 6. Scatter plots of the relationship between the 120 WorldView-2 classification points and MESMA with the IES endmember selection technique for green vegetation (GV), char, non-photosynthetic-vegetation (NPV), and soil.

3.5.2. Comparison with GeoCBI Plot Data

Due to the similarity between MESMA unmixed models used in this study, we limit our comparison to only the cover fractions generated from MESMA using the IES library and the full AVIRIS bands. First, the relationship between individual cover fractions and GeoCBI was assessed (Figure 7). The relationship between the GV fraction and GeoCBI appeared to be inversely linear (a linear regression produced an intercept of 2.93, a slope of −1.89 and r^2 value of 0.644). In contrast, the relationship between NPV and GeoCBI is clearly non-linear. NPV fractions are generally highest at GeoCBI values of 1.5 to 2.5 suggesting that NPV is high at moderate severities, but low at the highest and lower severity levels. The char fraction appears to be more of a binary relationship only appearing in the model at the highest GeoCBI levels (above 2.75) and is at or near zero for all other GeoCBI levels.

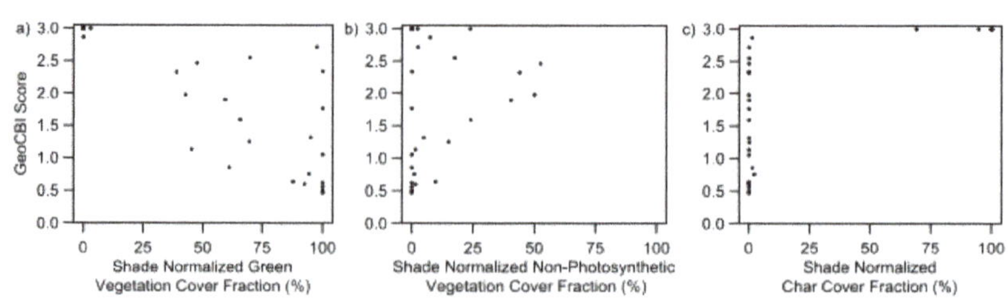

Figure 7. Scatter plots of the relationship between GeoCBI, a ground plot derived method of measuring fire severity and estimated cover fractions for green vegetation, non-photosynthetic vegetation, and char.

3.5.3. Comparison with Field Tree Status Data

We also compared the MESMA cover fractions of GV, NPV, and char with the percentage of canopy level trees with mostly green, brown, and black needles (Figure 8). Similar to the GeoCBI values, the char percentage and percentage of black trees appeared to have a near binary relationship in the plots. If the percentage of field identified black trees within the plot was under 50%, then typically no char fraction was modeled for the pixel, if it was above 50%, then the char fraction was almost always modeled as close to 100%. The relationship between the brown trees and NPV appeared to generally be more linear, although the modeled NPV cover was never greater than 50% even with near 100% brown trees. The relationship between the green trees and green vegetation may be positively linear, but the amount of scattering makes interpretation difficult. It is important to remember that few AVIRIS pixels are purely tree cover, most are a mixture of tree and substrate, so a pixel with 100% brown trees may indeed be made up of 50% other materials. Finally, it is important to note that the percent of trees in the plot may not be an ideal reference for the cover fraction as it does not account for variation in tree density, and this may partially contribute to the observed biases.

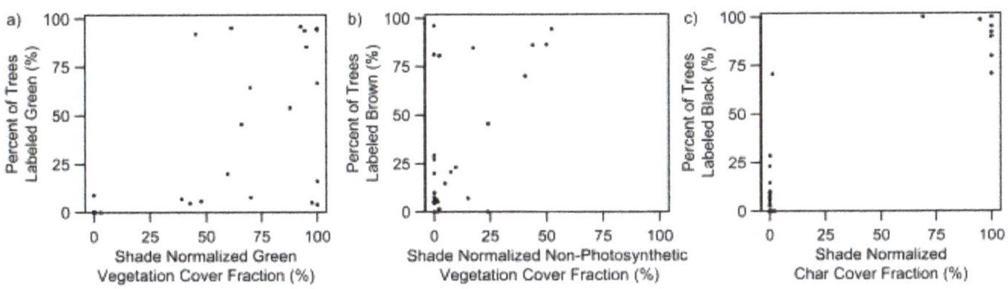

Figure 8. Scatter plots of the relationship between percent of black, brown, and green trees, and estimated cover fractions for green vegetation, non-photosynthetic vegetation, and char.

4. Discussion

4.1. Potential Bias and Uncertainty in the Cover Types

Whereas previous studies found the relationship between the char fraction and GeoCBI to be linear [18,58] this was not the case in this study. The GV fraction retrieval by MESMA, was found to be linear, however. GV's high accuracy is likely due to both the spectral separability and the temporal persistence of the class. In contrast to GV, at the spectral resolution of WorldView-2 soil, NPV, and ash were all somewhat difficult to separate. Human interpretation likely reduced inter-class confusion in classifying the WorldView-2 imagery, but this was still a source of error associated with the WorldView-2 image classification. In particular, separation of the soil and ash areas was particularly difficult with the WorldView-2 imagery. This is partially because of insufficient spectral information and partially because even at 2 m pixel resolutions, many pixels were likely a mixture of mostly ash with some soil. This mixture would impact the spectral signal, and therefore be reflected in the MESMA generated cover fractions, but these areas were generally just classified as ash in the manual interpretation of the WorldView-2 image. Another possible factor in diminished char is the amount of time between when the fire burned the area of the GeoCBI plots, which was in late August and early September (evaluated from fire spread map in [59]), and image acquisition, which was 28 September for the WorldView images and 17 November for AVIRIS. During the time period between the fire occurring and AVIRIS image acquisition, there were four rain events totaling 64.8 mm at a weather station in Groveland, CA (data accessed from http://www.ncdc.noaa.gov/cdo-web/datasets#GHCNDMS) within 20 km from where the field plots were collected. Combined with wind, rain likely attenuated the ash portion of the char signal. The impact of weather, combined with other previously mentioned factors, likely account for why char was more prevalent and soil less prevalent in the WorldView-2 classification than the MESMA cover fraction (Figure 6).

4.2. Evaluation of MESMA Techniques' Performance

The balance of library complexity with accuracy must be considered if MESMA is to be used on spaceborne imaging spectroscopy at regional to global scales for ecological monitoring [58]. Schaff et al. [31] and Roth et al. [32] have shown IES endmember selection produces high classification accuracy in the two endmember MESMA case. However, IES selected the most endmembers in its final spectral library in our study and when compared to other methods in [32]. In this study, regardless of the technique used to generate cover fractions, the more spectrally unique classes (GV and ash) had approximately similar performances when compared with the WorldView-2 cover fractions. Using uSZU In-CoB to generate MESMA cover fractions produced relatively high r^2 values when compared to WorldView-2, even though the number of endmembers used and processing times were significantly less. That a variant of In-CoB would perform as well as IES is not consistent with [32]. One possible explanation for this is that when using endmembers for classification of vegetation types, as was done in [32], some degrees of mixture within endmembers, particularly with soil and NPV, is necessary for accurate classification across a landscape. If endmembers within the endmember classes of the starting library are to some degree mixed, in order maximize kappa IES will tend to select mixed pixels. However, when using MESMA for spectral unmixing to estimate cover fractions, as was done in this study, the purest endmembers will obtain the highest accuracy.

The margin of computational efficiency created by uSZU, a 30–50% reduction in processing times, was similar to Somers and Asner [47]. However, unlike [47], we did not observe a clear trend of increased accuracy (Table 6). In Figure 5 and Table 3, it is clear that the techniques that used uSZU generally mapped less NPV than techniques that did not. Spectral regions that are critical for the separation of soils and NPV due to their association with lignin and cellulose absorption, such as the 2300–2400 nm region [60], did not get selected by uSZU (Figure 4). Many of the bands which were selected appear associated with atmospheric noise. In Veraverbeke et al. [58] bands were reduced to similar numbers as the uSZU, however they were not selected in a way that was designed to maximize differences between endmember classes, resulting in a much larger decrease in accuracy than was observed here. Future studies should investigate other band selection techniques and the optimal bands for the simultaneous separation of GV, NPV, soils, and ash.

4.3. Endmember Sources and Endmember Selection

Beyond pre-processing, another challenge to SMA becoming a global means of objective comparison is determining a proper base spectral library. Studies have shown the timing of the acquisition of the base spectra and the image is important [61], as is the spatial scale that the spectra were acquired [35,51]. It is notable that spectra collected in a different spatial location than the images and at different spatial scales than the images, were frequently selected to model cover fractions. This suggests that a common spectral library could be developed to map fire across at least a regional scale. It also suggests that even in environments not composed of as many complex materials as the urban [35,62,63], a diversity of spatial scales may be beneficial to cover fraction mapping.

Although, impurity of endmembers collected in non-laboratory settings is inevitable, it is a particular challenge at the spatial scale of AVIRIS or proposed spaceborne imaging spectrometers. It is notable that the endmember classes which it is most difficult to get pure pixels from images, NPV and soil (that was not exposed rock), also appeared least accurate. Although spectra collected in the field were included in our spectral library, these may not perfectly scale to the canopy level AVIRIS observations [51]. AVIRIS-Next Generation (AVIRIS-NG; [64]) poses a potential means for overcoming some of these challenges [55]. The superior spatial and spectral resolutions allows for collecting and evaluating endmembers that are closer to being pure while still being collected at the canopy-scale of AVIRIS.

4.4. SMA as a Novel Means for Assessing Fire Severity

Since properly identified cover fractions have physical meaning, they have the potential to be an objective and global means of assessing fire severity. Cover fractions, if shown to be accurate and comparable to field derived fire severity estimates, have the potential to overcome many of the criticisms of NBR-based indices [10,19,65,66]. With the potential launch of several spaceborne imaging spectrometers, and considering the demonstrated higher correlations of SMA with imaging spectroscopy data compared to broad-band data [58], the use of SMA for fire may increase in the future. Since the variation of soil and GV's spectral profile across space is one of the key reasons for NBR's subjectivity, if SMA is going to become an important means of assessing burn severity, endmember variability must be accounted for, and MESMA is one of the most reliable technique for doing so [22]. In this study, only a single fire is included in the study area; however, if SMA is to become a commonly used tool for fire severity assessment, a global assortment of geographic locations and their spectral variability will need to be tested and compared. As imaging spectroscopy data becomes more common, both through more airborne and spaceborne acquisitions, global assessment will become viable.

One possibility for assessing change using SMA is using a differenced SMA (dSMA) approach. By using dSMA the entirety of the information provided by MESMA classes could be used, potentially providing a robust and more ecologically meaningful method of evaluating fire severity. However, this approach would have an inherent disadvantage of all differenced imagery comparisons, in that relatively analogous pre-fire imagery would be needed. Since all imaging spectroscopy data with signal-to-noise ratios and spectral and spatial resolutions similar to AVIRIS are currently acquired from airborne platforms, pre-fire data is rare; however proposed spaceborne imaging spectrometers such as HyspIRI and ever-increasing computational power would make this analysis possible at regional to global scales. Future studies evaluating dSMA usefulness in fire severity evaluation will require careful planning in field validation plot placement to assure that plots are placed in areas that pre-fire were heterogeneous in terms of type and percentages of vegetation cover. Given the relative accuracy of mapping the GV cover class, using a differenced GV cover fraction (dGV) is likely to have high correlation with the field measures of burn severity. In addition, the high correlation observed for the char cover type, suggests that adding a post-fire char cover to a dGV assessment could further enhance the discrimination of burn severity.

5. Conclusions

We demonstrated the utility of imaging spectroscopy combined with MESMA for fire severity mapping over a large fire in California's Sierra Nevada. While currently limited to airborne acquisitions, future spaceborne missions will allow large-scale application of these techniques. One aspect that will be important for processing imaging spectroscopy and MESMA globally are techniques that maximize accuracy while minimizing calculation time. In this study, we performed MESMA using a spectral library that included several different collection methods and multiple different spatial scales. We demonstrated that band reduction can significantly reduce computational time with only small differences in performance. We also evaluated the performance of several endmember selection techniques and found that these can also be optimized between performance and calculation time. Further research is needed that evaluates cover fractions relation to fire severity comparing multiple fires across regions. However, the identification cover fractions represents a potentially objective and physically meaningful evaluation of fire severity using remote sensing.

Acknowledgments: We would like to acknowledge the US Forest Service, Region-5 Remote Sensing Lab staff, in particular Nathan Amboy for their assistance in the WorldView-2 image acquisition and pre-processing. We also would like to acknowledge Simon Hook and Linley Kroll for supporting the field data collection. We would like to acknowledge everyone at JPL that contributed with the collection and pre-processing of the AVIRIS imagery.

Author Contributions: Sander Veraverbeke, Susan Ustin, Zachary Tane conceived and designed the experiments; Ángeles Casas, Sander Veraverbeke, and Zachary Tane performed the experiments; Dar Roberts and Zachary Tane analyzed the data; Carlos Ramirez, Dar Roberts, Sander Veraverbeke, and Zachary Tane wrote the paper.

Conflicts of Interest: The authors declare no conflict of interest.

References

1. Dennison, P.E.; Brewer, S.C.; Arnold, J.D.; Moritz, M.A. Large Wildfire Trend in the Western United States, 1984–2011. *Geophys. Res. Lett.* **2014**, *41*, 2928–2933. [CrossRef]
2. Miller, J.D.; Safford, H.D.; Crimmins, M.; Thode, A.E. Quantitative evidence for increasing forest fire severity in the Sierra Nevada and southern Cascade Mountains, California and Nevada, USA. *Ecosystems* **2009**, *12*, 16–32. [CrossRef]
3. Steel, Z.L.; Safford, H.D.; Viers, J.H. The fire frequency-severity relationship and the legacy of fire suppression in California forests. *Ecosphere* **2015**, *6*, 8. [CrossRef]
4. Key, C.H.; Benson, N.C. *Landscape Assessment: Sampling and Analysis Methods*; General Technical Report RMRS-GTR-164-CD; USDA Forest Service: Fort Collins, CO, USA, 2006; pp. 1–55.
5. Miller, J.D.; Thode, A.E. Quantifying burn severity in a heterogeneous landscape with a relative version of the delta Normalized Burn Ratio (dNBR). *Remote Sens. Environ.* **2007**, *109*, 66–80. [CrossRef]
6. Cocke, A.E.; Fulé, P.Z.; Crouse, J.E. Comparison of burn severity assessments using Differenced Normalized Burn Ratio and ground data. *Int. J. Wildl. Fire* **2005**, *14*, 189. [CrossRef]
7. Van Wagtendonk, J.W.; Root, R.R.; Key, C.H. Comparison of AVIRIS and Landsat ETM+ detection capabilities for burn severity. *Remote Sens. Environ.* **2004**, *92*, 397–408. [CrossRef]
8. Smith, A.M.S.; Eitel, J.U.H.; Hudak, A.T. Spectral analysis of charcoal on soils: Implications for wildland fire severity mapping methods. *Int. J. Wildl. Fire* **2010**, *19*, 976–983. [CrossRef]
9. Epting, J.; Verbyla, D.; Sorbel, B. Evaluation of remotely sensed indices for assessing burn severity in interior Alaska using Landsat TM and ETM+. *Remote Sens. Environ.* **2005**, *96*, 328–339. [CrossRef]
10. Lentile, L.B.; Smith, A.M.S.; Hudak, A.T.; Morgan, P.; Bobbitt, M.J.; Lewis, S.A.; Robichaud, P.R. Remote sensing for prediction of 1-year post-fire ecosystem condition. *Int. J. Wildl. Fire* **2009**, *18*, 594–608. [CrossRef]
11. Adams, J.B.; Smith, M.O.; Johnson, P.E. Spectral mixture modeling: A new analysis of rock and soil types at the Viking Lander 1 Site. *J. Geophys. Res. Solid Earth* **1986**, *91*, 8098–8112. [CrossRef]
12. Roberts, D.A.; Smith, M.O.; Adams, J.B. Green vegetation, nonphotosynthetic vegetation, and soils in AVIRIS data. *Remote Sens. Environ.* **1993**, *44*, 255–269. [CrossRef]
13. Rogan, J.; Franklin, J. Mapping Wildfire Burn Severity in Southern California Forests and Shrublands Using Enhanced Thematic Mapper Imagery. *Geocarto Int.* **2001**, *16*, 91–106. [CrossRef]
14. Quintano, C.; Fernández-Manso, A.; Roberts, D.A. Multiple Endmember Spectral Mixture Analysis (MESMA) to map burn severity levels from Landsat images in Mediterranean countries. *Remote Sens. Environ.* **2013**, *136*, 76–88. [CrossRef]
15. Hudak, A.T.; Morgan, P.; Bobbitt, M.J.; Smith, A.M.S.; Lewis, S.A.; Lentile, L.B.; Robichaud, P.R.; Clark, J.T.; Mckinley, R.A. The Relationship of Multispectral Satellite Imagery. *Fire Ecol.* **2007**, *3*, 64–90. [CrossRef]
16. Kokaly, R.F.; Rockwell, B.W.; Haire, S.L.; King, T.V.V. Characterization of post-fire surface cover, soils, and burn severity at the Cerro Grande Fire, New Mexico, using hyperspectral and multispectral remote sensing. *Remote Sens. Environ.* **2007**, *106*, 305–325. [CrossRef]
17. Robichaud, P.R.; Lewis, S.A.; Laes, D.Y.M.; Hudak, A.T.; Kokaly, R.F.; Zamudio, J.A. Postfire soil burn severity mapping with hyperspectral image unmixing. *Remote Sens. Environ.* **2007**, *108*, 467–480. [CrossRef]
18. Veraverbeke, S.; Hook, S. Evaluating spectral indices and spectral mixture analysis for assessing fire severity, combustion completeness and carbon emissions. *Int. J. Wildl. Fire* **2013**, *22*, 707–720. [CrossRef]
19. Lentile, L.B.; Holden, Z.A.; Smith, A.M.S.; Falkowski, M.J.; Hudak, A.T.; Morgan, P.; Lewis, S.A.; Gessler, P.E.; Benson, N.C. Remote sensing techniques to assess active fire characteristics and post-fire effects. *Int. J. Wildl. Fire* **2006**, *15*, 319–345. [CrossRef]
20. Veraverbeke, S.; Lhermitte, S.; Verstraeten, W.W.; Goossens, R. The temporal dimension of differenced Normalized Burn Ratio (dNBR) fire/burn severity studies: The case of the large 2007 Peloponnese wildfires in Greece. *Remote Sens. Environ.* **2010**, *114*, 2548–2563. [CrossRef]
21. Roberts, D.A.; Gardner, M.; Church, R.; Ustin, S.L.; Scheer, G.; Green, R.O. Mapping chaparral in the Santa Monica Mountains using multiple endmember spectral mixture models. *Remote Sens. Environ.* **1998**, *65*, 267–279. [CrossRef]

22. Somers, B.; Asner, G.P.; Tits, L.; Coppin, P. Endmember variability in Spectral Mixture Analysis: A review. *Remote Sens. Environ.* **2011**, *115*, 1603–1616. [CrossRef]
23. Green, R.O.; Eastwood, M.L.; Sarture, C.M.; Chrien, T.G.; Aronsson, M.; Chippendale, B.J.; Faust, J.A.; Pavri, B.E.; Chovit, C.J.; Solis, M.; et al. Imaging spectroscopy and the Airborne Visible/Infrared Imaging Spectrometer (AVIRIS). *Remote Sens. Environ.* **1998**, *65*, 227–248. [CrossRef]
24. Jia, G.J.; Burke, I.C.; Goetz, A.F.H.; Kaufmann, M.R.; Kindel, B.C. Assessing spatial patterns of forest fuel using AVIRIS data. *Remote Sens. Environ.* **2006**, *102*, 318–327. [CrossRef]
25. Guanter, L.; Kaufmann, H.; Segl, K.; Foerster, S.; Rogass, C.; Chabrillat, S.; Kuester, T.; Hollstein, A.; Rossner, G.; Chlebek, C.; et al. The EnMAP spaceborne imaging spectroscopy mission for earth observation. *Remote Sens.* **2015**, *7*, 8830–8857. [CrossRef]
26. Stefano, P.; Angelo, P.; Simone, P.; Filomena, R.; Federico, S.; Tiziana, S.; Umberto, A.; Vincenzo, C.; Acito, N.; Marco, D.; et al. The PRISMA hyperspectral mission: Science activities and opportunities for agriculture and land monitoring. *Int. Geosci. Remote Sens. Symp.* **2013**, *2567*, 4558–4561.
27. Lee, C.M.; Cable, M.L.; Hook, S.J.; Green, R.O.; Ustin, S.L.; Mandl, D.J.; Middleton, E.M. An introduction to the NASA Hyperspectral InfraRed Imager (HyspIRI) mission and preparatory activities. *Remote Sens. Environ.* **2015**, *167*, 6–19. [CrossRef]
28. Boardman, J.; Kruse, F.; Green, R.O. Mapping Target Signatures via Partial Unmixing of AVIRIS Data. In Proceedings of the Fifth Annual JPL Airborne Earth Science Workshop, Volume 1: AVIRIS Workshop, Pasadena, CA, USA, 23–26 January 1995; pp. 23–26.
29. Tompkins, S.; Mustard, J.F.; Pieters, C.M.; Forsyth, D.W. Optimization of endmembers for spectral mixture analysis. *Remote Sens. Environ.* **1997**, *59*, 472–489. [CrossRef]
30. Dennison, P.E.; Roberts, D.A.; Thorgusen, S.R.; Regelbrugge, J.C.; Weise, D.; Lee, C. Modeling seasonal changes in live fuel moisture and equivalent water thickness using a cumulative water balance index. *Remote Sens. Environ.* **2003**, *88*, 442–452. [CrossRef]
31. Schaaf, A.N.; Dennison, P.E.; Fryer, G.K.; Roth, K.L.; Roberts, D.A. Mapping Plant Functional Types at Multiple Spatial Resolutions Using Imaging Spectrometer Data. *GISci. Remote Sens.* **2011**, *48*, 324–344. [CrossRef]
32. Roth, K.L.; Dennison, P.E.; Roberts, D.A. Comparing endmember selection techniques for accurate mapping of plant species and land cover using imaging spectrometer data. *Remote Sens. Environ.* **2012**, *127*, 139–152. [CrossRef]
33. Dennison, P.E.; Roberts, D.A. Endmember selection for multiple endmember spectral mixture analysis using endmember average RMSE. *Remote Sens. Environ.* **2003**, *87*, 123–135. [CrossRef]
34. Dennison, P.E.; Halligan, K.Q.; Roberts, D.A. A comparison of error metrics and constraints for multiple endmember spectral mixture analysis and spectral angle mapper. *Remote Sens. Environ.* **2004**, *93*, 359–367. [CrossRef]
35. Roberts, D.A.; Quattrochi, D.A.; Hulley, G.C.; Hook, S.J.; Green, R.O. Synergies between VSWIR and TIR data for the urban environment: An evaluation of the potential for the Hyperspectral Infrared Imager (HyspIRI) Decadal Survey mission. *Remote Sens. Environ.* **2012**, *117*, 83–101. [CrossRef]
36. Keshava, N.; Mustard, J.F. Spectral unmixing. *IEEE Signal Process. Mag.* **2002**, *19*, 44–57. [CrossRef]
37. Veganzones, M.A.; Grana, M. Endmember Extraction Methods: A Short Review. In Proceedings of the International Conference on Knowledge-Based and Intelligent Information and Engineering Systems, Zagreb, Croatia, 3–5 September 2008; pp. 400–407.
38. Parente, M.; Plaza, A. Survey of geometric and statistical unmixing algorithms for hyperspectral images. In Proceedings of the 2010 2nd Workshop on Hyperspectral Image and Signal Processing: Evolution in Remote Sensing (WHISPERS), Reykjavik, Iceland, 14–16 June 2010.
39. Pal, M.; Foody, G.M. Feature Selection for Classification of Hyperspectral Data by SVM. *IEEE Trans. Geosci. Remote Sens.* **2010**, *48*, 2297–2307. [CrossRef]
40. Miao, X.; Gong, P.; Swope, S.; Pu, R.; Carruthers, R.; Anderson, G.L.; Heaton, J.S.; Tracy, C.R. Estimation of yellow starthistle abundance through CASI-2 hyperspectral imagery using linear spectral mixture models. *Remote Sens. Environ.* **2006**, *101*, 329–341. [CrossRef]
41. Green, A.A.; Berman, M.; Switzer, P.; Craig, M.D. A Transformation for Ordering Multispectral Data in Terms of Image Quality with Implications for Noise Removal. *IEEE Trans. Geosci. Remote Sens.* **1988**, *26*, 65–74. [CrossRef]

42. Li, J. Wavelet-based feature extraction for improved endmember abundance estimation in linear unmixing of hyperspectral signals. *IEEE Trans. Geosci. Remote Sens.* **2004**, *42*, 644–649. [CrossRef]
43. Asner, G.P.; Lobell, D.B. A biogeophysical approach for automated SWIR unmixing of soils and vegetation. *Remote Sens. Environ.* **2000**, *74*, 99–112. [CrossRef]
44. Somers, B.; Delalieux, S.; Verstraeten, W.W.; van Aardt, J.A.N.; Albrigo, G.L.; Coppin, P. An automated waveband selection technique for optimized hyperspectral mixture analysis. *Int. J. Remote Sens.* **2010**, *31*, 5549–5568. [CrossRef]
45. Zhao, C.H.; Cui, S.L.; Qi, B. A sparse multiple endmember spectral mixture analysis algorithm of hyperspectral image. In Proceedings of the International Conference on Signal Processing, Hangzhou, China, 19–23 October 2014; pp. 687–692.
46. Peterson, S.H.; Roberts, D.A.; Beland, M.; Kokaly, R.F.; Ustin, S.L. Oil detection in the coastal marshes of Louisiana using MESMA applied to band subsets of AVIRIS data. *Remote Sens. Environ.* **2015**, *159*, 222–231. [CrossRef]
47. Somers, B.; Asner, G.P. Multi-temporal hyperspectral mixture analysis and feature selection for invasive species mapping in rainforests. *Remote Sens. Environ.* **2013**, *136*, 14–27. [CrossRef]
48. Thompson, D.R.; Gao, B.C.; Green, R.O.; Roberts, D.A.; Dennison, P.E.; Lundeen, S.R. Atmospheric correction for global mapping spectroscopy: ATREM advances for the HyspIRI preparatory campaign. *Remote Sens. Environ.* **2015**, *167*, 64–77. [CrossRef]
49. Settle, J.; Campbell, N. On the errors of two estimators of sub-pixel fractional cover when mixing is linear. *IEEE Trans. Geosci. Remote Sens.* **1998**, *36*, 163–170. [CrossRef]
50. Baldridge, A.M.; Hook, S.J.; Crowley, J.K.; Marion, G.M.; Kargel, J.S.; Michalski, J.L.; Thomson, B.J.; De Souza Filho, C.R.; Bridges, N.T.; Brown, A.J. Contemporaneous deposition of phyllosilicates and sulfates: Using Australian acidic saline lake deposits to describe geochemical variability on Mars. *Geophys. Res. Lett.* **2009**, *36*, 1–6. [CrossRef]
51. Roberts, D.A.; Ustin, S.L.; Ogunjemiyo, S.; Greenberg, J.; Dobrowski, S.Z.; Chen, J.; Hinckley, T.M. Spectral and Structural Measures of Northwest Forest Vegetation at Leaf to Landscape Scales. *Ecosystems* **2004**, *7*, 545–562. [CrossRef]
52. Roberts, D.A.; Dennison, P.E.; Gardner, M.E.; Hetzel, Y.; Ustin, S.L.; Lee, C.T. Evaluation of the potential of Hyperion for fire danger assessment by comparison to the airborne visible/infrared imaging spectrometer. *IEEE Trans. Geosci. Remote Sens.* **2003**, *41*, 1297–1310. [CrossRef]
53. Kruse, F.; Lefkoff, A.B.; Boardman, J.W.; Heidebrecht, K.B.; Shapiro, A.T.; Barloon, P.J.; Goetz, A.F.H. The spectral image processing system (SIPS)—Interactive visualization and analysis of imaging spectrometer data. *Remote Sens. Environ.* **1993**, *44*, 145–163. [CrossRef]
54. Cohen, J. A Coefficent of Agreement for Nominal Scales. *Educ. Psychol. Meas.* **1960**, *20*, 37–46.
55. Roberts, D.A.; Alonzo, M.; Wetherley, E.B.; Dudley, K.L.; Dennison, P.E. Multiscale Analysis of Urban Areas Using Mixing Models. In *Integrating Scale in Remote Sensing and GIS*; CRC Press: Boca Raton, FL, USA, 2017; pp. 247–282, ISBN 9781315373720.
56. Adams, J.B.; Gillespie, A.R. *Remote Sensing of Landscapes with Spectral Images: A Physical Modeling Approach*; Cambridge University Press: Cambridge, UK, 2006; ISBN 0521662214.
57. De Santis, A.; Chuvieco, E. GeoCBI: A modified version of the Composite Burn Index for the initial assessment of the short-term burn severity from remotely sensed data. *Remote Sens. Environ.* **2009**, *113*, 554–562. [CrossRef]
58. Veraverbeke, S.; Stavros, E.N.; Hook, S.J. Remote Sensing of Environment Assessing fire severity using imaging spectroscopy data from the Airborne Visible/Infrared Imaging Spectrometer (AVIRIS) and comparison with multispectral capabilities. *Remote Sens. Environ.* **2014**, *154*, 153–163. [CrossRef]
59. Peterson, D.A.; Hyer, E.J.; Campbell, J.R.; Fromm, M.D.; Hair, J.W.; Butler, C.F.; Fenn, M.A. The 2013 Rim Fire: Implications for predicting extreme fire spread, pyroconvection, smoke emissions. *Bull. Am. Meteorol. Soc.* **2015**, *96*, 229–247. [CrossRef]
60. Daughtry, C.S.T. Discriminating Crop Residues from Soil by Shortwave Infrared Reflectance. *Agron. J.* **2001**, *93*, 125–131. [CrossRef]
61. Dudley, K.L.; Dennison, P.E.; Roth, K.L.; Roberts, D.A.; Coates, A.R. A multi-temporal spectral library approach for mapping vegetation species across spatial and temporal phenological gradients. *Remote Sens. Environ.* **2015**, *167*, 121–134. [CrossRef]

62. Franke, J.; Roberts, D.A.; Halligan, K.; Menz, G. Hierarchical Multiple Endmember Spectral Mixture Analysis (MESMA) of hyperspectral imagery for urban environments. *Remote Sens. Environ.* **2009**, *113*, 1712–1723. [CrossRef]
63. Herold, M.; Gardner, M.E.; Roberts, D. Spectral resolution requirements for mapping urban areas. *IEEE Trans. Geosci. Remote Sens.* **2003**, *41*, 1907–1919. [CrossRef]
64. Hamlin, L.; Green, R.O.; Mouroulis, P.; Eastwood, M.; Wilson, D.; Dudik, M.; Paine, C. Imaging Spectrometer Science Measurements for Terrestrial Ecology: AVIRIS and New Developments. In Proceedings of the 2011 IEEE Aerospace Conference, Big Sky, MT, USA, 5–12 March 2011; pp. 1–8.
65. Cansler, C.A.; McKenzie, D. How robust are burn severity indices when applied in a new region? Evaluation of alternate field-based and remote-sensing methods. *Remote Sens.* **2012**, *4*, 456–483. [CrossRef]
66. Morgan, P.; Keane, R.E.; Dillon, G.K.; Jain, T.B.; Hudak, A.T.; Karau, E.C.; Sikkink, P.G.; Holden, Z.A.; Strand, E.K. Challenges of assessing fire and burn severity using field measures, remote sensing and modelling. *Int. J. Wildl. Fire* **2014**, *23*, 1045–1060. [CrossRef]

remote sensing

MDPI

Article

A New Algorithm for the On-Board Compression of Hyperspectral Images

Raúl Guerra *, Yubal Barrios, María Díaz, Lucana Santos, Sebastián López and Roberto Sarmiento

Institute for Applied Microelectronics (IUMA), University of Las Palmas de Gran Canaria (ULPGC), 35001 Las Palmas de Gran Canaria, Las Palmas, Spain; ybarrios@iuma.ulpgc.es (Y.B.); mdmartin@iuma.ulpgc.es (M.D.); lsfalcon@iuma.ulpgc.es (L.S.); seblopez@iuma.ulpgc.es (S.L.); roberto@iuma.ulpgc.es (R.S.)

* Correspondence: rguerra@iuma.ulpgc.es; Tel.: +34-928-451-220

Received: 1 February 2018; Accepted: 6 March 2018; Published: 9 March 2018

Abstract: Hyperspectral sensors are able to provide information that is useful for many different applications. However, the huge amounts of data collected by these sensors are not exempt of drawbacks, especially in remote sensing environments where the hyperspectral images are collected on-board satellites and need to be transferred to the earth's surface. In this situation, an efficient compression of the hyperspectral images is mandatory in order to save bandwidth and storage space. Lossless compression algorithms have been traditionally preferred, in order to preserve all the information present in the hyperspectral cube for scientific purposes, despite their limited compression ratio. Nevertheless, the increment in the data-rate of the new-generation sensors is making more critical the necessity of obtaining higher compression ratios, making it necessary to use lossy compression techniques. A new transform-based lossy compression algorithm, namely *Lossy Compression Algorithm for Hyperspectral Image Systems* (HyperLCA), is proposed in this manuscript. This compressor has been developed for achieving high compression ratios with a good compression performance at a reasonable computational burden. An extensive amount of experiments have been performed in order to evaluate the goodness of the proposed HyperLCA compressor using different calibrated and uncalibrated hyperspectral images from the AVIRIS and Hyperion sensors. The results provided by the proposed HyperLCA compressor have been evaluated and compared against those produced by the most relevant state-of-the-art compression solutions. The theoretical and experimental evidence indicates that the proposed algorithm represents an excellent option for lossy compressing hyperspectral images, especially for applications where the available computational resources are limited, such as on-board scenarios.

Keywords: hyperspectral compression; lossy compression; on-board compression; orthogonal projections; Gram–Schmidt orthogonalization; parallel processing

---

1. Introduction

The algorithms for compressing hyperspectral images, as any other state-of-the-art compression algorithm, take advantage of the redundancies in the image samples to reduce the data volume. Hyperspectral image compression algorithms may take into consideration the redundancies in the spatial and spectral domains for reducing the amount of data with or without losing information. Lossless compression algorithms have been traditionally preferred to preserve all the information present in the hyperspectral cube for scientific purposes despite their limited compression ratio. Nevertheless, the increment in the data-rate of the new-generation sensors is making more critical the necessity of obtaining higher compression ratios, making it necessary to use near-lossless and/or lossy compression techniques.

The general approach for compressing hyperspectral images consists of a spatial and/or spectral decorrelator, a quantization stage and an entropy coder, which tries to use shorter codewords for representing the symbols. The decorrelator can be transform-based or prediction-based. In the transform-based approaches, transforms like the *Discrete Wavelet Transform* (DWT) [1] or the *Karhunen–Loève Transform* (KLT) [2,3] are applied to decorrelate the data, while, in the prediction-based approaches, the samples are predicted from neighbouring (in the spectral or spatial directions) samples, and the predictions errors are encoded. While lossless compression is more efficiently performed by prediction-based methods, transform-based approaches are generally preferred for lossy compression.

In this scenario, the transform-based lossy compression approaches, based on the KLT transform for decorrelating the spectral information, have been proven to yield the best results in terms of rate-distortion as well as in preserving the relevant information for the ulterior hyperspectral analysis [4–7]. In particular, the *Principal Component Analysis* (PCA), which is equivalent to the KLT transform in this context, used for decorrelating and reducing the amount of spectral information, coupled with the JPEG2000 [8] for decorrelating the spatial information and performing the quantization stage and the entropy coding, stands out due to its lossy compression results, which have been demonstrated to be comparatively better than the results provided by other state-of-the-art approaches [5–7]. Indeed, the PCA algorithm has been widely used as a spectral decorrelator not only for compression, but also for other hyperspectral imaging applications such as classification or unmixing, increasing the accuracy of the obtained results [9,10].

Despite their optimal decorrelation features, the KLT approaches, including the PCA algorithm, have important disadvantages that prevent their use in several situations. These disadvantages include an extremely high computational cost, intensive memory requirements, high implementation costs and a non-scalable nature, which make these approaches not suitable for applications under latency/power/memory constrained environments, such as on-board compression. These limitations, as well as the promising compressions results achieved with the KLT approaches, have motivated the appearance of research works that aim to reduce the complexity of the transform by using divide-and-conquer strategies [11]. Nevertheless, the compression performance of these approaches is lower than the performance of the general KLT approach, or, in particular, than the performance of the PCA transform, while their computational complexity is still very high.

A new transform-based algorithm for performing lossy hyperspectral images compression, named *Lossy Compression Algorithm for Hyperspectral image systems* (HyperLCA), has been developed in this work with the purpose of providing a good compression performance at a reasonable computational burden. This compression process consists of three main compression stages, which are a spectral transform, a preprocessing stage and the entropy coding stage.

The compression process within the HyperLCA algorithm has been specifically designed for being able to independently compress blocks of pixels of the hyperspectral image without requiring any specific spatial alignment. The goal is to satisfy the requirements of the compression applications that must be executed under tight resources and latency constraints. The possibility of independently compressing blocks of pixels as they are captured avoids the necessity of storing big portions of the image until being able to compress them, reduces the amount of required resources for compressing the collected data, speeds up the process and provides parallelization and error-resilience. One example of application where this strategy may provide important advantages is the remote sensing on-board compression, especially when using pushbroom or whiskbroom sensors.

The most relevant contribution of this work consists in the HyperLCA spectral transform, which allows performing the spectral decorrelation and compression of the hyperspectral data. This transform is able to achieve high compression rate-distortion ratios with a low computational burden and high level of parallelism. The HyperLCA transform selects the most different pixels of the hyperspectral data to be compressed, and then compresses the image as a linear combination of these pixels. The number of selected pixels directly determines the compression ratio achieved in the compression process, and, hence, the compression ratio achieved by the HyperLCA transform can be perfectly fixed as an input parameter. The subsequent preprocessing and entropy coding proposed stages slightly increase the compression ratio achieved by the HyperLCA transform at a very low computational cost and without introducing additional losses of information. A further advantage of this methodology is that, after selecting each of the pixels used for compressing the image, the information that can be represented by the selected pixel is automatically subtracted from the image. Accordingly, the information remaining in the image corresponds with the information that would be lost in the compression–decompression process if no more pixels were selected. This fact enables the possibility of easily providing a stopping condition according to different quality measures such as the *Signal-to-Noise Ratio* (SNR) or the *Maximum Single Error* (MaxSE).

The HyperLCA algorithm has been developed also considering how the lossy compression–decompression process affects the ulterior hyperspectral imaging applications. Most of the lossy compression approaches typically behave as low pass filters, which may produce a reduction of the noise present in the image, positively affecting the results when processing the decompressed hyperspectral images in some applications [12]; but, at the same time, the low pass filter can also remove the most atypical elements of the image, which are crucial for several applications, such as anomaly detection, classification, unmixing, or target detection [13–17]. In this scenario, the HyperLCA algorithm provides important advantages with respect to the state-of-the-art solutions. Despite being a lossy compression approach, the HyperLCA algorithm is able to perfectly preserve the most different pixels in the data set through the compression–decompression process and also compresses the rest of the pixels introducing minimal spectral distortions, as it will be demonstrated in this paper.

This paper is organized as follows. Section 2 describes the different compression stages of the HyperLCA algorithm. Section 3 shows the followed methodology for evaluating the goodness of the proposed compressor while Section 4 contains the results obtained in the different accomplished experiments. Finally, Section 5 summarizes the conclusions that have been dragged from this work.

2. Hyperspectral Image Compression within the HyperLCA Algorithm

The HyperLCA algorithm is a lossy transform-based compressor specifically designed for providing a good compression performance at a reasonable computational burden for hyperspectral remote sensing applications. The compression process within the HyperLCA algorithm consists of three main compression stages, which are a spectral transform, a preprocessing stage and the entropy coding stage. Figure 1 graphically shows these three compression stages.

The spectral transform stage, carried on by the HyperLCA spectral transform, performs the spectral decorrelation and compression of the hyperspectral data. The HyperLCA transform is able to achieve high compression rate-distortion ratios with a low computational burden and high level of parallelism. The result of the HyperLCA transform consists of three different sets of vectors, as shown in Figure 1. First of all, a vector with N_b components that corresponds with the average pixel of the image or centroid pixel, c, where N_b refers to the number of bands of the hyperspectral image. Secondly, the HyperLCA transform also provides a set of vectors of N_b components that contains real pixels of the hyperspectral image selected by the transform for being the most different pixels in the data set. This set of pixels vectors is referred to as *Pixels* in the rest of the manuscript. Finally, the HyperLCA transform provides a set of vectors of N_p components, where N_p refers to the number of pixels of the image, which allows linearly combining the selected pixels, *Pixels*, for recovering the real hyperspectral image. This set of vectors is referred as *V vectors* in the rest of the manuscript.

After performing the HyperCLA transform, the HyperLCA algorithm executes a preprocessing stage followed by an entropy coding stage for slightly increasing the compression ratio achieved by the HyperLCA transform at a very low computational cost and without introducing further losses of information. These two compression stages independently process each of the different *Pixels* and *V vectors* as well as the centroid pixel, c. The main goal of the preprocessing stage is to make a very simple prediction of the different vectors' values, based on the previous value of the specific vector under analysis, and map the prediction error using positive values closer to zero in order to achieve a higher performance in the entropy coding stage. Before the prediction and error mapping, the *V vectors* are also scaled to positive integer values that perfectly fit the dynamic range produced by the number of bits, N_{bits} defined by the user, as shown in Figure 1. Once that each individual vector is independently preprocessed, its values are entropy encoded using a Golomb–Rice coder [18].

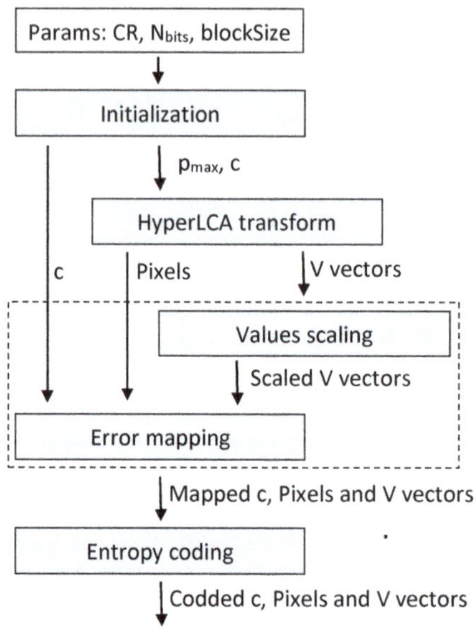

Figure 1. Diagram of the HyperLCA algorithm compression stages.

The HyperLCA algorithm has different characteristics that represent important advantages for remote sensing hyperspectral imaging applications. First of all, the compression process within the HyperLCA algorithm has been specifically designed for being able to independently compress blocks of pixels of the hyperspectral image without requiring any specific spatial alignment. The goal is to satisfy the remote sensing on-board compression requirements, especially when using pushbroom or whiskbroom sensors for collecting the images, allowing independently compressing the blocks of pixels as they are captured, as shown in Figure 2. This strategy avoids the necessity of storing large amounts of data until being able to compress them, reduces the amount of required resources for compressing the collected data, speeds up the process and provides parallelization and error-resilience. Additionally, the process performed by the HyperLCA algorithm for compressing each single block of pixels is highly parallel and has a low computational complexity in relation to other state-of-the-art transform-based approaches.

Secondly, the HyperLCA transform selects the most different pixels from the data set. These pixels are preprocessed and coded without losing information and hence they are perfectly preserved through the compression–decompression process. This is one of the most important differences of the HyperLCA algorithm with respect to other state-of-the-art lossy compression approaches, in which the most different pixels are typically lost in the compression. This fact represents a very important advantage for hyperspectral applications such as anomalies detection, target detection, tracking or classification, where it is really important to preserve the anomalous pixels.

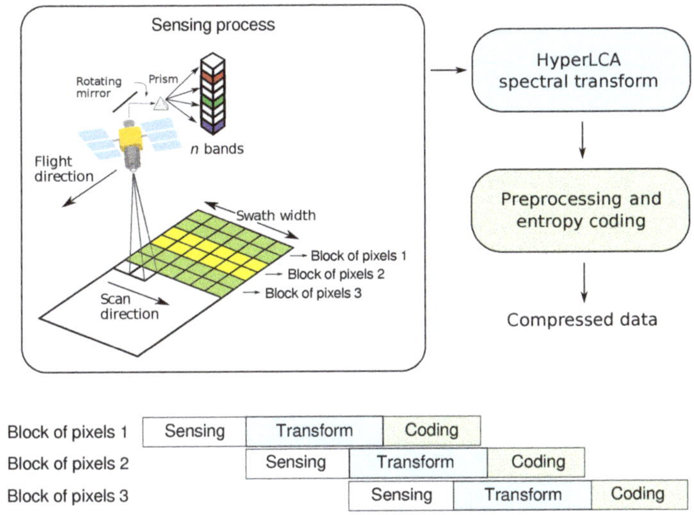

Figure 2. Diagram of the HyperLCA algorithm for independently compressing each block of pixels of the hyperspectral image.

Finally, the losses in the compression–decompression process as well as most of the compression ratio obtained are achieved in the spectral transformed stage, carried on by the HyperLCA transform. This provides two more important advantages. On one side, after selecting each of the pixels used for compressing the image (extracting one *Pixel* vector and its corresponding *V vector*), the information that can be represented by the selected pixel is automatically subtracted from the image. Accordingly, the information remaining in the image corresponds with the information that would be lost in the compression–decompression process if no more pixels were selected. This fact enables the possibility of easily providing a stopping condition according to different quality measures such as the *Signal-to-Noise Ratio* (SNR) or the *Maximum Single Error* (MaxSE). By doing so, if the stopping condition is satisfied, the process finishes and no more *Pixels* or *V vectors* are extracted, else, one new *Pixel* vector is extracted, its corresponding *V vector* is calculated and the stopping condition is checked again. This procedure also enables a progressive decoding of the compressed bitstream. The image can be reconstructed using just the first *Pixels* and *V vectors* in the same order that they are received, and progressively add the information of the subsequent *Pixels* and *V vectors* contained in the bitstream if a higher quality is desired.

On the other side, the number of pixels selected by the HyperLCA transform directly determines the compression ratio achieved in the transform stage, and, hence, the minimum compression ratio to be achieved by the HyperLCA transform can be perfectly fixed as an input parameter, ensuring that the compression ratio achieved in the overall process will be always higher (higher means smaller compressed data sets). The compression ratio is defined as *CR = (bits real image)/(bits compressed data)* in this manuscript, and is used by the HyperLCA transform for determining the maximum number of *V vectors* and *Pixels* to be extracted, $p_{\max}$. Once $p_{\max}$ *Pixels* and *V vectors* have been extracted, the HyperLCA transform finishes, even if the stopping condition based on a quality metric has not been satisfied.

Sections 2.1–2.8 deeply explain each of the different stages of the HyperLCA algorithm for compressing hyperspectral images.

2.1. HyperLCA Input Parameters

The HyperLCA algorithm has three main input parameters that need to be defined.

- *Compression ratio, CR*, which indicates the maximum amount of data desired in the compressed image. The HyperLCA algorithm will provide a compressed image at least *CR* times smaller than the original one.
- *Number of pixels per block, blockSize*. This is the number of pixels, with all their hyperspectral bands, to be independently compressed by the HyperLCA algorithm as a single block.
- *Bits for compressed image vectors*, N_{bits}. This is the number of bits used for scaling and coding the values of the $V = [v_1, v_2, ..., v_{p_{\max}}]$ vectors, as described in Section 2.5. The HyperLCA algorithm has been tested in this work using $N_{\text{bits}} = 16$, $N_{\text{bits}} = 12$ and $N_{\text{bits}} = 8$.

2.2. HyperLCA Initialization

The HyperLCA algorithm needs to accomplish two main operations before performing the HyperLCA transform according to the characteristics of the image to be compressed and parameters introduced by the user.

2.2.1. Determining the Number of Pixels to be Extracted by the HyperLCA Transform

The compression ratio achieved by the HyperLCA transform is directly determined by the number of extracted vectors and the number of bits used for representing the compressed vectors $V = [v_1, v_2, ..., v_{p_{\max}}]$. Accordingly, the maximum number of vectors to be extracted by the HyperLCA transform for each block of pixels, $p_{\max}$, is previously calculated according to the number of bands and the dynamic range of the image to be compressed, and the input parameters, *CR*, *blockSize* and N_{bits} as shown in Equation (1), where $N_{p_{\text{block}}}$ is the number of pixels per block (*blockSize*), *CR* is the minimum compression ratio desired, N_{bits} is the number of bits used for representing the compressed vectors $V = [v_1, v_2, ..., v_{p_{\max}}]$, and *DR* and N_b refer to the dynamic range and number of bands of the hyperspectral image to be compressed, respectively:

$$p_{\max} \leq \frac{DR \cdot (N_b \cdot (N_{p_{\text{block}}} - 1))}{CR \cdot (DR \cdot N_b + N_{\text{bits}} \cdot N_{p_{\text{block}}})}. \tag{1}$$

The process can be simplified by using the same precision for representing the compressed vectors $V = [v_1, v_2, ..., v_{p_{\max}}]$, N_{bits}, than for representing the image to be compressed, *DR*, and the Equation (1) would result in Equation (2)

$$p_{\max} \leq \frac{N_b \cdot (N_{p_{\text{block}}} - 1)}{CR \cdot (N_b + N_{p_{\text{block}}})}. \tag{2}$$

2.2.2. Calculating the Centroid Pixel for Each Block of Pixels

The HyperLCA transform requires the previous calculation of the average or centroid pixel, c, for every block of pixels to be processed. With independence of the data precision used for calculating c, it is rounded to the closest integer value before starting the HyperLCA transform stage. This has two main purposes. First of all, it eases the preprocessing and entropy coding stages, since these two stages need to work with integer values, as described in Section 2.5. Additionally, rounding c to integer values before performing the transformation stage ensures using the exact same vector c in both the HyperLCA transform and inverse transform, which increases the overall accuracy of the compression–decompression process.

2.3. HyperLCA Transform

The HyperLCA transform sequentially selects the most different pixels of the hyperspectral data set. The set of selected pixels is then used for projecting the hyperspectral image, obtaining a spectral decorrelated and compressed version of the data. The compression achieved within this process directly depends on the number of selected pixels. Selecting more pixels provides better decompressed images but lower compression ratios, understanding the compression ratio as the relation between the size of the real image and the compressed one (the higher, the better). Since the pixels are sequentially selected, the sooner the algorithm stops, the higher the compression ratio will be. The proposed method takes advantage of this fact for allowing the user to determine a minimum desired compression ratio, which is used for calculating the maximum number of pixels to be extracted. By doing so, the proposed method ensures the achievement of a compression ratio that will always be higher than the compression ratio specified by the user.

Moreover, each time that a pixel is selected, the information of the image that can be represented using the selected pixel is subtracted from the image. Accordingly, the remaining information directly corresponds with the information that would be lost if no more pixels were selected. This can be used for stopping the sequential extraction of pixels once a specific accuracy level is achieved in the spectral transform step, according to any desired evaluation measurement, without reaching the maximum number of pixels to be extracted, as described in Section 2.8. This way, the proposed method guarantees that if an accurate enough compression of the hyperspectral image is obtained before extracting all the required pixels according to the minimum compression ratio specified by the user, the algorithm will stop, providing a higher compression ratio at a lower computational burden.

Finally, both the extracted pixels as well as information of the image compressed as a linear combination of these pixels are set as the outputs of the HyperLCA transform. This is due to the fact that both are needed for recovering the original hyperspectral image.

The pseudocode that describes the HyperLCA transform for spectrally decorrelating and reducing the hyperspectral data set is presented in Algorithm 1, where matrix M contains N_p real hyperspectral pixels $[r_1, r_2, ..., r_{N_p}]$, placed in columns. Each of these pixels is a vector of N_b components, N_b being the number of bands of the hyperspectral image. The input parameter $p_{\max}$, defined in Section 2.1, determines the maximum number of pixels to be extracted. The input vector c corresponds with the average pixel of the image, also called centroid, in integer values, as described in Section 2.2. Additionally, the set of vectors $P = [p_1, p_2, ..., p_{p_{\max}}]$ and $V = [v_1, v_2, ..., v_{p_{\max}}]$ contain the vectors that are extracted by the HyperLCA transform as the compressed information. Specifically, P will store the $p_{\max}$ real hyperspectral pixels selected by the HyperLCA transform as the most different pixels of the data set (*Pixels*), and vectors contained in V will store the information of the image that can be represented by the extracted pixels (*V vectors*). Each of these vectors has N_p components, one per pixel. These vectors are used by the inverse transform for recovering the real image.

First of all, the hyperspectral data, M, is centered and stored in M_c, in line 3. This is done by subtracting the centroid pixel to all the pixels of the image. The amount of information present in matrix M_c decreases as more pixels are extracted. However, the real image matrix M is not modified.

Secondly, the pixels are sequentially extracted in lines 4–15. In this process, the brightness of each pixel within the M_c image is first calculated in lines 5–7. The extracted pixels are selected as those pixels from M that correspond with the highest brightness in matrix M_c, as shown in line 9. Then, the orthogonal projection vectors q and u are accordingly obtained as shown in lines 10 and 11.

After that, the information that can be spanned by the defined orthogonal vectors u and q is stored in the projected image vector v_p and subtracted from the M_c image in lines 12–13. The process finishes when the $p_{\max}$ pixels p_j have been selected and the information of the image that they can span has been stored in the $p_{\max}$ v_j vectors, or if an additional stopping condition is previously accomplished, as described in Section 2.8.

This methodology provides one important advantage for compressing and decompressing images for hyperspectral imaging applications. As described in lines 8 and 9 of Algorithm 1, the pixels to be transferred are selected as those with the largest amount of remaining information. By doing this, it is guaranteed that the most different pixels within the data set are perfectly preserved through the spectral decorrelation and compression steps of the compression–decompression process. This fact makes this compression approach especially suitable for applications in which some pixels may be very different from the rest of the pixels, such as anomaly detection, target detection, tracking or classification.

Algorithm 1: HyperLCA transform.

Inputs: $M = [r_1, r_2, ..., r_{N_p}]$, $p_{\max}$, c

$P = [p_1, p_2, ..., p_{p_{\max}}]$; {Extracted pixels.}
$V = [v_1, v_2, ..., v_{p_{\max}}]$; {Projected image vectors.}
$M_c = [x_1, x_2, ..., x_{N_p}]$ {Centralized version of M}
for $i = 1$ **to** $p_{\max}$ **do**
    for $j = 1$ **to** N_p **do**
        $b_j = {x_j}^t \cdot x_j$;
    end
    $j_{\max} = \arg\max(b_j)$;
    $p_i = r_{j_{\max}}$;
    $q = x_{j_{\max}}$;
    $u = x_{j_{\max}} / ((x_{j_{\max}})^t \cdot x_{j_{\max}})$;
    $v_i = u^t \cdot M_c$;
    $M_c = M_c - v_i \cdot q$;
    {Additional stopping condition.}
end

Outputs: $P = [c, p_1, p_2, ..., p_{p_{\max}}]$, $V = [v_1, v_2, ..., v_{p_{\max}}]$

2.4. HyperLCA Inverse Transform

The inverse HyperLCA transform linearly combines the centroid pixel, c, the $P = [p_1, p_2, ..., p_{p_{\max}}]$ extracted pixels and $V = [v_1, v_2, ..., v_{p_{\max}}]$ extracted vectors for reconstructing the hyperspectral image, obtaining the decompressed hyperspectral image, M'. The pseudocode that describes the inverse HyperLCA transform is presented in Algorithm 2. As shown in lines 2–4 of this pseudocode, all the pixels of M' are firstly initialized as the centroid pixel, c. After doing so, the centroid pixel, c, is subtracted to the pixels selected by the HyperLCA transform, as shown in lines 5–7. Finally, the decompressed image, M', is obtained by sequentially adding the result of each of the v_i vectors projected using the corresponding orthogonal vector q, as shown in lines 9–11. Lines 12–14 of the pseudocode show how the information that can be represented by the pixels that have been already used (p_i) is subtracted from the pixels that will be used in the next iterations ($p_{j=i+1}$ to $p_{j=p\max}$) of the inverse HyperLCA transform.

2.5. HyperLCA Preprocessing

This compression stage of the HyperLCA algorithm is executed after the HyperLCA transform for adapting the output data for being entropy coded in a more efficient way. This compression stage is divided into two different parts.

2.5.1. Scaling the V Vectors

The HyperLCA transform provides two different sets of vectors as a result. These two different sets of vectors, which have different characteristics, are needed for reconstructing the image using the inverse HyperLCA transform, and, hence, both should be entropy coded and transferred. The entropy coder proposed in this manuscript works with integer values. Accordingly, these two sets of vectors must be represented as integers.

Algorithm 2: Inverse HyperLCA transform.

```
Inputs: P = [p_1, p_2, ..., p_{p_max}], V = [v_1, v_2, ..., v_{p_max}], p_max, c
M' = [d_1, d_2, ..., d_{N_p}]; {Decompressed image.}
for k = 1 to N_p do
    d_k = c;
end
for i = 1 to p_max do
    p_i = p_i - c;
end
for i = 1 to p_max do
    q = p_i;
    u = p_i / (p_i^t · p_i);
    M' = M' + q · v_i;
    for j = i + 1 to p_max do
        p_j = p_j - (u^t · p_i) · q;
    end
end
Outputs: M' = [d_1, d_2, ..., d_{N_p}]
```

The first set of vectors obtained contains the centroid pixel, c, used for initializing the process, as well as the pixels selected by the HyperLCA transform, $P = [p_1, p_2, ..., p_{p_{\max}}]$. The pixels selected by the transform are already integers, since they are directly pixels of the image. The centroid pixel to be transferred has also been rounded to integers' values before starting the HyperLCA transform, as described in Section 2.2. Accordingly, adapting this set of vectors for the entropy coder is a straightforward step.

On the other hand, the second set of vectors obtained by the HyperLCA transform, $V = [v_1, v_2, ..., v_{p_{\max}}]$, contains the projection of the image pixels into the space spanned by the different orthogonal projection vector, u, calculated in each iteration. The value obtained when projecting one pixel vector over one specific vector u will be determined by the angle between both vectors and their magnitude, according to Equation (3), which describes the scalar product between two vectors:

$$u^t \cdot x_i = \frac{||u|| \cdot ||x_i||}{\cos(\theta).} \tag{3}$$

According to this equation, larger modules and smaller angles (more parallel vectors) will produce higher scalar product values. In the HyperLCA transform, we have selected the vector u as $u = x_{j\max} / ((x_{j\max})^t \cdot x_{j\max})$, $x_{j\max}$ being the brightest pixel in the image (the pixel with the largest magnitude). Hence, if we perform the scalar product of all the image pixels and the u vector as shown in Line 12 of Algorithm 1, $v_i = u^t \cdot M_c$, the image pixel $x_{j\max}$ will be the one producing the largest result, which will be exactly 1, as shown in Equations (4)–(8):

$$v_{j\max} = u^t \cdot x_{j\max} = (\frac{{x_{j\max}}^t}{((x_{j\max})^t \cdot x_{j\max})}) \cdot x_{j\max}, \tag{4}$$

$$v_{j\max} = u^t \cdot x_{j\max} = (\frac{{x_{j\max}}^t}{||x_{j\max}||^2}) \cdot x_{j\max}, \tag{5}$$

$$v_{j\max} = u^t \cdot x_{j\max} = (\frac{{x_{j\max}}^t}{||x_{j\max}||}) \cdot (\frac{x_{j\max}}{||x_{j\max}||}), \tag{6}$$

$$v_{j\max} = u^t \cdot x_{j\max} = \frac{||(\frac{x_{j\max}}{||x_{j\max}||})|| \cdot ||(\frac{x_{j\max}}{||x_{j\max}||})||}{\cos(\theta)}, \tag{7}$$

$$v_{j_{\max}} = u^t \cdot x_i = \frac{1 \cdot 1}{\cos(0)} = 1. \tag{8}$$

Since the rest of the image pixels $x_{j \neq j_{\max}}$ have smaller magnitudes than $x_{j_{\max}}$ and wider θ angles with respect to the vector u, the values $v_{j \neq j_{\max}}$ will be between -1 and 1. Accordingly, it can be said that all the values of the vectors $V = [v_1, v_2, ..., v_{p_{\max}}]$ will be between -1 and 1, as shown in Equation (9).

$$\forall v_j \in V : -1 < v_j \leq 1 \tag{9}$$

According to this particular property of the vectors $V = [v_1, v_2, ..., v_{p_{\max}}]$ we can easily scale the v_j values for representing them in integer values using all the dynamic range in order to avoid loosing too much precision in the conversion. The v_j values are scaled in the HyperLCA algorithm according to the input number of bits , N_{bits}, defined by the user for this purpose, as shown in Equation (10):

$$v_{j_{\text{scaled}}} = (v_j + 1) \cdot (2^{N_{\text{bits}}-1} - 1). \tag{10}$$

By doing this, it is guarantied that the maximum value obtained for each scaled v_j vector is always $2^{N_{\text{bits}}} - 1$, and its minimum value is always positive and very close to zero. After rescaling the v_j values, they are rounded to the closer integer values so they can be entropy coded.

2.5.2. Represent the Data with Positive Values Closer to Zero

The entropy coding stage takes advantage of the redundancies within the data to be coded, assigning shorter word length to the most common values. In order to achieve higher compression ratios in this stage, the centroid pixel, c, the selected pixels $P = [p_1, p_2, ..., p_{p_{\max}}]$, and the already scaled and converted to integer vectors $V = [v_1, v_2, ..., v_{p_{\max}}]$, are lossless preprocessed in the HyperLCA algorithm, making use of the prediction error mapper described in the CCSDS recommended standard for lossless multispectral and hyperspectral image compression [19]. This compression stage independently processes each individual vector in order to represent its values using only positive integer values that are closer to zero than the original values of the vector, using the same dynamic range of the values.

Let us assume that we have a vector $Y = [y_1, y_2, ...]$ whose components y_j are represented using *n-bits* integers values. In the HyperLCA compressor, this vector Y may be either the centroid pixel, c, a pixel vector p_i or a v_i vector. Due to the spectral redundancies between contiguous bands, when the vector Y corresponds with a pixel vector p_i or the centroid pixel c, we can assume that the difference between the y_j and y_{j-1} components of the vectors is closer to zero than the y_j value itself. Similarly, we can make the same assumption when the vector Y corresponds with a v_i vector, due to the spatial redundancies of the image. However, in order to prevent bit overflowing in this operation, the number of bits used for representing the prediction error, $\triangle_j = y_j - y_{j-1}$, needs to be increased to *(n+1)-bits*, and the values will be in the range $(-2^n + 1, 2^n - 1)$. In order to solve this issue, the prediction error, $\triangle_j = y_j - y_{j-1}$, is mapped using the aforementioned prediction error mapper [19].

The overall process works as follows. First of all, the possible minimum and maximum $(y_{\min}, y_{\max})$ values are calculated as $(-2^{n-1}, 2^{n-1} - 1)$ when the vector Y contains negative integer values and $(0, 2^n - 1)$ when it does not. Then, θ_j is calculated as $\theta_j = minimum(y_{j-1} - y_{\min}, y_{\max} - y_{j-1})$. Finally, the prediction error $\triangle_j = y_j - y_{j-1}$ is mapped according to the $\triangle_j$ and θ_j values as shown in Equation (11):

$$Y_{j_{\text{mapped}}} = \begin{cases} 2\triangle_j, & 0 \leq \triangle_j \leq \theta_j, \\ 2\|\triangle_j\| - 1, & -\theta_j \leq \triangle_j < 0, \\ \theta_j + \|\triangle_j\|, & \text{otherwise}. \end{cases} \tag{11}$$

By doing this, the values of the centroid pixel, c, the selected pixels $P = [p_1, p_2, ..., p_{p_{\max}}]$, and the already scaled and converted to integer vectors $V = [v_1, v_2, ..., v_{p_{\max}}]$, are represented using positive values closer to zero, and using the same amount of bits.

2.6. HyperLCA Entropy Coding

After the preprocessing stage of the HyperLCA compressor, the extracted vectors (centroid, Pixels and V vectors) are independently coded using a Golomb–Rice coding strategy. Each single vector is coded as follows:

- First of all, the compression parameter, M, is calculated as the average value of the vector.
- Secondly, the lowest power of 2 higher than M, 2^b, is calculated as $b = \log_2(M) + 1$.
- Then, each value of the vector is divided by the calculated parameter, M, obtaining the division quotient, q, and the reminder, r.
- Finally, each value of the vector is coded according to the q and r values obtained, as a code word composed by the quotient code followed by the reminder code.
    - The quotient code is obtained by coding the q-value using unary code ($q + 1$ bits are required).
    - The reminder code is obtained from the r-value. If M is a power of 2, the reminder code is obtained by coding r as plain binary using b bits. If M is not a power of 2, and $r < 2^b - M$, the reminder code is obtained by coding r in plain binary using $b - 1$ bits. In any other situation, the reminder code is obtained by coding $r + 2^b - M$ in plain binary using b bits.

The fact of independently coding each vector provides different advantages. On one side, it is possible to use the average value of each vector as the compression parameter, M, which provides almost the best coding performance for the Golomb–Rice method [20], without incrementing too much the complexity of the coder. Nevertheless, in those situations in which calculating the average value of the vector represents an important disadvantage due to its complexity, other solutions could be used, such as the median value, as it is done in other compression algorithms, without compromising its good performance.

On the other side, the fact of independently coding each vector allows coding them in the same order that they are obtained in the previous compression stages of the HyperLCA algorithm. This eases the parallelization of the process, making it possible to pipeline the inputs and outputs of the different compression stages for a single block of pixels, and also reducing the amount of memory required.

2.7. HyperLCA Bitstream Generation

Finally, the outputs of the previous compression stages are packed in the order that they are produced, generating the compressed bitstream. By doing so, the computational requirements for accomplishing this last step of the HyperLCA compressor are minimal. Additionally, this order also eases the decompression process, since the compressed data is used for reconstructing the image in the exact same order that it is produced by the compressor. Figure 3 graphically shows the produced bitstream structure.

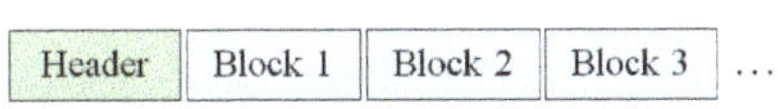

Figure 3. General structure of the bitstream generated by the HyperLCA algorithm

According to Figure 3, the first part of the bitstream is a header that contains the global information about the hyperspectral image and the parameters used in the compression process within the HyperLCA algorithm, which are needed for decompressing the image. Then, the compressed information of each individual block of pixels is packed in the exact same way and sequentially added to the bitstream, as shown in Figure 3.

The information contained in the header of the generated bitstream is:

- Size of the hyperspectral image, N_c, N_r and N_b, representing the number of columns, rows and bands, coded as plain binary using 16 bits each.
- The number of pixels per block, *blockSize*, used for compressing the image within the HyperLCA algorithm, coded as plain binary using 16 bits.
- The maximum number of *V vectors* and *Pixels* extracted for each block of pixels, p_{max}, coded as plain binary using 8 bits.
- The number of bits needed for covering the entire dynamic range of the hyperspectral image, *DR*, as well as the number of bits used for scaling the *V vectors*, N_{bits}, coded as plain binary using 8 bits each.
- One extra bit, *SC*, which indicates if one additional stopping condition, based on a quality metric, has been used ($SC = 1$) or not ($SC = 2$).

Figure 4 graphically describes the header structure. In this figure, *bS* represents the number of pixels per block, *blockSize*. According to the defined number of bits used for the different data contained in the header, 89 bits are required.

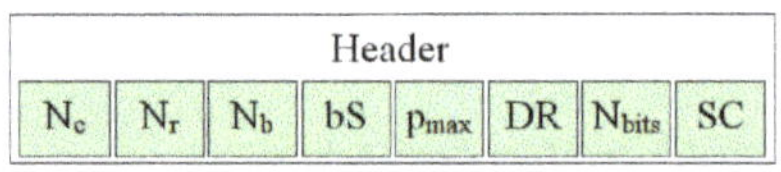

Figure 4. Structure of the bitstream header.

The information of each individual block of pixels may be packed in two ways that contain a minimal difference, as shown in Figure 5. The first one, graphically described in Figure 5a, is used when no additional stopping condition is used, and p_{max} *Pixels* and *V vectors* are extracted for each individual block. In this situation, the amount of compressed elements in each block is perfectly determined by the p_{max}, N_p and N_b values that are already included in the bitstream header. On the contrary, when using an additional stopping condition for stopping the algorithm if the desired quality is achieved before extracting p_{max} *Pixels* and *V vectors*, the number of *Pixels* and *V vectors* extracted for each block of pixels may be different and smaller than p_{max}, producing higher compression ratios. In this situation, the number of *Pixels* and *V vectors* used for compressing each block of pixels, $p \leq p_{\text{max}}$, needs to be included when coding each block of pixels, as shown in Figure 5b. This $p \leq p_{\text{max}}$ value is coded as plain binary using 8 bits.

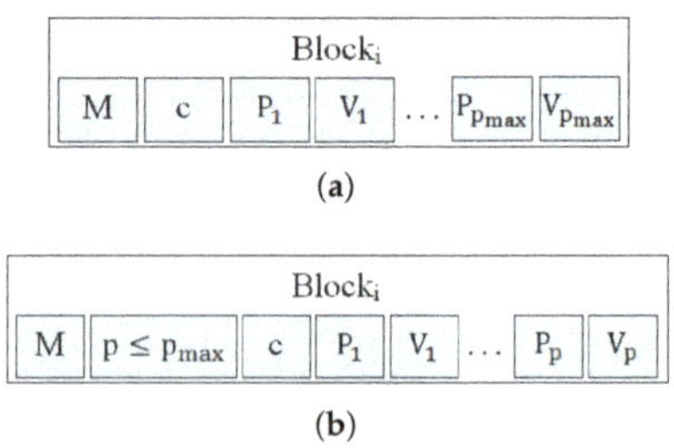

Figure 5. Structure of the bitstream for each block of pixels compressed by the HyperLCA algorithm. (**a**) without using extra stopping conditions; (**b**) using extra stopping conditions based on quality metrics.

2.8. Possible Stopping Conditions within the HyperLCA Algorithm

As previously described, all the losses of the compression–decompression process within the HyperLCA algorithm are produced in the spectral transformation and compression stage, carried on by the HyperLCA transform. As aforementioned, the HyperLCA transform sequentially selects the most different pixels of the image, and uses them for representing the hyperspectral data. Each time that a pixel is selected, the information that can be represented using that specific pixel is subtracted from the image, M_c, as described in Line 13 of the Algorithm 1. Accordingly, matrix M_c contains the information that would be lost if no more pixels were selected. This information can be used for setting an stopping condition in the sequential pixels selection process, based on the amount of losses desired within the compression–decompression process. For such purpose, two different approaches can be followed:

- *Global error approaches* measure the global or average error in the decompressed image with respect to the original one. Despite these approaches providing a good idea of the accuracy of the compression–decompression process, they do not necessarily yield an accurate reconstruction of the anomalous pixels.
- *Single error approaches* focus on measuring the largest single errors in the decompressed image. Despite these approaches not necessarily providing a very good idea of the overall compression–decompression performance, they are more suitable for verifying the accuracy of the reconstruction of the anomalous pixels.

Although many different metrics can be efficiently applied to the proposed HyperLCA transform, we will focus on the RMSE, the SNR and the MaxSE metrics since these are widely used state-of-the-art metrics for evaluating the compression–decompression performance. While the RMSE and SNR are global evaluation metrics that are useful for verifying an average good compression–decompression performance, the MaxSE, despite its simplicity, is more suitable for ensuring that the anomalous pixels are preserved through the compression–decompression process. Any of the possible stopping conditions based on these metrics would be placed in line 14 of the Algorithm 1, and would prevent the extraction of more pixels if the losses achieved are small enough according to the quality metric used.

2.8.1. RMSE Based Stopping Condition

The *Root Mean Square Error* (RMSE) is a frequently used metric for measuring the average differences between the real hyperspectral image and the decompressed one. A low RMSE represents low average compression–decompression errors. This metric is defined as:

$$\text{RMSE} = \frac{1}{N_p \cdot N_b} \cdot \sqrt{\sum_{i=1}^{N_b} \sum_{j=1}^{N_p} (M_{i,j} - M'_{i,j})^2}, \tag{12}$$

where M and M' refer to the real hyperspectral image and to the decompressed one, and N_p and N_b refer to the number of pixels and the number of bands in the hyperspectral image, respectively. Since M_c contains the information that cannot be represented with the already selected pixels, it could be directly calculated within the HyperLCA algorithm as:

$$\text{RMSE} = \frac{1}{N_p \cdot N_b} \cdot \sqrt{\sum_{i=1}^{N_b} \sum_{j=1}^{N_p} (M_{c_{i,j}})^2}. \tag{13}$$

2.8.2. SNR Based Stopping Condition

The *Signal-to-Noise Ratio* (SNR) generally compares the level of the desired signal to the level of background noise. It is defined as the ratio of signal power to the noise power, often expressed in decibels. In the compression scenario, the SNR metric measures the ratio between the real hyperspectral image power and the compression–decompression loses power, as shown in Equation (14), where M and M' refer to the real hyperspectral image and to the decompressed one, and N_p and N_b refer to the number of pixels and the number of bands in the hyperspectral image, respectively:

$$\mathrm{SNR} = 10 \cdot \log_{10}\left(\frac{\sum_{i=1}^{N_b}\sum_{j=1}^{N_p}(M_{i,j})^2}{\sum_{i=1}^{N_b}\sum_{j=1}^{N_p}(M_{i,j} - M'_{i,j})^2}\right). \quad (14)$$

As aforementioned, $(M_{i,j} - M'_{i,j})$ directly corresponds with $M_{c_{i,j}}$. Hence, the SNR can be calculated within the proposed HyperLCA transform every time that a new pixel is extracted as:

$$\mathrm{SNR} = 10 \cdot \log_{10}\left(\frac{\sum_{i=1}^{N_b}\sum_{j=1}^{N_p}(M_{i,j})^2}{\sum_{i=1}^{N_b}\sum_{j=1}^{N_p}(M_{c_{i,j}})^2}\right). \quad (15)$$

As a further optimization, $\sum_{i=1}^{N_b}\sum_{j=1}^{N_p}(M_{i,j})^2$ can be calculated just once at the beginning of the compression process, since it evaluates just the power of the real hyperspectral image, M.

Finally, it is important to remark that the HyperLCA transform is thought to be independently applied to blocks of pixels of the image, not to the entire image at once. Accordingly, these stopping conditions must be also independently applied to each block of pixels, and, hence, the number of pixels shown in this equations, N_p, should be substituted by the number of pixels per block, defined as $blockSize$.

2.8.3. MaxSE Based Stopping Condition

The *Maximum Single Error* (MaxSE) evaluates the maximum absolute difference between the real hyperspectral image and the compressed-decompressed one. The MaxSE that would be obtained in the compression–decompression process within the HyperLCA algorithm if no more pixels were extracted can be directly calculated, in each iteration, as the maximum absolute value of M_c.

2.9. Computational Complexity of the HyperLCA Compressor

The HyperLCA algorithm has been specifically designed for being able to independently compress blocks of pixels of the hyperspectral image without requiring any specific spatial alignment. This fact eases the image compression, especially when using pushbroom or whiskbroom sensors for collecting them, allowing for independently compressing the blocks of pixels as they are captured. This strategy avoids the necessity of storing big portions of the image until being able to compress them, reduces the amount of required resources for compressing the collected data, speeds up the process and provides parallelization and error-resilience. Besides the obvious gains that this fact brings in terms of lower computational complexity, the HyperLCA algorithm has the advantage that it uses simple mathematical operations that can be easily parallelized, avoiding complex matrix operations like, for instance, computing the eigenvalues and eigenvectors, operations that are present in many KLT based compression approaches.

The proposed HyperLCA compressor consists of different compression stages. The first and most computationally demanding one is the HyperLCA transform. This transform spectrally decorrelates each block of pixels of the image and reduces its number of spectral components, according to the specified compression ratio and/or quality measure stopping condition. The amount of data resulting from this transform is already much smaller than the original image. This, together with the simplicity of the subsequent compression stages, makes the computational complexity of these compression stages negligible in relation with the computational burden of the HyperLCA transform. Hence, this section focuses on analysing the computational complexity of the HyperLCA transform.

In order to simplify the analysis of the computational complexity of the HyperLCA transform, the description shown in Algorithm 1 has been followed, considering that no stopping condition based on quality metrics is used, and, hence, the maximum number of pixels to be extracted, $p_{\max}$, is always reached. In particular, the total number of operations done by the HyperLCA transform has been estimated. Despite the fact that the HyperLCA transform can be executed using integer or floating point values, the number of floating point operations (FLOPs) has been considered for simplicity.

Additionally, since the HyperLCA transform is independently applied to each block of pixels, the estimation of the number of FLOPs has been done for a single block of $N_{p_{\text{block}}}$ pixels. The amount of FLOPs required for processing the entire image directly scales with the number of blocks.

As shown in Algorithm 1, there are three sets of operations that are applied to all the pixels in each of the $p_{\max}$ iterations of the HyperLCA transform, and represents the majority of its required FLOPs. These are:

- The calculation of the amount of remaining information in each pixel by calculating its brightness (lines 5–7 of Algorithm 1). The calculation of the brightness of one pixel corresponds with the inner product between two vectors of N_b components, which results in $2 \cdot N_b$ FLOPs. This applied to all the pixels of the block produces a total of $2 \cdot N_b \cdot N_{p_{\text{block}}}$ FLOPs.
- The calculation of the information that can be spanned by the selected pixel, obtaining the corresponding *V vector* (line 12 of Algorithm 1). This also corresponds with one inner product between two vectors of N_b components for each pixel of the image, resulting in $2 \cdot N_b \cdot N_{p_{\text{block}}}$ FLOPs.
- The subtraction of the information that can be spanned by the selected pixel (line 13 of Algorithm 1). This consists of first multiplying the $1xN_{p_{\text{block}}}$ vector, v_i, with the N_bx1 vector, q_i, obtaining a $N_bxN_{p_{\text{block}}}$, and then subtracting this matrix to the image matrix, M_c. Both steps require $N_b \cdot N_{p_{\text{block}}}$ FLOPs, resulting in a total of $2 \cdot N_b \cdot N_{p_{\text{block}}}$.

The total number of FLOPs required by these three sets of operations is $6 \cdot N_b \cdot N_{p_{\text{block}}}$. Since these operations are done once per each of the $p_{\max}$ iterations of the HyperLCA transform, $6 \cdot p_{\max} \cdot N_b \cdot N_{p_{\text{block}}}$ FLOPs are required for completely processing one block of $N_{p_{\text{block}}}$ pixels of N_b. Additionally, the $p_{\max}$ value depends on the different HyperLCA input parameters, as it is described in Section 2.2. Its exact value can be calculated as shown in Equation (1). Accordingly, the amount of FLOPs to be done by the HyperLCA transform for a single block can be directly estimated from the input parameters as shown in Equation (16):

$$
\begin{gathered}
FLOPs_{\text{block}} = 6 \cdot N_b \cdot N_{p_{\text{block}}} \cdot p_{\max}, \\
FLOPs_{\text{block}} = 6 \cdot N_b \cdot N_{p_{\text{block}}} \cdot Integer(\frac{DR \cdot (N_b \cdot (N_{p_{\text{block}}} - 1))}{CR \cdot (DR \cdot N_b + N_{\text{bits}} \cdot N_{p_{\text{block}}})}).
\end{gathered} \tag{16}
$$

Finally, since the HyperLCA transform is independently applied to each block of pixels, the amount of FLOPs required for processing the entire image can be estimated by multiplying the number of FLOPs required for processing one block by the number of blocks to be compressed. The resulting amount of FLOPs required for processing the entire image withing the HyperLCA transform is shown in Equation (17)

$$
\begin{gathered}
FLOPs_{\text{Image}} = 6 \cdot N_b \cdot N_p \cdot p_{\max} \\
FLOPs_{\text{Image}} = 6 \cdot N_b \cdot N_p \cdot Integer(\frac{DR \cdot (N_b \cdot (N_{p_{\text{block}}} - 1))}{CR \cdot (DR \cdot N_b + N_{\text{bits}} \cdot N_{p_{\text{block}}})}).
\end{gathered} \tag{17}
$$

Different conclusions can be dragged from these expressions. Firstly, it can be observed that higher CR leads to lower $p_{\max}$ values, which results in less number of FLOPs required for compressing the image. Additionally, less data is produced by the HyperLCA transform as the CR increases, and, so, less data is to be compressed in the subsequent compression stages. These two facts make the HyperLCA compressor more efficient as the compression ratio increases. Secondly, for the same CR and N_b values, smaller block sizes $N_{p_{\text{block}}}$ produce smaller $p_{\max}$ values, decreasing the number of FLOPs required for compressing the entire image. This makes the HyperLCA compressor more efficient when smaller blocks are used. Additionally, the number of FLOPs required for processing each single block directly depends on the $N_{p_{\text{block}}}$ and the $p_{\max}$ value, and, hence, the computational complexity (in terms of FLOPs) for independently processing one single block exponentially decreases by decreasing the number of pixels per block, $N_{p_{\text{block}}}$. According to these facts, the HyperLCA compressor would be especially efficient (in terms of FLOPs) for processing hyperspectral images when it is desired to achieve high compression ratios and when using small blocks of pixels. All these facts present important advantages especially when using pushbroom or whiskbroom sensors for collecting the images. The strategy followed by the HyperLCA compressor, based on these facts, avoids the necessity of storing big portions of the image until being able to compress them, reduces the amount of required resources for compressing the collected data, speeds up the process and provides parallelization and error-resilience.

Finally, the operations done by the HyperLCA inverse transform, shown in Algorithm 2, are almost the same as the operations done by the HyperLCA transform, but are used for adding information to the image instead of subtracting it. In general, the operations required for decompressing the image using the HyperLCA algorithm are almost the same as the operations used for compressing it, but applied in reverse order. Due to this reason, the number of FLOPs required for decompressing each block of pixels of the image, as well as the entire image, can be also estimated as shown in Equations (16) and (17), respectively.

2.10. Advantages of the HyperLCA Compression Algorithm

The HyperLCA algorithm has several advantages with respect to other state-of-the-art solutions for lossy compressing hyperspectral images, which will be demonstrated in Section 4 of the current paper. The main advantages of the HyperLCA compressor are detailed next:

- High compression ratios and decent rate–distortion compression performance.
- Specially designed for preserving the most different pixels of the data set, which are important for several hyperspectral imaging applications such as anomalies detection, target detection, tracking or classification.
- The minimal desired compression ratio can be perfectly fixed in advance.
- Additional stopping conditions can be used for stopping the compression process if the desired quality is achieved at a higher compression ratio than the specified minimum compression ratio.
- Allows a progressive decoding of the compressed bitstream according to different quality metrics.
- Low computational complexity and high level of parallelism in relation with other transform-based compression approaches. This eases the hardware implementation of the HyperLCA algorithm for applications under tight latency constraints, also reducing the amount of required hardware resources.
- Designed for independently compressing blocks of pixels of the image without requiring any spatial alignment between the pixels. This allows starting the compression process just after capturing a block of pixels. This is specially suitable when using pushbroom or whiskbroom hyperspectral sensors.
- Error resilience. The HyperLCA algorithm independently compresses and codes each block of pixels, and, so, if there is an error in any block of pixels, that error will not affect any other block.
- The range of values that can be produced by the HyperLCA transform can be perfectly known in advance, which makes it simple to use integer values for representing its results, or for accomplishing all its operations, introducing minimal compression–decompression losses. This eases the integration of the HyperLCA transform with the subsequent compression stages and makes it possible to achieve more efficient hardware implementations.

All these advantages make the HyperLCA algorithm a very suitable option for applications under tight latency constraints or with limited available resources, such as the compression of hyperspectral images on board satellites, where it is critical to consider the amount of power, time and computational resources used.

3. Material and Methods

In order to evaluate the goodness of the proposed HyperLCA algorithm for compressing hyperspectral images in relation with other state-of-the-art approaches for this task, different experiments have been done. The hyperspectral images and quality metrics used in these experiments are detailed next.

3.1. Hyperspectral Images Used

A heterogeneous test bench has been selected in order to evaluate the behavior of the proposed compressor under different circumstances. On one side, we have selected three different images from the well known Airbone Visible/Infrarred Imaging Spectrometer (AVIRIS) [21]. This sensor captures 224 spectral bands in the wavelength range of 400–2500 nm. The first one is the Yellow Stone (Sc 0) radiance image. The other two images used, collected with the AVIRIS sensor, correspond to reflectance images. These are the Lunar Lake image and the Moffet Field one. These images have been cropped to portions of 512 × 512 pixels. Figure 6 graphically shows a false color representation of the portions of the AVIRIS hyperspectral images used.

(a) (b) (c)

Figure 6. RGB representation of the used hyperspectral images collected by the AVIRIS sensor. (**a**) Yellowstone; (**b**) Lunar Lake; (**c**) Moffet Field.

On the other side, in order to evaluate the behavior of the HyperLCA algorithm in a more challenging scenario, three different uncalibrated images from the Hyperion sensor have been used. The Hyperion sensor produces 242 spectral bands between 355.59 and 2577.08 nm [22]. The Hyperion images that have been used in the experiments are the Erta Ale image, the Lake Monona image and the Mt. St. Helens one. These images have been cropped to portions of 512 × 256 pixels. Figure 7 shows a grey scale representation of these images. As it can be observed in this figure, these images contain a high amount of striping noise, which makes the compression of these images more challenging.

Finally, the impact of the HyperLCA compression process in the ulterior hyperspectral imaging applications has also been evaluated. Specifically, classification, anomaly detection and spectral unmixing applications have been considered. For measuring the effect of the compression process in these applications, some well known images, typically used in these fields due to the existence of their corresponding ground truths, have been selected. These images are preprocessed as it is usually done in the corresponding targeted hyperspectral imaging applications.

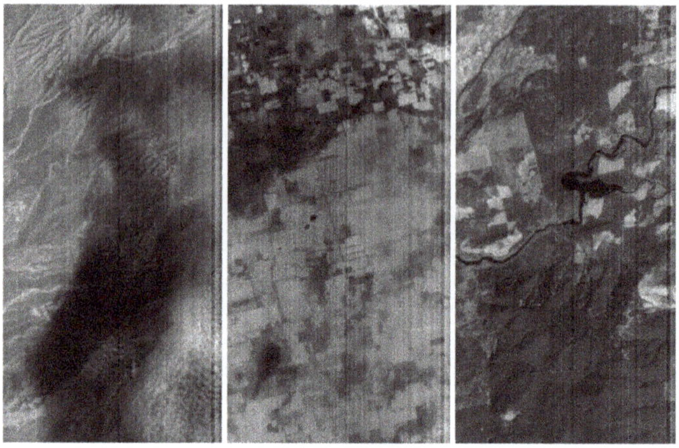

(a) (b) (c)

Figure 7. Grey scale representation of the used hyperspectral images collected by the Hyperion sensor. (**a**) Erta Ale; (**b**) Lake Monona; (**c**) Mt. St. Helens.

First, two different hyperspectral data sets have been used for evaluating the effect of the HyperLCA compression process in classification applications. The first one was collected in 1996 by the AVIRIS sensor over Indian Pines in northwestern Indiana. The selected scene is a total of 145 × 145 pixels with a spatial resolution of 20 m. Bands [1–4], [108–112], [154–167], and 224 have been removed due to water absorption, resulting in a total of 200 hyperspectral bands. The Indian Pines scene contains two-thirds agriculture, and one-third forest or other natural perennial vegetation. There are two major dual-lane highways, a rail line, as well as some low-density housing, other build structures, and smaller roads. The ground truth available is designated into sixteen classes that are briefly summarized in Table 1. The second data set used for evaluating the impact of the HyperLCA compression process in classifications' applications was captured over the city of Pavia, Italy, by the Reflective Optics Spectrographic Imaging System (ROSIS-03) airborne instrument. The ROSIS-03 sensor has 115 data channels with a spectral coverage ranging from 430 to 860 nm. Twelve channels have been removed due to noise. The data have been corrected atmospherically but not geometrically. The scene consists of 640 × 340 pixels with a spatial resolution of 1.3 m per pixel that covers the Engineering School at the University of Pavia and consists of nine different classes, briefly summarized in Table 1. Figures **??** and 8 show a false color representation of the two described hyperspectral images as well as their corresponding ground truths.

The impact of the HyperLCA compression in anomaly detection applications has been evaluated using two different hyperspectral data sets. The first one was captured over the Rochester Institute of Technology (RIT) by the Wildfire Airborne Sensor Program (WASP) Imaging System [23]. This system covers the visible, short, mid and long-wave infrared regions of the spectrum. The sensor was comprised by a high-resolution colour camera that covers the visible spectrum, a short wave infrared imager that covers from 900 nm to 1800 nm, a mid wave infrared imager that covers from 3000 nm to 5000 nm and a long wave infrared imager that covers from 8000 nm to 9000 nm. In particular, a portion of the overall image, taken over a parking lot, with a size of 180 × 180 pixels and 120 spectral bands, is used in this study. In this scene, the anomalies are fabric targets, which consist of 72 pixels and account for 0.22% of the image. The second data set used was captured by the AVIRIS sensor over the World Trade Center (WTC) area in New York City on 16 September 2001 [24]. A portion of the entire data set, with a size of 200 × 200 pixels and 224 spectral bands, has been used for the tests, where the anomalies are thermal hot spots that consist of 83 pixels and account for 0.21% of the image scene. Figure 9 shows a false color representation of the two described hyperspectral images as well as their corresponding ground truths.

Table 1. Ground truth classes for the Indian Pines and Pavia University scenes and their respective number of samples.

Indian Pines			Pavia University		
Label	**Class**	**Samples**	**Label**	**Class**	**Samples**
1	Alfalfa	46	1	Asphalt	6631
2	Corn-notill	1428	2	Meadows	18,649
3	Corn-mintill	830	3	Gravel	2099
4	Corn	237	4	Trees	3064
5	Grass-pasture	483	5	Painted metal sheets	1345
6	Grass-trees	730	6	Bare Soil	5029
7	Grass-pasture-mowed	28	7	Bitumen	1330
8	Hay-windrowed	478	8	Self-Blocking Bricks	3682
9	Oats	20	9	Shadows	947
10	Soybean-notill	972			
11	Soybean-mintill	2455			
12	Soybean-clean	593			
13	Wheat	205			
14	Woods	1265			
15	Buildings-Grass-Trees-Drives	386			
16	Stone-Steel-Towers	93			

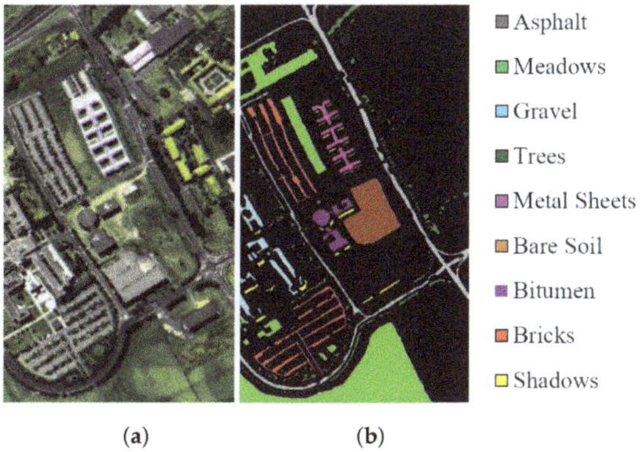

Figure 8. RGB representation of the Pavia University (PU) hyperspectral image and its corresponding ground truth for classification applications. (**a**) PU Image; (**b**) PU Ground truth.

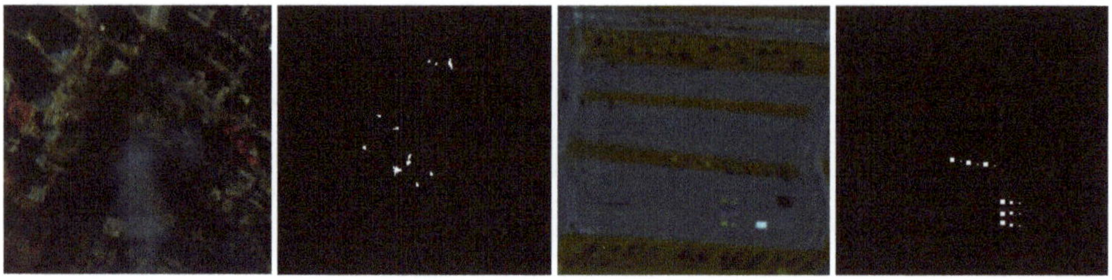

(a) (b) (c) (d)

Figure 9. RGB representation of the Rochester Institute of Technology (RIT) and World Trade Center (WTC) hyperspectral images and their corresponding ground truths for anomaly detection applications. (**a**) WTC Image; (**b**) WTC Ground truth; (**c**) RIT Image; (**d**) RIT Ground truth.

Finally, the Cuprite data set has been used for evaluating the impact of the HyperLCA compression process in unmixing applications. This image was taken by the AVIRIS sensor over the region of Cuprite, Nevada, USA in the summer of 1997. The selected scene consists of a total of 350 $\times$ 350 pixels with a spatial resolution of 4 m. Several bands have been removed due to water absorption and low SNR, resulting in a total of 188 spectral bands. The site is well understood mineralogically, and has several exposed minerals of interest including alunite, buddingtonite, calcite, kaolinite and muscovite. The ground truth of this data set consists of the spectral signature of these five minerals, each of them represented in the same 188 spectral bands that the hyperspectral image. Figure 10 graphically shows these five spectral signatures as well as a false color representation of the Cuprite hyperspectral image.

3.2. Evaluation Metrics

Lossy compression approaches for hyperspectral images are typically evaluated in terms of the rate-distortion relation achieved. However, it has been proven that low average distortions in the compression–decompression process do not necessarily ensure good results when the decompressed images are used in specific hyperspectral applications [5]. In general, lossy compression behaves as a low-pass filter, reducing the noise present in the image. This may improve the results obtained in some applications when using the decompressed images. Nevertheless, the low pass filter can also remove the most atypical elements of the image, which are crucial for several applications, such as anomaly detection, classification, unmixing, or target detection [13–17]. Due to this reason, three different evaluation metrics have been used for measuring the goodness of the compression–decompression process within the HyperLCA algorithm. On one side, the *Signal-to-Noise Ratio* (SNR) has been calculated as already described in Section 2.8.2 for measuring the average looses introduced by the compressor. High SNR values indicate good average compression performance.

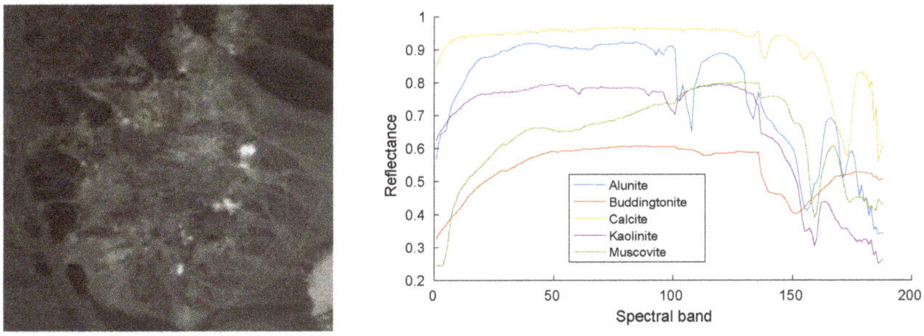

(**a**) (**b**)

Figure 10. RGB representation of the Cuprite hyperspectral image as well as the spectral signatures corresponding to the five minerals that are known to be present in this scene. These signatures conform the ground truth typically used for evaluating the endmembers extraction algorithms. (**a**) cuprite image; (**b**) reference spectral signatures.

On the other side, the *Maximum Single Error* (MaxSE), already described in Section 2.8.3, has also been used. The maximum single reconstructions errors are typically produced in the most different elements, and, hence, it can be assumed that lower MaxSE values indicate that the most different pixels in the data set are being better preserved through the compression–decompression process [25].

Finally, most of the hyperspectral imaging applications make use of the spectral information of the image for identifying or distinguishing between the different materials present in the scene. Due to this reason, in order to not affect the results of the subsequent hyperspectral applications, it is crucial to preserve the spectral signatures of the pixels through the compression–decompression process, introducing the minimal possible amount of spectral distortions. In order to evaluate the spectral distortions produced by the compression–decompression process within the HyperLCA algorithm, the *Spectral Angle* (SA) [26], has been used. Lower SA values indicate lower spectral distortions. In particular, the average and maximum spectral distortions have been measured in the experiments by calculating the average and maximum spectral angles between the pixels of the real images and the compressed-decompressed images.

It is important to remark that, although the HyperLCA compressor independently compresses each block of pixels of the image, all these metrics have been calculated for the entire compressed-decompressed images once each single block is decompressed. This allows for making fair comparisons with the other compression approaches used.

4. Results and Discussion

This section discloses the results obtained in all the uncovered experiments with the purpose of evaluating the goodness of the proposed HyperLCA algorithm for compressing hyperspectral images.

4.1. Effect of the Input Parameters of the HyperLCA Compressor

The HyperLCA algorithm has two main input parameters that may affect its compression performance: the number of pixels per block in which the image is divided, *blockSize*, and the number of bits used for scaling the extracted *V vectors*, N_{bits}. Different experiments have been done in order to evaluate the behavior of the HyperLCA compressor when using different values for these parameters. In particular, the number of pixels per block, *blockSize*, has been set to 256, 512 and 1024 pixels, and considering the dynamic range of the AVIRIS and Hyperion sensors, the N_{bits} parameter has been set to 16, 12 and 8 bits for the AVIRIS images and 12 and 8 for the Hyperion sensor images. Figure 11a graphically shows the average rate-distortion obtained, in terms of SNR, when compressing the Lunar Lake hyperspectral image collected by the AVIRIS sensor, according to these parameters for different compression ratios. Figure 11b graphically shows the MaxSE obtained for the same image and compression ratios, according to the different values of *blockSize* and N_{bits}. Similarly, Figure 12a,b display the same information but for the Erta Ale hyperspectral image collected by the Hyperion sensor.

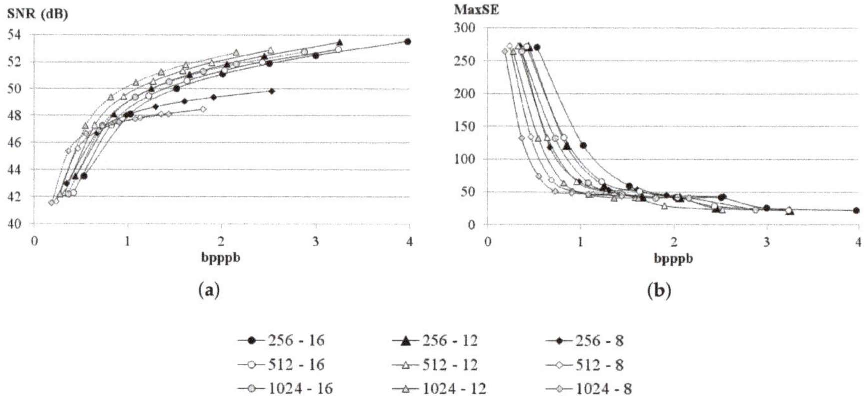

Figure 11. HyperLCA compression results for the Lunar Lake AVIRIS image using different *blockSize* and N_{bits} values.

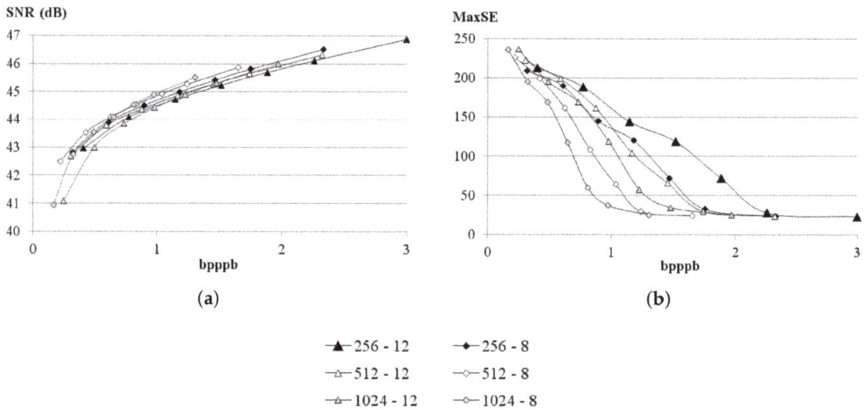

Figure 12. HyperLCA compression results for the Erta Ale Hyperion image using different *blockSize* and N_{bits} values.

The achieved compression has been represented in Figures 11 and 12 in terms of *Bits per Pixel per Band* (bpppb). The *bpppb* indicates the number of bits of the compressed data in relation with the number of bits of the original image. Accordingly, lower *bpppb* indicates higher compression ratios, *RC*.

According to the results shown in Figures 11 and 12 it can be observed that the HyperLCA compressor is able to achieve very high compression ratios with a good rate-distortion relation and low MaxSE values with all the tested combinations of the *blockSize* and N_{bits} parameters, for both the Lunar Lake AVIRIS image and the Erta Ale Hyperion image. Nevertheless, the best results have been produced when using blocks of 1024 pixels (*blockSize* = 1024). Additionally, the N_{bits} value that has produced the best results is, in average, 12 bits for the Lunar Lake AVIRIS image and 8 bits for the Erta Ale Hyperion image. Hence, in order to verify that these parameters values also produce good compression results for the complete set of images of the AVIRIS and Hyperion sensors, the rest of the images of the data set have been also compressed using the HyperLCA algorithm, using $blockSize = 1024$ and $N_{bits} = 12$ for the AVIRIS images and $blockSize = 1024$ and $N_{bits} = 8$ for the Hyperion sensor ones. Figure 13 graphically shows the obtained results. According to these results, the HyperLCA compressor is able to produce relatively high rate-distortion ratios for all the images in the data set, and for very high compression ratios. It is also worth mentioning that the compression performance of the HyperLCA algorithm is very solid for the three different uncalibrated images, collected by the Hyperion sensor, that contain a high amount of striping noise and typically represent a bigger challenge for the compression algorithms. Figure 13 also shows that the MaxSE values tend to very low values for both sensors in relation with their dynamic ranges, also demonstrating the good compression performance of the HyperLCA algorithm.

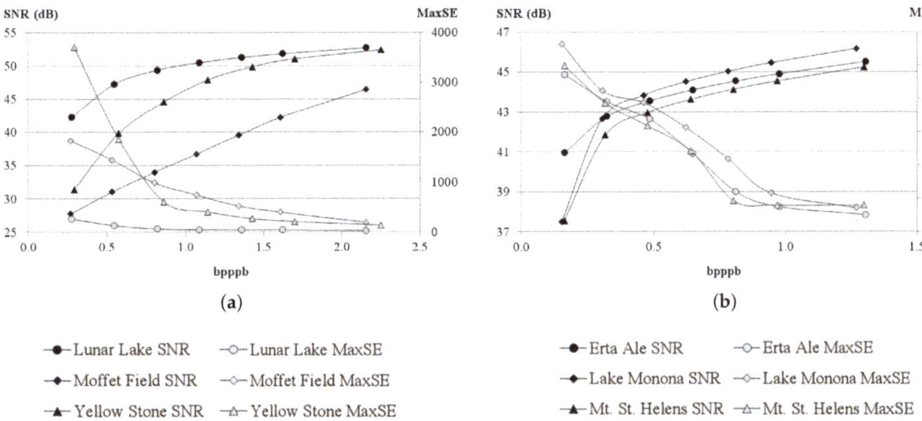

Figure 13. HyperLCA compression results for the images collected by the AVIRIS and Hyperion sensors. (**a**) AVIRIS images; (**b**) Hyperion images.

4.2. Evaluation of the HyperLCA Algorithm against Other Transform-Based Approaches

Most of the advantages of the HyperLCA compression algorithm, described in Section 2.10, are inherited from the HyperLCA transform, explained in Section 2.3. This transform allows to efficiently perform the spectral decorrelation and compression of the hyperspectral image. As aforementioned, the HyperLCA transform has been specially developed for being able to preserve the most different pixels through the compression–decompression process, since these pixels are very important for different hyperspectral imaging applications such as anomalies detection, target detection or classification. In order to verify the goodness of the HyperLCA transform against other state-of-the-art transforms used for the same purpose, the *Principal Component Analysis* (PCA) has been considered. Despite its computational complexity, the PCA produces some of the best transform-based compression results, in terms of rate-distortion as well as in preserving the relevant information for the ulterior hyperspectral analysis [4–7].

For making a fair comparison of both hyperspectral transforms, the different hyperspectral images of the data set have been spectrally decorrelated and reduced using the HyperLCA and the PCA transforms, for different compression ratios, without applying any further compression stage. Figures 14–17 graphically show the obtained results. It is also worth to mention that while the HyperLCA transform has been computed in order to produce integer values that can be directly processed by the upcoming compression stages, the PCA transform has been computing in Matlab using double precision floating point.

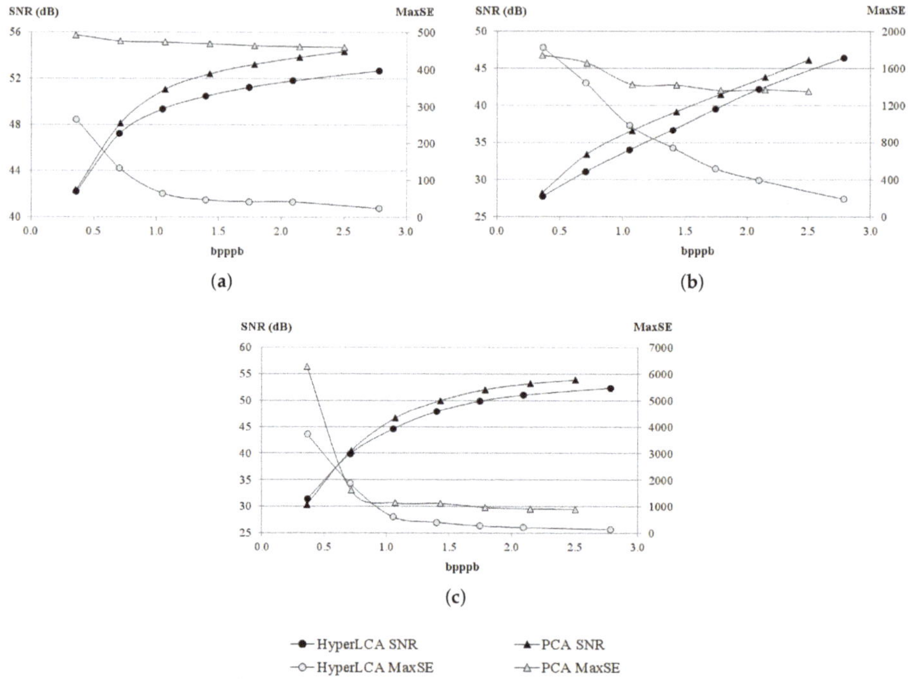

Figure 14. Comparison of the results produced by the HyperLCA and PCA transforms using the data set images collected by the AVIRIS sensor, in terms of SNR and MaxSE. (**a**) Lunar Lake; (**b**) Moffet Field; (**c**) Yellowstone.

Figures 14 and 15 graphically show the results produced by the HyperLCA and PCA transforms for the different data set images collected by the AVIRIS sensor. Similarly, Figures 16 and 17 show the results obtained with these transforms when processing the images collected by the Hyperion sensor. According to the results shown in Figure 14, both transforms produce similar results in terms of average rate-distortion ratio, the SNR produced by the PCA for these images being slightly better than the SNR produced by the HyperLCA transform. On the contrary, the MaxSE values obtained by the HyperLCA transform for these images are much lower than the MaxSE values obtained by the PCA transform, which indicates that the most different elements are much better preserved when using the HyperLCA transform. Figure 15 corroborates these conclusions. The average rate-distortion relation obtained, evaluated in terms of MeanSA, is similar for both transforms, being slightly better for the PCA transform. However, the MaxSA values obtained in the experiments indicate that there are spectral signatures with higher distortions when using the PCA transform than when using the HyperLCA transform. On the other hand, according to the results obtained when processing the images collected by the Hyperion sensor, shown in Figures 16 and 17, the HyperLCA transform clearly outperforms the PCA transform in terms of SNR, MaxSE and MaxSA, but not in MeanSA.

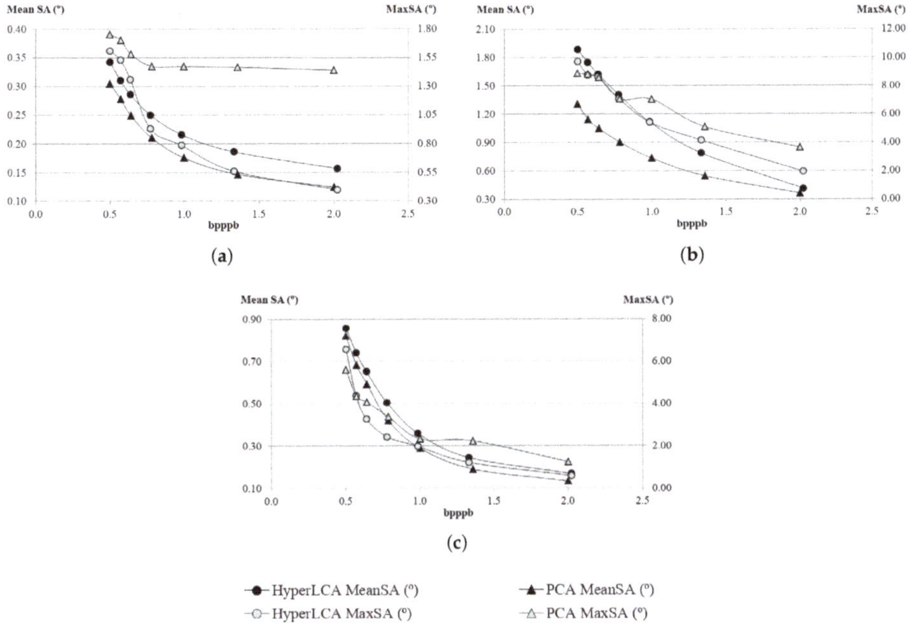

Figure 15. Comparison of the results produced by the HyperLCA and PCA transforms using the data set images collected by the AVIRIS sensor, in terms of SA. (**a**) Lunar Lake; (**b**) Moffet Field; (**c**) Yellowstone.

All these results demonstrate the goodness of the HyperLCA transform for spectrally decorrelating and reducing hyperspectral data sets, and suggest that its results are also good enough when processing images with high levels of noise. The obtained results also verify that the HyperLCA transform is capable of keeping the most different elements of the data set through the compression–decompression process introducing minimal spectral distortions, which makes it very useful when the compressed-decompressed images are to be used for hyperspectral applications such as anomalies detection, target detection, or classification.

Additionally, it can also be observed that the results of the HyperLCA transform, shown in Figures 14 and 16, display similar curves to the results provided by the entire HyperLCA compressor (HyperLCA transform, preprocessor and coder), but at lower compression ratios (higher *bpppb* values). This is due to the fact that the HyperLCA preprocessing and coding stages slightly increase the compression ratios achieved by the HyperLCA transform without introducing further compression losses.

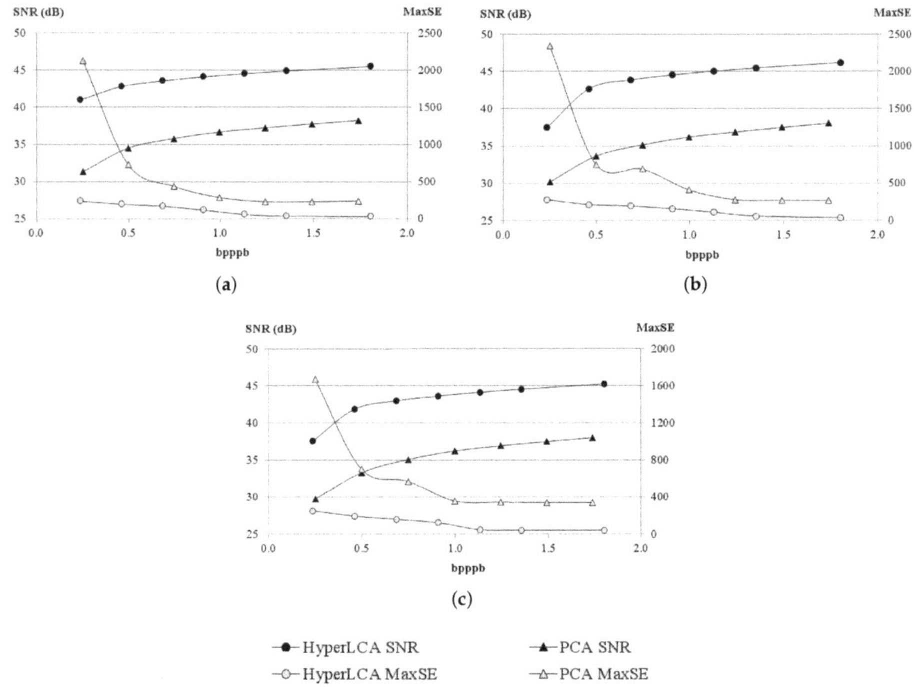

Figure 16. Comparison of the results produced by the HyperLCA and PCA transforms using the data set images collected by the Hyperion sensor, in terms of SNR and MaxSE. (**a**) Erta Ale; (**b**) Lake Monona; (**c**) Mt. St. Helens.

4.3. Evaluation of the Impact Produced by the HyperLCA Compression Process in the Ulterior Hyperspectral Imaging Applications

As already mentioned, the HyperLCA algorithm is a lossy compressor for hyperspectral images, especially designed for achieving high compression ratios at a reasonable computational burden. However, it is important to have in mind that the compressed-decompressed data sets have to be useful for the ulterior hyperspectral imaging applications, and, hence, some more requirements need to be fulfilled rather than just achieving a high compression rate–distortion relation. Due to this reason, the HyperLCA algorithm has been specifically designed for preserving the most different pixels of the data set through the compression–decompression process, since these pixels are very important for applications such as anomaly detection, spectral unmixing or classification, as well as for introducing few spectral distortions, since the spectral information is extremely important in most of the hyperspectral imaging applications. Different experiments have been uncovered in this section for evaluating the impact of the HyperLCA compression process in hyperspectral imaging classification, anomaly detection and spectral unmixing applications.

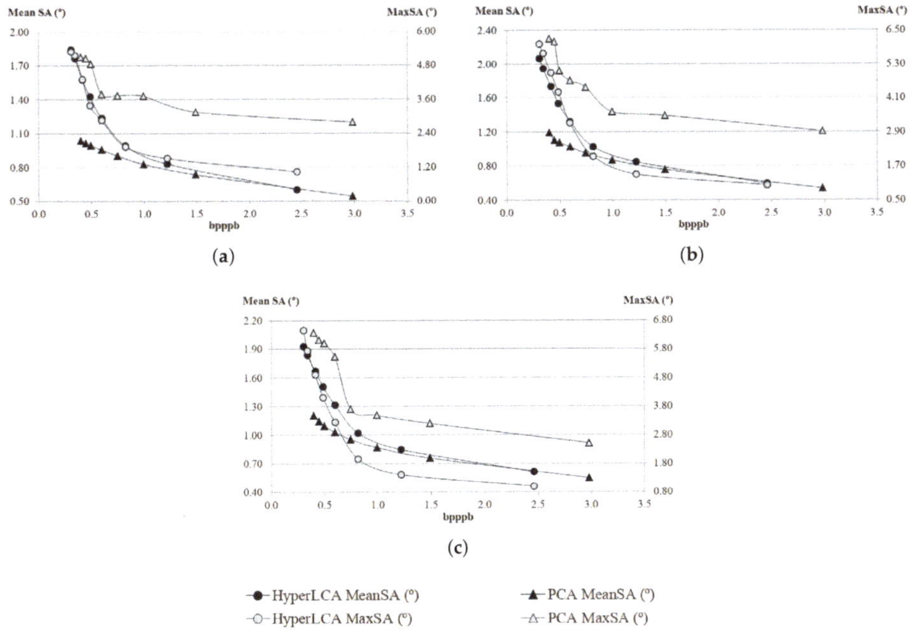

Figure 17. Comparison of the results produced by the HyperLCA and PCA transforms using the data set images collected by the Hyperion sensor, in terms of SA. (**a**) Erta Ale; (**b**) Lake Monona; (**c**) Mt. St. Helens.

From the entire HyperLCA compression process, all the losses of information are produced by the HyperLCA transform. Due to this reason, the impact of the compression–decompression process using just the HyperLCA transformation stage in the mentioned hyperspectral applications is evaluated and compared with the impact produced by the PCA transform for the same images and applications. It is worth to mention here that the PCA transform is used in many hyperspectral applications, such as unmixing or classification, for spectrally decorrelating the information and reducing the number of spectral components of the data set, with the goal of reducing the redundant information, increasing the separability of the different elements of interest and improving the application results [9,10,27]. Accordingly, the PCA transform can be considered as a good reference to compare with, regarding the impact of the HyperLCA transform in the ulterior hyperspectral applications.

4.3.1. Evaluation of the Impact Produced by the HyperLCA Compression Process in Hyperspectral Imaging Classification

For fairly evaluating the effect of the HyperLCA compressor in hyperspectral imaging classification applications, the Indian Pines and Pavia University data sets, described in Section 3.1, have been spectrally decorrelated and reduced using the HyperLCA and the PCA transforms, for different compression ratios, without applying any further compression stage. Figure 18 graphically shows the obtained performance using both transforms, according to the different evaluation metrics described in Section 3.2. These results have been obtained using blocks of 725 pixels ($blockSize = 725$) for the Indian Pines image, which corresponds to five lines of the image, and blocks of 680 pixels ($blockSize = 680$) for the Pavia University data set, which corresponds to two lines of the image. Additionally, the N_{bits} value has been set to 8 bits for both data sets. After doing so, all the reconstructed images as well as the original images (without applying any transformation) have been classified using the well known *Suport Vector Machine* (SVM) classifier, which is one of the most widely used classifiers for hyperspectral imaging applications [28]. The SVM classifier has been trained for performing linear prediction following the one versus one classification model, using the LIBSVM tool [29].

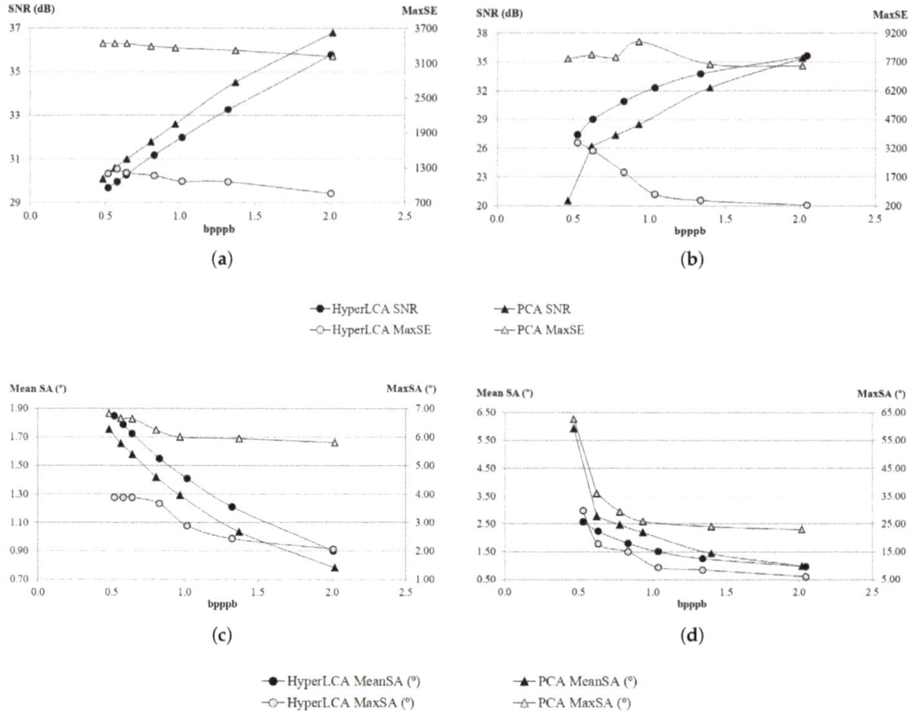

Figure 18. Comparison of the results produced by the HyperLCA and PCA transforms using the Indian Pines and Pavia University data sets, according to the different evaluation metrics. (**a**) Indian Pines; (**b**) Pavia University; (**c**) Indian Pines; (**d**) Pavia University.

The followed methodology is described next:

- Fifteen samples per class [28] have been randomly selected from the original image (without applying any transformation) for training the SVM classifier. The indexes of the selected samples have been stored for the next steps. After doing so, the rest of the labeled samples have been classified using the generated SVM model, and the Overall Accuracy (OA) has been measured [28]. This is the performance obtained with SVM classifier and the specified configuration for the real data set.
- The 15 samples per class corresponding to the stored indexes are extracted for each of the reconstructed images, and used for training the SVM classifier for each reconstructed data set, as it was previously done for the original image. These 15 samples may have distortions with respect to the original samples, introduced by the different transforms. The remaining labeled samples of each reconstructed data set have been classified using its corresponding SVM model, generated with its corresponding samples and the exact same configuration that was previously used. Finally, the OA has been measured for each compression ratio and for each of the different applied transforms. This is the performance obtained with the SVM classifier and the specified configuration for each of the reconstructed data sets.

- Finally, each of the classification maps obtained for each of the reconstructed images have been compared with the classification map obtained with its corresponding original image, calculating the percentage of coincidences (PC).
- These three steps have been repeated 10 times for each data set for calculating the average results.

The results obtained by following this process are graphically shown in Figure 19 for both the Indian Pines and Pavia University data sets. Each graph displays the OA obtained for each spectral transform and compression ratio (left vertical axis), as well as the PC (right vertical axis). Lower PC values indicate a higher impact of the spectral transform in the classification process (a lossless compression would produce $PC = 100\%$). However, the produced impact does not necessarily need to be negative, since the lost information may be removing part of the noise present in the image, which could improve the classification results, or relevant information, which would decrease the classification performance [12]. When the OA values obtained with the reconstructed images are higher than the OA values obtained with the original image, it indicates a positive impact of the transform. On the contrary, when the OA values obtained with the reconstructed images are lower than the OA values obtained with the original image, it indicates a negative impact. The horizontal dashed line displayed in both graphs shows the OA obtained with the original Indian Pines and Pavia University images.

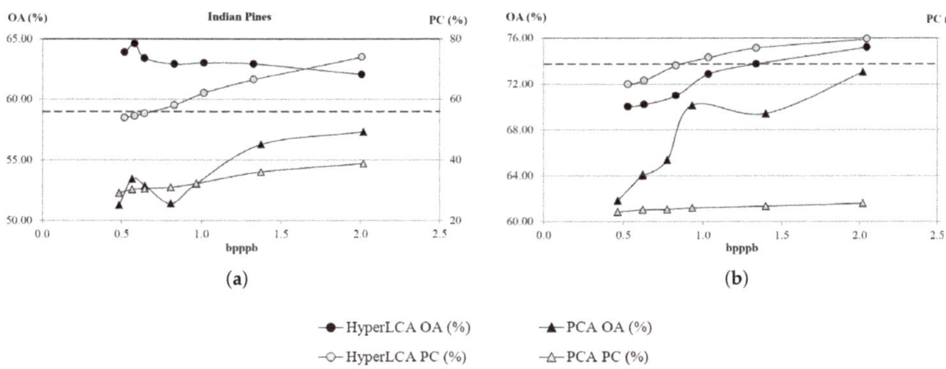

Figure 19. Comparison of the effect of the HyperLCA and PCA transforms in the SVM classification results using the Indian Pines and Pavia University data sets. (**a**) Indian Pines; (**b**) Pavia University.

Different conclusions can be dragged from the results shown in Figure 19. First of all, according to the obtained PC, it can be observed that the HyperLCA transform produces a lower impact in the classifier than the PCA transform, for both data sets, with independence of the compression ratio achieved. It can also be seen that the PC obtained using both transforms decreases as the compression ratio increases (smaller *bpppb*), which makes sense considering that the amount of losses introduced by the transforms increase with the compression ratio.

Regarding the OA obtained in the experiments, it can be observed that better classification results are obtained when using the HyperLCA transform than when using the PCA transform, when using the SVM classifier with the specified configuration and the two described data sets. It can also be observed that the OA obtained for the images processed using the PCA transform are always lower than the OA obtained for the original images, and decreases when the compression ratio increases. This suggests that part of the important information present in the original image is lost when using the PCA transform, and these losses are higher when the compression ratio increases. On the contrary, the OA obtained for the images processed using the HyperLCA transform tends to be higher than the OA obtained for the original images. This suggests that the information lost when using the HyperLCA transform is not relevant for the SVM classification process, and that these losses help to maximize the differences between the samples of the different classes.

According to these results, it could be considered that the HyperLCA transform has a smaller and more positive impact than the PCA transform in the ulterior hyperspectral imaging classification.

4.3.2. Evaluation of the Impact Produced by the HyperLCA Compression Process in Hyperspectral Anomaly Detection

For fairly evaluating the effect of the HyperLCA compressor in hyperspectral anomaly detection applications, the Rochester Institute of Technology (RIT) and the World Trade Center (WTC) data sets, described in Section 3.1, have been spectrally decorrelated and reduced using the HyperLCA and the PCA transforms, for different compression ratios, without applying any further compression stage. Figure 20 graphically shows the obtained performance using both transforms, according to the different evaluation metrics described in Section 3.2. These results have been obtained using blocks of 1080 pixels ($blockSize = 1080$) for the RIT image, which corresponds to six lines of the image, and blocks of 1000 pixels ($blockSize = 1000$) for the WTC data set, which corresponds to five lines of the image. Additionally, the N_{bits} value has been set to 8 bits for both data sets. After doing so, all the reconstructed images as well as the original images (without applying any transformation) have been processed using the well known *Orthogonal Subspace Projection Reed-Xiaoli* (OSPRX) detector for identifying the anomalous pixels. The OSPRX algorithm is one of the commonly used detectors for anomaly detection applications and provides good detection results [30–32].

The followed methodology is described next:

- The RIT and WTC original images (without applying any transformation) have been first processed using the OSPRX detector. This detector requires specifying the number of bands to use for representing the image information considered as background. This number has been set to 4 in the experiments, since this value produces very good anomaly detection results for both data sets [33]. By doing so, two anomaly maps have been obtained, one per image.
- The accuracy of the detection results using the original images has been evaluated using the *Receiver Operating Characteristics* (ROC) curves, and more specifically, the area under these curves (AUC) [30]. The ideal AUC value is 1. Lower AUC values indicate poorer detection performance. The obtained AUC values are 0.9837 and 0.9983 for the RIT and WTC data sets, respectively.
- The same process has been followed for each of the reconstructed images, also setting to 4 the input number of bands used by the OSPRX detector. By doing so, one anomaly map and AUC value are obtained for each image, transform and compression ratio achieved.
- Finally, each of the anomaly maps generated for each of the reconstructed images has been compared with the anomaly maps obtained with the original images, calculating the Mean Square Error (MSE) between the anomaly maps.

The results obtained by following this process are graphically shown in Figure 21 for both the RIT and WTC data sets. Each graph displays the AUC values obtained for each spectral transform and compression ratio (left vertical axis), as well as the MSE obtained for the different anomaly maps, corresponding to the reconstructed images (right vertical axis). Higher MSE values indicate a bigger difference between the anomaly maps generated with the reconstructed images and the one generated with the original image, which means a higher impact of the HyperLCA and PCA transforms in the anomaly detection process. This impact could be positive if just part of the noise present in the image is removed by the transforms, which could improve the anomaly detection results, or negative if relevant information, such as the anomalous pixels, is lost [12]. When the AUC values obtained with the reconstructed images are closer to 1 than the AUC value obtained for the original image, it indicates a positive impact of the transform. On the contrary, when the AUC values obtained with the reconstructed images are closer to 0 than the AUC values obtained with the original image, it indicates a negative impact. The horizontal dashed line displayed in both graphs shows the AUC value obtained with the original RIT and WTC images.

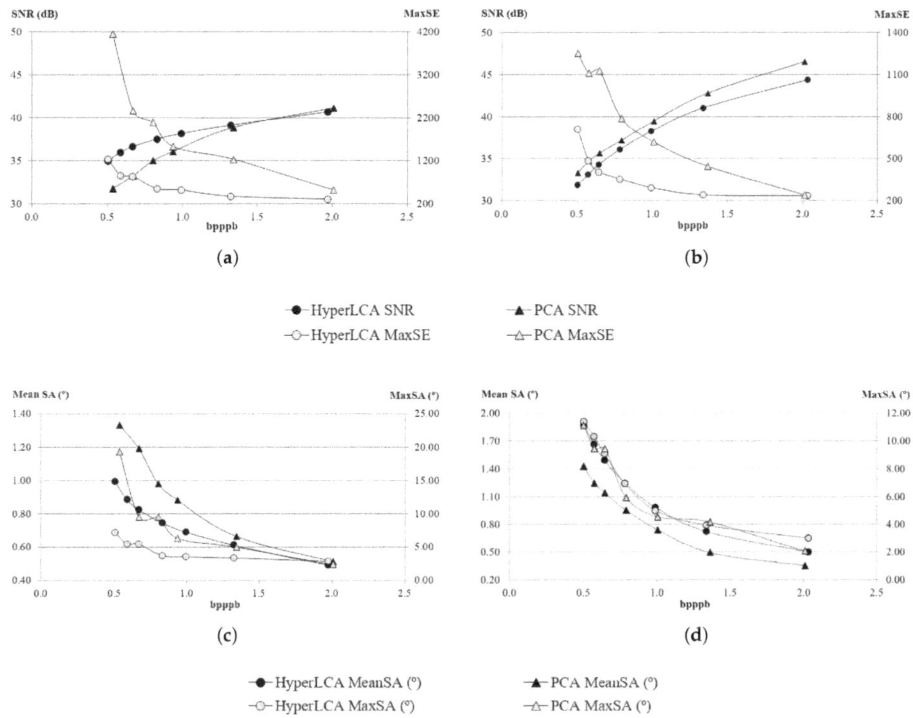

Figure 20. Comparison of the results produced by the HyperLCA and PCA transforms using the Rochester Institute of Technology (RIT) and the World Trade Center (WTC) data sets, according to the different evaluation metrics. (**a**) Rochester Institute of Technology; (**b**) World Trade Center; (**c**) Rochester Institute of Technology; (**d**) World Trade Center.

Different conclusions can be dragged from the results shown in Figure 21. First of all, according to the obtained MSE, it can be observed that the HyperLCA transform produces a lower impact in the anomaly detection process within the OSPRX detector than the PCA transform, for both data sets, with independence of the compression ratio achieved. It can also be seen that the MSE obtained using both transforms increases with the compression ratio (smaller *bpppb*), which makes sense considering that the amount of losses introduced by the transforms also increases with the compression ratio. It is also appreciable that the MSE values obtained are very close to 0. This is due to the fact that the anomaly maps should provide values close to 0 for the background pixels and values close to 1 for the anomalous pixels. Accordingly, most of the values obtained in the anomaly maps are very close to 0. Since the amount of anomalous pixels represents less than the 0.25% in both images, anomaly maps with big errors in the anomalous pixels would still produce very small MSE values. Hence, it is important to have in mind that low MSE values when comparing the obtained anomaly maps do not necessarily mean a low impact of the transforms in the anomaly detection process, and that this metric can be used just as a comparative between both transforms.

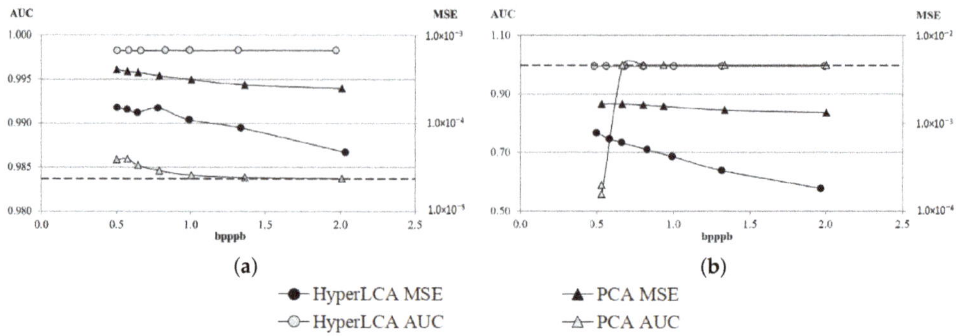

Figure 21. Comparison of the effect of the HyperLCA and PCA transforms in the OSPRX detector results using the Rochester Institute of Technology (RIT) and World Trade Center (WTC) data sets. (**a**) Rochester Institute of Technology; (**b**) World Trade Center.

Regarding the AUC values obtained in the experiments, it can be observed that better anomaly detection results are obtained when using the HyperLCA transform than when using the PCA transform, when using the OSPRX detector with the specified configuration and the two described data sets. It can also be observed that the AUC values obtained when using the compressed-decompressed RIT image, using both transforms, are better than the results obtained when using the original image. This suggests that the important information for the anomaly detection is preserved by both transforms and that just part of the noise or not useful information (for anomaly detection and the OSPRX detector) has been removed. It is also interesting that, for the RIT data set, the OSPRX performance increases with the compression ratio (higher AUC values are obtained for lower *bpppb*), for both spectral transforms. However, for the WTC data set, the OSPRX performance obtained with the compressed-decompressed image using the PCA transform significantly decreases for the highest compression ratios. This suggests that at such high compression ratios, the anomalous pixels have been lost through the PCA compression process.

According to these results, it could be considered that both spectral transforms are able to preserve the anomalous pixels through the compression–decompression process, having a positive impact in the anomaly detection process when using the OSPRX detector, at least for a wide range of compression ratios and the two selected data sets. It could also be concluded that the results obtained by the OSPRX detector are better when the HyperLCA transform has been applied rather than when the PCA transform has been applied.

4.3.3. Evaluation of the Impact Produced by the HyperLCA Compression Process in the Endmembers Finding Process for Spectral Unmixing Applications

The effect of the HyperLCA compressor in endmembers finding applications has been evaluated using the Cuprite data set, described in Section 3.1. This hyperspectral image has been spectrally decorrelated and reduced using the HyperLCA and the PCA transforms, for different compression ratios, without applying any further compression stage. Figure 22 graphically shows the obtained performance using both transforms, according to the different evaluation metrics described in Section 3.2. The HyperLCA results have been obtained using blocks of 700 pixels ($blockSize = 700$), which corresponds to two lines of the image, and setting the N_{bits} value to 8 bits.

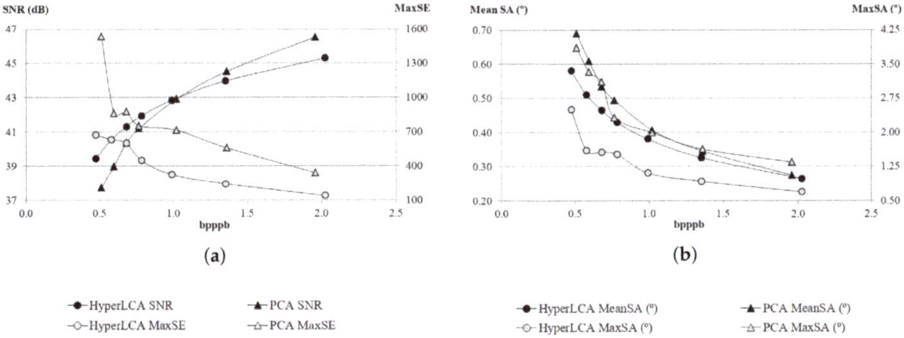

Figure 22. Comparison of the results produced by the HyperLCA and PCA transforms using the Cuprite data set, according to the different evaluation metrics.

After spectrally decorrelating and reducing the Cuprite image for the different compression ratios, using both transforms, all the reconstructed images as well as the original one (without applying any transformation) have been analysed in order to identify the pixels of the data set with the smallest differences in relation to the five pure spectral signatures of the ground truth: alunite, buddingtonite, calcite, kaolinite and muscovite. This allows identifying the pixels that would be the best candidates of each image to be selected as endmembers by spectral linear unmixing algorithms based on pure pixels [11], such as the *Vertex Component Analysis* (VCA) [34], the *Orthogonal Subspace Projection* (OSP) [35] or N-Finder [36] algorithms, among others. The best performance that the mentioned unmixing algorithms could achieve when processing the different compressed-decompressed images, as well as the original one, can be evaluated attending to the spectral angle obtained for the pixels identified as the best candidates to be endmembers. Lower spectral angle values indicate better achievable performance. The MeanSA and MaxSA values obtained for these pixels in the Cuprite image, when using both transforms and different compression ratios are graphically shown in Figure 23. The MeanSA and MaxSA values obtained for the original Cuprite image are also displayed in this graph as dashed lines.

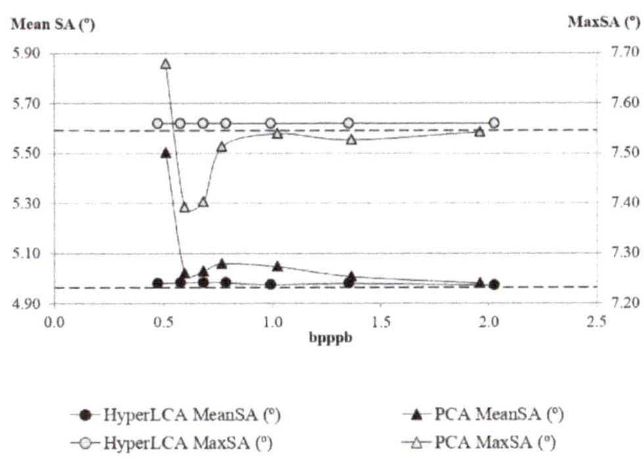

Figure 23. Comparison of the effect of the HyperLCA and PCA transforms in the unmixing results, using the Cuprite data set.

The results displayed in Figure 23 show that the effect of the HyperLCA transform in the SA values obtained does not vary too much with the compression ratio achieved. This suggests that the

HyperLCA transform could have a low impact in the results of the unmixing applications. On the other hand, the SA values obtained when using the PCA transform present larger variations for the different compression ratios, which suggests that the PCA transform has a higher impact on the achievable results of the ulterior unmixing applications. Nevertheless, the results obtained when using any of the tested transforms are very similar to the results obtained when using the Original Cuprite image, indicating that the impact produced by these transforms in the ulterior spectral unmixing applications could be negligible.

4.4. Evaluation of the HyperLCA Algorithm against the State-of-the-Art Solutions for on-Board Applications

Finally, as described in Section 1, the HyperLCA algorithm has been designed with the purpose of offering a hardware-friendly transformed based approach for lossy compressing hyperspectral images. The main goal is to provide a new solution for applications with limited available resources, such as compression on-board satellites. In particular, this new solution focuses on obtaining high compression ratios at a reasonable rate-distortion relation, without compromising too much the ulterior hyperspectral imaging applications.

This section is devoted to evaluating the characteristics of the HyperLCA compressor against the state-of-the-art solutions for on-board hyperspectral imaging compression, and, more specifically, against the actual standards proposed by the Consultative Committee for Space Data Systems [19] for space applications. The goal is to verify that the HyperLCA algorithm provides some new advantages for some specific situations and/or applications that are not totally covered by the existing standards.

The CCSDS has published two standards specially focused on hyperspectral imaging compression (also applicable to multispectral image compression). These are the CCSDS122.1-B-1 *Spectral Preprocessing Transform for Multispectral and Hyperspectral Image Compression* [37] and the CCSDS123.0-B-1 *Lossless Multispectral and Hyperspectral Image Compression* [38].

The CCSDS122.1-B-1 proposes a compressor that consists of two main functional parts: a spectral transform and a set of 2D encoders. The proposed 2D coders extend the (two-dimensional) CCSDS Image Data Compression standard, CCSDS 122.0-B-2 [39]. The process consists of first spectrally decorrelating the data set using one of the three proposed spectral transforms: the *Integer Wavelet Transform* (IWT) [40], the *Pairwise Orthogonal Transform* (POT) [41,42] or the *Arbitrary Affine Transform* (AAT), and, then, independently compressing each of the decorrelated image bands using the mentioned two-dimensional coder. The CCSDS122.1-B-1 also allows using the Identity Transform instead of any of the mentioned transforms. This transform is defined for the sake of providing a compressed data structure to encode a 3D image without requiring the implementation of a more complex transform stage. The compressor proposed in the CCSDS122.1-B-1 standard may produce both lossless or lossy hyperspectral image compression. However, the losses are mostly introduced by the 2D CCSDS 122.0-B-2 image coder, the spectral transform being mainly executed as lossless transforms. The AAT is the only one of the mentioned transforms that is not guaranteed to produced lossless results. In this sense, the HyperLCA compressor proposes a different strategy, the HyperLCA transform being the one taking control of the introduced losses while the subsequent HyperLCA compression stages produce lossless compression. This represents an additional computational advantage for the HyperLCA compressor, since the amount of data to be coded by the compression stages that goes after the HyperLCA transform is much smaller than the original image. On the contrary, when following the strategy proposed in the CCSDS122.1-B-1, the amount of information to be coded after applying the spectral transforms is exactly the same size as the original image. The equivalent approach using the transforms proposed in the CCSDS122.1-B-1 standard would be to preserve just a certain number of the decorrelated bands for reconstructing the hyperspectral image, and lossless compressing these bands using the CCSDS 122.0-B-2 coder. The information contained in the bands that are not considered would correspond with the losses produced in the compression process. However, even doing it this way, it would not exactly correspond with the strategy followed by the HyperLCA algorithm, since, in this approach, the spectral transform would being applied to the entire image at once, while the

HyperLCA transform is independently applied to a subset of pixels of the image (typically one or some lines of the image).

Nevertheless, the transform-based compression approach proposed by the CCSDS122.1-B-1 is the most similar to the HyperLCA compression strategy from all the CCSDS standards. As it is described in this standard, the Karhunen Loève Transform (KLT) is "the transform that provides perfect decorrelation", and the POT is an approximation of the (KLT) at a fraction of its computational cost. The standard also describes that, in general, the POT provides better coding performance than the IWT, but requires more computational resources and has a more complex implementation. Additionally, these assertions are empirically demonstrated in [5,7,41–43]. These results show that the compression performance of the KLT transform (PCA) clearly surpasses the compression performance provided by the rest of the transforms contained in the aforementioned standard. Accordingly, the results shown in the previous sections of this manuscript, where the results provided by the HyperLCA transform are widely compared with the results provided by the PCA transform, for different images and compression ratios, should be enough for verifying the good behavior of the HyperLCA transform for lossy compressing hyperspectral images, particularly when high compression ratios are desired. It is also important to remark here that the HyperLCA compressor has been specifically developed for lossy compressing hyperspectral images and, hence, its efficiency is not guaranteed for compressing multispectral images (with few bands), or for lossless compressing hyperspectral images.

On the other hand, the CCSDS123.0-B-1 proposes a 3D lossless prediction-based compression algorithm for hyperspectral images, which is hardly optimized for producing an efficient lossless compression at a relatively low computational cost. However, as any other lossless compressor, its achievable compression ratio is limited [44–46]. The authors consider that the use of a lossy transform-based approach, such as the HyperLCA compressor, which may be more complex that the compressor described in this standard, is only justified when it is important to achieve high compression ratios that are not achievable by the CCSDS123.0-B-1 compressor. Hence, the maximum compression ratios achieved by this compressor for the data sets described in Section 3.1 are shown in Table 2. Table 2 also shows different rate-distortion relations achieved by the HyperLCA compressor for the same images. The HyperLCA results shown in this table have been obtained using 1024 pixels per block ($blockSize = 1024$) and setting the N_{bits} parameter to 12 and 8 for the images collected by the AVIRIS and Hyperion sensors, respectively. The CCSDS123.0-B-1 results have been obtained using the WhiteDwarf software [47], developed by the European Space Agency (ESA). Additionally, an RGB representation of the compressed-decompressed AVIRIS images corresponding to Table 2 are displayed in Figure 24. Similarly, a grey scale representation of the compressed-decompressed Hyperion images corresponding to Table 2 are displayed in Figure 25. Figure 24 demonstrates that no appreciable spatial errors are introduced in the images collected by the AVIRIS sensor when compressing them using the HyperLCA algorithm, since the RGB images obtained for the different compression ratios seem very similar. However, there are some spatial differences in the compressed-decompressed images obtained for the Hyperion sensor, as can be observed in the Figure 25, especially for the highest compression ratios shown in Table 2. It seems that the striping noise is slightly removed, but also that some lines of the images are blurred (just for the highest compression ratios). Nevertheless, these results seem pretty good considering the quality of the Hyperion images and the achieved compression ratios.

The obtained results verify that the HyperLCA compressor is able to achieve much higher compression ratios (lower bpppb) than the CCSDS123.0-B-1 compressor at a reasonable good rate-distortion relation. This, together with the other features of the HyperLCA algorithm, makes this proposal a suitable option for those applications, where it is desired to perform lossy compression with the purpose of achieving high compression ratios (less than 2 *bpppb*).

Table 2. CCSDS123.0-B-1 compression results.

Sensor	Image	CCSDS123.0-B-1	HyperLCA Compressor			
		bpppb	bpppb	SNR	bpppb	SNR
AVIRIS	Lunar Lake	4.03	0.55	47.23	1.62	51.80
	Moffet Field	4.26	0.53	33.76	1.75	39.53
	Yellow Stone	6.37	0.58	39.84	1.70	51.04
Hyperion	Erta Ale	4.29	0.32	42.76	1.30	45.51
	Lake Monona	4.35	0.30	42.65	1.27	46.17
	Mt. St. Helens	4.27	0.32	41.83	1.30	45.24

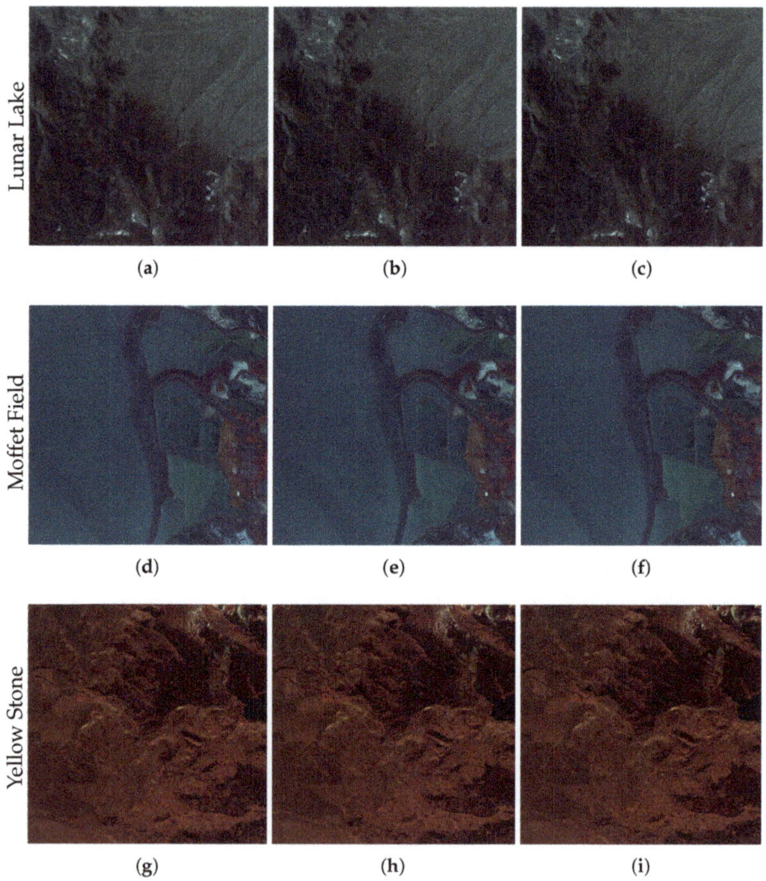

Figure 24. RGB representation of the different compressed-decompressed images collected by the AVIRIS sensor, according to the information displayed in Table 2. These images have been generated by displaying the bands number 50, 20 and 10 as the red, green and blue bands, respectively. (**a**) lossless; (**b**) 0.55 bpppb; (**c**) 1.62 bpppb; (**d**) lossless; (**e**) 0.53 bpppb; (**f**) 1.75 bpppb; (**g**) lossless; (**h**) 0.58 bpppb; (**i**) 1.70 bpppb.

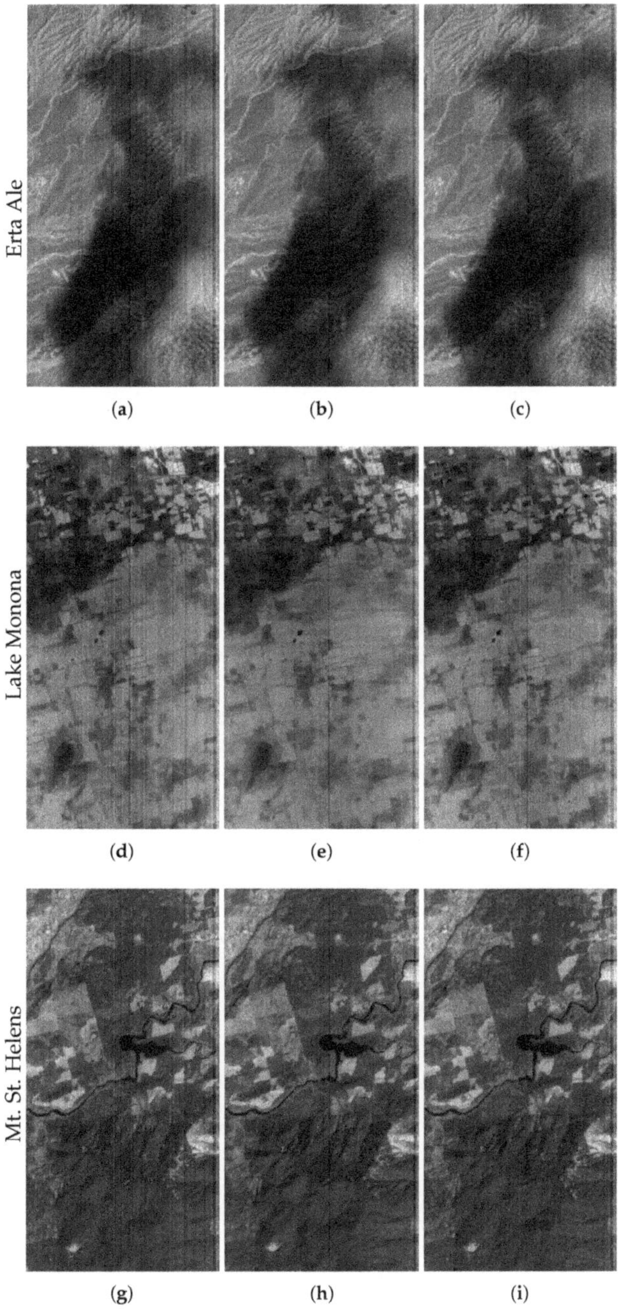

Figure 25. Grey scale representation of the different compressed-decompressed images collected by the Hyperion sensor, according to the information displayed in Table 2. These images have been generated by displaying the bands number 150 from the 242 bands of this sensor as an intensity image. (**a**) lossless; (**b**) 0.32 bpppb; (**c**) 1.30 bpppb; (**d**) lossless; (**e**) 0.30 bpppb; (**f**) 1.27 bpppb; (**g**) lossless; (**h**) 0.32 bpppb; (**f**) 1.30 bpppb.

5. Conclusions

In this manuscript, a new transform-based algorithm for performing lossy hyperspectral images compression, named *Lossy Compression Algorithm for Hyperspectral image systems* (HyperLCA), has been proposed. The main goal of this compressor is to provide a good compression performance at a reasonable computational burden, especially for very high compression ratios that are hardly achievable by lossless compression approaches. The proposed HyperLCA algorithm has different advantages. First of all, it is able to achieve high compression ratios while preserving the most different elements of the data set, which are crucial for many hyperspectral images applications such as anomaly detection, target detection or classification. Secondly, the compression ratio to be achieved can be perfectly fixed in advance. Additionally, some extra stopping conditions, based on quality metrics, can be added in order to stop the compression if the desire minimal quality is achieved at higher compression ratios than the specified. The possibility of adding these kinds of stopping conditions also enables a progressive decoding of the compressed bitstream. Furthermore, the HyperLCA algorithm is able to independently compress blocks of pixels of the image, increasing its error resilience and making it specially suitable for applications that use pushbroom or whiskbroom sensors. Finally, the HyperLCA also has many computational advantages, including low mathematical complexity and a high level of parallelism, which differentiate it from other state-of-the-art transform-based compression approaches and make it a more viable option for applications under tight latency constraints or applications with limited computational resources, such as hyperspectral compression on-board satellites.

An extensive amount of experiments have been performed in order to evaluate the goodness of the proposed HyperLCA compressor using different calibrated and uncalibrated hyperspectral images from the AVIRIS and Hyperion sensors. The results provided by the proposed HyperLCA compressor have been evaluated and compared against those produced by some of the most relevant state-of-the-art compression solutions. Additionally, the effect produced by compressing-decompressing the image using the HyperLCA transform in anomaly detection, hyperspectral imaging classification and spectral unmixing applications has also been evaluated. All the obtained results allow to conclude that the proposed HyperLCA compressor represents a very suitable option for lossy compressing hyperspectral images, especially when it is important to achieve high compression ratios while at the same time preserving the most different elements of the data set, and when it is important to perform the compression under tight latency and computational constraints.

Acknowledgments: This research is partially funded by the European Commission through the ECSEL Joint Undertaking (ENABLE-S3 project, No. 692455) and the Ministry of Economy and Competitiveness (MINECO) of the Spanish Government (ENABLE-S3 project, No. PCIN-2015-225 and REBECCA (Resilient EmBedded Electronic systems for Controlling Cities under Atypical situations) project, No. TEC2014-58036-C4-4-R.

Author Contributions: Raúl Guerra, Yubal Barrios, María Díaz, Lucana Santos, Sebastián López and Roberto Sarmiento were all involved in the study, design and writing of this paper. Raúl Guerra and María Díaz carried out the development of the HyperLCA compressor. Lucana Santos provided reference software and knowledge about the state-of-the-art solutions for on-board compression. Yubal Barrios performed the experiments. Sebastián López and Roberto Sarmiento supervised the work and revised the manuscript.

Conflicts of Interest: The authors declare no conflict of interest.

References

1. Penna, B.; Tillo, T.; Magli, E.; Olmo, G. Progressive 3-D coding of hyperspectral images based on JPEG 2000. *IEEE Geosci. Remote Sens. Lett.* **2006**, *3*, 125–129.
2. Chang, L.; Cheng, C.M.; Chen, T.C. An efficient adaptive KLT for multispectral image compression. In Proceedings of the 4th IEEE Southwest Symposium on Image Analysis and Interpretation, Austin, TX, USA, 2–4 April 2000; pp. 252–255.

3. Hao, P.; Shi, Q. Reversible integer KLT for progressive-to-lossless compression of multiple component images. In Proceedings of the 2003 International Conference on Image Processing, Barcelona, Spain, 14–17 September 2003; Volume 1.
4. Amrani, N.; Serra-Sagristà, J.; Laparra, V.; Marcellin, M.W.; Malo, J. Regression wavelet analysis for lossless coding of remote-sensing data. *IEEE Trans. Geosci. Remote Sens.* **2016**, *54*, 5616–5627.
5. Penna, B.; Tillo, T.; Magli, E.; Olmo, G. Transform coding techniques for lossy hyperspectral data compression. *IEEE Trans. Geosci. Remote Sens.* **2007**, *45*, 1408–1421.
6. Du, Q.; Zhu, W.; Fowler, J.E. Anomaly-based hyperspectral image compression. In Proceedings of the IGARSS 2008. IEEE International Conference on Geoscience and Remote Sensing Symposium, Boston, MA, USA, 6–11 July 2008; Volume 2.
7. Du, Q.; Fowler, J.E. Hyperspectral image compression using JPEG2000 and principal component analysis. *IEEE Geosci. Remote Sens. Lett.* **2007**, *4*, 201–205.
8. Taubman, D.; Marcellin, M. *JPEG2000 Image Compression Fundamentals, Standards and Practice: Image Compression Fundamentals, Standards and Practice*; Springer Science & Business Media: New York, NY, USA, 2012; Volume 642.
9. Bioucas-Dias, J.M.; Plaza, A.; Camps-Valls, G.; Scheunders, P.; Nasrabadi, N.M.; Chanussot, J. Hyperspectral remote sensing data analysis and future challenges. *IEEE Geosci. Remote Sens. Mag.* **2013**, *1*, 6–36.
10. Bioucas-Dias, J.M.; Plaza, A.; Dobigeon, N.; Parente, M.; Du, Q.; Gader, P.; Chanussot, J. Hyperspectral unmixing overview: Geometrical, statistical, and sparse regression-based approaches. *IEEE J. Sel. Top. Appl. Earth Obs. Remote Sens.* **2012**, *5*, 354–379.
11. Blanes, I.; Serra-Sagrista, J.; Marcellin, M.W.; Bartrina-Rapesta, J. Divide-and-conquer strategies for hyperspectral image processing: A review of their benefits and advantages. *IEEE Signal Process. Mag.* **2012**, *29*, 71–81.
12. Santos, L.; López, S.; Callico, G.M.; López, J.F.; Sarmiento, R. Performance evaluation of the H. 264/AVC video coding standard for lossy hyperspectral image compression. *IEEE J. Sel. Top. Appl. Earth Obs. Remote Sens.* **2012**, *5*, 451–461.
13. Aiazzi, B.; Alparone, L.; Baronti, S. Quality issues for compression of hyperspectral imagery through spectrally adaptive DPCM. In *Satellite Data Compression*; Springer: New York, NY, USA, 2012; pp. 115–147.
14. Lee, C.; Lee, S.; Lee, J. Effects of lossy compression on hyperspectral classification. In *Satellite Data Compression*; Springer: New York, NY, USA, 2012; pp. 269–285.
15. García-Vílchez, F.; Muñoz-Marí, J.; Zortea, M.; Blanes, I.; González-Ruiz, V.; Camps-Valls, G.; Plaza, A.; Serra-Sagristà, J. On the impact of lossy compression on hyperspectral image classification and unmixing. *IEEE Geosci. Remote Sens. Lett.* **2011**, *8*, 253–257.
16. Du, Q.; Ly, N.; Fowler, J.E. An operational approach to PCA+ JPEG2000 compression of hyperspectral imagery. *IEEE J. Sel. Top. Appl. Earth Obs. Remote Sens.* **2014**, *7*, 2237–2245.
17. Chang, C.I. *Hyperspectral Data Processing: Algorithm Design and Analysis*; John Wiley & Sons: Hoboken, NJ, USA, 2013.
18. Howard, P.G.; Vitter, J.S. Fast and efficient lossless image compression. In Proceedings of the DCC'93 Data Compression Conference, Snowbird, UT, USA, 30 March–2 April 1993; pp. 351–360.
19. Consultative Committee for Space Data Systems (CCSDS). Blue Books: Recommended Standards. Available online: https://public.ccsds.org/Publications/BlueBooks.aspx (accessed on 1 February 2018).
20. Kiely, A. Selecting the Golomb Parameter in Rice Coding; IPN Progress Report; 2004; Volume 42. Available online: https://www.researchgate.net/publication/252469081_Selecting_the_Golomb_Parameter_in_Rice_Coding (accessed on 1 February 2018).
21. Vane, G.; Green, R.O.; Chrien, T.G.; Enmark, H.T.; Hansen, E.G.; Porter, W.M. The airborne visible/infrared imaging spectrometer (AVIRIS). *Remote Sens. Environ.* **1993**, *44*, 127–143.
22. U.S. Geological Survey and NASA. Earth Observing 1, Hyperion Website. Available online: https://eo1.usgs.gov/sensors/hyperion (accessed on 1 February 2018).
23. Herweg, J.A.; Kerekes, J.P.; Weatherbee, O.; Messinger, D.; van Aardt, J.; Ientilucci, E.; Ninkov, Z.; Faulring, J.; Raqueño, N.; Meola, J. Spectir hyperspectral airborne rochester experiment data collection campaign. In Proceedings of the SPIE International Society for Optics and Photonics Defense, Security, and Sensing, Baltimore, MD, USA, 23–27 April 2012; p. 839028.

24. Plaza, A.; Du, Q.; Chang, Y.L.; King, R.L. High performance computing for hyperspectral remote sensing. *IEEE J. Sel. Top. Appl. Earth Obs. Remote Sens.* **2011**, *4*, 528–544.
25. Motta, G.; Rizzo, F.; Storer, J.A. *Hyperspectral Data Compression*; Springer Science & Business Media: New York, NY, USA, 2006.
26. Wan, K.X.; Vidavsky, I.; Gross, M.L. Comparing similar spectra: From similarity index to spectral contrast angle. *J. Ame. Soc. Mass Spectrom.* **2002**, *13*, 85–88.
27. Jablonski, J.A.; Bihl, T.J.; Bauer, K.W. Principal component reconstruction error for hyperspectral anomaly detection. *IEEE Geosci. Remote Sens. Lett.* **2015**, *12*, 1725–1729.
28. Ghamisi, P.; Plaza, J.; Chen, Y.; Li, J.; Plaza, A.J. Advanced Spectral Classifiers for Hyperspectral Images: A review. *IEEE Geosci. Remote Sens. Mag.* **2017**, *5*, 8–32.
29. Chang, C.C.; Lin, C.J. LIBSVM. A Library for Support Vector Machines. Available online: https://www.csie.ntu.edu.tw/~cjlin/libsvm/ (accessed on 1 February 2018)
30. Borghys, D.; Achard, V.; Rotman, S.; Gorelik, N.; Perneel, C.; Schweicher, E. Hyperspectral anomaly detection: A comparative evaluation of methods. In Proceedings of the 2011 XXXth URSI General Assembly and Scientific Symposium, Istanbul, Turkey, 13–20 August 2011; pp. 1–4.
31. Matteoli, S.; Diani, M.; Corsini, G. A tutorial overview of anomaly detection in hyperspectral images. *IEEE Aerosp. Electron. Syst. Mag.* **2010**, *25*, 5–28.
32. Borghys, D.; Kåsen, I.; Achard, V.; Perneel, C. Hyperspectral anomaly detection: Comparative evaluation in scenes with diverse complexity. *J. Electr. Comput. Eng.* **2012**, *2012*, 5. doi:10.1155/2012/162106.
33. María, D.; Guerra, R.; López, S.; Sarmiento, R. An Algorithm for an Accurate Detection of Anomalies in Hyperspectral Images with a Low Computational Complexity. *IEEE Trans. Geosci. Remote Sens.* **2018**. doi:10.1109/TGRS.2017.2761019.
34. Nascimento, J.M.; Bioucas Dias, J.M. Vertex component analysis: A fast algorithm to unmix hyperspectral data. *IEEE Trans. Geosci. Remote Sens.* **2005**, *43*, 898–910.
35. Chang, C.I. Orthogonal subspace projection (OSP) revisited: A comprehensive study and analysis. *IEEE Trans. Geosci. Remote Sens.* **2005**, *43*, 502–518.
36. Winter, M.E. N-FINDR: An algorithm for fast autonomous spectral end-member determination in hyperspectral data. In Proceedings of the SPIE's International Symposium on Optical Science, Engineering, and Instrumentation, Denver, CO, USA, 27 October 1999; pp. 266–275.
37. Consultative Committee for Space Data Systems (CCSDS). Spectral Preprocessing Transform for Multispectral and Hyperspectral Image Compression. Blue Book. Issue 1. September 2017. Available online: https://public.ccsds.org/Pubs/122x1b1.pdf (accessed on 1 February 2018).
38. Consultative Committee for Space Data Systems (CCSDS). Lossless Multispectral Andf Hyperspectral Image Compression. Blue Book. Issue 1. May 2012. Available online: https://public.ccsds.org/Pubs/123x0b1ec1.pdf (accessed on 1 February 2018).
39. Consultative Committee for Space Data Systems (CCSDS). Image Data Compression. Blue Book. Issue 2. September 2017. Available online: https://public.ccsds.org/Pubs/122x0b2.pdf (accessed on 1 February 2018).
40. Cohen, A.; Daubechies, I.; Feauveau, J.C. Biorthogonal bases of compactly supported wavelets. *Commun. Pure Appl. Math.* **1992**, *45*, 485–560.
41. Blanes, I.; Serra-Sagristà, J. Pairwise orthogonal transform for spectral image coding. *IEEE Trans. Geosci. Remote Sens.* **2011**, *49*, 961–972.
42. Blanes, I.; Hernández-Cabronero, M.; Auli-Llinas, F.; Serra-Sagristà, J.; Marcellin, M.W. Isorange pairwise orthogonal transform. *IEEE Trans. Geosci. Remote Sens.* **2015**, *53*, 3361–3372.
43. Huang, B. *Satellite Data Compression*; Springer Science & Business Media: New York, NY, USA, 2011.
44. Pizzolante, R.; Carpentieri, B. Visualization, band ordering and compression of hyperspectral images. *Algorithms* **2012**, *5*, 76–97.
45. Auge, E.; Santalo, J.; Blanes, I.; Serra-Sagrista, J.; Kiely, A. Review and implementation of the emerging CCSDS recommended standard for multispectral and hyperspectral lossless image coding. In Proceedings of the 2011 First International Conference on Data Compression, Communications and Processing (CCP), Cilento Coast, Italy, 21–24 June 2011; pp. 222–228.

46. Karaca, A.C.; Güllü, M.K. Lossless compression of ultraspectral sounder data using recursive least squares. In Proceedings of the 2017 8th International Conference on Recent Advances in Space Technologies (RAST), Istanbul, Turkey, 19–23 June 2017; pp. 109–112.
47. ESA. Data Compression Evaluation Tool for Standard CCSDS Compression Algorithms Used in ESA Missions. Available online: https://essr.esa.int/project/whitedwarf (accessed on 1 February 2018).

remote sensing

MDPI

Article

Hyperspectral Pansharpening Based on Intrinsic Image Decomposition and Weighted Least Squares Filter

Wenqian Dong, Song Xiao *, Yunsong Li * and Jiahui Qu

State Key laboratory of Integrated Service Networks, Xidian University, Xi'an 710071, China; dongwq_xd@163.com (W.D.); qujh_xd@163.com (J.Q.)

* Correspondence: xiaosong@mail.xidian.edu.cn (S.X.); ysli@mail.xidian.edu.cn (Y.L.); Tel.: +86-029-8820-4033 (S.X.); +86-029-8820-2721 (Y.L.)

Received: 19 December 2017; Accepted: 9 March 2018; Published: 12 March 2018

Abstract: Component substitution (CS) and multiresolution analysis (MRA) based methods have been adopted in hyperspectral pansharpening. The major contribution of this paper is a novel CS-MRA hybrid framework based on intrinsic image decomposition and weighted least squares filter. First, the panchromatic (P) image is sharpened by the Gaussian-Laplacian enhancement algorithm to enhance the spatial details, and the weighted least squares (WLS) filter is performed on the enhanced P image to extract the high-frequency information of the P image. Then, the MTF-based deblurring method is applied to the interpolated hyperspectral (HS) image, and the intrinsic image decomposition (IID) is adopted to decompose the deblurred interpolated HS image into the illumination and reflectance components. Finally, the detail map is generated by making a proper compromise between the high-frequency information of the P image and the spatial information preserved in the illumination component of the HS image. The detail map is further refined by the information ratio of different bands of the HS image and injected into the deblurred interpolated HS image. Experimental results indicate that the proposed method achieves better fusion results than several state-of-the-art hyperspectral pansharpening methods. This demonstrates that a combination of an IID technique and a WLS filter is an effective way for hyperspectral pansharpening.

Keywords: hyperspectral pansharpening; panchromatic; intrinsic image decomposition; weighted least squares filter

1. Introduction

Hyperspectral pansharpening aims to combine the preponderance and complementary information of the hyperspectral (HS) and panchromatic (P) images for image analysis and various applications [1]. Spatial information and spectral information plays an important role in remote sensing image analysis. Unfortunately, due to the limitation of sensor and theoretical aspects, most satellites cannot provide a remote sensing image with both high spatial and spectral resolution [2,3]. However, hyperspectral images with high spectral and spatial resolution have been in demand. Therefore, it is important to introduce hyperspectral pansharpening techniques to improve the spatial resolution of hyperspectral images.

Many methods dedicated to hyperspectral pansharpening have been proposed in the last two decades [4,5]. These hyperspectral pansharpening methods can be grossly divided into four groups: component substitution (CS), multiresolution analysis (MRA), matrix factorization, and Bayesian. In recent years, there has been increasing interest in Bayesian methods and matrix factorization methods. Bayesian methods usually model the HS and the P images as the degraded high-resolution HS images and then restore the HS images through solving optimization problems, such as Sparse

Representation [6,7], Bayesian HySure [8], and Bayesian Naive Gaussian prior (Bayesian Naive) [9]. Matrix factorization methods utilize the linear mixture model, and use it for the fusion optimization model. The coupled non-negative matrix factorization (CNMF) [10] method is a representative among the matrix factorization methods. Bayesian and matrix factorization methods have shown considerable potential in improving the quality of the fused images. However, for purpose of estimating a good solution, researchers have also made efforts to solve the ill-posed inverse problem, which is time consuming and computational expensive [11,12]. From a perspective of practical applications, it is a difficult problem.

The CS and MRA methods are easy and fast to implement [13,14]. The component substitution (CS) methods include algorithms, such as intensity hue saturation (IHS) [15,16], Gram–Schmidt (GS) [17], and principal component analysis (PCA) [18–20]. The primary concept of CS methods is that the HS image can be separated into spectral and spatial components, and the P image is a good substitution for the separated spatial component. The final fused image is obtained by the inverse spectral transformation [21]. The fusion step of the CS methods is summarized as

$$H_F^k = H^k + \alpha^k(P - S) \tag{1}$$

where, $k = 1, 2, \ldots, \lambda$, λ is the number of the HS image bands, α^k is the kth injection gain, matrix P is the panchromatic image, matrix S is the spatial component of the HS image, $P - S$ is generally called the detail map, matrix H is the interpolated HS image, H_F^k and H^k are the kth band of the fused image and the interpolated HS image, respectively. The injection gain is a gain used for merging the detail map and the interpolated HS image into a fused HS image. The CS methods have simple and fast implementation [22]. However, the spectral distortion is serious due to the spectral mismatch between the P image and the replaced component [23].

The multiresolution analysis (MRA) has algorithms such as modulation transfer function (MTF) generalized Laplacian Pyramid (MTF-GLP) [24], smoothing filter-based intensity modulation (SFIM) [25] and MTF-GLP with high pass modulation (MGH) [26]. The spatial filtering is performed on the P image to extract the high-frequency spatial details. The fused HS image is obtained by injecting the extracted spatial details into each band of the interpolated HS image. Following, a formulation of MRA methods is defined as [4]

$$H_F^k = H^k + \beta_k(P - P_L) \tag{2}$$

where, $k = 1, 2, \ldots, \lambda$, λ denotes the number of the HS image bands, β_k denotes the kth injection gain, and P_L denotes low-frequency component of the P image. The advantages of the MRA methods are good performance, temporal coherence, spectral consistency and acceptable computational complexity. In addition, the MRA methods can be easily adopted when the source of the high spatial frequencies is another multispectral/hyperspectral image [27]. However, blurry images may occur when the shapes of low-pass filters adopted have problems [24].

To overcome the problems of the CS and MRA methods, the CS-MRA hybrid frameworks were proposed [20,28,29]. These methods focus on fusing the P image and the spatial component of the HS image by multiscale transforms. The final fused image is obtained by the inverse spectral transformation. The performance of the CS-MRA hybrid methods has shown improvement compared with that of the CS or MRA methods. However, the fused images obtained by these hybrid methods suffer from spectral distortion of different degrees, since the structure of the HS image is not fully considered. To overcome the drawbacks of the CS-MRA hybrid methods, we propose a new CS-MRA hybrid framework based on intrinsic image decomposition and weighted least squares filter. Specifically, we filter the sharpened P image by the weighted least squares (WLS) filter to obtain the high-frequency component of the P image at first. Subsequently, the MTF-based deblurring method is performed on the interpolated HS image. The intrinsic image decomposition (IID) is applied to the deblurred interpolated HS image to extract the illumination component of the HS image. The detail map is generated by merging the high frequency information of P image with the illumination component of the HS image. Finally, the detail map is injected into the deblurred interpolated HS image to obtain the fused HS image.

The following are the major contributions of the proposed method using IID and WLS:

(a) It uses the IID technique which separates the deblurred HS image into the reflectance and illumination components to extract the spatial information from the HS image.
(b) Unlike the traditional CS and MRA methods where the spatial details are just extracted from the P image, the detail map in the proposed method depends on both the HS image and the P image. The spectral distortion caused by the spectral mismatch problem can be reduced.
(c) The WLS filter preserves the spatial details on edges in a better manner compared to traditional low pass filters, since it can make the best compromise between sharpening and blurring. Therefore, the WLS filter is adopted to extract the high-frequency component of the P image in the proposed method.
(d) Most CS and MRA methods are based on the assumption that each band of the HS image shares the same detail map. We assume that different detail map is required by different bands of the HS image. The detail map is generated according to the ratio of the information between different bands of the HS image.

This paper is organized as follows. We describe the related work in Section 2. The proposed method is discussed in Section 3. Section 4 displays the experimental results and discussion. Section 5 concludes the paper.

2. Related Work

2.1. Intrinsic Image Decomposition

Based on the principle of human visual perception, intrinsic image decomposition (IID) aims to decompose an image into reflectance and illumination components [30]. The reflectance component depends on the material of objects in an image. There is abundant edge and structure information in an image. However, these information is not directly related to the reflectance component. Edge and structure information is mainly preserved in the illumination component. The intrinsic image decomposition process can be expressed as:

$$I = SR \tag{3}$$

where matrix I represents an input image, matrices S and R represent the illumination and reflectance components, respectively. From the aforementioned equation, it is obvious that estimating S and R based on I is a difficult problem. In order to solve this problem, many solutions have been presented in recent years [31–34]. Among these methods, user intervention [34], extra constraints [30], and heuristic cues have shown good results to estimate the intrinsic image from the input image. Retinex algorithm is a widely used image enhancement method based on scientific experiments and analysis [35]. Compared with the traditional methods that can only enhance a certain characteristic, the retinex algorithm can balance the dynamic compression, contrast improvement and constant color. It is important that the detail information immersed in the light region or shadow can be effectively displayed by the retinex algorithm. However, it is difficult to weigh up the relationship between the detail contrast of the image and the color reservation by the single scale retinex algorithm. Hyperspectral pansharpening aims at enhancing the spatial information while preserving the spectral information. Multiscale retinex algorithm indeed shows the good performance in achieving this objective [36]. Motivated by the aforementioned finding, the multiscale retinex based-IID algorithm is really a good technique for hyperspectral pansharpening.

2.2. Weighted Least Squares Filter

The WLS filter which is an edge-preserving filter has become a highly active research topic in various image processing. Since the WLS filter do not blur strong edges in the process of image decomposition, the ringing artifacts can be avoided. Compared with other edge-preserving filters, the WLS filter can make a best compromise between sharpening and blurring [37]. The WLS filter can be used for estimating the low frequency image of the input image. Specifically, the WLS filter assumes that the filtered image f is as close as possible to the input image g, and should be as smooth as possible everywhere, except across the edges. Farbman et al. [37] proposed that the filtered image f can be obtained by seeking the minimum of the following equation

$$\sum_{p}((f_p - g_p)^2 + \gamma(\omega_{x,p}(g)(\frac{\partial f}{\partial x})_p^2 + \omega_{y,p}(g)(\frac{\partial f}{\partial y})_p^2)) \tag{4}$$

where p refers to the pth pixel. The first term $(f_p - g_p)^2$ ensures the minimum distance between the filtered image f and the input image g, $\omega_{x,p}(g)$ and $\omega_{y,p}(g)$ which depend on g are smoothness weights. γ is a regular term parameter that balances the two terms.

3. Proposed Method

The schematic diagram of the proposed method is shown in Figure 1. It consists of three major parts: (1) Extracting spatial details of the P image; (2) Extracting spatial detail of the H image; (3) Generating the detail map; and (4) Obtaining the fused H image.

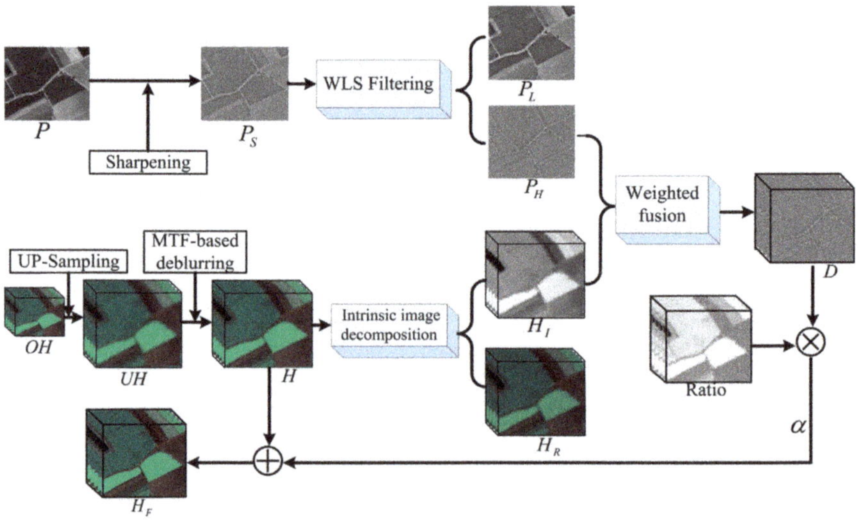

Figure 1. The schematic diagram of the proposed method.

3.1. Extracting Spatial Details of the P Image with Weighted Least Squares Filter

The P image contains plentiful spatial information. To enhance the spatial details of the P image and reduce noise, we use the Laplacian of Gaussian (LOG) enhancement algorithm to sharpen the spatial information of the P image. LOG algorithm can reduce image noise and sharpen structural details. First, Gaussian convolution filtering is performed on the P image to remove the noise. Laplace operator is then used for enhancing the spatial details of the denoised P image. The enhanced P image is finally obtained by combining the P image with the Laplacian filtered image. This process can be expressed as

$$P_S = P + a[P * f_L(x,y)] \tag{5}$$

where, P_S is the enhanced P image, $f_L(x,y)$ is the kernel function of LOG operator, a is a constant. a is related to the central coefficient of the kernel $f_L(x,y)$. In this paper, the central coefficient of the kernel is a negative value, and we set a to -1. $f_L(x,y)$ is defined as

$$f_L(x,y) = \frac{\partial^2}{\partial x^2} f_g(x,y) + \frac{\partial^2}{\partial y^2} f_g(x,y) \tag{6}$$

where, $f_g(x,y) = \frac{1}{\sqrt{2\pi\sigma^2}} \exp(-\frac{x^2+y^2}{2\sigma^2})$ is the Gaussian convolution function, and σ is the standard deviation.

To extract the spatial details of the enhanced P image, the WLS filter is applied to the P_S image. Based on the principle of the WLS filter, the low-frequency image P_L which should be as close as possible to the P_S image and be as smooth as possible except the edges can be obtained by the following optimization equation.

$$\underset{P_L}{\operatorname{argmin}}(\|P_L - P_S\|^2 + \gamma(\omega_x(\frac{\partial P_L}{\partial x})^2 + \omega_y(\frac{\partial P_L}{\partial y})^2)) \tag{7}$$

The first term $\|P_L - P_S\|^2$ ensures the minimum distance between P_L and P_S. γ is a regularization parameter to balance the first term and the second term. The second term $\omega_x(\partial P_L/\partial x)^2 + \omega_y(\partial P_L/\partial y)^2$ aims to smooth the P_L image by minimizing the derivatives of P_L with respect to x and y. ω_x and ω_y are smoothness weights. We rewrite Equation (7) by using matrix notation

$$(P_L - P_S)^T(P_L - P_S) + \gamma(P_L^T Z_x^T W_x Z_x P_L + P_L^T Z_y^T W_y Z_y P_L) \tag{8}$$

where matrices W_x and W_y are diagonal matrices, the matrices Z_x and Z_y are the discrete differentiation operators, and the smoothness weights ω_x and ω_y are contained in matrices Z_x and Z_y, respectively. The following linear equation can be obtained by taking the derivative of Equation (8)

$$[(I + \gamma(Z_x^T W_x Z_x + Z_y^T W_y Z_y))]P_L = P_S \tag{9}$$

where $Z_x^T W_x Z_x + Z_y^T W_y Z_y$ is a five-point spatially inhomogeneous Laplacian matrix. We define ω_x and ω_y in the same manner as in [38]:

$$\omega_{x,p(P_S)} = (\left|\frac{\partial l}{\partial x}(p)\right|^\alpha + \varepsilon)^{-1}, \omega_{y,p(P_S)} = (\left|\frac{\partial l}{\partial y}(p)\right|^\alpha + \varepsilon)^{-1} \tag{10}$$

where l is the log-luminance channel of the P_S image, ε is a constant that is close to zero, the parameter α controls the gradients of P_S, and is set to 2.0 in this paper. By the above analysis, the low-frequency image P_L can be estimated. Considering the MRA formulation (Equation (2)), we can obtain the high-frequency image P_H by subtracting the low-frequency image P_L from the enhanced P image P_S. This process can be expressed as

$$P_H = P_S - P_L \tag{11}$$

where P_L and P_H denote the low-frequency image and high-frequency image of the P_S image, respectively. We consider that most of the spatial details of the enhanced P image are contained in the high-frequency image.

Figure 2 shows the example of spatial detail processing of the P image. In this experiment, the respective parameters are adjusted to the optimal values. From Figure 2b, it can be observed that the Laplacian of Gaussian (LOG) enhancement algorithm indeed plays an important role in the spatial details enhancement. Figure 2c shows the high-frequency image of the enhanced P image obtained by the proposed method. To illustrate the effectiveness of the WLS filter in spatial extraction, Figure 2c is compared with Figure 2d which is obtained by Gaussian filtering on the enhanced image. Similarly, to illustrate the effectiveness of the LOG algorithm in spatial enhancement, Figure 2c is compared with Figure 2e which is obtained by WLS filtering on the original P image. A comparison of Figure 2c with Figure 2d,e shows that there is fine distinction between three filtered images. Figure 2d,e show that some low frequency components are mixed with the high-frequency image, especially in spherical regions. Therefore, the LOG enhancement algorithm indeed displays a good performance in spatial detail enhancement, and the WLS filter is really suitable for extracting the high frequency component of the P image.

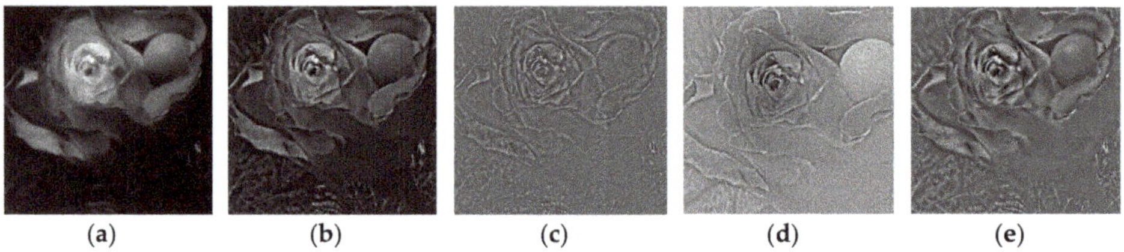

Figure 2. Example of spatial detail processing of the P image. (**a**) Original P image; (**b**) Enhanced P image. High-frequency images: (**c**) Sharpening + WLS filtering; (**d**) Sharpening + Gaussian filtering; (**e**) WLS filtering.

3.2. Extracting Spatial Detail of the HS Image with Intrinsic Image Decomposition

Upsampling is done on the original hyperspectral image (OH) to obtain the same size as the P image.

$$UH^k = \uparrow OH^k \tag{12}$$

for $k = 1, 2, \ldots, \lambda$, where λ is the number of the HS image bands, $\uparrow$ is the upsampling operation, UH is the interpolated HS image, OH^k and UH^k are the kth band of the original HS image and the interpolated HS image, respectively.

A preprocessing step that performs MTF-based deblurring [39] on the interpolated HS image. The MTF-based deblurring method is performed in the frequency domain as follows

$$\mathcal{H}^k = \frac{\mathcal{M}_k^*}{|\mathcal{M}^k|^2 + \frac{1}{SNR}} \mathcal{UH}^k \tag{13}$$

where, $\mathcal{H}$ is the Fourier transform of the deblurred interpolated HS image, $\mathcal{H}^k$ is the kth band of $\mathcal{H}$, $\mathcal{UH}^k$, and $\mathcal{M}^k$ are the Fourier transform of UH^k and the Fourier transform of the blurring kernel for the kth band, respectively, $1/SNR$ is the noise-to-power ratio, and $\mathcal{M}_k^*$ is the complex conjugate of $\mathcal{M}^k$. $1/SNR$ is a constant that is close to zero. The deblurred interpolated HS image H is obtained by the inverse Fourier transform of $\mathcal{H}$. Subsequently, according to [39], we adopt the de-ringing technique to decrease ringing artifacts caused by non-periodic boundaries.

According to the CS based method, the deblurred HS image H can be separated into spectral and spatial information. IID is introduced to separate the spectral and spatial information. Based on the principle of the IID, each band of the HS image is composed of two components, i.e., the illumination component and the reflectance component. For the HS image, the reflectance component is identified as the spectral information, while the illumination component is the spatial information. Each band of the deblurred HS image can be represented as follows

$$H^k(x,y) = H_R^k(x,y) \times H_I^k(x,y) \tag{14}$$

where (x,y) is spatial coordinate. $H_R^k(x,y)$ which depends on the intrinsic nature of the objects represents the reflectance component of the kth band of an HS image. $H_I^k(x,y)$, which is related to the structure information of the objects, represents the illumination component of the kth band of an HS image. Based on the single scale retinex algorithm, the output can be estimated by the difference between the input and the average of its neighborhood, which can be described by the following equation

$$h_R^k(x,y) = \log(H^k(x,y)) - \log(H^k(x,y) * F(x,y)) \tag{15}$$

where $h_R^k(x,y) = \log(H_R^k(x,y))$, H^k is the kth band of the input image, h_R^k is the kth band of the output image, F is the Gaussian surround function, and symbol $*$ is convolution. Illumination estimation is denoted by the convolution $H^k(x,y) * F(x,y)$. The Gaussian surround function $F(x,y)$ can be given as follows

$$F(x,y) = C\exp[-(x^2+y^2)/2\sigma^2] \tag{16}$$

$$\iint F(x,y)dxdy = 1 \tag{17}$$

where σ, the scale factor of the Gaussian kernel, controls the color information and the spatial resolution of the image. σ cannot be determined and theoretically modeled. Generally, the larger the scale parameter σ, the better the color fidelity and the lower the spatial resolution of the output image. To make the compromise between the extraction of spatial details and the preservation of spectral information, multiscale retinex is used for separating the reflectance component from the illumination component of the deblurred HS image. This process can be given by the following formula

$$h_{MR}^k(x,y) = \sum_{n=1}^{N} \omega_n h_R^k = \sum_{n=1}^{N} \omega_n \log(H^k(x,y)) - \log(H^k(x,y) * F_n(x,y)) \tag{18}$$

$$F_n(x,y) = C_n \exp[-(x^2+y^2)/2\sigma_n^2] \tag{19}$$

where N represents the number of scales, and ω_n represents the weighting factor. According to experimental experience, N is set to 3, and ω_n is set to 1/3. σ_1, σ_2, and σ_3 are set as 20, 40, and 80, respectively. Obtaining the scale factor σ_n, C_n which denotes the normalization factor can be determined by Equations (17) and (19). Then, the reflectance component of the deblurred HS image is easy to be obtained

$$H_R^k(\mathrm{x},\mathrm{y}) = \exp(h_{MR}^k(\mathrm{x},\mathrm{y})) \tag{20}$$

According to Equation (14), the illumination component of the deblurred HS image is calculated by

$$H_I^k = \frac{H^k}{H_R^k} \tag{21}$$

We consider that most of the spatial details of each band of the deblurred HS image are contained in H_I^k.

In [29], the spatial details of each band of the HS image are extracted by average filtering. The right picture of Figure 3b shows the detail map obtained by average filtering. The right picture of Figure 3c shows the detail map obtained by the adopted IID technique. The detail map shown in Figure 3b contains part of the spatial information of the HS image. The detail map obtained by the IID technique contains most of the spatial information of the HS image. The IID is indeed an effective technique for extracting spatial information of the HS image.

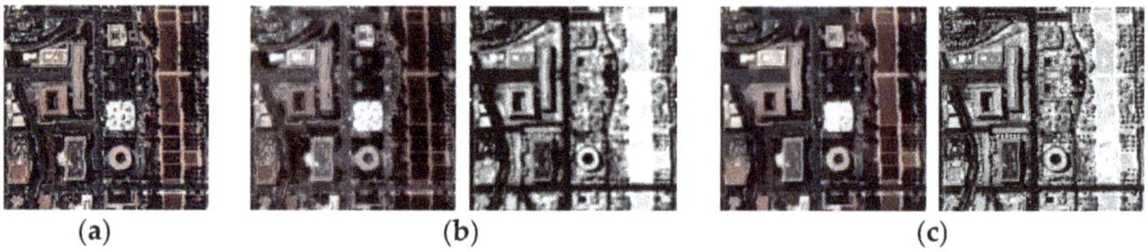

Figure 3. Spatial details extracted by average filtering and intrinsic image decomposition methods. (a) HS image; (b) Average filtering method (left: Spectral component of HS image; right: Spatial component of HS image); (c) Intrinsic image decomposition method (left: Reflectance component of HS image; right: Illumination component of HS image).

3.3. Generating the Detail Map

The CS methods suffer from the spectral distortion, since the detail map only depends on the P image. To reduce the spectral distortion, the detail map in this paper depends on both the P image and the HS image. The detail map of each band of the HS image is determined by the illumination component of the HS image and the high-frequency component of the P image. Specifically, after obtaining the high-frequency component of the P image and the illumination component of the kth band of the HS image, the detail map can be generated by

$$D^k = \zeta P_H + (1-\zeta) H_I^k \tag{22}$$

for $k = 1, 2, \ldots, \lambda$, where ζ is a weight coefficient, and D^k is the initial detail map. Since the P image contains much more spatial information compared with the HS image, ζ is set to 0.9. Many traditional CS and MRA based methods assume that the same detail map is in demand by different bands of the HS image. This hypothesis always produces spectral and spatial distortion. We consider that a different detail map is required by different bands of the HS image. Assuming that the amount of spatial detail required for different bands is proportional to the ratio of the information of that band. It is helpful for reducing the spectral distortion to keep this ratio unchanged. Thus, we define the following formula

$$D_F^k = \frac{UH^k}{(1/\lambda)\sum_{k=1}^{n} UH^k} D^k \tag{23}$$

for $k = 1, 2, \ldots, \lambda$, where λ is the bands number of the HS image, D_F^k denotes the final detail map which is required by the kth band of the HS image.

3.4. Obtaining the Fused HS Image

According to Equation (1), we define a constraint parameter α to control the final detail map. The fused HS image can be obtained by injecting the final detail map into the deblurred interpolated HS image for each band.

$$H_F^k = H^k + \alpha D_F^k \tag{24}$$

for $k = 1, 2, \ldots, \lambda$, where H_F is the fused HS image, and H_F^k is the kth band of the fused HS image. Since the amount of the injected details is regulated by the parameter α, the spectral and spatial distortion can be restrained.

4. Results

In the experiments, the test P and HS images are cropped from four different hyperspectral remote sensing datasets, i.e., the Salinas dataset [12], the Pavia University dataset [12], the Washington DC dataset, and Hyperion dataset [40]. Several widely used evaluation indexes are adopted to estimate the effectiveness of the proposed hyperspectral pansharpening method. Six representative hyperspectral pansharpening methods are utilized for comparison, i.e., principal component analysis (PCA) [18], Guided filter PCA (GFPCA) [41], HySure [8], coupled nonnegative matrix factorization (CNMF) [10], MTF-GLP with High Pass Modulation (MGH) [26] and Sparse Representation [7]. The PCA method is a most representative among the CS-based methods. The MGH method is a successful MRA-based hyperspectral pansharpening method. The CNMF method is one of the matrix factorization algorithms. The Sparse Representation and HySure methods which were presented recently belong to the Bayesian category. The GFPCA method has been awarded the "Best Paper Challenge" in the 2014 IEEE data fusion contest. Therefore, in the experiments, these six methods are compared with the proposed method.

4.1. Dataset Description

(1) The Salinas dataset: This dataset is composed of the urban and rural scene. The HS image was collected by the Airborne Visible/Infrared Imaging Spectrometer (AVIRIS) over Salinas Valley, California [12]. The collected HS image is characterized by 224 bands in the spectral range of 0.4–2.5 μm. The water absorption and noise corrupted bands are removed, and 204 bands are used for experimentation. In the experiments, the P image covers the visible spectral range with the spatial resolution of 20 m. The dimensions of the HS and P images are 40×40 and 200×200, respectively.

(2) The Pavia University dataset: This dataset was acquired by the Reflective Optics System Imaging (ROSIS) over Pavia, Italy [12]. ROSIS provides the dataset which covers the spectral range of 0.4–0.9 μm, and the dataset is characterized by 115 bands. After the water absorption and the noise corrupted bands are removed, 103 bands are used for experimentation. The P image has a spatial resolution of 1.3 m in the visible spectral range. The size of the HS and P images in the experiment are 40×40 and 200×200, respectively.

(3) The Washington DC dataset: This dataset was collected by the Spectral Information Technology Application Center of Virginia over the Washington DC Mall. The dataset consists of 210 bands in the spectral range of 0.4–2.4 μm. Some bands have been removed since the atmosphere is opaque, and 191 bands are used for the experiment. In the experiments, the P image has a spatial resolution of 0.8 m in the visible spectral range. The dimensions of the HS and P images are 40×40 and 200×200, respectively.

(4) The Hyperion dataset: The EO-I spacecraft which is operated by NASA carries two instruments: Hyperion and Advanced Land Imager (ALI) [12]. Hyperion provides the HS image which is characterized by 242 bands in the spectral range of 0.4–2.5 μm. The spatial resolution of the HS image is 30 m. ALI instrument is capable of providing the P image which covers the spectral range of 0.48–0.69 μm with the spatial resolution of 10 m.

4.2. Quality Measures

Generally, the performance of a hyperspectral pansharpening method can be assessed by the subjective effect and the objective indexes. The similarity of the colors between the reference HS image and the fused HS image can be determined by the subjective evaluation. Objective indexes are used for comparing the fusion quality accurately. This paper is limited to the five most widely used indexes, i.e., cross correlation (CC) [42], spectral angle mapper (SAM) [43], root mean squared error (RMSE), erreur relative globale adimensionnelle de synthèse (ERGAS) [44], and universal image quality index (UIQI) [45]. The CC is the spatial measure, and SAM is the spectral measure. RMSE, ERGAS, and UIQI are the global spectral and spatial measures. The formal definitions of these indexes are provided below. In the definitions, matrix $FH = [h_1, \ldots, h_m] \in R^{\lambda \times m}$ denotes the fused HS image with λ bands and m pixels. $RH \in R^{\lambda \times m}$ represents the reference HS image. RH_l and FH_l represent the lth columns of RH and FH, respectively. RH^j and FH^j represent the jth rows of RH and FH, respectively. $X, Y \in R^{1 \times m}$ denote two single band images, and X_i denotes the ith element of X.

(1) Cross correlation: The *CC* measures the degree of the geometric distortion. It is defined as follows:

$$CC(RH, FH) = \frac{1}{\lambda}\sum_{j=1}^{\lambda} CCS(RH^j, FH^j) \tag{25}$$

The *CCS* characterizes the geometric distortion of a single-band image as follows:

$$CCS(X, Y) = \frac{\sum_{i=1}^{m}(X_i - \mu_X)(Y_i - \mu_Y)}{\sqrt{\sum_{i=1}^{m}(X_i - \mu_X)^2 \sum_{i=1}^{m}(Y_i - \mu_Y)^2}} \tag{26}$$

where μ_X and μ_Y are the means of X and Y, respectively. The optimal value of *CC* is 1.

(2) Spectral angle mapper: The *SAM* measures the spectral distortion between the fused image *FH* and the reference image *RH*, which is defined as:

$$SAM(RH, FH) = \frac{1}{m}\sum_{l=1}^{m} \arccos\left(\frac{\langle RH_l, FH_l \rangle}{\|RH_l\| \|FH_l\|}\right) \tag{27}$$

The *SAM* is a spectral measure. The smaller the *SAM* value is, the better the fusion performance is.

(3) Root mean squared error: The *RMSE* which measures the standard difference between the two matrices *RH* and *FH*, is defined as

$$RMSE(RH, FH) = \frac{\sqrt{trace[(RH - FH)^T (RH - FH)]}}{\sqrt{m * \lambda}} \tag{28}$$

The optimal value of *RMSE* is 0.

(4) Erreur relative globale adimensionnelle de synthèse: The *ERGAS*, which is a global measure, is defined as

$$ERGAS(RH, FH) = 100c\sqrt{\frac{1}{\lambda}\sum_{j=1}^{\lambda}\left(\frac{RMSE_j}{\mu_j}\right)^2} \tag{29}$$

where, $RMSE_j = (\sqrt{trace[(RH^j - FH^j)^T (RH^j - FH^j)]}/\sqrt{m})$, c represents the ratio of the linear resolution between the P and HS images, and μ_j is the mean value of the jth band of the reference image. The optimal value of *ERGAS* is 0.

(5) Universal image quality index: The *UIQI*, which evaluates the similarity of the reference image *RH* and the fused image *FH*, is defined as

$$UIQI(RH, FH) = \frac{4\sigma_{RF}^2 \mu_R \mu_F}{(\sigma_R^2 + \sigma_F^2)(\mu_R^2 + \mu_F^2)} \tag{30}$$

where, $\mu_R, \sigma_R^2, \mu_F, \sigma_F^2$ are the sample means and standard deviations of the reference image *RH* and the fused image *FH*, and σ_{RF}^2 is the covariance of the two images. The ideal value of the *UIQI* value is 1.

4.3. Analysis of the Influence of Parameter α

In the experiments, α is the parameter which determines the quantity of the final injected spatial details and influences the fusion performance directly. To select an optimal parameter α, the proposed method is performed on the Salinas dataset with different α settings. We apply five quality measures to investigate the effects of the parameter α on the fusion performance. For clarity, the five quality measures are normalized to [0, 1] by the min-max normalization method and are displayed in one figure. Figure 4 shows the performance of the proposed method with different α settings. It can be observed that the CC and UIQI values are increasing from 0 to 0.1 when α is increased from 0 to 0.1. The CC value obtains the biggest value when α equals to 0.1. In addition, the values of SAM, RMSE, and ERGAS all are decreasing when α is increased from 0 to 0.1. While they will increase when α equals to 0.1. Therefore, we can draw a conclusion that when $\alpha = 0.1$, the performance of the proposed method is the best. We have also performed the performance of the proposed method on various hyperspectral remote sensing images. We found that $\alpha = 0.1$ also give the best performance there. Therefore, the parameter α is set as 0.1 in this paper.

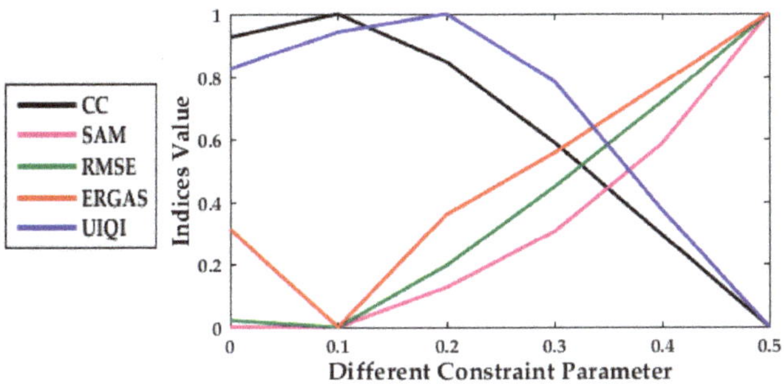

Figure 4. Performance of the proposed method with different α settings.

4.4. Experiments on Simulated Hyperspectral Remote Sensing Datasets

The Salinas dataset, Pavia University dataset, and Washington DC dataset are all simulated datasets. For the simulated dataset, a reference high spatial resolution HS image is given. The simulated P image and the simulated low spatial resolution HS image can be obtained by the Wald's protocol [44]. We can use the reference high spatial resolution HS image as the reference image to evaluate the performance of the fused image.

4.4.1. Salinas Dataset

The color displays of the fused HS images obtained by different methods are shown in Figure 5b–h. As reported in some articles, the PCA method generates serious spectral distortion. The fused image obtained by the GFPCA method looks blurry, since the injected spatial information seems to be not sufficient. There is less spectral distortion generated by the GFPCA method compared with the PCA method. The edges in the fused images obtained by the HySure and MGH methods appear too sharp due to the artifacts occurred around the edges. The CNMF and Sparse Representation methods can well preserve the spectral information of the original HS image. However, the edges in the vegetation and roof areas are not clear in the fused images obtained by these two methods. The halo artifacts and the blurring problems can be eliminated by the proposed method. It can be seen that the proposed method performs well in both spectral and spatial aspects.

To further compare the visual quality of the fused images obtained by different fusion methods, Figure 6 is given to show the difference images between the fused HS images and the reference HS image. Here, the difference image is generated by subtracting the reference image from the corresponding fused image, on a pixel by pixel strategy. It is observed that the difference image between the reference image and the fused image obtained by the proposed method shows the light blue color for almost the entire image. In other words, the proposed method causes the smallest value difference between the reference HS image and the fused image compared with other methods, which further proves that the outstanding fusion performance can be obtained by the proposed method. The quality metrics of different methods for the Salinas dataset are shown in Table 1. We consider the five quality metrics together to evaluate the performance of different pansharpening methods. It can be seen that, for the Salinas dataset, the proposed method gives the optimal quality indexes in terms of all the quality metrics. This means that the proposed method can perform well in both spectral and spatial aspects.

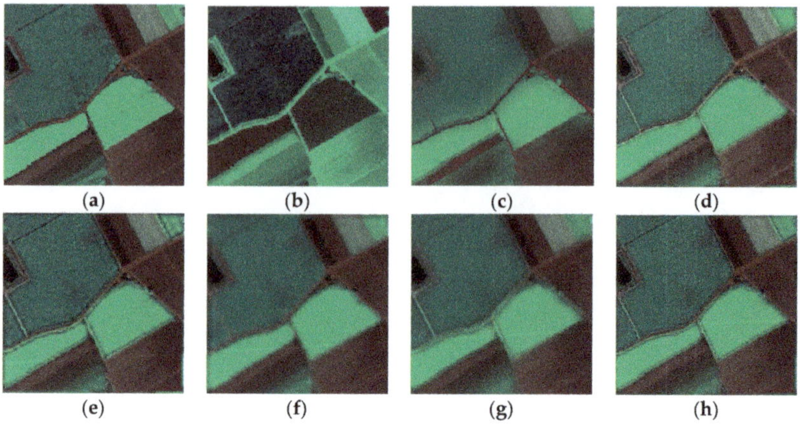

Figure 5. Visual comparison of different hyperspectral pansharpening methods for Salinas dataset. (**a**)Reference; (**b**) PCA; (**c**) GFPCA; (**d**) HySure; (**e**) MGH; (**f**) CNMF; (**g**) Sparse Representation; (**h**) Proposed.

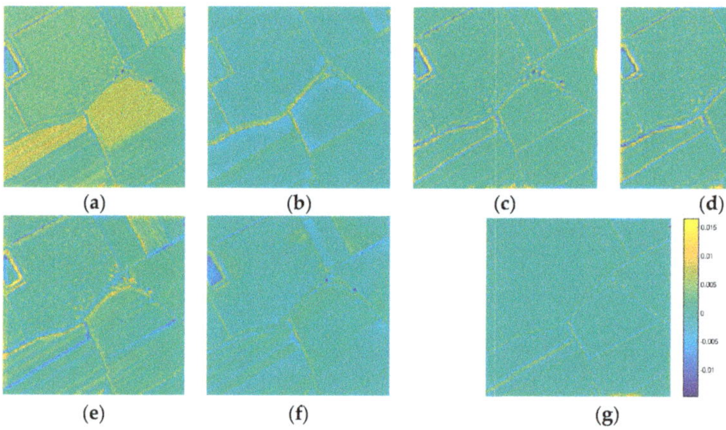

Figure 6. Visual comparison of difference images (light blue means small differences) between each fused HS image and the reference HS image (Salinas dataset). (**a**) PCA; (**b**) GFPCA; (**c**) HySure; (**d**) MGH; (**e**) CNMF; (**f**) Sparse Representation; (**g**) Proposed.

Table 1. Quality metrics of different methods for Salinas datasets.

Index	Method						
	PCA	**GFPCA**	**Hysure**	**MGH**	**CNMF**	**Sparse**	**Proposed**
CC	0.5341	0.9335	0.9485	0.9569	0.9385	0.9443	**0.9586**
SAM	10.5815	3.5553	2.1523	2.2694	2.3919	2.7381	**1.9475**
RMSE	0.0661	0.0262	0.0224	0.0231	0.0220	0.0209	**0.0188**
ERGAS	6.6333	2.8157	2.2176	2.3604	2.0145	2.0087	**1.1626**
UIQI	0.9008	0.9776	0.9877	0.9838	0.9867	0.9877	**0.9884**

4.4.2. Pavia University Dataset

Figure 7a shows the reference HS image of the Pavia University dataset. Figure 7b–h show the fused images obtained by different pansharpening methods. By visually comparing these fused images with the reference one, a similar conclusion as the above experiments can be drawn. The spatial and spectral quality of the fused image obtained by the PCA method is not desired. The spectral distortion caused by the PCA method is most visible, especially in the vegetation areas. The GFPCA method improves with respect to the spectral aspect. However, the spatial quality of the fused image obtained by the GFPCA method needs further improvement. The HySure and MGH methods produce halo artifacts around edges, although such artifacts make the edges appear sharper. The CNMF method introduces spectral distortion, since the color of the fused image obtained by the CNMF method is not match to that of the reference image in roof area. By contrast, the fused images produced by the Sparse Representation and the proposed method are the closest to the reference one.

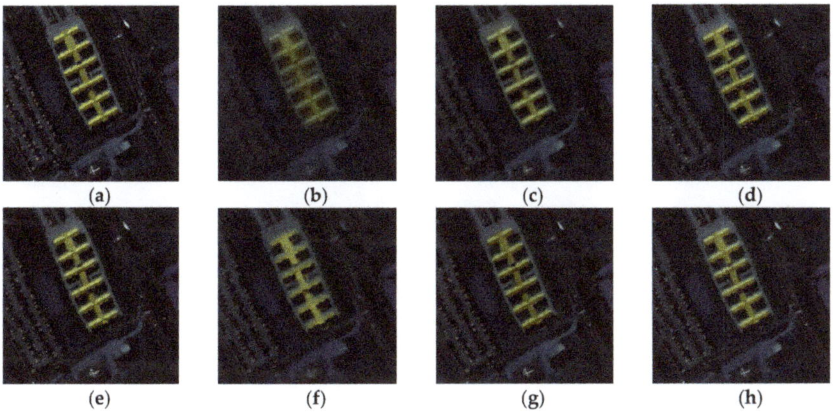

Figure 7. Visual comparison of different hyperspectral pansharpening methods for Paiva University dataset. (**a**) Reference; (**b**) PCA; (**c**) GFPCA; (**d**) HySure; (**e**) MGH; (**f**) CNMF; (**g**) Sparse Representation; (**h**) Proposed.

The visual quality of fused images obtained by different methods can be measured by the difference images between the fused HS images and the reference HS image. Figure 8 is given to show the difference images with outstanding defects and flat background between the fused HS images and the reference HS image. The difference image of the proposed method is almost all light blue with few yellow mixed. Based on the comparison of difference images, the proposed method indeed displays the best performance in visual quality. Table 2 shows the objective quality assessment of different methods for the Pavia University dataset. It can be clearly seen that the proposed method shows the best objective performance in most measurement terms including CC, SAM, RMSE, and ERGAS. The UIQI value of the proposed method is second largest. This further demonstrates that the proposed method can obtain the state-of-the-art fusion performance.

Table 2. Quality metrics of different methods for Pavia University dataset.

Index	Method						
	PCA	GFPCA	HySure	MGH	CNMF	Sparse	Proposed
CC	0.8967	0.8203	0.9404	0.9308	0.8723	0.9012	**0.9440**
SAM	6.2287	9.2413	6.5623	6.2832	7.2820	8.4505	**6.1135**
RMSE	0.0489	0.0664	0.0385	0.0475	0.0548	0.0484	**0.0330**
ERGAS	6.8652	6.9343	3.4842	4.6084	5.8609	4.6880	**3.4226**
UIQI	0.7824	0.7356	**0.8381**	0.8133	0.7908	0.7739	0.8310

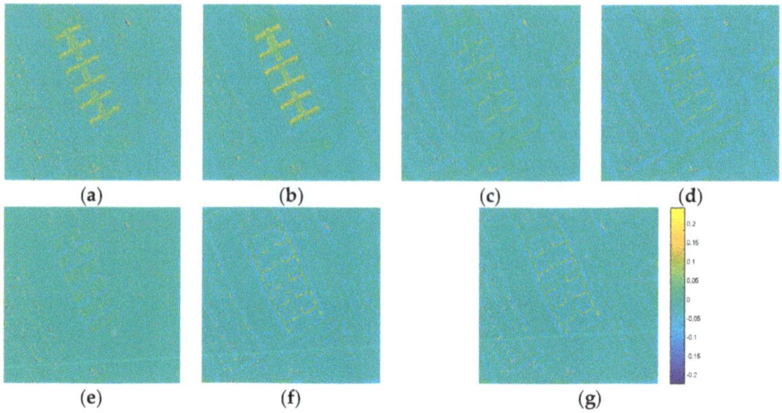

Figure 8. Visual comparison of difference image (light blue means small differences) between each fused HS image and the reference HS image (Paiva University dataset). (**a**) PCA; (**b**) GFPCA; (**c**) HySure; (**d**) MGH; (**e**) CNMF; (**f**) Sparse Representation; (**g**) Proposed.

4.4.3. Washington DC Dataset

Figure 9 shows the visual comparison of the fused images obtained by different fusion methods for the Washington DC dataset. The reference HS image is displayed in Figure 9a. Figure 9b–h show the fused images obtained by different fusion methods. It is apparent that the fused image obtained by the PCA method suffers from the spectral and spatial distortion. The fused image obtained by the GFPCA method shows improvement, but the spatial quality is not improved obviously. A visual comparison shows that the MGH method performs well in spectral aspect, but the fused image obtained by the MGH method looks blurry in some areas. The reason is that the injected spatial details is insufficient. The CNMF and Sparse Representation methods are close to the reference image in spectral aspect. However, the spatial quality of the CNMF and Sparse Representation methods in the edge regions is not desired. The fused results obtained by the HySure and the proposed fusion methods have superior performance in spectral and spatial aspects.

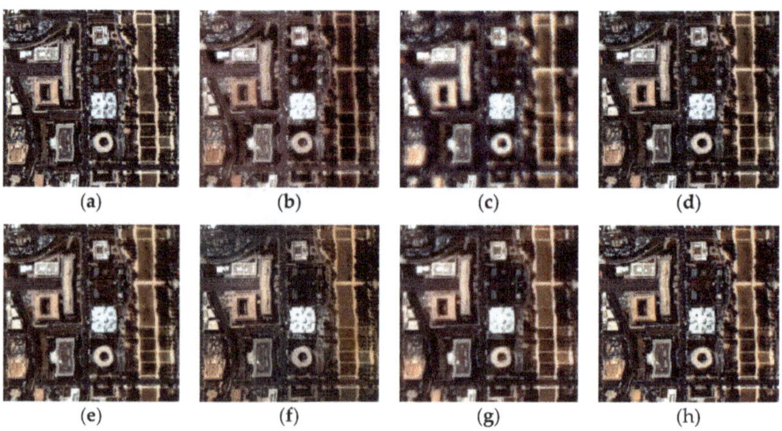

Figure 9. Visual comparison of different hyperspectral pansharpening methods for Washington DC dataset. (**a**) Reference; (**b**) PCA; (**c**) GFPCA; (**d**) HySure; (**e**) MGH; (**f**) CNMF; (**g**) Sparse Representation; (**h**) Proposed.

Figure 10 shows the visual comparison of difference images between the fused HS images and the reference HS image for the Washington DC dataset. The proposed method indeed performs best in achieving the objective that the fused HS image should be as close as possible to the HS image acquired by the high-resolution sensors. The quality metrics of different methods for the Washington DC dataset are shown in Table 3. It can be seen that, for the Washington DC dataset, the proposed method gives the smallest quality indexes for RMSE and ERGAS, and the optimal quality indexes for CC and UIQI. Although the objective assessment of the proposed method is not always the best, it achieves a very stable performance in terms of five widely used quality metrics. This means that the proposed method can perform well in terms of providing the spatial details while preserving the spectral information.

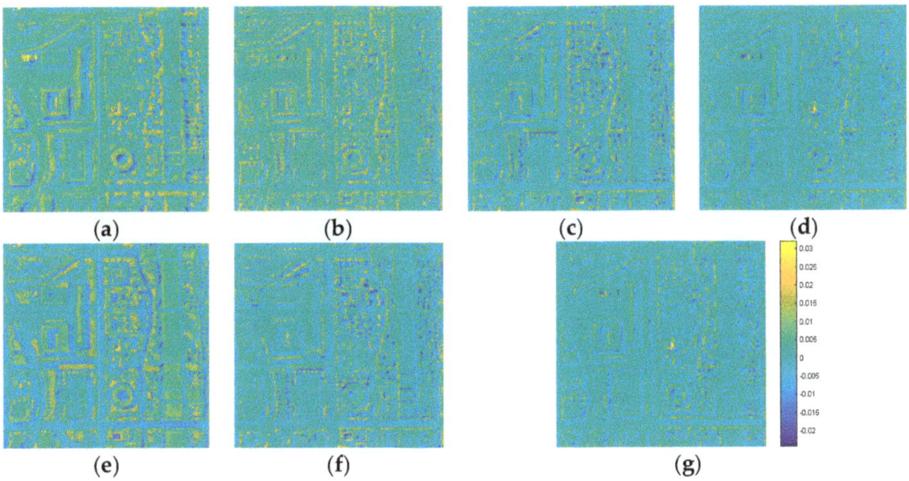

Figure 10. Visual comparison of difference image (light blue means small differences) between each fused HS image and the reference HS image (Washington DC dataset). (**a**) PCA; (**b**) GFPCA; (**c**) HySure; (**d**) MGH; (**e**) CNMF; (**f**) Sparse Representation; (**g**) Proposed.

Table 3. Quality metrics of different methods for Washington DC dataset.

Index	Method						
	PCA	GFPCA	HySure	MGH	CNMF	Sparse	Proposed
CC	0.7724	0.7829	0.8782	0.8771	0.7720	0.8179	**0.8940**
SAM	8.8123	10.7545	**7.3913**	7.9453	8.6588	10.0145	7.7436
RMSE	0.0123	0.0139	0.0084	0.0132	0.0120	0.0119	**0.0081**
ERGAS	33.4070	39.5999	27.1998	35.3178	31.2027	30.8796	**25.2085**
UIQI	0.9227	0.8971	0.9541	0.9350	0.9391	0.9517	**0.9584**

4.5. Experiments on Real Hyperspectral Remote Sensing Datasets

The Hyperion dataset which is the real hyperspectral dataset is utilized to evaluate the performance of the proposed method in real applications. For the real HS image, fusion is performed at the full scale for the subjective evaluation. The dimensions of the test P image are 210×150, and the size of the experimental HS image is 70×50. Figure 11a,b show the interpolated HS image and the P image, respectively. Figure 11c–i displays the results of different pansharpening methods. By visually comparing these fused HS images with the original HS image, it is clear that the blocking artifacts exist in the fused image obtained by the PCA method. The result obtained by the GFPCA method also looks blurry in this experiment. The HySure, MGH, and CNMF methods preserve the spectral information effectively, but the spatial quality of these methods is poor. The spectral distortion of the Sparse Representation method is visible in some areas. The proposed method can well preserve the spectral information and greatly improve the spatial quality of the original HS image.

The high spatial resolution HS image of the real dataset is generally not available. Fusion is performed at the degraded scale for the objective evaluation. Specifically, according to the literatures [46], we degrade the original HS and *P* images, and fuse the degraded HS and *P* images. The original HS image is used as the reference. The fused image is compared with the original HS image to evaluate the objective performance. Table 4 shows the quality metrics of different methods for the Hyperion dataset. The proposed method shows the best objective performance in terms of all the measurements terms including CC, SAM, RMSE, ERGAS, and UIQI.

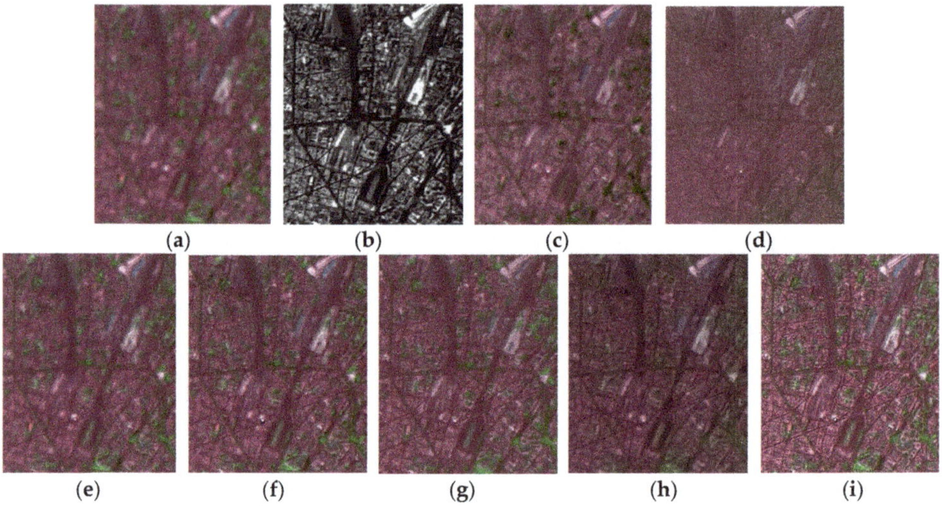

Figure 11. Visual comparison of different hyperspectral pan-sharpening methods for Hyperion dataset. (**a**) Interpolated HS image; (**b**) *P* image; (**c**) PCA; (**d**) GFPCA; (**e**) HySure; (**f**) MGH; (**g**) CNMF; (**h**) Sparse Representation; (**i**) Proposed.

Table 4. Quality metrics of different methods for Hyperion dataset.

Index	Method						
	PCA	**GFPCA**	**HySure**	**MGH**	**CNMF**	**Sparse**	**Proposed**
CC	0.7574	0.7430	0.5984	0.9087	0.8754	0.8273	**0.9123**
SAM	3.7563	4.6248	12.1445	2.7090	3.0467	4.5656	**2.6637**
RMSE	0.0353	0.0385	0.1012	0.0225	0.0249	0.0362	**0.0220**
ERGAS	8.5544	9.9020	22.0946	5.6159	6.6400	9.0590	**5.4775**
UIQI	0.9835	0.9797	0.8738	0.9922	0.9910	0.9775	**0.9927**

To verify the validity of the proposed method on the real HS images, the experiment is performed on another Hyperion image. The test *P* image is of size 300 × 300 pixels, and the size of the test HS image is 100 × 100. Figure 12a,b show the interpolated HS image and the *P* image, respectively. The fused images obtained by different methods are displayed in Figure 12c–i. The color of the fused image obtained by the PCA method is not match to that of the original HS image in some areas. The GFPCA method produces the serious spatial distortion, although it performs better in the spectral aspect compared with the PCA method. The fused images obtained by the HySure and Sparse Representation methods appear too sharp due to the artifacts occurred around the edges.

The color of the fused images obtained by the MGH, CNMF and the proposed methods is close to that of the original HS image, which indicates the superiority of these pansharpening methods in spectral preservation. However, the spatial quality of the CNMF method in some edges is not desired. By contrast, the fused images produced by the proposed method and the MGH method obtain the outstanding fusion performance in terms of spectral and spatial aspects. Table 5 shows the objective quality evaluation of each method for the Hyperion dataset. The proposed method performs best in terms of most of the indexes. The MGH method obtains the best ERGAS index. Although the objective performance of the proposed method is not always the best, it has a stable performance. Based on the analysis of the visual comparison and objective evaluation, we can draw a conclusion that the proposed method obtains the excellent performance for the real hyperspectral dataset in terms of the objective and subjective evaluations.

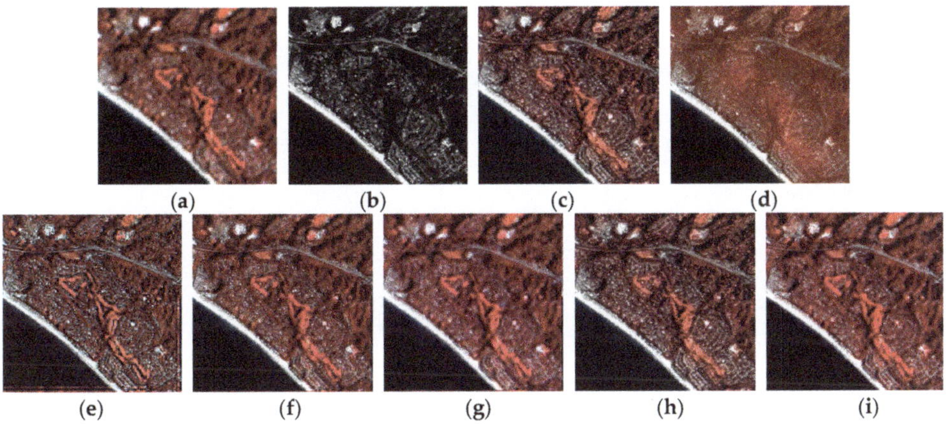

Figure 12. Visual comparison of different hyperspectral pan-sharpening methods for Hyperion dataset. (**a**) Interpolated HS image; (**b**) *P* image; (**c**) PCA; (**d**) GFPCA; (**e**) HySure; (**f**)MGH; (**g**) CNMF; (**h**) Sparse Representation; (**i**) Proposed.

Table 5. Quality metrics of different methods for Hyperion dataset.

Index	Method						
	PCA	**GFPCA**	**HySure**	**MGH**	**CNMF**	**Sparse**	**Proposed**
CC	0.7077	0.8234	0.7543	0.9341	0.9219	0.8781	**0.9344**
SAM	4.7224	5.7334	6.8596	3.0803	4.3548	5.4651	**3.0230**
RMSE	0.0399	0.0253	0.0396	0.0185	0.0189	0.0206	**0.0174**
ERGAS	12.5797	10.0371	15.8318	**6.2645**	6.7987	8.6556	6.5841
UIQI	0.7886	0.7324	0.7837	0.9418	0.9025	0.8537	**0.9446**

4.6. Discussion

Experimental results demonstrate that the proposed method outperforms the other six hyperspectral pansharpening methods. The proposed method has a good performance in the spectral fidelity, since it always obtains an optimal SAM index. The HySure method has an excellent performance on the simulated images, but it performs badly on the real HS images. The proposed method performs well on the simulated and real hyperspectral images.

The superiority of the proposed method is owing to the employment of the weighted least squares filter and the intrinsic image decomposition. The WLS filter can make a proper compromise between sharpening and blurring, which improves the spatial quality of the fused image. The IID is an effective technique to separate the HS image into the reflectance and illumination components, which plays an important role in reducing the spectral distortion.

A simple yet effective fusion rule is introduced in this paper α determines the quantity of the finial injected spatial details and influences the fusion performance directly. We have tested the performance of the proposed method on various remote sensing images and many real satellite images with different α settings. We found that $\alpha = 0.1$ always give the best performance. In future work, we plan to improve the performance of the proposed method by adaptively selecting parameter.

5. Conclusions

Hyperspectral pansharpening is an important subdivision of remote sensing image processing. A novel hyperspectral pansharpening method based on IID and WLS filter has been presented in this paper. The proposed method first obtains the spatial information of the P image with a weighted least squares filter, in which the LOG enhancement algorithm was used for the spatial enhancement. Then, the illumination component which is considered the spatial information of the HS image is estimated with the intrinsic image decomposition technique. The fused image can be obtained by injecting the detail map into each band of the deblurred interpolated HS image. The final injected spatial information takes full account of the P and the HS images. The impact of data independence can be eliminated. The existing problem of the CS and MRA-based fusion methods can be well solved by the combination of an IID technique and a WLS filter. Experiments conducted on six synthetic and real hyperspectral datasets demonstrated that the proposed method performs better than the state-of-the-art fusion methods as well as the CS and MRA-based fusion methods in terms of visual inspection and objective analysis.

Acknowledgments: This work was supported by NSFC (No. 61372069), National Defense Pre-research Foundation, SRF for ROCS, SEM (JY0600090102), "111" project (No. B08038) and the Fundamental Research Funds for the Central Universities.

Author Contributions: Wenqian Dong drafted the manuscript; Wenqian Dong, Song Xiao, and Yunsong Li conceived and designed the experiments; Wenqian Dong and Jiahui Qu performed the experiments; Wenqian Dong and Song Xiao analyzed the data; Yunsong Li and Jiahui Qu modified the manuscript.

Conflicts of Interest: The authors declare no conflict of interest.

Abbreviations

P	Panchromatic
HS	Hyperspectral
WLS	Weighted least squares
IID	Intrinsic image decomposition
OH	Original hyperspectral image
UH	Interpolated hyperspectral image
H	Deblurred interpolated hyperspectral image
D	Detail map
CC	Cross correlation
SAM	Spectral angle mapper
RMSE	Root mean squared error
ERGAS	Erreur relative globale adimensionnelle de synthèse
UIQI	Universal image quality index
PCA	Principal component analysis
GFPCA	Guided filter PCA
CNMF	Coupled nonnegative matrix factorization
MTF	Modulation transfer function
MGH	MTF-generalized Laplacian Pyramid with high pass modulation

References

1. Kempeneers, P.; Sedano, F.; Seebach, L.; Strobl, P.; San-MiguelAyanz, J. Data fusion of different spatial resolution remote sensing images applied to forest-type mapping. *IEEE Trans. Geosci. Remote Sens.* **2011**, *49*, 4977–4986. [CrossRef]
2. Yokoya, N.; Grohnfeldt, C.; Chanussot, J. Hyperspectral and multispectral data fusion: A comparative review of the recent literature. *IEEE Geosci. Remote Sens. Mag.* **2017**, *5*, 29–56. [CrossRef]
3. Zhang, H.; Huang, B. A new look at image fusion methods from a Bayesian perspective. *Remote Sens.* **2015**, *7*, 6828–6861. [CrossRef]
4. Vivone, G.; Alparone, L.; Chanussot, J.; Mura, M.D.; Garzelli, A.; Licciardi, G.A.; Restaino, R.; Wald, L. A critical comparison among pansharpening algorithms. *IEEE Trans. Geosci. Remote Sens.* **2015**, *53*, 2565–2586. [CrossRef]
5. Loncan, L.; Almeida, L.B.; Dias, J.M.; Briottet, X.; Chanussot, J.; Dobigeon, N.; Fabre, S.; Liao, W.; Licciardi, G.A.; Simões, M.; et al. Hyperspectral pansharpening: A review. *IEEE Geosci. Remote Sens. Mag.* **2015**, *3*, 27–46. [CrossRef]
6. Wei, Q.; Dobigeon, N.; Tourneret, J.Y. Bayesian fusion of multiband images. *IEEE J. Sel. Top. Signal Process.* **2015**, *9*, 1117–1127. [CrossRef]
7. Wei, Q.; Bioucas-Dias, J.; Dobigeon, N.; Tourneret, J. Hyperspectral and multispectral image fusion based on a sparse representation. *IEEE Trans. Geosci. Remote Sens.* **2015**, *53*, 3658–3668. [CrossRef]
8. Simões, M.; Dias, J.B.; Almeida, L.; Chanussot, J. A convex formulation for hyperspectral image superresolution via subspace-based regularization. *IEEE Trans. Geosci. Remote Sens.* **2015**, *53*, 3373–3388. [CrossRef]
9. Wei, Q.; Dobigeon, N.; Tourneret, J.Y. Fast fusion of multiband images based on solving a sylvester equation. *IEEE Trans. Image Process.* **2015**, *24*, 4109–4121. [CrossRef] [PubMed]
10. Yokoya, N.; Yairi, T.; Iwasaki, A. Coupled nonnegative matrix factorization unmixing for hyper-spectral and multispectral data fusion. *IEEE Trans. Geosci. Remote Sens.* **2012**, *50*, 528–537. [CrossRef]
11. Ghassemian, H. A review of remote sensing image fusion methods. *Inf. Fusion* **2016**, *32*, 75–89. [CrossRef]
12. Mookambiga, A.; Gomathi, V. Comprehensive review on fusion techniques for spatial information enhancement in hyperspectral imagery. *Multidimens. Syst. Signal Process.* **2016**, *27*, 863–889. [CrossRef]
13. Tu, T.M.; Su, S.C.; Shyu, H.C.; Huang, P.S. A new look at IHS-like image fusion methods. *Inf. Fusion* **2001**, *2*, 117–186. [CrossRef]
14. Hallabia, H.; Kallel, A.; Hamida, A.; Hegarat-Mascle, S. High spectral quality pansharpening approach based on MTF-matched filter banks. *Multidimens. Syst. Signal Process.* **2016**, *4*, 1–31. [CrossRef]
15. Carper, W.; Lillesand, T.M.; Kiefer, P.W. The use of Intensity-Hue-Saturation transformations for merging SPOT panchromatic and multispectral image data. *Photogramm. Eng. Remote Sens.* **1990**, *56*, 459–467.
16. Anderson, J.A.; Sides, S.C.; Chavez, P.S. Comparison of three different methods to merge multiresolution and multispectral data: Landsat TM and SPOT panchromatic. *Photogramm. Eng. Remote Sens.* **1991**, *57*, 295–303.
17. Laben, C.; Brower, B. Process for Enhacing the Spatial Resolution of Multispectral Imagery Using Pan-Sharpening. U.S. Patent 6,011,875, 1 April 2000.
18. Chavez, P.S.; Kwarteng, A.Y.A. Extracting spectral contrast in Landsat thematic mapper image data using selective principal component analysis. *Photogramm. Eng. Remote Sens.* **1989**, *55*, 339–348.
19. Shettigara, V.A. Generalized component substitution technique for spatial enhancement of multispectral images using a higher resolution data set. *Photogramm. Eng. Remote Sens.* **1992**, *58*, 561–567.
20. Shah, V.P.; Younan, N.; King, R.L. An efficient pan-sharpening method via a combined adaptive PCA approach and contourlets. *IEEE Trans. Geosci. Remote Sens.* **2008**, *56*, 1323–1335. [CrossRef]
21. Liu, J.M.; Liang, S.L. Pan-sharpening using a guided filter. *Int. J. Remote Sens.* **2016**, *37*, 1777–1800. [CrossRef]
22. Aiazzi, B.; Baronti, S.; Selva, M. Improving component substitution pansharpening through multivariate regression of MS + Pan data. *IEEE Trans. Geosci. Remote Sens.* **2007**, *45*, 3230–3239. [CrossRef]
23. Choi, J.; Yu, K.; Kim, Y. A new adaptive component-substitution based satellite image fusion by using partial replacement. *IEEE Trans. Geosci. Remote Sens.* **2011**, *49*, 295–309. [CrossRef]
24. Aiazzi, B.; Alparone, L.; Baronti, S.; Garzelli, A.; Selva, M. MTF-tailored multiscale fusion of high-resolution MS and pan imagery. *Photogramm. Eng. Remote Sens.* **2006**, *72*, 591–596. [CrossRef]

25. Liu, J.G. Smoothing filter based intensity modulation: A spectral preserve image fusion technique for improving spatial details. *Int. J. Remote Sens.* **2000**, *21*, 3461–3472. [CrossRef]
26. Vivone, G.; Restaino, R.; Mura, M.D.; Licciardi, G.; Chanussot, J. Contrast and error-based fusion schemes for multispectral image pansharpening. *IEEE Geosci. Remote Sens. Lett.* **2014**, *11*, 930–934. [CrossRef]
27. Selva, M.; Aiazzi, B.; Butera, F.; Chiarantini, L.; Baronti, S. Hyper-sharpening: A first approach on SIM-GA data. *IEEE J. Sel. Top. Appl. Earth Obs. Remote Sens.* **2015**, *8*, 3008–3024. [CrossRef]
28. Ghahremani, M.; Ghassemian, H. Remote-sensing image fusion based on Curvelets and ICA. *Int. J. Remote Sens.* **2015**, *36*, 4131–4143. [CrossRef]
29. Qu, J.H.; Li, Y.S.; Dong, W.Q. Hyperspectral pansharpening with guided filter. *IEEE Geosci. Remote Sens. Lett.* **2017**, *14*, 2152–2156. [CrossRef]
30. Tappen, M.F.; Freeman, W.T.; Adelson, E.H. Recovering intrinsic images from a single image. *IEEE Trans. Pattern Anal. Mach. Intell.* **2005**, *27*, 1459–1472. [CrossRef] [PubMed]
31. Shen, J.; Yang, X.; Li, X.; Jia, Y. Intrinsic image decomposition using optimization and user scribbles. *IEEE Trans. Cybern.* **2013**, *43*, 425–436. [CrossRef] [PubMed]
32. Jiang, X.; Schofield, A.J.; Wyatt, J.L. Correlation-based intrinsic image extraction from a single image. In Proceedings of the European Conference on Computer Vision, Heraklion, Greece, 5–11 September 2010; pp. 58–71.
33. Powell, M.W.; Sarkar, S.; Goldgof, D.B.; Ivanov, K. A methodology for extracting objective color from images. *IEEE Trans. Syst. Man Cybern. B Cybern.* **2004**, *34*, 1964–1978. [CrossRef] [PubMed]
34. Bousseau, A.; Paris, S.; Durand, F. User-assisted intrinsic images. *ACM Trans. Graph.* **2009**, *28*, 1. [CrossRef]
35. Jung, C.; Sun, T.; Jiao, L. Eye detection under varying illumination using the Retinex theory. *Neurocomputing* **2013**, *113*, 130–137. [CrossRef]
36. Jobson, D.J.; Rahman, Z.; Woodell, G.A. A multi-scale Retinex for bridging the gap between color images and the human observation of scenes. *IEEE Trans. Image Process.* **1997**, *6*, 965–976. [CrossRef] [PubMed]
37. Farbman, Z.; Fattal, R.; Lischinski, D.; Szeliski, R. Edge-preserving decompositions for multi-scale tone and detail manipulation. *ACM Trans. Graph.* **2008**, *27*, 1–10. [CrossRef]
38. Lischinski, D.; Farbman, Z.; Uyttenaaele, M.; Szeliski, R. Interactive local adjustment of tonal values. *ACM Trans. Graph.* **2006**, *25*, 646–653. [CrossRef]
39. Palsson, F.; Sveinsson, J.R.; Ulfarsson, M.O.; Benediktsson, J.A. MTF-Based Deblurring Using a Wiener Filter for CS and MRA Pansharpening Methods. *IEEE J. Sel. Top. Appl. Earth Obs. Remote Sens.* **2016**, *9*, 2255–2269. [CrossRef]
40. Chakrabarti, A.; Zickler, T. Statistics of real-world hyperspectral images. In Proceedings of the IEEE Conference on Computer Vision and Pattern Recognition, Colorado Springs, CO, USA, 20–25 June 2011; pp. 193–200.
41. Liao, W.; Huang, X.; Coillie, F.; Gautama, S.; Pizurica, A.; Philips, W.; Liu, H.; Zhu, T.; Shimoni, M.; Moser, G.; et al. Processing of multiresolution thermal hyperspectral and digital color data: Outcome of the 2014 IEEE GRSS data fusion contest. *IEEE J. Sel. Top. Appl. Earth Obs. Remote Sens.* **2015**, *8*, 2984–2996. [CrossRef]
42. Zhou, J.; Civco, D.L.; Silander, J.A. A wavelet transform method to merge Landsat TM and SPOT panchromatic data. *Int. J. Remote Sens.* **1998**, *19*, 743–757. [CrossRef]
43. Yang, Y.; Tong, S.; Huang, S.; Lin, P. Multifocus image fusion based on NSCT and focused area detection. *IEEE Sens. J.* **2015**, *15*, 2824–2838.
44. Wald, L.; Ranchin, T.; Mangolini, M. Fusion of satellite images of different spatial resolutions: Assessing the quality of resulting images. *Photogramm. Eng. Remote Sens.* **1997**, *63*, 691–699.
45. Wang, Z.; Bovik, A.C. A universal image quality index. *IEEE Signal Process. Lett.* **2002**, *9*, 81–84. [CrossRef]
46. Selva, M.; Santurri, L.; Baronti, S. On the Use of the Expanded Image in Quality Assessment of Pansharpened Images. *IEEE Geosci. Remote Sens. Lett.* **2018**, *15*, 1–5. [CrossRef]

MDPI
St. Alban-Anlage 66
4052 Basel
Switzerland
Tel. +41 61 683 77 34
Fax +41 61 302 89 18
www.mdpi.com

Remote Sensing Editorial Office
E-mail: remotesensing@mdpi.com
www.mdpi.com/journal/remotesensing

www.ingramcontent.com/pod-product-compliance
Lightning Source LLC
Chambersburg PA
CBHW061450220926
51850CB00004B/54
* 9 7 8 3 0 3 9 2 1 5 2 2 5 *